RELATIVISTIC ASTROPHYSICS

Related Titles from the AIP Conference Proceedings Subseries on Astronomy and Astrophysics

587 Gamma 2001: Gamma-Ray Astrophysics 2001
Edited by Steven Ritz, Neil Gehrels, and Chris R. Shrader, October 2001, 0-7354-0027-X

579 Radio Detection of High Energy Particles: First International Workshop; RADHEP 2000
Edited by David Saltzberg and Peter Gorham, July 2001, 0-7354-0018-0

566 Observing Ultrahigh Energy Cosmic Rays from Space and Earth: International Workshop
Edited by Humberto Salazar, Luis Villaseñor, and Arnulfo Zepeda, May 2001,
0-7354-0002-4

558 High Energy Gamma-Ray Astronomy: International Symposium
Edited by Felix A. Aharonian and Heinz J. Völk, April 2001, 1-56396-990-4

556 Explosive Phenomena in Astrophysical Compact Objects: First KIAS Astrophysics Wkshp
Edited by Heon-Young Chang, Chang-Hwan Lee, Mannque Rho, and Insu Yi, March 2001,
1-56396-987-4

555 Cosmology and Particle Physics: CAPP 2000
Edited by Ruth Durrer, Juan Garcia-Bellido, and Mikhail Shaposhnikov, March 2001,
1-56396-986-6

526 Gamma-Ray Bursts: 5th Huntsville Symposium
Edited by R. Marc Kippen, Robert S. Mallozzi, and Gerald J. Fishman, June 2000,
CD-ROM included, 1-56396-947-5

516 26th International Cosmic Ray Conference: ICRC XXVI, Invited, Rapporteur, and Highlight Papers
Edited by Brenda L. Dingus, David B. Kieda, and Michael H. Salamon, May 2000, 1-56396-939-4

515 GeV-TeV Gamma Ray Astrophysics Workshop: Towards a Major Atmospheric Cherenkov Detector VI
Edited by Brenda L. Dingus, Michael H. Salamon, and David B. Kieda, May 2000, 1-56396-938-6

510 The Fifth Compton Symposium
Edited by Mark L. McConnell and James M. Ryan, March 2000, 1-56396-932-7

493 General Relativity and Relativistic Astrophysics: Eighth Canadian Conference
Edited by C. P. Burgess and R. C. Myers, November 1999, 1-56396-905-X

478 COSMO-98: Second International Workshop on Particle Physics and the Early Universe
Edited by David O. Caldwell, May 1999, 1-56396-853-3

433 Wkshp. on Observing Giant Cosmic Ray Air Showers from $>10^{20}$ eV Particles from Space
Edited by John F. Krizmanic, Jonathan F. Ormes, and Robert E. Streitmatter, June 1998,
1-56396-788-X

To learn more about these titles, or the AIP Conference Proceedings Series, please visit the webpage **http://www.aip.org/catalog/aboutconf.html**

RELATIVISTIC ASTROPHYSICS

20th Texas Symposium

Austin, Texas   10–15 December 2000

EDITORS
J. Craig Wheeler
Hugo Martel
The University of Texas at Austin, Texas

Melville, New York, 2001
AIP CONFERENCE PROCEEDINGS ■ VOLUME 586

Editors:

J. Craig Wheeler
Hugo Martel

Department of Astronomy
University of Texas
Austin, TX 78712
USA

E-mail: wheel@astro.as.utexas.edu
hugo@simplicio.as.utexas.edu

The articles on pp. 420–425, 472–477, 632–634, and 702–704 were authored by U. S. Government employees and are not covered by the below mentioned copyright.

Authorization to photocopy items for internal or personal use, beyond the free copying permitted under the 1978 U.S. Copyright Law (see statement below), is granted by the American Institute of Physics for users registered with the Copyright Clearance Center (CCC) Transactional Reporting Service, provided that the base fee of $18.00 per copy is paid directly to CCC, 222 Rosewood Drive, Danvers, MA 01923. For those organizations that have been granted a photocopy license by CCC, a separate system of payment has been arranged. The fee code for users of the Transactional Reporting Service is: 0-7354-0026-1/01/$18.00.

© 2001 American Institute of Physics

Individual readers of this volume and nonprofit libraries, acting for them, are permitted to make fair use of the material in it, such as copying an article for use in teaching or research. Permission is granted to quote from this volume in scientific work with the customary acknowledgment of the source. To reprint a figure, table, or other excerpt requires the consent of one of the original authors and notification to AIP. Republication or systematic or multiple reproduction of any material in this volume is permitted only under license from AIP. Address inquiries to Office of Rights and Permissions, Suite 1NO1, 2 Huntington Quadrangle, Melville, N.Y. 11747-4502; phone: 516-576-2268; fax: 516-576-2450; e-mail: rights@aip.org.

L.C. Catalog Card No. 2001094694
ISBN 0-7354-0026-1
ISSN 0094-243X
Printed in the United States of America

CONTENTS

Preface xvii
Committees xix

CHAPTER 1:
BRANE-WORLD THEORIES: COSMOLOGY AND ASTROPHYSICS

The Current State of String Theory *(Plenary Talk)* 3
    G. T. Horowitz
Brane-World Astronomy *(Plenary Talk)* 11
    C. J. Hogan
Interaction of Bulk Matter with Brane-World Black-Holes 22
    A. Chamblin
A Universe in a Global Monopole 28
    K. Benson and I. Cho
Dynamics of Cosmological Models in the Brane-World Scenario 34
    A. Campos and C. F. Sopuerta
Initial Conditions for Hybrid Inflation on the Brane 37
    L. E. Mendes and A. R. Liddle

CHAPTER 2:
EARLY UNIVERSE, INFLATION, BARYOGENESIS,
AND PARTICLE PRODUCTION

New Constraints on Inflation from the CMB 43
    W. H. Kinney, A. Melchiorri, and A. Riotto
Natural Chaotic Inflation in Supergravity and Leptogenesis 49
    M. Yamaguchi, M. Kawasaki, and T. Yanagida
Diffractive Propagation of Light Waves in a Class of
Anisotropically-Expanding Cosmological Models 55
    B. Bochner
If the Parameters of the Vacuum Were Different, What Kind of
Universe Would We Have? Self-organization of the Vacuum 58
    V. Burdyuzha, Y. Ponomarev, and G. Vereshkov
A Cosmological Model with Particle Creation 64
    S. Chatterjee
Entropy and Adiabatic Components in Scalar Field Perturbations 68
    C. Gordon

CHAPTER 3:
DARK MATTER AND THE CDM PARADIGM

The Dark Matter Crisis *(Plenary Talk)* 73
    B. Moore

Recent Results from Super-Kamiokande *(Plenary Talk)* 83
    M. Nakahata for the Super-Kamiokande Collaboration
**Direct Search for Dark Matter Particles Deep Underground by
DAMA Experiment** *(Plenary Talk)* 95
    P. Belli, R. Bernabei, R. Cerulli, F. Montecchia, M. Amato,
    G. Ignesti, A. Incicchitti, D. Prosperi, C. J. Dai, H. H. He,
    H. H. Kuang, and J. M. Ma
Status of the CDMS Search for Dark Matter WIMPs *(Plenary Talk)* 107
    B. Cabrera, R. Abusaidi, D. S. Akerib, P. D. Barnes, Jr., D. A. Bauer,
    A. Bolozdynya, P. L. Brink, R. Bunker, D. O. Caldwell, J. P. Castle,
    C. Chang, R. M. Clarke, P. Colling, M. B. Cristler, A. Cummings,
    A. Da Silva, A. K. Davies, R. Dixon, B. L. Dougherty, D. Driscoll,
    S. Eichblatt, J. Emes, R. J. Gaitskell, S. R. Golwala, D. Hale, E. E. Haller,
    D. Holmgren, J. Hellmig, M. E. Huber, K. D. Irwin, J. Jochum,
    F. P. Lipshultz, A. Lu, C. Maloney, V. Mandic, J. M. Martinis, P. Meunier,
    S. W. Nam, H. Nelson, B. Neuhauser, M. J. Penn, T. A. Perera,
    M. C. Perillo Isaac, B. Pritychenko, R. R. Ross, T. Saab, B. Sadoulet,
    J. Sander, D. N. Seitz, P. Shestople, T. Shutt, A. Smith, G. W. Smith,
    R. W. Schnee, A. H. Sonnenschein, A. L. Spadafora, W. Stockwell,
    J. D. Taylor, S. White, S. Yellin, and B. A. Young
Microlensing: from Dark Matter to Planets *(Plenary Talk)* 119
    A. Gould
CDM: Numerical Predictions on Small Scales 130
    A. V. Kravtsov
**Warm Dark Matter, Small Scale Crisis, and the High Redshift
Universe** 136
    Z. Haiman, R. Barkana, and J. P. Ostriker
The Evolution of Dark-Matter Dominated Cosmological Halos 143
    M. Alvarez, P. R. Shapiro, and H. Martel
The Equilibrium Structure of Cosmological Halos 146
    I. T. Iliev and P. R. Shapiro
Self-Interacting Dark Matter with Flavor Mixing 149
    M. V. Medvedev
Off-Axis Cluster Mergers 152
    P. M. Ricker and C. L. Sarazin

CHAPTER 4:
COSMIC MICROWAVE BACKGROUND

Images of the Early Universe from the BOOMERanG Experiment
(Plenary Talk) 157
    P. de Bernardis, P. A. R. Ade, J. J. Bock, J. R. Bond, J. Borrill,
    A. Boscaleri, K. Coble, B. P. Crill, G. De Gasperis, G. De Troia,
    P. C. Farese, P. G. Ferreira, K. Ganga, M. Giacometti, E. Hivon,
    V. V. Hristov, A. Iacoangeli, A. H. Jaffe, A. E. Lange, L. Martinis,
    S. Masi, P. Mason, P. D. Mauskopf, A. Melchiorri, L. Miglio,
    T. Montroy, C. B. Netterfield, E. Pascale, F. Piacentini, G. Polenta,
    D. Pogosyan, S. Prunet, S. Rao, G. Romeo, J. E. Ruhl, F. Scaramuzzi,
    D. Sforna, and N. Vittorio

AMiBA: Array for Microwave Background Anisotropy 172
    K. Y. Lo, T. H. Chiueh, R. N. Martin, K.-W. Ng, H. Liang,
    U.-L. Pen, C.-P. Ma, M. Kesteven, R. Sault, R. Subrahmanyan,
    W. Wilson, and J. Peterson
First Results from the CBI 178
    B. S. Mason, T. J. Pearson, A. C. S. Readhead, M. Shepherd, J. Sievers,
    P. Udomprasert, J. K. Cartwright, and S. Padin
Galactic Synchrotron versus CMB Polarized Signal in the
Radiowavelength Region 184
    M. Tucci, E. Carretti, S. Cecchini, S. Cortiglioni, R. Fabbri, and
    E. Pierpaoli
Revisiting Kinetic Sunyaev Zel'dovich Effect 190
    P. Zhang and U.-L. Pen
The Role of CMB Polarization in Constraining Primordial
Non-Adiabatic Fluctuations 196
    M. Bucher, K. Moodley, and N. Turok
Interpreting CMB Anisotropy Observations: Trying to Tell the Truth
with Statistics 202
    E. Gawiser
Separation of E and B Patterns in the Local CMB Polarization Map 208
    T. Chiueh and C.-J. Ma
New Method of Extracting non-Gaussian Signals in the CMB 211
    J.-H. P. Wu
MAXIMA: Millimeter-wave Anisotropy Experiment Imaging Array 214
    C. Winant, M. Abroe, P. Ade, A. Balbi, D. Barbosa, J. Bock, J. Borrill,
    A. Boscaleri, J. Collins, P. de Bernardis, P. Ferreira, S. Hanany, V. Hristov,
    A. H. Jaffe, B. Johnson, A. E. Lange, A. T. Lee, P. D. Mauskopf,
    C. B. Netterfield, S. Oh, E. Pascale, B. Rabii, P. L. Richards, R. Stompor,
    G. F. Smoot, and J. H. P. Wu

CHAPTER 5:
LARGE-SCALE STRUCTURE, GALAXY FORMATION, AND EVOLUTION OF THE INTERGALACTIC MEDIUM

Cosmological Reionization *(Plenary Talk)* 219
    P. R. Shapiro
Gravitational Lensing: Recent Progress and Future Goals
(Plenary Talk) 233
    G. Soucail
Measuring Large-Scale Structure with the 2dF Galaxy Redshift
Survey *(Plenary Talk)* 245
    J. A. Peacock and the 2dFGRS Team
The Origin of the Magnetic Fields of the Universe:
The Plasma Astrophysics of the Free Energy of the Universe 259
    S. A. Colgate, H. Li, and V. I. Pariev
Visualization of Gravitational Lenses 262
    F. Frutos Alfaro

Explosions and Outflows during Galaxy Formation 265
    H. Martel and P. R. Shapiro
Cosmological Parameter Survey Using the Gravitational Lensing
Method 268
    P. Premadi, H. Martel, R. Matzner, and T. Futamase
Cosmological Matter Perturbations with Causal Seeds....................271
    J.-H. P. Wu
Cluster Abundance and Large Scale Structure 274
    J.-H. P. Wu

CHAPTER 6:
PROBES OF THE ACCELERATING UNIVERSE

The Quintessential Universe *(Plenary Talk)*................................279
    P. J. Steinhardt
Constraining the Properties of Dark Energy 297
    D. Huterer and M. S. Turner
Clusters in the Precision Cosmology Era 303
    Z. Haiman, J. J. Mohr, and G. P. Holder
A Local Void and the Accelerating Supernovae..........................310
    K. Tomita
Value of the Cosmological Constant: Theory versus Experiment 316
    M. Carmeli and T. Kuzmenko
Feasibility of Reconstructing the Quintessential Potential
Using SNIa Data 319
    T. Chiba and T. Nakamura
Constraints on Ω_m and Ω_Λ from Future Cluster Surveys 322
    G. Holder, Z. Haiman, and J. Mohr
Emergence of Discrete Structure Scales in Q-component Matter
Background 325
    M. P. Leubner
Conformal Gravity and a Naturally Small Cosmological Constant 328
    P. D. Mannheim
The Vacuum Energy from a New Perspective 331
    A. R. Mondragon and R. E. Allen

CHAPTER 7:
NUCLEOSYNTHESIS IN THE BIG BANG AND FIRST STARS

The Early Formation, Evolution and Age of the Neutron-Capture
Elements in the Early Galaxy...337
    J. J. Cowan, C. Sneden, and J. W. Truran
Coalescing Neutron Star Binaries and the First Stars 343
    S. Rosswog, C. Freiburghaus, F.-K. Thielemann, and M. B. Davies

Neutrino Heating in an Inhomogeneous Big Bang Nucleosynthesis
Model 349
    J. F. Lara
Primordial Nucleosynthesis with Hadron Injection from Low-mass
Primordial Black Holes 355
    J. Yokoyama and K. Kohri

CHAPTER 8:
SUPERMASSIVE BLACK HOLES AND GALAXY EVOLUTION

Supermassive Black Holes in Galactic Nuclei *(Plenary Talk)* 363
    J. Kormendy and K. Gebhardt
The X-ray Background and the Census of Quasars *(Plenary Talk)* 382
    A. J. Barger
Discovery of a 2 kpc Binary Quasar 394
    G. A. Shields, V. Junkkarinen, E. A. Beaver, E. M. Burbidge, R. D. Cohen,
    F. Hamann, and R. W. Lyons
Black Holes and Galaxy Metamorphosis................................. 400
    J. K. Holley-Bockelmann

CHAPTER 9:
ACCELERATION AND COLLIMATION OF COSMIC JETS

High Energy Phenomena in Blazars *(Plenary Talk)* 409
    L. Maraschi
Jet Power and Jet Suppression: The Role of Disk Structure and
Black Hole Rotation 420
    D. L. Meier
Poynting Jets from Accretion Disks 426
    R. V. E. Lovelace, H. Li, G. V. Ustyugova, M. M. Romanova, and
    S. A. Colgate
Magnetorotational Explosion: Results of 2D Simulations................. 433
    S. G. Moiseenko, G. S. Bisnovatyi-Kogan, and N. V. Ardeljan
Jet Formation from Rotating Magnetized Objects 439
    G. S. Bisnovatyi-Kogan, N. V. Ardelyan, and S. G. Moiseenko
Do Active Galactic Nucleus Jets Consist of a Pair Plasma?.............. 445
    K. Hirotani
Collimated Energy-Momentum Extraction from Rotating Black Holes
in Quasars and Microquasars Using the Penrose Mechanism 448
    R. K. Williams
A New Optically Thin Accretion Disk Model with Powerful
Electron-Positron Outflow 454
    T. Yamasaki, F. Takahara, and M. Kusunose

CHAPTER 10:
SUPERNOVAE

**Aspherical Supernova Explosions: Hydrodynamics,
Radiation Transport, and Observational Consequences** *(Plenary Talk)*.........459
    P. Höflich, A. Khokhlov, and L. Wang
**General Relativistic Simulations of Stellar Core Collapse and
Postbounce Evolution with Boltzmann Neutrino Transport**..................472
    M. Liebendörfer, O. E. B. Messer, A. Mezzacappa, and W. R. Hix
Hypernovae and Gamma Ray Bursts..478
    P. A. Mazzali, K. Nomoto, K. Maeda, and T. Nakamura
Simulations of Astrophysical Fluid Instabilities...........................484
    A. C. Calder, B. Fryxell, R. Rosner, L. J. Dursi, K. Olson, P. M. Ricker,
    F. X. Timmes, M. Zingale, P. MacNeice, and H. M. Tufo
Large Lepton Mixing and SN 1987A..487
    M. Kachelrieß, R. Tomàs, and J. W. F. Valle
Quenching Processes in Flame-Vortex Interactions.........................490
    M. Zingale, J. C. Niemeyer, F. X. Timmes, L. J. Dursi, A. C. Calder,
    B. Fryxell, D. Q. Lamb, P. MacNeice, K. Olson, P. M. Ricker, R. Rosner,
    J. W. Truran, and H. M. Tufo

CHAPTER 11:
PULSARS, MAGNETARS, AND SUPERNOVA REMNANTS

Gamma-Ray Bursts from Extragalactic Magnetar Flares...................495
    R. C. Duncan
**Long-Term *RXTE* Monitoring of the Anomalous X-ray Pulsar
1E 1048.1-5937**...501
    V. M. Kaspi, F. P. Gavriil, D. Chakrabarty, J. R. Lackey, and M. P. Muno
Testing Neutron Star Thermal Evolution Theories.........................507
    S. Tsuruta and M. A. Teter
Ubiquity: Relativistic Winds from Young Rotation-Powered Pulsars..........513
    E. V. Gotthelf
Interaction of Evolved Pulsars and Magnetars with the ISM................519
    M. M. Romanova, O. D. Toropina, Y. M. Toropin, and R. V. E. Lovelace
New Millisecond Pulsars in Globular Clusters.............................526
    N. D'Amico, A. Possenti, R. N. Manchester, J. Sarkissian, A. G. Lyne,
    and F. Camilo
Gamma-Ray Emission from Pulsar Magnetospheres.......................532
    K. Hirotani and S. Shibata
**Cyclotron-annihilation Imprints of Magnetized Vacuum in
MeV Emission from Neutron Stars**..538
    A. A. Belyanin, V. V. Kocharovsky, Vl. V. Kocharovsky
Large Redshifts from Compact Objects....................................541
    S. Chatterjee
Bardeen-Petterson Effect and QPOs in Low-Mass X-Ray Binaries...........544
    P. C. Fragile, G. J. Mathews, and J. R. Wilson

Galactic Population of Radio and Gamma-Ray Pulsars....................547
    P. L. Gonthier, M. S. Ouelette, S. O'Brien, J. Berrier, and A. K. Harding
Exact Solutions with w-Modes: Trapping of Gravitational Waves
Inside Neutron Stars..550
    M. Ishak, L. Chamandy, and K. Lake
Simulations of Glitches in Pulsars 553
    M. B. Larson and B. Link
On Companion-Induced Off-Center Supernova-Like Explosions 556
    C.-H. Lee, I. Yi, and H. K. Lee
Self-Similar Hot Accretion Flow onto a Neutron Star......................559
    M. V. Medvedev
Isolated Neutron Stars—Optical Non-Thermal Phenomenology 562
    A. Shearer, A. Gordon, P. O'Connor, and R. Butler
"Inverse Mapping" of Non-Thermal Optical Emission from Isolated
Neutron Stars 565
    P. O'Connor, A. Shearer, A. Golden, and S. Eikenberry
Magnetic Domain in Magnetar-Matters and Soft Gamma Repeaters..........569
    I.-S. Suh and G. J. Mathews

CHAPTER 12:
GAMMA-RAY BURSTS

Gamma-Ray Bursts—When Theory Meets Observations
(Plenary Talk)..575
    T. Piran
GRB and Environment Interaction 587
    P. Mészáros
A Survey of the Host Galaxies of Gamma-Ray Bursts 593
    S. Holland
Construction and Preliminary Application of the
Variability → Luminosity Estimator 599
    D. E. Reichart and D. Q. Lamb
Gamma-Ray Bursts as a Probe of Cosmology 605
    D. Q. Lamb and D. E. Reichart
Gamma-Ray Bursts Statistical Properties and Limitations on the
Physical Model 611
    G. S. Bisnovatyi-Kogan
Modification of Relativistic Jets in the Presence
of Neutron Component..614
    E. V. Derishev, V. V. Kocharovsky, and Vl. V. Kocharovsky
Cascaded sub-TeV Emission from Gamma-Ray Bursts:
Origin and Observational Tests 617
    E. V. Derishev, V. V. Kocharovsky, and Vl. V. Kocharovsky
Effects of Self-Compton Cooling on the Synchrotron Spectrum
of GRBs 620
    Vl. V. Kocharovsky, E. V. Derishev, V. V. Kocharovsky, and P. Mészáros

An Analytic Model of Gamma-Ray Burst Pulse Profiles 623
    D. Kocevski and E. P. Liang
GRBs from Unstable Poynting-Dominated Outflows 626
    M. Lyutikov
Generation of Magnetic Fields and Jitter Radiation in GRBs.
I. Kinetic Theory 629
    M. V. Medvedev
The Kinematics of the Lag-Luminosity Relationship 632
    J. D. Salmonson
Near-infrared Polarimetric Observations of the Afterglow of
GRB 000301C 635
    B. Stecklum, O. Fischer, S. Klose, R. Mundt, and C. Bailer-Jones
The Trans-Relativistic Blast Wave Model for SN 1998bw and
GRB 980425 638
    J. C. Tan, C. D. Matzner, and C. F. McKee

CHAPTER 13:
MAGNETIC ACCRETION INTO BLACK HOLES

X-rays and Accretion Discs as Probes of the Strong Gravity of
Black Holes *(Plenary Talk)* 643
    A. C. Fabian
Convection in Radiatively Inefficient Black Hole Accretion Flows
(Plenary Talk) 656
    I. V. Igumenshchev and M. A. Abramowicz
Magnetohydrodynamic Simulations of Black Hole Accretion 668
    C. S. Reynolds, P. J. Armitage, and J. Chiang
Magnetic Stresses at the Inner Edges of Accretion Disks Around
Black Holes 674
    J. H. Krolik
Magnetohydrodynamic Turbulence in Warped Accretion Discs 681
    U. Torkelsson, G. I. Ogilvie, A. Brandenburg, J. E. Pringle, Å. Nordlund,
    and R. F. Stein
Formation and Dynamics of Neutron Haloes in Disk Accreting
Black Holes 687
    A. A. Belyanin and E. V. Derishev
Shot Noise in USA Lightcurves of XTE J1118+480 and Cygnus X-1 690
    W. B. Focke, E. D. Bloom, B. Giebels, G. Godfrey, K. T. Reilly,
    P. Saz Parkinson, G. Shabad, K. S. Wood, P. S. Ray, R. M. Bandyopadhyay,
    M. T. Wolff, G. Fritz, P. Hertz, M. P. Kowalski, M. N. Lovellette,
    D. Yentis, and J. D. Scargle
Magnetic Flux Through a Slightly Charged Kerr Black Hole 693
    C. H. Lee, H. Kim, and H. K. Lee
Relativistic Slim Disk Model for NLS1s 696
    T. Manmoto and S. Mineshige

Estimation of Relativistic Accretion Disk Parameters from Iron
Line Emission 699
    V. I. Pariev, B. C. Bromley, and W. A. Miller
Probing Black Holes with Constellation-X 702
    K. A. Weaver

CHAPTER 14:
NUMERICAL RELATIVITY AND BLACK HOLE COLLISIONS

Recent Developments in Classical Relativity *(Plenary Talk)* 707
    B. G. Schmidt
Binary Neutron Star Mergers in Fully General Relativistic
Simulations *(Plenary Talk)* 717
    M. Shibata and K. Uryū
Numerical Evolution of the Kruskal Spacetime Using the Conformal
Field Equations 729
    B. G. Schmidt
Semiglobal Numerical Calculations of Asymptotically Minkowski
Spacetimes 734
    S. Husa
Simulations of Black Hole Binaries: Providing Initial Data 740
    P. Marronetti and R. A. Matzner
The Lazarus Project: Plunge Waveforms from Inspiralling
Black Holes 746
    J. Baker, B. Brügmann, M. Campanelli, C. Lousto, and R. Takahashi
Critical Phenomena Associated with Boson Stars 751
    S. H. Hawley and M. W. Choptuik
Gravitational Waves from Rotational Core Collapse in the
Conformally Flat Spacetime Approximation 757
    H. Dimmelmeier, J. A. Font, and E. Müller
Application of the Galerkin Method to the Problem of
Stellar Stability 760
    A. V. Dorodnitsyn and G. S. Bisnovatyi-Kogan
Light Cone Consistency in Bimetric General Relativity 763
    J. B. Pitts and W. C. Schieve
Dynamical Bar Instability in Relativistic Rotating Stars 766
    M. Saijo, M. Shibata, T. W. Baumgarte, and S. L. Shapiro
The Ultimate Future of the Universe, Black Hole Event Horizon
Topologies, Holography, and the Value of the Cosmological Constant 769
    F. J. Tipler

CHAPTER 15:
GRAVITY WAVE SIGNATURES

Post-Newtonian SPH Simulations of Binary Neutron Stars 775
    J. A. Faber and F. A. Rasio

Gravitational Radiation Evolution of Accreting Neutron Stars 781
    R. V. Wagoner, J. F. Hennawi, and J. Liu
Gravitational Wave Background from Coalescing Compact Stars in
Eccentric Orbits 787
    A. G. Kuranov, V. B. Ignatiev, K. A. Postnov, and M. E. Prokhorov
Detecting Eccentric Globular Cluster Binaries with LISA 793
    M. Benacquista
Coalescing Binary Neutron Star Systems 796
    A. C. Calder, F. D. Swesty, and E. Y. M. Wang
Quasi-Normal Modes in Schwarzschild anti-de Sitter Spacetimes 799
    V. Cardoso and J. P. S. Lemos
Constraining Post-Newtonian Parameters with Gravitational Waves 802
    J. S. Graber
KiloHertz QPO and Gravitational Wave Emission as the Signature of
the Rotation and Precession of a LMXB Neutron Star Near Breakup 805
    J. G. Jernigan
A Semi-analytic Model for the Radiation Reaction Luminosity for
Post-Newtonian Binary Neutron Star Mergers 808
    F. D. Swesty and A. C. Calder
Detection of Gravitational Waves from Gravitationally Lensed
Systems 811
    T. Wickramasinghe and M. Benacquista

CHAPTER 16:
ULTRA-HIGH ENERGY COSMIC RAYS

The Origin of Ultra High Energy Cosmic Rays: What We Know Now
and What the Future Holds *(Plenary Talk)* 817
    A. A. Watson
Recent Results from the High Resolution Fly's Eye Detector 827
    D. Bergman for the HiRes Collaboration
Probing Interactions beyond the Electroweak Scale with
Ultra High Energy Cosmic Radiation 832
    G. Sigl
Ultrahigh Energy Neutrinos as Probe for Weak-scale String Theories? 838
    M. Kachelrieß
Ultra-High-Energy Cosmic Rays from Relic Topological Defects 844
    K. D. Olum and J. J. Blanco-Pillado
Cen A as the Source of Ultrahigh Energy Cosmic Rays 850
    T. Piran and G. R. Farrar
On a Mechanism of Highest-Energy Cosmic Ray Acceleration 856
    C. Litwin and R. Rosner
Cerenkov Light from Cosmic Rays: a Comparison of Different
Parametrizations 862
    R. Barná, A. Butkevich, V. D'Amico, D. De Pasquale, A. Italiano,
    A. Trifiró, and M. Trimarchi

Pressure Imbalance of FRII Radio Source Lobes: a Role of Energetic
Proton Population...865
    M. Ostrowski and M. Sikora
GeV γ-ray Astronomy with STACEE-64................................868
    R. A. Scalzo, L. M. Boone, C. E. Covault, P. Fortin, D. Gingrich,
    D. S. Hanna, J. A. Hinton, R. Mukherjee, R. A. Ong, S. Oser, K. Ragan,
    D. R. Schuette, C. G. Theoret, and D. A. Williams

CHAPTER 17:
MISCELLANEOUS TOPICS

Neutrino Factory Detector and Long Baseline Oscillations 873
    E. J. Fenyves and R. F. Burkart
Magnetic Support and Alfvén Heating in Quasar Broad-Line Region
Clouds..876
    D. R. Gonçalves, A. C. S. Friaça, and V. Jatenco-Pereira
Quantum Singularity of Spacetimes with Dislocations and
Disclinations...879
    D. A. Konkowski and T. M. Helliwell
From Stellar Entropy to Black Hole Entropy........................882
    M. Mbonye, F. C. Adams, and M. J. Perry
GEST 885
    S. H. Rhie

CHAPTER 18:
CONFERENCE SUMMARY

Conference Summary..893
    S. Weinberg

APPENDICES
APPENDIX A: AFTER DINNER REMARKS

Texas Symposium 20..911
    W. Schild
20th Texas Symposium 913
    E. L. Schucking

APPENDIX B: REPRINT OF THE REPORT OF
THE FIRST TEXAS SYMPOSIUM IN AUSTIN

Relativistic Astrophysics: A Report on the Second Texas Symposium 915
    I. Robinson, A. Schild, and E. L. Schucking

APPENDIX C: THE FORMATION OF A BLACK HOLE

Recipe for Black Hole Formation... 922
    H. Martel and J. C. Wheeler

List of Attendees... 925
Author Index.. 933

PREFACE

For the 20th Symposium in the year 2000, the Texas Symposium on Relativistic Astrophysics returned to Austin and to its Texas roots. We took advantage of that numerology and the Austin locale to design the logo. In a further bit of amusing numerology, the 2nd Texas Symposium, in 1964, was also the first to be held in Austin. We reprint here the proceedings of that second symposium to show how that meeting highlighted the exciting topics of the day, for instance the dreams of a physical chemist named Ray Davis to detect solar neutrinos and the early stirrings of X-ray astronomy. In the intervening 36 years, there has been immense progress. The Texas Symposia have become a hallmark of modern astrophysics by celebrating that progress with biennial reports from the forefront of relativistic astrophysics. What a ride it has been. We believe that this symposium continued that spirit with discussions of currently cutting-edge topics from brane worlds to dark matter and dark energy to gamma-ray bursts.

We were honored to have two of the original organizers of the Texas Symposia, Ivor Robinson and Englebert Schucking, in attendance and, especially, Winnie Schild, representing her husband, Al, who built relativity at Texas and was the founding father of the symposia. We are delighted to have the after-dinner remarks of Winnie and Englebert as part of this record.

The Symposium has always been a place to highlight progress since the last meeting and more recent results with plenary talks by especially distinguished people. We are grateful to the plenary speakers at this meeting for a truly stimulating program. There is also a need to give a venue to a wider array of presentations and we accommodated that with a series of parallel sessions in the afternoon and with posters that were up for the whole length of the meeting. In addition, we scheduled "student lunches" so that young people, graduate students and postdocs, could sit with some of the biggest names in the business and chat. By the feedback we got, this was a very successful enterprise and we recommend it to all meeting organizers.

A great number of people supply the effort to make a meeting like this work. The International Scientific Advisory Committee provided excellent help to choose and organize the program. The Local Organizing Committee helped both with the program and with meeting arrangements.

Natasha Papousek got us off the ground with early organization and the launching of the web page. Amy Hendrick came on board in mid-stream and did a wonderful job of ensuring everything worked. As all who attended know, Amy was the backbone who made the meeting happen. Sheryl Anderson helped with the web page. Kim Orr provided a crucial boost with pre-meeting details and the smooth functioning of the meeting itself as did Jennifer Bieniek.

You cannot have a Texas Symposium without a snappy logo for the web page and T-shirts. We are greatly indebted to Tim Jones, Senior Artist at StarDate Magazine, for his inspired design incorporating key elements of the meeting theme and numerology featuring Dill, the Relativistic Armadillo.

Peter Höflich loaded our borrowed computers with Linux, set them up, and made sure they worked in our "internet cafe." Steve Maran volunteered his time to bring his great expertise to our press relations. An energetic crew of undergraduate members of the University of Texas Astronomy Student's Association and graduate students and postdocs from the Departments of Astronomy, Physics, and Computer Science helped in innumerable ways. We are also especially grateful to the volunteers from the Austin Astronomical Society who helped with registration and myriad other tasks. There were also many people at the University of Texas who helped behind the scenes with setting up accounts and other aid for the meeting.

The 20th Texas Symposium was made possible by generous financial support from the National Science Foundation, the National Aeronautics and Space Administration, the Department of Energy, and the Office of the Vice President for Research, the College of Natural Sciences, the Department of Astronomy and McDonald Observatory of the University of Texas at Austin. We are grateful to Provost Sheldon Ecklund Olson, Vice President for Research Juan Sanchez, Director of McDonald Observatory Frank Bash and Chair of the Department of Astronomy Chris Sneden for their support and for opening remarks at the meeting.

We are especially grateful for the personal generosity of David Chappel, Donald R. Counts, M. D., Richard C. Evans, Robert Cole Grable, C. Hastings Johnson and Robert Jorrie.

J. Craig Wheeler
Hugo Martel

TWENTIETH TEXAS SYMPOSIUM ON RELATIVISTIC ASTROPHYSICS

INTERNATIONAL ORGANIZING COMMITTEE

J. Audouze, J. D. Barrow, P. G. Bergmann, J. Ehlers, E. J. Fenyves, J. Frieman, T. Montmerle, L. Mestel, Y. Ne'eman, I. Ozvath, F. Pacini, I. Robinson, R. Ruffini, B. Sadoulet, E. Schucking, G. Setti, J. Silk, M. M. Shapiro, L. C. Shepley, J. Stachel, A. Trautman, V. Trimble, J. Trümper, S. Weinberg, J. A. Wheeler, J. C. Wheeler

INTERNATIONAL SCIENTIFIC ADVISORY COMMITTEE

Lars Bildsten, Roger Blandford, Blas Cabrera, Claude Canizares, Catherine Cesarsky, Marc Davis, Nathalie DeRuelle, Wendy Freeman, Margaret Geller, Steve Kahn, Vicki Kaspi, Edward Kolb, Ramesh Narayan, Ken'ichi Nomoto, Igor Novikov, Jerry Ostriker, Franco Pacini, Tsvi Piran, Martin Rees, Bernie Schutz, Paul Steinhardt, Meg Urry

LOCAL ORGANIZING COMMITTEE

Frank Bash, Robert Duncan, Peter Höflich, John Kormendy, Cecil Martinez, Richard Matzner, Hugo Martel, Sandra Preston, Roy Schwitters, Paul Shapiro, Lawrence Shepley, Gregory Shields, Steven Weinberg, J. Craig Wheeler (Chair)

WORKSHOP ORGANIZERS

Robert Duncan, Eanne Flanagan, Karl Gebhardt, Andrew Jaffe, Julien Krolik, Pablo Laguna, Donald Lamb, Andrei Linde, David Meier, Ken'ichi Nomoto, Angela Olinto, Brian Schmidt, Christopher Sneden, David Spergel, Paul Steinhardt, Robert Wagoner

CHAPTER 1

BRANE-WORLD THEORIES: COSMOLOGY AND ASTROPHYSICS

The Current State of String Theory

Gary T. Horowitz*

*Physics Department, University of California, Santa Barbara, CA 93111

Abstract. String theory is a promising candidate for a quantum theory of gravity and a unified theory of all forces and particles. I will briefly summarize the current status of this theory and discuss some recent results. The most important results involve describing quantum states of black holes in terms of strings, which has provided a fundamental explanation of black hole entropy and Hawking radiation. Further investigation of this connection between black holes and strings has led to a completely new nonperturbative formulation of string theory.

I INTRODUCTION

Two of the greatest achievements of physics in the last century were Einstein's general theory of relativity and quantum theory. Each of these theories has been extremely well tested and has been very successful. However, they are mutually incompatible. Thus our basic understanding of nature is not only incomplete – but inconsistent. One clearly needs a new theory, quantum gravity, which incorporates the principles of both of these theories and reduces to them in appropriate limits.

At first sight, the problem of constructing a quantum theory of gravity sounds easy since there are no experimental constraints! The task is simply to find any theory which unifies general relativity and quantum theory. However, on second thought, the problem sounds extremely difficult. General relativity teaches us that gravity is just a manifestation of the curvature of space and time. So quantum gravity must involve the quantization of space and time, something we have no previous experience with.

Surprisingly, even though there are no experimental constraints, there is a constraint on quantum gravity which was found in the early 1970's by studying black holes. Motivated by the close analogy between the laws of black hole mechanics and ordinary thermodynamics, Bekenstein proposed that black holes have an entropy proportional to their horizon area A [1]. Then Hawking showed that if matter is treated quantum mechanically (but gravity remains classical), black holes emit thermal radiation with a temperature $T = \hbar\kappa/2\pi$ where κ is the surface gravity of the black hole [2]. This confirmed Bekenstein's ideas and fixed the coefficient:

$$S_{bh} = \frac{A}{4G\hbar} \tag{1}$$

This is an enormous entropy – much larger than the entropy in the matter that collapsed to form the black hole. In a more fundamental statistical description, the entropy should be a measure of the log of the number of accessible states. So a constraint on any candidate quantum theory of gravity is to show that the number of quantum states associated with a black hole is indeed $e^{S_{bh}}$. We will see that there has been considerable progress in satisfying this constraint recently.

Over the years, there has been much discussion of possible consequences of quantum gravity. Let me comment on a few of the most popular:

- *Quantum gravity will smooth out spacetime singularities*

This is false as stated. Quantum gravity cannot smooth out all singularities. Some timelike singularities must remain, such as the one in the $M < 0$ Schwarzschild solution. This can be seen from the following simple argument [3]. Since quantum gravity must reduce to general relativity for weak fields (and large number of quanta), there must be solutions which look like $M < 0$ Schwarzschild at large radii, for any M. At small radii, the curvature becomes strong and the solution may be significantly modified. But if it is not singular in some sense, it would represent a state in the theory with negative total energy. Since the energy is unbounded from below, there would be no stable vacuum state.

- *Quantum gravity will allow the topology of space to change*

This is almost certainly true. There are semi-classical calculations of pair creation of magnetically charged black holes in a background magnetic field [4]. One can reliably calculate this rate for black holes with size much larger than the Planck scale. It is extremely small, but it does change the topology of space from R^3 to R^3 with an $S^2 \times S^1$ wormhole attached. This is because the black holes are created with their horizons identified. It is interesting to note that if one compares the rate of black hole pair creation to the rate of magnetic monopole creation, one sees an enhancement in the black hole case of $e^{S_{bh}}$ in line with the expectation that there are $e^{S_{bh}}$ different species of black holes [5]. We will see further evidence for topology change shortly.

- *Black hole evaporation will violate quantum mechanics: pure states will evolve to mixed states*

Since black holes can be formed from matter in a pure state and the radiation emitted is thermal (in the semiclassical approximation) it was thought that black hole evaporation would cause pure states to evolve into mixed states. There is recent evidence (discussed below) that this is false. In a more exact treatment there are likely to be correlations between the radiation emitted at early time and later time which ensure that the evolution is unitary.

- *Space and time will not be fundamental, but derived properties.*

This is likely to be true, but the key question is, what replaces them?

II STRING THEORY

Over the past fifteen years, there has been significant progress in constructing a quantum theory of gravity called string theory[1]. (There is another approach to quantum gravity [7] which will not be discussed here.) String theory is not only a promising candidate for a quantum theory of gravity, but may also be a unified theory of all forces and particles.

String theory starts with the idea that all elementary particles are not pointlike, but excitations of a one dimensional string. If one quantizes a free relativistic string in flat spacetime one finds a infinite tower of modes of increasing mass. There is a massless spin two mode which is identified with a linearized graviton. Next, one postulates a simple splitting and joining interaction between strings and finds, remarkably, that this reproduces the perturbative expansion of general relativity. One then adds fermionic degrees of freedom to the string so the theory is supersymmetric. This makes the theory better behaved and calculations easier to control. The consistent quantization of a string turns out to impose a constraint on the spacetime dimension. In most cases, one needs ten spacetime dimensions. Contact with observations obviously requires that six of these dimensions are unobservable. The simplest possibility is the old Kaluza-Klein idea that these six dimensions are wrapped up in a small compact manifold[2]. The natural size of this compact manifold is the string length, a new dimensionful parameter in the theory set by the string tension. The string length, ℓ_s, is related to the Planck length, ℓ_p, by a power of the (dimensionless) string coupling g. In ten dimensions, $\ell_p = g^{1/4}\ell_s$.

Since the string represents fluctuations about the background spacetime it is propagating in, the above description is strictly perturbative. String theory has now progressed far beyond this perturbative beginning. The current status of string theory is roughly the following.

1. Classical theory is well understood. The classical equations resemble general relativity with an infinite series of correction terms involving higher powers and derivatives of the curvature. Since the correction terms become important only when the curvature is of order the string scale, any solution to general relativity with curvature smaller than ℓ_s^{-2} is an approximate classical solution to string theory. To obtain exact solutions, it is useful to note that the classical field equations arise from the vanishing of the conformal anomaly in a certain two dimensional field theory. Several classes of exact solutions are known (often using special properties to ensure that the higher order correction terms vanish). One class, based on supersymmetry, includes compact Ricci flat six manifolds known as Calabi-Yau spaces which can be used to compactify spacetime down to four dimensions. Other classes include exact plane waves and group manifolds.

---

[1] A good general reference for string theory including many of the results discussed below is [6].
[2] Other possibilities are also being explored [8,9].

2. Quantum perturbation theory is well understood and well behaved. In particular, it is finite order by order in the loop expansion. Thus string theory provides a perturbatively finite quantum theory of gravity. However, the perturbation theory does not determine the quantum theory uniquely. Nonperturbative effects are important. This is evident in the fact that the loop expansion does not converge.

3. Some nonperturbative properties are known. These are mostly through clever use of supersymmetry, which guarantees that certain properties which are valid at weak coupling must continue to hold at strong coupling. In particular, extended objects (called branes) play an important role in the theory. At the perturbative level, there are five different string theories in ten dimensions which differ in the amount of supersymmetry and fundamental gauge groups they contain. There is convincing evidence that the strong coupling limit of one theory is equivalent to the weak coupling limit of another theory. These are known as duality symmetries. In fact, it is now believed that all of these theories can be obtained from a single eleven dimensional theory[3].

4. There exists a complete nonperturbative formulation of the theory for certain boundary conditions. This will be discussed further below.

Up until five years ago, the status of string theory was essentially just the first two points above. Point three has an interesting consequence. Since string theory includes gravity and strong coupling implies strong gravitational fields, one might have expected the strong coupling limit of string theory to have large fluctuations of space and time, i.e., spacetime foam. But this does not seem to be the case. It appears that the strong coupling limit of the theory can be described in terms of a weakly coupled theory in new variables. There is no evidence for spacetime foam.

III RESULTS SO FAR

Now I turn to discuss some results in string theory.

A) *Singularities:* The first thing to note is that the definition of a singularity is different in string theory than in general relativity, even classically. In general relativity, we usually define a singularity in terms of geodesic incompleteness which is based on the motion of test particles. In string theory, we must use test strings. So a spacetime is considered singular if test strings are not well behaved[4]. It turns out that some spacetimes which are singular in general relativity are completely nonsingular in string theory. A simple example is the quotient of Euclidean space by a discrete subgroup of the rotation group. The resulting space, called an orbifold, has a conical singularity at the origin. Even though this leads to geodesic

---

[3] This is often called M-theory, but I will continue to refer to this approach as string theory.
[4] Strictly speaking, one should also require that the other extended objects in the theory –branes – have well behaved propagation.

incompleteness in general relativity, it is completely harmless in string theory. This is essentially because strings are extended objects.

The orbifold has a very mild singularity, but even curvature singularities can be harmless in string theory. As mentioned above, string theory has exact solutions which are the product of four dimensional Minkowski space, and a compact Calabi-Yau space. A given Calabi-Yau manifold usually admits a whole family of Ricci flat metrics. So one can construct a solution in which the four large dimensions stay approximately flat and the geometry of the Calabi-Yau manifold changes slowly from one Ricci flat metric to another. In this process the Calabi-Yau space can develop a curvature singularity. In many cases, this can be viewed as arising from a topologically nontrivial S^2 or S^3 being shrunk down to zero area. It has been shown that when this happens, string theory remains completely well defined. The evolution continues through the geometrical singularity to a nonsingular Calabi-Yau space on the other side [10,11].

The reason this happens is roughly the following. There are extra degrees of freedom in the theory associated with branes wrapped around topologically nontrivial surfaces. As long as the area of the surface is nonzero, these degrees of freedom are massive, and it is consistent to ignore them. However when the surface shrinks to zero volume these degrees of freedom become massless, and one must include them in the analysis. When this is done, the theory is nonsingular.

The above singularities are all in the extra spatial dimensions. However other singularities which involve time in a crucial way have also been shown to be harmless. Putting many branes on top of each other produces a gravitational field which often has a curvature singularity at the location of the brane. It has been shown that one can understand physical processes near this singularity in terms of excitations of the branes.

Despite all this progress, we still do not yet have an understanding of the most important types of singularities: those arising from gravitational collapse or cosmology.

B) *Topology change:* It has been shown unambiguously that the topology of space can change in string theory. In fact, when one evolves through a singular Calabi-Yau space as described above, the topology of the manifold changes [10,12]. A simpler example of topology change is the following. Consider one direction in space compactified to a circle. If one identifies points under a shift $\theta \to \theta + \pi$, one obtains a circle of half the radius. If one identifies points under a reflection about a diameter, one obtains a line segment. It turns out that for a circle whose radius is the string scale, one can show these two Z_2 actions are equivalent in string theory[5]. There is no way for strings to distinguish them. So one can start with one direction compactified on a large circle, slowly shrink it down to the string scale, replace it with a line segment, and then slowly expand the line segment. As far as strings are concerned, the evolution is completely nonsingular.

---

[5] Usually, compactifying on a circle produces a $U(1)$ gauge field. At this special radius, there is an enhanced $SU(2)$ symmetry and these two Z_2 actions are conjugate subgroups.

C) *Black hole entropy:* By far the most important result is that it has been shown that string theory can satisfy the black hole constraint mentioned earlier. One starts by considering extreme black holes, i.e., charged black holes with the minimum possible mass for the given charge. Charged black holes are not of much interest astrophysically since they are expected to rapidly neutralize themselves. However, theoretically, they are very interesting since they are quantum mechanically stable. The Hawking temperature of an extreme black hole is zero. For a large class of extreme and near extreme charged black holes, one can count the number of quantum string states at weak coupling with the same mass and charge as the black hole. The answer turns out to agree exactly with the prediction made by Bekenstein and Hawking [13,14]. It is important to note that it is not just one number being reproduced. One can consider black holes with several different types of charges and angular momentum. String theory correctly reproduces the entropy as a function of all of these parameters.

The calculations are quite remarkable since one starts at weak coupling where gravitational effects are turned off and spacetime is flat. One considers configurations of branes and strings with appropriate charges and counts the number of states with a given energy. One then increases the string coupling. The gravitational field of the branes and strings becomes stronger and they eventually form a black hole. But general relativity tells us that there is only one black hole for a given mass and charge. So all of these different quantum states produce the same black hole at strong coupling. This is the origin of the thermodynamic properties of black holes. If one compares the number of states at weak coupling to the Bekenstein-Hawking entropy of the resulting black hole one finds complete agreement. It is possible to do this calculation exactly only for extremal and near extremal black holes. For more general black holes, one can show that the log of the number of string states is proportional to the area [15–17], but the coefficient is difficult to calculate.

In some cases, one can calculate the spectrum of Hawking radiation in string theory and show that it agrees with the semiclassical calculation. This is remarkable since the spectrum is not exactly thermal, but has grey body factors arising from spacetime curvature outside the horizon. These are correctly reproduced in string theory, even though the string calculation is done in flat space. The calculations look completely different, but the results agree.

D) *Nonperturbative formulation:* By studying these black hole results more closely, people were led to a new and more complete formulation of string theory. Recall that anti de Sitter spacetime is the maximally symmetric solution to Einstein's equation with negative cosmological constant. We have:

AdS/CFT Conjecture (Maldacena [18]): String theory on spacetimes which asymptotically approach the product of anti de Sitter (AdS) and a compact space, is completely described by a conformal field theory (CFT) "living on the boundary at infinity".

In particular, string theory with[6] $AdS_5 \times S^5$ boundary conditions is described by a four dimensional supersymmetric $SU(N)$ gauge theory. Since the gauge theory is defined nonperturbatively, this is a nonperturbative and (mostly) background independent formulation of string theory. A background spacetime metric only enters in the boundary conditions at infinity. The gauge theory is believed to describe all physics in spacetimes with these boundary conditions, including black holes, accretion disks, etc. The cosmological constant of the AdS space is a free parameter here and can be taken to be very small. It can be thought of as a long distance cut-off. Even though these boundary conditions are not physically the right ones, this conjecture is still of great interest since (if correct) it provides our first nonperturbative quantum theory of gravity.

At first sight this conjecture seems unbelievable. How could an ordinary field theory describe all of string theory? I don't have time to describe the impressive body of evidence in favor of this correspondence which has accumulated over the past few years. In the past three years, more than a thousand papers have been written on various aspects of this conjecture. A good review is [19].

This conjecture provides a "holographic" description of quantum gravity in that the fundamental degrees of freedom live on a lower dimensional space. The idea that quantum gravity might be holographic was first suggested by 't Hooft [20] and Susskind [21] motivated by the fact that black hole entropy is proportional to its horizon area. It also confirms earlier indications that string theory has fewer fundamental degrees of freedom than it appears in perturbation theory. This conjecture provides an answer to the longstanding question raised in the introduction: If space and time are not fundamental, what replaces them? Here the answer is that there is an auxiliary spacetime metric which is fixed by the boundary conditions at infinity. The gauge theory uses this metric, but the physical spacetime metric is a derived quantity. The dictionary relating spacetime concepts in the bulk and field theory concepts on the boundary is very incomplete, and still being developed.

This conjecture has an interesting consequence. Consider the formation and evaporation of a small black hole in a spacetime which is asymptotically $AdS_5 \times S^5$. By the AdS/CFT correspondence, this process is described by ordinary unitary evolution in the gauge theory. So black hole evaporation does not violate quantum mechanics. This is the basis for my earlier comment that the belief that black hole evaporation is not unitary is probably false.

Even though there has been enormous progress in string theory, much remains to be done. This includes finding the analog of the AdS/CFT conjecture for more realistic boundary conditions, and computing the entropy of neutral black holes exactly. Work on these and other questions is continuing.

This work was supported in part by NSF Grant PHY-0070895.

---

[6]) This is the product of five dimensional anti de Sitter spacetime and a five sphere. Recall that spacetimes in string theory have ten dimensions.

REFERENCES

1. Bekenstein, J., *Phys. Rev. D*, **7**, 2333 (1973).
2. Hawking, S., *Commun. Math. Phys.*, **43**, 199 (1975).
3. Horowitz, G., and Myers, R., *Gen. Rel. Grav.*, **27**, 915 (gr-qc/9503062) (1995).
4. Garfinkle, D., and Strominger, A., *Phys. Lett. B*, **256**, 146 (1991).
5. Garfinkle, D., Giddings, S., and Strominger, A., *Phys. Rev. D*, **49**, 958 (gr-qc/9306023) (1994).
6. Polchinski, J., *String Theory*, Cambridge University Press, (1998).
7. Rovelli, C., *Living Reviews*, 1 (gr-qc/9710008) (1998).
8. Arkani-Hamed, N., Dimopoulos, S., and Dvali, G. *Phys. Lett. B*, **429**, 263 (hep-ph/9803315) (1998).
9. Randall, L., and Sundrum, R., *Phys. Rev. Lett.*, **83**, 4690 (hep-th/9906064) (1999).
10. Aspinwall, P., Greene, B., and Morrison, D., *Nucl. Phys. B*, **416**, 414 (hep-th/9309097) (1994).
11. Strominger, A., *Nucl. Phys. B*, **451**, 96 (hep-th/9504090) (1995).
12. Greene, B., Morrison, D., and Strominger, A., *Nucl. Phys. B*, **451**, 109 (1995).
13. Strominger, A., and Vafa, C., *Phys. Lett. B*, **379**, 99 (hep-th/9601029) (1996).
14. Peet, A., preprint (hep-th/0008241) (2000).
15. Susskind, L., preprint (hep-th/9309145) (1993)
16. Horowitz, G., and Polchinski, J., *Phys. Rev. D*, **55**, 6189 (hep-th/9612146) (1997).
17. Damour, T., and Veneziano, G., *Nucl. Phys. B*, **568**, 93 (hep-th/9907030) (2000).
18. Maldacena, J., *Adv. Theor. Phys.*, **2**, 231 (hep-th/9711200) (1998).
19. Aharony, O., Gubser, S., Maldacena, J., Ooguri, H., and Oz, Y., *Phys. Rept.*, **323**, 183 (hep-th/9905111) (2000).
20. 't Hooft, G., preprint (gr-qc/9310026) (1993).
21. Susskind, L., *J. Math. Phys.*, **36**, 6377 (hep-th/9409089) (1995).

Brane-World Astronomy

Craig J. Hogan

Astronomy and Physics Departments, University of Washington, Seattle, WA 98195-1580, USA

Abstract. Unified theories suggest that space is intrinsically 10 dimensional, even though everyday phenomena seem to take place in only 3 large dimensions. In "Brane World" models, matter and radiation are localized to a "brane" which has a thickness less than $\approx (\text{TeV})^{-1}$ in all but the usual three dimensions, while gravity propagates in additional dimensions, some of which may extend as far as submillimeter scales. A brief review is presented of some of these models and their astrophysical phenomenology. One distinctive possibility is a gravitational wave background originating in the mesoscopic early universe, at temperatures above about 1 TeV and on scales smaller than a millimeter, during the formation of our 3-dimensional brane within a 10-dimensional space.

I UNIFIED THEORY

The Standard Model of strong, weak and electromagnetic interactions includes all the forms of mass-energy so far observed in nature, other than gravity. It is based on a relativistic quantum field theory of interacting fermion and boson fields, with forces arising from Yang-Mills vector gauge fields, propagating in a 3+1-dimensional spacetime. Gravity is formulated in a completely different way, using General Relativity, as a classical theory of dynamical spacetime itself: "Spacetime tells mass-energy how to move, and mass-energy tells spacetime how to curve."

Even though there is no direct inconsistency or disagreement of these theories with experiment or with each other, there is widespread dissatisfaction with the inelegance of this dualistic situation. It is suspected by those who believe in the unity of the natural world that there might be a single unified theory, derivable from simple principles of symmetry, which will appear in an appropriate mathematical limit as the Standard Model fields propagating in a General Relativistic spacetime.

Major steps have been taken recently in the construction of a unified theory; one can cite several triumphs of a theoretical nature, such as the ability to count precisely the quantum mechanical degrees of freedom of spacetime itself in certain special black hole spacetimes [1]. We may now also have observed [2–4] the first real-world phenomenon which specifically calls upon new quantum-gravity unification physics outside of "Standard Model Plus GR": the Cosmological Constant (or "Dark Energy"). Thus there is real hope that the Theory of Everything may become a real, testable physical theory. However it is not clear how to complete the most important step, connecting the fundamental theory to the real world which is so well described by "Standard Model Plus GR." The current best candidate for a Theory of Everything, "M theory", is formulated in ten spatial dimensions instead of three, and has no direct, distinctive connections with any real-world experiment.

Recent developments in M theory suggest that there may be an intermediate level of structure associated with dimensional reduction, which has spawned a wide variety of proposed designs for new "Brane World" models [5–12]: In these models, the fields of the Standard Model are confined to an approximately three-dimensional wall or "brane" imbedded in an extended ten-dimensional space or "bulk", which is described by adding extra dimensions to General Relativity. The brane has a thickness smaller than the TeV scale of current particle experiments, while the extra dimensions of the bulk can be as large as the 10^{-2}eV scale of current gravity experiments. This paper is a brief overview of brane worlds and some new effects they might produce in astrophysics.

Brane world models aim to short-cut the connection between fundamental theory and phenomena. They introduce a kind of "effective theory" as a conceptual bridge— a parametrized model broadly motivated by structures in the fundamental theory, which can be used to calculate new phenomena at low energy. Although it is not clear that this strategy will work in the long run, it has certainly broken a logjam in thinking and has spawned many intriguing new theoretical predictions and experimental tests.

II EXTRA DIMENSIONS

Direct detailed data from accelerators confirm the 3+1-dimensional behavior of Standard Model quantum fields directly to the current experimental energies of about 100 GeV, and with some modest extrapolation to about 1 TeV. That is, any effects of a fourth spatial dimension had better not appear in particle interactions unless it is on a length scale much smaller than $(1 \text{ TeV})^{-1}$.

The ideas [13,14] of the logarithmic running of couplings in standard supersymmetric grand unification (SUSY GUTs) suggest an extension to much higher energies. Renormalization group calculations allow extrapolation of the observed strong, weak and electromagnetic couplings as a function of energy; the three curves intersect at the supersymmetric grand unification scale, around 10^{16} GeV. This nontrivial intersection, which has been confirmed by increasingly precise accelerator experiments, is often cited as evidence for the unification scheme. In SUSY GUTs, the Standard Model structure including 3+1-D field theory is preserved up to this much higher energy scale.

The limits on the dimensionality of gravity are much weaker. Gauss' law tells us that the gravitational force falls off as r^{1-N_s} where N_s is the total number of spatial dimensions. Experiments [15] (motivated in part by brane world models) now confirm Newton's inverse square law (and hence $N_s = 3$) from astronomical scales down to hair's-width distances of about 250μm. This submillimeter scale is is however still vastly larger than those probed by the Standard Model fields; 0.3 mm corresponds to an energy of about 0.003 eV.

The "Cosmological Constant Problem" may be related to effects of extra dimensions. One way to state this problem (there are many) begins with the observation that the zero point fluctuations of Standard Model quantum fields on a given scale E, if they couple to gravity, correspond to a gravitating vacuum energy density of magnitude E^4. The observed vacuum energy (or Dark Energy) density is about equal to the critical density of the universe, which is about $(10^{-2}\text{eV})^4$. But the success of the Standard Model requires the presence of the field fluctuations on scales at least up to TeV scales. (Above that energy, it is possible for fermion and boson contributions to cancel exactly due to supersymmetry. Below that energy, we know that the system is not supersymmetric in today's vacuum). As presently formulated, the theory requires an offset of the zero energy level of the vacuum magically tuned to a precision of 17×4 orders of magnitude.

If for some reason the coupling of gravity to zero point modes were strongly suppressed above 10^{-2}eV, the gravitating energy of the vacuum would come out about right. The corresponding length scale, 0.1mm, lies just below the current experimental tests for gravitational coupling, but will become accessible with the next generation experiments. (Note that the predictions of the brane worlds with extra dimensions on these scales, which are discussed below, naiively have the wrong sign to solve the cosmological constant problem, since with $N_s > 3$ the gravitational force increases faster than r^{-2} on small scales).

A coincidence worth mentioning is that the gravitational timescale associated with matter at an energy density of $\rho \approx (\text{TeV})^4$ is $(G\rho)^{-1/2} \approx 1mm/c$. This coincidence is important in cosmology because it means that the uncertainty associated with the possible geometrical effects of gravity propagating in extra dimensions start at about the same place as the uncertainties associated with the possible new physics beyond the TeV scale. If there is indeed new physics at the 0.1mm scale responsible for the cosmological constant, then this also would explain the coincidence between the age of the universe when the cosmological constant starts to dominate the mass-energy (that is, about now), and the typical lifetimes of stars; that is, there may be a derivable reason why $\rho_{vacuum} \approx M_{Planck}^4 (M_{proton}/M_{Planck})^6$.

Aside from the cosmological constant, there is the corresponding Hierarchy Problem in particle physics itself: if the fundamental scale is 10^{19} GeV, how is the "light" TeV scale preserved in all orders of all interactions? (Related to this: what is the origin of Large Numbers of astrophysics, which derive from the large ratio m_{Planck}/m_{proton}?) The traditional approach is to invoke supersymmetry above the TeV scale to preserve the hierarchy, and to explain the large numbers as due to the logarithmically running couplings. In some of the new schemes however, the Planck scale is a kind of illusion; physics is fundamentally different above the ≈ 10 TeV scale, which is the only fundamental scale in the theory. The large numbers in these schemes arise from taking modest numbers to large powers; the gravitational force in ten dimensional space falls off as r^{-9}!

III EXTRA DIMENSIONS IN UNIFIED THEORY

The candidate Theory of Everything, sometimes called M theory (or supersymmetric superstring theory, or matrix theory, etc., depending on the limit and the context), consistently includes both quantum mechanics and general relativity, and possibly includes the Standard Model. It provides a framework for computing statistically the entropy of certain black holes from first principles. A hallmark of the theory is a formal melding and blurring of the distinction between string (and particle) degrees of freedom, and geometrical degrees of freedom. Powerful dualities are exploited to show the equivalence of different formulations and between large and small scales and strong and weak couping limits. In spite of the fluid character of the ideas, one central property seems to become more firmly established with time: the theory exists only in 10 fundamental spatial dimensions.

The central idea for dealing with the other 7 dimensions is "compactification": we are unaware of the extra dimensions because they are much smaller than the three normal large space dimensions. For example, a three dimensional tube can appear two dimensional if its walls are thin enough, and even one dimensional if it is long and thin enough. As the above remarks indicate, the size of the extra dimensions for Standard Model field propagation must be smaller than $(\text{TeV})^{-1}$, and smaller than $(10^{16}\text{GeV})^{-1}$ if SUSY GUT ideas are right.

The idea of compactification was investigated by Kaluza and Klein in the 1920's as a way to unify gravity and electromagnetism. They showed the geometrodynamics of an additional very small dimension could appear at low energies as an electromagnetic field tensor. Extra dimensions lead to new predicted degrees of freedom in fields— for example a field which is massless in 3+1 dimensions creates a "Kaluza- Klein tower" of new massive excitations corresponding to harmonics of states propagating around the short new directions. Altough traditionally the extra dimensions and new effects associated with M-theory have generally been assumed to happen close to the Planck scale, this is not neccesarily the case; the most notable result of brane world models is that the extra dimensions may be very large, possibly even infinite in extent.

M-theory is known to contain structures that offer suggestive clues to compactification. Features called branes appear which have lower dimensionality than the whole space. They form sites where the fundamental objects, one dimensional strings, can terminate, suggesting that in a low energy theory the gauge interactions may be confined to a lower dimensional surface embedded in the ten dimensional space. Branes are often thought of as classical, defect- or soliton-like structures resembling cosmic domain walls, with a surface tension and an internal vacuum energy larger than that in the surrounding higher dimensional background space. (Since the latter can be negative, the background space can be a higher-dimensional Anti-de Sitter solution even though our universe has a positive energy density). New types of excitations, corresponding to degrees of freedom such as displacement of the brane in the higher dimensions, correspond to propagating modes and new types of particles that might be observed.

One of the most spectacular discoveries in unification theory is Maldacena's AdS5/CFT correspondence: $N = 4$ supergravity in extended 5D Anti-deSitter space is exactly equivalent to a conformal field theory on its 4D boundary, which can be regarded as just ordinary Minkowski space. Here is a concrete example of a quantum gravity theory with all the richness associated with fields in five dimensions, all the details of which map onto the behavior of a conformal field theory in the standard 4D spacetime.

A related earlier idea called "holography" was inspired by the thermodynamics and information content of black holes. The conjecture is that all 3D fields are actually encoded by some theory acting on a 2D surface. We know that black hole entropy is given by a constant times the surface area of its (two-dimensional) event horizon. This means that a finite (indeed countable) amount of data on a two-dimensional surface (roughly, a few bytes per Planck area) must suffice to specify everything going on in the three-dimensional volume of space within it. The holography conjecture is that this applies to the whole universe— that three-dimensional space in some sense is an illusion, that the actual behavior is in some more fundamental sense two dimensional.

IV BRANE WORLDS

Brane worlds start with the idea that the familiar Standard Model fields are confined to a wall or "brane". This structure has three large dimensions but a thickness TeV^{-1} or smaller in the other dimensions. Gravity on the other hand can propagate much farther into one or more larger dimensions (called the "bulk"). The brane can be thought of as a stable classical defect embedded in a highly symmetric space of more dimensions, usually Anti-deSitter. Within this framework there are many options.

In some models, one or two extra dimensions can be of surprisingly large size [5–7], as they are only constrained by the direct experimental gravitational probes of the order of a few hundred microns. Most of the "large" extra dimensions however must be much smaller than this. Elaborations of brane worlds have been explored; for example, some have multiple branes which interact gravitationally. In others, there are different branes for different Standard Model fermion fields, with bosons allowed to travel in the bulk between them.

In one interesting class of brane-world models ("nonfactorizable geometries", [8–11]), the extra dimensions can be even larger, but the larger embedding space is highly curved, which traps gravitons in a bound state close to a brane. (Such geometries are said to have a "warp factor" (!)). The curvature radius of higher-dimensional (e.g. Anti-deSitter) space is again on a mesoscopic scale, which may be as large as ≈ 0.2 mm. Macroscopic black holes can be pictured as thin pancakes stuck to the brane, with only three large dimensions. The AdS space is Poincaré invariant and is itself a stable solution, so the setup is dynamically self-consistent, the kind of structure which might develop naturally from a defect in fields in higher

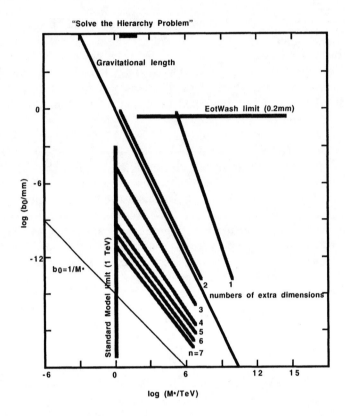

FIGURE 1. Summary of parameters for simple brane world models. It is assumed that there is a single unification scale M_* and that there are n extra dimensions of equal size b. The series of lines labeled $n = 1$ to 7 correspond to viable models with the right gravitational coupling at low energies. The "Gravitational length" line, degenerate with $n = 2$, denotes the Schwarzschild radius of a black hole with mean density M_*^4. The thresholds of direct current particle and gravity experiments are shown. Models with only one fundamental scale, which may "solve the hierarchy problem", lie not far beyond the reach of current accelerator constraints.

dimensions.

The apparent (usual) Planck mass in 3+1D, M_{Planck}, is related to the true fundamental scale M_* by $M_{Planck}^2 \approx M_*^2(M_*^n V_n)$ where V_n is the volume of n extra dimensions (larger than M_*^{-1}) in which gravity propagates. Thus if there is one extra dimension much larger than the others, the mm limits from gravity experiments require a unification scale $M_* = 10^6$TeV or larger. If there are two extra large dimensions the mm limits give M_* close to the TeV scale. If we try to solve the hierarchy problem with a single M_* not too far above the TeV scale, this can be accomplished for $n \geq 2$ by choosing suitable extra dimension sizes; for 7 equally large extra dimensions, we might have gravity propagating in ten dimensions, seven of which have size $b \approx 10^{-10}$ mm. The range of options for a simple model with n extra bulk dimensions of the same size b_0 is illustrated in figure 1.

An interesting result is that the unification implied by the running-together of Standard Model coupling constants can still work in brane-world scenarios, but the three gauge couplings come together at a much reduced energy [16,17]. With the addition of extra dimensions *for the gauge fields* (as well as gravity), the renormalization of the fields produces a power-law dependence of coupling on energy (like gravity always had), so that they run together in a rather modest range of energy. For example, if the brane has a width in a single extra dimension of TeV^{-1} then at higher energy the Standard Model couplings rapidly converge and meet in a point at about 20 TeV. This is regarded as less elegant than the parameter-free running-together of SUSY at the 10^{16} GUT scale but it may be the way nature works. These schemes thus hold out the attractive possibility of a unification scheme, even including gravity, with just one scale; it is even possible that we might find full quantum gravity effects accessible at the level of the next-generation accelerators. The famous "Desert" and the Planck scale, which have shaped so much discussion in the past, may be mirages.

V BRANE ASTRONOMY

The new fields and particles of these models might appear at accelerators in various manifestations. Some of these appear as "normal" new particle effects, such as excitation of Nambu-Goldstone modes of brane oscillations which would show up with the same signatures of missing energy and momentum as a weakly interacting scalar particle, or radion modes which might appear with signatures resembling (if not identical with) a Higgs scalar. Other possibilities include quirky signatures, such as multiple, evenly-spaced events produced as a particle traveling in the bulk punches periodically through the 3-brane. Null results in laboratory searches for measurable departures from Newton's inverse square law at short distances are an important constraint for $n = 1$ or 2.

As usual, astrophysical environments reach farther into parameter space. New weakly coupled species in these models are constrained in the same way axions are, using arguments based on energy losses from supernovae and red giants. The

"Kaluza-Klein tower" states can be particularly interesting. Massive KK modes of the graviton are a generic effect, and their cosmological production is an important constraint. They are produced thermally in the early universe, and only avoid causing an overclosure catastrophe in some cases because they can be very weakly coupled to the thermal particles on the brane. By the same token, for the right parameters they are a cold dark matter candidate. KK ladders of massive sterile neutrinos are a possible candidate for warm dark matter, and may display unusual nonthermal energy distributions induced by species oscillations. Brane worlds may bring important new insights into the cosmological constant [18] and inflation [19]. It is even possible for gravitational waves to travel faster than light since they can take a "short cut" across the bulk.

A more speculative phenomenon, which potentially reaches even farther into parameter space, is the classical production of gravitational waves in the early universe, which survive to the present as a nonthermal stochastic background [20]. Brane-world models suggest new sources of stochastic backgrounds: their new geometrical degrees of freedom can be coherently excited by symmetry breaking in the early universe, leading to gravitational radiation today at redshifted frequencies appropriate for new observatories such as LIGO and LISA [21,22]. New extra-dimensional effects remain important until the Hubble length $H^{-1} \approx M_{Planck}/T^2$ is comparable to the size or curvature radius b of the extra dimensions [7,23,24], or until the temperature falls below the new unification scale, whichever happens last.

Of particular interest are two new geometrical degrees of freedom common to many of these models: "radion" modes controlling the size or curvature of the extra dimensions [23,25], and new Nambu-Goldstone modes corresponding to inhomogeneous displacements of the brane in the extra dimensions [7,12]. Cosmological symmetry breaking can create large-amplitude, coherent classical excitations on scales of order H^{-1} as the configuration of the extra dimensions and the position of the brane settle into their present state.

The scalar modes of this distortion have long ago disappeared since they are on a very small scale (i.e., less than 1 mm times the redshift, or about the size of the solar system today), but the tensor modes might be observable. Extra dimensions with scale between 10 Å and 1 mm, which enter the 3+1-D era at cosmic temperatures between 1 and 1000 TeV, produce backgrounds with energy peaked at observed frequencies in the LISA band, between 10^{-1} and 10^{-4} Hz. The background is detectable above instrument and astrophysical foregrounds if initial metric perturbations are excited to a fractional amplitude of $\approx 10^{-3}$ or more. As shown in Figure 2, brane world models which "solve the hierarchy problem" naturally produce backgrounds in the range of frequencies encompassed by LISA for all the viable cases, $n = 2$ to 7. Ground based detectors (LIGO, VIRGO, TAMA, GEO), probing higher frequencies, reach extra dimensions down to 10^{-15} mm and unification energies up to 10^{13} GeV.

Thus it is possible that gravitational wave astrophysics might "see" outside of the four dimensions of ordinary spacetime, and trace the details of how the three

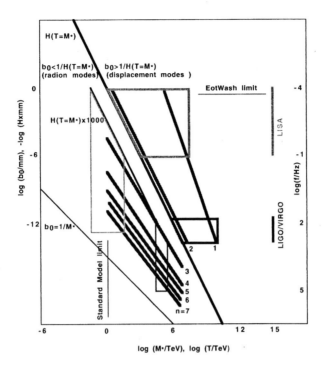

FIGURE 2. Summary of the new parameter space of extra dimensions that will be probed by gravitational-wave interferometers. Boxes indicate the corresponding regions of these parameters which may give rise to detectable mesoscopic gravitational radiation backgrounds in the LISA and LIGO bands. Heavy-line boxes show the displacement mode parameters, lighter-line boxes show the radion mode parameters. These regions extend well beyond those already constrained by gravitational experiments, direct particle production, or other astrophysical constraints. Theories which "solve the hierarchy problem" have M_* close to the Standard Model limit, and all of the viable ones ($2 \leq n \leq 7$) could possibly produce an observable background of one type or the other in the LISA band.

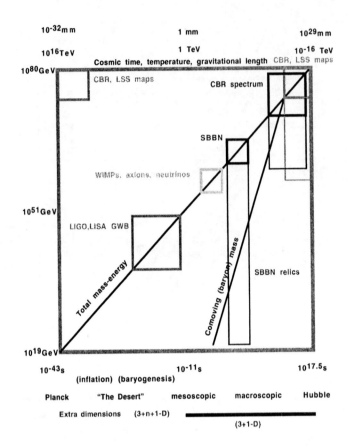

FIGURE 3. Summary of evidence about cosmic history, showing scale (in terms of total mass-energy) versus cosmic time/temperature. Boxes are labled by the technique used to constrain events in each domain of time and scale, including the microwave background anisotropy and spectrum, cosmological nucleosynthesis, and dark matter production. The box labeled LIGO,LISA GWB shows the hitherto unexplored region of mesoscopic phenomena which will either be opened up or constrained by gravitational wave astronomy— the universe earlier than 1 TeV, when it may have had more than three spatial dimensions.

spatial dimensions settled into their present shape, in brane worlds that cannot be tested by any other known technique (see Figure 2). The background spectrum also contains information about a regime of cosmic history not preserved by any other relic (Figure 3).

I am grateful for useful conversations with E. Adelberger, D. Kaplan, A. Nelson, C. Stubbs, and S. Weinberg.

REFERENCES

1. G. Horowitz, in Proc. 20th Texas Symposium, (this volume, 2001)
2. A. G. Riess et al. 1998, AJ, 116, 1009
3. S. Perlmutter et al. , ApJ, 517, 565 (1999)
4. N. Suntzeff, in Proc. 20th Texas Symposium, (this volume, 2001)
5. N. Arkani-Hamed, S. Dimopoulos, and G. Dvali, Phys. Lett. B429, 263 (1998)
6. I. Antoniadas, N. Arkani-Hamed, S. Dimopoulos, and G. Dvali, Phys. Lett. B436, 257 (1998)
7. N. Arkani-Hamed, S. Dimopoulos, S., and G. Dvali, Phys. Rev. D59, 086004 (1999)
8. L. Randall and R. Sundrum, Phys.Rev.Lett. 83, 3370 (1999)
9. L. Randall and R. Sundrum, Phys. Rev. Lett. 83, 460 (1999)
10. J. Lykken, J. L. Randall, J. High En. Phys. 0006, 014 (2000)
11. A. Karch and L. Randall, hep-th/0011156
12. R. Sundrum, Phys.Rev. D59, 085009 (1999) hep-ph/9805471
13. F. Wilczek, Rev. Mod. Phys. 71, S85 (1999)
14. S. Weinberg, in Proc. 20th Texas Symposium, (this volume, 2001)
15. Hoyle, C. D., Heckel, B., Adelberger, E., Schmidt, U., Gundlach, J., Swenson, E., & Kapner, D., in preparation (2000)
16. K. R. Dienes, E. Dudas and T. Gherghetta, hep-ph/9807522.
17. K. R. Dienes, E. Dudas and T. Gherghetta, Nucl. Phys. B537, 47 (1999) [hep-ph/9806292].
18. E. Flanagan et al., "A Brane World Perspective on the Brane World and the Hierarchy Problems", hep-th/0012129
19. R. Maartens, "Geometry and Dynamics of the Brane-World", gr-qc/0101059
20. M. Maggiore, "Gravitational Wave Experiments and Early Universe Cosmology", Phys. Rep. 331, 283 (2000)
21. C. J. Hogan, Phys. Rev. Lett., 85, 2044 (2000)
22. C. J. Hogan, "Scales of the Extra Dimensions and Their Gravitational Wave Backgrounds", Phys. Rev. D., submitted, astro-ph/0009136 (2000)
23. C. Csaki, M. Graesser, L. Randall, and J. Terning, Phys.Rev. D62, 045015, hep-ph/9911406 (2000)
24. N. Arkani-Hamed, S. Dimopoulos, N. Kaloper, and J. March-Russell, 2000, Nucl.Phys. B567, 189, hep-ph/9903224 (1999)
25. W. D. Goldberger and M. Wise, "Phenomenology of a Stabilized Modulus," hep-ph/9911457 (1999)

Interaction of bulk matter with brane-world black holes

A. Chamblin

MIT, Bldg. 6-304, 77 Mass. Ave., Cambridge, MA 02139

Abstract.
We study bulk timelike geodesics in the presence of a brane-world black hole, where the brane-world is a two-brane moving in a $3+1$-dimensional asymptotically adS_4 spacetime. We show that for a certain range of the parameters measuring the black hole mass and bulk cosmological constant, there exist stable timelike geodesics which orbit the black hole and remain bound close to the two-brane.

INTRODUCTION: BRANE-WORLD BLACK HOLES

Recently, phenomenological considerations have led people to consider the possibility that there may exist extra dimensions of space which are quite large. If this is correct then the universe would be a three-dimensional brane moving in some higher dimensional spacetime. This sort of picture is naturally realized in the Hořava-Witten corner of the M-theory moduli space, and it is also the basic assumption underlying the second model of Randall and Sundrum [1], where the universe itself is a thin "distributional" static flat domain wall or three-brane separating two regions of five-dimensional anti-de-Sitter spacetime.

Of course, we need to check that this idea that we live on a brane agrees with what we already know. For instance, if matter trapped on a brane undergoes gravitational collapse then a black hole will form. Such a black hole will have a horizon that extends into the dimensions transverse to the brane: it will be a higher dimensional object. Consistency demands that this higher dimensional object reproduce the usual astrophysical properties of black holes and stars.

Motivated by this reasoning, various authors have studied the problem of gravitational collapse on the brane. In particular, in [2] the authors proposed the existence of a 'black cigar' solution, and conjectured that this black cigar solution is the unique stable vacuum solution in five dimensions which describes the endpoint of gravitational collapse in a Randall-Sundrum brane-world. However, though they were able to discuss the gross features of this solution, they were not able to write the metric down explicitly. Since this work, other papers have appeared generalizing the analysis to include the effects of charge, and also the effects of more than one large extra dimension [3].

At present, about the only thing we know about the black cigar metric is that it is a generalization to higher dimensions of a solution in four dimensions known as the *adS-C-metric*. This is because the adS-C-metric naturally describes the pair creation of black holes by the breaking of a cosmic string in $3+1$ dimensions, and the black cigar metric should describe the fragmentation (due to the Gregory-Laflamme instability [4]) of a *black* string solution in $4+1$ dimensions.[1] Motivated by this observation, the authors of [5] studied the adS-C-metric, with a hope of learning something about the geometrodynamics of brane-worlds in one less dimension. We now turn to their construction.

THE CONSTRUCTION OF EMPARAN, HOROWITZ AND MYERS

In [5] the authors studied a lower-dimensional version of the Randall-Sundrum model, consisting of a two-brane moving in a $3+1$-dimensional asymptotically adS_4 background. They explicitly showed how the adS-C-metric describes a black hole bound to the two-brane, and therefore were able to conclude that adS-C describes the final state of gravitational collapse on the brane-world. At first glance, this seems to go against the conventional wisdom that gravitational collapse in $2+1$ dimensions (without a cosmological constant) will always produce a conical deficit, and not a horizon. However, this reasoning is flawed because matter which is trapped on the brane will still generate curvature in the full $3+1$ dimensional spacetime. Consequently, collapse on the brane will produce a black hole in $3+1$ dimensions, although at large distances along the brane $2+1$-gravity is reproduced in the sense that there is still a deficit angle in asymptotia. Following [5], we will only consider the following special case of the adS-C-metric:

$$ds^2 = \frac{l^2}{(x-y)^2}\left[-(y^2+2\frac{m}{l}y^3)dt^2 + \frac{dy^2}{(y^2+2\frac{m}{l}y^3)} + \frac{dx^2}{G(x)} + G(x)d\varphi^2\right] \quad (1)$$

where G is the cubic polynomial

$$G(x) = 1 - x^2 - 2\frac{m}{l}x^3$$

This metric is a solution of the bulk Einstein equations with a negative cosmological constant

$$\Lambda = -6/l^2$$

The global structure of this solution was more than adequately discussed in [5]; here, we simply recall the salient points.

First, the acceleration parameter A, which measures the rate at which the black hole has to accelerate in order to remain bound to the brane-world, has been fine tuned relative to the bulk cosmological constant:

$$A = 1/l \quad (2)$$

Of course, this is just a reflection of the fact that we are working in the original second Randall-Sundrum scenario, where the brane tension is fine tuned relative to the bulk cosmological constant so that the worldvolume of the brane is Minkowski. Any black

---

[1] The key point here is that in $3+1$ dimensions a self-gravitating Nambu string generates a conical deficit in spacetime, whereas in $4+1$ dimensions it actually generates an extended one-dimensional horizon, or 'black string'.

hole which comoves with a Minkowski brane in adS has to accelerate at the rate (2.2). On the other hand, if the brane is de Sitter the acceleration will be even greater. It would be interesting to study this more general case where the black hole is bound on a de Sitter brane-world.

Next, it is worth pointing out that 'infinity' corresponds to the set of points specified by $x = y$. This is because these points are infinitely far away from any other points. Not much is known about this infinity, except that it is still a timelike surface where presumably a holographic theory may reside. The structure of this set is not relevant for the construction of [5], since it is precisely this part of the spacetime that we 'throw away' when we cut-and-paste to obtain the Randall-Sundrum domain wall.

The coordinates used in (2.1) are probably not familiar to most people. Basically, the angle φ rotates around the axis of symmetry of the black hole (i.e., it rotates around the cosmic string which is pulling the black hole towards infinity). The coordinate x is like the cosine of the other angular variable, and the coordinate $-y$ is a radial variable. In this paper, we are going to be interested in geodesics which 'co-accelerate' with the black hole [6] as it is pulled towards infinity. That is, we want to see if there can exist geodesics which stay near the equatorial plane of the black hole ($x = 0$, where the brane-world intersects the black hole), so that x = constant, and which simultaneously orbit the black hole at some constant radius y = constant.

As usual, there is a horizon wherever the metric component g_{tt} has a root. The root at $y = 0$ corresponds to the acceleration horizon for the black hole, or equivalently, to the bulk adS Cauchy horizon. The horizon at $y = -1/2m$ is the black hole horizon. There is a curvature singularity at $y = -\infty$.

The parameter m measures the mass of the black hole, and as discussed in [5] once we specify m and l, the size of the deficit angle $\Delta\varphi$ is fixed. This deficit angle measures the tension of the 'cosmic string' which is accelerating the black hole towards the boundary at infinity. In the construction of Emparan et al, we throw away the part of the spacetime containing the conical deficit and we will have nothing more to say about this portion of the spacetime.

Finally, we should mention that we will assume the metric is Lorentzian signature $(-+++)$ throughout, and consequently we must have $G(x) \geq 0$. Also, we will require that G possesses three distinct real roots, so that $0 < \frac{m}{l} < \frac{1}{3\sqrt{3}}$ as in [5].

ANALYSIS OF THE GEODESICS

In [6], Pravda and Pravdova studied geodesic motion in the C-metric, which is a solution of the vacuum Einstein equations with vanishing cosmological constant. The following analysis is a straightforward generalization of their work to the adS-C-metric.

The lagrangian associated with the metric (2.1) is given as

$$L = \frac{l^2}{(x-y)^2}\left(+\frac{1}{G}(\frac{dx}{d\tau})^2 + \frac{1}{F}(\frac{dy}{d\tau})^2 + G(\frac{d\varphi}{d\tau})^2 - F(\frac{dt}{d\tau})^2\right) \tag{3}$$

where $F(y) = y^2 + 2\frac{m}{l}y^3$, τ is the proper time experienced by a free falling timelike observer (and the affine parameter for null geodesics), $L = -1$ for timelike geodesics and $L = 0$ for null geodesics.

The line element (2.1) is boost and rotation symmetric, that is to say, ∂_t and ∂_φ are Killing vectors. As described in [6], these Killing vectors define constants of the motion J and E such that for any geodesic $\lambda(\tau) = (t(\tau), y(\tau), x(\tau), \varphi(\tau))$ we have the relations

$$\frac{d\varphi(\tau)}{d\tau} = \frac{J}{l^2} \frac{[x(\tau) - y(\tau)]^2}{G(x(\tau))} \quad (4)$$

and

$$\frac{dt(\tau)}{d\tau} = \frac{E}{l^2} \frac{[x(\tau) - y(\tau)]^2}{F(y(\tau))} \quad (5)$$

Timelike geodesics

If we substitute equations (3.2) and (3.3) into (3.1), and let $L = -1$, we obtain the equation

$$\frac{Fl^4}{E^2(x-y)^4} \left[\frac{1}{G}(\frac{dx}{d\tau})^2 + \frac{1}{F}(\frac{dy}{d\tau})^2 \right] = \frac{E^2 - V^2}{E^2} \quad (6)$$

where V is an effective potential for the motion given as

$$V = F^{1/2} \left(\frac{l^2}{(x-y)^2} + \frac{J^2}{G} \right)^{1/2} \quad (7)$$

Clearly, (3.4) is precisely what we want, given that we are interested in the stability of timelike geodesics that orbit at $x = $ constant, $y = $ constant. In particular, we want to know if there can exist geodesics with $x = $ constant and $y = $ constant in a region where the potential V has a local minimum. As in [6], we find that such a local minimum for V can exist, provided the mass m and length scale l satisfy the necessary inequality:

$$\frac{m}{l} < C \quad (8)$$

where C is some constant which has to be determined numerically. The precise value of this constant is difficult to estimate, although our results imply that $C \sim O(10^{-3})$.

In other words, if a black hole does *not* satisfy (3.6) then there is definitely no local minimum for V and a slight perturbation will cause the particle to either fall away from the brane into the bulk or into the black hole. Conversely, if (3.6) is satisfied then it is possible to find geodesics with x and y constant. Generically, the stable orbits occur for a range of y outside of the black hole horizon: $-\frac{l}{2m} < y_{min} < y < y_{max} < 0$.

We are not able to find any geodesic which orbits *exactly* at $x = 0$, although we are able to find such orbits with x arbitrarily small. This is perhaps unsurprising, given the symmetry of the setup and the fact that we are trying to balance bulk acceleration vs. black hole acceleration. It would be interesting to know if it is ever possible to satisfy $x = 0$ for a stable orbit, perhaps by including the effects of charge or rotation.

Null geodesics

As in [6], it is straightforward to show that there exists a *null* geodesic ($L = 0$) which stays bound to the brane (at $x(\tau) = 0$), and which orbits the black hole at a radius

$$y(\tau) = -\frac{l}{3m}$$

Thus, we see explicitly how the adS length scale l will affect the radius at which there may exist a null geodesic orbit. As in [6], however, this orbit is unstable and the slightest perturbation will cause the photon to either fall into the black hole or off of the brane-world and into the bulk.

CONCLUSION: DO SOME BLACK HOLES HAVE 'HALOS'?

We have shown that a brane-world black hole can be used to 'trap' bulk degrees of freedom, in the sense that there can exist stable bulk geodesic orbits which remain bound arbitrarily close to the brane as they orbit the black hole. Of course, we have only really seen that this is true for a two-brane moving in an adS_4 background, but for the sake of argument let us suppose that a similar result holds for the full black cigar metric.

It would then follow that for any black hole which satisfies a bound of the form (3.6) there exists a sort of 'halo' surrounding the black hole where particles can orbit the black hole in such a way that the total acceleration 5-vector is nearly vanishing. Crudely, what is happening is that the acceleration due to the black hole is compensating for the acceleration due to the adS bulk, in such a way that in the halo region particles can orbit in inertial frames. Naively, one might conclude that this would imply that the clock of a brane-world observer would run *faster* inside of the halo than it would outside of the halo [7]. This is because an observer orbiting far outside of the halo region will have a non-vanishing acceleration $A_{outside}$ (the acceleration required to stay on a horospherical brane in adS), but an observer orbiting inside of the halo can arrange for her total acceleration to be arbitrarily small:

$$A_{halo} \sim 0$$

However, this reasoning is incorrect because it presupposes that acceleration makes sense on a spacetime that has a metric discontinuity. More precisely, the assumption that the brane-world is infinitely thin is merely an idealization, which implies that the acceleration experienced by observers who comove with the brane-world must jump discontinuously as one moves through the brane. To avoid this discontinous jump, the spacetime must somehow be 'smoothed out' in the region of the domain wall. Such a smoothing will imply that there is a region in the center of the domain wall where the force experienced by comoving observers is zero. Only by moving a certain critical distance from the brane will observers begin to feel the acceleration which drags them into the bulk of adS. In this way we can avoid the sort of paradoxical behaviour discussed above.

On a more speculative note, it is tempting to draw an analogy between the halo where bulk degrees of freedom might reside and the galactic halo of 'dark matter'.

Unfortunately, such a comparison seems unlikely given that the bound (3.6) is an *upper* bound on the mass of the black hole.

On the other hand, it certainly does make sense to think of these trapped bulk degrees of freedom in terms of the effective theory on the brane-world, as discussed in [8]. From this perspective, it is clear that a bulk particle which is trapped in the halo region will have a holographic image on the brane with some spread - a kind of cloud. What our results imply is that the bulk particles orbiting in the halo will generate an orbiting cloud of dark matter, which can interact with brane-world matter through the exchange of gravitons and SYM gauge bosons. One might think that in order to pursue this line of reasoning we need the full black cigar metric; however, it is likely that some linearized analysis, such as that presented in [9], will do the trick. Research on this and related issues is currently underway.

Acknowledgements

I thank R. Emparan, S. Giddings, A. Guth, E. Katz and N. Lambert for useful discussions, and the faculty and staff of the High Energy Theory Group at the University of Pennsylvania for hospitality while this work was completed. This work was supported in part by funds provided by the U.S. Department of Energy (D.O.E.) under cooperative research agreement DE-FC02-94ER40818.

REFERENCES

1. L. Randall and R. Sundrum, Phys. Rev. Lett. **83**: 4690-4693, (1999); hep-th/9906064.
2. A. Chamblin, S. W. Hawking and H. S. Reall, Phys. Rev. **D61**, 065007(2000); hep-th/9909205.
3. N. Dadhich, R. Maartens, P. Papadopoulos and V. Rezania, hep-th/0003061; A. Chamblin, C. Csaki, J. Erlich and T. J. Hollowood, Phys. Rev. **D62**, 044012(2000); A. Chamblin, H.S. Reall, H. Shinkai and T. Shiromizu, hep-th/0008177 (to appear in Physical Review D).
4. R. Gregory and R. Laflamme, Phys. Rev. Lett. 70, 2837 (1993); hep-th/9301052.
5. R. Emparan, G. T. Horowitz and R. C. Myers, JHEP **0001**, 07(2000); hep-th/9911043.
6. V. Pravda and A. Pravdova, gr-qc/0010051.
7. Sean Carroll, *Lecture Notes on General Relativity*, gr-qc/9712019 (see also http://pancake.uchicago.edu/ carroll/notes/).
8. S.B. Giddings and E. Katz, hep-th/0009176.
9. S.B. Giddings, E. Katz and L. Randall, JHEP 0003, 023 (2000); hep-th/0002091.

A Universe in a Global Monopole

Katherine Benson* and Inyong Cho*

Department of Physics, Emory University, Atlanta, Georgia 30322-2430

Abstract.
We investigate brane physics in a universe with an extra dimensional global monopole and negative bulk cosmological constant. To counter divergence of the graviton zero mode, we employ a physical cut-off. This cut-off yields 4D gravity on a brane at the monopole core; it also solves the hierarchy problem, by inducing the observed hierarchy between particle and Planck scales in the effective 4D universe. Our model has a discrete spectrum of massive Kaluza Klein modes, easily made consistent with 4D gravity on the brane. Extra-dimensional matter fields also induce 4D matter fields on the brane, with the same Kaluza Klein spectrum of excited states.

INTRODUCTION

Since Kaluza and Klein's work of the 1920's [1], theorists have invoked hidden extra dimensions to resolve foundational questions in field theory. Such extra dimensions were assumed to be compactified and small, negligibly affecting 4D gravitational force laws.

Recently, Arkani-Hamed, Dimopoulos and Dvali [2] suggested that extra dimensions could be large — as large 1 mm without endangering 4D gravitational force laws. They noted that large extra dimensions can solve the hierarchy problem, since large extradimensional volume transforms a natural higher-dimensional Planck mass into a hierarchically large effective 4D Planck mass.

Soon after, Randall and Sundrum proposed that extra dimensions can be infinite, if the metric is nontrivially "warped." [3] Specifically they embedded a 4D matter brane in a warped 5D spacetime with a negative bulk cosmological constant. Their model localizes massless gravitons on the brane, reproducing the observed $1/r^2$ gravitational force, with mild corrections due to a continuum of excited Kaluza-Klein gravitons. It also induces the observed hierarchy between particle and effective Planck scales, through integration over the fifth dimension to establish an effective 4D gravitational action.

Randall and Sundrum proposed the brane, and its intrinsically 4D matter fields, from string theoretic motivations. Later work realized the 4D subuniverse more naturally, as associated with topological defects formed by a matter condensate in the extra dimensions. The defect solution determines the warped metric, and binds both gravitons and matter fields to a 4D internal space at the defect's core. Such binding for matter fields is well-established (see, for example, [4]). Solutions for warped extradimensional defect metrics appeared in [6]; more complete models, showing bound $1/r^2$ 4D gravity plus corrections, and solved hierarchy problems, appeared in [5]

We here investigate the effective 4D universe induced by an extra dimensional global monopole. The monopole forms in the three transverse dimensions of a 7D universe with negative cosmological constant, generating a warped metric solution distinct from those considered previously. Each point in the extra transverse space corresponds to a 3D brane. The spacetime of this global monopole is singularity free in its geometry [6], with extra dimensional space stretching without bound.

Because our model has infinite extradimensional volume, the gravity zero mode is not normalizable. To normalize it, and achieve 4D gravity on the brane, we introduce a cut-off. We regard this cut-off as a natural element in our dynamical theory, measuring the typical separation between global monopoles formed in the dimension-reducing phase transition. We choose it so that the effective Planck scale flows upward to solve the hierarchy problem.

Independent of cut-off, our model yields a discrete graviton mass spectrum. This yields acceptable gravitational corrections, for either large or small model-dependent mass gap. Like gravitons, 7D matter fields become localized on the brane. In fact, our induced 4D matter fields determine a tower of excited Kaluza Klein states, with mass gap and wave functions identical to the graviton modes.

Below we present our model, discuss the graviton zero mode and 4D gravity on the brane, then summarize results for massive Kaluza Klein modes and matter fields.

THE MODEL

We consider a global monopole formed in a universe with 3 extra dimensions. The (4+3) dimensional metric, static and spherically symmetric in the extra three dimensions, takes the form

$$ds^2 = g_{MN}dx^M dx^N = B(r)\bar{g}_{\mu\nu}dx^\mu dx^\nu + A(r)dr^2 + r^2 d\Omega^2, \quad (1)$$

where $\bar{g}_{\mu\nu}$ is the apparent 4D-metric and we take the mainly + sign convention.[1]

The action is

$$S = \int d^7x \sqrt{-g} \left(\frac{\mathcal{R} - 2\Lambda}{16\pi G_N} + \mathcal{L}_m \right), \quad (2)$$

where G_N is the 7D gravitational constant $G_N = 1/8\pi M_N^{2+n}$, Λ is the 7D cosmological constant, and $\mathcal{L}_m$ is the Lagrangian of the monopole field $\phi^a = \phi(r)\hat{x}^a$, a scalar triplet with potential $V = \frac{\lambda}{4}(\phi^a\phi^a - \eta^2)^2$ and symmetry-breaking scale η.

With our conventions Einstein's equation is

$$R_{MN} - \frac{1}{2}g_{MN}\mathcal{R} + \Lambda g_{MN} = \kappa^2 T_{MN},$$

where $\kappa^2 = 8\pi G_N$ and T_{MN} is the monopole energy-momentum tensor $T_{MN} = \partial_M \phi^a \partial_N \phi^a + g_{MN}\mathcal{L}_m$, diagonal within the longitudinal subspace. This gives three distinct equations:

---

[1] Note that the capital Roman index runs over all seven dimensions, while the Greek index runs over our four longitudinal dimensions and the small Roman index runs over the extra three tranverse dimensions.

$$-G^\mu_\mu = \frac{1}{A}\left[-\frac{3}{2}\frac{B''}{B} + \frac{3}{4}\frac{A'B'}{AB} - 3\frac{B'}{Br} + \frac{A'}{Ar} + \frac{A-1}{r^2} + \frac{1}{4}\frac{A}{B}\bar{R}^{(4)}\right] = \kappa^2\left[\frac{\phi'^2}{2A} + \frac{\phi^2}{r^2} + V(\phi)\right] + \Lambda,$$

$$-G^r_r = \frac{1}{A}\left[-\frac{3}{2}\left(\frac{B'}{B}\right)^2 - 4\frac{B'}{Br} + \frac{A-1}{r^2} + \frac{1}{2}\frac{A}{B}\bar{R}^{(4)}\right] = \kappa^2\left[-\frac{\phi'^2}{2A} + \frac{\phi^2}{r^2} + V(\phi)\right] + \Lambda,$$

$$-G^{\theta_i}_{\theta_i} = \frac{1}{A}\left[-2\frac{B''}{B} - \frac{1}{2}\left(\frac{B'}{B}\right)^2 + \frac{A'B'}{AB} - 2\frac{B'}{Br} + \frac{1}{2}\frac{A'}{Ar} + \frac{1}{2}\frac{A}{B}\bar{R}^{(4)}\right] = \kappa^2\left[\frac{\phi'^2}{2A} + V(\phi)\right] + \Lambda, \quad (3)$$

where no sums are implied and $\bar{R}^{(4)}$ is the 4D Ricci scalar induced by $\bar{g}_{\mu\nu}$. Henceforward, we assume a flat 4D brane ($\bar{R}^{(4)} = 0$), which implies that the induced effective 4D cosmological constant vanishes.

The scalar field obeys its field equation

$$\frac{\phi''}{A} + \frac{1}{A}\left(-\frac{1}{2}\frac{A'}{A} + 2\frac{B'}{B} + \frac{2}{r}\right)\phi' - \frac{2}{r^2}\phi - \lambda\phi(\phi^2 - \eta^2) = 0. \quad (4)$$

It assumes a monopole configuration, with boundary conditions $\phi(0) = 0$, $\phi(\infty) = \eta$.

Near the origin, the scalar field has linear dependence on r, $\phi(r) \approx \phi_0 r$. The gravitational fields have asymptotic form

$$A(r) \approx 1 + \left[\frac{1}{2}\kappa^2\phi_0^2 + \frac{1}{15}(\kappa^2 V(0) + \Lambda) + \frac{\bar{R}^{(4)}}{6}\right]r^2,$$

$$B(r) \approx 1 + \left[-\frac{2}{15}(\kappa^2 V(0) + \Lambda) + \frac{\bar{R}^{(4)}}{12}\right]r^2, \quad (5)$$

due to regularity conditions at the origin. Here ϕ_0 is determined by Eq. (4), and B is arbitrary up to a constant multiplication, chosen to set $B(0) = 1$.

Asymptotics at large r depend on Λ. In this paper we take $\Lambda < 0$ and consider subPlanckian monopoles only ($\kappa\eta < 1$). Here the geometry approaches that of anti-de Sitter space asymptotically, with no coordinate singularities. Other geometries for extra dimensional global defects were discussed in Ref. [6].

Asymptotically, as r grows and $|\phi| \to \eta$, the local curvature $\hat{R}^7$ becomes dominated by the cosmological constant Λ, rather than the monopole stress-energy. For $r > \kappa\eta/\sqrt{|\Lambda|}$, the Einstein equations (3) give asymptotic solutions $A \approx -15/\Lambda r^2$, $B \approx br^2$ under the ansatz $A = cB^{-1}$. Note that A has the correct signature only for negative Λ, and b is fixed by setting $B = 1$ at the origin. For this asymptotic metric, the scalar field equation (4) yields asymptotic form for the monopole ϕ as $\phi(r) = \eta\left(1 - 15/4|\Lambda|r^2\right)$. These asymptotic limits well characterize our numerical solutions for A, B, and ϕ.[7]

GRAVITON ZERO MODE AND PLANCK MASS

We solve for the graviton, arising as a perturbation h_{MN} to the monopole background metric whose 4D part $\bar{g}_{\mu\nu}$ is flat. This gives metric

$$\begin{aligned}ds^2 &= [B(r)\eta_{\mu\nu}+h_{\mu\nu}]dx^\mu dx^\nu + A(r)dr^2 + r^2 d\Omega^2, \\ &= B(r)(\eta_{\mu\nu}+\bar{h}_{\mu\nu})dx^\mu dx^\nu + A(r)dr^2 + r^2 d\Omega^2. \end{aligned} \quad (6)$$

Here $\bar{h}_{\mu\nu}$ gives the apparent 4D graviton field. Note we consider only 4D fluctuations $h_{\mu\nu}$, with $h_{\mu j} = h_{ij} = 0$. We also apply transverse-traceless and harmonic gauge on h_{MN}, $h_M^M = 0$ and $h_{MN|}{}^N = 0$, where the vertical bar | in the subscript denotes the covariant derivative with respect to the background metric. Einstein's equations for h_{MN} then reduce to

$$h_{MN|A}{}^A + 2R_{MANB}^{(B)} h^{AB} = 0 \quad (7)$$

with $R_{MANB}^{(B)}$ determined by the unperturbed background metric $g_{MN}^{(B)}$.

Separating variables, we take $h_{\mu\nu} = \hat{e}_{\mu\nu} e^{ip_\rho x^\rho} R(r) Y_{lm}(\theta,\varphi)$. Note that the apparent 4D graviton $\bar{h}_{\mu\nu}$ has 4D mass $-p_\mu p^\mu = m^2$, and apparent radial wave function $\bar{R}(r) = R(r)/B(r)$, from Eq. (6). Applying Eq. (7) gives for this radial wave function $\bar{R}(r)$

$$\left[-\frac{B}{A}\frac{d^2}{dr^2} + \frac{B}{A}\left(\frac{1}{2}\frac{A'}{A} - 2\frac{B'}{B} - \frac{2}{r}\right)\frac{d}{dr} + \frac{l(l+1)B}{r^2}\right]\bar{R}(r) = m^2 \bar{R}(r). \quad (8)$$

This can be rewritten in Sturm-Liouville form:

$$\frac{1}{A^{1/2}Br^2}\frac{d}{dr}\left(A^{-1/2}B^2 r^2 \frac{d\bar{R}}{dr}\right) + \left(m^2 - \frac{l(l+1)B}{r^2}\right)\bar{R} = 0. \quad (9)$$

This is a Sturm-Liouville equation singular at both the origin and infinity, where regular boundary conditions must hold:

$$\bar{R} \text{ bounded}, \quad A^{-1/2}B^2 r^2 \bar{R}\bar{R}' \to 0 \quad \text{for } r \to 0, \infty. \quad (10)$$

We seek solutions $\bar{R}$ for allowed values of the eigenvalue m^2, taking $l=0$ for simplicity.

We consider first the graviton zero mode, $m=0$. In this case Equation (9) is integrable, with solution

$$\bar{R}_0(r) = \bar{R}_0 + \bar{R}_1 \int_0^r \frac{\sqrt{A}}{B^2 r^2} dr,$$

where $\bar{R}_0$ and $\bar{R}_1$ are constants. At $r \approx 0$, $A \approx 1$ and $B \approx 1$. The second term in the solution becomes proportional to $1/r$, which is irregular at $r=0$. Thus the only regular solution is $\bar{R}(r) = \bar{R}_0$, a constant. This solution is not normalizable. From the Sturm-Liouville form of the radial equation (9), the normalization weight for this zero mode solution is $A^{1/2}Br^2$. Thus the graviton zero mode has normalization

$$\bar{R}_0^2 \int dr\, r^2 B\sqrt{A}, \quad (11)$$

which diverges. Physically, the origin of this divergence lies in the infinite volume of the extra dimensions.

This normalization of the zero mode is directly related to localization of 4D gravity on the brane. Specifically, the induced 4D Planck mass is proportional to the graviton zero mode normalization. We see this by considering the 4D effective field theory induced by our monopole solution. For general $\bar{g}_{\mu\nu}$, our seven-dimensional theory has in its action the gravitational term

$$16\pi G_N S_g = \int d^7x\sqrt{-g}\mathcal{R} = \int d^4x\sqrt{-\bar{g}}\int r^2 dr d\Omega B^2 \sqrt{A}\left[\frac{\bar{R}^{(4)}}{B} + \mathcal{R}\left(\bar{g}_{\mu\nu} = \eta_{\mu\nu}\right)\right]. \tag{12}$$

Here we have evaluated the 7D Ricci scalar $\mathcal{R}$, in terms of $\bar{R}^{(4)}$, the 4D Ricci scalar induced by $\bar{g}_{\mu\nu}$, and $\mathcal{R}\left(\bar{g}_{\mu\nu} = \eta_{\mu\nu}\right)$, the 7D Ricci scalar, evaluated for flat $\bar{g}_{\mu\nu}$. Viewing the $\bar{R}^{(4)}$ term as an effective 4D gravitational action, we read off its effective coupling constant

$$M_{Pl}^2 = 4\pi M_N^{2+n}\int dr r^2 B\sqrt{A}, \tag{13}$$

after the substitution $G_N = 1/8\pi M_N^{2+n}$. This is directly proportional to the normalization of the graviton zero mode in Eq. (11); thus divergent normalization of the graviton zero mode leads to a divergent 4D Planck mass.

In the models of $n \leq 2$, the integral of Eq. (13) is finite due to the exponentially decreasing warp factor. However, in our model, the warp factor is $B \sim r^2$ at large r and the integral diverges. Therefore, in order to have normalizable graviton zero mode and finite 4D Planck mass $M_{Pl} \sim 10^{18}$GeV, we introduce a cut-off. Such a cut-off should arise dynamically, as the typical separation between monopoles formed when dimensional symmetry breaks. It could be imposed formally, by considering a monopole and anti-monopole pair, with the presence of the antimonopole cutting off the divergent monopole solution.

We estimate the cut-off radius r_* inducing the observed hierarchy; that is, inducing 4D $M_{Pl} \sim 10^{18}$GeV from a 7D Planck mass of TeV scale, $M_N \sim$TeV. The main contribution to the integral in Eq. (13) comes from the region outside the core where $r^2B\sqrt{A}$ is much larger than the unity. In this region, from the result of the previous section, $B\sqrt{A} = b\sqrt{a}r$. Then Eq. (13) becomes

$$M_{Pl}^2 = 4\pi b\sqrt{a}M_N^{2+n}\int^{r_*} dr r^3 \simeq \pi b\sqrt{a}M_N^5 r_*^4.$$

This gives the cut-off radius $r_* = \left(\frac{\text{TeV}}{b|\Lambda|^{-1/2}}\right)^{\frac{1}{4}} 10^{-10}$cm, where $b|\Lambda|^{-1/2}$ is a mass scale set by the free parameter Λ. A range of choices for Λ remains observationally allowed. Regarding r_* as the size of the extra dimensions in our model, gravity takes its usual 4D form so long as the separation of probe particles in the brane remains larger than r_*. Therefore, the cut-off radius r_* cannot not exceed 1 mm, the current observational resolution of gravitational measurement.

THE KALUZA-KLEIN MASS SPECTRA

In this section we summarize corrections due to massive Kaluza Klein gravitons, note the apparent 4D matter spectra, and conclude. More detailed analysis appears in [7].

Massive gravitons obey the radial graviton equation (9), with boundary conditions (10). This Sturm-Liouville problem gives the zero mass graviton responsible for $1/r^2$ 4D gravity; it also yields a spectrum of positive mass eigenvalues. We solve for those eigenvalues in [7], obtaining the approximate positive mass spectrum

$$m_j^2 = -C + \frac{\pi^2}{2Z^2}\left(j + \frac{5}{4}\right)^2 \qquad j = 1, 2, \ldots \qquad (14)$$

where $Z = \int_0^\infty dr \sqrt{A/2B}$ is finite because the integrand falls as $1/r^2$ at large r. Z is highly model-dependent; however, for small Z, the massive Kaluza Klein modes have large mass gap and negligibly affect 4D gravity. We also show that for large Z, in the continuum Kaluza Klein limit where $C \to 0$, the 4D gravitational potential has only $1/r^6$ corrections, which are adequately suppressed. We find 7D matter fields also bind to the brane at the monopole core, with 4D mass spectrum $m_{KG,j}^2 = m_j^2 + (1 - a/2Z^2)M_{KG}^2$, where M_{KG} is the 7D mass and $j \geq 1$.

In summary, we investigated the brane physics induced by a global monopole formed in three extra dimensions. This background induces a metric with divergent extra-dimensional volume. As usual for such models, the graviton zero-mode is not normalizable without a cutoff. We view a cutoff as physically natural, due to formation of adjacent defects during the dimension-reducing phase transition, and choose the cut-off radius to induce a hierarchical 4D Planck mass ($\sim 10^{18}$GeV) from a unified 7D Planck mass ($\sim$TeV). For reasonable parameters this cut-off radius remains consistent with current observations. Massive KK gravitons do not harm 4D gravity on the brane if they are either high or low enough in mass, a model-dependent question. 7D matter fields also localize on the brane, with 4D mass spectrum like that of the gravitons. Their effective field theory remains an interesting question.

We thank Gia Dvali, Alex Vilenkin, Takahiro Tanaka, and Gaume Garriga and Chi-Ok Hwang for helpful discussions. This work was supported by the University Research Committee of Emory University.

REFERENCES

1. T. Kaluza, Preus. Acad. Wiss. K **1**, 966 (1921); O. Klein, Zeit. Phys. **37**, 895 (1926).
2. N. Arkani-Hamed, S. Dimopoulos and G. Dvali, Phys. Lett. B **429**, 263 (1998); Phys. Rev. D **59**, 086004 (1999).
3. L. Randall and R. Sundrum, Phys. Rev. Lett. **83**, 3370 (1999); *ibid.* 4690 (1999).
4. V. Rubakov and M. Shaposhnikov, Phys. Lett. B **125**, 136 (1983); *ibid.* 139 (1983); G. Dvali and M. Shifman, Phys. Lett. B **396**, 64 (1997).
5. A. Cohen and D. Kaplan, Phys. Lett. B **470**, 52 (1999); T. Gherghetta and M. Shaposhnikov, Phys. Rev. Lett. **85**, 240 (2000); T. Gherghetta, E. Rossel and M. Shaposhnikov, hep-th/0006251.
6. I. Olasagasti and A. Vilenkin, Phys. Rev. D **62**, 044014 (2000).
7. K. Benson and I. Cho, hep-th/0104067, submitted to Phys. Rev. D.

Dynamics of Cosmological Models in the Brane-world Scenario

Antonio Campos* and Carlos F. Sopuerta*

*Relativity and Cosmology Group,
Portsmouth University, Portsmouth PO1 2EG, Britain*

Abstract. We present the results of a systematic investigation of the qualitative behaviour of the Friedmann-Lemaître-Robertson-Walker (FLRW) and Bianchi I and V cosmological models in Randall-Sundrum brane-world type scenarios.

Recently, Randall and Sundrum have shown that for non-factorizable geometries in five dimensions the zero-mode of the Kaluza-Klein dimensional reduction can be localized in a four-dimensional submanifold [1]. The picture of this scenario is a five-dimensional space with an embedded three-brane where matter is confined and Newtonian gravity is effectively reproduced at large distances.

Here, we summarize the qualitative behaviour of FLRW and Bianchi I and V cosmological models in this scenario (see [2] for more details). In particular, we have studied how the dynamics changes with respect to the general-relativistic case. For this purpose we have used the formulation introduced in [3]. From the Gauss-Codazzi relations the Einstein equations on the brane are modified with two additional terms. The first term is quadratic in the matter variables and the second one is the electric part of the five-dimensional Weyl tensor. In this communication we will consider the effects due to the first term. The study including both corrections has been carried out in [4]. We also assume that the matter content is described by a perfect fluid with energy density, ρ, and pressure, p, related by a linear barotropic equation of state, $p = (\gamma - 1)\rho$ with $\gamma \in [0, 2]$.

When the brane dynamics is described by a FLRW model we find five generic critical points: the flat FLRW models (F); the Milne universe (M); the de Sitter model (dS); the Einstein universe (E); and the non-general-relativistic Binétruy-Deffayet-Langlois (BDL) model (m) [5]. The dynamical character of these critical points and the structure of the state space depend on the equation of state, or in other words, on the parameter γ. This means that we have bifurcations for some values of γ, namely $\gamma = 0, \frac{1}{3}, \frac{2}{3}$. The bifurcation at $\gamma = \frac{1}{3}$ is a genuine feature of the brane world and is characterized by the appearance of an infinite number of non-general-relativistic critical points. The Einstein Universe critical point appears

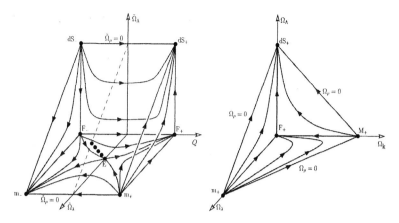

FIGURE 1. State space for the FLRW models with $\gamma \in (\frac{1}{3}, \frac{2}{3})$ and non-negative (left) and non-positive (right) spatial curvature. The variables $(\Omega_\rho, \Omega_k, \Omega_\Lambda, \Omega_\lambda)$, and their analogs with tilde, are fractional contributions of the energy density, spatial curvature, cosmological constant and brane tension, respectively, to the universe expansion [2]. Replacing Ω_k by Ω_σ (the shear contribution) and M_+ by K_+, the drawing on the right is also the state space for Bianchi I models with $\gamma \in (1, 2)$. For clarity, only trajectories on the invariant planes have been drawn. The dynamics of a general trajectory can be inferred from them. The subscript "+" ("−") refers to the expanding (contracting) character of the models. The planes $\tilde{\Omega}_\lambda = \Omega_\lambda = 0$ correspond to the state space of general relativity.

for $\gamma \geq \frac{1}{3}$, in contrast with the general-relativistic case, where it appears for $\gamma \geq \frac{2}{3}$. Actually, for $\frac{1}{3} < \gamma < 2$ we do not have an isolated critical point corresponding to the Einstein universe but a line of critical points, as can be seen in the state space shown in Figure 1. Another important feature of these scenarios is that the dynamical character of some of the points changes. For instance, the expanding and contracting flat FLRW models, which in general relativity are repeller and attractor for $\gamma > \frac{2}{3}$, are now saddle points for all values of γ. The new non-general-relativistic critical point, the BDL solution [5], describes the dynamics near the initial Big-Bang singularity and, for recollapsing models, near the Big-Crunch singularity. More precisely, the dynamical behaviour near these singularities is governed by a scale factor $a(t) = t^{1/(3\gamma)}$ which differs from the standard evolution in general-relativistic cosmology, where $a(t) = t^{2/(3\gamma)}$. Finally, the general attractor for ever expanding universes is, as in general relativity, the de Sitter model. For recollapsing universes, which now appear for $\gamma > \frac{1}{3}$, the contracting BDL model is the general attractor. However, if we only consider the invariant manifold representing general relativity, the contracting Friedmann universe is the general attractor for $\gamma > 2/3$. On the other hand, for zero cosmological constant and $\gamma < 2/3$ the expanding Friedmann universe is also an attractor.

For the homogeneous but anisotropic Bianchi I and V cosmological models, which contain the flat and negatively curved FLRW models respectively, we find an additional critical point, namely the Kasner vacuum spacetimes (K). In the Bianchi I case the state space can be represented by the same type of drawings used for the non-positive spatial curvature sector of the FLRW evolution (see Figure 1). A representative set of diagrams for Bianchi V models is given in [2]. For Bianchi I models we have found a new bifurcation at $\gamma = 1$ and for Bianchi V models at $\gamma = \frac{1}{3}, 1$, in addition to the general relativity bifurcations at $\gamma = 0, 2$ and $\gamma = 0, \frac{2}{3}, 2$, respectively. Some of the dynamical features explained above for the FLRW are shared by the these Bianchi models. However, the most interesting point here is the possibility of studying the dynamics of anisotropy in brane-world scenarios. Specifically, we have seen [2] that, although now we can have intermediate stages in which the anisotropy grows, expanding models isotropize as it happens in general relativity. This is expected since the energy density decreases and hence, the effect of the extra dimension becomes less and less important. The situation near the Big Bang is more interesting. In the brane-world scenario anisotropy dominates only for $\gamma < 1$, whereas in general relativity dominates for all the physically relevant values of γ.

To conclude, let us summarize the main features of the dynamics of cosmological models on the brane. First, we have found new equilibrium points, the BDL models [5], representing the dynamics at very high energies, where the extra-dimension effects become dominant. Thus, we expect them to be a generic feature of the state space of more general cosmological models in the brane-world scenario. Second, the state space presents new bifurcations for some particular equations of state. Third, the dynamical character of some of the critical points changes with respect to the general-relativistic case. Finally, for models in the range $1 < \gamma \leq 2$, that is for models satisfying all the ordinary energy conditions and causality requirements, we have seen that the anisotropy is negligible near the initial singularity. This naturally leads to the questions of whether the oscillatory behaviour approaching the Big Bang predicted by general relativity is still valid in brane-world scenarios. We are currently investigating this issue by considering Bianchi IX cosmological models [6].

Acknowledgments: This work has been supported by the European Commission (contracts HPMF-CT-1999-00149 and HPMF-CT-1999-00158).

REFERENCES

1. L. Randall and R. Sundrum, Phys. Rev. Lett **83**, 4690 (1999).
2. A. Campos and C. F. Sopuerta, Phys. Rev. D **63**, 104012 (2001).
3. T. Shiromizu, K. Maeda, and M. Sasaki, Phys. Rev. D **62**, 024012 (2000).
4. A. Campos and C. F. Sopuerta, submitted to Phys. Rev. D (hep-th/0105100).
5. P. Binétruy, C. Deffayet, and D. Langlois, Nucl. Phys. **B565**, 269 (2000).
6. A. Campos and C. F. Sopuerta, *in preparation*.

Initial conditions for hybrid inflation on the brane

Luís E. Mendes* and Andrew R. Liddle*

Astronomy Centre, University of Sussex, Brighton BN1 9QJ, United Kingdom

Abstract. In hybrid inflation models, typically only a tiny fraction of possible initial conditions give rise to successful inflation, even if one assumes spatial homogeneity. We show that hybrid inflation on the brane is not afflicted by this problem.

INTRODUCTION

Particle physics motivated model building of inflation has undergone a renaissance in the last few years, with the realization that the hybrid inflation model introduces a natural framework within which to implement supersymmetry and supergravity-based models of inflation [1, 2, 3]. The important new ingredient brought to the picture by supergravity is that the potential should only be believed for field values below the reduced Planck mass,[1] whereas previously only the constraint that the total energy density be below the Planck scale was imposed.

In conventional models of inflation where the inflationary epoch ends by leaving the slow-roll regime, it is problematic to obtain sufficient inflation subject to this condition on the field values. In the hybrid inflation model [4], inflation ends via an instability triggered by a second field, obviating the need for the troublesome fast-rolling phase, provided the vacuum energy part of the potential dominates over the $m^2\phi^2/2$ term. However Tetradis [5] has shown that in order for inflation to start, the fields must be initially located in a very narrow band around the valley of the potential in the direction of the inflaton, otherwise the fields will quickly oscillate around the bottom of the valley, and pass beyond the instability point in the potential without inflation, eventually settling in one of the minima of the potential along the axis of the second field.

In this contribution we show that the initial conditions problem is solved for hybrid inflation on the brane.

---

[1] The reduced Planck mass is given by $M_{Pl} = m_{Pl}/\sqrt{8\pi} \approx 0.2 m_{Pl}$, where m_{Pl} is the true Planck mass $G^{-1/2}$, G being Newton's constant.

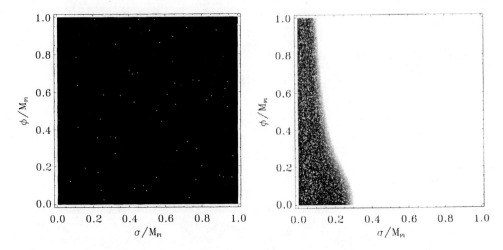

FIGURE 1. A plot of the space (ϕ_i, σ_i) of initial conditions for hybrid inflation. White regions correspond to initial conditions which give rise to successful inflation ($N_{efold} \geq 70$), while black regions correspond to initial conditions which quickly lead the fields to the minima along the σ axis with no successful inflation. The left panel shows the standard case. Although hard to see, the successful region is a very narrow strip at the $\sigma = 0$ axis, along with some scattered points off-axis. This plots was obtained with $\lambda = \lambda' = 1$, $m = 5 \times 10^{-6} M_{Pl}$ and $M = 2 \times 10^{-2} M_{Pl}$, which satisfies the COBE normalization. On the right panel we show the results for the brane scenario with $\lambda_b = 6 \times 10^{-8} M_{Pl}^4$, $m = 2 \times 10^{-6} M_{Pl}$, and $M = 10^{-2} M_{Pl}$: a much larger fraction of the space of initial conditions gives rise to successful inflation.

HYBRID INFLATION ON THE BRANE

We assume that the Universe is described by a flat Friedmann–Robertson–Walker geometry. We consider the original hybrid inflation potential, given by

$$V(\phi,\sigma) = \frac{1}{4}\lambda\left(\sigma^2 - M^2\right)^2 + \frac{1}{2}m^2\phi^2 + \frac{1}{2}\lambda'\phi^2\sigma^2, \tag{1}$$

where ϕ is the inflaton and σ the field which triggers the end of inflation. We impose the restrictions $0 < \lambda, \lambda' < 1$. Our numerical simulations (left panel in Fig. 1) confirm the results obtained by Tetradis [5].

In the brane scenario [6, 7] however, the Friedman equation takes the non-standard form [8]

$$H^2 = \frac{1}{3M_{Pl}^2}\rho\left(1 + \frac{\rho}{2\lambda_b}\right). \tag{2}$$

where λ_b is the brane tension. For high energies the quadratic term in ρ dominates and contributes to the increase of the friction term in the equations of motion for the scalar fields ϕ and σ. For lower energies however, we want to recover the usual behaviour $H^2 \sim \rho$. In particular, by the beginning of nucleosynthesis we require the linear term in the Friedman equation to be dominant, in order to reproduce the usual BBN scenario. This will constrain the brane tension $\lambda_b > 1\,\text{MeV}^4$.

Imposing the COBE normalization on the spectrum of density perturbations, M and m must obey [9]

$$\lambda^{1/4} M < 3.8 \times 10^{-1} (\lambda\lambda')^{-1/6} \lambda_b^{1/6} M_{Pl}^{1/3}; \qquad (3)$$

$$m < 1.1 \times 10^{-3} \left(\frac{\lambda}{\lambda'^2}\right)^{1/3} \lambda_b^{1/6} M_{Pl}^{1/3}. \qquad (4)$$

Due to the increased friction produced by the the quadratic term in the energy density, we expect the scalar fields to suffer more damping during the pre-inflationary stages of evolution, therefore preventing them from rolling too fast to one of the global minima before inflation begins. This is exactly the behaviour we observe in our simulations shown in the right panel of Fig. 1. As we decrease the value of λ_b, the onset of inflation becomes even easier as this further increases the friction term in the scalar field equations of motion. There is also a smoother transition between regions where successful inflation occurs (at least 70 e-foldings of expansion) and those where inflation cannot start [9].

It is debatable whether one should be too concerned that only a limited region of initial condition space leads to sufficient hybrid inflation in the models we've considered, as one can readily argue, either probabilistically or anthropically, that the Universe we live in would originate from one of these regions. Nevertheless, it is a relevant question as to whether or not the problem is mitigated once the scenario is generalized beyond the standard picture and we have studied a very promising alternative based on the brane scenario. It is encouraging to know that besides being a possible solution for the hierarchy problem, the brane scenario also mitigates some of the fine-tuning problems that plague conventional inflation.

ACKNOWLEDGMENTS

L.E.M. is supported by FCT (Portugal) under contract PRAXIS XXI BPD/14163/97. We thank Ed Copeland, David Lyth, Anupam Mazumdar and Arttu Rajantie for useful discussions. We acknowledge the use of the Starlink computer system at the University of Sussex. Part of this work was conducted on the SGI Origin platform using COSMOS Consortium facilities, funded by HEFCE, PPARC and SGI.

REFERENCES

1. Copeland, E.J., Liddle, A.R., Lyth, D.H., Stewart, E.D., and Wands, D. *Phys. Rev. D* **49**, 6410–6433 (1994).
2. Lazarides, G., and Tetradis, N., *Phys. Rev. D* **58**, 123502 (1998).
3. Lyth, D.H., and Riotto, A., *Phys. Rept.* **314**, 1–146 (1999).
4. Linde, A., *Phys. Lett B* **259**, 38–47 (1991).
5. Tetradis, N., *Phys. Rev. D* **57**, 5997–6002 (1998).
6. Gogberashvili, M., *Europhys. Lett.* **49**, 396–399 (2000).
7. Randall, L., and Sundrum, R., *Phys. Rev. Lett.* **83**, 4690–4693 (1999).
8. Binétruy, P., Deffayet, C., and Langlois, D., *Nucl. Phys.* **B565**, 269–287 (2000).
9. Mendes, L.E., and Liddle, A.R., *Phys. Rev. D* **62**, 103511 (2000).

CHAPTER 2

EARLY UNIVERSE, INFLATION, BARYOGENESIS, AND PARTICLE PRODUCTION

New constraints on inflation from the CMB

William H. Kinney*, Alessandro Melchiorri† and Antonio Riotto**

*Institute for Strings, Cosmology and Astroparticle Physics, Columbia University, New York, NY
†Dipartmento di Fisica, Universitá la Sapienza, Rome, Italy
**Scuola Normale Superiore, Piazza dei Cavalieri 7,Pisa I-56126, Italy

Abstract. The recent data from the Boomerang and MAXIMA-1 balloon flights have marked the beginning of the precision era of Cosmic Microwave Background anisotropy (CMB) measurements. We discuss the observational constraints from the current CMB anisotropy measurements on the simplest inflation models, characterized by a single scalar field ϕ, in the parameter space consisting of scalar spectral index n_S and tensor/scalar ratio r. The data are consistent with the simplest assumption of a scale invariant power spectrum, but the specific error contours in the $r - n$ plane depend on prior assumptions, particularly the baryon density of the universe and the reionization history. Models with significant "red" tilt ($n < 1$) and appreciable tensor fluctuations are disfavored in all cases.

THE INFLATIONARY MODEL SPACE

Inflation[1] is the leading theory for the origin of density fluctuations in the universe. It is therefore particularly relevant to investigate constraints on inflationary models arising from observation of the Cosmic Microwave Background (CMB). In this talk[2], we will confine ourselves to consideration of inflation models with a single dynamical degree of freedom, which we will take to be a scalar field ϕ. In single-field models of inflation, the inflaton "field" need not be a fundamental field at all, of course. Also, some "single-field" models require auxiliary fields. Hybrid inflation models[3, 4, 5], for example, require a second field to end inflation. What is significant is that the inflationary epoch be described by a single dynamical order parameter, the inflaton field. The metric perturbations created during inflation are of two types: scalar, or *curvature* perturbations, which couple to the stress-energy of matter in the universe and form the "seeds" for structure formation, and tensor, or gravitational wave perturbations. Both scalar and tensor perturbations contribute to CMB anisotropy. Scalar fluctuations can also be interpreted as fluctuations in the density of the matter in the universe. Most (but not all) inflationary models predict that these fluctuations are generated with approximately power law spectra,

$$P_S(k) \propto k^{n_S-1}, \\ P_T(k) \propto k^{n_T}. \quad (1)$$

The spectral indices n_S and n_T are assumed to vary slowly or not at all with scale:

$$\frac{dn_{S,T}}{d\log k} \simeq 0. \quad (2)$$

Some particular inflation models predict running of the spectral index with scale or even sharp features in the power spectrum. In this talk, we restrict ourselves to the more generic power-law case. In principle we have four independent observables with which to test inflation. One can be removed by considering the overall amplitude of CMB fluctuations: if the contribution of tensor modes to the CMB anisotropy can be neglected, normalization to the COBE four-year data gives[6, 7] $P_S^{1/2} = 4.8 \times 10^{-5}$. We then have three relevant observables: the tensor scalar ratio defined as a ratio of quadrupole moments,

$$r \equiv \frac{C_2^{\text{Tensor}}}{C_2^{\text{Scalar}}} \qquad (3)$$

and the two spectral indices n_S and n_T. However, r, n_S and n_T are not independent. The tensor spectral index and the tensor/scalar ratio are related as

$$n_T = -\frac{1}{6.8} r, \qquad (4)$$

known as the *consistency relation* for inflation. The relevant parameter space for distinguishing between inflation models is then the $r - n_S$ plane.

Even restricting ourselves to a simple single-field inflation scenario, the number of models available to choose from is large [8]. A generic single-field potential can be characterized by two independent mass scales: a "height" Λ^4, corresponding to the vacuum energy density during inflation, and a "width" μ, corresponding to the change in the field value $\Delta \phi$ during inflation:

$$V(\phi) = \Lambda^4 f\left(\frac{\phi}{\mu}\right). \qquad (5)$$

Different models have different forms for the function f. The height Λ is fixed by normalization, so the only free parameter is the width μ. It is convenient to define a general classification scheme, or "zoology" for models of inflation. We divide models into three general types: *large-field*, *small-field*, and *hybrid*, with a fourth classification, *linear* models, serving as a boundary between large- and small-field. Figure 1 shows the parameter space divided into regions for small-field, large-field and hybrid models. To illustrate the different behaviors of the different types of inflation models, we choose a set of potentials and plot them on the $r - n_S$ plane. The generic large-field potentials we consider are polynomial potentials $V(\phi) = \Lambda^4 (\phi/\mu)^p$, and exponential potentials, $V(\phi) = \Lambda^4 \exp(\phi/\mu)$. The generic small-field potentials we consider are of the form $V(\phi) = \Lambda^4 [1 - (\phi/\mu)^p]$, which can be viewed as a lowest-order Taylor expansion of an arbitrary potential about the origin. Linear models, $V(\phi) \propto \phi$, live on the boundary between large-field and small-field models. We consider generic potentials for hybrid inflation of the form $V(\phi) = \Lambda^4 [1 + (\phi/\mu)^p]$. The field value at the end of inflation is determined by some other physics, so there is a second free parameter characterizing the models. Because of this extra freedom, hybrid models fill a broad region in the $r - n_S$ plane. This enumeration of models is certainly not exhaustive. There are a number of single-field models that do not fit well into this scheme, for example logarithmic potentials $V(\phi) \propto \ln(\phi)$ typical of supersymmetry [8].

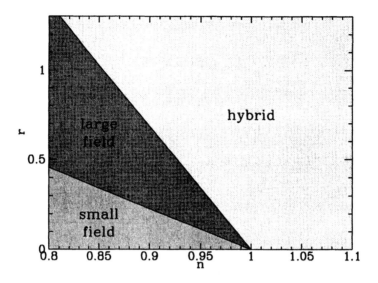

FIGURE 1. The parameter space divided into regions for small-field, large-field and hybrid models. The linear case is the dividing line between large- and small-field.

RESULTS

In this section we discuss the ability of current CMB data to place constraints on the inflationary parameter space. In particular, we wish to determine which regions of the inflationary model space are ruled out by the Boomerang and MAXIMA-1 data[9, 10, 11, 12], and which regions are consistent with the observed data. We perform a likelihood analysis over multiple cosmological parameters and project likelihood contours onto the $r - n_S$ plane. The choice of parameters to be varied in the analysis is crucial, and different assumptions result in different constraints on the inflationary model space. In addition to n_S and r, we also vary the Hubble constant h, the reionization optical depth τ_c, the amplitude of fluctuations, C_{10}, (in units of C_{10}^{COBE}), the cold dark matter density Ω_M, the vacuum energy density Ω_Λ, and the baryon density Ω_M, subject to the constraint of a flat universe $\Omega_{total} = \Omega_B + \Omega_{CDM} + \Omega_\Lambda = 1$, consistent with the prediction of inflation. Figure 2 shows the 1σ ($\delta\chi^2 = 2.3$), 2σ ($\delta\chi^2 = 6.0$), and 3σ ($\delta\chi^2 = 9.2$) contours on the $r - n_S$ plane for the case of no reionization and no BBN prior. The various inflation models described in the last section and in Figure 1 are now plotted as labeled lines on the graph. (Note that the entire region to the right of the case of the exponential potential is consistent with hybrid inflation models). The lines are obtained by varying a given parameter within a class of models, e.g. for large-field models with a power-law potential ϕ^p what varies is the power p. The constraints on model parameters also depend on the number of e-folds N of inflation assumed, which we take to be between 50 and 70.

The best model is nearly scale invariant, and the tensor/scalar ratio r is only weakly

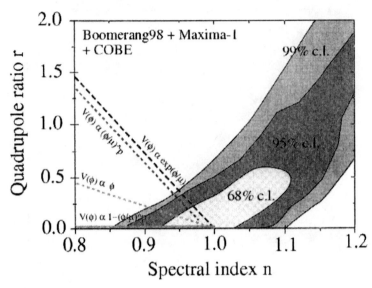

FIGURE 2. CMB constraints and inflation models for $\tau_c = 0$ and no BBN prior. The allowed contours are quite large but still exclude a significant portion of the inflationary model space.

constrained. Marginalizing over r (n_s) we obtain also the $1-\sigma$ constraint : $0.94 < n_s < 1.07$ ($r < 0.65$). Hybrid and small-field models are consistent with the data, but strongly tilted large-field models are in conflict with the CMB constraint to high significance. The 3σ contour on the $V(\phi) \propto \phi^p$ models result in a constraint that $p < 5$ for $N = 50$ and $p < 8$ for $N = 70$. In the power-law inflation case, $V(\phi) \propto \exp(\phi/\mu)$, the 3σ constraint corresponds to a lower limit $\mu > 0.75 M_{Pl}$. The 1σ contour is marginally consistent with $p = 2$ for $N = 50$ and requires $p < 3$ for $N = 70$ in the case of a polynomial potential, and constrains $\mu \geq M_{Pl}$ in the power-law inflation case.

Figure 3 shows a similar plot when BBN prior consistent with low deuterium abundance[13], $0.016 \leq \Omega_b h^2 \leq 0.021$, is included, still assuming negligible reionization. The contours in r are significantly tighter than in the case with no BBN prior, and the best fit model is shifted toward tilted models: $0.88 < n_S < 0.98$ and $r \leq 0.17$, with the scale invariant case just consistent with the data at 2σ. Power law inflation and polynomial potentials are ruled out to 68% confidence, as well as nearly all of the rest of the large-field region of the $n_s - r$ plane. The only models within 1σ are linear and small-field inflation. To 3σ, the case of a polynomial potential is constrained to $p < 6$ for $N = 50$ and $p < 8$ for $N = 70$.

Figure 4 shows the constraints when a strong constraint on Ω_b from BBN is assumed (left to right) and when the optical depth τ_c is increased. Fixing the optical depth to $\tau_c \sim 0.2$ shifts the likelihood towards "blue" tilted models ($n_S \geq 1$) making the scale invariant models consistent with the data even when the BBN constraint are assumed ($1.02 < n_S < 1.28$ for no constraint, $0.95 < n_S < 1.05$ for both BBN constraints, all at 68% c.l.). Increasing this parameter up to $\tau_c = 0.4$ moves the 68% likelihood contours

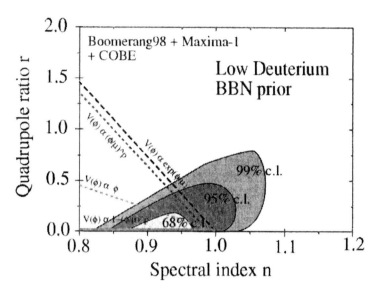

FIGURE 3. CMB constraints and inflation models for $\tau_c = 0$ and a strong BBN prior, $0.016 \leq \Omega_b h^2 \leq 0.021$. The error contours are significantly smaller than when no prior is assumed on the baryon density.

in the hybrid models region ($1.15 < n_S < 1.30$, $1.06 < n_S < 1.14$ for low deuterium and $1.00 < n_S < 1.08$ for high deuterium). These models are disfavored by the dataset itself, being the best fit model at $\Delta\chi^2 \sim 3$ from the corresponding best fit model with $\tau_c = 0$. Note in particular that large-field models with tilt $n_s < 0.9$ are strongly disfavored in all cases. Removing the consistency relation and fixing $n_t = 0$ doesn't affect this conclusion and has small effect on the overall result: for no BBN constraint and $\tau_c = 0$ we found $r < 0.67$ and $0.96 < n_s < 1.14$ (marginalized over r).

Perhaps the most important point to be drawn from the data at this stage is that for the first time, the error bars are smaller than the plot! This is the beginning of our ability to use the CMB to quantitatively constrain models of the very early universe. The quality of CMB data is rapidly becoming better, and we can expect much more accurate estimation of cosmological parameters in the very near future, including those which will enable us to "kill off" many of the proposed models of inflation.

REFERENCES

1. A. Guth, Phys. Rev. D **23**, 347 (1981).
2. For a more detailed treatment and extensive relevant references, see W. H. Kinney, A. Riotto, and A. Melchiorri, Phys. Rev. D **63**, 023505 (2001), astro-ph/0007375.
3. A. Linde, Phys. Lett. **B259**, 38 (1991).
4. A. Linde, Phys. Rev. D **49**, 748 (1994), astro-ph/9307002.
5. E. J. Copeland, A. R. Liddle, D. H. Lyth, E. D. Stewart, and D. Wands, Phys. Rev. D **49**, 6410 (1994).
6. E. F. Bunn and M. White, Astrophys. J. **480**, 6 (1997).
7. D. H. Lyth, Report No. hep-ph/9609431.

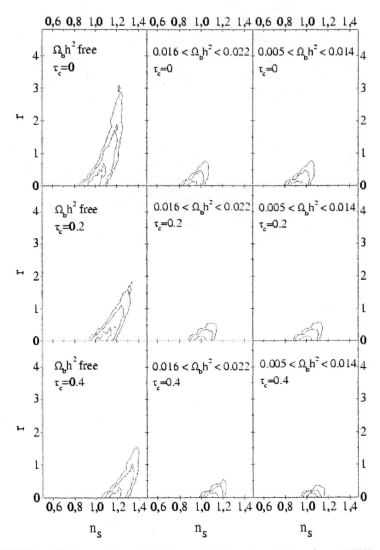

FIGURE 4. CMB constraints and inflation models for various choices of τ_c and BBN priors.

8. For a review, see D. H. Lyth and A. Riotto, Phys. Rept. **314**, 1 (1999), hep-ph/9807278.
9. P. de Bernardis *et al.*, Nature **404**, 955 (2000).
10. A. E. Lange *et al.*, astro-ph/0005004.
11. S. Hanany *et al.*, astro-ph/0005123.
12. A. Balbi *et al.*, astro-ph/0005124.
13. S. Burles, K. M. Nollett, J. W. Truran, and M. S. Turner, Phys. Rev. Lett. **82**, 4176 (1999), astro-ph/9901157.

Natural Chaotic Inflation in Supergravity and Leptogenesis

Masahide Yamaguchi[1], M. Kawasaki[2], and T. Yanagida[3]

[1,2,3] *Research Center for the Early Universe, University of Tokyo, Tokyo, 113-0033, Japan*
[3] *Department of Physics, University of Tokyo, Tokyo 113-0033, Japan*

Abstract. We propose a chaotic inflation model in supergravity. In this model, the form of Kähler potential is determined by a symmetry, that is, the Nambu-Goldstone like shift symmetry, which guarantees the absence of the exponential factor in the potential for the inflaton field. Though we need introduction of small parameters, the smallness of parameters is justified also by symmetries. That is, the zero limit of small parameters recovers symmetries, which is natural in 't Hooft's sense. The leptogenesis scenario via the inflaton decay in this chaotic inflation model is also discussed. We find that the lepton asymmetry enough to explain the present baryon number density is produced for low reheating temperatures avoiding the overproduction of gravitinos.

I INTRODUCTION

Chaotic inflation and supergravity have not been naturally realized simultaneously. The main reason is that the minimal supergravity potential has an exponential factor, $\exp(\frac{\varphi_i^* \varphi_i}{M_G^2})$, which prevents any scalar fields φ_i from having values larger than the gravitational scale $M_G \simeq 2.4 \times 10^{18}$ GeV. However, the inflaton φ is supposed to have a value much larger than M_G at the Planck time to cause the chaotic inflation. Thus, the above effect makes it very difficult to incorporate the chaotic inflation in the framework of supergravity. In fact, all of the existing models for chaotic inflation use rather specific Kähler potential, and one needs a fine tuning in the Kähler potential since there is no symmetry reason for having such specific forms of Kähler potentials. Thus, it is very important to find a natural chaotic inflation model without any fine tuning.

In this talk, we propose a natural chaotic inflation model where the form of Kähler potential is determined by a symmetry. With this Kähler potential the inflaton φ may have a large value $\varphi \gg M_G$ to begin the chaotic inflation. Our models, in fact, need two small parameters for successful inflation. However, we emphasize that the smallness of these parameters is justified by symmetries and hence the model is natural in 't Hooft's sense.

As an application of the above new type of chaotic inflation model, we discuss the leptogenesis. Recent experimental results on the atmospheric neutrinos strongly indicate that neutrinos have small masses of the order of 0.01-0.1 eV. Such small masses are naturally explained by the seesaw mechanism, which predicts superheavy right-handed neutrinos. The presence of Majorana masses of right-handed neutrinos naturally leads to the leptogenesis because it violates the lepton number conservation. The decay of superheavy Majorana neutrinos produces the lepton number asymmetry, in particular $B - L$ asymmetry, if C and CP symmetries are broken, which is converted into baryon asymmetry through the sphaleron effects. Therefore, we discuss a leptogenesis scenario in the above mentioned chaotic inflation model.

II NATURAL CHAOTIC INFLATION MODEL IN SUPERGRAVITY

For the inflaton chiral supermultiplet $\Phi(x,\theta)$, we assume that the Kähler potential $K(\Phi,\Phi^*)$ is invariant under the shift of Φ,

$$\Phi \to \Phi + i\, CM_G, \qquad (1)$$

where C is a dimensionless real parameter. Hereafter, we set M_G to be unity. Thus, the Kähler potential is a function of $\Phi + \Phi^*$, i.e. $K(\Phi,\Phi^*) = K(\Phi + \Phi^*)$. It is now clear that the supergravity effect $e^{K(\Phi+\Phi^*)}$ discussed above does not prevent the imaginary part of the scalar components of Φ from having a value larger than the gravitational scale. So, we identify it with the inflaton field φ (see eq.(8)). As long as the shift symmetry is exact, the inflaton φ never has a potential and hence it never causes inflation. Therefore, we need some breaking term in the superpotential. Here, we discuss the form of the superpotential. First of all, we assume that in addition to the shift symmetry, the superpotential is invariant under the $U(1)_R$ symmetry, which prohibits a constant term in the superpotential. Then, the above Kähler potential is invariant only if the R-charge of Φ is zero. Therefore, the superpotential comprised of only the Φ field is not invariant under the $U(1)_R$ symmetry, which compels us to introduce another supermultiplet $X(x,\theta)$ with its R-charge equal to two.

We now introduce a suprion field Ξ describing the breaking of the shift symmetry, and extend the shift symmetry including the suprion field Ξ as follows,[1]

$$\begin{aligned}\Phi &\to \Phi + i\,C \\ \Xi &\to \frac{\Phi}{\Phi + i\,C}\Xi.\end{aligned} \qquad (2)$$

---

[1] If Ξ transforms as $\Xi \to \frac{\Phi^n}{(\Phi+iC)^n}\Xi$ $(n \geq 2)$, we have $W = X\Xi\Phi^n$, which may cause φ^{2n} chaotic inflations.

That is, the combination $\Xi\Phi$ is invariant under the shift symmetry. Then, the general superpotential invariant under the shift and $U(1)_R$ symmetries is given by

$$W = X\left\{\Xi\Phi + \alpha_3(\Xi\Phi)^3 + \cdots\right\} + \delta_1 X\left\{1 + \alpha_2(\Xi\Phi)^2 + \cdots\right\}, \quad (3)$$

where we have assumed R-charge of Ξ vanish. The shift symmetry is softly broken by inserting the vacuum value $\langle\Xi\rangle = m$. The mass parameter m is fixed at a value much smaller than unity representing the magnitude of breaking of the shift symmetry (2). We see that higher order terms with α_i of the order of unity become irrelevant for the dynamics of the chaotic inflation. Thus, we neglect them in the following discussion [2] unless explicitly mentioned. We should note that the complex constant δ_1 is also of the order of unity in general. But, as shown later, the absolute magnitude of δ_1 must be at most of the order of m, which is much smaller than unity. Therefore, we introduce the Z_2 symmetry, under which both the Φ and X fields are odd. Then, the smallness of the constant δ_1 is associated with the small breaking of the Z_2 symmetry. That is, we introduce a suprion field Π with odd charge under the Z_2 symmetry. The vacuum value $\langle\Pi\rangle = \delta_1$ breaks the Z_2 symmetry. Though the above superpotential is not invariant under the shift and the Z_2 symmetries, the model is completely natural in 't Hooft's sense because we have enhanced symmetries in the limit m and $\delta_1 \to 0$. We use, in the following analysis, the superpotential,

$$W \simeq mX\Phi + \delta_1 X. \quad (4)$$

The Kähler potential invariant under the shift and $U(1)_R$ symmetries is give by

$$K = \delta_2(\Phi + \Phi^*) + \frac{1}{2}(\Phi + \Phi^*)^2 + XX^* + \cdots. \quad (5)$$

Here $\delta_2 \sim |\delta_1|$ is a real constant representing the breaking effect of the Z_2 symmetry. The terms $\delta_3 m_3 \Phi + \delta_3^* m_3^* \Phi^*$ and $(m_4\Phi)^2 + (m_4^*\Phi^*)^2$ may appear, where δ_3 and m_4 are complex constants representing the breaking of the Z_2 and the shift symmetries($|\delta_3| \sim |\delta_1|$ and $|m_3| \sim |m_4| \sim m$). But, these terms are extremely small so that we have omitted them in the Kähler potential (5). We have also omitted a constant term because it only changes the overall factor of the potential, whose effect can be renormalized into the constant m and δ_1. Here and hereafter, we use the same characters for scalar with those for corresponding supermultiplets.

III DYNAMICS OF CHAOTIC INFLATION

The Lagrangian density $\mathcal{L}(\eta, \varphi, X)$ neglecting higher order terms is given by

---

[2] Among all complex constants, only a constant becomes real by use of the phase rotation of the X field. Below we set m to be real.

$$\mathcal{L}(\eta,\varphi,X) = \frac{1}{2}\partial_\mu\eta\partial^\mu\eta + \frac{1}{2}\partial_\mu\varphi\partial^\mu\varphi + \partial_\mu X\partial^\mu X^* - V(\eta,\varphi,X), \tag{6}$$

with the potential $V(\eta,\varphi,X)$,

$$\begin{aligned} V(\eta,\varphi,X) &= m^2 e^{-\frac{\delta_2^2}{2}} \exp\left\{\left(\eta + \frac{\delta_2}{\sqrt{2}}\right)^2 + |X|^2\right\} \\ &\times \left[|X|^2\left\{1 + 2\left(\eta + \frac{\delta_2}{\sqrt{2}}\right)(\eta + \delta_R) + \left(\eta + \frac{\delta_2}{\sqrt{2}}\right)^2\left[(\eta + \delta_R)^2 + (\varphi + \delta_I)^2\right]\right\}\right. \\ &\left. + \frac{1}{2}\left\{(\eta + \delta_R)^2 + (\varphi + \delta_I)^2\right\}(1 - |X|^2 + |X|^4)\right]. \end{aligned} \tag{7}$$

Here, the complex field Φ and the complex constant δ'_1 are decomposed into a real and an imaginary parts,

$$\Phi = \frac{1}{\sqrt{2}}(\eta + i\varphi), \qquad \delta'_1 = \frac{1}{\sqrt{2}}(\delta_R + i\delta_I). \tag{8}$$

Then, due to the exponential factor, the potential can be approximated as,

$$V(\eta,\varphi,X) \simeq \frac{1}{2}m^2\tilde{\varphi}^2 + m^2|X|^2, \tag{9}$$

where $\tilde{\varphi} \equiv \varphi + \delta_I$ and we have taken $me^{-\delta_2^2/4} \simeq m$ since $\delta_2 \ll 1$. Thus, the term proportional to $\tilde{\varphi}^2$ becomes dominant and the chaotic inflation takes place. After a little calculation, we find that the condition $\delta_2 \sim |\delta_1| \lesssim m \sim 10^{-5}$ must be satisfied for a successful inflation.

The density fluctuations produced by this chaotic inflation is estimated as

$$\frac{\delta\rho}{\rho} \simeq \frac{1}{5\sqrt{3}\pi}\frac{m}{2\sqrt{2}}\left\{(\varphi + \delta_I)^2 + X^2\right\}. \tag{10}$$

Since we can easily show that $X \ll \varphi + \delta_I$, the amplitude of the density fluctuations is actually determined only by the φ field. Then, the normalization at the COBE scale gives $m \simeq 10^{13}$ GeV $\simeq 10^{-5}$.

The reheating may take place by introducing the following superpotential:

$$W = \delta_4 X H_u H_d, \tag{11}$$

where $\delta_4 = g\langle\Pi\rangle$ is a constant associated with the breaking of the Z_2 symmetry. For $g = \mathcal{O}(1)$, $\delta_4 \sim |\delta_1| \lesssim m \sim 10^{-5}$ as shown above. H_u and H_d are a pair of Higgs doublets. Then, the reheating temperature $T_{\rm RH}$ reads

$$T_{\rm RH} \lesssim 10^9 \text{ GeV}\left(\frac{\delta_4}{10^{-5}}\right)\left(\frac{m}{10^{13}\text{GeV}}\right)^{1/2}, \tag{12}$$

which is low enough to avoid the overproduction of gravitinos.

IV LEPTOGENESIS VIA THE INFLATON DECAY IN CHAOTIC INFLATION

In this section, we discuss the leptogenesis scenario via the inflaton decay in the above chaotic inflation model. For our purpose, we extend the Z_2 symmetry into a Z_4 symmetry. The charges of the Z_4 symmetry for various supermultiplets are given in Table 1. Then, we introduce the following superpotential invariant under the $U(1)_R$ and the Z_4 symmetries:

$$W = \lambda_i m \Phi N_i N_i + \gamma_i \Pi N_i N_i, \tag{13}$$

where λ_i and γ_i are constants and Π is the suprion field introduced before, whose vacuum value $\langle \Pi \rangle$ leads to the breaking of the Z_4 symmetry and must be less than $m \sim 10^{-5}$. Here, we set $\langle \Pi \rangle \sim m \sim 10^{-5}$. The Majorana masses of right-handed neutrinos M_i is given by $M_i = \gamma_i \langle \Pi \rangle$. For $\gamma_3 = \mathcal{O}(1)$, $M_3 \sim 10^{-5} \sim 10^{13}$GeV. The inflaton φ and the orthogonal field η can decay into right handed scalar neutrinos N_i through the above Yukawa interactions if $M_i < m/2$. Both decay rates are similar and given by

$$\Gamma_\varphi \simeq \Gamma_\eta \simeq \lambda^2 \frac{m^3}{32\pi} \sim 10\lambda^2 \text{ GeV}, \tag{14}$$

with $\lambda^2 \equiv \Sigma \lambda_i^2$, where i runs for $M_i \ll m$. Then, the reheating temperature T_{RH} is given by

$$T_{\text{RH}} \sim 10^9 \lambda \text{ GeV}. \tag{15}$$

The produced N_i immediately decay into leptons l_j and Higgs doublets H_u through the Yukawa interactions of Higgs supermultiplets.

Then, the ratio of lepton number to entropy density can be estimated as

$$\frac{n_L}{s} \simeq \frac{3}{2} \epsilon_1 B_r \frac{T_R}{m}$$
$$\sim -10^{-9} \delta_{\text{eff}} B_r \left(\frac{T_R}{10^9 \text{ GeV}}\right) \left(\frac{M_1}{10^{11} \text{ GeV}}\right) \left(\frac{10^{13} \text{ GeV}}{m}\right), \tag{16}$$

where B_r is the branching ratio of the inflaton decay into N_1. For $M_3 \sim M_2 \sim m \sim 10^{13}$GeV, the decay into N_3 and N_2 are prohibited kinematically or suppressed by the phase space and hence $B_r = \mathcal{O}(1)$ for $\lambda_1 = \mathcal{O}(1)$. In this case, we obtain $T_{\text{RH}} \sim 10^9$GeV, which results in $n_L/s \sim -10^{-9} \delta_{\text{eff}}$. Thus, our model of leptogenesis works well for $\gamma_2 \simeq \gamma_3 = \mathcal{O}(1)$, $\delta_{\text{eff}} = \mathcal{O}(1)$, and $\lambda_1 = \mathcal{O}(1)$.

V DISCUSSION AND CONCLUSIONS

In this paper we propose a natural chaotic inflation model with the shift symmetry in supergravity. In particular, the forms of the Kähler potential and the superpotential have been discussed. In order to suppress higher order terms of the inflaton field in the superpotential, the shift symmetry is extended into that including the suprion field Ξ with the combination $\Xi\Phi$ invariant. Also, the linear term of X in the superpotential is suppressed by introducing the Z_2 symmetry. We have found that if the magnitude of the breaking of the Z_2 symmetry is equal or smaller than that of the shift symmetry, a desired chaotic inflation can take place.

We have also discussed the leptogenesis via the inflaton decay in this chaotic inflation model. The inflaton φ can decay into right-handed neutrinos through the Yukawa interactions suppressed by the breaking of the shift symmetry, which leads low reheating temperature enough to avoid the overproduction of gravitinos. Right-handed neutrinos acquire their masses associated with the breaking of a Z_4 symmetry which is an extension of the Z_2 symmetry, whose magnitude is consistent with the result from the Superkamiokande experiment. Then, we have found that for a wide range of parameters, the lepton asymmetry enough to explain the observed baryon number density is produced.

Acknowledgments

M.Y. is grateful to R. Brandenberger, K. Hamaguchi, A. A. Starobinsky, E. D. Stewart, and J. Yokoyama for many useful discussions and comments. M.K. and T.Y. are supported in part by the Grant-in-Aid, Priority Area "Supersymmetry and Unified Theory of Elementary Particles"(#707). M.Y. is partially supported by the Japanese Society for the Promotion of Science.

REFERENCES

1. M. Kawasaki, M. Yamaguchi, and T. Yanagida, Phys. Rev. Lett. **85**, 3572 (2000); hep-ph/0011104, to appear in Phys. Rev. D.
2. See also references in the above two references.

	Φ	X	Ξ	Π	N	H_u	H_d	5^*	10
Q_R	0	2	0	0	1	0	0	1	1
Z_4	2	2	0	2	1	2	2	1	1

TABLE 1. The charges of various supermultiplets of $U(1)_R \times Z_4$. Here, R-charge of $H_u H_d$ is assigned to be 0. All supermultiplets of quarks and leptons have the Z_4 charge 1 and Higgs supermultiplets H_u and H_d carry the Z_4 charge 2.

Diffractive Propagation of Light Waves in a Class of Anisotropically-Expanding Cosmological Models

Brett Bochner

Department of Physics and Astronomy, Hofstra University, Hempstead, NY 11549

Abstract. We investigate the propagation of light for metrics in which the approximation of geometrical (ray) optics breaks down due to nonadiabatically rapid (in space or time) variations of the curved-spacetime geometry. This may lead to wavelength-dependent effects such as diffraction and dispersion. The rapidly-expanding cosmological spacetimes are good candidate metrics for such effects. We find that conformal invariance prevents such violations of adiabaticity for the spatially-flat Friedmann Robertson-Walker metric; but if one breaks conformal invariance via the anisotropic (but homogeneous) Kasner metrics, then the propagation of light waves will be shown to deviate from the predictions of simple ray-tracing.

INTRODUCTION

The main focus of this work involves the propagation of light waves in regimes where there are substantial deviations from the predictions of geometrical (ray) optics, the principle according to which wavefronts must propagate along null geodesics in general (vacuum) spacetime metrics [1].

The geometrical optics approximation breaks down when strongly varying gravitational fields induce effects upon the light waves that are significant on length scales much smaller than the light wavelength, and/or on time-scales much shorter than the light frequency. In such cases, the wavefronts may no longer be constrained to propagate along null geodesics, and in order to determine the correct wave propagation behaviors one must explicitly consider the generally covariant wave equations for fields propagating through the relevant curved-spacetime metric(s).

There is a growing literature (see, e.g., [2]) studying such effects, for example examining the existence of "tails" in the propagation of (vector, scalar, or tensor) fields which represents power propagating *within* the future light cone, rather than precisely on it. For the research that we outline here, we have derived explicit solutions for the diffractive propagation of light waves in various metrics that may have significance for the study of the very early universe. Cosmological metrics are particularly likely to violate the geometrical optics approximation, due to their highly nonadiabatic behavior in the vicinity of the initial Big Bang singularity [1]. The resulting effects may include dispersion and scattering, as well as parametric amplification of the wave amplitudes, the classical analog of particle creation induced by the rapidly evolving gravitational fields [1,3].

METHODS AND RESULTS

The Formalism

We solve for the propagation of electromagnetic fields in curved spacetime by deriving and solving Maxwell's Equations for each metric under study; this procedure will be discussed in detail in an upcoming paper by this author [4]. The crucial simplifying assumption for these calculations is the approximation of small-amplitude fields that have a negligible feedback upon the metric, and thus behave as "test particles" sampling a pre-determined background cosmology. We have therefore taken care to ensure that our solutions for the fields are nonsingular and well-behaved; furthermore, we note that our solutions will likely be invalid for times very close to the Big Bang, at which point the energy densities of the fields will be exceedingly large, regardless of how small they are for later times, once they have been diluted by the expanding universe.

Two additional simplifying assumptions made for these calculations are that we solve for electromagnetic *plane waves* which are formally infinite in extent, and that are "monochromatic" (besides the effects of gravitational redshifting). The former assumption should be valid (and not conflict with our test particle approximation) as long as the beam width is finite, yet significantly bigger than the light wavelength. The latter assumption does not prohibit us from considering the realistic transport of electromagnetic energy, as long as one can assemble finite wave packets from the individual frequency components without significantly misrepresenting how the composite wave packets would propagate. Based on considerations of the form of our solutions, and of the relative simplicity of the (vacuum) metrics being considered, this requirement seems likely to be satisfied [4].

Results for the Spatially-Flat, Friedman Robertson-Walker Metric

For reasons of cosmological relevance and calculational ease, we first consider the case of electromagnetic wave propagation in the isotropic, homogeneous, spatially-flat Friedman Robertson-Walker (FRW) metric, with a cosmological scale factor given by $R(t) \propto \sqrt{t}$ during the early, radiation-dominated epoch of the universe [1].

This case yields remarkably simple results: there is absolutely no deviation in the propagation of electromagnetic fields from the simple predictions of ray optics. This result stems from the inability of conformally flat spacetimes like the FRW metrics to nonadiabatically shock fields which obey conformally invariant wave equations (such as the generally covariant Maxwell Equations), thus preventing all effects such as diffraction, parametric resonance, particle creation, etc., from occurring [1,3-5].

To violate geometrical optics, one must somehow break conformal invariance in the physical system under study. We have chosen to do so by considering the class of homogeneous, anisotropically-expanding cosmologies known as Kasner metrics, which, as semi-stable epochs representing episodes of the "Mixmaster Universe", have been used to model the evolution of asymmetric or chaotic early universes not yet relaxed into the isotropic FRW form [1].

Results for the Anisotropically-Expanding Kasner Metrics

The Kasner metrics [1] are (vacuum) solutions to the Einstein equations in which each spatial axis has an independent scale factor given by $R_i(t) = t^{p_i}$, where the indices are real numbers related via $p_x + p_y + p_z = p_x^2 + p_y^2 + p_z^2 = 1$.

By explicitly solving the wave equations in different temporal regimes for different Kasner cases -- specifically $(p_x, p_y, p_z) = (1,0,0)$ and $(p_x, p_y, p_z) = (2/3, 2/3, -1/3)$ -- and also by considering the form of the wave equations for general Kasner indices, we were able to infer our main result: the coordinate phase velocity for an electromagnetic wavefront with wavenumbers (k_x, k_y, k_z) should generally be given as follows:

$$\vec{v}_{\text{coord}} = \sqrt{\frac{k_x^2}{t^{2p_x}} + \frac{k_y^2}{t^{2p_y}} + \frac{k_z^2}{t^{2p_z}}} \cdot \frac{-(k_x, k_y, k_z)}{[k_x^2 + k_y^2 + k_z^2]} \neq \vec{v}_{\text{coord}}^{\text{ray}} = \frac{-(k_x/t^{p_x}, k_y/t^{p_y}, k_z/t^{p_z})}{\sqrt{k_x^2 + k_y^2 + k_z^2}} . \quad (1)$$

(Note that we have set $c = 1$.) The above expression is understood to be approximate, and only valid in certain parameter regimes, e.g., $t \ll 1$ for $(p_x, p_y, p_z) = (1,0,0)$.

CONCLUSIONS

From Eq. 1 and supporting results, some of the conclusions we draw are as follows: (1) Light wave propagation in Kasner metrics deviates significantly from the simple coordinate phase velocity predictions of ray optics, leading to dispersion even in vacuum. (2) The behaviors of all components of the coordinate phase velocity vector are dominated by the same power of t, for any given temporal regime. (3) Closed-form expressions for propagating waves in highly nonadiabatic, non-FRW cosmologies can often be explicitly derived. Further conclusions about these effects, both abstract and regarding specific cosmological implications, are discussed at length in [4].

ACKNOWLEDGEMENTS

I would like to thank Hofstra University for supporting my attendance and presentation of research at the 20th Texas Symposium.

REFERENCES

1. Misner, C. W., Thorne, K. S., and Wheeler, J. A., *Gravitation*, W. H. Freeman, San Francisco, 1973, pp. 568-583, 703-735, 800-816.
2. Sonego, S., and Faraoni, V., *J. Math. Phys.* **33**, No. 2, 625-632 (1992).
3. Parker, L., *Phys. Rev. Lett.* **21**, No. 8, 562-564 (1968); Parker, L., *Phys. Rev.* **183**, No. 5, 1057-1068 (1969); Zel'dovich, Ya. B., and Starobinskii, A. A., *Sov. Phys. JETP* **34**, No. 6, 1159-1166 (1972).
4. Bochner, B., submitted to *Phys. Rev.* D15 (Jan. 2001).
5. Finelli, F., "Cosmological Magnetic Fields By Parametric Resonance?", astro-ph/0007290 (2000).

If the Parameters of the Vacuum Were Different, What Kind of Universe Would We Have? Self-organization of the Vacuum

V. Burdyuzha*, Yu. Ponomarev*, G. Vereshkov[†]

*Astro Space Centre of the Lebedev Physical Institute of Russian Academy of Sciences, 117810 Moscow, Russia
[†] Institute of Physics, Rostov State University, Rostov/Don, Russia

Abstract. We have researched the Universe vacuum properties (vacuum selforganization) that is the problem of the cosmological constant. This problem can't be solved in terms of the current quantum field theory which operates with Higgs and nonperturbative vacuum condensates and takes into account the changes of these condensates during relativistic phase transitions. The problem can't be completely solved also in terms of the conventional global quantum theory: Wheeler-DeWitt quantum geometrodynamics does not describe the evolution of the Universe in time (relativistic phase transitions in particular). We have investigated this problem in the context of energies density of different vacuum subsystems characteristic scales of which pervaid all energetic scale of the Universe. The transformation of the cosmological constant in dynamical variable is inevitably. The change of vacuum parameters brings to catastrophic consequences for the Universe (the antropic principle).

Recent observations have shown that the contribution of dark energy density Ω_Λ in the total density of the Universe $\Omega_0 = \Omega_\Lambda + \Omega_{CDM} + \Omega_b + \Omega_\gamma + \Omega_\nu$ dominates densities other components (see the review of Bahcall et al., 1999). This dark energy is the reason of accelerated expansion of the Universe in the modern epoch (Perlmutter et al., 1999). The dark energy may be the vacuum energy (the cosmological constant or Λ-term) with equation of state $w \equiv p/\varepsilon = -1$, which is the constant in time during practically all life-time of the Universe[1] and quintessence, which is the energetic component of total density of the Universe associated a scalar field

---

[1]) The vacuum energy might change by jumps in result of relativistic phase transitions in very early epoches of the Universe evolution ($T > 150\ MeV$). During time the density of matter ($\Omega_b + \Omega_{CDM}$ decreases as $1/a^3$ and the curvature decreases as $1/a^2$. Today the vacuum energy exceeds both these components (photon density Ω_γ and neutrino density Ω_ν are 5 orders values smaller than density energy of vacuum). Besides vacuum energy "overcome" curvature that is the Universe could form from "nothing".

(Armendariz-Picon et al., 2000; Hebecker and Wetterich, 2000; Rubakov, 2000). In this case the equation of state is the function of redshift (Moor et al., 2001).

Probably vacuum has local and global properties. The local properties of vacuum reflects properties of elementary particles (the connection between constants has put in characteristics of vacuum). Probably the local selforganization has also taken place. The global properties of vacuum are Universe vacuum properties. We have more detail investigated the global selforganization of vacuum, that is the cosmological constant problem (in cosmology terms: Λ-term and the cosmological constant are synonyms)

Today we can more exactly define the physical meaning of Λ-term which must contain the energy-momentum tensor (EMT) of gravitational vacuum $T_{\mu\nu(g)}$ and EMT of quantum fields $T_{\mu\nu(QF)}$

$$R_{\mu\nu} - \frac{1}{2}g_{\mu\nu}R = \kappa(T_{\mu\nu(g)} + <T_{\mu\nu(QF)}>) = \kappa(g_{\mu\nu}\Lambda_g + g_{\mu\nu}\Lambda_{QF}).$$

Here $\kappa = (10^{19} \ GeV)^{-2}$ is the gravitational constant in the system units where $\hbar = c = 1$; $<T_{\mu\nu(QF)}>$ is EMT of quantum fields averaged on some martix of density, which contains the information about state of plasma and vacuum of elementary particles. For $|R_\mu^\nu| \ll \kappa^{-1}$ the averaged EMT quantum fields is

$$T_{\mu\nu(g)} = g_{\mu\nu}\Lambda_g, \quad <T_{\mu\nu(QF)}> = <0|T_{\mu\nu(QF)}|0> = g_{\mu\nu}\Lambda_{QF}. \quad (1)$$

Here Λ_g is the second fundamental constant of the gravitation theory taking into account a gravitational vacuum condensate. Besides the quantum geometrodynamics introduces in the gravitational vacuum structure an additional subsystem consisting of the condensate of worm holes which is quantum fluctuations of space-time topology on Planck scales.

Therefore the observable Λ-term must contain two items

$$\Lambda = \Lambda_g + \Lambda_{QF} \quad (2)$$

which have practically exactly compensated each other since the observable value of Λ-term is near zero.

As known with the cooling of the cosmological plasma during relativistic phase transitions (RPT) vacuum condensates with a negative energy density were produced (Burdyuzha et al.,1997). In the Standard model (SM) two objects with the same status but with different physical properties take place: a Higgs vacuum condensate and a nonperturbative vacuum condensate. These condensates have the asymptotic equation of state $p_{vac} = -\epsilon_{vac} = const$. Thus RPT series were accompanied by the generation of negative contributions in initial Λ-term (if it was positive). Other words in the heterogenic vacuum careful coordination of different states of the vacuum subsystems the characteristic scales of which pervaid all energetic scales of the Universe has occured. The possibility of this coordination is also prompted by the mathematical formalism of strong nonlinear theory which describes RPT (a special solution of nonlinear equations). Therefore Λ-term was changed during the Universe evolution that is the phenomenon of vacuum selforganization has taken place.

SM of elementary particles physics belongs to class of renorm models of quantum field theory the vacuum of which contains three subsystems:

1) zeroth weakly correlated vibrations of quantum fields;
2) zeroth strongly correlated vibrations of quantum fields producing nonperturbative vacuum condensates;
3) quasiclassic (quasihomogeneous and quasistationary) fields usual named Higgs condensates.

We will discuss here the cosmological constant problem in context of densities energies of different vacuum subsystems. The problem is that each vacuum subsystem has a huge density energy, however the total value of vacuum energy in the Universe today is near zero as observational data confirm (Bahcall et al.,1999). Thus the phenomenon of selforganization of vacuum is evident although the mechanism of selforganization of nonperturbative condensates does not understand till now well.

In the interval of temperatures $150\ Mev < T < 100\ GeV$ the vacuum in the Universe was in the state of spontaneously breaking $SU(2)$ symmetry (for these temperatures the quark-gluon subsystem was in the state of de-confainment that is the quark-gluon vacuum condensate was absent).

$$\Lambda_{SM} = -\frac{m_H^2 m_w^2}{2g^2} - \frac{1}{128\pi^2}(m_H^4 + 3m_z^4 + 6m_w^4 - 12m_t^4) \qquad (3)$$

Here the first term is the energy density of quasiclassical Higgs condensate, the second term is the change of energy density of boson and fermion fields by quasiclassical field of a condensate. Excepting t-quark other fermions are very light and they involve a negligible small contribution in the formula (3). The numerical values of all constants except for Higgs boson mass are known from experiments As it follows from (3) $\Lambda_{SM} < 0$ that is the compensation of Λ-term during evolution of the Universe is inevitable. The methodics of Λ_{SM} calculation proposes that Λ_{SM} is a limiting value arising during dynamical evolution of a system. Besides this methodics allows to find the condiition of stability of this system which in this case is:

$$x^2 + x(\frac{1}{2a} - \frac{4ab}{9}) - \frac{2b}{3} > 0,$$

$$x < \frac{1}{a} + \frac{4ab}{9} \qquad (4)$$

Here $x = m_H^2/m_w^2; a = 3g^2/128\pi^2; b = \frac{12m_t^4 - 3m_z^4 - 6m_w^4}{m_w^4}, g^2 = 0.43$ is the gauge constant $SU_L(2)$ group; $m_w = 80\ GeV, m_z = 92\ GeV, m_t = 192\ GeV$. Thus from (3-4) it follows also that the mutual compensation of positive and negative contributions in vacuum density energy in SM is prohibited by the condition of stability.

The density energy of nonperturbative QCD vacuum is

$$\epsilon_{vac} = -\frac{9}{32} <0 \mid \frac{\alpha_s}{\pi} G_{\mu\nu}G^{\mu\nu} \mid 0> + \frac{1}{4} <0 \mid m_u \bar{u}u \mid 0> +$$
$$+ <0 \mid m_d \bar{d}d \mid 0> + <0 \mid m_s \bar{s}s \mid 0> = -8.2\lambda_{QCD}^4. \tag{5}$$

Here $m_\nu = 4.2\ MeV; m_d = 7.5\ MeV; m_s = 150\ MeV$ are masses of light quarks satisfying to the condition $m_q \leq \lambda_{QCD}; \lambda_{QCD} = 160\ MeV$. For $T < \lambda_{QCD} = 160\ MeV$ a nonperturbative quark-gluon vacuum is the state of dion condensate with negative energy density (the classic prototype of dions is nonlinear solutions of Yang-Mills equations similar to solitons, instantons, monopoles). That is today vacuum in the Universe has the confinement phase and the modern value of Λ-term can be calculated using formula

$$\Lambda_o = \Lambda'_g + \epsilon_{vac}. \tag{6}$$

Here ϵ_{vac} is the energy density of quark-gluon vacuum (see (5)) and Λ'_g is the constant taking phenomenologically into account all vacuum structures on energy scales more than ϵ_{vac}.

The calculation of vacuum energy density in superstring and supergravitational models is "less reliable" although it was recognized that vacuum energy density is formed by contributions from IR part of spectra of zeroth vibrations that is in total agreement with general considerations. Here the absence of a sum effect of vacuum polarization does not mean the absence of polarisations of separate parties (vacuum subsystems). This statement has been proved by experimentally.

Quite evidently that the coordination of vacuum subsystems was realized during cosmological evolution that is here we have all indicators of vacuum selforganization. It is pertinent to recall the anthropic principle (probably the selforganization of vacuum has provided the life of a organic type in the Universe).

Here it is necessary to do some notes relatively Λ-term in quantum geometrodynamics. As S.Hawking has shown the more probable state of the Universe was the state with $\Lambda_{eff} = 0$ since

$$P(\Lambda_{eff}) \sim \exp\left(\frac{3\pi}{\kappa^2 \Lambda_{eff}}\right). \tag{7}$$

This is the approch in which the additional fields of 3-form and the wave function of the Universe were involved and which may solve the Λ-term problem. But this was the theoretical phantom since RPT were in the Universe and Λ-term was changed during evolution. More radical step in the investigation of Λ-term problem was made by S.Coleman, which took into attention the realistic effect of quantum fluctuations of space-time topology (worm holes). In this approach the more probable the state of the Universe

$$P(\Lambda_{eff}) \sim \exp\left[\exp\left(\frac{3\pi}{\kappa^2 \Lambda_{eff}}\right)\right] \tag{8}$$

has had the more sharp peak than the distribution of S.Hawking. Certainly, the approach of S.Coleman does not solve of Λ-term problem since RPT were not included also, but this approach is more deep on the physical consideration. Wormholes and dions are topological nontrivial configurations of Yang-Mills fields being formed as quantum fluctuations on energetic scales $\lambda_{QGD} = \kappa^{-1/2} = 10^{19}\ GeV$ and $\lambda_{QCD} \sim 0.16\ GeV$. Thus a unified picture of nonperturbative vacuum is appeared, which contains a great number of subsystems the characteristic scales of which pervaid all energetic scale. The equality to zero of total Λ-term is the condition of selforganization of these subsystems.

As Higgs as and nonperturbative vacuum condensates in the modern quantum field theory are macroscopic mediums having quasiclassical properties. The periodic collective motions in those are perceived as pseudogoldstone bosons. The production of condensates is accompanied by decreasing of vacuum symmetry in comparison with symmetry of theory equations. This fact leads to the suggestion that in vacuum structures an additional one must be included which can be named a gravitational vacuum condensate (GVC). The information about GVC put in numerical constants characterizing its and in frequency spectrum of GVC pseudogoldstone excitations. Pseudogoldstone excitations must superweakly interact with usual particles that is they must possess properties of dark matter.

Thus our understanding of Λ-term problem on late stages of the Universe evolution is reduced to

$$\Lambda = \Lambda_{QF} + \Lambda_g'' + \frac{9\pi^2}{2\kappa^2} \lambda_n \qquad (9)$$

Here λ_n defines the spectrum of GVC possible states. In formula (9) every item has the physical content. The general for all items is that they were created during evolution of the Universe. Higgs and nonperturbative vacuum fluctuations being arisen during RPT form Λ_{QF}, the quantum fluctuations of space-time topology form Λ_g''; the evolution of gravitation vacuum condensate forms $\Lambda_{GVC} \equiv \frac{9\pi^2}{2\kappa^2} \lambda_n$. Probably, the Λ-term problem is the key to physics of the XXI century.

All values, which describe vacuum are strictly connected each others and any their changes lead to catastrophic consequences for the Universe. $\Lambda \approx 0$ is the best suited to cosmology. The Universe having large negative Λ-term never become macroscopic. If Λ-term value is large and positive then the production of complex nuclear, chemical and biological structures is impossible. Our Universe with observed hierarchy of large scale structures can exist for $\Lambda \approx 0$ only. The concrete calculations of vacuum parameters change of any vacuum subsystems are very complicated since, for example, near 90% nucleons mass depends on parameters of a quark-gluon condensate. Today we can only treat that Λ-term must be positive and have extremally small value.

REFERENCES

1. Armendariz-Picon, C., Mukhanov, S., and Steinhardt, P., *Phys. Rev. Lett.*, **85**, 4438 (2000).
2. Bahcall, N., Ostriker, J., Perlmutter, S., and Steinhard, P., *Science*, **284**, 1481 (1999).
3. Burdyuzha, V., Lalakulich, O., Ponomarev, Yu., and Vereshkov, G., *Phys. Rev. D.*, **55**, 7340R (1997).
4. Coleman, S., *Nucl. Phys. B*, **310**, 643 (1988).
5. Hawking, S., Phys. Lett. B, **134**, 403 (1984)
6. Hebecker, A., and Wetterich, C., *Phys. Rev. Lett.*, **85**, 3339 (2000).
7. Moor, I., Brustein, R., and Steinhardt, P., *Phys. Rev. Lett.*, **86**, 6 (2001).
8. Perlmutter, S., Turner, M., and White, M., *Phys. Rev. Lett.* **83**, 670 (1999).
9. Rubakov, V., *Phys. Rev. D.*, **61**, 061501 (2000).

A Cosmological Model with Particle Creation

Sujit Chatterjee

Department of Physics, New Alipore College, Calcutta 700 053, India

Abstract. A higher dimensional cosmological model is proposed where an expanding universe evolves from the vacuum fluctuation and matter creation takes place out of the gravitational energy. Choosing a particular form of the matter creation function $N(t)$ as an initial conditions it can be shown that starting from an inflationary era the cosmos enters the higher dimensional Friedmann-like phase after a time scale when the matter creation stops.

INTRODUCTION

The idea that a closed universe might be due to a vacuum fluctuation was first suggested independently by Tryon [1] and Fomin [2]. Following this idea Johri [3] recently proposed a new cosmological scenario where the fire ball phase and the subsequent cosmic expansion are assumed to be a consequence of high energy vacuum fluctuation as opposed to the standard big bang model. Here the role of big bang is taken over by the creation of particles in inducing expansion and the genesis of matter in the early universe.

It is true that inflationary model adequately dealt with the twin problems of "horizon" and "flatness" but it fails to account for the colossal amount of vacuum energy required to inflate the universe by about 28 orders of magnitude while its energy density remains constant. Although Guth [4] and also Hawking [5] have conjectured that gravitational energy might be the source of this huge energy no mechanism has been put forward by them. Prigogine [6] also offered an explanation of the creation of matter particles from the gravitational energy taking the universe as an open thermodynamical system.

On the other hand the higher dimensional spacetime is now an active field of research in its attempts to unify gravity with all other forces of nature [7,8]. The idea is particularly relevant in the cosmological context of the early universe before the extra dimensions underwent the compactification process to an unobservably small planckian scale. Recent impetus to higher dimensional theories has also come from its applications to the currently fashionable field of M-cosmology. In the present work we have thought it fits to examine the whole scenario described in Johri's work in the framework of higher dimensions.

HIGHER DIMENSIONAL MODEL

The line-element appropriate to a spherically symmetric distribution in $n+2$-dimensional space time is given by

$$ds^2 = dt^2 - R^2(dr^2 + r^2 dX_n^2) \tag{1}$$

where $dX_n^2 = d\theta_1^2 + \sin^2\theta_1 d\theta_2^2 \ldots \sin^2\theta_1 \sin^2\theta_2 \sin^2\theta_{n-1} d\theta_n^2$. The effective energy-momentum tensor in the presence of matter creation field is

$$T_{ij} = (\rho + p + p_c)v_i v_j - (p + p_c)g_{ij} \tag{2}$$

where the new term p_c comes from the particle creation and is given by

$$p_c = -\left(\frac{H}{N}\right)\left(\frac{dN}{dV}\right) < 0 \tag{3}$$

and $H = \dot{R}/R$, $V = R^{n+1}$, and dN is the number of particles created from the gravitational field energy (see Prigogine for details). The Einstein's equations for the metric (1) via equations (2) and (3) give [9]

$$\frac{n(n+1)}{2}\frac{\dot{R}^2}{R^2} = 8\pi\rho \tag{4}$$

$$n\frac{\ddot{R}}{R} + \frac{n(n+1)}{2}\frac{\dot{R}^2}{R^2} = 8\pi(p + p_c). \tag{5}$$

From Bianchi identity it further follows that

$$\dot{\rho} + (\rho + p)\theta = p_c\theta = (p + \rho)\frac{\dot{N}}{N} \tag{6}$$

where

$$\theta = (n+1)\dot{R}/R. \tag{7}$$

Invoking an equation of state $p = k\rho$, equation (6) gives

$$N(t) = N_0 R^{n+1} \rho^{1/(1+k)}. \tag{8}$$

The particle creation function $N(t)$ serves as an initial condition in our model and different choices of $N(t)$ give rise to different cosmological histories. Motivated by Anthropic principle, however we take

$$N(t) = e^{(n+1)\beta/t} t^{-(n+2)/(1+k)}; \qquad \beta > 0. \tag{9}$$

We take the initial state of the universe to be radiation dominated and the appropriate equation of state is

$$P = \frac{\rho}{n+1}. \tag{10}$$

For economy of space we skip here all intermediate mathematical steps and write the final equations (via eqns. [8]-[10]) as

$$\frac{2}{n+2}\frac{\dot{H}}{H} + H = \frac{\beta n}{2} t^{-(n+2)/2} - \frac{1}{t}. \tag{11}$$

This equation yields an immediate solution as

$$H = \frac{\beta n}{2} t^{-(n+2)/2} \tag{12}$$

such that the scale factor is

$$R = a e^{-\beta t^{-n/2}}. \tag{13}$$

Further from (12) it follows that

$$\frac{\ddot{R}}{R} = \frac{\beta n^2}{4} t^{-(n+4)/2} \left(\beta t^{-n/2} - \frac{n+2}{n} \right). \tag{14}$$

We also get from (9)

$$\frac{\dot{N}}{N} = (n+1) \left(\frac{n\beta}{2} t - \frac{n+2}{n} - \frac{1}{t} \right). \tag{15}$$

Thus we find that $\ddot{R} > 0$ until $t < [\beta n/(n+2)]^{2/n}$ indicating that universe accelerates with particle production until $t = [\beta n/(n+2)]^{2/n}$ and after that it starts decelerating. Moreover the particle creation stops at $t = (n\beta/2)^{2/n}$. From equation (11) it further follows that at this large stage

$$\frac{2}{n+2}\frac{\dot{H}}{H} + H = 0 \tag{16}$$

such that

$$R \sim t^{2/(n+2)}. \tag{17}$$

This is FRW-like solution in $(n+2)$-dimensions as found earlier by the present author [9].

To conclude it may be stated that we here present a multidimensional cosmological model which at the particle creation evolves naturally into the Friedmann era and the end of the particle creation function acts as a sort of initial condition in our model.

ACKNOWLEDGMENT

The author wishes to thank the University Grants Commission, India for financial support.

REFERENCES

1. Tryon, E., *Nature*, **246**, 396(A) (1973)
2. Fomin, P., *Dokl. Akad. Nauk. Ukr.*, **A9**, 831 (1975)
3. Johri, V., *Hadronic Journal*, **22**, 1 (1999)
4. Guth, A., *The Oscar Klein Memorial Lectures*, vol. 2 (World Scientific) (1992)
5. Hawking, S. *Black Holes and Baby Universes*, (Bantam Book) (1993)
6. Prigogine, I., *Gen. Rel. Grav.*, **21**, 767 (1989)
7. Wesson, P., *Space-Time-Matter* (World Scientific) (1999)
8. Chatterjee, S., Beesham, A., Bhui, B., and Ghosh, T., *Phys. Rev. D.*, **57**, 6544 (1998)
9. Chatterjee, S., and Bhui, B., *M.N.R.A.S.*, **247**, 57 (1990)

Entropy and adiabatic components in scalar field perturbations

Christopher Gordon

Relativity and cosmology group, University of Portsmouth, Britain.

INTRODUCTION

In [1] we developed a general formalism to study the evolution of both curvature and isocurvature perturbations in a wide class of multi-field inflation models by decomposing field perturbations into perturbations along the background trajectory in field space (the adiabatic field perturbation), and orthogonal to the background trajectory (the entropy field). We allowed an arbitrary interaction potential for the fields. I summarize the main formulas of [1] and discuss some of their consequences.

PERTURBATION EQUATIONS FOR MULTIPLE SCALAR FIELDS

We consider two interacting scalar fields, ϕ and χ. In order to clarify the role of adiabatic and entropy perturbations, their evolution and their inter-relation, we define new adiabatic and entropy fields by a rotation in field space. The "adiabatic field", σ, represents the evolution along the classical trajectory, such that

$$d\sigma = (\cos\theta)d\phi + (\sin\theta)d\chi, \qquad (1)$$

where

$$\cos\theta = \frac{\dot{\phi}}{\sqrt{\dot{\phi}^2 + \dot{\chi}^2}}, \quad \sin\theta = \frac{\dot{\chi}}{\sqrt{\dot{\phi}^2 + \dot{\chi}^2}}. \qquad (2)$$

This definition, plus the original equations of motion for ϕ and χ, give

$$\ddot{\sigma} + 3H\dot{\sigma} + V_\sigma = 0, \qquad (3)$$

where

$$V_\sigma = (\cos\theta)V_\phi + (\sin\theta)V_\chi. \qquad (4)$$

As illustrated in Fig. 1, $\delta\sigma$ is the component of the two-field perturbation vector along the direction of the background fields' evolution. Conversely, fluctuations orthogonal to

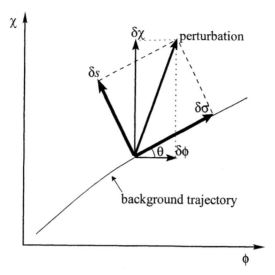

FIGURE 1. An illustration of the decomposition of an arbitrary perturbation into an adiabatic ($\delta\sigma$) and entropy (δs) component. The angle of the tangent to the background trajectory is denoted by θ. The usual perturbation decomposition, along the ϕ and χ axes, is also shown.

the background classical trajectory represent non-adiabatic perturbations, and we define the "entropy field", s, such that

$$ds = (\cos\theta)d\chi - (\sin\theta)d\phi. \qquad (5)$$

Perturbations in $\delta\sigma$, with $\delta s = 0$, describe adiabatic field perturbations, and this is why we refer to σ as the "adiabatic field".

The derivative of the comoving curvature can also be expressed neatly in terms of the new variables:

$$\dot{\mathcal{R}} = \frac{H}{\dot{H}}\frac{k^2}{a^2}\Psi + \frac{2H}{\dot{\sigma}}\dot{\theta}\delta s, \qquad (6)$$

where

$$\dot{\theta} = -\frac{V_s}{\dot{\sigma}}. \qquad (7)$$

The new source term on the right-hand-side of this equation, compared with the single-field case, is proportional to the relative entropy perturbation between the two fields, δs. Clearly, there can be significant changes to $\mathcal{R}$ on large scales if the entropy perturbation is not suppressed and if the background solution follows a curved trajectory, i.e., $\dot{\theta} \neq 0$, in field space. This can then produce a change in the comoving curvature on arbitrarily large scales (i.e., even in the limit $k \to 0$).

Equations of motion for the adiabatic and entropy field perturbations can be derived from the perturbed scalar field equations to give equation for the entropy perturbation as

$$\ddot{\delta s} + 3H\dot{\delta s} + \left(\frac{k^2}{a^2} + V_{ss} + 3\dot{\theta}^2\right)\delta s = \frac{\dot{\theta}}{\dot{\sigma}}\frac{k^2}{2\pi G a^2}\Psi. \tag{8}$$

Where Ψ is the longitudinal gauge curvature perturbation [2]. On large scales the inhomogeneous source term becomes negligible, and we have a homogeneous second-order equation of motion for the entropy perturbation, decoupled from the adiabatic field and metric perturbations.

By contrast, we cannot neglect the metric back-reaction for the adiabatic field fluctuations, or the source terms due to the entropy perturbations. Working in the spatially flat gauge, defining

$$Q_\sigma = \delta\sigma_Q = \delta\sigma + \frac{\dot{\sigma}}{H}\Psi, \tag{9}$$

we can write the equation of motion for the adiabatic field perturbation as

$$\ddot{Q}_\sigma + 3H\dot{Q}_\sigma + \left[\frac{k^2}{a^2} + V_{\sigma\sigma} - \dot{\theta}^2 - \frac{8\pi G}{a^3}\left(\frac{a^3 \dot{\sigma}^2}{H}\right)^{\cdot}\right]Q_\sigma$$
$$= 2(\dot{\theta}\delta s)^{\cdot} - 2\left(\frac{V_\sigma}{\dot{\sigma}} + \frac{\dot{H}}{H}\right)\dot{\theta}\delta s. \tag{10}$$

When $\dot{\theta} = 0$, this reduces to the single-field, but for a curved trajectory in field space, the entropy perturbation acts as an additional source term in the equation of motion for the adiabatic field perturbation, even on large scales. This can lead to *correlations* between the entropy and adiabatic components [1]. The coupling of the entropy and adiabatic component is also discussed in [3].

CONCLUSIONS

We have summarized the formalism in [1]. We decomposed arbitrary field perturbations into a component parallel to the background solution in field space, termed the *adiabatic* perturbation, and a component orthogonal to the trajectory, termed the *entropy* perturbation. We have rederived the field equations in terms of these rotated fields in Eqs. (8) and (10). These show that there can only be significant change in the large-scale comoving curvature perturbation if there is a non-negligible entropy perturbation, *and* if the background trajectory in field space is curved.

REFERENCES

1. C. Gordon, D. Wands, B. A. Bassett and R. Maartens, to appear in Phys. Rev. D. [astro-ph/0009131].
2. J. M. Bardeen, Phys. Rev. D **22**, 1882 (1980).
3. J. Hwang and H. Noh, astro-ph/0009268.
4. D. Langlois, Phys. Rev. **D59**, 123512 (1999) [astro-ph/9906080].

CHAPTER 3

DARK MATTER AND THE CDM PARADIGM

The dark matter crisis

Ben Moore

Department of Physics, Durham University, UK.

Abstract I explore several possible solutions to the "missing satellites" problem that challenges the collisionless cold dark matter model.

INTRODUCTION

Most dark matter candidates cannot be distinguished by observations on large scales. Although the observed universe appears consistent with hierarchical structure formation, this still leaves a wide range of potential dark matter candidates. For example, the clustering properties of galaxies, abundances of rich clusters or even halo masses and sizes are all very similar in universes with matter density dominated by cold dark matter, warm dark matter or collisional dark matter. We therefore seek tests of the nature of the dark matter that are sensitive to its interaction properties and small scale power which manifests itself on non-linear scales.

Cold dark matter (CDM) halos form via a complicated sequence of hierarchical mergers that lead to a global structure set primarily by violent relaxation. Numerical simulations have played an important role in determining the shape and scaling of CDM halo profiles that have subsequently lead to new observational tests of the model. A single functional form can fit CDM halos from a mass scale of $10^7 M_\odot$ – $10^{15} M_\odot$, where the density at a fixed fraction of the virial radius is higher in lower mass halos and the central profiles have steep singular cusps (c.f. Dubinski & Carlberg 1991, Warren etal 1992, Navarro etal 1996, Fukushige & Makino 1997, Moore etal 1998, Jing 2000 etc).

Galaxy clusters form via a similar process as individual galaxies, however most of the galactic fragments that formed clusters have survived the hierarchical growth, whereas on galaxy scales we find little trace of the merging hierarchy. Only a dozen satellites orbit the Milky Way, whereas a thousand satellites orbit within the Coma cluster. Numerical simulations of CDM halo formation have revealed that the abundance of dark matter subhalos within a galaxy is the same as found within a scaled galaxy cluster (Moore etal 1999, Klypin etal 1999). There are two solutions to this problem: (i) CDM is incorrect and the nature of the dark matter suppresses the formation of substructure halos, (ii) CDM is correct and the dark matter satellites of the Milky Way are present but only a few percent of them formed stars. Here I focus on the latter possibility.

1 DARK MATTER SUBSTRUCTURE

Figure 1. The distribution of dark matter with a CDM "Local Group" candidate. This is a binary pair of dark matter halos at a redshift z=0, separated by 1 Mpc and infalling at 100km/s. The large halos have virial masses of $\approx 2 \times 10^{12} M_\odot$, and with a particle mass of $10^6 M_\odot$ they are resolved with over 10^6 particles and 0.5 kpc force resolution. The grey scale represents the local density of dark matter – there are over 2000 dark matter satellites with circular velocity larger than 10 km/s.

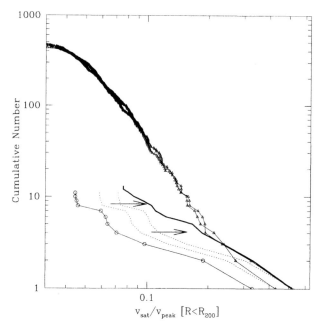

Figure 2. The cumulative number of satellites within the virial radius of the Milky Way (open circles) and within the two CDM halos from Figure 1 (open triangles). Here I have taken $v_{peak} = 210$ km/s for the Milky Way, however the CDM contribution to the Milky Way can be constrained to lie in the range 130 – 160 km/s once the baryonic component has been considered (Moore etal 2001). The dotted curves show the effects of this correction. The arrows show a correction for converting central velocity dispersion to v_{sat}. The thick solid curve shows the distribution of CDM satellites that could form stars before the universe is re-ionised.

Interpreting the observations

Figure 2 shows the cumulative distribution of subhalos within the high resolution Local Group halos of Figure 1. The open circles show the observed distribution for the Milky Way satellites, where I have normalised the distribution using $v_{peak} = 210$ km/s. However, baryons dominate the central region of the Galaxy and subtracting the contribution from the disk and bulge gives the maximum allowed CDM halo that has $v_{peak} = 160$ km/s (Moore etal 2001). A minimum value of $v_{peak} = 130$ km/s is required for the Galactic CDM halo to be massive enough to cool the observed mass of baryons. Figure 2 shows the effect of this correction.

Simon White has pointed out that the velocity dispersion of the dSph's are measured well within the cores of their dark matter halos (White 2000). We originally assumed isotropic orbits and isothermal potentials to derive v_{sat} from observations of the 1d central velocity dispersion (Moore etal 1999). CDM halos have central density profiles flatter than r^{-2} therefore one expects the velocity dispersion to drop in the inner region. M87 provides a good example of this. This galaxy lies at the center of the Virgo cluster and has a central velocity dispersion of ≈ 350 km/s whereas the cluster has a global value that is a factor of two larger. If we assume that the dSph's are similar to M87, then the correction should scale roughly as the concentration parameter. Since $c_{M87}/c_{dSph} \approx 0.4 - 0.5$, then we expect the maximum correction to v_{peak} to be an increase of 50% over our quoted values. This correction is indicated by the arrows in Figure 2 which brings the observed data into good agreement with a crude model for re-ionisation discussed later.

II FEEDBACK

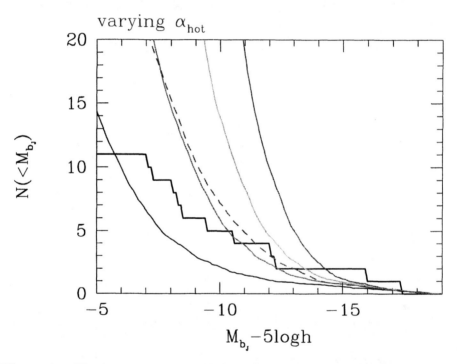

Figure 3. The histogram shows the cumulative distribution of absolute magnitudes of the 11 Galactic satellites. The curves show predictions from the Durham semi-analytic models of galaxy formation (Cole etal 2000) varying the parameter α that controls the efficiency of feedback. This can be tuned to give the correct distribution of satellite luminosities which in this case would lie somewhere in between the blue and pink curves.

Feedback is an essential component of galaxy formation within CDM models. It is invoked primarily to flatten the luminosity function given the steep mass function of CDM halos. For example, the faint end of the luminosity function in the Local Group is flat over a range that is about 10 magnitudes fainter than M_*. By varying the efficiency of feedback with halo mass, it is possible to get a reasonably flat luminosity function as Figure 3 demonstrates. The parameter α controls how much gas is ejected from dark matter halos of a given circular velocity and allows one to form systematically less stars in smaller mass halos.

The problem with a uniform feedback scheme is that the mass to light ratios of galaxies will increase rapidly for fainter galaxies. Thus we find that a satellite halo with absolute magnitude $M_B = -10$ is predicted to have a circular velocity of 40 km/s, roughly three times that observed for the dSph's. Figure 4 shows the "Tully-Fisher" relation for the 11 Galactic satellites compared with the curves predicted from the semi-analytic models employed in Figure 3. The circular velocities are overestimated by a factor of 3–4.

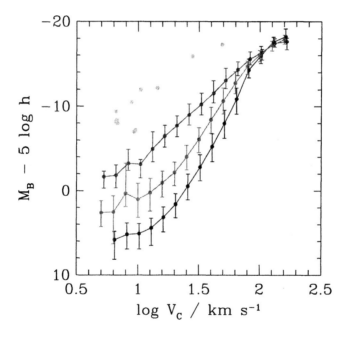

Figure 4. The Tully-Fisher relation for the Galactic satellites (green points) where the velocity dispersions of the spheroidals have been converted to circular velocity assuming isotropic orbits and isothermal potentials. The curves show the Tully-Fisher relation predicted by the semi-analytic models from Figure 3 where we varied the efficiency of feedback to match the numbers of dwarf galaxies.

III RE-IONISATION

We have just demonstrated that feedback does not solve the overabundance of CDM satellites – clearly some form of stochastic biasing is required. A solution was proposed by Bullock etal (2000) in which only those dark matter halos that have virialised prior to re-ionisation can cool gas and form stars. Once the IGM has been reheated then the smallest CDM halos cannot capture or cool gas and they remain completely dark.

In Figure 5 we mark all the progenitor halos that satisfy the condition for cooling gas prior to $z=10$, which we will take as the redshift of re-ionisation. We mark particles red if they lie within a region of overdensity larger than 1000. The locations of these particles are subsequently tracked to $z=0$ and marked in the right panel of Figure 5. Roughly 100 satellites satisfy the density criteria at a redshift $z=10$ and ≈ 80 of these physically merger together to form the very central region of the final galaxy halos. The remaining 20 survive intact and can be found orbiting within the virial radius of the two halos (see Figure 6). The mean radius of the surviving satellites is ≈ 80 kpc, which is a factor of 2.5 smaller than the half mass radius of the final halos.

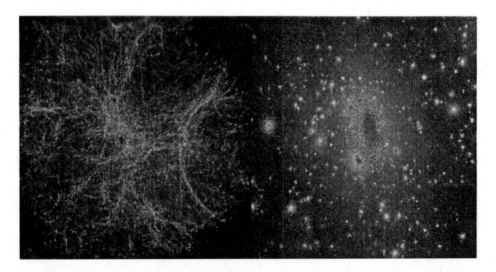

Figure 5. The left panel shows the Local Group simulation at $z=10$. Marked in red are all those particles that lie in regions with an overdensity larger than 1000. The right panel shows one of the high resolution halos at $z=0$ and the locations of the red particles marked at $z=10$.

Figure 6. The smoothed distribution of "starlight" in the Local Group at the present day. I plot only those stars that could form in dark matter halos prior to re-ionisation at z=10. The distribution of these stars is highly biased. Roughly a dozen dark matter dominated satellites orbit within each of the parent halos and they have a spatial distribution that matches the real Local Group. Most of the population II/III stars lie at the very centers of the halos surrounding M31 and the Galaxy. Their half light radius is just a few kiloparsecs (c.f. White & Springel 1999) and their luminosity density falls as r^{-3} (c.f. Figure 7).

The final cumulative distribution of satellites within one of the simulated halos is shown in Figure 2 and provides a good match to the corrected observational data points. Several puzzles remain. Why don't we find any satellites in the Galactic halo with velocity dispersion less than ~ 7 km/s? Is cooling that inefficient below ≈ 10 km/s such that we do not find any dark matter dominated systems containing just a handful of stars?

The star formation histories of the Local Group satellites presents a further puzzle. Most of the satellites show evidence for several bursts of star formation, some continuing to the present day. Both re-ionisation and the "essential" feedback have been extremely inefficient at removing gas from these tiny halos that have masses $\approx 10^8 M_\odot$.

IV RESOLUTION ISSUES

Central halo profiles

Have we really converged on the unique central structure of CDM halos? This is a hard numerical calculation to perform since we are always relaxation dominated on small scales. In a hierarchical universe the first halos to collapse will contain just a few particles and have relaxation times much shorter than a Hubble time.

Figure 7 shows the final density profile of one of the high resolution Local Group CDM halos. We also plot the final density profile of those particles that were located within highly non-linear regions at redshifts z=10 and z=20. The central 5 kpc of the halo is dominated by those particles that were in virialised halos at z=10. Most of these halos contained just a few particles and their internal structure is completely dominated by resolution effects. Until we can adequately resolve objects collapsing at z=10, we cannot claim to have converged upon the slope of the density profile at 1–2% of the virial radius.

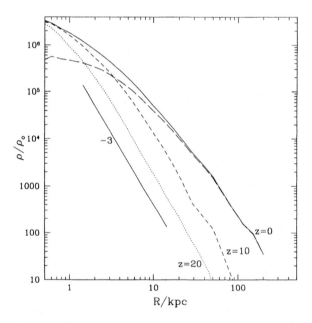

Figure 7. The solid curve shows the density profile of the high resolution halo shown in Figure 5. The dotted and dashed curves show the density profiles of those particles that lie in regions of overdensity larger than 50 at z=20 and z=10 respectively. The long-dashed curve shows the difference between the solid and short-dashed curves. The radial density profile of the marked particles at z=0 has a gradient of -3 which is similar to that of the Galactic spheroid.

Beam Smearing

Rotation curves of dwarf galaxies first highlighted potential problems with the structure of CDM halos (Moore etal 1994, Flores etal 1994, Burkert 1995). The quality of these data were recently questioned by several authors including van den Bosch & Swaters (2000) used rotation curves from 19 dwarf galaxies to claim that CDM halos are *consistent* with the data. However, to make this statement these authors had to throw away half of the galaxies and adopt unphysical (zero) mass to light ratios. Furthermore, seven of the remaining nine galaxies require concentration parameters in the range c=3–5 which cannot be obtained in any reasonable ΛCDM model. One could rephrase the conclusions of these authors by stating that only 2 galaxies from a sample of 19 are consistent with CDM!

Finally, I show the H_α and HI rotation curves of the nearby dwarf NGC3109 (Blais-Oullette etal 2001). These data clearly show that beam smearing is not an issue for the nearby dwarf galaxies. Furthermore, only a constant density core can fit these data. CDM profiles with central cusps < -1 are ruled out for any value of the concentration parameter. If CDM is correct then we are forced to conclude that galaxies such as NGC3109, NGC5585, IC2574, etc, are somehow strange and that their disk kinematics are somehow not measuring the mass distribution.

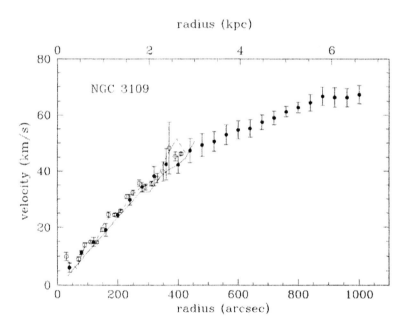

Figure 8. The rotation curve of NGC3109 (Blais-Oullette etal 2000) measured in HI (filled circles) and H_α (open circles). Beam smearing is clearly not an issue with nearby dwarf galaxies.

Acknowledgments I would like to thank Carlton Baugh for constructing Figure 3 and Figure 4. BM is supported by the Royal Society.

References

Blais-Ouellette, S., Carignan, C. & Amram, P. 2000, AJ, in press, astro-ph/0006449.
Bullock, J.S., Kravtsov, A.V. & Weinberg, D.H. 2000, ApJ, astro-ph/0007295.
Bullock, J.S., Kravstov, A.V. & Weinberg, D.H. 2000, ApJ, 548, 33.
Burkert, A. 1995, Ap.J.Lett., 447, L25.
Cole, S. Lacey, C., Baugh, C. Frenk, C.S. 2000, MNRAS, astro-ph/0007281.
Dubinski, J. & Carlberg, R. 1991, Ap.J., 378, 496.
Flores, R.A. & Primack, J.R. 1994, Ap.J.Lett., 457, L5.
Fukushige, T & Makino, J. 1997, Ap.J.Lett., 477, L9.
Ghigna, S., Moore, B., Governato, F., Lake, G., Quinn, T. & Stadel, J. 1998, M.N.R.A.S., 300, 146.
Jing, Y.P. 2000, ApJ, 535, 30.
Klypin, A., Kravtsov, A.V., Valenzuela, O., & Prada, F. 1999, ApJ, 522, 8.
Moore, B., 1994, Nature, 370, 620.
Moore, B., Ghigna, S., Governato, F., Lake, G., Quinn, T., Stadel, J., & Tozzi, P. 1999, ApJLett, 524, L19.
Moore, B., Quinn, T., Governato, F., Stadel, J., & Lake, G. 1999, MNRAS, 310, 1147.
Moore, B., Calcaneo, C., Quinn, T., Governato, F., Stadel, J., & Lake, G. PhysRevD, submitted.
Navarro, J.F., Frenk, C.S. & White, S.D.M. 1996, Ap.J., 462, 563.
van den Bosch, F.C. & Swaters, R.A. 2000, AJ, submitted.
Warren, S.W., Quinn, P.J., Salmon, J.K. & Zurek, H.W. 1992, Ap.J., 399, 405.
White, S.D.M. 2000, ITP conference, url: online.itp.ucsb.edu/online/galaxy_c00
White, S.D.M. & Springel, V. 1999, ESO Workshop, eds A. Weiss, T. Abel & V. Hill, astro-ph/9911378.

Recent Results from Super-Kamiokande

Masayuki Nakahata for Super-Kamiokande collaboration

Kamioka Observatory, Institute for Cosmic Ray research,
University of Tokyo, Higashi-Mozumi, Kamioka-cho, Yoshiki-gun,
Gifu, Japan, 506-1205, email=nakahata@suketto.icrr.u-tokyo.ac.jp

Abstract. Recent results from Super-Kamiokande on atmospheric neutrinos, solar neutrinos and supernova neutrinos are presented. The atmospheric neutrino data shows an evidence for ν_μ oscillations in the ratio of ν_μ/ν_e and zenith angle distributions of μ-like events. The neutrino oscillation parameters obtained from the atmospheric neutrino data are $\Delta m^2 = (1.7-4) \times 10^{-3} \text{eV}^2$ and $\sin^2 2\theta > 0.89$ for $\nu_\mu \leftrightarrow \nu_\tau$ oscillations. The zenith angle distributions of higher energy atmospheric neutrino data show that the pure $\nu_\mu \leftrightarrow \nu_\tau$ oscillations are favored over the pure $\nu_\mu \leftrightarrow \nu_{sterile}$ oscillations. The analyses of solar neutrino data on day/night difference in flux and spectral shape of ^{8}B solar neutrinos constrained the oscillation parameters for ν_e oscillations. The small mixing angle solution and just-so solutions are disfavored with about 95 % C.L. The flux constrained analysis of day/night and energy spectrum favours large mixing solutions. The huge Super-Kamiokande detector is able to detect ~4000 neutrino events, if a supernova happens at the center of our galaxy. No neutrino burst has been observed at SK yet. Progress of the search for relic supernova neutrinos is also discussed.

INTRODUCTION

Atmospheric and solar neutrinos are extremely good sources for investigating property of neutrinos. If neutrinos have mass and their weak interaction eigenstates are mixture of mass eigenstates, neutrinos change their species, known as neutrino oscillations, on the way from the place where they are produced to detectors on the earth. The probability of neutrino oscillations is given by the following formula:

$$P(\nu_\alpha \to \nu_\beta) = \sin^2 2\theta \sin^2(1.27 \frac{L}{E_\nu} \Delta m^2),$$

where θ is the mixing angle of neutrinos, L is the distance from a neutrino source to a detector in unit of kilometer, E_ν is the energy of neutrinos in unit of GeV, and Δm^2 is the difference of mass squares in unit of eV^2. As seen in the equation, one needs longer distance for investigating smaller Δm^2. The observation of atmospheric and solar neutrinos provides us an opportunity to investigate quite small neutrino masses.

In this report, recent results on those neutrinos from Super-Kamiokande are presented. Analyses of supernova neutrinos are also discussed.

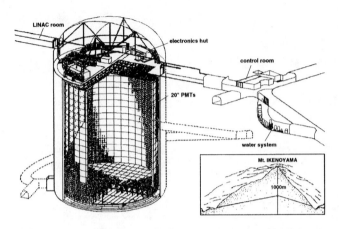

FIGURE 1. Schematic view of the Super-Kamiokande detector.

SUPER-KAMIOKANDE DETECTOR

Super-Kamiokande (SK) is a water Cherenkov detector located 1000 m underground in the Kamioka Mine in Japan. Schematic view of the detector is shown in Fig.1. The water tank consists of inner and outer detectors. The inner detector is composed of 11,146 20-inch diameter photomultiplier tubes (PMTs) placed facing inward on the surface of a 33.8 m diameter by 36.2 m high cylinder. The tubes are spaced on a 0.707 m grid, and enclose 32,000 metric tons of water. A 4π solid-angle outer detector surrounds the inner detector and viewed by 1,885 8-inch PMTs facing outwards.

ATMOSPHERIC NEUTRINOS

Atmospheric neutrinos are produced by the interaction of primary cosmic rays in the atmosphere. The hadronic showers resulting from the cosmic ray interaction produces electron and muon neutrinos through $\pi^+ \to \mu^+ + \nu_\mu$ followed by $\mu^+ \to e^+ + \bar{\nu}_\mu + \nu_e$ (and their charge conjugates). The ratio ($\equiv \nu_\mu/\nu_e$) of the flux of $\nu_\mu + \bar{\nu}_\mu$ to the flux of $\nu_e + \bar{\nu}_e$ is expected to be about two from the decay chain of $\pi^\pm$ and the uncertainty of the ratio is less than 5 % (Gaisser and Stanev 1995; Barr et al. 1989; Agrawal et al. 1996; Honda et al. 1990; Honda et al. 1995), whereas the absolute flux of each neutrino has an uncertainty of ~25 %. Hence, the ν_μ/ν_e ratio is a robust value to test neutrino oscillations. The measurements of the ν_μ/ν_e ratio are reported as $R \equiv (\mu/e)_{DATA}/(\mu/e)_{MC}$, where μ and e are the number of muon-like (μ-like) and electron-like (e-like) events observed in the detector for both data and Monte Carlo simulation. This ratio largely cancels experimental and theoretical uncertainties (Fukuda et al. 1998a; Fukuda et al. 1998b). $R = 1$ is expected if the physics in the Monte Carlo simulation accurately models the data. Much robuster way to test neutrino oscillations is the zenith angle of neutrinos(Fukuda et al. 1998d). The zenith angle distribution is expected to be up/down symmetric for neutrinos above a few GeV.

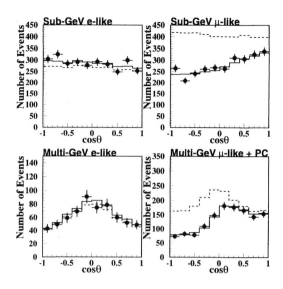

FIGURE 2. Zenith angle distribution of atmospheric neutrino events for sub-GeV e-like (left-top), sub-GeV μ-like (right-top), multi-GeV e-like (left-bottom), and multi-GeV μ-like + PC (right-bottom). cosθ=-1 corresponds to upward-going. The dashed histograms show the expectations from the Monte Carlo simulation for null oscillations and the solid histograms show the best fit assuming neutrino oscillations ($\Delta m^2 = 2.4 \times 10^{-3} eV^2$ and $\sin^2 2\theta = 1.00$).

The atmospheric neutrino events are observed as fully contained events (FC), partially contained events(PC), and upward-going muons (muons produced by neutrino interaction in the rock). The analysis of contained events of 1289 days' SK data shows that the R value is $0.638 \pm 0.017(stat.) \pm 0.051(sys.)$ for the lower energy sample (sub-GeV, $E_{vis} < 1.33$ GeV) and $0.675^{+0.034}_{-0.032}(stat.) \pm 0.080(sys.)$ for higher energy sample (multi-GeV, $E_{vis} > 1.33$ GeV). The zenith angle distributions of FC and PC events are shown in Fig.2 together with the Monte Carlo simulations. The shape of the observed zenith angle distribution of e-like events agrees to the simulation well, but the large distortion of the zenith angle distribution is observed in the μ-like events. The distorted zenith angle distribution is beautifully reproduced by assuming neutrino oscillations.

The upward-going muons are subdivided into through-going and stopping types depending on whether the muons are energetic enough to go through the detector or not. The averaged neutrino energy is ~100 GeV for upward-through-going muons and ~10 GeV for upward-stopping muons. SK observed 1416 upward-through-going muons and 345 upward-stopping muons during 1268 and 1247 days of live time, respectively. The zenith angle distribution of the upward-going muons are shown in Fig.3. The expected distributions without neutrino oscillations do not reproduce the data well. But, the distributions with best fit neutrino oscillation parameters fit quite well(Fukuda et al. 1999a; Fukuda et al. 1999b).

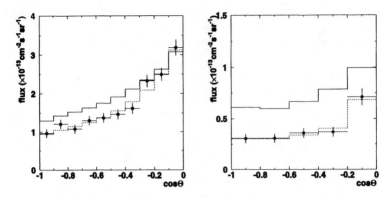

FIGURE 3. Zenith angle distributions of upward-through going muons(left) and upward-stopping muons(right). The solid histograms show expectations without neutrino oscillations and dashed histograms show best fit with neutrino $\nu_\mu \leftrightarrow \nu_\tau$ oscillations.

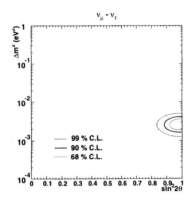

FIGURE 4. Allowed regions of neutrino oscillation parameters obtained by combining analyses of contained events, upward-through-going muons and upward-stopping muons.

An allowed region of neutrino oscillation parameters is obtained by combining contained and upward-going muon events as shown in Fig.4. The obtained neutrino oscillation parameters are $\Delta m^2 = (1.7-4) \times 10^{-3} \text{eV}^2$ and $\sin^2 2\theta > 0.89$ for $\nu_\mu \leftrightarrow \nu_\tau$ oscillations.

The neutrino oscillation analysis above assumed $\nu_\mu \leftrightarrow \nu_\tau$ oscillations. The possibility of $\nu_\mu \leftrightarrow \nu_s$(sterile neutrinos) oscillation is discussed using two methods, one using the MSW effect for oscillations and another using a neutral current (NC) enriched sample. The forward scattering amplitude of ν_μ and ν_τ are same, but that of ν_s is different. The difference in the amplitude causes suppression of neutrino oscillations for higher energy neutrinos. Figure 5(left) shows zenith angle distribution of high energy PC events (E>5GeV, <E>=~25GeV) together with expected distributions for $\nu_\mu \leftrightarrow \nu_\tau$

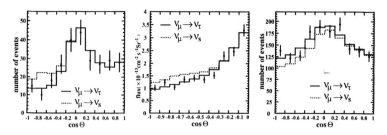

FIGURE 5. Zenith angle distribution of higher energy PC events(left), upward-going muons(middle) and NC enriched sample (right). The solid and dashed histograms show expectations for $\nu_\mu \leftrightarrow \nu_\tau$ and $\nu_\mu \leftrightarrow \nu_s$ oscillations, respectively, assuming $\Delta m^2 = 3 \times 10^{-3} \text{eV}^2$ and $\sin^2 2\theta = 1$.

and $\nu_\mu \leftrightarrow \nu_s$ oscillations assuming $\Delta m^2 = 3 \times 10^{-3} \text{eV}^2$ and $\sin^2 2\theta = 1$. The same figure for upward-through-going muons is shown in Fig.5(middle). The second method to discriminate $\nu_\mu \leftrightarrow \nu_\tau$ and $\nu_\mu \leftrightarrow \nu_s$ oscillations is using NC events. In case of $\nu_\mu \leftrightarrow \nu_\tau$ oscillations, the event rate of NC interaction should be same as null oscillations. On the other hand, the NC event rate is smaller for $\nu_\mu \leftrightarrow \nu_s$. A NC enriched sample was made requiring multi-ring event pattern and the most energetic ring to be e-like. The estimated fraction of NC event is 29% in the sample. The observed zenith angle distribution of the sample is shown in Fig.5(right). Those distributions show that the pure $\nu_\mu \leftrightarrow \nu_\tau$ hypothesis fits better than the pure $\nu_\mu \leftrightarrow \nu_s$ hypothesis. Scanning all possible Δm^2 and $\sin^2 2\theta$ ranges, $\nu_\mu \leftrightarrow \nu_s$ oscillations are disfavored with more than 99% C.L.(Fukuda et al. 2000).

If ν_μ's oscillate into ν_τ's, τ lepton events could be observed by $\nu_\tau + N \rightarrow \tau + X$ interactions. Figure 6(left) shows the comparison of ν_μ charged current (CC) and ν_τ CC interactions. Figure 6(right) shows expected τ lepton production rate as a function of Δm^2. $\sin^2 2\theta = 1$ is assumed in the figure. We expect about 20 τ events/year for $\Delta m^2 = 3 \times 10^{-3} \text{eV}^2$ and $\sin^2 2\theta = 1$. Searches for the τ appearance have been performed using signatures of τ decays, such as (1) higher multiplicity of Cherenkov rings, (2) more $\mu \rightarrow e$ decay signals, (3) more spherical event patterns, and etc. Three independent methods are applied to the atmospheric neutrino data to enhance the fraction of τ events. The first method uses energy flow and event shape analysis. The second one uses likelihood of τ signatures described above. And the third one is a neural network method putting Cherenkov ring information. The zenith angle distributions obtained by those methods are shown in Fig.7 together with expectations with and without τ production. All those three methods favor τ appearance with about 2σ significance. Note that events selected by those three methods are highly correlated and they are not statistically independent.

The atmospheric neutrinos at lower energy range is affected by the geomagnetic cutoff of primary cosmic rays. The primary cosmic rays coming from east is suppressed by the geometry of earth magnetic orientation and the direction of the cosmic rays. Figure 8 shows azimuthal angular distributions for e-like and μ-like events in the zenith angle range of $-0.5 < \cos\theta < 0.5$ and energy range from 400 MeV to 3 GeV. The

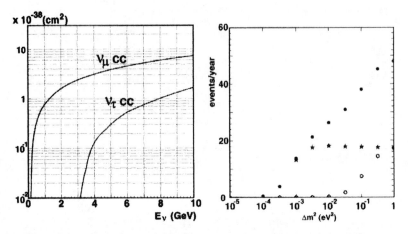

FIGURE 6. Left: Charged current cross section of ν_μ and ν_τ interactions as a function of neutrino energy. Right: expected event rate of τ lepton production a function of Δm^2. Filled circles show the total production rate. Open circles and stars show the rate for upward-going ($\cos\theta < -0.2$) and downward-going ($\cos\theta > 0.2$) events, respectively. $\sin^2 2\theta = 1$ is assumed.

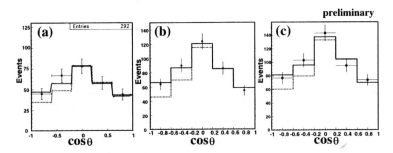

FIGURE 7. Zenith angle distributions after applying cuts to enhance τ events using (a) energy flow and event shape, (b) likelihood method, and (c) neural network method. Solid and dotted histograms are expectations with and without τ production, respectively.

shape of the observed azimuthal distributions are quite well reproduced by expected angular distributions which take into account the geomagnetic effect. Thus, the expected azimuthal effect(east-west effect) is actually observed at SK.

SOLAR NEUTRINOS

Solar neutrinos have been observed by Homestake, Kamiokande, GALLEX, SAGE, and SK(Cleveland et al. 1995; Davis 1994; Fukuda et al. 1996; Hirata et al. 1990; Anselmann et al. 1994; Anselmann et al. 1995; Abdurashitov et al. 1994,Fukuda et al. 1998c). All these experiments observed significantly smaller solar neutrino flux than the expectations from standard solar models (SSMs) (Bahcall, Pinsonneault &

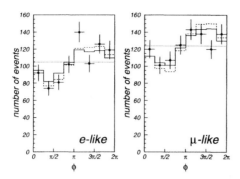

FIGURE 8. Azimuthal angular distributions of horizontal coming events, left:*e*-like, right:*μ*-like. $\phi=1/2\pi$ corresponds to neutrinos coming from east. Solid and dashed histograms show expectations based on the flux calculations by Handa et al. and Gaisser et al., respectively.

Basu 2000; Bahcall, Basu & Pinsonneault 1998; Turck-Chieze & Lopes 1993), known as "the solar neutrino problem". It is suggested that explaining the results of all five experiments in the framework of SSM has a difficulty, even if the input parameters of the SSM are changed. The most natural way to explain all those results is neutrino oscillations(Wolfenstein 1978; Mikheyev & Smirnov 1985). Using the flux observed by those experiments, three separate regions on the oscillation parameter plane (see Fig.11) remain as possible solutions with 95 % C.L. In order to solve the solar neutrino problem, solar model independent evidence for neutrino oscillations must be established and unique solution of oscillation parameters must be obtained. For this purpose, high precision energy spectrum measurement and time variation studies, such as day/night difference of the flux, are quite important. For ^{8}B solar neutrino measurements, SK has been taking data by neutrino-electron scattering(Fukuda et al. 1998c; Fukuda et al. 1999c; Fukuda et al. 1999d) with a quite high event rate.

SK has obtained 1258 live days' data between 31 May 1996 and 6 October 2000. The number of observed solar neutrino events above 5.0 MeV is about 18500 events corresponding to 14.7 events/day. The obtained flux of ^{8}B solar neutrinos is $(2.32\pm0.03(\text{stat.})^{+0.08}_{-0.07}(\text{syst.})) \times 10^6/\text{cm}^2/\text{s}$. Comparing with the expectation from the SSM by Bahcall et. al (BP2000) (Bahcall, Pinsonneault & Basu 2000), the *Data/SSM* is $0.451\pm0.005(\text{stat.})^{+0.016}_{-0.013}(\text{syst.})$.

The obtained data was divided into daytime and nighttime samples and their fluxes are $(2.28\pm0.04(\text{stat.})^{+0.08}_{-0.07}(\text{syst.})) \times 10^6/\text{cm}^2/\text{s}$ for daytime and $(2.36 \pm 0.04(\text{stat.})^{+0.08}_{-0.07}(\text{syst.})) \times 10^6/\text{cm}^2/\text{s}$ for nighttime. The difference of the flux between daytime(D) and nighttime(N) is given as

$$\frac{N-D}{\frac{1}{2}(D+N)} = 0.033 \pm 0.022^{+0.013}_{-0.012}.$$

The nighttime data is subdivided into 5 bins according to the zenith angle of the sun

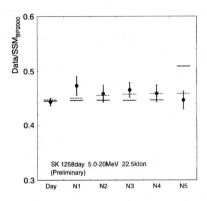

FIGURE 9. Solar neutrino fluxes for daytime and nighttime samples, the histograms are expectations from neutrino oscillations (see text).

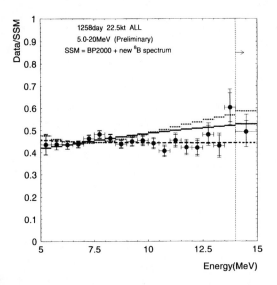

FIGURE 10. Recoil electron energy spectrum of solar neutrinos. Ratio of the observed spectrum to the expected spectrum is shown. The solid, dashed and dotted histograms show typical small mixing angle, large mixing angle and just-so solutions. $\nu_e \leftrightarrow \nu_\mu(\nu_\tau)$ oscillations are assumed.

(θ_z), N1 ($0 < \cos(\theta_z) \leq 0.2$), N2 ($0.2 < \cos(\theta_z) \leq 0.4$),..., N5 ($0.8 < \cos(\theta_z) \leq 1.0$). Neutrinos pass through the mantle for N1 to N4 and through the core of the earth in case of N5. The flux of each data set is shown in Fig. 9. The solid and dotted histograms in Fig.9 show expected flux variations for examples of the small mixing angle solution ($\sin^2 2\theta = 0.008$, $\Delta m^2 = 7.9 \times 10^{-6}$ eV^2) and the large mixing angle solution ($\sin^2 2\theta = 0.7$, $\Delta m^2 = 6.3 \times 10^{-5}$ eV^2) assuming $\nu_e \leftrightarrow \nu_\mu(\nu_\tau)$ oscillations. About 1-10% increase in flux during nighttime is expected for the large mixing angle solution. An increase in

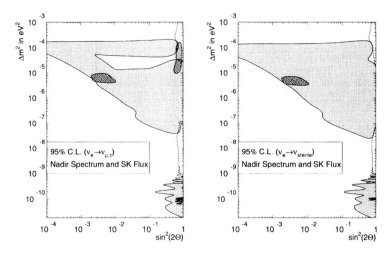

FIGURE 11. Excluded regions on neutrino oscillation parameters with 95 % C.L. obtained by the shape of the energy spectrum and day/night variation for $\nu_e \leftrightarrow \nu_\mu$ (left) and $\nu_e \leftrightarrow \nu_s$ (right) oscillations (gray regions). The cross hatched regions show allowed regions obtained by using the flux of Gallium, Chlorine and SK experiments. The dotted regions show allowed regions constrained with absolute ^{8}B solar neutrino flux in the SSM.

N5 or $\sim$1.0% negative day/night effect is expected in small mixing angle solution. The observed energy spectrum of solar neutrino events normalized by the SSM prediction is shown in Fig.10. For the expected ^{8}B neutrino spectrum shape, we have adopted the recent ^{8}B α decay measurement by Ortiz et al. 2000. The shape comparison between the observed spectrum and the expectation gives a χ^2 of 19.1 with 18 d.o.f., corresponding to 39 % C.L. Typical small mixing angle, large mixing angle, and just-so solutions are also shown in the figure.

Neutrino oscillation parameters are discussed using the shape of the energy spectrum and the day/night variation. The data are sub-divided into 8 energy bins and 7 day/night bins (day and 6 night bins). The flux of each energy-day/night bin is compared with expectations with neutrino oscillations. Fig.11(left) shows the excluded region overlaied with the allowed regions obtained by the fluxes of solar neutrino experiments (Gallium, Chlorine, and SK experiments) for $\nu_e \leftrightarrow \nu_\mu(\nu_\tau)$ oscillations. Small mixing angle solution and just-so solutions are disfavored with about 95 % C.L. If we assume the ^{8}B solar neutrino flux calculation in SSM is correct, large mixing solutions are allowed with 95 % C.L. by SK data as shown by a dotted curve in Fig.11(left). Similarly, Fig.11(right) shows the excluded regions for $\nu_e \leftrightarrow \nu_s$(sterile) oscillations. All possible solutions for $\nu_e \leftrightarrow \nu_s$ oscillations are disfavored with about 95 % C.L.

Thus, the precise measurement of energy spectrum and day/night variations by SK has constrained the neutrino oscillations parameters. The data on solar neutrinos is increasing not only in SK but also other solar neutrino experiments, such as SNO(McDonald 2000), and the solar neutrino problem should be solved in quite near future.

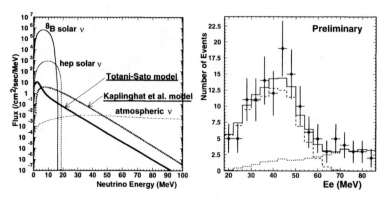

FIGURE 12. Left: expected relic SN neutrino spectrum calculated in Totani & Sato model and Kaplinghat et al. model. For a comparison, solar ^{8}B and hep neutrino spectra and atmospheric neutrino spectrum are shown. Right: Observed energy spectrum from 18 MeV to 90 MeV in SK. The dotted and dashed histograms show atmospheric neutrino backgrounds induced by ν_e and ν_μ interactions, respectively. The solid histogram shows sum of the backgrounds.

SUPERNOVA NEUTRINOS

Massive stars (> 8 times solar mass) become supernova(SN) at the end of their life. Supernova is a process of gravitational collapse of the central core of stars and almost 99% of the gravitational binding energy is released by neutrinos in type-II supernovae. Actually, such supernova neutrinos were observed by Kamiokande(Hirata et al. 1988) and IMB detectors(Bionta et al. 1987), when SN1987a in Large Magellanic Cloud was exploded. SK is able to detect more than ~4,000 supernova neutrino events, if it happens at the center of our galaxy. The contents of the events are about 4000 $\bar{\nu}_e + p \rightarrow e^+ + n$ events, 300 $\nu e^- \rightarrow \nu e^-$ events and ~400 gamma-ray events due to neutral current excitations of oxygen nuclei (Langanke, Vogel & Kolbe 1996; Beacom and Vogel 1998). The $\bar{\nu}_e + p \rightarrow e^+ + n$ events should provide precise energy spectrum and time profile and $\nu e^- \rightarrow \nu e^-$ events show the directionality of supernovae. Neutrino oscillations may play important role also in supernova neutrinos(Dighe and Smirnov 2000).

SK has been searching for supernova bursts for 3.95 years from the beginning of the experiment, but no candidate has been found yet. Upper limit on the galactic supernova rate is obtained to be <0.58 SN/year with 90% C.L. by the SK observation. Taking into account also the Kamiokande data, the limit becomes < 0.28 SN/year (90 % C.L.).

Neutrinos from past supernovae might be observed as relic neutrinos at SK. Figure 12(left) shows expected relic SN neutrino spectrum calculated by Totani-Sato model (Totani & Sato 1995; Totani, Sato & Yoshii 1996) and Kaplinghat et al. model (Kaplinghat, Steigman & Walker 2000). Possible backgrounds against the search for the relic SN neutrinos are spallation products induced by cosmic rays and atmospheric neutrinos. Those backgrounds limit the search energy window to about 18 - 40 MeV for the moment. Figure 12(right) shows the energy spectrum of observed events from 18 MeV to 90 MeV. The events around 40 MeV are mainly due to $\mu \rightarrow e$ decay electrons from atmospheric ν_μ interactions. The events above 50 MeV are mainly atmospheric ν_e interactions. The measured energy spectrum was fitted with those background contributions as shown in the figure. No significant excess over the background is seen at lower energy part of the spectrum and we set an upper limit of relic SN neutrino flux to be < 130 /cm^2/sec with 90 % C.L.(preliminary) assuming the spectrum shape given by Totani-Sato model. This limit is about 2-3 times larger than the estimation by Totani-Sato model (44 /cm^2/sec) and Kaplinghat et al. model(54 /cm^2/sec), but future progress in lowering the analysis energy threshold would give better limit of the flux.

CONCLUSIONS

ν_μ oscillations are established by atmospheric neutrinos in SK. And $\nu_\mu \leftrightarrow \nu_\tau$ oscillations are favored over $\nu_\mu \leftrightarrow \nu_s$ oscillations with 99 % C.L. Preliminary analyses of τ appearance search show about 2σ excess of τ-like events. The deficit of solar neutrinos observed by radiochemical experiments and SK is most likely due to neutrino oscillations. The day/night and spectrum analyses at SK shows small mixing angle and just-so solutions of the global flux analysis for $\nu_e \leftrightarrow \nu_\mu(\nu_\tau)$ is disfavored with about 95 % C.L. Taking into account absolute flux of ^{8}B neutrinos, large mixing solutions are favored. Similar analysis for $\nu_e \leftrightarrow \nu_s$ shows all possible solutions in the global analysis are disfavored with about 95 % C.L. SK is quite sensitive to supernova neutrinos. If a galactic supernova happens, it should give us fruitful information on astrophysics and particle physics. Non-observation of galactic supernova so far in SK and old Kamiokande sets an limit of the galactic SN rate to < 0.28 SN/year (90 % C.L.). No significant excess of events due to relic supernova nuetrinos are seen in the energy range above 18 MeV. Preliminary upper limit of the relic SN flux was given to be < 130 /cm^2/sec with 90 % C.L.

REFERENCES

1. Abdurashitov, J.N., et al. 1994, Phys. Lett. B328, 234
2. Agrawal, V., et al. 1996, Phys. Rev. D53, 1313
3. Anselmann, P., et al. 1994, Phys. Lett. B327, 377
4. Anselmann, P., et al. 1995, Phys. Lett. B342, 440
5. Bahcall, J.N., Basu, S., & Pinsonneault, M. 1998, Phys. Lett. B433, 1
6. Bahcall, J.N., Pinsonneault, M., & Basu, S. 2000, astro-ph/0010346.
7. Barr, G., et al. 1989, Phys. Rev. D39, 3532
8. Beacom, J.F., & Vogel, P. 1998, Phys. Rev. D58, 053010
9. Bionta, R.M., et al. 1987, Phys. Rev. Lett.58, 1494
10. Cleveland, B.T., et al. 1995, Nucl. Phys. B(Proc. Suppl.) 38, 47
11. Davis, R. 1994, Prog. Part. Nucl. Phys. 32, 13
12. Dighe, A.S., & Smirnov, A.Y. 2000, Phys. Rev. D62, 033007
13. Fukuda, Y., et al. 1996, Phys. Rev. Lett. 77,1683
14. Fukuda, Y., et al. 1998a, Phys. Lett. B433, 9
15. Fukuda, Y., et al. 1998b, Phys. Lett. B436, 33
16. Fukuda, Y., et al. 1998c, Phys. Rev. Lett. 81, 1158
17. Fukuda, Y., et al. 1998d, Phys. Rev. Lett. 81, 1562
18. Fukuda, Y., et al. 1999a, Phys. Rev. Lett. 82, 2644

19. Fukuda, Y., et al. 1999b, Phys. Lett. B467, 185
20. Fukuda, Y., et al. 1999c, Phys. Rev. Lett. 82, 1810
21. Fukuda, Y., et al. 1999d, Phys. Rev. Lett. 82, 2430
22. Fukuda, Y., et al. 2000, Phys. Rev. Lett. 85, 3999
23. Gaisser, T.K., & Stanev, T. 1995, Proc. 24th Int. Cosmic Ray Conf. (Rome) Vol.1 694
24. Honda, M., et al. 1990, Phys. Lett. B248, 193
25. Honda, M., et al., 1995, Phys. Rev. D52, 4985
26. Hirata, K.S., et al. 1988, Phys. Rev. D38, 448
27. Hirata, K.S., et al. 1990, Phys. Rev. Lett. 65, 1297; Hirata, K.S., et al. 1991, Phys. Rev. D44, 2241;1992, D45, 2170E
28. Kaplinghat, M., Steigman & G., Walker, T.P., 2000, Phys. Rev. D61, 103507.
29. Langanke, K., Vogel, P., &Kolbe, E. 1996, Phys. Rev. Lett. 76, 2629
30. McDonald, A. 2000, presentation in Neutrino 2000 conference, Sudbury, Canada, June
31. Mikheyev, S.P., & Smirnov, A.Y. 1985, Sov. Jour. Nucl. Phys. 42, 913
32. Ortiz, C.E. et al. 2000, Phys. Rev. Lett. 85, 2909
33. Totani, T. & Sato, K. 1995, Astropart. Phys. 3, 367
34. Totani, T., Sato, K., Yoshii, Y., 1996, Ap.J. 460, 303.
35. Turck-Chieze, S., & Lopes, I. 1993, Ap. J. 408,347
36. Wolfenstein, L. 1978, Phys. Rev. D17, 2369

Direct search for Dark Matter particles deep underground by DAMA experiment

P. Belli[1], R. Bernabei[1], R. Cerulli[1], F. Montecchia[1], M. Amato[2], G. Ignesti[2], A. Incicchitti[2], D. Prosperi[2], C.J. Dai[3], H.H. He[3], H.H. Kuang[3], J.M. Ma[3]

(1) Dip. di Fisica and INFN, sez. Roma2, Universita' di Roma "Tor Vergata", I-00133 Rome, Italy email: rita.bernabei@roma2.infn.it

(2) Dip. di Fisica and INFN, sez. Roma, Universita' di Roma "La Sapienza", I-00185 Rome, Italy.

(3) IHEP, Chinese Academy, P.O. Box 918/3, Beijing 100039, China

Abstract. DAMA is searching for rare processes by developing and using several kinds of radiopure scintillators: NaI(Tl), liquid Xenon and $CaF_2(Eu)$. Here, in particular, results achieved during four annual cycles with a model independent analysis and with model dependent ones in the investigation of the so-called WIMP annual modulation signature are summarized.

I INTRODUCTION

DAMA is devoted to the search for rare processes by developing and using low radioactivity scintillators. Its main aim is the search for relic particles (WIMPs: Weakly Interacting Massive Particles) [1–10], whose existence has been pointed out both by experimental observations and by theoretical considerations. Furthermore, due to the radiopurity achieved in the detectors and in the set-ups, several searches for other possible rare processes are also performed, such as e.g. for $\beta\beta$ processes, for charge-non-conserving processes, for Pauli exclusion principle violating processes and for nucleon instability [11].

The WIMPs are embedded in the galactic halo; thus our solar system, which is moving with respect to the galactic system, is continuously hit by a WIMP "wind". The quantitative study of this "wind" allows both to obtain information on the Universe evolution and to investigate Physics beyond the Standard Model. The WIMPs are mainly searched for by elastic scattering on target nuclei, which constitute a scintillation detector. In particular, the $\simeq 100$ kg NaI(Tl) set-up [7] has been realized to investigate the so-called WIMP "annual modulation signature". In fact, since the Earth rotates around the Sun, which is moving with respect to the galactic system, it would be crossed by a larger WIMP flux in June (when its rotational velocity is summed to the one of the solar system with respect to the Galaxy) and by a smaller one in December (when the two velocities are subtracted). The fractional difference between the maximum and the minimum of the rate is expected to be of order of $\simeq 7\%$. The $\simeq 100$ kg highly radiopure NaI(Tl) DAMA set-up [7] can effectively exploit such a signature because of its well known technology, of its high intrinsic radiopurity, of its mass, of the deep underground experimental site and of its suitable control of the operational parameters. The annual modulation signature is very distinctive as we have already pointed out [3,5,7–10,12]. In fact, a WIMP-induced seasonal effect must simultaneously satisfy all the following requirements: the rate must contain a component modulated according to a cosine function (1) with one year period (2) and a phase that peaks around $\simeq 2^{nd}$ June (3); this modulation must only be found in a well-defined low energy range, where WIMP induced recoils can be present (4); it must apply to those events in which just one detector of many actually "fires", since the WIMP multi-scattering probability is negligible (5); the modulation amplitude in the region of maximal sensitivity must be $\lesssim 7\%$ (6). Only systematic effects able to fulfil these 6 requirements could fake this signature; therefore for some other effect to mimic such a signal is highly unlikely.

Here the results obtained by analysing the data of four annual cycles investigating the annual modulation signature by means of the $\simeq 100$ kg NaI(Tl) set-up are summarized (see Table 1).

The whole set-up and its main performances have been discussed in ref. [7].

TABLE 1. Released data sets [1,3,5,6,8,10].

period	statistics $(kg \cdot d)$
DAMA/NAI-1	4549
DAMA/NaI-2	14962
DAMA/NaI-3	22455
DAMA/NaI-4	16020
Total statistics	57986
+ DAMA/NaI-0	upper limit on recoils by PSD

At beginning of year 2000 the data collected in two further annual cycles, DAMA/NaI-3 and DAMA/NaI-4 (total statistics 38475 kg·day) have been investigated. Cumulative analyses of a total statistics of 57986 kg·day have been carried out [8–10] properly including the physical constraint which arises from the measured upper limit on the recoil rate [1,9]. Both model independent and model dependent analyses have been performed. We remark that various uncertainties exist in every model dependent calculations concerning WIMP direct detection search (the same is for exclusion plot as well as for allowed regions in the cross section on proton versus WIMP mass plane). They are present both on some general features such as e.g. the real behaviour of the WIMP velocity distribution (which we have considered so far isothermal and Maxwellian), on the particle model and on the values taken for all the parameters needed in the calculations which are instead affected by uncertainties.

II ANNUAL MODULATION SIGNATURE AND RELATED STUDIES

Model independent analysis

A model independent analysis of the data of the four annual cycles offers an immediate evidence of the presence of an annual modulation of the rate of the single hit events in the lowest energy interval (2 – 6 keV) as shown in Fig. 1. There each data point has been obtained from the raw rate measured in the corresponding time interval, after subtracting the constant part.

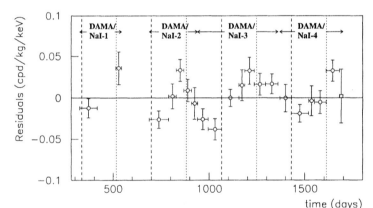

FIGURE 1. Model independent residual rate for single hit events, in the 2–6 keV cumulative energy interval, as a function of the time elapsed since January 1-st of the first year of data taking. The expected behaviour of a WIMP signal is a cosine function with minimum roughly at the dashed vertical lines and with maximum roughly at the dotted ones.

The χ^2 test on the data of Fig. 1 disfavors the hypothesis of unmodulated behaviour (probability: $4 \cdot 10^{-4}$), while fitting these residuals with the function $A \cdot \cos\omega(t - t_0)$, one gets: i) for the period $T = \frac{2\pi}{\omega} = (1.00 \pm 0.01)$ year when t_0 is fixed at the 152.5^{th} day of the year (corresponding to $\simeq$ 2 June); ii) for the phase $t_0 = (144 \pm 13)$ days, when T is fixed at 1 year. In the two cases A is: (0.022 ±

0.005) cpd/kg/keV and (0.023 ± 0.005) cpd/kg/keV, respectively. Similar results, but with slightly larger errors, are found in case all the parameters are kept free.

We have extensively discussed the results of the investigations of the known sources of possible systematics when releasing the data of each annual cycle; moreover, a dedicated paper [9] has been released on possible systematics, where in particular the data of the DAMA/NaI-3 and DAMA/NaI-4 running periods have been considered in quantitative evaluations. No known systematic effect or side reaction able to mimic a WIMP induced effect has been found; because of limited space available here, we invite the reader to refer to ref. [9] for details.

In conclusion, a WIMP contribution to the measured rate is candidate by the result of the model independent approach independently on the nature and coupling with ordinary matter of the involved WIMP particle.

Model dependent analyses

To investigate the nature and coupling with ordinary matter of a possible candidate, a suitable energy and time correlation analysis is necessary as well as a complete model framework. We remark that a model framework is identified not only by the general astrophysical, nuclear and particle physics assumptions, but also by the set of values used for all the parameters needed in the model itself and in related quantities (for example WIMP local velocity, v_0, form factor parameters, etc.).

At present the lightest supersymmetric particle named neutralino is considered the best candidate for WIMP; in supersymmetric theories both the squark and the Higgs bosons exchanges give contribution to the coherent (SI) part of the neutralino cross section, while the squark and the Z^0 exchanges give contribution to the spin dependent (SD) one. Note, in particular, that the results of the data analyses [8,10] summarized here and in the following hold for the neutralino, but are not restricted only to this candidate.

The differential energy distribution of the recoil nuclei in WIMP-nucleus elastic scattering can be calculated [1,13] by means of the differential cross section of the WIMP-nucleus elastic processes

$$\frac{d\sigma}{dE_R}(v, E_R) = \left(\frac{d\sigma}{dE_R}\right)_{SI} + \left(\frac{d\sigma}{dE_R}\right)_{SD} =$$
$$= \frac{2G_F^2 m_N}{\pi v^2} \{[Zg_p + (A-Z)g_n]^2 F_{SI}^2(E_R) +$$
$$+ 8\frac{J+1}{J}(a_p <S_p> + a_n <S_n>)^2 F_{SD}^2(E_R)\} \quad (1)$$

where: G_F is the Fermi coupling constant; m_N is the nucleus mass; v is the WIMP velocity in the laboratory frame; E_R is the recoil energy; Z is the nuclear charge and A is the atomic number; $g_{p,n}$ ($a_{p,n}$) are the effective WIMP-nucleon couplings for SI (SD) interactions; $<S_{p,n}>$ are the mean values of the nucleon spin in the nucleus and J is the nuclear spin; $F_{SI}^2(E_R)$ and $F_{SD}^2(E_R)$ are the SI and SD form factors, respectively. Therefore, the differential cross section and, consequently, the expected energy distribution depends on the WIMP mass, m_W, and on four unknown parameters of the theory: $g_{p,n}$ and $a_{p,n}$.

A generalized SI WIMP-nucleon cross section can be written as $\sigma_{SI} = \frac{4}{\pi}G_F^2 m_{W_p}^2 g^2$; there the coupling g is a function of g_p and g_n and – in a first approximation – is independent on the used target nucleus since $\frac{Z}{A}$ can be considered nearly constant for the nuclei typically used in WIMP direct searches; it is: $g = \frac{g_p+g_n}{2} \cdot \left[1 - \frac{g_p-g_n}{g_p+g_n}\left(1 - \frac{2Z}{A}\right)\right]$.

As regards the SD interaction we have introduced the useful notations [10]: $\bar{a} = \sqrt{a_p^2 + a_n^2}$, $tg\theta = \frac{a_n}{a_p}$ and $\sigma_{SD} = \frac{32}{\pi}\frac{3}{4}G_F^2 m_{W_p}^2 \bar{a}^2$, where σ_{SD} is a suitable SD WIMP-nucleon cross section.

In the following, the results obtained by analysing the data in some of the possible model frameworks are summarized.

I. WIMPs with dominant SI interaction in a given model framework

Often the spin-independent interaction with ordinary matter is assumed to be dominant since e.g. most of the used target-nuclei are practically not sensitive to SD interactions as on the contrary ^{23}Na and ^{127}I are. Therefore, first model dependent analyses of the data have been performed by considering a candidate in this scenario, that is neglecting the term $\left(\frac{d\sigma}{dE_R}\right)_{SD}$ in eq. (1).

A full energy and time correlation analysis – properly accounting for the physical constraint arising from the measured upper limit on recoils [1,9] – has been carried out in the framework of a given model for spin-independent coupled candidates with mass above 30 GeV. A standard maximum likelihood method has been used[1]. Following the usual procedure we have built the y log-likelihood function, which depends on the experimental data and on the theoretical expectations; then, y is minimized and parameters' regions allowed at given confidence level are derived. Note that different model frameworks (see above) vary the expectations and, therefore, the cross section and mass values corresponding to the y minimum, that is also the allowed region at given C.L.. In particular, the inclusion of the uncertainties associated to the models and to every parameter in the models themselves as well as other possible scenarios enlarges the allowed region as discussed e.g. in ref. [6] for the particular case of the astrophysical velocities. Also in the case considered here the minimization procedure has been repeated by varying v_0 from 170 km/s to 270 km/s to account for its present uncertainty. For example, in the model framework considered in ref. [8] $m_W = (72^{+18}_{-15})$ GeV and $\xi\sigma_{SI} = (5.7\pm1.1)\cdot 10^{-6}$ pb correspond to the position of y minimum when $v_0 = 170$ km/s, while $m_W = (43^{+12}_{-9})$ GeV and $\xi\sigma_{SI} = (5.4 \pm 1.0) \cdot 10^{-6}$ pb when $v_0 = 220$ km/s ($\xi = \frac{\rho_{WIMP}}{0.3 GeV cm^{-3}}$). Fig. 2 shows the regions allowed at 3σ C.L. in that model framework, when the uncertainty on

---
[1] Substantially the same results are obtained with other analysis approaches such as e.g. the Feldman and Cousins one.

v_0 is taken into account (solid contour) and when possible bulk halo rotation is considered (dashed contour). For simplicity, no other uncertainties on the used parameters have been considered there (some of them have been included in the approach summarized in the next subsection [10]), which obviously further enlarge the allowed region with the respect both to $\xi\sigma_{SI}$ and m_W quantities.

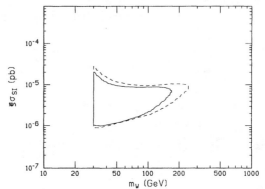

FIGURE 2. Regions allowed at 3σ C.L. on the plane $\xi\sigma_{SI}$ versus m_W for a WIMP with dominant SI interaction and mass above 30 GeV in the model framework considered in [8]: i) when v_0 uncertainty (170 km/s $\leq v_0 \leq$ 270 km/s; continuous contour) has been included; ii) when also a possible bulk halo rotation as in ref. [6] (dashed contour) is considered. See text. As widely known, the inclusion of present uncertainties on some other astrophysical, nuclear and particle physics parameters would enlarge these regions (varying consequently the position of the minimum for the y log-likelihood function); full estimates are in progress.

A quantitative comparison between the results of the model independent analysis and of this model dependent analysis has been discussed in ref. [8].

In conclusion, the observed effect investigated in terms of a WIMP candidate with dominant SI interaction and mass above 30 GeV in the model framework considered in ref. [8], supports allowed WIMP masses up to 130 GeV (1 σ C.L.) and even up to 180 GeV (1 σ C.L.) if possible dark halo rotation is included, while lower $\xi\sigma_{SI}$ would be implied by the inclusion of known uncertainties on parameters and model features.

Theoretical implications of these results in terms of a neutralino with dominant SI interaction and mass above 30 GeV have been discussed in ref. [13-15], while the case for an heavy neutrino of the fourth family has been considered in ref. [16].

II. WIMPs with mixed coupling in given model framework

Since the ^{23}Na and ^{127}I nuclei are sensitive to both SI and SD couplings – on the contrary of ^{nat}Ge and ^{nat}Si which are sensitive mainly to WIMPs with SI coupling (only 7.8 % is non-zero spin isotope in ^{nat}Ge and only 4.7% of ^{29}Si in ^{nat}Si) – the analysis of the data has been extended considering the more general case of eq. (1) [10]. This implies a WIMP having not only a spin-independent, but also a spin-dependent coupling different from zero, as it is also possible e.g. for the neutralino (see above) [2].

Following the usual procedure we have built the y log-likelihood function, which depends on the experimental data and on the theoretical expectations in the given model framework. Then, y has been minimized – properly accounting also for the physical constraint set by the measured upper limit on recoils [1] – and parameters' regions allowed at given confidence level have been obtained. In particular, the calculation has been performed by minimizing the y function with respect to the $\xi\sigma_{SI}$, $\xi\sigma_{SD}$ and m_W parameters for each given θ value. In the present framework the uncertainties on v_0 have been included; moreover, the uncertainties on the nuclear radius and the nuclear surface thickness parameter in the SI form factor, on the b parameter in the used SD form factor (see later) and on the measured quenching factors [1] of these detectors have also been considered [10].

Fig. 3 shows slices (colored areas) for some m_W of the region allowed at 3 σ C.L. in the ($\xi\sigma_{SI}$, $\xi\sigma_{SD}$, m_W) space at fixed θ value. Only the case of four particular couplings are shown here for simplicity: i) $\theta = 0$ ($a_n = 0$ and $a_p \neq 0$ or $|a_p| \gg |a_n|$); ii) $\theta = \pi/4$ ($a_p = a_n$); iii) $\theta = \pi/2$ ($a_n \neq 0$ and $a_p = 0$ or $|a_n| \gg |a_p|$); iv) $\theta = 2.435$ rad ($\frac{a_n}{a_p} = -0.85$, pure Z^0 coupling). The case $a_p = -a_n$ is nearly similar to the case iv). The dashed lines given for the case $m_W = 50$ GeV represent the limit curves calculated for simplicity in the case of $v_0 = 220$ km/s from the data of the DAMA liquid Xenon experiment [4]; regions above these dashed lines could be considered excluded at 90% C.L. Since the ^{129}Xe nucleus – on the contrary of the ^{23}Na and of the ^{127}I – has the neutron as odd nucleon, only the case $\theta \simeq \pi/2$ would be affected; similar results are obtained for all the other WIMP masses. In this context, we have to recall that, as widely known, the comparison of results achieved by different experiments – even more when different target nuclei and/or different techniques have been used – is affected by intrinsic experimental and theoretical uncertainties. Moreover, no direct quantitative comparison can be performed between results obtained in direct and in indirect searches since it strongly depends on assumptions and on the considered model framework[3]. In particular, it does not exist a biunivocal correspondence between the observables in the two kinds of experiments: WIMP-nucleus elastic scattering cross section (direct detection case) and flux of muons from neutrinos (indirect detection case). In fact, elastic cross sections on proton spanning several orders of magnitude can give the same muon flux for different e.g. neutralino configurations and viceversa (see e.g. Fig. 3 of ref. [14]).

As already pointed out, when the SD contribution goes to zero (y axis in Fig. 3),

---

[2] For the sake of completeness, we note that the possibility qualitatively derived in ref. [17] (after our ref. [18] where the possibility of a SD contribution was mentioned) that a SD component could be denied by data of direct or indirect searches is excluded by correct and cautious approaches as mentioned in ref. [10] and summarized here. We also note that, moreover, the regions quoted there have been evaluated with a largely arbitrary approach.

[3] In addition, large uncertainties are present in the evaluation of the results of the indirect searches themselves.

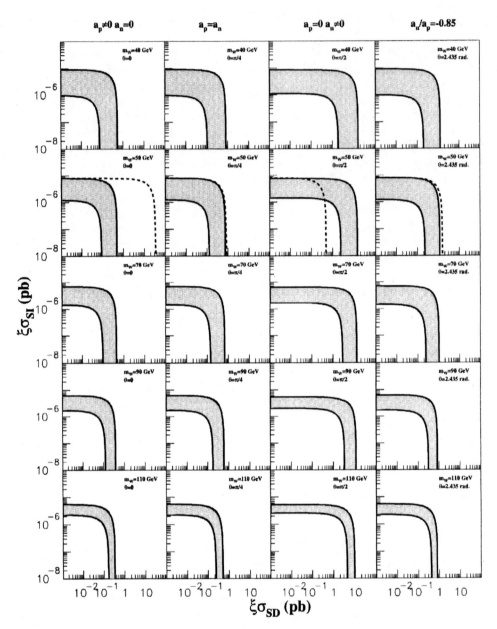

FIGURE 3. Slices (colored areas) for some m_W of the region allowed at 3 σ C.L. in the ($\xi\sigma_{SI}$, $\xi\sigma_{SD}$, m_W) space at fixed θ value in the model framework considered in ref. [10]. Only four particular couplings are reported here for simplicity: i) $\theta = 0$; ii) $\theta = \pi/4$ iii) $\theta = \pi/2$; iv) $\theta = 2.435$ rad.

an interval not compatible with zero is obtained for $\xi\sigma_{SI}$. Similarly, when the SI contribution goes to zero (x axis in Fig. 3), finite values for the SD cross section are obtained. Large regions are allowed for mixed configurations also for $\xi\sigma_{SI} \lesssim 10^{-5}$ pb and $\xi\sigma_{SD} \lesssim 1$ pb; only in the particular case of $\theta = \frac{\pi}{2}$ (that is $a_p = 0$ and $a_n \neq 0$) $\xi\sigma_{SD}$ can increase up to $\simeq 10$ pb, since the ^{23}Na and ^{127}I nuclei have the proton as odd nucleon. Moreover, in ref. [10] we have also shown that: i) finite values can be allowed for $\xi\sigma_{SD}$ even when $\xi\sigma_{SI} \simeq 3 \cdot 10^{-6}$ pb as in the region allowed in the pure SI scenario considered in the previous subsection; ii) regions not compatible with zero in the $\xi\sigma_{SD}$ versus m_W plane are allowed even when $\xi\sigma_{SI}$ values much lower than those allowed in the dominant SI scenario previously summarized are considered; iii) minima of the y function with both $\xi\sigma_{SI}$ and $\xi\sigma_{SD}$ different from zero are present for some m_W and θ pairs; the related confidence level ranges between $\simeq 3\ \sigma$ and $\simeq 4\ \sigma$ [10].

In conclusion, this analysis has shown that the DAMA data of the four annual cycles, analysed in terms of WIMP annual modulation signature, can also be compatible with a mixed scenario where both $\xi\sigma_{SI}$ and $\xi\sigma_{SD}$ are different from zero. The pure SD and pure SI cases in the model framework considered here are implicitly given in Fig. 3 for the quoted m_W and θ values.

Further investigations are in progress on these model dependent analyses to account for other known parameters uncertainties and for possible different model assumptions. As an example we recall that for the SD form factor an universal formulation is not possible since the internal degrees of the WIMP particle model (e.g. supersymmetry in case of neutralino) cannot be completely separated from the nuclear ones. In the calculations presented here we have adopted the SD form factors of ref. [19] estimated by considering the Nijmengen nucleon-nucleon potential. Other formulations are possible for SD form factors and can be considered with evident implications on the obtained allowed regions.

III DAMA/NAI-0 TO 4 ANNUAL MODULATION RESULT VERSUS CDMS-I RESULT

As previously mentioned, intrinsic experimental and theoretical uncertainties exist in the comparison of results achieved by different experiments and, even more, when different techniques are used as in the case of DAMA [3,5,6,8–10] and of CDMS [20]. In fact, DAMA is searching for a distinctive signature by using a large mass NaI(Tl) set-up, while CDMS is quoting counting rate by using a hybrid bolometer/ionizing technique for background rejection[4]. Moreover, when – as in

---

[4] For the sake of completeness, we comment that every kind of background rejection technique (such as PSD in scintillators, heat/ionizing and heat/light signals in bolometers) – even under the unrealistic assumption of an ideal electromagnetic background rejection – cannot allow to unambiguously identify a WIMP presence. In fact, e.g., recoils can be induced by other processes in competition with the WIMP-nucleus scattering. Moreover, tails of other kinds of populations (such as e.g. the so-called surface electrons in bolometers and the noise), the internal end-range

TABLE 2. Few numbers on the $\simeq 100$ kg NaI(Tl) DAMA and CDMS-I.

	$\simeq 100$ kg DAMA	CDMS-I
exposure	57986.0 kg · day	10.6 kg · day
depth	1400 m	10 m
number of events in the observed effect	total modulated amplitude $\simeq 2000$ events	13 evt in Ge, 4 evts in Si 4 multiple-evts in Ge + MonteCarlo on neutron flux

the DAMA and CDMS case – the involved target nuclei are different, no absolute comparison can be pursued at all and only model dependent comparisons can be considered with further uncertainties. In Table 2 few numbers are given to offer a prompt view on the two results. The CDMS-I result requires more informations on measured quenching factors, on sensitive volumes for the two different signals, on determination of the windows for rejection, on efficiencies, on energy calibrations, etc.. Moreover, the stability of these quantities during the about 100 days of running, the discussion of the performed significant data selection, the quantitative control of systematic uncertainties in the various hardware and software handlings must be demonstrated. In addition the quoted result arises from the joint analyses of two different experiments (Si and Ge) and, practically, by a MonteCarlo subtraction. The following discussion is made assuming – as it is not the real case – that all these informations have correctly been given.

In the Si experiment (used exposure $\simeq 1.5$ kg · day of $\simeq 3.3$ kg · day available) a large amount of events survived the ionizing/heat discrimination in the whole energy region allowed for recoil candidates. Thereafter, by the so-called athermal pulse shape discrimination, 4 events remained and were classified as "mostly neutrons", while all the others as so-called "surface electrons". The amount and the Y (ratio between ionizing and heat charges) and energy distributions of the latter ones give hint that the 4 "mostly neutrons" events could be indeed — all or partially — ascribed to the tail of the huge population of the "surface electrons" surviving the ionizing/heat discrimination. This possibility significantly affects the conclusions of ref. [20].

In the Ge experiment (used exposure $\simeq 10.6$ kg · day of $\simeq 48$ kg · day available for

---

alphas and fission fragments also generate signals indistinguishable from WIMP induced recoils; the corresponding contribution cannot be estimated and subtracted in any reliable manner at the needed level of precision. Thus, every result based on background rejection techniques will always be a conjecture on the real nature of the observed events. Moreover, the possibility of a reliable investigation of model independent signatures (such as the annual modulation one) on discriminated events (whatever technique will be used) is ruled out by the uncertainties existing in the background rejection itself.

3 of 4 Ge detectors), 13 recoil candidates survive the ionization/heat discrimination. This number of events is – under the mentioned assumptions – largely compatible with the DAMA allowed region estimated in ref. [8] in a particular model framework for a SI candidate with mass above 30 GeV. In fact, simple calculations assuming ideal values for the CDMS-I physical parameters show that, in that particular model framework [8], CDMS-I should measure from $\simeq$ 15 events down to < 1, even before considering the role of the parameters and model uncertainties which would enlarge the region allowed by DAMA and would vary the exclusion plot quoted by CDMS-I. In addition we note that the interpretation on the real nature of the considered 13 CDMS-I candidates strongly depends on the MonteCarlo estimates of the neutron background, which is constrained by the hypotised nature of the 4 Si candidates and of 4 multi-hit events. A similar procedure is strongly uncertain since it is affected by the latter assumptions and by the assumptions on the original neutron energy spectrum and on the neutron transport calculations. This can be verified by considering that the result of such a calculation gives in ref. [20] about 30 expected neutrons to be compared with the 13 quoted recoil candidates; this, in particular, suggests either an overestimate of the neutron background or an underestimate of the real events (related to real value and stability e.g. of quenching factors, efficiencies, sensitive volumes, discrimination windows, etc.). Changes of the given exclusion plot are straight forward. Furthermore, a CDMS-I representative has quoted [21] that analysing these data to determine their compatibility with DAMA, the result gives an upper limit for presence of WIMPs in CDMS-I Ge data of 8 events at 90% C.L., evidently compatible with the DAMA allowed region even in the particular model framework for SI coupled WIMPs of ref. [8]. Moreover, for the sake of completeness, we note that in ref. [20] the complete DAMA result has not been considered in that model dependent comparison.

In conclusion, a proper analysis of the experimental parts and a proper model dependent comparison – considering in addition the uncertainties on models and on needed parameters – shows that the claim of CDMS-I is largely arbitrary. This is even more evident when considering the further compatibility of the two results in the more complete model framework of ref. [10], since Ge and Si are poorly sensitive to SD interactions on the contrary of Na and I.

IV CONCLUSION

An annual modulation in the low energy rate with proper features has been observed in a model independent way in the data of four annual cycles. Model dependent analyses have shown that these data are compatible with a WIMP both with pure SI interaction and with mixed SI/SD couplings as well as with pure SD interaction. Other model scenarios are possible and are under consideration. As an example, we mention the role played by non-standard halo models on the allowed region; a similar study has qualitatively been discussed e.g. in ref. [22] and is quantitatively under analysis on the experimental data.

Moreover, the data of the 5th annual cycle are already at hand, while – after a full upgrading of the electronics and of the data acquisition system – the set-up is now running to collect the data of a 6th annual cycle. Finally, the exposed mass will be increased in near future up to $\simeq 250$ kg to achieve higher experimental sensitivity, higher C.L. values and to disentangle different possible model scenarios.

REFERENCES

1. R. Bernabei et al., Phys. Lett. B389, (1996), 757.
2. R. Bernabei et al., Phys. Lett. B387 (1996), 222 and Phys. Lett. B389 (1996), 783; Il N. Cim. C19 (1996), 537; Astrop. Phys. 7 (1997),73; Phys. Rev. Lett. 83 (1999), 4918; Il N. Cim. A112 (1999), 1541; Nucl. Phys. B563 (1999), 97.
3. R. Bernabei et al., Phys. Lett. B424, (1998), 195.
4. R. Bernabei et al., Phys. Lett. B436 (1998), 379.
5. R. Bernabei et al., Phys. Lett. B450 (1999), 448.
6. P. Belli et al., Phys. Rev. D61 (2000), 023512.
7. R. Bernabei et al., Il N. Cim. A112 (1999), 545.
8. R. Bernabei et al., Phys. Lett. B480, (2000), 23.
9. R. Bernabei et al., Eur. Phys. J. C18 (2000), 283.
10. R. Bernabei et al., ROM2F/2000-35 & INFN/AE-01/01 preprint at www.lngs.infn.it
11. R. Bernabei et al., Astrop. Phys. 5 (1996), 217; Il N. Cim. A110 (1997), 189; Phys. Lett. B408 (1997), 439; Astrop. Phys. 10 (1999), 115; Phys. Rev. C60 (1999), 065501; Phys. Lett. B460 (1999), 236; Phys. Lett. B465 (1999), 315; Phys. Rev. D61 (2000), 117301; Phys. Lett. B493 (2000), 12.
12. P. Belli et al.: in the volume "3K-Cosmology", AIP pub., 65 (1999).
13. A. Bottino et al., Phys. Lett. B402, 113 (1997); Phys. Lett. B423, 109 (1998); Phys. Rev. D59 (1999), 095004; Phys. Rev. D59 (1999), 095003; Astropart. Phys. 10 (1999), 203; Astropart. Phys. 13, 215 (2000); Phys. Rev. D62 (2000) 056006; hep-ph/0010203; hep-ph/0012377.
14. A. Bottino et al., Phys. Rev. D62 (2000) 056006.
15. R.W. Arnowitt and P. Nath, Phys. Rev. D60 (1999), 044002; E. Gabrielli et al., hep-ph/0006266;
16. D. Fargion et al., Pis'ma Zh. Eksp. Teor. Fiz. 68, (JETP Lett. 68, 685) (1998); Astropart. Phys. 12, 307 (2000).
17. P. Ullio et al., hep-ph/0010036
18. R. Bernabei et al., ROM2F/2000/32 to appear on the Proceed. of Int. Conference PIC20, Lisbon June 2000.
19. M.T. Ressell et al., Phys. Rev. C56, 535 (1997).
20. CDMS collaboration, Phys. Rev. Lett. 84 (2000) 5695.
21. T. Shutt, seminar given at LNGS, march, 2000.
22. A. M. Green, Phys. Rev. D63, 43005 (2001).

Status of CDMS Search for Dark Matter WIMPs

B. Cabrera[8], R. Abusaidi[8], D.S. Akerib[1], P.D. Barnes, Jr.[9], D.A. Bauer[10],
A. Bolozdynya[1], P.L. Brink[8], R. Bunker[10], D.O. Caldwell[10], J.P. Castle[8],
C. Chang[8], R.M. Clarke[8], P. Colling[8], M.B. Cristler[2], A. Cummings[9],
A. Da Silva[9], A.K. Davies[8], R. Dixon[2], B.L. Dougherty[8], D. Driscoll[1],
S. Eichblatt[2], J. Emes[3], R.J. Gaitskell[9], S.R. Golwala[9], D. Hale[10],
E.E. Haller[3], D. Holmgren[2], J. Hellmig[9], M.E. Huber[11], K.D. Irwin[4],
J. Jochum[9], F.P. Lipschultz[7], A. Lu[10], C. Maloney[10], V. Mandic[9],
J.M. Martinis[4], P. Meunier[9], S.W. Nam[4], H. Nelson[10], B. Neuhauser[7],
M.J. Penn[8], T.A. Perera[1], M.C. Perillo Isaac[9], B. Pritychenko[9],
R.R. Ross[3,9], T. Saab[8], B. Sadoulet[3,9], J. Sander[10], D.N. Seitz[9],
P. Shestople[7], T. Shutt[5], A. Smith[3], G.W. Smith[9], R.W. Schnee[1],
A.H. Sonnenschein[10], A.L. Spadafora[9], W. Stockwell[9], J.D. Taylor[3],
S. White[9], S. Yellin[10], B.A. Young[6]
(CDMS Collaboration)

[1]*Department of Physics, Case Western Reserve University, Cleveland, OH 44106, USA*
[2]*Fermi National Accelerator Laboratory, Batavia, IL 60510, USA*
[3]*Lawrence Berkeley National Laboratory, Berkeley, CA 94720, USA*
[4]*National Institute of Standards and Technology, Boulder, CO 80303, USA*
[5]*Department of Physics, Princeton University, Princeton, NJ 08544, USA*
[6]*Department of Physics, Santa Clara University, Santa Clara, CA 95053, USA*
[7]*Department of Physics & Astronomy, San Francisco State University, San Francisco, CA 94132, USA*
[8]*Department of Physics, Stanford University, Stanford, CA 94305, USA*
[9]*Center for Particle Astrophysics, University of California, Berkeley, Berkeley, CA 94720, USA*
[10]*Department of Physics, University of California, Santa Barbara, Santa Barbara, CA 93106, USA*
[11]*Department of Physics, University of Colorado, Denver, CO 80217, USA*

Abstract. We report on the latest results from the CDMS (cryogenic dark matter search) experiment. The experiment uses superconducting particle detectors, operated below 100 mK, to search for dark matter in the form of weakly interacting massive elementary particles or WIMPs. These detectors are either Si or Ge crystals, where the electron-hole production and the phonon production are measured for each event, allowing the discrimination of electron recoils (most backgrounds due to gammas and betas) from nuclear recoils (due to WIMPs and neutrons). We have recently reported new limits from the Stanford shallow site experiment (CDMS-I) which explore supersymmetric models where the lightest supersymmetric particle is often an excellent WIMP candidate. We will also report on the Soudan deep site facility for the CDMS-II experiment which is under construction, and on the status of the CDMS-II detector fabrication.

1. INTRODUCTION

There is strong theoretical justification for believing that the dark matter around galaxies, including our own, is in the form of a non-baryonic elementary particle.[1-2] The success of cold dark matter in generating large scale structure and galaxy formation in numerical simulations has shifted attention away from neutrinos to WIMPs or axions. Both of these candidates are well motivated by particle physics and make excellent candidates for the dark matter. The case for nonbaryonic dark matter remains strong even with the supernova data suggesting the existence of dark energy, and with the MACHO data suggesting that up to 20% of our galactic halo is baryonic. The lightest supersymmetric particle (LSP) has exactly the right properties to comprise the WIMP dark matter in a large class of possible supersymmetric (SUSY) theories. We search for WIMPs using low temperature detectors, which are operated at temperatures below 0.1 K and utilize the remarkable properties of insulating crystals and superconductors to open a new window for dark matter searches.

If WIMPs are the dark matter in the universe, the Earth is moving through a sea of these as yet undiscovered particles. The resulting interaction rate in a terrestrial experiment depends on the local density and velocity of the dark matter particles. Galactic modeling suggests that the dark matter particles have a local density $\rho_0 \approx 0.3 \text{GeV/cm}^3$ and a nearly Maxwellian velocity distribution with a velocity dispersion $v_0 \approx 220$ km/sec.[2-3] The calculated flux of WIMPs incident on a terrestrial experiment is substantial: $\Phi_\chi \approx (10^7/m_\chi) \text{GeV cm}^{-2} \text{ s}^{-1}$, where m_χ is the WIMP mass. The rate of WIMP interactions per unit recoil energy is:

$$\frac{dR}{dQ} = \frac{\sigma_0 \rho_0}{\sqrt{\pi} v_0 m_\chi m_{r\chi N}^2} F^2(Q) T(Q), \qquad (1)$$

where σ_0 is the WIMP elastic scattering cross section, m_N is the target nucleus mass with reduced mass $m_{r\chi N} = m_\chi m_N / (m_\chi + m_N)$, and Q is the recoil energy. In general, the cross section σ_0 contains contributions from both spin-independent (SI) and spin-dependent (SD) interactions. In most theories the coherent SI component dominates, and $F(Q)$ is approximated well by the Woods-Saxon equation[2]

$$F(Q) = 3 \frac{j_1(Qr_n/\hbar c)}{Qr_n/\hbar c} \exp\left[-(Qs/\hbar c)^2/2\right], \qquad (2)$$

over the entire range of target nuclei, where the nuclear radius is r_n.

To compare the coherent rates from different target materials for scalar interactions, the cross section $\sigma_{0\,scalar}$ may be written as

$$\sigma_{0\,scalar} = \frac{4 m_\chi^2 m_N^4}{\pi (m_\chi + m_N)^2} \left(\frac{f_n}{m_n}\right)^2, \qquad (3)$$

where $f_n \cong f_p$ is the WIMP-nucleon coupling and $m_n = 0.9315$ GeV is the average nucleon mass. We define the zero momentum transfer fundamental WIMP-nucleon cross section σ_{0Wn} as

$$\frac{\sigma_{0Wn}}{m_{r\chi n}^2} = \frac{4}{\pi} f_n^2 = \frac{\sigma_{0scalar}}{A^2 m_{r\chi N}^2}. \qquad (4)$$

where the atomic mass $A \cong m_N/m_n$, and $m_{r\chi n}^2$ is the reduced mass for a single nucleon.

To determine the optimal detector material, one must carefully consider four factors: (1) for a given WIMP mass, the characteristic recoil energy rE_0 is equal to the WIMP kinetic energy, about $25°\text{keV}\times°m_\chi/(50\text{ GeV})$ when $m_\chi \approx m_N$, and approaches $2m_N\langle v^2 \rangle$ for $m_\chi \gg m_N$, hence increasing linearly with atomic number A; (2) for some materials, the observed recoil energy Q_{obs} may be considerably less than Q; the thermal detectors used by CDMS have the advantage that $Q_{obs} \sim Q$; on the other hand, scintillator detectors have $Q_{obs} = \varepsilon\, Q$ where the measured value of ε is about 0.3 for Na, 0.09 for I, 0.08 for Ca, 0.12 for F, and 0.2 for Xe; (3) the effect of the form factor $F(Q)$ can be appreciable at higher Q values and for heavier nuclei; and (4) the coherent interaction cross section, which is proportional to A^2, favors heavy nuclei for small recoil energies but is suppressed at higher recoil energy by the form factor.

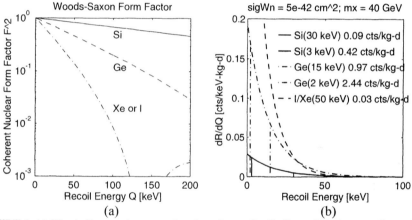

FIGURE 1. (a) Woods-Saxon coherent nuclear form factors for Si, Ge, and Xe (same as I) are plotted as a function of recoil energy . (b) Recoil spectra are plotted for the same materials, with the integrated events above detector thresholds tabulated.

In Fig 1a, the form factor has been plotted as a function of the recoil energy Q. Over the interesting energy range, there is little suppression for light nuclei like Si, more suppression for Ge, and severe suppression for heavier nuclei like xenon and iodine. In order to illustrate the role of these various experimental factors [Fig 1b], let us take a 40°GeV WIMP with $\sigma_{0Wn} \sim 5\times 10^{-42}$ cm^2, near the largest event rate allowed by minimal supersymmetric models. At our presently demonstrated thresholds for a 100 kg-d exposure with Si (3 keV threshold) and Ge (2 keV threshold), we would obtain 42 events in Si and 244 events in Ge. In contrast, we would expect only 3 events in Xe (or I) above a threshold of 50 keV recoil energy, which corresponds to 10 keV of observed energy in Xe (and 4 keV in I).

As shown in Fig 2, it is also important to operate detectors made of different materials. For example, the predicted rates for WIMPs are about a factor of five higher in Ge than in Si [Fig 2a]. However, Si is more sensitive than Ge to a neutron background [Fig 2b]. Thus the combination of Si and Ge data allows the separation of a WIMP signal from a neutron background on a statistical basis. In addition, because of the kinematic dependence of the WIMP recoil spectrum, in the absence of a neutron background, the comparison of the Si and Ge spectra would allow a determination of the WIMP mass.

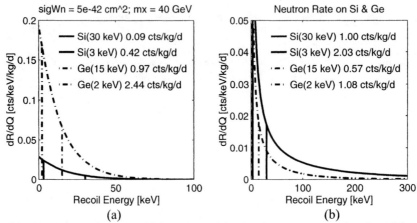

FIGURE 2. (a) Recoil spectra for WIMPs in Si and Ge showing the rates as a function of threshold. (b) Recoil spectra for neutrons in Si and Ge.

The CDMS detectors are capable of searching for a very broad class of WIMPs. As an example, we consider one of the best motivated WIMP candidates predicted by current theories of particle physics beyond the realm of the Standard Model, namely the lightest neutral supersymmetric particle (neutralino) of the minimal supersymmetric standard model (MSSM) with conserved R-parity. This neutralino is stable, and is therefore a very natural non-baryonic dark matter candidate. Indeed, over a large region of MSSM parameter space, neutralino dark matter is inevitable.

The neutralino-nuclear scattering cross section has two components: (i) an effective axial-vector interaction which gives a spin-dependent (SD) cross section which is non-zero only for nuclei with net spin, and (ii) scalar interactions which give spin-independent (SI) cross sections which involve the squares of the nuclear neutron and proton numbers. The relative strengths of the SD and SI parts of the scattering cross section depend on the neutralino composition. In general the neutralino will be a mixed state, and will have both SD and SI interactions, where the SI interaction is due to squark exchange and Higgs exchange and depends on the zino-higgsino mixture.

For a given point in the MSSM parameter space, and hence a given neutralino composition, the SD and SI interactions can be calculated. The envelope of predicted neutralino-nucleon cross sections that lead to cosmologically-interesting relic densities is shown in Fig 3 as a function of neutralino mass. Accelerator experiments give a lower bound to the allowed WIMP mass of about 35 GeV. Fig 3 also shows an estimate of the expected ultimate sensitivities for the Stanford and Soudan phases of CDMS. The Stanford curve is based on a kg-scale Si or Ge experiment with an exposure of 100 days (CDMS I), ambient backgrounds of approximately 3 events/(keV—kg—day), and intrinsic background rejection of 99% based on detector signals that distinguish WIMPs from the dominant electron-recoil backgrounds. Following background subtraction, the resulting sensitivity is 0.01 events/(keV kg day) at the Stanford site, good enough to allow CDMS I to discover relic neutralino WIMPs over a significant part of the allowed parameter space.

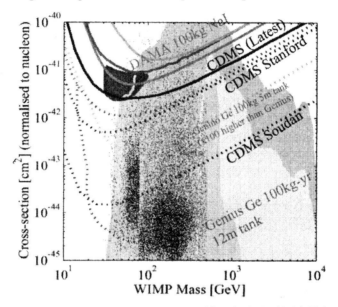

FIGURE 3. Goals for CDMS-I at Stanford site (0.01 events/(keV kg day) with 2 keV threshold for Ge), for CDMS-II at Soudan (0.0003 events/(keV kg day) with 2 keV threshold for Ge). The best current limits are shown for comparison along with possible SUSY model predictions (the peak corresponds to 5 events/kg/day for Ge). Also shown are new DAMA and CDMS results.

With a total exposure of 10,000 kg-days, CDMS II in the Soudan mine, will be capable of searching for WIMPs in general, and neutralinos in particular, over a substantially larger region of parameter space, as shown in Fig 3. Also shown for comparison, are the reported results from the DAMA collaboration which claims a suggested signal by exploiting a ~ 5% annual modulation in the WIMP-induced recoil spectrum due to the earth s solar orbit.[8-9] Also shown are proposals for large arrays of conventional Ge detectors operated at 77K (Genino and Genius).[10]

2. RECENT RESULTS FROM CDMS EXPERIMENT

For CDMS-I we chose a shallow site at Stanford, both for convenient access and because we recognized the site was at sufficient depth (17 mwe) that a competitive dark matter sensitivity could be achieved: assuming an active muon veto (>99.9% efficient) and detectors capable of discriminating against gammas (>99.5% efficient). For the Stanford Underground Facility (SUF), we modified an existing tunnel, and converted it into an underground experimental facility which provides 10.5 m of dirt overburden. The facility is located within End Station III in HEPL (Hansen Experimental Physics Laboratory) and is within 300 yards of the Stanford Physics Department. The location has proven ideal for bringing our advanced detector technology into operation.

After several years of running detectors and optimizing their performance, we have reached the point where neutrons are the dominant background of concern at SUF. All other sources (alpha, beta or gamma radioactivity in the laboratory structure or the apparatus immediately surrounding our detectors) can be avoided by using local shielding, careful selection of materials, background discrimination using the ratio of ionization to phonons in the detectors, and rejection of surface events. Three sources of neutrons must be considered: secondaries from the hadronic component of cosmic rays; neutrons from cosmic-ray muon interactions with nearby nuclei; and fission products from radioactivity in the ground and tunnel walls, mostly from uranium. The depth of our tunnel is sufficient to reduce the first source of neutrons to insignificance. An extensive series of measurements and Monte Carlo studies have been performed to understand backgrounds in SUF.

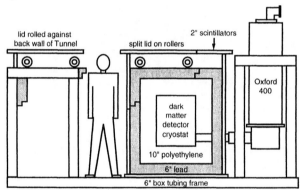

FIGURE 4. Schematic of dark matter detector experiment showing the lid of the Icebox, the polyethylene, lead, and plastic scintillators

Fig 4 shows a schematic design of the CDMS experimental apparatus, which has been successfully installed and brought into operation in the SUF. An Oxford Instruments 400 dilution refrigerator is attached by a 2 m long extension to the cryogenic environment, referred to as the Icebox, which houses the detector modules. Starting from the outside, the shielding structure around the Icebox consists of an active plastic scintillator enclosure which is about two inches thick, followed by a six inch thick Pb layer and then a ten inch thick polyethylene layer. The plastic scintillator is used as an active muon veto and allows good separation of muons from external background gammas. The scintillators of the muon veto are composed of 1.75" thick NE110 sheets with wave shifter bars attached to the edges. At least two PMTs are used per paddle, with the sum signal readout in coincidence.

In the low background counting experiments, most of the detector events associated with the passage of muons can be recognized by their occurrence within a few microseconds of a muon event in the scintillators. For each detector event, the pre- and post- muon hit history over a few msec are recorded. The Pb shield is used to attenuate the outside gammas from environmental radioactivity to an acceptably low rate. The polyethylene shield is used to moderate the external neutron flux incident on the detector down to a level such that any resultant nuclear recoils are below the energy threshold of the detectors.

2.1. Summary of SUF Run 18 with 100 g Si Detector

A schematic of the 100 g Si Fast Large Ionization and Phonon (FLIP) detector is shown in Fig 5a. A new design for Si and Ge crystals with similar phonon sensors and improved charged electrodes is called ZIPs (Z-sensitive Ionization and Phonon) and ZIP detectors are in production for CDMS-II. The FLIP phonon sensors for Run 18 are fabricated on the top surface using photolithographic processes. Each sensor is divided into 37 units each 5 mm square which themselves contain 12 individual sensor elements connected in parallel. Aluminum phonon collector fins cover 82% of the top surface of the Si and also provide the ground electrode for the charge measurement. The W outer ionization electrode is patterned (10% area coverage) to minimize athermal phonon absorption. The transition temperature of the W meanders is ~80 mK but the Si crystal is maintained at a base temperature of ~30 mK. The sensor temperature is maintained within the superconducting to normal transition by the Joule heating via negative electrothermal feedback associated with the voltage bias. The intrinsic stability of the voltage bias allows every one of the several hundred parallel W meanders to self-bias within their transition region, even if there is a gradient in the W transition temperature across the surface of the detector surface.[4]

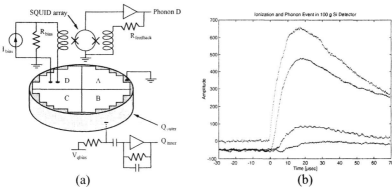

FIGURE 5. (a) Schematic diagram of 100 g Si FLIP, showing the four phonon W/Al QET sensors and the two ionization circuits. The Si crystal is 7.6 cm in diameter and 1 cm thick, and the phonon sensors cover 82% of the upper surface. (b) Ionization (top) and phonon pulses in three of the sensors from a single event in the 100g Si detector.

Discrimination of electron recoils from nuclear recoils on an event by event basis at low energy is the special feature of CDMS detectors. Typical "low background" measurements are dominated by alpha, beta, and gamma emissions from nearby radionuclide contaminants or cosmogenic sources. These backgrounds result in electron recoils. In contrast, weakly WIMPs and neutrons predominantly scatter off nuclei. Therefore, an important advantage for a dark matter WIMP detector is the ability to discriminate between electron and nuclear recoils. Since nuclear recoils dissipate a significantly smaller fraction of their energy into electron-hole pairs, an excellent discrimination technique is to measure simultaneously the phonons and ionization produced in each event.[5] This discrimination is accomplished through the use of independent sensors of phonons and charge-carriers, as with FLIP, BLIP (see below) and ZIP detectors.

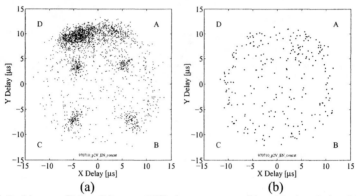

FIGURE 6. Position sensitivity of the new QET phonon sensors. Plotted is the relative delay between the four phonon sensors for collimated 60 keV x-rays from an 241Am source (left) and for background above 60 keV (right). The collimator holes were placed 1 inch apart. This phonon sensor clearly demonstrates position sensitivity on the order of a few mm at 60 keV.

Fig 5b show the pulses from a 60 keV gamma event in a Si FLIP detector. The ionization signal starts first, followed by the phonon signal from the sensor directly above the event, and then by the more distant phonon sensors. The measured time differences allow the event location to be reconstructed as shown in Fig 6. To better characterize this performance, a Si FLIP detector was cooled with an ^{241}Am source, which was collimated into four spots on the backside of the crystal, i.e., opposite the four phonon sensors. The spots were separated by one inch and offset toward sensors C and D. Data in Fig 6a clearly shows the four spots separated by approximately 12 s in a plot of relative delays between the phonon sensors. Fig 6b shows the more uniform illumination from background above 60 keV.

SUF Run 18 (November, 1997-July, 1998) produced excellent data from the 100 g Si FLIP detector. In particular there was a three month period of April-June, 1998 where continuous operation was maintained. Risetime discrimination was used to reject surface beta events.[6] These data resulted in a 1.6 kg-d exposure after cuts and were used as an important input to the SUF neutron background measurement. Fig 7 shows data from SUF Run 18. In the upper left, we show the response of the detector to a neutron calibration run using a ^{252}Cf source. A well defined nuclear recoil band is seen. The upper right shows a gamma calibration run with a ^{60}Co source. Now the gamma band is defined and we are able to measure the tail of the gamma distribution into the nuclear recoil band. The discrimination is better than 99.8% above 10 keV. The background data for 1.0 kg-d is shown in the two lower plots. First (lower left) are the muon coincident data, clearly showing a neutron band from muon induced neutrons. These data demonstrate the stability of the nuclear recoil band and continuously calibrate the detector throughout the run. Finally, the lower right plot shows the muon anticoincident data. A large number of gammas are still seen, coming mostly from residual radioactivity. Along the nuclear recoil band, where we expect to see a WIMP signal, we find two events (four for the entire 1.6 kg-d data set). From the calibrations done with a ^{14}C beta source, we measured the discrimination efficiency of the risetime cut against surface electron events. We obtained better than 95% rejection, so that of the four events, only 0.25 of an event should be from a misidentified beta. The events could be WIMPs, but as we argue below, neutrons produced from muon interactions in the rock outside the muon veto are more likely.

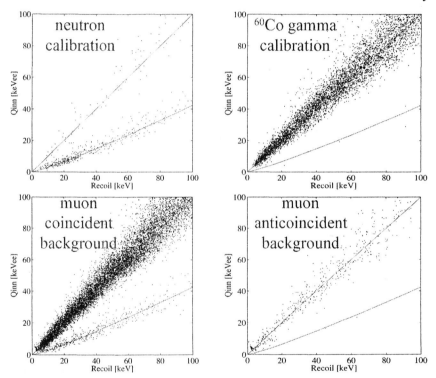

FIGURE 7. Data from 100 g Si FLIP detector in Run 18 showing neutron calibration (upper left), gamma calibration data (upper right), all muon coincident events over two months (lower left) and all muon anticoincident events over same time. Only two nuclear recoil events were seen in this data set of 1.0 kg-d (four over 1.6 kg-d).

2.2. Summary of SUF Run 19 with four 165 g Ge Detectors

SUF Run 19 (November, 1998 - September, 1999) produced a second set of excellent data using four 165 g Ge detectors utilizing NTD Ge thermister phonon sensors. These detectors were fabricated with the new amorphous Si electrodes (used in the new ZIP detectors) and were close packed with no material between them to allow vetoing of betas based on interactions with two surfaces. A schematic of the detector is shown in Fig 8a, the geometry of the four detectors is shown in Fig 8b.

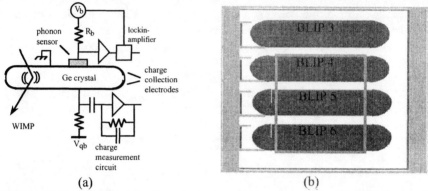

FIGURE 8. (a) The BLIP (Berkeley Large Ionization and Phonon) detectors with NTD Ge thermistors that measure temperature rise of crystal on msec timescale. (b) The geometry for the four close-packed detectors showing the fiducial volume.

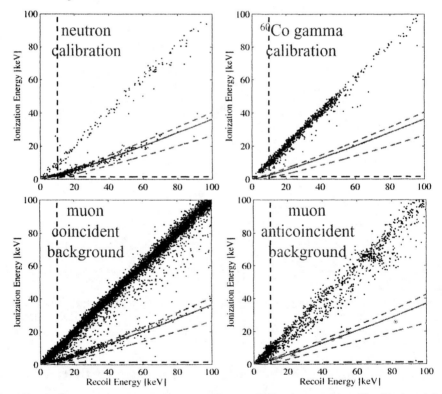

FIGURE 9. Data from four Ge BLIP detectors in Run 19 showing neutron calibration (upper left), gamma calibration data (upper right), muon coincident events (lower left) and all muon anticoincident events over the run. Only seventeen nuclear recoil events were seen in this data set of 10.6 kg-d of which four interacted with two detectors.

The results from this successful run are shown in Fig 9. Neutron calibration data (upper left) define the nuclear recoil band, and gamma calibration data (upper right) define the electron recoil band. Muon coincident data (lower left) clearly show the muon-induced neutrons and demonstrate conclusively that throughout the run the detectors were sensitive to nuclear recoils, and that the nuclear recoil band remained stable. The nuclear recoil band shown has a 90% acceptance. Finally, all of the muon anticoincident data (lower right) show a gamma band and a significant number of surface electrons, but the improved electrodes have pushed this beta band away from the nuclear recoil band and allowed our nuclear recoil measurements to be made. In the nuclear recoil band, there were 17 events over 10.6 kg-d of which 4 interacted with two detectors. The concentration of events in the gamma band is from Pb flourescence x-rays. The 13 single events could be WIMPs, but as we show below, neutrons are much more likely.

2.3. WIMP Limits from SUF Run 18 and SUF Run 19

The data from Run 18 with the Si detector and Run 19 with the four Ge detectors have set a new WIMP limit. Assuming coherent LSP interactions, this limit is not compatible with the result from the DAMA collaboration which uses NaI scintillator and claims to see annual modulation. Fig 10 summarizes our data, with 4 single events in the Si 1.6 kg-d, 13 single events in the Ge 10.6 kg-d and 4 multiple events in the Ge 10.6 kg-d data set. If the 13 single events were from WIMPs, one would expect zero Ge multiple events and 0.25 Si single events. On the other hand, if the 13 Ge events were produced by unvetoed neutrons, one would expect about 2.3 singles in Si and about 1.3 multiples in Ge. Clearly the Poisson statistics are poor, but a neutron model fits much better than a WIMP model. Fig 11a shows the new 90% C.L. set by our CDMS data and the comparison with the DAMA annual modulation signal. Our neutron-subtracted spectrum is shown in Fig 11b and is clearly not compatible with the DAMA central value. We expect to begin operating CDMS ZIP detectors in the Soudan mine by early 2002.

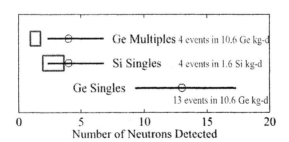

FIGURE 10. Summary of Run 18 Si and Run 19 Ge data, which are not dominated by WIMPs, but are in agreement with a neutron background. Shown are measured values (circles), predicted neutrons (boxes) from Ge singles and error bars (Poisson).

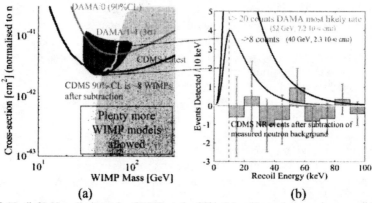

FIGURE 11. (left) New limit set by CDMS at the 90% C.L. These data are incompatible with the DAMA annual modulation claim and largely rule out that region of phase space. (right) The CDMS neutron subtracted spectrum compared with the central and three standard deviation values from the DAMA claim

3. ACKNOWLEDGMENTS

We thank the engineering and technical staffs at our respective institutions for invaluable support. This work is supported in part by the Department of Energy under contracts DE-FG03-90ER40569, DE-AC03-76SF00098, DE-FG03-91ER40618, by Fermilab, operated by the Universities Research Association, Inc., under Contract No. DE-AC02-76CH03000 with the Department of Energy, and by the Center for Particle Astrophysics, an NSF Science and Technology Center operated by the University of California, Berkeley, under Cooperative Agreement No. AST-91-20005, by the National Science Foundation under Grant No. PHY-9722414.

4. REFERENCES

1. Primack, J. R.., Seckel, D., and Sadoulet, B., *Annu. Rev. Nucl. Part. Sci.* **38**, 751 (1988).
2. Jungman, G., Kamionkowski, M., and Griest, K., *Phys. Rep.* **267**, 195 (1996).
3. Lewin, J.D., and Smith, P. F., *Astropart. Phys.* **6**, 87 (1996).
4. Irwin, K. D., et al., *Rev. Sci. Instr.* **66**, 5322 (1995).
5. Shutt, T., et al., *Phys. Rev. Lett.* **69**, 3531 (1992).
6. Clarke, R. M., et al., *Appl. Phys. Lett.* **76**, 2958 (2000).
7. Abusaidi, R., et al., *Phys. Rev. Lett.* **84**, 5699 (2000).
8. Bernabei, R., et al., *Phys. Lett.* **B 389**, 757 (1996).
9. Bernabei, R., et al., *Phys. Lett.* **B 480**, 23 (2000).
10. Baudis, L., et al., *Phys. Rev.* **D 59**, 22001 (1999).

Microlensing: from Dark Matter to Planets

A. Gould

Dept of Astronomy, Ohio State University, Columbus, OH 43210

Abstract.
When microlensing surveys were first undertaken a decade ago, their focus was overwhelmingly on dark matter. They have indeed fulfilled their original promise by ruling out massive compact halo objects (Machos) as the dominant form of dark matter over 8 decades of mass. However, microlensing has now developed a number of other important applications including searching for planets and probing the atmospheres of stars. More applications are promised in both the near and distant future.

INTRODUCTION

Following Paczyński's (1986) [1] suggestion, several groups began microlensing searches for dark matter in the form of massive compact halo objects (Machos) [2–4]. Now, a decade later, these searches have borne fruit. They have shown unequivocally that Machos in the mass range $10^{-7} M_\odot < M < 10 M_\odot$ do not constitute more than a modest fraction of the Milky Way's dark matter, and therefore that this unseen material must be something else: neutralinos, axions, or perhaps something more exotic.

But in the meantime, microlensing has greatly expanded its horizons. It is now a tool to search for planets and to probe the atmospheres of stars. In the relatively near future, it will give us an accurate census of the stellar mass function, including sub-luminous and non-luminous objects such as brown dwarfs, white dwarfs, neutron stars, and black holes. And in the still more distant future, microlensing could allow us to image the central engines of quasars with < 1 AU precision.

MICROLENSING BASICS

General relativity tells us that a spherical mass M deflects light by an angle $\alpha = 4GM/bc^2$, where b is the impact parameter. Hence, if the observer, and the mass (called a "lens") are perfectly aligned, and if the mass is compact enough not to block the light, then the source will be imaged into a ring of angular size θ_E, called the "Einstein radius"

$$\theta_E = \sqrt{\kappa M \pi_{\rm rel}}, \qquad \kappa \equiv \frac{4G}{c^2 {\rm AU}} \sim \frac{8\,{\rm mas}}{M_\odot}, \tag{1}$$

where $\pi_{\rm rel}$ is the relative parallax between the lens and the source. If the source is displaced from the lens by an angle $u\theta_E$, then the axial symmetry is broken and there are two images on opposite sides of the lens, at positions $u_\pm \theta_E$. Since surface brightness is conserved, the magnification tensor of these images is given by the derivatives of the source position with respect to the image position, and the scalar magnifications $A_\pm$ are just the determinant of this tensor,

$$u_\pm = \frac{u \pm \sqrt{u^2+4}}{2}, \qquad A_\pm = \frac{u_\pm^2}{u_+^2 - u_-^2}. \tag{2}$$

Thus, the total magnification is $A(u) = A_+ + A_- = (u^2+2)/[u\sqrt{u^2+4}]$. A microlensing event is then recognizable by the apparent time variation of the source.

$$F(t) = F_s A[u(t)] + F_b, \qquad u(t) = \sqrt{u_0^2 + \frac{(t-t_0)^2}{t_E^2}}, \tag{3}$$

where t_0 is the time of the event peak, $u_0 = u(t_0)$, t_E is the characteristic or "Einstein" timescale of the event, F_s is the unmagnified flux of the source, and F_b is the flux from any sources that are not being mangified but are entering the aperture.

The fact that microlensing events are describable by only 5 parameters is both a blessing and a curse: a blessing because it allows one to pick out microlensing events from much more common variable stars, and a curse because only one of these 5 parameters,

$$t_E = \frac{\theta_E}{\mu_{\rm rel}}, \tag{4}$$

tells us anything about the physical chacteristics of the lens. Moreover, as can be seen from equations (1) and (4), this lone piece of information comes entangled with $\pi_{\rm rel}$ and $\mu_{\rm rel}$, the lens-source relative parallax and proper motion.

An important microlensing quantity is τ, the "optical depth". This is the probability that a give source star falls inside an Einstein ring at a given time. The optical depth depends only on the mass density distribution along the line of sight and not on either the velocity distribution or the mass function. Hence, it is an extremely robust way to characterize dark matter in the form of Machos. If these comprise the dark matter, then $\tau \sim 5 \times 10^{-7}$ toward the LMC.

DARK MATTER SEARCHES

The main results on Macho dark matter come from the MACHO and EROS collaborations. Each searched for events toward the Large and Small Magellanic Clouds (LMC and SMC) in two regimes, "short" and "long" events, with the dividing line being $t_E \sim 1$ week.

Short Events

The short-event search is easier to interpret because no events were found. EROS conducted a search with very dense time sampling (several per hour) that could have not only detected but also characterized the events if they were found [5]. MACHO did not have this strategy, but happened quite often to acquire data in two colors for the same field at two closely spaced epochs on the same night. They were thus able to search for short timescale photometric excursions. Had they found something, they could not have figured out what it was, but since they did not, they could put strong constraints on short events [6]. MACHO and EROS then combined their results. Each of these studies had good sensitivity to events with timescales t_E ranging from about an hour to a few days. The expected kinematics of Machos (distributions of π_{rel} and μ_{rel}) together with equations (1) and (4) imply mass estimates [7]

$$M \sim \left(\frac{t_E}{65 \,\text{days}}\right)^2 M_\odot \qquad (5)$$

with a FWHM of a factor ~ 100. Hence using their combined null result, MACHO and EROS were able to constrain Machos to be less than 20% of the Milky Way halo over $10^{-7} M_\odot < M < 10^{-3} M_\odot$, with the limits being about twice that good over most of the range [8].

Long Events

The situation is slightly more complicated for the longer events. After 5.7 years of observations, MACHO had found about 15 candidates with timescales centered on 40 days [9]. The derived optical depth $\tau \sim 1.2^{+0.4}_{-0.3} \times 10^{-7}$, corresponds to 20% of the halo being in Machos, while (from eq. 5) the timescale corresponds to masses $M \sim 0.4 M_\odot$. The optical depth expected from known populations of stars toward this line of sight is about a factor 5 smaller, so this presented a very puzzling result indeed: the optical depth is too high to be stars, too low to account for all the dark matter, and the timescales are too long for the dark objects to be dark if they are made of hydrogen (since then they should burn).

The EROS survey [10] had a somewhat shorter duration, but covered overall a larger area, so that the total sensitivity was about 2/3 that of MACHO. They found a total of 5 candidates which, if interpreted as a detection of Machos would clearly lead to a lower estimate of the halo Macho fraction than MACHO's. However, EROS has chosen not make this interpretation. While one of their candidates looks quite nice, they argue that there is strong evidence that the lens is in the SMC, so it is not a Macho. None of the remainder of their candidates looks unequivocally like microlensing, and they note that several previous candidates have "gone away" because they varied again several years after their first detection. This means that they are almost certainly variable stars and not microlensing. Hence, EROS has taken the view that they will count their detections only toward an upper limit on the Macho fraction and not as a measure of it. Their resulting 95% confidence upper limit at $M = 0.4 M_\odot$ is $f < 30\%$. This is consistent with the MACHO detection ($f \sim 20\%$) at the same point.

Finally, MACHO has carried out a search for very long events, 75 days $< t_E <$ 300 days which, from equation (5), corresponds to $1.3 M_\odot < M < 20 M_\odot$. They find none and so are able to extend their limits to higher masses, ruling out halo fractions $f > 40\%$ all the way up to $\sim 10 M_\odot$.

Let me briefly summarize the situation. Machos are ruled out as dark matter for halo fractions $f > 20\%$ over the entire mass range $10^{-7} M_\odot < M < 10^{-1} M_\odot$, and for $f > 40\%$ to $10 M_\odot$. All the available evidence is consistent with a Macho fraction $f \sim 20\%$ at $M \sim 0.4 M_\odot$: MACHO claims to have detected such objects and EROS can't rule them out.

What can be done to resolve this issue? MACHO has stopped observing, although they still have about 2 years of data in the can. EROS will probably stop in 2002. The OGLE collaboration, which has almost a decade of experience with microlensing toward the Galactic bulge (see below), has initiated observations of the LMC that are of substantially higher quality than those of the previous generation and so could obtain significantly better statistics.

Pixel Lensing

However, there is also a completely independent approach, pixel lensing, which is being applied to observations of M31 by two groups, VATT/Columbia/MEGA and AGAPE [11,12]. Very few stars in M31 are resolved, and this at first sight seems to rule out microlensing searches because there are no stars to "watch" to see if they brighten in an event. However, one can still "watch" pixels (which each contain many stars), and if one of the stars it contains brightens, the pixel will also. In fact, although the surface of M31 is riddled with highly irregular surface-brightness fluctuations, if one subtracts two successive images, all of these go away and one is left with nearly perfect point-spread functions (PSFs), either positive or negative, at each position where a star has varied [11].

The problems of carrying out this work are nevertheless formidable, the main one being that M31 is 16 times further away than the LMC, so stars are 250 times fainter. This means that one needs to have bigger telescopes, or to restrict oneself to brighter (and hence rarer) sources, or to settle for worse photometry. In practice, one gets a bit of all three. The last problem, in particular, makes it difficult to definitively distinguish microlensing from other forms of stellar variability, a problem that we saw was fundamentally limiting the interpretation of LMC results.

However, the potential advantages are also large. The disk of M31 is inclined at about $78°$ toward the line of sight, and this implies that if its halo contains Machos, the optical depth toward the far disk is much higher than toward the near disk [13]. This would be an unambiguous signature: no form of variable, no matter how rare and perverse, could duplicate this asymmetry.

Both groups have reported detection of candidates [12,14], with a particularly nice one coming recently from AGAPE [15].

BULGE LENSING

Puzzles

The vast majority of microlensing events detected to date have been toward the Galactic bulge. When such observations were first proposed by Paczyński [16] and Griest et al., it was believed that the lenses would be disk stars with $\tau = 5 \times 10^{-7}$. However, soon after the detection of the first bulge event by OGLE when the initial event rate seemed to be much higher than expected, Kiraga & Paczyński [18] realized that lensing in this direction would be dominated by bulge lenses. MACHO made several estimates of the optical depth based on their first year of data, and argued that the most robust of these $\tau = 3.9^{+1.8}_{-1.2} \times 10^{-6}$ was the one based on clump giants [19]. Since these are bright, it is quite easy to determine how many stars are being effectively monitored, whereas for fainter stars "amplification bias" can produce events from stars below the detection threshold, thereby artificially enhancing the event rate [20]. Gould [21] and Kuijken [22] argued that it was impossible to produce this high τ with any axisymmetric Galaxy, and indeed the high rate has often been pointed to as evidence that we live in a barred galaxy. However, Binney et al. [23] have shown that reasonable non-axisymmetric models also cannot support such a high τ. MACHO has subsequently analyzed a much larger sample using image differencing, and still finds a high optical depth, although not quite as high $\tau = 3.2 \pm 0.5 \times 10^{-6}$. So this remains one important puzzle.

However, another puzzle concerns the distribution of events. Han & Gould [24] showed that the early MACHO and OGLE data had a strong excess of short events relative to what would be expected if the bulge mass function (MF) were similar to that of the solar neighborhood, perhaps implying an excess of low-mass stars and/or brown dwarfs. Part of this problem was explained by Han [20] as amplification bias, but much of it remains in the new MACHO analysis [25] which tries to take account of this bias.

SIM Measurement of the Bulge Mass Function

In brief, bulge lensing presents many interesting puzzles. What can be done to resolve them? The basic problem, as with LMC microlensing, is that one cannot tell much about events on an individual basis. One measures t_E and this is related to the mass through equations (4) and (1), but this relation involves the unknowns π_{rel} and μ_{rel}. For purposes of picking out a general mass scale (out of 8 orders of magnitude being probed), it is enough to make rough estimates of the quantities and derive a crude relation such as equation (5). However, toward the bulge, it is already known that a large fraction of microlensing is due to stars whose masses generally span about an order of magnitude. Since the FWHM of mass estimates like equation (5) is a factor 100, it is not much help in this case. One would like precise individual masses. It is enough to look at equation (1) to realize that there is a quantity conjugate to θ_E, called π_E, the "microlensing parallax"

$$\pi_E = \sqrt{\frac{\pi_{rel}}{\kappa M}}. \tag{6}$$

Similar to ordinary parallaxes, π_E tells how far an event is displaced in the Einstein ring if the *observer* changes position by 1 AU. Clearly, if both θ_E and π_E were measured, the mass would follow immediately,

$$M = \frac{\theta_E}{\kappa \pi_E}. \tag{7}$$

However, of the almost 1000 events discovered to date, there have been only about half a dozen measurements of each of these quantities [26–35] and no events for which both were measured.

Fortunately, the combined efforts of several workers have shown that the *Space Interferometry Mission (SIM)*, which will be a high-precision astrometry satellite, can measure both. First, Boden et al. [36] and Paczyński [37] showed that θ_E could be measured from the apparent motion of the *centroid* of the two images relative to source position

$$\delta \theta_c = \left(\frac{A_+ u_+ + A_- u_-}{A_+ + A_-} - u \right) \theta_E = \frac{u}{u^2 + 2} \theta_E. \tag{8}$$

The separation of the two images $2\theta_E \sim 600\,\mu as$ is generally too small to resolve with any existing or planned instruments, but the centroid can often be measured to much better precision than the resolution, and in particular *SIM* is expected to get down to $\sim 4\,\mu as$.

Gould & Salim [38] then pointed out that since *SIM* is to be launched into an Earth trailing orbit, π_E can be measured from the difference in the *photometric* event as seen from the Earth and *SIM*. Although *SIM* was not designed originally to do photometry, it measures positions by *counting* photons over the central fringe, and the sum of these counts constitutes a photometric measurement. Gould & Salim showed that 5% mass measurements were feasible with only 5 hours of *SIM* time. Thus it should be possible to get a complete census of all objects, dark or luminous, in the Galactic bulge.

MICROLENSING SEARCH FOR PLANETS

Mao & Paczyński [39] pointed out a decade ago that it should be possible to search for planets using microlensing: if a planet of the lens star lies close to the path of one of the two images, it will further perturb that image, leading to a mini "planetary event" superposed on the regular event and shorter by a factor $q = \sqrt{m/M}$ where m is the mass of the planet. Naively, one would expect that the cross section for such planetary events would be a factor q smaller than for the main event, so that for a system like Jupiter and Sun, the probability would be only $\sqrt{m_j/M_\odot} \sim 3\%$. However, Gould & Loeb [40] showed that if the planet-star separation was of order the Einstein radius, then the probability would be much larger. Since for a solar-like star sitting half-way to the Galactic center, the Einstein radius is about 4 AU, this enhancement is very relevant to the search for solar-system analogues. In fact, the probability that the solar system at 4 AU would be detected is about 20%. Gould & Loeb advocated setting up world-wide telescope networks to search for planets: world-wide because planetary perturbations last $\sim \sqrt{m/m_j}$ days, so that a single observatory could not see enough of the event to properly characterize it. After OGLE, MACHO, and EROS developed to capacity to alert on microlensing events in real time, such observations became feasible.

Primarily owing to the energy, imagination, and enthusiasm of two people, Penny Sackett and Dave Bennett, two such world-wide networks were established, PLANET [41] and Microlensing Planet Search (MPS) [42].

After 5 years of searching, there has been one claimed robust detection of a planet [43] but these data were shown to be better explained by an ordinary rotating binary [29]. In particular, PLANET searched 43 events for even faint traces of a planet and, finding none, put the first limits on planets of bulge M stars (the typical lens in the microlensing event): lens than 1/3 of such stars have Jupiter-mass planets with projected separations between 1.5 and 4 AU [44].

A much more ambitious program would be feasible from space. A proposed mission GEST would be able to detect 50,000 Jupiters and 100 Earths (assuming 1 per star) [45].

PROBING STELLAR ATMOSPHERES

Microlensing can be used to probe stellar atmospheres in two ways, one prosaic, the other elegant. In the first, the microlensing event is simply used as a giant telescope that can augment the power of the Earth-bound telescope. One can thereby obtain high S/N spectra of objects (such as dwarf stars at 8 kpc in the Galactic bulge) that would otherwise be too faint. In this way, Minnitti et al. [46] made the first lithium abundance measurement for a bulge dwarf, and Bennetti et al. [47] were able to type and measure the radial velocity ($-400\,\mathrm{km\,s^{-1}}$) of a $V = 20$ K0 subgiant using a 3.6 m telescopes at low resolution.

A more powerful application, however, is to use the microlens to *resolve* the surface of the star. Since one sees farther into the atmosphere when looking at the center than the limb, spatial resolution effectively permits one to investigate the atmosphere as a function of depth.

So far, I have discussed only point lenses. Unless a point lens actually passes over the face of a star, it does not resolve it: there is differential magnification, but when this is averaged over stellar annuli (on which the atmospheric properties are generally constant), this differential magnification almost completely cancels out. However, since even a bulge giant has an angular radius of $\theta_* \sim 6\mu$as, while (as mentioned above) typcially $\theta_E \sim 300\mu$as, the probability of such a crossing in an event is extremely low $\sim \theta_*/\theta_E \sim 1\%$. Moreover, it is extremely difficult to predict these transits, so if big telescopes are required (see below) it is extremely difficult to arrange the observations.

Most work on atmospheres has therefore made use of binary lenses. These generate caustics, concave polygonal curves in the source plane that define regions of infinite magnification. Although recognizable binaries constitute only 5–10% of all microlensing events [27], the caustics in these cases cover a large fraction of the Einstein ring. Hence, binary caustic crossings are the most common way that stars are resolved.

Limb Darkening

There are two ways to exploit resolution of the an atmosphere: photometric and spectroscopic. The first requires only 1 m class telescopes and so is well suited to microlensing search teams and follow-up teams who are well equipped with such telescopes and are more-or-less constantly monitoring microlensing events anyway. Four such measurements have been made, two of K giants and one of a G/K sub-giant, all in the Galactic bulge, and one of a metal-poor A star in the SMC [28–31]. No star of any of these types has ever had its limb-darkening measured by any other technique.

Spatial/Spectral Resolution

A more ambitious undertaking is to get spectra during a caustic crossing. This could potentially yield detailed information on the temperature, pressure, and ionization status of the star as a function of atmospheric depth. Such information is easily obtained for the Sun, and can thus be reasonably inferred for other dwarf stars of near-solar type. However, models of giant atmospheres are poorly constrained by observational data. Here, microlensing has begun to make a big contribution.

Valls-Gabaud [48] and Heyrovský et al. [49] were early advocates of this technique, while Gaudi & Gould [50] were the first to recognize that it would become practical only if focused on binary lenses. Alcock et al. [26] and Lennon et al. [51] made the first such measurements, but did not manage to obtain a detailed picture of the atmosphere.

Part of the problem is simply being able to predict when a caustic crossing will occur, to allow requests for big-telescope time at the appropriate location. Generally it is not difficult to predict the crossing to within a day, a few days in advance, but the crossing itself usually lasts only a few hours, so one cannot be very confident that a given telescope will be useful.

A major breakthrough came with EROS BLG-2000-5, in which a K3 giant source traversed a binary-lens caustic. Making use of alerts by the EROS and MPS collaborations, the PLANET team was able to accurately predict that the second caustic crossing for this event would last 4 days. As a result, spectra were obtained from two large telescopes on several successive nights: low-resolution spectra from VLT on 4 nights [52] and high-resolution spectra from Keck on 2 nights [53].

To date, all that have been published from these two remarkable sets of spectra are the Hα lines, but these appear quite interesting. The high-resolution spectra give a very precise (12σ) measurement of the drop in EW between the two nights, while the low resolution spectra show a 24% drop in equivalent width relative to pre-caustic value just ~ 3.5 hrs before the end of the crossing. The latter measurement would require that the outer few percent of the star be in emission, which would be very exciting if confirmed.

MICROLENS RESOLUTION OF QUASARS

When I began working on microlensing, I came up with many ideas that I thought looked nice on paper but would not be realized in practice for many decades. As a result, I almost did not publish a number of them. Then several of them were implemented in two or three years. I concluded that the observers needed stronger challengers. This prompted Scott Gaudi and me to come up with following project to image the central engines of quasars [54].

The first step is to find a nearby ($\sim 15\,\mathrm{pc}$) dwarf that is perfectly aligned with a distant quasar. To find such a pair, one must go roughly 40 AU from the Sun. That is routine with the aid of sling-shots from Jupiter, but the challenge is to stop the spacecraft at this spot.

The dwarf must be a binary, but most are. The Einstein radius at this distance is about $0.''01$, or 0.15 AU. So the binary companion will typically be at least 100 Einstein radii out. This means that it will induce an extremely small caustic.

The trick then is to align the quasar inside this tiny caustic and close to one of the cusps. There will then be 5 images of the quasar, one close to the companion (which we will ignore), one moderately magnified near the Einstein ring and on the opposite side from the cusp (which we will also ignore), and three highly magnified images along the Einstein ring on the same side of the cusp. How highly magnfied? It can easily be a factor of a million, so that $10^8 M_\odot$ black hole (~ 1 AU in radius), at 1 Gpc, would be expanded from a nas to a mas, easily big enough to resolve with a free-flyer optical interferometer with mirrors spread out over a few hundred meters.

Unfortunately, this is still only 1-dimensional resolution: the image is magnified, but only along the axis parallel to the Einstein ring. In the perpendicular direction it is actually demagnified by a factor 2.

Enter the 3 images. Since the magnification tensor is so anisotropic, the entire field is effectively mapped into a line (along the Einstein ring). This means that each "point" along the image maps back to a curve (which turns out to be pretty close to being straight) in the source plane, so a finite resolution element corresponds to a band. Since each part of the source plane is imaged into three highly magnified images, there are three such bands that intersect every point.

Let us consider what would happen if we combined the light from two resolution elements (i.e., two bands in the source plane) and put it through a spectrograph. In so far as the light from the two bands was coming from *different* regions of space it would add incoherently, but the light from the overlapping region would interfere. For each point within the region, there is a certain delay Δt between the two images. For each wavelength λ the interference will be constructive if $c\Delta t/\lambda$ is an integer, destructive if a half-integer. If the light is highly concentrated within the overlap region, then these fringes will be strong, while if a broad range of time delays is sampled, they will be weak. By combining all three sets of pairs of images, one can at least partially deconvolve the image on scales that are much smaller than 1 AU.

CONCLUSION

Microlensing is a young subject, but it is already proving to be a very powerful tool to explore a variety of astrophysical questions. It has moved from dark matter to planets to stellar atmospheres, and promises even more novel applications in the future.

ACKNOWLEDGMENTS

This work was supported by NSF grant AST 97-27520 and by a grant from Le Ministère Nationale de l'Éducation et de la Recherche et de la Technologie.

REFERENCES

1. Paczyński, B., *Astrophysical Journal*, **304**, 1 (1986)
2. Alcock, C., et al., *Nature*, **365**, 621 (1993)
3. Aubourg, E., et al., *Nature*, **365**, 623 (1993)
4. Udalski, A., Szymanski, M., Kaluzny, J., Kubiak, M., Krzeminski, W., Mateo, M.,Preston, G.W., and Paczynski, B., *Acta Astronomica*, **43**, 289 (1993)
5. Renault, C. et al., *Astronomy & Astrophysics*, **332**, 1 (1998)
6. Alcock, C., et al., *Astrophysical Journal*, **471**, 774 (1996)
7. Griest, K., *Astrophysical Journal*, **366**, 41 (1991)
8. Alcock, C., et al., *Astrophysical Journal*, **499**, L9 (1998)
9. Alcock, C., et al., *Astrophysical Journal*, **542**, 281 (2000)
10. Milsztajn, A., and Lasserre, T., "Not enough stellar mass Machos in the Galactic halo" in *XIX International Conference on Neutrino Physics and Astrophysics, Sudbury, Canada*, June 2000, in press (astro-ph/0011375)
11. Crotts, A.P.S., and Tomaney, A.B., *Astronomical Journal*, **112**, 2872 (2000)
12. Ansari, R., et al., **344**, L49 (1999)
13. Crotts, A.P.S., *Astrophysical Journal*, **399**, L43, (1992)
14. Crotts, A., Uglesich, R., Gould, A., Gyuk, G., Sackett, P., Kuijken, K, Sutherland, W., and Widrow, L. "First Results from MEGA" in *Microlensing 2000: A New Era of Microlensing Astrophysics*, edited by J.W. Menzies and P.D. Sackett, ASP conference proceedings, San Francisco, in press (2001) (astro-ph/0006282)
15. Aurière, M., et al., *Astrophysical Journal*, submitted (2001) (astro-ph/0102080)
16. Paczyński, B., *Astrophysical Journal*, **371**, L63 (1991)
17. Griest, K., et al., *Astrophysical Journal*, **372**, L79 (1991)
18. Kiraga, M., and Paczyński, B., *Astrophysical Journal*, **430**, 101 (1994)
19. Alcock, C., et al., *Astrophysical Journal*, **479**, 119 (1997)
20. Han, C., *Astrophysical Journal*, **484**, 555 (1997)
21. Gould, A., *Astrophysical Journal Letters*, submitted, (1994) (astro-ph/9408060)
22. Kuijken, K., *Astrophysical Journal*, **486**, L19 (1997)
23. Binney, J., Bissantz, N., and Gerhard, O., *Astrophysical Journal*, **537**, L99 (2000)
24. Han, C., and Gould, A., ApJ, **467**, 540 (1996)
25. Alcock, C., et al., *Astrophysical Journal*, **541**, 734 (2000)
26. Alcock, C., et al., *Astrophysical Journal*, **491**, 436 (1997)

27. Alcock, C., et al., *Astrophysical Journal*, **541**, 270 (2000)
28. Albrow, M.D., et al., *Astrophysical Journal*, **522**, 1011 (1999)
29. Albrow, M.D., et al, *Astrophysical Journal*, **534**, 894 (2000)
30. Albrow, M.D., et al., *Astrophysical Journal*, **549**, 000 (2001) (astro-ph/0004243)
31. Afonso, C. et al., *Astrophysical Journal*, **532**, 340 (2000)
32. Alcock, C., et al., *Astrophysical Journal*, **454**, L125 (1995)
33. Bennett, D.P., et al., *Bull. Am. Ast. Soc.*, **191**, 8303 (1997)
34. Mao, S., *Astronomy & Astrophysics*, **350**, L19 (1999)
35. Soszyński, I., et al., *Astrophysical Journal*, in press (2001) (astro-ph/0012144)
36. Boden, A.F., Shao, M., and Van Buren, D., *Astrophysical Journal*, **502**, 538 (1998)
37. Paczyński, B., *Astrophysical Journal*, **494**, L23 (1998)
38. Gould, A., and Salim, S., *Astrophysical Journal*, **524**, 794 (1999)
39. Mao, S. and Paczyńsk, B., *Astrophysical Journal*, **374**, L37 (1991)
40. Gould, A., and Loeb, A., *Astrophysical Journal*, **396**, 104 (1992)
41. Albrow, M.D., et al., *Astrophysical Journal*, **509**, 687 (1998)
42. Rhie, S.-H., et al., *Astrophysical Journal*, **535**, 378 (2000)
43. Bennett, D.P., et al., *Nature*, **402**, 57 (2000)
44. Albrow, M.D., et al., *Astrophysical Journal*, submitted (2001) (astro-ph/0008078)
45. Bennett, D.P., *Bull. Am. Ast. Soc.*, **197**,1108 (2000)
46. Minniti, D., Vandehei, T., Cook, K. H., Griest, K., and Alcock, C., *Astrophysical Journal*, **499**, L175 (1998)
47. Benetti, S., Pasquini, L., and West, R.M., *Astronomy & Astrophysics*, **294**, L37 (1995)
48. Valls-Gabaud, D., *Monthly Notices*, **294**, 747 (1998)
49. Heyrovský, D., Sasselov, D., and Loeb, A., *Astrophysical Journal*, submitted (1999) (astro-ph/9902273)
50. Gaudi, B.S., and Gould, A., **513**, 619 (1999)
51. Lennon, D.J., Mao, S., Fuhrmann, K., and Gehren, T., *Astrophysical Journal*, **471**, L23 (1996)
52. Albrow, M. et al., *Astrophysical Journal Letters*, in press (2001)
53. Castro, S.M., Pogge, R.W., Rich, R.M., DePoy, D.L., and Gould, A., *Astrophysical Journal Letters*, in press (2001)
54. Gaudi, B.S., and Gould, A., **477**, 152 (1997)

CDM: Numerical Predictions on small scales

Andrey V. Kravtsov[1]

Department of Astronomy, The Ohio State University,
140 W. 18th Ave., Columbus, OH 43210, U.S.A.

Abstract. I discuss recent results from the numerical CDM simulations concerning the density profiles of galactic halos and the abundance of satellite dark matter halos around galaxies. The models predict 1) halo density distributions that may be too concentrated compared to the density distributions implied by observed galactic rotation curves and 2) ≳5 − 10 times more satellites than are actually observed around the Milky Way and M31. This may indicate that there are either physical mechanisms at work which complicate the straightforward comparison between observations and dissipationless simulations (e.g., processes that greatly reduce the starformation efficiency in most satellite halos and render them invisible), or that essential physics is missing in the CDM models.

INTRODUCTION

During the last decade there has been an increasingly growing interest in testing the predictions of variants of cold dark matter (CDM) models at subgalactic (≲100 kpc) scales. Tests at such small scales are useful because, when combined with tests on intermediate (galaxy distribution) and large (CMB fluctuations) scales, they provide the greatest leverage on the shape of the power spectrum. Such tests are also interesting due to increasing evidence that predictions of the CDM models on small scales may be at odds with observations. In this contribution I present some of the recent results concerning two of these predictions: the density profiles and abundance of the small-mass halos in Cold Dark Matter (CDM) models.

DENSITY PROFILES OF THE CDM HALOS

A systematic study of halo density profiles for a wide range of halo masses and cosmologies was done by [1] (hereafter NFW), who showed that CDM predicted cuspy density profiles of DM halos with logarithmic slope of the density distribution changing gently from -3 in the outer regions of the halos, to the asymptotic slope of -1 in the innermost regions. Subsequently, Moore et al. ([2]) argued that limited mass resolution has a significant impact on the central density distribution of halos. They suggested that at least several million particles per halo are required to reliably model the density profiles at

---

[1] Hubble Fellow

scales $\lesssim 0.01 R_{\rm vir}$. Based on these results, Moore et al. advocated a density profile that behaves similarly ($\rho \propto r^{-3}$) to the NFW profile at large radii, but is steeper at small r: $\rho \propto r^{-1.5}$.

The exact slope of the central density distribution is a matter of ongoing debate from both the theoretical (e.g., [3]; [4]; [5]) and observational (e.g., R. Swaters, this volume) side. The numerical results are uncertain due to the high mass and force resolution required, which limits the analysis to only a handful of individual halos. Figure 1 illustrates these difficulties. The left panels of the figure show density profiles of three DM halos of similar mass ($\approx 1.2 \times 10^{12} h^{-1} M_\odot$) at two different epochs ($z = 0$ and $z = 1$). The halos are different in their environments and merger histories (halo D is rather isolated, while halos B and C are located about 0.5 Mpc from each other). The halos contain more than 10^6 particles within their virial radii and were simulated with a force resolution of $0.4 h^{-1}$ kpc (see [5] for details).

The right panels show fractional deviations from the best fit NFW and Moore et al. profiles as a function of the halo radius. We found that for halos B and C the errors in the Moore et al. fits were systematically smaller than those of the NFW fits, though the differences were not dramatic. But the Moore et al. profile was not a good fit for halo D. Remarkably, the same conclusions hold for the halo profiles at $z = 0$. The analysis does not show that one analytic profile is better then the other as a description of the density distribution in simulated halos. Instead, we see a certain degree of scatter in the shapes of halo density profiles. The scatter can be parameterized by the *halo concentration* which is defined as the ratio of the halo virial radius to the radius where the logarithmic slope of the density profile is -2. For a given cosmological model the concentration is a weak function of halo mass (NFW). As noted above, halos B, C, and D shown in Figure 1, have very similar masses and therefore should have similar concentrations if their profiles would have similar shapes. We find that this is not the case; halos B, C, and D have concentrations of 15.6, 11.2, and 11.9, respectively. This is consistent with results from systematic studies by [6] and [7], which used larger halo samples derived from the lower-resolution simulations.

Interpretation of the observed galaxy rotation curves is complicated by the effects of beam smearing, asymmetries in gas distribution and motions, non-gravitationally induced gas motions, etc. Although significant advances are being made in both simulations and observations, the situation is likely to remain uncertain in the near future. Thus, it seems reasonable to focus on constraints on the halo concentration because it is less sensitive to the details of the uncertain central density distribution. The concentration is also probably the most robustly predicted property of CDM halos.

Figure 2 shows the concentration as a function of halo mass for a large sample of halos in the LCDM model analyzed by [6]. The thick solid line represents the average $c(M)$ relation, while the dashed lines represent 1σ scatter in the concentrations of individual halos of a given mass. The normalization and shape of the $c(M)$ relation is sensitive to the normalization and shape of the primordial perturbation spectrum (NFW; [8]). The origin of scatter is less certain, but preliminary analysis of the merger histories of the halo sample used in Figure 2 (Wechsler et al., in preparation) indicates that it probably reflects differences in the halo merger histories. Note that the results are presented for the overall halo population. The scatter for observed spiral galaxies is expected to be smaller because spiral galaxies should correspond to a much narrower range of merger

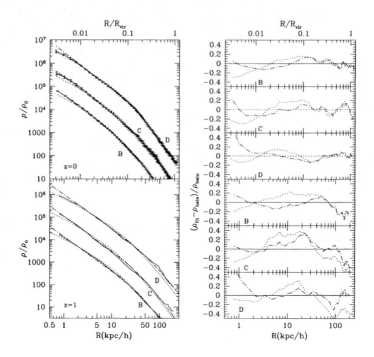

FIGURE 1. Fits of the NFW (dotted curves) and Moore et al. (dot-dashed curves) profiles to the density distributions of three DM halos B, C, D (solid curves) at $z = 0$ (*top left*). For clarity, the profiles of halos C and B were shifted down by factors of 10 and 100, respectively. *Top right* panel shows the fractional deviations of the fitted profile from the simulated halo profile as a function of scale. *The bottom panels* show the same but for the halos B, C, and D at $z = 1$. See [5] for details.

histories than the overall halo population (which would also include elliptical galaxies).

ABUNDANCE OF DWARF GALAXIES

Dwarf galaxies observed in the Local Group (LG) and its vicinity are the smallest objects which are believed to have formed from the primordial density fluctuations. As such, they can be used to probe the power spectrum on the corresponding mass scale. The volume of the Local Group is currently the best studied volume-limited sample of dwarf galaxies. Although several low-luminosity dwarfs were discovered in the Local Group during the past decade, it is unlikely that the current surveys miss a large fraction of the *luminous* dwarf galaxies.

A straightforward comparison between the abundance of small-mass ($\sim 10^7 - 10^9 h^{-1} M_\odot$) DM halos predicted for the Local Group mass systems in CDM simulations and the number of satellites actually observed in the Local Group shows that CDM models predict $\gtrsim 5 - 10$ (depending on the mass range) more DM satellites

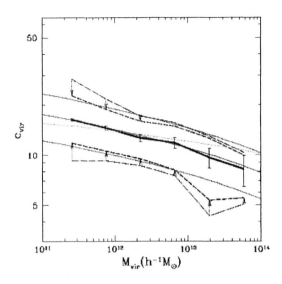

FIGURE 2. Concentration of isolated halos in the ΛCDM cosmology at $z = 0$ as a function of halo virial mass. The thick solid line represents the average over halos in a given mass range, while the thick dashed lines indicate the 1σ scatter of concentrations of individual halos about this average. The thin solid lines show the relation predicted using an analytical toy-model for halo concentration (see [6] for details).

than is observed in the LG ([9]; [10]). The discrepancy occurs for systems with circular velocities $v_c \lesssim 30$ km/s and is increasingly larger for smaller systems. For instance, the cumulative velocity function of DM satellites located within $R < 500 h^{-1}$ kpc from their host in simulations can be well described as: $N(> v_c) \approx 320(v_c/10 \text{ km s}^{-1})^{-3}$, while the observed cumulative VF is: $N(> v_c) \approx 12(v_c/10 \text{ km s}^{-1})^{-1}$. One can see that the functions are different in both their amplitude and power-law slope. The two functions intersect at $v_c \sim 40$ km s^{-1} where the predicted abundance of satellites with larger velocities (e.g., DM halos of masses similar to the Magellanic Cloud) are in agreement with observations.

What does the discrepancy at $v_c \lesssim 40$ km s^{-1} tell us? Two types of explanations have been proposed: 1) physical processes preventing gas collapse onto or removing gas from the small-mass DM halos ([11]; [12]); 2) incorrect properties of the dark matter assumed in the CDM models. The latter solutions include proposals to abandon the assumption that DM is cold (warm dark matter or WDM), that DM is collisionless (self-interacting DM; SIDM; [13]), or both ([14]). The implications of these models and observational constraints are currently being investigated, although these investigations are in the preliminary exploratory stage.

While it is not clear what physical processes could alleviate problems with the central

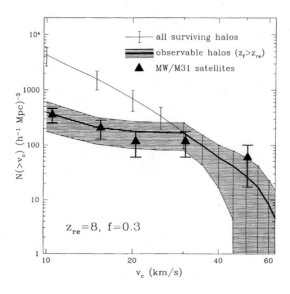

FIGURE 3. Cumulative velocity function of all dark matter subhalos surviving at $z = 0$ around Milky Way mass host halo (*thin solid line*) and "observable" satellites (i.e., halos that accrete a large fraction of their mass before reionization) (*thick solid line with shading*). The observed velocity function of satellite galaxies around the Milky Way and M31 is shown by triangles. See [11] for details.

density profiles of the CDM halos, in the case of dwarf galaxy abundance, feedback processes are likely to be important. Recall that the discrepancy is observed for relatively small-mass systems. It would be difficult for the shallow potential wells of such objects to accrete gas heated to temperatures of $\gtrsim 10^4$ K by the background ionizing radiation (e.g., [15]) or to retain the accreted gas after the energy injection due to supernova feedback ([16]) or after the supernova-driven wind from the neighboring systems passes through such system ([12]).

Figure 3 illustrates possible effects of gas heating by the ionizing background (see [11] for details). The figure compares the abundance of all DM satellites around a Milky Way-mass host predicted in a ΛCDM model to the abundance of observed satellites in the Local Group. The shaded strip shows the velocity functions of DM halos which accrete a substantial fraction of their mass (0.3 in this case) prior to the assumed epoch of reionization ($z_{re} = 8$) for $v_c \lesssim 30$ km s^{-1}. We assume that the accretion of gas onto halos with $v_c \lesssim 30$ km s^{-1} is inefficient after the universe is reionized and the ambient gas is heated. In this model, only halos that accreted a substantial amount of gas before reionization have a chance for significant starformation and become sufficiently luminous to be observable. The accretion of satellites onto the MW type host is followed using extended Press-Schechter merger trees. The model is admittedly simplistic. However, as Figure 3 shows, it indicates that heating by the ionizing background can fairly easily (and naturally) account for the observed paucity of small-mass satellites around the Milky Way.

CONCLUSIONS

The CDM models do not pass the test on small subgalactic scales naturally. They seem to predict halo profiles that are too dense to be consistent with observed galactic rotation curves. They also predict hundreds of satellites with circular velocities of $v_c \gtrsim 10$ km s^{-1} in the Local Group type systems, while only two dozen are observed. At present it is not clear how severe these problems are. Although the results of numerical simulations are converging, comparison of observed rotation curves to the predicted density profiles is complicated by uncertain gasdynamic processes and non-gravitationally induced gas motions in galaxies. Further progress can be made using constraints on the halo concentrations which are less uncertain and are easier to deduce from the rotation curves. The paucity of observed, small-mass dwarf systems and their spatial distribution can give us insights into the physical processes that govern the formation of dwarf galaxies. It seems plausible that this particular discrepancy can be naturally explained by the fragility of small-mass systems with respect to even relatively mild gas heating. Alternatively, if such small-scale problems of CDM continue to accumulate, these problems may give us insight into the nature and properties of dark matter particles, invaluable information which could prove cosmology to be the "poor man's accelerator" once again.

ACKNOWLEDGMENTS

I would like to thank my collaborators, Anatoly Klypin, James Bullock, Joel Primack and Avishai Dekel, for allowing me to present the results of our joint projects. The work presented here was partially supported by NASA through Hubble Fellowship grant.

REFERENCES

1. Navarro, J. F., Frenk, C. S., and White, S. D. M., *ApJ*, **490**, 493–508 (1997).
2. Moore, B., Governato, F., Quinn, T., Stadel, J., and Lake, G., *ApJ*, **499**, L5–L9 (1998).
3. Ghigna, S., Moore, B., Governato, F., Lake, G., Quinn, T., and Stadel, J., *ApJ*, **544**, 616–628 (2000).
4. Jing, Y. P., and Suto, Y., *ApJ*, **529**, L69–L72 (2000).
5. Klypin, A., Kravtsov, A., Bullock, J., and Primack, J. (2001), submitted, astro-ph/0006343.
6. Bullock, J. S., Kolatt, T. S., Sigad, Y., Somerville, R. S., Kravtsov, A. V., Klypin, A. A., Primack, J. R., and Dekel, A., *MNRAS*, **321**, 559+ (2001).
7. Jing, Y. P., *ApJ*, **535**, 30–36 (2000).
8. Eke, V., Navarro, J., and Steinmetz, M. (2001), submitted, astro-ph/0012337.
9. Klypin, A., Kravtsov, A. V., Valenzuela, O., and Prada, F., *ApJ*, **522**, 82–92 (1999).
10. Moore, B., Ghigna, S., Governato, F., Lake, G., Quinn, T., Stadel, J., and Tozzi, P., *ApJ*, **524**, L19–L22 (1999).
11. Bullock, J. S., Kravtsov, A. V., and Weinberg, D. H., *ApJ*, **539**, 517–521 (2000).
12. Scannapieco, E., Ferrara, A., and Broadhurst, T., *ApJ*, **536**, L11–L14 (2000).
13. Spergel, D. N., and Steinhardt, P. J., *Physical Review Letters*, **84**, 3760–3763 (2000).
14. Hannestad, S., and Scherrer, R. (2000), submitted, astro-ph/0003046.
15. Thoul, A. A., and Weinberg, D. H., *ApJ*, **465**, 608+ (1996).
16. Dekel, A., and Silk, J., *ApJ*, **303**, 39–55 (1986).

Warm Dark Matter, Small Scale Crisis, and the High Redshift Universe

Zoltan Haiman*[1], Rennan Barkana[†] and Jeremiah P. Ostriker*

Princeton University Observatory, Princeton, NJ 08544
[†]*Canadian Institute for Theoretical Astrophysics, 60 St. George Street #1201A, Toronto, Ontario, M5S 3H8, Canada*

Abstract. Warm Dark Matter (WDM) models have recently been resurrected to resolve apparent conflicts of Cold Dark Matter (DM) models with observations. Endowing the DM particles with non–negligible velocities causes free–streaming, which suppresses the primordial power spectrum on small scales. The choice of a root-mean-square velocity dispersion $v_{\rm rms,0} \sim 0.05$ km/s at redshift $z=0$ (corresponding to a particle mass $m_X \sim 1$ keV if the WDM particles are fermions decoupling while relativistic) helps alleviate most, but probably not all, of the small–scale problems faced by CDM. An important side–effect of the particle velocities is the severe decrease in the number of collapsed halos at high redshift. This is caused both by the loss of small–scale power, and by the delay in the collapse of the smallest individual halos (with masses near the effective Jeans mass of the DM). The presence of early halos is required in order (1) to host either early quasars or galaxies that can reionize the universe by redshift $z = 5.8$, and (2) to allow the growth of the supermassive black hole believed to power the recently discovered quasar SDSS 1044-1215 at this redshift. We quantify these constraints using a modified Press-Schechter formalism, and find $v_{\rm rms,0} \lesssim 0.04$ km/s (or $m_X \gtrsim 1$ keV). If future observations uncover massive black holes at $z \gtrsim 10$, or reveal that reionization occurred at $z \gtrsim 10$, this could conclusively rule out WDM models as the solution to the small–scale crisis of the CDM paradigm.

INTRODUCTION

The currently favored model of hierarchical galaxy formation in a universe dominated by cold dark matter (CDM) has been very successful in matching observations of the density distribution on large scales. However, recently some small-scale shortcomings of this model have appeared, as summarized elsewhere in these proceedings (see also [1] for a recent review). Although the observational significance of the discrepancies is still disputed, and astrophysical solutions involving feedback may still be possible, the accumulating tension with observations has focused attention on solutions involving the particle properties of dark matter. Proposals have

---

[1] Hubble Fellow

included self-interacting dark matter [2], adding a repulsive interaction to gravity [3,4], the quantum–mechanical wave properties of ultra–light dark matter particles [5], and a resurrection of warm dark matter (WDM) models [6–8]. By design, a common feature of models that attempt to solve the apparent small-scale problems of CDM is the reduction of fluctuation power on small scales. In the CDM paradigm, structure formation proceeds bottom-up: the smallest objects collapse first, and they subsequently merge together to form larger objects. It then follows that the loss of small-scale power modifies structure formation most severely at the highest redshifts; in particular, the number of self–gravitating objects at high redshift is reduced.

A strong reduction in the abundance of high-redshift objects could be in conflict with observations. First, the lack of a Gunn-Peterson trough in the spectrum of the bright quasar SDSS 1044-1215 at redshift $z = 5.8$ [9] implies that the hydrogen in the intergalactic medium (IGM) was highly ionized prior to this redshift. The most natural explanation for reionization is photo-ionizing radiation produced by an early generation of stars or quasars [10]. The sources of reionization reside in halos that have masses in the range corresponding to dwarf galaxies — the mass scale on which power needs to be reduced relative to CDM models. Second, if SDSS 1044-1215 is unlensed and radiating at or below the Eddington limit, its unusual intrinsic brightness implies that it is powered by an exceptionally massive black hole (BH). The growth of this BH, out of a stellar remnant seed, requires [11] that a host halo be present at a sufficiently high redshift ($z \gg 5.8$).

In this contribution, we briefly review the status of WDM models in resolving the problems of CDM, and then examine new constraints that arise on WDM models from the high redshift universe. We focus on WDM models, although similar constraints would apply to other modifications of the CDM paradigm that reduce the small-scale power. The cosmological parameters we adopt, based on present large scale structure data [12], are $(\Omega_0, \Omega_\Lambda, \Omega_b, h, \sigma_8, n) = (0.3, 0.7, 0.045, 0.7, 0.9, 1)$. The details of the study described here can be found in [13].

WARM DARK MATTER MODELS

The WDM is assumed to be composed of particles of about ~ 1 keV mass (compared to ~ 1 GeV in CDM, or ~ 10 eV in Hot Dark Matter models). The thermal velocities of the particles cause free streaming out of overdense regions, smoothing out small–scale fluctuations, leading to a small-scale cutoff in the linear power spectrum. In addition, the thermal velocities act similarly to pressure, and inhibit the growth of low–mass perturbations. One example of WDM is fermionic particles that decouple in the early universe while relativistic and in thermal equilibrium [14]. To produce a given contribution Ω_X to the cosmological critical density, the required particle mass m_X is then determined by $m_X n_X \propto \Omega_X h^2$, where the present number density n_X of WDM particles follows from their chosen r.m.s. velocities. This yields a relation between particle mass and r.m.s. velocity dispersion [8],

$$v_{\rm rms}(z) = 0.0437\,(1+z)\left(\frac{\Omega_X h^2}{0.15}\right)^{1/3}\left(\frac{g_X}{1.5}\right)^{-1/3}\left(\frac{m_X}{1\text{ keV}}\right)^{-4/3}\text{ km s}^{-1}, \qquad (1)$$

where g_X is the effective number of degrees of freedom of WDM. The comoving cutoff scale R_c, where free–streaming reduces the power spectrum to half of its value in CDM, is given by

$$\begin{aligned}R_c &= 0.201\left(\frac{\Omega_X h^2}{0.15}\right)^{0.15}\left(\frac{g_X}{1.5}\right)^{-0.29}\left(\frac{m_X}{1\text{ keV}}\right)^{-1.15}\text{ Mpc} \\ &= 0.226\left(\frac{\Omega_X h^2}{0.15}\right)^{-0.14}\left(\frac{v_{\rm rms,0}}{0.05\text{ km/s}}\right)^{0.86}\text{ Mpc}\,.\end{aligned}\qquad (2)$$

The length scale R_c corresponds to a characteristic mass scale M_c,

$$M_c = 1.74\times 10^8\left(\frac{\Omega_0 h^2}{0.15}\right)\left(\frac{R_c}{0.1\text{ Mpc}}\right)^3 M_\odot\,. \qquad (3)$$

The predictions of WDM models can be expected to differ from CDM on scales below R_c or M_c. In addition, it is useful to define an "effective Jeans mass" for WDM, at which the pressure corresponding to the r.m.s. particle velocities balances gravity,

$$\begin{aligned}M_J &= 3.06\times 10^8\left(\frac{g_X}{1.5}\right)^{-1}\left(\frac{\Omega_X h^2}{0.15}\right)^{1/2}\left(\frac{m_X}{1\text{ keV}}\right)^{-4}\left(\frac{1+z_i}{3000}\right)^{3/2} M_\odot \\ &= 4.58\times 10^8\left(\frac{\Omega_X h^2}{0.15}\right)^{-1/2}\left(\frac{v_{\rm rms,0}}{0.05\text{ km/s}}\right)^3\left(\frac{1+z_i}{3000}\right)^{3/2} M_\odot\,.\end{aligned}\qquad (4)$$

We have verified these scalings using spherically symmetric, one–dimensional collapse simulations [13]. The growth of perturbations on scales below M_J is slowed down by the "pressure" of WDM. Although this effect is irrelevant in most discussions of WDM models at lower redshifts (due to the smallness of M_J), we find that it is important in the context of reionization. The inhibited growth of perturbations in the linear regime results in a delay in the final virialization epoch of the individual, low–mass halos, that first condense out in the universe.

SUMMARY OF CURRENT OBSERVATIONS

The current status of WDM models in resolving various problems of CDM is summarized in Table 1. Based on analytical arguments, as well as numerical simulations, WDM particles with a mass in the range $0.6\text{ keV} \lesssim m_X \lesssim 1.5\text{ keV}$ can strongly suppress the number of satellite halos of the Milky Way, and produce low halo concentration parameters for both dwarfs, and Milky Way-sized galaxies

TABLE 1. Scorecard of WDM models.

CDM Problem[a]	WDM mass	Status	References
Milky Way satellites	0.6–1.5 keV	√	[6,8,16]
Halo concentration	0.6–1.5 keV	√	[6,8,17]
Inner density profile	<0.6 keV ?	?	[6,17]
No dwarfs in voids	<1.5 keV	?	[8,19]
Angular momentum	0.5–0.8 keV	x?	[7,20]

[a] Note that WDM particles with $m_X \lesssim 750$ eV would contradict the power spectrum of the Lyα forest [15]. A similar limit follows from the maximum observed phase space density in cores of dwarf spheroidals [17,18].

[6,8]. It remains unclear whether WDM can fully resolve some other, related problems. One of the major successes of the CDM paradigm is the interpretation of the statistics of the Lyα forest. An analysis of the Lyα forest power spectrum at redshift $z \approx 3$ finds that WDM models with $m_X < 0.75$ keV would spoil this success [15]. A similar limit from the maximum observed phase space density in dwarf spheroidal galaxies. To solve some of the problems with CDM, lower masses might be needed. In existing simulations for $m_x > 0.6$ keV, the inner slope of the density profiles of normal disk galaxies do not appear to flatten sufficiently to produce the observed, nearly constant cores, although these simulations do not yet probe the relevant innermost regions reliably [6]. A similar situation arises with two other CDM problems. The solution of the long–standing angular–momentum problem in galaxy formation might be solved if "subclumps" in the protogalaxy's halo were eliminated (CDM models produce too small disks, thought to be attributable to angular momentum transfer from the infalling gas to the DM subclumps). Although WDM does suppress halo substructure, the required WDM mass is estimated to be $0.5 - 0.8$ keV [7]. Furthermore, a simulation that artificially turns off angular momentum transfer between gas and DM halo subclumps reveals that the specific angular momentum distribution still contradicts observations [20]. Finally, the absence of dwarf galaxies in voids, a difficulty of CDM [19], is found to be eased in a large–scale simulation of an $m_X = 1.5$ keV WDM model [8], but a smaller m_X appears necessary to match the observations in detail.

CONSTRAINTS FROM REIONIZATION

As emphasized in section § 1, WDM models reduce the abundance of high–redshift halos, which could lead to conflicts with observations. Note that in general, the differences between WDM and CDM models are amplified at high redshifts. In a study based on a modified version of the extended Press–Schechter theory, we have quantified constraint on the velocities and masses of WDM particles from reionization, and from the presence of a supermassive BH in the bright $z = 5.8$ quasar SDSS 1044-1215. The main features of our models are as follows:

1. The effect of free streaming on the power spectrum was included using a fit to the numerical transfer function, obtained from a Boltzmann code.

2. The effect of WDM velocities on the growth of perturbations was included using spherically symmetric, one–dimensional hydrodynamical simulations.

3. The WDM halo mass function was computed using Monte Carlo–generated halo "merger trees". A moving barrier ($\delta_c \geq 1.69$) was adopted to generalize (an improved) Press–Schechter theory to the case of mass–dependent collapse times. Our semi–analytic halo mass function was demonstrated to agree with numerical simulations at low redshift over the halo masses of interest [8].

4. We assume that a fraction f_* of baryons turns into stars in halos with virial temperatures $T_{\rm vir} \geq 10^4$K (necessary for efficient cooling [21]). We assume a Scalo (1998) stellar IMF (producing ≈ 4000 ionizing photons per baryon), and that a fraction $f_{\rm esc}$ of the ionizing photons escape into the IGM. We parameterize our models by the product $\epsilon_* \equiv f_* f_{\rm esc} (= 0.01$ in our standard model), consistent with the ionizing background inferred from the proximity effect at redshift $z \approx 3$. [22]

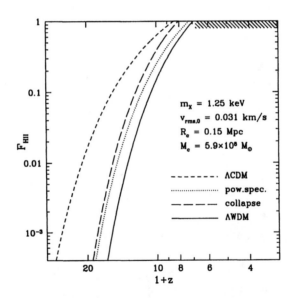

FIGURE 1. The filling factor of ionized hydrogen, $F_{\rm H\,II}$, as a function of redshift z in our standard model ($C = 10, \epsilon_* = 0.01$). The uppermost curve corresponds to ΛCDM, and the lowest curve to a WDM particle mass of $m_X = 1.25$ keV. The middle pair of curves shows separate contributions to the delay in reionization from the suppression of the power spectrum (dotted curve) and the "effective pressure" of WDM (long dashed curve).

5. The mean clumping of the ionized gas in the IGM is parameterized by a single constant $C \equiv \langle n_H^2 \rangle / \bar{n}_H^2 (= 10$ in our standard model).

Our main results on the constraints from reionization are shown in Figures 1 and 2. Figure 1 shows the filling factor of ionized hydrogen in our standard model, with $m_X = 1.25$ keV, $\epsilon_* = 0.01$, and $C = 10$. In this model, reionization occurs at $z = 5.8$. The contributions to the decrease of the filling factor from the suppression of the power spectrum and the effective pressure of WDM are comparable (dotted and long–dashed curves, respectively). In Figure 2, we demonstrate how the reionization redshift depends on m_X, and on our two model parameters, ϵ_* and C.

CONCLUSIONS

If high-redshift galaxies produce ionizing photons with an efficiency similar to their $z = 3$ counterparts ($\epsilon_* \sim 0.01$), reionization by redshift $z = 5.8$ places a limit of $m_X \gtrsim 1.25$ keV ($v_{\rm rms,0} \lesssim 0.03$ km/s) on the mass of the WDM particles. This limit is somewhat stronger than the limit inferred from the statistics of the

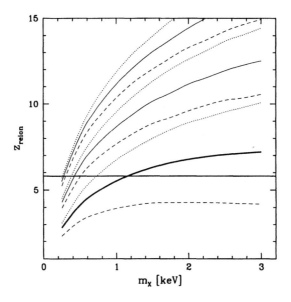

FIGURE 2. Reionization redshift as a function of WDM particle mass. Our standard model for the star-formation efficiency, escape fraction, and stellar IMF, is shown by the thick solid line. The three solid curves show models with the same clumping factor ($C = 10$) but different ionization efficiencies ($\epsilon_* = 0.01, 0.1, 1$, bottom to top). The three dotted curves correspond to the same models except with $C = 1$; the three dashed curves correspond to the same models with $C = 30$. In our standard model, the requirement $z_{\rm reion} > 5.8$ yields the constraint $m_X > 1.25$ keV.

Lyα forest (which yields $m_X \gtrsim 0.75$ keV; [15]), although our limit may weaken to $m_X \gtrsim 0.75$ keV ($v_{\rm rms,0} = 0.060$ km/s) given the uncertainty in current measurements of the stellar contribution to the ionizing intensity at $z = 3$ [22] (which we can use to normalize our models). We also find that the existence of a $\approx 4 \times 10^9$ M$_\odot$ supermassive black hole at $z = 5.8$, powering the quasar SDSS 1044-1215 (assuming it is unlensed and radiating at or below the Eddington limit), yields the somewhat weaker, but independent constraint $m_X \gtrsim 0.5$ keV (or $v_{\rm rms,0} \lesssim 0.10$ km/s), if this BH acquired most of its mass accreting at the Eddington rate. Finally, we also find that WDM models with $m_X \lesssim 1$ keV ($v_{\rm rms,0} \gtrsim 0.04$ km/s) produce a low-luminosity cutoff in the high-z galaxy luminosity function which is detectable with the *Next Generation Space Telescope*. Such an observation would directly break the degeneracy in the reionization redshift between low ionizing-photon production efficiency and small WDM particle mass. The constraints derived here will tighten considerably as observations probe still higher redshifts, offering increasingly stringent tests of models with diminished small–scale power, exemplified by WDM.

ZH acknowledges support from a Hubble Fellowship, and RB acknowledges support from CITA and from Institute Funds (IAS, Princeton).

REFERENCES

1. Sellwood, J., and Kosowsky, A., in *Gas and Galaxy Evolution*, eds. Hibbard, Rupen and van Gorkom, in press, astro-ph/0009074 (2001).
2. Spergel, D. N., and Steinhardt, P. J., *Phys. Rev. Lett.* **84**, 3760 (2000).
3. Goodman, J., *New Astronomy* **5**, 103 (2000).
4. Peebles, P. J. E., *ApJ* **534**, 127 (2000).
5. Hu, W., Barkana, R., and Gruzinov, A., *Phys. Rev. Lett.* **85**, 1158 (2000).
6. Colín, P., Avila-Reese, V., and Valenzuela, O., *ApJ* **542**, 622 (2001).
7. Sommer-Larsen, J., and Dolgov, A., *Ap. J.* submitted, astro-ph/9912166 (1999).
8. Bode, P., Ostriker, J. P., and Turok, N., *Ap. J.*, subm. astro-ph/0010389 (2001).
9. Fan, X., et al., *Astron. J.* **1210**, 1167 (2000).
10. Barkana, R., and Loeb, A., *Physics Reports*, in press, astro-ph/0010468 (2001).
11. Haiman, Z., and Loeb, A., *ApJ* in press, astro-ph/0011529 (2001).
12. Bahcall, N., et al., *Science* **284**, 1481 (1999).
13. Barkana, R., Haiman, Z., and Ostriker, J. P., *Ap. J.*, subm. astro-ph/0102000 (2001).
14. Kolb, E. W., and Turner, M. S., *The Early Universe*, Addison-Wesley (1990).
15. Narayanan, V. K., Spergel, D. N., Davé, R., and Ma, C., *Ap. J.* **543**, L103 (2000).
16. White, M., and Croft, R., *Ap. J.* **539**, 497 (2000).
17. Hogan, C. J., and Dalcanton, J. J., *Phys. Rev. D.* **62**, (2000).
18. Dalcanton, J. J., and Hogan, C. J., *Phys. Rev. D.* subm. astro-ph/0004381 (2001).
19. Peebles, P. J. E. 2001, preprint astro-ph/0101127
20. Kravtsov, A. V., these proceedings.
21. Haiman, Z., Abel, T., and Rees, M. J., *ApJ* **534**, 11 (2000).
22. Bajtlik, S., Duncan, R. C., and Ostriker, J. P., *ApJ* **327**, 570 (1988).

The Evolution of Dark-Matter Dominated Cosmological Halos

Marcelo Alvarez[*], Paul R. Shapiro[*] and Hugo Martel[*]

[*]*Department of Astronomy, University of Texas, Austin, TX 78712*

Abstract. Adaptive SPH and N-body simulations were carried out to study the evolution of the equilibrium structure of dark matter halos that result from the gravitational instability and fragmentation of cosmological pancakes. Such halos resemble those formed by hierarchical clustering from realistic initial conditions in a CDM universe and, therefore, serve as a test-bed model for studying halo dynamics. The dark matter density profile is close to the universal halo profile identified previously from N-body simulations of structure formation in CDM, with a total mass and concentration parameter which grow linearly with scale factor a. When gas is included, this concentration parameter is slightly larger than the pure N-body result. We also find that the dark matter velocity distribution is less isotropic and more radial than found by N-body simulations of CDM.

CDM Simulations vs. Observed Halos N-body simulations of structure formation from Gaussian-random-noise density fluctuations in a cold dark matter (CDM) universe have revealed that dark matter halos possess a universal density profile that diverges as $r^{-\gamma}$ near the center, with $1 \leq \gamma \leq 2$ [6] [7]. There exists a discrepancy between these singular density profiles found in N-body simulations and current observations of the rotation curves of nearby dwarf galaxies [6] and of strong lensing of background galaxies by the galaxy cluster CL0024+1654 [10] [8], which suggest that dark-matter dominated halos of all scales have flat density cores, instead.

Halo Formation by Pancake Instability and Fragmentation The model we use to examine the formation of dark-matter dominated halos is that of cosmological pancake instability and fragmentation, previously discussed in detail by [12]. Halos formed by such a pancake instability have density profiles very similar to those formed hierarchically in CDM models (e.g. NFW profile [7]), providing a convenient alternative to more complicated simulations with more realistic initial conditions [1] [5] [11]. The ASPH/P^{3}M simulations considered here were described by [1]; this paper extends that analysis to evolutionary trends in the dark matter halo structure.

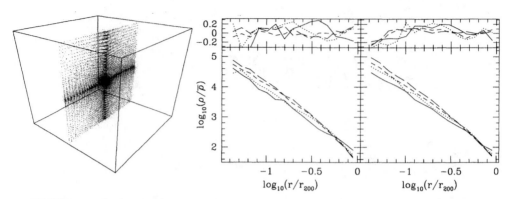

FIGURE 1. (left) Dark matter density field at $a/a_c = 3$. (middle) Density profile of the dark matter halo as simulated without gas at four different scale factors, $a/a_c = 3$ (solid), 4 (dotted), 5 (short dash), and 7 (long dash). Shown above are fractional deviations $(\rho_{NFW} - \rho)/\rho_{NFW}$ from best-fit NFW profile for each epoch. (right) Same as middle, but for DM halo simulated with gas+DM.

Main Results:

- For a/a_c between 3 and 7, the halo can be fit by an NFW profile, with mass within r_{200} growing linearly with scale factor a, when simulated either with or without gas: $M_{200}(x) \simeq 0.07x$, where $x \equiv a/a_c$, and a_c is the scale factor at primary pancake collapse (see Figs. 1 & 2). This mass evolution resembles that of self-similar spherical infall [2], despite the anisotropy associated with pancake collapse and filamentation and periodic boundary conditions.

- After $a/a_c = 3$, the concentration parameter $c_{NFW} \equiv r_s/r_{200}$, determined by best-fitting an NFW density profile [7] to our simulation halos, grows roughly linearly with scale factor a: $c_{NFW}(x) \simeq 1.33x - 0.18$ (without gas), $c_{NFW}(x) \simeq 1.49x - 0.37$ (with gas) (i.e. the linear slope is steeper in the case with gas included). Fluctuations in c_{NFW} around this trend are smaller when gas is included. This evolution we find for c_{NFW} is reminiscent of that reported for halos in CDM N-body simulations [3]. However, the latter applies to halos of a given mass which are observed at different epochs, and, therefore, reflects the statistical correlation of halo mass with collapse epoch in the CDM model, while our result follows an individual halo.

- The anisotropy parameter $\beta \equiv 1 - \langle v_t^2 \rangle / (2 \langle v_r^2 \rangle)$, where $v_t(v_r)$ are tangential (radial) velocities, is shown in Figure 2. Pancake halos are somewhat radially biased, with $\beta \geq 0.6$, about twice the value reported for halos in CDM N-body simulations [4] [9]. With no gas included, the average anisotropy in the halo does not change very much with time, while the inclusion of gas leads to a slight drop after $a/a_c = 5$.

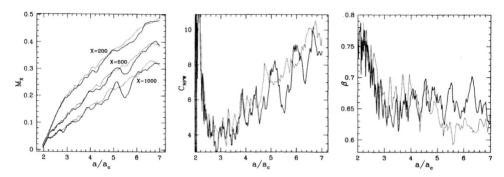

FIGURE 2. (left) Evolution of halo dark matter integrated mass M_X as simulated with (dotted) and without (solid) gas, within spheres of average overdensity $X \equiv \langle \rho \rangle / \bar{\rho}$ (in computational units, where $M_{box} = \lambda_p^3 \bar{\rho} = 1$). (middle) Evolution of halo concentration parameter for the dark matter halo as simulated with (dotted) and without (solid) gas. (right) Anisotropy parameter β averaged over all dark matter halo particles within a sphere of average overdensity 200, as simulated with (dotted) and without (solid) gas.

This work was supported by NASA ATP grants NAG5-7363 and NAG5-7821, NSF grant ASC-9504046, and Texas Advanced Research Program grant 3658-0624-1999.

REFERENCES

1. Alvarez, M., Shapiro, P.R., & Martel, H., *RevMexAA (S.C.)*, in press (2000) (astro-ph/0006203)
2. Bertschinger, E., *ApJS*, **58**, 39 (1985)
3. Bullock, J.S. *et al.*, *MNRAS*, **321**, 559 (2001)
4. Eke, V.R., Navarro, J.F., Frenk, C.S., *ApJ*, **503**, 569 (1998)
5. Martel, H., Shapiro, P.R., & Valinia, A., in preparation (2001)
6. Moore, B., Quinn, T., Governato, F., Stadel, J. & Lake, G., *MNRAS*, **310**, 1147 (1999)
7. Navarro, J.F., Frenk, C.S., and White, S.D.M., *ApJ*, **490**, 493 (1997)
8. Shapiro, P.R., Iliev, I.T., *ApJL*, **542**, 1 (2000)
9. Thomas, P.A. *et al.*, *MNRAS*, **296**, 1061 (1998)
10. Tyson, J.A., Kochanski, G.P., and dell'Antonio, I.P., *ApJL*, **498**, 107 (1998)
11. Valinia, A., Ph.D. Thesis, University of Texas (1996)
12. Valinia, A., Shapiro, P.R., Martel, H., & Vishniac, E.T., *ApJ*, **479**, 46 (1997)

The Equilibrium Structure of Cosmological Halos

Ilian T. Iliev* and Paul R. Shapiro[†]

*IA-UNAM, Mexico
[†]The University of Texas at Austin

Abstract. We have derived an analytical model for the postcollapse equilibrium structure of cosmological halos as nonsingular truncated isothermal spheres (TIS) and compared this model with observations and simulations of cosmological halos on all scales. Our model is in good agreement with the observations of the internal structure of dark-matter-dominated halos from dwarf galaxies to X-ray clusters. It reproduces many of the average properties of halos in CDM simulations to good accuracy, including the density profiles outside the central region, while avoiding the possible discrepancy at small radii between observed galaxy and cluster density profiles and the singular density profiles predicted by N-body simulations of the CDM model. While much attention has been focused lately on this possible discrepancy, we show that the observed galaxy rotation curves and correlations of halo properties nevertheless contain valuable additional information with which to test the theory, despite this uncertainty at small radii. The available data allows us to constrain the fundamental cosmological parameters and also to put a unique constraint on the primordial density fluctuation power spectrum at large wavenumbers (i.e. small mass scale).

The TIS Model. Our model is described in detail in [14] for an EdS universe and generalized to a low-density universe, either matter-dominated or flat with $\Lambda > 0$ in [7]. An initial top-hat density perturbation collapses and virializes, which leads to a nonsingular TIS in hydrostatic equilibrium, a solution of the Lane-Emden equation (appropriately modified for $\Lambda \neq 0$). Using the anzatz that the resulting TIS sphere is the one with the minimum-energy, out of the family of possible solutions, we find that a top-hat perturbation collapse leads to a unique, nonsingular TIS, yielding a universal, self-similar density profile for the postcollapse equilibrium of cosmic halos. Our solution has a unique length scale and amplitude set by the top-hat mass and collapse epoch, with a density proportional to the background density at that epoch. The density profiles for gas and dark matter are assumed to be the same.

Rotation Curves of Dark-Matter Dominated Galactic Halos. The TIS profile matches the observed mass profiles of dark-matter-dominated dwarf galaxies, which are well-fit by the empirical density profile of [1], with a finite density core.

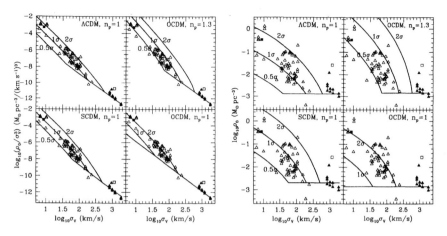

FIGURE 1. Galaxy and cluster halo phase-space density $Q \equiv \rho_0/\sigma_V^3$ (left panels) and halo central density ρ_0 (right panels) versus the halo velocity dispersion σ_V: empty triangles = galaxy data from [9]; filled square = data for dwarf galaxy Leo I from [10]; filled triangles = cluster data from [2] (the filled triangles with the same σ_V correspond to different mass estimates for the same cluster); empty square = galaxy cluster CL 0024 (σ_V is from [3], central density obtained in [13] by fitting TIS profile to the strong lensing data of [15]); curves = TIS + Press-Schechter (PS) prediction for four popular cosmological models, as labelled (n_p is the tilt of the power spectrum). The models in the two upper panels are COBE- and cluster-normalized, with $\Omega_0 = 0.3$ and $\lambda_0 = 0.7$ and 0, respectively. Results are for fluctuations of different amplitudes $\nu \equiv \delta_{\rm crit}/\sigma(M)$, where $\delta_{\rm crit}$ is the value of the linear density contrast with respect to the background density for a tophat fluctuation, extrapolated to the time when the actual nonlinear density inside the tophat reaches infinity, as labelled with $\nu - \sigma$. Curves for each ν connect to the curve for $z_{\rm coll} = 0$, for those $\nu - \sigma$ fluctuations which have not yet collapsed by $z = 0$.

The TIS profile gives a nearly perfect fit to the Burkert profile, providing it with theoretical underpinning and a cosmological context [6].

We have also combined the TIS halo model with the Press-Schechter (PS) formalism, which predicts the typical collapse epoch for objects of a given mass in the CDM model, to explain the observed correlation of $v_{\rm max}$ and $r_{\rm max}$ for dwarf spiral and LSB galaxies. The observational data indicates preference for the currently-favored, flat ΛCDM universe ($\Omega_0 = 1 - \lambda_0 = 0.3$, $h = 0.7$). For more details on our methods and results, see [6].

The Correlations of Halo Core and Maximum Phase-Space Densities with Velocity Dispersion. A comparison of observed halo properties with other correlations predicted by this TIS+PS approach can test the CDM model while constraining the fundamental cosmological parameters and the power-spectrum shape at small mass scales (i.e. large wavenumbers k). The core densities ρ_0 and maximum phase-space densities $Q \equiv \rho_0/\sigma_V^3$ for dark-matter dominated halos are predicted to be correlated with their velocity dispersions σ_V as shown in Figure 1. For cold, collisionless DM, Q is expected to be almost independent of the effects of baryonic dissipation [12]. The data on halos from dwarf spheroidal to clusters is consistent with these predictions, with preference for the flat, ΛCDM model. There have been recent claims that ρ_0 =const for all cosmological halos, independent of their mass, and that such behavior is expected for certain types of SIDM [4,8]. This claim, however, does not seem to be supported by the current data (Figure 1).

Galaxy Clusters: TIS vs. CDM Simulations We have shown previously [5,13,14] that the TIS halo model predicts to great accuracy the internal structure of X-ray clusters found by gas-dynamical/N-body simulations of cluster formation in the CDM model at $z = 0$. The TIS prediction for the redshift evolution of the halo mass-temperature and mass-velocity dispersion relations for galaxy clusters also matches to high accuracy ($\sim$ few percent) the empirical relations derived in [11] from CDM gas/N-body simulations and by the Virgo Consortium from their Hubble volume N-body simulations [Evrard, private communication].

Acknowledgments

This research was supported by NSF grant INT-0003682 from the International Research Fellowship Program and the Office of Multidisciplinary Activities of the Directorate for Mathematical and Physical Sciences to ITI and grants NASA ATP NAG5-7363 and NAG5-7821, NSF ASC-9504046, and Texas Advanced Research Program 3658-0624-1999 to PRS.

REFERENCES

1. Burkert, A. ApJ, **447**, L25 (1995).
2. Dalcanton, J.J., and Hogan, C.J., ApJ, submitted (astro-ph/0004381)
3. Dressler, A., Smail, I., Poggianti, B.M., Butcher, H., Couch, W.J., Ellis, R.S., and Oemler, A., Jr. ApJS, **122**, 51 (1999).
4. Firmani, C., D'Onghia, E., Chincarini, G., Hernandes, X., and Avila-Reese, V. MNRAS, 321, 713 (2000).
5. Iliev I.T., and Shapiro P.R., in "The Seventh Texas-Mexico Conference on Astrophysics: Flows, Blows, and Glows," eds. W. Lee and S. Torres-Peimbert, RevMexAA (Serie de Conferencias), in press (2001) (astro-ph/0006184)
6. Iliev, I.T., and Shapiro, P.R., ApJ, **546**, L5 (2001a).
7. Iliev, I.T., and Shapiro, P.R., MNRAS, in press (2001b) (astro-ph/0101067).
8. Kaplinghat, M., Knox, L., and Turner, M.S., preprint (astro-ph/0005210)
9. Kormendy, J. & Freeman K.C. 2001, in preparation.
10. Mateo, M., Olszewski, E.W., Vogt, S.S., and Keane, M.J. ApJ, **116**, 2315 (1998)
11. Mathiesen, B.F., Evrard, A.E. ApJ, 546, 100 (2001)
12. Sellwood, J.A. ApJ, 540, 1L (2000)
13. Shapiro P.R., and Iliev I.T. ApJ, **542**, L1 (2000)
14. Shapiro, P.R., Iliev, I.T., and Raga, A.C., MNRAS **307**, 203 (1999).
15. Tyson, J.A., Kochanski, G.P., and Dell'Antonio, I.P. ApJ, **498**, L107 (1998).

Self-Interacting Dark Matter with Flavor Mixing

Mikhail V. Medvedev

Canadian Institute for Theoretical Astrophysics, University of Toronto, Toronto, Ontario, M5S 3H8, Canada

Abstract. The crisis of the cold dark matter and problems of the self-interacting dark matter models is resolved by postulating flavor mixing of dark matter particles. Flavor-mixed particles segregate in the gravitational field to form dark halos composed of heavy mass eigenstates. Since these particles are mixed in the interaction basis, elastic collisions convert some of heavy eigenstates into light ones which leave dense central regions of the halo. This annihilation-like process will soften dense central cusps of halos. The proposed model accumulates most of the attractive features of self-interacting and annihilating dark matter models, but does not suffer from their severe drawbacks. This model is natural; it does not require fine tuning.

INTRODUCTION

Dark matter constitutes most of the mass in the Universe, but its nature and properties remain largely unknown. A model of structure formation in a universe with cold dark matter (CDM) is in excellent agreement with observations on large scales ($\gg$ Mpc) which, thus, supports the hypothesis that dark matter particles are heavy and weakly interacting with baryonic matter and photons, while on small (galactic and sub-galactic scales) it appears to be in conflict with recent observations. This fact suggests that the CDM model is a good first approximation, but it has to be corrected on a galactic scale.

The simplest and most popular models are the self-interacting dark matter (SIDM) [1] and the annihilating dark matter (ADM) [2]. There are some problems with each. In a simplest SIDM, the scattering cross-section of dark matter (DM) particles, σ_{si}, must be such that to ensure that the flat core forms just by now; larger σ_{si} result in core collapse and cuspy cores while for smaller σ_{si} the core flattening time is larger than the Hubble time. Moreover, it seems difficult to obtain flattened cores in galaxies and in clusters by the same time because of very different dynamical scales. The ADM model suffers from the severe "annihilation catastrophe" in the early universe (unless some *ad hoc* assumptions made). Here we present a natural model which has all attractive features of these both models but does not suffer from their drawbacks.

SIDM MODEL WITH QUANTUM MIXING

For the sake of simplicity, let us assume that there exist two flavors, the strongly self-interacting, $|\mathcal{I}\rangle$, and the non-self-interacting or "sterile", $|\mathcal{S}\rangle$, ones. Each of these interaction eigenstates is a superposition of two mass eigenstates, the "heavy", $|M\rangle$, and the "light", $|m\rangle$, ones. We write

$$\begin{pmatrix} |M\rangle \\ |m\rangle \end{pmatrix} = \begin{pmatrix} \cos\vartheta & \sin\vartheta \\ -\sin\vartheta & \cos\vartheta \end{pmatrix} \begin{pmatrix} |\mathcal{I}\rangle \\ |\mathcal{S}\rangle \end{pmatrix}, \qquad (1)$$

where ϑ is the mixing angle. Throughout the paper, we assume for simplicity that $M \gg m$, so that heavy states are *non-relativistic* and light states are *relativistic*.

The concept of flavor eigenstates arises when interactions of particles are considered. In the field-free theory the particle fields in the mass basis have a physical meaning instead. In general, such mass eigenstates have different momenta and energies, $E_{M,m}^2 = |\mathbf{p}_{M,m}|^2 c^2 + \{M, m\}^2 c^4$. Between the interactions, the mass eigenstates propagate independently, with different velocities $\mathbf{v}_{M,m} = \mathbf{p}_{M,m} c^2 / E_{M,m}$. Thus at times

$$t \gg t_s \sim \delta x / |\mathbf{v}_m - \mathbf{v}_M| \sim \delta x / c, \qquad (2)$$

where δx is the spatial width of a wave-packet, these states are separated from each other and their wave functions no longer overlap, as illustrated in Fig. 1a. The separation time above is negligibly small compared to a galactic dynamical time scale. In a gravitational field different eigenstates also segregate by mass. Non-relativistic $|M\rangle$-states form halos and the large-scale structure (this is CDM) while relativistic $|m\rangle$-states leave the halo and behave similar to hot DM.

As dark matter halos form, the density in the central parts increases and, at some point, self-interactions of the dark matter particles become important. Let us consider the elementary act of *elastic* scattering of two DM particles, i.e., $|M\rangle$ eigenstates. The initial wave function of two interacting particles in the center of mass frame, according to equation (1), is

$$\Psi_i = \left(e^{ikz} \pm e^{-ikz}\right)|M\rangle = e^{\pm ikz}\left(\cos\vartheta\,|\mathcal{I}\rangle + \sin\vartheta\,|\mathcal{S}\rangle\right), \qquad (3)$$

where "+" corresponds to Ψ_i symmetric to interchange of particles (integer total spin) and "−" – to an antisymmetric Ψ_i (half-integer total spin), the exponents represent two waves, propagating to the right and to the left, and $e^{\pm ikz} \equiv \left(e^{ikz} \pm e^{-ikz}\right)$ is the short-hand notation. For scattering, the interaction basis is appropriate, rather than the mass basis, hence the expansion above. During the scattering event, only $|\mathcal{I}\rangle$-component is changing, since $|\mathcal{S}\rangle$-component does not interact. The wave function after scattering at large distances thus becomes

$$\Psi_s \approx \left(e^{\pm ikz} + \frac{f_\pm(\theta)}{r} e^{ikr}\right)\cos\vartheta\,|\mathcal{I}\rangle + e^{\pm ikz}\sin\vartheta\,|\mathcal{S}\rangle$$

$$= \left(e^{\pm ikz} + \cos^2\vartheta \frac{f_\pm(\theta)}{r} e^{ikr}\right)|M\rangle - \cos\vartheta\sin\vartheta\frac{f_\pm(\theta)}{r} e^{ikr}|m\rangle, \qquad (4)$$

where the combination $f_\pm(\theta) = f(\theta) \pm f(\pi - \theta)$ arises because particles are indistinguishable and $f(\theta)$ is the amplitude of scattering of flavor states. The radial part of the wave function represents a diverging scattered wave. One can clearly see that an initial heavy eigenstate acquires, upon scattering, a light eigenstate admixture. In other words, a heavy eigenstate may be converted into a light eigenstate. This process is illustrated in Fig. 1b. The differential cross-section of a process is the scattering amplitude squared. Integrating over θ we have the relation between the scattering and conversion cross-sections:

$$\sigma_{conv} = \tan^2 \vartheta \, \sigma_{si}. \tag{5}$$

The rest is straightforward. The light particles escape from the halo core and decrease its density. This prevents core collapse even if the scattering time is much smaller then the Hubble time. In this respect, the proposed model resembles an ADM model. However, no annihilation occurs in the dense early Universe because forward conversions of M's into m's are balanced by the reverse ones of m's into M's. A more detailed discussion of the process of freeze-out of mixed DM particles and the resultant constraints on the model will be presented elsewhere.

REFERENCES

1. Spergel, D. N., and Steinhardt, J. P., *Phys. Rev. Lett.* **84**, 3760 (2000)
2. Kaplinghat, M., Knox, L., and Turner, M. S., *Phys. Rev. Lett.* **85**, 3335 (2000)
3. Medvedev, M. V., *Phys. Rev. Lett.* , submitted (astro-ph/0010161) (2001)

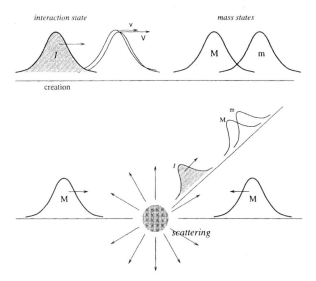

FIGURE 1. Illustration of separation of mass states (a) and of the scattering of two heavy states which after all leads to formation of light states (b).

Off-Axis Cluster Mergers[‡]

P. M. Ricker[*] and C. L. Sarazin[†]

[*]*Department of Astronomy and Astrophysics, University of Chicago, Chicago IL 60637*
[†]*Department of Astronomy, University of Virginia, Charlottesville VA 22903*

Abstract. We present a parameter study of offset mergers between clusters of galaxies, examining the effect of a cuspy dark matter profile on luminosity and temperature evolution and turbulence.

INTRODUCTION

The intracluster gas is repeatedly stirred during a cluster's life by merger shocks [10] and galaxy wakes [20, 18]. The Reynolds number associated with these motions is typically of order $10^2 - 10^3$ [19]. Turbulence may contribute to a number of the proposed secondary consequences of mergers, including the formation of constant-density gas cores, the destruction of cooling flows, and *in situ* acceleration of cosmic-ray particles.

Offset cluster mergers are a likely source of vorticity and turbulence in the intracluster medium. Such mergers are expected to result from tidal torques produced during the linear phase of structure formation [5, 11, 22].

To distinguish the effects of offset mergers from those of the many other simultaneous influences on the ICM, it is necessary to study them in isolation. Offset mergers have been studied by only a few groups [17, 13, 16, 21]; all have used dark matter density profiles with constant-density cores, despite the fact that simulations of hierarchical structure formation produce halos with cuspy central density profiles [9, 8, 6]. We have performed simulations of offset mergers with total density profiles given by the NFW model [9] and β-model gas profiles [1] in order to study the mechanisms by which entropy is generated and redistributed in mergers, and to study the properties of merger-generated turbulence. Further details of this work appear in [14].

We have assumed a Hubble constant $H_0 = 100h$ km s^{-1} Mpc^{-1} with $h = 0.6$.

SIMULATIONS

We have used COSMOS, an Eulerian shock-capturing N-body/hydrodynamics code [15]. The gas temperature and dark matter velocity dispersion profiles were set using

---

[‡] PMR is supported by the ASCI Flash Center at the University of Chicago under DOE grant B341495. Runs were carried out at the Pittsburgh Supercomputing Center and the San Diego Supercomputer Center.

the assumption of hydrostatic equilibrium.

We considered mergers between clusters of equal mass ($2 \times 10^{14} M_\odot$) and mass ratio 1 : 3, holding the smaller mass constant. The temperatures of the two clusters were chosen to be typical of rich clusters at low redshift [2, 3]. The masses were then constrained using the virial mass-temperature relation [4]. Other model parameters were constrained using the X-ray luminosity-temperature relationship for nearby clusters [7], the requirement of convective equilibrium, and the requirement that the central cooling time exceed H_0^{-1}. The sound crossing time for each cluster was $t_{sc} \equiv R/c_s \approx 2$ Gyr.

For each mass ratio, we considered impact parameters between zero and five times the larger NFW scale radius. Each run was followed for ~ 15 Gyr, more than the age of the universe for $\Omega = 1$ and $h = 0.6$. For the hydrodynamics and gravitational potential calculations, we used a nonuniform mesh with 256×128^2 zones, with a minimum zone spacing $\sim 20 h^{-1}$ kpc. For the dark matter, each run also used 128^3 particles of equal mass, distributed between the merging clusters in proportion to their total dark mass.

RESULTS

The morphological changes, relative velocities, and temperature jumps we observe agree well with previous studies of collisions between clusters modeled using the King profile [17]. We observe a larger jump in X-ray luminosity ($\sim 4 - 10\times$) than previous studies, and we argue that this increase is most likely a lower limit due to our spatial resolution. We emphasize that luminosity and temperature jumps due to mergers may have an important bearing on constraints on Ω derived from the observation of hot clusters at high redshift. We will address this issue further in a subsequent paper [12].

Shocks play an important dissipative role in mergers, but they are relatively weak in the highest-density regions. As a result they do not directly raise the entropy of the cluster cores. Instead, shocks create entropy in the outer parts of the clusters, and this high-entropy gas is mixed with the core gas during later stages of the merger. Mixing is initiated by ram pressure: the core gas is displaced from its potential center and becomes convectively unstable. The resulting convective plumes initiate large-scale turbulent motions with eddy sizes up to several 100 kpc. This turbulence is pumped by oscillations in the gravitational potential, which in turn are driven by the more slowly relaxing collisionless dark matter. Even after nearly a Hubble time these motions persist as subsonic turbulence in the cluster cores, providing $20 - 30\%$ of the support against gravity. The dark matter oscillations are also reflected in the extremely long time following a merger required for the remnant to come to virial equilibrium. Because of the increase in core entropy, if a constant-density gas core is initially present, it is not destroyed by a merger, even if the dark matter density profile has a central cusp.

REFERENCES

1. Cavaliere, A., & Fusco-Femiano, R., *Astron. Astrophys.* **49**, 137-144 (1976)
2. Edge, A. C., et al., *Mon. Not. R. Astron. Soc.* **245**, 559-569 (1990)
3. Edge, A. C., Stewart, G. C., & Fabian, A. C., *Mon. Not. R. Astron. Soc.* **258**, 177-188 (1992)

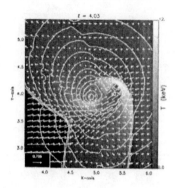

FIGURE 1. Interaction region in a 1:3 offset merger shortly after first core passage. Contours spaced by 1.78× indicate log gas density. Shading indicates gas temperature in keV. Arrows denote gas velocity; the fiducial arrow has length 1200 km s^{-1}. Axis scaling is in units of h^{-1} Mpc.

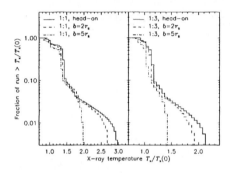

FIGURE 2. Cumulative distribution of X-ray temperatures $T_X(t)$ [14]. For $\sim 8\%$ of each run in the 1:1 cases, and 3% in the 1:3 cases, the system has a temperature greater than 1.5 times its initial value.

4. Evrard, A. E., Metzler, C. A., & Navarro, J. F., *Astroph. J.* **469**, 494-507 (1996)
5. Hoyle, F., in *Problems of Cosmical Aerodynamics*, Central Air Documents Office, Dayton, 1949
6. Jing, Y. P., & Suto, Y., *Astroph. J. Lett.* **529**, 69-72 (2000)
7. Markevitch, M., *Astroph. J.* **504**, 27-34 (1998)
8. Moore, B., et al., *Astroph. J. Lett.* **499**, 5-8 (1998)
9. Navarro, J. F., Frenk, C. S., & White, S. D. M. (NFW), *Astroph. J.* **490**, 493-508 (1997)
10. Norman, M. L., & Bryan, G. L., in *Ringberg Workshop on M87*, eds. K. Meisenheimer & H.-J. Röser, Springer, New York, 1998, p. 106
11. Peebles, P. J. E., *Astroph. J.* **155**, 393-401 (1969)
12. Randall, S., Sarazin, C. L., & Ricker, P. M., in preparation (2001)
13. Ricker, P. M., *Astroph. J.* **496**, 670-692 (1998)
14. Ricker, P. M., & Sarazin, C. L., *Astroph. J.*, submitted (2001)
15. Ricker, P. M., Dodelson, S., & Lamb, D. Q., *Astroph. J.* **536**, 122-143 (2000)
16. Roettiger, K., & Flores, R., *Astroph. J.* **538**, 92-97 (2000)
17. Roettiger, K., Stone, J., & Mushotzky, R., *Astroph. J.* **493**, 62-72 (1998)
18. Sakelliou, I., *Mon. Not. R. Astron. Soc.* **318**, 1164-1170 (2000)
19. Sarazin, C. L., *X-ray Emission from Clusters of Galaxies*, Cambridge U. P., Cambridge, 1988
20. Stevens, I. R., Acreman, D. M., & Ponman, T. J., *Mon. Not. R. Astron. Soc.* **310**, 663-676 (1999)
21. Takizawa, M., *Astroph. J.* **532**, 183-192 (2000)
22. White, S. D. M., *Astroph. J.* **286**, 38-41 (1984)

CHAPTER 4

COSMIC MICROWAVE BACKGROUND

Images of the Early Universe from the BOOMERanG experiment

P. de Bernardis[1], P.A.R.Ade[2], J.J.Bock[3], J.R.Bond[4], J.Borrill[5,6],
A.Boscaleri[7], K.Coble[8], B.P.Crill[9], G.De Gasperis[10], G.De Troia[1],
P.C.Farese[8], P.G.Ferreira[11], K.Ganga[9,12], M.Giacometti[1], E.Hivon[9],
V.V.Hristov[9], A.Iacoangeli[1], A.H.Jaffe[6], A.E.Lange[9], L.Martinis[13],
S.Masi[1], P.Mason[9], P.D.Mauskopf[14,15], A.Melchiorri[1], L.Miglio[16],
T.Montroy[8], C.B.Netterfield[16], E.Pascale[7], F.Piacentini[1], G.Polenta[1],
D.Pogosyan[4], S.Prunet[4], S.Rao[17], G.Romeo[17], J.E.Ruhl[8], F.Scaramuzzi[13],
D.Sforna[1], N.Vittorio[10]

[1] *Dipartimento di Fisica, Università di Roma La Sapienza, P.le A. Moro 2, 00185 Roma, Italy,*
[2] *Department of Physics, Queen Mary and Westfield College, Mile End Road, London, E1 4NS, UK,*
[3] *Jet Propulsion Laboratory, Pasadena, CA, USA,*
[4] *CITA University of Toronto, Canada,*
[5] *NERSC-LBNL, Berkeley, CA, USA,*
[6] *Center for Particle Astrophysics, University of California at Berkeley, 301 Le Conte Hall, Berkeley CA 94720, USA,*
[7] *IROE - CNR, Via Panciatichi 64, 50127 Firenze, Italy,*
[8] *Department of Physics, University of California at Santa Barbara, Santa Barbara, CA 93106, USA,*
[9] *California Institute of Technology, Mail Code: 59-33, Pasadena, CA 91125, USA,*
[10] *Dipartimento di Fisica, Università di Roma Tor Vergata, Via della Ricerca Scientifica 1, 00133 Roma, Italy,*
[11] *Astrophysics, University of Oxford, Keble Road, OX1 3RH, UK,*
[12] *PCC, College de France, 11 pl. Marcelin Berthelot, 75231 Paris Cedex 05, France,*
[13] *ENEA Centro Ricerche di Frascati, Via E. Fermi 45, 00044 Frascati, Italy,*
[14] *Physics and Astronomy Dept, Cardiff University, UK,*
[15] *Dept of Physics and Astronomy, U.Mass. Amherst, MA, USA,*
[16] *Departments of Physics and Astronomy, University of Toronto, Canada,*
[17] *Istituto Nazionale di Geofisica, Via di Vigna Murata 605, 00143, Roma, Italy,.*

THE CMB AND THE CURVATURE OF THE UNIVERSE

The CMB is the fundamental tool to study the properties of the early universe and of the universe at large scales. In the framework of the Hot Big Bang model, when we look to the CMB we look back in time to the end of the plasma era, at a redshift ~ 1000, when the universe was ~ 50000 times younger, ~ 1000 times hotter and $\sim 10^9$ times denser than today. The image of the CMB can be used to study the physical processes there, to infer what happened before, and also to study the background geometry of our Universe.

The photons of the CMB travel in space for ~ 15 billion years before reaching our microwave telescopes. Originally visible and near infrared light, they are converted in a faint glow of microwaves by the expansion of the Universe. Since CMB photons travel so long in the Universe, their trajectories are affected significantly by any large-scale curvature of space. The image of the CMB can thus be used to study the large scale geometry of space. The scale of the acoustic horizon at recombination is the "standard ruler" needed for these studies: density fluctuations larger than the horizon are frozen, while fluctuations smaller than the horizon can oscillate, arriving at recombination in a compressed or rarefied state, and thus producing a characteristic pattern of hot and cold spots in the CMB (see e.g. [1] [2] [3] [4] [5]).

In fact, the temperature fluctuations of the CMB are related to the density fluctuations at recombination through three physical processes: the photon density fluctuations δ_γ accompanying the fluctuation of density; the gravitational redshift/blueshift of photons coming from overdense/underdense regions with gravitational potential ϕ; the Doppler shift produced by scatter of photons by electrons moving with velocity v with the perturbation. In formulas (see e.g.[6]):

$$\frac{\Delta T}{T}(\vec{n}) \approx \frac{1}{4}\delta_{\gamma r} + \frac{1}{3}\phi_r - \vec{n}\frac{\vec{v}_r}{c}$$

where n is the line of sight vector and the subscript r labels quantities at recombination. In the CMB temperature distribution we see a snapshot of the status of the density perturbations at recombination: the characteristic size of the horizon in the density fluctuations is thus directly translated in a characteristic size in the spots of the CMB.

Recombination of hydrogen happens when the temperature drops below $\sim 3000K$, i.e. about 300000 years after the Big Bang. The causal horizon is about 300000 light years there. Since the distance travelled by CMB photons is $\sim$ 15 billion light years, and lenghts in the Universe increase by a factor 1000 meanwhile, the angle subtended now by the horizons at recombination is expected to be close to one degree, in a Euclidean Universe. The typical observed angular size of the hot and cold spots strongly depends on the average mass-energy density parameter Ω. The presence of mass and energy acts as a magnifying ($\Omega > 1$) or demagnifying ($\Omega < 1$) lens, producing horizon-sized spots larger or smaller than $\sim 1^o$ in the two cases, respectively.

The image of the CMB is described in statistical terms: the temperature field is expaned in multipoles

$$\frac{\Delta T}{T}(\vec{n}) = \sum_{\ell=1}^{\infty} \sum_{m=-\ell}^{\ell} a_{\ell m} Y_{\ell m}(\vec{n})$$

The $a_{\ell m}$ are random variables with zero average and ensemble variance $< a_{\ell m} a^*_{\ell' m'} > = c_\ell \delta_{\ell \ell'} \delta_{mm'}$. The c_ℓ's represent the angular power spectrum of the CMB. If we compute the power spectrum of the image of the CMB, we expect to see a peak at multipoles $\ell_1 \sim$ 200 corresponding to these degree-size spots. The location of the peak will be mainly driven by Ω, thus allowing a measurement of this elusive cosmological parameter. The dependance of ℓ_1 from Ω is not simple in the presently favoured cosmological model with significant vacuum energy Ω_Λ [7] [8]. Full spectral data must be compared to spectral models computed from a set of cosmological parameters, and degeneracies must be taken into account [9], [10]. It remains confirmed, however, that the main driver for the location of the first acoustic peak is the value of Ω. This can thus be retrieved with good accuracy from the power specturm of an image of the CMB with sub-degree resolution.

After recombination, the same density fluctuations driving the acoustic oscillations grow, and form the jerarchy of structures we see in the Universe today. Thus, the image of CMB anisotropies represents also a fundamental tool in the study of the formation of structures in the Universe.

The detailed shape of the CMB power spectrum c_ℓ has been computed in a number of scenarios with very high detail. A wide literature is available on the subject as well as publically available software ([11], [12]) to accurately compute the power spectrum given a set of cosmological parameters in the framework of adiabatic inflationary models. More work remains to be done for the general case (for example including isocurvature modes, see e.g. [13]).

The main feature, i.e. a harmonic series of peaks following the first one described above, can be understood as follows. The acoustic horizon increases with time, and at some point becomes larger than a given perturbation size. At this point, all the perturbations present in the Universe with that size will start to oscillate. This can be seen as a cosmic synchronization process, which initiates the oscillation of small perturbations before the oscillation of large ones: the phase of the perturbations at the recombination depends on their intrinsic size. Perturbations with size close to the horizon at recombination start last, and have just enough time to arrive to the maximum compression, producing the degree-size spots in the CMB, i.e. the first "acoustic" peak in the angular power spectrum of the CMB anisotropy. Perturbations with smaller intrinsic size have entered the acoustic horizon before, and arrive at recombination after a full compression and a return to the average density. They will produce a CMB temperature fluctuation smaller than the previous ones, because among the three physical processes producing temperature fluctuations from density ones, only the Doppler effect is effective. This corresponds to a dip in the angular power spectrum of the CMB at multipoles larger than the first acoustic peak. Even smaller perturbations have enough time to compress, return to the average density, and then arrive to the maximum rarefaction at recombination, producing a second peak in power spectrum, at multipoles about twice of those of the first one. Repeating this reasoning, we expect a harmonic series of acoustic peaks, up to very small sizes, smaller than the thickness of the last scattering surface. Photons diffusion, and the fact that many small-size positive and negative perturbations are aligned on the same line of sight, damps the temperature fluctuations measurable in the CMB at very high multipoles.

The shape of the power spectrum of the CMB, c_ℓ, depends on several cosmological parameters in addition to Ω. Increasing the physical density of baryons $\Omega_b h^2$ favours compressions against rarefactions in the acoustic oscillations. Compression peaks are thus enhanced with respect to rarefaction ones. The relative amplitude of the second peak (rarefaction) with respect to the amplitude of the first peak (compression) is thus a good measurement of $\Omega_b h^2$. The power spectrum of primordial density fluctuations $P(k)$ controls the general shape of the power spectrum of the CMB. In the inflationary scenario $P(k) = A k^n$ with $n \sim 1$. The value of n also drives the amplitude of the higher order peaks relative to the amplitude of the first one. If only the first and second peak are observed, increasing n has about the same effect as decreasing $\Omega_b h^2$. A measurement of the third peak removes this degeneracy.

CMB AND COSMIC INFLATION

In the previous section we have assumed the presence of density perturbations in the primeval plasma. In 1992 the DMR instrument on the COBE satellite has shown that these perturbations do exist at least at large scales[14]. The angular resolution of DMR was 7°, corresponding to a sensitivity to multipoles between 1 and ~ 20 in the angular power spectrum of the CMB. The power spectrum measured by DMR has a characteristic power spectrum[15] $c_\ell \sim 1/\ell/(\ell+1)$. This is consistent with a Harrison-Zeldovich power spectrum of density fluctuations $P(k) = Ak^n$ with $n = (1.2 \pm 0.3)$.

Such a spectrum is expected in the simplest inflationary scenarios (see e.g. [16] [17] [18] [19] [20]), where microscopic quantum fluctuations in the very early universe ($\sim 10^{-36}$ s after the big bang) are inflated to cosmological scales by the exponential growth of space during an early phase transition at the grand-unification era. The density fluctuations generated in this way are gaussian and adiabatic (see e.g. [21]). Inflation explains the homogeneity of the CMB at large scales: regions causally disconnected at the recombination epoch had been in close causal contact in the very early Universe, before the superluminal inflation of space. Inflation naturally produces a flat geometry of space, stretching any inital curvature, due to the huge expansion factor. The power law spectrum of primordial fluctuations is reminiscent of the initial quantum fluctuations: constraining n through the measurement of the power spectrum is thus a way to test the inflationary hypothesis. However, some inflation variants feature values of $n \sim 0.9$ or even less (see e.g. [22]). The adiabatic inflationary scenario is shown in cartoon form in fig.1.

The alternative scenario for the generation of density fluctuations is based on topological defects (see e.g. [23]). In this scenario non gaussian isocurvature fluctuations are favored. The peak at ℓ_1 is either not present of shifted to higher ℓs. The analysis of the power spectrum and of the image of the CMB (and in particular of its gaussianity properties) can thus distinguish between the two alternative scenarios for the generation of density fluctuations.

The problem of recovering the set of cosmological parameters from the measured power spectrum c_ℓ has been widely studied in view of the satellite missions MAP and Planck, which promise to measure the power spectrum with $\sim 1\%$ accuracy on a very wide range of multipoles, allowing an accurate determination of the main cosmological parameters (see e.g. [25]).

MEASURING THE IMAGE OF THE CMB

The contrast of the image of the CMB is very low (~ 10 ppm). Moreover, atmospheric and Galactic signals can be much larger than the CMB anisotropy, depending on wavelength, observed sky region and location of the observer. Trying to map the CMB anisotroy from the total brightness coming from the sky in a ground-based exper-

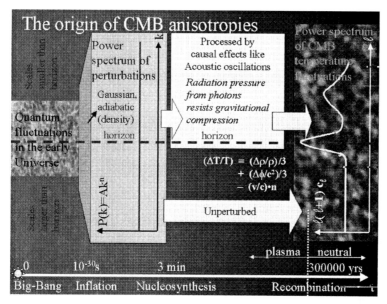

FIGURE 1. Adiabatic inflationary scenario for the generation of CMB anisotropy from quantum fluctuations in the very early universe. Inflation boosts the microscopic fluctuations producing adiabatic gaussian density fluctuations at all cosmological scales. Fluctuatios larger than the causal horizon are effectively frozen, and produce the large scale anisotropy in the CMB measured by the COBE-DMR. Fluctuations entering the horizon before recombination are processed by the cosmological plasma and oscillate as sound waves, thus producing a characteristic power spectrum of anisotropy in the CMB at sub-horizon scales. The angular scale under which these are seen today depends on the curvature of space.

iment is like trying to map distant galaxies in the visual band during daytime. Detector and istrumental noise is an additional problem in these measurements.

For all these reasons, 27 years were needed to produce the first detection of CMB anisotropy after its the discovery. In 1992 the DMR instrument on board of the COBE satellite detected for the first time anisotropy of the CMB at large scales [14]. The resolution of the instrument ($\sim 7^o$ FWHM) was not sufficient to resolve the degree-scale spots useful to measure the curvature of the Universe. Following this detection, many experiments were carried out with degree and sub-degree resolution. The first clear detections of a peak at $\ell \sim 200$ in the angular power spectrum of the CMB anisotropy arrived with the use of new detectors. HEMT based microwave amplifiers [26] [27] were used with the MAT telescope operating at 5000m of altitude in Chile [28] [29]. Spider web bolometers[30] were used at higher frequencies in the test flight of the balloon borne experiment BOOMERanG [31], which was flown in a short flight in Texas in 1997, while TOCO was taking data. Both experiments produced convincing evidence for the peak at $\ell \sim 200$. But the most exciting breakthrough was the recent measurement of wide, resolved images of the CMB, obtained by the BOOMERanG-LDB[32] and by the MAXIMA-1[33] experiments. In the following we will focus on the results from the BOOMERanG experiment.

BOOMERANG

BOOMERanG (Balloon Observations Of Millimeter Extragalactic Radiation and Geophysics) is a 1.3m off-axis scanning telescope using fast, ultra-sensitive bolometric detectors in four frequency bands. The telescope scans the sky at nearly constant speed, of the order of $1^o/s$, along $L_{scan} = 60^o$ wide scans at constant elevation (either 40°, 45°, or 50°). The main beam is similar to a gaussian with $\sigma_{beam} \sim 5'$ at 150 GHz. Different multipoles of the CMB anisotropy are converted by the scan into different audio frequencies in the detectors[34], thus avoiding the effects of 1/f noise and other low-frequency disturbances [35]. This allows the detection of the angular power spectrum over a wide range of multipoles ($\ell_{max} \sim 1/\sigma_{beam}$; $\ell_{min} \sim 1/L_{scan}$) in a single experiment. Two different scan speeds (1 dps and 2 dps) have been used during the measurements to detect possible systematic effects due to the transfer function and noise spectrum of the instrument. The instrument operates in the stratosphere, avoding the large-scale signals and noise produced by atmospheric emission. The full payload is rotated around the vertical axis to avoid scan synchronous instrumental signals. The azimuth of the center of the scan tracks the azimuth of the selected region to be mapped, and sky rotation produces a nicely crosslinked pattern of scans, very useful to reconstrunct the sky map from the time-ordered data. The mesurement is repeated several times in different days, while the instrument drifts by hundreds of kilometers in its stratospheric circumnavigation of Antarctica. This allows the repetition of the measurements under very different experimental conditions, which is the best way to check for systematic effects contaminating the data. For example, every day the ground environment (ice, sea, rocky areas) is different, and so is the ground spillover in the sidelobes of the telescope. A sensitive null test can be obtained by comparing the maps obtained in different days. If the maps are the same, ground spillover contamination can be excluded.

BOOMERanG maps the sky simultaneously at 90, 150, 240 and 410 GHz. Comparison of the maps measured at different frequencies is a powerful tool to test for foregrounds contamination. The two low frequency bands are mainly sensitive to CMB anisotropies, while the two higher frequency bands can be used to monitor atmosperic and interstellar contaminations. 16 detectors have been distributed in the focal plane as shown in fig.2. The location of different detectors in the focal plane has been optimized in order to have robust confirmation of structures in the sky at different time scales.

The technical details of the instrument are reported in [36] and in [37]. The main characteristics are as follows:

- Telescope: off-axis gregorian with cryogenic secondary and tertiary
- Primary mirror: off-axis, aluminum, 45° off-axis, 1.3m diameter, f/1.
- 90 GHz detectors: 2, 18' FWHM, best $NET_{CMB} \sim 150 \mu K \sqrt{s}$
- 150 GHz detectors: 6, 10' FWHM, best $NET_{CMB} \sim 150 \mu K \sqrt{s}$
- 240 GHz detectors: 4, 14' FWHM, best $NET_{CMB} \sim 210 \mu K \sqrt{s}$
- 410 GHz detectors: 4, 12' FWHM, best $NET_{CMB} \sim 3000 \mu K \sqrt{s}$
- Attitude control: azimuth flywheels, passive pendulation damper
- Azimuth scans at 1 to $2^o/s$
- Attitude reconstruction: differential GPS, digital sundial sun sensors, laser gyroscopes

FIGURE 2. The cryogenic focal plane of the BOOMERanG experiment. In the bottom right panel the entrance of the 8 photometers are shown and labeled with the corresponding frequency (in GHz).

The instrument was flown by NASA-NSBF from Dec.29, 1998 to Jan.8, 1999, at 39 Km of altitude, circumnavigating Antarctica at a latitude $\sim -78^o S$. 57 million 16 bit samples were acquired for each detector during the flight.

THE MAPS OF THE MICROWAVE SKY

Maximum likelihood sky maps were constructed from the time-streams using the MADCAP package [38] and a recursive estimator of instrumental noise [39]. The maps cover ~ 1800 square degrees in one of the best (lowest foreground) sky regions in the southern hemisphere. The structures at large angular scales ($\gtrsim 10^o$) have been filtered out to remove the effect of 1/f noise and instrumental drifts. These maps, smoothed to a resolution of 22' FWHM, have been published in [32]. In fig.3 we show sum ($\Delta T_{240} + \Delta T_{150}$) and difference ($\Delta T_{240} - \Delta T_{150}$) maps obtained from the 150 and 240 GHz channels, expressed in CMB temperature fluctuation units. Degree-size structures uniformly covering the surveyed area are the dominant feature of the sum map. The structures disappear in the difference map. This is already a proof of the cosmological nature of the structures.

It is very important to compare the structures visible in the maps at 90, 150 and 240 GHz [32]. The structures are very similar in shape, and the relative amplitude of the fluctuations is perfectly consistent with the derivative of a 2.73K blackbody, as expected for CMB anisotropy. A simple scatter plot of these signals expressed in CMB temperature fluctuations units has best fit slopes very consisten with 1, confirming the visual impres-

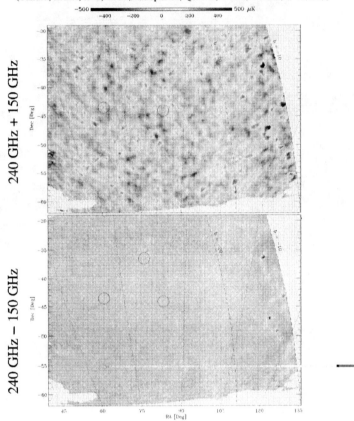

FIGURE 3. The sum (top panel) and difference (bottom panel) maps obtained from the 150 and 240 GHz channels of BOOMERanG. CMB temperature units are used for both the channels. For this reason, CMB fluctuations are enhanced in the sum map, while are removed in the difference map. Only non-CMB structures remain in the difference map. The three circles surround three AGNs present in the map.

sion above[40]. Masi et al. [41] find that the contamination from the main local foreground (i.e. thermal emission from diffuse interstellar dust) is less than 1% of the mean square fluctuation detected at 150 GHz. Moreover, the brightness fluctuations generated by unresolved point-like radio sources can be estimated from existing catalogues ([42]) and is found to be negligible ($\sim 160(\ell/1000)^2 \mu K^2$). From these results we conclude that the dominant feature in the maps is CMB anisotropy, copying the pattern of the acoustic horizons at the last scattering surface. The amplitude of the fluctuations ($\Delta T_{rms} \sim 80 \mu K$) is consistent with the level of CMB anisotropy computed in the inflationary adiabatic scenario, normalized to the COBE-DMR detection.

The MAXIMA experiment has also produced a map of $\sim 0.25\%$ of the sky at 150 GHz, by observing a high latitude sky region in the Northern Hemisphere [33]. BOOMERanG and MAXIMA nicely complement each other. In fact BOOMERanG covers a large region of the sky, producing data with smaller cosmic variance at $\ell \lesssim 300$, but the precision in the pointing reconstruction is presently limited to 3 arcminutes. For this reason the maps have been smoothed to 22.5' FWHM and the power spectrum is limited to $\ell \lesssim 600$. MAXIMA instead covers a smaller sky region, producing larger errors at multipoles $\lesssim 300$, but has already achieved sub-arcmin precision in the pointing reconstrunction, and the resolution of the map is limited only by the intrinsic resolution on the telescope at ~ 10' FWHM. The power spectrum correspondingly extends up to $\ell \sim 800$. The map from MAXIMA is statistically very similar to the map of BOOMERanG, and this agreement is a very important independent confirmation of the results for both experiments.

THE POWER SPECTRUM OF THE CMB

The angular power spectrum[32] has been computed from the maps of BOOMERanG in two different ways: using a simple spherical harmonics transform [43] and using the MADCAP [38] maximum likelihood algorithm. The two methods produce very consistent results. In fig.4 we report the angular power spectrum of the center region of the BOOMERanG map ($\sim 1\%$ of the sky). together with the data from COBE-DMR and from the recently published MAXIMA, CBI [44] and BIMA [45] experiments.

There are several features immediately evident from these data. The level and shape of fluctuations is consistent with a constant level at the large scales sampled by COBE-DMR, plus a peak due to acoustic horizon effects at scales around one degree. The second peak (and the higher order ones) has not been detected yet, but there is significant power detected at multipoles between 300 and 600. There is a damped tail at multipoles $\gtrsim 1000$, as expected by photon diffusion and line of sight averaging effects at recombination. This behaviour is in remarkable agreement with the simple theory presented in section 2 and we can start to say that the big picture is correct. A word of caution is needed, however. The presence of a second peak is consistent with the current data, but it has not been observed yet. Several experiment promise a detection soon, including the combined analysis of 12 channels in BOOMERanG with refined pointing reconstruction, the second flight of MAXIMA or from the interferometers (DASI [46], CBI [47], VSA [48] etc.) currently getting data. Such detection will be the final proof that the Universe underwent a hot phase with acoustic oscillations, and that the structure we see in the Universe today was grown after the hot plasma phase through gravitational instability from the same primordial fluctuations.

Several experimental problems are immediately evident from the power spectrum shown in fig.4. The data at $50 \lesssim \ell \lesssim 300$ are limited in precision by cosmic variance. This will improve with larger surveys, as in the Archeops [49] and TOPHAT [50] balloon experiments and in the full sky survey of the MAP experiment. Significant calibration errors are still present in the same data sets. A reliable standard of calibration is still

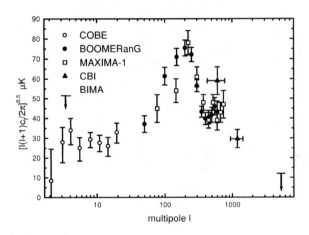

FIGURE 4. Recently published measurements of the angular power spectrum of the cosmic microwave background

non-trivial at these wavelengths, and the Dipole calibration used in BOOMERanG can be affected by scan synchronous effects difficult to monitor to better than 10% [51]. The data at $300 \lesssim \ell \lesssim 600$ are still detector noise limited. In the case of BOOMERanG, this will improve with the simultaneous analysis of all the 12 detectors sensitive to CMB anisotropy (the published power spectrum has been computed from the best detector alone).

The BOOMERanG results have been obtained from a preliminary pointing solution. Jitter in the telescope pointing is a relevant concern, and we are currently working on the development of a refined pointing solution. The main effect of pointing jitter is to spread the equivalent beam of the telescope. In fig.5 we compute the effect of changing the equivalent beam on the most inportant features detected in the spectrum: the location of the peak, its amplitude, and the ratio between the general level of fluctuations at $\ell > 300$ and the amplitude of the peak. It is evident from the figure that the measurement of the location of the peak (controlling the measurement of Ω) is very robust, while the other two quantities (controlling for example the measurement of $\Omega_b h^2$) are less robust.

The present data from the interferometers at $\ell \gtrsim 600$ suffer for lack of spectral resolution and are still single frequency. Moreover, only differential maps have been produced, due to the presence of significant ground spillover in the absolute maps. This is going to improve a lot with the analysis of the larger dataset already acquired and with future system developments.

The significant mismatch between the lowest multipoles sampled by interferometric measurements and the highest multipoles measured by BOOMERanG and MAXIMA is reduced to less than 2 σ once the beam error and the partial overlap of the ℓ-bands are taken into account. It is, however, a possible indication of a calibration systematic still

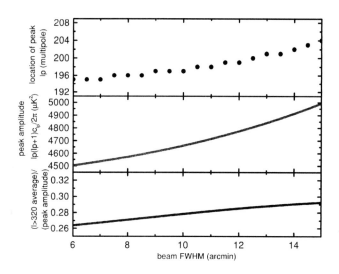

FIGURE 5. Effect of the beam FWHM on the most important characteristics of the power spectrum measured by BOOMERanG. The nominal FWHM is 10 arcmin.

unaccounted for.

The final measurements will arrive with the MAP (launch in a few months) and Planck (launch in 2007) satellite experiments. In the following we focus on what we can already say about the curvature.

COSMOLOGY FROM CMB ANISOTROPY

The position, amplitude and width of the peak evident in fig.4 are consistent with the general adiabatic inflationary scenario, while the simplest models based on topological defects do not fit the data as well.

Using a naive quadratic fit estimator for the MAXIMA and BOOMERanG data we find that the peak is located at multipole $\ell = (197 \pm 6)$ for BOOMERanG (1σ, bandcenters from 50 to 300, data from [32]) , $\ell = (233 \pm 33)$ for MAXIMA (1σ, bandcenters from 55 to 300, data from [33]) , $\ell = (195 \pm 8)$ for the combination of the two datasets (1σ, bandcenters from 65 to 296, data from [52]). These results can shift a little bit higher if one assumes skewed power spectra similar to the adiabatic scenario ones [53].

This location of the maximum is consistent with a flat geometry of space, but it is not univocally related to the density parameter Ω if we allow for a non vanishing cosmological constant [8]. Moreover, the location of the maximum alone is not the best way to constrain Ω, and there is much more information contained in the datasets. Jaffe et al.[52] carried out a full Bayesian analysis on the combined BOOMERanG, MAXIMA and COBE-DMR datasets, constraining simultaneously the parameters Ω, $\Omega_b h^2$, n_s, $\Omega_c h^2$.

It must be stressed that this kind of analysis assumes an adiabatic inflationary model. Moreover, it is very important to specify in detail the prior distributions assumed for each of the parameters. This is especially important in the case of CMB power spectrum measurements, since an important geometrical degeneracy is present [9], so different combinations of the parameters produce very similar power spectra. The 95% confidence intervals for Ω, taking into account all the degeneracies, range from $(0.88 - 1.12)$ to $(0.97 - 1.35)$ depending on the assumed priors and parametrizations [52, 10, 24]. This strongly suggests a flat geometry of the Universe, and at least implies, with 95% confidence, a curvature length

$$R = \frac{c}{H_o} \frac{1}{\sqrt{|\Omega - 1|}}$$

larger than 8.3 to 2.9 times the Hubble length.

Needless to say, having demonstrated that $\Omega \sim 1$ has very important cosmological consequences. According to several independent measurements and methods, the matter density parameter Ω_M is significantly smaller than unity and close to 30% [54, 55, 56, 57, 58]. This means that about 70% of the mass and energy present in the Universe must be in a different form: not ordinary matter, not dark matter.

The presence of dark energy (a form of repulsive energy with negative pressure) is an interesting hypothesis to solve the puzzle. Its presence has been independently proposed [7, 59, 60, 61, 62] to drive the acceleration of the expansion rate of the Universe hypotised to explain the observation of distant supernovae [63, 64]. This hypothesis is still widely debated, the main interpretation problem remaining the lack of a convincing particle physics model for the dark, negative pressure form of energy required.

In the same adiabatic perturbations framework, BOOMERanG and MAXIMA constrain the slope of the initial power spectrum of density perturbations n_s in the range 1.01 ± 0.17 (95% confidence) [52]. In addition to the Ω measurement, tese results for n_s are also consistent and support the simplest inflationary scenarios [65, 19, 66].

The third parameter constrained by BOOMERanG and MAXIMA is the density of baryons. In the power spectrum this parameter controls the relative amplitude of the first peak to the second one. From the power spectrum of fig.4, this ratio is constrained to be $\sim 2\sigma$ larger than expected for a standard primordial nucleosynthesis (BBN) ($\Omega_b h^2 = (0.019 \pm 0.002)$ from [67], $\Omega_b h^2 = (0.020 \pm 0.002)$ from [68]). This suggests a high physical density of baryons. Depending on the assumed priors, 95% intervals for $\Omega_b h^2$ ranging from $(0.019 - 0.045)$ to $(0.026 - 0.048)$ are obtained from the CMB power spectrum data. These results are only marginally overlapping the BBN one, but given the orthogonality of the methods and of the systematic errors, I would rather speak of good overall consistency of the model. It must be stressed, in fact, that the density of baryons enters in the two observable in completely different ways. It controls the primordial abundance of light elements due to its effect in the nuclear reactions happening in the first minutes after the big bang. It also affects the density oscillations (acoustic waves) producing CMB anisotropies about 300000 years after the big bang. Completely different physical phenomena at completely different regimes in the early history of the Universe require the density of baryons to be the same to within a factor 1.5: this should be considered a wonderfull success of cosmology. New physics will be required only if future, more precise measurements will confirm and increase the present $\sim 2\sigma$ disagreement. Also, note that frequentist methods point to a statistical consistency of the two results [69].

BEYOND THE POWER SPECTRUM

At this point the next step is obvious: we should check if the detected temperature fluctuations are gaussian distributed, another prediction of inflation. This is very difficult to test, since instrumental effects can mask small cosmic non gaussianities, and can produce subtle non gaussianities as well. Moreover, different methods probe different kinds of non gaussianity (see e.g. [70] and references therein). A reality check, simply using the 1 point distribution of the BOOMERanG map, has been published in [40]: the detected pixel temperature fluctuations (normalized to the square root of the sum in quadrature of sky variance and instrument noise variance in that pixel) are indeed very precisely gaussian distributed. This is only a starting point in the demonstration of the gaussian character of the CMB, since the central limit theorem helps a lot in this kind of test. However, it is at least reassuring the fact that we do not detect deviations from gaussianity in the data at high latitudes at 150 GHz, while deviations are evident in the data at lower galactic latitudes at 150 GHz, and at all the latitudes in the dust dominated map at 410 GHz. A detailed non gaussianity analysis is underway. The MAXIMA team has already published a first analysis of gaussianity of the CMB map [71]. The main difficulty is, of course, to separate with high confidence small instrumental effects and foreground contaminations from real non-gaussian signatures in the CMB, if any. Realistic montecarlo simulations are the only way to assess the statistical significance of detections of (or upper limits to) non-gaussianity.

CMB Photons are last scattered by electrons at the recombination epoch. ItŠs a Thomson scattering. If the distribution of incoming radiation has a quadrupole moment, the scattered radiation has some degree of linear polarization. The degree of linear polarization is of the order of $\lesssim 10\%$ of the anisotropy: the expected signal is thus of the order of a few μK rms or less (see e.g. [72]). Despite a long lasting experimental effort (see e.g. [73, 74, 75, 76, 77, 78]) the polarization of the CMB has not been detected yet. The best upper limits to date are of the order of 6×10^{-6} in $\Delta T/T$ at angular scales around one degree. In the standard scenario [79] we expect acoustic peaks in the polarization power spectrum at $\ell \gtrsim 200$. The presence of tensor perturbations generated by inflation produces a curl component B in the CMB polarization field which adds to the curl-free E component described above. Measuring the power spectrum of polarization of the CMB is very important for several reasons. First, it provides four power spectra to measure: in addition to the $<TT>$ power spectrum of the anisotropy T, it is possible to measure the $<TE>$ anisotropy-polarization cross-spectrum; the pure polarization power spectrum $<EE>$ is more challenging, while the $<BB>$ component is even smaller and is non-zero only if gravity-waves were present at last scattering. Its detection would prove directly the physics of inflation. The $<EB>$ and $<TB>$ spectra are zero by parity. The BOOMERanG instrument has been modified including a new focal plane with polarization senitive bolometers: its new long duration flight (called B2K) is scheduled for the end of 2001. The forecast sensitivity to polarization has been computed in [80]. Many other attempts to measure CMB polarization are in progress(see e.g. [81] and references therein), and a first detection is expected soon .

ACKNOWLEDGMENTS

The BOOMERANG project has been supported by the CIAR and NSERC in Canada, by Programma Nazionale Ricerche in Antartide, Universitá "La Sapienza", and Agenzia Spaziale Italiana in Italy, by PPARC in the UK, and by NASA, NSF OPP and NERSC in the U.S. We received superb field and flight support from NSBF and the USAP personnel in McMurdo.

REFERENCES

1. Peebles, P.J.E, and Yu J.T., 1970, Ap.J. 162, 815
2. Sunyaev, R.A. & Zeldovich, Ya.B., 1970, Astrophysics and Space Science 7, 3
3. Doroshkevich A.G., Zeldovich Ya.B., Sunyaev R.A., 1978, Soviet Astronomy, 22, 523
4. Silk J. & Wilson M. L. 1980, Physica Scripta, 21, 708
5. Hu W., Sugiyama N. & Silk J., 1997, Nature, 386, 37
6. Martínez-González E., Sanz J.L. & Silk J., 1990, ApJ, 355, L5
7. Weinberg S. , Rev. Mod. Phys. 61 1 1989
8. Weinberg S. astro-ph/0006276
9. G. Efstathiou and J. R. Bond, Mon. Not. R. Astron. Soc. **304**, 75 (1999).
10. Melchiorri A. and Griffiths L.M., astro-ph/0011147
11. Seljak, U. & Zaldarriaga, M. 1996, Ap.J. 469, 437
12. Lewis A., Stewart E., Lasenby A., astro-ph/9911176
13. Bucher, M., Moodley, K. & Turok, N. 2000, astro-ph/0007360
14. Smoot G.F. *et al.* , 1992, ApJ, 396, L1
15. Bennett C.L. *et al.* , 1996, ApJ, 464, L1
16. Guth A.H., 1982, Phil. Trans. R. Soc., A307, 141
17. Linde A., 1982, Phys.Lett., 108B, 389
18. Linde A., 1983, Phys.Lett., 129B, 177
19. Albrecth A. & Steinhardt P.J., 1982, Phys.Rev.Lett., 48, 1220
20. Guth A.H. 1997, The Inflationary Universe: The quest for a new theory of cosmic originis, Addison-Wesley, New York.
21. Kolb E.B. & Turner M., 1990,'The Early Universe', Addison-Wesley, New York
22. Kinney W., Melchiorri A., Riotto A., PRD in press, astro-ph/0007375 (2000)
23. Vilenkin A. & Shellard E.P.S., 1994, 'Cosmic Strings and other Topological Defects', Cambridge University Press
24. Bond. R., et al., The Quintessential CMB, Past and Future, Proc. of the CAPP2000 meeting, Verbier, astro-ph/0011379
25. J. R. Bond, G. Efstathiou and M. Tegmark, Mon. Not. R. Astron. Soc. **291**, L33 (1997) and references therein
26. Pospieszalski, M., IEEE-MTT-S Digest, 1369, (1992)
27. Pospieszalski, M., IEEE-MTT-S Digest, 1121, (1995)
28. Torbet E., et al., Ap.J., **521**, L79-L82 , (1999)
29. Miller A., et al., Ap.J., **524**, L1-L4 , (1999)
30. Mauskopf P., et al., Applied Optics, **36**, 765-771, (1997)
31. Mauskopf P., et al., Ap.J., **536**, L59-L62 (2000)
32. de Bernardis P., et al., Nature **404**, 955 (2000).

33. Hanany S., et al., Ap.J., **545**, L5-L9 (2000)
34. Delabrouille J. et al., MNRAS, **298**, 445-450 (2000)
35. de Bernardis P. & Masi S., 1998, in 'Fundamental parameters in cosmology', Proc. of the XXXIIIrd rencontres de Moriond (France), Trân Thanh Vân J., Giraoud-Héraud Y., Bouchet F., Damour T. & Mellier Y. eds., Editions Frontiéres, p.209
36. Piacentini F., et al., 2001, astro-ph/0105148
37. Crill B., et al., 2001, in preparation.
38. Borrill J. in 3K cosmology, Roma 1998, AIP CP 476, 277, (1999).
39. Prunet, S. et al., 2000, in "Energy densities in the Universe", Bartlett J., Dumarchez J. eds., Editions Frontieres, Paris - astro-ph/0006052
40. P. de Bernardis, et al., ŞFirst results from the BOOMERanG experiment Ť, Proc. of the CAPP2000 meeting, Verbier, July 2000, astro-ph/00011469
41. Masi S., et al. 2001, ApJ Letters in press, astro-ph/0101539
42. WOMBAT collaboration, 1998, see http://astron.berkeley.edu/wombat/foregrounds/radio.html.
43. Hivon E. et al., astro-ph/0105302
44. Padin et al., 2001, astro-ph/0012211
45. Dawson K.S. et al., 2000, astro-ph/0012151
46. http://astro.uchicago.edu/dasi/
47. http://astro.caltech.edu/ tjp/CBI/
48. http://www.mrao.cam.ac.uk/telescopes/vsa/index.html
49. http://www.archeops.org/
50. http://www.topweb.gsfc.nasa.gov/
51. P. de Bernardis, G. De Troia, L. Miglio "Calibration of balloon-borne CMB experiments" NEW ASTRONOMY REVIEWS, 43, 281-287, 1999.
52. Jaffe, A., et al., 2001, PRL, 86, 3475-3479
53. Knox L., Page L., Phys.Rev.Lett. 85 (2000) 1366-1369
54. Donahue M., Voit M, astro-ph/9907333
55. Bahcall N.A. et al, *Ap.J.* **541**, 1 (2000)
56. Blakesee J.P., et al., astro-ph/9910340
57. Juszkiewicz R., et al., *Science* **287**, 109 (2000)
58. Wittman et al., Nature, May 11 2000
59. Ostriker J.P., and Steinhardt P.J., *Nature*, **377**, 600 (1995)
60. Caldwell R.R., Dave R., Steinhardt P.J., *Phys. Rev. Lett.*, **80**, 1582 (1998)
61. Armendariz C., Mukhanov V., Steinhardt P.J., astro-ph/0004134 (2000)
62. Amendola L., astro-ph/0006300
63. Riess A.G. et al, *Ap.J.* **116**, 1009 (1998)
64. Perlmutter S. et al, *Ap.J.* **517**, 565 (1999)
65. Linde A.D., *Phys.Lett.* **108B**, 389, (1981)
66. Watson, G.S., astro-ph/0005003, (2000)
67. Tytler, D. et al. 2000, Physica Scripta submitted, astro-ph/0001318
68. Burles, S., Nollett, K.M. & Turner, M.S. 2000, astro-ph/0010171
69. Gawiser E., astro-ph/0105010 (in this book)
70. R. B. Barreiro, astro-ph/9907094
71. J.H.P.Wu et al., 2001, Tests for Gaussianity of the MAXIMA-1 CMB Map , astro-ph/0104248
72. Kaiser N., 1983, M.N.R.A.S., 101, 1169
73. Caderni N., et al., 1978, PRD, 17.
74. Lubin P., Smoot G., Ap.J., 245, 1
75. Lubin P., et al., Ap.J., 273, L51
76. Wollack et al., 1993, Ap.J., 419, L49
77. Netterfield B., et al., 1996, Ap.J., 445, L69.
78. Pisano G., New Astronomy Reviews, 2000, 43, 329-339
79. A. Melchiorri, N. Vittorio, astro-ph/9610029
80. Tegmark M. et al. Astrophys.J. 530 (2000) 133-165
81. Staggs, S. et al., astro-ph/9904062

AMiBA: Array for Microwave Background Anisotropy

K. Y. Lo *†, T. H. Chiueh †*, R. N. Martin *, Kin-Wang Ng *,
H. Liang **, Ue-li Pen ‡*, Chung-Pei Ma §*, M. Kesteven ††,
R. Sault †† and R. Subrahmanyan ††, W. Wilson ††, J. Peterson #

* Academia Sinica Institute of Astronomy & Astrophysics (ASIAA),
† Physics department, National Taiwan University (NTU),
** Physics department, University of Bristol,
‡ Canadian Institute of Theoretical Astrophysics (CITA),
§ Physics department, University of Pennsylvania,
†† Australia Telescope National Facility,
# Physics department, Carnegie-Mellon University

Abstract. As part of a 4-year Cosmology and Particle Astrophysics (CosPA) Research Excellence Initiative in Taiwan, AMiBA – a 19-element dual-channel 85-105 GHz interferometer array is being specifically built to search for high redshift clusters of galaxies via the Sunyaev-Zeldovich Effect (SZE). In addition, AMiBA will have full polarization capabilities, in order to probe the polarization properties of the Cosmic Microwave Background. AMiBA, to be sited on Mauna Kea in Hawaii or in Chile, will reach a sensitivity of $\sim 1\,\mathrm{mJy}$ or $9\mu K$ in 1 hour. The project involves extensive international scientific and technical collaborations. The construction of AMiBA is scheduled to start operating in early 2004.

I INTRODUCTION

The Academia Sinica Institute of Astronomy & Astrophysics and the National Taiwan University Physics department in Taipei, Taiwan are jointly developing experimental and theoretical Cosmology.

The idea of a ground-based millimeter-wave interferometric array (MINT) to study the primary anisotropy of the Cosmic Microwave Background (CMB) was first suggested by Lyman Page of Princeton University at a workshop on Cosmology held in Taiwan in December 1997. Subsequently, the ASIAA focussed on designing an instrument specifically for observing the Sunyaev-Zel'dovich Effect (SZE) to study clusters of galaxies and to search for high redshift clusters to take advantage of the distance-independent nature of the SZE. The resulting specifications consist of a high sensitivity 19-element 90 GHz interferometer that can achieve one arcminute resolution.

In response to the Research Excellence Initiative of the Ministry of Education and the National Science Council in Taiwan, a proposal on Cosmology and Particle Astrophysics (CosPA) was jointly submitted by the NTU Physics department, the ASIAA, the National Central University and the National Tsinghua University during the Spring of 1999, with the goals of developing cosmological research and Optical/Infrared Astronomy. CosPA consists of five inter-related projects: (1) construction and use of the Array for Microwave Background Anisotropy (AMiBA); (2) theoretical work in Cosmology; (3) access to large Optical/Infrared (OIR) facilities; (4) improving the infra-structure at the Lulin observatory site in the Jade Mountains in Taiwan; (5) a feasibility study of cold dark matter detection.

During December of the same year, the 4-year US$15M CosPA proposal was funded in full. In February 2000, a science and engineering specification meeting on AMiBA was held in order to define an instrument that will have unique scientific capabilities when completed in late 2003. The important conclusion was reached to achieve full polarization capability for AMiBA in order to probe the polarization properties of the CMB, in addition to observing the secondary anisotropy of the CMB.

II SCIENCE GOALS

There are three principal science goals for AMiBA: (1) a survey for high z clusters via the SZE; (2) in search of the missing baryons in large scale structures via the SZE; and (3) the polarization of the CMB.

A High z Cluster Survey

The formation history of clusters depends on Ω_m, the matter density and Λ, the cosmological constant (e.g. Barbosa et al. 1996), as well as σ_8, a measure of the initial fluctuation amplitude (Fan & Chiueh 2000). What is needed observationally to define the history is a survey of high z clusters over a sufficiently large area of the sky, so that the cosmic variance does not affect the results significantly. Because the SZE is distance independent, surveying for the SZ decrement in the CMB is very well suited to search for clusters at high z (Sunyaev & Zel'dovich 1972; Birkinshaw 1998).

To optimize the sensitivity of the survey, AMiBA is designed for maximal sensitivity, by maximizing the number of elements, adopting dual channel for the receivers and to have a 20 GHz bandwidth. The choice of the 90 GHz range is to minimize the foreground and point source confusion, and to minimize the scale of the array.

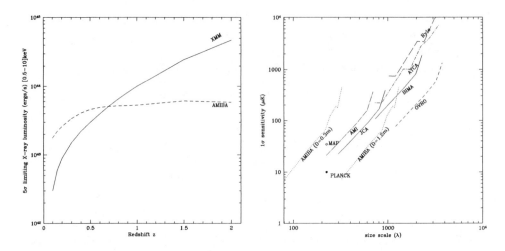

FIGURE 1. LEFT: A comparison of the limiting sensitivity of AMIBA (dotted curve) and XMM (Solid curve) for detecting clusters at various redshifts. AMIBA is more sensitive than XMM at detecting clusters beyond a redshift of $z \geq 0.7$. Limiting X-ray luminosity in ergs s^{-1} for a 5σ detection in 20 ksec for both telescopes for a typical cluster of core radius $r_c = 250$ kpc, shape parameter $\beta = 2/3$ and gas temperature $T_e = 8$ keV, assuming an average galactic neutral hydrogen column density of $N(H) = 6 \times 10^{20}$ cm^{-2} for the X-ray observation. (XMM sensitivities provided by Monique Arnaud) RIGHT: A comparison of brightness sensitivities (in 1hr) of various ground based aperture synthesis telescopes for the SZE either in operation or planning. The telescopes plotted is by no means a complete list of telescopes capable of detecting the SZE. The brightness sensitivity in μK over 1hr's observation is plotted against the uv-spacing (or size scale) expressed in terms of the number of wavelengths (multiplied by 2π gives l). A hexagonal close packed configuration is assumed for AMIBA, and a scaled version of a standard BIMA D-array configuration is assumed for the non-platform arrays BIMA, OVRO, AMI and JCA.

To detect clusters more massive than 2.5×10^{14} M$_\odot$ at $z \geq 0.7$, the AMiBA will be more sensitive compared to X-ray detection by satellites such as the XMM (fig. 1). A comparison of the sensitivity of AMiBA with other existing and planned instruments that can be applied to such a survey is also shown in fig. 1. We plan to survey for 50 square degrees of sky at a speed of 1.2 square degree per month.

B Super-clusters, Filaments - Missing Baryons ?

In addition to the hot gas in highly collapsed objects such as rich clusters, a weaker SZ effect should be produced when CMB photons scatter off warm baryons in lower-density environments such as filaments and inter-cluster regions in super-clusters. There is a growing consensus that a significant fraction (from 1/3 to 1/2) of the present-day baryons from big bang nucleosynthesis may be in the form of warm to hot gas with $10^5 < T < 10^7$ K which has mostly eluded detection thus far (cf. Fukugita et al 1998). Both gravitational and non-gravitational (such as supernova feedback) heating mechanisms have been discussed for this gas component (Dave et al. 2000; Cen & Ostriker 1999; Pen 1999; Wu et al. 1999). The gravitational case is due to shock heating as intergalactic gas flows along dark matter in filaments and the large scale structure. The dark matter in these regions are at moderate overdensities with a mean of 10 to 30, and the expected SZ distortion is of order 10 μK. Detection of signals from sensitive, non-targeted SZ surveys over large regions of the sky may be feasible, but it will be challenging to separate out the warm gas component from the primary anisotropy and the hot gas in clusters. We are currently carrying out more detailed studies to assess this exciting possibility.

C Polarization of the CMB

The CMB polarization contains a wealth of information about the early Universe. It provides a sensitive test of the reionization history as well as the presence of non-scalar metric perturbations, and improves the accuracy in determining the cosmological parameters (Zaldariaga et al 1997). The degree of polarization is of order a few μK at $l \sim 1000$ (Bond and Efstathiou 1984).

So far, the current upper limit on the CMB linear polarization is 16 μK (Netterfield et al 1995). A handful of new experiments, adopting low-noise receivers as well as long integration time per pixel, are underway or being planned (Staggs et al 2000). The MAP mission, launched in 2001, will be sensitive to the temperature-polarization correlation. The balloon-borne Boomerang and Maxima experiments have scheduled flights in 2001 to measure polarization with an angular resolution of $l < 800$ and pixel sensitivity of a few μK. The ESA space mission Planck will have sensitivity to CMB polarization, but the mission is not scheduled for launch until 2007.

An interferometer array is very attractive for CMB observations in that it directly measures the power spectrum. In addition, many systematic problems that are inherent in single-dish experiments, such as ground and near field atmospheric pickup, and spurious polarization signal, can be reduced or avoided in interferometry (cf. White et al 1999). Balloon-borne experiments are usually plaqued by pointing accuracy.

The AMiBA, with dual-channel receivers and 4 correlators, will be able to measure all four Stokes parameters simultanesously, so that the array will be much more sensitive to detecting CMB polarization than existing arrays, such as the Very Small Array (VSA), Degree Angular Scale Interferometer (DASI), Cosmic Background Imager (CBI). The AMiBA, when used with the 0.3 m apertures, will be sensitive to CMB polarization over the range $700 < l < 2000$. The S/N ratio in polarization in 24 hr is about 4 at $l \sim 700$, and about 2 at $l \sim 1150$.

Frequency (ν)	85-105 GHz
Bandwidth ($\delta\nu$)	20 GHz
Polarisations (N_p)	2-linear (XX,YY)
Receiver type	HEMT, cooled to 15 K
System Temperature	70 K
Number of Antennas (N)	19
Size of antennas (D)	2 sets; 1.2m, 0.3m
Number of Baselines	171
Primary beam	$11'$, $44'$ FWHM
Synthesized beam (full range)	$2' - 19'$
Fequency bands	8 chunks over 20 GHz
Flux Sensitivity	1.1 mJy, 17 mJy in 1 hr
Brightness sensitivity	$9\mu K$ in 1hr
Mount	Hexapod Mount: 3 rotational axes used
Platform	CFRP structure - 3 fold symmetry

III ORGANIZATION OF AMiBA

The AMiBA project is a collaboration principally between ASIAA/NTU and the Australia Telescope National Facility (ATNF), with important participation by scientists elsewhere. K. Y. Lo is the PI, with Robert Martin as the project manager, T. H. Chiueh as the project scientist, Paul Shaw (NTU/ASIAA) as the project administrator, Michael Kesteven (ATNF) as the system scientist. The other science and engineering team members include Ron Ekers, R. Sault, M. Sinclair, Ravi Subrahmanyan and W. Wilson from the ATNF, M. T. Chen, Y. J. Hwang, Kin-wang Ng from the ASIAA, T. D. Chiueh, T. Chu, and H. Wang from NTU, Haida Liang (Bristol), Chung Pei Ma (Penn/ASIAA), Ue-li Pen (CITA/ASIAA), Jeff Peterson (CMU), and John Payne (NRAO).

IV SOME AMiBA TECHNICAL DETAILS

The dual channel 85-105 GHz receievers will be based on the MIC InP HEMT amplifiers supplied by the National Radio Astronomy Observatory, with similar specifications built for the MAP project (Popieszalski 2000). The local oscillator system will be based on photonic devices with fiber-optic transmission lines, which will minimize component counts and make the distribution more stable. The 20 GHz bandwidth poses considerable technical challenges that are being met by a 17-lag analog correlator. There will be four correlators built to provide full polarization capabilities for AMiBA. The 19 apertures will be mounted on one 6 to 8 meter platform supported on a hexapod mount.

As the project is also funded to develop the research capabilities of the universities in Taiwan, there are parallel development projects on the InP and GaAs MMICs that are aimed at satisfying the requirements of the AMiBA. However, these development efforts are not placed on the critical paths of the AMiBA construction.

V SCHEDULE OF EVENTS

A preliminary design review meeting was held in July 2000 in Taipei, where the decision was made to build a prototype by September 2001 to test the basic concepts and specifications. After the proving of concepts, the full system will be built to be completed in late 2003 and to start observations in early 2004.

To further review the science goals and to keep up with the latest development in this rapidly evolving field, an international workshop in Taiwan on AMiBA-related science goals is being planned for June 2001.

REFERENCES

1. Barbosa, D., Bartlett, J. G., Blanchard, A. and Oukbir, J., *A&A* **314**, 13 (1996).
2. Birkinshaw, M., *Phys. Rept.* **310**, 97 (1999).
3. Bond, J. R., Efstathiou, G., *ApJ* **285**, L45 (1984).
4. Cen, R., Ostriker, J., *ApJL* **519**, L109 (1999).
5. Dave, et al., *ApJ (submitted), astro-ph/0007217*, (2000).
6. Fan, Z., Chiueh, T., *ApJ (in press), astro-ph/0011452*, (2000).
7. Fukugita, M., Hogan, C. J., Peebles, P. J. E. 1998, *ApJ* **53**, 518 (1998).
8. Netterfield, C. B. et al., *ApJL* **474**, L69 (1995).
9. Pen, U-L, *ApJL* **510**, 1L (1999).
10. Popieszalski, M. et al., *IEEE MTT-S Symp. Digest (in press)*, (2000).
11. Staggs, S. T., Gundersen, J. O., Church, S. E., *astro-ph/9904062*, (1999).
12. Sunyaev, R.A., Zel'dovich, Ya. B., *Comm. Astrophys. Sp. Phys.* **4**, 173 (1972).
13. White, M., Carlstrom, J. E., Dragovan, M., Holzapfel, W. L., *ApJ* **514**, 12 (1999).
14. Zaldarriaga, M., Spergel, D. N., Seljak, U., *ApJ* **488**, 1 (1997).

First Results from the CBI

B. S. Mason, T.J. Pearson, A.C.S. Readhead, M. Shepherd, J. Sievers, P. Udomprasert, J.K. Cartwright, S. Padin

105-24 Caltech, Pasadena CA 91125

Abstract.
The Cosmic Background Imager (CBI) is an instrument designed to measure intrinsic anisotropies in the cosmic microwave background (CMB) on angular scales from about 3 arc minutes to one degree (spherical harmonics of $\ell \sim 4250$ to $\ell \sim 400$). The CBI is a 13 element interferometer mounted on a 6 meter platform operating in ten 1-GHz frequency bands from 26 to 36 GHz. We present a review of the capabilities of the instrument and a discussion of observations which have been taken over the past year from the Atacama desert of Chile. We also present first results from the CBI which show a strong cutoff in the power spectrum between $\ell = 600$ and $\ell = 1200$ which is consistent with the photon-diffusive damping predicted by most models of structure formation in the early universe. We discuss future topics which the CBI will address.

INTRODUCTION

Anisotropies in the Cosmic Microwave Background (CMB) contain a wealth of information about fundamental cosmological parameters [1], as well as providing a direct link to theories of high-energy physics [2]. The most straightforward models for anisotropies in the CMB (see, *e.g.*,[3]) feature two key physical scales: a length scale associated with acoustic oscillations of density fluctuations in the primordial plasma, and an exponential damping of these fluctuations caused by photon diffusion. Both of these physical scales depend upon the values of the cosmological parameters at and before the time of recombination, and the relation of these scales to observable angular scales on the sky is determined by the angular diameter distance between the present and $z = 1100$. If this basic picture is correct, the observable anisotropies are capable of strongly constraining the parameters of cosmology.

Inspired by the possibility of such accurate determinations of classical cosmological parameters, many experiments have sought and detected the most prominent large-scale CMB anisotropies due to the first Doppler peak (*e.g.*, [4, 5, 6, 7]). In these proceedings we report CBI measurements of intrinsic anisotropies on scales from $\ell = 600$ to $\ell = 1200$. These measurements confirm the existence of a strong cutoff in the power spectrum and provide an independent constraint on the total energy density of the universe, Ω_{tot}.

FIGURE 1. The CBI in the configuration in which the results reported in these proceedings were obtained.

THE INSTRUMENT

The CBI is an interferometric array of 13 0.9-meter diameter antennas mounted on a 6-meter steerable platform, and operating in 10 1-GHz bands between 26 and 36 GHz ($\sim$ 1 cm). Configurations available to the CBI yield synthesized beamwidths ranging from $4'$ to $15'$ (FWHM). Interferometry confers the significant advantage, relative to total power or beam-switched single-dish methods, of providing a *direct* measurement of C_ℓ on a scale determined by the baseline length. The range of baselines available to the CBI correspond to $400 < \ell < 4250$. By changing the array configuration of the CBI, the instrument's sensitivity can be optimized for varying ranges of ℓ. The primary beam width $44'$ (FWHM at 30 GHz) implies a resolution $\delta\ell \sim 420$ (FWHM); this can be significantly improved by mosaicked observations. One configuration of the CBI is shown in Figure 1.

The key design challenges in the project were eliminating cross-talk in a compact array and developing a wide-band correlator. Receiver noise scattering between adjacent antennas (cross-talk) causes false signals at the correlator output and this could limit the sensitivity of the instrument. We developed a shielded Cassegrain antenna with low scattering to reduce the cross-talk [8]. The antennas have machined, cast aluminum primaries which sit at the bottom of deep cylindrical shields. The upper rims of the shields are rolled with a radius of a few wavelengths to reduce scattering from the shield itself. The secondaries are made of carbon fiber epoxy, to minimize weight, and supported on transparent polystyrene feed legs. Cross-talk between the antennas is < -110 dB in any CBI band.

The antennas are mounted on a rigid tracking platform supported by an altazimuth mount that is fully steerable to elevations $> 42°$. The antenna platform can be rotated about the optical axis. In normal observations, the platform tracks the parallactic angle so that observations are made at fixed (u,v) points: i.e., the baseline orientations are fixed relative to the sky. Additional discrete steps in the orientation of the platform are used to change the baseline orientations and thus sample more (u, v) points.

The correlator [9] is an analog filter bank correlator with ten 1-GHz bands. A fast phase-switching scheme, in which the receiver local oscillators are inverted in Walsh function cycles, is used to reject cross-talk and low-frequency pickup in the signal processing system. The system is calibrated by nightly observations of celestial sources; CMB data are referenced onto this scale by comparison with an internal source of correlated noise which is injected before the first stage of each receiver. Variations in the calibration of the CBI are at the 1% and 1 deg level. Further discussion of the instrument can be found in [10].

Construction of the CBI was begun in August 1995 in Pasadena, CA on the Caltech campus, and completed in January 1999. After a period of test observations in Pasadena, the telescope was disassembled and shipped in August 1999 to its site high in the Chilean Andes. This site, at an altitude of 5000 meters, was chosen in order that our sensitivity not be limited by atmospheric water vapor emissions. First light in Chile was achieved November 1, 1999, and routine observations have been taken from January 2000 to the present.

OBSERVING STRATEGY AND FIRST RESULTS

Our observing strategy is dictated by the fact that, on the shortest CBI baselines, the ground produces significant correlated signals. To remove these, pairs of (LEAD/TRAIL) fields are observed over identical ranges in azimuth and elevation. The difference between these data cancels ground-based signals. This differencing also controls other possible systematics due, *e.g.*, to correlator offsets or antenna cross-talk. From January through April 2000, observations of two pairs of such fields (C0844−0310 and C1442−0350) were taken in a test configuration providing a maximally uniform distribution of baseline lengths and easy access to all receivers. In all ~ 160 hours of integration were obtained. The data are calibrated with reference to Jupiter at 32 GHz, using $T_J = 152 \pm 5\,\text{K}$ [11]. Since Jupiter does not have a simply thermal spectrum, this calibration is bootstrapped to other frequencies using TauA, which has a known power-law spectrum between 26 and 36 GHz.

The difference image of the C0844 field is shown in Figure 2. The observed signal is confined to the telescope main beam, indicating that it is of celestial origin. We have searched for contaminating signals by dividing the data by epoch and by zenith angle and differencing the resulting datasets, yielding a doubly-differenced dataset which in the absence of contaminating signals should be consistent with thermal noise. The zenith angle tests showed no measureable excess signal; from this we conclude that our LEAD/TRAIL differencing leaves no significant residual ground signal in the data. There is a slight noise excess in the data when divided by epoch which amounts to a $< 1.5\%$ contamination in C_ℓ. This could be due to slight changes in the instrument calibration over long periods of time, but at present this excess is not understood in detail.

In order to remove discrete radio sources, the dominant foreground to CMB measurements at our frequency and resolution, dedicated observations of sources selected from

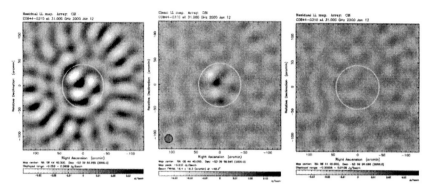

FIGURE 2. Differenced image of the 08^h field observed on the 100 and 104 cm CBI baselines (left), an image with the point spread function deconvolved (center), and the residuals after subtraction of the deconvolved signal (right). The large circles in the map centers indicate the 5% power level of the telescope primary beam (2 FWHM $\sim 88'$); the small circle in the lower left of the center map indicates the synthesized beamwidth of $\sim 16'$ FWHM

the NVSS[12] are conducted at 30 GHz with the OVRO 40-meter telescope. Sources detected by the 40-meter are subtracted directly from the CBI data. The bright source subtraction affects our results on these scales by less than 2%, and the longer baseline data show no evidence for bright missed sources.

Our data give a robust detection of CMB anisotropy on scales from $\ell = 300$ to $\ell = 1500$. A maximum-Likelihood analysis of the visibility data gives flat bandpowers ($\delta T_{band} \equiv [\ell(\ell+1)\overline{C}_\ell/(2\pi)]^{1/2} \times T_{cmb}$) of $\delta T_{band} = 58.7^{+7.7}_{-6.3} \mu K$ for $\ell = 603^{+180}_{-166}$ and $\delta T_{band} = 29.7^{+4.8}_{-4.2} \mu K$ for $\ell = 1190^{+261}_{-224}$. These results are shown in Figure 3.

Currently only weak constraints on the spectral index are available and only in the lower ℓ bin. In this range we find a temperature spectral index $\beta = 0.0 \pm 0.4 (1\sigma)$, where $\beta = 0$ corresponds the a thermal blackbody spectrum. For the realistic case that the foreground powers $[\ell(\ell+1)C_\ell]$ are falling as $\ell^{-0.2}$, then at the 2σ limit a free-free foreground would contribute 15% to the observed power in the first bin, and a synchrotron foreground would contribute 11%. Our data are also consistent with no foreground contribution to the observed power level. Furthermore, the C0844−0310 and C1442−0350 fields show different levels of emission in our foreground templates (IRAS and synchrotron maps), but the power levels observed by the CBI at 32 GHz in these two fields are consistent.

A cosmological maximum-Likelihood analysis of these data yields $\Omega_{tot} \leq 0.4$ or $\Omega_{tot} \geq 0.7$ (90% confidence).

More discussion of this analysis can be found in [13].

FURTHER SCIENCE WITH THE CBI

Observations of two $5° \times 3°$ mosaic fields and one $5° \times 6°$ field have been completed and the analysis of these data is in progress. When complete, these results will improve our

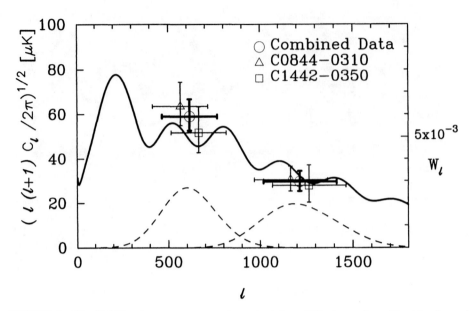

FIGURE 3. The CMBR anisotropy spectrum determined from CBI observations. The triangles and squares show results on the 08^h and 14^h differenced fields; the circles show the results of a joint maximum likelihood analysis of both differenced fields. The individual 08^h and 14^h field results are offset in l for clarity. The window functions for each bin are shown as dashed lines. The solid curve represents a flat universe with $H_0 = 75$ km s^{-1} Mpc^{-1}, $\Omega_b h^2 = 0.019$, and $\Omega_{cdm} = 0.2$.

resolution in ℓ by a factor of 3 – 4. We have also acquired data to improve the spectral index constraints on the signal out to $\ell \sim 1200$.

The CBI has one cross-polarized antenna. This will enable stringent limits to be placed on the CMB polarization in the vicinity of the second Doppler peak in the polarization power spectrum. Polarization data were acquired on the CBI fields from January through December 2000 and the analysis of these data is in progress.

The range of angular scales available to the CBI is also well-suited to measuring the Sunyaev-Zeldovich Effect (SZE) in nearby galaxy clusters. A campaign to determine H_0 from observations of the SZE in a sample of 20 $z < 0.1$ clusters is underway with the CBI. This campaign has the feature of selecting targets from an orientation-unbiased sample [14]. The large sample size is important for reducing the effects of intrinsic CMB anisotropy, and for understanding possible X-ray modelling systematics associated with clusters at a range of dynamical states. Previous results from a similar program with the OVRO 5-meter telescope have been presented in [15].

ACKNOWLEDGMENTS

We are grateful for the contributions to this project of our collaborators: Russ Keeney, Steve Miller, Walter Schaal, and John Yamasaki (Caltech); John Carlstrom and Erik

Leitch (University of Chicago); Bill Holzapfel (University of California, Berkeley); Steven Myers (National Radio Astronomy Observatory); Marshall Joy (NASA's Marshall Space Flight Center); Angel Otarola (European Southern Observatory); and Leonardo Bronfman, Jorge May, Simon Casassus, and Pablo Altamirano (University of Chile). The CBI project has been supported by the National Science Foundation under grants AST-9413935 and AST-9802989, and we are also grateful for the generous support of Maxine and Ronald Linde, Cecil and Sally Drinkward, and our colleagues at the California Institute of Technology, especially the Provost, the President, and the Chairman of the Division of Physics, Mathematics, and Astronomy. We are grateful to CONICYT for permission to operate the CBI in the Chajnantor Scientific Preserve in Chile. JS and PSU acknowledge support from National Science Foundation Graduate Student Fellowships.

REFERENCES

1. White, M., Scott, D., and Silk, J., *Ann. Rev. A. & A.*, **32**, 319–370 (1994).
2. Kamionkowski, M., and Kosowski, A., *Ann. Rev. Nucl. Part. Sci.*, **49**, 77–123 (1999).
3. Hu, W., Sugiyama, N., and Silk, J., *Nature*, **386**, 37–43 (1997).
4. Miller, A. D., Caldwell, R. R., Herbig, T., Page, L., Torbet, E., Tran, H., Devlin, M., and Puchalla, J., *ApJ*, **524**, L1–L4 (1999).
5. Leitch, E. M., Readhead, A. C. S., Pearson, T. J., Myers, S. T., and Gulkis, S., *ApJ*, **532**, 37–56 (2000).
6. de Bernardis, P., et al., *Nature*, **404**, 955–959 (2000).
7. Hanany, S., et al., *ApJ*, **545**, L5–L9 (2000).
8. Padin, S., Cartwright, J. K., and Joy, M., *IEEE Trans. Antennas & Propagation*, **48**, 836–838 (2000).
9. Padin, S., Cartwright, J. K., Shepherd, M. C., Yamasaki, J. K., and Holzapfel, W. L., *IEEE Trans. Instrum. Meas.* (submitted).
10. Padin, S., Shepherd, M. C., Cartwright, J. K., Readhead, A. C. S., Pearson, T. J., Schaal, W. L., Yamasaki, J. K., Sievers, J., Keeney, R. G., Udomprasert, P. S., Mason, B. S., Holzapfel, W. L., Joy, M., Myers, S. T., Carlstrom, J. E., and Otarola, A. (in preparation).
11. Mason, B. S., Leitch, E. M., Myers, S. T., Cartwright, J. K., and Readhead, A. C. S., *AJ*, **118**, 2908–2918 (1999).
12. Condon, J. J., Cotton, W. D., Greisen, E. W., Yin, Q. F., Perley, R. A., Taylor, G. B., and Broderick, J. J., *AJ*, **115**, 1693–1716 (1998).
13. Padin, S., Cartwright, J. K., Mason, B. S., Pearson, T. J., Readhead, A. C. S., Shepherd, M. C., Sievers, J., Udomprasert, P. S., Holzapfel, W. L., Myers, S. T., Carlstrom, J. E., Leitch, E. M., Joy, M., Bronfman, L., and May, J., *ApJL* (in press).
14. Udomprasert, P. S., Mason, B. S., and Readhead, A. C. S., "The Sunyaev-Zel'dovich Effect with the Cosmic Background Imager", in *Constructing the Universe with Clusters of Galaxies*, edited by F. Durret and D. Gerbal, in press.
15. Mason, B. S., Myers, S. T., and Readhead, A. C. S., *ApJL* (in press).

Galactic Synchrotron versus CMB Polarized Signal in the Radiowavelength Region

M. Tucci[*], E. Carretti[†], S. Cecchini[†], S. Cortiglioni[†], R. Fabbri[**] and E. Pierpaoli[‡]

[*]*Dipartimento di Fisica G. Occhialini, Università di Milano, Bicocca, Italy*
[†]*Istituto Te.S.R.E.-C.N.R., Bologna, Italy*
[**]*Dipartimento di Fisica, Università di Firenze, Italy*
[‡]*Department of Physics and Astronomy, University of British Columbia, Vancouver, Canada*

Abstract. Galactic synchrotron will be an important contaminant for forthcoming experiments on CMB polarisation. We investigate the angular power spectra of polarized Galactic synchrotron in several surveys (Parkes at 2.4 GHz, Effelsberg at 2.7 GHz and 1.4 GHz), showing that partial sky coverage analysis may lead to largely different results. Reasonable extrapolations to tens of GHz allow us to find a region in the (ν, l) plane where the CMB polarized signal is expected to prevail largely.

INTRODUCTION

The measurement of the cosmic microwave background polarisation (CMBP) is a challenge for a large number of planned experiments, both from ground and from space (see Staggs et al. 1999 for a review). The increasing interest on CMBP is due to its ability to substantially improve the accuracy with which some cosmological parameters can be measured, breaking the degeneracy between certain parameter combinations.

One of the main difficulties in CMBP measurements is the presence of polarized foreground components. In particular, Galactic synchrotron emission is expected to be the dominant component of the linearly polarized sky emission in a wide frequency range up to several tens of GHz. In this work we compute the intensity and polarisation angular power spectra of Galactic synchrotron emission. This kind of analysis is becoming relatively common also for foregrounds, because it allows us to quantify the level of contamination in the CMB observations at different angular scales, and thus to forecast the dominant contribution at a given range of spherical-harmonic index l.

The intensity power spectrum of Galactic synchrotron has been computed on degree scales ($l \lesssim 200$) and has been found to be reasonably approximated by a power law, $C_{Il} \propto l^{-\alpha_I}$, with $\alpha_I = 2 \div 3$ (for a summary of the works in literature see Tucci et al. 2000). The situation about the polarized component is much less clear, due to the small number of available surveys. The only one covering a large area of the sky (about 40%) at latitudes out of the Galactic plane is provided by Brouw & Spoelstra (1976); nevertheless, the analysis of such sparse and undersampled data is problematic. However, in the next years we expect a great improvement in our knowledge on the synchrotron emission: for example, the SPOrt experiment (Cortiglioni et al. 1999) will be able to provide a nearly full–sky map of polarized synchrotron at 22 and 32 GHz. The first estimation of the synchrotron polarisation spectrum has been derived by Tucci et al. (2000) from the 2.4 GHz Parkes survey (Duncan et al. 1995, 1997, hereafter D97), over a strip 127° wide of the southern Galactic plane. They found that the spectra can be well fitted by power laws with slope $\alpha_{E,B} \sim 1.4 \div 1.5$. These values, if compared to theoretical models of the CMB polarisation spectra, indicate a different behaviour of the synchrotron emission at small scales with respect to the CMB. In this contribution, we extend the previous analysis to the northern Galactic plane, by using the data from Effelsberg at 2.7 GHz (Duncan et al. 1999, hereafter D99) and to intermediate latitudes using the data from Effelsberg at 1.4 GHz (Uyaniker et al. 1999, hereafter U99).

INTENSITY AND POLARISATION POWER SPECTRA

In our analysis we use three high resolution surveys, that allow us to investigate the synchrotron angular spectra at small scales: the first two cover the Galactic plane in a longitude range between $\ell = [238°, 5°]$ (D97) and $\ell = [5°, 74°]$ (D99) with a declination range of 10°. The FWHM resolutions are 10.4' and 5.1' respectively, sufficient to achieve angular scales up to $l \sim 10^3$. The nominal rms noise in D97 is 8 mK for total power and 5.3 mK for polarisation (5.3 and 2.9 mK in some more sensitive areas), while in D99 rms is 9 mK. The third one is the U99 survey, consisting of five different regions at intermediate latitudes: ($45° \leq \ell \leq 55°, 4° \leq b \leq 20°$), ($65° \leq \ell \leq 95°, 5° \leq b \leq 15°$), ($70° \leq \ell \leq 100°, -15° \leq b \leq 5°$), ($140° \leq \ell \leq 153°, 3.5° \leq b \leq 10°$), ($190° \leq \ell \leq 210°, 3.8° \leq b \leq 15°$). The rms noise is about 15 mK for total intensity and about 8 mK for linear polarisation; the angular resolution is 9.35'.

Because of the limited sky coverage, power spectra can be suitably obtained from Fourier analysis instead of the standard spherical harmonics approach (Seljak 1997). This technique has been already applied by Tucci et al. (2000) on D97 data (the results are shown in Fig. 1), and technical details can be found there. We extract from each survey square patches of $10° \times 10°$ and for each of them we perform Fourier analysis of the Stokes parameters I, Q and U. The estimators for the power spectrum of the total intensity can be derived by

$$C_{Il} = \left\{ \frac{\Omega}{N_l} \sum_l I(\mathbf{l})I^*(\mathbf{l}) - w_I^{-1} \right\} b^{-2}(l), \qquad (1)$$

and for the E– and B–modes of polarisation by

$$C_{El} = \left\{ \frac{\Omega}{N_l} \sum_l |Q(\mathbf{l})\cos(2\phi_l) + U(\mathbf{l})\sin(2\phi_l)|^2 - w_P^{-1} \right\} b^{-2}(l), \qquad (2)$$

$$C_{Bl} = \left\{ \frac{\Omega}{N_l} \sum_l |-Q(\mathbf{l})\sin(2\phi_l) + U(\mathbf{l})\cos(2\phi_l)|^2 - w_P^{-1} \right\} b^{-2}(l), \qquad (3)$$

where $b(l)$ is the window function of the instrument and $w_{I,P}^{-1}$ represents the noise contribution.

Fig. 1 shows our results for C_{Il} (dotted lines) and for C_{El} (solid lines; results for B–mode are similar to E–mode) from the D97 and D99 surveys. We observe a large variation in the normalization of the intensity spectra, more than three orders of magnitude. The same feature is not found in polarisation spectra. This can be explained by observing how the total power emission fastly decreases when moving away from the Galactic center. In spite of this the polarized component remains much more uniform by changing along the longitude and the latitude (see Fig. 4, 6 in Duncan et al. 1995 and Fig. 9 in D97). In particular, in the polarized emission D97 notices a "background component", nearly constant over all the survey at about 20 mK, indipendent of the latitude.

Most of the curves in Fig. 1 are reasonably approximated by a power law, $C_{Xl} = A_X l^{-\alpha_X}$ ($X = I, E, B$). In Table 1 we report the results of spectral fits in the range $100 \leq l \leq 800$ for all submaps; we also report the values of the average slope, $\langle \alpha \rangle$, as well as the fit parameters for the average spectrum, $\langle C_l \rangle$. The distribution of the index α highlights some differences between total intensity and polarisation spectra: the values of α_I range between 0.4 and 2.2, while the slope of the polarisation spectra remains relatively close to the mean value 1.4. Moreover, in total intensity low–emission regions show very flat spectra, while no meaningful differences are found between high– and low–polarized emission regions.

Fig. 2 and Table 2 report the results obtained from 5 intermediate–latitude patches of U99. The intensity spectra are found to be extremely flat ($\alpha_I < 1$) and low, indicating that the diffuse emission drops just out the Galactic plane. On the contrary, the polarisation spectra do not show a decrease in amplitude with respect to the other two surveys; this means that the polarized background observed in D97 and D99 extends at least up to $\ell \sim 15°$. In U99 we observe large differences in the slope of polarisation spectra from region to region: there are three patches with $\alpha_{E,B} > 2$ and two with $\alpha_{E,B} \sim 1$. However, some of these regions cannot be considered as typical; for example, the area $140° \leq \ell \leq 153°, 4° \leq b \leq 10.4°$ lies within the so called "fan region", and the two regions centered at $\ell = 80°$ show rather complex structures.

DISCUSSION

Our conclusions regard the angular spectrum slope of the synchrotron emission up to $l \sim 10^3$. The total intensity spectrum is found to have a very moderate slope, in contrast to the values reported in literature at large scales. This is easily interpreted as due to Galactic plane sources. On the other hand, the results on polarisation spectra are not so clear: along the Galactic plane the spectrum slope is about $1.4 \div 1.5$, with no systematic differences between high and low emission submaps. We previously suggested that this slope might describe the global properties of synchrotron emission. However, in the U99 survey the intermediate–latitude polarisation slope ranges from 1 to ~ 2.5, proving that at small scales significant changes on the spectra slope can occur in different sky regions. The knowledge of such local behaviours will be necessary in order to separate the foregrounds from the cosmological signal using high resolution data.

Finally, in order to evaluate the impact of synchrotron on CMB studies, we extrapolate our results on the polarisation to the "cosmological" frequencies. Such extrapolation is

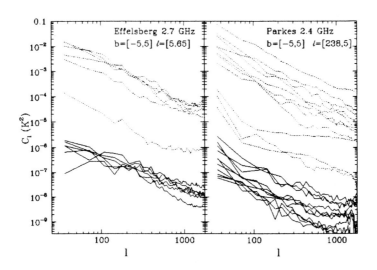

FIGURE 1. The angular power spectra for total intensity (dotted lines) and polarisation (solid lines) from D97 (left box) and D99 (right box) surveys

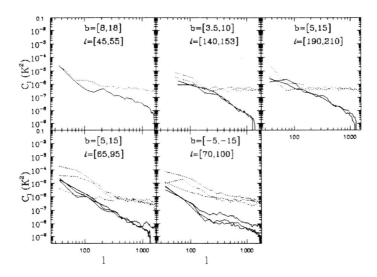

FIGURE 2. The angular power spectra for total intensity (dotted lines) and polarisation (solid lines) from the 5 areas of Effelsberg 1.4 GHz survey (U99)

TABLE 1. Power spectrum parameters from D97 ($\ell = [360°, 250°]$) and D99 ($\ell = [10°, 60°]$); the first column is the Galactic longitude of the submap center

ℓ (deg.)	$A_T(K^2)$	α_T	$A_E(K^2)$	α_E	α_B
360	19.5	1.65 ±0.12	0.63×10^{-4}	1.28 ±0.11	1.33 ±0.12
350	3.5	1.64 ±0.11	0.92×10^{-6}	$0.85^{+0.13}_{-0.06}$	1.12 ±0.12
340	17.5	2.00 ±0.11	0.48×10^{-4}	1.31 ±0.11	1.52 ±0.12
330	0.21	1.46 ±0.09	0.22×10^{-3}	1.57 ±0.12	$1.74^{+0.15}_{-0.09}$
320	0.48	1.76 ±0.11	0.40×10^{-4}	1.49 ±0.11	1.40 ±0.12
310	5.1	1.89 ±0.12	0.28×10^{-5}	1.11 ±0.11	$0.96^{+0.12}_{-0.09}$
300	0.07	$1.22^{+0.12}_{-0.08}$	0.15×10^{-4}	1.56 ±0.12	1.88 ±0.12
290	1.8	1.43 ±0.11	0.32×10^{-3}	2.04 ±0.12	1.90 ±0.12
280	0.48×10^{-4}	$0.44^{+0.08}_{-0.12}$	0.44×10^{-4}	1.56 ±0.12	1.78 ±0.11
270	0.02	1.02 ±0.11	2.6×10^{-3}	1.79 ±0.12	1.25 ±0.12
260	0.20×10^{-2}	1.48 ±0.12	0.96×10^{-5}	1.48 ±0.12	1.32 ±0.10
250	0.14×10^{-3}	$0.96^{+0.12}_{-0.08}$	0.13×10^{-4}	1.58 ±0.11	1.45 ±0.11
$\langle \alpha \rangle$		1.37 ±0.44		1.44 ±0.30	1.46 ±0.29
$\langle C_l \rangle$	2.2	1.60 ±0.13	0.12×10^{-3}	1.53 ±0.11	1.43 ±0.12
10	3.40	1.60 ±0.12	0.21×10^{-4}	0.92 ±0.10	$1.13^{+0.08}_{-0.12}$
20	124.4	2.18 ±0.07	0.39×10^{-3}	$1.44^{+0.13}_{-0.07}$	$1.52^{+0.07}_{-0.15}$
30	11.5	1.79 ±0.10	0.88×10^{-3}	1.50 ±0.10	1.63 ±0.10
40	0.10	1.21 ±0.12	0.25×10^{-3}	$1.36^{+0.11}_{-0.06}$	$1.67^{+0.05}_{-0.10}$
50	0.31	$1.28^{+0.21}_{-0.15}$	0.80×10^{-4}	1.24 ±0.09	1.52 ±0.13
60	0.65×10^{-3}	$1.00^{+0.12}_{-0.15}$	0.71×10^{-3}	1.68 ±0.07	1.82 ±0.10
$\langle \alpha \rangle$		1.71 ±0.43		1.40 ±0.23	1.57 ±0.19
$\langle C_l \rangle$	10.3	1.82 ±0.11	0.31×10^{-3}	1.39 ±0.11	1.55 ±0.12

TABLE 2. Power spectrum parameters for the U99 survey; first and second columns are the longitude and the latitude of the submap center, respectively.

ℓ (deg.)	b (deg.)	$A_T(K^2)$	α_T	$A_E(K^2)$	α_E	α_B
50	13	0.53×10^{-5}	$0.37^{+0.13}_{-0.10}$	0.16×10^{-3}	1.22 ±0.08	1.19 ±0.08
143	7	0.35×10^{-6}	$0.^{+0.02}$	0.22	$2.55^{+0.14}_{-0.17}$	$2.70^{+0.03}_{-0.25}$
150	7	0.50×10^{-6}	$0.^{+0.02}$	0.80×10^{-1}	$2.38^{+0.06}_{-0.17}$	$2.10^{+0.23}_{-0.16}$
195	10	0.45×10^{-6}	$0.^{+0.02}$	0.35×10^{-1}	2.28 ±0.08	2.32 ±0.08
205	10	0.72×10^{-6}	$0.^{+0.11}$	0.70×10^{-2}	$1.99^{+0.08}_{-0.12}$	1.98 ±0.10
70	10	0.99×10^{-4}	0.82 ±0.15	0.73×10^{-1}	2.41 ±0.11	2.39 ±0.12
80	10	0.89×10^{-3}	1.13 ±0.13	0.90×10^{-3}	1.71 ±0.15	1.79 ±0.19
90	10	0.6×10^{-6}	$0.^{+0.13}$	0.11	$2.48^{+0.02}_{-0.12}$	2.23 ±0.13
75	-10	0.16×10^{-5}	0.16 ±0.13	0.23×10^{-5}	0.87 ±0.08	1.50 ±0.12
85	-10	0.12×10^{-4}	0.87 ±0.12	0.36×10^{-5}	1.00 ±0.11	1.17 ±0.12
95	-10	0.20×10^{-4}	0.81 ±0.11	0.47×10^{-4}	1.20 ±0.09	$0.63^{+0.07}_{-0.02}$

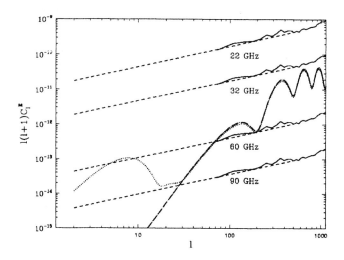

FIGURE 3. The angular spectra of polarized synchrotron from D97 extrapolated to 22, 32, 60, 90 GHz (solid lines) with the linear fits (dashed lines), compared to the CMB E-spectrum for a standard CDM model (dot–dashed line) and for a CDM with a secondary ionization optical depth $\tau_{ion} = 0.1$ (dotted line).

reliable if the effects of Faraday depolarization are negligible, as suggested by data on the rotation measure (Spoelstra 1984). Fig. 3 shows our extrapolations to 22, 32, 60 and 90 GHz, taking the average synchrotron spectrum from the D97 survey ($\alpha_E = 1.53, A_E = 0.12 \times 10^{-3}$ K^2) and a spectral index of 3. Although a slope of 1.5 cannot be extended to the entire sky, it can however be considered a lower limit, since small–scale structures are expected to be weaker at high latitudes than along the Galactic plane. The Fig. 3 shows that at 60 GHz CMB prevails for $l > 100$ (in spite of the above Galactic plane normalization). Higher values for α_E would increase again the differences between the two components and the CMB signal would prevail even more strongly.

REFERENCES

1. Brouw, W.N., & Spoelstra, T.A., 1976, A&AS, 26, 129.
2. Cortiglioni, S., et al., 1999, in: 3 K Cosmology, Maiani, L., Melchiorri, F. & Vittorio, N. (Eds.), AIP Conference Proc. 476, 186; and SPOrt home page: http://tonno.tesre.bo.cnr.it/~sport/.
3. Duncan, A.R., Haynes, R.F., Jones, K.L., & Stewart, R.T., 1997, MNRAS, 291, 279 (D97).
4. Duncan, A.R., Stewart, R.T., Haynes, R.F., & Jones, K.L., 1995, MNRAS, 277, 36.
5. Duncan, A.R., Reich, P., Reich, W. & Fürst, E., 1999, A&A, 350, 447 (D99).
6. Seljak, U., 1997, ApJ, 482, 6
7. Spoelstra, T.A., 1984, A&A, 135, 238
8. Staggs, S.T., Gundersen, J.O., Church, S.E., 1999, in: Microwave Foregrounds, de Olivera-Costa, A. & Tegmark, M. (eds.), ASP Conf. Ser. Vol. 181, 299
9. Tucci, M., Carretti, E., Cecchini, S., Fabbri, R., Orsini, M., & Pierpaoli, E., 2000, NewA, 5, 181.
10. Uyaniker, B., et al., 1999, A&AS, 138, 31 (U99).

Revisiting kinetic Sunyaev-Zel'dovich effect

Pengjie Zhang* and Ue-Li Pen[†]

*Astronomy Department, Univ. of Toronto, Toronto, Canada, M5S 3H8
[†]Canadian Institute for Theoretical Astrophysics, Cananda, M5S 3H8

Abstract. We propose an analytical method to calculate the vorticity power spectrum of the intergalactic gas peculiar momentum in the non-linear regime and thus to calculate the kinetic Sunyaev Zel'dovich effect. Our results show that the KSZ power spectrum peaks around $l = 3000$ with $\Delta T \simeq 1.5 \mu K$.

INTRODUCTION

Upcoming CMB experiments such as AMIBA [1] and Planck [2] will be able to measure the Sunyaev Zel'dovich effect [3, 4] (thermal SZ and kinetic SZ effect, respectively) to the arcminute scale. In order to extract as much information as possible from such experiments, we need to understand such effects from the theoretical point of view. Recently, we have proposed an analytical method to model the thermal SZ effect and presented a variation method to extract the redshift information of the intergalactic medium [5]. Here, we propose an analytical method for the kinetic SZ effect. Recent simulations [6, 7] suggest that equi-partition of the intergalactic gas is reached in the momentum space in the highly non-linear regime. We adopt this result and the stable clustering approximation to obtain the analytical formula of the gas momentum power spectrum. Combined with a second-order calculation of the momentum vorticity in the linear and mildly non-linear regime, we obtain the vorticity power spectrum of the gas momentum in full range and we subsequently calculate the kinetic Sunyaev Zel'dovich effect on the CMB.

METHOD

The bulk flow of free electrons in the intergalactic medium generates the temperature fluctuation in the CMB through Compton scattering. This is the so called kinetic Sunyaev-Zel'dovich effect given by the expression [4]:

$$\Theta(\hat{n}) \equiv \Delta T / T_{CMB} = \int n_e \sigma_T \mathbf{v}/c \cdot \hat{n} a C(r) dr \equiv \int \chi_e \bar{n}_e \sigma_T \mathbf{p} \cdot \hat{n} \exp[-\tau(z)] a C(r) dr. \quad (1)$$

Here, $n_e = \chi_e \bar{n}_e (1 + \delta_e)$ is the free electron number density with the χ_e as the ionization fraction. Hereafter we will set $\chi_e = \tilde{\chi}_e$, namely, omit the patchy ionization effect as indicated both in the simulation [8] and in the theory [9]. We further set $\chi_e = 1$ until the reionization redshift z_i. σ_T is the Thompson cross section. $\hat{n}$ is the direction on the sky and $aC(r)dr$ is the physical distance interval at r in terms of the comoving radial coordinate and $C(r) \equiv 1/\sqrt{1-(r/A)^2}$. $\tau(z)$ is the Thompson optical depth. $\mathbf{v}$ is the electron peculiar velocity. $\mathbf{p} \equiv (1+\delta_e)\mathbf{v}/c \equiv \mathbf{p}_B + \mathbf{p}_E$ is the dimensionless electron momentum. $\mathbf{p}_E$ is the irrotational part of $\mathbf{p}$ satisfying $\nabla \times \mathbf{p}_E = 0$. Its contribution to the CMB is negligible. The dominant contribution to CMB is from $\mathbf{p}_B$, the rotational part of $\mathbf{p}$, for which $\nabla \cdot \mathbf{p}_B = 0$. We define the Fourier transform of $\mathbf{p}_{B,E}(\mathbf{k})$ by

$$\mathbf{p}_{B,E}(\mathbf{k}) = \int \mathbf{p}_{B,E}(\mathbf{x}) \exp(i\mathbf{k}\cdot\mathbf{x}) d^3x. \tag{2}$$

Some notations are: $\xi_{B,E}(r) \equiv \langle \mathbf{p}_{B,E}(\mathbf{x}) \cdot \mathbf{p}_{B,E}(\mathbf{x}+\mathbf{r}) \rangle$ and $p_{B,E}^2(k) \equiv \langle |\mathbf{p}_{B,E}(\mathbf{k})|^2 \rangle$ is the Fourier transformation of $\xi_{B,E}(r)$. Then, $\langle p_{B,i}(\mathbf{k}) p_{B,j}(-\mathbf{k}) \rangle = (\delta_{ij} - k_i k_j/k^2) p_B^2(k)/2$. $k_i k_j$ terms correspond to $\nabla_i \nabla_j \xi_{B,E}(r)$ in real space. After integrating along the line of sight, they become surface terms and negligible since $\nabla \xi_{B,E} \to 0$ when $r \to 0, \infty$. Adopting Limber's approximation we get the CMB angular correlation function:

$$w(\theta) \simeq \sigma_T^2 \cos\theta \int drC(r)[a\bar{n}_e\bar{\chi}_e]^2 \exp[-2\tau(z)] \int \frac{1}{2}\xi_B(\sqrt{d_A^2(1+z)^2\theta^2+y^2}) dy \tag{3}$$

$$\simeq \sigma_T^2 \int drC(r)[a\bar{n}_e\bar{\chi}_e]^2 \exp[-2\tau(z)] \int \frac{1}{2}\xi_B\sqrt{d_A^2(1+z)^2\theta^2+y^2} dy.$$

Here, d_A is the angular distance. The last approximation introduces errors less than 0.2% for $\theta \le 1^0$. Thus, the usual expression of the CMB power spectrum [10] still holds:

$$C_L = 16\pi^2/(2L+1)^3 [\bar{n}_e(0)\sigma_T c/H_0]^2 \tag{4}$$
$$\times \int_0^{z_i} (1+z)^4 \frac{1}{2}\Delta_B^2(k,z)|_{k=L/[d_A(1+z)]} \bar{\chi}^2(z) \exp[-2\tau(z)] C(x) x(z) \frac{dx(z)}{dz} dz.$$

Here, $\Delta_B^2(k,z) \equiv k^3 p_B^2(k)/(2\pi^2)$ and $x(z) \equiv r(z)/(c/H_0)$.

The power spectrum of $\mathbf{p}_B$ in the linear and mildly non-linear regime was originally discussed by Vishniac [11] using second order perturbation theory. After that, some efforts [9, 12] have been taken for the fact that the universe is highly non-linear. Recent simulations [6, 7] show that on small scales, equal partition in $\mathbf{p}$ space is reached. Since $\mathbf{p}_B$ has two degrees of freedom, while $\mathbf{p}_E$ has one, equi-partition means that $p_B^2 = 2p_E^2$ in the highly non-linear region and gives a way to calculate p_B^2. The mass conservation law states that:

$$\dot{\delta}_e + \nabla\cdot\mathbf{p}c/a = 0. \tag{5}$$

After some algebra, we get:

$$p_E^2(k) \equiv \langle \mathbf{p}_E(\mathbf{k})\cdot\mathbf{p}_E(-\mathbf{k})\rangle = \frac{a^2}{c^2}\langle|\dot{\delta}_e(k)|^2\rangle/k^2. \tag{6}$$

Under the stable clustering assumption, $\delta_e(\mathbf{x}) = a^3/a_i^3 \delta_e^i(a/a_i\mathbf{x})$ (We choose the comoving coordinate origin as the center of each stable clustered object) [1]. Here, 'i' means an

---

[1] More general form is $\delta_e(\mathbf{x}) = a^3/a_i^3 \delta_e^i[\mathbf{x_c}+a/a_i(\mathbf{x}-\mathbf{x_c})]$. Here, $\mathbf{x_c}$ is the comoving coordinate of the center of each stable clustered object. Then, $\dot{\delta}_e = 3H\delta_e + H\nabla\delta_e\cdot(\mathbf{x}-\mathbf{x_c})$. In Fourier space, the

arbitrary initial epoch. Then, $\dot{\delta}_e = H\nabla \cdot (\delta_e \mathbf{x})$ Here, $H(z) = \dot{a}/a$ is the Hubble constant. Then, $\langle |\dot{\delta}_e(k)|^2 \rangle = H^2[k^2 d^2 P_g(k)/dk^2 - 3/2k dP_g/dk]$. Here, P_g is the gas density power spectrum. On small scales, baryonic matter does not trace the dark matter due to the effects of gas pressure. The relation of δ_e and δ can be described by a window function. Hui and Gnedin [13] and Gnedin [14] introduced a Gaussian window function such that the baryon power spectrum $P_g(k) = \exp(-2k^2/k_F^2)P(k) = W_g^2(k)P(k)$. $P(k)$ is the dark matter density power spectrum. $k_F = k_F(z)$, the so called filtering scale, is generally a function of time and relies on the total thermal history of the universe. Hereafter we will adopt the value of k_F in Gnedin [14]. Then,

$$\Delta_B^2(k) = \frac{a^2 H^2}{c^2} \frac{k}{2\pi^2}[k^2 \frac{d^2 P_g(k)}{dk^2} - \frac{3}{2}k\frac{dP_g}{dk}]. \quad (7)$$

The momentum power spectrum in the intermediate region between the regions, where second order perturbation and stable clustering apply, respectively, is difficult to obtain analytically, so we simply choose it as the second order result before intersection and the stable clustering result after intersection. The power spectrum obtained in this way should serve as the upper limit of the real one and the asymptotic behavior should be exact. The results are shown in Fig. 1 and 2.

DISCUSSION

Our result is intrinsically different to former analytical results and simulations [9, 12, 15, 16]. Our CMB power spectrum peaks at lower amplitude ($\Delta T \simeq 1.5 \mu k$, as compared to their result $\Delta T \simeq 2.7 \mu k$) and larger angular scale ($l \simeq 3000$, as compared to their result $l \sim 10^4$). Our vorticity power spectrum and the resulted C_l are both smaller than the corresponding Vishniac results. This is due to the stable clustering assumption, which ensures that the motion of highly non-linear substructure of virialized objects cancel most of its contribution to the vorticity power spectrum. Then the highly non-linear regime is not the most important contribution to the kinetic SZ effect, as shown in Fig. 1. So C_l peaks at lower amplitude and larger angular scalea. Other analytical methods, such as Hu [12], state that the highly non-linear regime is the most important to the vorticity power spectrum and thus the C_l is bigger than the Vishniac effect and peaks at smaller angular scales.

Though we still need more detailed simulation and analytical work to solve these intrinsic discrepancies, we can show that the behavior of the vorticity power spectrum in the non-linear regime in our model happens in other cases. For example, in the second order perturbation theory, the density variance $\Delta^2(k) \propto \Delta_{linear}^4(k)$. But the stable clustering states that in the highly non-linear regime, $\Delta^2(k) \propto \Delta_{linear}^3(k)$ and is thus smaller than the second order result at sufficiently smaller scales.

---

contribution of the last term $\nabla \delta_e \cdot \mathbf{x_c}$ is $\propto \sum_j \mathbf{x}_c(j) \cdot \int_{S_j} \delta_e \exp(i\mathbf{k} \cdot \mathbf{x}) d^2\mathbf{S} - i\mathbf{k} \cdot \sum_j \mathbf{x}_c(j) \int_{V_j} \delta_e \exp(i\mathbf{k} \cdot \mathbf{x}) d^3 x = 0$. Here, "$j$" is the label of each stable clustered object. So, the power spectrum of $\langle |\dot{\delta}_e|^2 \rangle$ remains the same.

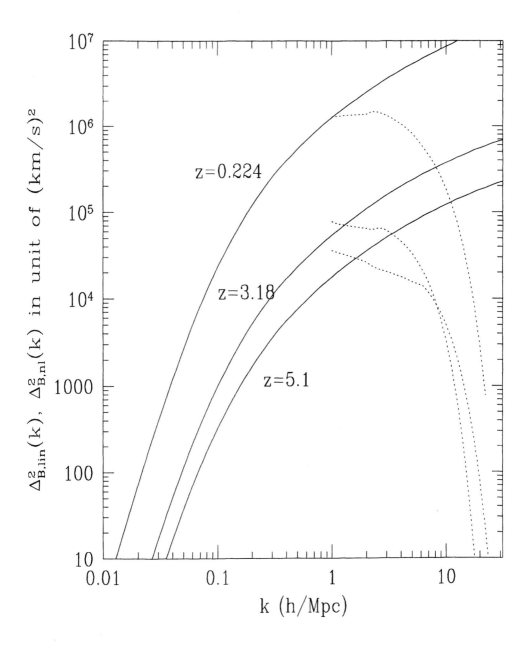

FIGURE 1. $p_B^2(k)$: The power spectrum of the rotational part of **p**. Solid lines are from second order perturbation. Dot lines are the result of the equi-partition and stable clustering approximation. Parameters of the gas window function is adopted from [14]. $\Omega_m = 0.37, \Omega_\Lambda = 0.63, \Omega_B h^2 = 0.02, h = 0.67, \sigma_8 = 0.9$ are adopted.

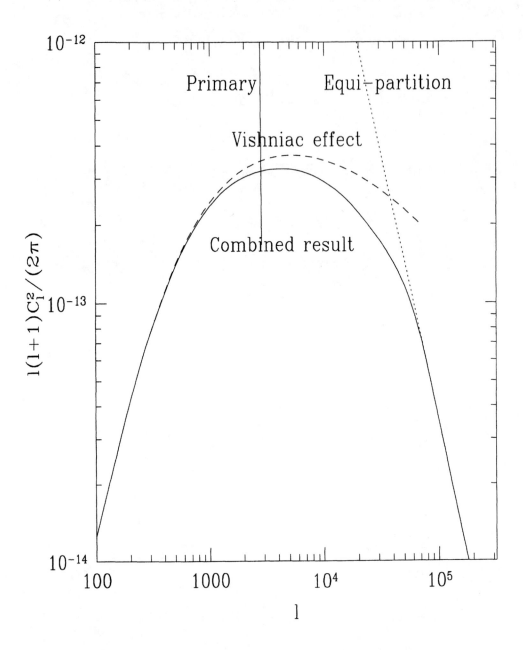

FIGURE 2. The CMB Kinetic SZ effect power spectrum. $\Omega_m = 0.37$, $\Omega_\Lambda = 0.63$, $\Omega_B h^2 = 0.02$, $h = 0.67$, $\sigma_8 = 0.9$, $z_i = 15$ are adopted.

REFERENCES

1. http://www.asiaa.sinica.edu.tw/AMIBA/
2. http://astro.estec.esa.nl/SA-general/Projects/Planck
3. Zel'dovich, Y.B. and Sunyaev, R.A., 1969, Astronomy & Astrophysics, 20, 189
4. Sunyaev., R. and Zel'dovich, Y., 1980, MNRAS, 190, 413
5. Zhang, P.J. and Pen, U.L., 2001, will appear on ApJ, 549. astro-ph/0007462
6. Scoccimarro, R., 2000, astro-ph/0008277
7. Pen, U.L., 2001, in preparation.
8. Gnedin, N.Y., 2000b, astro-ph/0008469
9. Valageas, P., Balbi, A. and Silk, J., 2000, astro-ph/0009040
10. Peacock, J.A., 1999, *Cosmological Physics*, Cambridge University Press
11. Vishniac, Ethan T., 1987, ApJ, 322, 597
12. Hu, W., 2000, ApJ, 529, 12
13. Hui, L., and Gnedin, N.Y., 1997, MNRAS, 292, 27
14. Gnedin, N.Y., 2000a, astro-ph/0002151
15. Gnedin, N.Y. and Jaffe, A., 2000, astro-ph/0008469
16. Springel, V., White, M. and Hernquist, L., 2000, astro-ph/0008133

The Role of CMB Polarization in Constraining Primordial Non-Adiabatic Fluctuations

Martin Bucher, Kavilan Moodley and Neil Turok

Department of Applied Mathematics and Theoretical Physics, Centre for Mathematical Sciences, University of Cambridge, Wilberforce Road, Cambridge CB3 0WA, U.K.

Abstract. We review the general primordial perturbation valid at early times in a universe with no new physics and consisting of four regular isocurvature modes in addition to an adiabatic growing mode. The prospects for testing the adiabaticity of the cosmic microwave background (CMB) anisotropy using MAP and PLANCK are then discussed. The role of CMB polarization in constraining the presence of primordial isocurvature modes is investigated.

With impressive measurements of the CMB temperature anisotropy already achieved, experimenters are now attempting to measure the anticipated but as-yet-undetected CMB polarization signal. An accurate mapping of the polarization spectrum, after the foregrounds are properly removed and systematics brought under control, will independently confirm the pattern of acoustic oscillations now being seen in the temperature spectrum [1], in addition to providing useful complementary constraints on the underlying cosmological model. In particular, polarization has been proposed as a sensitive probe of the baryon density once the temperature fluctuations are measured [2], and can constrain the epoch of reionization [3] from the shape of the spectrum on the largest angular scales. If the experiments can attain the required sensitivity to measure the B-mode polarization, we can also check whether the anisotropies were of a purely scalar origin or whether vector and tensor modes [4] contributed as well. This contribution highlights the importance of a polarization measurement in testing the hypothesis that the CMB fluctuations originated from adiabatic perturbations, by distinguishing the isocurvature signature on angular scales larger than a degree.

The main motivation for previous studies of isocurvature models has been to seek an alternative to the adiabatic cold dark matter (CDM) model. Both the baryon isocurvature model [5] and the CDM isocurvature model [6] did not fare well, however, in trying to simultaneously explain the CMB anisotropy spectrum and data from large-scale structure surveys. Other work in this direction [7] has

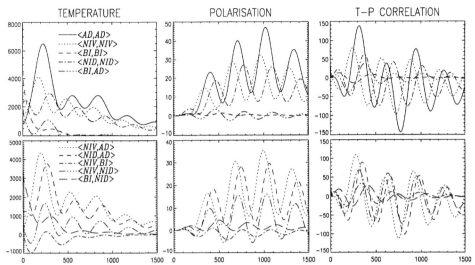

FIGURE 1. CMB spectra for adiabatic and isocurvature auto-correlation and cross-correlation modes with cosmological parameters $h = 0.65$, $\Omega_b = 0.06$, $\Omega_c = 0.25$, $\Omega_\Lambda = 0.69$ and $\tau = 0.1$. All modes have scale-invariant spectra and are normalised to have equal power as the fiducial model.

concentrated on trying to detect or constrain admixtures of the adiabatic and CDM isocurvature modes observationally, though cross-correlations between these modes were not considered.

Our present motivation for studying isocurvature perturbations is rather to test the adiabatic hypothesis by considering non-adiabatic fluctuations and checking if the observational data can be used to discover whether these modes were significantly excited or whether they in fact contributed negligibly as is often assumed. Constraints must also be placed on the contributions from non-orthogonal directions that arise from cross-correlations between the different modes. While most attention has focussed on studying an adiabatic spectrum of perturbations which results from single scalar-field inflationary models, it is important to note that multi-field inflationary models generically produce adiabatic and isocurvature perturbations [8]. Cross-correlations between modes arise naturally in models of this type [9].

In previous work [10] we solved the perturbation equations describing a universe filled with baryons, photons, neutrinos and cold dark matter to obtain the perturbation modes that are valid at early times and on superhorizon scales. Apart from the gauge and decaying modes which are not physically relevant, there exist the familiar adiabatic, baryon isocurvature and CDM isocurvature modes. If we also consider perturbations in the neutrino fluid after it has decoupled there exist two additional regular modes, a neutrino isocurvature density mode and a neutrino isocurvature velocity mode, that were identified previously in the literature [11] but the consequences of which were not studied. The isocurvature modes are characterised by the different components (eg. photons, baryons, neutrinos etc.) having

	PLANCK-T ADIA	PLANCK-TP ADIA	PLANCK-T ADIA + 3 ISO	PLANCK-TP ADIA + 3 ISO
$\delta h/h$	9.93	3.69	40.13	4.36
$\delta\Omega_b/\Omega_b$	19.37	7.26	68.85	8.61
$\delta\Omega_k$	4.92	1.83	20.56	2.18
$\delta\Omega_\Lambda/\Omega_\Lambda$	2.74	1.21	5.93	1.49
$\delta n_s/n_s$	0.73	0.37	3.92	0.70
τ_{reion}	8.25	0.41	35.35	0.56
$\langle NIV, NIV \rangle$	..	..	43.45	1.14
$\langle BI, BI \rangle$	..	..	53.29	4.23
$\langle NID, NID \rangle$	..	..	19.18	2.37
$\langle NIV, AD \rangle$	..	..	121.59	4.69
$\langle BI, AD \rangle$	..	..	58.75	8.97
$\langle NID, AD \rangle$	..	..	114.39	5.77
$\langle NIV, BI \rangle$	..	..	46.91	3.67
$\langle NIV, NID \rangle$	..	..	80.01	2.97
$\langle BI, NID \rangle$	..	..	100.97	4.60

TABLE 1. 1-σ error bars (in percent) on cosmological parameters and isocurvature mode amplitudes as determined by the PLANCK experiment (i) assuming adiabatic perturbations ('ADIA') and (ii) when 3 isocurvature modes are also included ('ADIA + 3 ISO'). 'T' refers to a temperature measurement alone and 'TP' refers to the total information available from combining a temperature and polarization measurement.

spatially varying abundances that combine to yield a zero net curvature perturbation, in contrast to the adiabatic mode which has a spatially uniform equation of state that results in a warping of the spatial hypersurfaces initially. The different features of the above modes, as well as possible mechanisms to generate them, are discussed in more detail in [10], which also contains the power series solutions that characterise the initial conditions.

The CMB spectra for each of the modes and their cross-correlations are plotted in Figure 1. We omit the CDM isocurvature spectra as they agree to a fraction of a percent with those of the baryon isocurvature mode. We would like to determine what fractions of the isocurvature spectra are allowed by the data. This is done by first generalising the familiar power spectrum of perturbations to a 5x5, positive-definite, symmetric matrix

$$\langle A_i(\mathbf{k}) \, A_j(\mathbf{k'}) \rangle = P_{ij}(\mathbf{k}) \cdot \delta^3(\mathbf{k} - \mathbf{k'}), \quad (i,j = 0,1,2,3,4)$$

the entries of which characterise the amplitudes of the auto-correlation and cross-correlation modes. We present constraints on the mode amplitude parameters (omitting the CDM isocurvature mode) and a set of six cosmological parameters by considering their variations about an adiabatic fiducial model, in the absence of the precision satellite data. The values of cosmological parameters chosen for the fiducial model are listed in the Figure 1 caption. A detailed description of the statistical method used to determine the constraints, which involves a calculation

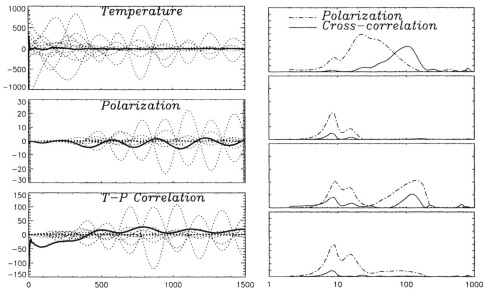

FIGURE 2. (a) The first panel shows the near perfect cancellation that characterises the most uncertain direction of a PLANCK temperature measurement, while the lower panels demonstrate that the "cancellation" is far less pronounced for polarization and T-P correlation measurements. The solid curve is a weighted sum of the corresponding parameter variations and is multiplied by 10 for clarity. (b) The panels correspond to the four flattest directions of a temperature measurement, and at each ℓ indicate the gain in information in these directions (given by the area under the curves) from the additional polarization and T-P correlation measurements, which reduces the uncertainties in the flattest directions from 239%, 60%, 36%, and 23% to 11.1%, 10.3%, 6.6%, and 4.6%, respectively.

of the Fisher matrix about the fiducial model, can be found in ref. [12]. In Table 1 we list the 1-σ error bars that indicate the accuracy with which the PLANCK experiment will measure the parameters under consideration. The constraints on cosmological parameters assuming purely adiabatic perturbations are included in the first two columns for comparison. We note that the errors on the mode amplitudes determined by a PLANCK temperature measurement are of order unity, as in the case of the MAP experiment (see ref. [13]). However, including polarization information significantly improves the accuracy with which PLANCK will measure all parameters – all errors are reduced to less than 10%.

We discuss this point in more detail. The upper panel of Figure 2(a) indicates how variations in the temperature spectrum with respect to the different parameters yield shapes that combine to provide a near perfect cancellation. This results in a very poorly determined direction in parameter space. The spectral decomposition of the Fisher matrix that characterises this degeneracy can be found in Table 8 in ref. [12]. We note here that several different parameter directions are

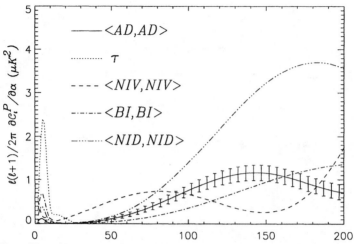

FIGURE 3. The polarization spectra of isocurvature modes and a late-time reionization are compared to the fiducial model at low ℓ, with the PLANCK error bars due to instrumental noise and cosmic variance also plotted, to emphasise the large signatures that these parameters have.

necessary to maintain this degeneracy. The lower panels show that the degeneracy in the flat temperature direction is broken when polarization and temperature-polarization correlation information is included – the ripples represent variations that can be detected by the PLANCK experiment. It may appear that this signal is accessible to PLANCK over a range of angular scales. However we see from Figure 2(b) that the additional information gained from a polarization measurement and the temperature-polarization correlation, which breaks the degeneracy in the flat direction, lies mainly on angular scales larger than a degree ($\ell \lesssim 100$).

We can ask whether this degeneracy breaking information originates from any specific polarization signatures and if so which parameters contribute most to the information gained. Figure 3 shows the fiducial adiabatic polarization spectrum compared to the three isocurvature spectra for $\ell \leqslant 200$ and the signal expected from a late reionization of the universe at an optical depth $\tau = 0.1$; variations in other parameters do not produce such marked deviations from the fiducial model. We observe that the neutrino velocity isocurvature mode ($\ell \approx 50$) and a late-time reionization event ($\ell \approx 10$) contribute most significantly to the gain in polarization information shown in Figure 2(b). The role that the polarization measurement plays in breaking the degeneracy between the optical depth and the amplitude of the temperature fluctuations (evident in Figure 3) has been studied before [14]. Here we highlight the neutrino isocurvature velocity mode signature that originates from an initial power spectrum which is scale-invariant in the photon velocity $P_v(k) \sim v_\gamma^2 \sim k^{-3}$. The polarization spectrum which scales as $C_\ell \sim (\nabla \cdot \mathbf{v}_\gamma)^2$, directly probes the photon dipole at last scattering and is thus sensitive to the enhanced power $\propto \ell^{-2}$ of the neutrino isocurvature velocity mode relative to the adiabatic mode.

What are the prospects for measuring the polarization signal at low-ℓ? PLANCK will provide a very sensitive map of the polarization sky over a wide range of angular scales which will allow us to strongly constrain or detect isocurvature modes, as we have demonstrated above. The MAP satellite will fly this summer but is not optimised for a polarization measurement and may detect it only marginally. The BOOMERANG and MAXIMA teams are planning a polarization flight to be launched later this year with the range of scales covered ($\ell \approx 100-1000$) allowing a possible constraint on the non-adiabatic signal. The ground-based experiments POLAR and Milano and the space-based experiment SPOrt, have focussed on angular scales $\gtrsim 10°$ to measure the polarization signal (for an overview of ground-based and space-based experiments see ref. [15]), but face the challenge of controlling systematics that will arise from long integration times. These are necessary to attain the sensitivity required to measure the small signal. The present work provides strong motivation for a polarization experiment of high sensitivity on angular scales larger than a degree.

REFERENCES

1. Netterfield, C. B. et al., 2001, astro-ph/0104460; Halverson, N. W. et al., 2001, astro-ph/0104489; Lee, A. T. et al., 2001, astro-ph/0104459.
2. Jaffe, A. H., Kamionkowski, M. & Wang, L., 2000, Phys. Rev. D, **61**, 083501.
3. Zaldarriaga, M., 1997, Phys. Rev. D, **55**, 1822; Ng, K. L. & Ng, K. W., 1996, Ap. J., **456**, 413.
4. Seljak, U. & Zaldarriaga, M., 1997, Phys. Rev. Lett., **78**, 2054; Kamionkowski, M., Kosowsky, A. & Stebbins, A., 1997, Phys. Rev. Lett., **78**, 2058.
5. Peebles, P. J. E., 1987, Nature, **327**, 210; Peebles, P. J. E., 1987, Ap. J. **315**, L73.
6. Bond, J. R. & Efstathiou, G., 1987, M.N.R.A.S., **22**, 33.
7. Enqvist, K., Kurki-Suonio, H. & Valiviita, J., 2000, Phys. Rev. D, **62**, 103003; Enqvist, K. & Kurki-Suonio, H., 2000, Phys. Rev. D, **61**, 043002; Pierpaoli, E., García-Bellido, J. & Borgani, S., 1999, J.H.E.P., **9910**, 15.
8. Polarski, D. & Starobinsky, A. A., 1994, Phys. Rev. D, **50**, 6123.
9. Langlois, D., 1999, Phys. Rev. D, **59**, 123512; Gordon, C., Wands, D., Bassett, B. A. & Maartens, R., Phys. Rev. D, **63**, 2001, 023506.
10. Bucher, M., Moodley, K. & Turok, N., 2000, Phys. Rev. D, **62**, 083508.
11. Rebhan, A. & Schwarz, D., 1994, Phys. Rev. D, **50**, 2541; Challinor, A. & Lasenby, A., 1999, Ap. J., **513**, 1.
12. Bucher, M., Moodley, K. & Turok, N., 2000, to appear in Phys. Rev. D, (astro-ph/0007360).
13. Bucher, M., Moodley, K. & Turok, N., 2000, to appear in Phys. Rev. Lett., (astro-ph/00012141).
14. Zaldarriaga, M., Spergel, D. N. & Seljak, U., 1997, Ap. J., **488**, 1.
15. Staggs, S. T., Gundersen, J. O. & Church, S. E., 1999, astro-ph/9904062.

Interpreting CMB Anisotropy Observations: Trying to Tell the Truth with Statistics

Eric Gawiser

Center for Astrophysics and Space Sciences, University of California, San Diego
La Jolla, CA 92093

Abstract. A conflict has been reported between the baryon density inferred from deuterium observations and that found from recent CMB observations by BOOMERanG and MAXIMA. Despite the flurry of papers that attempt to resolve this conflict by adding new physics to the early universe, we will show that it can instead be resolved via a more careful usage of statistics. Indeed, the Bayesian analyses that produce this conflict are by their nature poorly suited for drawing this type of conclusion. A properly defined frequentist analysis can address this question directly and appears not to find a conflict. Finally, a conservative accounting of systematic uncertainties in measuring the deuterium abundance could reduce what is nominally a 3σ conflict to 1σ.

INTRODUCTION

The recent BOOMERanG [1,2] and MAXIMA-1 [3,4] observations of Cosmic Microwave Background (CMB) anisotropy provide the first high-quality, high-resolution observations to cover the angular scales over which the first two acoustic peaks are expected in the angular power spectrum. If used by themselves, these data are sufficient to determine the location and rough amplitude of the first acoustic peak, providing evidence that the universe is near critical density. A simultaneous fit to numerous cosmological parameters is impossible, however, because of strong degeneracies amongst those parameters in determining the shape of the CMB angular power spectrum at the precision of the observations. As a result, a Bayesian framework has been used in which numerous other cosmological observations are used as priors. These priors come from large-scale structure, Type Ia Supernovae, direct determinations of Hubble's constant, and the baryon density inferred from combining observations of the deuterium-to-hydrogen abundance ratio with the standard predictions of Big Bang Nucleosynthesis (BBN).

A funny thing happened on the way to precision cosmology. While the first acoustic peak is clearly defined by CMB anisotropy data, thereby providing evidence for a flat universe, the second acoustic peak is either at surprisingly low amplitude or missing entirely. Within the parameter space of the standard adiabatic CDM paradigm, this can be produced either by a red-tilted ($n < 1$) primordial power spectrum or by a baryon density higher than that inferred from deuterium observations plus BBN ($\Omega_b h^2 = 0.021 \pm 0.002$) [5]. In order to test the latter idea, [6] performed a Bayesian analysis on the combined BOOMERanG, MAXIMA-1, and COBE-DMR [7] data without using the BBN baryon density as a prior, and they found $\Omega_b h^2 = 0.032 \pm 0.004$. Hence there appears to be a conflict between the CMB and BBN values for the baryon density at roughly the 3σ level.

One way to resolve this conflict is to postulate additional physics in the early universe that alters Big Bang Nucleosynthesis such that the observed deuterium abundance is consistent with the higher value of $\Omega_b h^2$ preferred by the CMB. This has been attempted using degenerate neutrinos due to a large lepton asymmetry [8,9], a decaying neutrino that likewise produces extra entropy during BBN [10,11], or inhomogeneous BBN [12]. Even if the precise priors on the nature of nonstandard BBN are allowed to vary, a robust need for new physics is claimed [13]. Another approach adds new physics to the earlier inflationary epoch in the form of an unexplained bump in the primordial power spectrum of density perturbations [14].

Injecting new physics into the first few minutes of the universe is a serious step and needs to be motivated by a strong observational signal. While the claimed 3σ conflict between CMB and BBN baryon densities seems to have been interpreted by many authors as a sufficient signal, a close examination of the statistics involved reveals that this conflict has been exaggerated and may not exist at all.

BAYESIAN ANALYSES

A Bayesian analysis seeks to answer the question, "Given what I knew before plus the data I have just obtained, what do I now think the truth is?" What was known before is incorporated in the form of a prior probability function. Basic probability theory gives us the starting point,

$$p(model, data) = p(data|model)p(model) = p(model|data)p(data) , \qquad (1)$$

which is equivalent to the statement that the probability of two things being true is equal to the probability that the first one is true given that the second one is true times the probability that the second one is indeed true. Bayes' theorem involves specializing this statement to the case of a set of models and the observed data and dividing by $p(data)$ to get

$$p(model|data) \propto p(data|model)p(model) . \qquad (2)$$

The probability that the data would be observed given a particular model is often easy to calculate and is referred to as the likelihood function. This means that as long as we know the prior probability of various models being true, $p(model)$, and can calculate the likelihood function, we can determine the posterior likelihood that each model is correct. We can either think of deuterium observations as part of the current data and do a joint likelihood analysis or we can account for the results of the deuterium observations when we choose a prior probability function for $\Omega_b h^2$. To ignore the deuterium observations entirely would imply that we do not consider them trustworthy.

Since the various models considered by [6] all lie within the adiabatic CDM parameter space, the prior $p(model)$ can be expressed as the product of the prior probability functions of various independent parameters, including the baryon density. When a uniform prior $0.0031 \leq \Omega_b h^2 \leq 0.2$ is used, these authors find a posterior likelihood described by $\Omega_b h^2 = 0.032 \pm 0.004$. If a prior consistent with BBN plus deuterium observations, $\Omega_b h^2 = 0.019 \pm 0.002$ [15], is used, they find a posterior likelihood described by $\Omega_b h^2 = 0.021 \pm 0.003$.

Although these authors conclude that the BBN plus deuterium value of the baryon density is "disfavored by the data," this is not the right interpretation to ascribe to the results of their Bayesian analysis. If they assume that the BBN plus deuterium value of the baryon density is correct by including it as a prior, they produce a posterior likelihood in good agreement, showing that while the CMB data may favor a higher value of $\Omega_b h^2$, BBN is a much stronger constraint. Indeed, they note that this prior is strong enough to alter the results on other parameters, for instance yielding a scalar spectral index of the primordial power spectrum of $n_s = 0.89 \pm 0.06$ rather than the $n_s = 1.03 \pm 0.08$ produced by the uniform prior on the baryon density. Starting and ending with a baryon density consistent with BBN plus deuterium is a self-consistent result.

When they instead use a uniform prior on the baryon density and ignore the implications of deuterium observations, these authors are starting from the assumption that the BBN value of the baryon density is not worthy of consideration. Producing a posterior likelihood for $\Omega_b h^2$ that is different from the BBN value is again self-consistent; in this case we start and end with the idea that $\Omega_b h^2$ may well be greater than 0.019. In the case of the uniform prior, the CMB data do show the ability to narrow a broad prior into a localized posterior. The correct conclusion to draw from this exercise is that it is quite important to decide a priori whether we believe the deuterium observations and what they imply for the baryon density, because it makes a significant difference in our posterior estimation of the truth.

One complication of Bayesian statistics is that if we do not know the correct priors we should vary our priors over the range of reasonable functions. If indeed both the BBN and the uniform prior on the baryon density are reasonable, then the correct conclusion about the posterior likelihood is that $\Omega_b h^2$ could be anywhere from 0.02 to 0.03. Because it requires a specific choice of prior assumptions and produces only relative likelihoods at the end, the simple Bayesian analysis is not well-suited to answering the question, "Are the CMB and BBN values for the baryon density in conflict?" This question could be pursued in a Bayesian format using prior assumptions on how likely such a conflict is, but this is far beyond the scope of the analyses that have been done.

FREQUENTIST ANALYSES

A frequentist analysis is a bit simpler to describe; each model is viewed as a separate hypothesis to be tested against the observations. One looks only at the likelihood function i.e. $p(data|model)$. Typically a misfit statistic such as χ^2 is used, and any models for which the chance of getting a better agreement with the data is greater than or equal to e.g. 95% are considered to be ruled out at the e.g. 95% confidence level. This is more akin to the basic scientific method taught to children; a frequentist analysis seeks to answer the question, "Which of these models are reasonably likely to produce the observed data?" While a best-fit model can still be found, one concentrates on discarding those models that are ruled out beyond some confidence level. Further discrimination requires better data. This frequentist approach has the added benefit of being able to rule out an entire parameter space if none of its models are a reasonably good fit to the data; in this case the hypothesis that the true model lies somewhere in this parameter space has been rejected. The Bayesian approach can be modified to compare one parameter space to another but it always makes a conclusion based on relative likelihood.

In the case of the recent CMB data, a frequentist goodness-of-fit analysis was performed by the MAXIMA team [4]. They find that the ΛCDM model with $\Omega_b h^2 = 0.021$ has $\chi^2 = 10/10$ when compared with the MAXIMA-1 data alone and $\chi^2 = 40/40$ compared with MAXIMA-1 plus COBE-DMR, but their best-fit model with $\Omega_b h^2 = 0.025$ has $\chi^2 = 8/10$ and $\chi^2 = 38/40$ respectively. This is consistent with the Bayesian result that a high baryon density has greater likelihood, but now we have a chance to assess the absolute goodness-of-fit of these models. The "best-fit" model is a slightly better fit than one expects but this is likely explained by having varied seven parameters to find it. Indeed, we should subtract up to seven degrees of freedom from the above results if these seven parameters successfully span the space of possible functions of seventh order. A simple way to state this effect is that even if ΛCDM is the true model we expect that by varying n of its parameters freely we will be able to drop the χ^2 by a value of n; these parameter variations are fitting the observational errors in the data rather than telling us more about the truth.

Of course, we would like to have a similar frequentist analysis of the full set of CMB data, particularly BOOMERanG, MAXIMA-1, and COBE-DMR. [16] analyze the combined BOOMERanG and MAXIMA-1 data using a relative likelihood analysis of their χ^2 values; since their best-fit model is close to $\chi^2/d.o.f. = 1$ this is nearly the correct frequentist approach. The problem is that they ignore the significant calibration uncertainties of BOOMERanG and MAXIMA-1 so this analysis is seriously flawed and will eliminate a large set of viable models.

An acceptable frequentist analysis of BOOMERanG, MAXIMA-1, and COBE-DMR data has been performed by [14] in the course of adding a bump to the primordial power spectrum in the ΛCDM model. The standard ΛCDM model, i.e. amplitude of bump equals zero, is ruled out at the 68% confidence level but not at 95% confidence. This means that ΛCDM is a reasonably good fit to the current set of high-quality CMB data, and it seems to eliminate the motivation for considering a primordial bump. These authors do not analyze models with higher baryon fractions but most likely would find an even better fit. One must then consider whether a model can be ruled out not for being a bad fit but simply because another model is a better fit. In general, this is a dangerous approach although it is implicit in the Bayesian formalism. If there is evidence to suggest that the observational errors have been overestimated[1] then the relative likelihood approach may be justified, but otherwise it is premature to discard models for which the data is a quite reasonable result.

SYSTEMATIC UNCERTAINTIES IN BARYON DENSITY FROM DEUTERIUM

An alternative manner in which the careful usage of statistics may resolve the apparent conflict between CMB and BBN values of the baryon density is via a fuller accounting of systematic uncertainties in the usage of the observed deuterium-to-hydrogen abundance ratio to infer the baryon density. This issue is explored by [17], who find that while the best-fit CMB baryon density of $\Omega_b h^2 = 0.03$ "cannot be accomodated," a very conservative consideration of systematic errors would allow $0.016 \leq \Omega_b h^2 \leq 0.025$. Although it is unlikely that the systematic errors in converting the observed deuterium-to-hydrogen abundance ratio into a baryon density are nearly this large, this does allow for the possibility that improved deuterium observations could reduce the claimed 3σ conflict between the CMB- and BBN-preferred baryon densities to 1σ.

Although the above range is quite conservative, the most recent high-redshift quasar absorption system in which [5] measured the deuterium-to-hydrogen abundance ratio, HS0105+1619, would by itself give a result of $\Omega_b h^2 = 0.023$. There are reasons to believe that this is the best measurement of deuterium yet performed; it has the highest hydrogen column density and therefore deuterium was seen in several Lyman-series transitions with a reduced chance of contamination from the Lyman alpha forest. Such contamination would increase the perceived abundance of deuterium, leading to an underestimate of the true baryon density. Indeed the HS0105+1619 deuterium-to-hydrogen abundance ratio is higher than the two previous detections of [15] by an amount greater than the observational errors. [5] are forced to add an empirical uncertainty to these points in order to account for their scatter. Unfortunately, it is also possible that mild levels of deuterium destruction due to star formation in the higher metallicity system HS0105+1619 have caused a systematic error in this system instead of the previous ones. Although deuterium destruction is not expected to be significant at the significantly sub-solar metallicity of this system, it is unclear where the true deuterium-to-hydrogen abundance ratio lies amongst the range of observed values.

---

[1] How often does this happen in observational astrophysics?

CONCLUSION

There are thus a number of ways in which a careful usage of statistics seems to eliminate the claimed 3σ conflict between the CMB- and BBN-preferred values of the baryon density. The first is that the Bayesian analyses used are actually producing consistent results; the proper conclusion to be drawn is that whether or not to include prior information from BBN is an important choice. Utilizing relative likelihood information and prior probability functions makes these analyses poorly suited to answering the question of whether a conflict exists between the CMB and BBN values for $\Omega_b h^2$. When a better-suited frequentist analysis is used, we find that the standard ΛCDM model with a BBN plus deuterium preferred value of $\Omega_b h^2 = 0.021$ is in reasonably good agreement with recent CMB observations. While a model with higher baryon fraction may be an even better fit this could simply be caused by having fit several free parameters; we need more precise observations to make a clear discrimination between these models. Additionally it is possible that systematic errors in measuring the deuterium-to-hydrogen abundance ratio are responsible for underestimating the BBN value of the baryon density. Given any one of these reasons, there is no longer a conflict between CMB anisotropy results and the value of $\Omega_b h^2$ preferred by observations of deuterium. Given all of them, it is clearly unnecessary to introduce additional physics to the early universe.

REFERENCES

1. de Bernardis, P. et al., *Nature* **404**, 955–959 (2000).
2. Lange, A. E. et al., *Phys. Rev. D* **63**, 042001 (2001).
3. Hanany, S. et al., *Astrophys. J., Lett.* **545**, L5–L9 (2000).
4. Balbi, A. et al., *Astrophys. J., Lett.* **545**, L1–L4 (2000).
5. O'Meara, J. M., Tytler, D., Kirkman, D., Suzuki, N., Prochaska, J. X., Lubin, D., & Wolfe, A. M., *Astrophys. J.* in press, astro-ph/0011179 (2001).
6. Jaffe, A. H. et al., preprint, astro-ph/0007333 (2000).
7. Bennett, C. L. et al., *Astrophys. J., Lett.* **464**, L1–L4 (1996).
8. Lesgourgues, J. & Peloso, M., preprint, astro-ph/0004412 (2000).
9. Esposito, S., Mangano, G., Melchiorri, A., Miele, G., & Pisanti, O., *Phys. Rev. D* **63**, in press, 043004 (2001).
10. Hansen, S. H. & Villante, F. L., *Physics Letters B* **486**, 1–5 (2000).
11. Kaplinghat, M. & Turner, M. S., *Phys. Rev. Lett.* **86**, 385 (2001).
12. Kurki-Suonio, H. & Sihvola, E., preprint, astro-ph/0011544 (2000).
13. Kneller, J. P., Scherrer, R. J., Steigman, G., & Walker, T. P., preprint, astro-ph/0101386 (2001).
14. Griffiths, L. M., Silk, J., & Zaroubi, S., preprint, astro-ph/0010571 (2000).
15. Burles, S. & Tytler, D., *Astrophys. J.* **507**, 732–744 (1998).
16. Padmanabhan, T. & Sethi, S. K., preprint, astro-ph/0010309 (2000).
17. Burles, S., Nollett, K. M., & Turner, M. S., preprint, astro-ph/0008495 (2000).

Separation of E and B Patterns in the Local CMB Polarization Map

Tzihong Chiueh and Cheng-Jiun Ma

Physics Department, National Taiwan University, Taipei, Taiwan

Abstract. The CMB polarization E and B tensor patterns possess different parities, much like those in the familiar curl-free and divergence-free vector patterns. It is possible to project the polarization tensor map into a vector map, which consists of the linear superposition of a curl-free and a divergence-free vector field. This map not only allows one to visualize what the E and B modes look like, but it also permits the $E - B$ separation by a topological means. The measured variance of surface brightness for the E and B modes by this topological strategy is presented.

INTRODUCTION

Photon polarization is a two-dimensional tensor **P**, with Stokes parameters Q, U and V as its tensor elements. The CMB photons are linearly polarized, thus yielding a zero Stokes V. One often conceives the linear polarization tensor field through the polarization "vector" **p**, which should more accurately be regarded geometrically as a segment rather than a vector. Its length characterizes the polarization strength and its orientation is either parallel or anti-parallel to the instantaneous electric field of incoming photons. This polarization "vector" is simply $\mathbf{p} = \sqrt{\mathbf{P}}$, which is a multi-valued function that accounts for the bi-directionality. For arbitrary functions of Q and U, the polarization "vector" **p** is generally non-analytical and contains branch points, at which both Q and U vanish linearly [1].

Unlike **p**, which is not a vector, a genuine "polarization" *vector* can in fact be constructed from Q and U, which not only retains the original topological characteristics of the tensor **P** but it is also analytical. To show this, we note that **P** in the CMB is nothing more than a second derivative acting on a random field F:

$$\mathbf{P} = Q + iU \equiv 4\frac{\partial^2 F}{\partial \bar{w}^2}, \tag{1}$$

where $\bar{w} \equiv x - iy$. Here, F is either a scalar or pseudo-scalar field. In general $F = f + ig$, with real f and g which correspond to the scalar (E-mode) and pseudo-scalar (B-mode) components respectively. These E and B modes result

FIGURE 1. Projected E-Mode Vector Field

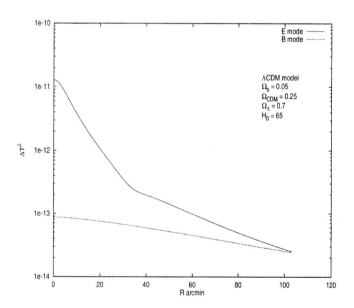

FIGURE 2. The Polarization Surface Brightness

from different origins, with the former from primordial scalar fluctuations and the latter from primordial gravitational waves and exotic vortex motion.

The important goal of future CMB polarization experiments is to detect each mode separately so as to identify its respective origins. The conventional approach for extracting the two modes is to project $\mathbf{P}$ in Eq.(1) into a scalar and a pseudo-scalar by taking appropriate second derivatives. In the complex representation, the unique projection is $4\partial^2 \mathbf{P}/\partial w^2$, which equals $\nabla^2\nabla^2(f+ig)$. The real part captures the E mode and imaginary part the B mode. However, such a procedure is not desirable for data analysis since the differentiation tends to amply the noise in the maps, let alone it takes derivatives twice.

A viable way for separating E and B modes is to take advantage of their distinct topological natures, and extracts them by topological means [2]. We first take a projection derivative on the polarization field to form a vector:

$$\mathbf{A} \equiv 2\frac{\partial \mathbf{P}}{\partial w} = 2\frac{\partial}{\partial \bar{w}}\nabla^2(f+ig) = \nabla(\nabla^2 f) - \nabla \times (\hat{z}\nabla^2 g)$$
$$= \hat{x}(\frac{\partial Q}{\partial x}+\frac{\partial U}{\partial y}) - \hat{y}(\frac{\partial U}{\partial y}-\frac{\partial Q}{\partial x}), \qquad (2)$$

where the real and imaginary parts are identified as the x and y components of the vector $\mathbf{A}$. The curl-free and divergence-free vectors are associated with the E and B modes, respectively. To extract E mode, we employ the Stokes theorem and take a contour integral along a closed loop of arbitrary shape: $\int_C d\mathbf{l} \times \mathbf{A} = \int d^2 S \nabla^2\nabla^2 f$, where $d^2 S$ is confined within the contour; to extract the B mode, we take $\int_C d\mathbf{l} \cdot \mathbf{A} = \int d^2 S \nabla^2\nabla^2 g$. By changing the contour size, one may probe either E or B mode of various scales. Fig.1 shows the $\mathbf{A}$ map for a pure E mode constructed from the CMBFAST in a two degree field, and the circle indicates a possible contour C. A pure B mode pattern can be envisaged by rotating all local vectors by 90 degrees. The surface brightness of filtered E and B modes can be measured by a circular contour C of radius R. Fig.2 plots the squared variance of the contour integral divided by the interior area, i.e., $\langle(\int_C (...)/\pi R^2)^2\rangle$ in unit of K^2.

Notice that we do need to take one spatial derivative on the Q and U maps to construct the $\mathbf{A}$ map, but it is followed by a line integration. The integral suppresses some of the amplified noise due to the differentiation, but likely not completely. An entirely different scheme for projecting out E and B modes without any differentiation has been demonstrated recently [3]; there, it requires the contour of integration to be strictly a circle.

REFERENCES

1. Naselsky, P.D. and Novikov, D.I. 1998, ApJ, 507,31.
2. Chiueh, T. 2000, astro-ph/0010433.
3. Chiueh, T. and Ma, C.J. 2001, astro-ph/0101205.

New Method of Extracting non-Gaussian Signals in the CMB

Jiun-Huei Proty Wu

Astronomy Department, University of California, Berkeley, CA 94720-3411, USA

Abstract. Searching for and characterizing the non-Gaussianity (NG) of a given field has been a vital task in many fields of science, because we expect the consequences of different physical processes to carry different statistical properties. Here we propose a new general method of extracting non-Gaussian features in a given field, and then use simulated cosmic microwave background (CMB) as an example to demonstrate its power. In particular, we show its capability of detecting cosmic strings.

With the cosmological principle as the basic premise, two currently competing theories for the origin of structure in the universe are inflation [1] and topological defects [2,3]. Although the recent CMB observations seem to have favored the former [4], the latter can still coexist with it. In particular, the observational verification of defects will have certain impact to the grand unified theory, since they are an inevitable consequence of the spontaneous symmetry-breaking phase transition in the early universe. In addition to the conventional study of the power spectra of cosmological perturbations, another way to distinguish these models is via the search for intrinsic NG—while the standard inflationary models predict Gaussianity, theories like isocurvature inflation [5] and topological defects [6] generate NG. Here we shall propose a new method of extracting the NG from a given field [7], and then apply it to the CMB, which is arguably the cleanest cosmic signals [8].

The new method aims to nothing but removing the 'Gaussian' components: the mean and the power. Using an n-dimensional field $\Delta(\mathbf{x})$ as an example, the method first Fourier transforms $\Delta(\mathbf{x})$ to yield $\tilde{\Delta}(\mathbf{k})$. Then the power spectrum can be estimated as $C_k = \langle|\tilde{\Delta}(\mathbf{k})|^2\rangle_k/V^n$, where V^n is the n-dimensional volume of the field and $k \equiv |\mathbf{k}|$. Next we define and calculate ($\forall \mathbf{k}$ with $C_k \neq 0$)

$$\tilde{\Delta}_P(\mathbf{k}) = \left[\tilde{\Delta}(\mathbf{k}) - \tilde{\Delta}(\mathbf{0})\delta(\mathbf{k})\right] C_k^{-1/2} P_k^{1/2}, \qquad (1)$$

where $\delta(\mathbf{k})$ is a Dirac Delta and P_k is a given function of k. Finally, $\tilde{\Delta}_P(\mathbf{k})$ is transformed back to the real space $\Delta_P(\mathbf{x})$. Now the field Δ_P has a mean $\overline{\Delta}$ equal to zero and a power spectrum renormalized to P_k. For the simplest case $P_k = 1$, the field Δ is 'whitened' in the Fourier space, and we shall use the superscript 'W' to

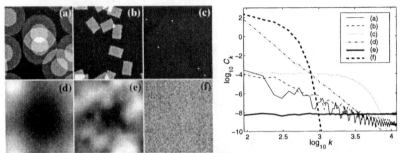

FIGURE 1. Six different components in a simulated CMB map, and their power spectra.

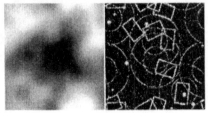

FIGURE 2. A simulated CMB map Δ (left), and the extracted NG signal $(\Delta^W)^2$ (right).

denote such whitened fields. In the real space, this means $\Delta = \Delta^W \otimes D + \overline{\Delta}$, where $\otimes$ denotes a convolution, and $D \equiv \int dk^n C_k^{1/2} e^{i\mathbf{k}\cdot\mathbf{x}}$. Thus Δ is now decomposed into two parts: the 'Gaussian components', $\overline{\Delta}$ and D, which carry the information in the mean and the power spectrum, and the 'NG component', Δ^W, which possesses all the remaining information. Therefore, if Δ is a Gaussian field, then all samples in Δ^W should appear uncorrelated as pure white noise. Otherwise Δ^W would contain all the non-Gaussian features [7]. We note that the above treatment can be easily converted to the conventional multipole transform for the CMB, although we shall continue to use the Fourier convention, which is appropriate for small CMB fields. In this case, we have $\ell \equiv k$ and $C_\ell \equiv C_k$. We also notice that the above new method is equivalent to the matrix manipulation $\mathbf{d}_P = \mathbf{P}^{1/2}\mathbf{C}^{-1/2}(\mathbf{d} - \overline{\mathbf{d}})$, where $\mathbf{d} \equiv \Delta$, and $\mathbf{P}$ and $\mathbf{C}$ are the two-point correlation matrices specified by P_k and C_k respectively. This is similar but different from the Wiener filtering.

We now test this formalism using simulations. Figure 1 shows six simulated CMB components: (a)–(d) (where (d) contains 5 diffuse points) are non-Gaussian, while (e) and (f) are Gaussian. They are then linearly summed to yield $\Delta = \sum_i \Delta_{(i)}$, with RMS ratios ((a)–(f)) 1 : 1 : 10 : 500 : 1000 : 0.2 (Figure 2 left). We then apply our method to obtain the extracted NG signal Δ^W (Figure 2 right).

In a second test, we simulate a CMB field of $(2°)^2$ (Figure 3(c)): $\Delta_s = [\Delta W_p] \otimes W_o + \Delta_{noi}$, where $\Delta = \Delta_{bg}$ (Gaussian background) $+\Delta_{SISW}$ (string-induced CMB [9]; Figure 3(b)) $+\Delta_{pnt}$ (point source; Figure 3(a)) with RMS ratios 5 : 1 : 2, Δ_{noi} is a 5% noise, and W_i with $i =$ 'p' and 'o' denote the primary and observing beams respectively. The whitened field Δ_s^W is shown in Figure 3 (d).

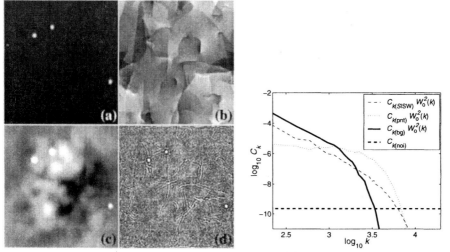

FIGURE 3. Simulated CMB, the extracted NG signal (d), and the power spectra (right).

With even more tests, the main observation remains the same: in a field $\Delta = \Delta_{(G)} + \Delta_{(NG)}$, regardless how stronger the $\Delta_{(G)}$ is, the NG features of $\Delta_{(NG)}$ can always show up in the whitened field Δ^W as long as $C_{k(NG)}$ dominates $C_{k(G)}$ within a certain range of k. In fact, this can be analytically proved [7]. In addition, the NG features of uncorrelated NG components do not mix up in the extracted NG field Δ^W, even if some of them dominate the others in power.

Finally we notice that according to equation (1), in principle we can design a 'window function' P_k to keep the power only on scales where the NG components of a field dominate. However, in general we do not know what these scales are and thus taking $P_k = 1$ is optimal. This may even enable us to find the NG signals of unknown physical processes. We acknowledge the support from NSF KDI Grant (9872979) and NASA LTSA Grant (NAG5-6552).

REFERENCES

1. Guth, A. H., *Phys.Rev.*, **D23**, 347 (1981).
2. For a review see Vilenkin, A., Shellard, E. P. S., *Cosmic strings and other topological defects*, Cambridge University Press, Cambridge, 1994.
3. Wu, J. H. P., Avelino, P. P., Shellard, E. P. S., Allen, B., *astro-ph/9812156* (2001).
4. Jaffe, A. H. et al., *Phys.Rev.Lett.* in press, *astro-ph/0007333* (2001).
5. Peebles, P. J. E., *ApJ.*, **510**, 523 (1999); Peebles, P. J. E., *ApJ.*, **510**, 531 (1999).
6. Avelino, P. P., Shellard, E. P. S., Wu, J. H. P., Allen, B., *ApJ.*, **507**, L101 (1998).
7. Wu, J. H. P., *astro-ph/0012206* (2000).
8. For a review, see Hu, W., Sugiyama, N., Silk, J., *Nature*, **386**, 37 (1997).
9. Wu, J. H. P., (in preparation)

MAXIMA: Millimeter-wave Anisotropy Experiment Imaging Array

C. Winant (1), M. Abroe (2), P. Ade (3), A. Balbi (4,5), D. Barbosa (6), J. Bock (7,8), J. Borrill (4,6), A. Boscaleri (9), J. Collins (4), P. de Bernardis (10), P. Ferreira (11), S. Hanany (2), V. Hristov (7), A. H. Jaffe (4), B. Johnson (2), A. E. Lange (7), A. T. Lee (4,6), P. D. Mauskopf (12), C. B. Netterfield (13), S. Oh (6), E. Pascale (9), B. Rabii (4), P. L. Richards (4,6), R. Stompor (4), G. F. Smoot (4,6), J. H. P. Wu (4)

1 Department of Physics, University of California at Berkeley, cwinant@physics.berkeley.edu
2 University of Minnesota, Twin Cities; 3 Queen Mary and Westfield College; 4 University of California, Berkeley; 5 University of Rome (II); 6 Lawrence Berkeley National Laboratory; 7 California Institute of Technology; 8 Jet Propulsion Laboratory; 9 IROE-CNR; 10 University of Rome (I); 11 University of Oxford; 12 University of Cardiff, Wales; 13 University of Toronto

Abstract.

We discuss the status of the data obtained from the first two flights of the MAXIMA balloon-borne experiment. MAXIMA is sensitive to CMB fluctuations on angular scales from 10 arcmin to 5 degrees. The instrument uses a 16 element bolometric array with 3 frequency bands centered at 150, 240, and 410 GHz.

An angular power spectrum of the CMB anisotropy has been obtained from the data of the first flight, MAXIMA-1, which shows a peak at $\ell \sim 220$ of $78 \pm 6\mu K$ and an amplitude varying between 40 μK and 50 μK from $400 < \ell < 785$ [1]. During a second flight (MAXIMA-2), in 1999, we mapped an additional 225 square degree region of the sky, nearly twice the area of MAXIMA-1.

Data from MAXIMA-1 are consistent with adiabatic inflationary models with the total energy density, $\Omega = 1.0^{+0.15}_{-0.3}$, the total baryon density, $\Omega_b h^2 = 0.03 \pm 0.01$, and the spectral index of the initial power spectrum, $n_s = 1.08 \pm 0.1$ [2]. Limits are quoted at the 95% confidence level. Data from MAXIMA have been combined with those from LSS observations, SNIa observations, COBE, and BOOMERanG, and have been used for cosmological parameter estimation [3]. MAXIMA data can be obtained at http://cfpa.berkeley.edu/maxima.

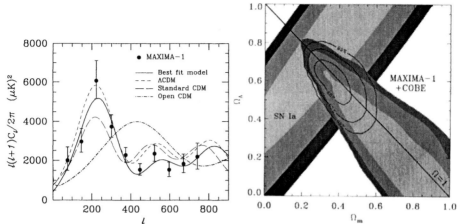

FIGURE 1. left: The angular power spectrum from MAXIMA-1. The solid line is the bestfit model of our analysis, having $\{\Omega_b, \Omega_{cdm}, \Omega_\Lambda, n_s, h\} = \{0.105, 0.595, 0.3, 1.08, 0.53\}$. The long-dashed line is the Standard CDM model $\{0.05, 0.95, 0.1, 1, 0.5\}$. The short-dashed line is the Λ-CDM model with $\{0.05, 0.35, 0.6, 1, 0.65\}$. The dot-dashed line is the Open CDM model with $\Omega = 0.35$. **right:** Constraints on the $\Omega_m - \Omega_\Lambda$ plane from the combined MAXIMA-1 and COBE/DMR sets. The borders of the regions correspond to the 0.32, 0.05, and 0.01 of the peak value of the likelihood. These are overlaid on the bounds obtained from high redshift supernovae data [8], [9]. The closed contours are the confidence levels from the obtained likelihood. Both figures are taken from Balbi et al [2].

Lee et al [4] gives a detailed description of the MAXIMA system. It is an off-axis Gregorian telescope mounted on an attitude controlled balloon platform. The receiver consists of 16 photometers: eight operating at 150 GHz, four at 240 GHz and four at 410 GHz. All have 10' FWHM beams. CMB radiation is detected with spider web bolometers [5] operated at 0.1K. The combined sensitivity of the eight 150 GHz photometers for MAXIMA-1 was 41 $\mu K\sqrt{sec}$.

MAXIMA-1 was launched at 00:58 UT on August 2, 1998, from the National Scientific Balloon Facility (NSBF) in Palestine, Texas. The CMB anisotropy was measured in two overlapping observations. The first observation lasted 1.7 hours and the second lasted 1.4 hours. Due to the rotation of the sky, the two scans were cross-linked at an average angle of 22°. Approximately 122 square degrees of the sky were observed in a region of low dust contrast.

MAXIMA-2 was launched at 00:07 UT June 17, 1999, also from the NSBF in Palestine, TX. Again, two overlapping CMB observations were made. The first observation lasted 2.4 hours and the second lasted 2.2 hours. The two scans were cross-linked at an average angle of 27°. Approximately 255 square degrees of the sky were observed in a region of low dust contrast, with roughly 50 degrees of overlap with the observations of MAXIMA-1.

For both flights, a full beam calibration of the 150 and 240 GHz photometers was obtained from the CMB dipole. A calibration uncertainty of 4% was obtained for MAXIMA-1. Beam contour maps and were obtained from observations of planets (Jupiter in MAXIMA-1, Mars in MAXIMA-2). Planet observations were also used as a secondary source of calibration.

First results from MAXIMA-1 have been published, including a CMB map and power spectrum [1]. The map consists of 15,000 5' by 5' pixels and is derived from four independent CMB detectors in the 150 GHz and 240 GHz bands. Data from all four detectors were found to be statistically consistent. Data from 410 GHz detectors indicate negligible levels of dust emission in the observed region.

From these data, a power spectrum has been estimated in 10 bins spanning the range $36 < \ell < 785$. The amplitude in each bin is found by a maximum likelihood analysis using a quadratic estimator technique [6]. The power spectrum shows a clear peak at $\ell \sim 220$. The MAXIMA-1 power spectrum constrains a variety of cosmological parameters, including the total density ($\Omega = 1.0^{+0.15}_{-0.3}$), the baryon density ($\Omega_b h^2 = 0.03 \pm 0.01$), and the spectral index of the initial power spectrum ($n_s = 1.08 \pm 0.1$), with limits quoted at the 95% confidence level [2].

Further constraints are obtained by combining MAXIMA data with LSS and SNIa observations, and with other CMB data. The results of MAXIMA and BOOMERanG are highly complimentary [3]. MAXIMA-1 provides strong constraints at high ℓ, while BOOMERanG provides greater resolution and sensitivity at intermediate ℓ [7].

Data from the MAXIMA-2 flight are being analyzed. The MAXIMA telescope is currently being upgraded for use as a polarimeter in a followup experiment, MAXIPOL.

MAXIMA is supported by NASA through grants NAG5-4454 and NAG5-3941, and by the Center for Particle Astrophysics, a National Science and Technology Center operated by the University of California, Berkeley, under Cooperative Agreement No. AST 9120005. The author is supported by NASA through grant NGT5-50195.

REFERENCES

1. Hanany, S., *et al.* 2000, ApJ, 545L, L5
2. Balbi, A., *et al.* 2000, ApJ, 545L, L1
3. Jaffe, A. H., *et al.* 2001, Phys.Rev.Lett. in press, *astro-ph/0007333*
4. Lee, A. T., *et al.* 1998, Proceedings of 3K Cosmology Conference, Rome, Italy, *astro-ph/9903249*
5. Bock, J. J., *et al.* 1998, ESA proceedings
6. Bond, J. R., Jaffe, A. H., and Knox, L. 1998, PhysRev, D57, 2117
7. Lange, A. E., *et al.* 2001, Phys. Rev. D., 63, 042001
8. Perlmutter, S., *et al.* 1999, ApJ, 517, 565
9. Riess, A. G., *et al.* 1998, AJ, 116, 1009

CHAPTER 5

LARGE-SCALE STRUCTURE, GALAXY FORMATION, AND EVOLUTION OF THE INTERGALACTIC MEDIUM

Cosmological Reionization

Paul R. Shapiro

Dept. of Astronomy, The University of Texas, Austin, TX 78712 USA

Abstract. The universe was reionized by redshift $z \approx 6$ by a small fraction of the baryons in the universe, which released energy following their condensation out of a cold, dark, and neutral IGM into the earliest galaxies. The theory of this reionization is a critical missing link in the theory of galaxy formation. Its numerous observable consequences include effects on the spectrum, anisotropy and polarization of the cosmic microwave background and signatures of high-redshift star and quasar formation. This energy release also created feedback on galaxy formation which left its imprint on the mass spectrum and internal characteristics of galaxies and on the gas between galaxies long after reionization was complete. Recent work suggests that the photoevaporation of dwarf galaxy minihalos may have consumed most of the photons required to reionize the currently-favored ΛCDM universe. We will review recent developments in our understanding of this process.

I INTRODUCTION

Observations of quasar absorption spectra indicate that the universe was reionized prior to redshift $z = 5$ (e.g. [68]). CMB anisotropy data on the first acoustic peak set limits on the electron scattering optical depth of a reionized IGM which imply $z_{\rm rei} \lesssim 40$ (model-dependent) [26]. Together with the lack of a hydrogen Lyα resonance scattering ("Gunn-Peterson") trough in the spectrum of an SDSS quasar discovered at $z = 5.8$ [19], these limits currently suggest that $6 \lesssim z_{\rm rei} \lesssim 40$. The origin and consequences of this reionization are among the major unsolved problems of cosmology. (For prior reviews and further references, see, e.g. [3,29,55].) Photons emitted by hitherto undetected massive stars or miniquasars formed within early galaxies are generally thought to be the reionization source.

This is consistent with the theoretical expectation in cosmological models like the Cold Dark Matter (CDM) model, in which the first objects to condense out of the background and possibly begin star formation were small, of subgalactic mass (i.e. $M \lesssim 10^6 M_\odot$) and began to form as early as $z \gtrsim 30$ (e.g. [25,30,57,71]). The initial collapse of such objects only led to star formation if radiative cooling was possible after collapse, usually involving H_2 molecules, and it is likely that the first objects to form stars released radiation (and possibly SN explosion energy as well) which exerted a strong feedback on the subsequent formation history of other objects. The details of this feedback and even the overall *sign* (i.e. negative or positive) are poorly understood (e.g. [8,14,17,20,28,31,32,36,55,57,59]). It appears that the first stars would have photodissociated H_2 long before enough UV was emitted to bring about reionization, so it is currently thought that if reionization was accomplished by stars, they formed in objects with virial temperature $T_{\rm vir} > 10^4$K, of mass $\gtrsim 10^8 M_\odot$, which were able to cool by atomic radiative cooling, even without H_2. This conclusion changes, however, if the first sources were miniquasars whose nonthermal spectra had a significant X-ray flux, since this would have created a *positive* feedback on the H_2.

If we adopt an optimum efficiency for massive star formation and radiation release by the collapsed baryon fraction in a standard CDM model which is flat, matter-dominated, and COBE-normalized (an unrealistic model which has too high an amplitude to satisfy X-ray cluster abundance constraints at $z = 0$, but conservatively overestimates $z_{\rm rei}$), $z_{\rm rei} \approx 50$ is possible [55]. But more suitable CDM models all tend to yield $z_{\rm rei} \lesssim 20$ (e.g. [4,13,15,25,30,57,74]).

II COSMOLOGICAL IONIZATION FRONTS

Ionization Fronts in a Clumpy Universe. The neutral, opaque IGM out of which the first bound objects condensed was dramatically reheated and reionized at some time between a redshift $z \approx 50$ and $z \approx 6$ by the radiation released by some of these objects. When the first sources turned on, they ionized their surroundings by propagating weak, R-type ionization fronts which moved outward supersonically with respect to both the neutral gas ahead of and the ionized gas behind the front, racing ahead of the hydrodynamical response of the IGM [54,56]. The problem of the time-varying radius of a spherical I-front which surrounds isolated sources in a cosmologically-expanding IGM was solved analytically by [54,56], taking proper account of the I-front jump condition generalized to cosmological conditions. They applied these solutions to determine when and how fast these I-front-bounded spheres would grow to overlap and, thereby, complete the reionization of the universe. The effect of density inhomogeneity on the rate of I-front propagation was described by a mean "clumping factor" $c_l > 1$, which slowed the I-fronts by increasing the average recombination rate per H atom inside clumps. This suffices to describe the average rate of I-front propagation as long as the clumps are either not self-shielding or, if so, only absorb a fraction of the ionizing photons emitted by the central source. In two recent calculations, this analytical prescription was adapted to N-body simulations of structure formation in the CDM model, to calculate the evolving size of the spherical H II regions with which to surround putative sources of ionizing radiation, to model the growth of the ionized volume filling factor leading to reionization [4,15].

Numerical radiative transfer methods are currently under development to solve this problem in 3D for the inhomogeneous density distribution which arises as cosmic structure forms, but so far are limited to an imposed density field without gas dynamics (e.g. [1,16,45,51]). A different approach, which replaces radiative transfer with a "local optical depth approximation," intended to mimic the average rate at which I-fronts expanded and overlapped during reionization within the context of cosmological gas dynamics simulation, has also been developed [22]. These recent attempts to model inhomogeneous reionization numerically are handicapped by their limited spatial resolution ($\gtrsim 1\,\text{kpc}$), which prevents them from resolving the most important density inhomogeneities. The dynamical response of density inhomogeneities to the I-fronts which encountered them and the effect which these inhomogeneities had, in turn, on the progress of universal reionization, therefore, require further analysis. Toward this end, we have developed a radiation-hydrodynamics code which incorporates radiative transfer and have focused our attention on properly resolving this small-scale structure. In what follows, we summarize the results of new radiation-hydrodynamical simulations of what happens when a cosmological I-front overtakes a gravitationally-bound density inhomogeneity – a dwarf galaxy minihalo – during reionization. According to [27], the photoevaporation of these sub-kpc-sized objects is likely to be the dominant process by which ionizing photons were absorbed during reionization, so this problem is of critical importance in determining how reionization proceeded.

Dwarf Galaxy Minihalos at High Redshift. The effect which small-scale clumpiness had on reionization depended upon the sizes, densities, and spatial distribution of the clumps overtaken by the I-fronts during reionization. For the currently-favored ΛCDM model ($\Omega_0 = 1 - \lambda_0 = 0.3$, $h = 0.7$, $\Omega_b h^2 = 0.02$, primordial power spectrum index $n_p = 1$; COBE-normalized), the universe at $z > 6$ was already filled with dwarf galaxies capable of trapping a piece of the global, intergalactic I-fronts which reionized the universe and photoevaporating their gaseous baryons back into the IGM (see Figure 1). Prior to their encounter with these I-fronts, "minihalos" with $T_{\text{vir}} < 10^4\,\text{K}$ were neutral and optically thick to hydrogen ionizing radiation, as long as their total mass exceeded the Jeans mass M_J in the unperturbed background IGM prior to reionization [i.e. $M_J = 5.7\times10^3(\Omega_0 h^2/0.15)^{-1/2}(\Omega_b h^2/0.02)^{-3/5}((1+z)/10)^{3/2} M_\odot$], as was required to enable baryons to collapse into the halo along with dark matter. Their "Strömgren numbers" $L_S \equiv 2R_{\text{halo}}/\ell_S$, the ratio of a halo's diameter to its Strömgren length ℓ_S inside the halo (the length of a column of gas within which the unshielded arrival rate of ionizing photons just balances the total recombination rate), were large. For a uniform gas of H density $n_{H,c}$, located a distance r_{Mpc} (in Mpc) from a UV source emmitting $N_{\text{ph},56}$ ionizing photons (in units of 10^{56}s^{-1}), the Strömgren length is only $\ell_S \approx (100\,\text{pc})(N_{\text{ph},56}/r_{\text{Mpc}}^2)(n_{H,c}/0.1\,\text{cm}^{-3})^{-2}$, so $L_S \gg 1$ for a wide range of halo masses and sources of interest. In that case, the intergalactic, weak, R-type I-front which entered each minihalo during reionization would have decelerated to about twice the sound speed of the ionized gas before it could exit the other side, thereby transforming itself into a D-type front, preceded by a shock. Typically, the side facing the source would then have expelled a supersonic wind backwards toward the source, which shocked the surrounding IGM as the minihalo photoevaporated.

The importance of this photoevaporation process has long been recognized in the study of interstellar clouds exposed to ionizing starlight (e.g. [5,6,38,39,41,43,47,53,69]). In the cosmological context, however, its importance has only recently been fully appreciated.

In proposing the expanding minihalo model to explain Lyα forest ("LF") quasar absorption lines, [9] discussed how gas originally confined by the gravity of dark-matter minihalos in the CDM model would have been expelled by pressure forces if photoionization by ionizing background radiation suddenly heated all the gas to an isothermal condition at $T \approx 10^4$K, a correct description only in the optically thin limit. The first discussion of the photoevaporation of a primordial density inhomogeneity overtaken by a cosmological I-front, including radiation-hydrodynamical simulations, however, was in [65,66]. [2] subsequently estimated the relative importance of this process for dwarf galaxy minihalos of different masses at different epochs in the CDM model, concluding that 50%–90% of the gas which had already collapsed into gravitationally bound objects when reionization occurred should have been photoevaporated.

Not only did this photoevaporation during reionization affect most of the collapsed baryons, but so common were minihalos with $T_{\rm vir} < 10^4 K$ during reionization that the typical reionizing photon is likely to have encountered one of them, according to [27]. A catalogue of all halos in a cubic volume 100 kpc (50 kpc) on a side at $z = 9$ in the currently-favored ΛCDM model based on N-body simulations reveals that minihalos with $T_{\rm vir} < 10^4 K$ (i.e. $M < 10^{7.6} M_\odot$) were separated on average by only $d \approx 7$ kpc (3.5 kpc) for $M \geq 10^{5.6} M_\odot$ ($M \geq 10^{4.7} M_\odot$), respectively, while their geometric cross sections together covered $f \sim 16\%$ (30%) of the area along every 100 kpc of an average line of sight [60] [see Figs. 1(b), (c)]. If the

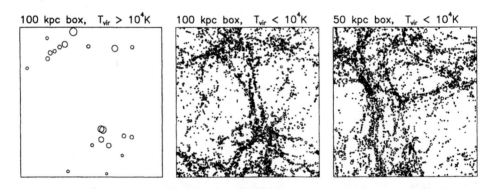

FIGURE 1. ΛCDM Halos at High Redshift: DM Halos at $z = 9$ are shown with sizes and locations determined by FOF algorithm applied to P^{3}M simulations (128^3 particles, 256^3 cells), projected onto face of simulation cube of proper size $L_{\rm box}$, as labelled. (a) (left) $M_{\rm halo} > 10^{7.6} M_\odot$ only (i.e. $T_{\rm vir} > 10^4 K$). (b) (middle) $10^{5.6} M_\odot < M_{\rm halo} < 10^{7.6} M_\odot$ only (i.e. $T_{\rm vir} < 10^4 K$) – THE MINIHALOS; (c) (right) like (b), but higher resolution simulation with same number of particles and cells in 1/8 volume (i.e. $M_{\rm halo,min} = 10^{4.7} M_\odot$).

sources of reionization, on the other hand, were larger-mass halos with $T_{\rm vir} > 10^4 K$ [like those in Fig. 1(a)], then these were well-enough separated that typical reionization photons were likely to have been absorbed by intervening photoevaporating minihalos. To demonstrate this in a statistically meaningful way with more dynamic range than the N-body results can yet provide, an analytical approximation to the detailed numerical results is required. We shall combine the well-known Press-Schechter (PS) prescription for deriving the average number density of halos of different mass at each epoch with the nonsingular truncated isothermal sphere (TIS) model of [34,58] for the size, density profile, and temperature of each halo as unique functions of the halo mass and collapse redshift for a given background universe. (The TIS model and further tests and applications of it are briefly mentioned elsewhere in this volume [35]). We illustrate the validity of these aproximations in Figure 2 by comparing them with the N-body simulation results depicted in Figure 1. The PS halo mass functions plotted in Figure 2(a) are shown in Figure 2(b) to reproduce the N-body results over a range of 10^5 in halo mass, for which $dn_{\rm halo}/dM \approx 10^{12}(M/M_\odot)^{-2}{\rm Mpc}^{-3}M_\odot$ (proper units) at $z = 9$. Likewise, Figure 2(c) shows that the TIS model correctly predicts the average virial ratio, $[GM_{200}/(\sigma_V^2 r_{200})]_{\rm TIS} = 2.176$, for halos of different mass according to the N-body results, over this same mass range.

This TIS+PS approximation allows us to determine which halo masses are subject to photoevaporation and how common they are, as functions of redshift. Each minihalo with $T_{\rm vir} < 10^4{\rm K}$ is opaque to H-ionizing photons with a geometric cross section $\sigma_{\rm halo} = \pi r_t^2$, where $r_t = $ TIS radius $\approx (0.75\,{\rm kpc})(M/10^7 M_\odot)^{1/3}[(1 + $

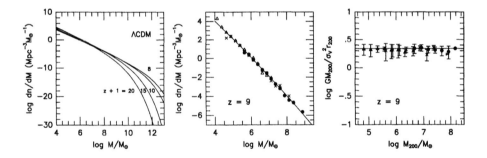

FIGURE 2. CDM Halos: N-body Results Vs. PS+TIS Approximations: (a) (left) Proper Number Density of Halos per Unit Halo Mass, $dn_{\rm halo}/dM$, in ΛCDM at different redshifts, as labelled, according to PS approximation; (b) (middle) Same as (a), with PS mass function at $z = 9$ (solid curve) compared with N-body results for halos in Fig. 1 for three different resolutions: $L_{\rm box} = 100\,{\rm kpc}$ (circles), $50\,{\rm kpc}$ (crosses), and $25\,{\rm kpc}$ (triangles); (c) Virial ratio, $GM_{200}/(\sigma_V^2 r_{200})$, versus halo mass M_{200}, where σ_V^2 is the halo DM velocity dispersion and M_{200} and r_{200} are the mass and radius of the sphere whose average density is 200 times the cosmic mean value, for all halos in Fig. 1 with at least 200 particles per halo [symbols same as in (b), with 1σ error bars]. Horizontal line is analytical prediction of the TIS model.

$z_{\rm coll})/10]^{-1}(\Omega_0 h^2/0.15)^{-1/3}$. Ionizing photons travelling through this universe will suffer absorption by these minihalos [i.e. those with mass $M_J \leq M \leq M(10^4 K)$ at each z] with a mean free path $\lambda_{\rm mfp}$, as shown in Figure 3(a). For comparison, the mean separation $<d_{\rm sep}>$ of halos of each mass is plotted in Figure 3(b). At $z = 9$, for example, $\lambda_{\rm mfp} = 160$ kpc, while halos of mass $M \gtrsim 10^8 M_\odot$ are separated on average by $<d_{\rm sep}> \approx 50(M/10^8 M_\odot)^{1/3}$kpc. As shown in Figure 3(c), their ratio, $<d_{\rm sep}>/\lambda_{\rm mfp}$, gives the fraction of the sky, as seen by a source halo of a given mass, which is covered by opaque minihalos located within the mean volume per source halo. If halos with $M \gtrsim 10^8 M_\odot$ are the reionization sources, their minihalo covering fraction is close to unity and increases with increasing source mass. This estimate will increase by a factor of a few if we take account of the statistical bias by which minihalos tend to cluster around the source halos [27].

An argument like this led [27] to argue that our photoevaporating minihalos are the chief consumers of the ionizing photons responsible for reionization. As a result, they suggest, the photoevaporation of these minihalos drives the number of ionizing photons per baryon required to reionize the universe up by an order of magnitude compared to previous estimates! A recent semi-analytical study of inhomogeneous reionization by [44], which neglected this effect, concluded that only one ionizing photon per hydrogen atom would have been sufficient to reionize most of the volume of the IGM by $z \approx 5$, an order of magnitude too low when compared to the new photoevaporation-dominated estimates. The gas clumping model they adopted apparently missed the smallest scales because it was adjusted to match numerical simulation results which could not resolve these scales. Simulations by [22], which agreed with [44], have the same problem since their resolution limit exceeded ~ 1 kpc, too large to resolve the minihalos which photoevaporate.

We are led to conclude that further study of the photoevaporation of cosmological minihalos during reionization is essential if we are to advance the theory of

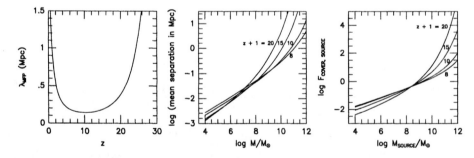

FIGURE 3. Minihalo Sinks and Sources: (a) (left) Proper mean free path $<n_{\rm halo}\sigma_{\rm halo}>^{-1}$ for absorption of photons by minihalos in ΛCDM at different z (i.e. if they photoevaporate at $z_{\rm ev} < z$); (b) (middle) Proper mean separation of halos $<d_{\rm sep}> \equiv (M \, dn_{\rm halo}/dM)^{-1/3}$ versus halo mass at different redshifts in ΛCDM, as labelled; (c) (right) Fraction of sky covered by minihalos located within the mean volume per source halo, $F_{\rm cover, source} = <d_{\rm sep}>/\lambda_{\rm mfp}$, versus source halo mass, at different redshifts.

reionization. The new "photoevaporation-dominated" reionization scenario suggests that significantly more than one photon per baryon may have been required, and this is difficult to understand on the basis of simple extrapolation (e.g. [21,42]) of either quasar or stellar photon production rates observed at $z < 5$ to $z > 5$. Perhaps this problem will be alleviated by appeal to the recently revised stellar output for zero metallicity stars [10,73], or to a higher escape fraction of ionizing photons from their source than the nominal $f_{\rm esc} \approx 0.1$ typically assumed (e.g. [18,40,52,70,76]). Alternatively, if minihalo sources, alone, reionized the IGM (e.g. with miniquasars instead of starlight), then their ionizing photons could have done so without encountering any other minihalos, since $F_{\rm cover,source} < 1$ for halos with $T_{\rm vir} < 10^4$K. Photons released, thereafter, would have encountered fewer opaque minihalos, as well, since baryons would not then have condensed out of a reheated IGM to form *new* minihalos, according to [57].

The good news is that observations should be able to distinguish these possibilities. In order for photoevaporating minihalos to have consumed most of the ionizing photons produced before reionization is complete, the covering fraction of the ionization sources by these photoevaporative flows must be of order unity. Hence, if we can observe the sources of universal reionization directly, we will generally be able to observe foreground photoevaporating minihalos in absorption toward these sources. Such observations will constrain and help diagnose the reionization process. As shown in Figure 4, photons emitted by *any* high z source before or during reionization will typically encounter large numbers of photoevaporating minihalos at $z > 6$. The number of minihalos probed per unit redshift interval, in fact, is just $(1 + z)^{-1}$ times the ratio of the horizon size $c/H(t)$ at z to $\lambda_{\rm mfp}$, a large number. It is important, therefore, for us to predict the effect which reionization will have on the gas in these minihalos.

The Photoevaporation of Dwarf Galaxy Minihalos Overtaken by Cosmological Ionization Fronts. We have performed radiation-hydrodynamical

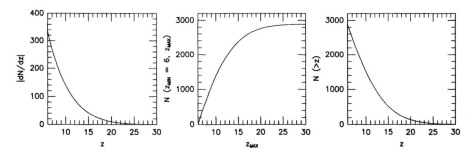

FIGURE 4. Minihalos Encountered By a Photon Emitted At High Redshift: (a) (left) Number of minihalos per unit redshift interval in ΛCDM encountered by a photon at redshift z which travels along the average LOS; (b) (middle) Total number of minihalos in ΛCDM along photon path which probes redshift interval $z_{\min} \le z \le z_{\max}$ for $z_{\min} = 6$ is plotted versus $z_{\max}$; (c) (right) Asymptotic number of halos along photon path which probes redshift interval (z, ∞).

simulations of the photoevaporation of a cosmological minihalo overrun by a weak, R-type I-front in the surrounding IGM, created by an external source of ionizing radiation [61–64]. The gas contained H, He, and a possible heavy element abundance of 10^{-3} times solar. Our simulations in 2D, axisymmetry used an Eulerian hydro code with Adaptive Mesh Refinement and the Van Leer flux-splitting algorithm, which solved nonequilibrium ionization rate equations (for H, He, C, N, O, Ne, and S) and included an explicit treatment of radiative transfer by taking into account the bound-free opacity of H and He [43,49,50]. The reader is referred to [65–67] for earlier results which considered uniform clouds and demonstrated the importance of a proper treatment of optical depth.

Here we compare some of our results of minihalo photoevaporation in a ΛCDM universe [64] for two types of source spectra: a quasar-like source with emission spectrum $F_\nu \propto \nu^{-1.8}$ ($\nu > \nu_H$) and a stellar source with a 50,000 K blackbody spectrum, with luminosity and distance adjusted to keep the ionizing photon fluxes the same in the two cases. In particular, if r_{Mpc} is the proper distance (in Mpc) between source and minihalo and $N_{\text{ph},56}$ is the H-ionizing photon luminosity (in units of 10^{56} s^{-1}), then the flux at the location of the minihalo would, if unattenuated, correspond initially to $N_{\text{ph},56}/r_{\text{Mpc}}^2 = 1$; thereafter, $r_{\text{Mpc}} \propto a(t)$, the cosmic scale factor.

Our initial condition before ionization is that of a $10^7 M_\odot$ minihalo in the ΛCDM universe which collapsed out and virialized at $z_{\text{coll}} = 9$, yielding a truncated, nonsingular isothermal sphere ("TIS") of radius $r_t = 0.75$ kpc in hydrostatic equilibrium with $T_{\text{vir}} = 4000$ K and dark-matter velocity dispersion $\sigma_V = 5.2$ km s^{-1}. The TIS

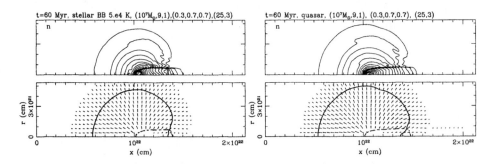

FIGURE 5. Photoevaporating Minihalo I. One time-slice, 60 Myr after I-front caused by ionizing source (located far to the left of computational box along the x-axis) overtakes a $10^7 M_\odot$ minihalo [centered at $(r, x) = (0, 1.06 \times 10^{22}$cm)] at $z = 9$ in the ΛCDM universe, for two types of source spectra: (a) (left) STELLAR CASE and (b) (right) QUASAR CASE. (upper panels) isocontours of atomic density, logarithmically spaced, in (r, x)–plane of cylindrical coordinates; (lower panels) velocity arrows are plotted with length proportional to gas velocity. An arrow of length equal to the spacing between arrows has velocity 25 km s^{-1}; minimum velocities plotted are 3 km s^{-1}. Solid line shows current extent of gas initially inside minihalo at $z = 9$. Dashed line is I-front (50% H-ionization contour).

profile has a central density and an average density which are 18,000 and 130 times the mean background density, respectively, with core radius $r_0 \equiv r_{King}/3 \sim r_t/30$. This hydrostatic sphere is embedded in a self-similar, spherical, cosmological infall according to [7].

The results of our simulations on an (r,x)-grid with 256×512 cells are illustrated by Figures 5–9. The minihalo shields itself against ionizing photons, traps the R-type I-front which enters the halo, causing it to decelerate inside the halo to close to the sound speed of the ionized gas and transform itself into a D-type front, preceded by a shock. The side facing the source expels a supersonic wind backwards towards the source, which shocks the IGM outside the minihalo. The wind grows more isotropic with time as the remaining neutral halo material is photoevaporated. Since this gas was initially bound to a dark halo with $\sigma_V < 10 \, \mathrm{km \, s^{-1}}$, photoevaporation proceeds unimpeded by gravity. Figures 5 and 6 show

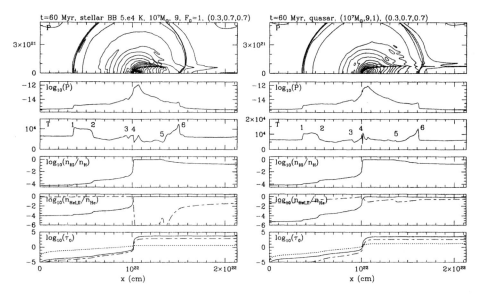

FIGURE 6. Photoevaporating Minihalo II. Same time-slice as Fig. 5. (a) (left) STELLAR CASE and (b) (right) QUASAR CASE. From top to bottom: (i) isocontours of pressure, logarithmically spaced, in (r,x)–plane of cylindrical coordinates; (ii) pressure along the $r=0$ symmetry axis; (iii) temperature; (iv) H I fraction; (v) He I (solid) and He II (dashed) fractions; (vi) bound-free optical depth measured from $x=0$ along the x-axis, at the threshold ionization energies for H I (solid), He I (dashed), He II (dotted). Key features of the flow are indicated by the numbers which label them on the temperature plots: 1 = IGM shock; 2 = contact discontinuity which separates shocked halo wind (between 2 and 3) from swept-up IGM (between 1 and 2); 3 = wind shock; between 3 and 4 = supersonic wind; 4 = I-front; 5 = boundary of gas initially inside minihalo at $z=9$; 6 = shock in shadow region caused by compression of shadow gas by shock-heated gas outside shadow.

the structure of the photoevaporative flow 60 Myrs after the global I-front first overtakes the minihalo, with key features of the flow indicated by the labels on the temperature plot in Figure 6. Figure 7(a) shows how the neutral mass of the gas initially within the original hydrostatic sphere gradually declines as the minihalo photoevaporates, within $t_{\rm ev} \approx 250$ (100) Myrs for the stellar (quasar) cases, respectively. The gradual decay of the opaque cross section of the minihalo as seen by the source is illustrated by Figure 7(b).

Some observational signatures of this process are shown in Figures 8 and 9. Figure 8 shows the spatial variation of the relative abundances of C, N, and O ions along the symmetry axis after 60 Myrs. While the quasar case shows the presence at 60 Myrs of low as well as high ionization stages for the metals, the softer spectrum of the stellar case yields less highly ionized gas on the ionized side of the I-front (e.g. mostly C III, N III, O III) and the neutral side as well (e.g. C II, N I, O I and II). The column densities of H I, He I and II, and C IV for minihalo gas of different velocities as seen along the symmetry axis at different times are shown in Figure 9. At early times, the minihalo gas resembles a weak Damped Lyα ("DLA") absorber with small velocity width ($\gtrsim 10\,\rm km\,s^{-1}$) and $N_{\rm HI} \gtrsim 10^{20}\rm cm^{-2}$, with a Ly$\alpha$-Forest("LF")-like red wing (velocity width $\gtrsim 10\,\rm km\,s^{-1}$) with $N_{\rm HI} \gtrsim 10^{16}\rm cm^{-2}$ on the side moving toward the source, with a He I profile which mimics

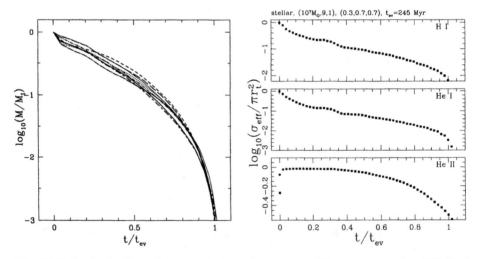

FIGURE 7. Evolution of Neutral Gas Content of Photoevaporating Minihalo. (a) (left) Fraction of mass M_I, the mass which is initially inside the minihalo when the intergalactic I-front overtakes it, which remains neutral versus time t (in units of the evaporation time $t_{\rm ev}$ at which this $M/M_I = 10^{-3}$), for a large range of cases with different assumed source spectra and fluxes and minihalo masses; (b) (right) fraction of minihalo initial geometric cross section πr_t^2 which is opaque to source photons that can ionize H I (top panel), He I (middle panel), or He II (bottom panel) versus time (in units of $t_{\rm ev}$), for the case with stellar source shown in Figs. 5, 6 ($t_{\rm ev} = 245\,\rm Myr$).

that of H I but with $N_{\text{He I}}/N_{\text{H I}} \sim [\text{He}]/[\text{H}]$, and with a weak C IV feature with $N_{\text{C IV}} \sim 10^{11}\,(10^{12})\xi\,\text{cm}^{-2}$ for the stellar (quasar) cases, respectively, displaced in this same asymmetric way from the velocity of peak H I column density, where $\xi \equiv [\text{C}]/[\text{C}]_\odot \times 10^3$. For He II at early times, the stellar case has $N_{\text{He II}} \approx 10^{18}\,\text{cm}^{-2}$ shifted by 10's of km/sec to the red of the H I peak, while for the quasar case,

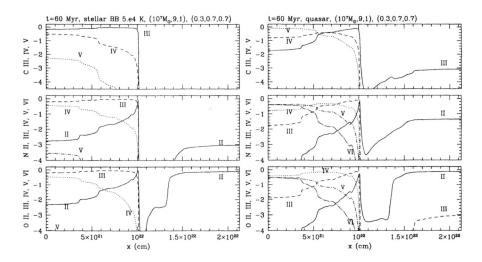

FIGURE 8. Observational diagnostics I: ionization structure of metals. C, N, and O ionic fractions along symmetry axis at $t = 60\,\text{Myr}$, for photoevaporating minihalo of Figs. 5, 6. (a) (left) STELLAR CASE; (b) (right) QUASAR CASE.

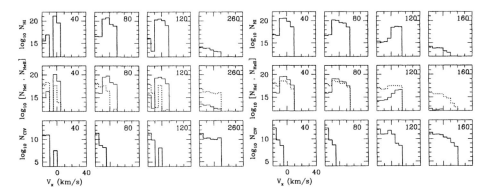

FIGURE 9. Observational diagnostics II. Absorption lines. Minihalo column densities (cm^{-2}) along symmetry axis for gas at different velocities, for photoevaporating minihalo of Figs. 5, 6. (a) (left) STELLAR CASE; (b) (right) QUASAR CASE. (Top) H I; (Middle) He I (solid) and He II (dotted); (Bottom) C IV (i.e. if $[\text{C}]/[\text{C}]_\odot = \xi \times 10^{-3}$, then plotted values are $N_{\text{C IV}}/\xi$). Each box labelled with time (in Myrs) since arrival of intergalactic I-front.

He II simply follows the H I profile, except that $N_{\text{He II}}/N_{\text{H I}} \approx 10$ in the red wing but $N_{\text{He II}}/N_{\text{H I}} \approx 10^{-2}$ in the central H I feature. After 260 (160) Myr, however, only a LF-like H I feature with column density $N_{\text{H I}} \sim 10^{14}\,\text{cm}^{-2}$ remains, with $N_{\text{He I}}/N_{\text{H I}} \sim 1/4\,(\lesssim 1/30)$, $N_{\text{He II}}/N_{\text{H I}} \sim 10^3\,(10^2)$, and $N_{\text{C IV}}/N_{\text{H I}} \sim 3(1) \times [\text{C}]/[\text{C}]_\odot$ for the stellar (quasar) cases, respectively.

As described above, intervening minihalos like these are expected to be ubiquitous along the line of sight to high redshift sources. With photoevaporation times $t_{\text{ev}} \gtrsim 100\,\text{Myr}$, this process can continue down to redshifts significantly below $z = 10$. For stellar sources in the ΛCDM model, these simulations show that photoevaporation of a $10^7\,M_\odot$ minihalo which begins at $z_{\text{initial}} = 9$ can take 250 Myr to finish, at $z_{\text{final}} = 6.8$, during which time such minihalos can survive without merging into larger halos. Observations of the absorption spectra of high redshift sources like those which reionized the universe should reveal the presence of these photoevaporative flows and provide a useful diagnostic of the reionization process.

Dwarf Galaxy Suppression and Reionization. As emphasized by [57], the reheating of the IGM which accompanied its reionization must have filtered the linear growth of baryonic fluctuations in the IGM, thereby reducing the baryon collapsed fraction and preventing baryons from condensing out further into the minihalos (see also [12,13,23,24,74]). A related effect, the suppression of baryon accretion onto dark matter halos, has also been studied, by 1D [33,36,37,72] and 3D simulations (e.g. [46,48,75]). The current conclusion is expressed in terms of the threshold circular velocity, v_c, below which baryonic infall and star formation were suppressed by photoionization: $v_c \sim 30\,\text{km\,s}^{-1}$ for complete suppression, partial suppression of infall found to extend even to $v_c \sim 75\,\text{km\,s}^{-1}$. This suppression of gas accretion onto low-mass halos by the feedback effect of reionization may naturally explain why there are so many fewer dwarf galaxies observed in the Local Group than are predicted by N-body simulations of the CDM model [11].

Acknowledgments: I thank A. Raga, I. Iliev, and H. Martel for their collaboration on the work presented here, supported by grants NASA ATP NAG5-7363 and NAG5-7821, NSF ASC-9504046, and Texas Advanced Research Program 3658-0624-1999.

REFERENCES

1. Abel, T., Norman, M. L., and Madau, P. *ApJ*, **523**, 66 (1999).
2. Barkana, R., and Loeb, A., *ApJ*, **523**, 54 (1999).
3. Barkana, R., and Loeb, A., preprint (2001) (astro-ph/0010468).
4. Benson, A. J., Nusser, A., Sugiyama, N., and Lacey, C. G., *MNRAS*, **320**, 153 (2001).
5. Bertoldi, F., *ApJ*, **346**, 735 (1989).
6. Bertoldi, F., and McKee, C. F., *ApJ*, **354**, 529 (1990).
7. Bertschinger, E., *ApJS*, **58**, 39 (1985).
8. Blanchard, A., Valls-Gabaud, D., and Mamon, G., *A&A*, **264**, 365 (1992).

9. Bond, J. R., Szalay, A. S., and Silk, J., *ApJ*, **324**, 627 (1988).
10. Bromm, V., Kudritzki, R. P., and Loeb, A., preprint (2000) (astro-ph/0007248).
11. Bullock, J. S., Kravtsov, A. V., and Weinberg, D. H., *ApJ*, **539**, 517 (2000).
12. Cen, R., and Ostriker, J. p., *ApJ*, **417**, 404 (1993).
13. Chiu, W. A., and Ostriker, J. P., *ApJ*, **534**, 507 (2000).
14. Ciardi, B., Ferrara, A., and Abel, T., *ApJ*, **533**, 594 (1998).
15. Ciardi, B. Ferrara, A., Governato, F. and Jenkins A., *MNRAS*, **314**, 611 (2000).
16. Ciardi, B., Ferrara, A., Marri, S., and Raimondo, G., T., *MNRAS*, submitted (2000) (astro-ph/0005181).
17. Couchman, H. M. P., and Rees, M. J., *MNRAS*, **221**, 53 (1986).
18. Dove, J. B., Shull, J. M., and Ferrara, A., *ApJ*, **531**, 846 (2000).
19. Fan, X., et al., *AJ*, **121**, 54 (2000).
20. Ferrara, A., *ApJ*, **499**, L17 (1998).
21. Giroux, M. L., and Shapiro, P. R., *ApJS*, **102**, 191 (1996).
22. Gnedin, N. Y., *ApJ*, **535**, 530 (2000).
23. Gnedin, N. Y., *ApJ*, **542**, 535 (2000).
24. Gnedin, N. Y., and Hui, L., *MNRAS*, **296**, 44 (1998).
25. Gnedin, N. Y., and Ostriker, J. P., *ApJ*, **486**, 581 (1997).
26. Griffiths, L. M., Barbosa, D., and Liddle, A. R., *MNRAS*, **308**, 854 (1999).
27. Haiman, Z., Abel, T., and Madau, P., preprint (2000) (astro-ph/0009125).
28. Haiman, Z., Abel, T., and Rees, M. J., *ApJ*, **534**, 11 (1999).
29. Haiman, Z., and Knox, L., in *Microwave Foregrounds* (ASP Conference Series) eds. A. de Oliveira-Costa and M. Tegmark, p. 227 (1999).
30. Haiman, Z., and Loeb, A., *ApJ*, **483**, 21 (1997).
31. Haiman, Z., Rees, M. J., and Loeb, A., *ApJ*, **467**, 522 (1996).
32. Haiman, Z., Rees, M. J., and Loeb, A., *ApJ*, **476**, 458 [Erratum: *ApJ*, **484**, 985] (1997).
33. Haiman, Z., Thoul, A. A., and Loeb, A., *ApJ*, **464**, 523 (1996).
34. Iliev, I. T., and Shapiro, P. R., *MNRAS*, in press (2001) (astro-ph/0101067).
35. Iliev, I. T., and Shapiro, P. R., in *20th Texas Symposium on Relativistic Astrophysics and Cosmology*, eds. J. C. Wheeler and H. Martel, in press (2001) (astro-ph/0104069).
36. Kitayama, T., and Ikeuchi S., *ApJ*, **529**, 615 (2000).
37. Kitayama, T., Tajiri, Y., Umemura, M., Susa, H., and Ikeuchi, S., *MNRAS*, **315**, 1 (2000).
38. Klein, R. I., Sandford, M. T., and Whitaker, R. W., *ApJ*, **271**, L69 (1983).
39. Lefloch, B., and Lazareff, B., *A&A*, **289**, 559 (1994).
40. Leitherer, C. et al., *ApJ*, **454**, L19 (1995).
41. Lizano, S., Cantó, J., Garay, G., and Hollenbach, D., *ApJ*, **468**, 739 (1996).
42. Madau, P., Haardt, F., and Rees, M. J., *ApJ*, **514**, 648 (1999).
43. Mellema, G., Raga, A. C., Canto, J., Lundquist, P., Balick, B., Steffen, W., and Noriega-Crespo, A., *A&A*, **331**, 335 (1997).
44. Miralda-Escudé, J., Haehnelt, M., and Rees, M. J., *ApJ*, **530**, 1 (2000).
45. Nakamoto, T., Umemura, M., and Susa, H., *MNRAS*, **321**, 593 (2001).
46. Navarro, J. F., and Steinmetz, M., *ApJ*, **478**, 13 (1997).

47. Oort, J. H., and Spitzer, L., *ApJ*, **121**, 6 (1955).
48. Quinn, T., Katz, N., and Efstathiou, G., *MNRAS*, **278**, L49 (1996)
49. Raga, A. C., Mellema, G., and Lundquist, P., *ApJS*, **109**, 517 (1997).
50. Raga, A. C., Taylor, S. D., Cabrit, S., and Biro, S., *A&A*, **296**, 833 (1995).
51. Razoumov, A., and Scott, D., *MNRAS*, **309**, 287 (1999).
52. Ricotti, M., and Shull, J. M., *ApJ*, **542**, 548 (2000).
53. Sandford, M. T., Whitaker, R. W., and Klein, R. I., *ApJ*, **260**, 183 (1982).
54. Shapiro, P. R., *PASP*, **98**, 1014 (1986).
55. Shapiro, P. R., in *The Physics of the Interstellar Medium and The Intergalactic Medium*, eds. A. Ferrara, C. F. McKee, C. Heiles, and P. R. Shapiro (ASP Conference Series, vol. 80), pp. 55–97 (1995).
56. Shapiro, P. R., and Giroux, M. L., *ApJ*, **321**, L107 (1987).
57. Shapiro, P. R., Giroux, M. L., and Babul, A., *ApJ*, **427**, 25 (1994).
58. Shapiro, P. R., Iliev, I., and Raga, A. C., *MNRAS*, **307**, 203 (1999)
59. Shapiro, P. R., and Martel, H., in *Dark Matter*, eds. S. S. Holt and C. L. Bennett (AIP Conference Proceedings 336), pp. 446–449 (1995).
60. Shapiro, P. R., Martel, H., and Iliev, I. T., to be submitted (2001).
61. Shapiro, P. R., and Raga, A. C., in *Astrophysical Plasmas: Codes, Models, and Observations*, eds. S. J. Arthur, N. Brickhouse, and J. J. Franco, *Rev.Mex.A.A. (SC)*, **9**, 292 (2000) (astro-ph/0002100)
62. Shapiro, P. R., and Raga, A. C., in *Cosmic Evolution and Galaxy Formation: Structure, Interactions, and Feedback* (ASP Conference Series, vol. 215), eds. J. Franco, E. Terlevich, O. Lopez-Cruz, and I. Aretxaga, pp. 1–6 (2000) (astro-ph/0004413).
63. Shapiro, P. R., and Raga, A. C., in *The Seventh Texas-Mexico Conference on Astrophysics: Flows, Blows, and Glows*, eds. W. Lee and S. Torres-Peimbert, *Rev.Mex.A.A. (SC)*, in press (2000) (astro-ph/0006367).
64. Shapiro, P. R., Raga, A. C., and Iliev, I. T., to be submitted (2001).
65. Shapiro, P. R., Raga, A. C., and Mellema, G., in *Structure and Evolution of the IGM from QSO Absorption Line Systems (13th IAP Colloquium)*, eds. P. Petitjean and S. Charlot (Paris: Editions Frontieres), pp. 41–46 (1997) (astro-ph/9710210)
66. Shapiro, P. R., Raga, A. C., and Mellema, G., in *Proceedings of the Workshop on H_2 in the Early Universe*, eds. F. Palla, E. Corbelli, and D. Galli, Memorie Della Societa Astronomica Italiana, **69**, pp. 463–469 (1998) (astro-ph/9804117)
67. Shapiro, P. R., Raga, A. C., and Mellema, G. 2000, *Nucl. Phys. B.*, **80**, CD-Rom 05/07 (2000)
68. Songaila, A., Hu, E. M., Cowie, L. L., and McMahon, R. G., *ApJ*, **525**, L5 (1999).
69. Spitzer, L., *Physical Processes in Interstellar Matter* (Wiley: New York) (1978).
70. Steidel, C. C., Pettini, M., and Adelberger, K. L., *ApJ*, **546**, 665 (2001).
71. Tegmark, M., Silk, J., Rees, M. J., Blanchard, A., Abel, T., and Palla, P., *ApJ*, **474**, 1 (1997).
72. Thoul, A. A., and Weinberg, D. H., *ApJ*, **465**, 608 (1996).
73. Tumlinson, J., and Shull, J. M., *ApJ*, **528**, L65 (2000).
74. Valageas, P., and Silk, J., *A&A*, **350**, 725 (1999).
75. Weinberg, D., Hernquist, L., and Katz, N. 1997, *ApJ*, **477**, 8 (1997).
76. Wood, K., and Loeb, A., *ApJ*, **545**, 86 (2000).

Gravitational Lensing: Recent Progress and Future Goals

Geneviève Soucail

Observatoire Midi-Pyrénées, 14 Avenue Belin, 31400 Toulouse, France

Abstract. After the enthusiasm raised by the discovery of multiple quasars and a few years later of giant luminous arcs in clusters of galaxies, gravitational lensing has proven to be a versatile tool to study a broad range of astrophysical problems. I will illustrate recent results obtained in several domains such as: the mass determination for galaxies and clusters of galaxies, the mapping of large scale structures, and the use of lenses to constrain the Hubble constant.

INTRODUCTION

Gravitational lensing is now well admitted as a powerful tool to address several cosmological questions in an original and efficient way. There are many sub-topics and this review does not have the ambition to cover all of them. On the contrary, I will (partly) review those concerning lensing by individual galaxies and by clusters of galaxies, extending the analysis up to the cosmological large scale structures.

Historically, the first gravitational lens was observed in 1979 as the double quasar Q0957+561, rapidly identified as a double image of a quasar lensed by a galaxy [1]. More generally, in the case of lensing by a galaxy, a typical system consists in a quasar or an AGN lensed by an intervening galaxy, splitting the source into 2 to 4 images, separated by a few arcseconds. In the case of lensing by clusters of galaxies, it was initially thought that the occurence of such systems would be very rare, and that the projected mass density was not high enough to become critical, even in the central regions. But after the first identification of the so-called "gravitational arcs" in clusters of galaxies as due to the lensing on some background galaxies [2,3], the systematic study of these systems became of great importance for the understanding of the mass distribution in clusters. Moreover, the large number of magnified sources observed that way opened a new vision of the distribution and physical propoerties of high redshift galaxies, especially thanks to the observations obtained with the Hubble Space Telescope (HST).

I LENSING BY GALAXIES

It is now well recognised that most of the lenses are elliptical field galaxies. One of the main output of the individual study of a gravitational lens is the mass determination of the deflector, generally constrained by lens modeling. Such modeling is in principle quite simple to operate, as the lens equation is a very simple derivate of the theory of General Relativity. The physical assumptions underlying the equation of the deviation of light rays are quite valid for galaxies or cluster lenses: **1.** thin lens approximation, (the light propagation follows the null geodesics in an expanding Universe outside the lens); **2.** small deviation angles; **3.** transparent lens (no absorption by dust, so gravitational lensing is achromatic); **4.** the gravitational potential of the lens is stationary during the light travel time.

Under these assumptions, the lens equation can be derived from the general statement of the Fermat's theorem: "the arrival time of a light ray (null geodesic) is stationary under first-order variation of the light paths between the source and the observer". The resulting equation is a mapping of the Image plane in the Source plane, depending on the two-dimensional newtonian potential of the lens.

In the case of the modeling of galaxy-size deflectors, typical models assume that the deflecting potential is an isothermal sphere (singular or with a core radius). This has the main advantage that it is analytically very simple, and its physical meaning is quite understood. The total mass is scaled with respect to the galaxy lens luminosity **L**, and the models are constrained by the accurate positions of the images, and of the lens if detected. In most cases, additional external shear is required to better fit the observational constraints.

A The generic example of Q0957+561

Although this double system was discovered long ago [1], it is still one of the most studied gravitational lens system, and I will illustrate the main scientific results which can be obtained with galaxy lenses through this example. Let us first summarise the configuration of the system: it consists in 2 images of a background quasar ($z_Q = 1.41$), separated by slightly more than 6" and lensed by an ellipticl galaxy at $z_S = 0.36$ (Fig. 1). The large separation value immediatly implies that a single galaxy cannot be the only deflector, and it was reconginsed early that the deflector was embedded in a larger mass structure, identified as a cluster of galaxies. Spectroscopic surveys of the area confirmed that this galaxy is indeed the central galaxy of a rather poor cluster. Moreover, although extensively searched, no central third image has ever been detected, even with the resolution of the HST. This pecular property is observed in most lenses, and it is understood by assuming that the central potential is nearly singular. Q0957+561 being a radio source quasar, it was deeply observed in radio, up to the resolution of the VLBI. The resoled structure presents a core and a jet, both detected in both images, increasing the number of constraints in any lens model. Since this system was identified as

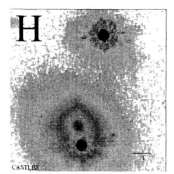 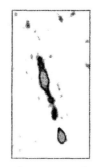 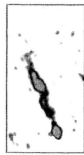

FIGURE 1. (Left): HST image of the lens configuration in Q0957+561. The 2 images of the quasar are splitted by the central galaxy (from CASTLES). (Right): resolved VLBI images of the two images of the quasar, with the core radio source and a jet escaping from it, used to constrain the lens models.

a gravitational lens, it was recognised as a potentially interesting target for the measurement of the time delay between the images. This time delay, predicted by the theory of General Relativity, is due to the differences in the optical paths of the light ray, including the geometrical delay and the gravitational one. It depends quite simply on the Hubble constant, so provided the lens potential is very well known, it is a direct and long range measure of H_0:

$$c\Delta t_{AB} = \frac{D_{OL} D_{OS}}{D_{LS}} (1+z_L) \left[\frac{1}{2}\left(\alpha^2(\theta_A) - \alpha^2(\theta_B)\right) - (\psi(\theta_A) - \psi(\theta_B))\right]$$
$$= \frac{c}{H_0} \times f(z_L, z_S, \Omega, \Lambda) \times \text{lens model}$$

Several groups struggled to monitor the light of the two images of Q0957+561 for many years, and finally yielded to a now admitted value of: $\Delta t_{AB} = 417 \pm 3$ days [4,5]. The present day best results on the lens modeling give $H_0 = 77^{+29}_{-24}$ km s^{-1} Mpc^{-1} [6], although the uncertainties are still rather large. They are mainly related to two intrinsic degeneracies in the knowledge of the lens. The first one, the so-called "mass-sheet degeneracy", means that if a potential $\psi(\vec{\theta})$ fits the observations, any new one defined as $\psi'(\vec{\theta}) = 1/2\kappa|\vec{\theta}|^2 + (1-\kappa)\psi(\vec{\theta})$ will fit equally well (the first term corresponds to the potential of a mass sheet). But the consequences on the value of H_0 are not equivalent: $\Delta t' = (1-\kappa)\Delta t$ or $H_0' = (1-\kappa)H_0$. So an absolute measure of the external shear is required to raise this degeneracy. Second, there is a degeneracy between the shape of the main lens galaxy G1 (ellipticity and orientation) and the external shear. Additional data are then required to increase the constraints. The recent identification of the host galaxy, multiple imaged into two arcs [7] can help breaking these degeneracies. Increasing the knowledge of the cluster in which the main lens is embedded is also a priority to get a complete picture of this gravitational lens.

B Surveys for gravitational lenses

A very promising programme, named CASTLES, has started with the HST. The aim is to image at optical and near-IR wavelengths a sample of 60 known gravitational lenses with the beautiful resolution attained with the HST (http://cfa-www.harvard.edu/castles/). Detailed modeling of each of these lenses, constrained by the position of the images of the lens (if detected), brings new insight in the physical properties of the lenses. The sample corresponds to an interesting "mass selected" sample [8]. In particular the analysis of these properties in terms of evolution of the fundamental plane of elliptical galaxies suggests a high redshift of formation for early-type galaxies ang that the mass distribution of the ellpitical galaxies is strongly dominated by dark matter [9].

Another promising project is the optical monitoring of multiple quasars for the measurement of the time delay between images. The search for the simplest and optimal lens is very important to reduce the uncertainties and the degeneracies in the lens models [10,6].

Finally, I can mention the JVAS/CLASS survey of systematic search for gravitational lenses from radiosources, as it corresponds to a large and well defined sample, very usefull for statistical studies. The statistics of occurences of lenses is a highly sensitive function of the cosmological parameters Ω and Λ (or the global geometry of the Universe) [11,12].

C Radio rings

Discovered about 10 years ago, these systems represent a spectacular manifestation of gravitational lensing. They correspond to distant radio sources lensed by a foreground galaxy with a nearly perfect alignment. If the source is resolved, the image forms a radio ring with a few arcsec in diameter. MG1131+0456 was the first complete Einstein ring observed [13], and since that first example, many cases were identified, mainly from systematic radio surveys for the search of gravitational lenses (the MIT Green Bank survey, MERLIN observations, the JVAS/CLASS survey). In most cases, the lens vanishes in radio so the ring is easily detectable, and deep optical images lead to the optical identification of the lens and sometimes the ring itself. Several tools were developped to reconstruct the source, from these cases of gravitational lensing on an extended source [14,15], using similar techniques for the source reconstruction as the radio deconvolution techniques. In most cases, the host galaxies correspond to bright radiogalaxies, with some stellar emission detected in the optical [9]. Their statistical study tends to explore the physical properties of high redshift radiogalaxies.

The gain in sensitivity and/or spatial resolution of all clases of radiotelescopes will increase significantly the sample, with a better understanding of the relation between the radio and optical properties of the faintest radiosources.

D galaxy/galaxy lensing

The lensing effect of bright and massive galaxies on the numerous background sources has long been recognised as a very promising effect for the understanding of the mass distribution of the haloes of galaxies. The first attempts to detect this effect started in the 80s [16,17]. But the expected intensity of the effect is so small that the results were quite controversial. The first reliable measures were obtained more recently with deep subarcsecond imaging [18]. Thanks to the high spatial resolution of the HST, an attempt to detect a lensing signal from the MDS data was also proposed [19] . All these result confirm the presence of a large halo of dark matter around field galaxies. In a very near future, one expects to see results coming from the deep wide field surveys in progress [20,21] with huge statistics.

In clusters of galaxies, the configuration is more favorable as all individual cluster galaxies benefit from the additional global shear due to the large scale lensing by the cluster itself. Some convincing results were proposed by several groups who analysed different cluster configurations, both from simulations of mass distributions and from real data [22–25]. The results suggest than about 10% of the dark matter in clusters is associated with the halos of cluster galaxies, and that their truncation radius is smaller than in the field. This truncation effect is interpreted as a signature of tidal stripping of galaxies in clusters.

II CLUSTERS OF GALAXIES AS GRAVITATIONAL LENSES

Gravitational lensing in clusters of galaxies is generally of different appearance than in galaxy lenses. In most cases, the source galaxy is extended so the images appear very distorted with or without multiple images. The extension of the image surface creates the magnification, with a global conservation of surface brightness. The level of details acheived is much higher than in galaxy lenses for at least two reasons: first, with extended sources both the magnification and the shear are measurable. Second, with clusters of galaxies of typical mass M $\sim 10^{14} M_\odot$ in the core, the size of the Einstein radius goes up to $R_E \sim 30''$, leaving the possibility to get spatial information on the deflector. As a first step, measuring the size of the Einstein radius gives a very simple determination of the projected mass inside it. This estimate is quite robust and has confirmed thanin most cases a significant amount of dark matter was not related to the galaxies, but distributed more widely in the clusters. Lens modeling is also efficient to constrain the mass profile in the central regions, with estimates of the core radius of the dark matter. Finally, the level of details is so high that different mass scales can be scanned, from M $\sim 10^{14} M_\odot$ and $r_c \sim 100$ kpc for the main deflector to M $\sim 10^6$ to $10^{11} M_\odot$ for the effects of individual galaxies (galaxy/galaxy lensing in clusters). Many clusters have been studied so far, and I will illustrate what can be learned from lens modeling on one of the most studied case.

A The generic example of Abell 370

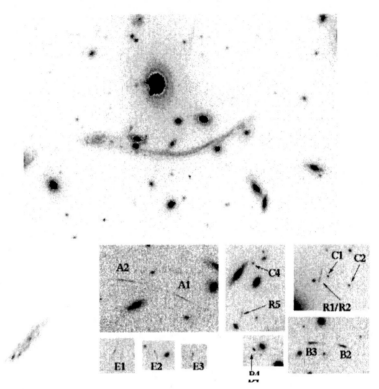

FIGURE 2. (Top): HST image of the giant arc in Abell 370, located close to one of the two giant elliptical galaxies dominating the cluster potential. (Bottom left): Reconstructed source of the giant arc, with typical morphology of a spiral galaxy. (Bottom right):faint arclets and multiple images candidates identified on HST images. Their redshifts are predicted by the lens model [26].

The cluster is a rich one at intermediate redshift ($z_L = 0.37$), dominated by two giant elliptical galaxies, and also known as a strong X-ray emitter. The giant arc in A370 was discovered in 1985 [2] and rapidly confirmed as a gravitational lensed galaxy [3] at redshift $z_S = 0.725$. More recently, HST imaging of the cluster allowed to identify the giant arc as a system of 3 merging images. In addition, several faint distorted structures were detected, increasing the number of multiple images in this cluster. Several attempts to model the mass distribution in the cluster have been proposed. The last one [26] took advantage of all the multiple images detected by HST (Figure 2) and was very successfull in predicting the redshifts of the faintest arclets. The mass distribution fitted by the model consists in 2 large clumps of mass centered on the two giant galaxies. This global mass distribution is quite compatible with the ROSAT/HRI map observed in the cluster [27]. Moreover, the "dark matter follows the light" rule is again verified with good accuracy if one considers the ellipticity and the orientation of the enveloppes of the two central giant galaxies. The core radii are very small (about $r_c \sim 100$ kpc) and correspond to a peaked mass profile. In addition, thanks to the accurate mass model, it is possible to reconstruct the morphological shape of the source galaxy of the giant arc, leading to a nearly face-on galaxy (Figure 2).

B Mass discrepancy in clusters

Masses determined from gravitational lensing in clusters of galaxies correspond to an independant mass estimator in clusters, which must be compared to other ones: distribution of the galaxies (luminosities, radial velocities) or of the intra-cluster gas (X-ray maps). All these estimators call for different physical assumptions which can be directly tested by comparing the results.

The question of the so-called "mass discrepancy in clusters of galaxies" was initially pointed out by Miralda-Escudé & Babul [28] who found a contradiction of a factor $\simeq 2$ between masses measured from weak lensing effects and masses derived from the properties of the X-ray gaz ! This appeared to be rather systematic: lensing masses were always in excess compared to X-ray masses. This question was adressed by many authors, and deeply discussed, although the limited quality of the data used in this analysis was recognised. More recently, Allen [29] re-analysed a sample of a few tens of X-ray clusters with ROSAT data and found a much better agreement between X-ray and lensing masses, especially for cooling flow clusters. These clusters are known to have in their history a long period of dynamical stability, as cooling flows are rapidly disrupted during any merging event from the cluster. Moreover, a very detailed comparison of the mass estimates have been re-processed recently, based on the few lensing X-ray clusters with high quality X-ray and lensing data, namely MS1008−1224 [30] and Cl0024+17 [31].

Mass discrepancy in the cluster Cl0024+1654

This cluster, located at intermediate redshift $z_L = 0.39$, is one of the most spectacular case of multiple arcs, with its system of four arcs. It is a rich cluster, with a high concentration of galaxies in the very center, but not dominated by a giant elliptical galaxy. The velocity dispersion of the galaxies was estimated by Schneider et al. [32] from a sample of abour 50 cluster members ($\sigma_{los} \simeq 1300$ km/s), giving a total virial mass of $M_{Virial} = 1 \pm 0.4 \, 10^{15} \, h_{50}^{-1} M_\odot$. From the lensing point of view, deep HST images confirmed the lensing nature of the 3 main arcs and identified one counter-arc. This system has a large angular radius ($r = 35''$) suggesting a high mass density in the center. Recently the redshift of the arc was finally determined from Keck spectroscopy [33], giving a total mass inside the arc radius:

$$M(r < r_{arc}) = 2.6 \pm 0.01 \times 10^{14} \, h_{50}^{-1} M_\odot$$

The weak lensing regime was also detected in this cluster, with the pioneering work of Bonnet et al. [34], up to the large radius of $3h_{50}^{-1}$ Mpc. A fit of the shear profile with standard mass profiles gave a total mass

$$M(r < 3Mpc) \simeq [2.4 - 4] \times 10^{15} \, h_{50}^{-1} M_\odot$$

with most uncertainties due to mass inversion and the assumptions on the mass profile.

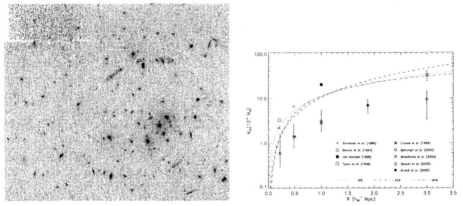

FIGURE 3. (Left): HST image of the center of the cluster Cl0024+1654, displaying a beautiful triple arc, with its counter-image. (Right): compilation of the mass determinations in the cluster, illustrating the question of the mass discrepancies. All the lower points are coming from X-ray mass determinations, except one corresponding to the "corrected" Virial mass from Czoske et al. [35].

Cl0024+1654 is also known as an X-ray emitter and was observed deeply in X-rays. Thanks to the good spatial resolution of the ROSAT/HRI data combined with spectral information from ASCA it was possible to subtract the individual sources superimposed on the cluster emission, fit the profile with a standard β-profile and then reconstruct the mass profile [31]. The global discrepancy between the different mass estimators reaches a factor 1.5 to 3 in this cluster (Fig. 3). To better understand the consequences, several arguments must be pointed out: first, as for any distant clusters, the uncertainties in the fit of the β-profile on X-ray data underestimates the total mass by a factor of 1.5 to 2 [36], because of the high background level of HRI data. Second, the shear mass is probably overestimated and needs to be re-analysed with recent and efficient mass reconstruction techniques [37,38]. This work is in progress now. Finally, if the X-ray measured temperature

($T_X \simeq 6$ keV) is correct, the cluster falls out of the correlation σ_{los}/T_X with the measured value of the velocity dispersion. A better and more detailed study of the internal dynamics of this cluster was recently presented [35] from a wide field spectroscopic survey of the galaxies. With more than 300 redshifts of cluster members, the new velocity histogram is clearly bi-modal and the velocity dispersion of the main mode drops to ~ 750 km/s, reducing the Virial mass by a factor of 3. The new understanding of the cluster is that we are witnessing the interaction between two massive structures well aligned on the line of sight. In all cases the apparent mass discrepancy can be solved with accurate and detailed studies on cluster lenses.

C Dark clumps and large scale structures

Gravitational lensing is affecting the light rays at every scale, and all mass scales can be scanned in principle. In practise when the shear effect becomes small, it is impossible to identify it individually: the induced ellipticity on the background galaxies is so small that it cannot be disentangled from their intrinsic ellipticity. Only a statistical approach must be used, but it allow to scan mass scales up to very large structures.

The first attempts to detect weak lensing signal was applied to the periphery of clusters of galaxies (see the above section), thanks to deep wide field imaging. More interestingly, several cases of the so-called "dark clumps" were detected that way: they correspond to regions of the sky with a positive shear detection, but with no or very faint optical counter-parts, implying a high M/L ratio for the identified structure. The most intriguing case is the dark clump detected in the field of A1942 [39], a few arcmin SW of the cluster center. Although a strong shear signal was detected, no optical counter-part was identified in the shear region, even with reasonably deep images. Near-IR images, considered to be more efficient to detect faint high redshift galaxies, also failed to detec any overdensity of galaxies [40]. This case remains presently the best candidate for a "true" dark clump.

The weak lensing technique, which is just starting to leave its infancy, is opening a completely new vision of the large scale mass distribution of structures. Very soon, one can expect to see complete mass maps in clusters and between them. First results on the detection of a bridge of matter between 2 clusters have already been presented [41], although still preliminary. With the advent of wide field panoramic cameras on large telescopes and with good image quality, this will be a major topic which will progress in the next few years. But it is important to remind that although the prospects are very exciting, the difficulties related to the detection of statistical weak lensing at a few % level is an observational challenge.

As a first step in the study of the large scale dsitribution of mass in the universe, the first detections of a "cosmic shear" signal were claimed last year. This corresponds to the detection of the cosmic shear statistical variance $\langle \gamma^2 \rangle$, measured on angular scales ranging from 0.1' to 30' [42–45]. It is essential to insist on the difficulty of this measure, which has to take rid of all the systematics which may induce some distorsion in the shape of the galaxies: PSF anisotropies, telescope tracking residuals, optical distorsion... But the concordance of the results obtained by independant teams on independant data give confidence in the reality of the measurement. The cosmological constraints of such measurement start to be considered. Preliminary results, based on the comparison between the data and both some analytical development and numerical simalutions of the non-linear groth of structures in cosmological models [46], show that the cosmic variance can be related to the cosmological parameters σ_8, Ω_M and the exponent of the power spectrum n, provided the redshift distribution of the sources is known (see the detailed discussion in Van Waerbeke et al. [47]), with an intrinsic degeneracy between σ_8 and Ω_M.

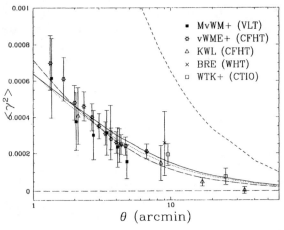

FIGURE 4. Summary of the "cosmic shear" measurements, with various cosmological model simulations superimposed. The straight lines correspond to different configurations of non-vanishing cosmological constant [45].

III CONCLUSIONS

This review is far from a complete one! I would like to point out the diversity of the cosmological topics which have been addressed these recent years by the studies of gravitational lenses. Gravitational lensing is now totally accepted as a very original and efficient tool to address many questions, most of them related to masses, at all scales. It is also a prospective tool for constraining the main cosmological parameters as it is a direct estimator of the global geometry of the Universe. Masses and cosmological parameters (H_0, Ω, Λ): indeed gravitational lensing is a wonderfull tool for Cosmology !!!

ACKNOWLEDGMENTS

I wish to thank the organisers of the 20th Texas Symposium of General Relativity for their hospitality, and in particular Craig Wheeler for the enthusiasm with which he supervised it. I also want to thank my collegues with which I got so many scientific interactions along these years: Richard Ellis, Bernard Fort, Jean-Paul Kneib, Yannick Mellier, Roser Pello, Peter Schneider, Ludovic Van Waerbeke and many others.

REFERENCES

1. Walsh, D., Carswell, R. F., and Weymann, R. J., *Nat*, **279**, 381–384 (1979).
2. Soucail, G., Fort, B., Mellier, Y., and Picat, J. P., *A&A*, **172**, L14–L16 (1987).
3. Soucail, G., Mellier, Y., Fort, B., Mathez, G., and Cailloux, M., *A&A*, **191**, L19–L21 (1988).
4. Kundic, T., Turner, E. L., Colley, W. N., Gott, J. R., Rhoads, J. E., Wang, Y., Bergeron, L. E., Gloria, K. A., Long, D. C., Malhotra, S., and Wambsganss, J., *ApJ*, **482**, 75 (1997).
5. Haarsma, D., Hewitt, J. N., Lehar, J., and Burke, B. F., *ApJ*, **510**, 64 (1999).
6. Bernstein, G., and Fischer, P., *AJ*, **118**, 14–34 (1999).
7. Keeton, C. R., Falco, E. E., Impey, C. D., Kochanek, C. S., Lehár, J., McLeod, B. A., Rix, H. ., Muñoz, J. A., and Peng, C. Y., *ApJ*, **542**, 74 (2000).
8. Lehár, J., Falco, E. E., Kochanek, C. S., McLeod, B. A., Muñoz, J. A., Impey, C. D., Rix, H. ., Keeton, C. R., and Peng, C. Y., *ApJ*, **536**, 584 (2000).
9. Kochanek, C. S., Falco, E. E., Impey, C. D., Lehár, J., McLeod, B. A., Rix, H. ., Keeton, C. R., Muñoz, J. A., and Peng, C. Y., *ApJ*, **543**, 131 (2000).
10. Fassnacht, C. D., Pearson, T. J., Readhead, A. C. S., Browne, I. W. A., Koopmans, L. V. E., Myers, S. T., and Wilkinson, P. N., *ApJ*, **527**, 498 (1999).
11. Helbig, P., *A&A*, p. 1 (1999).
12. Macias-Perez, J. F., Helbib, P., Quast, R., Wilkinson, A., and Davies, R., *A&A*, **353**, 419 (2000).
13. Chen, G. H., Kochanek, C. S., and Hewitt, J. N., *ApJ*, **447**, 62 (1995).
14. Kochanek, C. S., and Narayan, R., *ApJ*, **401**, 461–473 (1992).
15. Wallington, S., Narayan, R., and Kochanek, C. S., *ApJ*, **426**, 60–73 (1994).
16. Webster, R. L., *MNRAS*, **213**, 871–888 (1985).
17. Tyson, J. A., Valdes, F., Jarvis, J. F., and Mills, A. P., *ApJ Lett.*, **281**, L59–L62 (1984).
18. Brainerd, T. G., Blandford, R. D., and Smail, I., *ApJ*, **466**, 623 (1996).
19. Rhodes, J., Refregier, A., and Groth, E. J., *ApJ*, **536**, 79–100 (2000).
20. Fischer, P., et al.,, and The SDSS Collaboration, *AJ*, **120**, 1198–1208 (2000).
21. Wilson, G., Kaiser, N., Luppino, G., and Cowie, L., *ApJ* (2001), submitted, astro-ph/0008504.
22. Natarajan, P., and Kneib, J., *MNRAS*, **287**, 833–847 (1997).
23. Natarajan, P., Kneib, J., Smail, I., and Ellis, R. S., *ApJ*, **499**, 600 (1998).

24. Geiger, B., and Schneider, P., *MNRAS*, **295**, 497 (1998).
25. Geiger, B., and Schneider, P., *MNRAS*, **302**, 118–130 (1999).
26. Bézecourt, J., Kneib, J. P., Soucail, G., and Ebbels, T. M. D., *A&A*, **347**, 21–29 (1999).
27. Fort, B., and Mellier, Y., *A&AR*, **5**, 239–292 (1994).
28. Miralda-Escude, J., and Babul, A., *ApJ*, **449**, 18 (1995).
29. Allen, S. W., *MNRAS*, **296**, 392–406 (1998).
30. Athreya, R., Mellier, Y., Van Waerbeke, L., Fort, B., Pelló, R., and Dantel-Fort, M., *A&A* (2001), submitted, astro-ph/9909518.
31. Soucail, G., Ota, N., Böhringer, H., Czoske, O., Hattori, M., and Mellier, Y., *A&A*, **355**, 433–442 (2000).
32. Schneider, D. P., Dressler, A., and Gunn, J. E., *AJ*, **92**, 523 (1986).
33. Broadhurst, T., Huang, X., Frye, B., and Ellis, R., *ApJ Lett.*, **534**, L15–L18 (2000).
34. Bonnet, H., Mellier, Y., and Fort, B., *ApJ*, **427**, L83–L86 (1994).
35. Czoske, O., Kneib, J., Soucail, G., Bridges, T., Mellier, Y., and Cuillandre, J., *A&A* (2001), submitted, astro-ph/0103123.
36. Bartelmann, M., and Steinmetz, M., *MNRAS*, **283**, 431 (1996).
37. Seitz, S., Schneider, P., and Bartelmann, M., *A&A*, **337**, 325–337 (1998).
38. Bridle, S. L., Hobson, M. P., Lasenby, A. N., and Saunders, R., *MNRAS*, **299**, 895–903 (1998).
39. Erben, T., van Waerbeke, L., Mellier, Y., Schneider, P., Cuillandre, J.., Castander, F. J., and Dantel-Fort, M., *A&A*, **355**, 23–36 (2000).
40. Gray, M., Ellis, R., Lewis, J., McMahon, R., and Firth, A., *MNRAS* (2001), submitted, astro-ph/0101431.
41. Kaiser, N., Wilson, G., Luppino, G., Kaufman, L., Gioia, I., Metzger, M., and Dahle, H., p, astro-ph/9809268.
42. Van Waerbeke, L. et al., *A&A*, **358**, 30–44 (2000).
43. Bacon, D., Refregier, A., and Ellis, R., *MNRAS*, **318**, 625 (2000).
44. Kaiser, N., Wilson, G., and Luppino, G., *ApJ* (2001), in press, astro-ph/0003338.
45. Maoli, R., Van Waerbeke, L., Mellier, Y., Schneider, P., Jain, B., Bernardeau, F., Erben, T., and Fort, B., *A&A*, **368**, 766–775 (2001).
46. Jain, B., Seljak, U., and White, S., *Astrophysics*, **530**, 547–577 (2000).
47. Van Waerbeke, L. et al., *A&A* (2001), submitted, astro-ph/0101511.

Measuring large-scale structure with the 2dF Galaxy Redshift Survey

J.A. Peacock and the 2dFGRS team*

Institute for Astronomy, University of Edinburgh,
Royal Observatory, Edinburgh EH9 3HJ, UK

Abstract. The 2dF Galaxy Redshift Survey is the first to measure more than 100,000 redshifts. This allows precise measurements of many of the key statistical measures of galaxy clustering, in particular redshift-space distortions and the large-scale power spectrum. This paper presents the current 2dFGRS results in these areas. Redshift-space distortions are detected with a high degree of significance, confirming the detailed Kaiser distortion from large-scale infall velocities, and measuring the distortion parameter $\beta = 0.43 \pm 0.07$. The power spectrum is measured to $\lesssim 10\%$ accuracy for $k > 0.02\,h\,\mathrm{Mpc}^{-1}$, and is well fitted by a CDM model with $\Omega_m h = 0.20 \pm 0.03$ and a baryon fraction of 0.15 ± 0.07.

I AIMS AND DESIGN OF THE 2DFGRS

The large-scale structure in the galaxy distribution is widely seen as one of the most important relics from an early stage of evolution of the universe. The 2dF Galaxy Redshift Survey (2dFGRS) was designed to build on previous studies of this structure, with the following main aims:

1. To measure the galaxy power spectrum $P(k)$ on scales up to a few hundred Mpc, bridging the gap between the scales of nonlinear structure and measurements from the the cosmic microwave background (CMB).

2. To measure the redshift-space distortion of the large-scale clustering that results from the peculiar velocity field produced by the mass distribution.

---

*) *The 2dF Galaxy Redshift Survey team:* Matthew Colless (ANU), John Peacock (ROE), Carlton M. Baugh (Durham), Joss Bland-Hawthorn (AAO), Terry Bridges (AAO), Russell Cannon (AAO), Shaun Cole (Durham), Chris Collins (LJMU), Warrick Couch (UNSW), Nicholas Cross (St Andrews), Gavin Dalton (Oxford), Kathryn Deeley (UNSW), Roberto De Propris (UNSW), Simon Driver (St Andrews), George Efstathiou (IoA), Richard S. Ellis (Caltech), Carlos S. Frenk (Durham), Karl Glazebrook (JHU), Carole Jackson (ANU), Ofer Lahav (IoA), Ian Lewis (AAO), Stuart Lumsden (Leeds), Steve Maddox (Nottingham), Darren Madgwick (IoA), Peder Norberg (Durham), Will Percival (ROE), Bruce Peterson (ANU), Will Sutherland (ROE), Keith Taylor (Caltech)

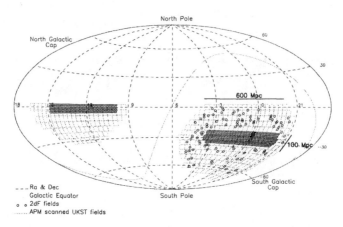

FIGURE 1. The 2dFGRS fields (small circles) superimposed on the APM catalogue area (dotted outlines of Sky Survey plates). There are approximately 140,000 galaxies in the 75° × 15° southern strip centred on the SGP, 70,000 galaxies in the 75° × 7.5° equatorial strip, and 40,000 galaxies in the 100 randomly-distributed 2dF fields covering the whole area of the APM catalogue in the south.

3. To measure higher-order clustering statistics in order to understand biased galaxy formation, and to test whether the galaxy distribution on large scales is a Gaussian random field.

The survey is designed around the 2dF multi-fibre spectrograph on the Anglo-Australian Telescope, which is capable of observing up to 400 objects simultaneously over a 2 degree diameter field of view. Full details of the instrument and its performance are given in Lewis et al. (2001). See also http://www.aao.gov.au/2df/.

The source catalogue for the survey is a revised and extended version of the APM galaxy catalogue (Maddox et al. 1990a,b,c). The extended version of the APM catalogue includes over 5 million galaxies down to $b_J = 20.5$ in both north and south Galactic hemispheres over a region of almost 10^4 deg^2 (bounded approximately by declination $\delta \leq +3$ and Galactic latitude $b \gtrsim 20$). This catalogue is based on Automated Plate Measuring machine (APM) scans of 390 plates from the UK Schmidt Telescope (UKST) Southern Sky Survey. The b_J magnitude system for the Southern Sky Survey is defined by the response of Kodak IIIaJ emulsion in combination with a GG395 filter. The photometry of the catalogue is calibrated with numerous CCD sequences and has a precision of approximately 0.2 mag for galaxies with $b_J = 17$–19.5. The star-galaxy separation is as described in Maddox et al. (1990b), supplemented by visual validation of each galaxy image.

The survey geometry is shown in Figure 1, and consists of two contiguous declination strips, plus 100 random 2-degree fields. One strip is in the southern Galactic hemisphere and covers approximately 75°×15° centred close to the SGP at $(\alpha,\delta)=(01^h,-30)$; the other strip is in the northern Galactic hemisphere and covers

75° × 7.5° centred at $(\alpha, \delta) = (12.5^h, +00)$. The 100 random fields are spread uniformly over the 7000 deg^2 region of the APM catalogue in the southern Galactic hemisphere. At the median redshift of the survey ($\bar{z} = 0.11$), $100 \, h^{-1}$ Mpc subtends about 20 degrees, so the two strips are $375 \, h^{-1}$ Mpc long and have widths of $75 \, h^{-1}$ Mpc (south) and $37.5 \, h^{-1}$ Mpc (north).

The sample is limited to be brighter than an extinction-corrected magnitude of $b_J = 19.45$ (using the extinction maps of Schlegel et al. 1998). This limit gives a good match between the density on the sky of galaxies and 2dF fibres. Due to clustering, however, the number in a given field varies considerably. To make efficient use of 2dF, we employ an adaptive tiling algorithm to cover the survey area with the minimum number of 2dF fields. With this algorithm we are able to achieve a 93% sampling rate with on average fewer than 5% wasted fibres per field. Over the whole area of the survey there are in excess of 250,000 galaxies.

II SURVEY STATUS

By the end of 2000, observations had been made of 161,307 targets in 600 fields, yielding redshifts and identifications for 141,402 galaxies, 7958 stars and 53 QSOs, at an overall completeness of 93%. Repeat observations have been obtained for 10,294 targets. Figure 2 shows the projection of the galaxies in the northern and southern strips onto (α, z) slices. The main points to note are the level of detail apparent in the map and the slight variations in density with R.A. due to the varying field coverage along the strips.

The adaptive tiling algorithm is efficient, and yields uniform sampling in the final survey. However, at this intermediate stage, missing overlaps mean that the sampling fraction has large fluctuations, as illustrated in Figure 3. This variable sampling makes quantification of the large scale structure more difficult, and limits any analysis requiring relatively uniform contiguous areas. However, the effective survey 'mask' can be measured precisely enough that it can be allowed for in low-order analyses of the galaxy distribution.

III REDSHIFT-SPACE CORRELATIONS

The simplest statistic for studying clustering in the galaxy distribution is the the two-point correlation function, $\xi(\sigma, \pi)$. This measures the excess probability over random of finding a pair of galaxies with a separation in the plane of the sky σ and a line-of-sight separation π. Because the radial separation in redshift space includes the peculiar velocity as well as the spatial separation, $\xi(\sigma, \pi)$ will be anisotropic. On small scales the correlation function is extended in the radial direction due to the large peculiar velocities in non-linear structures such as groups and clusters – this is the well-known 'Finger-of-God' effect. On large scales it is compressed in the radial direction due to the coherent infall of galaxies onto mass concentrations – the Kaiser effect (Kaiser 1987).

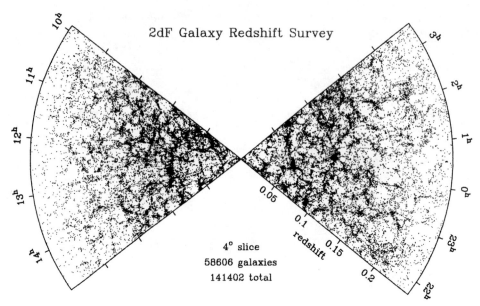

FIGURE 2. The distribution of galaxies in part of the 2dFGRS, drawn from a total of 141,402 galaxies: slices 4° thick, centred at declination $-2.5°$ in the NGP and $-27.5°$ in the SGP. Not all 2dF fields within the slice have been observed at this stage, hence there are weak variations of the density of sampling as a function of right ascension. To minimise such features, the slice thickness increases to 7.5° between right ascension 13.1^h and 13.4^h. This image reveals a wealth of detail, including linear supercluster features, often nearly perpendicular to the line of sight. The interesting question to settle statistically is whether such transverse features have been enhanced by infall velocities.

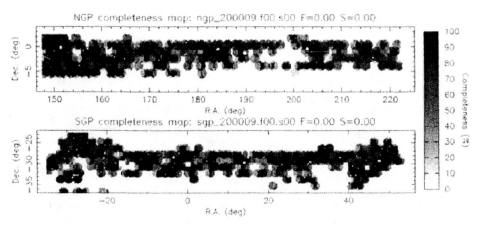

FIGURE 3. The completeness as a function of position on the sky. The circles are individual 2dF fields ('tiles'). Unobserved tiles result in low completeness in overlap regions. Rectangular holes are omitted regions around bright stars.

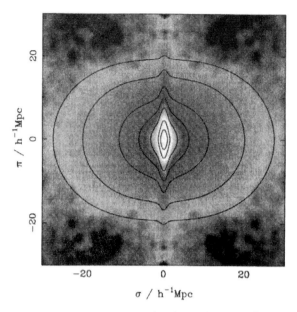

FIGURE 4. The galaxy correlation function $\xi(\sigma, \pi)$ as a function of transverse (σ) and radial (π) pair separation is shown as a greyscale image. It was computed in $0.2\,h^{-1}$ Mpc boxes and then smoothed with a Gaussian having an rms of $0.5\,h^{-1}$ Mpc. The contours are for a model with $\beta = 0.4$ and $\sigma_p = 400\,\mathrm{km\,s^{-1}}$, and are plotted at $\xi = 10, 5, 2, 1, 0.5, 0.2$ and 0.1.

To estimate $\xi(\sigma, \pi)$ we compare the observed count of galaxy pairs with the count estimated from a random distribution following the same selection function both on the sky and in redshift as the observed galaxies. We apply optimal weighting to minimise the uncertainties due to cosmic variance and Poisson noise. This is close to equal-volume weighting out to our adopted redshift limit of $z = 0.25$. We have tested our results and found them to be robust against the uncertainties in both the survey mask and the weighting procedure. The redshift-space correlation function for the 2dFGRS computed in this way is shown in Figure 4. The correlation-function results display very clearly the two signatures of redshift-space distortions discussed above. The 'fingers of God' from small-scale random velocities are very clear, as indeed has been the case from the first redshift surveys (e.g. Davis & Peebles 1983). However, this is the first time that the large-scale flattening from coherent infall has been seen in detail.

The degree of large-scale flattening is determined by the total mass density parameter, Ω, and the biasing of the galaxy distribution. On large scales, it should be correct to assume a linear bias model, so that the redshift-space distortion on large scales depends on the combination $\beta \equiv \Omega^{0.6}/b$. On these scales, linear distortions should also be applicable, so we expect to see the following quadrupole-to-monopole

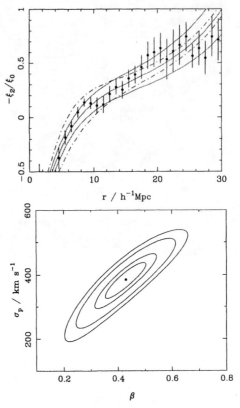

FIGURE 5. (a) The compression of $\xi(\sigma,\pi)$ as measured by its quadrupole-to-monopole ratio, plotted as $-\xi_2/\xi_0$. The solid lines correspond to models with $\sigma_p = 400\,\mathrm{km\,s^{-1}}$ and (bottom to top) $\beta = 0.3, 0.4, 0.5$, while the dot-dash lines correspond to models with $\beta = 0.4$ and (top to bottom) $\sigma_p = 300, 400, 500\,\mathrm{km\,s^{-1}}$. (b) Likelihood contours for β and σ_p from the model fits. The inner contour is the one-parameter 68% confidence ellipse; the outer contours are the two-parameter 68%, 95% and 99% confidence ellipses. The central dot is the maximum likelihood fit, with $\beta = 0.43$ and $\sigma_p = 385\,\mathrm{km\,s^{-1}}$.

ratio in the correlation function:

$$\frac{\xi_2}{\xi_0} = \frac{3+n}{n} \frac{4\beta/3 + 4\beta^2/7}{1 + 2\beta/3 + \beta^2/5} \qquad (1)$$

where n is the power spectrum index of the fluctuations, $\xi \propto r^{-(3+n)}$. This is modified by the Finger-of-God effect, which is significant even at large scales and dominant at small scales. The effect can be modelled by introducing a parameter σ_p, which represents the rms pairwise velocity dispersion of the galaxies in collapsed structures, σ_p (see e.g. Ballinger et al. 1996). Full details of the fitting procedure are given in Peacock et al. (2001).

Figure 5a shows the variation in ξ_2/ξ_0 as a function of scale. The ratio is positive on small scales where the Finger-of-God effect dominates, and negative on large scales where the Kaiser effect dominates. The best-fitting model (considering only the quasi-linear regime with $8 < r < 25\,h^{-1}$ Mpc) has $\beta \simeq 0.4$ and $\sigma_p \simeq 400\,\mathrm{km\,s^{-1}}$; the likelihood contours are shown in Figure 5b. Marginalising over σ_p, the best estimate of β and its 68% confidence interval is

$$\beta = 0.43 \pm 0.07 \qquad (2)$$

This is the first precise measurement of β from redshift-space distortions; previous studies have shown the effect to exist (e.g. Hamilton, Tegmark & Padmanabhan 2000; Taylor et al. 2000; Outram, Hoyle & Shanks 2000), but achieved little more than 3σ detections.

IV COSMOLOGICAL PARAMETERS AND THE POWER SPECTRUM

The detailed measurement of the signature of gravitational collapse is the first major achievement of the 2dFGRS; we now consider the quantitative implications of this result. The first point to consider is that there may be significant corrections for luminosity effects. The optimal weighting means that our mean luminosity is high: it is approximately 1.9 times the characteristic luminosity, L^*, of the overall galaxy population (Folkes et al. 1999). Benoist et al. (1996) have suggested that the strength of galaxy clustering increases with luminosity, with an effective bias that can be fitted by $b/b^* = 0.7 + 0.3(L/L^*)$. This effect has been controversial (see Loveday et al. 1995), but the 2dFGRS dataset favours a very similar luminosity dependence. We therefore expect that β for L^* galaxies will exceed our directly measured figure. Applying a correction using the given formula for $b(L)$, we deduce $\beta(L = L^*) = 0.54 \pm 0.09$. Finally, the 2dFGRS has a median redshift of 0.11. With weighting, the mean redshift in the present analysis is $\bar{z} = 0.17$, and our measurement should be interpreted as β at that epoch. The extrapolation to $z = 0$ is model-dependent, but probably does not introduce a significant change (Carlberg et al. 2000).

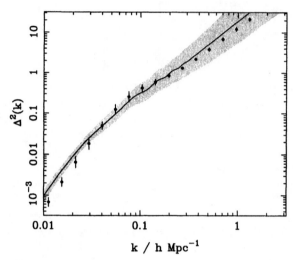

FIGURE 6. The dimensionless matter power spectrum at zero redshift, $\Delta^2(k)$, as predicted from the allowed range of models that fit the microwave-background anisotropy data, plus the assumption that $H_0 = 70\,\mathrm{km\,s^{-1} Mpc^{-1}} \pm 10\%$. The solid line shows the best-fit model from Jaffe et al. (2000) [power-spectrum index $n = 1.01$, and density parameters in baryons, CDM, and vacuum of respectively 0.065, 0.285, 0.760]. The effects of nonlinear evolution have been included, according to a revised version of the procedure of Peacock & Dodds (1996). The shaded band shows the 1σ variation around this model allowed by the CMB data. The solid points are the real-space power spectrum measured for APM galaxies. The clear conclusion is that APM galaxies are consistent with being essentially unbiased tracers of the mass on large scales.

Our measurement of $\Omega^{0.6}/b$ would thus imply $\Omega = 0.36 \pm 0.10$ if L^* galaxies are unbiased, but it is difficult to justify such an assumption. In principle, the details of the clustering pattern in the nonlinear regime allow the $\Omega - b$ degeneracy to be broken (Verde et al. 1998), but for the present it is interesting to use an independent approach. Observations of CMB anisotropies can in principle measure almost all the cosmological parameters, and Jaffe et al. (2000) obtained the following values for the densities in collisionless matter (c), baryons (b), and vacuum (v): $\Omega_c + \Omega_b + \Omega_v = 1.11 \pm 0.07$, $\Omega_c h^2 = 0.14 \pm 0.06$, $\Omega_b h^2 = 0.032 \pm 0.005$, together with a power-spectrum index $n = 1.01 \pm 0.09$. Our result for β gives an independent test of this picture, as follows.

The only parameter left undetermined by the CMB data is the Hubble constant, h. Recent work (Mould et al. 2000; Freedman et al. 2000) indicates that this is now determined to an rms accuracy of 10%, and we adopt a central value of $h = 0.70$. This completes the cosmological model, requiring a total matter density parameter $\Omega \equiv \Omega_c + \Omega_b = 0.35 \pm 0.14$. It is then possible to use the parameter limits from the CMB to predict a conservative range for the mass power spectrum at $z = 0$, which is shown in Figure 6. A remarkable feature of this plot is that the mass power spectrum appears to be in good agreement with the clustering observed in the APM survey (Baugh & Efstathiou 1994). For each model allowed by the CMB, we can predict both b (from the ratio of galaxy and mass spectra) and also β (since a given CMB model specifies Ω). Considering the allowed range of models, we then obtain the prediction $\beta_{\rm CMB+APM} = 0.57 \pm 0.17$. A flux-limited survey such as the APM will have a mean luminosity close to L^*, so the appropriate comparison is with the 2dFGRS corrected figure of $\beta = 0.54 \pm 0.09$ for L^* galaxies. These numbers are in very close agreement. In the future, the value of β will become one of the most direct ways of confronting large-scale structure with CMB studies.

V THE 2DFGRS POWER SPECTRUM

Of course, one may question the adoption of the APM power spectrum, which was deduced by deprojection of angular clustering. The 3D data of the 2dFGRS should be capable of improving on this determination, and we have made a first attempt at doing this, shown in Figure 7. This power-spectrum estimate uses the FFT-based approach of Feldman, Kaiser & Peacock (1994), and needs to be interpreted with care. Firstly, it is a raw redshift-space estimate, so that the power beyond $k \simeq 0.2\,h\,{\rm Mpc}^{-1}$ is severely damped by fingers of God. On large scales, the power is enhanced, both by the Kaiser effect and by the luminosity-dependent clustering discussed above. Finally, the FKP estimator yields the true power convolved with the window function. This modifies the power significantly on large scales (roughly a 20% correction). We have made an approximate correction for this in Figure 7 by multiplying by the correction factor appropriate for a $\Gamma = 0.25$ CDM spectrum. The precision of the power measurement appears to be encouragingly high, and the systematic corrections from the window are well specified.

The next task is to perform a detailed fit of physical power spectra, taking full account of the window effects. The hope is that we will obtain not only a more precise measurement of the overall spectral shape, as parameterized by Γ, but will be able to move towards more detailed questions such as the existence of baryonic features in the matter spectrum (Meiksin, White & Peacock 1999). We summarize here results from the first attempt at this analysis (Percival et al. 2001).

The likelihood of each model has been estimated using a covariance matrix calculated from Gaussian realisations of linear density fields for a $\Omega_m h = 0.2$, $\Omega_b/\Omega_m = 0.15$ CDM power spectrum, for which $\chi^2_{\rm min} = 34.4$, given an expected value of 28. The best fit power spectrum parameters are only weakly dependent on this choice. The likelihood contours in Ω_b/Ω_m versus $\Omega_m h$ for this fit are shown in Figure 8. At each point in this surface we have marginalized by integrating the Likelihood surface over the two free parameters, h and the power spectrum amplitude. The result is not significantly altered if instead, the modal, or Maximum Likelihood points in the plane corresponding to power spectrum amplitude and h were chosen. The likelihood function is also dependent on the covariance matrix (which should be allowed to vary with cosmology), although the consistency

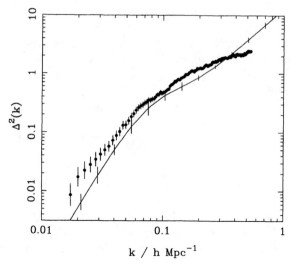

FIGURE 7. The 2dFGRS redshift-space power spectrum, estimated according to the FKP procedure. The solid points with error bars show the power estimate. The window function correlates the results at different k values, and also distorts the large-scale shape of the power spectrum An approximate correction for the latter effect has been applied. The line shows the real-space power spectrum estimated by deprojection from the APM survey.

of result from covariance matrices calculated for different cosmologies shows that this dependence is negligibly small. Assuming a uniform prior for h over a factor of 2 is arguably over-cautious, and we have therefore added a Gaussian prior $h = 0.7 \pm 10\%$. This corresponds to multiplying by the likelihood from external constraints such as the HST key project (Freedman et al. 2000); this has only a minor effect on the results.

Figure 8 shows that there is a degeneracy between $\Omega_m h$ and the baryonic fraction Ω_b/Ω_m. However, there are two local maxima in the likelihood, one with $\Omega_m h \simeq 0.2$ and $\sim 20\%$ baryons, plus a secondary solution $\Omega_m h \simeq 0.6$ and $\sim 40\%$ baryons. The high-density model can be rejected through a variety of arguments, and the preferred solution is

$$\Omega_m h = 0.20 \pm 0.03; \qquad \Omega_b/\Omega_m = 0.15 \pm 0.07. \qquad (3)$$

The 2dFGRS data are compared to the best-fit linear power spectra convolved with the window function in Figure 9. This shows where the two branches of solutions come from: the low-density model fits the overall shape of the spectrum with relatively small 'wiggles', while the solution at $\Omega_m h \simeq 0.6$ provides a better fit to the bump at $k \simeq 0.065 \, h \, \mathrm{Mpc}^{-1}$, but fits the overall shape less well.

Perhaps the main point to emphasize here is that the results are not greatly sensitive to the assumed tilt of the primordial spectrum. We have used the CMB results to motivate the choice of $n = 1$, but it is clear that very substantial tilts are

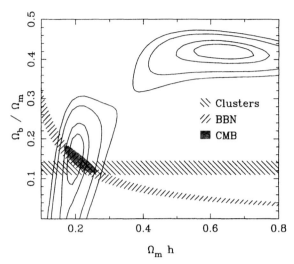

FIGURE 8. Likelihood contours for the best-fit linear power spectrum over the region $0.02 < k < 0.15$. The normalization is a free parameter to account for the unknown large scale biasing. Contours are plotted at the usual positions for one-parameter confidence of 68%, and two-parameter confidence of 68%, 95% and 99% (i.e. $-2\ln(\mathcal{L}/\mathcal{L}_{\max}) = 1, 2.3, 6.0, 9.2$). We have marginalized over the missing free parameters (h and the power spectrum amplitude) by integrating under the Likelihood surface. A prior on h of $h = 0.7 \pm 10\%$ was assumed. This result is compared to estimates from x-ray cluster analysis (Evrard 1997), big-bang nucleosynthesis (O'Meara et al. 2001) and recent CMB results (Jaffe et al. 2000). The CMB results assume that $\Omega_b h^2$ and $\Omega_{\mathrm{cdm}} h^2$ were independently determined from the data.

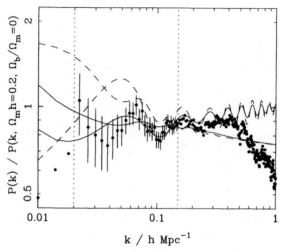

FIGURE 9. The 2dFGRS data compared with the two preferred models from the Maximum Likelihood fits convolved with the window function (solid lines). Error bars show the diagonal elements of the covariance matrix, for the fitted data that lie between the dotted vertical lines. The unconvolved models are also shown (dashed lines). The $\Omega_m h \simeq 0.6$, $\Omega_b/\Omega_m = 0.42$, $h = 0.7$ model has the higher bump at $k \simeq 0.05\,h\,\mathrm{Mpc}^{-1}$. The smoother $\Omega_m h \simeq 0.20$, $\Omega_b/\Omega_m = 0.15$, $h = 0.7$ model is a better fit to the data because of the overall shape.

required to alter our conclusions significantly: $n \simeq 0.8$ would be required to turn zero baryons into the preferred model.

VI CONCLUSIONS

The 2dFGRS is now the largest 3D survey of the local universe, by a factor of over 5 compared to any published survey. When it is complete, we expect to have obtained definitive results on a number of key issues relating to galaxy clustering. For details of the current status of the 2dFGRS, see http://www.mso.anu.edu.au/2dFGRS. In particular, this site gives details of the 2dFGRS public release policy, in which we intend to release approximately the first half of the survey data by mid-2001, with the complete survey database to be made public by mid-2003.

At present, the 2dFGRS data allow the galaxy power spectrum to be measured to high accuracy (10–15% rms) over about a decade in scale at $k < 0.15\,h\,\mathrm{Mpc}^{-1}$. We have carried out a range of tests for systematics in the analysis and a detailed comparison with realistic mock samples. As a result, we are confident that the 2dFGRS result can be interpreted as giving the shape of the linear-theory matter power spectrum on these large scales, and that the statistical errors and covariances between the data points are known.

By fitting our results to the space of CDM models, we have been able to reach a number of interesting conclusions regarding the matter content of the universe:

(1) The power spectrum is close in shape to that of a $\Omega_m h = 0.2$ model, to a tolerance of about 20%.

(2) Nevertheless, there is sufficient structure in the $P(k)$ data that the degeneracy between Ω_b/Ω_m and $\Omega_m h$ is weakly broken. The two local likelihood maxima have $(\Omega_m h, \Omega_b/\Omega_m) \simeq (0.2, 0.15)$ and $(0.6, 0.4)$ respectively.

(3) Of these two solutions, the preferred one is the low-density solution. The evidence for detection of baryon oscillations in the power spectrum is presently modest, with a likelihood ratio of approximately 3 between the favoured model and the best zero-baryon model. Conversely, a large baryon fraction can be very strongly excluded: $\Omega_b/\Omega_m < 0.28$ at 95% confidence, provided $\Omega_m h < 0.4$.

(4) These conclusions do not depend strongly on the value of h, but they do depend on the tilt of the primordial spectrum, with $n \simeq 0.8$ being required to make a zero-baryon model the best fit.

(5) The sensitivity to tilt emphasizes that the baryon signal comes in good part from the overall shape of the spectrum. Although the eye is struck by a single sharp 'spike' at $k \simeq 0.065\,h\,\mathrm{Mpc}^{-1}$, the correlated nature of the errors in the $P(k)$ estimate means that such features tend not to be significant in isolation. We note that the convolving effects of the window would require a very substantial spike in the true power in order to match our data exactly. This is not possible within the compass of conventional models, and the conservative conclusion is that the apparent spike is probably enhanced by correlated noise. A proper statistical treatment is essential in such cases.

It is interesting to compare these conclusions with other constraints. According to Jaffe et al. (2000), the current CMB data require $\Omega_m h^2 = 0.17 \pm 0.06$, $\Omega_b h^2 = 0.032 \pm 0.005$, together with a power-spectrum index of $n = 1.01 \pm 0.09$, on the assumption of pure scalar fluctuations. If we take $h = 0.7 \pm 10\%$, this gives

$$\Omega_m h = 0.24 \pm 0.09; \quad \Omega_b/\Omega_m = 0.19 \pm 0.07, \qquad (1)$$

in remarkably good agreement with the estimate from the 2dFGRS

$$\Omega_m h = 0.20 \pm 0.03; \quad \Omega_b/\Omega_m = 0.15 \pm 0.07. \qquad (2)$$

Latest estimates of the Deuterium to Hydrogen ratio in QSO spectra combined with big-bang nucleosynthesis theory predict $\Omega_b h^2 = 0.0205 \pm 0.0018$ (O'Meara et al. 2001), which disagrees with the CMB measurement at about the 2σ level. The confidence interval estimated from the 2dFGRS power spectrum overlaps both regions. X-ray cluster analysis predicts a baryon fraction $\Omega_b/\Omega_m = 0.127 \pm 0.017$ (Evrard 1997) which is again within 1σ of our preferred value.

The above limits are all shown on Figure 9, and paint a picture of qualitative consistency: it appears that we live in a universe that has $\Omega_m h \simeq 0.2$ with a baryon fraction of approximately 15%. It is hard to see how this conclusion can be seriously in error. Although the CDM model is claimed to have problems in matching galaxy-scale observations, it clearly works extremely well on large scales. Any new model that cures the small-scale problems will have to look very much like $\Omega_m = 0.3$ ΛCDM on large scales.

REFERENCES

Ballinger W.E., Peacock J.A., Heavens A.F., 1996, MNRAS, 282, 877
Baugh C.M., Efstathiou G., 1994, MNRAS, 267, 323
Benoist C., Maurogordato S., da Costa L.N., Cappi A., Schaeffer R., 1996, ApJ, 472, 452
Carlberg R.G., Yee H.K.C., Morris S.L., Lin H., Hall P.B., Patton D., Sawicki M., Shepherd C.W., 2000, ApJ, 542, 57
Davis M., Peebles, P.J.E., 1983, ApJ, 267, 465
Folkes S.J. et al., 1999, MNRAS, 308, 459
Feldman H.A., Kaiser N., Peacock J.A., 1994, ApJ, 426, 23
Freedman W.L. et al., 2000, astro-ph/0012376
Hamilton A.J.S., Tegmark M., Padmanabhan N., 2000, MNRAS, 317, L23
Jaffe A. et al., 2000, astro-ph/0007333
Kaiser N., 1987, MNRAS, 227, 1
Lewis I., Taylor K., Cannon R.D., Glazebrook K., Bailey J.A., Farrell T.J., Lankshear A., Shortridge K., Smith G.A., Gray P.M., Barton J.R., McCowage C., Parry I.R., Stevenson J., Waller L.G., Whittard J.D., Wilcox J.K., Willis K.C., 2001, MNRAS, submitted
Loveday J., Maddox S.J., Efstathiou G., Peterson B.A., 1995, ApJ, 442, 457
Maddox S.J., Efstathiou G., Sutherland W.J., Loveday J., 1990a, MNRAS, 242, 43P
Maddox S.J., Sutherland W.J., Efstathiou G., Loveday J., 1990b, MNRAS, 243, 692
Maddox S.J., Efstathiou G., Sutherland W.J., 1990c, MNRAS, 246, 433
Meiksin A.A., White M., Peacock J.A., 1999, MNRAS, 304, 851
Mould J.R. et al., 2000, ApJ, 529, 786
Outram P.J., Hoyle F., Shanks T., 2000, astro-ph/0009387
Peacock J.A., Dodds S.J., 1996, MNRAS, 280, L19
Peacock J.A. et al., 2001, Nature, 410, 169
Percival W.J. et al., 2001, MNRAS, submitted
Schlegel D.J., Finkbeiner D.P., Davis M., 1998, ApJ, 500, 525
Taylor A.N., Ballinger W.E., Heavens A.F., Tadros H., 2000, astro-ph/0007048
Verde L., Heavens A.F., Matarrese S., Moscardini L., 1998, MNRAS, 300 747

The Origin of the Magnetic Fields of the Universe: The Plasma Astrophysics of the Free Energy of the Universe

Stirling A. Colgate*, Hui Li* and Vladimir I. Pariev*[†‡]

*Theoretical Astrophysics Group, T6, Los Alamos National Laboratory,
Los Alamos, New Mexico 87545, USA
[†]Steward Observatory, 933 N. Cherry Ave., Tucson, AZ 85721, USA
[‡]P.N. Lebedev Physical Institute, Leninsky Prospect 53, Moscow 117924, Russia

Abstract. The largest accessible free energy in the universe is almost certainly the binding energy of the massive central black hole (BH) of nearly every galaxy. We have calculated one mechanism that produces this characteristic mass, $10^8\,M_\odot$, by initiating a Rossby vortex dominated accretion disk at a critical thickness, $\sim 100\,\mathrm{g\,cm^{-2}}$, in the development of the flat rotation curve of nearly every galaxy. We have simulated how an α–Ω dynamo should work due to star-disk collisions and plume rotation. The back reaction of this saturated dynamo may convert almost all the accretion energy into a single force-free magnetic field helix. This helix and field energy is then distributed as a quasi-static, hydrodynamically stable, Poynting flux configuration, filling the intergalactic space with a magnetized plasma. This energy and flux also explains the Faraday rotation maps of AGN in clusters. This energy density is $\sim 10^3$ times the virial energy of a galactic mass of baryonic matter in the combined gravity of dark and baryonic matter on the galaxy scale and before and during galaxy formation. This extra galactic energy density should affect subsequent galaxy formation. This possibly explains why the large extra galactic mass of gas in both clusters and the walls has not subsequently formed further galaxies. Also the reconnection of this magnetic field during a Hubble time provides enough energy to maintain the extra galactic cosmic ray spectrum.

The total energy released by the growth of supermassive black holes at the center of nearly every galaxy is large and can be comparable or even larger than that emitted by stars in the universe. Recent observations suggest that radiation from Active Galactic Nuclei (AGNs or quasars) might account for only $\sim 10\%$ of this energy [1]. Where did the rest of the energy go? We propose that a major fraction of this energy has been converted into magnetic energy and stored in the large scale magnetic fields primarily external to each galaxy, in galaxy clusters and "walls". We have envisioned a sequence of key physical processes that describes this energy flow [2–4].

The interpretation of Faraday rotation measure maps of jets and radio lobes of radio galaxies within galaxy clusters has revealed ordered or coherent regions, $L_{\mathrm{mag}} \sim 50 - 100\,\mathrm{kpc}$ ($\sim 3 \times 10^{23}\,\mathrm{cm}$), that are populated with large, $\sim 30\mu\mathrm{G}$ magnetic fields [5–7]. The magnetic energy of these coherent regions is $L_{\mathrm{mag}}^3 (B^2/8\pi) \sim 10^{59} - 10^{60}$ ergs, and the total magnetic energy over the whole cluster ($\sim 1\,\mathrm{Mpc}$ across) is expected to be even larger. Understanding the origin and role of these magnetic fields is a major challenge to plasma astrophysics.

A sequence of physical processes that are responsible for the production, redistribution and dissipation of these magnetic fields is proposed. These fields are associated with single Active Galactic Nucleus (AGN) within the cluster and therefore with all galaxies during their AGN (Active Galactic Nucleus or Quasar) phase, simply because only the central supermassive black holes ($\sim 10^8 M_\odot$) formed during the AGN phase have an accessible energy of formation, $\sim 10^{61}$ ergs, that can account for the magnetic field energy budget. An $\alpha - \Omega$ dynamo process has been

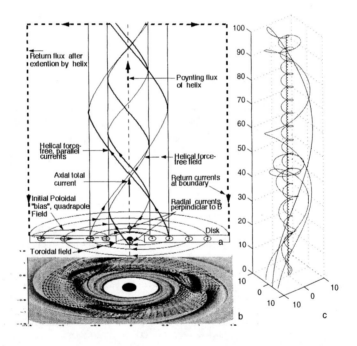

FIGURE 1. The helix generated by the dynamo in the disk. The foot points of the external quadrupole field lines, *(a)*, generated by the dynamo, are attached to the surface of the disk *(b)*. In the conducting low density plasma, $\beta \ll 1$, these field lines are wrapped into a force-free helix. This helix extends from the disk carrying flux and energy, a Poynting flux. *(c)*.

proposed that operates in an accretion disk around a black hole. Such an accretion disk is formed in the central part of the collapsing cloud of barionic matter, which forms the galaxy. When the surface density of barionic ellipsoid exceeds the value $\sim 100\,\mathrm{g\,cm^{-2}}$ large enough for a disk to contain its heat against radiative cooling, the mechanism of outward angular momentum transport by Rossby vortices sets in and allows the barionic gas to reach the center and to form a $\sim 10^8$ solar masses black hole. We have found a global nonaxisymmetric instability to exist in thin disks that can excite large scale vortices in the disk. Such vortices are shown to transport angular momentum outward efficiently [8].

The disk rotation naturally provides a large winding number, $\sim 10^{11}$ turns, sufficient to make both large gain and large flux. The helicity of the dynamo can be generated by the differential plume rotation derived from star-disk collisions. This helicity generation process has been demonstrated in the laboratory and the dynamo gain was simulated numerically. A liquid sodium analog of the dynamo is being built. Unlike dynamos driven by hydrodynamic or hydromagnetic instabilities in the accretion disk, star-disk collision driven dynamo has no problems with α quenching and can result in the magnetic fields stronger than equipartition value with the thermal pressure in the disk. Speculations are that the back reaction of the saturated dynamo will lead to the formation of a force-free magnetic helix, which will carry the energy and flux of the dynamo away from the accretion disk and redistribute the field within the clusters and walls. Such a helix is shown schematically in Figure 1a and calculated with the force-free field, Grad–Shafranov equations with a cylindrical boundary, Figure 1c. The magnetic reconnection of a small fraction of this energy logically is the source of the AGN luminosity, and the remainder of the field energy should then dominate the free energy of the present-day universe. The reconnection of this intergalactic field during a Hubble time is the only sufficient source of energy necessary to produce an extragalactic cosmic ray energy spectrum as observed in our Galaxy, and at the same time allow this spectrum to escape to the voids faster than the time of the Gruneisen-Zatsepin-Kuzmin loss due to black body radiation photons.

REFERENCES

1. Richstone, D.O., Ajhar, E.A., Bender, R. et al., *Nature* **395**, 14 (1998).
2. Colgate, S.A., and Li, H., *Astrophys. Space Sci.* **264**, 357 (1999)
3. Colgate, S.A., and Li, H., *IAUS 195, ASP Conf. Series 334*, eds. P.C.H. Martens and S. Tsurta, 1999
4. Colgate, S.A., Li, H., Pariev, V.I., *Physics of Plasmas*, in press (2001)
5. Taylor, G.B., and Perley, R.A., *Astrophys. J.* **416**, 554 (1993).
6. Clarke, T.E., Kronberg, P.P., and Böhringer, H. *Astrophys. J. Letters* **547**, L111 (2001).
7. Kronberg, P.P., *Prog. Phys.* **57**, 325 (1994).
8. Li, H., Colgate, S.A., Wendroff, B., and Liska, R., *Astrophys. J.*, in press (2001).

Visualization of Gravitational Lenses

Francisco Frutos Alfaro
Max Planck Institut für Aeronomie
Max Planck Straße 2
37191 Katlenburg-Lindau
Germany
frutos@linmpi.mpg.de

INTRODUCTION

Gravitational Lens phenomena are presently playing an important role in astrophysics. These lenses deflects the light rays coming from distant objects and allow the origination of multiple images of the same source object. This effect opens the possibility of determining not only the parameters of the lens (mass) but also the computation of Hubble's constant provided that the time delay of a pair of images is known. Using C, Xforms, Mesa (free version of Open GL) and Imlib a computer program to visualize this phenomenon has been developed. This program has been applied to generate sequences of images of a source object and its corresponding images. It has also been used to visually test different models of gravitational lenses.

THE VISUALIZATION PROGRAM

Description of the program

The program was written in C and the *Mesa graphic libraries* (free version of the Open GL) have been used. These graphic subroutines are available for Unix and Linux systems. The *XForms Library* was used to design the control panel program (see fig. 1). The *Image Library* permits to load an image file on the program. These libraries can be found at the following addresses:
http://world.std.com/~xforms/ftp/ftp.html
http://www.mesa3d.org/
ftp://ftp.enlightenment.org/pub/enlightenment/imlib
The program creates a window: the *Control Panel* (see fig. 1). The user is able to control all items on it just by clicking. A second window, the *Image window*, appears when the Image Window button is clicked on this panel (see fig. 1). On this window the events (images, ray plot, etc.) are displayed. All variations of

the parameters on the control panel are shown on-line on this window. The Help button on the panel gives the user a concise program guide.

A version of this program employing the SGI *Graphic Libraries* and the *Forms Library* is also available. The author prepared websites for downloading this program:
http://lia.efis.ucr.ac.cr/~frutos/
http://www.tat.physik.uni-tuebingen.de/~frutto/

APPLICATIONS

From both the didactical and scientific point of view a program to visualize the gravitational lenses is useful. This versatile program works quickly and interactively with the mouse. With this computer program the user has a tool to visualize and to visually model gravitational lenses. Two of many applications of the program are:
- *Sequences of images*
- *Easy visual modeling*

The user can produce sequences of images for a chosen gravitational lens model (see fig. 2). Through the variation of model parameters he or she can investigate the structure of the images. The user can also attempt to visually model observed gravitational lenses. The observed position data can be input and the model parameters can be easily varied in order to approximate the observed images (see fig. 3). So the user can quickly obtain model parameter estimations. Some observed lenses have already been modeled and the user can compare those results with the output from a chosen model of the control panel.

FUTURE WORK

The program can be improved by the inclusion of some additional subroutines:
- *Contour subroutine for the isochrones (time delay)*
- *Light curves subroutine (dependence of brightness with time)*
- *Subroutine for computing the image magnification*
- *Subroutine to calculate critical curves and caustics*
- *Fitting subroutine*
- *Root finder subroutine*
- *Subroutine to load images of observed gravitational lenses*
- *Subroutine with more complex (elliptical) models*
- *Subroutine for superposition of models in different lens planes*
- *Subroutine with cosmic string lens models*
- *Subroutine for non-parametric reconstruction*
- *Kaiser-Squires Subroutine*

The author is actually working on the implementation of some of the abovementioned improvements.

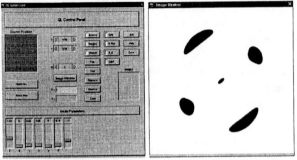

FIGURE 1. Control Panel and Image Window

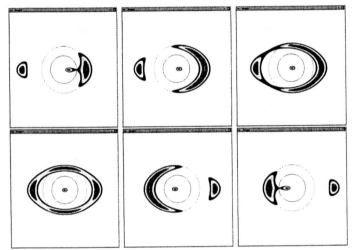

FIGURE 2. An elliptical lens

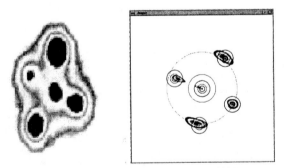

FIGURE 3. Gravitational lens 2237 + 0305

Explosions and Outflows during Galaxy Formation

Hugo Martel* and Paul R. Shapiro*

*Department of Astronomy, University of Texas, Austin, TX 78712

Abstract. We consider an explosion at the center of a halo which forms at the intersection of filaments inside a cosmological pancake, a convenient test-bed model for galaxy formation. ASPH/P^{3}M simulations reveal that such explosions are anisotropic. The energy and metals are channeled into the low density regions, away from the pancake. The pancake remains essentially undisturbed, even if the explosion is strong enough to blow away all the gas located inside the halo and reheat the IGM surrounding the pancake. Infall quickly replenishes this ejected gas and gradually restores the gas fraction as the halo continues to grow. Estimates of the collapse epoch and SN energy-release for galaxies of different mass in the CDM model can relate these results to scale-dependent questions of blow-out and blow-away and their implication for early IGM heating and metal enrichment and the creation of gas-poor dwarf galaxies.

INTRODUCTION

The release of energy that occurs during galaxy formation can have important consequences. We present 3D gas dynamical simulations of the effect of energy release by supernovae (SNe) on the evolution of the halo in which the explosions take place, the surrounding large-scale structure, and the IGM.

Structure formation from Gaussian random noise is highly anisotropic, favoring pancakes and filaments over quasi-spherical objects. However, pancake fragmentation results in the formation of quasi-spherical halos [1,2,6,8] with density profiles similar to the universal profile [7] found in CDM simulations. This suggests that pancake fragmentation may be used to study galaxy formation. This provides a good compromise between simulations of structure formation in CDM models, with limited resolution, and those of isolated virialized objects, which ignore cosmological initial and boundary conditions. We use our ASPH/P^{3}M method to simulate the formation of a halo at the intersection of two filaments in the plane of a cosmological pancake which collapses at scale factor $a = a_c$, in an $\Omega_0 = 1$ universe with baryon fraction $\Omega_B = 0.03$ (see [3,4] for further description). These simulations also describe early galaxy formation in a flat, ΛCDM model. The explosion is induced at scale factor $a_{\exp}/a_c \cong 2$, when the central gas density first exceeds

$\rho_{\text{gas}}/\langle\rho_{\text{gas}}\rangle = 1000$, by boosting the thermal energy of this central gas to the level $E_{\text{exp}} = \chi \mathcal{E}_{\text{halo}}$, where $\mathcal{E}_{\text{halo}}$ is the total thermal energy of gas in the halo, for four cases: $\chi = 0, 10, 100,$ and 1000.

RESULTS

Figure 1 shows the pancake-filament-halo structure at $a/a_c = 3$, with and without explosions. For all cases, the dark matter distribution is essentially the same as the gas distribution for $\chi = 0$. The pancake and filaments ensure that a highly anisotropic explosion results, channelling the energy and mass ejection along the symmetry axis. The ability of explosions to eject metal-enriched gas depends on χ.

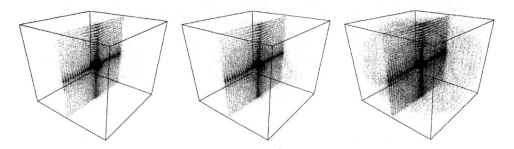

FIGURE 1. Gas distribution at $a/a_c = 3$. (left) no explosion case ($\chi = 0$); (middle) explosion case with $\chi = 100$; (right) explosion case with $\chi = 1000$.

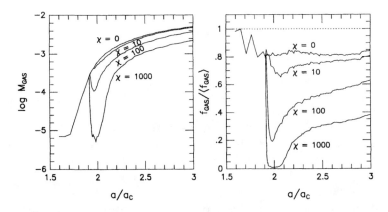

FIGURE 2. (Left) gas mass M_{gas} (in computational units $M_{\text{box}} = \bar{\rho}\lambda_p^3 = 1$) in the central halo (sphere of mean overdensity 200), versus a/a_c, for different explosion intensities, as labelled. (Right) gas mass fraction f_{gas} inside halo, divided by the universal gas mass fraction $\langle f_{\text{gas}}\rangle = \Omega_B/\Omega_0$, versus a/a_c.

For $\chi = 10$, relatively little gas is blown out of the halo, and is quickly re-accreted. For $\chi > 100$, most of the halo gas is blown out. However, as shown in Figure 2, infall along the pancake plane continues despite the explosion and replenishes the ejected halo gas very efficiently, although the final gas fraction f_{gas} is below the cosmic mean value by an amount which increases with increasing χ. Even the most energetic explosions fail to disturb the pancake and filaments.

Each dimensionless, scale-free simulation for a given χ can be applied to any particular halo mass M_{halo} at any epoch by adjusting the values of λ_p (the wavelength of the primary pancake) and collapse redshift z_c when converting to physical units. We can also use Milky Way (MW) star formation efficiencies and IMF to estimate a typical value $\chi_*(\lambda_p, z_c)$ of the explosion parameter χ, for each λ_p and z_c. $\chi \gtrsim 100$ is required to eject gas and metals into the IGM. This matches MW efficiencies only for $M_{halo} \lesssim 10^7 M_\odot$, for both cluster-normalized SCDM (i.e. $\lambda_p < 0.2\,\text{Mpc}$) and COBE-normalized ΛCDM ($\Omega_0 = 0.3$, $\lambda_0 = 0.7$, $h = 0.7$) (i.e. $\lambda_p < 0.3\,\text{Mpc}$). However, gas replenishment by continued infall may enable a single halo to contribute multiple outbursts. Halos $< 10^{10} M_\odot$ form early enough to cause heavy element distribution prior to $z = 3$. Only the smallest mass objects, however, can expect to do so if limited to MW efficiencies. Such small mass halos are also the ones which are most likely to form early enough that they may explode *before* the reionization of the IGM is complete, after which the enhanced IGM pressure inhibits gas ejection. Otherwise efficiencies greatly exceeding MW values are required.

A final determination of the success or failure of heavy element distribution by SN explosions in low-mass halos forming at high redshift will depend sensitively on the efficiencies of star formation and SN energy release, and the relative timing of these explosions versus universal reionization. Our results suggest that heavy element distribution at the observed level of $\approx 10^{-3}$ solar in the IGM could have been accomplished most efficiently by low-mass objects prior to the completion of universal reionization. (NASA ATP grants NAG5-7363 and NAG5-7821, NSF grant ASC-9504046, and Texas Advanced Research Program grant 3658-0624-1999).

REFERENCES

1. Alvarez, M., Shapiro, P. R., and Martel, H., *Rev.Mex.A.A.(SC)*, in press (2001a) (astro-ph/0006203)
2. Alvarez, M., Shapiro, P. R., and Martel, H., in press (this volume) (2001b)
3. Martel, H., and Shapiro, P. R., *Nucl.Phys.B*, **80**, CD-Rom 09/16 (2000) (astro-ph/9904121)
4. Martel, H., and Shapiro, P. R., *Rev.Mex.A.A. (SC)*, in press (2001a) (astro-ph/0006309)
5. Martel, H., and Shapiro, P. R., in preparation (2001b).
6. Martel, H., Shapiro, P. R., and Valinia, A., in preparation (2001)
7. Navarro, J. F., Frenk, C. S., and White, S. D. M., *Ap.J.*, **490**, 493 (1997)
8. Valinia, A., Shapiro, P. R., Martel, H., and Vishniac, E. T., *Ap.J.*, **479**, 46 (1997)

Cosmological Parameter Survey using the Gravitational Lensing Method

Premana Premadi*, Hugo Martel‡, Richard Matzner∥, and Toshifumi Futamase¶

*Department of Astronomy & Bosscha Observatory, Bandung Institute of Technology, Indonesia
‡Department of Astronomy, University of Texas, Austin, TX 78712
∥Department of Physics and Center for Relativity, University of Texas, Austin. TX 78712
¶Astronomical Institute, Tohoku University, Sendai, Japan

Abstract. We study light propagation in inhomogeneous universes, for 43 different *COBE*-normalized CDM models, with various values of Ω_0, λ_0, H_0, and σ_8. We provide statistics of the magnification and multiple imaging of distant sources. The results of these experiments might be compared with observations, and eventually help constraining the possible values of the cosmological parameters.

INTRODUCTION

The evolution of a homogeneous, isotropic, expanding universe is described in terms of three parameters: the Hubble constant H_0, the density parameter Ω_0, and the cosmological constant λ_0. The large-scale structure of the universe represents the deviations from this overall homogeneity and isotropy, and can be quantified using the rms density fluctuation at scale $8h^{-1}$ Mpc, σ_8. Gravitational lenses can be used to estimate the values of these parameters. However, if the cosmological model has several free parameters, a full survey of the cosmological parameter space is required in order to determine all cosmological parameters *simultaneously*. We present a study of light propagation in inhomogeneous universe that surveys the full 4-parameter phase-space formed by Ω_0, λ_0, H_0, and σ_8.

To simulate light propagation in inhomogeneous universes, we use a newly developed version of the *Multiple Lens-Plane Algorithm* [1,2]. Our algorithm uses a P³M code with 64^3 particles to simulate the formation and evolution of large-scale structure in the universe, inside a computational volume of comoving size $L_{\text{box}} = 128$ Mpc, combined with an empirical Monte Carlo method for locating galaxies inside the computational volume, based on the underlying distribution of dark matter. Each galaxy is modeled by a truncated, non-singular isothermal sphere. We consider TCDM models normalized to *COBE*, with $T_{\text{CMB}} = 2.7$ K and $\Omega_{\text{B0}} = 0.015h^{-2}$. The independent parameters in this parameter space are Ω_0,

λ_0, H_0, and σ_8, while the tilt n of the power spectrum is a dependent parameter (adjusted to produce the desired value of σ_8). We survey this parameter space by considering 43 different cosmological models. For each model, we performed numerous ray-tracing experiments.

THE MAGNIFICATION DISTRIBUTIONS

Figure 1 shows the magnification distribution $P(\mu)$ for various models, and its dependence upon the various parameters. (1) As σ_8 increases, the peak of the distribution decreases, the low edge of the distribution moves to even lower values, but the right edge is hardly affected. A larger σ_8 implies that (i) the underdense regions are more underdense and the overdense regions are more overdense, and (ii) the fraction of the universe occupied by underdense regions increases. In the case of demagnification, these two effects act in the same direction, while in the case of magnification, they act in opposite directions, and almost perfectly cancel each other, making the distributions at values of $\mu > 1$ independent of σ_8. (2) The distributions are independent of H_0. This results from competing effects, between the dependences upon H_0 of cosmological distances and mean background density. (3) The dependence upon Ω_0 is difficult to interpret, because it involves several competing effects. The importance of lensing increases with Ω_0, resulting in a shift of the distribution toward lower values. (4) The presence of a cosmological constant λ_0 increases the effect of magnification by increasing the cosmological distances. The cosmological constant results in a widening of the distribution, and a shift toward lower magnifications.

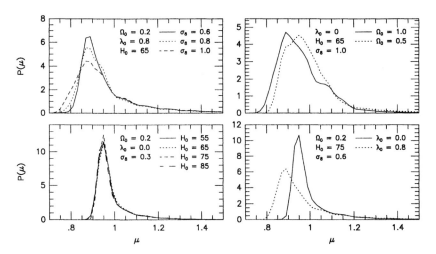

FIGURE 1. Magnification distributions for various models, showing the effect of varying σ_8 (1st panel), H_0 (2nd panel), Ω_0 (3rd panel), and λ_0 (4th panel).

THE DISTRIBUTION OF IMAGE SEPARATIONS

Figure 2 shows histograms of the angular separations of double images, for models with $\Omega_0 = 0.2$, $\lambda_0 = 0.8$. Several trends are apparent. We are considering sources with an angular diameter of 1″, hence the smallest possible image separation 0.5″. Most histograms show a distributions that rises sharply from 0.5″ to 1″, and then drops slowly at larger separations, with a high-tail that extends to separations of order 4″ − 6″. We find no obvious correlation between the shape of the histograms and the value of σ_8, indicating that double images are caused primarily by individual galaxies, and not by the large-scale structure.

SUMMARY

(1) Magnification is caused primarily by the distribution of background matter, with negligible contribution from galaxies. Consequently, it is sensitive to the value of σ_8. (2) Multiple images are caused mainly by galaxies. Consequently, their properties are independent of σ_8. (3) The dependences upon H_0 and Ω_0 are complex, because of competing effects. Determining λ_0 and σ_8 from observations seems much more promising than determining Ω_0 and H_0.

REFERENCES

1. Jaroszyński, M., *M.N.R.A.S.*, **255**, 655 (1992)
2. Premadi, P., Martel, H., and Matzner, R., *Ap.J.*, **493**, 10 (1998)

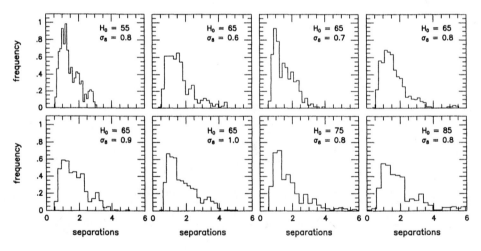

FIGURE 2. Histograms of the distribution of image separations in arc seconds, for various models with $\Omega_0 = 0.2$, $\lambda = 0.8$. The values of H_0 (in units of $\mathrm{km\,s^{-1}Mpc^{-1}}$), σ_8, and the number of double images are indicated in each panel. (with H_0 in units of $\mathrm{km\,s^{-1}Mpc^{-1}}$).

Cosmological Matter Perturbations with Causal Seeds

Jiun-Huei Proty Wu

Astronomy Department, University of California, Berkeley, CA 94720-3411, USA

Abstract. We investigate linear matter density perturbations in models of structure formation with causal seeds. Under the fluid approximation, we obtain the analytic solutions using Green-function technique. Some incorrect solutions in the literature are corrected here. Based on this, we analytically prove that the matter density perturbations today are independent of the way the causal seeds were compensated into the background contents of the universe when they were first formed. We also find that the compensation scale depends not only on the dynamics of the universe, but also on the properties of the seeds near the horizon scale. It can be accurately located by employing our Green-function solutions.

With the cosmological principle as the basic premise, there are currently two main paradigms for structure formation of the universe—inflation [1] and topological defects [2]. Although recent observations of the cosmic microwave background (CMB) seem to have favored inflation [3], defects can still coexist with it and their observational verification will have certain impact to the grand unified theory. In the literature the power spectra of models with causal seeds like defects have been investigated using the full Einstein-Boltzmann equations. However, the study of the phase information of these perturbations still remains difficult due to the limited computation power [4]. Although there have been some detailed treatments for models with causal seeds [5,6], we shall present a simpler formalism to provide not only a physically transparent way to understand the evolution of their density perturbations, but also a computationally economical scheme to investigate their phase information [7]. This formalism is parallel to those presented in [8] and [9], but we give modifications to include the cosmological constant Λ, as well as some other improvements and corrections.

In a flat Friedmann-Robertson-Walker (FRW) model with an evolving weak source field of energy-momentum tensor $\Theta_{\mu\nu}(\mathbf{x},\eta)$, the full evolution equations of linear perturbations can be obtained by considering the stress-energy conservation of the fluids and the source, as well as the linearly perturbed Einstein equations [7]. With the photon-baryon tight coupling approximation, a closed set of equations for the density perturbations in the synchronous gauge are:

$$\ddot{\delta}_r - \frac{4}{3}\ddot{\delta}_c + \frac{\dot{R}}{1+R}(\dot{\delta}_r - \frac{4}{3}\dot{\delta}_c) - \frac{1}{3(1+R)}\nabla^2\delta_r = 0, \tag{1}$$

$$\ddot{\delta}_c + \frac{\dot{a}}{a}\dot{\delta}_c - \frac{3}{2}\left(\frac{\dot{a}}{a}\right)^2[\Omega_c\delta_c + (2+R)\Omega_r\delta_r] = 4\pi G\Theta_+, \tag{2}$$

where $R = 3\rho_B/4\rho_r$, $\Theta_+ = \Theta_{00} + \Theta_{ii}$, a is the scale factor, a dot represents the derivative with respect to the conformal time η, and the subscripts 'c', 'B' and 'r' denote cold dark matter (CDM), baryons, and radiation respectively.[1] By splitting the perturbations into the initial (I) and subsequent (S) parts as $\delta_N(\mathbf{x},\eta) = \delta_N^I(\mathbf{x},\eta) + \delta_N^S(\mathbf{x},\eta)$ where $N = c, r$, and employing the zero entropy fluctuation condition on super-horizon scales as part of the initial condition, we solve the above equations for $\Lambda = 0$ in the Fourier space to yield

$$\tilde{\delta}_N^I(\mathbf{k},\eta) = \tilde{\mathcal{G}}_3^N(k;\eta,\eta_i)\tilde{\delta}_c(\mathbf{k},\eta_i) + \tilde{\mathcal{G}}_4^N(k;\eta,\eta_i)\dot{\tilde{\delta}}_c(\mathbf{k},\eta_i), \tag{3}$$

$$\tilde{\delta}_N^S(\mathbf{k},\eta) = 4\pi G \int_{\eta_i}^{\eta} \tilde{\mathcal{G}}_4^N(k;\eta,\hat{\eta})\tilde{\Theta}_+(\mathbf{k},\hat{\eta})\,d\hat{\eta}, \tag{4}$$

where η_i is the initial conformal time, and the full expressions of $\tilde{\mathcal{G}}_i^N$, including the baryonic effects, are presented in [7]. We notice that to solve for δ_c we need only two Green functions ($\tilde{\mathcal{G}}_3^c$ and $\tilde{\mathcal{G}}_4^c$), instead of five as presented in [8], some of which are incorrect due to incorrect initial conditions. Figure 1 (left) shows the asymptotic behaviors of these two Green functions on the super-horizon ($k\hat{\eta} \ll 1$) and sub-horizon ($k\hat{\eta} \gg 1$) scales: $\tilde{T}_i^c(k;\hat{\eta}) \equiv \lim_{\eta/\eta_{eq}\to\infty} \tilde{\mathcal{G}}_i^c(k;\eta,\hat{\eta})a_{eq}/a$, where the subscript 'eq' denotes the epoch of radiation-matter density equality. A simple and accurate extrapolation scheme can then be used to obtain solutions in the non-flat or $\Lambda \neq 0$ cosmologies [7]. All the solutions are numerically verified to high accuracy.

One important problem for structure formation with causal seeds is to investigate how the source energy was compensated into the radiation and matter background when the seeds were formed at η_i, and how the resulting perturbations today depend on this. First consider the pseudo energy $\tau_{00} = \Theta_{00} + (3/8\pi G)(\dot{a}/a)^2(\Omega_c\delta_c + \Omega_r\delta_r) + (\dot{a}/a)\dot{\delta}_c/4\pi G$. Since causality requires $\tau_{00} = 3\delta_r/4 - \delta_c = 0$ on superhorizon scales, it follows that the initial source energy Θ_{00} can be compensated into between δ_N and $\dot{\delta}_N$ with different portions. With our Green-function solutions, it can be straightforwardly shown that no matter how Θ_{00} was compensated into the background at η_i, the resulting δ_c^I and thus δ_c today will be the same. This was first numerically observed in [9], and here we can provide an analytic proof.

Finally we use our Green-function solutions to study the compensation mechanism and the scale on which it operates. First it can be shown that [7] $\tilde{\tau}_{00}(\mathbf{k},\eta_0) = (1 - T(k))\tilde{\Theta}_{00}(\mathbf{k},\eta_0) + \int_{\eta_i}^{\eta_0}\left[T'(k;\hat{\eta})(\dot{a}(\hat{\eta})/a(\hat{\eta}))\tilde{\Theta}_+(\mathbf{k},\hat{\eta}) + T(k)\dot{\tilde{\Theta}}_{00}(\mathbf{k},\hat{\eta})\right]d\hat{\eta}$, where $T(k)$ is the standard CDM transfer function in the inflationary models, and

---

[1] Here we have ignored neutrinos, whose effects on the current study is negligible.

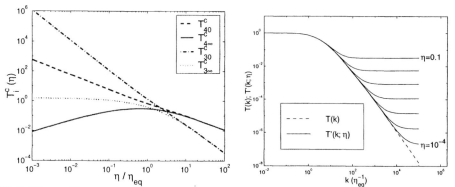

FIGURE 1. The asymptotic behaviors of the CDM Green-function solutions on the super-horizon and sub-horizon scales (left), as well as on all scales (right).

$T'(k;\hat{\eta}) = \tilde{\mathcal{G}}_4^c(k;\eta_0,\hat{\eta})/\tilde{\mathcal{G}}_{40}^c(k;\eta_0,\hat{\eta})$ (see Figure 1 right). On super-horizon scales, it is clear that $T(k)$ is unity by definition so that only the integral survives, and that the quantity inside the square brackets is nothing but $\tilde{\Theta}_{0i,i}(\mathbf{k},\hat{\eta})$ due to the source stress-energy conservation and the fact $T = T'$. Since $\tilde{\Theta}_{0i,i}$ has a k^4 fall-off power spectrum due to causality, it follows immediately that the pseudo-energy τ_{00} also has a k^4-decay power spectrum outside the horizon. On sub-horizon scales, on the other hand, although $(1 - T(k))$ is approximately unity, the usual sub-horizon power-law decay in $\tilde{\Theta}_{00}$ (as in the case of topological defects) will make the first term negligible, while the quantity inside the integral is no longer simply $\tilde{\Theta}_{0i,i}$. As a result, we see that the compensation scale, above which the power of both δ_c and τ_{00} decays as k^4, is determined not only by the functions T and T', but also by the properties of the source near the horizon scale. Once the detailed behavior of the source near the horizon scale is known, we can accurately locate the compensation scale using our formalism [7]. We acknowledge the support from NSF KDI Grant (9872979) and NASA LTSA Grant (NAG5-6552).

REFERENCES

1. Guth, A. H., *Phys.Rev.*, **D23**, 347 (1981).
2. For a review see Vilenkin, A., Shellard, E. P. S., *Cosmic strings and other topological defects*, Cambridge University Press, Cambridge, 1994.
3. Jaffe, A. H. et al., *Phys.Rev.Lett.* in press, *astro-ph/0007333* (2001).
4. Avelino, P. P., Shellard, E. P. S., Wu, J. H. P., Allen, B., *ApJ.*, **507**, L101 (1998).
5. Hu, W., Seljak, U., White, M., Zaldarriaga, M., *Phys.Rev.*, **D57**, 3290 (1998).
6. Hu, W., White, M., *Phys.Rev.*, **D56**, 596 (1997).
7. Wu, J. H. P., *astro-ph/0012205* (2000).
8. Veeraraghavan, S., Stebbins, A., *ApJ.*, **365**, 37 (1990).
9. Pen, U., Spergel, D., Turok, N., *Phys.Rev.*, **D49**, 692 (1994).

Cluster Abundance and Large Scale Structure

Jiun-Huei Proty Wu

Astronomy Department, University of California, Berkeley, CA 94720-3411, USA

Abstract. We use the presently observed number density of large X-ray clusters and linear mass power spectra to constrain the shape parameter (Γ), the spectral index (n), the amplitude of matter density perturbations on the scale of $8h^{-1}$Mpc (σ_8), and the redshift distortion parameter (β). The non-spherical collapse as an improvement to the Press-Schechter formula is accounted for. An analytical formalism for the formation redshift of halos is also derived.

One of the most important constraints on models of structure formation is the observed abundance of galaxy clusters. Because they are the largest virialized objects in the universe, their abundance can be simply predicted by and thus used to constrain the linear perturbation theory. In the light of the new observations and the improvement in modeling cluster evolution, we revisit this application [1], which has been extensively explored in the literature.

Based on the maximum-likelihood analysis, we first use the observed linear mass power spectra $P(k)$ by Peacock & Dodds [2] (PD, combination of galaxy surveys) and by Hamilton, Tegmark, & Padmanabhan [3] (HTP, based on PSCz [4]; see Figure 1), to estimate the spectral index n, the shape parameter Γ, and the amplitude of perturbations σ_8 in the parameterization of the standard model $P(k) \propto \sigma_8^2 k^n T^2(k/\Gamma)$ [1]. The results are shown in Figure 1 and Table 1, with $\sigma_{8(I)} = 0.78 \pm 0.26$ for IRAS galaxies. The degeneracy between Γ and n is clear, motivating us to find the theoretically expected 'degenerated' shape parameter $\Gamma' = 0.247\Gamma \exp(1.4n) = 0.220^{+0.036}_{-0.031}$, which has a much more constrained likelihood. These results are consistent with the current constraints from CMB [5].

TABLE 1. Best fits of different data sets (all errors at 95% confidence level).

	n	Γ	Γ'	χ^2/degrees of freedom (conf. level)
HTP	$0.91^{+1.09}_{-0.91}$	$0.18^{+0.74}_{-0.18}$	$0.160^{+0.085}_{-0.051}$	15.4/19 (70%)
PD	$0.99^{+0.81}_{-0.86}$	$0.23^{+0.55}_{-0.16}$	$0.229^{+0.042}_{-0.033}$	6.95/9 (64%)
HTP+PD	$0.84^{+0.67}_{-0.67}$	$0.27^{+0.42}_{-0.16}$	$0.220^{+0.036}_{-0.031}$	24.6/30 (74%)

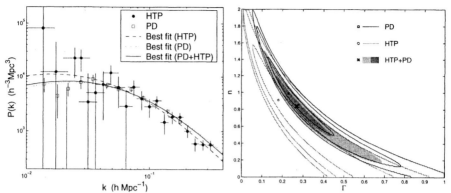

FIGURE 1. Matter power spectra of different observations and their best fits (left). The 68%, 95%, and 99% likelihood contours (inner out) in the (Γ, n) parameter space (right).

TABLE 2. Values in the fits of $\sigma_{8(i)}$ and $\beta_{I(j)}$.

i	PS	ST	LS	ST+LS	j	PS	ST+LS
c_1	0.54	0.50	0.455	0.477	d_1	0.693	0.613
c_2	0.45	0.37	0.31	0.34	d_2	0.26	0.24

Following a similar formalism as in Ref. [6], we then derive the probability distribution function $p_{z(i)}(z)$ of cluster formation redshift z [1] for different models of mass function $n_i(M)$, where i =PS (Press & Schechter [7]), ST (Sheth & Tormen [8]), or LS (Lee & Shandarin [9]), the last two of which incorporate non-spherical collapse. With specified $n_i(M)$, $p_{z(i)}(z)$, σ_8, and the previously estimated n and Γ (or Γ'), we can project the present cluster abundance of a given mass M into the space of formation redshift z, and then use the virial mass-temperature relation to associate this abundance with the virial temperature T that corresponds to the given M and z. An integration over T and z will give us a prediction of cluster abundance, which is a function of σ_8. A comparison of this with the observation [10] will give the normalization of σ_8. Combined with the $\sigma_{8(I)}$ estimated earlier, it further yields the constraint on the redshift distortion parameter $\beta_I \approx \Omega_m^{0.6} \sigma_8/\sigma_{8(I)}$, which quantifies the confusion between the Hubble expansion and the local gravitational collapse [11]. Our results can be fitted by $\sigma_{8(i)}(\Omega_{m0}, \Omega_{\Lambda 0}) = c_1 \Omega_{m0}^{\alpha}$, where i =PS, ST, LS or ST+LS, and $\alpha \equiv \alpha(\Omega_{m0}, \Omega_{\Lambda 0}) = -0.3 - 0.17\Omega_{m0}^{c_2} - 0.13\Omega_{\Lambda 0}$ (see Figure 2 left), and $\beta_{I(j)}(\Omega_{m0}, \Omega_{\Lambda 0}) = d_1 \Omega_{m0}^{d_2 - 0.16(\Omega_{m0} + \Omega_{\Lambda 0})}$, where j =PS or ST+LS (see Figure 2 right). The parameter values of these fits are given in table 2.

It is clear that the σ_8 and β_I resulted from non-spherical-collapse models (ST and LS) are systematically lower than those based on the PS formalism, mainly owing to the larger mass function on cluster scales. A detailed investigation of the uncertainties in our final results shows that the main contributor is the uncertainty in the normalization of the virial mass-temperature relation. Therefore further improvement in this normalization will provide us with more stringent constraint

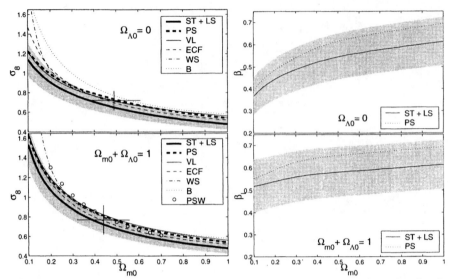

FIGURE 2. The cluster-abundance-normalized σ_8 and β_I, in comparison with results of σ_8 from the literature (see [12] for the abbreviations used in the figure legend).

on both σ_8 and β_I. In addition, since we saw significant corrections in the resulting σ_8 and β_I when switching from the PS formalism to the more accurate non-spherical-collapse models, we urge the use of these models in all relevant studies, especially when we are entering the regime of precision cosmology. We acknowledge the support from NSF KDI Grant (9872979) and NASA LTSA Grant (NAG5-6552).

REFERENCES

1. Wu, J. H. P., *astro-ph/0012207* (2000).
2. Peacock, J. A., Dodds, S. J., *MNRAS.*, **267**, 1020 (1994).
3. Hamilton, A. J. S., Tegmark, M., Padmanabhan, N., *MNRAS.*, **317**, L23 (2000).
4. Saunders, W. et al., astro-ph/0001117 (2000).
5. Jaffe, A. H. et al., *Phys.Rev.Lett.* in press, *astro-ph/0007333* (2001).
6. Lacey, C., Cole, S., *MNRAS.*, **262**, 627 (1993); *MNRAS.*, **271**, 676 (1994).
7. Press, W. H., Schechter, P., *ApJ.*, **187**, 452 (1974).
8. Sheth, R. K., Tormen, G., *MNRAS.*, **308**, 119 (1999).
9. Lee, J., Shandarin, S. F., *ApJ.*, **500**, 14 (1999).
10. Henry, J. P., *ApJ.*, **534**, 565 (2000).
11. Kaiser, N., *MNRAS.*, **227**, 1 (1987).
12. Viana, P.T.P., Liddle, A.R., *MNRAS.*, **303**, 535 (1999; VL); Eke, V.R., Cole, S., Frenk, C.S., *MNRAS.*, **282**, 263 (1996; ECF); Wang, L., Steinhardt, P.J., *ApJ.*, **508**, 483 (1998; WS); Borgani, S., Rosati, P., Tozzi, P., Norman, C., *ApJ.*, **517**, 40 (1999; B); Pierpaoli, E., Scott, D., White, M., *astro-ph/0010039* (2000; PSW).

CHAPTER 6

PROBES OF THE ACCELERATING UNIVERSE

The Quintessential Universe

Paul J. Steinhardt[†]

[†]*Department of Physics, Princeton University, Princeton, NJ 08540*

Abstract. Recent evidence suggests that most of the energy in the universe consists of some form of dark energy that is gravitationally self-repulsive and that is causing the expansion rate of the universe to accelerate. We review the evidence, including recent measurements of the cosmic microwave background by the DASI, BOOMerang and MAXIMA groups. The dark energy may consist of vacuum energy density (or, equivalently, a cosmological constant), or quintessence, a time-evolving, spatially inhomogeneous component with negative pressure. A key problem is to explain the initial conditions required to have the energy density nearly coincident with the matter density today. A possible solution is "k-essence," a form of quintessence with an attractor-like solution which leads to cosmic acceleration today for a very wide range of initial conditions without fine-tuning and without invoking an anthropic argument.

INTRODUCTION

In the 20th century, three revolutionary discoveries have totally changed our understanding of the evolution of the universe. First, in 1927, Hubble and Slipher established that the universe is expanding. The discovery of the cosmic microwave background in the 1960's and the successful explanation of primordial nucleosynthesis have confirmed that the observable universe was hotter and denser in the past and has been expanding cooling as predicted by Einstein's theory of general relativity over the last 14 billion years. Second, beginning with the work of Fritz Zwicky in the 1930's, evidence has accumulated that most of the mass of the universe consists of extraordinary, non-baryonic matter. Ordinary matter consists of less than five percent of the total energy of the universe. Third, over the last decade of the 20th century, we have learned that most of the energy of the universe is not matter at all. Rather, the universe is dominated today by some form of dark energy with a peculiar property: it is gravitationally self-repulsive. As a result, the expectation for the future has been dramatically revised. Based on the conventional assumption that the universe contains only matter and radiation – the forms of energy we can readily detect – the expectation for the future had been that the expansion rate of the universe would slow continuously due to the gravitational self-attraction of matter. The only major issue seemed to be whether the universe would expand forever or ultimately recollapse to a big crunch. Now, we know that cosmic dark energy exists that opposes the self-attraction of matter and is causing the expansion of the universe to accelerate.

The evidence for dark energy became apparent by the mid-90's [1,2]. First, improved observations confirmed that the total mass density is probably less than half of the critical density [3–5]. At the same time, combined measurements of the cosmic microwave background (CMB) temperature fluctuations and the distribution of galaxies on large scales began to suggest that the universe may be flat [1], consistent with the standard inflationary prediction. The only way to have a low mass density and a flat universe, as expected from the inflationary theory, is if an additional, nonluminous, "dark" energy component dominates the universe today. The dark energy would have to resist gravitational collapse, or else it would already have been detected as part of the clustered energy in the halos of galaxies. But, as long as most of the energy of the universe resists gravitational collapse, it is impossible for structure to form in the universe. The dilemma can be resolved if the hypothetical dark energy was negligible in the past and then over time became the dominant energy in the universe. According to general relativity this requires that the dark energy have *negative* pressure. This argument [1] would rule out almost all of the usual suspects, such as cold dark matter, neutrinos, radiation, and kinetic energy, because they have zero or positive pressure.

With the recent measurements of distant exploding stars, supernovae, the existence of negative-pressure dark energy has gained broad consideration. Using Type Ia supernovae as standard candles to gauge the expansion of the universe, observers have found evidence that the universe is accelerating [6,7]. A dark energy with significant negative pressure [1,8] will in fact cause the expansion of the universe to speed up, so the supernova observations provide empirical evidence of a dark energy with strongly negative pressure [6,9–11].

The news has brought the return of the cosmological constant, first introduced by Einstein for the purpose of allowing a static universe with the repulsive cosmological constant delicately balancing the gravitational attraction of matter [12]. In its present incarnation, the cosmological constant is out of balance, causing the expansion of the universe to accelerate. It can be viewed as a vacuum energy assigned to empty space itself, a form of energy with negative pressure.

Cosmologists are familiar with other hypothetical forms of dark energy with negative pressure that can accelerate the universe. In inflationary cosmology, a cosmic scalar field, known as the inflaton, pervades space and causes acceleration at a far greater rate than seen today. The same general notion can be applied to explain the current dilemma. A different field, with much tinier energy, coined "quintessence" [13], could account for the acceleration suggested to be observed today. Unlike a cosmological constant, quintessence energy changes with time and naturally develops inhomogeneities that can produce variations in the distribution of mass and the CMB temperature [13,14].

The dark energy component is characterized by its equation-of-state, $w = p/\rho$, where p is the pressure and ρ is the energy density. See Section III for a fuller description. A cosmological constant has w equal to precisely -1. For quintessence, w can take any value less than zero (by definition, it is a negative pressure component) and, in general, w varies with time (although, in the simplest models, w is nearly constant).

I EVIDENCE FOR DARK ENERGY

The current evidence for dark energy can be usefully summarized in a a "cosmic triangle" plot [15], which incorporates all the major constraints on cosmological models. Here we have included the results announced in Spring, 2001, based on measurements of anisotropies in the cosmic microwave background by the DASI, BOOMerang and MAXIMA groups. [16–20]

According to Einstein's theory of general relativity, the evolution of the universe is determined by the Friedmann equation:

$$H^2 = \frac{8\pi G}{3}\rho - \frac{k}{a^2}. \quad (1)$$

On the left-hand-side is the Hubble parameter $H = H(t)$, which measures the expansion rate of the universe as a function of time. The right-hand side contains the factors which determine the expansion rate. The first factor is the energy density ρ (times Newton's gravitational constant G). The energy density $\rho = \rho(t)$ can have several different subcomponents: a mass density associated with ordinary and dark matter, the kinetic energy of the particles and radiation, vacuum energy and quintessence. The second term on the right-hand side describes the effect of curvature. The curvature constant k can be positive, negative or zero. The scale factor, $a = a(t)$, measures how much the universe stretches as a function of time. The terms closed, open and flat refer, by definition, to the cases of positive, negative, and zero curvature. It has been common to use the same terms to describe whether the universe will ultimately recollapse, expand forever, or lie on the border between expansion and recollapse. This second usage does not generally apply if there is vacuum density or quintessence, a point which often causes confusion. (For example, a closed universe with vacuum energy may never recollapse into a big crunch.)

For simplicity, we will consider a universe composed today of baryonic (ordinary) and dark (exotic) matter, curvature and dark energy, where the dark energy is either a cosmological constant, or quintessence (but not a combination of both). The fractional contributions to the right-hand side of the Friedmann equation are given the symbols $\Omega_{mass} \equiv 8\pi G\rho_{mass}/(3H^2)$, $\Omega_{dark\,energy} \equiv 8\pi G\rho_{dark\,energy}/(3H^2)$. and $\Omega_{curv} \equiv -k/(aH)^2$, respectively. Dividing both sides of Eq. (1) by H^2 yields a simple sum rule

$$1 = \Omega_{mass} + \Omega_{curv} + \Omega_{dark\,energy}. \quad (2)$$

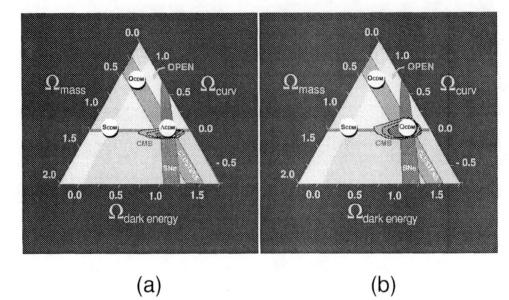

FIGURE 1. Triangle plots, as introduced in Ref. 15, represent the three key cosmological parameters – Ω_{mass}, $\Omega_{dark\,energy}$, and Ω_{curv} – where each point in the triangle satisfies the sum rule $\Omega_{mass} + \Omega_{dark\,energy} + \Omega_{curv} = 1$. The tightest constraints from measurements at low red shift (clusters, including the mass-to-light method, baryon fraction, and cluster abundance evolution), and intermediate red shift (supernovae) are shown by the shaded vertical and diagonal bands (both representing 1-σ uncertainties). The CMB contours are 1-, 2-, and 3-σ constraints based on COBE, TOCO, DASI, BOOMerang and MAXIMA measurements through Spring, 2001. The CMB fit in triangle (a) presumes the dark energy is vacuum energy, whereas triangle (b) presumes quintessence and marginalizes over the equation-of-state parameter w.

The sum rule can be represented by an equilateral triangle (Fig. 1). Lines of constant Ω_{mass}, Ω_{curv} and $\Omega_{dark\,energy}$ run parallel to each of the edges of the equilateral triangle. Every point lies at an intersection of lines of constant Ω_{mass}, Ω_{curv} and $\Omega_{dark\,energy}$ such that the sum rule is satisfied. Although Ω_{mass} is non-negative, the curvature and cosmological constant can be positive or negative.

If the curvature is zero, i.e., the universe is flat, then the sum rule reduces to $\Omega_{mass} + \Omega_{dark\,energy} = 1$, corresponding to the blue line (marked "flat") in the figure. Open models lie above the line and closed models lie below the line. Three important models are indicated: the best fit standard cold dark matter (SCDM) model with $\Omega_{mass} = 1$ and $\Omega_{dark\,energy} = \Omega_{curve} = 0$; the best-fit open cold dark matter (OCDM) model, with $\Omega_{mass} = 1/3$, $\Omega_{curve} = 2/3$ and $\Omega_{dark\,energy} = 0$; and the best-fit flat model with vacuum energy (ΛCDM, Fig. 1(a)) or quintessence ($QCDM$, Fig. 1(b)), corresponding to $\Omega_{mass} = 1/3$, $\Omega_{curv} = 0$ and the remainder in dark energy. All the models (and our analysis) assume the standard inflationary prediction that the density fluctuations are gaussian and adiabatic (*i.e.*, radiation, ordinary and dark matter fluctuate spatially in the same manner) [21–24], which agree with current observations.

Superposed on the triangle plot are the three major observational constraints on Ω_i: (a) the number of galaxy clusters observed today and compared with the number observed at red shift $z = 1/2$; (b) measurements of the luminosity-distance red shift relation for Type IA supernovae; and (c) the most recent (Spring 2001) measurements of the cosmic microwave background (CMB) anisotropy. For the latter measurements, we impose the modest constraints that the Hubble constant $H_0 \equiv 100h$ km/s/Mpc lies between $0.5 < h < 0.9$ and that the baryon density lies between $0.015 < \Omega_b h^2 < 0.025$, limits which are both much broader than the reported uncertainties. For the first time, the CMB constraints have closed to occupy a much smaller region than either of the other two constraints.

Together, the three measurements show redundantly that the SCDM and OCDM models are strongly rejected. Flatness is strongly favored. Measurements are not yet able to distinguish clearly between a cosmological constant and quintessence. However, it is noticeable that the CMB contours for ΛCDM are shifted to the closed side of the triangle, whereas the fit for QCDM is centered on flat models. Indeed, the best-fit QCDM model is flat, whereas the best-fit ΛCDM model is closed. (The predictions for the two models lie almost exactly on top of one another, illustrating a degeneracy in which a flat QCDM model is indistinguishable from a closed ΛCDM model. Other degeneracies are discussed in 25.) The preferred value for w based on CMB measurements is about $w = -2/3$. However the error bars are too broad to reach firm conclusions.

Some groups have reported more stringent measurements of w based on supernovae measurements. [26,27] However, as pointed out by Maor et al., [28,29] these analyses suffer from serious flaws that cause them to underestimate their errors.

At this point, it is fair to say that there is compelling, redundant evidence for dark energy based on the CMB, or any combination of CMB, supernovae and cluster measurements. My own estimate suggests $\Omega_{mass} = 1/3$, but the value could lie anywhere between 0.15 and 0.40, depending on which measurements one considers to be most reliable. By any count, the dark energy corresponds to $\Omega_{dark\,energy} = 1 - \Omega_{mass} \approx 2/3$, sufficient to dominate the energy density of the universe and trigger cosmic acceleration.

II QUINTESSENCE

Although both the cosmological constant and quintessence are viable candidates for the dark energy, we focus on the case of quintessence for reasons described in the following sections. By definition [13], quintessence is a dynamical, evolving, spatially inhomogeneous component with negative pressure. The term derives from the ancient term for the fifth element after earth, air, fire and water, an all-pervasive, weakly interacting component. In the current context, quintessence refers to a weakly interacting component in addition to the four already known to

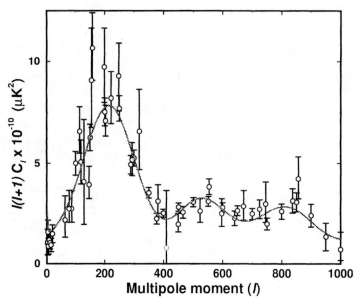

FIGURE 2. The CMB power spectrum plot comparing the best-fit cold dark matter models to COBE, TOCO, DASI, BOOMerang and MAXIMA measurements (as of Spring, 2001). The best-fit ΛCDM model is slightly closed: $\Omega_{total} = 1.05$. $\Omega_{mass} = 0.35$, $h = 0.56$, $\Omega_b h^2 = 0.022$, and scalar spectral index $n_s = 0.92$. The best-fit quintessence model is flat and has: $\Omega_{mass} = 0.30$, $h = 0.60$, $\Omega_b h^2 = 0.022$, and $n_s = 0.91$. The power spectra for the two models, one flat and one closed, lie on top of one another and are indistinguishable over the angular range shown.

influence the evolution of the universe: baryons, leptons, photons, and dark matter. More specifically, quintessence refers to a component with negative pressure p and an equation-of-state ($w \equiv p/\rho$) with $0 \geq w > -1$. (A cosmological constant has w precisely equal to -1.) Unlike a cosmological constant, the quintessential pressure and energy density evolve in time, and w may also. Furthermore, because the quintessence component evolves in time, quintessence is necessarily spatially inhomogeneous.

The quintessential concept is related to the notion of a time-dependent cosmological constant, which dates back to the early days of relativistic cosmology, when it was ascribed political as well as scientific significance [30]. In these and some later treatments, an energy component was assumed whose density decreases more slowly than the matter density but which is smoothly distributed. However, it is important to note that a time-dependent cosmological constant, in the sense of a spatially uniform but temporally varying energy, is inconsistent with the equivalence principle [13]. Quintessence, which has non-negligible spatial fluctuations on large scales (≥ 100 Mpc) is the closest approximation that is physically consistent.

The general description of quintessence incorporates many proposals for the missing energy that have been proposed over the past decade or so [13,31–39]. For example, under some conditions, light, nonabelian cosmic defects, strings or walls, evolve into a frozen network whose equation-of-state corresponds to $w = -1/3$ and $w = -2/3$, respectively [32]. The example considered here and in most papers is the energy density associated with a scalar field Q rolling down the potential $V(Q)$, which can have an equation-of-state anywhere between zero and -1. The equations are nearly identical to the equations that describe the inflaton field in inflationary cosmology [40,41]. The major difference is that the energy scale associated with quintessence is much lower so that it remains a subdominant contribution to the Friedmann equations that describe the expansion of the universe until recently. In this framework, quintessence can be viewed as the onset of a low-energy, late inflationary epoch.

The pressure of the scalar field, $p = \frac{1}{2}\dot{Q}^2 - V(Q)$ is negative if the field rolls slowly enough that the kinetic energy density is less than the potential energy density. The ratio of kinetic-to-potential energy is determined by equation-of-motion for the scalar field:

$$\ddot{Q} + 3H\dot{Q} + V(Q) = 0. \tag{3}$$

Depending on the detailed form of $V(Q)$, the equation-of-state w can vary between 0 and -1. For most potentials, w evolves slowly over time. The field is assumed to couple only gravitationally to matter. The Q-energy density decreases with time as $1/a^{3(1+w_Q)}$, so negative pressure corresponds to a density which decreases more slowly than $1/a^3$.

Spatial fluctuations in Q evolve over time due to the gravitational interaction between Q and clustering matter [13]. The perturbations are important because they can leave a distinguishable imprint on the CMB and large-scale structure.

To determine how the perturbations evolve, specifying w is insufficient. One must know the response of the component to perturbations. This can be defined by specifying the sound speed c_s as a function of wavenumber k or, alternatively, by specifying the equations-of motion (from which the perturbative equations can be derived). Note that it is possible, in principle, to have two fluids with the same w but different c_s, which would lead to distinct observational predictions.

For a scalar field, the equation-of-motion for the perturbations δQ in synchronous gauge is:

$$\ddot{\delta Q} + 3H\dot{\delta Q} + (k^2 + a^2 V'(Q))\delta_Q = -\frac{1}{2}\dot{h}_k \dot{Q}, \qquad (4)$$

where the dot represents the derivative with respect to conformal time, the prime represents the derivative with respect to Q, and h_k is the kth fourier mode of the perturbed metric. The source term in Eq. (4) has several important properties. First, any realistic cosmological model has clustering matter, so $\dot{h}_k$ must be non-zero. Also, except for the limit of a constant $V(Q)$ (which corresponds to a true cosmological constant), $\dot{Q}$ is non-zero. Hence, the source term must be non-zero overall for any $p > \rho$ ($w > -1$). This is significant because it ensures that Q cannot be smoothly spread. Even if δQ is non-zero initially, the source term causes perturbations to grow. (If the universe consisted initially of a uniform distribution of quintessence, $\dot{h}_k$ could remain zero and Q could remain smoothly spread. However, once there is a mixture of components and at least one clusters, smoothness is no longer possible.) A further consequence of the source term is that the perturbations in Q observed today are extremely insensitive to the initial conditions for δQ [13]. Assuming that $\delta \rho_Q/\rho_Q$ is comparable to the perturbations in other energy components, the transient solution to the perturbation equation is negligible today compared to the particular solution set by the source term.

III WHY QUINTESSENCE?

Why consider quintessence if its effect on the expansion of the universe is similar to cosmological constant? The principle reasons are: (a) quintessence has different implications for fundamental physics; (b) quintessence may fit the observational data better than cosmological constant; and (c) quintessence may solve cosmological problems which a cosmological constant cannot.

Distinguishing quintessence from a cosmological constant is important both for cosmology and for fundamental physics. A vacuum density or cosmological constant (Λ) is static and spatially uniform. Its value is set once and for all in the very early universe. Hence, Λ is tied directly to quantum gravity physics near the Planck scale. Quintessence is new dynamics at ultra-low energies (energy scale ~ 1 meV today), perhaps a harbinger of a whole spectrum of new low-energy phenomena.

At present, both models with cosmological constant and cold dark matter (ΛCDM) and models with a mixture of quintessence and cold dark matter (QCDM) agree with current data. [10] However, QCDM fits marginally better, as shown in Figs. 1 and 2.

For quintessence composed of scalar fields, there is the added observational constraint that the coupling to ordinary matter be sufficiently suppressed to evade fifth force and other constraints on light fields. [42] Also, it should be emphasized that the quintessence explanation for the dark energy does not explain at all the longstanding problem of the cosmological constant. A mechanism must be assumed that cancels the vacuum energy altogether or at least to a level where it is negligible small.

Whatever form the dark energy takes, two new cosmological problems arise. First, the component must be comparable in density today to the critical density, 10^{-47} GeV^4. This small value is forced by the observation that Ω_Q and Ω_m are comparable today. We will refer to the puzzle of explaining this tiny energy as the "fine-tuning problem."

A second problem arises when the cosmological model is extrapolated back in time to the very early universe, at the end of inflation, say. The curvature, the cosmological constant, and, in general, quintessence decrease at different rates than the matter density. At the end of inflation, it appears that the ratio of missing energy to matter and radiation energy must be specially fixed to an extraordinarily small value (of order 10^{-100} for the case of vacuum density) in order to have the ratio evolve to be of order unity today. Accounting for the special ratio in the early universe will be referred to as the "coincidence problem." [43] The coincidence problem is a generalization of the flatness problem pointed out by Dicke and Peebles. [44]

IV A QUINTESSENTIAL SOLUTION TO THE FINE-TUNING AND COSMIC COINCIDENCE PROBLEMS

So vexing are the fine-tuning and the cosmic coincidence problems that many cosmologists and physicists have given renewed attention to anthropic models [47]. They often pose the problems as: why should the acceleration begin shortly after structure forms in the universe and sentient beings evolve? If the dark energy component consists of vacuum density (Λ) or quintessence [13] in most forms that have been discussed to date, the answer is either pure coincidence or the anthropic principle.

An alternative approach has been proposed which suggests a dynamical explanation which does not require the fine-tuning of initial conditions or mass parameters and which is decidedly non-anthropic. The proposal, known as k-essence, is a form of quintessence model with non-linear kinetic couplings which enable unusual classical dynamical attractor behavior [48,49]. In this scenario, cosmic acceleration and human evolution are related because both phenomena are linked to the onset of matter-domination. The k-essence component has the property that it only behaves as a negative pressure component after matter-radiation equality, so that it can only overtake the matter density and induce cosmic acceleration after the matter has dominated the universe for some period, at about the present epoch. And, of course, human evolution is linked to matter-domination because the formation of planets, stars, galaxies and large-scale structure only occurs after the matter-dominated epoch begins. A further property of k-essence is that, because of the dynamical attractor behavior, cosmic evolution is insensitive to initial conditions, and so the fine-tuning problem is obviated.

The existence of attractor solutions is reminiscent of quintessence models based on evolving scalar fields with exponential [50] "tracker" [51,52] potentials. In these models, an attractor solution causes the energy density in the scalar field to track the equation-of-state of the dominant energy component, be it radiation or matter. An advantage is that the cosmic evolution is insensitive to the initial energy density of the quintessence field, and, for many models, the scenario can begin with the most natural possibility, equipartition initial conditions. (For the case of vacuum energy or cosmological constant, the vacuum energy must be set 120 orders of magnitude less than the initial matter-radiation density.) However, so long as the field tracks any equation-of-state, it cannot overtake the matter-density and induce cosmic acceleration. Indeed, for a purely exponential potential, the field never overtakes the matter density and dominates the universe. Hence, this is an unacceptable candidate for the dark energy component. In tracker models [51,52], the curvature of the potential ultimately dips to a critically small value once the field passes a particular value, $\bar{Q}$ such that the field Q becomes frozen and begins to act like a cosmological constant. The value of the potential energy density at $Q = \bar{Q}$ determines when quintessence overtakes the matter density and cosmic acceleration begins. The overall scale of the potential must be finely adjusted in order for the component to overtake the matter-density at the present epoch. So, while tracker models allow equipartition initial conditions (and, hence, resolve the coincidence problem), they require the same fine-tuning as models with cosmological constant.

The distinctive feature of the k-essence models we consider is that k-essence only tracks the equation-of-state of the background during the radiation-dominated epoch. A tracking solution during the matter-dominated epoch is physically forbidden. Instead, at the onset of matter-domination, the k-essence field energy density ε drops several orders of magnitude as the field approaches a new attractor solution in which it acts as a cosmological constant with pressure p approximately equal to $-\varepsilon$. That is, the equation-of-state, $w \equiv p/\varepsilon$, is nearly -1. The k-essence energy density catches up and overtakes the matter-density, typically several billions of years after matter-domination, driving the universe into a period of cosmic acceleration. As it overtakes the energy density of the universe, it begins to approach yet another attractor solution which, depending on details, may correspond to an accelerating universe with $w < -1/3$ or a decelerating or even dust-like solution with $-1/3 < w \leq 0$. In this scenario, we observe cosmic acceleration today because the time for human evolution and the time for k-essence to overtake the matter density are both severals of billions of years due to independent but predictive dynamical reasons. The scenario is depicted in Fig. 3.

The attractor behavior required for avoiding the cosmic coincidence problem can be obtained in models with non-standard (non-linear) kinetic energy terms. In string and supergravity theories, non-standard kinetic terms appear generically in the effective action describing the massless scalar degrees of freedom. Normally, the non-linear terms are ignored because they are presumed to be small and irrelevant. This is a reasonable expectation since the Hubble expansion damps the kinetic energy density over time. However, one case in which the non-linear terms cannot be ignored is if there is an attractor solution which forces the non-linear terms to remain non-negligible. This is precisely what is being considered here. Hence, we wish to emphasize that k-essence models are constructed from building blocks that are common to most quantum field theories and, then, utilize dynamical attractor behavior (that often arises in models with non-linear kinetic energy) to produce novel cosmological models.

The Lagrangian density for k-essence models takes form

$$\mathcal{L} = -\frac{1}{6}R + \frac{1}{\varphi^2}\tilde{p}_k(X) + \mathcal{L}_m \tag{5}$$

where R is the Ricci scalar, $X \equiv \frac{1}{2}(\nabla\varphi)^2$, $\mathcal{L}_m$ is the Lagrangian density for dust and radiation and we use units where $8\pi G/3 = 1$. The energy density of the k-field φ is $\rho_k = (2X\tilde{p}_{k,X} - \tilde{p}_k)/\varphi^2$; the pressure is $p_k = \tilde{p}_k/\varphi^2$; and the speed of sound of k-essence is $c_s^2 = p_{k,X}/\rho_{k,X}$. [53,54]

The attractor behavior can be explained most easily by changing variables from X to $y = 1/\sqrt{X}$ and rewriting the k-field Lagrangian as:

$$\mathcal{L}_k = \tilde{p}_k(X)/\varphi^2 \equiv g(y)/\varphi^2 y. \tag{6}$$

In this case, the energy density and pressure are $\rho_k = -g'/\varphi^2$ and $p_k = g/\varphi^2 y$, where prime indicates derivative with respect to y. The equation-of-state is

$$w_k \equiv p_k/\rho_k = -g/yg' \tag{7}$$

and the sound speed is $c_s^2 = p_k'/\rho_k' = (g - g'y)/g''y^2$.

In order to have a sensible, stable theory, we require $\rho_k > 0$ and $c_s^2 > 0$. These conditions are satisfied if $g' < 0$ and $g'' > 0$ in the region where p_k' is positive. Therefore, a general, convex, decreasing function $g(y)$, such as shown in Fig. 4, satisfies these necessary conditions. Using the Friedmann equation: $H^2 = \rho_{tot} = \rho_k + \rho_m$, where ρ_m is the energy density of matter (radiation and dust), and the energy conservation equations, $\dot{\rho}_i = -3\rho_i(1+w_i)$ for k-essence ($i \equiv k$) and radiation ($i \equiv R$) (replaced by matter/dust ($i \equiv D$) after matter-domination), we obtain the following equations of motion

$$\frac{dy}{dN} = \frac{3}{2}\frac{(w_k(y) - 1)}{r'(y)}\left[r(y) - \sqrt{\frac{\rho_k}{\rho_{tot}}}\right] \tag{8}$$

$$\frac{d}{dN}\left(\frac{\rho_k}{\rho_{tot}}\right) = 3\frac{\rho_k}{\rho_{tot}}\left(1 - \frac{\rho_k}{\rho_{tot}}\right)(w_R - w_k(y)), \tag{9}$$

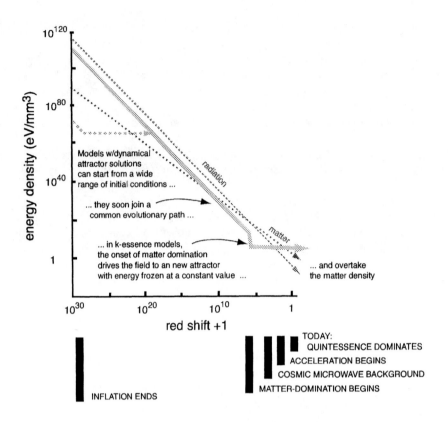

FIGURE 3. A plot showing the matter, radiation, and quintessence energy density as a function of red shift for the case of k-essence models. The k-essence models are special cases of "tracker" models with dynamical attractor solutions that funnel a wide range of initial conditions into a common evolutionary track (upper left). The distinctive property of k-essence is that the behavior shifts at the onset of matter domination to an attractor solution that acts like a cosmological constant (bottom). The bars below the graph indicate important events in cosmic history. The late conversion to a cosmological constant behavior explains why cosmic acceleration has begun only recently, a modest period (in terms of temperature scale) after matter domination.

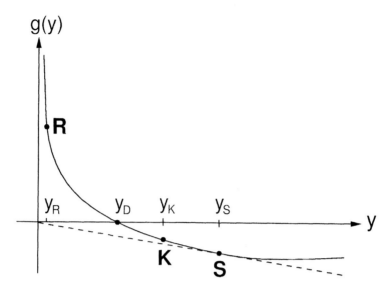

FIGURE 4. A plot of $g(y)$ vs. y (see Eq. (6) for definition) indicating the points discussed in the text. **R** corresponds to the attractor solution during the radiation-dominated epoch; **S** is the de Sitter attractor at the onset of matter-domination; and **K** is the attractor as k-essence dominates. For our range of $g(y)$, there is no dust-like attractor solution at $y = y_D$.

where $N \equiv \ln a$ and

$$r(y) \equiv \left(-\frac{9}{8}g'\right)^{1/2} y(1+w_k) = \frac{3}{2\sqrt{2}} \frac{(g-g'y)}{\sqrt{-g'}}. \quad (10)$$

These are the master equations describing the dynamics of k-essence models. Once some general properties of $g(y)$ are specified, the attractor behavior follows from these coupled equations.

We are seeking a tracker solution $y(N)$ in which the k-essence equation-of-state is constant and exactly equal to the background equation-of-state, $w_k(y(N)) = w_R$, and the ratio ρ_k/ρ_{tot} is fixed. Generically, this requires $y(N)$ be a constant y_{tr} and therefore $\rho_k/\rho_{tot} = r^2(y_{tr})$. The last condition can only be satisfied if $r(y_{tr})$ is less than unity. In ranges where $r(y)$ exceeds unity, there are no attractor solutions.

A radiation attractor corresponds to positive pressure, so it can be located only at $y < y_D$. Hence, we must have $g(y)$ such that there is a point $y_R < y_D$ where $r(y_R) < 1$ and $w_k(y_R) = 1/3$. During the radiation-dominated epoch, the ratio of k-essence to the total density remains fixed on this attractor and equal to $(\rho_k/\rho_{tot}) = r^2(y_R)$.

In Fig. 4, the pressure $p_k = g/\varphi^2 y$ is positive above the y-axis, and negative below the y-axis. The dust equation-of-state $p_k = 0$ can only be obtained at $y = y_D$ where $g(y)$ goes through zero. However, this point can be an attractor only if the second condition, $r(y_D) < 1$, is satisfied. If it so happens that $r(y_D) > 1$, then there is no dust attractor in the matter-dominated epoch. This is precisely what we want for our scenario, and this is possible for a broad class of functions g.

If g possesses a radiation attractor but no dust attractor, what happens at dust-radiation equality? To answer this question let us study the solutions of the master equations Eqs. (8-10) when the energy density of k-essence is much smaller than the matter energy density. If $\rho_k/\rho_m \ll 1$, one can neglect the last term in the equation (8) and it is obvious that $y(N) \simeq y_S$, where y_S satisfies the equation $r(y_S) = 0$, is an approximate solution of the equations of motion. The point S satisfies $g(y_S) = g'(y_S)y_S$, so the tangent of g at y_S passes through the origin, as shown in Fig. 4. Since $r \propto (1 + w_k)$, the equation of state of k-essence at y_S corresponds to $w_k(y_S) \approx -1$; this solution is the de Sitter attractor denoted by S in Fig. 4. From Fig. 4, it is clear that y_S nearly always exists for convex decreasing functions g. As a result, if ρ_k during the radiation dominated epoch is significantly less than the radiation density which it tracks, which is both typical and required to satisfy nucleosynthesis constraints, then k-essence proceeds to the de Sitter attractor immediately after dust-radiation equality. [49]

As the transition to dust-domination occurs, ρ_k first drops to a small, fixed value, as can be simply understood. Suppose that $(\rho_k/\rho_{tot})_R = r^2(y_R) = \alpha < 10^{-2}$ during the radiation dominated epoch, where the bound is set by nucleosynthesis constraints. From the equation of state, Eq. (7), we have the relation: $g(y_R) = -g'(y_R)y_R/3$. The condition $r(y_D) \geq 1$ is required in order to have no dust attractor solution. Combining these relations, we obtain:

$$\frac{g'_R y_R^2}{g'_D y_D^2} \leq \frac{9}{16}\alpha < 10^{-2} \tag{11}$$

On the other hand, it is apparent from Fig. 4 that $-g'_R > -g'_D$, so $y_R \ll y_D$ if $\alpha \ll 1$. In particular, the tangent at y_D falls below $g(y_R)$, so $g'_D(y_R - y_D) \approx -y_D g'_D \leq g(y_R) = -y_R g'_R/3$. Using this relation, we obtain

$$\frac{y_R}{y_D} \leq \frac{3}{16}\alpha < 2 \cdot 10^{-3} \text{ and } \frac{g'_D}{g'_R} \leq \frac{\alpha}{16} < 7 \cdot 10^{-4}. \tag{12}$$

Since $\rho_k = -g'/\varphi^2$ and $|g'(y_S)| \leq |g'(y_D)|$ we conclude that after radiation domination, when the k-field reaches the vicinity of the **S**-attractor, the ratio of energy densities in k-essence and dust does not exceed $(\rho_k/\rho_{tot})_R \times g'_D/g'_R$; that is, $\rho_k/\rho_{dust} < \alpha^2/16 < 7 \cdot 10^{-6}$. Hence, provided $(\rho_k/\rho_{tot})_R \leq 10^{-2}$ at dust-radiation equality, the k-essence field loses energy density on its way to the **S**-attractor down to a value below 7×10^{-6}.

By definition, the **S**-attractor is one in which $w \approx -1$ and the energy density is nearly constant. Hence, once ρ_k has reached its small but non-zero value, it freezes. In the further evolution of the universe, the matter density decreases, but the k-essence energy density remains constant, eventually overtaking the matter density of the universe. Note that, as ρ_k approaches ρ_D, the condition $\rho_k/\rho_m \ll 1$

is necessarily violated and a new attractor solution is found for the case where k-essence itself dominates the background energy density. This attractor is denoted **K** in Fig. 4.

To prove that that the **K**-attractor exists, we once again consider the master equations, Eqs. (8-10), in the limit where $\rho_k/\rho_{tot} \to 1$. If y_K satisfies the equation $r(y_K) = 1$, then $y(N) \simeq y_K$ is an approximate solution of the equations of motion. When dust is not a tracker, there always exists a unique attractor y_K in the interval $y_D < y < y_S$. [49] To prove this, note that, within this interval, the function $r(y)$ has a negative derivative. Recall that $r(y_S) = 0$ (definition of **S**-attractor) and $r(y_D) > 1$ (to avoid a dust attractor). Since $r(y)$ is a monotonically decreasing, continuous function, there exists a unique point y_K ($y_D < y_K < y_S$) where $r(y)$ becomes equal to unity. At $y > y_D$ the pressure of k-essence is negative. Hence, generically the **K** attractor, located near y_K, describes a universe dominated by a negative pressure component which induces power-law cosmic acceleration. As acceleration proceeds, ρ_k increasingly dominates and $y \to y_K$.

Following along using Fig. 4, the dynamics can be summarized as follows: k-essence is attracted to $y = y_R$ during the radiation dominated epoch; at matter-domination, the energy density drops sharply as k-essence skips past $y = y_D$, because there is no dust attractor, and heads towards $y \approx y_S$. The energy density ρ_k freezes and, after a period, overtakes the matter density. As it does so, y relaxes towards y_K. In this scenario, our current universe would be making the transition from y_S to y_K. All this occurs for generic $g(y)$ satisfying broad conditions on its first and second derivatives. If the ratio of ρ_k to the radiation density is near the maximum allowed by nucleosynthesis (roughly equipartition initial conditions), the scenario predicts that the ρ_k dominates by the present epoch. These results can be confirmed by numerical calculations, as illustrated in Fig. 3. A full analytical and numerical treatment of attractor dynamics is given in Ref. 49

V CONCLUSIONS

In the next decade, the focus of cosmologists will be to determine if the dark energy consists of vacuum energy or quintessence. Quintessence has observable consequences. Because its value of the equation-of-state w differs from that of vacuum energy, it produces a different rate of cosmic acceleration. More precise measurements of supernovae over a longer span of distances may separate the two cases, although the measurement can be confounded by parameter degeneracies [28,29]. Astronomers have proposed two new observatories–the orbiting Supernova Acceleration Project (SNAP) and the earth-based Large Scale Synoptic Telescope (LSST)– to resolve the issue. Differences in acceleration rate also produce small differences in the angular size of hot and cold spots in the cosmic microwave background, as the Microwave Anisotropy Probe (MAP) and Planck satellites should be able to detect. Fig. 2 illustrates how the microwave background can potentially distinguish flat models with quintessence versus cosmological constant, and the measurements will improve greatly over the coming decade.

The distribution of galaxies is yet another test. The average density of galaxies is thought to be uniform throughout space. That is, for a fixed range of distances, astronomers should find the same number of galaxies whether the range lies nearby or far away. But astronomers seldom measure the distance to a galaxy directly; they infer it from the galaxy's red shift. The conversion from red shift to distance follows a simple linear relation (the Hubble law) if the distance is small, but a nonlinear relation if the distance is great. In an accelerating universe, the number of galaxies in a fixed range of red shifts should decrease as astronomers probe more deeply into space. The steeper the decrease, the more negative the value of w. An earth-based project known as the Deep Extragalactic Evolutionary Probe (DEEP) will look for this effect.

Over the longer term future, cosmologists, as well as scientists and humanists, will be left to ponder the profound implications of the revolutionary discoveries about our universe made just as the millennium was coming to a close. Certainly they lead to a strange new interpretation of our place in cosmic history. First came the inflationary theory of the universe, which proposes an extended period of accelerated expansion during the first instants after the big bang. Space was nearly devoid of matter and a quintessence-like quantum field with negative pressure constituted most of the energy of the universe. During that period, the universe expanded by a greater factor than it has during the 15 billion years since inflation ended. At the end of inflation, the field decayed to a hot gas of quarks, gluons, electrons, light and dark energy.

The next stage–our epoch–has been one of steady cooling, condensation and the evolution of intricate structure. But this period is coming to an end. Cosmic acceleration has returned. The universe as we know it, with shining stars, galaxies and clusters, appears to be a brief interlude. As the cosmic acceleration takes hold over the next tens of billions of years, the matter and energy in the universe will become more and more diluted and space will stretch too rapidly to enable new structures to form. If the acceleration is caused by vacuum energy, then the cosmic story is complete, and the universe we see today is the pinnacle of cosmic structure formation.

But if the acceleration is caused by quintessence, the story is less certain. The universe might accelerate forever, or the quintessence could decay into new forms of matter and radiation, repopulating the universe with new structure. Because the dark energy density today is so small, one might suppose that the matter and radiation derived from its decay would have too little energy to anything of interest. Under some circumstances, however, quintessence could decay through the nucleation of bubbles. The bubble interior would be a void, but the bubble wall would be the site of vigorous activity. As the bubble wall moved outward, it would collect all of the energy derived from the decay of quintessence. Occasionally, two bubbles would collide in a fantastic fireworks display. In the process, massive particles such as protons and neutrons might arise, and perhaps stars and planets. To future inhabitants, the universe would look highly inhomogeneous, with life confined to distant islands surrounded by vast voids. Would they figure out that their origin was the homogeneous and isotropic universe we see about us today? Would they ever know that the universe had once been alive?

Experiments are already testing the quintessence hypothesis and perhaps they will decide among the future possibilities for the universe. The ultimate answer will be obtained when quintessence is incorporated into the hoped-for unified theory of fundamental forces, perhaps string theory. Our place in cosmic history hinges on the interplay between the science of the very big and the very small.

REFERENCES

1. J. P. Ostriker and P. J. Steinhardt, *Nature* **377**, 600 (1995).
2. L. Krauss and M.S.Turner, *Gen. Rel. Grav.* **27**, 1135 (1995).
3. N.A. Bahcall, L.M. Lubin and V. Dorman, *Ap.J.* **447**, L81 (1995).
4. R.G. Carlberg *et al.*, *Ap.J.* **462**, 32 (1996).
5. N.A. Bahcall and X. Fan, PNAS **95**, 5956 (1998); *Ap.J.* **504**, 1 (1998).
6. S. Perlmutter, *et al.*, LBL-42230 (1998), astro-ph/9812473; S. Perlmutter, *et al.*, *Ap.J.* (in press), astro-ph/9812133.
7. A.G. Riess, *et al.*, Astron. J. **116**, 1009 (1998).
8. J.P.E. Peebles, *Ap.J.* **284**, 439 (1984).
9. P. M. Garnavich, *et al.*, *Ap.J.* **509**, 74 (1998).
10. L. Wang, R. Caldwell, J.P. Ostriker and P.J. Steinhardt, *Ap.J.* (submitted), astro-ph/9901388, (1999).
11. S. Perlmutter, M. S. Turner and M. White, *Ap.J.* (submitted), astro-ph/9901052, (1999).
12. A. Einstein, *Sitz. Preuss. Akad. Wiss.* **142**, (1917).
13. R. R. Caldwell, R. Dave and P. J. Steinhardt, *Phys. Rev. Lett.* **80**, 1582 (1998).
14. N. Weiss, *Phys. Lett. B* **197**, 42 (1987); B. Ratra and J.P.E. Peebles, *Ap.J.*, 325, L17 (1988); C. Wetterich, *Astron. Astrophys.* **301**, 32 (1995); J.A. Frieman, *et al. Phys. Rev. Lett.* **75**, 2 077 (1995); K. Coble, S. Dodelson, and J. Frieman, *Phys. Rev. D* **55**, 1851 (1995); P.G. Ferreira and M. Joyce, *Phys. Rev. Lett.* **79**, 4740 (1997); *Phys. Rev. D* **58**, 023503 (1998).
15. N. Bahcall, J.P. Ostriker. S. Perlmutter, and P.J. Steinhardt, *Science* **284**, 1481-1488, (1999).
16. N.W. Halverson, *et al.*, astro-ph/0104489; C. Pryke, *et al.*, astro-ph/0104490.
17. C.B. Netterfield, *et al.*, astro-ph/0104460.
18. R. Stompor, *et a.*, astro-ph/0105062.
19. A.D. Miller *et al*, *Ap.J.* **524**, L1 (1999); E. Torbet *et al*, *Ap. J.* **521**, L79 (1999).
20. G.F. Smoot, *et al.*, *Ap.J.* **396**, L1 (1992); C. L. Bennett *et al.*, *Ap.J.* **464**, L1 (1996).
21. J. Bardeen, P. J. Steinhardt and M. S. Turner, *Phys. Rev. D* **28**, 679 (1983).
22. A. H. Guth and S.-Y. Pi, *Phys. Rev. Lett.* **49**, 1110 (1982).
23. A. A. Starobinskii, *Phys. Lett. B* **117**, 175 (1982).
24. S. W. Hawking, *Phys. Lett. B***115**, 295 (1982).

25. G. Huey, L. Wang, R. Dave, R.R. Caldwell and P.J. Steinhardt, *Phys. Rev.* **D59**, 063005 (1999).
26. D. Huterer and M.S. Turner, *Phys. Rev. D* **60**, 081301 (1999);
27. J. Weller and A. Albrecht, *Phys. Rev. Lett.* **86**, 1939 (2001).
28. I. Maor, R. Brustein, and P.J. Steinhardt, *Phys. Rev. Lett.* **86**, 6, (2001).
29. I. Maor, R. Brustein, and P.J. Steinhardt, in preparation.
30. See Ref. 33 in J.M. Overduin and F.I. Cooperstock, *Phys. Rev.* **D58** 043506 (1998); M. Bronstein, Physikalische Zeitschrift Sowjet Union 3, 73 (1933).
31. M.S. Turner, and M. White, Phys. Rev. D **56**, R4439 (1997).
32. D. Spergel and U.-L. Pen, *Astrophys. J.* **491**, L67, (1997).
33. N. Weiss, *Phys. Lett. B* **197**, 42 (1987).
34. B. Ratra, and P.J.E. Peebles, Phys. Rev. D **37**, 3406 (1988); P.J.E. Peebles and and B. Ratra, ApJ **325**, L17 (1988).
35. C. Wetterich, *Astron. Astrophys.* **301**, 32 (1995).
36. J.A. Frieman, *et al. Phys. Rev. Lett.* **75**, 2077 (1995).
37. K. Coble, S. Dodelson, and J. Frieman, *Phys. Rev. D* **55**, 1851 (1995).
38. P.G. Ferreira and M. Joyce, *Phys. Rev. Lett.* **79**, 4740 (1997); *Phys. Rev. D* **58**, 023503 (1998).
39. E. J. Copeland, A.R. Liddle, and D. Wands, *Phys. Rev.* **D57**, 4686 (1998).
40. A. D. Linde, *Phys. Lett.* **108B**, 389 (1982).
41. A. Albrecht and P. J. Steinhardt, *Phys. Rev. Lett.* **48**, 1220 (1982).
42. S. Carroll, *Phys. Rev. Lett.***81**, 3067 (1998).
43. P. Steinhardt, in 'Critical Problems in Physics," ed. by V.L. Fitch and D.R. Marlow (Princeton U. Press, 1997).
44. R.H. Dicke and P.J.E. Peebles, in **General Relativity: An Einstein Centenary Survey**, ed. by S.W. Hawking & W. Israel (Cambridge U. Press, 1979).
45. I. Zlatev, L. Wang, and P.J. Steinhardt, *Phys. Rev. Lett.* **82**, 896 (1999).
46. P. Steinhardt, L. Wang, and I. Zlatev, *Phys. Rev. D* **59**, 123504 (1999).
47. See, for example, S. Weinberg, astro-ph/0005265, and references therein.
48. C. Armendariz-Picon, V. Mukhanov, and P.J. Steinhardt, *Phys. Rev. Lett.* **85**, 4438 (2000).
49. C. Armendariz-Picon, V. Mukhanov, and P.J. Steinhardt, *Phys. Rev. D* **63**, 103510 (2001).
50. C. Wetterich, *Nucl. Phys. B* **302**, 668 (1988), and *Astron. Astrophys.* **301**, 32 (1995); P. G. Ferreira and M. Joyce, *Phys. Rev. Lett.* **79**, 4740 (1997).
51. I. Zlatev, L. Wang, P. Steinhardt, *Phys. Rev. Lett.* **82**, 895 (1998).
52. P. Steinhardt, L. Wang, I. Zlatev, *Phys. Rev. D* **59**, 123504 (1999).
53. C. Armendariz-Picon, T. Damour, and V. Mukhanov, *Phys. Lett.* B**458**, 209 (1999).
54. J. Garriga, V. Mukhanov, *Phys. Lett. B* **458**, 219 (1999).

Constraining the properties of dark energy

Dragan Huterer* and Michael S. Turner[†]

*Physics Department, University of Chicago, Chicago, IL
[†]Physics and Astronomy and Astrophysics Departments, University of Chicago, Chicago, IL

Abstract. The presence of dark energy in the Universe is inferred directly from the accelerated expansion of the Universe, and indirectly, from measurements of cosmic microwave background (CMB) anisotropy. Dark energy contributes about 2/3 of the critical density, is very smoothly distributed, and has large negative pressure. Its nature is very much unknown. Most of its discernible consequences follow from its effect on evolution of the expansion rate of the Universe, which in turn affects the growth of density perturbations and the age of the Universe, and can be probed by the classical kinematic cosmological tests. Absent a compelling theoretical model (or even a class of models), we describe dark energy by an effective equation of state $w = p_X/\rho_X$ which is allowed to vary with time. We describe and compare different approaches for determining $w(t)$, including magnitude-redshift (Hubble) diagram, number counts of galaxies and clusters, and CMB anisotropy, focusing particular attention on the use of a sample of several thousand type Ia supernova with redshifts $z < 1.7$, as might be gathered by the proposed SNAP satellite. Among other things, we derive optimal strategies for constraining cosmological parameters using type Ia supernovae. While in the near term CMB anisotropy will provide the first measurements of w, supernovae and number counts appear to have the most potential to probe dark energy.

INTRODUCTION

Three major lines of evidence point to the existence of a smooth energy component in the universe. Various measurements of the matter density indicate $\Omega_M \simeq 0.3 \pm 0.1$ (e.g., [1]). Recent cosmic microwave background (CMB) results strongly favor a flat (or nearly flat) universe, with the total energy density $\Omega_0 \simeq 1.1 \pm 0.07$ [2]. Finally, there is direct evidence coming from type Ia supernovae (SNe Ia) that the universe is accelerating its expansion, and that it is dominated by a component with strongly negative pressure, $w = p_X/\rho_X < -0.6$ [3, 4]. Two out of of three of these arguments would have to prove wrong in order to do away with the smooth component.

Even before the direct evidence from SNe Ia, there was a dark energy candidate: the energy density of the quantum vacuum (or cosmological constant) for which $p = -\rho$. However, the inability of particle theorists to compute the energy of the quantum vacuum – contributions from well understood physics amount to 10^{55} times critical density – casts a dark shadow on the cosmological constant. Another important issue is the coincidence problem: dark energy seems to start dominating the energy budget, and accelerating the expansion of the universe, just around the present time. A number of other candidates have been proposed: rolling scalar field (or quintessence), and network of frustrated topological defects, to name a couple. While these and other models have some motivation and attractive features, none are compelling.

In this work we discuss the cosmological consequences of dark energy that allow its

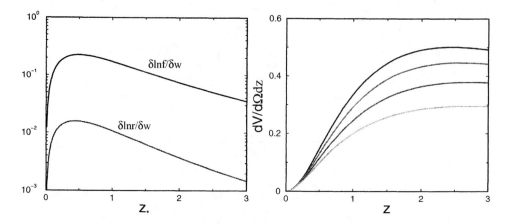

FIGURE 1. Left panel: relative sensitivity of the comoving distance $r(z)$ and the comoving volume element $f(z) \equiv dV/d\Omega dz$ to a localized change in the value of w at redshift z_*. Right panel: comoving volume element $dV/d\Omega dz$ vs. redshift for constant $w = -1, -0.8, -0.6, -0.4$ (from top to bottom).

nature to be probed. We also discuss relative merits of various cosmological probes, focusing particular attention to a large, well-calibrated sample of SNe Ia that would be obtained by the proposed space telescope SNAP[1]. We parameterize dark energy by its scaled energy density Ω_X and equation of state w, and assume a fiducial cosmological model with $\Omega_X = 1 - \Omega_M = 0.7$ and $w = -1$, unless indicated otherwise.

COSMOLOGICAL CONSEQUENCES OF THE DARK ENERGY

Dark energy is smooth, and if does clump at all, it does so at very large scales only ($k \sim H_0$). All of its consequences therefore follow from its modification of the expansion rate

$$H(z)^2 = H_0^2 \left[\Omega_M (1+z)^3 + \Omega_X \exp[3 \int_0^z (1+w(x)) d\ln(1+x)] \right].$$

For a given rate of expansion today H_0, the expansion rate in the past, $H(z)$, was smaller in the presence of dark energy. Therefore, dark energy increases the age of the universe. The comoving distance $r(z) = \int dz/H(z)$ also increases in the presence of dark energy. The same follows for the comoving volume element $dV/d\Omega dz = r^2(z)/H(z)$.

At small redshifts $r(z)$ is insensitive to w for the simple reason that *all* cosmological models reduce to the Hubble law ($r = H_0^{-1} z$) for $z \ll 1$:

---

[1] http://snap.lbl.gov

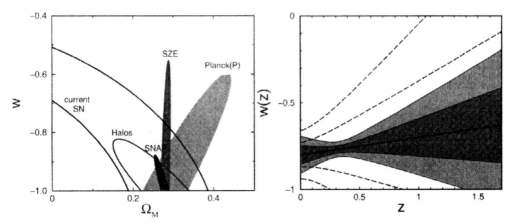

FIGURE 2. Left panel: present and projected future constraints (68% C.L.) on constant w. The Sunyaev-Zeldovich effect (SZE) constraint roughly corresponds to the estimate from [5], while for the halo counts we assumed a random uncertainty of 10% in each of eight redshift bins. Right panel: projected constraints (68% and 95% C.L.) on time-varying w using simulated data expected from SNAP. The fiducial model is $w(z) = -0.8 + 0.1z$ and Ω_M is considered to be known. The broken lines show the effect of assuming a Gaussian uncertainty of 0.05 in Ω_M.

$$r(z) \to H_0^{-1}\left[z - \frac{3}{4}z^2 - \frac{3}{4}\Omega_X w z^2 + \cdots\right] \quad \text{for } z \ll 1. \tag{1}$$

At redshift greater than about five, the sensitivity of $r(z)$ to a change in w levels off because dark energy becomes an increasingly smaller fraction of the total energy density, $\rho_X/\rho_M \propto (1+z)^{3w}$. Note also that the volume depends upon dark energy parameters through $H(z)$, which contains one integral less than the distance $r(z)$. This is the reason why number-count surveys (which effectively measure $dV/d\Omega dz$) are potentially a strong probe of dark energy.

Constraints on (constant) w

Supernovae are perhaps the strongest probes of dark energy, as the relevant observable $r(z)$ depends only on Ω_M, Ω_X and w; moreover, SNe probe the optimal redshift range. A powerful supernova program, such as SNAP, with about 2000 SNe distributed at $0.2 < z < 1.7$, would be able to determine w with an accuracy $\sigma_w = 0.05$.

Number-count surveys are potentially also a very strong probe of dark energy. To estimate the expected number density of objects (galaxies, galaxy clusters etc.), one typically uses the Press-Schechter formalism [6]. For example, number counts of galaxy clusters either from an X-ray survey or from Sunyaev-Zeldovich survey can serve as a probe of dark energy [5]. The number of these objects depends upon the growth of structure as well as on cosmological volume, and the former makes the overall constraint complementary to that of SNe Ia. Another example is using halos of fixed rotational speed [7] to infer the dark energy dependence of the volume element. In order to obtain constraints on dark energy using number-count surveys, control over modeling and systematic errors will be critical.

CMB anisotropy, which mainly probes the epoch of recombination ($z \sim 1000$) is weakly sensitive to the presence of dark energy because dark energy becomes inconsequential at such high redshift. The small dependence comes through the distance to the surface of last scattering which slightly increases in the presence of dark energy, moving the acoustic peaks to smaller scales. However, this effect is small:

$$\frac{\Delta l_1}{l_1} = -0.084 \Delta w - 0.23 \frac{\Delta \Omega_M h^2}{\Omega_M h^2} + 0.09 \frac{\Delta \Omega_B h^2}{\Omega_B h^2} + 0.089 \frac{\Delta \Omega_M}{\Omega_M} - 1.25 \frac{\Delta \Omega_0}{\Omega_0},$$

which shows that the location of the first peak is least sensitive to w. The left panel of Fig. 2 further illustrates this: even the Planck experiment with polarization information would be able to achieve only $\sigma_w \approx 0.25$ (after marginalization over all other parameters). Nevertheless, CMB constraints are of crucial importance because they complement SNe and other probes, in particular providing the measurement of the total energy density Ω_0.

The Alcock-Paczynski shape test and the age of the Universe are also sensitive to the presence of dark energy, but they seem somewhat less promising: the former because of the small size of the effect (around 5%); and the latter because the errors in the two needed quantities, H_0 and t_0, are not likely to become small enough in the near future. Finally, large-scale structure surveys have weak direct dependence upon dark energy simply because this component is smooth on observable scales.

Probing $w(t)$

There is no *a priori* reason to assume constant w, as some models (in particular quintessence) generically produce time-varying w. Here we discuss how to best constrain $w(t)$ (or $w(z)$). We find that $w(z)$ will be much more difficult to constrain than constant w due to additional degeneracies. In order to illustrate prospects for constraining $w(z)$, we use SNAP's projected dataset with 2000 SNe and assume that, by the time this difficult task is seriously attempted, Ω_M and Ω_X will be pinned down accurately by a combination of CMB measurements and large-scale structure surveys.

One of the simplest ways to characterize $w(z)$ is to divide the redshift range into B redshift bins and assume constant equation of state ratio w_i in each. The resulting constraint, however, is poor for $B \geq 3$ (and ideally one would want many more bins). A better way to proceed is to assume linearly varying $w(z)$ expanded around a suitably chosen redshift w_1 [8]

$$w(z) = w_1 + w'(z - z_1).$$

We choose z_1 so as to de-correlate w_1 and w'. The resulting constraint is shown in the right panel of Fig. 2. The constraint is best at $z \approx 0.4$ and deteriorates at lower and higher redshifts. Despite the relatively large uncertainty in the slope ($\sigma_{w'} = 0.16$),

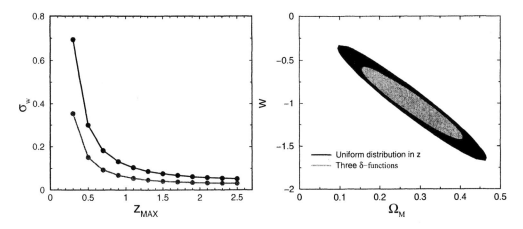

FIGURE 3. Left panel: upper curve shows the constraint upon (constant) w using the (nearly uniform) distribution of 2000 SNe out to z_{max}. Lower curve shows uncertainties using the same number of SNe with mathematically optimal distribution in redshift. Right panel: constraints on Ω_M and w using 100 SNe with uniform (dark) and optimal (light) distribution in redshift.

this analysis may be useful in constraining dark energy models. The parameterization $w(z) = w_1 - \alpha \ln[(1+z)/(1+z_1)]$ yields comparable constraints.

OPTIMAL STRATEGIES

Given the importance and difficulty of probing dark energy, it is worthwhile to consider how to optimize data sets in order to obtain tighter constraints. We address the following question: what is the optimal redshift distribution of SNe Ia in order to best constrain the cosmological parameters?

We choose to minimize the area of uncertainty ellipsoid for P parameters, which (in the Fisher matrix formalism [9]) is equivalent to maximizing $\det F$, where F is the Fisher matrix. Making a few simplifying but reasonable assumptions, in [10] we show that P parameters are always best determined if SNe are distributed as $P+1$ delta-functions in redshifts, two of which are always at the lowest and highest redshifts available. For example, to best constrain Ω_X and w (assuming a flat universe), SNe should be located at $z = 0$, $z \approx 2/5 z_{max}$ and $z = z_{max}$ in equal numbers, where z_{max} is the maximum redshift available. In Fig. 3 we show the merits of optimal distribution. Note, however, that there are other considerations that may favor a more uniform distribution of SNe in redshift – for example, ability to perform checks for dust and evolution.

CONCLUSIONS

We discussed prospects for constraining dark energy. Dark energy is best probed at redshifts roughly between 0.2 and 1.5, where it is dynamically important. Probes of the low-redshift Universe (supernovae and number counts) seem the most promising, as they only depend upon three cosmological parameters (Ω_M, Ω_X and w), which will be effectively reduced to two (Ω_X and w) when precision CMB measurements determine $\Omega_0 = \Omega_M + \Omega_X$ to better than 1%. CMB is weakly sensitive to the presence of dark energy, but is very important as a complementary probe. Alcock-Paczynski test and the age of the universe seem less promising.

Constraining the equation of state w is the first step in revealing the nature of the dark energy. We show that future surveys will be able to determine constant w quite accurately. A high-quality sample of 2000 supernovae out to redshift $z \sim 1.7$ could determine w to a precision of $\sigma_w = 0.05$. A similar accuracy might be achieved by number counts of galaxies and clusters of galaxies out to $z \sim 1.5$, provided systematics are held in check (e.g., in the case of galaxies, the evolution of the comoving number density needs to be known to better than 5%). Time-varying w will be considerably more difficult to pin down because of additional degeneracies that arise in this case. Nevertheless, interesting constraints may be achieved using high-quality data and assuming a slowly-varying $w(z)$. Consequently, use of complementary measurements and further development of techniques to constrain dark energy will be of crucial importance.

ACKNOWLEDGMENTS

We would like to thank the organizers for the wonderful conference and the "black hole" drink recipe. This work was supported by DOE at Chicago.

REFERENCES

1. Turner, M. S., *Physica Scripta*, **85**, 210 (2000).
2. Jaffe, A. H., et al., *astro-ph/0007333* (2000).
3. Riess, A., et al., *Astron. J.*, **116**, 1009 (1998).
4. Perlmutter, S., et al., *Astrophys. J.*, **517**, 545 (1999).
5. Haiman, Z., Mohr, J. J., and Holder, G. P., *astro-ph/0002336* (2000).
6. Press, W. H., and Schechter, P. L., *Astrophys. J.*, **187**, 425 (1974).
7. Newman, J., and Davis, M., *Astrophys. J. Lett.*, **534**, 11 (2000).
8. Cooray, A. R., and Huterer, D., *Astrophys. J. Lett.*, **513**, 95 (1999).
9. Tegmark, M., Taylor, A. N., and Heavens, A. F., *Astrophys. J.*, **480**, 22 (1997).
10. Huterer, D., and Turner, M. S., *astro-ph/0012510* (2000).

Clusters in the Precision Cosmology Era

Zoltán Haiman[*,1], Joseph J. Mohr[†] and Gilbert P. Holder[△]

[*] *Princeton University Observatory, Princeton, NJ 08544*
[†] *Departments of Astronomy and Physics, University of Illinois, Urbana, IL 61801*
[△] *Department of Astronomy and Astrophysics, University of Chicago, Chicago, IL 60637*

Abstract. Over the coming decade, the observational samples available for studies of cluster abundance evolution will increase from tens to hundreds, or possibly to thousands, of clusters. Here we assess the power of future surveys to determine cosmological parameters. We quantify the statistical differences among cosmologies, including the effects of the cosmic equation of state parameter w, in mock cluster catalogs simulating a 12 deg^2 Sunyaev-Zel'dovich Effect (SZE) survey and a deep 10^4 deg^2 X-ray survey. The constraints from clusters are complementary to those from studies of high–redshift Supernovae (SNe), CMB anisotropies, or counts of high–redshift galaxies. Our results indicate that a statistical uncertainty of a few percent on both Ω_m and w can be reached when cluster surveys are used in combination with any of these other datasets.

INTRODUCTION

Because of their relative simplicity, galaxy clusters provide a uniquely useful probe of the fundamental cosmological parameters. The formation of the large-scale dark matter potential wells of clusters is likely independent of complex gas dynamical processes, star formation, and feedback, and involve only gravitational physics. The observed abundance of nearby clusters implies robust constraints on the amplitude σ_8 of the power spectrum on cluster scales to an accuracy of $\sim 25\%$ [1,2]. In addition, the redshift–evolution of the observed cluster abundance constrains the matter density Ω_0 [3–5].

In order to be useful for these cosmological studies, the masses of galaxy clusters have to be known. Existing studies utilized the presently available tens of clusters with mass estimates [6,7], and as a result, were limited in their scope. The present samples, however, will likely soon be replaced by catalogs of thousands of intermediate redshift and hundreds of high redshift ($z > 1$) clusters. At a minimum, the analysis of the European Space Agency *X-ray Multi-mirror Mission (XMM)* archive for serendipitously detected clusters will yield hundreds, and perhaps thousands of new clusters with temperature measurements [8]. Dedicated X-ray and

---

[1)] Hubble Fellow

SZE surveys could likely surpass the *XMM* sample in areal coverage, number of detected clusters or redshift depth.

The imminent improvement of distant cluster data motivates us to estimate the cosmological power of future surveys. In particular, we study the constraints provided by a 12 deg^2 SZE survey [9], or by a deep 10^4 deg^2 X-ray survey. Our primary goals are (1) to quantify the accuracy to which various cosmological models can be distinguished from a standard Λ Cold Dark Matter (ΛCDM) cosmology; and (2) to contrast constraints from clusters to those available from CMB anisotropy measurements, from high-redshift SNe, or from high-redshift galaxy counts [10–12].

The galaxy cluster abundance provides a natural test of models that include a dark energy component with an equation of state parameter $w \equiv p/\rho \neq -1$ [13–15]. The value of w directly affects the linear growth of fluctuations, and the angular diameter distance (and hence the SZE decrement and the X-ray luminosity) to individual clusters. We restrict our analysis to a flat universe, and focus on the following four parameters: the matter density Ω_m; the equation of state parameter w (assumed to be constant); the Hubble constant $h \equiv H_0/100$ km s^{-1} Mpc^{-1}; and the amplitude of the power spectrum on $8h^{-1}$Mpc scales, σ_8. A broader range of parameters, including open/closed universes, and evolving $w(z)$, will be examined in future work [16]. Details of the study described here can be found in [17].

MODELING DETAILS

General Approach

To quantify the power of a future cluster survey to distinguish cosmologies, we utilize the following approach:

1. We pick a fiducial cosmological (ΛCDM) model, with the parameters $(\Omega_\Lambda, \Omega_m, h, \sigma_8, n) = (0.7, 0.3, 0.65, 0.9, 1)$, based on present large scale structure data [18]. We assume this model describes the "real" universe.

2. In the fiducial model, we compute the abundance of clusters dN_{fid}/dz as a function of redshift in a specific (SZE or X-ray) survey. This simulates the dataset that will be available in the future for cosmological tests.

3. We vary parameters of our model, and recompute the cluster abundance dN_{test}/dz as a function of redshift in this new "test" cosmology. In each test cosmology, we set the value of σ_8 by requiring the local cluster abundance at redshift $z \approx 0$ to match the value in the fiducial cosmology.

4. We compute the likelihood of observing the redshift distribution dN_{fid}/dz if the true distribution were dN_{test}/dz. We utilize both the normalization and shape of the distributions, by combining the standard Poisson and Kolmogorov–Smirnov tests.

5. We repeat steps 3 and 4 for a wide range of values of w, Ω_m, and h.

Predicting Cluster Abundance and Evolution

The fundamental ingredient of this approach is the cluster abundance, given a cosmology and the parameters of a survey. In this study, we utilize the "universal" halo mass function found in a series of recent large–scale cosmological simulations [19]. Following these simulations, we assume that the comoving number density of clusters at redshift z with mass M is given by

$$\frac{dn}{dM}(z, M) = 0.315 \frac{\rho_0}{M} \frac{1}{\sigma_M} \frac{d\sigma_M}{dM} \exp\left[-|0.61 - \log(D_z \sigma_M)|^{3.8}\right], \quad (1)$$

where σ_M is the present day r.m.s. density fluctuation on mass–scale M [20], D_z is the linear growth function, and ρ_0 is the present–day mass density. The directly observable quantity is the number of clusters with mass above $M_{\min}$ at redshift $z \pm dz/2$ in a solid angle $d\Omega$:

$$\frac{dN}{dz d\Omega}(z) = \left[\frac{dV}{dz d\Omega}(z) \int_{M_{\min}(z)}^{\infty} dM \frac{dn}{dM}\right] \quad (2)$$

where $dV/dzd\Omega$ is the cosmological volume element, and $M_{\min}(z)$ is the limiting mass of the survey, as discussed below. Equations 1 and 2 reveal that the cluster abundance depends on cosmology through several quantities: (1) the growth function D_z; (2) the volume element $dV/dzd\Omega$; (3) the power spectrum σ_M; (4) the mass density ρ_0; and (5) the survey mass threshold $M_{\min}$. The first four of these dependencies are well–determined, once the parameters of the cosmological model and the power spectrum are specified (in particular, note that the abundance is exponentially sensitive to the growth function). The scaling of the limiting mass with cosmology depends on the specific survey.

Cluster Surveys

In this study, we examine two specific surveys, in which clusters are detected through either their SZ decrements, or X–ray fluxes. In practice, the only survey details we utilize in our analyses are the virial mass of the least massive, detectable cluster (as a function of redshift and cosmological parameters), and the solid angle of the survey. The SZE survey we consider is that proposed by J. Carlstrom and collaborators [9]. This interferometric survey will detect clusters more massive than $\sim 2 \times 10^{14} M_\odot$, nearly independent of their redshift, and will cover an area of 12 deg^2 in a year. The relatively small solid angle of the survey will allow cluster redshifts to be determined by deep optical and near infrared followup observations. The X–ray survey we consider is similar to a proposed Small Explorer mission, called the Cosmology Explorer, spear-headed by G. Ricker and D. Lamb. The survey depth is 3.6×10^6 cm^{2}s at 1.5 keV, and the coverage is 10^4 deg^2 (approximately half the available unobscured sky). We focus on clusters which produce 500 detected source counts in the 0.5:6.0 keV band, sufficient to reliably estimate the emission weighted mean temperature. The X-ray survey could be combined with the Sloan Digital Sky Survey (SDSS) to obtain redshifts for the clusters.

The most important aspect of both surveys is the limiting halo mass $M_{\min}(z, \Omega_m, w, h)$, and its dependence on redshift and cosmological parameters. For an interferometric SZE survey, the relevant observable is the cluster visibility V, which is proportional to the total SZE flux decrement. The detection limit as a function of redshift and cosmology for this survey has been studied using mock observations of simulated galaxy clusters [21]. In the X-ray survey, $M_{\min}$ follows from the cluster X-ray luminosity – virial mass relation [22]. Illustrative examples of the mass limits in both surveys are shown in Figure 1, both for ΛCDM and for a $w = -0.5$ universe. The SZE mass limit is nearly independent of redshift, and changes little with cosmology. As a result, the cluster sample can extend to $z \approx 3$. In comparison, the X-ray mass limit is a stronger function of w, and it rises rapidly with redshift. For the X-ray survey considered here the number of detected clusters beyond $z \approx 1$ is negligible. The total number of clusters in the SZE survey is ~ 200, while in the X-ray survey, it is $\sim 2,000$.

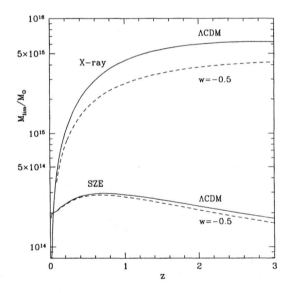

FIGURE 1. Limiting cluster virial masses for detection in a 10^4 deg^2 X-ray survey (upper pair of curves) and in a 12 deg^2 SZE survey (lower pair of curves). The solid curves show the mass limit in our fiducial flat ΛCDM model, with $w = -1$, $\Omega_m = 0.3$, and $h = 0.65$, and the dotted curves show the masses in the same model except with $w = -0.5$.

RESULTS AND CONCLUSIONS

Our results are summarized by the likelihood contours in the $\Omega_m - w$ plane, shown in Figures 2 and 3 for the SZE and X-ray surveys, respectively. Figure 2 shows three different cross-sections of constant total probability in the SZE survey, at fixed values of h (0.55, 0.65, and 0.80) in the investigated 3-dimensional Ω_m, w, h parameter space. These diagrams demonstrate that the constraints on Ω_m are at the $\sim 10\%$ level, while w remains essentially unconstrained ($w \lesssim -0.2$). Nevertheless, the narrowness of the contours in Figure 2 implies that the SZE survey can yield accurate constraints on w if combined with other data. We find that the differences among cosmologies in the SZE survey are driven nearly entirely by the growth function D_z. This results from the cluster sample extending to high redshifts ($z > 1$), where the growth functions in different cosmologies diverge rapidly. This makes the SZE sample especially useful. For comparison, galaxy counts at $z \sim 1$ probe mostly the cosmological volume, and constitute and independent test from clusters [12]. Note that our constraints scale weakly with h: this arises from the weak dependence of the power spectrum on h.

Figure 3 shows constraints in the X-ray survey for a fixed $h = 0.65$. The increased number of clusters translates to significantly narrower contours compared to the SZE survey. The orientation of the contours remains similar, but we find that the cosmological sensitivity arises nearly entirely from the mass limit $M_{\min}$. Indeed, the X-ray flux is more sensitive to cosmology than the SZE decrement (cf. Fig. 1), and the X-ray sample extends only out to $z \sim 1$, where the growth functions are less

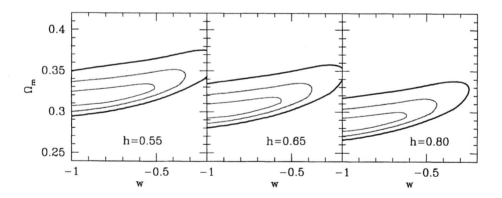

FIGURE 2. Contours of 1, 2, and 3σ likelihood for different models when they are compared to a fiducial flat ΛCDM model with $\Omega_m = 0.3$ and $h = 0.65$, using the SZE survey. The three panels show three different cross-sections of constant total probability at fixed values of h (0.55, 0.65, and 0.80) in the investigated 3-dimensional Ω_m, w, h parameter space.

divergent. Also shown in Figure 3 are constraints expected from CMB anisotropies and from high–z SNe. The dashed curves correspond to the CMB constraints (a $\pm 1\%$ determination of the position of the first Doppler peak); the dotted curves to the constraints from SNe (a $\pm 1\%$ determination of the luminosity distance to $z=1$). As these curves show, the constraints from the CMB and SNe data are complementary to the direction of the parameter degeneracy in cluster abundance studies, making the cluster surveys especially valuable.

Our findings suggest that cluster surveys will lead to tight constraints on a combination of Ω_m and w, especially valuable because of their high complementarity to constraints from CMB anisotropies, magnitudes of high–z SNe, or counts of high–z galaxies. In combination with either of these data, clusters can determine both Ω_m and w to a few percent accuracy. We have focused primarily on the statistics of cluster surveys: further work is needed to clarify the role of systematic uncertainties, arising from the cosmology–scaling of the mass limits and the the cluster mass function, as well as our neglect of issues such as galaxy formation in the lowest mass clusters.

We thank J. Carlstrom and the COSMEX team for providing access to proposed instrument characteristics. ZH acknowledges support from a Hubble Fellowship.

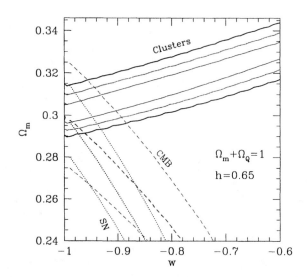

FIGURE 3. Likelihood contours for a fixed $h = 0.65$ in the X–ray survey. Also shown are combinations of w and Ω_m that keep the spherical harmonic index ℓ of the first Doppler peak in the CMB anisotropy data constant to within $\pm 1\%$ (dashed lines); and combinations that keep the luminosity distance to redshift $z=1$ constant to the same accuracy.

REFERENCES

1. White, S.D.M., Efstathiou, G., Frenk, C.S., *Mon. Not. Roy. Ast. Soc.* **262**, 1023 (1993).
2. Viana, P.T.P. and Liddle, A.R., *Mon. Not. Roy. Ast. Soc.* **281**, 323 (1996).
3. Bahcall, N.A. and Fan, X., *Ap. J.* **504**, 1 (1998).
4. Blanchard, A. and Bartlett, J.G., *Ast. and Astroph.* **332**, L49 (1998).
5. Viana, P.T.P. and Liddle, A.R., *Mon. Not. Roy. Ast. Soc.* **303**, 535 (1999).
6. Gioia, I.M., et al., *Ap. J. Supp.* **72**, 567 (1990).
7. Vikhlinin, A., et al., *Ap. J.* **503**, 77 (1998).
8. Romer, A.K., Viana, P.T.P., Liddle, A.R. and Mann, R.G., *Ap. J.* **547**, 594 (2001).
9. Carlstrom, J.E., et al., *Physica Scripta* **T**, 148, (2000).
10. Schmidt, B.P. et al. *Ap. J.* **507**, 46 (1998).
11. Perlmutter, S. et al. *Ap. J.* **517**, 565 (1999).
12. Newman, J. A., and Davis, M. *Ap. J.* **534**, L11 (2000).
13. Turner, M. S., and White, M., *Phys. Rev. D.* **56**, 4439 (1997).
14. Caldwell, R.R., Dave, R. and Steinhardt, P.J., *Ast. Space Sci.* **261**, 303 (1998).
15. Huterer, D., and Turner, M. S., *Phys. Rev. D.* submitted, astro-ph/0012510
16. Holder, G.P., Haiman, Z., Mohr, J.J., Turner, M. S., & Carlstrom, J.E., in prep; see also these proceedings.
17. Haiman, Z., Mohr, J.J., and Holder, G.P., *Ap. J.*, in press, astro-ph/0002336 (2001).
18. Bahcall, N.A., et al., *Science* **284**, 1481 (1999).
19. Jenkins, A. et al., *Mon. Not. Roy. Ast. Soc.* **321**, 372 (2001).
20. Eisenstein, D.J. and Hu, W. *Ap. J.* **504**, L57 (1998).
21. Holder, G.P., et al., *Ap. J.* **544**, 629 (2000).
22. Bryan, G.L. and Norman, M.L. *Ap. J.* **495**, 80 (1998).

A Local Void and the Accelerating Supernovae

Kenji Tomita

Yukawa Institute for Theoretical Physics, Kyoto University, Kyoto, 606-8502, Japan

Abstract. The magnitude-redshift relation in a cosmological model with a local void is investigated, corresponding to the recent observational evidences that we are in a local void (an underdense region) on scales of 200 – 300 Mpc. It is already evident that the accelerating behavior of high-z supernovae can be explained in this model, because the local void plays a role similar to the positive cosmological constant. In this note the dependence of the behavior on the gaps of cosmological parameters in the inner (low-density) region and the outer (high-density) region is shown.

INTRODUCTION

One of the most important cosmological observations at present is the $[m, z]$ relation for high-z supernovae (SNIa), which play a role of standard candles at the stage reaching epochs $z \sim 1$. So far the observed data of SNIa have been compared with the theoretical relation in homogeneous and isotropic models, and many workers have made efforts to determine their model parameters (Perlmutter et al. 1999, Riess et al. 1998, 2000, Riess 2000).

Here is, however, an essentially important problem to be taken into consideration. It is the homogeneity of the Universe. According to Giovanelli et al. (1999) the Universe is homogeneous in the region within $70h^{-1}$ Mpc (the Hubble constant H_0 is $100h$ km s^{-1} Mpc^{-1}). On the other hand, recent galactic redshift surveys (Marinoni et al. 1999, Marzke et al. 1998, Folkes et al. 1999, Zucca et al. 1997) show that in the region around $200 - 300h^{-1}$ Mpc from us the distribution of galaxies may be inhomogeneous. This is because the galactic number density in the region of $z < 0.1$ or $< 300h^{-1}$ Mpc from us was shown to be by a factor > 1.5 smaller than that in the remote region of $z > 0.1$. Recently a large-scale inhomogeneity suggesting a wall around the void on scales of $\sim 250h^{-1}$ Mpc has been found by Blanton et al. (2000) in the SDSS commissioning data (cf. their Figs. 7 and 8). Similar walls on scales of $\sim 250h^{-1}$ Mpc have already been found in the Las Campanas and 2dF redshift surveys near the Northern and Southern Galactic Caps (Shectman et al. 1996, Folkes et al. 1999, Cole et al. 2000). These results mean that there is a local void with the radius of $200 - 300h^{-1}$ Mpc and we live in it.

Moreover, the measurements by Hudson et al. (1999) and Willick (1999) for a systematic deviation of clusters' motions from the global Hubble flow may show some inhomogeneity on scales more than $100h^{-1}$ Mpc. This fact also may suggest some inhomogeneity in the above nearby region.

If the local void exists really, the Hubble constants also must be inhomogeneous, as well as the density parameters, and the theoretical relations between observed quantities are different from those in homogeneous models. At present, however, the large-scale

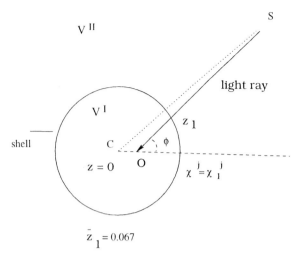

FIGURE 1. Model with a spherical single shell. Redshifts for observers at O and C are z and $\bar{z}$.

inhomogeneity of the Hubble constant has not been observationally established yet because of the large error bars in the various measurements (cf. Tomita 2001).

In my previous papers (Tomita 2000a and 2000b), I showed various models with a local void and discussed the bulk flow, CMB dipole anisotropy, distances and the $[m,z]$ relation in them in the limited parameter range. It was found that the accelerating behavior of supernovae can be explained in these models without cosmological constant.

In this note I describe first a simplified cosmological model with a local void, and next I show the dependence of the $[m,z]$ relation on model parameters such as the ratios of density parameters and Hubble constants in the inner (low-density) and outer (high-density) regions. The constraints to the parameters are discussed in comparison between the above relations in the present models and the relations in homogeneous models.

DISTANCES IN MODELS WITH A LOCAL VOID

Inhomogeneous models we consider consist of inner (low-density) region V^I and outer (high-density) region V^{II} which are separated by a single shell. It is treated as a spherical singular shell and the mass in it compensates the mass deficiency in V^I. So V^I and the shell are regarded as a local void and the wall, respectively. The Hubble constants and density parameters are expressed as (H_0^I, H_0^{II}) and $(\Omega_0^I, \Omega_0^{II})$, where we assume that $H_0^I > H_0^{II}$ and $\Omega_0^I < \Omega_0^{II}$. The distances of the shell and the observer O (in V^I) from the centre O (in V^I) are assumed to be 200 and $40 h_I^{-1}$ as a standard case. This shell corresponds to the redshift $\bar{z}_1 = 0.067$ (see Figure 1).

In a previous paper we derived the full-beam distances ($\overline{CS}$) between the centre C and a source S, and the distances ($\overline{OS}$) between an observer O and S. The two distances are nearly equal in the case when $\overline{CS}$ or $\overline{OS}$ is much larger than $\overline{CO}$. Since we notice this

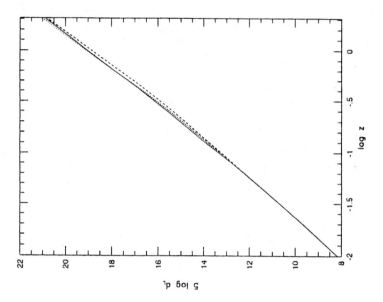

FIGURE 2. The $[m, z]$ relation in cosmological models with a local void. The solid line denotes the case with a standard parameter set given in (3). The dotted and dash lines stand for homogeneous models with $(\Omega_0, \lambda_0) = (0.3, 0.7)$ and $(0.3, 0.0)$, respectively, for comparison.

case alone in the following, we treat the light paths $\overline{CS}$ for simplicity.

Here we treat the following equation for the angular-diameter distance to consider the clumpiness along paths into account (Dyer and Roeder 1973, Schneider et al. 1992, Tomita 1999):

$$\frac{d^2(d_A^j)}{d(z^j)^2} + \left\{\frac{2}{1+z^j} + \frac{1}{2}(1+z^j)\left[\Omega_0^j(1+3z^j) + 2 - 2\lambda_0^j\right]\right.$$
$$\left. \times F^{-1}\right\}\frac{d(d_A^j)}{dz^j} + \frac{3}{2}\Omega_0^j\alpha(1+z^j)F^{-1}d_A^j = 0, \quad (1)$$

where $j = $ I and II, z^j is the redshift in the region V^j, α is the clumpiness parameter, and

$$F \equiv (1 + \Omega_0^j z^j)(1+z^j)^2 - \lambda_0^j z^j(2+z^j). \quad (2)$$

The luminosity distance d_L is related to the angular-diameter distance d_A by $d_L = (1+z)^2 d_A$.

PARAMETER DEPENDENCE OF THE MAGNITUDE-REDSHIFT RELATION

As for homogeneous models it is well-known from the comparison with observational data that the flat case with nonzero cosmological constant of $(\Omega_0, \lambda_0) = (0.3, 0.7)$ can represent the accelerating behavior of high-z SNIa, while an open model with $(0.3, 0)$ cannot explain their data for $z \approx 1.0$ (Perlmutter et al. 1999, Riess et al. 1998, 2000, Riess 2000). In this note the relations in these two homogeneous models are used as a measure for inferring how the relations in inhomogeous models with various parameters can reproduce the observational data. That is, we deduce that the model parameters are consistent with the observational data, if at the interval $0.5 < z < 1.0$ the curve in the $[m, z]$ relation is near that in the homogeneous model $(0.3, 0.7)$ comparing with the difference between those for $(0.3, 0.7)$ and $(0.3, 0)$.

For the $[m, z]$ relation in an inhomogeneous model, we first treat the case with the following *standard* parameters to reproduce the accelerating behavior in the above homogeneous model $(0.3, 0.7)$ in similar way:

$$\begin{aligned}
(\Omega_0^I, \Omega_0^{II}) &= (0.3, 0.6), \\
H_0^I &= 71, \quad H_0^{II}/H_0^I = 0.82, \\
\alpha &= 1.0, \quad z_1 = 0.067, \text{ and } \lambda_0^I = \lambda_0^{II} = 0.
\end{aligned} \quad (3)$$

The radius of the local void is $r_1 \equiv (c/H_0^I)z_1 = 200(h_1)^{-1}$ Mpc. In Figure 2, the relation is shown for $z = 0.01 - 2.0$ in comparison with that in two homogeneous models with parameters: $(\Omega_0, \lambda_0) = (0.3, 0.7)$ and $(0.3, 0), H_0 = 71$ and $\alpha = 1.0$. For $z < z_1$ the relation is equal to that in the open model $(0.3, 0.0)$. In the Figure we used $5 \log d_L$ for the magnitude m in the ordinate axis.

It is found that the behavior in the case of $(\Omega_0^I, \Omega_0^{II}) = (0.3, 0.6)$ with $\lambda_0^I = \lambda_0^{II} = 0$ accords approximately with that in the flat, homogeneous model with $(\Omega_0, \lambda_0) = (0.3, 0.7)$ for $z_1 < z < 1.0$. Accordingly they is similarly fit for the observed data of SNIa.

Next, to examine the parameter dependence of the $[m, z]$ relation, we take up two cases with following parameters different from the above standard case:

$$\begin{aligned}
\Omega_0^{II} &= 0.45 \text{ and } 0.80 \quad (\text{for } \Omega_0^I = 0.3) \\
H_0^{II}/H_0^I &= 0.80 \text{ and } 0.87 \quad (\text{for } H_0^I = 71).
\end{aligned} \quad (4)$$

In Figure 3, the cases with $H_0^{II}/H_0^I = 0.80, 0.82$, and 0.87 are shown in a model with $(\Omega_0^I, \Omega_0^{II}) = (0.3, 0.6)$, $z_1 = 0.067$, $\alpha = 1.0$, and $\lambda_0^I = \lambda_0^{II} = 0$. In the cases with $H_0^{II}/H_0^I = 0.82, 0.87$, the relations are found to be consistent with the relation in the above flat, homogeneous model for $z = 0.5 - 1.0$, but in the case with $H_0^{II}/H_0^I = 0.80$ or < 0.80, the $[m, z]$ relation is difficult to explain the observed data.

In Figure 4, the cases with $\Omega_0^{II} = 0.45, 0.6$, and 0.8 are shown in a model with $\Omega_0^I = 0.3$, $H_0^{II}/H_0^I = 0.82$, $z_1 = 0.067$, $\alpha = 1.0$, and $\lambda_0^I = \lambda_0^{II} = 0$. In the case with $\Omega_0^{II} = 0.45$ and 0.6, the relation are found to be consistent with those in the above flat, homogeneous model for $z = 0.5 - 1.0$, but in the case with $\Omega_0^{II} = 0.8$ we have less consistency.

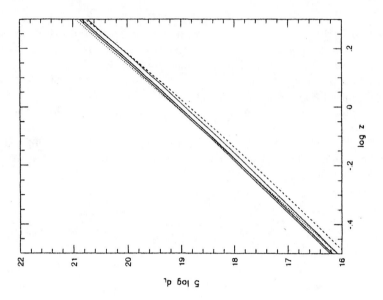

FIGURE 3. The $[m,z]$ relation in cosmological models with a local void. The solid lines denote the cases with $H_0^{II}/H_0^I = 0.87, 0.82$, and 0.80 from the top to the bottom. The other parameters are same as those in a standard parameter set given in (3). The dotted and dash lines stand for homogeneous models, as in Figure 2.

CONCLUDING REMARKS

As for the $[m,z]$ relation in cosmological models with a local void, we studied the parameter dependence of their accelerating behavior, and found that the local void with $r_1 \sim 200h^{-1}$ Mpc, $H_0^{II}/H_0^I \sim 0.82$, and $\Omega_0^{II} \sim 0.6$ is appropriate for explaining the accelerating behavior of SNIa without cosmological constant. More details can be seen in a recent paper (astro-ph/0011484). On the basis of these results we can determine in the next step what values of the model parameters are best in the direct comparison with the observational data of SNIa.

In the near future the void structure on scales of $\sim 200h^{-1}$ Mpc will be clarified by the galactic redshift survey of SDSS in the dominant part of whole sky. Then, observational cosmology will be developed taking into account that we are in the local void.

REFERENCES

1. Blanton, M. R., Dalcanton, J., Eisenstein, J., Loveday, J., Strauss, M. A., SubbaRau, M., Weinberg, D. H., Anderson, Jr., J. E., et al., astro-ph/0012085
2. Cole, S., Norberg, P., Baugh, C. M., Frenk, C. S., BlandHawthorn, J., Bridges T., Cannon, R., Colless, M., astro-ph/0012429
3. Dyer, C. C., Roeder, R. C., 1973, ApJ, 180, L31

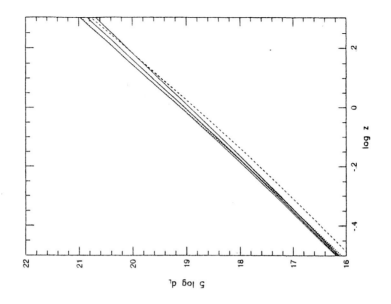

FIGURE 4. The $[m,z]$ relation in cosmological models with a local void. The solid lines denote the cases with $\Omega_0^{II} = 0.45, 0.6$, and 0.8 from the top to the bottom. The other parameters are same as those in a standard parameter set given in (3). The dotted and dash lines stand for homogeneous models, as in Figure 2.

4. Folkes, S., Ronen, S., Price, I., Lahav, O., Colless, M., Maddox, S., Deeley, K., Glazebrook, K, et al., 1999, MNRAS, 308, 459
5. Giovanelli, R., Dale, D. A., Haynes, M. P., Hardy, E., Campusano, L., 1999, ApJ, 525, 25
6. Hudson, M. J., Smith, R. J., Lucey, J. R., Schlegel, D. J., Davies, R. L., 1999, ApJ, 512, L79
7. Marinoni, C., Monaco, P., Giuricin, G, Costantini, B., 1999, ApJ, 521, 50
8. Marzke, R. O., da Costa, L. N., Pellegrini, P. S., Willmer, C. N. A., Geller, M. J., 1998, ApJ, 503, 617
9. Perlmutter, S., Aldering, G., Goldhaber, G., Knop, R. A., Nugent, P., Groom, D. E., Castro, P. G., Deustua, S. et al., 1999, ApJ, 517, 565
10. Riess, A. G., Filippenko, A. V., Challis, P., Clocchiatti, A., Diercks, A., Garnavich, P. M., Gilliland, R. L., et al., 1998, AJ, 116, 1009
11. Riess, A. G., Filippenko, A. V., Li, W., Schmidt, B., 2000, AJ, 118, 2668
12. Riess, A. G., 2000, PASP, 112, 1284
13. Shectman, S. A., Landy, S. D., Oemler, A., Tucker, D. L., Lin, H., Kirshner, R. P., Schecter, P. L., 1996, ApJ, 470, 172
14. Schneider, P., Ehler, J., Falco, E. E., 1992, Gravitational Lenses (Berlin: Springer)
15. Tomita, K., 1999, Prog. Theor. Phys. Suppl. 133, 155; astro-ph/9904351
16. Tomita, K., 2000a, ApJ, 529, 26; astro-ph/9905278
17. Tomita, K., 2000b, ApJ, 529, 38; astro-ph/9906027
18. Tomita, K., 2001, Prog. Theor. Phys., 105, No.3; astro-ph/0005031
19. Willick, J. A., 1999, ApJ, 522, 647
20. Zucca, E., Zamorani, G., Vettolani, G., Cappi, A., Merighi, R., Mignoli, M., MacGillivray, H., Collins, C. et al., 1997, A&A, 326, 477

Value of the Cosmological Constant: Theory versus Experiment

Moshe Carmeli and Tanya Kuzmenko

Department of Physics, Ben Gurion University, Beer Sheva 84105, Israel

Abstract. The numerical value of the cosmological constant is calculated using a recently suggested cosmological model and found to be $\Lambda = 2.036 \times 10^{-35} \text{s}^{-2}$. This value of Λ is in excellent agreement with the measurements recently obtained by the *High-Z Supernova Team* and the *Supernova Cosmology Project*.

The problem of the cosmological constant and the vacuum energy associated with it is of high interest these days. There are many questions related to it at the quantum level, all of which are related to quantum gravity. Why there exists the critical mass density and why the cosmological constant has this value? Trying to answer these questions and others were recently the subject of many publications [1–13].

In this paper it is shown that the recently suggested cosmological model [14] predicts the value $\Lambda = 2.036 \times 10^{-35} \text{s}^{-2}$ for the cosmological constant. This value of Λ is in excellent agreement with the measurements recently obtained by the *High-Z Supernova Team* and the *Supernova Cosmological Project* [15–21].

The Einstein gravitational field equations with the added cosmological term are [22]:

$$R_{\mu\nu} - \frac{1}{2}g_{\mu\nu}R + \Lambda g_{\mu\nu} = \kappa T_{\mu\nu}, \tag{1}$$

where Λ is the cosmological constant, the value of which is supposed to be determined by experiment. In Eq. (1) $R_{\mu\nu}$ and R are the Ricci tensor and scalar, respectively, $\kappa = 8\pi G$, where G is Newton's constant and the speed of light is taken as unity.

Recently the two groups (the *Supernovae Cosmology Project* and the *High-Z Supernova Team*) concluded that the expansion of the universe is accelerating [15–21]. Both teams obtained

$$\Omega_M \approx 0.3, \quad \Omega_\Lambda \approx 0.7, \tag{2}$$

and ruled out the traditional $(\Omega_M, \Omega_\Lambda)=(1, 0)$ universe. Their value of the density parameter Ω_Λ corresponds to a cosmological constant that is small but, nevertheless, nonzero and positive,

$$\Lambda \approx 10^{-52} \text{m}^{-2} \approx 10^{-35} \text{s}^{-2}. \tag{3}$$

In Ref. 14 a four-dimensional cosmological model was presented. The model predicts that the universe accelerates and hence it is equivalent to having a positive value for cosmological constant in it. In the framework of this model the zero-zero component of Einstein's equations is written as

$$R_0^0 - \frac{1}{2}\delta_0^0 R = \kappa \rho_{eff} = \kappa \left(\rho - \rho_c^{BC}\right) \tag{4}$$

where $\rho_c^{BC} = 3/\kappa\tau^2$ is the critical mass density and τ is Hubble's time in the zero-gravity limit.

Comparing Eq. (4) with the zero-zero component of Eq. (1), one obtains the expression for the cosmological constant,

$$\Lambda = \kappa \rho_c^{BC} = 3/\tau^2. \tag{5}$$

To find out the numerical value of τ we use the relationship between $h = \tau^{-1}$ and H_0 given in Ref. 14 [Eq. (5.23)]:

$$H_0 = h\left[1 - \left(1 - \Omega_M^{BC}\right) z^2/6\right], \tag{6}$$

where z is the redshift and $\Omega_M^{BC} = \rho_M/\rho_c^{BC}$ where $\rho_c^{BC} = 3h^2/8\pi G$ [14]. (Notice that ρ_c^{BC} is different from the standard ρ_c defined with H_0.) The redshift parameter z determines the distance at which H_0 is measured. We choose $z = 1$ (Fig. 11 in Ref. 14) and take for

$$\Omega_M^{BC} = 0.245 \tag{7}$$

(roughly corresponds to 0.3 in the standard theory), Eq. (6) then gives

$$H_0 = 0.874h. \tag{8}$$

At the value $z = 1$ the corresponding Hubble constant H_0 according to the latest results from HST can be taken [23] as $H_0 = 72$km/s-Mpc, thus $h = (72/0.874)$km/s-Mpc or

$$h = 82.380 \text{km/s-Mpc}. \tag{9}$$

What is left is to find the value of Ω_Λ^{BC}. We have $\Omega_\Lambda^{BC} = \rho_c^{ST}/\rho_c^{BC}$, where $\rho_c^{ST} = 3H_0^2/8\pi G$ and $\rho_c^{BC} = 3h^2/8\pi G$. Thus $\Omega_\Lambda^{BC} = (H_0/h)^2 = 0.874^2$, or

$$\Omega_\Lambda^{BC} = 0.764. \tag{10}$$

As is seen from Eqs. (7) and (10) one has

$$\Omega_M^{BC} + \Omega_\Lambda^{BC} = 1.009 \approx 1, \qquad (11)$$

which means the universe is flat.

As a final result we calculate the cosmological constant according to Eq. (5). One obtains

$$\Lambda = 3/\tau^2 = 2.036 \times 10^{-35} s^{-2}. \qquad (12)$$

Our results confirm those of the supernovae experiments and indicate on existance of the dark energy as has recently received confirmation from the Boomerang cosmic microwave background experiment [24,25], which showed that the universe is flat.

REFERENCES

1. Weinberg, S., Talk given at Dark Matter 2000, February 2000. (astro-ph/0005265)
2. Witten, E., Talk given at Dark Matter 2000, February 2000.
3. Rubakov, V.A., and Tinyakov, P.G., *Phys. Rev.* **D61**, 087503 (2000).
4. Weinberg, S., *Phys. Rev.* **D61**, 103505 (2000). (astro-ph/0002387)
5. Zlatev, I., Wang, L., and Steinhardt, P.J., *Phys. Rev. Lett.* **82**, 896 (1999).
6. Estrada, J., and Masperi, L., *Mod. Phys. Lett.* **A13**, 423 (1998). (hep-ph/9710522)
7. Adler, S.L., *Gen. Rel. Grav.* **29**, 1357 (1997). (hep-th/9706098)
8. Carroll, S.M., The cosmological constant (preprint). (astro-ph/0004075)
9. Fujii, Y., *Phys. Rev.* **D62**, 064004 (2000). (gr-qc/9908021)
10. Cohn, J.D., *Astrophys. Sp. Sci* **259**, 213 (1998). (astro-ph/9807128)
11. Sahni, V., and Starobinsky, A., *Intern. J. Mod. Phys.* **D9**, 373 (2000).
12. Goliath, M., and Ellis, G.F.R., *Phys. Rev.* **D60**, 023502 (1999). (gr-qc/9811068)
13. Roos, M., and Harun-or-Rashid, S.M., *Astron. Astrophys.* **329**, L17 (1998).
14. Behar, S., and Carmeli, M., *Intern. J. Theor. Phys.* **39**, 1375 (2000). (astro-ph/0008352)
15. Garnavich, P.M., et al., *Astrophys. J.* **493**, L53 (1998). (astro-ph/9710123)
16. Schmidt, B.P., et al., *Astrophys. J.* **507**, 46 (1998). (astro-ph/9805200)
17. Riess, A.G., et al., *Astronom. J.* **116**, 1009 (1998). (astro-ph/9805201)
18. Garnavich, P.M., et al., *Astrophys. J.* **509**, 74 (1998). (astro-ph/9806396)
19. Perlmutter, S., et al., *Astrophys. J.* **483**, 565 (1997). (astro-ph/9608192)
20. Perlmutter, S., et al., *Nature* **391**, 51 (1998). (astro-ph/9712212)
21. Perlmutter, S., et al., *Astrophys. J.* **517**, 565 (1999). (astro-ph/9812133)
22. Carmeli, M., *Classical Fields: General Relativity and Gauge Theory* (Wiley, New York, 1982).
23. Freedman, W.L., et al., Talk given at the 20th Texas Symposium, Austin, Texas 10-15 December 2000. (astro-ph/0012376)
24. De Bernardis, P., et al., *Nature* **404**, 955 (2000). (astro-ph/0004404)
25. Bond, J.R., et al., in *Proc. IAU Symposium* 201 (2000). (astro-ph/0011378)

Feasibility of Reconstructing the Quintessential Potential Using SNIa Data

Takeshi Chiba* and Takashi Nakamura[†]

*Department of Physics, Kyoto University, Kyoto 606-8502, Japan
[†] Yukawa Institute for Theoretical Physics, Kyoto University, Kyoto 606-8502, Japan

Abstract. We investigate the feasibility of the method for reconstructing the equation of state and the effective potential of the quintessence field from SNIa data. We introduce a useful functional form to fit the luminosity distance with good accuracy (the relative error is less than 0.1%). We assess the ambiguity in reconstructing the equation of state and the effective potential which originates from the uncertainty in Ω_M. We find that the equation of state is sensitive to the assumed Ω_M, while the shape of the effective potential is not. We also demonstrate the actual reconstruction procedure using the data created by Monte-Carlo simulation. Future high precision measurements of distances to thousands of SNIa could reveal the shape of the quintessential potential.

INTRODUCTION

Evidence for dark energy is now compelling. Quintessence is an alternative to Λ, and a lot of models have been proposed so far. However, there is currently no clear guidance from particle physics as to which quintessence models may be suitable. Then it should be the observations that decide which model is correct or not.

In view of the future prospect for high-z SNIa search, such as SNAP, we investigate in detail the feasibility of the method for reconstructing the equation of state and the effective potential of the quintessence field from SNIa data [1,2].

RECONSTRUCTION PROGRAM

Parameterizing the Luminosity Distance: We present a fitting function for $r(z) = d_L(z)/(1+z)$ which has the following properties: (1)the good convergence (the relative error is hopefully less than 0.1% because the distance error expected from SNAP will be less than a few percent) for $0 < z < 10$; (2) the correct asymptotic behavior for $z \gg 1$ ($H(z) \propto (1+z)^{3/2}$).

In analogy with Pen's powerful fitting formula for $r(z)$ for a flat FRW universe with Λ [3], we propose to fit $r(z)$ in the following functional form [4]:

$$H_0 r(z) = \eta(1) - \eta(y), \qquad \eta(y) = 2\alpha \left[y^{-8} + \beta y^{-6} + \gamma y^{-4} + \delta y^{-2} + \sigma \right]^{-1/8}, \qquad (1)$$

where $y = 1/\sqrt{1+z}$. The use of variables y and r is detailed in [5].

Reconstructing the Equation of State and the Effective Potential:
In order to demonstrate the effectiveness of the fitting function Eq.(1), we reconstruct the effective potential of the quintessence field. We consider an inverse power law potential, $V(\phi) = M^4 \phi^{-4}$. For other types of potential, see [4].

The results are shown in Fig. 1. The dotted curves are numerically reconstructed $w(y)$ and $V(\phi)$ with $\Omega_M = 0.30$ being assumed, while the solid curves are the original ones up to $y = 0.6$. Non-dimensional variables, $\kappa^2 V/H_0^2$ and $\kappa \phi$ with $\kappa^2 = 8\pi G$, are shown there. We fix the present value of $\kappa \phi$ to unity.

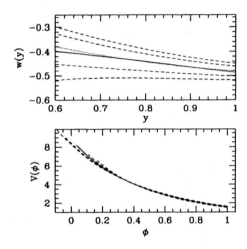

FIGURE 1. The reconstructed equation of state and the effective potential. Solid curve is the original equation of state or effective potential. Dotted curves are for $\Omega_M = 0.30$. The curves assuming $\Omega_M = 0.25, 0.27, 0.33, 0.35$ (from top to bottom for $w(y)$ or from left to right for V) are shown by dashed ones.

Since the smaller y, the larger the error in $w(y)$, the range of ϕ significantly depends on the assumed Ω_M. The range of ϕ is larger (smaller) for smaller (larger) Ω_M. However, it is interesting that the whole shape of $V(\phi)$ is less sensitive to the uncertainty in Ω_M, although w and the range of ϕ are dependent on the assumed value of Ω_M. If we assume smaller Ω_M, the resulting potential energy is larger. On the other hand, ϕ decreases more rapidly back in time. The opposite is the case for larger Ω_M. Both effects make the reconstructed shape of $V(\phi)$ converge to the true one.

Reconstructing the Effective Potential from Simulated Data: We simulate the actual reconstruction procedure using numerically generated data. We

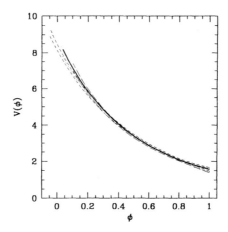

FIGURE 2. One sigma intervals for the reconstructed quintessential potential assuming luminosity distance error of 0.5% with $N = 30$. Solid curves are the original potentials. Dotted curves are for $\Omega_M = 0.30$, short dashed curves for $\Omega_M = 0.27$, long dashed curves for $\Omega_M = 0.33$

consider $N = 30$ data with variance $\sigma = 0.005$. The error is the distance error of the binned data expected from observations of 6000 supernovae by SNAP. We only consider statistical uncertainties. The 68% confidence intervals of the reconstructed potentials are shown in Fig. 2. See Ref. [4] for the details.

To summarize, we have studied the feasibility of reconstructing the equation of state of dark energy and the effective potential of the quintessence field from SNIa data by taking into account the uncertainty in Ω_M as well as the error in the luminosity distance. We have found that w and the range of ϕ are dependent on the assumed value of Ω_M, while the whole shape of $V(\phi)$ is less sensitive to the uncertainty in Ω_M. If Ω_M could be constrained to some 10% accuracy by other observations, which may not be unrealistic expectation [6], then future high precision measurements of distances to thousands of SNIa could reveal the shape of the quintessential potential.

REFERENCES

1. T. Nakamura and T. Chiba, Mon. Not. Roy. Astron. Soc. **306**, 696 (1999).
2. A. Starobinsky, JETP Lett. **68**, 757 (1998); D. Huterer and M.S. Turner, Phys. Rev. D **60**, 081301 (1999).
3. U.-L. Pen, Astrophys.J.Suppl. **120**, 49 (1999).
4. T. Chiba and T. Nakamura, Phys. Rev. D **62**, 121301 (2000).
5. T. Nakamura and T. Chiba, astro-ph/0011157, to be published in ApJ.
6. Z. Haiman, J. Mohr, and G.P. Holder, astro-ph/002336.

Constraints on Ω_m and Ω_Λ from Future Cluster Surveys

Gilbert Holder*, Zoltan Haiman†, Joseph Mohr‡

* *Department of Astronomy and Astrophysics, University of Chicago, Chicago IL 60637*
† *Princeton University Observatory, Princeton NJ 08544*
‡ *Departments of Astronomy and Physics, University of Illinois, Urbana IL 61801*

Abstract. We show that deep cluster surveys can put strong constraints on Ω_m and Ω_Λ, in a manner that is nearly independent of the Hubble constant. Degeneracies between parameters are very different from either CMB or supernovae constraints, allowing very accurate joint constraints on parameters. Redshift information extending past $z \sim 0.5$ will be crucial for constraining the vacuum energy.

INTRODUCTION

Galaxy clusters can allow strong constraints on cosmological parameters [1]. Cluster surveys are mainly probing the amplitude of the power spectrum as a function of redshift, rather than distance or volume, allowing unique constraints on cosmology and providing a sensitive test of structure formation. As we show below, surveys out to $z \sim 0.5$ are very powerful probes of Ω_m while deep surveys ($z \geq 1$) allow a determination of the vacuum energy density, Ω_Λ, as well. Upcoming cluster surveys using the Sunyaev-Zel'dovich effect will be capable of finding large numbers of clusters, some with redshifts extending well past $z \sim 1$. In this work, we use the expected survey yields to estimate the precision to which the matter density and vacuum energy density can be reasonably measured.

CALCULATING CLUSTER SURVEY YIELDS

To calculate the expected number of clusters per square degree as a function one must consider several elements: the comoving volume per unit redshift per unit solid angle, $\frac{dV}{dzd\Omega}$; the minimum observable mass as a function of redshift and cosmology, $M_{lim}(z)$; and the comoving number density of halos above the mass threshold as a function of redshift and cosmology, $n(> M_{lim}, z)$.

We assume constant mass limits for this work, a suitable approximation for non-targeted SZ surveys [2,3], and use the Jenkins et al. (2000) mass function, obtained from large cosmological simulations.

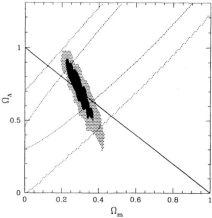

FIGURE 1. Confidence levels (solid 68%, shaded 95%) on energy densities, marginalized over σ_8. Dashed contours show the current constraints from the Supernova Cosmology Project (see http://supernova.lbl.gov), while the diagonal line indicates a flat universe.

As a fiducial model, we take a flat, low-density model, with $\Omega_m = 0.3, \Omega_\Lambda = 0.7, h = 0.65, \sigma_8 = 1.0$. We keep h fixed for all models and vary the other three parameters. To isolate the effects the growth of the power spectrum, we use the power spectrum of the fiducial model for all models. Relaxing this assumption will slightly affect the constraints on Ω_m and introduce an h dependence, but the qualitative results are unchanged.

We construct a likelihood space by a Monte Carlo method, allowing Ω_m, Ω_Λ and σ_8 to find their best-fit values for many mock cluster catalogs drawn from the fiducial model. For this work, we then marginalize over σ_8 to construct confidence limits in the $\Omega_m - \Omega_\Lambda$ plane.

We first look at a relatively low value for the constant mass limit, $M_{lim} = 1 \times 10^{14} h^{-1} M_\odot$, appropriate for a deep SZ survey that would cover approximately 12 deg^2 (e.g., one square degree per month for a year). Results are shown in figure 1. Constraints can be seen to be complementary to other probes. The relatively small number of (~ 300) clusters spread over only a few square degrees should be relatively easy to follow up to get redshift information.

We also used a relatively high value, $M_{lim} = 8 \times 10^{14} h^{-1} M_\odot$, appropriate for an SZ survey like PLANCK, but covering a large fraction of the sky. Results are shown in figure 2. From the nearby sample (shaded region), it can be seen that, even though this is a large fraction of the total cluster sample, it is difficult to constrain the vacuum energy. The growth function is just not very sensitive to vacuum energy out to this redshift, and therefore there is very little difference in the comoving number density. It is the added information from clusters past $z \sim 0.5$ that carry almost all information on the cosmological constant.

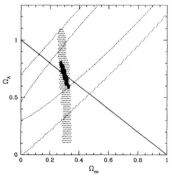

FIGURE 2. Confidence levels on energy densities for a mass limit of $8 \times 10^{14} M_\odot$. The shaded region shows the expected 95% confidence region using only clusters with $z < 0.5$, while the solid region shows the 95% confidence region for the entire sample. It is the high z differences in the growth function that allow constraints on Ω_Λ.

CONCLUSIONS AND FUTURE WORK

Cluster surveys in the next few years will be very powerful probes of cosmological parameters, and in a manner that will serve as a very useful check of other methods. As well, they will provide a view of the structure formation history of the universe. The results shown here are optimistic in that systematic errors will almost certainly be very important in any interpretation of real surveys, but they are indicative of the potential science yields of non-targeted SZ surveys.

To get meaningful constraints on the vacuum energy density, redshift information will be required for clusters with $z > 0.5$, making considerations of follow-up observations very important. In some respects, this might argue for a preference for deep surveys on a relatively small patch of sky, as follow-up observations could be done on a smaller scale.

Future work will look at more realistic mass limits as a function of survey strategy, as well as exploring the σ_8 direction of the degeneracy.

REFERENCES

1. P. Viana and A. Liddle, *Mon. Not. Roy. Ast. Soc.* **303**, 535 (1999).
2. G. P. Holder, J. J. Mohr, J. E. Carlstrom, E. A. E., and E. M. Leitch, *Ap. J.* **544**, 629 (2000).
3. Z. Haiman, J. J. Mohr, and G. P. Holder, *Ap. J.* , submitted: astro-ph/0002336 (2000).
4. A. Jenkins et al., *Mon. Not. Roy. Ast. Soc.* **321**, 372+ (2001).

Emergence of Discrete Structure Scales in Q-component Matter Background

Manfred P. Leubner

Institute for Theoretical Physics
University of Innsbruck, A-6020 Innsbruck, Austria

Abstract. The sequence of observed discrete cosmic structure scales ranging from hadrons to globular cluster, galaxies and superclusters is shown to be a consequence of a specific quantization rule unifying the scenario of observed mass and length scales. Constraints discretizing the increase of gravitational entropy for a set of merging particles are discussed yielding a unique condition for a hierarchical clustering process. Furthermore, the missing energy density in the universe appears to require the introduction of a new mass scale significantly below the electroweak scale. An appropriate mass scale turns out as a natural ingredient of the theory and is shown to define the lower limit of the hierarchy of structure scales, providing a negative pressure quintessence component matter background. Since the resulting three lowest order mass scales are to be considered as Q-component, the typical particle physics scale and the atomic scale of condensed matter also the 'great desert' between the latter two, a content of GUT theories, is found as natural outcome. The accelerating universe is discussed within the FRW environment and distinct bounds on permissible density perturbations are provided.

The question why condensed matter is made up of atoms and why the universe is composed by stellar systems, globular clusters or galaxies, but nothing between was originally addressed by Chandrasekar [1]. Suggestions for a cosmic quantization of gravitational interaction were proposed for astrophysical objects [2,3] and recently Leubner [4] provided a hierarchical clustering technique in dimensionless notation for the problem of discrete structure formation.

Let a clustering procedure result in a sequence of hierarchically nested sets of higher order clusters $(...\{G_{i-1}\},\{G_i\},\{G_{i+1}\},\{G_{i+2}\},...)$ where the members of any specific generation admit nearly equal richness. Substructures G_i continuously merge into higher order systems G_{i+n} finally approaching the root, the universe, a cluster with one element. A separation into internal interactions between the individuals of each member of a set of equally bound clusters and their global external interactions with partners of nearly equal richness provides the basis to formulate a quantization condition for higher order merging. This approach reflects just what nature demonstrates, since mutual interaction of substructures of spatially separated bound macro-systems is not realized.

The question how many substructures merge in average forming sets of nearly equal richness of a higher order generation is studied from gravitational entropy increase due to clumping of structure. We verify the only feasible configuration where the entropy increase within any two levels of cluster formation $(i, i+1)$ is successively at each higher order merging process subject to an extremum.

Let for instance a universe consist of N_i galaxies and ask how many galaxies n_i merge in average into one cluster, yielding a total of N_{i+1} clusters of galaxies. Then the solution is found within this concept from the requirement [4,5]

$$N_{i+1} n_i (n_i - 1)/2 + N_{i+1}(N_{i+1} - 1)/2 \Longrightarrow extremum, \tag{1}$$

a condition driving the gravitational entropy increase of the entire system at each higher order merging process towards a maximum. The evaluation of equation (1) yields $N_{i+1} = (N_i^2/2)^{1/3}$ which can be rewritten with proper subscripts in terms of a fundamental recurrence relation defining the mass hierarchy as $m_{i+1} = (2m_i^2 M)^{1/3}$. Here M denotes the mass parameter of a flat universe which can be introduced via Hubble's parameter. Similar relations are found for the mean size and mean distance of discrete mass concentrations from a basic principle of statistics [6] along with two invariants originating from gravitational pressure balance. The results are summarized in Fig. 1 where Σ_{crit} is in analogy to $\rho_{crit} = 3H^2/8\pi G$ the critical surface density and the specific parameter values are provided in [4]. Stars refer to the actual position of an inhomogeneity in the $m - r$ space with a related ρ_{crit}.

The initial condition $N_0 \simeq 10^{121}$ can be calculated and corresponds to the current value of the Beckenstein-Hawking entropy of the universe inside a Hubble radius [7], equivalent to the ratio of the cosmological constant set at Planck scale and it's present value Λ_0 [8]. The associated mass scale $m_0 \simeq 2 \times 10^{-33} eV$ is identified from the requirements of a **quintessence** potential [9] as effective mass of the order of the expansion rate H_0 of the present universe with Planck's length as related interaction scale [10]. The gravitational quantization condition predicts consecutively a series of higher order distinct mass/radius structure scales where the typical **particle physics scale** is followed by a **condensed matter** state with density of the order of unity, both separated by a 'great desert gap' favored from GUT approaches. Successively higher order structure scales must be assigned to **planetesimals**, known to play a key role in modern theories of solar system evolution [11] and constituting the building blocks of **stellar systems**. Continuing to large scales it can be verified easily [12] that the predictions for permissible inhomogeneities are reproduced in the universe by representative members of **globular star clusters, galaxies, galaxy clusters** and **superclusters**. Superclusters as largest well defined structures are known to be surrounded by low density regions of similar scale leading to an overall agreement of this concept with observations. The sequence converges to the **universe** as the number of clusters $N_i \Rightarrow 1$ where nine fundamental inhomogeneity scales are clearly identified between $10^{122} \geq N_i \geq 1$.

The formalism implies for the quintessence matter background density $\rho_1 = 1/\sqrt{2}\rho_{crit}$ yielding a density parameter $\Omega = 0.7\Omega_Q + 0.3\Omega_M = 1$ for the dark energy

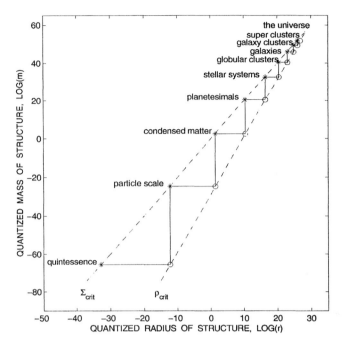

FIGURE 1. The hierarchy of discrete structure scales

to matter separation [13]. The time evolution of discrete structure scales can be simulated within the FRW- environment omitted here due to lack of space.

The concept of structure unification derived in view of a quantized gravitational entropy principle appears to be reproduced on all scales of the cosmic scenario.

REFERENCES

1. Chandrasekar S., *Nature* **139**, 757 (1937).
2. Sivaram C., *Astrophys. Space Sci.* **207**, 317 (1993).
3. Terazawa H., *Mod. Phys. Lett.* **A 11**, 2971 (1996).
4. Leubner M. P., *Nucl. Phys. B (Proc. Suppl.)* **80**, 09/10 (2000).
5. Leubner M. P., *Gravitation and Cosmology Suppl.* **6**, 144 (2000).
6. Carazza B., *Found. Phys. Lett.* **9**, 79 (1996).
7. Barrow J. D., *Mod. Phys. Lett.* **A 14**, 1067 (1999).
8. Sivaram C., *Mod. Phys. Lett.* **A 14**, 2363 (1999).
9. Yasunori N., T. Watari and T. Yanagida, *Phys. Lett.* **B 484**, 103 (2000).
10. Kuchiev M. Yu., *Class. Quant. Grav.* **15**, 1895 (1998).
11. Bailey M., in *Asteroids, Comets, Meteors*, IAU Symposium **160**, 1994.
12. Fabian A. C., ed., *Clusters and Superclusters of Galaxies*, Kluwer AP, 1992.
13. Wang L., R. R. Caldwell, J. P. Ostriker, and P. J. Steinhardt, *ApJ* **530**, 17 (2000).

Conformal gravity and a naturally small cosmological constant[1]

Philip D. Mannheim

Department of Physics, University of Connecticut, Storrs, CT 06269
mannheim@uconnvm.uconn.edu

Abstract.
With attempts to quench the cosmological constant Λ having so far failed, we instead investigate what could be done if Λ is not quenched and actually gets to be as big as elementary particle physics suggests. Since the quantity relevant to cosmology is actually Ω_Λ, quenching it to its small measured value is equally achievable by quenching not Λ but G instead, with the G relevant to cosmology then being much smaller than that measured in a low energy Cavendish experiment. A gravitational model in which this explicitly takes place, viz. conformal gravity, is presented, with the model being found to provide for a completely natural, non fine tuned accounting of the recent high z accelerating universe supernovae data, no matter how big Λ itself actually gets to be. Thus to solve the cosmological constant problem we do not need to change or quench the energy content of the universe, but rather only its effect on cosmic evolution.

The recent discovery [1,2] that the current era deceleration parameter $q(t_0)$ is close to $-1/2$ has made the already extremely disturbing cosmological constant problem even more vexing than before. Specifically, with $q(t_0)$ being given in standard gravity by $q(t_0) = (n/2 - 1)\Omega_M(t_0) - \Omega_\Lambda(t_0)$ [where $\Omega_M(t) = 8\pi G\rho_M(t)/3c^2H^2(t)$ is due to ordinary matter (i.e. matter for which $\rho_M(t) = A/R^n(t)$ where $A > 0$ and $3 \leq n \leq 4$), and where $\Omega_\Lambda(t) = 8\pi G\Lambda/3cH^2(t)$ is due to a cosmological constant], we see that not only must $c\Lambda$ be non-zero, it must be of order $3c^2H^2(t_0)/8\pi G = \rho_C(t_0)$ in magnitude, i.e. it must be quenched by no less than 60 orders of magnitude below its natural value as expected from fundamental particle physics. Additionally, since such a quenched $c\Lambda$ would then be of order $\rho_M(t_0)$ as well (the so-called cosmic coincidence), our particular cosmological epoch would then only be achievable in standard gravity if the macroscopic Friedmann evolution equation were to be fine-tuned at very early times to incredible precision. Any still to be found fundamental microscopic physics mechanism which might in fact quench $c\Lambda$ by the requisite sixty orders of magnitude would thus still leave standard gravity with an additional macroscopic coincidence to explain.

---

[1] astro-ph/9901219 v2, February 26, 2001.

Since no mechanism has yet been found which might actually quench $c\Lambda$ and since its quenching might not necessarily work macroscopically anyway, we shall thus turn the problem around and ask what can be done if $c\Lambda$ is not in fact quenched and is in fact as big as elementary particle physics suggests. To this end we note immediately that it would still be possible to have $q(t_0)$ be of order one today (the measurable consequence of $c\Lambda$) if instead of quenching $c\Lambda$ we instead quench G, with the cosmological G then being replaced by an altogether smaller G_{eff}. Since observationally $\rho_M(t_0)$ is known to not be bigger than $\rho_C(t_0)$, any successful such cosmological quenching of G (successful in the sense that such relativistic quenching not modify standard non-relativistic physics) would immediately leave us with a non-quenched $c\Lambda$ which would then not suffer from any cosmic coincidence problem.

Given these remarks it is thus of interest to note that it is precisely a situation such as this which obtains in the conformal gravity theory which has recently been advanced [3–9] as a candidate alternative to the standard gravitational theory. Conformal gravity is a fully covariant gravitational theory which, unlike standard gravity, possesses an additional local scale invariance, a symmetry which when unbroken sets any fundamental cosmological constant and any fundamental G to zero [3]. Unlike standard gravity conformal gravity thus has a great deal of control over the cosmological constant, a control which is found to be of relevance even after the conformal symmetry is spontaneously broken by the non-vanishing of a scalar field vacuum expectation value S_0 below a typical critical temperature T_V. In fact in the presence of such breaking the standard attractive G phenomenology is found to still emerge at low energies [5], while cosmology is found [4] to instead be controlled by the effective $G_{eff} = -3c^3/4\pi\hbar S_0^2$, a quantity which by being negative immediately entails cosmic repulsion [7], and which, due to its behaving as $1/S_0^2$, is made small by the very same mechanism which serves to make Λ itself large.

Other than the use of a changed G the cosmic evolution of conformal gravity is otherwise the same as that of the standard one, viz. [7–9]

$$\dot{R}^2(t) + kc^2 = -3c^3\dot{R}^2(t)(\Omega_M(t) + \Omega_\Lambda(t))/4\pi\hbar S_0^2 G \equiv \dot{R}^2(t)(\bar{\Omega}_M(t) + \bar{\Omega}_\Lambda(t))$$
$$q(t) = (n/2 - 1)\bar{\Omega}_M(t) - \bar{\Omega}_\Lambda(t) \quad (1)$$

(Eq. (1) serves to define $\bar{\Omega}_M(t)$ and $\bar{\Omega}_\Lambda(t)$). Moreover, unlike the situation in the standard theory where values for the relevant evolution parameters (such as the sign of Λ) are only determined phenomenologically, in conformal gravity essentially everything is already a priori known. With conformal gravity not needing dark matter to account for non-relativistic issues such as galactic rotation curve systematics [6], $\rho_M(t_0)$ can be determined directly from luminous matter alone, with galaxy luminosity accounts giving a value for it of order $0.01\rho_C(t_0)$ or so. Further, with $c\Lambda$ being generated by vacuum breaking in an otherwise scaleless theory, since such breaking lowers the energy density, $c\Lambda$ is unambiguously negative, with it thus being typically given by $-\sigma T_V^4$. Then with G_{eff} also being negative, $\bar{\Omega}_\Lambda(t)$ is necessarily positive, just as needed to give cosmic acceleration. Similarly, the sign of the spatial 3-curvature k is known from theory [9] to be negative,[2] something which has been independently confirmed from a study of galactic rotation curves [6]. Finally, since G_{eff} is negative, the cosmology is singularity free and thus expands from a finite maximum temperature T_{max}, a temperature which for $k < 0$ is necessarily greater than T_V [7–9] (so that a large T_V entails an even larger T_{max}).

Given only that Λ, k and G_{eff} are all negative, the temperature evolution of the theory is then completely determined for arbitrary T_{max} and T_V, to yield [7–9]

$$\bar{\Omega}_\Lambda(t) = (1 - T^2/T_{max}^2)^{-1}(1 + T^2 T_{max}^2/T_V^4)^{-1}, \quad \bar{\Omega}_M(t) = -(T^4/T_V^4)\bar{\Omega}_\Lambda(t) \qquad (2)$$

at any T. Thus, from Eq. (2) we see that simply because $T_{max} \gg T(t_0)$, i.e. simply because the universe is as old as it is, it immediately follows that $\bar{\Omega}_\Lambda(t_0)$ has to lie somewhere between zero and one today no matter how big (or small) T_V might be. Then, since $T_V \gg T(t_0)$, $\bar{\Omega}_M(t_0)$ has to be completely negligible,[3] so that $q(t_0)$ must thus necessarily lie between zero and minus one today notwithstanding that T_V is huge. Moreover, the larger T_V gets to be, the more $\bar{\Omega}_\Lambda(t_0)$ will be reduced below one, with it taking a value close to one half should $T(t_0)T_{max}/T_V^2$ be close to one. With $\bar{\Omega}_M(t_0)$ being negligible today, $\bar{\Omega}_\Lambda(t_0)$ is therefore given as $1 + kc^2/\dot{R}^2(t_0)$, a quantity which necessarily lies below one if k is negative. Thus in a $k < 0$ conformal gravity universe, once the universe has cooled enough, $\bar{\Omega}_\Lambda(t)$ will then be forced to have to lie between zero and one no matter how big Λ may or may not be. The contribution of Λ to cosmology is thus seen to be completely under control in conformal gravity, with the theory thus leading us right into the $\bar{\Omega}_\Lambda(t_0) \simeq 1/2$, $\bar{\Omega}_M(t_0) = 0$ region, a region which, while foreign to standard gravity, is nonetheless still fully compatible with the reported supernovae data fits. Hence to solve the cosmological constant problem we do not need to change or quench the energy content of the universe, but rather only its effect on cosmic evolution. This work has been supported in part by the Department of Energy under grant No. DE-FG02-92ER40716.00.

REFERENCES

1. A. G. Riess et. al., Astronom. J. **116**, 1009 (1998).
2. S. Perlmutter et. al., Astrophys. J. **517**, 565 (1999).
3. P. D. Mannheim, Gen. Relativ. Gravit. **22**, 289 (1990).
4. P. D. Mannheim, Astrophys. J. **391**, 429 (1992).
5. P. D. Mannheim and D. Kazanas, Gen. Relativ. Gravit. **26**, 337 (1994).
6. P. D. Mannheim, Astrophys. J. **479**, 659 (1997).
7. P. D. Mannheim, Phys. Rev. D **58**, 103511 (1998).
8. P. D. Mannheim, astro-ph/9910093 (1999).
9. P. D. Mannheim, Founds. Phys. **30**, 709 (2000).

---

[2] At the highest temperatures the zero energy density required of a (then) completely conformal invariant universe is maintained by a cancellation between the positive energy density of ordinary matter and the negative energy density due to the negative curvature of the gravitational field.
[3] $\bar{\Omega}_M(t_0)$ is suppressed by G_{eff} being small, and not by $\rho_M(t_0)$ itself being small.

The Vacuum Energy from a New Perspective

Antonio R. Mondragon and Roland E. Allen

*Center for Theoretical Physics, Texas A&M University,
College Station, Texas 77843, USA*

Abstract. It is commonly believed that the vacuum energy problem points to the need for (1) a radically new formulation of gravitational physics and (2) a new principle which forces the vacuum stress-energy tensor (as measured by gravity) to be nearly zero. Here we point out that a new fundamental theory contains both features: (1) In this theory the vierbein is interpreted as the "superfluid velocity" associated with the order parameter Ψ_s for a GUT-scale Higgs condensate. (2) The vacuum stress-energy tensor $\mathcal{T}_{\mu\nu}^{vac}$ is exactly zero in the vacuum state, because the action is extremalized with respect to variations in Ψ_s. With inhomogeneously-distributed matter present, $\mathcal{T}_{\mu\nu}^{vac}$ is shifted away from zero.

The vacuum energy is one of the deepest issues in theoretical physics, and no conventional theory – including superstring/M theory – has offered a convincing solution to the problem of why the vacuum stress-energy tensor (as measured by gravity) is vastly smaller than expected but still nonzero [1-3].

In this paper we consider an unconventional theory which contains a radically new formulation of gravitational physics [4-6]. The gravitational vierbein is interpreted as the "superfluid velocity" of a GUT-scale condensate Ψ_s which forms in the very early universe:

$$g^{\mu\nu} = \eta^{\alpha\beta} e^{\mu}_{\alpha} e^{\nu}_{\beta} \quad , \quad e^{\mu}_{\alpha} = v^{\mu}_{\alpha} \quad , \quad v^{\mu} = v^{\mu}_{\alpha} \sigma^{\alpha} \quad \text{with} \quad \mu, \alpha = 0, 1, 2, 3 \quad (1)$$

$$v^{\mu} = \eta^{\mu\nu} v_{\nu} \quad , \quad m v_{\mu} = -i U^{-1} \partial_{\mu} U \quad , \quad \Psi_s = n_s^{1/2} U \eta_s \quad , \quad \eta_s^{\dagger} \eta_s = 1. \quad (2)$$

Here $\eta^{\mu\nu} = \mathrm{diag}\,(-1,1,1,1)$ is the Minkowski metric tensor, the σ^{α} are the identity matrix and three Pauli matrices, U is a 2×2 unitary matrix, η_s is a constant 2-component vector, and n_s is the condensate density. (After a Kaluza-Klein reduction from a higher-dimensional theory, the initial group of this order parameter is $SO(10) \times SU(2) \times U(1)$. For the purposes of this paper, however, the gauge group $SO(10)$ can be ignored, leaving the simpler description of (1) and (2).) In the present theory, Ψ_s is not static but instead exhibits $SU(2) \times U(1)$ rotations as a function of position and time. This condensate also supports Planck-scale $SU(2)$ instantons (in a Euclidean picture), which are analogous to the $U(1)$ vortices in an ordinary superfluid. In the present theory, the Einstein-Hilbert action and the curvature of spacetime result from these instantons [4]. Quantum gravity has a natural cutoff at the energy scale $m \sim$ the Planck energy m_P, but at lower energies one recovers the Einstein field equations

$$\frac{\delta S_{total}}{\delta g^{\mu\nu}} = \frac{\delta S_{vac}}{\delta g^{\mu\nu}} + \frac{\delta S_{fields}}{\delta g^{\mu\nu}} + \frac{\delta S_{EH}}{\delta g^{\mu\nu}} = 0 \quad (3)$$

or
$$R_{\mu\nu} - \frac{1}{2}g_{\mu\nu}{}^{(4)}R = \frac{1}{2}\ell_P^2 T_{\mu\nu}^{total} = \frac{1}{2}\ell_P^2 \left(T_{\mu\nu}^{vac} + T_{\mu\nu}^{fields}\right) \quad (4)$$

where
$$T_{\mu\nu}^{vac} = -\frac{2}{\sqrt{-g}}\frac{\delta S_{vac}}{\delta g^{\mu\nu}}, \quad T_{\mu\nu}^{fields} = -\frac{2}{\sqrt{-g}}\frac{\delta S_{fields}}{\delta g^{\mu\nu}} \quad (5)$$

and ℓ_P is the Planck length defined in Ref. 4.

In the vacuum state, there is no stress-energy tensor for matter and radiation: $T_{\mu\nu}^{fields} = 0$. We additionally assume that there there is no contribution from topological defects in the vacuum state: $\delta S_{EH}/\delta g^{\mu\nu} = 0$. This assumption will be discussed in more detail elsewhere [6], but it will be seen below that it leads to a consistent solution (whereas such a solution could not be obtained in conventional physics). We then have

$$\delta S_{vac} = \delta S_{total} = 0 \quad \text{in the vacuum state} \quad (6)$$

for arbitrary variations of the order parameter Ψ_s. Variations in v_α^μ, however, are a special case of functional variations in Ψ_s. It follows that the vacuum stress-energy tensor is exactly zero:

$$T_{\mu\nu}^{vac} = 0 \quad \text{in the vacuum state.} \quad (7)$$

It is interesting to see in more detail how (7) can be achieved. According to the quantum Bernoulli equation (3.20) of Ref. 4, we have

$$\frac{1}{2}m\eta_s^\dagger \eta_{\mu\nu} v^\mu v^\nu \eta_s + V + P + V_{vac} = \mu \quad (8)$$

$$V = bn_s, \quad P = -\frac{1}{2m}n_s^{-1/2}\eta^{\mu\nu}\partial_\mu\partial_\nu n_s^{1/2} \quad (9)$$

where μ is a fundamental energy which plays the role of a chemical potential here and which is comparable to m_P. We have added a term V_{vac} which represents the contribution of all other vacuum fields to δS_{vac} when Ψ_s is varied. Let us rewrite (8) as

$$-\frac{1}{2m}n_s^{-1/2}\eta^{\mu\nu}\partial_\mu\partial_\nu n_s^{1/2} + bn_s = \mu - V_{vac} - \frac{1}{2}m\eta_{\mu\nu}e_\alpha^\mu e_\beta^\nu \eta_s^\dagger \sigma^\alpha \sigma^\beta \eta_s. \quad (10)$$

After V_{vac} and e_α^μ are specified, the condensate density n_s adjusts itself to satisfy (10), (6), and (7). One might express this result as follows: In the vacuum state, the vacuum stress-energy tensor is tuned to exactly zero through adjustments of the condensate density. Notice that the extremalization (3) in conventional physics requires a contribution from the Einstein-Hilbert action S_{EH} even when $\delta S_{fields}/\delta g^{\mu\nu} = 0$, but in the present theory this extremalization in the vacuum state can be accomplished with S_{vac} alone.

In a more general state with matter, radiation, and topological defects present, it is the *total* action which is extremalized in (3). Then $\delta S_{vac}/\delta g^{\mu\nu}$ is shifted away from zero:

$$\mathcal{T}_{\mu\nu}^{vac} \neq 0 \quad \text{with matter and radiation present.} \tag{11}$$

There are two primary aspects of the vacuum energy problem [1, 2]: (i) Why is the vacuum stress-energy tensor many orders of magnitude smaller than predicted by conventional physics? This question is addressed in (7). (ii) Why is the vacuum stress-energy tensor not exactly zero? This is addressed in (11).

Although these are the "big" questions, one can add two more in the present context: (iii) Why was the vacuum energy density small compared to the density of matter and radiation during the period of big-bang nucleosynthesis? (iv) Why is it comparable to the density of matter now? To fully answer these questions will require a detailed treatment of how the vacuum energy is affected by the presence of matter and radiation. Suppose, however, that the dominant mechanism is a Casimir-like effect, in which the vacuum energy is modified by the boundary conditions imposed on the vacuum fields when there is an inhomogeneous distribution of matter. In a radiation-dominated universe, the energy density will be relatively homogeneous, and such an effect should be relatively small. In the present epoch, on the other hand, there is an extremely inhomogeneous distribution of matter. This plausibility argument indicates that the vacuum energy should play an important role only in the present epoch, and that the vacuum energy density (as measured by the stress-energy tensor) will be comparable to the density of inhomogeneously-distributed matter.

Acknowledgement

This work was supported by the Robert A. Welch Foundation.

References

[1] S. Weinberg, *Rev. Mod. Phys.* 61, 1 (1989).

[2] S. Weinberg, talk at the 4th International Symposium on Sources and Detection of Dark Matter in the Universe (DM2000) and astro-ph/0005265.

[3] E. Witten, talk at DM2000 and hep-ph/0002297.

[4] R. E. Allen, *Int. J. Mod. Phys. A* 12, 2385 (1997) and hep-th/9612041.

[5] R. E. Allen, to be published and hep-th/0008032.

[6] R. E. Allen, to be published and talk given at the conference on Problems with Vacuum Energy (Copenhagen, August 24-26, 2000).

CHAPTER 7

NUCLEOSYNTHESIS IN THE BIG BANG AND FIRST STARS

The Early Formation, Evolution and Age of the Neutron-Capture Elements in the Early Galaxy

John J. Cowan*, Christopher Sneden† and James W. Truran¶

*Department of Physics and Astronomy, University of Oklahoma, Norman, OK 73019
†Department of Astronomy, University of Texas, Austin, TX 78712
¶Department of Astronomy and Astrophysics, University of Chicago, Chicago, IL 60637

Abstract. Abundance observations indicate the presence of rapid-neutron capture (i.e., r-process) elements in old Galactic halo and globular cluster stars. These observations demonstrate that the earliest generations of stars in the Galaxy, responsible for neutron-capture synthesis and the progenitors of the halo stars, were rapidly evolving. Abundance comparisons among several halo stars show that the heaviest neutron-capture elements (including Ba and heavier) are consistent with a scaled solar system r-process abundance distribution, while the lighter such elements do not conform to the solar pattern. These comparisons suggest two r-process sites or at least two different sets of astrophysical conditions. The large star-to-star scatter observed in the neutron-capture/iron ratios at low metallicities – which disappears with increasing [Fe/H] – suggests an early, chemically unmixed and inhomogeneous Galaxy. The stellar abundances indicate a change from the r-process to the slow neutron capture (i.e., s-) process at higher metallicities in the Galaxy. The detection of thorium in halo and globular cluster stars offers a promising, independent age-dating technique that can put lower limits on the age of the Galaxy.

INTRODUCTION

In this paper we briefly review some of the important abundance trends for the slow- or rapid-neutron capture elements. We focus on how these neutron-capture elements can be employed to (1) study the nature of the progenitors and the nucleosynthesis history in the early Galaxy, (2) explore the chemical evolution of the Galaxy, and (3) obtain radioactive age estimates for the oldest stars, which in turn puts limits on the age of the Galaxy and the universe.

NEUTRON-CAPTURE ABUNDANCES

Extensive abundance studies have been made of the ultra-metal-poor ($\equiv$ UMP, [Fe/H] = -3.1) but neutron-capture rich halo star CS 22892–052 [1–3]. In Figure 1 we show the n-capture abundances as determined by Sneden et al. [3] compared to a scaled solar system r-process abundance distribution. As has been noted previously, the upper end of the stellar abundance distribution (*i.e.*, Ba and above) is in very close agreement with the solar system r-process curve. It has been seen only recently, and so far only in this star, that the lighter n-capture abundances between Zr and Ba (*e.g.*, Nb, Pd, and Ag) do not lie on the same solar curve. This lends support to previous suggestions that there may be two sites for the r-process, one for the heavier and one for the lighter n-capture elements [4]. It is unclear whether both of those sites might be supernovae occurring at different frequencies [5] or neutron star binaries [6] or some combination of those. Alternatively, it has been proposed that both ends of the abundance distribution could be synthesized in different regions of the same neutron-rich jet of a core-collapse supernova [7].

ABUNDANCE TRENDS IN THE GALAXY

Observations of n-capture abundances in a wide range of Galactic, including metal-poor halo and disk, stars have now been made over a range of metallicities. These data demonstrate several interesting abundance trends in the Galaxy.

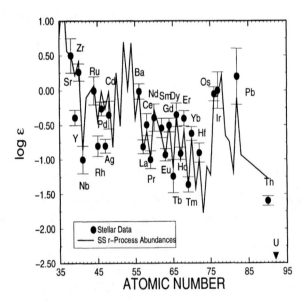

FIGURE 1. Neutron-capture abundances in CS 22892–052 compared with a scaled solar system r-process distribution (solid line).

Scatter in the Early Galaxy

Earlier work by Gilroy et al. [8] first demonstrated that the stellar abundances of r-process elements with respect to iron, particularly Eu/Fe, showed a large scatter at low metallicities. This scatter appeared to diminish with increasing metallicity. A more extensive study by Burris et al. [9] confirmed the very large star-to-star scatter in the early Galaxy, while studies of stars with higher metallicities – mostly disk stars – [10,11] show little scatter. In Figure 2 we plot the data from a number of surveys [9–13], along with detailed abundance determinations from several single stars [3,14]. These studies, which cover a metallicity range $-3.5 \leq$ [Fe/H] $\leq +0.5$ and include large numbers of stars, had attempted to minimize observational errors. The star-to-star scatter illustrated in this figure can be explained as the result of individual nucleosynthetic events (e.g., supernovae) [8,9] and strongly suggests an early, unmixed, chemically inhomogeneous Galaxy. (See also [15] for further discussion.) It should be noted that while the absolute levels of n-capture/Fe abundances vary widely, the relative abundances are similar in all of the very metal-poor halo stars.

One other important trend is notable in Figure 2. At higher metallicities, particularly for [Fe/H] $\simeq -1$, the values of [Eu/Fe] tend downward. This demonstrates clearly the effect of increasing iron production, presumably from Type Ia supernovae, at higher Galactic metallicities [9]. At very low metallicities high mass (and

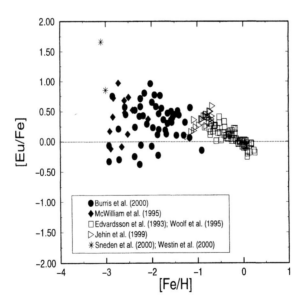

FIGURE 2. [Eu/Fe] vs. metallicity for Galactic halo and disk stars. The dotted line indicates the solar value.

rapidly evolving) Type II supernovae contribute to Galactic iron production. The onset of the bulk of iron production from Type Ia supernovae (with longer evolutionary timescales due to lower mass progenitors) occurs only at higher [Fe/H] and later Galactic times.

Chemical Evolution of the r- and s-Process

The abundances observed for the elements in CS 22892–052 and other UMP halo stars demonstrate the early onset of the r-process in the Galaxy. These results (see Figure 1) also confirm earlier predictions [16] that elements synthesized by the s-process in the solar system (*e.g.*, Ba) were formed solely in the r-process early in the history of the Galaxy. Further confirmation of this early Galactic dominance of the r-process is seen in Figure 3, where we plot [Ba/Eu] as a function of [Fe/H]. We have utilized a combination of data sets [9–11,17–19], including some shown in Figure 2, to produce this new figure. It is clearly seen in Figure 3 that at the lowest metallicities the stellar Ba/Eu ratios cluster around the solar system (pure) r-process value. Eu is almost exclusively an r-process element, but Ba is produced predominantly in solar system material by the s-process in low mass (1-3 $M_\odot$) AGB stars [9]. At the lowest metallicities early in the history of the Galaxy, the halo stars show the products only of r-process nucleosynthesis from rapidly evolving (with short stellar evolutionary timescale) progenitors typical of, for example, Type II supernovae. As the metallicity grows larger, the ratio of Ba/Eu rises due to the

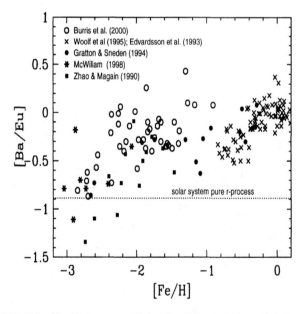

FIGURE 3. [Ba/Eu] vs. metallicity for Galactic halo and disk stars.

increased production of Ba (in the *s*-process) but not the *r*-process element Eu. The transition between a pure *r*-process production of Ba and production dominated by the main *s*-process occurs between $-3 <$ [Fe/H] < -2, with most stars consistent with the (total) solar value of [Ba/Eu] for metallicities [Fe/H] ≥ -2. The delay in the onset of the *s*-process with respect to the *r*-process is consistent with the longer stellar evolutionary timescales typical of low-mass (1–3 $M_\odot$) stars thought to be the site for *s*-process synthesis. It is interesting to note that the onset of the bulk of the main *s*-process in the Galaxy at [Fe/H] $\simeq -2$ occurs at a lower metallicity, and likely an earlier time, than the bulk of the iron production from Type Ia supernovae at [Fe/H] $\simeq -1$, as discussed above. We note further that while Ba has been commonly used for these abundance studies, in the future La may prove to be a more reliable indicator of the Galactic evolution of the *s*-process [15].

CHRONOMETRIC AGES

The detection of the radioactive element thorium in halo stars such as CS 22892–052 (see Figure 1) has provided the exciting opportunity of directly determining stellar ages. This technique relies upon comparing the observed stellar abundances with estimates of the initial abundance of the radioactive element. To minimize systematic errors, ratios of Th to Eu (produced almost exclusively in the *r*-process) are usually employed for these age determinations. Cowan *et al.* [20] and Westin *et al.* [14] obtained an average (minimum) age of 13.8 Gyr for the UMP stars CS 22892–052 and HD 115444 by comparing their observed Th/Eu abundances with the solar system ratio. This value represents a lower limit on their ages since Th has partially decayed over the last 4.5 Gyr. Improvements in these age estimates were then made by determining the initial (zero-decay) values of Th/Eu in the same calculations that reproduced the observed stable *n*-capture abundance distributions for both of these stars. The stable lead and bismuth solar system isotopic abundances were then employed to determine the most reliable mass formulae for predicting the properties of nuclei far from stability, critical for the theoretical *r*-process abundance calculations. Utilizing these constraints to obtain zero-decay Th/Eu abundances, an average age for CS 22892–052 and HD 115444 of 15.6 Gyr, with an estimated uncertainty of $\simeq 4$ Gyr, was obtained [20,14].

In addition to studies of halo stars there have been recent observations of the globular cluster M 15 [21]. Detailed abundance determinations confirm that the heavier *n*-capture abundances are consistent with the scaled solar system *r*-process curve. Further, the detection of thorium in several of the globular stars has allowed a chronometric age estimate of 14 Gyr to be determined for this system [21].

Uranium, another long-lived radioactive element, can also serve as a chronometer. This element has not been detected to date in CS 22892–052 or in HD 115444, but has been found for the first time in the star CS 31082–001 [22]. Combining both chronometers Th and U, in conjunction with several stable heavy elements, an age of 12.5 ± 3 Gyr has been estimated for this star. While this technique still has some observational and theoretical uncertainties associated with it [23], it offers great promise. In particular such chronometric age estimates of UMP stars are independent of, and consequently avoid the large uncertainties in, Galactic chemical evolution models. Finally, we note that the radioactive age determinations of these oldest halo stars put meaningful constraints on both Galactic and cosmological age estimates.

ACKNOWLEDGMENTS

We thank all of our colleagues who have collaborated with us on various studies of n-capture elements in halo stars. This research has received support from NSF grants AST-9986974 to J.J.C., AST-9987162 to C.S. and from DOE contract B341495 to J.W.T., and from the Space Telescope Science Institute grant GO-8342.

REFERENCES

1. Cowan, J. J., *et al.*, *ApJ* **439**, L51 (1995).
2. Sneden, C., *et al.*, *ApJ* **467**, 819 (1996).
3. Sneden, C., *et al.*, *ApJ* **533**, L139 (2000).
4. Wasserburg, G. J., *et al.*, *ApJ* **466**, L109 (1996).
5. Wasserburg, G. J., and Qian, Y.-Z., *ApJ* **529**, L21 (2000).
6. Rosswog, S., *et al.*, *A&A* **341**, 499 (1999).
7. Cameron, A. G. W., *Nuclear Physics A*, in press (2001).
8. Gilroy, K. K., *et al.*, *ApJ* **327**, 298 (1988).
9. Burris, D. L., *et al.*, *ApJ* **544**, 302 (2000).
10. Edvardsson, B., *et al.*, *A&A* **275**, 101 (1993).
11. Woolf, V. M., *et al. ApJ* **453**, 660 (1995).
12. McWilliam, A., *et al.*, *AJ* **109**, 2757 (1995).
13. Jehin, E., *et al.*, *A&A* **341**, 241 (1999).
14. Westin, J., *et al.*, *ApJ* **530**, 783 (2000).
15. Sneden, C., *et al.*, in *Proceedings of Cosmic Evolution*, in press (2001) (astro-ph/0101439).
16. Truran, J. W., *A&A* **97**, 391 (1981).
17. Gratton, R., and Sneden, C., *A&A* **287**, 927 (1994).
18. McWilliam, A., *AJ* **115**, 1640 (1998).
19. Zhao, G. and Magain, P. *A&A* **238**, 242 (1990).
20. Cowan, J. J., *et al.*, *ApJ* **521**, 194 (1999).
21. Sneden, C., *et al.*, *ApJ* **536**, L85 (2000).
22. Cayrel, R., *et al.*, *Nature* **409**, 691.
23. Cowan, J. J., *et al.*, in *Proceedings of Cosmic Evolution*, in press (2001) (astro-ph/0101438).

Coalescing Neutron Star Binaries and the first Stars

S. Rosswog[*], C. Freiburghaus[†], F.-K. Thielemann[†] and M.B. Davies[*]

[*]*University of Leicester, Leicester LE1 7RH, UK*
[†]*Universität Basel, 4056 Basel, CH*

Abstract.
There is growing evidence from the observation of metal-poor stars that the rapid neutron capture elements show a unique pattern beyond Barium which excellently fits the (scaled) solar abundance pattern. However, the actual production site of r-process nuclei is still subject of lively debates. Recent nucleosynthesis calculations for the ejecta of neutron star mergers, one of the suggested production sites, have shown that the observed abundance pattern can be reproduced very accurately, *if* the electron fraction of the ejected material is close to 0.1. Constraints on this scenario from observations are discussed.

INTRODUCTION

Around one half of the nuclei beyond the iron peak are produced in rapid neutron capture reactions, the so-called r-process. Starting from preexisting seed nuclei a sequence of rapid (as compared to the relevant beta-decay time scales) neutron captures drives the nuclei away from the valley of beta-stability into the very neutron rich regions of the nuclear chart. This process only ceases when either the beta-decay time scales become comparable to neutron capture time scale and/or the neutron fluxes decrease. Then these neutron rich nuclei decay towards the valley of beta stability. Although the r-process involves nuclear physics far from stability and far from being accessible in laboratory experiments the basic r-process mechanism has been understood already the fifties [1, 2, 3]. But still, the astrophysical event where these elements are formed, is despite intense research for four decades still subject to lively debates.

The basic requirements for a successful r-process are neutron number densities in excess of 10^{20} cm^{-3} and temperatures above $\sim 10^9$ K. These conditions suggest an explosive event in a neutron rich environment, possibly related to neutron stars. An obvious candidate event is the explosion of a massive star, a type II supernova. Especially the high entropy wind scenario yielded initially promising results (e.g. [4, 5, 6]). However, more detailed studies revealed deficits in reproducing the mass range $80 < A < 120$ and the abundance peak around $A = 195$ could be reproduced only at the expense of entropies much higher than those found in state-of-the-art supernova calculations [7, 8]. Several cures have been suggested, among them mechanisms that rely on a mixing between active and sterile neutrinos (e.g. [9]) which would drive matter more neutron rich thereby reducing the need for the enormous entropies.

Other attempts include more compact neutron stars that would lead to higher entropies and shorter expansion time scales, both favorable conditions for a successful r-process ([10] and references therein). However, this would require a very soft equation of state and very massive neutron stars, conditions which are difficult to achieve.

Although current supernova models face severe problems in producing r-process material in the correct amounts it has to be said that the models are not yet mature enough to draw final conclusions on this question. There is growing observational evidence for strong asymmetries in the explosion process of a SNII. It cannot be excluded that in the possibly emerging jets in a SN explosion favorable conditions for the r-process can be achieved (Höflich, these proceedings). However, this topic needs further investigation.

NEUTRON STAR MERGERS AS POSSIBLE R-PROCESS SITES

An alternative possible r-process site is related to the decompression of initially cold neutron star material during a neutron star merger (NSM) or a coalescence of a neutron star and a low mass black hole. This scenario has been suggested almost three decades ago ([11, 12]; see also [13] and [14]) since it would provide in a natural way the high neutron densities and the rapid decompression.

The problem that arises with this scenario is that there is almost no field of (astro)physics that does not enter this process at some stage: the coalescence process is intrinsically 3D, gravity is governed by the highly relativistic self-gravitating neutron star fluid, nuclear physics is of highest importance for the dynamical evolution and stability of the merged configuration, neutrino emission and transport processes are crucial for a determination of the composition of the hot nuclear matter, magnetic fields might come into play at some stage and so on and so on.

This short discussion should have made clear that this scenario bears very high potentials but is also afflicted with considerable uncertainties. At the present stage each modelling of this scenario has to make compromises in order to make this problem tractable, either sacrificing microphysics in favour of sophisticated gravity treatment (e.g. [15]) or relying on more or less Newtonian approaches thereby taking the most important microphysical processes into account [16, 17, 18, 19].

We have performed careful calculations using the smoothed particle method to solve the equations of fluid motion in 3D. The system of hydrodynamics equations is closed with an equation of state for hot and dense nuclear matter ([20]; LS-EOS). The self-gravity treatment of the fluid is Newtonian, gravitational forces are calculated efficiently using a binary tree [21]. Forces emerging from the emission of gravitational waves are taken into account in the quadrupole approximation [19]. For details and test calculations we refer to [18, 19].

The range of possible outcomes depending on the physical parameters of the inspiralling binary has been carefully investigated. For example, six different spin configurations and combinations of three different neutron star masses (1.3, 1.4 and 1.6 $M_\odot$) have been examined. Further the sensitivity to the EOS has been investigated by comparing the results of the physical EOS to polytropic and pseudo-polytropic equations of state. In all cases where the LS-EOS has been used between $\sim 4 \cdot 10^{-3}$ and $\sim 4 \cdot 10^{-2}$ $M_\odot$ were found to be dynamically ejected from the system (further material might be ejected by

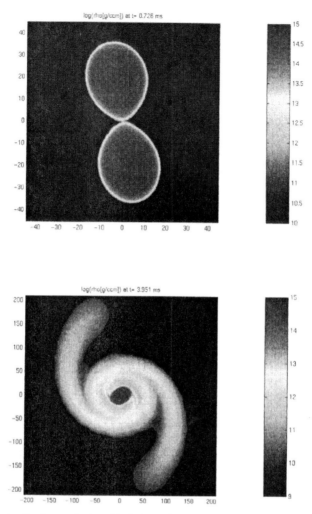

FIGURE 1. Density distribution in the orbital plane of a corotating system with twice 1.6 $M_\odot$, lengths are given in km.

means of the strong emerging neutrino wind). If folded with the estimated merger rate [22, 23, 24, 25] one finds that these amounts could substantially enrich the universe with heavy elements.

To investigate the nature of the ejected material Freiburghaus et al. [26] performed fully dynamical r-process calculations (including fission cycling) along the ejecta-trajetory of a blob of matter located in one of the rapidly expanding spiral arms (from an initially corotating binary system). Initially all beta-decays are Pauli-blocked and thus the material cools adiabatically during the rapid expansion. Along the expansion trajectory of the ejected matter blob all relevant nuclear reactions including beta-decays have been

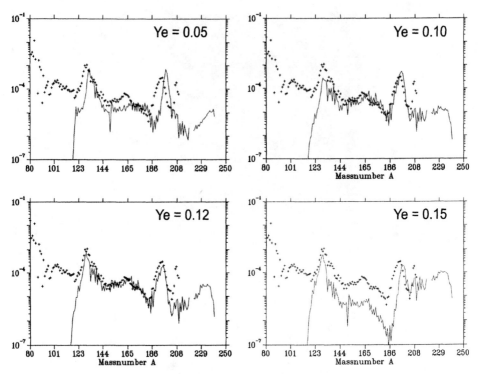

FIGURE 2. Resulting abundances as a function of Y_e, the solid line indicates the results of the calculations. The drop below $A \approx 130$ is caused by fission cycling.

followed [26]. The temperature history within the ejecta is determined by the interplay of cooling due to the rapid expansion and the heat input from the decaying nuclei. To ensure that results are not influenced by possibly artificially high numerical temperatures, we also performed test calculations starting exclusively with neutrons and protons at an artificially low temperature ($T = 10^{-2}$ MeV). In this case the almost immediate recombination of the nucleons into α-particles heated up the material to r-process-like temperatures on extremely short time scales [26].

In all investigated cases r-process took place, and the initial electron fraction of the material, Y_e, turned out to be the decisive parameter that determined the resulting abundance distribution. Due to restrictions in the EOS, the limited numerical resolution and the lack of a detailed neutrino transport the electron fraction Y_e was treated as a free parameter. Y_e was varied in a range characteristic for neutron star material, $0.05 < Y_e < 0.2$. The main result is that *if* Y_e is close to 0.1, the solar r-process pattern for nucleon numbers $A > 130$ is excellently reproduced, lighter nuclei are (due to fission cycling) strongly underabundant in comparison with their solar values.

DISCUSSION

Since this scenario involves so many different physical processes that rely on extrapolations of terrestrial physics several question marks remain for all results obtained so far (this, however, is equally true for supernova calculations, which still refuse to explode in most simulations). For example the equation of state of the high density part of a neutron star is still only poorly known. However, its seems that the stiffness of the EOS in this regime determines the amount of ejected material [19]. For soft equations of state we could not resolve any mass loss, for the stiff LS-EOS the event ejects perhaps (depending on the real merger rate) even more material than needed in order to explain the observed r-process abundances (which, however may be "corrected" by the stronger relativistic gravity forces, see below). Recent test calculations with the EOS of Shen et el. [27] support the ejecta results found with the LS-EOS.

As mentioned above Y_e is not determined self-consistently from the current calculations due to the lack of numerical resolution and neutrino transport. Here further investigations are needed. The probably weakest approximation of the current approach is the Newtonian treatment of gravity. It is to be expected that that the stronger general relativistic forces may reduce the amount of ejected material. In the light of most recent estimates concerning the merger rates [25] even an amount as small as a few times 10^{-5} $M_\odot$ per event might be of interest for the enrichment of the universe with heavy elements. This point will have to be clarified in future calculations including a sophisticated treatment of gravity *and* the relevant microphysical input.

It is a very interesting question whether the constraints coming from the observation of metal-poor (i.e. low iron abundance) stars can be met by the neutron star merger scenario. Recently it has been argued on grounds of elemental abundance ratios in metal-poor stars against the possibility of NSMs as sources of r-process material. However, we doubt, for several reasons, the soundness of this line of reasoning. For a further discussion of this point we refer to the literature [28, 29].

Observations [30, 31] show that the emergence of r-process material is slightly delayed with respect to iron. This could be explained either in terms of only 7-8 $M_\odot$ producing r-process material or -perhaps more naturally- due to the inspiral of a neutron star binary. The large amounts ejected per event would lead to some kind of "clumpyness" in the early distribution of r-process material. This is actually observed [32, 33].

There is growing evidence from observation (for the star CS 22892-052, [34, 32, 35]; for HD 115444 [36]; three giants in M15, [37]) for a unique r-process pattern beyond Ba whose relative abundances agree excellently with scaled solar system abundances. Lighter nuclei show deviations from the (scaled) solar values. Since the same high mass pattern is observed in stars of very different ages the production scenario has to produce the same relative abundances every time independent of the initial metallicity. This behaviour points to a scenario with a rather restricted parameters. This would find a natural explanation in terms of the neutron star binary scenario since the observed neutron star masses are restricted to a very narrow range [38] and one can expect the systems to be close to irrotational systems, where stellar spins are negligible in comparison with the orbital angular momentum [39]. In summary, observations seem to point to a low frequency event undergoing fission cycling as the source of the high-mass r-process that produces an abundance pattern independent of the age of the Galaxy. The merger of two neutron stars suggests itself as a strong candidate.

REFERENCES

1. Cameron, A. G. W., *ApJ*, **121**, 144 (1955).
2. Cameron, A. G. W., *Chalk River Rept.*, **CRL-41** (1957).
3. Burbidge, G., Burbidge, R., Fowler, W., and Hoyle, F., *Rev. Mod. Phys.*, **29**, 547 (1957).
4. Woosley, S. E., and Hoffman, R. D., *ApJ*, **395**, 202 (1992).
5. Meyer, B. S., Mathews, G. J., Howard, W. M., Woosley, S. E., and Hoffman, R. D., *ApJ*, **399**, 656 (1992).
6. Howard, W. M., Goriely, S., Rayet, M., and Arnould, M., *ApJ*, **417**, 713 (1993).
7. Takahashi, K., Witti, J., and Janka, H.-T., *A&A*, **286**, 857 (1994).
8. Freiburghaus, C., Rembges, J., Rauscher, T., Kolbe, E., Thielemann, F.-K., Kratz, K.-L., and Cowan, J., *ApJ*, **516**, 381 (1999).
9. McLaughlin, G., Fetter, J., Balantekin, A., and Fuller, G., *preprint*, **astro-ph/9902106** (1999).
10. Wanajo, S., Kajino, T., Mathews, G., and Otsuki, K., *astro-ph/0102261* (2001).
11. Lattimer, J. M., and Schramm, D. N., *ApJ, (Letters)*, **192**, L145 (1974).
12. Lattimer, J. M., and Schramm, D. N., *ApJ*, **210**, 549 (1976).
13. Meyer, B. S., *ApJ*, **343**, 254 (1989).
14. Eichler, D., Livio, M., Piran, T., and Schramm, D. N., *Nature*, **340**, 126 (1989).
15. Shibata, M., Baumgarte, T., and Shapiro, S., *Phys. Rev. D*, **61**, 064001 (2000).
16. Ruffert, M., Janka, H., and Schäfer, G., *A & A*, **311**, 532 (1996).
17. Ruffert, M., Janka, H., Takahashi, K., and Schäfer, G., *A & A*, **319**, 122 (1997).
18. Rosswog, S., Liebendörfer, M., Thielemann, F.-K., Davies, M., Benz, W., and Piran, T., *A & A*, **341**, 499–526 (1999).
19. Rosswog, S., Davies, M., Thielemann, F.-K., and Piran, T., *A & A*, **360**, 171 (2000).
20. Lattimer, J. M., and Swesty, F. D., *Nucl. Phys.*, **A535**, 331 (1991).
21. Benz, W., Bowers, R., Cameron, A., and Press, W., *ApJ*, **348**, 647 (1990).
22. Narayan, R., Piran, T., and Shemi, A., *ApJ*, **379**, L17 (1991).
23. van den Heuvel, E., and Lorimer, D., *MNRAS*, **283**, L37 (1996).
24. Fryer, C. L., Woosley, S. E., and Hartmann, D. H., *ApJ*, **526**, 152–177 (1999).
25. Kalogera, V., and Belczynski, K., *astro-ph/0101047* (2001).
26. Freiburghaus, C., Rosswog, S., and Thielemann, F.-K., *ApJ*, **525**, L121 (1999).
27. Shen, H., Toki, H., Oyamatsu, K., and Sumiyoshi, K., *Nuclear Physics*, **A 637**, 435 (1998).
28. Qian, Y.-Z., *ApJ*, **536**, L67 (2000).
29. Rosswog, S., Freiburghaus, C., and Thielemann, F.-K., "Nucleosynthesis Calculations for the Ejecta of Neutron Star Mergers", in *Proceedings of Nulcei in the Cosmos VI*, 2001.
30. Mathews, G. J., Bazan, G., and Cowan, J., *ApJ*, **391**, 719 (1992).
31. MCWilliam, A., *ARA&A*, **35**, 503–556 (1997).
32. Sneden, C., McWilliam, A., Preston, G. W., Cowan, J. J., Burris, D. I., and Armosky, B. J., *ApJ*, **467**, 819 (1996).
33. Cowan, J. J., McWilliam, A., Sneden, C., and Burris, D. L., *ApJ*, **480**, 246 (1997).
34. Cowan, J. J., Burris, D., Sneden, C., McWilliam, A., and Preston, G., *ApJ*, **439**, L51 (1995).
35. Sneden, C., Cowan, J., Ivans, I., Fuller, G., Burles, S., Beers, T., and Lawler, J., *ApJ*, **533**, L139 (2000).
36. Westin, J., Sneden, C., Gustafsson, B., and Cowan, J., *ApJ*, **530**, L783 (2000).
37. Sneden, C., Johnson, J., Kraft, R., Smith, G., Cowan, J., and Bolte, M., *ApJ*, **536**, L85 (2000).
38. Thorsett, S., and Chakrabarti, *ApJ*, **7512**, 288 (1999).
39. Bildsten, L., and Cutler, C., *ApJ*, **400**, 175 (1992).

Neutrino Heating in an Inhomogeneous Big Bang Nucleosynthesis Model

Juan F. Lara*

*Center for Relativity, University of Texas at Austin
Austin, Texas 78712

Abstract.
The effect of the heating of neutrinos by scattering with electrons and positrons and by e^-e^+ annihilation on nucleosynthesis is calculated for a spherically symmetric baryon inhomogeneous model of the universe. The model has a high baryon density core and a low density outer region. The heating effect is calculated by solving the Boltzmann Transport Equation for the distribution functions of electron and muon/tau neutrinos. For a range of baryon-to-photon ratio $\ln(\eta_{10}) = [0, 1.5]$ and $r_i = [10^2, 10^8]$ cm the heating effect increases the mass fraction $X_{^4He}$ by a range of $\Delta X_{^4He} = [1,2] \times 10^{-4}$. The change of the value of $X_{^4He}$ appears similiar to the change caused by an upward shift in the value of η_{10}. But the change to deuterium is a decrease in abundance ratio $Y(d)/Y(p)$ on the order of 10^{-3}, one order less than the decrease due to a shift in η_{10}.

I INTRODUCTION

When discrepencies arise between observations of light isotope abundances and the predictions of standard (homogeneous and isotropic) Big Bang Nucleosynthesis (BBN) models research has turned to BBN models with inhomogeneous baryon distributions (IBBN models) to attempt to resolve the discrepencies. Articles in the 1970's and 1980's looked for IBBN models with an overall baryon density that both equalled the critical density and satifised observational constraints on the light elements. [1] These IBBN models could not satisfy observational constraints to isotope abundances for baryon densities other than densities demanded by standard models. In the late 1990's observations of ^{4}He and deuterium placed conflicting constraints on the baryon density in standard models. (KKS, 1999) [2] found a range of agreement for the outer fringes of the 2σ ranges of the observations using an IBBN model code. Most recently (KS, 2000) [3] used an IBBN model to bring baryon density constraints from cosmic microwave background observations in agreement with constraints from ^{4}He and deuterium observations, though not with ^{7}Li observations.

Nucleosynthesis models have become more descriptive as they include smaller effects that faster computers can calculate accurately. One effect is the neutrino heating effect. Electrons and positrons pass a fraction of their energy to neutrinos through annihilation and scattering. That energy transfer can change the rates of neutron-proton interconversion, and then the results of nucleosynthesis. Hannestad and Madsen [4] derive a means of solving the Boltzmann Transport Equation for the neutrino distribution functions, presenting results for a standard BBN model. This author will discuss the neutrino heating effect in an IBBN model. This article will show how the heating effect slightly alters neutron distribution and nucleosynthesis, and show the effect's dependence on IBBN parameters.

II THE MODEL

The IBBN code for this article corresponds to a model with a spherically symmetric baryon distribution. The model is divided into a core and 31 inner shells with high baryon density, and 32 outer shells with low baryon density. A run starts at electromagnetic plasma temperature $T = 100$ GK. The distance scale r_i, the radius of the model when $T = 100$ GK, can be varied. During a run the number density $n(i,s)$ of isotope species i in shell s is determined by the equation [5]

$$\frac{dn(i,s)}{dt} = \frac{1}{n_b(s)} \sum_{j,k,l} N_i \left(-\frac{n^{N_i}(i,s) n^{N_j}(j,s)}{N_i! N_j!}[ij] + \frac{n^{N_k}(k,s) n^{N_l}(l,s)}{N_k! N_l!}[kl] \right)$$
$$-3\dot{a} n(i,s) + \frac{1}{r^2} \frac{\partial}{\partial r}\left(r^2 D_n \frac{\partial n(i,s)}{\partial r}\right) \quad (1)$$

The first term corresponds to nuclear reactions and beta decays within shell s, the second term to the expansion of the universe, and the third term to diffusion of isotope i between shells. In this model only neutrons diffuse. As T falls neutrons diffuse from the high density shells to the low density shells, until the neutrons are homogeneously distributed. The weak reactions $n + \nu_e \leftrightarrow p + e^-$, $n + e^+ \leftrightarrow p + \bar{\nu}_e$, and $n \leftrightarrow p + e^- + \bar{\nu}_e$ convert protons in the high density shells to neutrons, which then diffuse into the low density shells where the weak reactions convert them back to protons. Nucleosynthesis occurs earlier in the high density shells, depleting neutrons. Neutrons from the low density shells then back diffuse into the high density shells until nucleosynthesis incorporates all neutrons into nuclei, mostly ^{4}He nuclei.

Figure (1a) shows the mass fraction $X_{^4\text{He}}$ of ^{4}He nuclei produced by the model, as a function of η and r_i (in centimeters). Contour lines correspond to specific values of $X_{^4\text{He}}$. Neutron diffusion occurs later as r_i increases. The shape of the contour lines is determined by when neutron diffusion occurs compared with the weak reaction rates mentioned above and with nucleosynthesis. As r_i increases the time of neutron diffusion coincides less and less with the time before the weak reactions fall out of equilibrium. More protons then remain in the high density

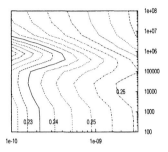
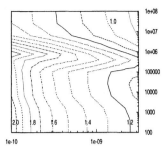

FIGURE 1. $X_{^4He}$ (1a) and $10^4 \Delta X_{^4He}$ (1b) due to neutrino heating.

shells. When $r_i = 25000$ cm the innermost shells retain all their protons, and nucleosynthesis occurs earlier enough that nearly all neutrons back diffuse into the high density shells to undergo nucleosynthesis. $X_{^4He}$ increases as r_i increases to 25000 cm. But for larger r_i neutrons cannot reach the innermost shells before nucleosynthesis, leading to decreasing production of ^{4}He. ^{4}He production bottoms out at $r_i = 790000$ cm when diffusion occurs at the same time as nucleosynthesis. For higher $r_i =$ the high density shells and the low density shells act separately from each other, with the high density shells producing large amounts of ^{4}He.

To calculate the neutrino heating effect, the model solves the Boltzmann Transport Equation [4]

$$\frac{df_1(p_1)}{dt} = \frac{1}{2E_1}\left(\frac{kT_N}{R}\right)^5 \int \frac{d^3p_2}{(2\pi)^3 2E_2} \int \frac{d^3p_3}{(2\pi)^3 2E_3}$$
$$\int \frac{d^3p_4}{(2\pi)^3 2E_4} S|M|^2 (2\pi)^4 \delta^4(p_1 + p_2 - p_3 - p_4)$$
$$\{[1 - f_1(p_1)][1 - f_2(p_2)]f_3(p_3)f_4(p_4)$$
$$- f_1(p_1)f_2(p_2)[1 - f_3(p_3)][1 - f_4(p_4)]\} \quad (2)$$

for the neutrino distribution functions $f_i(p_i)$. The distribution functions are used to calculate the neutrino energy densities, and $f_{\nu_e}(p_e)$ for electron neutrinos is used to calculate the neutron-proton conversion reaction rates. Increased $f_{\nu_e}(p_e)$ makes the conversion reactions produce fewer neutrons for nucleosynthesis, but decreased electron energy density ρ_e makes those reactions produce more neutrons at the same time. Decreased ρ_e decreases temperature T, making nucleosynthesis occur earlier, when more neutrons are present. This last effect, the clock effect, ultimately determines the increase in $X_{^4He}$ due to neutrino heating.

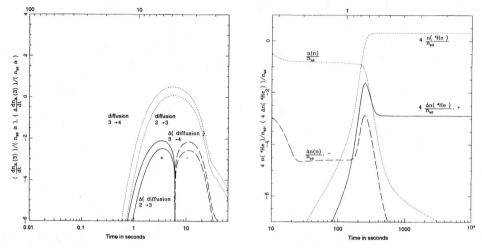

FIGURE 2. In Figure (2a) the dotted lines are the diffusion rates to and from shell 3, and the solid (increase) and dashed (decrease) lines are the changes of the rates due to neutrino heating. Figure (2b) shows $n(n)$ and $n(^4\text{He})$ (dotted lines) and the changes in these number densities due to heating. n_{bO} is a normalizing factor and $\dot\alpha$ is the expansion rate of the universe.

III RESULTS

The diffusion coefficient D_n is calculated from collisions between neutrons and electrons/positrons (D_{ne}) and neutrons and protons (D_{np}). [6]

$$\frac{1}{D_n} = \frac{1}{D_{ne}} + \frac{1}{D_{np}} \tag{3}$$

$$D_{ne} = \frac{3}{8}\sqrt{\frac{\pi}{2}} \frac{c}{n_e \sigma_{ne}} \frac{K_2(z)}{\sqrt{z} K_{5/2}(z)} \left(1 - \frac{n_n}{n_t}\right) \tag{4}$$

$$z = \frac{m_e}{kT}$$

Neutrino heating affects D_{ne} by lowering the electron number density n_e and lowering the electromagnetic plasma temperature T at a given time. The neutrinos and the electrons/positrons are homogeneously distributed throughout this model. So D_{ne} is the same for all the shells.

Figure (2a) shows the neutron diffusion rates into and out of Shell 3, a high density inner shell, for the case of $\eta_{10} = 3.0$ and $r_i = 25000$ cm. Figure (2a) also shows the change in the rates due to heating. This change is at first positive, and then becomes negative. This change corresponds to a shift in the diffusion rate to a time earlier by 0.1 %. All other shells also have this earlier time shift. Figure (2b) shows the number densities of neutrons (n_n) and ^{4}He nuclei ($n_{^4\text{He}}$) at around $T = 0.8$ GK, when nucleosynthesis occurs, for shell 3 and the same values of η_{10}

and r_i. Figure (2b) also shows the changes in n_n and $n_{^4He}$ due to neutrino heating. The clock effect mentioned above is seen in the humps in the graphs for the changes Δn_n and $\Delta n_{^4He}$. Δn_n and $\Delta n_{^4He}$ are larger between the time when neutrons are converted to ^{4}He nuclei with the heating effect on and when neutrons are converted to ^{4}He nuclei without heating.

Figure (1b) shows $10^4 \Delta X_{^4He}$, the change due to neutrino heating. For the ranges of η and r_i observed $\Delta X_{^4He}$ remains within a range of $[1.1, 2.0] \times 10^{-4}$. $X_{^4He}$ is determined by the baryon number densities in the shells, and those densities also determine the magnitude of the clock effect. That seems to account for the contour lines of $\Delta X_{^4He}$ tracking the contour lines of $X_{^4He}$ itself. The heating effect on $X_{^4He}$ seems to be similar to the effect one would get by shifting the value of η upwards by about 1 to 2 %. The earlier shift in neutron diffusion due to heating would have the effect of the lines in Figure (2a) being stretched to higher values of r_i. So for a given value of r_i and η the change $\Delta X_{^4He}$ will have a value corresponding to $X_{^4He}$ at that point in the graph plus a shift corresponding to a lower value of r_i. The change is bigger for $r_i \leq 25000$ cm, when $X_{^4He}$ decreases with decreasing r_i, and is smaller for 25000 cm $< r_i \leq 790000$ cm when $X_{^4He}$ increases with decreasing r_i.

Figure (3a) shows the overall abundance ratio $Y(d)/Y(p)$ between deuterium and protons. Figure (3b) shows $10^3 \Delta \log[Y(d)/Y(p)]$, the change due to neutrino heating. The nuclear reaction rates that create and destroy deuterium depend on the electromagnetic plasma temperature T and on shell s's baryon energy density $\rho_b(s)$. Through T neutrino heating shifts the nuclear reaction rates to a slightly earlier time. So for $r_i < 10^5$ cm less deuterium remains at the end of a run. This decrease in $Y(d)/Y(p)$ due to heating turns out to be one order of magnitude less than the decrease that would come from an increase in η_{10} mentioned above. For $\eta_{10} = 3.0$ and $r_i = 25000$ cm one has to increase η by 1.53 % to increase $X_{^4He}$ by as much as neutrino heating does. But that upward shift would decrease $\log[Y(d)/Y(p)]$ by around 9.19×10^{-3}, instead of the decrease of 1.12×10^{-3} done by heating. For $r_i > 10^5$ cm the low density outer shells produce considerable amounts of deuterium. In that range the heating effect increases deuterium production. The neutrino heating effect on ^{4}He can then be distinguished from a shift in η_{10} by looking at how $Y(d)/Y(p)$ differs from the results without heating.

IV CONCLUSIONS AND FUTURE RESEARCH

The neutrino heating effect increases the mass fraction of ^{4}He by $[1.1, 2.0] \times 10^{-4}$ for the observed range of baryon-to-photon ratio $\eta_{10} = [1, 32]$ and distance scale $r_i = [10^2, 10^8]$ cm. That's the same order of magnitude as in the standard case. The increase in the mass fraction resembles the increase that would come from increasing η_{10} by about 1 % to 2 %. The heating effect also decreases deuterium production. But the decrease in deuterium is an order of magnitude less than the decrease that would come from the increase of η_{10} mentioned above. Further, the effect increases deuterium production for $r_i \geq 10^5$ cm.

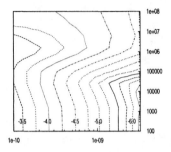 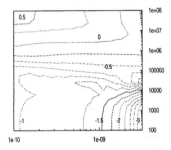

FIGURE 3. log[Y(d)/Y(p)] (3a) and $10^3 \Delta \log[Y(d)/Y(p)]$ (3b) due to neutrino heating.

A larger future article can show the earlier time shift of neutron diffusion for values of η_{10} and r_i other than 3.0 and 25000 cm, and for shells other than shell 3 as well. These graphs would show how this time shift is characteristic of the whole observed range of η_{10} and r_i. The article can also look at deuterium production for r_i lower than and higher than 10^5 cm, to explain in more detail how the change due to neutrino heating changes signs. The article can also comment on production of other isotopes, such as ^{7}Li. One can determine if the change due to heating has a distance scale dependence similar to the dependence for deuterium.

One should note that CMB observations have placed parameters on the baryon-to-photon ratio ($\Omega_B h^2 = 0.032^{+0.009}_{-0.008}$ where $\Omega_B h^2 = (3.650 \pm 0.008) \times 10^{-3} \eta_{10}$) [7]. A future article should particularly look at the IBBN code's behavior in regions corresponding to these parameters.

REFERENCES

1. Malaney, R.A., and Mathews, G.J., *Phys. Repts* **229**, 147 (1993) has an exhaustive list of such articles.
2. Kainulainen, K., Kurki-Suonio, H., and Sihvola, E., *Phys. Rev.* **D59**, 083505 (1999).
3. Kurki-Suonio, H., and Sihvola, E., astro-ph/0011544.
4. Hannestad, S., and Madsen, J., *Phys. Rev.* **D52**, 1764 (1995).
5. Mathews, G.J., Meyer, B.S., Alcock, C.R., and Fuller, G.M., *ApJ* **358**, 36 (1990).
6. Kurki-Suonio, H., Aufderheide, M.B., Graziani, F., Mathews, G.J., Banerjee, B., Chitre, S.M., and Schramm, D.N., *Phys. Let.* **B289**, 211 (1992).
7. Burles, S., Nollet, K.M., and Turner, M.S., astro-ph/0008495.

Primordial Nucleosynthesis with Hadron Injection from Low-mass Primordial Black Holes

Jun'ichi Yokoyama* and K. Kohri[†]

Department of Earth and Space Science, Osaka University, Toyonaka, 560-0043, Japan
[†]*Yukawa Institute for Theoretical Physics, Kyoto University, Kyoto 606-8502, Japan*

Abstract. We investigate the influence of hadron injection from evaporating primordial black holes (PBHs) in the early stage of the primordial nucleosynthesis era ($t \simeq 10^{-3} - 10^4$ sec). The emitted quark-antiquark pairs or gluons immediately fragment into a lot of hadrons and scatter off the thermal plasma which is constituted by photons, electrons and nucleons. For the relatively low mass holes we point out that the dominant effect is the inter-conversion between ambient proton and neutron through the strong interaction induced by the emitted hadrons. Even after the freeze-out time of the week interactions between neutron and proton, more neutrons are produced and the synthesized light element abundances could be drastically changed. Comparing the theoretical predictions with the observational data, we constrain the PBH's density and their lifetime. We obtain the upper bound for PBH's initial mass fraction, $\beta \lesssim 10^{-20}$ for $10^8 \text{g} \lesssim M \lesssim 10^{10} \text{g}$, and $\beta \lesssim 10^{-22}$ for $10^{10} \text{g} \lesssim M \lesssim 3 \times 10^{10} \text{g}$.

INTRODUCTION

Primordial black holes (PBHs) are formed in the hot early Universe if overdensity of order of unity exists and a perturbed region enters the Hubble radius [1]. They serve as a unique probe of primordial density fluctuations on small scales. If their existence is obervationally confirmed, we can determine model parameters of the inflation model with high accuracy [2]. Even the nonexistence of PBHs over some mass ranges, however, would provide useful informations on the primordial spectrum of density fluctuations. In this sense it is very important to obtain accurate constraints on the abundance of PBHs on each mass scale. High energy particles emitted from evaporating black holes [3, 4] induce a number of interesting processes in the cosmic evolution, and various constraints have been imposed from primordial big-bang nucleosynthesis (BBN) [5, 6, 7, 8, 9], microwave background radiation [10], and gamma-ray background radiation [11].

Here we reanalyze the effects of evaporating primordial black holes on BBN in order to improve constraints on their mass spectrum. Contrary to the previous papers, we incorporate the modern view of high energy physics of the energy scale relevant to PBHs evaporating in the nucleosynthesis era. That is, hadrons are not emitted in the form of nucleons or mesons but as a quark-gluon jet according to the quantum chromodynamics (QCD). Thus we should adopt the Elementary Particle Picture of Carr [4] and assume that elementary particles in the standard model are emitted from PBHs and they generate jets. We adopt a simple and conventional view that PBHs are produced with a single mass, M, when the horizon mass is equal to M, namely at $t = 2.4 \times 10^{-29} M_{10}$ sec $\equiv t_{form}$, and obtain an improved constraints on their initial abundance at each mass scale.

Once quark-antiquark pairs or gluons are emitted from a PBH, a lot of hadrons, *e.g.* pions, kaons and nucleons (protons and neutrons) are produced through their hadronic fragmentation. They inter-convert the ambient protons and neutrons each other through the strong interaction. If the inter-conversion rate between neutrons and protons becomes large again after the freeze-out time of the weak interactions in the standard BBN (SBBN) or $t \simeq 1$ sec, the protons which are more abundant than the neutrons at that time are changed into neutrons. That is, this results in an excess of neutrons compared with SBBN. Therefore, the hadron injection significantly influences the freeze-out value of the neutron-to-proton ratio and the final abundances of ^{4}He, D and ^{7}Li are drastically changed. In this case it is expected that both ^{4}He and D tend to become more abundant than in SBBN. If PBHs are so massive that they continue to emit particles even after ^{4}He are produced, then spallation of ^{4}He due to high energy particles will tend to increase the final D. Here we concentrate on the low-mass PBHs and consider only the former effects.

Reno and Seckel [12] investigated the detail of the physical mechanism and the influences of the hadron injection from long-lived massive decaying particles on BBN. They constrained the parent particle's lifetime and the number density comparing the theoretical prediction of the light element abundances with the observational data. Here we basically follow their treatment and apply it to the hadron injection originated in the PBH evaporation.

EVAPORATION AND JETS

First we briefly summarize basic results of PBH evaporation. The temperature and the lifetime of evaporating black hole are given by [3]

$$T_{\rm BH} = 1.06 M_{10}^{-1} {\rm TeV}, \quad \tau_{\rm BH} = 435 M_{10}^3 \, {\rm sec}, \tag{1}$$

respectively, where M_{10} is the black hole mass normalized by 10^{10}g and we have assumed the particle contents of the standard particle physics model. Thus a PBH with mass $M = 10^8 - 10^{10}$g evaporates in the nucleosynthesis era $t = 1 - 10^3$ sec. The average energy of emitted particles is given by $\overline{E} = 4.4 T_{\rm BH}$ and lies in the range $10^2 - 10^4$ GeV.

The emitted high-energy quarks and gluons fragment into further quarks and gluons until they cluster into the observable hadrons when they have traveled a distance $\Lambda_{\rm QCD}^{-1} \sim 10^{-13}$cm. The hadron jet thus produced would be similar to that produced in e^+e^- annihilation. The emission rate of the hadron jet is estimated by

$$J(t) = \frac{n_{\rm BH}(t)}{\overline{E}(t)} \frac{{\rm d}M(t)}{{\rm d}t}, \tag{2}$$

where $n_{\rm BH}(t)$ is the number density of the PBHs. Hadrons in these jets are thermalized before they interact with nucleons in most cases and we can use the thermally averaged cross sections $\langle \sigma v \rangle_{N \to N'}^{H_i}$ for the strong interaction between hadron H_i and the ambient nucleon N, where N denotes proton p or neutron n. For a hadron interaction process $N + H_i \to N' + \cdots$, the strong interaction rate is estimated by

$$\Gamma^{H_i}_{N \to N'} = n_N \langle \sigma v \rangle^{H_i}_{N \to N'}$$
$$\simeq 10^8 \sec^{-1} f_N \left(\frac{\eta_i}{10^{-9}}\right) \left(\frac{\langle \sigma v \rangle^{H_i}_{N \to N'}}{10\,\mathrm{mb}}\right) \left(\frac{T_\nu}{2\,\mathrm{MeV}}\right)^3, \tag{3}$$

where n_N is the number density of the nucleon species N, η_i is the initial baryon to photon ratio ($= n_B/n_\gamma$ at $T \gtrsim 10\,\mathrm{MeV}$), $n_B = n_p + n_n$ denotes the baryon number density, $f_N \equiv n_N/n_B$, and T_ν is the neutrino temperature.

Thus all we need to consider are particles with lifetime larger than $O(10^{-8})$ sec. The corresponding mesons are π^+, π^-, K^+, K^-, and K_L and the baryons are p, $\bar{p}$, n, and $\bar{n}$. We have therefore calculated the final yield of these particles per two jets, n^{H_i}, out of Table 38.1 of [13]. The results are:

$$n^{\pi^+} = n^{\pi^-} = 14.1, \quad n^{K^+} = n^{K^-} = 1.67, \quad n^{K_L} = 1.19, \quad n^p = n^{\bar{p}} = n^n = n^{\bar{n}} = 0.772.$$

For these particles, the cross section $\langle \sigma v \rangle^{H_i}_{N \to N'}$ lies in the range $1 - 40\,\mathrm{mb}$.

HADRON INJECTION AND BBN

As we mentioned in the previous section, since we concentrate on the effects of low-mass PBHs which evaporate before $^4\mathrm{He}$ formation, the extra conversion between n and p is the dominant effect. Then the time evolution equations for the number density of a nucleon $N(= p, n)$ is represented by

$$\frac{dn_N}{dt} + 3H(t)n_N = \left[\frac{dn_N}{dt}\right]_{SBBN} - B_h J(t) \sum_{N'} (K_{N \to N'} - K_{N' \to N}), \tag{4}$$

where $H(t)$ is the cosmic expansion rate, $[dn_N/dt]_{SBBN}$ denotes the contribution from the standard weak interaction rates [14] and nuclear interaction rates, B_h is the hadronic branching ratio, $J(t)$ denotes the emission rate of the hadron jet per unit time and $K_{N \to N'}$ denotes the average number of the transition $N \to N'$ per one hadron jet emission.

Though PBHs generally emit not only quarks and gluons but also all lighter particle species than the temperature of PBH, the emitted neutrinos, photons and the other electro-magnetic particles do not influence the light element abundances significantly for the relatively short lifetime. For $\tau_{\mathrm{BH}} \lesssim 10^4$ sec the injection of photon or the other electro-magnetic particles do not induce the photo-dissociation of the light elements. On the other hand, the emitted neutrinos scatter off the background neutrinos and produce the electron-positron pairs. Although they also induce the electro-magnetic cascade, we should not be worried about photo-dissociation for the same reason. Hence we concentrate on the effects of hadron injection in BBN epoch. Then the resultant upper bound on the abundance of PBHs turns out to be proportional to B_h^{-1}. Below we analyze the case $B_h = 1$, because its magnitude is not precisely known, although we expect

$B_h = O(0.1 - 1)$. Anyway the constraint for other cases with $B_h \neq 1$ can easily be obtained from the above scaling law.

The average number of the transition $N \to N'$ per jet is expressed by

$$K_{N \to N'} = \sum_{H_i} \frac{n^{H_i}}{2} R^{H_i}_{N \to N'}, \tag{5}$$

where H_i runs the hadron species which are relevant to the nucleon inter-converting reactions, and $R^{H_i}_{N \to N'}$ denotes the probability that a hadron species H_i induces the nucleon transition $N \to N'$. The transition probability is estimated by

$$R^{H_i}_{N \to N'} = \frac{\Gamma^{H_i}_{N \to N'}}{\Gamma^{H_i}_{dec} + \Gamma^{H_i}_{abs}}, \tag{6}$$

where $\Gamma^{H_i}_{dec} = \tau^{-1}_{H_i}$ is the decay rate of the hadron H_i, $\Gamma^{H_i}_{abs}$ is the total absorption rate of H_i. For K_L, which is not stopped, the decay rate is approximately estimated by $\Gamma^{K_L}_{dec} = m_{K_L}/E_{K_L} \tau^{-1}_{K_L}$ where E_{K_L} is the averaged energy of the emitted K_L.

OBSERVATIONAL CONSTRAINTS

Since the effects of the hadron emission tend to increase D and ^{4}He in our analysis, we can constrain mass and abundance of PBHs by comparing the yield with the observational upperbounds of the abundance of these light elements. In order to get the conservative bound for the hadron injection induced by PBH evaporation, we adopt the following mild observational bounds here,

$$Y_p \leq 0.252 \ (2\sigma), \tag{7}$$
$$D/H \leq 4.0 \times 10^{-5} \ (2\sigma) \quad \text{for LowD} \tag{8}$$
$$D/H \leq 5.7 \times 10^{-4} \ (2\sigma) \quad \text{for HighD}, \tag{9}$$
$$3.3 \times 10^{-11} \leq {}^7\text{Li/H} \leq 9.2 \times 10^{-10} \ (2\sigma), \tag{10}$$

where we summed all the errors in quadrature [15, 16, 17, 18]. We use two classes of the observed D values, Low D and High D, because it is still under discussion which value reflects the primordial one.

RESULTS

We now compare the theoretical predictions of the light-element abundances in the hadron injection scenario with the observational constraints. We have three free parameters, the baryon to photon ratio η, the PBH's lifetime τ_{BH}, and the initial number density of the PBH normalized by the entropy density, s, $Y_{BH} \equiv n_{BH}/s$. We can relate Y_{BH} to the

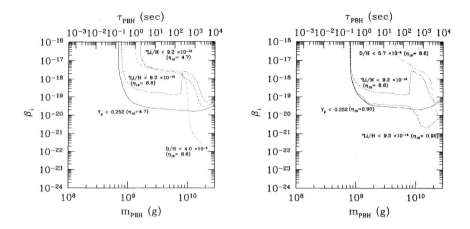

FIGURE 1. The most conservative upper bounds for ^{4}He (solid line), D (dashed line) and ^{7}Li (dotted line) as mild as possible in the parameter region $\eta_{10} = 4.7 - 8.6$, where $\eta_{10} \equiv \eta \times 10^{10}$. Left figure shows the "LowD" case and the right figure shows the "HighD" case.

initial mass fraction β by the following equation:

$$\beta = 5.4 \times 10^{21} \left(\frac{\tau_{BH}}{1 \text{ sec}} \right)^{\frac{1}{2}} Y_{BH}. \tag{11}$$

We start the BBN calculation at the cosmic temperature $T = 100$ MeV. Since η is the value at present time, the initial value η_i should be set to an appropriate value which turns out to the present η after the possible entropy production due to PBH evaporation and the photon heating due to e^+e^- annihilation.

Since the hadron injection tends to increase the produced D and ^{4}He abundances in the present situation, the parameter range of η is necessarily restricted to a narrow region if we adopt a set of the observational upper bounds for D and ^{4}He. For the "LowD" case, the constraints (7), (8) and (10) imply that the baryon-to-photon ratio is restricted to lie in the range $\eta = (4.7 - 8.6) \times 10^{-10}$, while for the "HighD" case it should satisfy $\eta = (0.90 - 8.6) \times 10^{-10}$. In Figures 1 we plot the most conservative bounds for each case for the PBH mass $M = 10^8 - 3 \times 10^{10}$g which corresponds to the lifetime $\tau_{BH} = 4.4 \times 10^{-4} - 10^4$ sec. The shorter lifetime, $\tau_{BH} \lesssim 10^{-2}$ sec, do not affect the freeze-out value of n/p and do not change any predictions of SBBN. For $\tau_{BH} \gtrsim 10^{-2}$ sec, the freeze-out value of n/p ratio is increased by the hadron-induced inter-converting interactions and the produced neutron increases the ^{4}He abundance because most of the free neutrons burned into ^{4}He through D. Then PBH abundance is strongly constrained by the upper bound of the observational ^{4}He abundance. For $\tau_{BH} \gtrsim 10^2$ sec, since the produced free D can no longer burn into ^{4}He, the extra free neutrons remain in D. Then β may also be constrained by the upper bound of the observational D/H depending on which observational value we adopt. On the other hand, ^{7}Li abundance traces D abundance for the longer lifetime $\tau_{BH} \gtrsim 10^2$ sec and is produced more than SBBN at the relatively high η ($\gtrsim 3 \times 10^{-10}$).

CONCLUSION

We have investigated the influence of hadron jets from evaporating PBHs in the early stage of the BBN epoch ($t \simeq 10^{-3} - 10^4$ sec). We obtain the following upper bounds for the initial mass fraction of PBH β as a function of the initial mass of PBH when they are formed.

$$\beta \lesssim 10^{-20} \quad \text{(for } 10^8 \text{g} \lesssim M \lesssim 10^{10} \text{g)} \tag{12}$$

$$\beta \lesssim 10^{-22} \quad \text{(for } 10^{10} \text{g} \lesssim M \lesssim 3 \times 10^{10} \text{g)}. \tag{13}$$

Here we adopt the case of "lowD" because the produced D is more severely constrained, and we take the most conservative bounds which are independent of the baryon to photon ratio [19].

This work was partially supported by the Monbukagakusho Grant-in-Aid for Scientific Research Nos. 11740146 (JY) and 10-04502(KK).

REFERENCES

1. Ya.B. Zel'dovich and I.D. Novikov, Soviet Astronomy **10**, 602 (1967); S.W. Hawking, Mon. Not. R. astr. Soc. **152**, 75 (1971); B.J. Carr, Astrophys. J. **201**, 1 (1975).
2. J. Yokoyama, Astron. Astrophys. **318**, 673 (1997); Phys. Rev. D **58**, 083510 (1998); Phys. Rep. **307**, 133 (1998).
3. S.W. Hawking, Nature **248**, 30 (1974); Comm. Math. Phys. **43**, 199 (1975).
4. B.J. Carr, Astrophys. J. **206**, 8 (1976).
5. B.V. Vainer and P.D. Nasel'skii, Astron. Zh. **55**, 231 (1978) [Sov. Astron. **22**, 138 (1978).]
6. S. Miyama and K. Sato, Prog. Theor. Phys. **55**, 1012 (1978).
7. Ya.B. Zel'dovich, A.A. Starobinskii, M.Yu. Khlopov, and V.M. Chechëtkin, Pis'ma Astron. Zh. **3**, 208 (1977) [Sov. Astron. Lett. **3**, 110 (1977).]
8. B.V. Vainer, O.V. Dryzhakova, and P.D. Nasel'skii, Pis'ma Astron. Zh. **4**, (1978) [Sov. Astron. Lett. **4**, 185 (1978).]
9. D. Lindley, Mon. Not. Roy. Astron. Soc. **193**, 593 (1980).
10. Ya.B. Zel'dovich, A.A. Starobinskii, JETP Lett. **24**, 571 (1976).
11. J.H. MacGibbon, Nature **329**, 308 (1987); J.H. MacGibbon and B.J. Carr, Astrophys. J. **371**, 447 (1991).
12. M. H. Reno and D. Seckel, Phys. Rev. **D37**, 3441 (1988)
13. C. Caso *et al.*, Particle Data Group, Euro. Phys. J. *C* **3**, 1 (1998).
14. For a review of BBN, see, *e.g.* K.A. Olive, G. Steigman, and T.P. Walker, astro-ph/9905320
15. B.D. Fields and K.A. Olive, Astrophys. J., **506**, 177 (1998).
16. S. Burles and D. Tytler, Astrophys. J. **507**, 732 (1998).
17. D. Tytler, S. Burles, L. Lu, X-M. Fan and A. Wolfe, Astron. J. **117**, 63 (1999).
18. P. Bonifacio and P. Molaro, Mon. Not. Roy. Astron. Soc. **285**, 847 (1997).
19. K. Kohri and J. Yokoyama, Phys. Rev. **D61**, 023501 (1999).

CHAPTER 8

SUPERMASSIVE BLACK HOLES AND GALAXY EVOLUTION

Supermassive Black Holes in Galactic Nuclei

John Kormendy* and Karl Gebhardt*

Department of Astronomy, RLM 15.308, University of Texas, Austin, TX 78712

Abstract.
We review the motivation and search for supermassive black holes (BHs) in galaxies. Energetic nuclear activity provides indirect but compelling evidence for BH engines. Ground-based dynamical searches for central dark objects are reviewed in Kormendy & Richstone (1995, ARA&A, 33, 581). Here we provide an update of results from the *Hubble Space Telescope* (HST). This has greatly accelerated the detection rate. As of 2001 March, dynamical BH detections are available for at least 37 galaxies.

The demographics of these objects lead to the following conclusions: (1) BH mass correlates with the luminosity of the bulge component of the host galaxy, albeit with considerable scatter. The median BH mass fraction is 0.13 % of the mass of the bulge. (2) BH mass correlates with the mean velocity dispersion of the bulge inside its effective radius, i.e., with how strongly the bulge stars are gravitationally bound to each other. For the best mass determinations, the scatter is consistent with the measurement errors. (3) BH mass correlates with the luminosity of the high-density central component in disk galaxies independent of whether this is a real bulge (a mini-elliptical, believed to form via a merger-induced dissipative collapse and starburst) or a "pseudobulge" (believed to form by inward transport of disk material). (4) BH mass does not correlate with the luminosity of galaxy disks. If pure disks contain BHs (and active nuclei imply that some do), then their masses are much smaller than 0.13 % of the mass of the disk.

We conclude that present observations show no dependence of BH mass on the details of whether BH feeding happens rapidly during a collapse or slowly via secular evolution of the disk. The above results increasingly support the hypothesis that the major events that form a bulge or elliptical galaxy and the main growth phases of its BH – when it shone like a quasar – were the same events.

MOTIVATION

Black holes (BHs) progressed from a theoretical concept to a necessary ingredient in extragalactic astronomy with the discovery of quasars by Schmidt (1963). Radio astronomy was a growth industry at the time; many radio sources were identified with well-known phenomena such as supernova explosions. But a few were identified only with "stars" whose optical spectra showed nothing more than broad emission lines at unfamiliar wavelengths. Schmidt discovered that one of these "quasi-stellar radio sources" or "quasars", 3C 273, had a redshift of 16 % of the speed of light. This was astonishing: the Hubble law of the expansion of the Universe implied that 3C 273 was one of the most distant objects known. But it was not faint. This meant that 3C 273 had to be enormously luminous – more luminous than any galaxy. Larger quasar redshifts soon followed. Explaining their energy output became the first strong argument for gravity power (Zel'dovich 1964; Salpeter 1964).

Studies of radio jets sharpened the argument. Many quasars and lower-power active galactic nuclei (AGNs) emit jets of elementary particles that are prominent in the radio and sometimes visible at optical wavelengths. Many are bisymmetric and feed lobes of emission at their ends (e.g., Fig. 1). Based on these, Lynden-Bell (1969, 1978) provided a convincing argument for gravity power. Suppose that we try to explain the typical quasar using nuclear fusion reactions, the most efficient power source that was commonly studied at the time. The total energy output of a quasar is at least the energy stored in its radio halo, $E \sim 10^{54}$ J. Via $E = mc^2$, this energy weighs 10^7 solar masses ($M_\odot$). But nuclear reactions have an efficiency of only 0.7 %. So the mass that was processed by the quasar in order to convert 10^7 $M_\odot$ into energy must have been 10^9 $M_\odot$. This waste mass became part of the quasar engine. Meanwhile, rapid brightness variations showed that quasars are tiny, with diameters $2R \lesssim 10^{13}$ m. But the gravitational potential energy of 10^9 $M_\odot$ compressed inside 10^{13} m is $GM^2/R \sim 10^{55}$ J. "Evidently, although our aim was to produce a model based on nuclear fuel, we have ended up with a model which has produced more than enough energy by gravitational contraction. The nuclear fuel has ended as an irrelevance" (Lynden-Bell 1978). This argument convinced many people that BHs are the most plausible quasar engines.

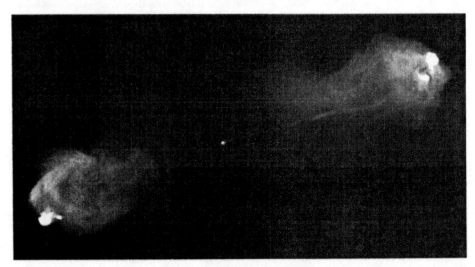

FIGURE 1. Cygnus A at 6 cm wavelength (Perley, Dreher, & Cowan 1984). The central point source is the galaxy nucleus; it feeds oppositely directed jets (only one of which is easily visible at the present contrast) and lobes of radio-emitting plasma. The resolution of this image is about 0.″4.

Jets also provide more qualitative arguments. Many are straight over $\sim 10^6$ pc in length. This argues against the most plausible alternative explanation for AGNs, namely bursts of supernova explosions. The fact that jet engines remember ejection directions for $\gtrsim 10^6$ yr is suggestive of gyroscopes such as rotating BHs. Finally, in many AGNs, jet knots are observed to move away from the center of the galaxy at apparent velocities of several times the speed of light, c. These can be understood if the jets are pointed almost at us and if the true velocities are almost as large as c (Blandford, McKee, & Rees 1977). Observations of superluminal motions provide the cleanest argument for relativistically deep potential wells.

By the early 1980s, this evidence had resulted in a well-established paradigm in which AGNs are powered by BHs accreting gas and stars (Rees 1984). Wound-up magnetic fields are thought to eject particles in jets along the rotation poles. Energy arguments imply masses $M_\bullet \sim 10^6$ to $10^{9.5}$ $M_\odot$, so we refer to these as supermassive BHs to distinguish them from ordinary-mass (several-$M_\odot$) BHs produced by the deaths of high-mass stars. But despite the popularity of the paradigm, there was no direct dynamical evidence for supermassive BHs. The black hole search therefore became a very hot subject. It was also dangerous, because it is easy to believe that we have proved what we expect to find. Standards of proof had to be very high.

THE SEARCH FOR SUPERMASSIVE BLACK HOLES

Kormendy & Richstone (1995) review BH search techniques and summarize the ground-based detections. Recent reviews (e. g., Richstone *et nuk.* 1998) concentrate on BH astrophysics. There has not been a comprehensive review of BH discoveries made with the *Hubble Space Telescope* (HST), so we provide a summary here.

There are ground-based BH detections in 10 galaxies, including the nearest (our Galaxy, M 31, and M 32) and the best (our Galaxy and NGC 4258) candidates. References are given in Kormendy & Richstone (1995) and Table 1. Of the 7 stellar-dynamical cases, 6 have been reobserved with HST. In all cases, the BH detection was confirmed and the ground-based BH mass agrees with the HST result to within a factor of ~ 2. The HST papers are: M 31: Statler *et al.* (1999), Bacon *et al.* (2001); M 32: van der Marel *et al.* (1998); NGC 3115: Kormendy *et nuk.* (1996a), Emsellem *et al.* (1999); NGC 3377: Richstone *et nuk.* (2001), NGC 4486B (Green *et al.* 2001), and NGC 4594: Kormendy *et nuk.* (1996b).

Our Galaxy is the strongest BH case, based on observations of velocities in the plane of the sky of stars in a cluster within $0\rlap{.}''5 = 0.02$ pc of the central radio source Sgr A* (Eckart & Genzel 1997; Genzel *et al.* 1997, 2000; Ghez *et al.* 1998, 2000). The fastest star is moving at 1350 ± 40 km s^{-1}. Acceleration vectors have been measured for three stars; they intersect, to within the still-large errors, at Sgr A*, supporting the identification of the radio source with the inferred central mass of $M_\bullet = (2.6 \pm 0.2) \times 10^6$ $M_\odot$ (Ghez *et al.* 2000). The stellar orbital periods could be as short as several decades, so we can look forward to seeing the Galactic center rotate in our lifetimes! Most important, the mass $M_\bullet$ is constrained to live inside such a small radius that alternatives to a supermassive black hole are ruled out by astrophysical constraints. Brown dwarf stars would collide, merge, and become luminous, and clusters of white dwarf stars, neutron stars, or stellar-mass black holes would evaporate too quickly (Maoz 1995, 1998; Genzel et al. 1997, 2000).

The next-best BH case is NGC 4258. In it, a water maser disk shows remarkably Keplerian rotation velocities inward to a radius of 0.2 pc (Miyoshi et al. 1995). The implied central mass, $M_\bullet = 4 \times 10^7 \, M_\odot$, again is confined to a small enough volume to exclude the above BH alternatives (Maoz 1998). Such arguments cannot yet be made for any other galaxy. Nevertheless, they increase our confidence that all of the dynamically detected central dark objects are BHs.

The BH search has now largely moved to HST. With the aberrated HST, BH work was based on indirect arguments that have serious problems (Kormendy & Richstone 1995). But with COSTAR, HST has become the telescope of choice for BH searches, and the pace of detections has accelerated remarkably.

The HST era is divided into two periods. Before the installation of the Space Telescope Imaging Spectrograph (STIS) in 1997, the main instrument used was the Faint Object Spectrograph (FOS). It was inefficient, because it used an aperture instead of a slit. Nevertheless, the first HST BH detections were made with the FOS (M 87: Harms et al. 1994; NGC 4261: Ferrarese et al. 1996; NGC 7052: van der Marel & van den Bosch 1998). It is often suggested that HST was required to make BH cases convincing. This is an exaggeration. HST beats ground-based resolution by a factor of 5, but the first BH detections made with HST were in Virgo cluster galaxies or in ones that are 2 – 4 times farther away. Virgo is ∼ 20 times farther away than M 31 and M 32. Therefore the ground-based BH discoveries in M 31 and M 32 had better spatial resolution (in pc) than the HST BH discoveries in the above galaxies. Of course, the distant BHs have higher masses. Therefore a better measure of relative resolution is the ratio of the radius $r_{\mathrm{cusp}} = GM_\bullet/\sigma^2$ of the BH sphere of influence to the resolution. Table 1 lists r_{cusp} for all BH detections. Since the PSF in the ground-based discovery observations had a radius of ∼ 0″.3 – 0″.5 while the FOS observations used a 0″.26 circular aperture (M 87 and NGC 7052) or a 0″.09 square aperture (NGC 4261), Table 1 shows that the FOS BH detections in M 87, NGC 4261, and NGC 7052 had comparable or lower relative resolution than the ground-based observations of M 31, M 32, NGC 3115, and NGC 4594. As HST spatial resolution improved (especially with STIS), BH cases have indeed gotten stronger. But the main thing that HST has provided is many more detections.

STIS has begun a new period in the BH search. With the efficiency of a long-slit spectrograph and CCD detector, the search has become feasible for most nearby galaxies that have unobscured centers and old stellar populations. It is still not easy; finding a 10^6-$M_\odot$ BH is difficult at the distance of the Virgo cluster and impossible much beyond. But the pace of discoveries has accelerated dramatically. At the 2000 Summer AAS meeting, 14 new BH detections were reported, and several more have been published since. As a result, about 37 BH candidates are now available. We say "about" because not all cases are equally strong: which ones to include is a matter of judgment. Table 1 provides a census.

TABLE 1
Census of Supermassive Black Holes (2001 March)

Galaxy	Type	$M_{B,\text{bulge}}$	$M_\bullet$ $(M_{\text{low}}, M_{\text{high}})$ $(M_\odot)$	σ_e (km/s)	D (Mpc)	r_{cusp} (arcsec)	Reference
Galaxy	Sbc	−17.65	2.6 (2.4−2.8) e6	75	0.008	51.40	See notes
M 31	Sb	−19.00	4.5 (2.0−8.5) e7	160	0.76	2.06	Dressler + 1988; Kormendy 1988a
M 32	E2	−15.83	3.9 (3.1−4.7) e6	75	0.81	0.76	Tonry 1984, 1987
M 81	Sb	−18.16	6.8 (5.5−7.5) e7	143	3.9	0.76	Bower + 2001b
NGC 821	E4	−20.41	3.9 (2.4−5.6) e7	209	24.1	0.03	Gebhardt + 2001
NGC 1023	S0	−18.40	4.4 (3.8−5.0) e7	205	11.4	0.08	Bower + 2001a
NGC 2778	E2	−18.59	1.3 (0.5−2.9) e7	175	22.9	0.02	Gebhardt + 2001
NGC 3115	S0	−20.21	1.0 (0.4−2.0) e9	230	9.7	1.73	Kormendy + 1992
NGC 3377	E5	−19.05	1.1 (0.6−2.5) e8	145	11.2	0.42	Kormendy + 1998
NGC 3379	E1	−19.94	1.0 (0.5−1.6) e8	206	10.6	0.20	Gebhardt + 2000a
NGC 3384	S0	−18.99	1.4 (1.0−1.9) e7	143	11.6	0.05	Gebhardt + 2001
NGC 3608	E2	−19.86	1.1 (0.8−2.5) e8	182	23.0	0.13	Gebhardt + 2001
NGC 4291	E2	−19.63	1.9 (0.8−3.2) e8	242	26.2	0.11	Gebhardt + 2001
NGC 4342	S0	−17.04	3.0 (2.0−4.7) e8	225	15.3	0.34	Cretton + 1999a
NGC 4473	E5	−19.89	0.8 (0.4−1.8) e8	190	15.7	0.13	Gebhardt + 2001
NGC 4486B	E1	−16.77	5.0 (0.2−9.9) e8	185	16.1	0.81	Kormendy + 1997
NGC 4564	E3	−18.92	5.7 (4.0−7.0) e7	162	15.0	0.13	Gebhardt + 2001
NGC 4594	Sa	−21.35	1.0 (0.3−2.0) e9	240	9.8	1.58	Kormendy + 1988b
NGC 4649	E1	−21.30	2.0 (1.0−2.5) e9	375	16.8	0.75	Gebhardt + 2001
NGC 4697	E4	−20.24	1.7 (1.4−1.9) e8	177	11.7	0.41	Gebhardt + 2001
NGC 4742	E4	−18.94	1.4 (0.9−1.8) e7	90	15.5	0.10	Kaiser + 2001
NGC 5845	E	−18.72	2.9 (0.2−4.6) e8	234	25.9	0.18	Gebhardt + 2001
NGC 7457	S0	−17.69	3.6 (2.5−4.5) e6	67	13.2	0.05	Gebhardt + 2001
NGC 2787	SB0	−17.28	4.1 (3.6−4.5) e7	185	7.5	0.14	Sarzi + 2001
NGC 3245	S0	−19.65	2.1 (1.6−2.6) e8	205	20.9	0.21	Barth + 2001
NGC 4261	E2	−21.09	5.2 (4.1−6.2) e8	315	31.6	0.15	Ferrarese + 1996
NGC 4374	E1	−21.36	4.3 (2.6−7.5) e8	296	18.4	0.24	Bower + 1998
NGC 4459	SA0	−19.15	7.0 (5.7−8.3) e7	167	16.1	0.14	Sarzi + 2001
M 87	E0	−21.53	3.0 (2.0−4.0) e9	375	16.1	1.18	Harms + 1994
NGC 4596	SB0	−19.48	0.8 (0.5−1.2) e8	136	16.8	0.22	Sarzi + 2001
NGC 5128	S0	−20.80	2.4 (0.7−6.0) e8	150	4.2	2.26	Marconi + 2001
NGC 6251	E2	−21.81	6.0 (2.0−8.0) e8	290	106	0.06	Ferrarese + 1999
NGC 7052	E4	−21.31	3.3 (2.0−5.6) e8	266	58.7	0.07	van der Marel + 1998
IC 1459	E3	−21.39	2.0 (1.2−5.7) e8	323	29.2	0.06	Verdoes Kleijn + 2001
NGC 1068	Sb	−18.82	1.7 (1.0−3.0) e7	151	15	0.04	Greenhill + 1996
NGC 4258	Sbc	−17.19	4.0 (3.9−4.1) e7	120	7.2	0.36	Miyoshi + 1995
NGC 4945	Scd	−15.14	1.4 (0.9−2.1) e6		3.7		Greenhill + 1997

Notes – BH detections are based on stellar dynamics (top group), ionized gas dynamics (middle) and maser dynamics (bottom). Column 3 is the B-band absolute magnitude of the bulge part of the galaxy. Column 4 is the BH mass $M_\bullet$ with error bars ($M_{\text{low}}, M_{\text{high}}$). Column 5 is the galaxy's velocity dispersion (see Figure 2). Column 6 is the distance (Tonry et al. 2001). Column 7 is the radius of the sphere of influence of the BH. References are the BH discovery papers (+ means et al.; for reviews of our Galaxy, see Kormendy & Richstone 1995; Yusef-Zadeh + 2000). BH masses are from the above papers except for our Galaxy (Genzel + 1997; Ghez + 1998), M 31 (Kormendy + 1999; Bacon + 2001), M 32 (van der Marel + 1998), NGC 3115 (Kormendy + 1996a; Emsellem + 1999), NGC 3377 (Gebhardt + 2001), NGC 4374 (Maciejewski + 2001); M 87 (Harms + 1994; Macchetto + 1997), and NGC 4486B (Green + 2001).

An important HST contribution has been to enable BH searches based on ionized gas dynamics (middle part of Table 1). The attraction of gas is simplicity – unlike the case of stellar dynamics, velocity dispersions are likely to be isotropic and projection effects are small unless the disks are seen edge-on. Especially important is the fact that gas disks are easy to observe even in giant ellipticals with cuspy cores. These galaxies are a problem for stellar-dynamical studies: they are expensive to observe because their surface brightnesses are low, and they are difficult to interpret because they rotate so little that velocity anisotropy is very important. It is no accident that most BH detections in the highest-luminosity ellipticals are based on gas dynamics. Without these, we would know much less about the biggest BHs.

At the same time, the uncertainties in gas dynamics are often underestimated. Most studies assume that disks are cold and in circular rotation. But the gas masses are small, and gas is easily pushed around. It would be no surprise to see velocities that are either slower or faster than circular. Faster-than-circular motions can be driven by AGN or starburst processes, while motions that are demonstrably slower than circular are observed in many bulges (see Kormendy & Westpfahl 1989). A separate issue is the large emission-line widths seen in many galaxies. If these are due to pressure support, then the observed rotation velocity is less than the circular velocity and $M_\bullet$ is underestimated if the line width is ignored. The situation is like that in any stellar system that has a significant velocity dispersion, and the cure is similar. In the context of a not-very-hot disk like that in our Galaxy, the correction from observed to circular velocity is called the "asymmetric drift correction", and in the context of hotter stellar systems like ellipticals, it is handled by three-integral dynamical models. For gas dynamics, the state of the art is defined by Barth et al. (2001), who discuss the line broadening problem in detail. They point out that asymmetric drift corrections may be large or they may be inappropriate if the line width is due to the internal microturbulence of gas clouds that are in individual, nearly circular orbits. We do not understand the physics of line broadening, so it adds uncertainty to BH masses. But it is likely that $M_\bullet$ will be underestimated if the line width is ignored. In contrast, Maciejewski & Binney (2001) emphasize that $M_\bullet$ can be overestimated by as much as a factor of three if we neglect the smearing effects of finite slit sizes. The best gas-dynamic $M_\bullet$ estimates (e. g., Sarzi et al. 2001; Barth et al. 2001) are thought to be accurate to $\sim 30\%$. In future, it will be important to take all of the above effects into account. It is not clear a priori whether they are devastating or small. The best sign that they are manageable is the observation that stellar- and gas-dynamical analyses imply the same $M_\bullet$ correlations (compare the squares and circles in Figure 2).

Gas-dynamical BH searches are limited mainly by the fact that suitable gas disks are rare. Sarzi et al. (2001) found gas disks with well-ordered, nearly circular velocities in only about 15 % of their sample of galaxies that were already known to have central gas. So $\lesssim 10\%$ of a complete sample of bulges is likely to have gas disks that are usable for BH searches. Nevertheless, within the next year, we should have gas-kinematic observations of 30 – 40 galaxies from a variety of groups. They will provide a wealth of information both on nuclear gas disks and on BHs.

THE $M_\bullet - M_{B,\mathrm{bulge}}$ AND $M_\bullet - \sigma_E$ CORRELATIONS

The list of BHs is now long enough so that we have finished the discovery era, when we were mainly testing the AGN paradigm, and have begun to use BH demographics to address a variety of astrophysical questions.

Two correlations have emerged. Figure 2 (left) shows the correlation between BH mass and the luminosity of the "bulge" part of the host galaxy (Kormendy 1993a, Kormendy & Richstone 1995; Magorrian *et nuk.* 1998) brought up to date with new detections. A least-squares fit gives

$$M_\bullet = 0.78 \times 10^8 \ M_\odot \left(\frac{L_{B,\mathrm{bulge}}}{10^{10} \ L_{B\odot}} \right)^{1.08}. \quad (1)$$

Since $M/L \propto L^{0.2}$, Equation (1) implies that BH mass is proportional to bulge mass, $M_\bullet \propto M_{\mathrm{bulge}}^{0.90}$.

Figure 2 (right) shows the correlation between BH mass and the luminosity-weighted velocity dispersion σ_e within the effective radius r_e (Gebhardt *et nuk.* 2000b; Ferrarese & Merritt 2000). A least-squares fit to the galaxies with most reliable $M_\bullet$ measurements (Gebhardt *et nuk.* 2001) gives

$$M_\bullet = 1.3 \times 10^8 \ M_\odot \left(\frac{\sigma_e}{200 \ \mathrm{km \ s^{-1}}} \right)^{3.65}. \quad (2)$$

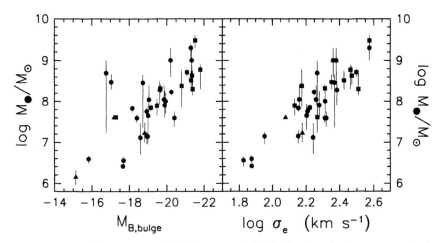

FIGURE 2. Correlation of BH mass with (left) the absolute magnitude of the bulge component of the host galaxy and (right) the luminosity-weighted mean velocity dispersion inside the effective radius of the bulge. In both panels, filled circles indicate $M_\bullet$ measurements based on stellar dynamics, squares are based on ionized gas dynamics, and triangles are based on maser disk dynamics. All three techniques are consistent with the same correlations.

The scatter in the $M_\bullet$ – $M_{B,\text{bulge}}$ relation is large: the RMS dispersion is a factor of 2.8 and the total range in $M_\bullet$ is two orders of magnitude at a given $M_{B,\text{bulge}}$. There are also two exceptions with unusually high BH masses. The more extreme case, NGC 4486B (Kormendy et nuk. 1997; Green et al. 2001), is still based on two-integral models. Its BH mass may decrease when three-integral models are constructed. Despite the scatter, the correlation is robust. One important question has been whether the $M_\bullet$ – $M_{B,\text{bulge}}$ correlation is real or only the upper envelope of a distribution that extends to smaller BH masses. The latter possibility now seems unlikely. Ongoing searches find BHs in essentially every bulge observed and in most cases would have done so even if the galaxies were significantly farther away.

In contrast, the scatter in the $M_\bullet$ – σ_e correlation is small, and the galaxies that were discrepant above are not discrepant here. Gebhardt et nuk. (2000b) find that the scatter is consistent with the measurement errors for the galaxies with the most reliable $M_\bullet$ measurements. So the $M_\bullet$ – σ_e correlation is more fundamental than the $M_\bullet$ – $M_{B,\text{bulge}}$ correlation. What does this mean?

Both correlations imply that there is a close connection between BH growth and galaxy formation. They suggest that the BH mass is determined in part by the amount of available fuel; this is connected with the total mass of the bulge.

Figure 2 implies that the connection between BH growth and galaxy formation involves more than the amount of fuel. Exceptions to the $M_\bullet$ – $M_{B,\text{bulge}}$ correlation satisfy the $M_\bullet$ – σ_e correlation. This means that, when a BH is unusually high in mass for a given luminosity, it is also high in σ_e for that luminosity. In other words, it is high in the Faber-Jackson (1976) $\sigma(L)$ correlation. One possible reason might be that the mass-to-light ratio of the stars is unusually high; this proves not to be the main effect. The main effect is illustrated in Figure 3. Ellipticals that have unusually high dispersions for their luminosities are unusually compact: they have unusually high surface brightnesses and small effective radii for their luminosities. Similarly, cold galaxies are fluffy: they have low effective surface brightnesses and large effective radii for their luminosities. Therefore, when a galaxy is observed to be hotter than average, we conclude that it underwent more dissipation than average and shrunk inside its dark halo to a smaller size and higher density than average. That is, it "collapsed" more than the average galaxy.

We can show this quantitatively by noting that the $M_\bullet$ – $M_{B,\text{bulge}}$ and $M_\bullet$ – σ_e relations are almost equivalent. The left panel in Figure 2 is almost a correlation of $M_\bullet$ with the mass of the bulge, because mass-to-light ratios vary only slowly with luminosity. Bulge mass is proportional to $\sigma_e^2 r_e / G$. So a black hole that satisfies the $M_\bullet$ – σ_e correlation will look discrepant in the $M_\bullet$ – $\sigma_e^2 r_e$ correlation if r_e is smaller or larger than normal. We conclude that BH mass is directly connected with the details of how bulges form.

This result contains information about when BHs accreted their mass. There are three generic possibilities. (1) BHs could have grown to essentially their present masses before galaxies formed and then regulated the amount of galaxy that grew around them (e. g., Silk & Rees 1998). (2) Seed BHs that were already present at the start of galaxy formation or that formed early could have grown to their present masses as part of the galaxy formation process. (3) Most BH mass may have been accreted after galaxy formation from ambient gas in the bulge. The problem is that the $M_\bullet$ correlations do not directly tell us which alternative dominates. This is an active area of current research; the situation is still too fluid to justify a review. All three alternative have proponents even on the Nuker team.

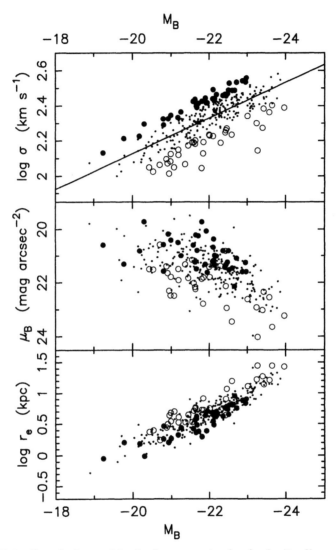

FIGURE 3. Correlations with absolute magnitude of velocity dispersion (upper panel), effective surface brightness (middle panel) and effective radius (lower panel) for elliptical galaxies from the Seven Samurai papers. Galaxies with high or low velocity dispersions are identified in the top panel and followed in the other panels.

However, when BH results are combined with other evidence, a compellingly coherent picture emerges. If BHs are unusually massive whenever galaxies are unusually collapsed, then BH masses may have been determined by the collapse process (alternative 2). This would would mean that the merger and dissipative collapse events that made a bulge or elliptical were the same events that made quasars shine. Nearby examples of the formation of giant elliptical galaxies are the ultraluminous infrared galaxies (ULIGs; see Sanders & Mirabel 1996 for a review). Sanders et al. (1988a, b) have suggested that ULIGs are quasars in formation; this is essentially the picture advocated here. Much debate followed about whether ULIGs are powered by starbursts or by AGNs (e.g., Filippenko 1992; Sanders & Mirabel 1996; Joseph 1999; Sanders 1999). Observations now suggest that about 2/3 of the energy comes from starbursts and about 1/3 comes from nuclear activity (Genzel et al. 1998; Lutz et al. 1998). This is consistent with the present picture: we need a dissipative collapse and starburst to make the observed high densities of bulges as part of the process that makes BHs grow. Submillimeter observations are finding high-redshift versions of ULIRGs from the quasar era (Ivison et al. 2000). Many are AGNs. Further evidence for a connection between ULIGs and AGN activity is reviewed in Veilleux (2000). ULIG properties strongly suggest that bulge formation, BH growth, and quasar activity all happen together.

WHICH GALAXIES CONTAIN BHs? WHICH DO NOT? $M_\bullet$ UPPER LIMITS

So far, BHs have been discovered in every galaxy that contains a bulge and that has been observed with enough resolution to find a BH consistent with the correlations in Figure 2. The canonical BH is about 0.13 % of the mass of the bulge; the scatter is more than a factor of 10. Table 2 lists the strongest BH mass limits. We fail to find BHs in pure disk and related galaxies. These are discussed in the next section.

TABLE 2
Limits on Supermassive Black Holes (2001 March)

Galaxy	Type	$M_{B,\text{nucleus}}$	$M_\bullet$ upper limit ($M_\odot$)	σ (km/s)	D (Mpc)	Reference
M 33	Scd	-10.21	1.0 e3	24	0.8	Gebhardt + 2001
NGC 205	Sph	-10	9.0 e4	15	0.72	Jones + 1996
NGC 4395	Sm		8.0 e4	30	2.6	Filippenko + 2001
IC 342	Scd	-14	5.0 e5	33	1.8	Böker + 1999

Note – These galaxies do not contain bulges; the absolute magnitude $M_{B,\text{nucleus}}$ and velocity dispersion σ refer to the nuclear star cluster. NGC 205 is a spheroidal galaxy; it does not fit into the traditional Hubble Sequence, but it is physically related to late-type galaxies (Kormendy 1985, 1987).

THE $M_\bullet - M_{B,\text{TOTAL}}$ CORRELATION: BHS DO NOT KNOW ABOUT DISKS

It is important to note that BH mass does not correlate with disks in the same way that it does with bulges. Figure 4 shows the correlations of BH mass with (left) bulge and (right) total luminosity. Figure 4 (right) shows that disk galaxies with small bulge-to-total luminosity ratios destroy the reasonably good correlation seen in Figure 4 (left). In addition, Figure 4 shows four galaxies that have strong BH mass limits but no bulges. They further emphasize the conclusion that disks do not contain BHs with nearly the same mass fraction as do bulges. In particular, in the bulgeless galaxy M 33, the upper limit on a BH mass from STIS spectroscopy is $M_\bullet \lesssim 1000\ M_\odot$. If M 33 contained a BH with the median mass fraction observed for bulges, then we would expect that $M_\bullet \sim 3 \times 10^7\ M_\odot$.

Figure 4 tells us that BH masses do not "know about" galaxy disks. Rather, they correlate with the high-density bulge-like component in galaxies.

These results do not preclude BHs in pure disk galaxies as long as they are small. Filippenko & Ho (2001) emphasize that some pure disks are Seyfert galaxies. They probably contain BHs. An extreme example is NGC 4395, the lowest-luminosity Seyfert known (Fig. 4). However, if its BH were radiating at the Eddington rate, then its mass would be only $M_\bullet \sim 100\ M_\odot$ (Filippenko & Ho 2001). So disks can contain BHs, but their masses are *much* smaller in relation to their disk luminosities than are bulge BHs in relation to bulge luminosities. It is possible that the small BHs in disks are similar to the seed BHs that once must have existed even in protobulges before they grew monstrous during the AGN era.

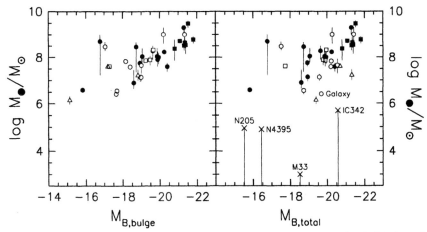

FIGURE 4. (left) $M_\bullet - M_{B,\text{bulge}}$ correlation from Figure 1. (right) Plot of $M_\bullet$ against the total absolute magnitude of the host galaxy. Filled symbols denote elliptical galaxies, open symbols denote bulges of disk galaxies. Crosses denote galaxies that do not contain a bulge: M 33 is from Gebhardt *et al.* (2001); IC 342 is from Böker *et al.* (1999), and NGC 4395 is from Filippenko & Ho (2001).

THE $M_\bullet - M_{B,\mathrm{bulge}}$ CORRELATION. II. BULGES VERSUS PSEUDOBULGES

So far, we have discussed elliptical galaxies and the bulges of disk galaxies as if they were equivalent. In terms of BH content, they are indistinguishable: they are consistent with the same $M_\bullet - M_{B,\mathrm{bulge}}$ and $M_\bullet - \sigma_e$ correlations. But a variety of observational and theoretical results show that there are two different kinds of high-density central components in disk galaxies. Both have steep surface brightness profiles. But, while classical bulges in (mostly) early-type galaxies are like little ellipticals living in the middle of a disk, the "pseudobulges" of (mostly) late-type galaxies are physically unrelated to ellipticals.

Pseudobulges are reviewed in Kormendy (1993b). Observational evidence for disklike dynamics includes (i) velocity dispersions σ that are smaller than those predicted by the Faber-Jackson (1976) $\sigma - M_B$ correlation, (ii) rapid rotation V that implies V/σ values above the "oblate line" describing rotationally flattened, isotropic spheroids in the V/σ - ellipticity diagram, and (iii) spiral structure that dominates the pseudobulge part of the galaxy. These observations and n-body simulations imply that high-density central disks can form out of disk gas that is transported toward the center by bars and oval distortions. They heat themselves, e. g. by scattering of stars off of bars (Pfenniger & Norman 1990). The observations imply that most early-type galaxies contain bulges, that later-type galaxies tend to contain pseudobulges, and that only pseudobulges are seen in Sc – Sm galaxies.

Andredakis & Sanders (1994), Andredakis, Peletier, & Balcells (1995), and Courteau, de Jong, & Broeils (1996) show that the "bulges" of many late-type galaxies have nearly exponential surface brightness profiles. It is likely that these profiles are a signature of pseudobulges, especially since blue colors imply that they are younger than classical bulges (Balcells & Peletier 1994).

HST observations strengthen the evidence for pseudobulges. Carollo *et al.* (1997, 1998a, b) find that many bulges have disk-like properties, including young stars, spiral structure, central bars, and exponential brightness profiles. It seems safe to say that no-one who saw these would suggest that they are mini-ellipticals living in the middle of a disk. They look more like late-type or irregular galaxies. To be sure, Peletier *et al.* (2000) find that bulges of early-type galaxies generally have red colors: they are old. True bulges that are similar to elliptical galaxies do exist; M 31 and NGC 4594 contain examples. But the lesson from the Carollo papers is that pseudobulges are more important than we expected. Like Kormendy (1993b) and Courteau *et al.* (1996), Carollo and collaborators argue that these are not real bulges but instead are formed via gas inflow in disks.

So there is growing evidence that the "bulges" in Fig. 2 are two different kinds of objects. Classical bulges are thought to form like ellipticals, in a dissipative collapse triggered by a merger. Pseudobulges are thought to form by secular evolution in disks. In both cases, gas flows inward and may feed BHs. One way to explore this is to ask whether bulges and pseudobulges have the same BH content.

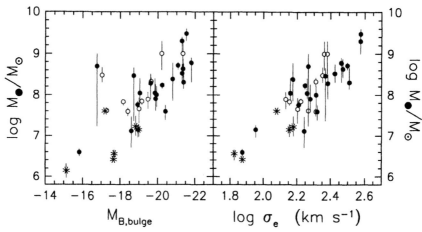

FIGURE 5. The $M_\bullet - M_{B,\text{bulge}}$ (left) and $M_\bullet - \sigma_e$ (right) correlations for elliptical galaxies (filled circles), bulges (open circles) and pseudobulges (stars).

The answer is shown in Figure 5. Pseudobulges have relatively low luminosities; this is plausible, since they are made from disks. But for their low luminosities, they have normal BH masses. The indentification of pseudobulges is still somewhat uncertain, and only a few have been observed. It will be important to check our result with a larger sample. However, it is consistent with the hypothesis that (pseudo)bulge formation and BH feeding are closely connected. Present data do not show any dependence of $M_\bullet$ on the details of whether BH feeding happens rapidly during a collapse or slowly via secular evolution of the disk.

If disks contain only small BHs while the pseudobulges that form from disks contain standard BHs with 0.13 % of the pseudobulge mass, then we conclude that these BHs must have grown to their present masses during pseudobulge formation.

The smallest BHs provide an argument that most BH growth did not happen after bulge formation. Some pseudobulges are still forming now; there is little time after bulge formation. Also, these objects do not contain fuel in the form of x-ray gas. And galaxies like M 32 contain little gas of any sort for late accretion.

BLACK HOLES AND GALAXY FORMATION: CONCLUSION

Galaxy formation is complicated, so any conclusions that we reach now are less secure than the observational results discussed above. However: *The observations suggest that the major events that form a bulge and the major growth phases of its BH – when it shone as an AGN – were the same events.* The likely formation processes are either a series of dissipative mergers that fuel starbursts and AGN activity (Sanders *et al.* 1988a, b) or secular inward flow of gas in disks that builds pseudobulges and simultaneously feeds their BHs (Kormendy *et nuk.* 2001).

THE FUTURE

The future is promising: (1) the census of BHs is expected to grow rapidly as HST reaches its full potential and as new techniques allow us to measure $M_\bullet$ in more distant galaxies; (2) the ongoing unification of the BH and galaxy formation paradigms is fundamental progress, and (3) x-ray satellites and gravitational wave detectors are expected to probe the immediate vicinity of the Schwarzschild radius.

Black Holes in Distant Galaxies

Measuring $M_\bullet$ by making dynamical models of observations that spatially resolve the central kinematics (Tables 1 and 2) are well tested techniques. Confidence is growing that the resulting BH masses are accurate to within $\sim 30\,\%$ in the best cases. This has allowed us to begin demographic studies of BHs in nearby galaxies. But the above techniques have a fundamental limitation. They cannot be applied unless the galaxies are close enough so that we can spatially resolve the region that is dynamically affected by the BH. Within a few more years, the most interesting galaxies that are accessible with HST resolution will have been observed, and new detections will slow down. Expected advances in spatial resolution will enable important but only incremental progress. The subject could use a breakthrough that allows us to measure BH masses in much more distant objects.

In this context, Figure 6 is encouraging news. It compares BH masses based on spatially resolved kinematics with masses derived by two other techniques, reverberation mapping (Blandford & McKee 1982; Netzer & Peterson 1997) and ionization models (Netzer 1990; Rokaki, Boisson, & Collin-Souffrin 1992). Both techniques have been available for some time, but it was not clear how much they could be trusted. Figure 6 shows that both techniques produce BH masses that are consistent with the $M_\bullet$ correlations discussed in earlier sections.

Reverberation mapping exploits the time delays measured between brightness variations in the AGN continuum and in its broad emission lines. These are interpreted as the light travel times between the BHs and the clouds of line-emitting gas. The result is an estimate of the radius r of the broad-line region. We also have a velocity V from the FWHM of the emission lines. Together, these measure a mass $M_\bullet \approx V^2 r/G$. However, a number of authors (Wandel 1999; Ho 1999; Wandel, Peterson, & Malkan 1999) have pointed out that reverberation mapping BH masses are systematically low in the $M_\bullet$ - $M_{B,\text{bulge}}$ correlation. Recently, Gebhardt et nuk. (2000c; see Figure 6, below, for an update) have shown that reverberation mapping BH masses agree with the $M_\bullet$ - σ_e correlation. This suggests that the problem uncovered in previous comparisons was that the bulge luminosities of the reverberation mapping galaxies were measured incorrectly or were inflated by young stars. Gebhardt et nuk. (2000c) and Figure 6 here suggest that reverberation mapping does produce reliable BH masses.

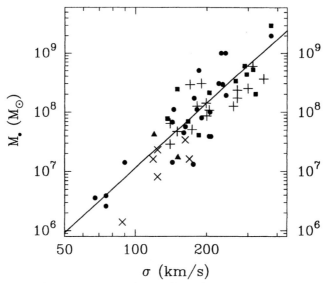

FIGURE 6. The $M_\bullet - \sigma_e$ correlation for galaxies with BH masses from detailed dynamical models applied to spatially resolved kinematics (filled symbols as in Figure 2), reverberation mapping (crosses), and ionization models (plus signs).

Similarly, ionization model BH masses – ones based on the observed correlation between quasar luminosity and the radius at which the broad-line-emitting gas lives – are largely untested and therefore uncertain. Laor (1998) and Gebhardt *et nuk.* (2001) now show that this technique also appears to produce $M_\bullet$ values with no systematic offset from other techniques (Figure 6).

These results are important because neither reverberation mapping nor ionization models require us to spatially resolve the central region affected by the BH. Both techniques can be applied to objects at arbitrarily large distances. Therefore BH masses can now be estimated for quasars out to redshifts of nearly $z = 6$. Ongoing surveys like 2dF and the Sloan Digital Sky Survey are producing thousands of quasar detections. BH masses should therefore be derivable for very large samples that span all redshifts from $z = 0$ to the most distant objects known. It will be important to check as well as possible that the physical circumstances that make the ionization models work so well are still valid far away. Nevertheless, it should be possible to directly measure the growth of BHs in the Universe.

The Effects of Black Holes on Galaxy Structure

There is other encouraging news, too. In the past, the BH search was decoupled from other galaxy studies. It was carried out largely in isolation to test the AGN paradigm. Furthermore, the early BH detections were mostly in inactive galaxies, so even the connection with AGN physics was indirect. This situation was reviewed in Kormendy & Richstone (1995). But now BH results are beginning to connect up with a variety of work of galaxy physics. The subject is large and our space is limited. We therefore mention briefly only three of the developing results.

1 – Triaxial elliptical galaxies evolve rapidly toward axisymmetry if the central gravitational potential well gets steep enough (Lake & Norman 1983; Gerhard & Binney 1985; Norman, May, & van Albada 1985; Valluri & Merritt 1998; Merritt & Quinlan 1998; Poon & Merritt 2001; Holley-Bockelmann et al. 2001; see Merritt 1999 for a review). This can be achieved either by increasing the central density of stars via gas infall and star formation or by the growth of a BH. In either case, chaotic mixing of stellar orbits redistributes stars in phase space and causes orbit shapes to evolve. Box orbits, which support the triaxial structure but which allow stars to pass arbitrarily close to the center, are destroyed in favor of orbits that suppport axisymmetric structure. To the extent that triaxiality promotes gas infall and BH feeding, the evolution may also turn off the feeding when the BH has grown to 1 or 2 % of the bulge mass. These processes help to explain the observed upper limit to the BH mass fraction.

2 – Some elliptical galaxies have "cuspy cores", i.e., density distributions that break at small radii from steep outer power laws to shallow inner power laws. These cores may be produced by the orbital decay of binary BHs (Begelman, Blandford, & Rees 1980; Ebisuzaki, Makino, & Okamura 1991; Makino & Ebisuzaki 1996; Quinlan 1996; Quinlan & Hernquist 1997; Faber et nuk. 1997; Milosavljević & Merritt 2001). The formation of BH binaries is a natural consequence of hierarchical galaxy mergers. The orbits then decay (i.e., the binaries get "harder") by flinging stars away. This BH scouring may reduce the stellar density enough to produce a break in the density profile.

3 – Three-integral dynamical models tell us the distribution of stellar orbits around a BH. Preliminary results (van der Marel et al. 1998; Cretton et al. 1999b; Gebhardt et nuk. 2000a, 2001; Richstone et nuk. 2001) show an important difference between core and power-law galaxies. In core galaxies, the central tangential velocity dispersion σ_t is larger than the radial dispersion σ_r. Large tangential anisotropy is consistent with the effects of BH binaries (Nakano & Makino 1999a, b) and BH scouring (Quinlan & Hernquist 1997). In contrast, coreless galaxies, which have featureless, almost power-law density profiles, are observed to have $\sigma_r \simeq \sigma_t$. This is more consistent with the adiabatic growth of single BHs via gas accretion (Quinlan, Hernquist, & Sigurdsson 1995 and references therein). Further studies of the relationship between BHs and properties of their host galaxies should provide much better constraints on the relationship between BHs and galaxy formation.

The above developments are an important sign of the developing maturity of this subject. Finding convincing connections between BH properties and the microphysics of galaxies contributes in no small measure to our confidence in the BH picture. The medium-term future of this subject is therefore very promising.

In the longer-term future, the most fundamental progress is expected to come from gravitational wave astronomy. We can look forward to the true maturity of work on supermassive BHs when the Laser Interferometer Space Antenna (LISA) begins to provide a direct probe of strong gravity.

ACKNOWLEDGMENTS

It is a pleasure to thank our Nuker collaborators for helpful discussions and for permission to use our BH detection results before publication. We are also most grateful to Gary Bower, Richard Green, Mary Beth Kaiser, and Charles Nelson for communicating STIS team BH detections before publication.

REFERENCES

1. Andredakis, Y. C., Peletier, R. F., & Balcells, M. 1995, MNRAS, 275, 874
2. Andredakis, Y. C., & Sanders, R. H. 1994, MNRAS, 267, 283
3. Bacon, R., et al. 2001, A&A, in press (astro-ph/0010567)
4. Balcells, M., & Peletier, R. F. 1994, AJ, 107, 135
5. Barth, A. J., et al. 2001, ApJ, in press (astro-ph/0012213)
6. Begelman, M. C., Blandford, R. D., & Rees, M. J, 1980, Nature, 287, 307
7. Blandford, R. D., & McKee, C. F. 1982, ApJ, 255, 419
8. Blandford, R. D., McKee, C. F., & Rees, M. J. 1977, Nature, 267, 211
9. Böker, T., van der Marel, R. P., & Vacca, W. D. 1999, AJ, 118, 831
10. Bower, G. A., et al. 1998, ApJ, 492, L111
11. Bower, G. A., et al. 2001a, ApJ, in press (astro-ph/0011204)
12. Bower, G. A., et al. 2001b, in preparation
13. Carollo, C. M., & Stiavelli, M. 1998a, AJ, 115, 2306
14. Carollo, C. M., Stiavelli, M., de Zeeuw, P. T., & Mack, J. 1997, AJ, 114, 2366
15. Carollo, C. M., Stiavelli, M., & Mack, J. 1998b, AJ, 116, 68
16. Courteau, S., de Jong, R. S., & Broeils, A. H. 1996, ApJ, 457, L73
17. Cretton, N., & van den Bosch, F. C. 1999a, ApJ, 514, 704
18. Cretton, N., de Zeeuw, P. T., van der Marel, R. P., & Rix, H.-W. 1999b, ApJS, 124, 383
19. Dressler, A., & Richstone, D. O. 1988, ApJ, 324, 701
20. Ebisuzaki, T., Makino, J., & Okamura, S. K. 1991, Nature, 354, 212
21. Eckart, A., & Genzel, R. 1997, MNRAS, 284, 576
22. Emsellem, E., Dejonghe, H., & Bacon, R. 1999, MNRAS, 303, 495
23. Faber, S. M., & Jackson, R. E. 1976, ApJ, 204, 668
24. Faber, S. M., et nuk. 1997, AJ, 114, 1771
25. Ferrarese, L., & Ford, H. C. 1999, ApJ, 515, 583
26. Ferrarese, L., Ford, H. C., & Jaffe, W. 1996, ApJ, 470, 444
27. Ferrarese, L., & Merritt, D. 2000, ApJ, 539, L9
28. Filippenko, A. V. (ed.) 1992, Relationships Between Active Galactic Nuclei and Starburst Galaxies (San Francisco: ASP)
29. Filippenko, A. V., & Ho, L. C. 2001, ApJ, submitted
30. Gebhardt, K., et nuk. 2000a, AJ, 119, 1157
31. Gebhardt, K., et nuk. 2000b, ApJ, 539, L13
32. Gebhardt, K., et nuk. 2000c, ApJ, 543, L5
33. Gebhardt, K., et nuk. 2001, in preparation (four papers)

34. Genzel, R., Eckart, A., Ott, T., & Eisenhauer, F. 1997, MNRAS, 291, 219
35. Genzel, R., *et al.* 2000, MNRAS, 317, 348
36. Genzel, R., *et al.* 1998, ApJ, 498, 579
37. Gerhard, O. E., & Binney, J. 1985, MNRAS, 216, 467
38. Ghez, A. M., Klein, B. L., Morris, M., & Becklin, E. E. 1998, ApJ, 509, 678
39. Ghez, A. M., *et al.* 2000, Nature, 407, 349
40. Green, R. F., *et al.* 2001, in preparation
41. Greenhill, L.J., Gwinn, C.R., Antonucci, R., Barvainis, R., 1996, ApJL, 472, L21
42. Greenhill, L.J, Moran, J.M., & Hernstein, J.R. 1997, ApJL, 481, L23
43. Harms, R.J. *et al.* 1994, ApJ, 435, L35
44. Ho, L. C. 1999, in Observational Evidence for Black Holes in the Universe, ed. S. K. Chakrabarti (Dordrecht: Kluwer), 157
45. Holley-Bockelmann, K., *et al.* 2001, ApJ, submitted
46. Ivison, R. J., *et al.* 2000, MNRAS, 315, 209
47. Jones, D. H., *et al.* 1996, ApJ, 466, 742
48. Joseph, R. D. 1999, A&SS, 266, 321
49. Kaiser, M. E., *et al.* 2001, in preparation
50. Kormendy, J. 1985, ApJ, 295, 73
51. Kormendy, J. 1987, in Nearly Normal Galaxies: From the Planck Time to the Present, ed. S. M. Faber (New York: Springer-Verlag), 163
52. Kormendy, J. 1988a, ApJ, 325, 128
53. Kormendy, J. 1988b, ApJ, 335, 40
54. Kormendy, J. 1993a, in The Nearest Active Galaxies, ed. J. Beckman, L. Colina, & H. Netzer (Madrid: CSIC), 197
55. Kormendy, J. 1993b, in IAU Symposium 153, Galactic Bulges, ed. H. Dejonghe & H. J. Habing (Dordrecht: Kluwer), 209
56. Kormendy, J., & Bender, R. 1999, ApJ, 522, 772
57. Kormendy, J., & Bender, R., Evans, A. S., & Richstone, D. 1998, AJ, 115, 1823
58. Kormendy, J., & Richstone, D. 1992, ApJ, 393, 559
59. Kormendy, J., & Richstone, D., 1995, ARA&A, 33, 581
60. Kormendy, J., & Westpfahl, D. J. 1989, ApJ, 338, 752
61. Kormendy, J., *et nuk.* 1996a, ApJ, 459, L57
62. Kormendy, J., *et nuk.* 1996b, ApJ, 473, L91
63. Kormendy, J., *et nuk.* 1997, ApJ, 482, L139
64. Kormendy, J., *et nuk.* 2001, in preparation
65. Lake, G., & Norman, C. 1983, ApJ, 270, 51
66. Laor, A. 1998, ApJ, 505, L83
67. Lutz, D., *et al.* 1998, ApJ, 505, L103
68. Lynden-Bell, D. 1969, Nature, 223, 690
69. Lynden-Bell, D. 1978, Physica Scripta, 17, 185
70. Macchetto, F., *et al.* 1997, ApJ, 489, 579
71. Maciejewski, W., & Binney, J. 2001, MNRAS submitted (astro-ph/0010379)
72. Magorrian, J., *et nuk.* 1998, AJ, 115, 2285
73. Makino, J., & Ebisuzaki, T. 1996, ApJ, 465, 527
74. Maoz, E. 1995, ApJ, 447, L91

75. Maoz, E. 1998, ApJ, 494, L181
76. Marconi, A., et al. 2001, ApJ, 549, 915
77. Merritt, D. 1999, PASP, 111, 129
78. Merritt, D., & Quinlan, G. D. 1998, ApJ, 498, 625
79. Milosavljević, M., & Merritt, D. 2001, ApJ, submitted (astro-ph/0103350)
80. Miyoshi, M., et al. 1995, Nature, 373, 127
81. Nakano, T., & Makino, J. 1999a, ApJ, 510, 155
82. Nakano, T., & Makino, J. 1999b, ApJ, 525, L77
83. Netzer, H. 1990, in Active Galactic Nuclei, Saas-Fee Advanced Course 20, ed. T. J.-L. Courvoisier & M. Mayor (Berlin: Springer), 57
84. Netzer, H., & Peterson, B. M. 1997, in Astronomical Time Series, ed. D. Maoz, A. Sternberg, & E. M. Leibowitz (Dordrecht: Kluwer), 85
85. Norman, C. A., May, A., & van Albada, T. S. 1985, ApJ, 296, 20
86. Peletier, R. F., et al. 2000, MNRAS, 310, 703
87. Perley, R. A., Dreher, J. W., & Cowan, J. J. 1984, ApJ, 285, L35
88. Pfenniger, D., & Norman, C. 1990, ApJ, 363, 391
89. Poon, M. Y., & Merritt, D. 2001, ApJ, 549, 192
90. Quinlan, G. D. 1996, NewA, 1, 35
91. Quinlan, G. D., & Hernquist, L. 1997, NewA, 2, 533
92. Quinlan, G. D., Hernquist, L., & Sigurdsson, S. 1995, ApJ, 440, 554
93. Rees, M. J. 1984, ARA&A, 22, 471
94. Richstone, D., et nuk. 1998, Nature, 395, A14
95. Richstone, D., et nuk. 2001, in preparation
96. Rokaki, E., Boisson, C., & Collin-Souffrin, S. 1992, A&A, 253, 57
97. Salpeter, E. E. 1964, ApJ, 140, 796
98. Sanders, D. B. 1999, A&SS, 266, 331
99. Sanders, D. B., & Mirabel, I. F. 1996, ARA&A, 34, 749
100. Sanders, D. B., et al. 1988a, ApJ, 325, 74
101. Sanders, D. B., et al. 1988b, ApJ, 328, L35
102. Sarzi, M., et al. 2001, ApJ, 550, 65
103. Schmidt, M. 1963, Nature, 197, 1040
104. Silk, J., & Rees, M. J. 1998, A&A, 331, L1
105. Statler, T. S., King, I. R., Crane, P., & Jedrzejewski, R. I. 1999, AJ, 117, 894
106. Tonry, J. L. 1984, ApJL, 283, L27
107. Tonry, J. L. 1987, ApJ, 322, 632
108. Tonry, J. L., et al. 2001, ApJ, 546, 681
109. Valluri, M., & Merritt, D. 1998, ApJ, 506, 686
110. van der Marel, R. P., Cretton, N., de Zeeuw, T., & Rix, H.-W. 1998, ApJ, 493, 613
111. van der Marel, R. P., & van den Bosch, F. C. 1998, AJ, 116, 2220
112. Veilleux, S. 2000, astro-ph/0012121
113. Verdoes Kleijn, G. A., et al. 2001, AJ, 120, 1221
114. Wandel, A. 1999, ApJ, 519, L39
115. Wandel, A., Peterson, B. M., & Malkan, M. A. 1999, ApJ, 526, 579
116. Yusef-Zadeh, F., Melia, F., & Wardle, M. 2000, Science, 287, 85
117. Zel'dovich, Ya. B. 1964, Soviet Physics – Doklady, 9, 195

The X-ray Background and the Census of Quasars

Amy J. Barger[*,†,‡]

[*]Institute for Astronomy, University of Hawaii, Honolulu, Hawaii 96822
[†]Department of Astronomy, University of Wisconsin, Madison, WI 53706
[‡]Hubble Fellow and Chandra Fellow at Large

Abstract. The *Chandra X-ray Observatory* detects X-rays emitted during the accretion of matter onto supermassive black holes, even when they are highly obscured. A cosmic census of AGN contributing to the X-ray background is nearing completion. Follow-up observations with ground-based telescopes provide new knowledge about distant supermassive black holes. Here we discuss deep optical, near-infrared, submillimeter, and radio data, as well as high quality optical spectra, of a complete sample of twenty 2 – 10 keV X-ray sources selected in a deep *Chandra* observation of the Hawaii Deep Survey Field SSA13. The data allow us to estimate the duration and times of black hole activity. Surprisingly, the duty cycle is about a billion years, and accretion is still occurring at cosmologically recent times. These conclusions challenge conventional wisdom that supermassive black hole growth is only associated with host galaxy formation or violent galaxy mergers. With the optical, submillimeter, and radio data, we estimate the bolometric luminosities of the black holes and infer their masses. We find that the emitted bolometric energy from accretion onto supermassive black holes is about a third of the emitted energy from all the stars in the optical.

INTRODUCTION

A major goal in observational cosmology is to understand the accretion and star formation histories of the Universe. A serious complication in achieving this goal is the effects of obscuration by gas and dust, which can hide much of the activity from optical view. However, with hard X-ray, submillimeter, and radio observations, we can "see" through the dust. Thus, we can make a cosmic census of all the galaxies and AGN in the Universe. How will we know when we have succeeded? The extragalactic background light (EBL) does the bookkeeping. The EBL is the emission from all extragalactic sources along the line-of-sight. Once we have accounted for the full EBL, our census is complete.

Whereas star forming galaxies dominate the optical background, active galactic nuclei (AGN) dominate the X-ray background (XRB). The engines of AGN are supermassive black holes at the centers of galaxies. As matter falls into a black hole, it radiates at all wavelengths. X-rays reveal the presence of these black holes. In obscured AGN and galaxies, low energy X-rays and light get absorbed by the surrounding gas and dust while the high energy X-rays can escape unabsorbed. The reprocessed light appears in the far-infrared (FIR), or, for high redshift sources, in the submillimeter.

In this proceeding I will first discuss the resolution of the hard X-ray background using deep imaging surveys with the *Chandra X-ray Observatory*. I will then discuss the deep ground-based follow-up work that we have been carrying out to learn about the nature of the hard X-ray sources. I will end by addressing what we have learned about supermassive black hole activity from our multi-wavelength dataset.

THE X-RAY BACKGROUND

The XRB photon intensity, $P(E)$, with units photons cm^{-2} s^{-1} keV^{-1} sr^{-1} can empirically be described by the power law $P(E) = AE^{-1.4}$ between energies of 1 and 15 keV (e.g., Marshall et al. 1980; Chen, Fabian, & Gendreau 1997). The focus in X-ray astronomy has traditionally been on two broad energy bands: the soft band (0.5 – 2 keV) and the hard band (2 – 10 keV). *ROSAT* resolved some 70 – 80 percent of the soft X-ray background (Hasinger et al. 1998). The optical counterparts to the X-ray sources were mostly AGN. The soft band X-ray spectra, however, were too steep ($\sim E^{-1.7}$) to explain the XRB spectrum, giving rise to the so-called "spectral paradox". Hard band observations were needed to detect the obscured AGN that were inferred to be present. *ASCA* and *BeppoSAX* were able to resolve some 20 – 30 percent of the hard XRB but with limited spatial resolution. The full resolution of the XRB into discrete sources at hard energies had to wait for the arcsecond imaging quality and high-energy sensitivity of *Chandra*.

Deep *Chandra* imaging surveys of three fields — SSA13 (100 ks; Mushotzky et al. 2000), the *Chandra* Deep Field South (CDF-S; 130 ks; Giacconi et al. 2001), and the Hubble Deep Field North (HDF-N; 225 ks; Garmire et al. 2001) — have already resolved some 80 percent of the hard XRB into discrete sources. Figure 1 shows the cumulative counts versus hard X-ray flux for the combined Mushotzky et al. (2000) and Giacconi et al. (2001) published samples. The conversion from observed counts to flux depends on the spectral slope of the source, and the two samples are statistically indistinguishable from one another once a consistent photon index of $\Gamma = 1.2$ is assumed. The two samples become almost identical if we allow for a 15% systematic difference in the absolute flux calibration between the front-side illuminated and back-side illuminated chips. The CDF-S and HDF-N fields have since each been observed for *one million* seconds. These data will essentially resolve all of the hard XRB and be one of the main legacies of *Chandra*.

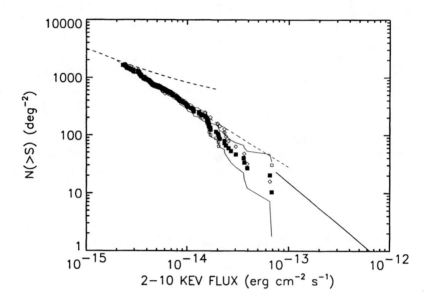

FIGURE 1. Combined hard X-ray counts (solid squares) and 1σ uncertainties (jagged solid lines) from Mushotzky et al. (2000; open diamonds) and Giacconi et al. (2001; open squares), where we have corrected the Giacconi et al. counts to an assumed photon index of $\Gamma = 1.2$ and also included a 15% systematic difference in the absolute flux calibration between the front-side illuminated and back-side illuminated chips. The fit to the *ASCA* counts from Ueda et al. (1999) is shown as the solid line, $N(>S) \propto S^{-1.5}$. The fit to the *Chandra* counts over the restricted flux range 2×10^{-15} to 2×10^{-14} erg cm^{-2} s^{-1} is shown as the lower dashed line, $N(>S) \propto S^{-1.1}$. The upper dashed line shows where the counts would exceed the $2 - 10$ keV background of Chen, Fabian, & Gendreau (1997) under the condition that the total intensity of the XRB is explained by the integral of sources assuming $N(>S) \propto S^{-1.1}$.

MULTIWAVELENGTH FOLLOW-UP OBSERVATIONS

Because of the excellent $< 1''$ X-ray positional accuracy of *Chandra*, counterparts to the X-ray sources in other wavebands can be securely identified. Deep optical imaging studies of SSA13 have revealed a heterogeneous sample of optical counterparts. At the current flux limits, almost half (9/20) of the hard X-ray sources have optically bright counterparts ($I < 23.5$) in the $z < 1.5$ redshift range (Barger et al. 2001). All of these can be spectroscopically identified. Many are early-type galaxies with near-L_* luminosities. Contrary to the situation for the faint *ROSAT* sources (Schmidt et al. 1998), the vast majority of the *Chandra* sources do not have broad optical or ultraviolet lines, and a substantial number (6/20) show no obvious spectroscopic signatures of AGN activity. I will refer to the latter as 'normal' galaxies to reflect the fact that they are not likely to have been identified as AGN in a spectroscopic survey of a magnitude-limited field galaxy sample. There are two plausible explanations for the lack of observed optical AGN characteristics in the 'normal' galaxies: (i) absorption due to dust and gas or (ii) an actual lack of ultraviolet/optical emission, as is the case in many low luminosity objects (Ho 1999). The line-of-sight hydrogen column densities that we infer from our X-ray spectra ($N_H = 2 \times 10^{22}$ cm^{-2} to 3×10^{23} cm^{-2}) are sufficiently large to obscure the optical AGN signatures, but the extent of the obscuration will depend on the geometry. It is an open question whether the numbers of 'normal' galaxies will increase with the deeper *Chandra* exposures.

About 35 percent (7/20) of the hard X-ray sources have optically faint ($I > 23.5$) counterparts. These galaxies are often extremely red. It has been speculated that they are higher redshift analogs of the $z < 1.5$ optically bright X-ray sources, since red spectral energy distributions (SEDs) would result in substantial optical fading at higher redshifts (Crawford et al. 2001; Barger et al. 2001).

Clusters act as lenses which magnify any background sources. Cowie et al. (2001) combined deep near-infrared (NIR) images of the massive clusters A2390 and A370 with deep optical data to map the SEDs of the *Chandra* X-ray sources lying behind the clusters. The three X-ray sources behind A2390 were found to have extremely red colors; their SEDs are consistent with the sources being evolved galaxies at $z > 1.4$. The photometric redshift of one has been confirmed at $z = 1.467$ from NIR spectroscopy. Mapping of optically faint hard X-ray sources may prove to be an extremely efficient way to locate luminous evolved galaxies at high redshift.

Light that is absorbed by gas and dust is re-radiated into the FIR, or, for high redshift sources, into the submillimeter. From *COBE* measurements of the EBL, the total re-radiated emission is comparable to the entire optical Universe (see Fig. 3). Deep 850-micron submillimeter observations with SCUBA have resolved 20 − 30 percent of the 850-micron background at fluxes greater than 2 mJy (e.g., Barger, Cowie, & Sanders 1999), revealing a new population of highly obscured, exceptionally luminous sources that appear to be the distant analogs of local ultra-luminous infrared galaxies (ULIGs). There is an ongoing debate on whether local ULIGs are dominantly powered by star formation or by heavily dust enshrouded AGN, and the same applies to the distant SCUBA sources. Barger et al. (1999) carried out a spectroscopic survey of a complete sample of submillimeter sources detected in a survey of massive lensing clusters. Only three of the 17 sources could be reliably identified spectroscopically; of these, two showed AGN signatures. These results leave open the possibility that many SCUBA sources may contain AGN. A key goal is to determine how much overlap there is between the hard X-ray and submillimeter populations in order to discern whether AGN are the primary power sources in these objects. So far only three submillimeter sources have also been detected in hard X-rays (Bautz et al. 2000; Fabian et al. 2000; Hornschemeier et al. 2000, 2001; Barger et al. 2001). This may reflect the fact that the 850-micron flux limits obtainable with SCUBA are quite close to the expected submillimeter fluxes from obscured AGN. An important corollary is that the bright submillimeter

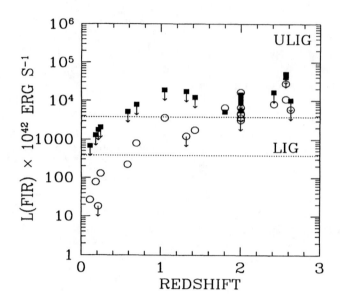

FIGURE 2. FIR luminosities versus redshift for the SSA13 hard X-ray sample. Open circles were calculated from the radio data and the FIR-radio correlation. Filled squares (mostly upper limits) were calculated from the submillimeter data (see Barger et al. 2001 for details). The spectroscopically unidentified sources are placed at $z = 2$. The one source detected in both the radio and the submillimeter has a millimetric redshift of 1.8. Downward pointing arrows indicate 3σ limits for either the radio inferred FIR luminosities or the submillimeter inferred FIR luminosities. The dashed horizontal lines indicate the luminosities at which a galaxy would be classified as either a ULIG (upper line) or a LIG (lower line).

sources are mainly star formers and can safely be used to map the star formation history at high redshifts (e.g., Barger, Cowie, & Richards 2000).

Because there is very little overlap between the submillimeter and hard X-ray populations, our submillimeter observations can only be used to infer upper limits on the FIR luminosities of the hard X-ray sources. However, we do have another means to estimate the FIR luminosities of the hard X-ray sources: radio observations. Galaxies and the intergalactic medium are transparent in the radio. Locally, a tight correlation is observed between the FIR and radio luminosities of star-forming galaxies and radio-quiet AGN (Condon 1992). The recent discovery that the majority of bright SCUBA sources are associated with optically faint microjansky radio sources implies that the correlation continues to hold at high redshifts (Barger, Cowie, & Richards 2000; Chapman et al. 2001). Only four of the twenty hard X-ray sources in SSA13 are not detected above the 15 microjansky (3σ) limit in the 20 cm data of Richards et al. (2001). Thus, we can use the radio luminosities to infer FIR luminosities for the hard X-ray sources in our sample (see Fig 2).

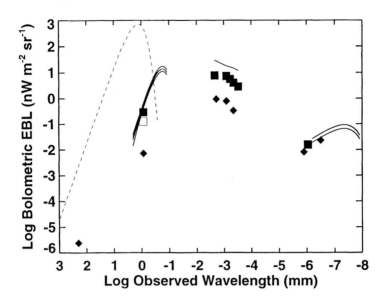

FIGURE 3. Contribution of the *Chandra* SSA13 hard X-ray selected sample to the EBL at each of the measured wavelengths (filled diamonds). The dashed line shows the cosmic microwave background. Solid curves show the measured total extragalactic light in each wavelength regime: submillimeter from Puget et al. (1996) and Fixsen et al. (1998); optical/NIR from Bernstein et al. (1999) and Wright & Reese (2000); X-ray from Marshall et al. (1980), both at the original flux and re-normalized to the more recent higher flux determinations. Filled squares show the integrated contribution of the source counts: submillimeter counts to 0.5 mJy from Blain et al. (1999) (submillimeter counts > 2 mJy from Barger, Cowie, & Sanders (1999) are shown by an open square); optical/NIR from Williams et al. (1996) and Gardner et al. (1993); 1 – 2 keV X-ray from Mushotzky et al. (2000).

It is interesting to note that three-quarters of the hard X-ray sources with radio detections in our sample would be classified as LIGs ($L_{FIR} > 10^{11}$ $L_\odot$; Sanders & Mirabel 1996) from their FIR luminosities, and half would be classified as ULIGs ($L_{FIR} > 10^{12}$ $L_\odot$).

HARD X-RAY CONTRIBUTION TO THE EBL

The summed contributions of the SSA13 *Chandra* hard X-ray sources to the EBL are shown versus wavelength in Fig. 3. These are compared with measurements of the EBL and with the integrated light from direct counts. The direct counts now lie close to the EBL at all wavelengths. The hard X-ray contribution to the submillimeter EBL is about 10%. It appears that the great bulk of the submillimeter

EBL must arise in sources which are too faint at 2 − 10 keV to be seen at the current threshold. The hard X-ray sample also contributes about 10% of the light at UV/optical wavelengths, but there is almost certainly strong contamination of the estimate of the AGN contribution by the light of the host galaxies, so this should be considered an upper limit.

SUPERMASSIVE BLACK HOLES

There is dynamical evidence for 10^6 M$_\odot$ to 10^9 M$_\odot$ black holes in local ellipticals and spirals. A tight correlation has recently been found between the mass of the black hole, M_{bh}, and the velocity dispersion, σ, of the host bulge, $M_{bh} \propto \sigma^4$ (e.g., Ferrarese & Merritt 2000; Gebhardt et al. 2000). A looser correlation had previously been found between the mass of the black hole and the luminosity of the host bulge, $M_{bh} \propto L_B$ (Kormendy & Richstone 1995; Magorrian et al. 1998). Assuming this local correlation holds at higher redshifts, and subject to the uncertainties in translating our observed B-band magnitudes into M_B(bulge) that would worsen the relation, we find that our 'normal' galaxies have approximate black hole masses in the range 5×10^7 M$_\odot$ to 10^9 M$_\odot$.

Chandra detects black holes in their active phases, so an interesting question to ask is, what is the duty cycle of X-ray activity? From our SSA13 sample, we know that the spectroscopically identified hard X-ray sources are common in bulge-dominated optically luminous galaxies (see Fig. 4). We can therefore invert our approach and ask, what fraction of an optically luminous spectroscopically identified field galaxy sample is X-ray active at any given time? The result is about 10 percent. If the fraction of galaxies showing such behavior reflects the fraction of time that each galaxy spends accreting onto its supermassive black hole, then we require each such galaxy to be active for around 1 Gyr. This duration of the X-ray activity is much longer than the theoretically estimated accretion time of 0.01 Gyr for black hole fuelling by mergers (Kauffmann & Haenelt 2000) and may suggest that the accretion is being powered by smaller mergers or by internal flows within the galaxies.

We can take our analysis one step further and construct redshift slices of supermassive black hole growth. In Fig. 5 we have combined data from three deep *Chandra* samples: SSA13, the HDF-N (Hornschemeier et al. 2001), and A370 (from a 66 ks exposure; Bautz et al. 2000). We find that about 18% of the X-ray energy production has occurred since $z = 1$, when the Universe was half its present age. (This does not include the brighter X-ray fluxes from sources detected in pre-*Chandra* surveys.) This result challenges the traditional view that black hole formation is only associated with the galaxy formation era.

The X-ray contribution to the bolometric AGN light is only 1 − 20%, but with our multi-wavelength data we can estimate the total luminosity associated with accretion onto supermassive black holes. We find that all but three of the hard X-ray sources in our sample are dominated by the FIR light rather than by the

UV/optical light. If we assume that the FIR light in the hard X-ray sources is reprocessed AGN light, uncontaminated by star formation in the galaxies, then we may use our data to compute the bolometric correction from the hard X-ray luminosity to the bolometric luminosity of the AGN. Because galaxy contamination of the optical light is a problem, we adopt the hard X-ray flux-weighted average ratio $L_{FIR}/L_{HX} = 33$ for a conservative minimum of the bolometric correction, recognizing that the true ratio could be larger by a factor of two or more.

We may now address the issue of the mass of the black holes. Our first approach, discussed above, was to estimate the masses of the 'normal' galaxy black holes using the local relation between black hole mass, M_{bh}, and the absolute magnitude of the bulge component of the host galaxy, M(bulge). These estimates are shown in Fig. 6 as open stars. A second approach is to estimate the total accreted mass required to account for the luminosity in each galaxy. The mass inflow rate is $\epsilon \Delta M_{bh} c^2 / \Delta t = L_{bol}$, where ϵ is the efficiency for re-radiation of the accretion energy. Taking the

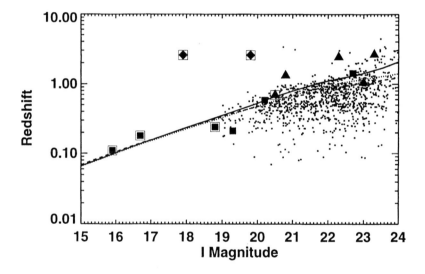

FIGURE 4. Redshift versus I magnitudes for the thirteen $I < 23.5$ hard X-ray sources with spectroscopic identifications (large symbols) and for the $I < 24$ field galaxies from the Hubble Deep Field and the Hawaii Deep Survey Fields SSA13 and SSA22 (small symbols). Quasars are denoted by filled diamonds, AGN by filled triangles, and optically 'normal' galaxies by filled squares. Sources with 5σ radio counterparts are enclosed in open squares. The hard X-ray sources are almost exclusively galaxies lying at or above L_*, as indicated by Coleman, Wu, & Weedman (1980) tracks for an early-type galaxy (solid), an early spiral galaxy (dashed), and an irregular galaxy (dotted) with $M_I = -22.5$.

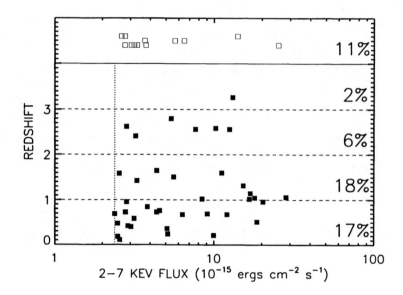

FIGURE 5. Redshift versus 2 – 7 keV flux for the combined SSA13, HDF-N (Hornschemeier et al. 2001), and A370 (Bautz et al. 2000) *Chandra* hard X-ray samples with fluxes above 2.4×10^{-15} ergs cm^{-2} s^{-1} (dashed vertical line). Spectroscopically unidentified sources are indicated with open symbols in the topmost box. Percentages in each redshift bin indicate the bin's contribution to the hard XRB from Chen, Fabian, & Gendreau (1997).

accretion time, Δt, to be 1 Gyr, the accretion masses for our 'normal' galaxies are 7×10^5 to 3×10^8 $M_\odot$. Thus, even for the maximum plausible efficiency, $\epsilon \sim 0.1$, we are seeing a substantial fraction of the growth of these supermassive black holes. A final approach is to impose the Eddington limit, $\eta \equiv L_{bol}/L_{Edd} < 1$, where η is the bolometric accretion luminosity in Eddington units, $L_{Edd} = 4\pi G M_{bh} c \mu_e m_p / \sigma_T = 1.4 \times 10^{38} M_{bh}/M_\odot$ erg s^{-1}. The value of η must be less than one, else material would not be able to accrete onto the black hole because of the radiation pressure. These minimum masses are indicated in Fig. 6 as solid squares or downward pointing open triangles, the latter symbols denoting sources with only radio flux upper limits. Clearly the 'normal' galaxies are accreting at rates much less than the Eddington rate.

CONCLUSIONS

The cosmic census of supermassive black holes in the Universe is nearing completion, and unexpected phenomena have been revealed. Luminous hard X-ray sources are found to be common in bulge dominated optically luminous galaxies

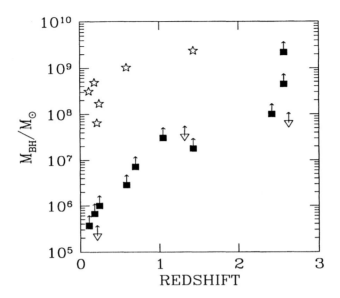

FIGURE 6. Black hole masses versus redshift for the SSA13 hard X-ray spectroscopic sample. Open stars have been estimated from the local empirical relation between black hole mass and absolute magnitude of the bulge component of the host galaxy, assuming that the relation holds at high redshift and subject to the uncertainties in translating our observed B-band magnitudes into M_B(bulge) that would worsen the relation. Solid squares with upward pointing arrows are the minimum black hole masses estimated from the Eddington luminosity limit and the bolometric luminosities. Open downward pointing triangles with upward pointing arrows are for sources that only had radio upper limits.

with about 10% of the population showing activity at any given time. Since the hard X-ray emission is associated with episodes of accretion onto the central supermassive black holes in these galaxies, the 10% represents the duty cycle of galaxies that are active at any given time. This implies that supermassive black hole accretion is slow (~ 1 Gyr) and still occurring at cosmologically recent times, which challenges the traditional view that black hole growth is only associated with host galaxy formation or violent galaxy mergers. We can estimate the integrated history of energy release in the Universe: we find that the emitted bolometric energy from accretion onto supermassive black holes is about one-third of the emitted energy from all the stars in the optical (2000–8000Å).

ACKNOWLEDGEMENTS

I thank Craig Wheeler and the other Texas Symposium organizers for inviting me to this stimulating meeting and for providing support. I acknowledge support from NASA through Hubble Fellowship grant HF-01117.01-A awarded by the Space Telescope Science Institute, which is operated by the Association of Universities for Research in Astronomy, Inc., for NASA under contract NAS 5-26555. I also acknowledge support from NSF through grant AST-0084847.

REFERENCES

1. Barger, A.J., Cowie, L.L., Sanders, D.B., *ApJ* **518**, L5 (1999).
2. Barger, A.J., Cowie, L.L., Smail, I., Ivison, R.J., Blain, A.W., Kneib, J.-P., *AJ* **117**, 2656 (1999).
3. Barger, A.J., Cowie, L.L., Richards, E.A., *AJ* **119**, 2092 (2000).
4. Barger, A.J., Cowie, L.L., Mushotzky, R.M., Richards, E.A., *AJ* **121**, 662 (2001).
5. Bautz, M.W., Malm, M.R., Baganoff, F.K., Ricker, G.R., Canizares, C.R., et al., *ApJ* **543**, L119 (2000).
6. Bernstein, R.A., et al., ASP Conference Series 170, Eds. J.I. Davies, C. Impey, S. Phillipps, Astronomical Society of the Pacific (San Francisco), p. 341 (1999).
7. Blain, A.W., Kneib, J.-P., Ivison, R.J., Smail, I., *ApJ* **512**, L87 (1999).
8. Chapman, S.C., Richards, E.A., Lewis, G., Wilson, G., Barger, A.J., *ApJL* in press, [astro-ph/0011066].
9. Chen, L.-W., Fabian, A.C., Gendreau, K.C., *MNRAS* **285**, 449 (1997).
10. Coleman, G.D., Wu, C.-C., Weedman, D.W., *ApJS* **43**, 393 (1980).
11. Condon, J.J., *ARA&A* **30**, 575 (1992).
12. Cowie, L.L., Barger, A.J., Bautz, M.W., Capak, P., Crawford, C.S., et al., *ApJL* in press, [astro-ph/0102306].
13. Crawford, C.S., Fabian, A.C., Gandhi, P., Wilman, R.J., Johnstone, R.M., *MNRAS* submitted, [astro-ph/0005242].
14. Fabian, A.C., Smail, I., Iwasawa, K., Allen, S.W., Blain, A.W., et al., *MNRAS* **315**, L8 (2000).
15. Ferrarese, L., Merritt, D., *ApJ* **539**, L9 (2000).
16. Fixsen, D.J., Dwek, E., Mather, J.C., Bennett, C.L., Shafer, R.A. *ApJ* **508**, 123 (1998).
17. Gardner, J.P., Cowie, L.L., Wainscoat, R.J. *ApJ* **415**, L9 (1993).
18. Garmire, G.P., et al., in preparation.
19. Gebhardt, K., et al., *ApJ* **539**, L13 (2000).
20. Giacconi, R., Rosati, P., Tozzi, P., Nonino, M., Hasinger, G., et al., *ApJ* in press, [astro-ph/0007240].
21. Hasinger, G., Burg, R., Giacconi, R., Schmidt, M., Trumper, J., Zamorani, G., *A&A* **329**, 482 (1998).
22. Ho, L.C., *ApJ* **516**, 672 (1999).
23. Hornschemeier, A.E., Brandt, W.N., Garmire, G.P., Schneider, D.P., Broos, P.S., et al., *ApJ* **541**, 49 (2000).
24. Hornschemeier, A.E., Brandt, W.N., Garmire, G.P., Schneider, D.P., Barger, A.J., et al., *ApJ* in press, [astro-ph/0101494].

25. Hughes, D.H., Serjeant, S., Dunlop, J., Rowan-Robinson, M., Blain, A., et al., *Nature* **394**, 241 (1998).
26. Kauffmann, G., Haehnelt, M., *MNRAS* **311**, 576.
27. Kormendy, J., Richstone, D., *ARA&A* **33**, 581 (1995).
28. Magorrian, J., et al., *AJ* **115**, 2285 (1998).
29. Marshall, F., et al., *ApJ* **235**, 4 (1980).
30. Mushotzky, R.M., Cowie, L.L., Barger, A.J., Arnaud, K.A., *Nature* **404**, 459 (2000).
31. Puget, J.-L., Abergel, A., Bernard, J.-P., Boulanger, F., Burton, W.B., et al., *A&A* **308**, L5 (1996).
32. Richards, E.A., et al., in preparation.
33. Sanders, D.B., Mirabel, I.F. 1996, *ARA&A* **34**, 749 (1996).
34. Schmidt, M., Hasinger, G., Gunn, J., Schneider, D., Burg, R., et al., *A&A* **329**, 495 (1998).
35. Ueda, Y., et al., *ApJ* **518**, 656 (1999).
36. Williams, R., et al., *AJ* **112**, 1335 (1996).
37. Wright, E.L., Reese, E.D., *ApJ* **545**, 43 (2000).

Discovery of a 2 Kpc Binary Quasar [1]

G. A. Shields*, V. Junkkarinen[†][2], E. A. Beaver[†], E. M. Burbidge[†],
R. D. Cohen[†], F. Hamann[‡], and R. W. Lyons[†]

*Department of Astronomy, University of Texas, Austin TX 78712
[†]Center for Astrophysics and Space Sciences, University of California, San Diego
La Jolla, CA 92093
[‡]Department of Astronomy, University of Florida, Gainsville FL 32611

Abstract.
LBQS 0103−2753 is a binary quasar with a separation of only 0.3 arcsec. The projected spacing of 2.3 kpc at the distance of the source ($z = 0.848$) is much smaller than that of any other known binary QSO. The binary nature is demonstrated by the very different spectra of the two components and the low probability of a chance pairing. LBQS 0103−2753 presumably is a galaxy merger with a small physical separation between the two supermassive black holes. Such objects may provide important constraints on the evolution of binary black holes and the fueling of AGN.

INTRODUCTION

Binary QSOs occur at a rate of about one per thousand QSOs [12,15]. Typical angular separations are 3″ to 10″. True binaries, as opposed to gravitational lenses, may be recognized by different radio emission or different emission-line or BAL characteristics in their component spectra [15,18]. Binary QSOs presumably represent galaxy encounters in which tidal interactions have led to an enhanced probability of nuclear activity [8]. Such events may offer important insights into the fueling of AGN and the timescales for the mergers of galaxies and their central black holes [3].

In the course of a study of QSOs with broad absorption lines (BALs), we have discovered that LBQS 0103−2753 is a binary QSO with a separation of only 0″.3 [14]. This corresponds to a physical separation of only 2 kpc. The closest previously known binaries have spacings of 15 kpc or more [7]. Such a close spacing suggests

---

[1]) Based on observations made with the NASA/ESA Hubble Space Telescope. STScI is operated by the Association of Universities for Research in Astronomy, Inc. under NASA contract NAS5-26555.
[2]) Visiting Astronomer, Cerro Tololo Inter–American Observatory, which is operated by the Association of Universities for Research in Astronomy, Inc., under contract with the National Science Foundation.

an advanced merger, with the two supermassive black holes well within the nascent bulge of the merged galaxy.

OBSERVATIONS

LBQS 0103−2753 is one of eight BALQSOs included in a study, by the authors, of the ultraviolet spectra of BALQSOs at moderate redshift with the Space Telescope Imaging Spectrograph (*STIS*) on the Hubble Space Telescope (*HST*). The object has an emission-line redshift $z_e = 0.848$ and an apparent magnitude $B_J = 18.1$, and its LBQS spectrum shows broad absorption in Mg II $\lambda 2798$ [17]. Figure 1 shows the STIS CCD acquisition image. The A/B flux ratio is 3.1, implying magnitudes $B_J = 18.2$ and 19.4 for components A and B, respectively. The FWHMs of the images are consistent with point sources, and the separation is $\Delta\theta = 0.295 \pm 0.011$ arcsec in position angle $+30.1° \pm 2.2°$ (A northeast of B).

A 52″ by 0.″2 slit was used for the spectroscopic observations (Figure 2) with the STIS NUV–MAMA using the G230L grating. The QSO pair was, by luck, almost perfectly aligned with the STIS slit, itself aligned along the image +y axis. The resolution at 2400 Å is 3.3 Å FWHM.

We obtained an optical spectrum of LBQS 0103−2753 (sum of both components) in August 1998 using the CTIO Blanco 4m telescope. The spectrum [14] shows a dramatic weakening of the Mg II Bal, which is unusual [13].

LBQS 0103−2753: A CLOSE BINARY QSO

LBQS 0103−2753 A has strong BALs of C IV $\lambda\lambda 1548, 1551$, Si IV $\lambda 1400$, N V $\lambda 1240$, and O VI $\lambda 1034$ that are absent in the spectrum of component B. In addition, the two components have very different emission-line properties. This

FIGURE 1. Acquisition image of LBQS 0103−2753 obtained with the STIS CCD. The brighter object is A, the BALQSO, and the fainter is B. Reproduced from Junkkarinen et al. (2001) with permission.

effectively rules out the possibility that the object is a lensed QSO. Junkkarinen et al. [14] argue that the likelihood also is small that LBQS 0103−2753 is a chance alignment of a wider binary of the 3″ to 10″ (∼40 kpc) variety.

Could LBQS 0103−2753 be a chance superposition of two unrelated QSOs? The emission-line redshifts of the two components, though similar, differ by more than the 600 km s^{-1} difference typical of binary QSOs [15]. The redshift of B is well determined at $z_B = 0.858$, but the BALs and emission-line profiles of component A make the emission-line redshift quite uncertain. The C IV line gives $z_A = 0.834$, which corresponds to a velocity difference from z_B of 3900 km s^{-1}. However, C IV emission lines are often blueshifted from the systemic redshift [9], especially in the case of weak, low contrast lines as in component A [6]. How likely is a chance coincidence of a QSO as bright as the apparent magnitude of component B within this redshift range and within an angular separation of 0″.3? We take a surface density of 10 QSOs per square degree between B = 18.2 and 19.4, for QSOs with redshifts $z < 2.2$ [4]. If we assume these are distributed roughly uniformly with redshift, then the odds of having a second QSO within 0″.3 of a given QSO, and with a redshift difference not more than $\Delta z = z_B - z_A = 0.024$, is $10^{-8.6}$. If ∼500 QSOs

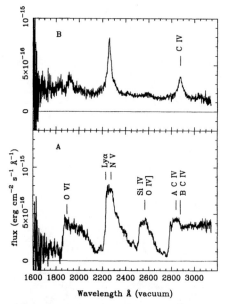

FIGURE 2. Spectra of LBQS 0103−2753 A and B obtained with HST/STIS using the NUV-MAMA and G230L grating. The upper panel is component B and the lower panel is component A. The measured C IV λ 1549 emission-line wavelengths for components A and B are both shown with vertical marks above component A's CIV λ 1549 emission line. Reproduced from Junkkarinen et al. (2001) with permission.

have been observed in a way that would reveal such a binary (see below), then the odds of having one chance alignment resembling LBQS 0103−2753, among all these targets, is only $10^{-5.9}$. This supports the conclusion that LBQS 0103−2753 is a true, physical binary.

DISCUSSION

The greatest number of opportunities to discover a pair like LBQS 0103−2753 appears to be offered by the *HST* snapshot survey for gravitationally lensed QSOs [16]. This involved 498 QSOs with $z > 1$ and $M_V < -25.5$ (H_0=100, q_0= 0.5). The snapshot survey could have detected a binary with the characteristics of LBQS 0103−2753 [1]. However, it did not find any binary QSOs with separations less than 1″.0. From the one example of LBQS 0103−2753, we therefore estimate a rate of roughly 1/500 for separations ∼0″.3. This in turn suggests that there are ∼10 unrecognized binaries in the 0″.3 range among the ∼10^4 known QSOs.

Junkkarinen et al. [14] suggest that LBQS 0103−2753 represents an advanced stage of a galactic merger. The cores of the two galaxies are merging, and the black holes are settling into an orbit of radius similar to the observed spacing of 2 kpc. The wider binaries observed, with physical spacing of ∼20 to 100 kpc, likely represent mergers at an earlier stage. Simulations of colliding disk galaxies [2] indicate that gas accumulates in the nucleus of each galaxy as the two orbit away from each other after the first close encounter. The duration of this first large orbital loop is ∼$10^{8.5}$ yr. Dynamical friction causes rapid decay of subsequent loops, and the central regions of the galaxies merge quickly. The final merger involves a massive concentration of gas in the nucleus capable of fueling powerful starbursts and AGN activity. Junkkarinen et al. [14] estimate the masses of the black holes and the properties of the host galaxies from the QSO's luminosities and the relationship between black hole mass and bulge properties of the host galaxy [10]. This leads to an estimated lifetime of the binary black hole orbit, at the observed 2 kpc spacing, of only ∼10^7 years. This in turn suggests that the probability of a merger being active as a binary QSO is larger during the 2 kpc stage than during the 40 kpc stage.

Ultraluminous infrared galaxies (ULIGS, $L_{ir} > 10^{11}$ $L_\odot$) typically involve galactic mergers [19,11]. Interestingly, the mean nuclear spacing of ∼2 kpc is similar to that of LBQS 0103−2753. The brighter ULIGs are believed to be powered predominantly by AGN. LBQS 0103−2753 is distinguished from typical ULIGs by the lack of heavy extinction, as well as an exceptionally high luminosity. Infrared observations of LBQS 0103−2753 would be valuable to determine whether massive amounts of dust are still present in the vicinity of the QSOs.

OTHER UNRESOLVED BINARY QSOS?

We have estimated that binaries as close as LBQS 0103−2753 occur with a frequency of $\sim 10^{-3}$. Given this low incidence, searches for $0''\!.3$ binaries would benefit from any criterion to narrow the list of candidates. Until our *HST* observations, LBQS 0103−2753 was not, to our knowledge, suspected of being a binary QSO. The great diversity of QSO spectra makes it difficult to specify spectroscopic criteria with which to identify candidate binaries. Objects with double peaked narrow line emission ([O III] $\lambda 5007$ or [O II] $\lambda 3727$) might include some candidates. Composite broad emission-line profiles of an unusual character might also be an indicator, e.g., the QSO B340 [20]. Ideally, one could identify two properties of QSOs that almost never occur together. The rare exceptions would then be candidates for unresolved binaries. One possibility might be BALQSOs with strong, unabsorbed soft X-ray fluxes; BALQSOs as a class show weak soft X-ray fluxes, relative to the optical continuum, evidently because of absorption [5].

Support for this work was provided by NASA through grants GO-07359, GO-07359.02 from the Space Telescope Science Institute, which is operated by the Association of Universities for Research in Astronomy under NASA contract NAS5-26555.

REFERENCES

1. Bahcall, J., Maoz, D., Doxsey, R., Schneider, D. P., Bahcall, N. A., Lahav, O., and Yanny, B. 1992, Ap. J., 387, 56
2. Barnes, J. E., and Hernquist, L. 1996, Ap. J., 370, L65
3. Begelman, M., Blandford, R., and Rees, M. 1980, Nature, 287, 307
4. Boyle, B. J., Shanks, T., and Peterson, B. A. 1988, MNRAS, 235, 935
5. Brandt, W. N., Laor, A., and Wills, B. J. 2000, Ap. J., 528, 637
6. Brotherton, M. S., Wills, B. J., Steidel, C. C., and Sargent, W. L. W. 1994, Ap. J., 423, 131
7. Brotherton, M. S., Gregg, M. D., Becker, R. H., Laurent-Muehleisen, S. A., White, R. L., and Stanford, S. A. 1999, Ap. J., 514, L61
8. Djorgovski, S. 1991, ASP Conf. 21, Space Distribution of Quasars, ed. D. Crampton (San Francisco: ASP), 349
9. Espey, B. R., Carswell, R. F., Bailey, J. A., Smith, M. G., and Ward, M. J. 1989, ApJ, 342, 666
10. Gebhardt, K., et al. 2000, Ap. J. (Letters), 539, L13
11. Genzel, R., and Cesarsky, C. 2000, Ann. Rev. Astr. Ap., 38, 761
12. Hewett, P. C., Foltz, C. B., Harding, M. E., and Lewis, G. F. 1998, Ast. J., 115, 383
13. Junkkarinen, V. T., Cohen, R. D., and Hamann, F. 1999, BAAS, 31, 951
14. Junkkarinen, V., Shields, G. A., Beaver, E. A., Burbidge, E. M., Cohen, R. D., Hamann, F., and Lyons, R. W. 2001, Ap. J. (Letters), 549, L155
15. Kochanek, C. S., Falco, E. E., and Muñoz, J. A. 1999, Ap. J., 510, 590
16. Maoz, D., et al. 1993, Ap. J., 409, 28

17. Morris, S. L., Weymann, R. J., Anderson, S. F., Hewett, P. C., Foltz, C. B., Chaffee, F. H., Francis, P. J., and MacAlpine, G. M. 1991, Ast. J., 102, 1627
18. Mortlock, D. J., Webster, R. L., Francis, P. J. 1999, MNRAS, 309 836
19. Sanders, D. B., and Mirabel, I. F. 1996, Ann. Rev. Astr. Ap., 34, 749
20. Shields, G. A., and McKee, C. F. 1981, Ap. J. (Letters), 246, L57

Black Holes and Galaxy Metamorphosis

J. Kelly Holley-Bockelmann

Case Western Reserve University
Cleveland, Ohio 44106

Abstract. Supermassive black holes can be seen as an agent of galaxy transformation. In particular, a supermassive black hole can cause a triaxial galaxy to evolve toward axisymmetry by inducing chaos in centrophilic orbit families. This is one way in which a single supermassive black hole can induce large-scale changes in the structure of its host galaxy – changes on scales far larger than the Schwarzschild radius ($O(10^{-5})$pc) and the radius of influence of the black hole ($O(1) - O(100)$pc).

INTRODUCTION

Observations are beginning to conclude that massive central black holes are a natural part of elliptical galaxy centers [1,2]. In fact, best-fit models of black hole demography indicate that approximately 97% of ellipticals harbor a massive black hole. Black hole mass seems to be correlated with the host bulge potential; current dynamical estimates of the best galactic black hole candidates have yielded masses on the order of $0.005 M_{\text{bulge}}$ [3–5]. There also seems to be a strong correlation between black hole mass and global velocity dispersion, implying that the Fundamental Plane exists even in the four-dimensional space described by ($\log M_{\text{BH}}, \log L, \log \sigma_e, \log R_e$) [6,7]. It appears, then, that black hole formation and the evolution of the host galaxy may be deeply connected. This proceeding explores one way in which black holes may drastically change the structure of an elliptical galaxy over large scales: by inducing axisymmetry in a triaxial galaxy. It appears, then, that a supermassive black hole's impact on the structure and subsequent evolution of its host galaxy is both dramatic and far-reaching.

BLACK HOLES AND TRIAXIAL GALAXIES

There is evidence, both observationally and theoretically, that elliptical galaxies are at least mildly triaxial in shape [8,9]. Even a mildly triaxial galaxy will generate entirely different orbit families than are present in a spheroid. In particular, there are a rich variety of regular box and boxlet orbits that are centrophilic and comprise the backbone of the galaxy (Figure 1). These centrophilic orbits can be driven chaotic with the introduction of a supermassive black hole [10–15]. And, since chaotic orbits will eventually fill all available phase space, the time-averaged shape of a chaotic orbit is spherical. Hence, the destruction of these centrophilic orbits breaks the backbone of the triaxial model and the system evolves toward axisymmetry. This effect has been shown in numerous computational and analytic studies [10–15]. In fact, the main controversies are whether the galaxy becomes axisymmetric locally or globally, and whether the transformation occurs in a few crossing times or over many Hubble times.

If the transformation is rapid and global, there are important implications for elliptical galaxy evolution. For example, one possible difference between an intrinsically bright elliptical (which is thought to be more triaxial) and a faint elliptical is that the faint elliptical, with its shorter crossing time, has had more interactions with the black hole and is thus more dynamically evolved [14]. Secondly, the black hole/bulge mass relation can be explained in terms of galaxy evolution [14]. In this scenario, spiral galaxies begin as gas-rich disks with a small triaxial bulge. Since triaxiality supports box orbits, gas can flow radially inward along these orbits, which can rapidly funnel matter into a black hole. The black hole grows until a critical black hole mass of $M_\bullet = 0.02 M_{\rm gal}$, which breaks triaxiality and strongly curtails the gas inflow. Subsequent disk-disk merging can create a elliptical galaxy, and black hole feeding ensues in this larger triaxial bulge until the critical black hole mass is achieved. In both types of galaxies, the process is the same: once the black hole mass fraction is large enough to disrupt box orbits, gas inflow is sharply diminished.

However, most self-consistent N-body simulations which have studied this effect have employed astrophysically unrealistic galaxy models [10,13]. For example, the models have been highly flattened, maximally triaxial models, while observations indicate that ellipticals are most likely mildly triaxial and not very flattened. Furthermore, the black hole was often grown within a flat inner density profile (a 'core') which, while it maximized the change in the inner potential, does not reflect the observations of cuspy inner density profiles in ellipticals.

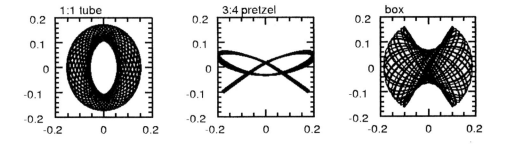

FIGURE 1. Planar Orbits in Triaxial Potentials.

My collaborators and I [15] have explored the issue of black hole-embedded triaxial galaxies with more realistic galaxy models. Our models exhibited a range of moderate triaxialities ($0.25 < T < 0.75$) with mild flattening and cuspy inner density profiles. They were generated by 'adiabatically squeezing' a cuspy Hernquist sphere into a stable triaxial figure, and then adiabatically growing a central black hole in this model [15]. The models were populated with $N = 512,000$ multimass particles to accurately reproduce the central density cusp, and the particles were advanced with a multiple timestep, high-order Hermite integrator in the SCFcode.

Figure 2 shows the change in the physical structure of a triaxial model as the black hole is grown. Initally, the model had axis lengths $a : b : c = 1 : 0.85 : 0.7$, and a central density cusp $\gamma = 1$. If this galaxy were to lie on the core and global Fundamental Planes, the density cusp dictates the absolute magnitude ($M_v \approx -21.6$), mass ($M_{\text{gal}} \approx 2 \times 10^{12} M_\odot$), core radius ($r_{\text{core}} \approx 150\text{pc}$) and effective radius ($r_{\text{eff}} \approx 4\text{kpc}$) of the corresponding galaxy.

Although the final state of this black hole-embedded model is decidedly not axisymmetric on global scales, it is certainly more round near the center. We developed an automated orbit analysis routine to determine if this central roundening was caused by black hole-induced centrophilic orbit destabilization. Figure 3 presents xy,xz surfaces of section for the initial state of the triaxial potential (ie no black hole). Despite the rich variety of resonant boxlets, boxes, and tubes, less than 0.2% of the orbits were strongly chaotic (see [15] for details).

After the black hole has grown, the orbital content is quite different. In the

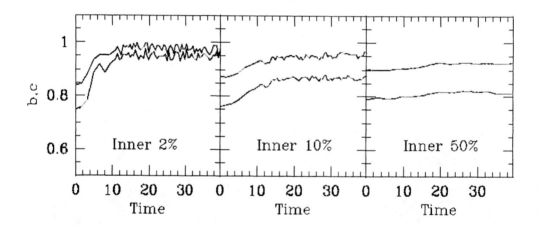

FIGURE 2. The intermediate and minor axes lengths as a function of time for particle sets binned by mass in the model. The axes lengths are iteratively calculated from the ellipsoidal density distribution using the moment of inertia tensor. The lack of evolution in the axes lengths in the last panel argues for a stable shape at the half mass radius, in spite of the black hole.

inner regions, nearly all centrophilic orbits have become stochastic, including the population of eccentric loops. However, for less bound orbits, there are only a scattered few strongly chaotic orbits. A common interpretation for the lack of chaos in these less bound orbits is that they have been integrated for far fewer dynamical times than the highly bound orbits. Thus these outermost orbits have not been exposed to the perturbative black hole potential enough for substantial chaos to set in. To test this explanation, we integrated a subset of the lesser bound box and boxlet orbits ($E = -0.40$) in the xy plane for ≈ 200 orbital times. Although the percentage of chaotic orbits in this subset increased from 4% to 71% over the experiment, many stable centrophilic orbits existed after ≈ 2 Hubble times, including a large fraction of non-resonant boxes. A possible explanation is that these weakly bound orbits spend very little of the orbit in the inflection region between spherical potential and the triaxial potential, and are thus not driven stochastic at all. This explanation is far different than scenarios which involve scattering by the black hole, but black hole scattering is difficult to envision as the cause of the chaos observed in our models, since there were many outer orbits that passed quite close to the black hole, yet remained stable after hundreds of dynamical times.

CONCLUSION

The black hole in our model induced axisymmetry out to nearly 100 parsecs, and resulted in a clearly observable change in the shape and structure of the galaxy. Since the transformation did not take place globally, it is tempting to say that the black hole mass/bulge mass relation observed in the current galaxy population is it not simply an artifact of gas inflow in a more triaxial-shaped progenitor population. However, it is not immediately clear how the more localized axisymmetry we observed would effect gas inflow and subsequent black hole feeding. While it it true in our globally triaxial model that gas inflow from outside the half mass radius would never be entirely cut off, the behavior of the gas once it hits the axisymmetric region requires detailed gas dynamical simulations. Nonetheless, it is clear that a central supermassive black hole causes dramatic and long-lasting changes in the host galaxy over scales well outside the region in which it dominates the potential.

REFERENCES

1. Gebhardt, K. *et al* 2000, ApJL, 543, 5.
2. Richstone, D. *et al* 1998, Nature, 395, 14.
3. Kormendy, J. & Richstone D. 1995, ARAA, 33, 581.
4. Magorrian, J. *et al* 1998, AJ, 115, 2285.
5. van der Marel, R. 1999, AJ, 117, 744.
6. Gebhardt, K. *et al* 2000, ApJL, 543, 13.
7. Ferrarese, L., Merritt, D. 2000, ApJL, 539, 9.

8. Bak, J. & Statler, T. 2000, AJ, 120, 110.
9. Dubinski, J. & Carlberg, R. 1991, ApJ, 378, 496.

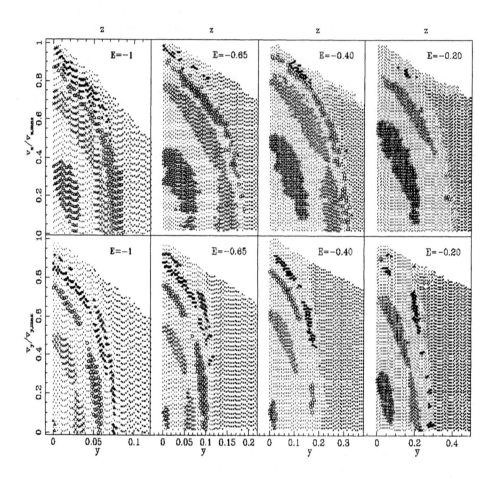

FIGURE 3. Surfaces of section for the initial triaxial model without a central black hole, plotted for orbital populations of differing binding energies. Top: Surfaces of section for orbits in the xz plane. Bottom: Surfaces of section for orbits in the xy plane. Orbits are coded by point type - strongly chaotic: large filled circles, loops: filled boxes, boxes: small points, bananas: horizontal lines, fish: open triangles, pretzels: X marks, 5:4 resonance: open pentagons, 6:5 resonance: filled hexagons, 7:6 resonance: asterisks. We zoom in on the x-axis of the plot to show as many box and boxlets as possible; the orbits outside the limits of the plot are all tubes. If plotted to the full extent of the x axis, the boxlet region would comprise $\sim$ 50% of the most bound panels, and only $\sim$ 10% of the least bound panels. Notice that there are no strongly chaotic regions in this stable triaxial figure. See http://burro.astr.cwru.edu/kelly/texas.talk/triax4.html for this plot in color.

10. Norman, C., May, A., van Albada, T. 1985, ApJ, 296, 20.
11. Gerhard, O., & Binney, J. 1985, MNRAS, 216, 467.
12. Fridman, T. & Merritt, D. 1997, ApJ, 114, 1479.
13. Merritt, D. & Quinlan, G. 1998, ApJ, 498, 625.
14. Valluri, M. & Merritt, D. 1998, ApJ, 506, 686.
15. Holley-Bockelmann, K., Mihos, C., Sigurdsson, S., Hernquist, L. 2001, ApJ, in press.

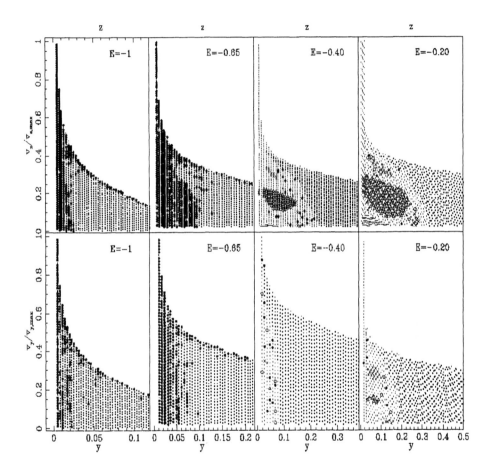

FIGURE 4. Surfaces of section for the final state of the triaxial model with a central black hole of mass $M_\bullet = 0.01 M_{\text{gal}}$. See Figure 3 for explanation of symbols. Notice that the box and boxlet space in the inner regions of this model is almost entirely taken over by chaotic orbits, while the outer region is nearly devoid of strongly chaotic orbits. See http://burro.astr.cwru.edu/kelly/texas.talk/triax5.html for this plot in color.

CHAPTER 9

ACCELERATION AND COLLIMATION OF COSMIC JETS

High Energy Phenomena in Blazars

L. Maraschi

Osservatorio Astronomico di Brera, via Brera 28, 20121 Milano, Italy

Abstract. Advances in the capabilities of X-ray, gamma-ray and TeV telescopes have brought new information on the physics of relativistic jets, which are responsible for the blazar "phenomenon". In particular the broad band sensitivity of the *Beppo*SAX satellite, extending up to 100 KeV has allowed unprecedented studies of their hard X-ray spectra. I summarize here some basic results and present a unified view of the blazar population, whereby all sources contain essentially similar jets despite diversities in other properties, like the presence or absence of emission lines in their optical spectra. Blazars with emission lines are of particular interest in that it is possible to estimate both the luminosity of the jet and the luminosity of the accretion disk. Implications for the origin of the power carried by relativistic jets, possibly involving rapidly spinning supermassive black holes are discussed. We suggest that emission line blazars are accreting at near critical rates, while BL lacs, where emission lines are weak or absent are highly subcritical.

INTRODUCTION

The original argument which led to hypothesize supermassive black holes as basic engines for the AGN phenomenon (Zeldovich & Novikov 1964; Salpeter 1964) was extremely "simple". A source of high luminosity which varies rapidly must be very "compact" : since its size R must be less than ct_{var} the photon density $L/(4\pi R^2 c)$ is extremely high. A source of high compactness (defined in adimensional terms as $l = L\sigma_T/Rmc^3$) must be very efficient as shown most clearly by Fabian (1979) who derived the well known limit $\Delta L/\Delta t \lesssim \eta \, 10^{43} erg/sec^2$, where η is the radiative efficiency. Accretion onto black holes can be orders of magnitude more efficient than ordinary nuclear reactions powering stars, with η up to 42% (e.g., Rees 1984) but still limited to values < 1.

Thus there is a maximum compactness that any source powered by accretion cannot exceed. However observationally some AGN violate even this limit. Early results concerned the excessive brightness temperatures inferred from the variability of compact radio sources. More recently, the observation of gamma-rays varying rapidly, from the same class of sources exhibiting the fast radio variability, provided new independent evidence of violation of the fundamental compactness limit (Maraschi et al. 1993). For this relatively small fraction of AGN, called blazars, relativistic motion of the emitting plasma has been invoked in order to reconcile the observed properties with basic physics (Blandford and Rees,1978).

It is now generally accepted that the blazar "phenomenon" (highly polarised and rapidly variable radio/optical continuum) is due to a relativistic jet pointing close to the line of sight. An additional step towards unification is to propose that jets are basically similar in all blazars , despite diversities in other properties , most notably the presence or absence of emission lines in their optical spectra (flat spectrum quasars (FSQ) vs. BL Lacs). This hypothesis was put forward by Maraschi & Rovetti (1994) on the basis of the "continuity" of the radio and X-ray luminosity functions of the two classes of blazars.

Here I will follow the point of view that Quasars with Flat Radio Spectrum (FSQs, which include OVVs and HPQs) and BL Lac objects belong to a single population, in the sense that the nature of the central engine is similar apart from differences in scaling. The plan is to start from a physical comprehension of the phenomenology common to all blazars, in particular of the broad band spectral energy distributions (SEDs) from radio to γ-rays, with the goal of understanding eventually the role of more fundamental parameters, like the central black hole mass, angular momentum and accretion rate, in determining the properties of the jets and of the associated accretion disks.

THE UNIFIED FRAMEWORK FOR THE SEDS OF BLAZARS.

It was noted early on that the SEDs of blazars exhibit remarkable systematic properties (Landau et al. 1986, Sambruna et al. 1996). The subsequent discovery by the Compton Gamma Ray Observatory of gamma-ray emission from blazars (a summary can be found in Mukherjee et al. 1997) was a major step forward, showing that in many cases the bulk of the luminosity was emitted in this band and questioning the importance of previous studies of the SEDs at lower frequencies.

A systematic investigation on the SEDs of the main complete samples of blazars (X-ray selected, radio-selected and Quasar-like, Fossati et al. 1998) including gamma-ray data showed that the systematic trends found previously indeed persisted, suggesting a continuity of spectral properties (spectral sequence). A suggestive plot where sources from different complete samples have been grouped in radio-luminosity decades (see Fossati et al. 1998 for a full descripton) is shown in Fig. 1.

All the SEDs show two broad components with peaks in the $10^{13} - 10^{18}$Hz and $10^{21} - 10^{25}$Hz ranges respectively. Both peaks appear to shift to higher frequencies with decreasing luminosity. We will call red and blue the objects respectively at the low and high frequency extremes of the sequence.

Beamed synchrotron and inverse Compton emission from a single population of relativistic electrons accounts well for the first and second peak respectively in the observed SEDs. Note that the relativistic particle spectrum must be "curved" in order to explain the peaks observed in the SEDs. The curvature is often modelled with a broken power law. The change in spectral index must be quite large to explain the emission peaks and the energy, γ_b, at which the change (or break) occurs corresponds to the energy of electrons which radiate at the peak.

Homogenous models fail however to reproduce the SEDs in the radio to mm range, where selfabsorption cuts off the contribution of the electron population accounting for the higher energy emission (see also Kubo et al. 1998). In fact it is well known that at low frequency the observed spectra are due to the superposition of different components located further down the jet, with selfabsorption turnover at lower and lower frequencies (e.g., Konigl 1989). These "external" regions, with scales of the order of parsecs resolved by VLBI observations, are not considered in the present discussion which refers to scales

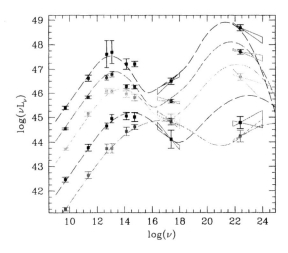

FIGURE 1. Average SEDs of the "merged" blazar sample binned according to radio luminosity, irrespective of the original classification. Empty asymmetric triangles represent uncertainties in spectral shapes as measured in the X-ray and γ-ray bands. The overlayed dashed curves are analytic approximations obtained assuming that the ratio of the two peak frequencies is constant and that the luminosity of the second peak is proportional to the radio luminosity (from Fossati et al.1998).

of the order of light days or smaller (in the observer's frame).

The homogenous model predicts that the synchrotron and IC emissions should vary in a correlated fashion, since they derive from the same electron population. In particular radiation at frequencies near the two peaks derives form electrons in the same energy interval (in the absence of Klein-Nishina effects) therefore variations in these two "corresponding" frequency ranges should be strongly correlated. This has been verified at least in some well studied objects (see below).

Ghisellini et al. 1998 derived the physical parameters of jets of different luminosities along the spectral sequence shown in Fig. 1 applying the above model and including seed photons of internal (SSC) as well as external (EC) origin for the inverse Compton process. The results suggest that i) the importance of external seed photons increases with increasing jet luminosity ii) the "critical" energy of the radiating electrons decreases with increasing (total) radiation energy density as seen in the jet frame. The latter dependence is physically plausible since the radiation energy density determines the energy losses of relativistic particles and may limit the energy attained by particles in shock acceleration processes.

In a broader perspective, if FSQs and BL Lacs contain "similar" jets (at least close to the nucleus) as suggested by the continuity of the SEDs, we still need to understand the differences in emission line properties. Also in this respect continuity could hold, in the sense that the accretion rate may decrease continuously along the sequence but the emission properties of the disk may not simply scale with the accretion rate.

STUDIES OF INDIVIDUAL OBJECTS.

It is impossible to review here all the important multifrequency studies of selected objects often triggered by flares (e.g. Bloom et al. 1996, Kubo et al. 1998, Tagliaferri et al. 2000, 2001, Sambruna 2000 and references therein,) We discuss below only three cases which are among those with the best data collection and can be considered representative of the behaviour of sources at the red and blue ends of the blazar sequence.

3C 279

This source is a prototype of red blazars and the first one to be discovered as a powerful gamma-ray emitter. Its spectral energy distribution illustrates well the presence of two main continuum components, the first one peaking in the IR, the second one in the gamma-ray range, attributed to the Synchrotron and inverse Compton mechanisms respectively. Two SEDs with simultaneous optical, X-ray and gamma-ray data obtained respectively in 1996 and 1997 are shown in Fig. 2. The 1997 X-ray data derive from observations with *Beppo*SAX (Hartman et al. 2001, Maraschi et al. in preparation). The two SEDs differ largely in brightness: that of 1996 is close to a historical maximum, while that of 1997 is rather faint. Clearly the two components of the SED vary in a correlated fashion (this is confirmed by a handful of other observations with comparable simultaneous multifrequency coverage) and the variability amplitude in the IR-optical branch is much less than at gamma-rays. This was predicted by the SSC model (Ghisellini & Maraschi 1996) but the physical parameters derived using a homogeneous SSC model are inconsistent with the rapid gamma-ray variations observed. Applying the EC model for gamma-ray production yields acceptable parameters. However the photon field surrounding the jet is not expected to vary rapidly except under special conditions (Ghisellini & Madau 1996, Bednarek 1998, Böttcher & Dermer 1998,). The large amplitude variability in gamma-rays can then only be attributed to a variation in the bulk flow velocity as illustrated in Fig 2.

The two spectral states have been reproduced using the same theoretical model (e.g. Tavecchio et al. 2000) varying only the bulk Lorentz factor Γ of the emitting plasma (Maraschi et al. 2001 in prep). The contributions of synchrotron photons and of external photons to the Inverse Compton emission are shown separately (dot dashed and dashed lines respectively. While this picture is probably still oversimplified, it fits nicely with a recently proposed scenario derived from the "internal shock" model developed for Gamma-Ray Bursts. In the latter (Spada et al. 2001, in press), most of the variability is attributed to the collisions of plasma sheaths moving along the jet with different Γs. This scenario is very promising for explaining the full range of variability of 3C 279.

For red blazars the study of the synchrotron component is difficult, because the peak falls in the poorly covered IR - FIR range. Furthermore the study of the gamma-ray component in the MeV-GeV region of the spectrum has been difficult in the last few years due to the loss of efficiency of EGRET and is now impossible after reentry of CGRO. Substantial progress will have to await the launch of new gamma-ray satellites, like AGILE planned by the Italian Space Agency (ASI) and GLAST by NASA.

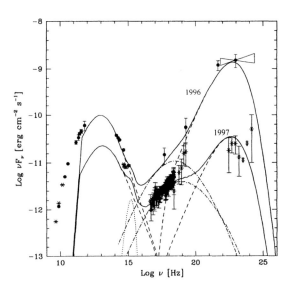

FIGURE 2. Quasi-simultaneous SEDs of the quasar 3C279 obtained at two epochs 1996 and 1997. The 1996 SED represents a historical maximum. The X-ray data of 1997 were obtained with *Beppo*SAX. The continuous lines represent synchrotron plus Inverse Compton models computed to reproduce the observations at the two epochs (see text). For both epochs subcomponents of the Inverse Compton emission are shown as dot-dashed (SSC) and dashed (EC) lines respectively. The dotted component is a Blackbody approximating the estimated emission from an accretion disk, assumed to be constant. The models for the two epochs differ mainly in the value of Γ_{bulk}

Mkn 501, Mkn 421

In the last years high energy observations have concentrated on blue blazars. For several sources of this class the Synchrotron component peaks in the X-ray band, where numerous satellites can provide good data. In few bright extreme BL Lac objects the high energy γ-ray component is observable from ground with TeV telescopes (for a general account see Catanese & Weekes 1999). In these particular cases the contemporaneous X-ray/TeV monotoring demonstrated well the correlation between the Synchrotron and the IC components. Dramatic TeV flares exhibited by Mkn 501 were accompanied by exceptional outbursts in X-rays where *in the brightest state the peak of the synchrotron spectrum reached 100 KeV* as observed by *Beppo*SAX (Pian et al. 1998). Similar behaviour was observed with RXTE (Sambruna et al. 2000, Catanese & Sambruna 2000).

Another important case is that of Mkn 421 for which a rapid flare (timescale of hours) was observed simultaneously in the TeV and X-ray bands by the Whipple observatory and by the *Beppo*SAX satellite in 1998. This observation probed for the first time the existence of correlation on short time scales (Maraschi et al. 1999). An associated intensive multiwavelength campaign (involving EUVE ASCA RXTE and the CAT, HEGRA and Whipple observatories) organized by Takahashi showed that TeV variations

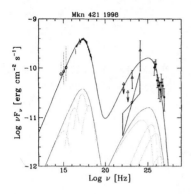
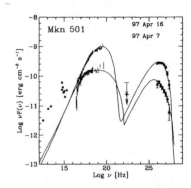

FIGURE 3. *Left:* Overall SED of Mkn 421 obtained from observations taken in 1998 April (from Maraschi et al. 1999). The X-ray and TeV data are exactly simultaneous. The solid line is the spectrum computed with the SSC model. Below the actual model, subcomponents due to electrons in 4 fixed energy intervals are shown both for the synchrotron and SC processes. *Right:* Overall SED of Mkn 501 observed simultaneously by *Beppo*SAX and CAT during the major flare in April 1997 (From Tavecchio et al. 2000, in press). The solid lines are the spectra computed with the SSC model. The models for the high and low state differ only in the value of $\gamma_b mc^2$, the energy of the electrons radiating at the peak of the synchrotron component. For both sources the observational constraints on the two peaks allow to obtain robust estimates of the physical parameters of the jet.

correlate with X-rays also on longer timescales and confirmed the trend of spectral hardening with increasing X-ray intensity (Takahashi et al. 1999).

When the position of the synchrotron and SSC peaks can be well determined observationally, as is possible in this type of sources, robust estimates of the physical parameters of the jet can be obtained (e.g. Tavecchio et al. 1998). This was done for both Mkn 421 and Mkn 501 as illustrated in Fig. 3 (Maraschi et al. 1999, Tavecchio et al., in press). For Mkn 421 subcomponents due to electrons in 4 fixed energy intervals are shown in Fig. 3 both for the synchrotron and SSC process, in order to give an intuitive view of the correlation between different energy ranges. Note for Mkn 501 the definite change in the TeV spectra measured by the CAT group (Djannati-Ataı et al. 1999) indicating a shift of the IC peak consistent with the change of the X-ray spectrum.

As a result of these model fits to different states with the simultaneous constraint on the X-ray and TeV spectra, we can confidently deduce that the flares are due to an increase of the critical electron energy γ_b rather than to a variation of the bulk Lorentz factor Γ as suggested for 3C279. These "modes" of variability may represent a significant difference between red and blue blazars.

JET POWER VS. ACCRETION POWER

We now turn to discuss luminous blazars with emission lines. These fall at the high-luminosity end of the sequence, with the Synchrotron peak in the FIR region. In these sources the beamed X-ray emission is believed to be produced through IC scattering between soft photons external to the jet (produced and/or scattered by the Broad Line Region) and *relativistic electrons at the low energy end of their energy distribution*. It is important to stress that the broad band sensitivity of *Beppo*SAX allowed to measure the X-ray spectra from 0.3 up to 100 KeV for a number of these objects. Note that for this type of sources the hard X-ray emission has luminosity comparable to that measured in gamma-rays, due to the fact that the EC peak falls in between the two ranges.

Measuring the X-ray spectra and adapting a broad band model to their SEDs yields reliable estimates of the total number of relativistic particles involved, which is dominated by those at the lowest energies. This is interesting in view of a determination of the total energy flux along the jet (e.g. Celotti et al. 1997, Sikora et al. 1997). The total "kinetic" power of the jet can be written as:

$$P_{\text{jet}} = \pi R^2 \beta c U \Gamma^2 \tag{1}$$

where R is the jet radius, Γ is the bulk Lorentz factor and U is the total energy density in the jet, including radiation, magnetic field, relativistic particles and eventually protons. If one assumes that there is 1 (cold) proton per relativistic electron, the proton contribution is usually dominant.

In high luminosity blazars the UV bump is often directly observed and/or can be estimated from the measurable emission lines, yielding direct information on the accretion process in the hypothesis that the UV emission derives from an accretion disk. *Thus the relation between accretion power and jet power can be explored*. This approach was started by Celotti et al. (1997) but their estimates of P_{jet} were obtained applying the SSC theory to VLBI radio data which refer to pc scales, much larger than the region responsible for the high energy emission ($10^{-2} - 10^{-3}$ pc).

We took advantage of *Beppo*SAX data for a number of emission line blazars deriving their jet powers as described above. The SEDs for three objects together with the models computed to represent the data are shown in Fig 4 (from Tavecchio et al. 2000). We have preliminary results for 6 other sources with similar characteristics, all observed with *Beppo*SAX. We further consider blazars with less prominent emission lines, for which we had previuos good quality *Beppo*SAX and multifrequency data, namely 3C 279, BL Lac, ON231 (Tagliaferri et al. 2001, Tagliaferri et al. 2000) plus Mkn 501 and Mkn 421 discussed above.

In all cases we estimated physical parameters by means of a homogeneous SSC+EC model and derived accordingly the kinetic power of the jet including 1 cold proton per electron, P_{jet}, as well as the total luminosity radiated by the jet in the observer frame (L_{jet}). The luminosity of the disk could be estimated for all objects except the latter three BL Lac, for which we could set only upper limits on the luminosity of their putative accretion disks. For 3C 279 and BL Lac, the presence of broad Ly_α and H_α respectively allowed to estimate the ionizing continuum (e.g. Corbett et al. 2000).

In Fig 5a the derived radiative luminosity L_{jet} and kinetic power of the jet $P\text{jet}$ are compared. The ratio between these two quantities gives directly the "radiative efficiency" of the jet, which turns out to be $\eta \simeq 0.1$, though with large scatter. The line traces the result of a least-squares fit: we found a slope ~ 1.3, indicating a decrease of the radiative efficiency with decreasing power.

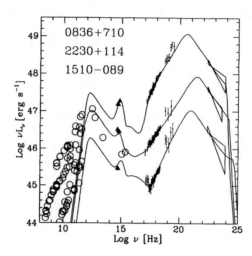

FIGURE 4. Overall SEDs of three powerful emission-lines Blazars (from Tavecchio et al. 2000). The continuous lines are theoretical models computed to account for the jet emission, plus a blackbody component. The objects are characterized by the presence of a strong UV-bump, allowing the determination of the luminosity of the accretion disk as well as the luminosity and power of the jet from the model fits to the non thermal emission.

In Fig. 3b we compare the luminosity of the jet, L_{jet}, which is a *lower limit* to P_{jet}, with the luminosity of the disk, L_{disk}.

A first important result is that on average the minimal power transported by the jet is *of the same order* as the luminosity released in the accretion disk. This result poses an important constraint for models elaborated to explain the formation of jets.

Two main classes of models consider either extraction of rotational energy from the black hole itself or magnetohydrodynamic winds associated with the inner regions of accretion disks. Let us parametrize the two possibilities as follows. Blandford & Znajek (1977) summarize the result of their complex analysis of extraction of rotational energy from a black hole in the well known expression:

$$P_{BZ} \simeq B_0^2 r_g^2 a^2 c \qquad (2)$$

Assuming maximal rotation for the black hole ($a = 1$), the critical problem is the estimate of the intensity reached by the magnetic field threading the event horizon, which must be provided by the accreting matter. Using a spherical free fall approximation with $B_0^2/8\pi \simeq \rho c^2$ one can write:

$$P_{BZ} \simeq g\dot{M}c^2 \qquad (3)$$

where $P_{acc} = \dot{M}c^2$ is the accretion power and g is of order 1 in the spherical case, but in fact it is a highly uncertain number since it also depends on the field configuration.

Several authors have recently discussed this difficult issue in the case of an accretion disk: the arguments discussed by Ghosh & Abramovicz (1997) (GA; see also Livio, Ogilvie & Pringle 1999) plus equipartition within an accretion disk described by the Shakura and Sunyaev (1973) model lead to $g \simeq 1$ when gas pressure dominates. However at high accretion rates, when radiation pressure dominates, the pressure and consequently the estimated magnetic field do not increase further with $\dot{M}$ but saturate at the transition value. The estimates of P_{BZ} derived by GA for various values of the mass of the central black hole are compared with the values of L_{jet} and L_{disk} in Fig 5b. The accretion rate which appears in the formulae of GA has been converted into a disk luminosity using an efficiency $\varepsilon \simeq 0.1$, while 100% radiative efficiency has been assumed for the jet. Clearly the model fails to explain the large power observed in the jets of bright quasars, even for BH masses ($M \sim 10^9 M_\odot$). Different hypotheses on the structure of the flow near the black hole, for instance frame dragging by the rotating hole may however increase g to values even larger than 1 (Meier 1999, Krolik 1999; Acceleration and collimation of cosmic jets (session 6), this volume).

As argued by Livio et al. the accretion flow itself may power jets through a hydromagnetic wind. However for consistency only some fraction $f\dot{M}c^2$ can be used to power the jet. Further recall that the luminosities observed from the jet and disk are related to their respective powers by efficiency factors $L_{jet} = \eta P_{jet}$; $L_{disk} = \varepsilon P_{acc}$.

Using the condition that $P_{jet} \leq (P_{BZ} + f P_{acc})$ together with the previous relations we finally find

$$L_{jet} \leq \frac{\eta(g+f)}{\varepsilon} L_{disk}. \qquad (4)$$

The data we have used suggest $L_{jet} \simeq L_{disk}$ at high luminosities and $L_{jet} > L_{disk}$ at intermediate and low luminosities.

At the high luminosity end the observed luminosities are extremely large. For a disk luminosity of 10^{47} erg s^{-1} a mass of 10^9 $M_\odot$ is implied if the disk is close to the Eddington luminosity, which corresponds to an accretion rate of 10 $M_\odot$ y^{-1} for $\varepsilon = 10^{-1}$. It seems then implausible that such disks could be low efficiency radiators. Assuming that $\eta = 10^{-1}$ as estimated above (Fig 5a), the near equality of L_{jet} and L_{disk} requires g or f or both to be of order 1.

On the other hand, a dominance of L_{jet} over L_{disk} at lower luminosities could be attributed to a lower value of $\varepsilon \ll 0.1$ which may be expected if the accretion rate is largely sub-Eddington (e.g., Blandford 1990). In the latter case the range in luminosities spanned by Fig 5b should be mainly a range in accretion rates rather than a range in black hole masses. For instance the minimum jet powers of three of the BL Lacs in Fig 5b are around 10^{44} erg/s which suggests $P_{jet} \simeq 10^{45}$ requiring a mass of 10^7 for critical accretion rate. Since the disk luminosity is less than 10^{42}, if the accretion rate is 1% Eddington the implied mass is again 10^9 $M_\odot$. This scenario is attractive (see also Cavaliere & Malquori 1999) and could be verified observationally if the mass of the central object can be determined independently. In fact for low luminosity objects the velocity dispersion close to the core of the galaxy, indicative of the central black hole mass (e.g. Ferrarese et al. 2000) should be measurable.

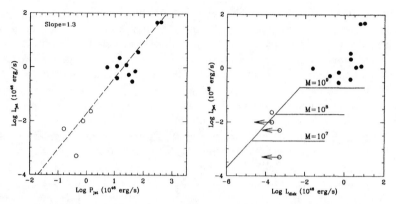

FIGURE 5. *Left:* Radiative luminosity vs. jet power for the sample of Blazars discussed in the text (open circles represent BL Lac objects). The dashed line indicates the least-squares fit to the data. *Right:* Radiative luminosity of jets vs disk luminosity. The solid lines represent the *maximum* jet power estimated for the Blandford & Znajek model for black holes with different masses (in Solar units).

CONCLUSIONS

The study of blazars yields unique information on the physical conditions and emission processes in relativistic jets. A unified approach is possible whereby the jets in all blazars are similar and their power sets the basic scale. While the phenomenological framework is suggested to be "simple" (e.g. "red" blazars are highly luminous, have low average electron energies and emit GeV gamma-rays while "blue" blazars have low luminosity, high average electron energies and emit TeV gamma-rays) we do not yet know what determines the emission properties of jets of different power nor what determines the jet power in a given AGN. We suggest that the basic parameter may be the accretion rate rather than the black hole mass that is all blazars contain very massive black holes and the lower luminosity ones are accreting at sub-Eddington rates. An observational verification of this hypothesis may come from black hole mass determinations in the nearest, lowest luminosity BL Lacs.

Blandford, R.D., & Znajek, R.L., 1977, MNRAS, 179, 433
Blandford, R.D., & Rees, M.J., 1978, Pittsburgh Conf. on BL Lac Objects, p. 341-347.
Blandford, R.D. 1990, Saas-Fee Advanced Course 20. Lecture Notes 1990. Swiss Society for Astrophysics and Astronomy, XII, 280 pp. 97.
Bloom, S. D. et al.1997, ApJ, 490, L145
Bednarek, W., 1998, A&A, 386, 123
Böttcher, M. & Dermer, C.D., 1998, ApJ, 501, 51
Catanese, M., & Weekes, T.C., 1999, PASP, 111, 1193
Catanese, M.& Sambruna, R. M. 2000, ApJ, 534, L39
Cavaliere, A. & Malquori, D. 1999, ApJ, 516, L9
Celotti, A., Padovani, P., & Ghisellini, G. 1997, MNRAS, 286, 415
Corbett, E. A., et al. 2000, MNRAS, 311, 485

Djannati-Ataı, A. et al. 1999, A&A, 350, 17
Fabian, A.C. 1979, Proc. Royal. Soc., 366, 449
Fossati, G., et al. 1998, MNRAS, 299, 433
Ghisellini, G., & Madau, P., 1996, MNRAS, 280, 67
Ghisellini, G., & Maraschi, L., 1996, ASP Conf. Series, 110, 436
Ghisellini, G., et al. 1998, MNRAS, 301, 451
Ghosh, P., & Abramowicz, M.A. 1997, MNRAS, 292, 887
Krolik, J.H., 1999, ApJ, 515, L73
Hartman, R.C. et al. 2001, ApJ, in press
Konigl, A., 1989, in "BL Lac Objects"..., 321
Kubo, H., et al. 1998, ApJ, 504, 693
Landau, R., et al. 1986, ApJ, 308, 78
Livio, M., Ogilvie, G. I. & Pringle, J. E. 1999, ApJ, 512, 100
Maraschi, L. & Rovetti, F. 1994, ApJ, 436, 79
Maraschi, L., et al. 1999, ApJ, 526, L81
Meier, D. L. 1999, ApJ, 522, 753
Mukherjee, R., et al. 1997, ApJ, 490, 116
Pian, E., et al. 1998, ApJ, 492, L17
Rees, M.J., 1984, ARA&A, 22, 471
Salpeter, E. E. 1964, ApJ, 140, 796
Sambruna, R.,M., Maraschi, L., &Urry, C.M., 1996, ApJ, 463, 444
Sambruna, R. M., 2000, AIP, 2000. AIP Conference Proceedings, Vol. 515, 19
Sambruna, R. M., et al. 2000, ApJ, 538, 127
Shakura, N. I. & Sunyaev, R. A. 1973, A&A, 24, 337
Sikora, M., et al. 1997, ApJ, 484, 108
Takahashi, T., et al. 1999, Astroparticle Physics, 11, 177
Tagliaferri, G. et al. 2000, A&A, 354, 431
Tagliaferri, G.et al. 2001, A&A, 368, 38
Tagliaferri, G. et al. 2001, to appear Proceedings of the Conference "X-ray Astronomy'99: Stellar Endpoints, AGN and Diffuse Background", 2000, Astrophysical Letters and Communications, in press
Tavecchio, F., Maraschi, L., & Ghisellini, G., 1998, ApJ, 509, 608
Tavecchio, F., et al. 2000, ApJ, 543, 535
Zeldovich Y.B., & Novikov, I.D. 1964 Usp. Fiz. Nauk 84, 377

Jet Power and Jet Suppression: The Role of Disk Structure and Black Hole Rotation

David L. Meier*

*Jet Propulsion Laboratory, California Institute of Technology [1]
Pasadena California 91109

Abstract. I review a model in which the strongest radio-emitting jets are produced by black hole systems only when the accretion flow is both geometrically thick *and* the black hole is rotating rapidly. The model accounts for the radio powers of the strongest extragalactic sources and explains the correlation between radio jet power and accretion disk state seen in galactic black hole candidate sources. It appears unlikely that an ADAF-accreting Schwarzschild black hole will be able to produce a strong jet, despite the possibility of an inner rotationally-supported disk near the last stable orbit; the poloidal magnetic field will not be strong enough in that region. Comparison of these theoretical results with a radio power/black hole mass relation recently derived for quasars suggests that more massive black holes (*e.g.*, bright quasars) may be rotating more rapidly than less massive ones (*e.g.*, Seyfert galaxies).

INTRODUCTION

There is now substantial evidence that, in binary black hole candidate X-ray sources, a radio-emitting jet outflow is produced only when the X-rays are in the hard state [1-6]. In addition, there is some evidence that those black hole systems that produce a strong jet (both stellar-mass holes and supermassive ones) contain a rapidly-rotating Kerr black hole. In particular, in the supermassive case, this "spin paradigm" has useful advantages for explaining the difference between radio loud and radio quiet quasi-stellar objects (QSOs) and why giant radio sources occur only in elliptical galaxies [7,8].

Meier [9] has shown that these properties are a natural consequence of black hole accretion and can be explained by combining modern disk accretion theory (including geometrically thick accretion flows), standard jet production theory (in which the jet power is a strong function of the poloidal magnetic field and the rotation rate of that field), and simple dynamo theory (in which the poloidal field strength is a strong function of disk thickness). While thin disks have rapid Keplerian rotation

---

[1] Under contract to the National Aeronautics and Space Administration.

rates, their poloidal magnetic fields will be weak, resulting in weak jets, regardless of the type of central black hole. This result is similar to an argument discussed by Livio, Olgivie, & Pringle [10]. On the other hand, thick accretion flows are expected to have strong poloidal magnetic fields, but suffer from slower rotation when the central hole is non-rotating (Schwarzschild). Only when the disk is thick *and* the geometry rotates rapidly (Kerr) are both strong fields and rapid rotation, and therefore strong jets, obtained.

THE SUPERMASSIVE BLACK HOLE CASE

We can estimate quantitatively the jet power in the different cases by combining the theories of jet production, accretion disk structure, and dynamos. The results are [9]

$$L_{jet}^{Schw-thin} = 5 \times 10^{41} \text{ erg s}^{-1} m_9^{0.9} \dot{m}_{-1}^{1.2} \tag{1}$$

$$L_{jet}^{Kerr-thin} = 5 \times 10^{42} \text{ erg s}^{-1} m_9^{0.9} \dot{m}_{-1}^{1.2} (1 + 1.1 j + 0.29 j^2) \tag{2}$$

$$L_{jet}^{Schw-thick} < 6 \times 10^{43} \text{ erg s}^{-1} m_9 \dot{m}_{-1} \tag{3}$$

$$L_{jet}^{Kerr-thick} > 1.3 \times 10^{45} \text{ erg s}^{-1} m_9 \dot{m}_{-1} j^2 \tag{4}$$

where the geometrically thin models are Shakura & Sunyaev disks [11] and geometrically thick models are advection-dominated accretion flows (ADAFs) [12], m_9 is the black hole mass $M_\bullet$ in units of $10^9 M_\odot$, $\dot{m}_{-1}$ is the accretion rate in units of $0.1 \dot{M}_{Edd} = 2.2 M_\odot \text{ yr}^{-1} m_9$, and $j = J/(GM_\bullet^2/c)$ is the dimensionless black hole angular momentum parameter. The upper limit in equation (3) occurs because, depending on the equation of state, the rotation rate of the thick advection-dominated flow assumed therein could be substantially less than the 40% Keplerian rate generally quoted. On the other hand, the lower limit in equation (4) is due to the use of a conservative value for the ADAF magnetic field: shear enhancement of the field in the Kerr metric is not taken into account.

We also can derive an observed relation between (maximum) jet power and black hole mass by analyzing the upper envelope of the radio power/optical magnitude diagram for radio galaxies [13], coupling this with the relation between black hole mass and galaxy optical magnitude [14], and using standard conversion factors to turn jet radio power into total power [15]. This gives [9]

$$L_{jet}^{obs-max} = 3 \times 10^{45} \text{ erg s}^{-1} m_9^{1.12} \tag{5}$$

While all of the above models predict an approximately linear relation between jet power and black hole mass, only the thick advection-dominated flow in a Kerr metric (equation 4) produces jet powers approaching the observed limit in equation (5). We have named this need for *both* black hole spin and thick accretion flows (in order to produce powerful jets) the "modified spin paradigm". Comparing equation (5) with equation (4), we see that choosing the upper envelope of the Ledlow-Owen radio-optical diagram is consistent with choosing black holes of a uniformly-high relative angular momentum ($j = 1$) and relative accretion rate.

M. Lacy (private communication), on the other hand, has recently derived, from the FIRST Bright Quasar Survey (FBQS), a similar relation for radio loud and radio quiet QSOs using detected quasars of all radio luminosities (Seyferts, radio quiet and radio loud QSOs alike) and not just the most powerful radio galaxies in a given galaxy optical magnitude class, as was done for equation (5). Furthermore, the black hole mass was derived from quasar line widths, rather than from the galaxy optical magnitude (which was not always measured). He finds a relation in which the average radio jet power is given by

$$L_{jet}^{obs-ave} \propto m_9^{1.9} \dot{m}_{-1} \qquad (6)$$

This relation is decidedly *non*-linear in the black hole mass and, at first sight, seems inconsistent with equation (5). However, we can once again compare with the theoretical relation (4) and conclude that it (and equation 5) are consistent with equation (6) if the *average* black hole spin increases with black hole mass as

$$j \propto m_9^{0.45}$$

That is, the results from the FBQS are consistent with those for the most powerful radio galaxies, and with the above jet production models, if more massive black holes in powerful radio galaxies and quasars rotate relatively faster on average than the less massive black holes in Seyfert galaxies, for example. This conclusion is consistent with the ideas of Wilson & Colbert [7] and Meier [8] which attribute a significant portion of the radio loudness of a quasar to the spin rate of the central black hole.

THE STELLAR MASS BLACK HOLE CASE

For reference we will repeat equations (1) - (4) below for the case of a stellar mass ($10 \, M_\odot$) black hole:

$$L_{jet}^{Schw-thin} = 3 \times 10^{34} \, \text{erg s}^{-1} \, m_1^{0.9} \, \dot{m}_{-1}^{1.2} \qquad (7)$$

$$L_{jet}^{Kerr-thin} = 3 \times 10^{35} \, \text{erg s}^{-1} \, m_1^{0.9} \, \dot{m}_{-1}^{1.2} (1 + 1.1j + 0.29j^2) \qquad (8)$$

$$L_{jet}^{Schw-thick} < 6 \times 10^{35} \, \text{erg s}^{-1} \, m_1 \, \dot{m}_{-1} \qquad (9)$$

$$L_{jet}^{Kerr-thick} > 1.3 \times 10^{37} \, \text{erg s}^{-1} \, m_1 \, \dot{m}_{-1} j^2 \qquad (10)$$

Again, powerful jets of order the accretion power ($1.3 \times 10^{38} \, \text{erg s}^{-1} \, m_1 \, \dot{m}_{-1}$) are produced only when the hole is rotating rapidly and the disk is geometrically thick. Furthermore, should the disk suddenly become geometrically thin (due to a state transition, for example), the sudden shift from equation (10) to equation (8) should suppress the jet power by a factor of $> 20 \, m_1^{0.1} \, \dot{m}_{-1}^{-0.2}$, which compares well with the observed jet suppression factor of > 35 for GX 339-4 [1] when it goes into the soft state and the jet disappears.

PREDICTIONS

In the case of high luminosity galactic superluminal sources such as GRS 1915+105, where a thermally unstable disk should go through a limit cycle and oscillate between geometrically thin cool disk and hot geometrically thick accretion flow [16], a jet should be ejected only in the hot portion of the cycle. Sufficient time resolution in both radio/infrared and X-ray observations will be necessary to test this prediction.

Furthermore, in the supermassive case, accretion states similar to those observed in galactic black hole candidate sources also should exist, *and* there should be a similar correlation between hard black hole state and jet production. When performing observations to test this prediction, care must be taken to compare short time scale radio observations (to determine jet power) and optical/X-ray observations (to determine disk state) rather than comparing long-term properties such as total radio power and average optical magnitude.

EFFECTS OF A CENTRIFUGAL BARRIER IN THICK ACCRETION FLOWS AROUND SCHWARZSCHILD HOLES

The above results for thick accretion flows (equations 3, 4, 9, and 10) assume that the self-similar solution for ADAFs (see, *e.g.*, [12]) continues inward into the black hole horizon. In this case, the rotation always remains sub-Keplerian (slow) and the power of any jet produced remains weak.

However, M. Abramowicz (private communication) has pointed out that an alternative scenario is possible near the last stable orbit at $R = 6\,GM/c^2$, if the inflow time scale exceeds the time to transport angular momentum outward (the viscous time). In that case, the angular momentum per unit mass in the inflow will remain approximately constant, eventually exceeding the Keplerian value, which decreases approximately as the square root of the cylindrical radius ($R^{0.5}$) and then increases dramatically interior to the last stable orbit. If this occurs, the geometrically thick inflow will be in rapid rotation which, when coupled to the strong magnetic field that is expected to be present, may drive a strong jet — *independent of the spin of the black hole*. If, in fact, Schwarzschild holes can produce jets as powerful as Kerr holes, then the above theoretical arguments for the spin paradigm, and its advantages for understanding radio loud and quiet QSOs, for example, would collapse. It therefore is important to assess whether such an inner super-Keplerian centrifugal barrier can exist and whether it can drive a strong jet.

Considerable light is shed on this problem by recent global three-dimensional simulations of thick accretion tori. Hawley [17] has shown that, in the presence of even a weak magnetic field, the magneto-rotational instability (MRI) grows rapidly in the super-Keplerian region, transferring angular momentum to larger radii.[2] On a dynamical time scale, the MRI reduces the inner region near the last stable orbit to a relatively thin, Keplerian disk inflow. In fact, because radiative dissipation was not included in these simulations, it is likely that the resulting inner Keplerian region will be even thinner geometrically than the simulations depict (and therefore have quite a weak poloidal magnetic field). Exterior to the last stable orbit the relative disk thickness will increase, giving way to a sub-Keplerian, ADAF-like inflow. Interior to the last stable orbit, the Keplerian speed diverges, causing the inflow to rapidly become permanently sub-Keplerian and spiral into the hole.

The net effect for a Schwarzschild hole is that a geometrically thick (strong magnetic field) super-Keplerian (rapidly-rotating) inner region, which would have been able to produce a strong jet, does not exist. Instead, a geometrically thin Keplerian flow sets up (inside the slowly-rotating outer geometrically thick inflow), whose jet-producing powers will be weak and given by equation (1). In short, Schwarzschild black holes still should produce weak jets, despite having a thin inner Keplerian centrifugal barrier near the last stable orbit.

For a Kerr hole with a geometrically thick inflow, the structure of the inner regions should be similar, with one important difference: the hole and surrounding geometry will be rotating. Although a thin Keplerian inner disk still will form near the last stable orbit (now at $R \sim GM_\bullet/c^2$ instead of $6GM_\bullet/c^2$), the inflow near and inside the ergosphere ($R \lesssim 2GM_\bullet/c^2$), where frame dragging is important, will be geometrically thick and have a strong poloidal magnetic field. For a rapidly-rotating hole ($j \to 1$) the effect of rotation of the reference frame on this strongly-magnetized inflow will be as great as if it had been rotating at the Keplerian rate. In this case, and only in this case (equation 4), will the power of the jet produced be similar to that seen in the strongest radio sources (equation 5).

ACKNOWLEDGMENTS

This research was carried out at the Jet Propulsion Laboratory, California Instituted of Technology, under contract to the National Aeronautics and Space Administration.

REFERENCES

1. Fender, R. et al., *Ap J* **519**, L165 (1999).

---

[2] At first one might suspect that, because the inner super-Keplerian region may be dynamically unstable to the Papaloizou-Pringle instability [18], it is this effect that destroys the inner geometrically-thick, rapidly-rotating region. However, the recent global simulations [17] have shown that the MRI grows so fast that the disk does not maintain its initial constant angular momentum structure long enough to develop the coherent structures associated with the PP instability.

2. Fender, R.P. 1999, in *Black Holes in binaries and galactic nuclei*, in press. Available as astro-ph/9911176.
3. Harmon, B.A. et al., *Nature* **374**, 703 (1995).
4. Harmon, B.A. et al. *Ap J* **477**, L85 (1997).
5. McCollough, M.L., et al. *Ap J* **517**, 951 (1999).
6. Brocksopp, C. et al. *M.N.R.A.S.* **309**, 1063 (1999).
7. Wilson, A.S. & Colbert, E.J.M., *Ap J* **438**, 62 (1995).
8. Meier, D.L. *Ap J* **522**, 753 (1999).
9. Meier, D.L. *Ap J* **548**, L9 (2001).
10. Livio, M., Ogilvie, G.I., & Pringle, J.E., *Ap J* **512**, 100 (1999).
11. Shakura, N.I. & Sunyaev, R.A., *Astron. Astroph.* **24**, 337 (1973).
12. Narayan, R., Mahadevan, R., & Quataert, E., in *The Theory of Black Hole Accretion Disks*, eds. Abramowicz, Bjornsson, & Pringle, Cambridge: Cambridge Univ. Press, 1998, pp. 148-182.
13. Ledlow, M.J. & Owen, F.N., *A J* **112**, 9 (1996).
14. Kormendy, J. & Richstone, D., *Ann. Rev. Astr. Ap.* **33**, 581 (1995).
15. Bicknell, G.V., *Ap J S* **101**, 29 (1995).
16. Szuszkiewicz, E. & Miller, J.C., *M.N.R.A.S.* **298**, 888 (1998).
17. Hawley, J.F., *Ap J* **528**, 462 (2000).
18. Papaloizou, J.C.B. & Pringle, J.E., *M.N.R.A.S.* **208**, 721 (1984).

Poynting Jets from Accretion Disks

R.V.E. Lovelace*, H. Li[†], G.V. Ustyugova[+], M.M. Romanova*, and S.A. Colgate[†]

*Department of Astronomy, Cornell University, Ithaca, NY 14853
[†] Theoretical Astrophysics, T-6, MS B288, Los Alamos National Laboratory, Los Alamos, NM 87545
[+] Keldysh Institute of Applied Mathematics, Russian Academy of Sciences, Miusskaya Square 4, Moscow 125047, Russia

Abstract.
   The powerful narrow jets observed to emanate from many compact accreting objects may arise from the twisting of a magnetic field threading a differentially rotating accretion disk which acts to magnetically extract angular momentum and energy from the disk. Two main regimes have been discussed, *hydromagnetic outflows*, which have a significant mass flux and have energy and angular momentum carried by both the matter and the electromagnetic field and, *Poynting outflows*, where the mass flux is negligible and energy and angular momentum are carried predominantly by the electromagnetic field. Here we consider a Keplerian disk initially threaded by a dipole-like magnetic field and we present solutions of the force-free Grad-Shafranov equation for the coronal plasma. We find solutions with Poynting jets where there is a continuous outflow of energy and toroidal magnetic flux from the disk into the external space. This behavior contradicts the commonly accepted "theorem" of Solar plasma physics that the motion of the footpoints of a magnetic loop structure leads to a stationary magnetic field configuration with zero power and flux outflows.

   In addition we discuss recent magnetohydrodynamic (MHD) simulations which establish that quasi-stationary collimated Poynting jets similar to our Grad-Shafranov solutions arise from the inner part of a disk threaded by a dipole-like magnetic field. At the same time we find that there is a steady uncollimated hydromagnetic outflow from the outer part of the disk. The Poynting jets represent a likely model for the jets from active galactic nuclei, microquasars, and gamma ray burst sources.

INTRODUCTION

Highly-collimated, oppositely directed jets are observed in active galaxies and quasars [1], and in old compact stars in binaries [2,3]. Further, well collimated emission line jets are seen in young stellar objects [4,5]. Different ideas and models have been put forward to explain astrophysical jets [6] with recent work favoring models where twisting of an ordered magnetic field threading an accretion disk acts to magnetically accelerate the jets [7]. There are two regimes: (1) the *hydromagnetic regime*, where energy and angular momentum are carried by both the electromagnetic field and the kinetic flux of matter, which is relevant to the jets from young stellar objects; and (2) the *Poynting flux regime*. where energy and angular from the disk are carried predominantly by the electromagnetic field, which is relevant to extra galactic and microquasar jets.

Here, we discuss recent theoretical and simulation work on Poynting jets. Stationary Poynting flux dominated jets have been found in axisymmetric MHD simulations of the opening of magnetic loops threading a Keplerian disk [8, 9]. Theoretical studies have developed models for Poynting jets from accretion disks [10-12].

THEORY OF POYNTING OUTFLOWS

Consider the coronal magnetic field of a differentially rotating Keplerian accretion disk for a given poloidal field threading the disk. The disk is perfectly conducting with a very small accretion speed. Further, consider "coronal" or force-free magnetic fields in the non-relativistic limit. Cylindrical (r, ϕ, z) coordinates are used and axisymmetry is assumed Thus the magnetic field has the form $\mathbf{B} = \mathbf{B}_p + B_\phi \hat{\phi}$, with $\mathbf{B}_p = B_r \hat{\mathbf{r}} + B_z \hat{\mathbf{z}}$. We can write $B_r = -(1/r)(\partial \Psi/\partial z), B_z = (1/r)(\partial \Psi/\partial r)$, where $\Psi(r, z) \equiv r A_\phi(r, z)$. In the force-free limit, the magnetic energy density $\mathbf{B}^2/(8\pi)$ is much larger than the kinetic or thermal energy densities; that is, the flow speeds are sub-Alfvénic, $\mathbf{v}^2 \ll v_A^2 = \mathbf{B}^2/4\pi\rho$, where v_A is the Alfvén velocity. In this limit, $0 \approx \mathbf{J} \times \mathbf{B}$; therefore, $\mathbf{J} = \lambda \mathbf{B}$ [13] and consequently,

$$\Delta^\star \Psi = -H(\Psi) \frac{dH(\Psi)}{d\Psi}, \qquad (1)$$

with $\Delta^\star \equiv \partial^2/\partial r^2 - (1/r)(\partial/\partial r) + \partial^2/\partial z^2$, which is the Grad-Shafranov equation for Ψ (see e.g. [10]).

We consider an *initial value problem* where the disk at $t = 0$ is threaded by a dipole-like poloidal magnetic field. The form of $H(\Psi)$ is then determined by the differential rotation of the disk: The azimuthal *twist* of a given field line going from an inner footpoint at r_1 to an outer footpoint at r_2 is fixed by the differential rotation of the disk. The field line slippage speed through the disk due to the disk's finite magnetic diffusivity is estimated to be negligible compared with the Keplerian velocity. For a given field line we have $rd\phi/B_\phi = ds_p/B_p$, where $ds_p \equiv \sqrt{dr^2 + dz^2}$ is the poloidal arc length along the field line, and $B_p \equiv \sqrt{B_r^2 + B_z^2}$. The total twist of a field line loop is

$$\Delta\phi(\Psi) = \int_1^2 ds_p \frac{-B_\phi}{rB_p} = -H(\Psi) \int_1^2 \frac{ds_p}{r^2 B_p}, \qquad (2)$$

with the sign included to give $\Delta\phi > 0$. For a Keplerian disk around an object of mass M the angular rotation is $\omega_K = \sqrt{GM/r^3}$ so that the field line twist after a time t is

$$\Delta\phi(\Psi) = \omega_0 \, t \left[\left(\frac{r_0}{r_1}\right)^{3/2} - \left(\frac{r_0}{r_2}\right)^{3/2} \right] = (\omega_0 \, t) \, \mathcal{F}(\Psi/\Psi_0) \quad (3)$$

where r_0 is the radius of the O-point, $\omega_0 \equiv \sqrt{GM/r_0^3}$, and $\mathcal{F}$ is a dimensionless function (the quantity in the square brackets).

The Grad-Shafranov equation (1) can be readily solved by the method of successive over-relaxation with $H(\Psi)$ determined by iteratively solving equations (1) - (3) (see [12]). Figure 1 shows a sample solution exhibiting a Poynting jet. The solution consists of a region near the axis which is *magnetically collimated* by the toroidal B_ϕ field and a region far from the axis, on the outer radial boundary, which is *anti-collimated* in the sense that it is pushed against the outer boundary. The field lines returning to the disk at $r > r_0$ are anti-collimated by the pressure of the toroidal magnetic field.

Most of the twist $\Delta\phi$ of a field line of a Poynting jet occurs along the jet from $z = 0$ to Z_m. Because $-r^2 d\phi/H(\Psi) = dz/B_z$, we have

$$\frac{\Delta\phi(\Psi)}{-H(\Psi)} = \frac{(\omega_0 t)\mathcal{F}(\Psi/\Psi_0)}{-H(\Psi)} \approx \frac{Z_m}{r^2 B_z}, \quad (4)$$

where $r^2 B_z(r,z)$ is evaluated on the straight part of the jet at $r = r(\Psi)$. In the core of the jet $\Psi \ll \Psi_0$, we have $\mathcal{F} \approx 3^{9/8}(\Psi_0/\Psi)^{3/4}$, and in this region we

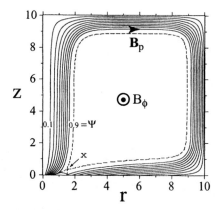

FIGURE 1. Poloidal field lines for Poynting jet case for twist $\omega_0 t = 1.84$ rad. and $(-H)_{max} = 1.13\Psi_0/r_0$ with Ψ =const contours measured in units of Ψ_0 which is the maximum value of Ψ. The outer boundaries at $r = R_m$ and $z = Z_m$ are perfectly conducting and correspond to an external plasma. This external plasma will expand in response to the magnetic pressure of the jet so that R_m and Z_m will increase with time. The dashed contour is the separatrix with the X-point indicated. The initial dipole-like poloidal magnetic field is characterized by the flux function on the disk surface $\Psi(r, z = 0) = (r_0^3/2^{3/2}) r^2 K [1-(r/r_0-1)^2/81]/(r_0^2/2+r^2)^{3/2}$, where r_0 is the radius of the O-point of the field in the disk. In this and subsequent plots Ψ is measured in units of $\Psi_0 = r_0^2 K/3^{3/2}$. Note that $B_z(0) \approx 10.4\Psi_0/r_0^2$.

can take $\Psi = C\Psi_0(r/r_0)^q$, and $H = -\mathcal{K}(\Psi_0/r_0)(\Psi/\Psi_0)^s$, where C, q, $\mathcal{K}$, and s are dimensionless constants. Equation (1) for the straight part of the jet implies $q = 1/(1-s)$ and $C^{2(1-s)} = s(1-s)^2\mathcal{K}^2/(1-2s)$. Thus we find $s = 1/4$ so that $q = 4/3$, $C = [9/32]^{2/3}\mathcal{K}^{4/3}$, and $\mathcal{K} = 3^{1/8}4(r_0\omega_0 t/Z_m)$.

In order to have a Poynting jet, we find that $\mathcal{K}$ must be larger than ≈ 0.5. For the case of uniform expansion of the top boundary, $Z_m = V_z t$, this condition is the same as $V_z < 9.2(r_0\omega_0)$. For Figure (1), $\mathcal{K} \approx 0.844$. The field components in the straight part of the jet are

$$B_\phi = -\sqrt{2}B_z = -\sqrt{2}\left(\frac{3}{16}\right)^{1/3} \mathcal{K}^{4/3} \left(\frac{\Psi_0}{r_0^2}\right) \left(\frac{r_0}{r}\right)^{2/3}. \qquad (5)$$

These dependences agree approximately with those found in numerical simulations of Poynting jets [9]. On the disk, $\Psi \approx 3^{3/2}\Psi_0(r/r_0)^2$ for $r < r_0/3^{3/4}$. Using this and the formula for $\Psi(r)$ gives the relation between the radius of a field line in the disk, denoted r_d, and its radius in the jet, $r/r_0 = 6.5(r_d/r_0)^{3/2}\mathcal{K}^{-1}$. Thus the power law for Ψ is applicable for $r_1 < r < r_2$, where $r_1 = 6.5r_0(r_i/r_0)^{3/2}/\mathcal{K}$ and $r_2 = 1.9r_0/\mathcal{K}$, with r_i the inner radius of the disk. The outer edge of the Poynting jet has a transition layer where the axial field changes from $B_z(r_2)$ to zero while (minus) the toroidal field increases from $-B_\phi(r_2)$ to $(-H)_{max}/r_2$. Using equations (5), which are only approximate at r_2, gives $(-H)_{max} \approx 1.2\mathcal{K}\Psi_0/r_0$, which agrees approximately with our Grad-Shafranov solutions.

MHD SIMULATIONS OF POYNTING JETS

Full, axisymmetric MHD simulations of the evolution of the coronal plasma of a Keplerian disk initially threaded by a dipole-like magnetic field are shown in Figure 2. For these simulations the outer boundaries at $r = R_m$ and $z = Z_m$ were treated as free boundaries following the methods of [14]. These simulations established that a quasi-stationary collimated Poynting jet arises from the inner part of the disk while a steady uncollimated hydromagnetic outflow arises in the outer part of the disk.

Quasi-stationary Poynting jets from the two sides of the disk within r_0 give an energy outflow per unit radius of the disk $d\dot{E}_B/dr = rv_K(-B_\phi B_z)_h$, where the h subscript indicates evaluation at the top surface of the disk. This outflow is $\sim r_0 d\dot{E}_B/dr \sim v_K(r_0)(\Psi_0/r_0)^2 \sim 10^{45}(\text{erg/s})r_{015}^{3/2}\sqrt{M_8}[B_z(0)/6\text{kG}]^2$ where r_{015} is in units of 10^{15}cm, and M_8 in units of $10^8 M_\odot$. This formula agrees approximately with the values derived from the simulations. The jets also give an outflow of angular momentum from the disk which causes disk accretion (without viscosity) at the rate $\dot{M}_B(r) \equiv -2\pi\Sigma v_r = -2(r^2/v_K)(B_\phi B_z)_h \sim 2\Psi_0^2/[r_0^2 v_K(r_0)]$, where Σ is the disk's surface mass density [14]. The Poynting jet has a net axial momentum flux $\dot{P}_z = (1/4)\int r dr (B_\phi^2 - B_z^2) \sim 0.5(\Psi_0/r_0)^2$, which acts to drive the jet outward through an external medium. Further, the Poynting jet generates toroidal magnetic flux at the rate $\dot{\Phi}_t \sim -12[v_K(r_0)/r_0]\Psi_0$.

For long time-scales, the Poynting jet is of course time-dependent due to the angular momentum it extracts from the inner disk ($r < r_0$). This loss of angular momentum leads to a "global magnetic instability" and collapse of the inner disk [13]. An approximate model of this collapse can be made if the inner disk mass M_d is concentrated near the O−point radius $r_0(t)$, if the field line slippage through the disk is negligible [13], Ψ_0 =const, and if $(-rB_\phi)_{max} \sim \Psi_0/r_0(t)$ (as found here). Then, $M_d dr_0/dt = -2\Psi_0^2(GMr_0)^{-1/2}$. If t_i denotes the time at which $r_0(t_i) = r_i$ (the inner radius of the disk), then $r_0(t) = r_i[1 - (t - t_i)/t_{coll}]^{2/3}$, for $t \leq t_i$, where $t_{coll} = \sqrt{GM} \; M_d r_i^{3/2}/(3\Psi_0^2)$ is the time-scale for the collapse of the inner disk. (Note that the time-scale for r_0 to decrease by a factor of 2 is $\sim t_i(r_0/r_i)^{3/2} \gg t_i$ for $r_0 \gg r_i$.) The power output to the Poynting jets is $\dot{E}(t) = (2/3)(\Delta E_{tot}/t_{coll})[1 - (t - t_i)/t_{coll}]^{-5/3}$, where $\Delta E_{tot} = GMM_d/2r_i$ is the total energy of the outburst. Roughly, $t_{coll} \sim 2$ day $M_8^2(M_d/M_\odot)(6 \times 10^{32} \text{Gcm}^2/\Psi_0)^2$ for a Schwarzschild black hole, where validity of the analysis requires $t_{coll} \gg t_i$. Such outbursts may explain the flares of active galactic nuclei blazar sources [16,17] and the one-time outbursts of gamma ray burst sources [18].

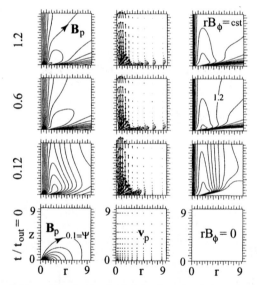

FIGURE 2. Time evolution of dipole-like field threading the disk from the initial configuration $t = 0$ (bottom panels) to the final quasi-stationary state $t = 1.2t_{out}$, where t_{out} is the rotation period of the disk at the outer radius R_m of the simulation region from [9]. The left-hand panels show the poloidal field lines which are the same as $\Psi(r, z)$ =const lines; Ψ is normalized by Ψ_0 (the maximum value of $\Psi(r, z)$) and the spacing between lines is 0.1. The middle panels show the poloidal velocity vectors $\mathbf{v}_p$. The right-hand panels show the constant lines of $-rB_\phi(r, z) > 0$ in units of Ψ_0/r_0 and the spacing between lines is 0.1.

CONCLUSIONS

Recent MHD simulation studies support the idea that an ordered magnetic field of an accretion disk can give powerful outflows or jets of matter, energy, and angular momentum. Most of the studies have been in the hydromagnetic regime and find asymptotic flow speeds of the order of the maximum Keplerian velocity of the disk, v_{Ki}. These flows are clearly relevant to the jets from protostellar systems which have flow speeds of the order of v_{Ki}. In contrast, observed VLBI jets in quasars and active galaxies point to bulk Lorentz factors $\Gamma \sim 10$ - much larger than the disk Lorentz factor. In the jets of gamma ray burst sources, $\Gamma \sim 100$. The large Lorentz factors as well as the small Faraday rotation measures point to the fact that these jets are in the Poynting flux regime. These jets may involve energy extraction from a rotating black hole [19, 20].

The authors (RVEL, MMR, and SAC) thank Drs. C. Wheeler and D.L. Meier with assistance in attending the symposium. This research was supported in part by NASA grants NAG5-9047 and NAG5-9735 and NSF grant AST-9986936. MMR received partial support from NSF POWRE grant AST-9973366.

REFERENCES

1. Bridle A.H., & Eilek, J.A. (eds) 1984, in *Physics of Energy Transport in Extragalactic Radio Sources*, Greenbank: NRAO.
2. Mirabel I.F., & Rodriguez L.F. 1994 Nature, **371**, 46.
3. Eikenberry S., Matthews K., Morgan E.H., Remillard R.A., & Nelson R.W. 1998, ApJ Lett., **494**, L61.
4. Mundt R. 1985, in *Protostars and Planets II*, D.C. Black and M.S. Mathews, eds. Univ. of Arizona Press, Tucson, 414.
5. Bührke T., Mundt R., & Ray T.P. 1988, A&A, **200** 99.
6. Bisnovatyi-Kogan, G.S. 1993, in *Stellar Jets and Bipolar Outflows*, eds. L. Errico & A. A. Vittone (Dordrecht: Kluwer), p. 369.
7. Meier, D.L., Koide, S., & Uchida, Y. 2001, Science, **291**, 84.
8. Romanova M.M., Ustyugova G.V., Koldoba A.V., Chechetkin V.M., & Lovelace R.V.E. 1998, ApJ, **500**, 703.
9. Ustyugova G.V., Lovelace, R.V.E., Romanova M.M., Li, H., & Cogate, S.A. 2000 ApJ, **541**, L21.
10. Lovelace R.V.E., Wang J.C.L., & Sulkanen M.E. 1987, ApJ, **315**, 504.
11. Colgate S.A. & Li H. 1998, in *Proc. of VII International Conference and Lindau Workshop on Plasma Astrophysics and Space Physics*, Lindau, Germany.
12. Li, H., Lovelace, R.V.E., Finn, J.M., & Colgate, S.A. 2001, in preparation.
13. Gold T., & Hoyle F. 1960, Month. Not. R.A.S., **120**, 7.
14. Ustyugova G.V., Koldoba A.V., Romanova M.M., Chechetkin V.M., & Lovelace R.V.E. 1999 ApJ, **516**, 221.
15. Lovelace R.V.E., Newman W.I., & Romanova M.M., 1997, ApJ, **424**, 628.
16. Romanova M.M., & Lovelace R.V.E. 1997, ApJ, **475**, 97.

17. Levinson A. 1998, ApJ, **507**, 145.
18. Katz J.I. 1997, ApJ, **490**, 633.
19. Blandford R.D., & Znajek R.L. 1977, MNRAS, **179**, 433.
20. Livio M., Ogilvie G.I., & Pringle J.E. 1999, ApJ, **512**, 100.

Magnetorotational explosion. Results of 2D simulations.

S.G.Moiseenko*, G.S.Bisnovatyi-Kogan* and N.V.Ardeljan[†]

*Space Research Institute, Profsoyuznaya 84/32, Moscow 117997, Russia
[†]Department of Computational Mathematics and Cybernetics, Moscow State University, Vorobjevy Gory, Moscow B-234 119899, Russia

Abstract.
Results of the magnetorotational explosion simulations are presented for differentially rotating magnetized cloud. Differential rotation leads to the amplification of the toroidal component of the magnetic field. The part of the rotational energy of the cloud is transformed to the kinetic energy of the radial motion. Part of the matter $\sim 7\%$ of the mass of the cloud (3.3% of the final gravitational energy of the cloud) gets radial kinetic energy which is larger than its potential energy and can be thrown away to the infinity, This part of the cloud carries about 30% of the initial angular momentum of the cloud. For the simulations we used specially developed completely conservative implicit 2D Lagrangian scheme on triangular grid with grid reconstruction.

INTRODUCTION

Many astrophysical objects have such important features like rotation and magnetic fields. Physical processes where magnetic field interacts with differential rotation are called magnetorotational. Magnetorotational phenomena are often take place in accretion discs, jets. Magnetrotational mechanism, suggested in [4], of supernova explosion is one of the most realistic for supernova type II.

Here we describe the investigation of the magnetorotational phenomena in the collapsing rotating magnetized cloud. Initially we suppose that the cloud is rigidly rotating uniform gas sphere. After collapse of the cloud and formation of the differentially rotating configuration the toroidal component of the magnetic field amplifies with the time.

First 2D calculations of magnetorotational phenomena have been performed in [7]. Authors used rather unrealistic configuration of magnetic field, as result they have got ejection along rotational axis. This mechanism was also investigated analytically in [8]. In [9] authors also get the ejection of the matter due to the magnetorotational mechanism, but in this case the shape of the initial magnetic field and the shape of ejection was qualitatively different.

Magnetorotational explosion of the rotating magnetized gas cloud in 2D approach has been investigated earlier in [1] for a divergency-free, but not force-free (balanced-free) magnetic field configuration. The effect of the ejection has been found there, but use of unbalanced initial field leads to artificial effects, which do not allow us to extend calculations as far as is needed.

In the calculations, described in this paper, the initial configuration of the magnetic field has been chosen in the following way: magnetic forces produced by the magnetic field are balanced with pressure and gravitational forces distribution.

BASIC EQUATIONS

For the simulations of the magnetorotational phenomena we consider set of MHD equations with self-gravitation and with infinite conductivity [6]

$$\frac{d\mathbf{x}}{dt} = \mathbf{u}, \quad \frac{d\rho}{dt} + \rho\,\mathrm{div}\,\mathbf{u} = 0,$$

$$\rho\frac{d\mathbf{u}}{dt} = -\mathrm{grad}\left(p + \frac{\mathbf{H}\cdot\mathbf{H}}{8\pi}\right) + \frac{\mathrm{div}(\mathbf{H}\otimes\mathbf{H})}{4\pi} - \rho\,\mathrm{grad}\,\Phi,$$

$$\rho\frac{d}{dt}\left(\frac{\mathbf{H}}{\rho}\right) = \mathbf{H}\cdot\nabla\mathbf{u}, \quad \Delta\Phi = 4\pi G\rho, \qquad (1)$$

$$\rho\frac{d\varepsilon}{dt} + p\,\mathrm{div}\,\mathbf{u} = 0, \quad \frac{1}{\rho} = \frac{T\mathfrak{R}}{p}, \quad \varepsilon = \frac{T\mathfrak{R}}{\gamma - 1},$$

where $\frac{d}{dt} = \frac{\partial}{\partial t} + \mathbf{u}\cdot\nabla$ is the total time derivative, $\mathbf{x} = (r,\varphi,z)$, $\mathbf{u}$ is velocity vector, ρ is density, p is pressure, $\mathbf{H} = (H_r, H_\varphi, H_z)$ is magnetic field vector, Φ is gravitational potential, ε is internal energy, G is gravitational constant, $\mathfrak{R}$ is universal gas constant, γ is adiabatic index, $\mathbf{H}\otimes\mathbf{H}$ is tensor of rank 2.

Axial symmetry ($\frac{\partial}{\partial\varphi}$) and symmetry to the equatorial plane ($z = 0$) are assumed.

The problem is solved in the restricted domain. Outside the domain, the density of the matter is zero, but poloidal components of magnetic field H_r, H_z can be non-zero.

To write the set of equations in dimensionless form we choose the following scale values:

$$\rho_0 = 1.492\cdot 10^{-17}\mathrm{g/cm}^3,\ r_0 = z_0 = x_0 = 3.81\cdot 10^{16}\mathrm{cm},$$

$$r = \tilde{r}x_0,\ z = \tilde{z}x_0,\ u = \tilde{u}u_0,\ u_0 = \sqrt{4\pi G\rho_0 x_0^2},$$

$$p = \tilde{p}p_0,\ \varepsilon = \tilde{\varepsilon}\varepsilon_0,\ T = \tilde{T}T_0,\ \Phi = \tilde{\Phi}\Phi_0,\ \Phi_0 = 4\pi G\rho_0 x_0^2,$$

$$t_0 = \frac{x_0}{u_0},\ p_0 = \rho_0 u_0^2 = \rho_0 x_0^2 t_0^{-2},\ T_0 = \frac{u_0^2}{\mathfrak{R}},$$

$$\varepsilon_0 = u_0^2 = x_0^2 t_0^{-2},\ H_0 = \sqrt{p_0} = x_0 t_0^{-1}\rho_0^{1/2}.$$

Here the values with index zero are the scale factors and the functions under a tilde are dimensionless functions.

COLLAPSE OF A ROTATING MAGNETIZED GAS CLOUD

Formulation of the problem

Consider a magnetized rotating gas cloud which is described by the set of equations (1). All graphs and figures below are in a nondimensional form. At the initial moment ($t = 0$) the cloud is a rigidly rotating uniform gas sphere with the following parameters:

$$r = 3.81 \cdot 10^{16} \text{cm}, \rho = 1.492 \cdot 10^{-17} \text{g/cm}^3, \eta = 5/3,$$

$$M = 1.73 M_\odot = 3.457 \cdot 10^{33} \text{g}, u^r = u^z = 0, \tag{2}$$

$$\beta_{r0} = \frac{E_{\text{rot}0}}{|E_{\text{gr}0}|} = 0.04, \tag{3}$$

$$\beta_{i0} = \frac{E_{\text{in}0}}{|E_{\text{gr}0}|} = 0.01, \tag{4}$$

$$\xi = \frac{E_{\text{mag}1}}{|E_{\text{gr}1}|} = 10^{-2}, 10^{-4}, 10^{-6}. \tag{5}$$

Here the subscript "0" corresponds to the initial moment ($t = 0$), subscript "1" to the moment of the beginning of the evolution of magnetic field ($t = t_1 > 0$).

Initial magnetic field configuration

The initial magnetic field must satisfy the condition of absence of magnetic charges $\text{div}\mathbf{H} = 0$. It also has to correspond to the boundary conditions of the problem.

The best choice would be dipole or quadrupole. While they satisfy initial and boundary conditions, they have singularities in the origin of coordinates ($r = 0, z = 0$). Using such magnetic fields in numerical simulations can lead to loss of accuracy of calculations.

To define the initial magnetic field we use the following method. We defined toroidal current j_φ in the central part of the core of the collapsed cloud by a formula:

$$j_\varphi = j_\varphi^u + j_\varphi^d, \tag{6}$$

$$j_\varphi^u = \left[\sin\left(\pi \frac{r}{0.3} - \frac{\pi}{2}\right) + 1\right]\left[\sin\left(\pi \frac{z}{0.3} - \frac{\pi}{2}\right) + 1\right] \times \left[1 - \left(\frac{r}{0.3}\right)^2 - \left(\frac{z}{0.3}\right)^2\right],$$

at $r^2 + z^2 < 0.3^2, z > 0,$

$$j_\varphi^d = -j_\varphi^u \text{ at } r^2 + z^2 < 0.3^2, z < 0.$$

After getting the defferentially rotating stationary solution for nonmagnetized cloud we use Bio-Savara law for calculation of the poloidal components of the magnetic field

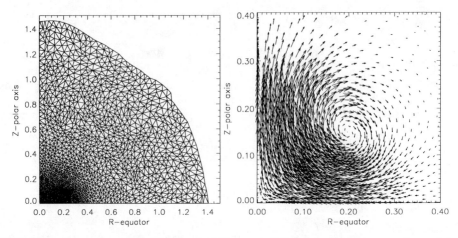

FIGURE 1. Lagrangian grid (left), and poloidal magnetic field distribution in the central part of the cloud (right) produced by current (6).

H_{r0}, H_{z0}. This magnetic field is divergency free, but it is not force-free and should be balanced at the initial moment. Then we use the following method: we "turn on" the poloidal magnetic field H_{r0}, H_{z0}, but "switch off" the equation for the evolution of the toroidal component H_φ in (1). Actually it means that we define $H_\varphi \equiv 0$, $\frac{dH_\varphi}{dt} \equiv 0$. From the physical point of view it means that we allow magnetic field lines to slip through the matter of the cloud in the toroidal direction. After "turning on " such a field, we let the cloud come to the steady state, where magnetic forces connected with the purely poloidal field are balanced by other forces.

The calculated balanced configuration has the magnetic field of quadrupole-like symmetry. For testing we run our code with this purely poloidal field for a large number of time steps ($\sim 10^3$), during which the parameters of the cloud did not change.

RESULTS

We describe results of simulations of the problem for initial $\xi(t_1) = 10^{-2}$. Results of simulations for $\xi(t_1) = 10^{-4}$, 10^{-6} are qualitatively similar to the first case. The amounts of the ejected mass and energy are the same, $\sim 7\%$ of mass and $\sim 3.3\%$ of energy. The main difference between these three variants is the duration of the process.

After number of oscillations the cloud at $t_1 = 18.45$ came to the steady differentially rotating configuration. At this time moment it consists of rapidly rotating dense core and slowly rotating prolate envelope.

At t_1 the initial poloidal magnetic field was "turned on".

The toroidal component of the magnetic field amplifies with the time and magnetic pressure in the transition zone between the core and the envelope grows with time. At the periphery of the core of the cloud compression wave forms and moves outwards on steeply decreasing density profile. Soon after the formation it transforms to the MHD

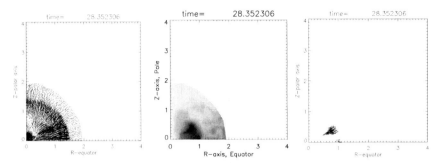

FIGURE 2. Velocity field (left), specific angular momentum distribution (center), ejected part of the cloud (right) at $t = 28.35$.

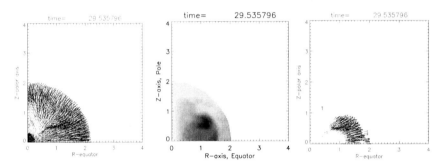

FIGURE 3. Velocity field (left), specific angular momentum distribution (center), ejected part of the cloud (right) at $t = 29.54$.

shock. When the shock propagates through the envelope of the cloud the $\sim 7\%$ of the total mass of the cloud gets radial kinetic energy larger than its potential energy and can be ejected, it carries about 3.3% of the final gravitational energy of the cloud. The core of the cloud at the final stage of the calculations rotates much slower than at the

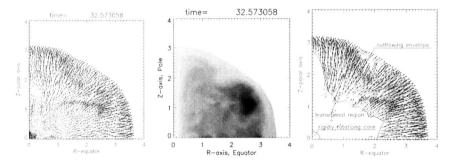

FIGURE 4. Velocity field (left), specific angular momentum distribution (center), ejected part of the cloud (right) at $t = 32.57$.

beginning of the evolution of the magnetic field. Part of the angular momentum of the cloud was transported from the inner parts of the cloud to the expanding ejected shell. 1D simulations of a similar problem, but for the magnetorotational supernova explosions made in [5] and [3] show that magnetic field transforms rotational energy of the star to the energy of radial motion. The central core of the star in 1D results started to rotate in the opposite direction to the initial one. Such magnetorotational oscillations have been investigated analytically in [5]. In our 2D simulations we do not get magnetorotational oscillations while the core of the cloud after ejection rotates slower and lost significant part of angular momentum due to magnetic breaking. The ejecting matter looks like an expanding shell. At the final stage of the calculations the cloud consists of outflowing envelope, a transitional region, and almost rigidly rotating core.

ACKNOWLEDGMENTS

This work was partially supported by RFBR grant 99-02-18180 and INTAS-ESA grant 120. The authors are grateful to Prof. J.C.Wheeler and the Organizing Committee for support and hospitality.

REFERENCES

1. Ardeljan, N. V., Bisnovatyi-Kogan, G. S., Moiseenko, S. G., *The local bubble and beyond Proc. of IAU colloquium No.166, Lecture notes in physics; Vol. 506*, 145-148 (1998).
2. Ardeljan, N. V., Bisnovatyi-Kogan, G. S., Moiseenko, S. G., *A&A*, **355**, 1181-1190 (2000).
3. Ardeljan, N. V., Bisnovatyi-Kogan, G. S., Popov Ju. P. *Astron. Zh.*, **56**, 1244-1255 (1979).
4. Bisnovatyi-Kogan, G. S., *Astron. Zh.*, **47**, 813-816 (1970).
5. Bisnovatyi-Kogan, G. S., Popov, Ju. P., Samokhin, A. A., *Astrophys. Space.Sci.*, **41**, 321-356 (1976).
6. Landay, L. D., Lifshitz, E. M., *Electrodynamics of Continuous Media*, 2nd ed., Pergamon Press, Oxford (1984).
7. Le Blanc, L. M., Wilson, J. R., *ApJ*, **161**, 541-551 (1970).
8. Meier, D. L., Epstein, R. I., Arnett, W. D., Schramm, D. N., *ApJ*, **204**, 869-878 (1976).
9. Ohnishi, T., *Tech. Rep. Inst. At. En. Kyoto Univ.*, No.198 (1983).

Jet Formation from Rotating Magnetized Objects

G.S.Bisnovatyi-Kogan*, N.V.Ardelyan[†] and S.G.Moiseenko*

*Space Research Institute, Profsoyuznaya 84/32, Moscow 117810, Russia
[†]Department of Computational Mathematics and Cybernetics, Moscow State University, Vorobjevy Gory, Moscow B-234 119899, Russia

Abstract. Jet formation is connected most probably with matter acceleration from the vicinity of rotating magnetized bodies. It is usually related to the mass outflows and ejection from accretion disks around black holes. Problem of jet collimation is discussed. Collapse of a rotating magnetized body during star formation or supernovae explosion may lead to a jet-like mass ejection for certain angular velocity and magnetic field distributions at the beginning of the collapse. Jet formation during magnetorotational explosion is discussed basing on the numerical simulation of collapse of magnetized bodied with quasi-dipole field.

INTRODUCTION

Matter outflow is observed in most astrophysical objects in the form of stellar wind, or collimated outflow (ejection) from young stellar objects, AGN and quasars, microquasars (galactic X-ray sources). Mechanisms of mass loss are connected with a radiative and/or electromagnetic acceleration. The last one, which is probably the best for producing collimated outflows, may work quasi-stationary, or may be connected with explosive events. Rotation is always present in compact objects, and formation of collimated jets results from the action of magneto-rotational phenomena.

PROBLEM OF COLLIMATION

During accretion and outflow of matter the relative dynamical action of magnetic field increases for a given element of matter. As was shown in[19], the conservation of a magnetic flux during stationary accretion implies a dependence $B_r \sim r^{-2}$. At constant mass flux $\dot{M} = 4\pi\rho v_r r^2$ and free-fall velocity $v_r \sim r^{-1/2}$, the density increases as $\rho \sim r^{-3/2}$, and kinetic energy density $E_k \sim r^{-5/2}$. The growth of the magnetic energy density is faster $E_m \sim B_r^2 \sim r^{-4}$. After equipartition $E_k \sim E_M$ is reached, it cannot be violated during subsequent accretion.

The outflow from rotating magnetized object contain azimuthal component of the magnetic field, which energy density decreases not faster than the specific kinetic energy of matter. For the outflow with constant velocity v_r in stationary spherical outflow the density $\rho \sim r^{-2}$, $B_\phi \sim r^{-1}$. In this conditions $E_k \sim E_m$, and the trajectories of mass outflow become more tightly spiraled, and finally the flow is becoming highly collimated with direction of the flow along the rotational axis [12], [10]. We may say here about the universal magnetic collimation in the outflows from rotating magnetized objects.

The outbursts may be collimated at the very beginning. When jets are separated from the object of its origin the problem appears of preservation of the jet against its spherization during a motion in a rarefied medium. One of the plausible mechanism of jet preservation is a magnetic pinch collimation produced by an axial electrical current, suggested in [4].

JET FORMATION BY MATTER OUTFLOW FROM MAGNETIZED ACCRETION DISK

Accretion of matter with a large scale magnetic field into a black hole leads to formation of an accretion disk with a strong poloidal magnetic field. Models of non-rotating accretion disks, supported by magnetic field had been constructed in [6], [7], see Figure 1.

Rotation of matter in such a disk is accompanied by generation of a strong electrical field, leading to particle acceleration and ejection of matter along magnetic field lines. Qualitative and phenomenological models had been considered in [3], [8], [14]. Analytical self-similar solutions for jets and winds produced by centrifugally and magnetically driven mechanisms had been obtained in [9], [15].

Extensive numerical simulations have been done for construction of more realistic models of quasi-stationary jet formation from magnetized accretion disks. In the paper [18] numerical simulations have been done of dynamics of magnetic loops in the coronae of accretion disks. It was obtained that in presence of differential rotation the loops are opened in the inner parts of the disk, and opening of loops is followed by magnetically driven outflow from the disk. The outflow may be transient, consisting of small-scale outbursts from different loop opening, or may be steady, corresponding to outflow along the open field lines at the inner part of the disk, and showing some collimation. Computations of magnetocentrifugally driven winds had been performed in [20]. The stationary regime of the outflow was obtained by solution of time-dependent MHD equations, with a split-monopole poloidal field configuration frozen into the disk. Close to the disk the outflow is driven by the centrifugal force, while at all larger distances the flow is driven by the magnetic force, which is proportional to $-\nabla(rB_\phi)^2$, where B_ϕ is the toroidal field. The collimation distance over which the flow becomes collimated is much larger than the size of the sumulation region, so the obtained outflows are approximately spherical.

MHD simulation have been performed in [21] of Pointing outflows, where the mass flux is negligible and energy and angular momentum are carried predominantly by electromagnetic field in the form of MHD waves. As a result of time-dependent simulations a quasi-stationary collimated Pointing jet arises from the inner part of the disk and a steady uncollimated hydromagnetic outflow from the outer disk for a case Keplerian disk initially threaded by a dipole like poloidal magnetic field.

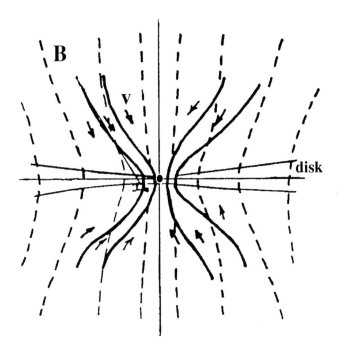

FIGURE 1. Schematic picture of magnetic field lines, obtained from the external uniform field by accretion of a non-rotating gas with infinite conductivity into a black hole. Dash lines are produced by the self-similar accretion flow with a sink in the equatorial plane. In reality the accretion gas is forming a disk in the equatorial plane, where matter is collected, and is moving into a black hole due to non-perfect conductivity. This motion is generating additional azymuthal currents in the disk, changing the external magnetic field. Solid lines give the self-consistent magnetic field structure with account of currents in the disk, from [7].

MAGNETOROTATIONAL EXPLOSIONS

Mechanism of magnetorotational explosion was suggested in [2], where amplification of the toroidal component of the magnetic field due to differential rotation leads to the transformation of a part of the rotational energy to the energy of the ejection. First numerical simulations devoted to the magnetorotational mechanism were made in [13], and investigated analytically by [17].

One of the most important parameters for this problem is relation of magnetic energy to the gravitational energy of the star: $\alpha = \frac{E_{mag}}{|E_{grav}|}$. 2-D simulations have been done in [1] for the initial values of the $\alpha = 10^{-2}, 10^{-4}, 10^{-6}$. The computations have been performed using implicit Lagrangian scheme with triangle reconstructive grids, specially designed for astrophysical problems. Details of the definition of the initial magnetic field are described in the paper [1].

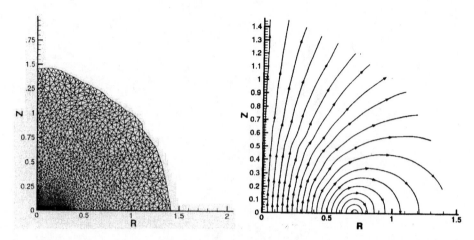

FIGURE 2. Lagrangian grid (left), and poloidal magnetic field distribution (right) in a stationary equilibrium differentially rotating star at $t_1 = 18.918451$, when the magnetic field was included into calculations.

Initial magnetic field chosen for our simulations has quadrupole-like kind of symmetry (i.e. its "z"- component is equal to zero at the equatorial plane). As a result an ejection predominantly in the equatorial plane was obtained. The toroidal magnetic field grows with time and produces MHD shock which moves to the boundary of the star. Part of the matter of the envelope of the star (about 7% of the mass of the star) has radial kinetic energy larger than its potential energy and can be ejected. This ejected matter carries about 3.3% of the total energy of the star (see also the previous contribution). When the initial magnetic field have a dipole-like symmetry jet formation is possible as was first shown in calculations of [13]. Calculations with the initial magnetic field structure, similar to [13] have been performed using the same program as in [1]. The results are presented in Figures 2-4. The initial model (Fig.2, left) was constructed as a differentially rotating star with a quasi-dipole magnetic field (Fig.2, right), produced by toroidal currents ring situated at about 0.5 of the equatorial radius, where density was less than 0.1 of the central density.

The time is given in non-dimensional units, one unit of time is roughly equal to the time of crossing of stellar radius with the parabolic speed of the initial model. Initial distributions of the density and angular velocity are represented in Figure 3.

Magnetic field twisting produced strong toroidal magnetic field with local energy density of the order of the one of the matter. Absence of the radial component of the magnetic field in the equatorial plane led to formation of toroidal field rings at higher latitudes. The excess of the magnetic pressure produced a matter compression across the axis, and ejection of the matter in the direction attached to the axis. Density distribution and velocity field in the last calculational point with clear indication of formation of the ejection are represented in Figure 4.

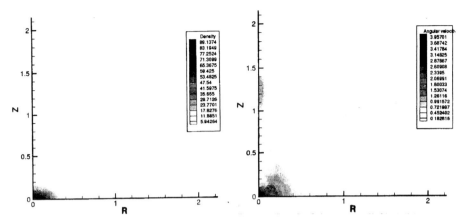

FIGURE 3. Density (left), and angular velocity (right) distributions at $t_1 = 18.918451$.

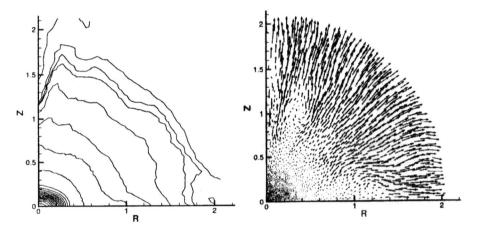

FIGURE 4. Density distribution (left) and velocity field (right) at $t = 30.947880$, when the mass ejection preferentially in the pole direction is formed.

MIRROR SYMMETRY BRAKING OF MAGNETIC FIELD IN DIFFERENTIALLY ROTATING STARS

There is a possibility of asymmetric MHD explosion, when we have asymmetric magnetic field amplification. One example of development of asymmetric picture was considered by [5]. When the collapsing and exploding star has initially toroidal and poloidal fields of different symmetry: dipole poloidal and symmetric poloidal, or quadrupole poloidal and antisymmetric toroidal fields, formation of an additional toroidal field from the existing poloidal due to differential rotation leads to spontaneous breaking of the symmetry. Due to the asymmetry of the toroidal field during the explosion the outbursts could also become asymmetric, giving observational asymmetric or even one-side jets often observed in extragalactic jets, and may be in microquasars [16], [11]. Another possibility of the asymmetry of the magnetic field was considered in [22], where disk with the mixture of dipole and quadrupole fields was considered. Asymmetric magnetic field in the accretion disk may lead to appearance of asymmetric jets in stationary and non-stationary variants. In transient accretion disks, formed by tidal capture or destruction of the nearby star by a black hole, asymmetric magnetic field may be formed, and magnetorotational explosion in such disk may lead to the one side ejection.

ACKNOWLEDGMENTS

This work was partially supported by RFBR grant 99-02-18180 and INTAS-ESA grant 120. The authors are grateful to Prof. J.C.Wheeler and the Organizing Committee for support and hospitality, and to O.D.Toropina for help.

REFERENCES

1. Ardeljan, N. V., Bisnovatyi-Kogan, G. S., Moiseenko, S. G., *A&A*, **355**, 1181-1190 (2000).
2. Bisnovatyi-Kogan, G. S., *Astron. Zh.*, **47**, 813-816 (1970).
3. Bisnovatyi-Kogan, G. S., Blinnikov, S. I., *Pisma Astron. Zh.*, **2**, 489-493 (1976).
4. Bisnovatyi-Kogan, G. S., Komberg, B. V., Friedman,A. M., *Astron. Zh.*, **46**, 465-471 (1969).
5. Bisnovatyi-Kogan, G. S., Moiseenko, S. G., *Astron. Zh.*, **69**, 563-571 (1992).
6. Bisnovatyi-Kogan, G. S., Ruzmaikin, A. A., *Astrophys. Sp. Sci.*, **28**, 45-58 (1974).
7. Bisnovatyi-Kogan, G. S., Ruzmaikin, A. A., *Astrophys. Sp. Sci.*, **42**, 401-425 (1976).
8. Blandford, R. D., *Mon. Not. RAS*, **176**, 465-481 (1976).
9. Blandford, R. D., Payne, D.G., *Mon. Not. RAS*, **199**, 883-903 (1982).
10. Bogovalov, S. V., *Pisma Astron. Zh.*, **24**, 381-392 (1998).
11. Fender, R.P., *astro-ph/9907050* (1999).
12. Heyvaerts, J., Norman, C. A., *ApJ*, **347**, 1055-1081 (1989).
13. Le Blanc, L. M., Wilson, J. R., *ApJ*, **161**, 541-551 (1970).
14. Lovelace, R. V. E., *Nature*, **262**, 649-652 (1976).
15. Lovelace, R. V. E., Berk, H. L., Contopulos, J., *ApJ*, **379**, 696-705 (1991).
16. Mirabel, I.F., Rodriguez, L.F., Cordier, B., Paul, J., Lebrun, F., *Nature*, **358** 215-217 (1992).
17. Meier, D. L., Epstein, R. I., Arnett, W. D., Schramm, D. N., *ApJ*, **204**, 869-878 (1976).
18. Romanova, M. M., Ustyugova, G. V., Koldoba, A. V., Chechetkin, V. M., Lovelace, R. V. E., *ApJ*, **500**, 703-713 (1998).
19. Schwartzman, V. F., *Astron. Zh.*, **48**, 479-488 (1971).
20. Ustyugova, G. V., Koldoba, A. V., Romanova, M. M., Chechetkin, V. M., Lovelace, R. V. E., *ApJ*, **516**, 221-235 (1999).
21. Ustyugova, G. V., Lovelace, R. V. E., Romanova, M. M., Li, H., Colgate, S.A., *ApJ*, **516**, 221-235 (2000).
22. Wang, J. C. L., Sulkanen, M. E., Lovelace, R. V. E., *ApJ*, **390**, 46-65 (1992)

Do Active Galactic Nucleus Jets Consist of a Pair Plasma?

Kouichi HIROTANI

National Astronomical Observatory, Osawa, Mitaka, Tokyo 181-8588, Japan
hirotani@hotaka.mtk.nao.ac.jp

Abstract. We investigate whether the parsec-scale jet of quasar 3C 345 is dominated by a normal (proton-electron) plasma or a pair (electron-positron) plasma. We first present a new method to compute the kinetic luminosity of a conical jet by using the core size observed at a single very long baseline interferometry frequency. The deduced kinetic luminosity gives electron densities of individual radio-emitting components as a function of the composition. We next constrain the electron density independently by using the theory of synchrotron self-absorption. Comparing the two densities, we can discriminate the composition. We then apply this procedure to the five components in the 3C 345 jet and find that they are pair-plasma dominated at 14 epochs out of the total 19 epochs, provided that the bulk Lorentz factor is less than 15 throughout the jet. The conclusion does not depend on the lower cutoff energy of radiating particles.

INTRODUCTION

Very long baseline interferometry (VLBI) is uniquely suited to the study of the matter content of active galactic nucleus (AGN) jets on parsec scales. Reynolds et al. (1996) analyzed historical VLBI data of the M87 jet at 5 GHz and concluded that the core is probably dominated by an electron-positron pair plasma. However, this method was applicable only for the VLBI observations of M87 core at epochs September 1972 and March 1973. Therefore, Hirotani et al. (2001) generalized Reynolds' condition and demonstrated that four components of the 3C 345 jet is likely dominated by a pair plasma, by deducing the kinetic luminosity, L_{kin}, from a reported core-position offset (Lobanov 1998). In this paper, we develop the method to infer L_{kin} and confirm this conclusion. We use a Hubble constant $H_0 = 65h$ km/s/Mpc and $q_0 = 0.5$. Spectral index α is defined such that $S_\nu \propto \nu^\alpha$.

KINETIC LUMINOSITY OF A CONICAL JET

We assume that the jet is flowing in a conical geometry with a small opening angle and that the proper electron number density and the magnetic field are proportional to r^{-2} and r^{-1}, respectively, where r refers to the distance from the AGN. Then $L_{\rm kin}$ is found to be independent of r and given by (Hirotani 2001)

$$L_{\rm kin} \propto C_{\rm kin}\Gamma(\Gamma-1)F^{2(5-2\alpha)/(7-2\alpha)}$$
$$\times \left[\frac{1}{\sin\varphi}\left(\frac{\delta}{1+z}\right)^{3/2-\alpha}\right]^{-4/(7-2\alpha)} \gamma_{\rm min}^{4(1+2\alpha)/(7-2\alpha)}, \quad (1)$$

where

$$F \propto \theta_{\rm d,core} \cdot \nu. \quad (2)$$

Here, $C_{\rm kin} = \pi^2 \langle \gamma_- \rangle m_e c^3/\gamma_{\rm min}$ for a pair plasma, while $C_{\rm kin} = \pi^2 m_p c^3/(2\gamma_{\rm min})$ for a normal plasma; m_p is the rest mass of a proton, $\Gamma \equiv 1/\sqrt{1-\beta^2}$ the bulk Lorentz factor, φ the viewing angle, $\gamma_{\rm min}$ the lower cutoff Lorentz factor of particles' random motion, δ the Doppler factor, z the redshift, $\theta_{\rm d,core}$ the angular core size.

The electron number density, N_e^*, can be obtained from $L_{\rm kin}$ as a function of the composition. Let us denote the density obtained under the assumption of a normal plasma as $N_e^*(\rm nml)$, which is typically 100 times less than that for a pair plasma.

SYNCHROTRON SELF-ABSORPTION CONSTRAINTS

Applying the SSA theory, we can alternatively compute N_e^* from the observed peak frequency ν_m, peak flux density S_m, and component anglar size, θ_d, as follows:

$$N_e^*({\rm SSA}) \propto \gamma_{\rm min}\theta_d^{4\alpha-7}\nu_m^{4\alpha-5}S_m^{-2\alpha+3}\left(\frac{\delta}{1+z}\right)^{2\alpha-3}, \quad (3)$$

where the constant of proportionarity is given in Hirotani (2001). If $0.01 < N_e^*({\rm nml})/N_e^*({\rm SSA}) < 1.0$ holds, then the possibility of a normal plasma dominance can be ruled out.

APPLICATION TO THE 3C 345 JET

Applying the method described in the previous sections to the 3C 345 core, and assuming $\Gamma = 15$, we obtain

$$L_{\rm kin} = 9.2 \times 10^{45} h^{-1.57} K^{0.57} \left(\frac{\gamma_{\rm min}}{100}\right)^{-1.56} {\rm ergs\ s}^{-1}, \quad (4)$$

where $K \sim 0.1$ holds when an energy equipartition is realized between radiating particles and the magnetic field and when $\alpha \sim -0.5$.

We present the ratio, $N_e^*({\rm nml})/N_e^*({\rm SSA})$ as a function of the projected distance in figure 1. The solid line represents $N_e^* = N_e^*(\rm nml)$. It follows from the figure that

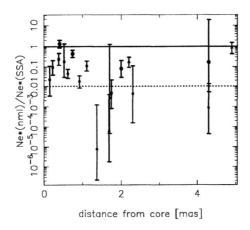

FIGURE 1. The ratio $N_e^*(\mathrm{nml})/N_e^*(\mathrm{SSA})$ as a function of the distance from the core. Above the solid, horizontal line, the dominance of a normal plasma is allowed. $\Gamma = 15$ is assumed.

neither a normal-plasma dominance nor the pair-plasma dominance with $\gamma_{\min} > 100$ are allowed within one-σ errors for the 14 epochs out of the 19 epochs. It is noteworthy that $N_e^*(\mathrm{nml})/N_e^*(\mathrm{SSA}) \propto (\gamma_{\min}/100)^{-2\alpha-1.56}$ for the spectral index of -1.15 for the core. Therefore, for a typical spectral index $\alpha \sim -0.75$ for the jet components, the dependence on $\gamma_{\min}$ virtually vanishes. That is, the conclusion does not depend on the assumed value of $\gamma_{\min} \sim 100$ for a normal plasma. Since $N_e^*(\mathrm{nml})/N_e^*(\mathrm{SSA})$ increases with increasing assumed value of Γ, we can conclude that *the jet is strongly suggested to be dominated by a pair plasma with $\gamma_{\min} \ll 100$, if we assume $\Gamma < 15$ throughout the jet.*

REFERENCES

1. Gould 1979, AA **76**, 306
2. Hardee, P. E., Norman, M. L. 1989, ApJ **342**, 680
3. Hewitt, A., Burbidge, G. 1993, ApJs **87**, 451
4. Hirotani, K., Iguchi, S., Kimura, M., & Wajima, K. 2001, ApJ **545**, 100
5. Hirotani, K. 2001, submitted to MNRAS (astro-ph/0009487)
6. Lobanov, A. P. 1998, AA **330**, 79
7. Reynolds, C. S., Fabian, A. C., Celotti, A., & Rees, M. J. 1996 MNRAS **283**, 873

Collimated Energy-Momentum Extraction from Rotating Black Holes in Quasars and Microquasars Using the Penrose Mechanism

Reva Kay Williams[*][†]

University of Florida, Gainesville, FL 32611
[†]*Bennett College, Greensboro, NC 27401*

Abstract. For almost four decades, since the discovery of quasars, mounting observational evidence has accumulated that black holes indeed exist in nature. In this paper, I present a theoretical and numerical (Monte Carlo) fully relativistic 4-D analysis of Penrose scattering processes (Compton and $\gamma\gamma \longrightarrow e^+e^-$) in the ergosphere of a supermassive Kerr (rotating) black hole. These model calculations surprisingly reveal that the observed high energies and luminosities of quasars and other AGNs, the collimated jets about the polar axis, and the asymmetrical jets (which can be enhanced by relativistic Doppler beaming effects) all are inherent properties of rotating black holes. From this analysis, it is shown that the Penrose scattered escaping relativistic particles exhibit tightly wounded coil-like cone distributions (highly collimated vortical jet distributions) about the polar axis, with helical polar angles of escape varying from 0.5° to 30° for the highest energy particles. It is also shown that the gravitomagnetic (GM) field, which causes the dragging of inertial frames, exerts a force acting on the momentum vectors of the incident and scattered particles, causing the particle emission to be asymmetrical above and below the equatorial plane, thus appearing to break the equatorial reflection symmetry of the Kerr metric. When the accretion disk is assumed to be a two-temperature bistable thin disk/ion corona (or torus), energies as high as 54 GeV can be attained by these Penrose processes alone; and when relativistic beaming is included, energies in the TeV range can be achieved, agreeing with observations of some BL Lac objects. When this model is applied specifically to quasars 3C 279 and 3C 273, their observed high energy luminosity spectra can be duplicated and explained. Moreover, this energy extraction model can be applied to any size black hole, irrespective of the mass, and therefore applies to microquasars as well. When applied specifically to microquasar GRS 1915+105 the results are consistent with observations.

INTRODUCTION

Astrophysical jets are one of the most poorly understood phenomena today. It is clear that they are present where gravitational accretion or contraction and magnetic fields exist. We observe these jets in quasars and microquasars due to supermassive and stellar size black holes, respectively. They are also present in contracting protostars. Perhaps, understanding the mechanism responsible for jets appearing from the energy source of black holes, where the gravitational field is dominant, we can understand their appearences associtated with photostars. At present there are two popular trains of thought associated with jets in black holes: one is that the jets are inherent properties of geodesic trajectories in the Kerr [1] metric of a rotating black hole, and thus, can be described by Einstein's general theory of relativity; and the other is that the accretion disk and its magnetic field through magnetohydrodynamics (MHD) are producing the jets. Perhaps it could be a combination of the two, with gravity controlling the flow near the event horizon, and MHD controlling the flow at distances farther away. In this paper, an anaylsis of the Penrose mechanism [2] is presented to describe gravitational-particle interactions close to the event horizon in the subparsec regime.

MODEL

The model consists of a supermassive $10^8 M_\odot$ Kerr (rotating) black hole plus particles from an assumed relativistic thin disk/ion corona (or torus): two-temperature [separate temperatures for protons ($\sim 10^{12}$ K) and electrons ($\sim 10^9$ K)] bistable accretion flow. See Williams [3] for a detailed description of the model. The Penrose mechanism is used to extract rotational energy by scattering processes inside the ergosphere ($r_{\text{erg}} \simeq 2M$, where $c = G = 1$). The "quasi-Penrose" processes investigated are (a) Penrose Compton scattering (PCS) of equatorial low energy radially infalling photons by equatorially confined and nonequatorially confined orbiting target electrons, at radii between the marginally bound ($r_{\text{mb}} \simeq 1.089M$) and marginally stable ($r_{\text{ms}} \simeq 1.2M$) orbits; (b) Penrose pair production (PPP) ($\gamma p \longrightarrow e^- e^+ p$) at r_{mb}; (c) PPP ($\gamma\gamma \longrightarrow e^- e^+$) by equatorial low energy radially infalling photons and high energy blueshifted (by factor $\simeq 52$) nonequatorially confined γ-rays at the *photon orbit* ($r_{\text{ph}} \simeq 1.074M$). Note, the target particles are initially in bound (marginally stable or unstable) trapped orbits.

METHOD

Monte Carlo computer simulations of up to $\sim 50,000$ scattering events of infalling accretion disk photons (normalized to a power-law distribution) are executed for each quasar. Energy and momentum (i.e., 4-momentum) spectra of escaping particles (γ, e^-, e^+), as measured by an observer at infinity, are obtained. The following ingredients are used: (1) General relativity [the Kerr metric spacetime geometry yields equatorially and nonequatorially confined ("spherical-like") particle orbits and escape conditions, conserved energy and angular momentum parameters, and transformations from the Boyer-Lindquist coordinate frame (BLF) to the local nonrotating frame (LNRF)]. Note, BLF is the observer at infinity; LNRF is the local Minkowski (flat) spacetime. (2) Special relativity [in the LNRF, physical processes (i.e., the scatterings) are done; Lorentz transformations between inertial frames are performed; and Lorentz invariant laws are applied]. (3) Cross sections (application of the Monte Carlo method to the cross sections, in the electron rest frame for PCS, in the proton rest frame for $\gamma p \longrightarrow e^- e^+ p$, and in the center of momentum frame for $\gamma \gamma \longrightarrow e^- e^+$, give the distributions of scattering angles and final energies).

OVERALL RESULTS

The energies attained are the following: (1) *PCS*: For the input (photon) energy range: 5 eV to 1 MeV, the corresponding output energy range: $\sim$ 15 keV to 14 MeV, is attained. (2) *PPP* ($\gamma p \longrightarrow e^- e^+ p$): No escaping pairs for radially infalling γ-rays ($\sim$40 MeV), and no energy boost: implying that the assumption (negligible recoil energy given to the proton) made in the conventional cross section and, perhaps, the geometry of the scattering must be modified. It had been predicted (Leiter and Kafatos 1978) that pairs with energies ($\sim$1 GeV) can escape. (3) *PPP* ($\gamma \gamma \longrightarrow e^- e^+$): Input (photon) energy range: 3.5 keV to 100 MeV, yields output ($e^- e^+$) energy range: $\sim$ 2 MeV to 10 GeV (for BB), and higher up to $\sim$ 54 GeV (for PL), where BB, PL $\equiv$ blackbody and power-law distributions, respectively, for the accretion disk protons that yield the neutral pion decays $\pi^0 \longrightarrow \gamma \gamma$ [4,5] to populate the photon orbit.

The luminosity spectrum due to Penrose processes for the specific case of quasar 3C 273 is plotted in Figure 1(a), along with the observed spectrum for comparison. The outgoing (escaping) luminosity spectrum produced by the Penrose scattered particles is given by [6]

$$L_\nu^{\text{esc}} \approx 4\pi d^2 F_\nu^{\text{esc}} \quad (\text{erg/s\,Hz})$$
$$\approx 4\pi d^2 h\nu^{\text{esc}} f_1 f_2 \cdots f_n (N_\nu^{\text{in}} - N_\nu^{\text{cap}}),$$

where d is the cosmological distance of the black hole source; F_ν^{esc} is the flux of escaping photons; N_ν^{in} and N_ν^{cap} are the emittance of incoming and captured photons, respectively; f_n defines the total fraction of the particles that undergoes scattering [$n = 2$ for PCS and $n = 5$ for PPP ($\gamma \gamma \longrightarrow e^- e^+$)]. The values of $f_1, \ldots, f_n$ are the fitting factors, which make the Penrose calculated luminosities agree with observations for the specific case of 3C 273, Note, if every particle scatters, and $f_1 = f_3 \sim 10^{-2}$ defines the fraction of the disk luminosity intersecting the scattering regime, then the remaining f_n's equal 1 [top curves on Figures 1(a), labeled with numbers for specific cases; see [6] for details].

THE GRAVITOMAGNETIC FIELD

The gravitomagnetic (GM) force field is the gravitational analog of a magnetic field. It is the additional gravitational force that a rotating mass produces on a test particle. The GM force is produced by the gradient of $\vec{\beta}_{\text{GM}} = -\omega \hat{\mathbf{e}}_\Phi$, where ω

is the frame dragging velocity and $\vec{\beta}_{GM}$ is gravitomagnetic potential [7]. Analysis of the equations governing the particles' trajectories show that the GM force, which acts on the momentum of a particle, alters the incoming and outgoing angles of the incident and scattered particles, resulting in asymmetrical distributions, thus, appearing to break the reflection symmetry of the Kerr metric above and below the equatorial plane [3,8,9]. Notice how the effects of the GM increase with increasing energies [compare Figs. 1(d) and 1(e)].

THE VORTICAL ORBITS PRODUCED

It is found that the Penrose scattered particles escape along vortical trajectories [10] collimated about the polar axis [11]. These distributions are fluxes of coil-like trajectories of relativistic jet particles escaping concentric the polar axis. The highest energy escaping particles have the largest P_Φ values, the near largest P_Θ values, and smallest P_r values. Note, P_r is negative (inward) for most of the PCS photons, and positive (outward) for the e^-e^+ pairs. The helical angles of escape $(\delta_i)_{esc} = |90° - \theta|$ of particle type i, for the highest energy scattered particles are $(\delta_{ph})_{esc} \simeq 1° - 30°$ for PCS and $(\delta_\mp)_{esc} \sim 25° - 0.5°$ for the e^-e^+ pairs. The above characteristics of the escaping particles imply strong collimation about the polar axis, giving rise to relativistic jets with particle velocities up to $\sim c$ [see Figures 1(b)–1(f))].

CONCLUSIONS

From this model to extract energy-momentum from a black hole we can conclude the following: PCS is an effective way to boost soft x-rays $\longrightarrow$ γ-rays (~ 14 MeV). PPP ($\gamma\gamma \longrightarrow e^-e^+$) is an effective way to produce relativistic e^-e^+ pairs up to $\sim$ 54 GeV: This is the probable mechanism producing the fluxes of relativistic pairs emerging from cores of AGNs. These Penrose processes can operate for any size rotating black hole, from quasars to microquasars [12]. Overall, the main features of quasars: (a) high energy particles (e^-, e^+, γ) coming from the central source; (b) large luminosities; (c) collimated jets; (d) one-sided (or uneven jets), can all be explained by these Penrose processes to extract energy from a black hole.

Moreover, it is shown here that the geodesic treatment of individual particle processes close to the event horizon in the subparsec regime, as governed by the black hole, is sufficient to described the motion of the particles. This finding is consistent with MHD through the statement made by de Felice and Zanotti [13], that the behavior of such individual particles on geometry-induced collimated trajectories is also that of the bulk of fluid elements in the guiding center approximation. In light of this, MHD should be incorporated into these calculations to describe the flow of the Penrose escaping particles away from the black hole, i.e., to further collimate and accelerate these jet particles out to the observed kpc distances.

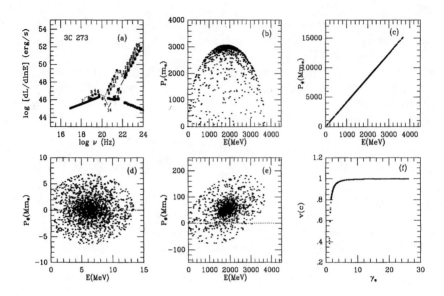

FIGURE 1. (a) Comparing the theoretical spectrum with observations for 3C 273. The calculated PCS and PPP ($\gamma\gamma \longrightarrow e^-e^+$) luminosity spectra are represented by the solid squares and large solid dots, respectively. The observed spectra is indicated by the solid line. The upper curves with the solid squares and solid dots superimposed on the dotted line and the dashed line, respectively, for PCS and PPP, are the spectra calculated from this model. Superimposed on the lower solid line of the observations are solid squares and solid dots that have been adjusted to agree with observations. These adjustments depend on the f_n's values (see text). (b) and (c) Penrose pair production ($\gamma\gamma \longrightarrow e^-e^+$) at $r_{ph} = 1.074M$: scatter plots showing momentum components (each point represents a scattering event). The radial momenta $(P_\mp)_r$ vs. $E_\mp$, the azimuthal momenta $(P_\mp)_\Phi$ ($\equiv L_\mp$) vs. $E_\mp$. For the infalling photons $E_{\gamma 1} = 0.03$ MeV, and for the target photons $E_{\gamma 2} = 3.893$ GeV. (d) and (e) PPP: $(P_\mp)_\Theta$ vs. $E_\mp$. For $E_{\gamma 1} = 0.03$ MeV, $E_{\gamma 2} = 13.54$ MeV; and $E_{\gamma 1} = 0.03$ MeV, $E_{\gamma 2} = 3.893$ GeV, respectively. (f) The velocity distribution vs. $\gamma_e (= E_\mp/m_e c^2)$ for the same case as (d) above.

REFERENCES

1. Kerr, R. P., *Phys. Rev. Letters* **11**, 237 (1963).
2. Penrose, R., *Rivista Del Nuovo Cimento: Numero Speciale* **1**, 252 (1969).
3. Williams, R. K., *Phys Rev. D* **51** 5387 (1995).

4. Eilek, J. A., *Astrophys. J*, **236**, 664 (1980).
5. Mahadevan, R., Narayan, R., and Krolik, J., *Astrophys. J.* **486**, 268 (1997).
6. Williams, R. K., *Astrophys. J.*, to appear (2001).
7. K. S. Thorne, R. H. Price, and D. A. Macdonald, *Black Holes: The Membrane Paradigm* (Yale University Press, New Haven, 1986).
8. Williams, R. K., *Phys. Rev. D*, to appear (2001).
9. Williams, R. K., in *The Proceedings of The Eighth Marcel Grossmann Meeting on General Relativity*, ed. T. Piran & R. Ruffini (Singapore: World Science), 416 (1999).
10. de Felice, F. and Carlotto, L., *Astrophys. J.*, **481**, 116 (1997).
11. Williams, R. K., *Astrophys. J.*, in preparation.
12. Williams, R. K. and Hjellming, R. M., *Astrophys. J.*, in preparation.
13. de Felice, F. and Zanotti, O., *General Relativity and Gravitation*, **8**, No. 32, 1449 (2000).

A New Optically Thin Accretion Disk Model with Powerful Electron-Positron Outflow

Tatsuya Yamasaki*[1], Fumio Takahara* and Masaaki Kusunose[†]

*Department of Earth and Space Science, Graduate School of Science, Osaka University, Toyonaka, Osaka, 560-0043, Japan
[†]Department of Physics, School of Science, Kwansei Gakuin University, Nishinomiya, 662-8501, Japan

Abstract. In the two-temperature optically thin accretion disks, it has been believed that pair processes are not important, in general. However by taking into account the effect of escape of created pairs from accretion disks, pair processes become important, and powerful outflows of escaping pairs are formed. When the mass accretion rate becomes as large as a tenth of the Eddington value, most of the viscously dissipated power is extracted by electron-positron outflow.

INTRODUCTION

Relativistic jets are observed in active galactic nuclei and Galactic black hole candidates. Although it is still unknown how these relativistic jets are formed, their physical properties have become clear by recent observations. They suggest that some of the jets are so energetic that their power amounts to about 10 % of the Eddington luminosity of the central objects. Further, they suggest that the jets are mainly composed of electron-positron pairs rather than electron-proton plasmas, at least within a parsec-scale distance from the central black holes (Takahara [1], Reynolds et al. [2], Wardle et al. [3]). The relativistic jets are most likely powered by accretion disks around the central black holes and formed by the ejection of electron-positron pairs produced in the accretion disks. However, only a few papers have addressed the issue relating the accretion disks with the pair ejection and outflow formation. In this paper, we study pair production/annihilation and ejection processes in accretion disks and investigate whether a large amount of accretion power can be extracted by escaping pairs or not.

Indeed in optically thin two-temperature accretion disk models, electron temperature becomes higher than 10^9K, and the pair processes are important. The pair

---

[1)] Research Fellow of the Japan Society for the Promotion of Science

processes in the optically thin accretion disks have been studied by Kusunose & Takahara [4], White & Lightman [5] and many others. In most of these papers, however, the effects of the pair ejection were neglected. Since the electron mass is much smaller than the proton mass, the produced pairs can be ejected more easily than protons. This is a plausible situation to form electron-positron outflow. In this paper, we investigate the pair production processes in the accretion disks, coupled with the pair ejection by solving the disk structure equations, including the vertical momentum equation of the electron-positron pairs.

FORMULATIONS

To investigate whether the powerful outflow of electron-positron can be formed in the accretion disks, it is necessary to know the structure of accretion disks. Since the main component of accretion disk flows in the radial direction and created pairs are ejected nearly in the vertical direction, this problem is two-dimensional one. However, when the temperature is not so high as the virial temperature, the velocity of inflowing gas is so slow that we can solve the radial structure locally. We solve the vertical motion of the electron-positron pairs coupled with vertical structure of accretion disks.

Solving these problems, we made several assumptions for simplicity. First, we assume the disk consists of two electrically neutral fluids, i.e., an electron-proton component and an electron-positron pair component. The former component stays in the disk and the latter is allowed to move vertically without any friction with the former one [the motion of the electon-positron component is assumed purely vertical]. Next, electrons and positrons are heated only by the Coulomb coupling with protons. Then, in the present case, number density of the gas is so low that the disk becomes two temperature. Further, we made non-relativistic formulation for simplicity. Although, several observations suggest that the ejection of jets takes place intermittently, we treat only steady case, since our main concern is whether enough power extraction is possible or not.

CONCLUSIONS

Our results show that roughly half of the viscously dissipated energy is transferred to the thermal and the kinetic energies of the ejected pairs when the normalized mass accretion rate $\dot{M}/\dot{M}_{\rm Edd}$ is as large as 0.1 ($\dot{M}_{\rm Edd}$ is the Eddington mass accretion rate). The other half of the dissipated energy is radiated from the disk. That is, accretion disks can efficiently transform the dissipated energy to the outflow energy. The present pair outflow formation model is plausible to explain the fact that the ejected jets have energy at least comparable to that radiated from the accretion disks. This is owing to another result that the pairs produced in accretion disks are ejected by their own gas pressure rather than the radiative force (cf. Yamasaki et al. [6]). This result means that the outflow energy is not supplied by the radiation field, but by the thermal energy of the electron-positron pairs which comes from protons through Coulomb collisions. It solves the difficulty explaining this observational fact in models that the pairs are produced and accelerated outside of the disks (In these models, most of the dissipated energy is inevitably lost by escape of photons).

This work was supported in part by a Grant-in-Aid from the Ministry of Education, Science, Sports and Culture of Japan.

REFERENCES

1. Takahara, F. 1995, in Towards a Major Atmospheric Cerenkov Detector III, ed. T. Kifune (Tokyo: Universal Academy), 131
2. Reynolds, C. S., Fabian, A. C., Celotti, A., & Rees, M. J. MNRAS, **283**, 873 (1996).
3. Wardle, J. F. C., Homan, D. C., Ojha, R., & Roberts, D. H. Nature, **39** 5, 457 (1998).
4. Kusunose, M., & Takahara, F. PASJ, **40**, 435 (1998).
5. White, T. R., & Lightman, A. P. ApJ, **340**, 1024 (1989).
6. Yamasaki, T., Takahara, F., & Kusunose, M. ApJL, **523**, 21 (1999).

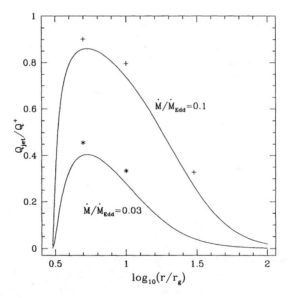

FIGURE 1. Radial distribution of $Q_{\rm jet}/Q^+$, where $Q_{\rm jet}$ is the energy extracted from a disk by pairs and Q^+ is the viscously dissipated energy; $r_{\rm g}$ is the Schwarzschild radius. The crosses denote the results for $\dot{M}/\dot{M}_{\rm Edd} = 0.1$ and the asterisks denote the results for $\dot{M}/\dot{M}_{\rm Edd} = 0.03$. The solid curves are the results obtained by calculations within one-zone approximation.

CHAPTER 10

SUPERNOVAE

Aspherical Supernova Explosions: Hydrodynamics, Radiation Transport & Observational Consequences

P. Höflich[1], A. Khokhlov[2], L. Wang[3]

[1] Department of Astronomy, University of Texas, Austin, TX 78681, USA
[2] Naval Research Lab, Washington DC, USA
[3] Lawrence Berkeley Lab, 1 Cyclotron Rd, Berkeley, CA 94720, USA

Abstract. Core collapse supernovae (SN) are the final stages of stellar evolution in massive stars during which the central region collapses, forms a neutron star (NS), and the outer layers are ejected. Recent explosion scenarios assume that the ejection is due to energy deposition by neutrinos into the envelope but detailed models do not produce powerful explosions. There is new and mounting evidence for an asphericity and, in particular, for axial symmetry in several supernovae which may be hard to reconcile within the spherical picture. This evidence includes the observed high polarization and its variation with time, pulsar kicks, high velocity iron-group and intermediate-mass elements observed in remnants, and direct observations of the debris of SN1987A. Some of the new evidence is discussed in more detail. To be in agreement with the observations, any successful mechanism must invoke some sort of axial symmetry for the explosion.

As a limiting case, we consider jet-induced/dominated explosions of "classical" core collapse supernovae. The discovery of magnetars revived the idea that a MHD-jet with appropriate properties may be formed at the NS. Our study is based on detailed 3-D hydrodynamical and radiation transport models. We demonstrate the influence of the jet properties and of the underlying progenitor structure on the final density and chemical structure. Our calculations show that low velocity, massive jets can explain the observations. Both asymmetric ionization and density/chemical distributions have been identified as crucial for the formation of asymmetric photospheres. Even within the picture of jet-induced explosion, the latter effect alone fails to explain early polarization in core collapse supernovae with a massive, hydrogen-rich envelope such as SN1999em. Finally, we discuss observational consequences and tests.

I OBSERVATIONAL EVIDENCE FOR ASYMMETRY

In recent years, there has been a mounting evidence that the explosions of massive stars (core collapse supernovae) are highly aspherical. (1) The spectra of core-collapse supernovae (e.g., SN87A, SN1993J, SN1994I, SN1999em) are significantly polarized indicating asymmetric envelopes with axis ratios up to 2 (Méndez et al. 1988; Höflich 1991; Jeffrey 1991; Wang et al. 1996; Wang et al. 2001b). The degree of polarization tends to vary inversely with the mass of the hydrogen envelope, being maximum for Type Ib/c events with no hydrogen (Wang et al. 2001b). For SN1999em (Fig. 1), Leonard et al. (2000) showed that the polarization and, thus, the asphericity increases with time. Both trends suggest a connection of the asymmetries with the central engine. For supernovae with a good time and wavelength coverage, the orientation of the polarization vector tends to stay constant both in time and with wavelength. This implies that there is a global symmetry axis in the ejecta (Leonard et al. 2001, Wang et al. 2001b). (2) Observations of SN 1987A showed that radioactive material was brought to the hydrogen rich layers of the ejecta very quickly during the explosion (Lucy 1988; Tueller et al. 1991). (3) The remnant of the Cas A supernova shows rapidly moving oxygen-rich matter outside the nominal boundary of the remnant and evidence for two oppositely directed jets of high-velocity material (Fesen & Gunderson 1997). (4). Recent X-ray observations with the CHANDRA satellite have shown an unusual distribution of iron and silicon group elements with large scale asymmetries in Cas A (Hughes et al. 2001). (5) After the explosion, neutron stars are observed with high velocities, up to 1000 km/s (Strom et al. 1995). (6) Direct HST-images from June 11,2000, are able to resolve the inner debris of SN1987A showing its prolate geometry with an axis ratio of ≈ 2 (Fig. 1). Both the ejecta and the inner ring around SN1987A show a common axis of symmetry. By combining the HST-images with spectral and early-time polarization data, Lifan Wang worked out the details of the chemical and density structure. By connecting the HST-images with the polarization data from earlier times, he demonstrated that the overall geometry of the entire envelope of SN1987A (including the H-rich envelope) was elongated all along the same symmetry axis, and that the distribution of the products of stellar burning (O, Ca, etc.) are concentrated in the equatorial plane (Wang et al. 2001a).

II MODELS FOR COLLAPSE SUPERNOVA

There is a general agreement that the explosion of a massive star is caused by the collapse of its central parts into a neutron star or, for massive progenitors, into a black hole. The mechanism of the energy deposition into the envelope is still debated. The process likely involves the bounce and the formation of the prompt shock (e.g. Van Riper 1978, Hillebrandt 1982), radiation of the energy in the form of neutrinos (e.g. Bowers & Wilson 1982), and the interaction of the neutrino with the material of the envelope and various types of convective motions (e.g. Herant et al. 1994, Burrows et al. 1995, Müller & Janka 1997, Janka & Müller 1996), rotation (e.g. LeBlanc & Wilson 1970, Saenz & Shapiro S.L. 1981, Mönchmeyer et al. 1991, Zwerger & Müller 1997) and magnetic fields (e.g. LeBlanc & Wilson 1970, Bisnovati-Kogan 1971).

Currently, the most favored mechanism invokes the neutrino deposition. The results depend critically on the progenitor structure, equation of state, neutrino

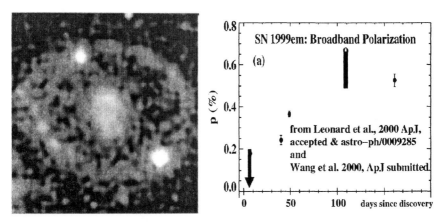

FIGURE 1. Observational evidence for asphericity in core collapse supernovae. The HST image of SN 1987A (left panel) shows the inner debris of the SN-ejecta with an axis ratio of ≈ 2 and the ring on June 11th, 2000 (from Wang et al. 2001). Note that the inner ring has been formed during the stellar evolution about 10,000 years before the explosion. The right panel shows the evolution with time of the linear polarization P in the plateau supernovae SN1999em. Although P increased with time, the polarization angle remained constant with time and wavelength indicating a common axis of symmetry in the expanding envelope (Leonard et al. 2001, Wang et al. 2001c).

physics, and implementation of the neutrino transport. Recent results indicate that spherical models fail to produce successful explosions even when using sophisticated, relativistic Boltzmann solvers for the neutrino transport and taking different flavors and neutrinos into account (Yamada et al. 1999, Ramp & Janka 2000, Mezzacappa et al. 2001). Multi-dimensional effects such as convection during the core collapse itself must be included but, still, it is an open question whether convection combined with the neutrino transport provides the solution to the supernova problem (Ramp et al. 1998 and references therein). In the current calculations, the size and scale of the convective motions seem to be too small to explain the observed asymmetries in the envelope. The angular variability of the neutrino flux caused by the convection has been invoked to explain the neutron star kicks (Burrows et al. 1995, Janka & Müller 1994). Calculations give kick velocity up to $\simeq 100$ km/s whereas NS with velocities of several 100 km/s are common. For a more detailed discussion of the current status and problems to explain the observed asymmetries, see Khokhlov & Höflich (2001).

It has long been suggested that the magnetic field can play an important role in the explosion (LeBlanc & Wilson 1970; Ostriker & Gunn 1971, Bisnovati-Kogan 1971, Symbalisty 1984). Simulations by LeBlanc & Wilson showed the amplification of the magnetic field due to rotation and the formation of two oppositely directed, high-density, supersonic jets of material emanating from the collapsed core. Their simulations assumed a rather high initial magnetic field $\sim 10^{11}$ Gauss and produced a very strong final fields of the order of $\sim 10^{15}$ Gauss which seemed to be unreasonable at the time. The recent discovery of pulsars with very high magnetic fields (Kouveliotou et al. 1998, Duncan & Thomson 1992) revived the interest in the role of rotating magnetized neutron stars in the explosion mechanism. It is not clear whether a high initial magnetic field required for the LeBlanc & Wilson mechanism is realistic. On the other hand it may not be needed.

The current picture of the core collapse process is unsettled. A quantitative model of the core collapse must eventually include all the elements mentioned above.

Due to the difficulty of modeling core collapse from first principles, a very different line of attack on the explosion problem has been used extensively and proved to be successful in understanding of the supernova problem, SN1987A in particular (Arnett et al. 1990, Hillebrand & Höflich 1991). The difference of characteristic time scales of the core (a second or less) and the envelope (hours to days) allows one to divide the explosion problem into two largely independent parts - the core collapse and the ejection of the envelope. By assuming the characteristics of the energy deposition into the envelope during the core collapse, the response of the envelope can be calculated. Thus, one can study the observational consequences of the explosion and deduce characteristics of the core collapse and the progenitor structure. This approach has been extensively applied in the framework of the 1D spherically symmetric formulation. The major factors influencing the outcome have been found to be the explosion energy and the progenitor structure. The same approach can be applied in multi-dimensions to investigate the effects of asymmetric explosions. In this paper we study the effects and observational consequences of an asymmetric, jet-like deposition of energy inside the envelope of SN.

III RESULTS FOR JET-INDUCED SUPERNOVAE

A Numerical Methods

3-D Hydrodynamics: The explosion and jet propagation are calculated by a full 3-D code within a cubic domain. The stellar material is described by the time-dependent, compressible, Euler equations for inviscid flow with an ideal gas equation with $\gamma = 5/3$ plus a component due to radiation pressure with $\gamma = 4/3$. The Euler equations are integrated using an explicit, second-order accurate, Godunov type, adaptive-mesh-refinement, massively parallel, Fully-Threaded Tree (FTT) program, ALLA (Khokhlov 1998).

1-D Radiation-Hydrodynamics: About 1000 seconds after the core collapse and in case of the explosion of red supergiants, the propagation of the shock front becomes almost spherical (see below). To be able to follow the development up to the phase of homologous expansion (≈ 3 to 5 days), the 3-D structure is remapped on a 1-D grid, and the further evolution is calculated using a one-dimensional radiation-hydro code that solves the hydrodynamical equations explicitly in the comoving frame by the piecewise parabolic method (Colella and Woodward 1984).

The radiation transport part is solved implicitly using the method of variable Eddington factors. Expansion opacities and LTE-equations of state are used (Höflich et al. 1998 and references therein).

3-D Radiation Transport for Ionization Structures and Light Curves: For given, arbitrary 3-D density, velocity and chemical distributions, we calculated the detailed 3-D ionization structure and LCs using the same assumptions as for our 1-D hydro code. The γ-ray transport is computed in 3-D using a Monte Carlo method (Höflich et al. 1992, 1993) which includes relativistic effects and a consistent treatment of both the continua and line opacities. Subsequently, for low energy photons, we solve the three-dimensional radiation transport using a hybrid scheme of Monte-Carlo and non-equilibrium diffusion methods. As a first step, we solve the time-dependent radiation transport equation in non-equilibrium, diffusion approximation for 3-D geometry including the scattering and thermalization terms for the source functions, and include the frequency derivatives into the formulation for the opacities and emissivities. We solve the same set of momentum equations as Höflich et al. (1993) but for 3-D geometry and an Eddington factor of 1/3. At large optical depths, this provides the solution for the full radiation transport problem. In a second step, to obtain the correction solution for the radiation transport equation at small optical depths, the difference between the solution of the diffusion and full radiation transport equation is calculated in a Monte Carlo scheme. We calculate the difference between the solutions for computational accuracy and efficiency. Consistency between the solution at the outer and inner region is obtained iteratively. The same Monte Carlo solver is used which has been applied to compute γ-ray and polarization spectra. For simplification, the relation between energy density and temperature is taken from the 1-D radiation hydrodynamics at the corresponding time (see below).

3-D Radiation for Spectra and Polarization: For several moments of time, detailed polarization and flux spectra for asymmetric explosions are calculated using our MC-code including detailed equations of state, and detailed atomic models for some of the ions. For details, see Höflich (1995), Höflich et al. (1995) and Wang et al. (1998).

B Jet propagation:

The Setup: The computational domain is a cube of size L with a spherical star of radius R_{star} and mass M_{star} placed in the center. The innermost part with mass $M_{\text{core}} \simeq 1.6 M_\odot$ and radius $R_{\text{core}} = 4.5 \times 10^8$ cm, consisting of Fe and Si, is assumed to have collapsed on a timescale much faster than the outer, lower-density material. It is removed and replaced by a point gravitational source with mass M_{core} representing the newly formed neutron star. The remaining mass of the envelope M_{env} is mapped onto the computational domain. At two polar locations where the jets are initiated at R_{core}, we impose an inflow with velocity v_j ρ_j. At R_{core}, the jet density and pressure are the same as those of the background material. For the

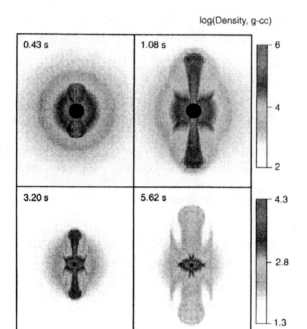

FIGURE 2. Logarithm of the density structure as a function of time for a helium core. The total mass of the ejecta is 2.6 $M_\odot$. The initial radius, velocity and density of the jet were taken to 1200 km 32,000 km/sec and $6.5E5g/cm^3$, respectively. The shown domains 7.9, 9.0, 36 and 45 $\times 10^9 cm$. The total energy is about 9E50 erg. After about 4.5 seconds, the jet penetrates the star. The energy deposited in the stellar envelope by the jet is about 4E50 erg, and the final asymmetry is of the order of two.

first 0.5 s, the jet velocity at R_{core} is kept constant at v_j. After 0.5 s, the velocity of the jets at R_{core} was gradually decreased to zero at approximately 2 s. The total energy of the jets is E_j. These parameters are consistent with, but somewhat less than, those of the LeBlanc-Wilson model.

The reference model: As a baseline case, we consider a jet-induced explosion in a helium star. Jet propagation inside the star is shown in Fig. 2. As the jets move outwards, they remain collimated and do not develop much internal structure. A bow shock forms at the head of the jet and spreads in all directions, roughly cylindrically around each jet. The jet-engine has been switched off after about 2.5 seconds. The material of the bow shock continues to propagate through the star. The stellar material is shocked by the bow shock. Mach shocks travel towards the equator resulting in a redistribution of the energy. The opening angle of the jet depends on the ratio between the velocity of the bow shock to the speed of sound. For a given star, this angle determines the efficiency of the deposition of the jet energy into the stellar envelope. Here, the efficiency of the energy deposition is about 40 %, and the final asymmetry of the envelope has an axis ratio of about two.

Influence of the jet properties: Fig. 3 shows two examples of an explosion with with a low and a very high jet velocity compared to the baseline case (Fig. 2). Fig. 3 demonstrates the influence of the jet velocity on the opening angle of the jet and, consequently, on the efficiency of the energy deposition. For the low velocity jet, the jet engine is switched off long before the jet penetrates the stellar envelope. Almost all of the energy of the jet goes into the stellar explosion. On a contrary, the fast jet (61,000 km/sec) triggers only a weak explosion of 0.9 foe although its total energy was $\approx 10\,foe$.

Influence of the progenitor: For a very extended star, as in case of 'normal' Type II Supernovae, the bow shock of a low velocity jet stalls within the envelope, and the entire jet energy is used to trigger the ejection of the stellar envelope. In our example (Fig. 5), the jet material penetrates the helium core at about 100 seconds. After about 250 seconds the material of the jet stalls within the hydrogen rich envelope and after passing about 5 solar masses in the radial mass scale of the spherical progenitor. At this time, the isobars are almost spherical, and an almost spherical shock front travels outwards. Consequently, strong asphericities are limited to the inner regions. After about 385 seconds, we stopped the 3-D run and remapped the outer layers into 1-D structure, and followed the further evolution in 1-D. After about 1.8E4 seconds, the shock front reaches the surface. After about 3 days, the envelope expands homologously. The region where the jet material stalled, expands at velocities of about 4500 km/sec.

Fallback: Jet-induced supernovae have very different characteristics with respect to fallback of material and the innermost structure. In 1-D calculations and for stars with Main Sequence masses of less than 20 $M_\odot$ and explosion energies in excess of 1 foe, the fallback of material remains less than 1.E-2 to 1.E-3 $M_\odot$ and an inner, low density cavity is formed with an outer edge of ^{56}Ni. For explosion

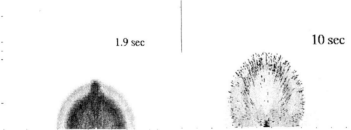

FIGURE 3. Same as Fig. 2 ($0.5 \leq log(\rho) \leq 5.7$) but for a jet velocity of 61,000 km/sec and a total energy of 10 foe at $\approx 1.9sec$ (left), and 11,000 km/sec and a total energy of 0.6 foe (right). The size of the presented domains are 5 (left) and 2 $10^{10}cm$ (right), respectively. For the high velocity jet, most of the energy is carried away by the jet. Only 0.9 foe are deposited in the expanding envelope. In case of a low velocity jet, the bow-shock still propagates through the star after the jet is switched off, and the entire jet energy is deposited in the expanding envelope.

energies between 1 and 2 foe, the outer edge of the cavity expands typically with velocities of about 700 to 1500 km/sec (e.g. Woosley 1997, Höflich et al. 2000). In contrast, we find strong, continuous fallback of $\approx 0.2 M_\odot$ of O-rich matter in the the 3-D hydro models, and no lower limit for the velocity of the expanding material (Fig. 4 of Khokhlov & Höflich 2001). This significant amount of fallback must have important consequences for the secondary formation of a black hole. The exact amount and time scales for the final accretion on the neutron star will depend sensitively on the rotation and momentum transport.

Chemical Structure: The final chemical profiles of elements formed during the stellar evolution such as He, C, O and Si are 'butterfly- shaped' whereas the jet material fills an inner, conic structure (Fig. 5, upper, middle panel).

The composition of the jets must reflect the composition of the innermost parts of the star, and should contain heavy and intermediate-mass elements, freshly synthesized material such as ^{56}Ni and, maybe, r-process elements because, in our examples, the entropy at the bow shock region of the jet was as high as a few hundred. In any case, during the explosion, the jets bring heavy and intermediate mass elements into the outer H-rich layers.

C Radiation Transport Effects

For the compact progenitors of SNe Ib/c, the final departures of the iso-density contours from sphericity are typically a factor of two. This will produce a linear polarization of about 2 to 3 % (Fig. 4) consistent with the values observed for Type Ib/c supernovae. In case of a red supergiant, i.e. SNe II, the asphericity is restricted to the inner layers of the H-rich envelope. There the iso-densities show an axis ratios of up to ≈ 1.3. The intermediate and outer H-rich layers remain spherical.

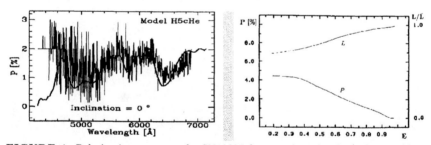

FIGURE 4. Polarization spectrum for SN1993J for an axis ratio of 1/2 for an oblate ellipsoid in comparison with observations by Trammell et al. (1993) are given in the left plot. On the right, the dependence of the continuum polarization (right) and directional dependence of the luminosity is shown as a function axis ratios for oblate ellipsoids seen from the equator (from Höflich, 1991 & Höflich et al. 1995).

This has strong consequences for the observations, in particular, for polarization measurements. The polarization should be larger in SNe Ib/c compared to classical SNe II which is consistent with the observations by Wang et al. (2001b). Early on, we expect no or little polarization in supernovae with a massive, hydrogen rich envelope but that the polarization will increase with time to about 1% (Höflich 1991), depending on the inclination the SN is observed. This is also consistent both with the long-term time evolution of SN1987A (Jefferies 1991).

Recently, the plateau supernova SN1999em has been observed with VLT and Keck providing the best time coverage up to now of any supernovae (Wang et al. 2001b; Leonard et al. 2001). The basic trend has been confirmed which we expected from the hydro. Indeed, P is very low early on, and it rises when more central parts are seen. However, there are profound differences which point towards an additional mechanism to produce aspherical photospheres.

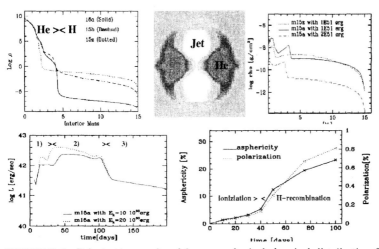

FIGURE 5. Polarization produced by an aspherical chemical distribution for a SN IIp model with $15M_\odot$ and an explosion energy $E_{exp} = 2 \times 10^{51} erg$. This model resembles SN 1999em (see above). The initial density profile is given for a star at the final stage of stellar evolution for metallicities Z of 0.02, 0.001 and 0 (models 15a, 15b, 15z, upper left panel, from Höflich et al. 2000 & Chieffi et al. 2001). The model for the Red Supergiant, 15a, has been used to calculate the jet-induced explosion 2). In the upper, middle panel, the chemical distribution of He is given at 250 sec for the He-rich layers after the jet material has stalled. The color-codes white, yellow, green, blue and red correspond to He mass fractions of 0., 0.18, 0.36, 0.72, and 1., respectively. The subsequent explosion has been followed in 1-D up to the phase of homologous expansion. In the upper, right panel, the density distribution is given at about 5 days after the explosion. The steep gradients in the density in the upper right and left panels are located at the interface between the He-core and the H-mantle. In the lower, left panel, the resulting bolometric LCs are given for $E_{exp} = 2 \times 10^{51} erg$ (dotted line) and, for comparison, for $1 \times 10^{51} erg$, respectively. Based on full 3-D calculations for the radiation & γ-ray transport, we have calculated the location of the recombination front as a function of time. The resulting shape of the photosphere is always prolate. The corresponding axis ratio and the polarization seen from the equator are shown (lower, right panel). Note the strong increase of the asphericity after the onset of the recombination phase between day 30 to 40 (see also SN 1999em in Fig. 1).

SN1999em is an extreme plateau supernovae with a plateau lasting for more than 100 days (IAUC 7294 to 7355). However, no detailed light curves have been published. Therefore, in Fig. 5, we show a theoretical LC which resembles SN 1999em with respect to the duration of the plateau and its brightness, and the typical expansion velocities. The light curves of SNe IIp show three distinct phases (Fig. 5). 1) Most of the envelope is ionized. This phase depends sensitively on the explosion energy, mixing of radioactive Ni, and the mass of the progenitor, e.g. either strong mixing or $E_{kin} \leq 1 foe$ will cause a steep and steady increase in the luminosity (and in B and V); 2) The emitted energy is determined by the receding (in mass) of the H recombination front which is responsible for both the release of stored, thermal and the recombination energy. At the recombination front, the opacity drops by about 3 orders of magnitude when it changes from electron scattering dominated to bound-free/ free-free. This provides a self-regulating mechanism for the energy release. If too little energy is released, the opacity drops fast causing an increase in the speed at which the photosphere is receding. In turn, this causes a larger energy release and vs. . Hydrogen recombines at a specific temperature at or just below the photosphere. Due to the flat density profiles of the expanding envelopes in the RSG case, the photospheric radius and, thus, the luminosity L stays almost constant. After the recombination front has passed through the H-rich envelope, the brightness drops fast. During phase 3), L is given by the instant energy release by radioactive decay of ^{56}Co. Obviously, the steep rise in P of SN1999em coincides with the transition from phase 1 to 2, pointing towards an-isotropic excitation as a new mechanism for producing aspherical photospheres and, consequently, polarization. To quantify the effect, we have calculated the temperature and ionization structure of a SN IIp (Fig. 5). Starting from a spherical model, the initial chemical distribution has been taken from our 3-D jet simulation. As mentioned above, the chemical profile is frozen out after about 250 sec, and the expansion becomes spherical. The further evolution can be followed with our 1-D radiation code. For several moments of time, we have calculated the ionization structure and continuum polarization based of the 3-D chemical structure and the spherical density distribution under the assumption that the distribution of the radioactive Ni coincides with the jet-material. Note that the ^{56}Ni layers extend throughout the He-core ($\approx 5 M_\odot$) in polar directions but they are confined to the very center along the equatorial plane (Fig. 5, upper, middle panel) leading to an increased transport of energy and, consequently, heating toward polar directions. Before the recombination phase, the opacity is dominated by Thomson scattering which does not depend on the temperature, and the shape of the photosphere remains almost spherical. However, during the recombination phase, the location of the photosphere depends sensitively on the heating, and the photosphere becomes prolate (Fig. 5, lower, right panel). There is a gradual increase in P and no jump in P because the optically thick H-rich layers below the photosphere redistribute the photons. The increase of P depends on the geometrical expansion and speed (in mass) of the receding recombination front.

IV CONCLUSIONS

We have numerically studied the explosion of Core Collapse supernovae caused by supersonic jets generated in the center of the supernova as a result of the core collapse into a neutron star. We simulated the process of the jet propagation through the star, the redistribution of elements, and radiation transport effects. A strong explosion and a high efficiency for the conversion of the jet energy requires low jet velocities or a low initial collimation of the jet. With increasing extension of the envelope, the conversion factor increases. Typically, we would expect higher kinetic energies in SNe II compared to SNe Ib/c if a significant amount of explosion energy is carried away by jets. Within the framework of jet-induced SN, the lack of this evidence suggests that the jets have low velocities.

The He, C, O and Si rich layers of the progenitor show characteristic, butterfly-shape structures. This morphology and pattern should be observable in supernova remnants, e.g. with the Chandra observatory, despite some modifications and instabilities when the expanding medium interacts with the interstellar material.

During the explosion, the jets bring heavy and intermediate mass elements including ^{56}Ni into the outer layers. Due to the high entropies of the jet material close to the center, this may be a possible site for r-process elements. Spatial distribution of the jet material will influence the properties of a supernova. In our model for a SN II, the jet material stalled within the expanding envelope corresponding to a velocity of $\approx 4500 km/sec$ during the phase of homologous expansion. In SN1987A, a bump in spectral lines of various elements has been interpreted as due to material excited by a clump of radioactive ^{56}Ni (Lucy 1988). Within our framework, this bump may be a measure of the region where the jet stalled. This could also explain the early appearance of X-rays in SN1987A which requires strong mixing of radioactive material into the hydrogen-rich layers (see above), and the overall distribution of elements and distribution of elements in the resolved HST-images of the inner debris of SN 1987A. We note that, if this interpretation is correct, the 'mystery spot' (Nisenson et al. 1988) would be unrelated. In contrast to 1-D simulations, we find in our models strong, continuous fallback over an extended period of time, and a lack of an inner, almost empty cavity. This significant amount of fallback during late times and the consequences for the secondary formation of a black hole shall be noted. Moreover, fallback and the low velocity material may alter the escape probability for γ-rays produced by radioactive decay of ^{56}Ni. In general, the lower escape probability is unimportant for the determination of the total ^{56}Ni production by the late LCs because full thermalization can be assumed in core collapse SN during the first few years. However, in extreme cases such as SN98BW (e.g. Schaefer et al. 2000), only a small fraction of gamma's are trapped. Effects of multi-dimensionality will strongly alter the energy input by radioactive material and disallow a reliable estimate for the total ^{56}Ni mass.

Qualitatively, the jet-induced picture allows reproduction of the polarization observed in core collapse supernovae. Both asymmetric ionization and density/chemical distributions have been identified as crucial. Even within the picture of jet-induced explosion, the latter effect alone cannot (!) account for the high polarization produced in the intermediate H-rich layers of core-collapse SN with a massive envelope such as SN1999em. The former effect operates only during a recombination phase, and can be expected to dominate the polarization in core-collapse supernovae with massive H-rich envelopes during the first 1 to 2 months. Complete time series of polarization measurements are needed to test this suggestion.

Finally, we want to emphasize the limits of this study and some of the open questions which will be addressed in the future. We have assumed that jets are formed in the course of the formation of a neutron star, and have addressed observational consequences and constrains. However, we have not calculated the jet formation, we do not know if they really form, and, if they form, whether they form in all core-collapse supernovae. Qualitatively, the observational properties of core collapse supernovae are consistent with jet-induced supernovae and support strongly that the explosion mechanism is highly aspherical, but more detailed comparisons with individual objects must be performed as soon as the data become available. We cannot claim that the jets are the only mechanism that can explain asphericity in supernovae, but any competing mechanism must involve some sort of axial symmetry on large scales with a profound impact on the explosion such as rapid rotation. It remains to be seen whether asymmetry and axial symmetry are the 'smoking gun' for our understanding of the SN-mechanism.

Acknowledgments: We want to thank our colleagues for helpful discussions, in particular, D. Baade, E.S. Oran, J.C. Wheeler, Inzu Yi A., C. Mayers, J.C. Wilson, A. Chieffi, M. Limongi, and O. Straniero. This work is supported in part by NASA Grant LSTA-98-022.

REFERENCES

1. Arnett W.D., Bahcall J.N., Kirshner, R.P., Woosley, S.E. 1990, ARAA 27, 62
2. Bisnovatyi-Kogan 1971, Soviet Astronomy AJ, 14, 652
3. Bowers R.L., Wilson J.R. 1982 ApJS 50, 115
4. Burrows A., Hayes J., Fryxell B. 1995, ApJ 450
5. Colella, P.; Woodward, P.R. 1984, J.Comp.Phys. 54, 174
6. Duncan, R. C. & Thompson, C. 1992, ApJ, 392, L9
7. Fesen, R. A. & Gunderson, K. S. 1996, ApJ, 470, 967
8. Herant M., Benz W., Hix W.R., Fryer C.L., Colgate S.A. 1994, ApJ 435, 339
9. Hillebrandt W., Höflich 1991, Nuclear Physics B 19, 113
10. Hillebrandt W. 1982, ApJ 103, 147
11. Höflich P., Straniero O., Limongi M. Dominguez I. Chieffi A. 2000, 7th TexMex-Conference, eds. W. Lee & S. Torres-Peimbert, UNAM-Publ., in press & astro-ph/005037
12. Höflich, P., Wheeler, J. C., and Thielemann, F.K. 1998, ApJ 495, 617
13. Höflich, P. 1995, ApJ 443, 89
14. Höflich P., Wheeler, J.C., Hines, D., Trammell S. 1995, ApJ 459, 307

15. Höflich, P., Müller E., Khokhlov A. 1993, A&A 268, 570
16. Höflich, P., Müller E., Khokhlov A. 1992, A&A 259, 243
17. Höflich, P. 1991 A&A 246, 481
18. Hughes J.P., Rakowski C.E., Burrows D.N., Slane P.O. 2001, AJ, in press & astro-ph/9910474
19. Janka H.T. & Müller E., 1994 A&A 290, 496
20. Jeffrey D.J., 1991, ApJ, 375, 264
21. Khokhlov A., Höflich P., Oran E.S., Wheeler J.C., P. Wang L., 1999, ApJ 524, L107
22. Khokhlov A.M., Höflich P. 2001, in: 1st KIAS Astrophysics, IAP-Publishing, ed. I. Yi, in press & astro-ph/0011023
23. Khokhlov, A.M. 1998, J.Comput.Phys., 143, 519
24. Kouveliotou, C., Strohmayer, T., Hurley, K., Van Paradijs, J., Finger, M. H., Dieters, S., Woods, P., Thompson, C. & Duncan, R. C. 1998, ApJ, 510, 115
25. Lucy L.B. 1988, Proc. of the 4th George Mason conference, ed. by M. Kafatos, Cambridge University Press, p. 323
26. LeBlanc, J. M. & Wilson, J. R. 1970, ApJ, 161, 541
27. Leonard D.C., Filippenko, A.V., Barth A.J., Matheson T. 2000, ApJ 536, 239
28. Mendez R.H., Clocchiatti A., Benvenuto O.G., Feinstein C. Marraco H. 1977, ApJ 334, 295
29. Mezzacappa A., Liebendoerfer M., Bronson Messer O.E. Hix R., Thielemann F-K, Bruenn S.W. 2001, Phy.Rev.Let., accepted
30. Mönchmeyer R., Schaefer G., Mueller E., Kates R.E. 1991 A&A 246, 417
31. Müller E., Janka H.T. 1997, A&A 317, 140
32. Nisenson P., Papaliolios C., Karovska M., Noyes R. 1988, ApJ 324, 35
33. Ostriker, J. P. & Gunn, J. E. 1971, ApJ, 164, L95
34. Rammp M., Müller E., Ruffert M. 1998, A&A 332, 969
35. Ramp M. and Janka, H.-T. 2000, ApJ 593, L33
36. Schaefer B. 2000, ApJ 533, 21
37. Saenz R.A., Shapiro S.L. 1981, ApJ 244, 1033
38. Symbalisty E.M.D. 1984, ApJ 285, 729
39. Strom R., Johnston H.M., Verbunt F., Aschenbach B. 1995, Nature, 373, 587
40. Symbalisty E.M.D. 1984, ApJ 285, 729
41. Trammell S., Hines D., Wheeler J.C. 1993, ApJ 414, 21
42. Tueller J., Barthelmy S., Gehrels N., Leventhal M., MacCallum C.J., Teegarden B.J. 1991, in: Supernovae, ed. S.E. Woosley, Springer Press, p. 278
43. Van Riper K.A. 1978, ApJ 221, 304
44. Wang L. et al. 2001a, *The Bipolar Ejecta of SN1987A*, ApJ, submitted
45. Wang L., Howell A., Höflich P., Wheeler J.C. 2001b, ApJ, in press
46. Wang L., Baade D., Höflich P., Wheeler J.C. 2001c, ApJ, submitted
47. Wang, L., Wheeler, J. C., Li, Z. W., & Clocchiatti, A. 1996, ApJ, 467, 435
48. Wang, L., Wheeler, J.C., Höflich, P. 1997, ApJ, 476, 27
49. Yamada S., Janka H.T., Suzuki H. 1999, A&A 344, 533
50. Zwerger T., Müller E. 1997, A&A 320, 209

General Relativistic Simulations of Stellar Core Collapse and Postbounce Evolution with Boltzmann Neutrino Transport

M. Liebendörfer[*], O. E. B. Messer[*], A. Mezzacappa[†] and W. R. Hix[*]

[*]*Department of Physics and Astronomy, University of Tennessee, Knoxville, TN-37996 and Physics Division, Oak Ridge National Laboratory, Oak Ridge, TN-37831*
[†]*Physics Division, Oak Ridge National Laboratory, Oak Ridge, TN-37831*

Abstract. We present self-consistent general relativistic simulations of stellar core collapse, bounce, and postbounce evolution for 13, 15, and 20 solar mass progenitors in spherical symmetry. Our simulations implement three-flavor Boltzmann neutrino transport and standard nuclear physics. The results are compared to our corresponding simulations with Newtonian hydrodynamics and $O(v/c)$ Boltzmann transport.

INTRODUCTION

A supernova explosion is a dramatic event that includes such a rich diversity of physics that, with current computer hardware, a self-consistent model based on numerical simulations can not possibly include all of them at once. After stellar core collapse, a compact object at the center of the event is formed, requiring a description in general relativity. Neutrinos radiating from this central object are strongly coupled to the matter at high densities before streaming out at lower densities. Multi-frequency radiation hydrodynamics must be used to quantify the energy that these neutrinos deposit in the material behind the shock. Moreover, evidence suggests that this heating drives convection behind the shock, and significant rotation and strong magnetic fields might also be present. Observations of neutron star kicks, mixing of species, inhomogeneous ejecta, and polarization of spectra support the presence of asymmetries in supernova explosions (e.g. Tueller et al. [1], Strom et al. [2], Galama et al. [3], Leonard et al. [4], and references therein). Motivated by such observations, various multi-dimensional explosion mechanisms have been explored (Herant et al.[5], Miller et al.[6], Herant et al.[7], Burrows et al.[8], Janka and Müller [9], Mezzacappa et al.[10], Fryer [11], Fryer and Heger [12]) and jet-based explosion scenarios have recently received new momentum (Höflich et al. [13], Khokhlov et al. [14], MacFadyen and Woosley [15], Wheeler et al. [16]). Many exciting aspects of jets are discussed in these proceedings. However, for the time being, one has to single out a subset of the known physics and to investigate the role each part plays in a restricted simulation. It is natural to start with ingredients that have long been believed to be essential for the explosion and to add modifiers in a systematic way until the observables can be reproduced (see e.g. Mezzacappa et al. [17]). Spherically symmetric supernova modeling has a long tradition and is nearing a definitive point in the sense that high resolution hydrodynamics, general relativity, complete Boltzmann neutrino transport, and reasonable nuclear and weak interaction physics are being combined to dispel remaining uncertainties.

Our simulations are based on the three-flavor Boltzmann solver of Mezzacappa and Bruenn [18, 19, 20] that was consistently coupled to hydrodynamics on an adaptive grid and extended to general relativistic flows (Liebendörfer [21]). We performed relativistic simulations for progenitors with different masses and compare the results to the counterpart simulations in Newtonian gravity with O(v/c) Boltzmann transport (Messer [22]). While the quantitative results are sensitive to the inclusion of all neutrino flavors and general relativity (Bruenn et al. [23]), we find the same qualitative outcome as did Rampp and Janka [24] in their independent simulations with one-flavor O(v/c) Boltzmann transport: Given current input physics, there are no explosions in spherical symmetry without invoking multidimensional effects (Mezzacappa et al. [17], Liebendörfer et al. [25]). This is consistent with the results of Wilson and Mayle [26], who found explosions only by including neutron finger convection.

CORE COLLAPSE AND NEUTRINO BURST

The simulations discussed herein are initiated from progenitors with main sequence masses of 13 $M_\odot$, 20 $M_\odot$ (Nomoto and Hashimoto [27]), and 15 $M_\odot$ (Woosley and Weaver [28]). We use in our models the equation of state of Lattimer and Swesty [29] and "standard" weak interactions (e.g. Bruenn [30]). The first phase in the simulations leads through core collapse to bounce and shock formation. When the shock passes the neutrinospheres approximatively 4 ms after bounce, an energetic neutrino burst is released from the hot shocked material, rendering it "neutrino-visible" to the outside world. The corresponding deleptonization behind the shock is dramatic. The energy carried off with the neutrino burst enervates the shock in both the NR and the GR cases. A pure accretion shock continues to propagate outwards as infalling material is dissociated and layered on the hot mantle. This stage is the definitive end of a "prompt", i.e. purely hydrodynamic, explosion.

The neutrino luminosities from the general relativistic simulations are shown in Fig. (1). The electron neutrino luminosity slowly rises during collapse and decreases as the core reaches maximal density. It remains suppressed for the ~ 4 ms the shock needs to propagate to the electron neutrinosphere. The most prominent feature is the electron neutrino burst, reaching 3.5×10^{53} erg/s at shock breakout, and declining afterwards.

The initial similarities in density and temperature in the inner parts of the three cores are responsible for the similar evolution of each core with respect to collapse, bounce, and the signature of the electron neutrino burst (Messer [22]). This is shown in Figs. (1) and (2). Any initial difference in the electron fraction, Y_e, in the inner part of each core cannot survive in regions that reach sufficient densities for neutrino trapping and equilibration to occur. If a mass element reaches neutrino trapping density, the total lepton fraction, Y_l, becomes the variable determining the state of the mass element. At equilibrium, the electron capture and neutrino absorption rates are related to one another through detailed balance, and the final Y_e becomes a function of the local temperature

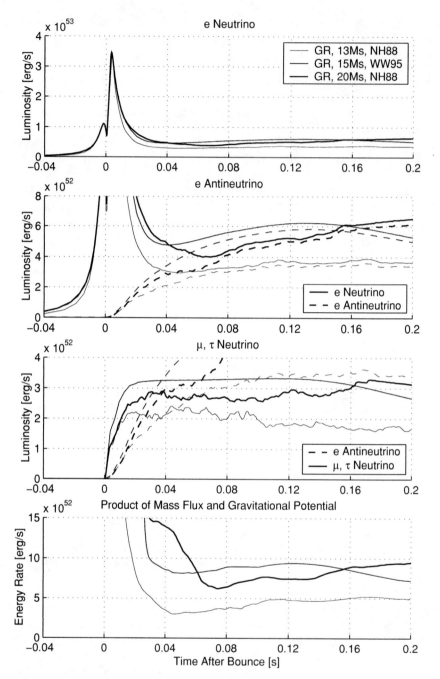

FIGURE 1. Luminosities for all flavors in the GR simulations. The last graph shows an energy rate calculated by the product of the gravitational potential at the neutrinosphere and the mass flux crossing it.

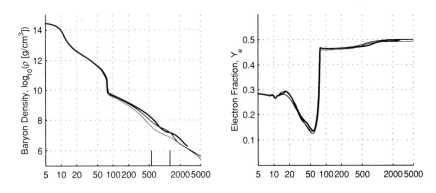

FIGURE 2. Similar core profiles in the GR cases at 10 ms after bounce.

and density through the electron chemical potential. Since the inner temperature and density profiles differ very little between the models, each settles to essentially the same equilibrium Y_e deep in the core. This determines the size of the homologous inner core and, consequently, the position of shock formation. Since this position sets the amount of mass that has to be dissociated when the shock ploughs to the neutrinospheres, it strongly influences the shock energy. Thus, a strikingly similar shock initiation is found for the three progenitors. However, the initially dissociated mass differs largely between the GR and NR cases because of the smaller enclosed mass at shock formation due to the GR effects in the gravitational potential (GR: 0.53 $M_\odot$, NR: 0.65 $M_\odot$). The luminosities in the GR and NR cases are very similar for the electron neutrino burst. Afterwards, the GR luminosities are generally 10% – 20% larger than the NR luminosities.

HEATING AND ACCRETION

In the standard picture, the ensuing evolution is driven by electron neutrino heating. Electron flavor neutrinos diffuse out of the cold unshocked core and are created in the accreting and compactifying matter around the neutrinospheres in a hot shocked mantle. If the heating is not sufficient to revive the shock within the first ~ 200 ms after bounce (as is the case in our simulations), the material in the heating region is drained from below (Janka [31]), and the conditions for heating deteriorate because of the shortened time the infalling matter spends in the heating region. We observe this process earlier in the general relativistic simulations than in the nonrelativistic simulations.

The shock trajectories for the different progenitors are shown in Fig. (3). The main difference between the GR and NR simulations stems from the difference in the size of the proto-neutron star, which is caused by the nonlinear GR effects at very high densities in the center of the star. At radii of order 100 km and larger, the gravitational potential becomes comparable in the GR and NR cases. However, large differences arise if the steep gravitational well is probed at different *positions* in the GR and NR simulations, as happens with accretion down to the neutrinospheres. The deeper neutrinospheres

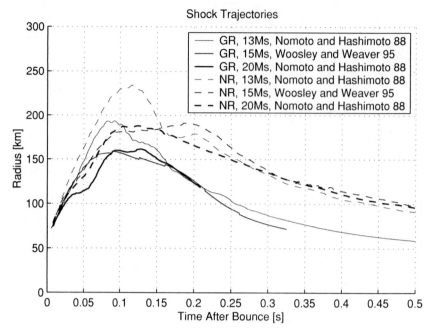

FIGURE 3. Shock trajectories for the GR and NR cases.

in the more compact GR case encounter material that has traversed a larger potential difference, producing more energetic accretion luminosities. After the shock has stalled, the surrounding layers (cooling/heating region, shock radius) adjust to a smaller radius and settle to a stationary equilibrium in the spirit of Burrows and Goshy [32]. In the GR simulation, this occurs at a smaller radius, deeper in the gravitational well, with higher infall velocities, leading to higher accretion luminosities with higher rms energies and higher heating rates (see also Bruenn et al. [23]).

This accretion-determined picture obtains additional support from the following observation: In the last graph in Fig. (1) we provide the mass flux at the electron neutrinosphere multiplied by the local gravitational potential. We can compare this energy rate to the luminosities of the electron flavor neutrinos that are shown in the second graph in Fig. (1). The striking similarity suggests that the luminosity is indeed mainly determined by the gravitational potential at the boundary of the proto-neutron star and the mass accretion rate. Moreover, we can directly relate the different accretion rates to differences in the density profiles of the initial progenitors. The energy accumulation rate at the neutrinospheres for the 15 $M_\odot$ model exceeds that of the 20 $M_\odot$ model between 60 ms and 160 ms after bounce. We have marked in Fig. (2) the region that crosses the neutrinospheres during this time. The density of the 15 $M_\odot$ progenitor exceeds the density of the 20 $M_\odot$ progenitor in exactly this region. Thus, the time dependent electron flavor luminosities reflect the position of the neutrinospheres and are directly modulated by the variations in the spatial density profiles of the progenitors.

ACKNOWLEDGMENTS

M.L. is supported by the National Science Foundation under contract AST-9877130 and, formerly, was supported by the Swiss National Science Foundation under contract 20-53798.98. O.E.M. is supported by funds from the Joint Institute for Heavy Ion Research and a Dept. of Energy PECASE award. A.M. is supported at the Oak Ridge National Laboratory, managed by UT-Battelle, LLC, for the U.S. Department of Energy under contract DE-AC05-00OR22725. W.R.H. is supported by NASA under contract NAG5-8405 and by funds from the Joint Institute for Heavy Ion Research. Our simulations were carried out on the ORNL Physics Division Cray J90 and the National Energy Research Supercomputer Center Cray J90.

REFERENCES

1. J. Tueller, S. Barthelmy, N. Gehrels, M. Leventhal, C. J. MacCallum, and B. J. Teegarden, in *Supernovae*, edited by S. E. Woosley, 278 (Berlin, Springer, 1991).
2. R. Strom., H. M. Johnston, F. Verbunt, and B. Aschenbach, Nature **373**, 587 (1995).
3. T. J. Galama et al., Nature **395**, 670 (1998).
4. D. C. Leonard, A. V. Filippenko, A. J. Barth, and T. Matheson, Astrophys. J. **536**, 239 (2000).
5. M. Herant, W. Benz, and S. A. Colgate, Astrophys. J. **395**, 642 (1992).
6. D. S. Miller, J. R. Wilson, and R. W. Mayle, Astrophys. J. **415**, 278 (1993).
7. M. Herant, W. Benz, R. W. Hix, C. L. Fryer, and S. A. Colgate, Astrophys. J. **435**, 339 (1994).
8. A. Burrows, J. Hayes, and B. A. Fryxell, Astrophys. J. **450**, 830 (1995).
9. H.-T. Janka and E. Müller, Astronomy and Astrophysics **306**, 167 (1996).
10. A. Mezzacappa, A. C. Calder, S. W. Bruenn, J. M. Blondin, M. W. Guidry, M. R. Strayer, and A. S. Umar, Astrophys. J. **495**, 911 (1998).
11. C. F. Fryer, Astrophys. J. **522**, 413 (1999).
12. C. F. Fryer and A. Heger, Astrophys. J. **541**, 1033 (2000).
13. P. Höflich, J. C. Wheeler, L. Wang, Astrophys. J. **521**, 179 (1999).
14. A. M. Khokhlov, P. A. Höflich, E. S. Oran, J. C. Wheeler, L. Wang, A. Yu. Chtchelkanova, Astrophys. J. Lett. **524**, 107 (1999).
15. A. I. MacFadyen and S. E. Woosley, Astrophys. J. **524**, 262 (1999).
16. J. C. Wheeler, I. Yi, P. Höflich, L. Wang, Astrophys. J. **537**, 810 (2000).
17. A. Mezzacappa, M. Liebendörfer, O. E. B. Messer, R. W. Hix, F.-K. Thielemann, and S. W. Bruenn, accepted for publication in Phys. Rev. Lett., astro-ph/0005366 (2001).
18. A. Mezzacappa and S. W. Bruenn, Astrophys. J. **405**, 669 (1993).
19. A. Mezzacappa and S. W. Bruenn, Astrophys. J. **405**, 637 (1993).
20. A. Mezzacappa and S. W. Bruenn, Astrophys. J. **410**, 740 (1993).
21. M. Liebendörfer, Ph.D. thesis, University of Basel, (Basel, Switzerland, 2000).
22. O. E. B. Messer, Ph.D. thesis, University of Tennessee, (Knoxville, USA, 2000).
23. S. W. Bruenn, K. R. DeNisco, and A. Mezzacappa, accepted for publication in Astrophys. J., astro-ph/0101400 (2001).
24. M. Rampp and H.-T. Janka, Astrophys. J. Lett. **539**, L33 (2000).
25. M. Liebendörfer, A. Mezzacappa, F.-K. Thielemann, O. E. B. Messer, R. W. Hix, S. W. Bruenn, submitted to Phys. Rev. D, astro-ph/0006418, (2001).
26. J. R. Wilson and R. W. Mayle, Physics Reports **227**, 97 (1993).
27. K. Nomoto and M. Hashimoto, Physics Reports **163**, 13 (1988).
28. S. E. Woosley and T. A. Weaver, Astrophys. J. Supplement **101**, 181 (1995).
29. J. Lattimer and F. D. Swesty, Nuclear Physics **A535**, 331 (1991).
30. S. W. Bruenn, Astrophys. J. Supplement Series **58**, 771 (1985).
31. H.-T. Janka, accepted for publication in Astronomy and Astrophysics, astro-ph/0008432 (2000).
32. A. Burrows and J. Goshy, Astrophys. J. Lett. **416**, L75 (1993).

Hypernovae and Gamma Ray Bursts

P. A. Mazzali[1], K. Nomoto[2,3], K. Maeda[2] and T. Nakamura[2]

[1] *Osservatorio Astronomico, via G.B. Tiepolo, 11, Trieste, Italy*
[2] *Department of Astronomy, University of Tokyo, Tokyo, Japan*
[3] *Research Center for the Early Universe, University of Tokyo, Tokyo, Japan*

Abstract. The properties of the very energetic Type Ic SNe 1997ef and 1998bw, the latter probably connected to GRB980425, are reviewed. It is shown that the SNe have explosion kinetic energies well in excess of normal SNe, hence they are called 'Hypernovae'. Evidence for asymmetry in the ejecta of SN 1998bw is discussed, and the results of a 2D calculation of an axisymmetric exposion are presented. The nebular line profiles of SN 1998bw computed assuming an axisymmetric exposion viewed from a direction close to that of the jet are shown to reproduce the observed profiles.

I INTRODUCTION

Recently, there have been a number of SNe with a possible GRB counterpart (see Nomoto et al. 2001 for references). Among these, the Type Ic Supernovae (SNe) 1998bw and 1997ef are characterised by a very large kinetic explosion energy, $E_K \gtrsim 10^{52}$ erg. This is more than one order of magnitude larger than in typical SNe, so these objects may be called "hypernovae". These SNe produced more ^{56}Ni than the average core collapse SN, and ejected material at very high velocities. The possible link to GRB's suggests that these explosion are not spherically symmetric.

II SN 1997EF

SN 1997ef was noticeable for its unique light curve and spectra. At early times, the spectra were dominated by broad oxygen and iron absorptions, but did not show any clear feature of hydrogen or helium (Garnavich et al. 1997; Filippenko et al. 1997), and the SN was classified as Type Ic (SN Ic). The broadness of the line features suggests that the SN may have had a large explosion energy.

In order to clarify whether SN 1997ef is indeed a hypernova, Iwamoto et al. (2000) constructed hydrodynamical models for an ordinary SN Ic and for a hypernova. For the ordinary SN Ic, a C+O star with a mass $M_{CO} = 6.0 M_\odot$ (the core of a 25 $M_\odot$ star) was exploded with a kinetic energy $E_K = 1.0 \times 10^{51}$ ergs and an ejecta mass $M_{ej} = 4.6 M_\odot$ (model CO60). For the hypernova model (CO100), a C+O star of

$M_{CO} = 10.0 M_\odot$ (the core of a 30 - 35 $M_\odot$ star) was exploded with $E_K = 8.0 \times 10^{51}$ ergs and $M_{ej} = 8.0 M_\odot$. The mass of ^{56}Ni was set to be 0.15 $M_\odot$ for both models to explain the observed peak brightness.

It is difficult to distinguish clearly between the ordinary SN Ic and the hypernova model from the light curve alone, since models with different values of M_{ej} and E_K can display a similar light curve if $M_{ej}^{3/4} E_K^{-1/4} = $ const. In fact, the synthetic V light curves for both models show a broad peak and a relatively slow tail, and reproduce the light curve of SN 1997ef reasonably well. However, these models can be expected to produce different spectra, because of the different E_K. Therefore, spectrum synthesis can distinguish between them.

Spectra computed with our Monte Carlo code (Mazzali 2000) using model CO60 for the early epochs show narrow lines, much narrower than the observations. This clearly indicates a lack of material at high velocity in model CO60, and suggests that the kinetic energy of this model is too small. Synthetic spectra computed with model CO100 have much broader lines and are in good agreement with the observations, and even better agreement is obtained if the outer density is artificially increased (Fig.1, left; Mazzali et al. 2000). We therefore conclude that SN 1997ef was a hypernova.

III SN 1998BW

SN 1998bw was discovered as the optical counterpart of GRB980425 (Galama et al. 1998). The absorption lines are so broad in SN 1998bw that they blend together, giving rise to broad absorption features. Velocities in the Si II 6355Å line are as high as 30,000 km s^{-1}. Also, the SN was very bright for a SN Ic: the observed peak luminosity, $L \sim 1.4 \times 10^{43}$ erg s^{-1}, is almost ten times higher than that of typical SNe Ib/Ic (Woosley et al. 1999).

Iwamoto et al. (1998) computed light curves and spectra for various C+O star models with different values of E_K and M_{ej}. The best model is the explosion of a 13.8 $M_\odot$ C+O star with a large kinetic energy ($E_K = 6 \times 10^{52}$ erg). Such a massive C+O star progenitor is the product of a main-sequence star of $\sim$ 40 $M_\odot$ which lost its H/He envelope via a stellar wind or binary interaction. A ^{56}Ni mass of $\sim$ 0.63$M_\odot$ is necessary to reproduce the light curve maximum. The broad lines are well reproduced with model CO138 (Fig. 1, right) because there is enough material at high velocities in this model.

Despite the success of our 'hypernova' model at early times, the model light curve declines too rapidly after t $\lesssim$ 60 days (Nakamura et al. 2001). This suggests that there might be a high density region in the ejecta of SN 1998bw where the γ-rays are efficiently trapped. Another peculiarity appeared in the late phase spectra. Measuring the velocity of the various elements from the width of their emission lines, Patat et al. (2001) noted that iron expands faster than oxygen, which is contrary to expectations.

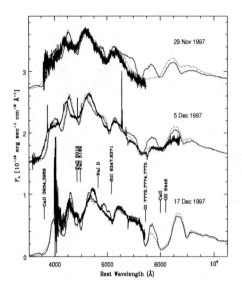 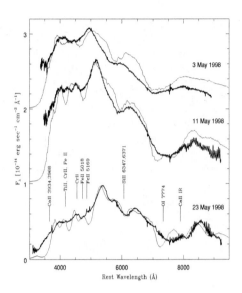

FIGURE 1. Left: Observed spectra of SN 1997ef (bold lines) and synthetic spectra computed using model CO100 (thin lines) and an artificial extension of the outer density (dotted lines). Right: Observed spectra of SN1998bw (full lines) and synthetic spectra computed using model CO138 (dashed lines).

We used a simple NLTE code to compute the nebular spectrum of SN 1998bw. We modelled 4 spectra of SN 1998bw, dating from four months to more than one year after maximum. In all spectra six dominant features can be identified: Mg I] at 4500Å, an Fe II] complex near 5250Å, Na I D at 5900Å, [O I] 6300Å, an O I]-Ca II] blend near 7200Å and the Ca II IR triplet near 8500Å. In all spectra the lines of [O I], Mg I] and partly also Ca II] IR are narrower than the Fe II] complex. Furthermore, the strength of the [O I] and Mg I] lines, and that of the O I]-Ca II] blend, grows with time relative to that of the other four features.

The outer velocity of the nebula can be determined by fitting the width of the emission features, including complex blends like Fe II], which is the broadest feature in the nebular spectra of SN 1998bw. The velocity of this feature is about 11,000 km s^{-1}.

For all the 'broad' models (Fig.2), the ^{56}Ni mass was about $0.65 M_\odot$. This is in good agreement with the light curve calculations, although it is perhaps a slight overestimate because the narrow observed lines are reproduced as broad lines. It is also encouraging that the ^{56}Ni mass is constant over the entire period, showing that the Fe II] feature is indeed due to γ-ray deposition in an expanding nebula.

The narrow-line spectra fit well the lines of Mg I], Na I D and [O I] (Fig.2). The width of the lines decreases only slowly with time, as was the case for the broad Fe II] feature: velocities range from 7500 to 5000 km s^{-1}. The synthetic

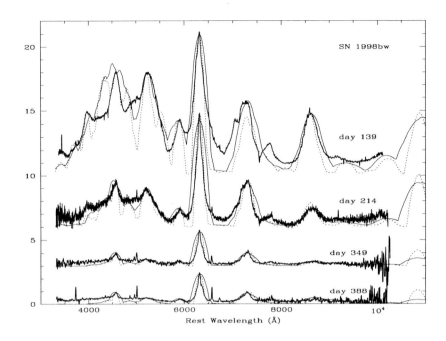

FIGURE 2. The nebular spectra of SN 1998bw (bold lines) and synthetic spectra computed to fit the broad lines (continuous lines) or the narrow lines (dashed lines).

Fe II] feature is in these models significantly too narrow. Therefore we chose to reproduce its peak only. Once again we found that the ^{56}Ni mass is very nearly constant in all four models, with a value of $\sim 0.35 M_\odot$. However, this is certainly an underestimate, as the total flux in the Fe II] feature is not reproduced.

The fact that the narrow lines of [O I] and Mg I] become stronger with time appears to indicate that there is a significant amount of these elements at low velocity. The different widths of the Fe II] feature and of [O I] 6300Å also suggests that the Fe/O ratio is larger at high velocity. All this is very difficult to explain with a spherically symmetric explosion, even taking into account mixing.

IV 2D MODELS OF AN ASYMMETRIC EXPLOSION

The need for a high density region and the velocity inversion might indicate that the explosion was aspherical (Höflich et al. 1999; MacFadyen & Woosley 1999). The outburst in SN 1998bw may have taken the form of a prolate spheroid, as also indicated by polarization measurements (Patat et al. 2001).

We computed 2D models to follow the explosive nucleosynthesis in an axisymmetric explosion (Maeda et al. 2001). The progenitor model was CO138. The

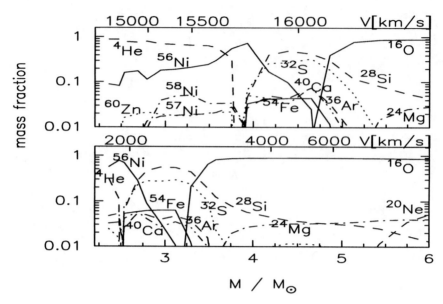

FIGURE 3. The isotopic compositions of the ejecta of the axisymmetric explosion in the direction of the jet (top) and perpendicular to the jet (bottom)

hydrodynamical simulation was started by depositing the energy as 50% thermal and 50% kinetic below the mass cut which divides the ejecta from the collapsing core. More kinetic energy was deposited in the direction of the jet than in other directions.

In the jet direction (z) the shock is stronger and the post-shock temperatures are higher, so that explosive nucleosynthesis takes place in a more extended, lower density region compared with the perpendicular direction (r). The results of the nucleosynthesis in the z and r directions are shown in Fig.3. A large amount of ^{56}Ni is produced in the jet direction and ejected at high velocity. In addition, elements produced by α-rich freezeout are enhanced because nucleosynthesis proceeds at higher entropies than in the region away from the jet. In the jet direction a larger amount of ^{4}He is left after shock decomposition. Hence, elements synthesized via α-particle capture, such as ^{44}Ti and ^{48}Cr (which decay into ^{44}Ca and ^{48}Ti, respectively) are more abundant (see also Nagataki et al. 1997). In contrast, in the r direction little ^{56}Ni is produced and the expansion velocities are lower than in the z-direction. Therefore, the velocity of Fe (mostly in the z-direction) can exceed that of O (in the r-direction), as observed in SN 1998bw. The density in the r direction may be high enough that γ-rays are trapped even at advanced phases, thus giving rise to the slowly declining tail. The distribution of ^{56}Ni and ^{16}O in (r,z) space is shown in Fig.4 (left). The estimated kinetic energy in the aspherical model is smaller ($E_K = 1 \times 10^{52}$erg), but still very large for a normal SN.

Maeda et al. (2001) computed nebular emission lines of Fe and O based on the

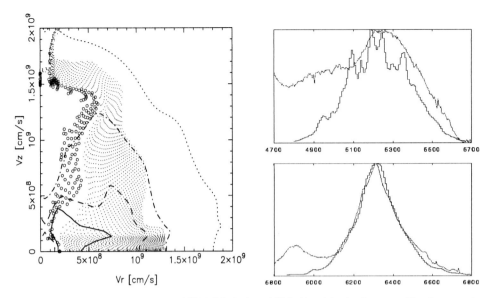

FIGURE 4. Left: Distribution of ^{56}Ni (circles) and ^{16}O (dots) in (r, z) space. The lines mark iso-density contours. Right: Observed spectra of SN1998bw (full lines) and synthetic emission lines of Fe (top) and O (bottom) computed using the aspherical model (dashed lines).

distribution shown in Fig.4 (left). If the explosion is viewed from about 15° from the jet axis, the profiles of the Fe blend near 5200Å and of O I] 6300Å are correctly reproduced, as shown in Fig.4 (right).

REFERENCES

1. Filippenko, A.V. et al. 1997, IAU Circ. No.6783, 6809
2. Galama, T.J. et al. 1998, Nature, 395, 670
3. Garnavich, P. et al. 1997, IAU Circ. No. 6778, 6786, 6798
4. Höflich, P., Wheeler, J.C., Wang, L., 1999, ApJ, 521, 179
5. Iwamoto, K. et al. 1998, Nature, 395, 672
6. Iwamoto, K. et al. 2000, ApJ, 534, 660
7. MacFadyen, A.I. & Woosley, S.E. 1999, ApJ, 524, 262
8. Maeda, K., et al., 2001, ApJ, submitted (astro-ph/0011003)
9. Mazzali, P.A., 2000, A&A, 363, 705
10. Mazzali, P.A. et al., 2000, ApJ, 545, 407
11. Nakamura, T., et al., 2001, ApJ, 550 (astro-ph/0007010)
12. Nomoto, K., et al., 2001, in "Supernovae and GRBs" (CUP) (astro-ph/0003077)
13. Patat, F. et al., 2001, ApJ, submitted
14. Woosley, S.E., Eastman, R.G., Schmidt, B.P. 1999, ApJ, 516, 788

Simulations of Astrophysical Fluid Instabilities

A. C. Calder[*,†], B. Fryxell[*,‡], R. Rosner[*,†,‡], L. J. Dursi[*,†],
K. Olson[*,‡,§], P. M. Ricker[*,†], F. X. Timmes[*,†], M. Zingale[*,†],
P. MacNeice[§], and H. M. Tufo[*]

[*] *Center for Astrophysical Thermonuclear Flashes*[1], *University of Chicago Chicago, IL 60637*
[†] *Department of Astronomy and Astrophysics, University of Chicago Chicago, IL 60637*
[‡] *Enrico Fermi Institute, The University of Chicago, Chicago, IL 60637*
[§] *NASA Goddard Space Flight Center, Greenbelt, MD 20771*

Abstract. We present direct numerical simulations of mixing at Rayleigh-Taylor unstable interfaces performed with the FLASH code, developed at the ASCI/Alliances Center for Astrophysical Thermonuclear Flashes at the University of Chicago. We present initial results of single-mode studies in two and three dimensions. Our results indicate that three-dimensional instabilities grow significantly faster than two-dimensional instabilities and that grid resolution can have a significant effect on instability growth rates. We also find that unphysical diffusive mixing occurs at the fluid interface, particularly in poorly resolved simulations.

INTRODUCTION

Many of the problems of interest in relativistic astrophysics involve fluid instabilities. The shock of a core-collapse supernova propagating through the outer layers of the collapsing star, for example, is subject to Rayleigh-Taylor instabilities occurring at the boundaries of the layers. A fluid interface is said to be Rayleigh-Taylor unstable if either the system is accelerated in a direction perpendicular to the interface such that the acceleration opposes the density gradient or if the pressure gradient opposes the density gradient [1,2]. Growth of these instabilities can lead to mixing of the layers. The early observation of ^{56}Co, an element formed in the core, in SN 1987A strongly suggested that mixing did indeed play a fundamental role in the dynamics. Following this observation, supernova modelers embraced multi-dimensional models with the goal of understanding the role of fluid instabilities in the core collapse supernova process [3]. Despite years of modeling

---

[1] This work is supported by the U.S. Department of Energy under Grant No. B341495 to the Center for Astrophysical Thermonuclear Flashes at the University of Chicago.

these events, many fundamental questions remain concerning fluid instabilities and mixing. In this manuscript, we present early results of our research into resolving fluid instabilities with FLASH, our simulation code for astrophysical reactive flows.

The FLASH code [4] is an adaptive mesh, parallel simulation code for studying multi-dimensional compressible reactive flows in astrophysical environments. It uses a customized version of the PARAMESH library [5] to manage a block-structured adaptive grid, placing resolution elements only where needed in order to track flow features. FLASH solves the compressible Euler equations by an explicit, directionally split version of the piecewise-parabolic method [6] and allows for general equations of state using the method of Colella & Glaz [7]. FLASH solves a separate advection equation for the partial density of each chemical or nuclear species as required for reactive flows. The code does not explicitly track interfaces between fluids, so a small amount of numerical mixing can be expected during the course of a calculation. FLASH is implemented in Fortran 90 and uses the Message-Passing Interface library to achieve portability. Further details concerning the algorithms used in the code, the structure of the code, verification tests, and performance may be found in Fryxell et al. [4] and Calder et al. [8].

RESULTS

From our single-mode Rayleigh-Taylor studies, we find significantly faster instability growth rates in three-dimensional simulations than in two-dimensional simulations. In addition, we find that obtaining a converged growth rate requires at least 25 grid points per wavelength of the perturbation, that grid noise seeds small scale structure, and that the amount of small scale structure increases with resolution due to the lack of a physical dissipation mechanism (such as a viscosity). Another result is that poorly-resolved simulations exhibit a significant unphysical diffusive mixing. Figure 1 shows the growth of bubble and spike amplitudes for two well-resolved simulations beginning from equivalent initial conditions. The three-dimensional result (left panel) shows faster growth than the two-dimensional result (right panel). Results of our single-mode studies will appear in Calder et al. [9].

Our single-mode studies serve as a prelude to multi-mode studies, which are works in progress; our single mode results strongly suggest that using sufficient resolution is essential in order to obtain physically-sensible results for these calculations. In the multi-mode case, bubble and spike mergers are thought to lead to an instability growth according to a t^2 scaling law, which for the case of a dense fluid over a lighter fluid in a gravitational field may be written as [10]

$$h_{b,s} = \alpha_{b,s} g A t^2 \qquad (1)$$

where $h_{b,s}$ is the height of a bubble or spike, g is the acceleration due to gravity, $A = (\rho_2 - \rho_1)/(\rho_2 + \rho_1)$ is the Atwood number where $\rho_{1,2}$ is the density of the lighter (heavier) fluid, and t is the time. α is a proportionality 'constant' that may be thought of as a measure of the efficiency of potential energy release. Experiments

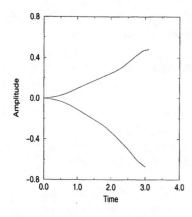
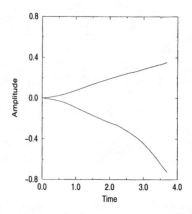

FIGURE 1. Bubble and spike amplitudes for two-dimensional (right) and three-dimensional (left) simulations of single-mode instabilities. The resolutions are 128 X 768 (2-d) and 128 X 128 X 768 (3-d). The amplitudes are measured by tracking the advection of each fluid. The initial conditions consisted of a dense fluid ($\rho = 2$) over a lighter fluid ($\rho = 1$) and $g = 1$. The initial perturbation consisted of a sinusoidal vertical velocity perturbation of 2.5% of the local sound speed with the horizontal components chosen so the initial velocity field was divergence-free.

and simulations indicate that α lies in the range of 0.03 to 0.06, and it is thought to depend on Atwood number, evolution time, initial conditions, and dimensionality. See Young et al. [11] and references therein for a discussion of experimental results. Results of our multi-mode studies will appear in publications of the Alpha Group, a consortium formed by Guy Dimonte in 1998 to determine if the t^2 scaling law holds for the growth of the Rayleigh-Taylor instability mixing layer, and if so, to determine the value of α [12].

REFERENCES

1. Taylor, G., *Proc. Roy. Soc.*, **A 201**, 192 (1950)
2. Chandrasekhar, S., *Hydrodynamic and Hydromagnetic Stability*, New York: Dover, 1961, ch. X, pp. 428-480.
3. Arnett, D., Fryxell, B., and Müller, E., *Ap. J.*, **341**, L63 (1989)
4. Fryxell, B. A., et al., *Ap. J. S.* **131**, 273 (2000)
5. MacNeice, P., et al., *Comp. Phys. Comm.*, **126**, 330 (2000)
6. Colella, P. and Woodward, P., *J. Comp. Phys.* **54**, 174 (1984)
7. Colella, P. and Glaz, H. M., *J. Comp. Phys.* **59**, 264 (1985)
8. Calder, A. C., et al., in Proc. Supercomputing 2000, IEEE Computer Soc., 2000
9. Calder, A. C. et al., in prep. (2001)
10. Youngs, D. L., *Lasers and Particle Beams*, **12**, no. 4, 725 (1994)
11. Young, Y.-N., et al., *J. Fluid Mech.*, in press (2001)
12. Dimonte, G. et al., in prep. (2001)

Large lepton mixing and SN 1987A

M. Kachelrieß*, R. Tomàs† and J.W.F. Valle†

TH Division, CERN, Geneva
†*Institut de Física Corpuscular - CSIC - Universitat de València*

Abstract. We reconsider the impact of $\bar{\nu}_e \leftrightarrow \bar{\nu}_{\mu,\tau}$ neutrino oscillations on the observed $\bar{\nu}_e$ signal of supernova SN 1987A. Performing a maximum-likelihood analysis using as fit parameters the released binding energy E_b and the average neutrino energy $\langle E_{\bar{\nu}_e}\rangle$, we find as previous analyzes that $\bar{\nu}_e \leftrightarrow \bar{\nu}_{\mu,\tau}$ oscillations with large mixing angles have lower best-fit values for $\langle E_{\bar{\nu}_e}\rangle$ than small-mixing angle (SMA) oscillations, which already turn out to be smaller than those found in simulations. In order to quantify the degree to which the experimental data favor the SMA over the large mixing solutions we use their likelihood ratios as well as a Kolmogorov-Smirnov test. We find within the range of SN parameters predicted by simulations regions in which the LMA-MSW solution is either only marginally disfavored or favored compared to the SMA-MSW solution. We conclude therefore that the LMA-MSW solution is not in conflict with the current understanding of SN physics. In contrast, the vacuum oscillation and the LOW solutions to the solar neutrino problem can be excluded at the 4σ level for most of the SN parameter ranges found in simulations.

INTRODUCTION

The consistency of the neutrino signal from SN 1987A with the different allowed solutions to the solar neutrino problem [1] has been recently reanalyzed in Ref. [2]. In order to quantify the agreement of different oscillation parameters Δm^2 and $\tan^2\theta$ with the experimental data we have performed a likelihood as well as a Kolgomorov-Smirnov test. Our analysis is based on the following assumptions:

a) use of a two-neutrino oscillations scenario, $\bar{\nu}_e \leftrightarrow \bar{\nu}_h = \cos\phi\bar{\nu}_\mu + \sin\phi\bar{\nu}_\tau$, motivated both by the results of the Chooz experiment and fits of atmospheric neutrino data as well as by the fact that the energy spectra of $\bar{\nu}_\mu$ and $\bar{\nu}_\tau$ are identical.

b) the SN emits the same amount of energy, $\sim \frac{1}{6}E_b$, in all neutrino flavors.

c) the time-averaged spectra of the neutrinos can be described by Maxwell-Boltzmann distributions with $\langle E_{\bar{\nu}_e}\rangle < \langle E_{\bar{\nu}_{\mu,\tau}}\rangle$.

Then the $\bar{\nu}_e$ fluence arriving to the detectors is given by

$$F_{\bar{\nu}_e} = P_{\bar{e}\bar{e}}F^0_{\bar{\nu}_e} + (1 - P_{\bar{e}\bar{e}})F^0_{\bar{\nu}_h},$$

where $F^0_{\bar{\nu}}$ stands for the time-integrated flux of neutrinos emitted by the SN and $P_{\bar{e}\bar{e}}$ is the survival probability of a $\bar{\nu}_e$ to arrive at the detector.

RESULTS

We have first tested with the likelihood function described in [4]

$$\mathcal{L}(\alpha) \propto \exp\left(-\int n(E,\alpha)\mathrm{d}E\right) \prod_{i=1}^{N_{\mathrm{obs}}} n(E_i,\alpha)$$

the hypothesis that a prescribed neutrino fluence $F_{\bar{\nu}_e}(E_{\bar{\nu}_e},\alpha)$ leads to the observed experimental data E_i with probability distribution $n(E,\alpha)$. The maximization of $\mathcal{L}(\alpha)$ gives an estimate of the values α_* which best represent the data set E_i.

The best-fit point for the LMA-MSW case is shifted to smaller values of $\langle E_{\bar{\nu}_e}\rangle$ for increasing values of $\tau \equiv \langle E_{\bar{\nu}_e}\rangle/\langle E_{\bar{\nu}_h}\rangle$ with respect to the SMA-MSW one. This shift is more pronounced for VO: already for $\tau = 1.4$ the 95.4% C.L. likelihood contour does not include $\langle E_{\bar{\nu}_e}\rangle \sim 12$ MeV. Moreover, the LMA-MSW solution does not improve significantly the compatibility of the Kamiokande and IMB data sets unless there is a fine tuning of Δm^2 and θ.

In order to decide how strong the LMA-MSW, LOW and VO solutions are (dis-)favored against the SMA-MSW solution we have considered the ratio

$$R(E_b, \langle E_{\bar{\nu}_e}\rangle, \tau) = \frac{\mathcal{L}_{LMA}(E_b, \langle E_{\bar{\nu}_e}\rangle, \tau)}{\mathcal{L}_{SMA}(E_b, \langle E_{\bar{\nu}_e}\rangle, \tau)}.$$

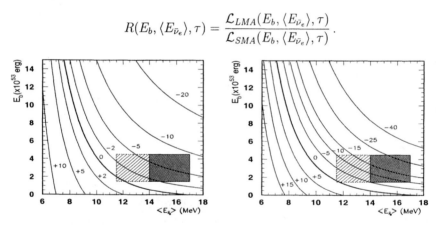

FIGURE 1. LMA-MSW (left) and VO (right) compared to SMA-MSW, both for $\tau = 1.4$.

In Fig. 1 we show $\ln(R)$ for $\tau = 1.4$, together with the typical range of parameters found in simulations, cross hatched from Ref. [3] and hatched from Ref. [5]. We find that for large values of $\langle E_{\bar{\nu}_{e,h}}\rangle$ and E_b all three large mixing solutions are practically excluded when compared to the SMA-MSW one. However the probability that LMA-MSW is compatible with the neutrino signal rises for lower values, and already for $\langle E_{\bar{\nu}_e}\rangle \sim 12$ MeV, as found in Ref. [5], the LMA-MSW solution can be even *favored* compared to SMA-MSW.

In order to determine the impact of the oscillations in the LOW region we show $\ln(R)$ as function of $\tan^2\theta$ and Δm^2 relative to the SMA hypothesis for $\tau = 1.7$ and

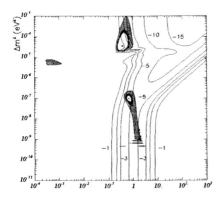 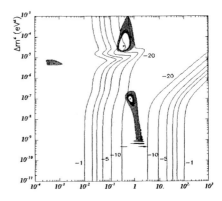

FIGURE 2. $\ln(R)$ relative to SMA-MSW for $\tau = 1.7$, and $E_b = 1.5 \times 10^{53}$ erg and $\langle E_{\bar{\nu}_e} \rangle = 12$ MeV, and $E_b = 3. \times 10^{53}$ erg and $\langle E_{\bar{\nu}_e} \rangle = 14$ MeV respectively.

$\langle E_{\bar{\nu}_e} \rangle = 12, 14$ MeV respectively, in Fig. 2. We find that below $10^{-7} - 10^{-8}$ eV^2 the dark side is as compatible with the SN 1987A neutrino signal as the light side. The absence of Earth matter effects below 10^{-5} eV^2 also shows that regeneration effect are negligible for the LOW and VO solutions, and "explains" why their status is worse than that of LMA.

We also use the one-dimensional Kolmogorov-Smirnov test to estimate the probability that the observed data points E_i agree with the spectral shape predicted by a given distribution $n_{\bar{\nu}_e}(E)$. We find that the LMA-MSW hypothesis is consistent with the data at the 5% ($\tau = 1.4$) for $\langle E_{\bar{\nu}_e} \rangle = 14$ MeV, while the compatibility increases to 25% for the same τ at the lower end of predicted energies, $\langle E_{\bar{\nu}_e} \rangle = 12$ MeV. In the vacuum case, though, the maxima of the probability as function of $\langle E_{\bar{\nu}_e} \rangle$ compared to the SMA-MSW is so shifted to lower energies that make it highly incompatible with the range of neutrino temperatures predicted by simulations.

In conclusion, we have found that the LMA-MSW solution is not significantly disfavored compared to the SMA-MSW solution by the neutrino signal of SN 1987A. In contrast, the VO and the LOW solutions to the solar neutrino problem can be excluded at the 4σ level for most of the SN parameter ranges found in simulations.

REFERENCES

1. For a discussion of the solar neutrino problem see the contribution of M. Nakahata in these proceedings.
2. M. Kachelrieß, R. Tomàs and J.W.F. Valle, *J. High Energy Phys.* **01030** (2001).
3. H.-T. Janka, in Proc. *Frontier Objects in Astrophysics and Particle Physics*, Vulcano 1992, eds. F. Giovaelli and G. Mannocchi.
4. B. Jegerlehner, F. Neubig and G. Raffelt, *Phys. Rev.* **D54**, 1194 (1996).
5. A. Burrows *et al.*, *Astrophys. J.* **539**, 865 (2000).

Quenching Processes in Flame-Vortex Interactions

M. Zingale[*,†], J. C. Niemeyer[‡], F. X. Timmes[*,†], L. J. Dursi[*,†],
A. C. Calder[*,†], B. Fryxell[*,†], D. Q. Lamb[*,†], P. MacNeice[||], K. Olson[||],
P. M. Ricker[*,†], R. Rosner[*,†], J. W. Truran[*,†], and H. M. Tufo[*]

[*] *Center for Astrophysical Thermonuclear Flashes[1], Chicago, IL 60637*
[†] *Department of Astronomy and Astrophysics, University of Chicago, Chicago, IL 60637*
[‡] *Max-Planck-Institut für Astrophysik, Garching, Germany 85748*
[||] *Goddard Space Flight Center, Greenbelt Maryland 20771*

Abstract. We show direct numerical simulations of flame-vortex interactions in order to understand quenching of thermonuclear flames. The key question is—can a thermonuclear flame be quenched? If not, the deflagration-detonation transition mechanisms that demand a finely tuned preconditioned region in the interior of a white dwarf are unlikely to work. In these simulations, we pass a steady-state laminar flame through a vortex pair. The vortex pair represents the most severe strain the flame front will encounter inside the white dwarf. We perform a parameter study, varying the speed and size of the vortex pair, in order to understand the quenching process. No quenching is observed in any of the calculations performed to date.

INTRODUCTION

The favored model for a Type Ia supernova begins as a flame deep in the interior of a Chandrasekhar mass white dwarf. At some point, the burning may undergo a deflagration-detonation transition (DDT). Some mechanisms for this transition require a preconditioned region in the star. As the flame propagates down the temperature gradient, the speed increases, and the transition to a detonation may occur [1,2]. For this to happen, the region must be free of any temperature fluctuations—any burning must be quenched.

A flame propagates by balancing thermal diffusion and nuclear burning. Heat diffuses from the hot ash to the cold fuel ahead of the burning front, raising its temperature to ignition. The width and speed of the burning front depends on how rapidly energy is released by nuclear burning, and how efficiently conduction can transport energy to the material ahead of the front. If a flame encounters a

---

[1] The Center for Astrophysical Thermonuclear Flashes is supported by the Department of Energy under Grant No. B341495 to the University of Chicago

large rate of strain, the burning may not be able to keep pace with the increase in flame surface, and the flame can quench. Experiments and simulations involving chemical flames have observed flame quenching (see for example [3]).

FLAME-VORTEX SIMULATIONS

To study quenching processes in thermonuclear flames, we ran a grid of flame-vortex interaction simulations with the FLASH Code [4]. FLASH solves the compressible Euler equations with a nuclear energy generation source term and a realistic equation of state. FLASH uses an adaptive mesh, which concentrates spatial resolution at the complex features of the flow, greatly improving the efficiency of the simulation. For these simulations, FLASH was extended to include explicit thermal diffusion. The one-dimensional laminar flame speeds calculated by FLASH were compared to those computed by Timmes and Woosley [5] and found to be in excellent agreement.

For a typical white dwarf density of 5×10^8 g cm^{-3} and a pure carbon environment, the flame thickness is $\sim 4 \times 10^{-4}$ cm and the speed is $\sim 6 \times 10^6$ cm s^{-1} [5]. The radius of a white dwarf is $\sim 10^8$ cm, meaning a full direct numerical simulation of a supernova would need to span 13 orders of magnitude. We are interested in understanding the microphysics of flames, which will be necessary for sub-grid modelling for large calculations.

The calculation presented here was set up by mapping a steady-state flame solution onto a grid with a superposed vortex given by the stream function [3]:

$$\Psi = C e^{\left((x-x_\text{ctr})^2 + (y-y_\text{ctr})^2\right)/2R_\text{vortex}^2} \tag{1}$$

There are two free parameters: the vortex size (R_vortex) and the amplitude (C). Additionally, the distance between the vortices must be chosen such that they do not overlap. The boundary conditions transverse to the direction of propagation are periodic, and they are zero-gradient along the direction of propagation.

We look at a single case here (see Fig. 1), where the vortex size is greater than the flame thickness ($R_\text{vortex} = 5 \times 10^{-4}$ cm). The figures show the flame shortly after encountering the vortex pair. We see that the flame wraps around the vortices, but at no point is the flame disrupted. Additionally, no unburned fuel survives behind the front.

CONCLUSION

We have simulated a wide range of flame-vortex interactions, and no quenching has been observed. In contrast to the chemical cases, there is no heat loss to the walls of the system that help the quenching process. In none of the simulations does unburned fuel persist behind the flame front. A full anaylsis of these simulations will appear in Zingale et al. [6].

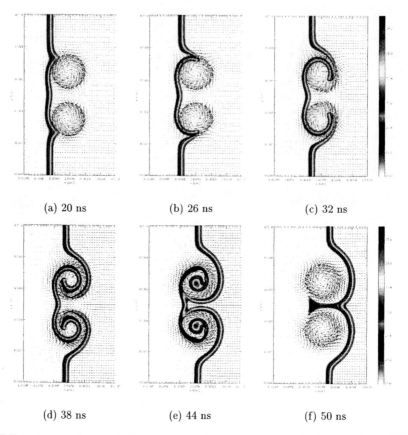

FIGURE 1. Time sequence of a flame encountering a vortex whose size is > the flame thickness. Log10 of nuclear energy generation rate is shown.

REFERENCES

1. Khokhlov, A. M., Oran, E. S., & Wheeler, J. C., *Ap. J*, **478**, 678 (1997)
2. Niemeyer, J.C. & Woosley, S. E., *Ap. J.* **475**, 740 (1997)
3. Poinsot, T., Veynante, D., and Candel, S., *J. Fluid Mech.*, **228**, 561 (1991)
4. Fryxell, B. A., *et al.*, *Ap. J. S.* **131**, 273 (2000)
5. Timmes, F. X. and Woosley, S. E., *Ap. J.* **396**, 649 (1992)
6. Zingale, M. *et al.* in prep. (2001)

CHAPTER 11

PULSARS, MAGNETARS, AND SUPERNOVA REMNANTS

Gamma-Ray Bursts from Extragalactic Magnetar Flares

Robert C. Duncan

Dept. of Astronomy, University of Texas, Austin, Texas 78712

Abstract.
The prototype for events that we call MFs—"March Fifth" events or "Magnetar Flares"—was observed on March 5, 1979. There is evidence that MFs are powered by catastrophic magnetic instabilities in ultra-magnetized neutron stars. These events begin with brief ($\Delta t \sim 0.1$–1 s), intense, hard spikes of gamma rays, probably emitted in concurrence with relativistic outflows; followed by long ($t \sim 100$ s) softer tails of hard X-rays, modulated on the stellar rotation period. Prototypical MFs could have been detected by BATSE out to ~ 13 Mpc, nearly reaching the Virgo cluster. The likely number of isotropic, standard-candle MFs detected by the BATSE experiment is ~ 12. These short-duration, fast-rising gamma-ray bursts could in principle be identified by their positional coincidences with nearby galaxies and the Supergalactic Plane. The ensuing soft tail emission would not have been detected by BATSE for sources more distant than the Andromeda Galaxy. Bayes' Theorem implies that there is a $\sim 99\%$ chance for at least 1 isotropic MF in the BATSE catalog, and a $\sim 16\%$ chance for more than 20.

It is possible that MFs also emit an intense, hard, *beamed* component during the intial spike phase. If this beamed component has opening angle $\psi = 8°\psi_8$ and peak luminosity comparable to the power of the isotropic component, then BATSE would detect such beamed sources out to redshift $z \sim 0.1\,\psi_8^{-1}$, at a full-sky rate of $\dot{N}_B \sim 100\,\psi_8^{-1}$ yr^{-1}. We speculate that such beamed MFs could account for the short, hard Class II gamma-ray bursts (GRBs) in the BATSE catalog, or some significant subset of them. If true, then Class II GRBs positions should correlate with the positions of galaxies and galaxy clusters within ~ 350 Mpc.

INTRODUCTION

The short-duration, hard-spectrum gamma-ray bursts (GRBs) in the BATSE catalog, often called Class II bursts, have no detected afterglows. As a consequence, they have not yet been subject to scrutiny for absorption lines, and the distance from Earth at which these bursts originate is uncertain. Here we will explore the possibility of a relatively nearby extragalactic origin, $z \lesssim 0.1$, for (at least) some Class II GRBs.

Kouveliotou et al. first showed that the GRB population is bimodal [1]. More recently, Murkherjee et al. found evidence that there exist *three* classes of BATSE GRBs [2]. But Hakkilla et al. (ref. [3]) argued that Class III properties can be produced from Class I by a combination of measurement error, hardness-intensity correlation, and a newly-identified BATSE bias, the fluence duration bias. Class II GRBs, which are short-duration, low-fluence, hard-spectrum events, do not seem to be related to Class I/III. It is Class II GRBs, or some subset of Class II, that we suggest are MFs.

Class II GRBs have many features in common with the hard spikes of MFs. About 40% of Class II bursts are single-peaked [4]. Most peaks exhibit a fast rise and slower decline, with hard-to-soft spectral evolution [4]. These properties are shared by the hard spikes of the March 5th and August 27th events [5–7], which also had similar spectral hardness [8,9]. Class II GRBs that are sufficiently bright to study with fine time resolution often show substructure down to time scales of ~ 10 ms and less [10,11], as did the 1979 March 5 event [12].

BATSE DETECTION OF MF'S

Consider a beamed source of gamma-rays, emitting uniformly into a beaming fraction $f = \Delta\Omega/4\pi$, of the unit sphere, with peak beam luminosity $L_{\rm bm}$. For a BATSE peak-flux detection threshold F_B, the sampling depth of BATSE is

$$D = \left(\frac{L_{\rm bm}}{4\pi f F_B}\right)^{1/2}. \quad (1)$$

If D is small enough that source density evolution and departures from Euclidean geometry are negligible, then BATSE's rate of event detection is

$$\dot{N}_B = \frac{4\pi}{3} D^3 n_* f \Gamma. \quad (2)$$

Here n_* is the density of L_* galaxies (with luminosities comparable to that of the Milky Way) and Γ is the rate of MFs within each L_* galaxy. Thus the (full-sky) BATSE detection rate is

$$\dot{N}_B = \frac{n_* \Gamma}{6(\pi f)^{1/2}} \left(\frac{L_{\rm bm}}{F_B}\right)^{3/2}. \quad (3)$$

We evaluate this using known parameter values. In particular, $F_B \approx 10^{-7}$ erg cm^{-2} s^{-1} (ref. [10]) and $n_* = 0.01\, h^3$ Mpc^{-3} from the local normalization of the Schechter luminosity function (e.g., ref. [13]). We will adopt a Hubble constant of $H_o = 65\, h_{65}$ km s^{-1} Mpc^{-1}, thus $n_* = 2.8 \times 10^{-3}\, h_{65}^3$ Mpc^{-3}.

Given 2 MFs in our Galaxy (or in nearby dwarf satellite galaxies) in the past $t_o \sim 20$ years of effective full sky coverage during which have had capability for detecting them (Hurley 1999, private communication), the MF rate per L_* galaxy is roughly

$$\Gamma = 0.1\,\Gamma_{0.1}\;\mathrm{yr}^{-1}. \tag{4}$$

The simplest idealization is that the gamma-rays in MF hard spikes are emitted isotropically, $f \approx 1$. Since the long, soft, oscillating tails of MFs have never been observed without hard spikes at their onsets, we infer that many or most MFs begin with a spike of quasi-isotropic emissions. (This is in addition to a possible high-intensity, beamed component. We have no direct evidence for beamed emission; however, with only 2 detected MFs, we would not expect to have observed any beams with $f \ll 1$.)

Setting $f = 1$ in the above formulae, we find a BATSE sampling depth for the isotropic component of MFs of

$$D = 13\,(L_{45}/2)^{1/2}\;\mathrm{Mpc}, \tag{5}$$

where we have scaled to the value of peak luminosity found for the 1979 March 5 event by Fenimore [8]. This falls just short of the Virgo cluster at $D[\mathrm{Virgo}] \approx 18\,h_{65}^{-1}$ Mpc. The full-sky BATSE detection rate is

$$\dot N_B = 2.6\,(L_{45}/2)^{3/2}\,\Gamma_{10}\,h_{65}^3\;\mathrm{yr}^{-1} \tag{6}$$

Since BATSE operated for 9.5 years, corresponding to 4.75 yrs of full-sky coverage, there should be

$$\langle \mathcal{N} \rangle \sim 12\,(L_{45}/2)^{3/2}\,\Gamma_{0.1}\,h_{65}^3 \tag{7}$$

extragalactic MFs in the BATSE catalog, detected via their quasi-isotropic hard spike emissions.

The most uncertain parameter in this estimate is Γ, the galactic rate of MFs. The Bayesian probability distribution for Γ, $\mathcal{P} \equiv dP/d\Gamma$, is

$$\mathcal{P}(\Gamma) = \frac{P(2|\Gamma)\,\mathcal{P}_{prior}(\Gamma)}{\int_0^\infty d\Gamma\;P(2|\Gamma)\,\mathcal{P}_{prior}(\Gamma)}, \tag{8}$$

where the probability for observing 2 MFs in time $t_o \approx 20$ yr, given Γ, is $P(2|\Gamma) = \frac{1}{2}(\Gamma t_o)^2 \exp(-\Gamma t_o)$ since this is a Poisson process. The appropriate prior distribution, when we don't know the order of magnitude of Γ a priori, is $(dP_{prior}/d\log\Gamma) = $ constant, or $\mathcal{P}_{prior} \propto \Gamma^{-1}$. Thus

$$\mathcal{P}(\Gamma) = \Gamma\,t_o^2\,\exp(-\Gamma t_o), \tag{9}$$

with a mean value $\langle \Gamma \rangle = 2 t_o^{-1} = 0.1\;\mathrm{yr}^{-1}$ as noted above [eq. (4)]. The probability for the galactic flare rate to exceed a cutoff value, $\Gamma > \Gamma_x$, is thus $P(\Gamma > \Gamma_x) = \exp(-\Gamma_x t_o)(1 + \Gamma_x t_o)$. If other sources of uncertainty can be neglected, then there is a 99% probability for at least one isotropic MF in the BATSE catalog, $\mathcal{N} > 1$. The probability is 80% for $\mathcal{N} > 5$; 52% for $\mathcal{N} > 10$; 16% for $\mathcal{N} > 20$; and 4.5% for $\mathcal{N} > 30$.

COULD ALL CLASS II GRB'S BE MF'S?

Suppose that there also exists a beamed component in the intial hard spikes of MFs, with a full opening angle ψ. The beaming fraction is $f = [1 - \cos(\psi/2)]$ assuming two beams, one at each magnetic pole. This is $f \approx \psi^2/8$ for $\psi \ll 1$, or $f = 2 \times 10^{-3} \psi_8^2$, where $\psi_8 \equiv (\psi/8°)$. From eqs. (1)–(3), the BATSE sampling depth for beamed MFs is

$$D = 350 \; \psi_8^{-1} \left(\frac{L_{\text{bm}}}{3 \times 10^{45} \text{ erg s}^{-1}} \right)^{1/2} \text{ Mpc}, \qquad (10)$$

and the (full-sky) rate of detection is

$$\dot{N}_B = 100 \; \Gamma_{0.1} \; h_{65}^3 \; \psi_8^{-1} \left(\frac{L_{\text{bm}}}{3 \times 10^{45} \text{ erg s}^{-1}} \right)^{3/2} \text{ yr}^{-1} \qquad (11)$$

This is plausibly in agreement with the rate of detection of Class II bursts by BATSE. Out of 796 bursts classified by Murkherjee et al., about 185 were Class II, or $\sim 23\%$. Since BATSE detects about 300 GRB per year with half-sky coverage (due to the fact that the *Compton Observatory* is in low Earth orbit), the full-sky detection rate of Class II bursts is roughly $\dot{N}_B(\text{Class II}) \approx 140 \text{ yr}^{-1}$.

However, there is no compelling reason to expect such a beamed emission component based upon the magnetar model [14–16,7].

Moreover, the number-intensity or V/V_{max} distribution of Class II GRBs seems to give evidence against a local extragalactic origin for these events. This is a serious concern; however, note that the evidence for $\langle V/V_{max} \rangle < 0.5$ in Class II bursts is much less compelling than in Class I/III. It is possible that selection effects cause a paucity of BATSE events just above threshold for peak flux on 256 and 64-ms time scales. (Flux averaged over 1.024 s does not give a good brightness measure for these short bursts.) Faint and poorly-measured bursts are systematically removed because of insufficient information to make class identifications (e.g., only 778 out of 1122 catalogued bursts were classified in ref. [3]); and the class identifications themselves are least reliable near threshold where measurements are most statistically dubious. It is also possible that the *bright* Class II bursts include a contamination of physically-distinct, non-Euclidean (low V/V_{max}) events; e.g., a tail of Class I bursts extending to short durations, or a subclass of short bursts from galactic SGRs or AXPs like the bright, hard-spectrum events with $T_{90} \sim 1$ s already identified from SGR 1900+14 [17]. Note that Tavani [18] found $\langle V/V_{max} \rangle = 0.458 \pm 0.044$ for short duration, soft spectrum bursts in the BATSE catalog (66 events with $T_{90} < 2.5$ s and $H_{32}^e < 3$), and Cline, Matthey & Otwinowski [19] found $\langle V/V_{max} \rangle = 0.52 \pm 0.06$ for all BATSE bursts with $T_{90} < 0.1$ s.

CONCLUSIONS: OBSERVATIONAL TESTS

Under the idealization that MF hard spikes are emitted quasi-isotropically, we find that the BATSE catalog contains ~ 12 extragalactic MFs [eq. (7)]. These events are expected to be fast-rising ($t_{rise} \lesssim 1$ ms), short-duration bursts, correlated in position with galaxies at $D_{gal} < 20$ Mpc. Insofar as the quasi-isotropic component of MF hard spikes have uniform peak luminosities, there will be a diminishment of the peak flux with D_{gal} according to $F_{peak} \simeq L_{\text{iso}}/(4\pi D_{gal}^2)$, affording a possible auxiliary check on associations. As a group, these "isotropic MFs" should tend to concentrate toward the supergalactic plane [20]. Hartmann, Briggs & Mannheim (ref. [21]) found no significant supergalactic anisotropy in the BATSE catalog, but their statistics were dominated by Class I bursts. At distances less than $D_{\text{vir}} \sim 18 h_{65}^{-1}$ Mpc the distribution of candidate MFs may show a significant Virgocentric dipole moment (e.g., Fig. 3.3 in ref. [13]); and a correlation with Virgo's discrete position on the sky if the sampling depth (eq. [5]) extends as far as D_{vir}.

This prediction of "isotropic MFs" in the BATSE catalog is based upon direct, reliable extrapolation from observations of the 1979 March 5 and 1998 August 27 events. More speculatively, in any model of MFs, a full-sky BATSE rate $\dot{N}_B$ is possible if there exists a *beamed emission component* with full opening angle

$$\psi = 8° \; \Gamma_{0.1} \; h_{65}^3 \left(\frac{L_{\text{bm}}}{3 \times 10^{45} \text{ erg s}^{-1}} \right)^{3/2} \left(\frac{\dot{N}_B}{100 \text{ yr}^{-1}} \right)^{-1}, \quad (12)$$

where L_{bm} is the total beam peak power. This equation assumes two polar beams; for a single beam, ψ goes up by $\sqrt{2}$.

If such beamed MFs accounted for many or all Class II GRBs, then the events would have positions correlated with galaxies and galaxy clusters within the BATSE sampling depth at redshift

$$z_{\text{B}} = 0.076 \; \psi_8^{-1} \; h_{65} \left(\frac{L_{\text{bm}}}{3 \times 10^{45} \text{ erg s}^{-1}} \right)^{1/2}. \quad (13)$$

Several studies have found correlations of GRB positions with galaxy clusters in the Abell, Corwin & Olowin (hereafter ACO) catalog [22], but only at a statistically marginal level [23–25]. These studies have focused primarily on bursts with small positional error boxes, which are mostly bright Class I events. Note that the MFs come from young neutron stars in star-forming regions. Such sources are not expected to concentrate strongly within rich galaxy clusters, which contain mostly early-type (gas-stripped) galaxies.

The positional correlations expected for MFs are difficult to study using BATSE data. Class II GRBs, with short durations and relatively low fluences, tend to have poorly-determined positions, often with BATSE error boxes of size $\sim 10°$. Future experiments will localize short-duration GRBs well enough to test for correlations with nearby galaxies and galaxy clusters, hopefully making the identification of extragalactic MFs possible.

ACKNOWLEDGMENTS

We thank R. Knill-Dgani for discussions. This work was supported by NASA grant NAG5-8381; by the Texas Advanced Research Program grant ARP-028.

REFERENCES

1. Kouveliotou, C. et al., *Ap.J.*, **413**, L101 (1993).
2. Murkherjee, S. et al., *Ap.J.*, **508**, 314 (1998).
3. Hakkila, J. et al., *Ap.J*, **538**, 165 (2000).
4. Gupta, V., Gupta, P.D., and Bhat, P.N. in *Gamma-Ray Bursts: 5th Huntsville Symposium,* eds. R.M. Kippen et al., New York: AIP Conf. Proc. No. 526, pp. 215-219 (2000).
5. Mazets, E.P. et al., *Nature,* **282**, 587 (1979).
6. Mazets, E.P. et al., *Astron. Lett.,* **25 (10),** 635 (1999).
7. Feroci, M., Hurley, K., Duncan, R.C. and Thompson, C., *Ap J,* **549,** in press (2001) [astro-ph / 0010494].
8. Fenimore, E., in *High-Velocity Neutron Stars and Gamma-Ray Bursts,* ed. R.E. Rothschild & R.E. Lingenfelter, New York: AIP Conf. Proc. No. , pp. 68-72 (1996).
9. Fenimore, E., Klebesadel, R. and Laros, J., *Ap.J.*, **460,** 964 (1996).
10. Meegan, C. et al., *ApJ Supp.*, **92,** 229 (1994).
11. Meegan, C. et al., *ApJ Supp.*, **106,** 65 (1996).
12. Barat, C. et al., *Astron. & Astroph.*, **126,** 400 (1983).
13. Peebles, P.J.E., *Principles of Physical Cosmology,* Princeton: Princeton University Press (1993).
14. Duncan, R.C. and Thompson, C., *Ap.J.*, **392,** L9 (1992).
15. Thompson, C. and Duncan, R.C., *M.N.R.A.S.,* **275,** 255 (1995).
16. Thompson, C. and Duncan, R.C., *Ap.J.*, submitted (2001).
17. Woods, P.M. et al., *Ap.J.*, **527,** L47 (1999).
18. Tavani, M., *Ap.J.*, **497,** L21 (1998).
19. Cline, D.B., Mathhey, C. and Otwinowski, S. *Ap.J.*, **527,** 827 (1999).
20. de Vaucouleurs, G., *Astronom.J.*, **58,** 29 (1953).
21. Hartmann, D.H., Briggs, M.S. and Mannheim, K., in *Gamma-Ray Bursts: 3rd Huntsville Symposium,* eds. M.A. Briggs et al., New York: AIP Conf. Proc. 384, pp. 397-403 (1996).
22. Abell, G.O., Corwin, H. and Olowin, R.P., *Ap.J.Supp.*, **70,** 1 (1989).
23. Marani, G.F., Nemiroff, R.J., Norris, J.P. and Bonnell, J.P., *Ap.J.*, **474,** 576 (1997).
24. Kolatt, T. and Piran, T., *Ap.J.*, **467,** L41 (1996).
25. Hurley, K. et al., *Ap.J.*, **479,** L113 (1997).

Long-Term *RXTE* Monitoring of the Anomalous X-ray Pulsar 1E 1048.1−5937

Victoria M. Kaspi[*][†], Fotis P. Gavriil[*], Deepto Chakrabarty[†], Jessica R. Lackey[†], Michael P. Muno[†]

[*]*Department of Physics, Rutherford Physics Building, McGill University, 3600 University Street, Montreal, Quebec, H3A 2T8, Canada*
[†]*Department of Physics and Center for Space Research, Massachusetts Institute of Technology, Cambridge, MA 02139*

Abstract. We report on long-term monitoring of the anomalous X-ray pulsar 1E 1048.1−5937 using the *Rossi X-ray Timing Explorer*. This pulsar's timing behavior is different from that of other AXPs. In particular, the pulsar shows significant deviations from simple spin-down such that phase-coherent timing has not been possible over time spans longer than a few months. We show that in spite of the rotational irregularities, the pulsar exhibits neither pulse profile changes nor large pulsed flux variations. We discuss the implications of our results for AXP models. We suggest that 1E 1048.1−5937 may be a transition object between the soft gamma-ray repeater and AXP populations, and the AXP most likely to one day undergo an outburst.

INTRODUCTION

The nature of anomalous X-ray pulsars (AXPs) has been a mystery since the discovery of the first example some 20 years ago. Although it is clear that AXPs are young neutron stars, it is not clear why they are observable. In particular, they show no evidence for possessing a binary companion, making conventional accretion problematic. Furthermore, given their spin periods and period derivatives, their rate of loss of rotational kinetic energy is orders of magnitude too small for these sources to be rotation-powered. One important clue is that two AXPs (and one AXP candidate) are clearly associated with supernova remnants. Although only five AXPs are known, their origin is likely to be of great importance to our understanding of the fate of massive stars and the basic properties of the young neutron star population. For an excellent recent review of these objects, see [16].

Currently there are two models to explain AXPs. One model proposes that AXPs are young, isolated, highly magnetized neutron stars or "magnetars" [5,21]. High magnetic fields ($10^{14} - 10^{15}$G) are inferred from their spin-down under the assumption of magnetic dipole braking, as well as by association with the soft gamma repeaters (SGRs) which show AXP-like pulsations in quiescence [10,11], and are thought to have high magnetic fields for independent reasons [21]. The second model of AXP emission is that they are powered by accretion from a fall-back disk of material remaining from the supernova explosion [4].

One way to distinguish between these classes of models may be through timing observations. In the magnetar model, relatively smooth spin-down should be expected, punctuated by occasional abrupt spin-up or spin-down events or "glitches," as well as low-level, long-time-scale deviations from simple spin-down, or "timing noise." Both phenomena are well known among young radio pulsars (e.g. [12]), although their physical origins in magnetars may be different given the much larger inferred magnetic field. However, according to the magnetar model, no extended spin-up should be seen. On the other hand, accretion power is usually associated with much noisier timing behavior, which can be correlated with spectral, luminosity, and pulse morphology changes. In addition, some accreting binary systems undergo extended ($\sim$years) episodes of spin-up, although these generally seem to alternate with long intervals of spin-down [2].

1E 1048.1−5937 is a 6.4 s AXP in the Carina region [20]. It exhibits no evidence for any binary companion, as no Doppler shifts of the pulse period are seen [17], and no optical counterpart to a limiting magnitude of $m_V \sim 20$ has been detected [14]. The pulsar's spectrum, like those of other AXPs, is well described with a two component model consisting of a soft black body with a power-law tail [18]. Occasional monitoring observations over more than 20 years show that the pulsar is spinning down, though significant deviations from a simple spin-down model have been noted [18,19,1]. The paucity of data thus far has made it impossible to unambiguously identify the origin of the deviations.

Here we report on our monthly *Rossi X-Ray Timing Explorer* (*RXTE*) monitoring of 1E 1048.1−5937 in which we have attempted long-term phase-coherent timing like that achieved for other AXPs [8]. The results described here are reported in more detail in [9].

OBSERVATIONS AND RESULTS

The observations we report on were made with *RXTE*'s Proportional Counter Array (PCA; [7]). Observations of 3–6 ks in length of 1E 1048.1−5937 were made on a monthly basis during 1996 November – 1997 December and 1999 January – 2000 August. In addition, we used archival observations from 1996; these generally had longer integration times than the other data sets. To minimize use of telescope time, our monitoring data consist of brief (usually 3 ks) snapshots of the pulsar. These snapshots suffice to measure pulse arrival times for a phase-coherent timing analysis to good precision. However, for any one epoch, the measured period has typical uncertainty ~ 3 ms, quite large by normal timing standards. Thus, our snapshot method of measuring pulse arrival times can determine spin parameters with extremely high precision only when phase coherence can be maintained. For details regarding this timing procedure, see [9]. The snapshot observations are always, however, useful for monitoring the source pulsed flux and pulse morphology (see below).

Timing

We maintained unambiguous phase coherence for 1E 1048.1−5937 in our monthly observations from 1999 January 23 through 1999 November 15. We required a fourth-order polynomial to characterize the 17 pulse arrival times obtained in this span. These results alone clearly imply that the rotational behavior of 1E 1048.1−5937 is quite different from that of AXPs 1E 2259+586 and RXS J170849.0−400910. Those AXPs exhibit much more stable rotation on comparable and even longer time scales, that is, terms of higher order than $\dot{\nu}$ are very small or negligible for those pulsars on time scales of over a year [8]. The span 1999 January through November represents the longest over which we can phase-connect timing data from 1E 1048.1−5937. Investigating archival *RXTE* data going back to 1997 for 1E 1048.1−5937, we find timing results that are similar those obtained in our recent monitoring program, namely we are able to maintain phase coherence only over few-month intervals.

We can compare our pulse ephemerides with measurements of pulse frequency made over the past 20 yr in order to look for long-term trends. Figure 1 shows the spin history of 1E 1048.1−5937 with previously measured spin frequencies plotted as points with their corresponding 1σ error bars. Data were taken from a variety of sources [18,19,1]. Our *RXTE* timing results are plotted as lines representing separate, short, phase-connected segments. The dotted line represents an extrapolation of the ν and $\dot{\nu}$ from the 1999 coherent fit. The lower plot shows the same data set with the linear term subtracted off. This ephasizes deviations from the simple linear trend.

Pulsed Flux and Pulse Morphology

In accreting systems in which the neutron star is undergoing spin-up, changes in torque should be correlated with changes in X-ray flux. AXPs are spinning down, however. Chatterjee et al. (2000) suggest that AXPs might be spinning down in the propeller regime due to accretion from a fall-back disk. In that case, although the physics of the propeller regime is not well understood, it is still likely that L_x should be correlated with torque [9].

Given the large field-of-view of the PCA and that the bright, nearby, but unrelated source η Carinae exhibited large flux changes over the course of our observations, direct flux measurements of 1E 1048.1−5937 could not be made with our *RXTE* data. Instead, we determined the pulsed component of the flux by using off-pulse emission as a background estimator. This renders our analysis insensitive to changes in the fluxes of other sources in the field-of-view.

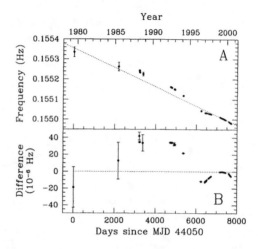

FIGURE 1. Spin history for 1E 1048.1−5937. The points represent past measurements of the frequency of the pulsar. The solid lines represent the *RXTE* phase-connected intervals. See [9] for details. Panel A shows the observed frequencies over time. The dotted line is the extrapolation of the ν and $\dot\nu$ of the 1999 phase-coherent ephemeris. Panel B shows the difference between the ephemeris indicated by the dotted line and the data points.

The results are shown in Figure 2. We find no large pulsed flux variations. The χ^2 strictly speaking does suggest some low-level variability; longer individual observations are clearly necessary to verify this is the case. However, as we discuss below, the pulsed flux is certainly much more stable than previous analyses have suggested [18].

We have also used the *RXTE* data to search for pulse profile changes, as many accretion-powered pulsars exhibit significant changes in their average pulse profiles. Such changes can be correlated with the accretion state, and hence accretion torque and timing behavior [2]. Furthermore, X-ray pulse profiles from the SGRs 1806−20 and 1900+14 have shown differences at different epochs depending on time since outburst [10,11]. However, we find no significant changes in the pulse profile morphology in any of the *RXTE* observations of 1E 1048.1−5937.

DISCUSSION AND CONCLUSIONS

Long-term *RXTE* monitoring of the AXP 1E 1048.1−5937 has shown it to be a much less stable rotator than other AXPs, yet its pulse profile and pulsed flux are stable. Previously, Oosterbroek et al. (1998) compiled flux data from a variety of different X-ray instruments that observed 1E 1048.1−5937. That compilation suggested that the pulsar shows variability by over a factor of ∼5 on time scales of a few years. The reality of those flux changes is not supported by our results. One

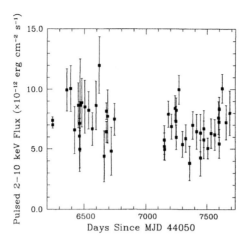

FIGURE 2. Pulsed flux time series in the 2–10 keV band for *RXTE* observations of 1E 1048.1−5937.

caveat is that we measure pulsed flux, while they report flux, so the results could be reconciled if the pulsed fraction is variable.

In the context of the magnetar model, we note that the timing behavior of 1E 1048.1−5937 is somewhat similar to that observed for the soft gamma repeaters SGR 1806−20 and 1900+14 [13,23,24]. However, as the stable flux time series (Fig. 2) for the AXP shows, it has not undergone any outbursts. This can perhaps be understood in terms of persistent seismic activity and small-scale crustal fractures [22] or low amplitude toroidal modes resulting in angular momentum loss following crustal twisting fractures [6].

1E 1048.1−5937 is unusual among AXPs for reasons other than just its timing behavior. In particular, it shows the highest ratio of blackbody to total flux (once energy band is accounted for), and the largest pulsed fraction. In addition, it has the lowest photon index for the power-law tail in its spectrum of any AXP, which makes it the closest to those measured in the X-ray band for SGRs 1806−20 and 1900+14. Further, the thermal component of 1E 1048.1−5937's spectrum has the highest temperature (0.64 keV) of any AXP. This temperature is comparable to that seen for SGR 1900+14 post-burst, 0.62 keV [24]. It therefore could be the case that 1E 1048.1−5937 is a transition object between the populations of AXPs and SGRs, and the AXP most likely to one day undergo an outburst.

In the context of accretion models, perhaps the best source with which to compare 1E 1048.1−5937 is 4U 1626−67, a 7.7 s accreting pulsar in a 42-min binary with a low-mass companion. Although 1E 1048.1−5937 is noisy by AXP timing standards, its noise is comparable in strength to that of 4U 1626−67 [3]. Still, we regard the case for 1E 1048.1−5937 as an accreting binary, even with a very low-mass companion, as weak, given the other evidence against this hypothesis, namely, its much softer spectrum than other accreting binaries, the absence of pulsed flux or pulse morphology changes correlated with the timing behavior, and the spin-down over some 20 yr. It is more difficult to dismiss the possibility that 1E 1048.1−5937 is accreting from a supernova fall-back disk, since there is not yet a consensus on the properties such a disk would have or on the expected timing and variability properties of the pulsar. However, one expectation is that such a disk would be a significant emitter in the optical and infrared, Future optical/IR observations following a more precise localization using the *Chandra X-ray Observatory* could test the fallback disk model.

REFERENCES

1. Baykal, A., Strohmayer, T., Swank, J., Alpar, A., & Stark, M. J. *MNRAS*, in press
2. Bildsten, L. et al. *ApJS*, 113, 367 (1997).
3. Chakrabarty, D. et al. *ApJ*, 474, 414, (1997).
4. Chatterjee, P., Hernquist, L., & Narayan, R. *ApJ*, 534, 373, (1999).
5. Duncan, R. C. & Thompson, C. *ApJ*, 392, L9 (1992).
6. Duncan, R. C., in *Explosive Phenomena in Astrophysics: First KIAS Astrophysics Workshop*, eds. Chang et al., New York: AIP, 2000, in press
7. Jahoda, K., Swank, J. H., Giles, A. B., Stark, M. J., Strohmayer, T., Zhang, W., & Morgan, E. H. *Proc. SPIE*, 2808, 59 (1996).
8. Kaspi, V. M., Chakrabarty, D., & Steinberger, J. *ApJ*, 525, L33, (1999).
9. Kaspi, V. M., Gavriil, F., Chakrabarty, D., Lackey, J. R., Muno, M. P. *ApJ*, in press.
10. Kouveliotou, C. et al. *Nature*, 393, 235 (1998).
11. Kouveliotou, C. et al. *ApJ*, 510, L115 (1999).
12. Lyne, A. G. *Pulsars: Problems and Progress, IAU Colloquium 160*, ed. S. Johnston, M. A. Walker, & M. Bailes, San Francisco: Astronomical Society of the Pacific, 1996, p. 73
13. Marsden, D., Rothschild, R. E., & Lingenfelter, R. E. *ApJ*, 520, 107, (1999).
14. Mereghetti, S., Caraveo, P., & Bignami, G. F. *A&A*, 263, 172, (1992).
15. Menou, K., Esin, A. A., Narayan, R., Garcia, M. R., Lasota, J.-P., & McClintock, J. E. *ApJ*, 520, 276, (1999).
16. Mereghetti, S. *Mem. della Soc. Ast. It.*, 69, 819 (1999).
17. Mereghetti, S., Israel, G. L., & Stella, L. *MNRAS*, 296, 689 (1998).
18. Oosterbroek, T., Parmar, A. N., Mereghetti, S., & Israel, G. L. *A&A*, 334, 925 (1998).
19. Paul, B., Kawasaki, M., Dotani, T., & Nagase, F. *ApJ*, 537, 319, (2000).
20. Seward, F. D., Charles, P. A., & Smale, A. P. *ApJ*, 305, 814, (1986).
21. Thompson, C. & Duncan, R. C. *ApJ*, 473, 322 (1996).
22. Thompson, C. & Blaes, O. *Phys. Rev. D* 57, (1998).
23. Woods, P. M. et al. *ApJ*, 535, L55 (2000).
24. Woods, P. M. et al. *ApJ*, in press

Testing Neutron Star Thermal Evolution Theories

Sachiko Tsuruta and Marcus A. Teter

Department of Physics, Montana State University, Bozeman, MT 59717, USA

Abstract. *Einstein Observatory* gave the first hope for detecting the radiation directly from the neutron star surface. ROSAT detected such radiation from at least four pulsars. With successful launch of Chandra and XMM-Newton combined with the lower-energy band (EUV, UV and optical) observations, we expect more of these detections and better determination of surface temperature and composition of these and more neutron stars. Here we demonstrate that careful comparison of neutron star thermal evolution theories with these observations will finally make it possible to distinguish among various competing theories. That will help us obtain better insight into the properties of dense matter, for instance, by finding whether the central core of observed neutron stars consists of nucleons, pions, or kaons. That will also help us determine the degree of superfluidity of these particles, the equation of state and stellar radius.

INTRODUCTION

The launch of the *Einstein Observatory* gave the first hope for detecting thermal radiation directly from the surface of neutron stars (NSs). However, the temperatures obtained by the *Einstein* were only the upper limits [1]. ROSAT first detected such surface radiation from at least four pulsars, PSR 0833-45 (Vela), PSR 0656+14, PSR 0630+18 (Geminga), and PSR 1055-52 [2]. The spectral data from ROSAT and ASCA showed that these sources consist of soft X-ray thermal radiation from the whole stellar surface and the harder X-ray component which is thermal, nonthermal or the combination of these two. The harder thermal component is considered to come from hot polar caps, while the nonthermal component is X-rays emitted in the magnetosphere [3]. Recently, the number of detections of surface thermal X-rays has increased to at least eight, by the addition of RX J0822-4300 [4a], 1E 1027.4-5209 [4b], RX J185635-3754 [5], and RX J0720.4-3125 [6]. These developments have proved to become serious `turning points' for the detectability of radiation directly from NSs. It has now become possible to seriously compare the observed temperatures, not just the upper limits, with NS thermal evolution theories. Possible detections of NS surface radiation have been reported also from the lower energy optical to EUV band observations by HST and EUVE, for PSR 0656+14, PSR 1929+10, PSR 0950+08, and RX J185635-3754 [5][7]. Most recently the prospect for testing NS thermal evolution theories with observation has increased tremendously, through the successful launch of Chandra and XMM-Newton in July 1999 and December 1999, respectively. Exciting results have already started coming out [8]. On the theoretical side, more detailed, careful investigations have been carried out [9][10].

THERMAL EVOLUTION MODELS

After a supernova explosion a newly formed NS first cools via various neutrino emission mechanisms before the surface photon radiation takes over [11]. The first NS thermal evolution calculations [12] showed that NSs should be visible as X-ray sources for about a million years. Among the important factors which seriously affect the nature of thermal evolution are: neutrino emission processes, superfluidity of the core particles, composition, and mass.

The conventional neutrino cooling mechanisms, such as the modified URCA, plasmon neutrino and bremsstrahlung processes, which were adopted in the earlier and subsequent cooling calculations, are generally called 'standard' cooling. On the other hand, the more exotic extremely fast cooling processes, such as the URCA process involving pions, kaons, and quarks, and the direct URCA process involving nucleons and hyperons, are called 'nonstandard' processes [3]. For convenience, in this report we adopt this terminology. Very recently, some additional neutrino processes have been suggested [13a]. We have examined each of these processes carefully, and found that only one of these, called the 'Cooper pair neutrino emission', is important. This process, neutrino emission due to the breaking and formation of Cooper pairs, takes place right after the stellar core temperature decreases below the critical superfluid temperature, T^{cr}, but it soon decreases exponentially[13b]. The net effect is to enhance, for some superfluid models, the neutrino emission right after the superfluidity sets in.

As the star cools after the explosion and the stellar core temperature reaches below the superfluid critical temperature, T^{cr}, the core particles become superfluid. That causes exponential suppression of both specific heat (and hence the internal energy) and all neutrino processes involving core particles (nucleons, pions, etc.) which become superfluid. The net effect is that the star cools more slowly (the surface temperature and luminosity are higher) during the neutrino cooling era (than in the absence of the suppression).

The composition of the core is predominantly neutrons (with several percent of protons and electrons) if the density is moderate ($< \sim 10^{15}$ gm/cm^3). For higher densities hyperons, pions, kaons, quarks, etc. may dominate the central core. Both observational and theoretical considerations support the view that the NS mass is close to 1.4 $M_\odot$ [14].

Figures 1a and 1b show typical thermal evolution curves (time history of total surface photon luminosity (which corresponds to the effective temperature) to be observed at infinity.

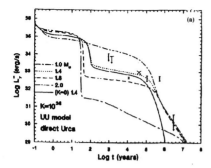 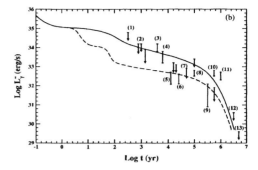

Figure 1. Thermal evolution curves for (a) UU Model with various stellar mass, and (b) FP Model for 0.8 $M_\odot$ (solid curve) and 1.4 $M_\odot$ (dashed curve) stars. The data points are: (1) Cas A, (2) Crab, (3) RX J0822-4300, (4) 1E 1207.4-5209, (5) the Vela pulsar, (6) 0002+6246, (7) PSR 2334+61, (8) PSR 0656+14, (9) Geminga, (10) PSR 1055-52, (11) RX J185635-3754, (12) PSR 1929+10, and (13) PSR 0950+08.

In Figure 1a we adopt a medium equation of state (EOS) model, UU Model, with an AO superfluid[15]. As an example of the nonstandard cooling, the nucleon direct URCA process is used. As the stellar mass increases the core density increases. A sufficiently low mass star, whose core density ρ_c is less than ρ_{crit}, cools with the standard process (ρ_{crit} is the critical density beyond which the nonstandard cooling process sets in). On the other hand, a more massive (i.e., $\rho_c > \rho_{crit}$) star cools by the nonstandard process. The superfluid energy gap (which is proportional to T^{cr}) depends on density. The energy gap for neutrons starts to appear near the nuclear density ρ_N (~ 2.4 x 10^{14} g/cm^3) and increases, but then it decreases again and disappears. The peak of the gap and the density where it disappears depend on the superfluid model. The AO model adopted in Figure 1a is an intermediate case where the gap disappears at ρ_{max} ~ 2 x 10^{15} g/cm^3. That means if the star is massive enough and hence the core density exceeds ρ_{max}, the superfluid disappears in the interior. In this figure, the 1 $M_\odot$ star is not dense enough to exceed ρ_{crit}, and hence it cools by the standard process. By the time the mass reaches 1.4 $M_\odot$, the density exceeds ρ_{crit}, and hence it cools by the nonstandard process. However, the density is less than ρ_{max} (where the superfluidity disappears). That means the central core particles are in the superfluid state and the superfluid suppression of the nonstandard cooling is effective. By the time the mass reaches 2 $M_\odot$, however, the stellar density exceeds ρ_{max}, and hence the core particles are not in the superfluid state. This is because the density is essentially constant within the stellar interior. Without the superfluid suppression, the star cools very fast. *The conclusion is that depending on the stellar mass the cooling curves lie anywhere between the standard curve and the very fast nonstandard curve without the superfluid suppression.* The solid curve shows cooling of a 1.4 $M_\odot$ star without heating. In all other curves a moderate degree of internal frictional heating is added. This figure includes all neutrino processes except the Cooper pair process.

Figure 1b shows our latest cooling curves where the Cooper pair neutrino emission, as well as all others, are included. FP Model, with a typical medium EOS [3], is adopted. The solid curve refers to the standard cooling of a 0.8 $M_\odot$ star which consists predominantly of neutrons because the stellar core density ρ_c is <u>below</u> ρ_{crit} for the neutron-pion phase transition. For this star we adopted the T72 superfluid model [16] which is considered to be the most accurate for the neutron matter. The superfluid suppression is minor because for the T72 model the gap disappears at ρ_{max} which is <u>less</u> than the stellar core density ρ_c. Consequently, the Cooper pair cooling also hardly affects the solid curve. On the other hand, the 1.4 $M_\odot$ star (dashed curve) cools by the pion cooling as the nonstandard case. This is because for this star ρ_c exceeds ρ_{crit}. As the superfluid model for pions we adopted a medium superfluid gap model, called the E1-0.5 Model [15]. For this superfluid model both the enhanced Cooper pair neutrino emission right below T^{cr} and subsequent suppression are significant.

COMPARING THERMAL EVOLUTION MODELS WITH OBSERVATION

In Figure 1b theoretical curves are compared with the most recent (as of February 2001) observational data. The bars are detections, while the downward arrows show the upper limits. Many sources appear to be consistent with the standard cooling. Both PSR 1055-52 (10) and RX J185635-3754 (11) lie somewhat higher than the standard curve. Only the lower limit to the age is known for RX J0720.4-3125 and hence this source is not shown in Figure 1b, but this source also lies slightly above the standard curve. Nevertheless we conclude that all these sources are probably consistent with the standard cooling when various factors are taken into account, such as the effects of heating, the 2D (two dimensional) heat flow under quantized magnetic fields, and age uncertainties [3]. However, we note that *several data points, e.g., the Vela pulsar (5), 0002+6246 (6), PSR 2334+61(7), PSR 0656+14(8), and Geminga(9), lie <u>below</u>* the standard curve. The age uncertainty, of a factor of ~ 2, should not affect this conclusion for younger sources such as the Vela pulsar because the cooling curves are essentially horizontal near these sources.

The conclusion reached above is most naturally explained as the effects of stellar mass and superfluidity of the core particles. In Figure 1b with the medium FP Model, a 0.8 $M_\odot$ star (solid curve) cools with the standard scenario while a star more massive, with 1.4 $M_\odot$(dashed curve), cools with the pion cooling with significant superfluid suppression. Then the hotter pulsars are consistent with the standard cooling of 0.8 $M_\odot$ star while the cooler ones agree with the cooling curve of a 1.4 $M_\odot$ star.

However, we noted that both observational and theoretical considerations suggest the stellar mass to be ~ 1.4 $M_\odot$[14]. In Figure 1b a majority of the detections appear to be consistent with the standard cooling while several other sources appear to be cooling with the pion cooling which is significantly modified by the superfluid suppression. The implication is that the transition from the standard to the pion cooling should take place between somewhat higher mass values, e.g.,~ 1.4 $M_\odot$ and ~ 1.6 $M_\odot$, rather than between 0.8 $M_\odot$ and 1.4 $M_\odot$. To satisfy this requirement, the EOS then should be somewhat <u>stiffer than the medium</u> FP Model adopted in Figure 1b. This is because a <u>stiffer</u> EOS model possesses <u>larger</u> mass for given ρ_{crit}. For the nucleon-pion transition $\rho_{crit} \sim 2.2\, \rho_N$ [3]. For a given EOS the stellar radius is fixed for a given mass. Then, for a 1.4 $M_\odot$ star with EOS somewhat stiffer than medium the stellar radius is somewhat larger than 10 km. For very soft EOS, such as BPS Model, the stellar density $\rho_c \sim \rho_{crit}$ (for the nucleon-pion transition) when $M \sim 0.2\text{-}0.3\, M_\odot$ [3]. That means very soft EOS should be excluded. Similar conclusion applies to other nonstandard scenarios if their ρ_{crit}s are similar to the pion case.

It has been found that the qualitative behavior of all nonstandard scenarios are similar [3][15]. Then, how can we distinguish among these competing nonstandard scenarios? Here we try to show that is still possible, in the following sense. First of all, we note that these nonstandard mechanisms alone are all *too fast* to be consistent with any observed detection data, even with heating [3]. That means *significant amount of superfluid suppression is required*. However, Takatsuka and Tamagaki [9][10] recently showed that T^{cr} for both kaon cooling and nucleon direct URCA are *too low*, ~ several x 10^7 K. Then observed pulsars are hotter (core temperature ~ 10^8 K) than T^{cr}, and hence *the core particles are not yet in the superfluid state*! That leaves, as possibly viable nonstandard mechanisms, the pion cooling, quark cooling, and the direct URCA involving hyperons. It has been already shown that T^{cr} for pion condensates should be higher [17], and therefore *pions in the observed pulsars should be already in the superfluid state*. The problem for quarks is that there are so many unknown factors that it would be virtually impossible to go beyond the regime of speculation. If anything, ρ_{crit} for nucleon-quark transition should be too high to be of interest to neutron star problems [18]. The superfluid calculations similar to those done for kaons, pions and nucleons are possible for hyperons, and these calculations are currently under way (Takatsuka, private communication).

Figure 1b shows that the pion cooling (dashed curve) with the moderate E1-0.5 superfluid gap model agrees with the observed temperature of cooler pulsars such as Vela. In this way, the degree of superfluidity can be estimated for pion condensates by comparison with observation. Similarly, Figure 1b suggests that the superfluid gap should be small for neutron matter, supporting the T72 model adopted, because otherwise the standard curve will lie too low to be consistent with the hotter NS data. Especially, due to the enhanced Cooper pair neutrino emission, the standard curve will lie too low if an intermediate superfluid gap model, such as AO or E1-0.5 model, is adopted, to be consistent with the detection data of PSR 1055-52(10) and RX J185635-3754 (11). This conclusion does not change even when heating and other factors are taken into account. The spectral data from observations require that the surface/atmospheric composition of both of these NSs is heavy elements such as Fe, not hydrogen, and hence these data points cannot become lower than indicated[5][8].

SUMMARY AND CONCLUSION

We have shown that the most up-to-date observed temperature data are consistent with the current NS thermal evolution theories if less massive stars (~ 1.4 $M_\odot$) cool by the standard cooling while more massive stars (e.g., 1.6 $M_\odot$) cool with the nonstandard cooling. Among various nonstandard cooling scenarios, the pion cooling is shown to be consistent with observation, while both kaon cooling and the direct URCA with nucleons are not. The comparison of cooling theory with observation suggests that the EOS should be somewhat stiffer than medium and the stellar radius somewhat larger than 10km. We have shown that recent observational and theoretical developments have already started distinguishing among various competing theories. We emphasize that more and better data expected soon from Chandra and XMM-Newton, when combined with further, careful theoretical work, should give still better insight to this exciting area of astrophysics.

ACKNOWLEDGMENTS

We thank many of our colleagues for stimulating discussions. We thank Dr. G. Pavlov for providing us some of the newer Chandra detection data points (with error bars). This work was supported in part by NSF grant PHY99-07949 and NASA grant NAG5-3159.

REFERENCES

1. e.g., Nomoto, K., and Tsuruta, S., *Ap. J. Letters* **305**, L19 (1986).
2. e.g., Ögelman, H., in *Lives of the Neutron Stars*, edited by M.A. Alpar and J. van Paradijs, Kluwer Acad. Publ., Dordrecht, 1994, p. 101.
3. Tsuruta, S., *Physics Reports* **292**, 1-130 (1998).
4a. Zavlin, V.E., Trümper, J., and Pavlov, G.G., *Ap. J.* **525**, 959 (1999).
4b. Zavlin, V.E., Pavlov, G.G., and Trümper, J., *Ap. J.* **331**, 821 (1998).
5. Walter, F.M., Wolk, S.J., and Neuhäuser, R., *Nature* **379**, 233 (1996).
6. Haberl, F. et al., *A & A* **326**, 662 (1997).
7. Pavlov, G.G., Stringfellow, G.S., and Córdova, F.A., *Ap. J.* **467**, 10 (1996).
8. Pavlov, G.G., Sanwall, D., Teter, M.A., Tsuruta, S., and Zavlin, V., to be submitted to Ap.J.(2001).
9. Takatsuka, T., and Tamagaki, R., *Prog. Theor. Phys.* **94**, 457-461 (1995).
10. Takatsuka, T., and Tamagaki, R., *Prog. Theor. Phys.* **97**, 345-350 (1997).
11. Tsuruta, S., *Physics Reports* **56**, 237 (1979).
12. Tsuruta, S., Ph.D. thesis, Columbia University, N.Y. (1964).
13a. e.g., Page, D., in *the Many Faces of neutron Stars*, edited by R. Buccheri et al., Kluwer Acad. Publ., Dordrecht, 1998, p. 538.
13b. e.g., Flowers, E., Ruderman, M., and Sutherland, P., *Ap.J.* **205**, 541-544 (1976); Yakovlev, D.G., Levenfish, K.P., and Shibanov, Yu. A. *Physics-Uspekhi* **42**, 737-778 (1999).
14. e.g., Brown, G.E., Weingartner, J.C., and Wijers, R.A.M.J., *Ap.J.* **463**, 297 (1996).
15. Umeda, H., Tsuruta, S., and Nomoto, K., *Ap. J.* **433**, 256 (1994).
16. Takatsuka, T., *Prog. Theor. Phys.* **48**, 1517 (1972).
17. Takatsuka, T., and Tamagaki, R., *Prog. Theor. Phys.* **67**, 1649 (1982).
18. Iwamoto, N., *Ann. Phys.* **141**, 1 (1982).

Ubiquity: Relativistic Winds from Young Rotation-Powered Pulsars

E. V. Gotthelf

Columbia Astrophysics Laboratory, 550 West 120th St, New York, NY 10027, USA

Abstract. Recent X-ray observations of young rotation-powered pulsars are providing an unprecedented detailed view of pulsar wind nebulae. For the first time, coherent emission features involving wisps, co-aligned toroidal structures, and axial jets are fully resolved in X-rays on arc-second scales. These structures, which are remarkably coherent and symmetric, are similar to features seen in the optical Crab nebula. In this report, we present the latest high resolution Chandra X-ray images of six young rotation-powered pulsars in supernova remnants. These data suggest an X-ray morphology perhaps common to all young rotation-powered pulsars and serve as a guide for developing the next generation of theoretical models for pulsar wind nebulae.

INTRODUCTION

Young Crab-like pulsars are thought to lose their rotational energy predominantly in the form of a highly relativistic particle wind. Evidence for this wind is manifest as a bright, centrally condensed synchrotron emission nebula, which Weiler & Panagia (1978) referred to as a "plerion". Models for these pulsar wind nebulae (PWNe) are based on the Crab Nebula, the first known and brightest example. The freely expanding pulsar wind is initially invisible as it travels though the surrounding self-evacuated region. Eventually, the wind encounters the ambient medium where it is reverse-shocked, resulting in the thermalization and re-acceleration of particles. Synchrotron radiation from these particles is most easily observed in the form of a bright radio nebula which acts as a calorimeter of the pulsar's current energy loss. Models were developed by Pacini and Salvati (1973), Reynolds & Chevalier (1984) and others and subsequently refined (e.g. Kennel & Coroniti 1984; Aron 1998). A review is found in Chevalier (1998).

High resolution Chandra X-ray images of the Crab pulsar show the limitations of the current models. These observations display coherent concentric toroidal structures and jet-like features apparently aligned along the spin axis, and in the direction of the pulsar's velocity vector. The remarkable structures and alignments clearly delineate the complex magneto-hydrodynamics associated with this pulsar which have yet to be fully explained. Herein we present several more examples of Crab-like pulsars in supernova remnants (SNRs) observed with Chandra, all of which show evidence of similarly orientated toroidal and possible jet-like features. Collectively, these observations suggest, for the first time, the fundamental relationship of these structures to the central engines in young rotation-powered pulsars. We briefly present each pulsar and discuss the implications of their individual and common morphology.

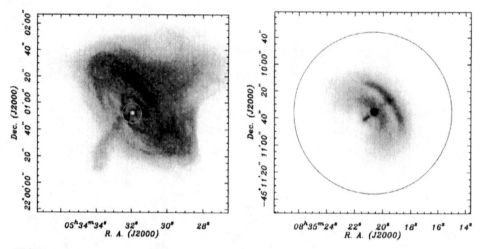

FIGURE 1. A scaled comparison of the relativistic wind nebulae surrounding two young pulsars observed by the Chandra Observatory, (left) the 1,000 yrs Crab pulsar and (right) the 12 kyrs Vela pulsar. These images are displayed with the same plate scale, but the Vela nebula is a factor of 16 times smaller than the Crab assuming distances of 2 kpc (Crab) and 250 pc (Vela); the circles represent the same physical size at the distance of the pulsar. Although Vela is an order of magnitude older and smaller, the two objects are found to be similar in shape and overall brightness distribution. From Helfand, Gotthelf, & Halpern (2001); see also Weisskopf (2000)

NEW CHANDRA OBSERVATIONS

The Chandra Observatory (Weisskopf, O'Dell, and van Speybroeck 1996) has targeted most of the known Crab-like pulsars using one or both of its two imaging focal plane detectors, the High Resolution Camera (HRC) and the CCD-camera (ACIS). Both cameras provide arc-sec imaging over a $\sim 0.5° \times 0.5°$ field-of-view. The HRC allows pulse-phase imaging to isolate the pulsars from their nebulae while ACIS provides moderate resolution spectroscopy. Table I presents a summary of the observational characteristics of the pulsars presented herein. The Chandra images for each of these SNRs are displayed in Fig. 1a-3b.

PSR J0534+2200, The Crab [Fig. 1a] – The remarkable Chandra observation of the Crab nebula has been reported in Weisskopf et. al 2000. As previously mentioned, images from this data set shows toroidal and jet-like X-ray structures aligned along the spin axis. These features were first hinted at by earlier X-ray observation (Aschenbach & Brinkmann 1975) but the quality of the new Chandra images now fully reveals the central engine powering the nebula, clearly delineating its geometry with respect to the optical nebula. The observed morphology displayed in Fig. 1a forms the basis for a comparison with other PWN pulsars. The Crab Nebula, however, is unique amongst young supernova remnants in that it has no discernible SNR shell; the reason for this is not yet satisfactorily explained.

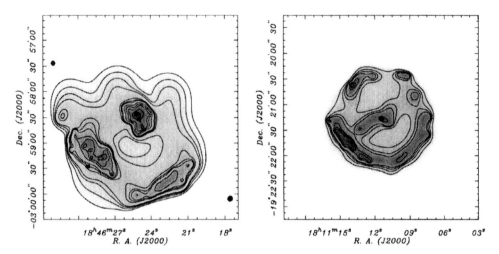

FIGURE 2. Two young supernova remnants observed by the Chandra Observatory containing recently discovered pulsars. (Left) – Kes 75 with its 324 ms pulsar PSR J1846−0258, which has a spin-down age of only 700 yrs. (Right) – G11.2-0.3 contains the 69 ms pulsar PSR J1811−1926, possibly the stellar remnant of the historic supernova of A.D. 386. The resolved pulsars in the center of both images are by far the brightest object in the image which is saturated to highlight the diffuse emission. These images are adaptively smoothed and their respective contour levels and plate scales set the same for comparison. From Helfand & Gotthelf (2001) and Kaspi et al. (2001)

PSR J0835−4510 in Vela XYZ [Fig. 1b] – The 89 ms Vela pulsar was observed twice with the HRC, ~ 3.5 and ~ 35 days following an extreme glitch in its rotation frequency (Helfand, Gotthelf, and Halpern 2001). Most surprisingly was the discovery of a coherent toroidal structure and axial jet remarkably similar to that observed in the Chandra observation of the Crab Nebula. Furthermore, as for the Crab, the axis of symmetry is also along the directions of the pulsar proper motion. Clearly resolved are two concentric arcs which may form tori. The physical size of these structures is 16 times smaller relative to the Crab (see Figure 1), perhaps commensurate with its lesser spin-down energy. Comparison of the two Vela observations shows that the brightness of the outer arc increases significantly while the flux of the pulsar remained relatively steady. If this increase is associated with the glitch, the inferred propagation velocity is $\gtrsim 0.7c$, similar to that seen in the brightening of the optical wisps of the Crab Nebula.

PSR J1846−0258 in Kes 75 [Fig. 2a] – Kes 75 is a young, distant (~ 19 kpc) Galactic shell-type remnant with a central core whose observed properties have long suggested a PWN similar to the Crab Nebula (Becker, Helfand & Szymkowiak 1983). Located within the core of Kes 75 resides the recently discovered PSR J1846−0258 (Gotthelf et al. 2000) – a pulsar with exceptional timing properties: its period, spin-down rate, and spin-down conversion efficiency, are each an order-of-magnitude greater than that of the Crab, most likely as a result of its extreme magnetic field (Helfand & Gotthelf 2001). Although the PWN is found to be noticeable elongated, the statistics of the current observations is insufficient to resolve any detailed Crab-like structure. The association of a shell-type remnant in Kes 75 with a centrally located, coeval pulsar provides strong

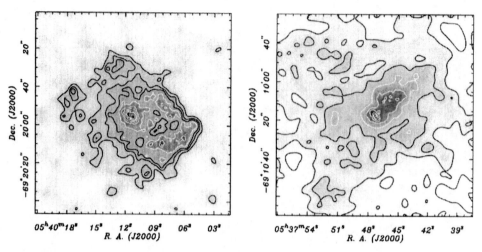

FIGURE 3. A global view around two young Crab-like pulsars located close to each other in the Large Magellenic Clouds. (Left) – the 50 ms pulsar PSR J0540−6919 in SNR 0540-69.3 has an incomplete SNR shell containing an elongated core. Phase resolved blow-up of the core (see Gotthelf & Wang 2000) shows weak evidence of a perpendicular jet. (Right) – the recently discovered 16 ms pulsar PSR J0537−6910 in the N157B nebula, the most rapidly spinning young pulsar known, has faint SNR emission, a bright PWN with an enormous tail of diffuse emission. These images are adaptively smoothed and their respective contour levels and plate scales set the same for comparison. From Gotthelf & Wang (2000) and Wang & Gotthelf (2001).

evidence that neutron stars are born in supernovae explosions. PSR J1846−0258 has the youngest characteristic age ($P/2\dot{P} \sim 700$ yrs) and is likely being spun down rapidly by torques from a large magnetic dipole, just above B_{QED} (see Table 1). The role of the magnetic field is important for understanding the transport and dissipation of particle and magnetic field energy in the relativistic pulsar wind.

PSR J1811−1926 in G11.2−0.3 [Fig. 2b] – The apparently young SNR G11.2−0.3 is a shell-type Galactic remnant containing the 65 ms pulsar PSR J1811−1926 (Torii et al 1998). Based on it location and thermal remnant age estimate, G11.2-0.3 has been proposed as the remnant of supernova SN 386, one of only a few historical supernovae known. The spin-down age of PSR J1811−1926 is, however, 24,000 yrs, a severe discrepancy which suggests the two sources could be unrelated. This is most puzzling, as the pulsar is located near the geometric center of the nearly complete X-ray and radio shell. This suggests the intriguing possibility that the pulsar was born spinning near its current rate or suffered an episode of rapid spin-down. If associated, the symmetry and completeness of the thermal shell strongly constrains the projected magnitude of any initial velocity imparted to the pulsar during its formation.

PSR J0540−6919 in SNR 0540-69.3 [Fig. 3a] – The Chandra calibration observation of the X-ray-bright 50 ms pulsar PSR J0540−6919 in the Large Magellanic Cloud (LMC) has conclusively demonstrated its Crab-like nature (Gotthelf & Wang 2000; Kaaret et al. 2000). Although much older than the Crab (based on its spin-down age), PSR J0540−6919 is contained within a central PWN with toroidal structure similar in

TABLE 1. Young Crab-like Pulsars in Supernova Remnants: Observational Characteristics

Remnant*	Pulsar	Distance (kpc)	P (ms)	$\dot{P}$ ×10^{-14} (s/s)	Age (kyrs)	$\dot{E}^\dagger$ ×10^{35} (ergs/s)	B_p/B_{QED}**
G11.2−0.3	PSR J1811−1926	5	69	4.4	23.0	63	0.04
Vela XYZ	PSR J0835−4510	0.25	89	12.0	12.0	67	0.08
Kes 75	PSR J1846−0258	19	324	710.0	0.7	82	1.10
SNR 0540-69.3	PSR J0540−6919	LMC	50	48.0	17.0	1516	0.11
Crab Nebula	PSR J0534+2200	2	33	40.0	1.3	4394	0.08
N157B Nebula	PSR J0537−6910	LMC	16	5.1	5.0	4916	0.02

* References for Chandra images − G11.2−0.3: Kaspi et al. (2001); Vela XYZ: Helfand, Gotthelf, & Halpern (2001); Kes 75: Helfand & Gotthelf (2001); SNR 0540-69.3: Wang & Gotthelf (2000), Kaaret et al. (2000); Crab Nebula: Weisskopf et. al (2000); N157B Nebula: Wang, Gotthelf, Chu, & Dickel (2001).

† Ranked by spin-down energy.

** The inferred magnetic field is normalized to the quantum critical field defined as $B_{QED} = m_e^2 c^3/e\hbar = 4.4 \times 10^{13}$ G.

size to that of the Crab. The PWN is just resolved at the distance of the LMC, but analyses of pulse phase-dependent images reveals the characteristic elliptical emission of a Crab-like torus, plus weak evidence for a jet emanating from the pulsar, perpendicular to the major axis of the torus as seen for the Crab and Vela systems.

PSR J0537−6910 in N157B [Fig. 3b] − This remnant is also located in the LMC, only 17 arcmins from SNR 0540-69.3, and contains the recently discovered 16 ms pulsar PSR J0537-6910, the most energetic and rapidly rotating young pulsar known. In addition to the pulsar itself and its surrounding compact PWN, the Chandra X-ray observations resolve a third distinct features, a region of large-scale diffuse emission trailing from the pulsar (Wang, Gotthelf, Chu, & Dickel 2001). This X-ray feature, the largest among all known Crab-like SNRs, is a comet-shaped bubble coexisting with enhanced radio emission and is oriented nearly perpendicular to the major axis of the PWN. It is likely powered by a toroidal pulsar wind of relativistic particles which is partially confined by the ram-pressure from the supersonic motion of the pulsar. Ram-pressure confinement also allows a natural explanation for the observed X-ray luminosity of the compact nebula and for the unusually small X-ray to spin-down luminosity ratio of $\sim 0.2\%$, compared to the Crab pulsar.

DISCUSSION

This preliminary look at the new Chandra images of rotation-powered pulsars hint at what might be learned from these observations. Common to these pulsars is their location in the centers of their respective SNR. Apparently these pulsars have not travelled far from their origin. Collectively, these images provide important constraints on the average birth-kick velocity imparted to young pulsars.

The conjecture that the Crab Nebula is the result of the historic supernova SN 1054 provides a direct link between neutron stars formation and supernovae explosions. The case may be better made, however, by pulsars residing in identifiable thermal SNR shells, apparently absent for the Crab. The 1000-yr young Crab may be in a pre-shell evolutionary stage; however, a counter example is provided by Kes 75, whose pulsar has a similar spin-down age coeval with its SNR. Furthermore, N157B and G11.2−0.3 show only weak shell emission, perhaps like the Crab, but their characteristic ages are much older. As often suggested, the spin-down age is likely unreliable for young pulsars. This notion is furthered advanced by the association of G11.2−0.3 with SN 386, which is inconsistent with the spin-down age of PSR J1811−1926. The standard assumptions used to associate the spin-down age with the "true" pulsar age may well be violated for young pulsars.

Perhaps ultimately more revealing is the common alignment of the symmetry axis for several of the PWN tori and jets along the direction of the pulsar's velocity vector, when known. Quite inscrutable is the ostensible chance sky alignment of the principle axis for the observed PWNs with position angle of ~ 45 degs. What is clear, however, is that the observed complex toroidal and jet structures are likely ubiquitous to young rotation-powered pulsars. Furthermore, imaging-spectroscopy of these feature are consistent with a highly energetic particle wind emitting non-thermal synchrotron radiation. Of course much theoretical work is needed to model these remarkable structures.

ACKNOWLEDGMENTS

This work is made possible by NASA LTSA grant NAG 5-7935.

REFERENCES

1. Arons, J. 1998, MmSAI, 69, 989
2. Aschenbach, B. & Brinkmann, W. 1975, A&A, 41, 147
3. Becker, R. H., Helfand, D. J. & Szymkowiak, A. E., 1983, ApJ, 268, L93
4. Chevalier, R. A. 1998, MmSAI, 69, 977
5. Chevalier, R. A. 2000, ApJ, 539, L45
6. Kaspi, V. M., Roberts, M. E., Vasisht, G., Gotthelf, E. V., & Kawai, N. 2001, ApJ, submitted
7. Helfand, D. J. & Gotthelf, E. V. 2001, ApJ submitted
8. Helfand, D. J., Gotthelf, E. V., & Halpern, J. P. 2001, ApJ, in press
9. Gotthelf, E. V., Vasisht, G., Boylan-Kolchin, M., & Torii, K. 2000, ApJL, 542, L37
10. Gotthelf, E. V. & Wang, Q. D. 2000, ApJL, 532, L117
11. Kaaret, P., et. al 2001, ApJ, 546, 1159
12. Kennel, C. F. & Coroniti, F. V. 1984, ApJ, 283, 710
13. Reese, M. J. & Gunn, J. E. 1974, NMRAS, 167, 1
14. Reynolds, S. P. & Chevalier, R. A. 1984, ApJ, 278, 630
15. Torii, K., Tsunemi, H., Dotani, T.,& Mitsuda, K. 1997 ApJ, 489, L145
16. Pacini, F. & Salvati, M. 1973, ApJ, 186, 249.
17. Wang, Q. D. & Gotthelf, E. V. 1998, ApJL, 509, L109
18. Wang, Q. D., Gotthelf, E. V., & Chu Y.-H. & Dickel, J. R. 2001, ApJ sumbitted.
19. Weiler, K. W. & Panagia, N. 1978, A&A, 70, 419
20. Weisskopf, M. C. O'Dell, S. L., van Speybroeck, L. P. 1996, Proc. SPIE 2805, Multilayer and Gazing Incidence X-ray/EUV Optics III, 2.
21. Weisskopf, M. C. et. al 2000, ApJ, 536, L81.

Interaction of Evolved Pulsars and Magnetars With the ISM

Marina M. Romanova*, Olga D. Toropina[†], Yuriy M. Toropin[††], Richard V.E. Lovelace*

Cornell University, Ithaca, NY 14853
[†] *Space Research Institute, Moscow, Russia 117810*
[††] *Keldysh Institute of Applied Mathematics, Moscow, Russia 125047*

Abstract.
After the pulsar stage of evolution, evolved (radio quiet) pulsars and magnetars are still strongly magnetized and rotating objects. They continue to generate a magnetically dominated (Poynting flux) wind from the region of the light cylinder. Part of the magnetic energy converts to energy of relativistic particles (pulsar wind). Expansion of the magnetic field and relativistic particles outward significantly increases the effective size of the magnetosphere of these objects and thus the cross-section for their interaction with the interstellar medium (ISM).

The interaction leads to the formation of a bow shock wave and a highly elongated magnetotail. The standoff distance of the bow shock is determined by the Poynting power outflow from the light cylinder of the pulsar. We performed axisymmetric MHD simulations of propagation of rotating magnetized stars through the ISM, and observed that a prominent bow shock stands in front of magnetosphere. The magnetosphere is stretched by the ISM forming elongated magnetotail. We observed that a rotating star has an expanded cross-section compared to a non-rotating one. In the case of a non-moving star, a disk-like equatorial MHD wind is observed. We conclude that the closest evolved pulsars and magnetars may be observable owing to interaction with the ISM. The origin of the shock wave near the object RX J1856.5-3754 is discussed.

Introduction

There are ~ 1000 pulsars observed in our Galaxy. Their evolution time-scale is $\sim 10^6 - 10^8$ yr. The estimated total number of isolated neutron stars in our Galaxy is $10^8 - 10^9$ (e.g. [1]), so that most of them are presently invisible. Some neutron stars (magnetars) are born with stronger magnetic field $B \sim 10^{14} - 10^{16}$ G [2-3]. The number of these objects is estimated to be $\sim 10\%$ of all new born magnetized neutron stars [4-5]. Magnetars pass their pulsar stage much faster, in $\sim 10^4$ yr.

At the pulsar stage of evolution, most of the star's spin-down energy goes into a

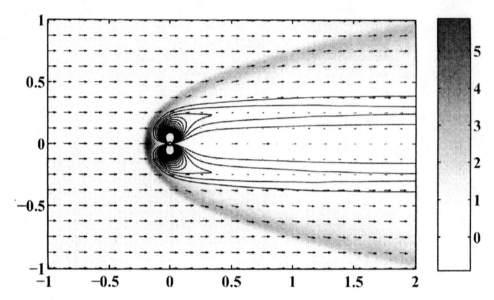

FIGURE 1. Poloidal magnetic field lines and velocity vectors for the case of a non-rotating magnetized star moving through the ISM with Mach number $\mathcal{M} = 10$. The gray scale represents density. The figure is based on axisymmetric MHD simulations on a (R,Z) grid of 129×385.

magnetically dominated (Poynting flux) MHD wind, which originates in the vicinity of the light cylinder R_L [6]. A fraction of the MHD wind energy is thought to go into relativistic particles [6]. Bow shocks and elongated structures have been observed around several pulsars, which were interpreted as interaction of pulsar wind of relativistic particles with the ISM (e.g. [7-8]). After the radiopulsar stage, evolved pulsars and magnetars are still strongly magnetized and rotating objects which continue to generate a MHD wind and possibly a wind of relativistic particles. In this paper we study the interaction of evolved pulsars and magnetars with the ISM. Our earlier numerical simulations have shown that the magnetospheres of moving stars interact with the ISM and form a bow shock wave and an elongated magnetotail (see Figure 1 and [9]). In this paper we show in addition simulations of a *rotating*, magnetized stars moving supersonically through the ISM. We analyze possible observational consequences.

Interaction of Magnetosphere or Pulsar Wind With the ISM

After the radiopulsar stage, at periods $P \gtrsim 1 - 3$ s, the light cylinder radius $R_L = cP/2\pi \gtrsim 4.8 \times 10^9 P_s$ cm ($P_s = P/1$ s) is much smaller than Alfvén radius R_A (which is discussed below). An MHD wind forms in the region of the light

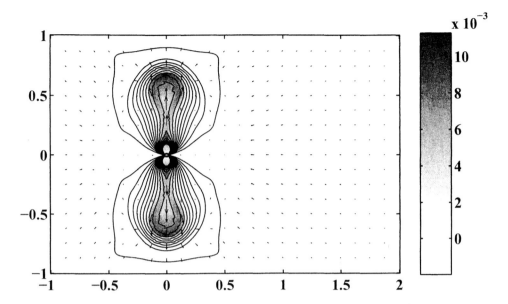

FIGURE 2. Poloidal magnetic field lines and velocity vectors of a non-moving star rotating with azimuthal velocity $v_\phi = v_K$ at the radius of the numerical star. The gray scale represents the magnitude of azimuthal velocity $|v_\phi|$.

cylinder and propagates outward [6]. The wind is at least initially magnetically (Poynting flux) dominated. The dominant component of the magnetic field is azimuthal, as in the Solar wind. For $r > R_L$, the azimuthal magnetic field decreases as $B \approx B_L(R_L/r)$ [6], where $B_L = B_*(R_*/R_L)^3 = 9B_{12}P_s^{-3}$ G is magnetic field at $r = R_L$ and B_* is the field at the star's surface. If a pulsar moves through the ISM with velocity V, then the effective Alfvén radius (or standoff distance of the bow shock) R_A is determined by the balance of the magnetic pressure of the wind and the ram pressure of the ISM, $B^2/4\pi = \rho V^2$. Thus,

$$R_A = (B_L^2 R_L^2 / 4\pi \rho V^2)^{1/2} = 9.4 \times 10^{14} B_{12} P_s^{-2} V_7^{-1} n^{-1/2} \text{ cm}, \tag{1}$$

where $B_{12} = B/10^{12}$ G, $V_7 = V/10^7$ cm/s, n is a number density of the ISM, and the radius of the neutron star is taken to be 10^6 cm. Note, that for a non-rotating star with the dipole magnetic field, Alfvén radius $R_{A0} = R_*(B_*^2/4\pi\rho v^2)^{1/6} = 1.8 \times 10^{11} R_6 (B_{12}^2/n V_7^2)^{1/6}$ cm is much smaller.

The standoff distance for moving *pulsars* can be also determined from observed spin-down luminosity $\dot{E}_{sd}$. The spin-down power approximately equals the electromagnetic energy released at the light cylinder $\dot{E}_{sd} \sim \dot{E}_{Poynt} \sim R_L^2 B_L^2 c$ [6]. (The angular momentum outflow, carried by this wind is $\dot{L}_z = \dot{E}_{Poynt} P/2\pi$). Thus, the standoff distance is (e.g. [7]) $R_A = (\dot{E}_{sd}/4\pi\rho V^2 c)^{1/2}$, which equals to that determined by equation (1). Both formulae describe the same fact that the standoff

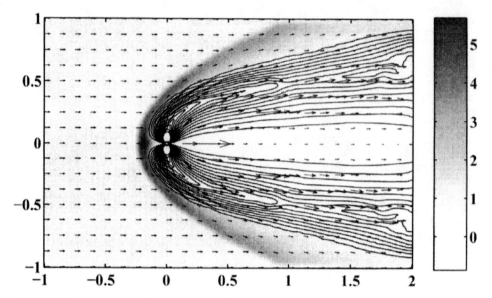

FIGURE 3. Magnetic field lines and velocity vectors for an aligned rotator moving with Mach number $\mathcal{M} = 10$ and rotating with $v_\phi = 3v_K$ (at the radius of the numerical star). The gray scale represents density.

distance is determined by total electromagnetic power released near the light cylinder [6]. A fraction $\alpha < 1$ of electromagnetic energy may be converted to energy of relativistic particles at $r > R_L$ owing to reconnection and annihilation of magnetic field [10] (or other processes). If most of energy is carried by relativistic particles ($\alpha = 1$), then the standoff distance of the bow shock is determined by interaction of pulsar wind with ISM (see e.g. [7]). If $\alpha \ll 1$, then the standoff distance is determined by interaction of expanded magnetosphere with the ISM. We emphasize, that the standoff distance does not depend on α and is determined by total spin-down power of the star.

External regions of the shock interact with ISM, while internal regions interact with pulsar wind. Interaction with ISM leads to heating of the bow shock matter at the rate:

$$\dot{E}_{es} \approx \frac{1}{2}\rho V^3 \pi R_A^2 \approx 2.3 \times 10^{27} B_{12}^2 P_s^{-4} V_7 \text{ erg/s}. \qquad (2)$$

Similar energy will be released in the magnetotail owing to reconnection of magnetic field lines [9]. Heating of matter at the internal shock by pulsar wind is at the rate:

$$\dot{E}_{is} = \alpha R_L^2 B_L^2 c \approx 5.6 \times 10^{31} \alpha B_{12}^2 P_s^{-4} \text{ erg/s}. \qquad (3)$$

The ratio of powers released at the internal shock to that at the external shock

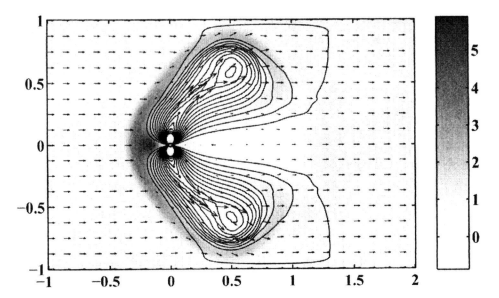

FIGURE 4. Magnetic field lines and velocity vectors for an align rotator moving with Mach number $\mathcal{M} = 3$ and rotating with $v_\phi = 3v_K$ (at the radius of the numerical star). The gray scale represents density.

is $\dot{E}_{is}/\dot{E}_{es} \approx 8\alpha(c/V)$. Heating by pulsar wind dominates at $\alpha > 0.125 V/c \approx 4 \times 10^{-5} V_7$.

Numerical Models

We studied the interaction of rotating moving magnetized star with the ISM by axisymmetric, non-relativistic MHD simulations. All three components of velocity and magnetic field $(v_r, v_\phi, v_z, B_r, B_\phi, B_z)$ were calculated, but $\partial/\partial\phi = 0$. Other equations are those for matter density ρ and energy-density ε (see detailed description of all equations in [11]). The radius of a neutron star $R_* \approx 10^6$ cm is much smaller than R_A, $R_A/R_* \sim 10^5 - 10^8$ and it is presently impossible to model numerically this range of sizes. Thus we place the actual star and part of its magnetosphere inside our numerical star. We model only the *external* part of the magnetosphere outside the numerical star. The surface of the neutron star rotates with subkeplerian velocity $v < v_K$, where $v_K = \sqrt{GM/R_*}$. However, external regions of (rigidly rotating) magnetosphere rotate with velocity larger than *local* keplerian velocity $v >> v_K(r)$. Thus, we rotate the surface of the numerical star R_{num} with velocity $v_\phi = \delta v_K$, where $v_K = \sqrt{GM/R_{num}}$, $\delta \sim 1 - 10$. This fast rotation, leads to outflow of matter and magnetic field, and to expansion of magnetosphere in the equatorial direction. Of course, a relativistic approach is needed to treat the actual very low density plasma outside the light cylinder of a pulsar. We find that when a star is not moving $V = 0$, the magnetosphere expands forming a torus-like structure in the equatorial region (Fig. 2). The azimuthal component of the velocity and the azimuthal magnetic field predominate in an equatorial wind. The observed disk-like outflows resemble the disk-like structure seen in X-ray image of Crab [12].

When a rotating star propagates through the ISM then the cross-section of interaction with ISM (Fig. 3) is substantially larger than that with non-rotating star (Fig. 1). Part of the star's magnetic flux is stretched behind the star forming a highly elongated magnetotail. Reconnection events are observed in the tail similar to those seen in case of non-rotating star [9]. However, when the star rotates, magnetic flux in the tail is larger, and the expected reconnection power is larger also. If the velocity of rotation v_ϕ is of the same order as propagation velocity V, then a more complicated shape of expanded magnetosphere is observed as shown at Fig. 4. Thus, different shapes of resulting expanded magnetosphere are expected depending on Mach number $\mathcal{M}$ and v_ϕ, from torus-shaped magnetosphere at $\mathcal{M} = 0$, to extremely stretched elongated magnetotails at $\mathcal{M} >> 1$.

Is RX J1856.5-3754 an Evolved Pulsar or Magnetar ?

Recently, a shock wave was discovered around the nearby $d \approx 60$ pc [13] isolated neutron star RX J1856.5-3754 in H_α line by Drs. M. van Kerkwijk and S. Kulkarni. Here, we check hypothesis that this object is a possible evolved pulsar or magnetar. The observed standoff distance between a star and a bow shock is $\theta \approx 1''$ [14] which corresponds to $R_A \approx 9 \times 10^{14}$ cm. This bow shock may be connected with interaction of pulsar wind with the ISM. Assuming number density of the ISM particles before the front $n \approx 0.6$ cm^{-3} [14], and velocity of the star $v \approx 100$ km/s [13], we get spin-down power of the star $\dot{E}_{sd} \approx 4\pi\rho V^2 c R_A^2 \approx 3.1 \times 10^{31}$ erg/s. Equating spin-down power $B^2 P^{-4} c$ to one derived from standoff distance, we get $B_{12}^2 P_s^{-4} \approx 0.5$, from which the broad spectrum of possible B ad P may be derived. For example, a) if $B = 10^{11}$ G, then $P \approx 0.4$ s, and life-time of the pulsar ($T \approx 10^8 B_{12}^{-2} P_s^2$ yr) is $T \approx 1.4 \times 10^9$ yr, b) if $B = 10^{12}$ G, then $P \approx 1.2$ s and $T \approx 1.4 \times 10^8$ yr, c) if $B = 10^{13}$ G, then $P \approx 3.8$ s and $T \approx 1.4 \times 10^7$ yr, d) if $B = 10^{14}$ G, then $P \approx 12$ s and $T \approx 1.4 \times 10^6$ yr. Objects with these parameters are in the region of active radiopulsars, which means that either radiobeam is not directed towards us, or radiopulses are very weak.

The derived spin-down power $\dot{E}_{sd}$ is comparable or smaller than X-ray luminosity of the central star $L_x \approx 5 \times 10^{31}$ erg/s. which means that the nature of this object may be close to one of anomalous X-ray pulsars, and possibly X-ray pulses may be discovered in the future. Thus, cases c) and d) corresponding to higher magnetic field and larger period are favorable at this point. If the total number of isolated neutron stars in our Galaxy is $\sim 10^9$, then there should be about $\sim 10^3$ in the volume with radius $R \sim 100$ pc. Thus, there should be about ten objects with age $T \sim 10^8$ yr, and one object with age $\sim 10^7$ yr. There is also a probability 0.1 to see an object with age $\sim 10^6$ yr. Thus, the probability is high that the object RX J1856.5-3754 is a pulsar with parameters b) or c). The probability that this object is a nearby magnetar (case d)) also can not be excluded.

This work was supported in part by NASA grant NAG5-9047 and NSF grant AST-9986936. MMR thanks NSF POWRE grant for partial support. RVEL was partially supported by grant NAG5-9735. Authors thank Drs. J. Cordes, R. Duncan, M. van Kerkwijk, and I. Wasserman for helpful discussions. MMR and RVEL thank the Texas Symposium organizers for financial support and warm hospitality.

REFERENCES

1. Narayan, R. and Ostriker, J.P., *ApJ* **352**, 222 (1990).
2. Duncan, R.C., and Thompson, C., *ApJ* **392**, L9 (1992).
3. Thompson, C., and Duncan, R.C., *MNRAS* **275**, 255 (1995).
4. Kulkarni, S.R. and Frail, D.A., *Nature* **365**, 33 (1993).
5. Kouveliotou, et al., *Nature* **368**, 125 (1994).
6. Goldreich, P. and Julian, W.H., *ApJ* **157**, 869 (1969).
7. Cordes, J.M., Romani, R.W., and Lundgren, S.C., *Nature* **362**, 133 (1993).
8. Wang, Q.D., Li, Z.-Y., and Begelman, M.C., *Nature* **364**, 127 (1993).
9. Toropina, O.D., Romanova, M.M., Toropin, Yu.M., and Lovelace, R.V.E., *ApJ*, submitted (2001).
10. Coroniti, F.V., *ApJ* **349**, 538 (1990).
11. Toropin, Yu.M., Toropina, O.D., Savelyev, V.V., Romanova, M.M., Chechetkin, V.M., and Lovelace, R.V.E., *ApJ* **517**, 906 (1999).
12. Weisskopf, M.C., et al., *ApJ* **536**, L81 (2000)
13. Walter, F.M., astro-ph/0009031 (2001).
14. van Kerkwijk, M., private communication (2001).

New Millisecond Pulsars in Globular Clusters

Nichi D'Amico[1], Andrea Possenti[1], Richard N. Manchester[2], John Sarkissian[3], Andrew G. Lyne[4] and Fernando Camilo[5]

[1] *Osservatorio Astronomico di Bologna, via Ranzani 1, 40127 Bologna, Italy*

[2] *Australia Telescope National Facility, CSIRO, PO Box 76, Epping, NSW 2121, Australia*

[3] *Australia Telescope National Facility, CSIRO, Parkes Observatory, PO Box 276, Parkes, NSW 2870, Australia*

[4] *University of Manchester, Jodrell Bank Observatory, Macclesfield, SK11 9DL, UK*

[5] *Columbia Astrophysics Laboratory, Columbia University, 550 West 120th Street, New York, NY 10027*

Abstract. A new search of globular clusters for millisecond pulsars is in progress at Parkes. In this paper we describe the motivation, the new hardware and software systems adopted, the survey plan and the preliminary results. So far, we have discovered ten new millisecond pulsars in four clusters for which no associated pulsars were previously known.

INTRODUCTION

Exchange interactions in the core of globular clusters result in the formation of binary systems containing neutron stars. In these systems, the neutron star is eventually spun up through mass accretion from the evolving companion [1–3], resulting in the formation of millisecond pulsars. These objects are valuable for studies of the dynamics of clusters, the evolution of binaries, and the interstellar medium [4–7]. But searches for millisecond pulsars are difficult because they are usually rather weak and their signals are distorted by propagation through the interstellar medium, and because the apparent spin period may be affected by doppler-shift changes due to binary motion.

After several discoveries, made mainly in the early 1990s, no additional pulsars were found in globular clusters, leaving the question open why some clusters (e.g. 47 Tucanae or M15) had large numbers of detectable pulsars, whereas other apparently similar clusters have few or none.

When, a few years ago, a new multibeam 20-cm receiver was installed at Parkes, we decided to initiate a new search of globular clusters for millisecond pulsars. This receiver has a system temperature of ~ 21 K and bandwidth of ~ 300 MHz, resulting in an unprecented sensitivity. In order to further improve our search capability, we have constructed at Jodrell Bank and Bologna a new high resolution filterbank system, made of 512×0.5 MHz adjacent pass-band filters. This makes possible to remove the effects of dispersion in the interstellar medium more efficiently than previous searches, and allows the detection of millisecond pulsars with dispersion measures (DMs) of more than 200 cm^{-3}pc. The combination of this new equipment with the relatively high frequency of the multibeam receiver and its sensitivity level gives a unique opportunity to probe distant clusters. Also, because globular clusters are known to contain short-binary period millisecond pulsars, and because this class of objects is a very interesting one, we have implemented a new multi-dimensional code to search over a range of accelerations resulting from binary motion, in addition to the standard search over a range of DMs. So far, we have discovered 10 millisecond pulsars in four clusters, none of which had previously known pulsars associated with them. Five of these pulsars are members of short-period binary systems, and four of them have relatively high DM values.

OBSERVATIONS AND RESULTS

We have selected about 60 clusters, on the basis of their central density and distance. Observations consist usually of a 2.3h integration on each target. The resulting nominal (8σ) sensitivity to a typical 3 ms pulsar with DM ~ 200 cm^{-3} pc is about 0.14 mJy, several times better than previous searches. Sampling the 512 channels every 125 μs, each observation produces a huge array, 32 Gsamples, or 4 Gbytes (packing the data at 1-bit/sample), and requires significant CPU resources for offline processing. In Bologna, we have implemented the new code on a local cluster of Alpha-500MHz CPUs and on the Cray-T3E 256-processor system at the CINECA Supercomputing Center.

In the off-line processing, each data stream is split into non-overlapping segments of 2100, 4200 or 8400 sec and these are separately processed. When the DM is not known precisely from the existence in a given cluster of a previously known pulsar, data are first de-dispersed over a wide range of dispersion measures centered on the value expected for each cluster on the basis of a model of the Galactic electron layer [8], and then transformed using a Fast Fourier Transform (FFT). The analyis method exploits the fact that even relatively highly accelerated binaries might have significant spectral power in the FFT. Time-domain data are fast-folded at periods corresponding to a significant number of spectral features above a threshold to form a series of 'sub-integration arrays' and these arrays are searched for the parabolic signatures of an accelerated periodicity. Parameters for final pulse profiles having significant signal-to-noise ratio are output for visual examination.

When a pulsar is detected and confirmed in a cluster, we usually reprocess the data. The raw data are de-dispersed at the single DM value of the newly discovered pulsar; then the resulting time series is interpolated to compensate for an acceleration and transformed using a FFT, with many trials to cover a large acceleration range. Since this analysis involves many FFTs, it is relatively slow, and has benn rarely used when a DM value (or a narrow DM range) was not available.

Millisecond pulsars in NGC 6752

NGC 6752 is believed to have a collapsed core and was already known to possess a large proportion of binary systems and dim X-ray sources. In this cluster, we have first discovered a 3.26 ms pulsars in a 21 h orbital period binary system, PSR J1910−59A [9](see Fig 1). This pulsar has a relatively low DM, 34 $cm^{-3}pc$ and scintillates markedly, similar to the pulsars in 47 Tucanae, so it is seen rarely. As has been already experienced on 47 Tucanae [10], amplification due to scintillation might occasionally help in the detection of additional rather weak millisecond pulsars in the same cluster. And in fact, devoting a large amount of observing time to this cluster, we have already found four additional previously unseen millisecond pulsars (Table 1).

An eclipsing millisecond pulsar in NGC 6397

NGC 6397 is a prime candidate for globular cluster searches. It is close and has a very dense and probably collapsed core and it contains at least four X-ray sources, but there was no known pulsar associated with NGC 6397 prior to this search. In this cluster we have found PSR J1740−53, a millisecond pulsar with a spin period of 3.65ms and an orbital period of 1.35 days [9]. This pulsar is eclipsed for more than 40 % of the orbital phase. Similar eclipses are observed in other binary pulsars [11,12]. But these systems are close binary systems (with orbital periods of a few hours) and have relatively light companions (minimum mass < 0.1 $M_\odot$). In contrast, J1740-53 is in a rather wide binary system, with an orbital period of 1.35 days, and has a heavier companion (> 0.18 $M_\odot$). It seems unlikely that a wind of sufficient density could be driven off a degenerate companion, and hence produce the observed eclipses. Therefore, follow-up observations of this pulsar will be useful to probe the eclipse mechanism in millisecond pulsars.

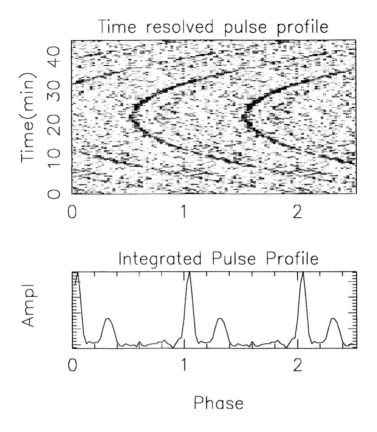

FIGURE 1. Integrated (bottom panel) and time-resolved (top panel) pulse profile for PSR J1910−58A in NGC6752. The curvature in the top panel show that the apparent period varies during the observation due to the pulsar's orbital motion

Milankersecond pulsars in NGC 6266

We have discovered three millisecond binary pulsars in NGC 6266, another relatively dense cluster. The first one, PSR J1701−30A [9], has a spin period of 5.24 ms, an orbital period, 3.8 days, and the mass function indicates a minimum companion mass of 0.19 $M_\odot$. This system is similar to several low-mass binary pulsars, associated with globular clusters or in the Galactic field. But the two other systems found, PSR J1701−30B and PSR J1701−30C, belong to the class of short-binaries. They have spin periods of 3.6 ms and 3.8 ms and orbital periods of 3.8h and 5.2h (Fig. 2).

An ultra-short binary in NGC 6544

This cluster is also one of the closest, and most concentrated globular clusters known. The pulsar discovered, PSR J1807−24 [9,13], has a spin period of 3.06 ms and it is binary, with an extremely short orbital period, 1.7 hours, the second

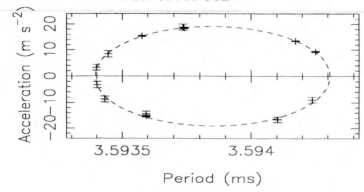

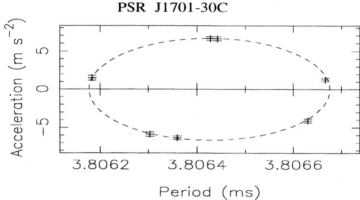

FIGURE 2. Observed accelerations plotted against barycentric period for PSR J1701−30B and PSR J1701−30C in NGC 6266. Dashed ellipses represent best fits circular orbits given in Table 1

shortest known. Even more interestingly, the projected semi-major axis of the orbit is only 12 light-ms. The corresponding minimum companion mass is only 0.0089 $M_\odot$ or about 10 Jupiter masses.

CONCLUSIONS

In Table 1 we report the preliminary parameters of the new millisecond pulsars discovered so far in four globular clusters. It is too early to draw new conclusions on the pulsar content of globular cluster, as there are many clusters in our sample that need to be searched. The present experiment raises an interesting issue from the observational point of view: besides sensitivity and powerful search algorithms,

TABLE 1. Parameters of the millisecond pulsars discovered

Cluster	Pulsar	Period (ms)	DM (cm^{-3} pc)	Orbital period (days)	Mass function (M$_\odot$)
NGC6266	J1701–30A	5.2415660(4)	114.4(3)	3.805(1)	0.0031
	J1701–30B	3.59386(2)	114.4(3)	0.159(3)	0.0009
	J1701–30C	3.80643(2)	114.4(3)	0.221(5)	0.0002
NGC6397	J1740–53	3.650328896926(9)	71.8(2)	1.35405971(2)	0.0027
NGC6544	J1807–24	3.0594487974(3)	134.0(4)	0.071092(1)	3.85×10^{-7}
NGC6752	J1910–59A	3.266186212(3)	34(1)	0.83711(1)	0.0029
	J1910–59B	8.35779(1)	34(1)	-	-
	J1910–59C	5.27732(2)	34(1)	-	-
	J1910–59D	9.03528(2)	34(1)	-	-
	J1910–59E	4.57177(2)	34(1)	-	-

a key strategy in a search for millisecond pulsars in globular clusters is to devote a large amount of observing time to each target. In fact, quite often these objects have properties such that they can be seen very rarely only: scintillation in low DM clusters, abnormally long eclipses, and unfavourable orbital phases in the case of ultra-short binaries might easily prevent the detection during a single observation. But the hidden systems are very often the very interesting ones.

REFERENCES

1. Smarr, L. L. & Blandford, R. 1976, ApJ, 207, 574.
2. Bhattacharya, D. & van den Heuvel, E. P. J. 1991, Phys. Rep., 203,1.
3. Kulkarni, S. R. & Anderson, S. B. 1996, in Dynamical Evolution of Star Clusters - Confrontation of Theory and Observations, ed. P. Hut & J. Makino, (Dordrecht: Kluwer), p. 181.
4. Phinney, E. S. 1992, Phil. Trans. Roy. Soc. A, 341, 39.
5. Hut, P. et al 1992, PASP, 104, 981.
6. Bhattacharya, D. & Verbunt, F. 1991, A&A, 242, 128.
7. Freire, P. C., Camilo, F., Lorimer, D. R., Lyne, A. G., Manchester, R. N., & D'Amico, N. 2001 MNRAS. In press.
8. Taylor, J. H. & Cordes, J. M. 1993, ApJ, 411, 674.
9. D'Amico, N., Lyne A.G., Manchester R.N., Possenti A., & Camilo, F. 2001, ApJL 548, L171.
10. Camilo, F., Lorimer, D. R., Freire, P., Lyne, A. G., & Manchester, R. N. 2000, ApJ, 535, 975.
11. Fruchter, A. S. et al. 1990, ApJ, 351, 642.
12. Stappers, B. W. et al 1996, ApJ, 465, L119.
13. Ransom, S.M., Greenhill, L.J., Herrnstein, J.R., Manchester, R.N., Camilo, F., Eikenberry, S.S. & Lyne, A.G. 2001, ApJL 546, L25.

Gamma-Ray Emission from Pulsar Magnetospheres

Kouichi Hirotani* and Shinpei Shibata[†]

National Astronomical Observatory, Mitaka, Tokyo 181-8588, Japan
hirotani@hotaka.mtk.nao.ac.jp
[†]*Department of Physics, Yamagata University, Yamagata 990-8560, Japan*
shibata@sci.kj.yamagata-u.ac.jp

Abstract. We investigate a stationary pair production cascade in the outer magnetosphere of a spinning neutron star. The charge depletion due to a global current, causes a large electric field along the magnetic field lines. Migratory electrons and/or positrons are accelerated by this field to radiate curvature gamma-rays, some of which collide with the X-rays to materialize as pairs in the gap. The replenished charges partially screen the electric field, which is self-consistently solved together with the distribution functions of particles and gamma-rays. If no current is injected across either of the boundaries of the accelerator, the gap is located around the conventional null surface, where the local Goldreich-Julian charge density vanishes. However, we first find that the gap position shifts outwards (or inwards) when particles are injected at the inner (or outer) boundary. Applying the theory to the Crab pulsar, we demonstrate that the pulsed TeV flux does not exceed the observational upper limit for moderate infrared photon density and that the gap should be located near to or outside of the conventional null surface so that the observed spectrum of pulsed GeV fluxes may be emitted via curvature process.

INTRODUCTION

The EGRET experiment on the Compton Gamma Ray Observatory has detected pulsed signals from seven rotation-powered pulsars (e.g., for Crab, Nolan et al. 1993, Fierro et al. 1998). The modulation of the γ-ray light curves at GeV energies testifies to the production of γ-ray radiation in the pulsar magnetospheres either at the polar cap (Harding, Tademaru, & Esposito 1978; Daugherty & Harding 1982, 1996), or at the vacuum gaps in the outer magnetosphere (Cheng, Ho, & Ruderman 1986a,b, hereafter CHR; Romani 1996; Zhang & Cheng 1997). Effective γ-ray production in a pulsar magnetosphere may be extended to the very high energy (VHE) region above 100 GeV as well. If the VHE emission originates the pulsar magnetosphere, a significant fraction of them can be expected to show pulsation. However, only the upper limits have been obtained for pulsed TeV radiation (e.g.,

Hillas et al. 1998; Lessard et al. 2000). Therefore, the lack of pulsed TeV emissions provides a severe constraint on the modeling of particle acceleration zones in a pulsar magnetosphere. For example, in the CHR picture, the magnetosphere should be optically thick for pair–production in order to reduce the TeV flux to an unobserved level by absorption. This in turn requires very high luminosities of infrared photons. However, the required IR fluxes are generally orders of magnitude larger than the observed values (Usov 1994). We are therefore motivated by the need to contrive an outer–gap model which produces less TeV emission with a moderate infrared luminosity.

High-energy emission from a pulsar magnetosphere, in fact, crucially depends on the acceleration electric field, $E_\parallel$, along the magnetic field lines. It was Hirotani and Shibata (1999, hereafter Papers I), and Hirotani (2000a, Paper V; 2000b, Paper VI) who first considered the spatial distribution of $E_\parallel$ together with particle and γ-ray distribution functions. By solving the Vlasov equations, they demonstrated that a stationary gap is formed around the conventional null surface at which the local Goldreich–Julian charge density,

$$\rho_{\rm GJ} = -\frac{\Omega B_z}{2\pi c}, \tag{1}$$

vanishes, where B_z is the component of the magnetic field along the rotation axis, Ω the angular frequency of the neutron star, and c the speed of light.

In the next section, we formulate the pair-production cascade. We then apply the theory to the Crab pulsar and present the expected γ-ray spectra in § 3.

BASIC EQUATIONS AND BOUNDARY CONDITIONS

Let us first consider the Poission equation for the electrostatic potential, Φ. Neglecting relativistic effects, and assuming that typical transfield thickness of the gap, $D_\perp$, is greater than or comparable with the longitudinal gap width, W, we can reduce the Poisson equation into the form (Paper VI)

$$E_\parallel = -\frac{d\Phi}{ds}, \qquad \frac{dE_\parallel}{ds} = 4\pi e \left(N_+ - N_- - \frac{\rho_{\rm GJ}}{e} \right) \tag{2}$$

where N_+ and N_- designate the positronic and electronic densities, respectively, e the magnitude of the charge on an electron, and s the length along the last-open fieldline.

We next consider the continuity equations for the particles. Assuming that both electrostatic and curvature-radiation-reaction forces cancel out each other, we obtain the following continuity equations

$$\pm B \frac{d}{ds}\left(\frac{N_\pm}{B}\right) = \frac{1}{c}\int_0^\infty d\epsilon_\gamma \, [\eta_{\rm p+} G_+ + \eta_{\rm p-} G_-], \tag{3}$$

where $G_\pm(s, \epsilon_\gamma)$ are the distribution functions of γ-ray photons having momentum $\pm m_e c \epsilon_\gamma$ along the poloidal field line. Since the electric field is assumed to be positive in the gap, e^+'s (or e^-'s) migrate outwards (or inwards). The pair production redistribution functions, $\eta_{p\pm}$, are defined as

$$\eta_{p\pm}(\epsilon_\gamma) = (1-\mu_c)\frac{c}{\omega_p}\int_{\epsilon_{th}}^{\infty} d\epsilon_x \frac{dN_x}{d\epsilon_x}\sigma_p(\epsilon_\gamma, \epsilon_x, \mu_c), \tag{4}$$

where σ_p is the pair-production cross section and $\cos^{-1}\mu_c$ refers to the collision angle between the γ-rays and the X-rays.

A combination of equations (3) gives the current conservation law,

$$\frac{\Omega}{2\pi}j_{tot} \equiv ce[N_+(s) + N_-(s)] = \text{constant for } s. \tag{5}$$

When $j_{tot} = 1.0$, the current density per unit flux tube equals the Goldreich–Julian value, $\Omega/(2\pi)$.

Assuming that the outwardly (or inwardly) propagating γ-rays dilate (or constrict) at the same rate with the magnetic field, we obtain (Paper VI)

$$\pm B\frac{\partial}{\partial s}\left(\frac{G_\pm}{B}\right) = -\frac{\eta_{p\pm}}{c}G_\pm + \frac{\eta_c}{c}N_\pm, \tag{6}$$

where η_c is explicitly defined by equations (49)–(50) in Paper VI.

We impose the following boundary conditions at the *inner* (starward) boundary $(s = s_1)$:

$$E_\parallel(s_1) = 0, \quad \Phi(s_1) = 0, \quad G_+(s,\epsilon_\gamma) = 0, \quad \text{and} \quad N_+(s) = \frac{\Omega}{2\pi ce}j_1. \tag{7}$$

The last condition yields (eq. [5]) $N_-(s_1) = (\Omega/2\pi ce)(j_{tot} - j_1)$.

At the *outer* boundary $(s = s_2)$, we impose

$$E_\parallel(s_2) = 0, \quad G_-(s_2,\epsilon_\gamma) = 0, \quad \text{and} \quad N_-(s_2) = \frac{\Omega}{2\pi ce}j_2. \tag{8}$$

The current density created in the gap is expressed as $j_{gap} = j_{tot} - j_1 - j_2$. We adopt j_{gap}, j_1, and j_2 as the free parameters.

Dividing the γ-ray energies into m bins, we obtain totally $6 + 2m$ boundary conditions for $4 + 2m$ unknowns. Thus two extra boundary conditions must be compensated by making the positions of the boundaries s_1 and s_2 be free. The two free boundaries appear because $E_\parallel = 0$ is imposed at *both* the boundaries and because j_{gap} is externally imposed. In other words, the gap boundaries shift, if j_1 and/or j_2 varies.

APPLICATION TO THE CRAB PULSAR

In this section, we apply the theory to the Crab pulsar. The rotational frequency and the magnetic moment are $\Omega = 188.1 \text{rad s}^{-1}$ and $\mu = 3.38 \times 10^{30} \text{G cm}^3$.

HEAO 1 observations revealed that the X-ray spectrum in the primary pulse phase is expressed by

$$\frac{dN_{\text{pl}}}{d\epsilon_x} = N_{\text{pl}} \epsilon_x{}^\alpha \quad (\epsilon_{\min} < \epsilon_x < \epsilon_{\max}), \tag{9}$$

with $\alpha = -1.81$ and $N_{\text{pl}} = 5.3 \times 10^{15} d^2 (r_0/\varpi_{\text{LC}})^{-2}$ (Knight 1982), where d refers to the distance in kpc. We adopt $\epsilon_{\min} = 0.1\text{keV}/511\text{keV}$ and $\epsilon_{\max} = 50\text{keV}/511\text{keV}$. Substituting this power-law spectrum into equation (4), we can solve the Vlasov equations by the method described in § 2.

To reveal the spatial distribution of the acceleration field, we consider four representative boundary conditions:
case 1 $(j_1, j_2) = (0, 0)$,
case 2 $(j_1, j_2) = (0.25, 0)$,
case 3 $(j_1, j_2) = (0.5, 0)$,
case 4 $(j_1, j_2) = (0, 0.25)$.
That is, for case 2 (or case 4), the positronic (or electronic) current density flowing into the gap per unit flux tube at the inner (or outer) boundary is 25% of the typical Goldreich-Julian value, $\Omega/2\pi$. We fix $j_{\text{gap}} = 0.01$ for all the four cases, because the solution forms a 'brim' to disappear (fig. 2 in Hirotani & Okamoto 1998) if j_{gap} exceeds a few percent. In what follows, we adopt 45° as the magnetic inclination, which is necessary to compute B at each point for the Newtonian dipole field.

The results of $E_\parallel(\xi)$ for the four cases are presented in figure 1. The abscissa designates the distance along the last-open field line and covers the range from $s = 0$ (neutron star surface) to $s = 1.2\varpi_{\text{LC}} = 1.91 \times 10^6$ m. The solid curve (case 1) shows that the gap is located around the conventional null surface. However, the gap shifts outwards as j_1 increases, as the dashed (case 2) and dash-dotted (case 3) curves indicate. On the other hand, when j_2 increases, the gap shifts inwards and the potential drop, $\Psi(s_2)$, reduces significantly. For example, we obtain $\Psi(s_2) = 7.1 \times 10^{12}$ V for case 4, whereas 1.7×10^{13} V for case 2.

Let us now consider the GeV and TeV emission from the gap. The GeV spectra is readily computed from the solved $G_\pm$. The TeV spectra is, on the other hand, obtained by inputting some infrared spectrum. For simplicity, assume that the seed, infrared photons are homogeneous and isotropic within the radius ϖ_{LC}. Interpolating the pulsed fluxes in radio, near IR, and optical bands from the Crab pulsar (Moffett and Hankins 1996; Percival et al. 1993; Eikenberry et al. 1997), we obtain (Paper V) $dN_{\text{IR}}/d\epsilon_{\text{IR}} = 1.5 \times 10^{17} d^2 \epsilon_{\text{IR}}{}^{-0.88}$, where $\epsilon_{\text{IR}} m_e c^2$ refers to the IR photon energy. Using this IR spectrum equation, we can compute the spectrum of the upscattered photons by the method described in § 4.4 in Paper V. It is noteworthy that the optical depth for the TeV photons to be absorbed by the same IR field is about 5 for the Crab pulsar (§ 4.2 in Hirotani & Shibata 2001).

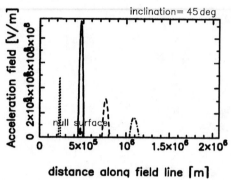

FIGURE 1. Distribution of $E_\parallel(s)$; the abscissa is in meters. The solid, dashed, dash-dotted, and dotted curves correspond to the cases 1, 2, 3, and 4, respectively (see text).

We present the γ-ray spectra for the four cases in figure 2, multiplying the cross sectional area of $D_\perp^2 = (6W)^2$. If $D_\perp$ increase twice, both the GeV and TeV fluxes increases four times. The figure shows that the TeV fluxes are kept below the observational upper limits (downarrows) as a whole for appropriate GeV fluxes. Therefore, we can conclude that the problem of the excessive TeV flux does not arise for a reasonable IR density for the Crab pulsar.

It is noteworthy that the GeV spectrum depends on j_1 and j_2 significantly. In particular, in case 4 (as the dotted lines show), the GeV emission significantly decreases and softens, because both the potential drop and the maximum of $E_\parallel$ reduce as the gap shifts inwards. As a result, it becomes impossible to explain the EGRET flux (open circles) around 10^{24} Hz, if the gap is located well inside of the null surface.

To sum up, we have developed a one-dimensional model for an outer-gap accelerator in the magnetosphere of a rotation-powered pulsar. When a magnetospheric current flows into the gap from the outer (or inner) boundary, the gap shifts inwards (or outwards). In particular, when a good fraction of the Goldreich-Julian current density is injected from the outer boundary, the gap is located well inside of the conventional null surface; the resultant GeV emission becomes very soft and weak. Applying this method to the Crab pulsar, we find that the gap should be located near to or outside of the conventional null surface, so that the observed GeV spectrum of pulsed GeV fluxes may be emitted via curvature process. By virtue of the absorption by the dense IR field in the magnetosphere, the problem of excessive TeV emission does not arise.

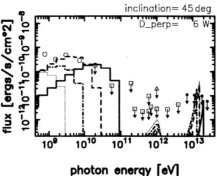

FIGURE 2. Gamma-ray spectra from the Crab pulsar magnetosphere. The thick (or thin) lines represent the flux of outwardly (or inwardly) propagating γ-rays. The solid, dashed, dash-dotted, and dotted lines correspond to the same cases as in figure 2.

REFERENCES

1. Cheng, K. S., Ho, C., Ruderman, M., 1986a ApJ, **300**, 500
2. Cheng, K. S., Ho, C., Ruderman, M., 1986b ApJ, **300**, 522
3. Daugherty, J. K., Harding, A. K. 1982, ApJ, **252**, 337
4. Daugherty, J. K., Harding, A. K. 1996, ApJ, **458**, 278
5. Eikenberry, S. S., Fazio, G. G., Ransom, S. M. et al. 1997, ApJ **477**, 465
6. Harding, A. K., Tademaru, E., Esposito, L. S. 1978, ApJ, **225**, 226
7. Fierro, J. M., Michelson, P. F., Nolan, P. L., Thompson, D. J., 1998, ApJ **494**, 734
8. Hillas, A. M., Akerlof, C. W., Biller, S. D. et al. 1998, ApJ 503, 744
9. Hirotani, K. 2000a ApJ **545**, in press (Paper V)
10. Hirotani, K. 2000b PASJ **52**, 645 (Paper VI)
11. Hirotani, K. Okamoto, I., 1998, ApJ, **497**, 563
12. Hirotani, K. Shibata, S., 1999, MNRAS **308**, 54 (Paper I)
13. Hirotani, K. Shibata, S., 2001, submitted to ApJ (astroph/0101498)
14. Knight F. K. 1982, ApJ **260**, 538
15. Lessard, R. W., Bond, I. H., Bradbury, S. M. et al. 2000, ApJ 531, 942
16. Moffett, D. A., Hankins, T. H. 1996, ApJ **468**, 779
17. Nolan, P. L., Arzoumanian, Z., Bertsch, D. L., et al. 1993, ApJ **409**, 697
18. Percival, J. W., et al. 1993, ApJ **407**, 276
19. Romani, R. W. 1996, ApJ, **470**, 469
20. Usov, V. V. 1994, ApJ **427**, 394
21. Zhang, L. Cheng, K. S. 1997, ApJ **487**, 370

Cyclotron-annihilation imprints of magnetized vacuum in MeV emission from neutron stars

A.A. Belyanin[†*], V.V. Kocharovsky[*†] and Vl.V. Kocharovsky[†]

[*]*Physics Department, Texas A&M University, College Station, TX 77843*
[†]*Institute of Applied Physics RAS, 46 Ulyanov Street, 603600 Nizhny Novgorod, Russia*

Abstract.
We analyze formation of annihilation-cyclotron spectral features due to gamma-ray propagation through magnetized vacuum around neutron stars, taking into account collective effects in virtual electron-positron plasma which go beyond the standard rate approach of QED. We show that detection of absorption lines at the first two annihilation-cyclotron resonances provides reliable and model-independent estimate of the magnetic field strength. We identify the best observation candidates for future INTEGRAL mission.

INTRODUCTION

Reliable determination of the magnetic field strength B on neutron stars is one of the challenge problems in modern astrophysics. For many neutron stars, the magnetic fields reach hundredths and tenths of the critical field $B_c = m^2c^3/e\hbar \simeq 4.4 \times 10^{13}$ G in which the cyclotron energy of the electron $\hbar\omega_B = \hbar eB/mc$ is equal to its rest energy 511 keV.

Current methods are based on the simplest models of electromagnetic spin-down and radiation of neutron stars (from radio to gamma-rays). They have limited region of applicability and allow one, at best, to estimate an order of magnitude of the magnetic field. Direct information on the magnetic field strength can be obtained only from observations of the cyclotron features in the spectrum, typically in the range of several tens keV. Even in this case an independent confirmation and detailed theoretical analysis are necessary for reliable interpretation.

In this paper we suggest to employ another observation strategy allowing direct determination of the magnetic field strength, which is based on the detection of annihilation-cyclotron features in the radiation spectrum, formed in the energy range above 1 MeV [1]. Detection of such features would be also the first direct manifestation of the effects of magnetized vacuum and the first strong-field test of quantum electrodynamics.

The best candidates for the detection of annihilation-cyclotron features are isolated pulsars. Several of the known pulsars have high enough flux at 1 MeV to be detected by INTEGRAL [2]. In addition, the INTEGRAL is expected to discover new gamma-ray sources from Galactic population of isolated neutron stars. Such sources are also good candidates for annihilation-cyclotron line detection.

We provide below an outline of the theoretical analysis of annihilation-cyclotron line formation. Detailed treatment is presented elsewhere [1]. We analyze here the simplest case of emission from the surface of a neutron star in vacuum, having dipole magnetic field with surface strength B_0. Then the problem is split into three parts:

(a) analysis of dispersion and absorption of the normal waves close to annihilation-cyclotron resonances in magnetized vacuum [3-6];

(b) calculation of the optical depth for absorption and inhomogeneous linewidths in a dipole magnetic field;

(c) analysis of the dependence of spectral features from the magnetic field strength.

ANNIHILATION-CYCLOTRON FEATURES

In the inhomogeneous magnetic field of a neutron star all annihilation-cyclotron resonances become broadened and their optical depth is considerably reduced as compared with the main annihilation resonance which frequency does not depend on the magnetic field. This can be illustrated by calculating the optical depths in the dipole magnetic field of a neutron star, performed for definiteness in the equatorial plane ($\sin\theta = 1$) and for given value of the surface magnetic field $b_0 = B_0/B_c \ll 1$ (more accurately, this is the field at the gamma-ray emitting surface at a distance $r_0 \sim 10-15$ km from the star center).

Maximum optical depth for the photons at the main annihilation resonance turns out to be equal to

$$\max \tau_0 \simeq \frac{3^{1/2}\alpha^{2/3}b_0^{5/3}r_0\sin\theta}{2^{1/3}16\lambda_c}\exp\left(-\frac{4}{3b_0}\right), \quad (1)$$

where $\alpha = e^2/\hbar c$, $\lambda_c = \hbar/mc$. Equating the right-hand side of (1) to unity, we find the minimum magnetic field in which the X-polarized absorption feature at frequency $\omega_0 = 2mc^2/\hbar\sin\theta$ can be observed: $B_x \simeq 2.2 \times 10^{12}$ G.

Maximum optical depth for the X-mode (having the electric field vector in the plane formed by the magnetic field and photon wave vector) at the next (first) annihilation-cyclotron resonance $\omega_1 = mc^2[1+(1+2B/B_c)^{1/2}]/\hbar\sin\theta$ is equal to

$$\max \tau_1 \simeq \frac{4\alpha r_0\sin\theta}{3\lambda_c}\exp\left(-2-\frac{2}{b_0}\right). \quad (2)$$

It is reached at $\kappa^2\sin^2\theta - 4 - 4b_0 \simeq b_0^2$, where $\kappa = \hbar k/mc$ is dimensionless photon wavenumber, and corresponds to the following minimum magnetic field which provides and absorption feature at ω_1: $B_{1x} \simeq 2.7 \times 10^{12}$ G. For the O-mode (with electric field orthogonal to the above plane) at the first resonance the optical depth is lower by only a factor of $b_0/2$, so the minimum field is practically unchanged.

Similar analysis can be extended to higher resonances. However, it can be shown that resonances higher than one are practically unresolvable.

DETERMINATION OF THE MAGNETIC FIELD STRENGTH

The formation of absorption lines at the main ($\omega = \omega_0$) and first ($\omega = \omega_1$) annihilation-cyclotron resonances and of the transparency window between them is possible only for a certain range of magnetic fields. When the field is too low, absorption features are absent. For very high fields, the optical depth between the resonances becomes greater than unity. From the latter restriction we can easily find the width of absorption line and the conditions for its observability:

$$b_0 < b_m = 2/[\log(\alpha b_m^{3/2} r_0 \sin\theta/(12 e^2 \lambda_c))] \sim 0.08,$$

i.e., $B_0 < B_m \simeq 3.6 \times 10^{12}$ G. When the above condition is satisfied, the absorption between the main and the first resonances is still weak. In larger fields the absorption in continuum washes out the transparency window between the main and the first resonances. In the latter case we will observe a high-energy cutoff at frequency ω_0 which allows us to estimate the value of $B_0/\sin\theta$.

Our calculations indicate that the most robust, reliable, and essential result of using the method of annihilation-cyclotron resonances for determination of the magnetic fields on neutron stars would be the detection of narrow (< 50 keV) one-photon absorption line at frequency $\omega_0 = 1.02/\sin\theta$ MeV, separated from absorption continuum in upper resonances by an emission line of width $\lesssim 40$ keV. In this case the magnetic field on the gamma-ray emitting surface of a neutron star could be restricted within $(2.7 - 3.6) \times 10^{12}$ G. For larger fields the optical depth in the whole continuum above the main annihilation resonance becomes greater than unity, and we will observe a single cutoff at frequencies above ω_0. Note that the θ-dependence of the center frequency of the line can lead to the dynamic drift of the line during the pulsar period, which can be observed in time-resolved measurements. We emphasize that the above estimates are only weakly (logarithmically) sensitive to the model of the magnetic field or its inhomogeneity scale, and for any resonable non-dipole model can vary only in the second digit.

The best candidates for detection of annihilation-cyclotron features are isolated neutron stars with high enough radiation flux between 1-3 MeV. Roughly, to be detected with confidence by INTEGRAL the flux in the lines should exceed 10^{-5} ph/cm^{2}s. Several known pulsars observed by COMPTEL satisfy this condition. These candidates include well- known Crab pulsar, PSR B1509-58, PSR B1951+32, PSR B0833-45. The detection of previously unknown isolated neutron stars with significant gamma-ray flux is also possible.

REFERENCES

1. Belyanin, A.A., Kocharovsky, V.V., and Kocharovsky, Vl.V., *Radiophys. Quant. Electron.* **44**, No. 1-2 (2001).
2. Schonfelder, V. et al. (The first COMPTEL source catalogue), astro-ph/0002366 (2000).
3. Shabad, A.E., *Ann. Phys.* **90**, 166 (1975); Peres Rojas, H., and Shabad, A.E., *Ann. Phys.* **138**, 1 (1982).
4. Melrose, D.B., and Stoneham, R.J., *Nuovo Cim.* **A32**, 435 (1976).
5. Belyanin, A.A., Kocharovsky, V.V., and Kocharovsky, Vl.V., *Phys. Lett.* **A149**, 258 (1990); *JETP* **72**, 70 (1991).
6. Belyanin, A.A., Kocharovsky, V.V., and Kocharovsky, Vl.V., *Quant. Semiclass. Opt.* **9**, 1 (1997).

Large Redshifts from Compact Objects

Sujit Chatterjee

Department of Physics, New Alipore College, Calcutta 700 053, India

Abstract. Einstein-Maxwell equations are solved for the case of a prolate spheroid of charged dust in equilibrium under its own gravitational attraction and electrical repulsion. Interesting to point out that this solution and another by Gautreau, Hoffman and Armenti Jr. representing an axially symmetric cluster of charged bodies in equilibrium may exhibit very large gravitational redshifts under certain circumstances.

INTRODUCTION

The large redshifts observed in the spectra of Quasistellar objects have been puzzling astronomers for a long time [1]. If the redshifts were assumed to be cosmological it requires colossal amounts of energy to be released from QSO's. However, subsequent observations favored non-cosmological redshifts [2]. In a pioneering work, Bondi [3] showed that for a realistic equation of state the gravitational redshift cannot exceed 0.62 for a spherical distribution of matter. Bonnor and Wickramasuriya [4] found solutions for Einstein-Maxwell equations for a charged oblate spheroid which may give arbitrary large gravitational redshifts in specific situations, where the gravitational attraction of the charged fluid is balanced by the electrical repulsion.

In the present work we have obtained an exact solution for a charged fluid in an axially symmetric distribution and showed that our solution along with another given by Gautreau et al. [5] can be made to yield very large gravitational redshifts even though the mass density is not necessary infinite.

PROLATE SPHEROID

We take the axially symmetric line-element as

$$ds^2 = e^{2\lambda}dt^2 - e^{-2\lambda}(dr^2 + dz^2 - r^2d\phi^2) \tag{1}$$

where $\lambda = \lambda(r, z)$. With the help of transformation equations

$$r = a\sinh u \cos\theta, \qquad z = a\cosh u \sin\theta \tag{2}$$

we convert the metric (1) to the prolate-spheroidal form as

$$ds^2 = e^{2\lambda}dt^2 - e^{-2\lambda}\left[a^2(\sinh^2 u + \cos^2 \theta)(ds^2 + d\theta^2) + a^2 \sinh^2 u \cos^2 \theta\, d\phi^2\right] \quad (3)$$

where $0 \leq \phi \leq 2\pi$ and $-\pi/2 \leq \theta \leq \pi/2$. Earlier Zipoy [6] gave a set of solutions in empty space in spheroidal coordinates. With the help of Bonnor's theorem [7] we can generate electrovac solutions outside a prolate spheroid $u = u_0$. Skipping all mathematical details the final electrovac solution in prolate symmetry is given by

$$\exp(-\lambda_E) = 1 + \frac{m}{a}\tanh^{-1}(\operatorname{sech} u), \quad u_0 \leq u < \infty. \quad (4)$$

With the help of the boundary condition at $u = u_0$,

$$\lambda_E = \lambda_I; \quad (\lambda_E)_{,k} = (\lambda_I)_{,k} \quad (5)$$

(here the suffix E and I mean external and internal) we obtain from the Einstein-Maxwell equations the interior metric as

$$\exp(-\lambda_I) = 1 + \frac{m}{a}\left[\tanh^{-1}(\operatorname{sech} u_0) + \frac{u_0^4 - u^4}{4u_0^3}\operatorname{cosech} u_0\right]. \quad (6)$$

If a photon is emitted from a point P inside the charged dust and is received by an observer at a large distance where the metric may be assumed to be Euclidean then the redshift is $e^{-\lambda_P} - 1$. Thus the redshift for light coming from $u = 0$ (i.e. from z-axis) is given by

$$\xi = [\exp(-\lambda_I)]_{u=0} - 1 = \frac{m}{a}\left[\tanh^{-1}(\operatorname{sech} u_0) + \frac{1}{4}u_0 \operatorname{cosech} u_0\right]. \quad (7)$$

Following Bonnor and Wickramasuriya we see that for both the two limiting cases (keeping the mass m constant) $u_0 \to 0$ and $a \to 0$ the redshift ξ becomes infinite.

MULTIPARTICLE SYSTEM

Gautreau, Hoffman and Armenti took an axially symmetric metric in Weyl canonical form as

$$ds^2 = e^{2\lambda}dt^2 - e^{2(\nu-\lambda)}(dr^2 + dz^2) - e^{-2\lambda}r^2 d\phi^2 \quad (8)$$

where $\lambda = \lambda(r,z)$, $\nu = \nu(r,z)$. They considered a multiparticle system in which the gravitational attraction between bodies is balanced by the Coulomb repulsion and found solutions as

$$\bar{\lambda} = \sum_{i=1}^{n}\bar{\lambda}_i(r,z); \quad \bar{\lambda}_i = \frac{1}{2}\ln\frac{L_i - A_i}{L_i + A_i} \quad (9)$$

where

$$L_i = \frac{1}{2}(\ell_{i+} + \ell_{i-}), \tag{10}$$

$$\ell_{i+}^2 = r^2 + (z - b_i + A_i)^2, \qquad \ell_{i-}^2 = r^2 + (z - b_i - A_i)^2 \tag{11}$$

where b_i is the position of the center of the ith particle and

$$A_i = (a_i^2 - q_i^2)^{1/2}; \qquad a_i = \text{mass}, \quad q_i = \text{charge}. \tag{12}$$

λ and $\bar{\lambda}$ are related by the equation

$$e^{-\lambda} = \frac{1}{2}\left[1 + c(c^2-1)^{-1/2}\right]\exp(-\bar{\lambda}) + \frac{1}{2}\left[1 - c(c^2-1)^{1/2}\right]\exp(\bar{\lambda}) \tag{13}$$

where c is an arbitrary constant. For a point on the axis, $r = 0$ we find

$$e^{-2\bar{\lambda}} = \frac{(z - b_1 + A_1)(z - b_2 + A_2) \cdots (z - b_n + A_n)}{(z - b_1 - A_1)(z - b_2 - A_2) \cdots (z - b_n - A_n)}. \tag{14}$$

For positive values of z each factor in the numerator is greater that the corresponding factor in the denominator. Hence by taking a large number of bodies $e^{-2\lambda}$ and hence the redshift becomes large.

To conclude it may be stated that here we have given examples of two axially symmetric equilibrium configurations which can exhibit arbitrarily large redshifts but the density is not always infinite.

ACKNOWLEDGMENT

The author wishes to thank the University Grants Commission, Delhi for financial support.

REFERENCES

1. Burbudge, G., and Burbidge, M., *Quasistellar Objects* (W. B. Freeman) (1967)
2. Sanders, R. E., *Nature*, **248**, 390 (1974)
3. Bondi, H., *Proc. Roy. Soc..*, **282A**, 303 (1964)
4. Bonnor, W., and Wickramasuriya, S., *Int. J. Theo. Phys.*, **5**, 371 (1972)
5. Gautreau, R., Hoffman, R., and Armenti Jr., A. *Nuov. Cim.* **78**, 71 (1972)
6. Zipoy, D., *J. Math. Phys.*, **7**, 1137 (1966)
7. Bonnor, W. *Z. Phys.*, **161**, 439 (1961)

Bardeen-Petterson Effect and QPOs in Low-Mass X-Ray Binaries

P. Chris Fragile*, Grant J. Mathews*, and James R. Wilson[†]*

Center for Astrophysics, University of Notre Dame, Notre Dame, IN 46556
[†]*Lawrence Livermore National Laboratory, Livermore, CA 94550*

Abstract. The Bardeen-Petterson effect around a rapidly-rotating compact object causes a tilted accretion disk to warp into the equatorial plane of the rotating body at a characteristic radius. We propose that accreting material passing through the transition region may generate quasi-periodic brightness oscillations (QPOs) such as have been observed in a number of low-mass X-ray binaries (LMXBs). We argue that the QPO frequency range predicted by this model is consistent with observed QPO frequencies in both black-hole and neutron-star LMXBs.

I INTRODUCTION

The Bardeen-Petterson effect [1] is the combined result of differential Lense-Thirring precession and internal viscosity in a tilted accretion disk around a rapidly-rotating black hole (BH) or neutron star (NS). (By "tilted" we mean that the angular momentum of the disk is not parallel to the spin axis of the BH or NS.) The Lense-Thirring precession causes the disk to "twist up." Damping of the fluid motion by viscosity limits this twisting. Close to the accreting object, where the Lense-Thirring precession is strongest, viscosity allows the misaligned angular momentum of the tilted disk to be transported outward. This allows the inner region to settle into the rotation plane of the BH or NS. Further out, the disk remains in its original plane because the Lense-Thirring precession rate drops off rapidly with increasing radial distance. The end result is an aligned inner accretion disk, a tilted outer accretion disk, and a transition region between the two.

The transition radius is expected to occur approximately where the rate of twisting up by differential precession is balanced by the rate at which warps of the disk are diffused or propagated away by viscosity. We can estimate the transition radius by equating the Lense-Thirring precession frequency $\Omega_{\rm LT}$ with the inverse of the infall time $t_{infall} = R/|\overline{V_R}|$ for the accreting material, i.e. $\Omega_{\rm LT} = t_{infall}^{-1}$. For a standard Keplerian disk, $\overline{V_R} = -3\nu/2R$ where ν is the kinematic viscosity. After rewriting the viscosity in terms of the standard Shakura-Sunyaev α parameter and the scale-height δ of the disk, we get the following estimate for the Bardeen-Petterson transition radius:

$$R_{\rm BP} = \left(\frac{4a_*}{3\alpha\delta^2}\right)^{2/3} R_{\rm GR} . \qquad (1)$$

Here a_* is the dimensionless specific angular momentum ($a_* = Jc/GM^2$ where J is the angular momentum) and $R_{\rm GR} = GM/c^2$, where M is the gravitational mass of the BH or NS. The corresponding Keplerian orbital frequency is therefore:

$$f_{Kep,{\rm BP}} = \frac{3c^3\alpha\delta^2}{8\pi GM a_*} . \qquad (2)$$

For the geometrically thin, cool disks being considered here, $\alpha \sim 0.1$ and $\delta \ll 1$.

II   X-RAY MODULATION MECHANISM

In [2] we argue that a large fraction ($\sim 1/2$) of all LMXBs may be candidates for observation of the Bardeen-Petterson effect. One manner in which this effect may generate observable signals is through quasi-periodic brightness oscillations (QPOs) in the X-ray spectra of such systems. This could occur, for instance, if the transition between the inner and outer disks is relatively abrupt. Then accretion may occur in the form of discrete gas clumps breaking off from the outer disk and colliding with the inner disk. In this case, there is likely to be significant shock heating at the point where this accreting material impacts the inner disk. As the shock-heated gas continues to orbit the BH or NS, it may generate periodic brightness oscillations at the orbital frequency appropriate for that radius. We expect this shock-heated region to lie just inside the Bardeen-Petterson transition radius, so that the QPO frequency should be very close to the Keplerian orbital frequency given in equation 2.

III   NEUTRON STAR X-RAY BINARIES

For neutron-star LMXBs, the parameter space that is relevant to this model is much narrower than for black-hole binaries. Both the mass and the angular momentum of NSs are tightly constrained by theory and observation. Models for the NS equation of state give the following upper limits [3,4]: $M_{NS} \lesssim 2.6 M_\odot$ and $a_{*,NS} \lesssim 0.7$. Observed radio pulsars in binary systems are all consistent with $1.35 M_\odot \leq M_{NS} \leq 1.45 M_\odot$ [5]. We can use equation 2, along with the theoretical upper limits for M_{NS} and $a_{*,NS}$, to set a lower limit for the Bardeen-Petterson orbital frequency in NSs: $f_{Kep,{\rm BP}} \gtrsim 1$ Hz. The corresponding upper limit is fixed by the orbital frequency at the inner edge of the accretion disk. For a putative 1.4 $M_\odot$ NS with $a_* = 0.1$, the orbital frequency at the marginally stable orbit is ≈ 1700 Hz. This range of frequencies (1 − 1700 Hz) is consistent with most of the observed QPOs in NS LMXBs. In high-luminosity Z-sources, it is consistent with the normal/flaring and horizontal branch oscillations and kHz QPOs. In less-luminous atoll sources, it is consistent with the low frequency QPOs, the peaked noise components, and the kHz QPOs.

However, in order for the Bardeen-Petterson effect to be a consistent interpretation, we should expect all the QPOs associated with this effect to occur at similar frequencies in all NS LMXBs since they are all characteristically very similar. We have already argued that the NS masses are similar ($M_{NS} \approx 1.4 M_\odot$). Furthermore, interpretation of the nearly-coherent oscillations seen during Type I X-ray bursts in ten NS LMXBs argue for a narrow range of angular momenta for the NSs in these systems [6]. The two remaining parameters, the viscosity and the disk aspect ratio, are presumably similar in all LMXBs since the accretion mechanism (Roche-lobe overflow) is assumed to be the same.

Thus, we feel that the Bardeen-Petterson effect is likely to be important in a large fraction of LMXBs and may be responsible for one class of QPOs in these systems. The interested reader is referred to [2] for a more thorough review of the model outlined here and an application of this model to the black-hole candidate LMXB GRO J1655-40.

IV ACKNOWLEDGMENTS

P. C. F. would like to thank the Arthur J. Schmidt Foundation for fellowship support at the University of Notre Dame. This work was supported in part by the National Science Foundation under grant PHY-97-22086 and the Department of Energy under Nuclear Theory grant DE-FG02-95-ER40934.

REFERENCES

1. Bardeen, J. M. and Petterson, J. A., *Astrophys. J.* **195**, L65–L67 (1975).
2. Fragile, P. C., Mathews, G. J., and Wilson, J. R., *Astrophys. J.* (2001), in press.
3. Salgado, M., Bonazzola, S., Gourgoulhon, E., and Haensel, P., *Astron. Astrophys.* **291**, 155–170 (1994).
4. Cook, G. B., Shapiro, S. L., and Teukolsky, S. A., *Astrophys. J.* **424**, 823–845 (1994).
5. Thorsett, S. E. and Chakrabarty, D., *Astrophys. J.* **512**, 288–299 (1999).
6. Strohmayer, T. E., *Adv. Space Res.* (2001), in press.

Galactic Population of Radio and Gamma-Ray Pulsars

Peter L. Gonthier*, Michelle S. Ouellette[†], Shawn O'Brien[‡], Joel Berrier* and Alice K. Harding[||]

Hope College, Department of Physics, 27 Graves Place, Holland, MI 49422-9000
[†]*Michigan State University, Physics and Astronomy Department, East Lansing, MI 48824-1116*
[‡]*University of Notre Dame, Department of Physics, Notre Dame, IN 46556*
[||]*NASA - Goddard Space Flight Center, LHEA, Greenbelt, MD 20771*

Abstract. We simulate the characteristics of the Galactic population of radio and gamma-ray pulsars using Monte Carlo techniques. At birth, neutron stars are spatially distributed with supernova-kick velocities in the Galactic disk and randomly dispersed in age back to 10^9 years. They are evolved in the Galactic gravitational potential to the present time. From a radio luminosity model, the radio flux is filtered through a selected set of radio-survey parameters. Using the features of recent polar cap acceleration models invoking space-charge-limited flow, a pulsar death region further attenuates the population of radio-loud pulsars, and gamma-ray luminosities are assigned. Assuming a featureless emission geometry of 1 steradian corresponding to an alignment of the radio and gamma-ray beams, our model predicts that EGRET should have seen 17 radio loud and 5 radio quiet, gamma-ray pulsars. GLAST, on the other hand, is expected to observe 132 radio loud and 212 radio quiet, gamma-ray pulsars of which 21 are expected to be identified as pulsed sources.

With the advent of the *Compton Gamma Ray Observatory* (CGRO), the number of γ-ray pulsars has grown to eight, with several additional candidates. We anticipate that many more pulsed sources will be added to the list with the future telescope, *Gamma-Ray Large Area Space Telescope* (GLAST) scheduled for launch in late 2005. Among the known γ-ray pulsars, only Geminga is radio-quiet or at least radio weak [1,2] remaining an enigma. Of the 271 sources listed in the Third EGRET Catalog [3], about 170 of these γ-ray point sources have not been identified with sources at other wavelengths. Recently Grenier & Perrot [4] and Gehrels et al. [5] suggested that some of these unidentified sources are correlated with the Gould Belt of massive stars from a nearby Galactic structure consisting of an expanding disk of gas with young stars ($\leq$ 30 million years) inclined about $20°$ to the Galactic plane. Harding & Zhang [6] suggest that some of the sources associated with the Gould Belt are indeed radio-quiet, off-beam γ-ray pulsars seen at large angles to the magnetic pole.

We develop a model to simulate the production of neutron stars within the Galaxy, evolving their trajectories, periods and period derivatives from their birth forward in time to the present. Assuming that the radio and γ-ray beams are aigned, we supply the radio and γ-ray characteristics to each neutron star and filter its properties through radio surveys and γ-ray thresholds (in and out of plane) associated with EGRET and expected for GLAST. These γ-ray thresholds correspond to the number of photons required for the instrument to identify the object as a point source. Higher thresholds would be required to obtain sufficient photons to identify the object independently as a pulsed source. In our simulations, the magnetic field of the pulsar is assumed to be constant throughout its lifetime. The birth rate of neutron stars is assumed to be constant during the history of the Galaxy with the simulation suggesting a birth rate of 0.7 neutron stars per century. The age of the pulsar is randomly selected from the present to 10^9 years in the past. Using the recent study of Zhang, Harding & Muslimov [7], we have introduced a pulsar death line predicted by a multipole magnetic field configuration near the stellar surface within a space-charge-limited flow model. We have taken the expressions describing the γ-ray luminosity from the work of Zhang & Harding [8] where a polar-cap model simulates the pair cascade region with curvature radiation or inverse Compton scattering of the primary particles and synchrotron radiation and inverse Compton scattering of subsequent higher generation pairs. In addition, the model uses the self-consistent acceleration model of Harding & Muslimov [9] to produce the primary particles.

In Figure 1, we compare in a $(\dot{P}P)$ plot of the select group of observed pulsars (1a) with the one of simulated pulsars (1b). The dotted lines are shown for the locus of constant magnetic field with the indicated strength. The dashed lines represent the indicated ages of pulsars assuming a dipole spin-down field. The solid lines show the pulsar death lines for dipole and multipole magnetic field distributions in the space-charge-limited-flow model (SCLF) [7]. The radio pulsars observed (1a) and those simulated and filtered through the select group of surveys (1b) are indicated with solid dots. The clear absence of high-field, high-period observed pulsars in Figure 1a as compared to those simulated in Figure 1b, might be suggestive of the decay of the magnetic field. In fact, the whole pear-shaped distribution of observed pulsars in Figure 1a could be explained by field decay. The excess of simulated high-field, high-period pulsars have ages of the order of 10^7 years. Therefore, a decay constant of this order is required for these high-field pulsars to have their fields decay by an order of magnitude. We anticipate exploring in a future study the effects of the decay of the magnetic field.

The model predicts that GLAST should observe 132 radio-loud, γ-ray pulsars compared to 17 predicted for EGRET detected as point sources. The model predicts that GLAST should observe 212 radio-quiet, γ-ray pulsars compared to 5 predicted for EGRET. The GLAST sensitivity for blind period searches is expected to be about the same as the EGRET point-source detection sensitivity (S. Ritz, priv. comm.). GLAST will therefore be expected to identify 21 of these 212 objects as pulsed sources. The results predicted for GLAST are interesting in that many

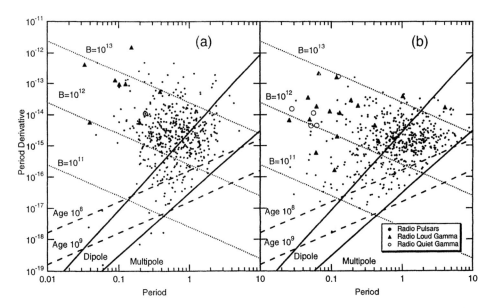

FIGURE 1. Distributions of observed pulsars (a) and simulated pulsars (b) as a function of the period derivative and period of the pulsars. Solid dots indicate radio pulsars, solid triangles represent radio-loud, γ-ray pulsars and open circles symbolize radio-quiet, γ-ray pulsars observed (a) and predicted (b) for EGRET.

more radio-quiet, γ-ray pulsars are expected to be observed than radio-loud, γ-ray pulsars. Given the flux thresholds we have used, these are objects observed as point sources. Thus the simulations suggest that due to the increased sensitivity, GLAST will be able to detect more pulsars that are further away than even those detected by the radio surveys used in this study.

REFERENCES

1. Kuzmin, A.D. & Losovsky, B. Ya., *IAU Circular*, 6559 (1997).
2. Malofeev, V.M. & Malov, O.I., *Nature*, **389**, 697 (1997).
3. Hartman, R. et al., *ApJ Supp.*, **123**, 79 (1999).
4. Grenier, I.A. & Perrot, C., *Proc. XXVI Int. Cosmic Ray Conf. Salt Lake City*, **3**, 476 (1999).
5. Gehrels, N. et al., *Nature*, **404**, 363 (2000).
6. Harding, A.K., & Zhang, B., *ApJ Letters*, in press (2001).
7. Zhang, B., Harding, A. K., & Muslimov, A. G., *ApJ*, **531**, L135 (2000).
8. Zhang, B., & Harding, A. K., *ApJ*, **532**, 1150 (2000).
9. Harding, A. K., & Muslimov, A. G., *ApJ*, **508**, 328 (1998).

Exact Solutions with w-modes: trapping of gravitational waves inside neutron stars

Mustapha Ishak[1], Luke Chamandy and Kayll Lake

Queen's University, Kingston, Ontario, Canada

Abstract. An explicit necessary condition for the occurrence of resonance scattering of axial gravitational waves, along with the internal trapping of null geodesics, is proposed for static spherically symmetric perfect fluid solutions to Einstein's equations. Some exact inhomogeneous solutions which exhibit this trapping are given with special attention to boundary conditions and the physical acceptability of the space times. In terms of the tenuity ($\alpha = R/M$ at the boundary) all the examples given lie in the narrow range $2.8 < \alpha < 2.9$. The tenuity can be raised to more interesting values (e.g. to neutron stars) by the addition of an envelope without altering the trapping by using the Darmois-Israel junction conditions. We are able to exhibit exact solutions which support w-modes and which can be used for numerical studies of gravitational waves incident on neutron stars.

INTRODUCTION

The line element in conventional form for static spherically symmetric space time is

$$ds^2 = \frac{dr^2}{1 - \frac{2m(r)}{r}} + r^2(d\theta^2 + sin(\theta)^2 d\phi^2) - e^{2\Phi(r)} dt^2 \qquad (1)$$

with the coordinates comoving in the sense that the fluid streamlines are given by $u^a = e^{-\Phi(r)} \delta^a_t$. In terms of the functions $\Phi(r)$ and $m(r)$ the central regularity conditions reduce to

$$\Phi'(0) = m(0) = m'(0) = 0, \qquad (2)$$

with $' \equiv d/dr$ and $\Phi(0)$ a constant fixed by the scale of t. Next, in terms of the perfect fluid decomposition ($T^a_b = (\rho(r) + p(r))u^a u_b + p(r)\delta^a_b$), solving for $\Phi'(r)$ from the r-component of the conservation equations and Einstein's equations ($\nabla_a T^a_r = 0$ and $G^r_r - 8\pi p(r) = 0$) we obtain the Tolman-Oppenheimer-Volkoff (TOV) equation

---

[1] ishak@hera.phy.queensu.ca

$$\Phi'(r) = \frac{-p'(r)}{\rho(r) + p(r)} = \frac{m(r) + 4\pi p(r)r^3}{r(r - 2m(r))}, \quad (3)$$

where, from the t component of the Einstein equations $(G_t^t = -8\pi\rho(r))$,

$$4\pi\rho(r) = \frac{m'(r)}{r^2}. \quad (4)$$

THE EXPLICIT CONDITION

We showed that a necessary and sufficient condition for the internal trapping of null geodesics (that is the existence of r_0 such that $r^\bullet = 0$ and $r^{\bullet\bullet} < 0$ at r_0) is given by [1]

$$\Phi'(r) > \frac{1}{r} \quad (5)$$

Using the TOV equation it follows that (5) can be given as

$$r < 3m(r) + 4\pi p(r)r^3 \quad (6)$$

a relation which makes the trapping of null geodesics a manifestly relativistic phenomenon.

THE FULL POTENTIAL

The w-modes are non-radial odd parity (axial) perturbations which do not couple to fluid motions in the star. In terms of the frequency ϖ and mode number $l \geq 2$ the governing equation is given by

$$(\frac{d^2}{dr_*^2} + \varpi^2)Z = V(r_*)Z \quad (7)$$

The potential is conveniently expressed in terms of r and is given by

$$V(r) = \frac{1}{B(r)^2}(l(l+1) + 4\pi r^2(\rho(r) - p(r)) - \frac{6m(r)}{r}). \quad (8)$$

A necessary condition for the occurrence of resonance scattering of axial gravitational waves by an isolated distribution of fluid is a local minimum in $V(r)$ within the boundary of the fluid [2]. It is the purpose of this paper to explore the occurences of this minimum in physically acceptable exact solutions. It is the shape of the function $V(r)$ which is of interest, and since the exterior vacuum has a well known local maximum at $r \sim 3.28M$ (for $l = 2$), the boundary conditions associated with the fluid - vacuum interface need careful attention.

EXACT SOLUTIONS WITH W-MODES

The solutions that passed the physical acceptability conditions and which exhibit trapping were found to be:

a) The Finch-Skea [3,6] exact solution

$$ds^2 = v^2 dr^2 + r^2 d\Omega^2 - A^2((C_2 - C_1 v)\cos(v) + (C_1 + C_2 v)\sin(v))^2 dt^2, \quad (9)$$

where $v \equiv \sqrt{1+\omega^2}$, $\omega^2 \equiv Cr^2$ and A, C_1 and C_2 are constants.

b) The Heintzmann [4,6] solution

$$ds^2 = \frac{dr^2}{\left(1 - \frac{3ar^2}{2}\frac{1+C(1+4ar^2)^{-1/2}}{1+ar^2}\right)} + r^2 d\Omega^2 - A^2(1+ar^2)^3 dt^2. \quad (10)$$

c) The Durgapal solution For $n = 4$ (Durgapal IV) [5,6]

$$ds^2 = \frac{(1+Cr^2)^2 dr^2}{\left(\frac{7-10Cr^2-C^2r^4}{7} + \frac{KCr^2}{(1+5Cr^2)^{\frac{2}{5}}}\right)} + r^2 d\Omega^2 - A(1+Cr^2)^4 dt^2. \quad (11)$$

d) The Durgapal solution For $n = 5$ (Durgapal V) [5,6]

$$ds^2 = \frac{(1+Cr^2)^3 dr^2}{\left(1 - \frac{Cr^2(309+54Cr^2+8C^2r^4)}{112} + \frac{KCr^2}{\sqrt[3]{1+6Cr^2}}\right)} + r^2 d\Omega^2 - A(1+Cr^2)^5 dt^2. \quad (12)$$

REFERENCES

1. Ishak M., Chamandy L., and Lake K. gr-qc/0007073.
2. S. Chandrasekhar and V. Ferrari, Proc. R. Soc. Lond. A **434**, 449 (1991).
3. M. Finch and J. Skea Class. Quantum Grav **6**, 467 (1989)
4. H. Heintzmann,Z. Phys. **228** 489 (1969).
5. M. Durgapal, J. Phys. A **15** 2637 (1982).
6. M. Delgaty and K. Lake, Comput.Phys.Commun. **115**, 395 (1998) (gr-qc/9809013).

Simulations of Glitches in Pulsars

Michelle B. Larson and Bennett Link

Dept. of Physics, Montana State University, Bozeman, MT. 59717

Abstract.
Pulsar glitches are thought to represent variable coupling between the neutron star and its superfluid interior. With the aim of distinguishing among different theoretical explanations for the glitch phenomenon, we study the response of a neutron star to two types of perturbations to the vortex array that exists in the superfluid interior: 1) thermal excitation of vortices pinned to inner crust nuclei by sudden heating of the crust, (e.g. a starquake), and 2) mechanical motion of vortices, (e.g. from crust cracking by superfluid stresses). The thermal glitch model adequately fits the timing data of the 1989 glitch in the Crab pulsar (Fig. 1), while both models fit the Vela "Christmas" glitch of 1988 (Fig. 2). The two models make different predictions for the generation of internal heat and subsequent enhancement of surface emission. The mechanical glitch model predicts a negligible temperature increase. For a pure and highly-conductive crust, the thermal glitch model predicts a surface temperature increase of $\sim 0.2\%$ in the Crab (Fig. 3) and $\sim 1.5\%$ in Vela (Fig. 4), occurring several weeks after the glitch. If the thermal conductivity of the crust is lowered by a high concentration of impurities, however, the surface temperature increase is $\sim 10\%$ for a thermal glitch, and occurs about a decade after the glitch. A thermal glitch in an impure crust is consistent with the surface emission measurements following the January 2000 glitch in the Vela pulsar (Fig. 4). Future surface emission measurements coordinated with radio observations will constrain glitch mechanisms and the conductivity of the crust. See Larson & Link (2001) for a detailed description of this work.

REFERENCES:
Dodson R.G., McCulloch P.M., Costa M.E., IAUC 7347, (2000)
Helfand D.J., Gotthelf E.V., Halpern J.P., *astro-ph* 0007310, (2000).
Larson M.B., and Link B., *MNRAS*, to be submitted (2001).
Lyne A.G., Smith F.G., Pritchard R.S., *Nature*, **359**, 706 (1992).
McCulloch P.M. et al., *Nature*, **346**, 822 (1990).
Pavlov G.G. et al., *AAS Mtg.*, **196**, 37.04, (2000).

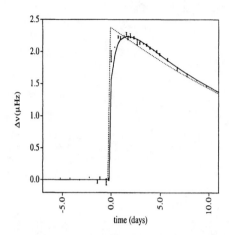

FIGURE 1. Modeling of the 1989 Crab glitch (Lyne, Smith & Prichard 1992), showing a thermal glitch (*solid line*), after an energy deposition of 1.5×10^{42} ergs, and a mechanical glitch (*dotted line*), resulting from the sudden motion of superfluid vortex lines. The mechanical glitch occurs much too quickly to explain the data.

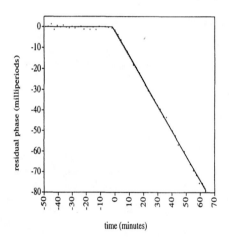

FIGURE 2. Modeling of the 1988 Vela "Christmas" glitch (McCulloch et al. 1990), showing a thermal glitch, after an energy deposition of 6.5×10^{42} ergs, and a mechanical glitch, resulting from the sudden motion of superfluid vortex lines. The simulations are indistinguishable on this scale; both models fit the data equally well.

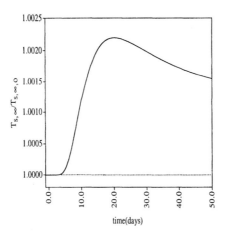

FIGURE 3. Surface temperature changes in the Crab pulsar after a thermal glitch (*solid line*) and a mechanical glitch (*dotted line*) in a pure crust model. The thermal glitch generates a maximum surface temperature increase of $\sim 0.2\%$. The temperature increase following a mechanical glitch is negligible.

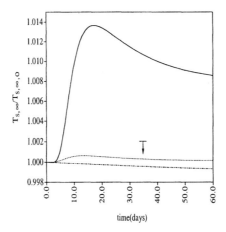

FIGURE 4. Surface temperature changes in the Vela pulsar after a thermal (*solid line*) and mechanical (*dotted line*) glitch in a pure crust model. The *dashed-dot* line shows the initial surface temperature response for an impure crust after a thermal glitch. The thermal conductivity in the impure crust is much lower than the thermal conductivity of the pure crust, delaying the arrival of the thermal wave at the surface by ~ 15 years (Larson & Link 2001). These curves correspond to a glitch of magnitude $\Delta\nu/\nu \simeq 3 \times 10^{-6}$, as was observed in the Vela pulsar in January 2000 (Dodson et al. 2000). An upper limit on the surface temperature obtained 35 days after that glitch is shown (Helfand et al. 2000, Pavlov et al. 2000). For the thermal glitch model to be consistent with this limit, the crust must contain a significant impurity fraction.

On Companion-Induced Off-Center Supernova-Like Explosions

Chang-Hwan Lee*, Insu Yi* and Hyun Kyu Lee[†]

Korea Institute for Advanced Study, Seoul 130-012, Korea
[†]*Dept. of Physics, Hanyang University, Seoul 133-791, Korea*

Abstract. We suggest that a neutron star with a strong magnetic field, spiraling into the envelope of a companion star, can generate a "companion induced SN-like off-center explosion". If the strongly magnetized neutron star ("magnetar") collapses into a black hole via the hypercritical accretion during the spiral-in phase, a rapidly rotating black hole with a strong magnetic field at the horizon results. The Blandford-Znajek power is sufficient to power a supernova-like event with the center of explosion displaced from the companion core. The companion core, after explosion, evolves into a C/O-white dwarf or a neutron star with a second explosion. The detection of highly eccentric black-hole, C/O-white dwarf binaries or the double explosion structures in the supernova remnants could be an evidence of the proposed scenario.

HYPERCRITICAL ACCRETION ONTO MAGNETARS

It has recently been reported that neutron stars with large luminosities could be powered by ultra-strong magnetic fields $\sim 10^{14} G$ ("magnetars") rather than rapid rotation as in rotation-powered pulsars (e.g. Kouveliotou et al. 1998, 1999). These "magnetars" are believed to be produced in supernova explosions or accretion induced collapses (Duncan & Thompson 1992).

On the other hand, the Blandford-Znajek process has been suggested as a powerful source for a variety of very energetic events ranging from active galactic nuclei (Blandford & Znajek 1977) to gamma-ray bursts (Lee, Wijers & Brown 2000). The extremely high luminosities or power outputs generally require ultra-strong magnetic fields around rapidly rotating black holes. The origin of such strong magnetic fields around the rapidly spinning black holes has not been clearly understood. It is interesting to point out that if magnetars directly collapse to black holes via rapid accretion, the resulting black holes could spin rapidly with magnetic fields left at the horizon(Lee, Lee & van Putten), which are large enough to make the Blandford-Znajek power interesting for very energetic events.

Based on this possibility, we propose that the hypercritical accretion onto a neutron star with a very strong magnetic field can produce a rapidly spinning black hole with an ultra-strong field near the horizon (Lee, Yi & Lee 2000). The hypercritical accretion has been invoked for the common envelope phase evolution of the massive stellar binaries (Bethe & Brown 1998). Unless the strong magnetic fields decay within time scales shorter than the duration of the accretion phase, ~ 1 year, it is highly likely that the black holes with very strong fields could be produced. The Blandford-Znajek power is sufficient to power a supernova-like event with the center of explosion displaced from the companion core. The companion core, after explosion, evolves into a C/O-white dwarf or a neutron star with a second explosion. The detection of highly eccentric black-hole, C/O-white dwarf binaries or the double explosion structure in the supernova remnants, with two different centers of explosion, could be an evidence of the proposed scenario. Especially, in the later case, the neutron star can have double kicks, resulting in very high kick velocities, > 1000 km s^{-1}. If the spiral-in of the neutron star leads to the merger with the companion core, the explosion at the core would appear as an explosion from a single star. The neutron star-neutron star binary system is expected when the companion collapses before significant accretion of the neutron star occurs.

BLACK HOLE FORMATION IN THE SPIRAL-IN PHASE

When the first born star with strong magnetic field accretes material from the companion envelope, the black hole should be formed inside the envelope. Since the black hole is formed in the process of the hypercritical accretion, the temperature near the surface of the neutron star should remain ~ 1 MeV until the collapse into a black hole. This temperature is required to generate neutrinos. In this case, the spin frequency for the r-mode instability is so high that we can neglect the effect (Lindblom & Owen 1999). If the Blandford-Znajek process works at the time of black hole formation, there is enough energy to generate GRBs. However, because of the optically thick envelope, gamma rays cannot come out through the hydrogen (or helium) envelope. Instead of GRBs, the energy will pile up in the envelope to generate supernova. The main difference from the normal SN scenarios is that the SN is not induced by the core collapse of the progenitors. The natural consequence is the strong asymmetry in the SN events. Since the companion mass is not the unique parameter, even the less massive stars (normal white-dwarf progenitors) can have SN events with high kick velocities.

The available energies powered by rotating black holes ($> 1.5\ M_\odot$) with Blandford-Znajek process can be $\sim 10^{51}$ erg. Since the binding energies ($0.6GM^2/R$) of the hydrogen envelope of giant stars ($10\ M_\odot$) are $\sim O(10^{48})$ erg, newly formed black holes have enough energy to blow the envelope off. If only 1% of this SN energy is available for the kinetic energy of the He core of the giant star, the available kick velocities are $\sim O(1000$ km s$^{-1})$.

Fast moving white dwarfs from SN remnants, if exists, will be the consequences of this "companion induced SN events". If the black holes are formed just before the normal SN events of the companions, the double SN are also possible leaving (bh,ns) as remnants. Especially, in the later case, the neutron star can have double kicks, resulting in very high kick velocities, > 1000 km s^{-1}.

According to our scenario, in the formation of (bh, ns) and $(bh, cowd)$ binaries by hypercritical accretion, about 10% of the neutron stars have very strong magnetic field and will have the companion (newly formed black holes) induced supernova-like explosions. As a result $\sim 10\%$ of progenitors of $(bh, cowd)$ binaries end up as eccentric $(bh, cowd)_e$ binaries instead of circular ones. Also $\sim 10\%$ of progenitors of (bh, ns) binaries have double explosions with two different centers of explosion. By assuming that half of the progenitors of $(bh, cowd)_e$ or (bh, ns) binaries survive the explosion induced by rotating black holes, we have the formation rates of eccentric $(bh, cowd)_e$ binaries or double explosions, $R \sim 10^{-5}$ yr^{-1} per galaxy. Since the cooling time of CO white dwarfs, i.e. the time required to reach luminosities $\log(L/L_\odot) = -4.5$ with $L_\odot =$ solar luminosity, is $1 - 10$ Gyr (Salaris et al. 1997), the number of eccentric $(bh, cowd)_e$ binaries with luminosity $\log(L/L_\odot) > -4.5$ is $\sim 10^4 - 10^5$ in Galaxy. Note that the number of non-pulsating eccentric $(ns, cowd)_e$ binaries with luminosity $\log(L/L_\odot) > -4.5$ is ~ 10 times more popular than $(bh, cowd)_e$ systems (Brown, Lee, Portegies Zwart, & Bethe, 2001).

DISCUSSION

There are a number of points in the proposed scenario which could be observationally testable. Firstly, if the Blandford-Znajek phase occurs while the neutron star collapses into black hole in the envelope, the resulting off-center explosion will give a high eccentricity, which should be testable in terms of the highly eccentric binary systems. Secondly, there will be little metal spread from the core since there will be no core collapse. The dominant effect would be blowing-off the hydrogen envelope preceded by the expansion of the envelope.

If the merger of the companion core with the neutron star occurs before the black hole formation, then the explosion will not cause an off-center explosion. Although this possibility exists, the estimated accretion time scale on which the neutron star can accrete enough mass to collapse to a black hole is comparable to the spiral-in time scale. This implies that the off-center explosion is possible.

CHL and IY is supported in part by 1999-2000 KIAS Research Fund. IY is supported in part by KRF Research Fund KRF 1998-001-D00365. HKL is supported in part by BK21 Program of Ministry of Education and by KOSEF Grant No. 1999-2-112-003-5.

REFERENCES

- Bethe, H. A. & Brown, G. E. 1998, ApJ, 506, 780
- Blandford, R.D. & Znajek, R.L. 1977, MNRAS, 179, 433
- Brown, G.E., Lee, C.-H., Portegies Zwart, S., & Bethe, H.A. 2001, ApJ, 547, 345
- Duncan, R. C. & Thompson, C. 1992, ApJ, 392, L9
- Kouvelioutou, C. et al. 1998, Nature, 393, 235
- Kouvelioutou, C. et al. 1999, ApJ, 510, L115
- Lindblom, L. & Owen, B.J. 1999, Phys. Rev. Lett., 80, 4843
- Lee, H.K., Lee. C.H., & van Putten, M.H.P.M., to appear in MNRAS, astro-ph/0009239
- Lee, C.-H., Yi, I., & Lee, H.K. 2000, astro-ph/0003266
- Lee, H.K. Wijers, R.A.M.J., & Brown, G.E. 2000, Phys. Rep., 325, 83
- Salaris, M., Dominguez, I., Garcia-Berro, E., Hernanz, M., Isern, J., & Mochkovitch, R. 1997, ApJ, 486, 413

Self-Similar Hot Accretion Flow onto a Neutron Star

Mikhail V. Medvedev

Canadian Institute for Theoretical Astrophysics, University of Toronto, Toronto, Ontario, M5S 3H8, Canada
Harvard-Smithsonian Center for Astrophysics, 60 Garden Street, Cambridge, MA 02138

Abstract. We present analytical and numerical solutions which describe a hot, viscous, two-temperature accretion flow onto a neutron star or any other compact star with a surface. We assume Coulomb coupling between the protons and electrons, and free-free cooling from the electrons. Outside a thin boundary layer, where the accretion flow meets the star, we show that there is an extended settling region which is well-described by two self-similar solutions: (1) a two-temperature solution which is valid in an inner zone $r \lesssim 10^{2.5}$ (r is in Schwarzschild units), and (2) a one-temperature solution at larger radii. In both zones, $\rho \propto r^{-2}$, $\Omega \propto r^{-3/2}$, $v \propto r^0$, $T_p \propto r^{-1}$; in the two-temperature zone, $T_e \propto r^{-1/2}$. The luminosity of the settling zone arises from the rotational energy of the star as the star is braked by viscosity; hence the luminosity is independent of $\dot{M}$. The settling solution is convectively and viscously stable and is unlikely to have strong winds or outflows. The flow is thermally unstable, but the instability may be stabilized by thermal conduction. The settling solution described here is not advection-dominated, and is thus different from the self-similar ADAF found around black holes. When the spin of the star is small enough, however, the present solution transforms smoothly to a (settling) ADAF.

INTRODUCTION

At mass accretion rates less than a few per cent of the Eddington rate, black holes (BHs) and neutron stars (NSs) are believed to accrete via a hot, two-temperature, radiatively inefficient, quasi-spherical, advection-dominated accretion flow, or ADAF [1,2]. While the properties of BH ADAFs are quite well known, hot flows onto NSs have not been investigated. Their properties, such as the luminosity, spectra, torque applied to a central object, etc., are expected to be different from the BH ADAFs because a NS has a surface while a BH has an event horizon [2,3]. Here we discuss the structure of a hot accretion flow around a NS [4]. We do not attempt a detailed analysis of the boundary layer region near the NS surface.

SELF-SIMILAR SETTLING SOLUTION

We consider a steady, rotating, axisymmetric, quasi-spherical, two-temperature accretion flow onto a star with a surface, and we use the height-integrated form of the viscous hydrodynamic equations. We assume the Shakura-Sunyaev-type viscosity parametrized by dimensionless α. We assume viscous heating of protons, Bremsstrahlung cooling of electrons and Coulomb energy transfer from the protons to the electrons. We neglect thermal conductivity and Comptonization. In the inner zone $r < 10^{2.5}$ (r is in Schwarzchild units, $R_S = 2GM/c^2$), the flow is two-temperature with the density, proton and electron temperatures, angular and radial velocities scalings as

$$\rho = \rho_0 \, r^{-2}, \quad T_p = T_{p0} \, r^{-1}, \quad T_e = T_{e0} \, r^{-1/2}, \quad \Omega = \Omega_0 \, r^{-3/2}, \quad v = v_0 \, r^0, \quad (1)$$

where $\rho_0, T_{p0}, T_{e0}, \Omega_0, v_0$ are functions of M, α and the star spin $s = \Omega_*/\Omega_K(R_*)$, and $\Omega_K(R) = (GM/R^3)^{1/2}$ is the Keplerian angular velocity. In the outer zone $r > 10^{2.5}$, we have $T_e = T_p \propto r^{-1}$ and the same other scalings. This self-similar solution is valid for the part of the flow below the radius r_s related to the mass accretion rate $\dot{m}$ (in Eddington units, $\dot{M}_{\rm Edd} = 1.4 \times 10^{18} m$ g/s, and here $m = M/M_\odot$):

$$\dot{m} < 2.2 \times 10^{-3} \alpha_{0.1}^2 s_{0.3}^2 r_{s,3}^{-1/2}, \quad (2)$$

where $r_{s,3} = r_s/10^3$, $\alpha_{0.1} = \alpha/0.1$, etc.. The numerical solution of the hydrodynamic equations with appropriate inner and outer boundary conditions is represented in Figure 1. It is in excellent agreement with the self-similar solition (1).

PROPERTIES OF THE SELF-SIMILAR SOLUTION

Spin-Up/Spin-Down of the Neutron Star — The angular momentum flux in the flow, $\dot{J}$, is negative which implies that the accretion flow removes angular momen-

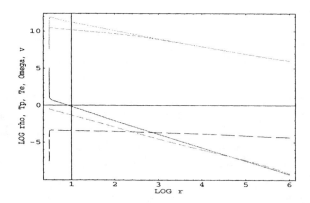

FIGURE 1. The radial profiles of density (*solid* curve), proton and electron temperatures (two *dotted* curves), angular velocity (*dashed* curve), and radial velocity (*long-dashed* curve).

tum from the star and spins it down. This behavior is quite different from that seen in thin disks [5,6], where for most choices of the stellar spin parameter s, the accretion disk spins up the star with a torque $\dot{J}_{\text{thin}} \approx \dot{M}\Omega_K(R_*)R_*^2$. In contrast, for the self-similar solution derived here, the torque is negative for nearly all values of s. Moreover, $\dot{J}$ is independent of $\dot{M}$. Equivalently, the dimensionless torque, $j = \dot{J}/\dot{J}_{\text{thin}}$, which is ~ 1 under most conditions for a thin disk, here takes on the value $j \simeq -43\dot{m}^{-1}\alpha^2 s^3 \left(1-s^2\right)^{3/2}$. This torque spins down the NS as $s = s_0/\sqrt{1+t/\tau}$, where the spin-down time is

$$\tau \simeq 2 \times 10^8 s_{0.1}^{-2} \alpha_{0.1}^{-2} (R_m/R_*)^{-3/2} \text{ yr} \quad \text{or} \quad \dot{P}_*/P_*^2 \simeq 2.7 \times 10^{-12} m_{1.4}^{-1} \alpha_{0.3}^2 s_{0.5}^3 \text{ s}^{-2}, \quad (3)$$

which is in excellent agrement with observational spin-down rates of some X-ray pulsars [7] (here R_m is the magnetospheric radius). Note, the spin-down rate is independent of $\dot{M}$!

Luminosity and Spectrum — The total luminosity has two contributions: from the settling flow and from the boundary layer:

$$L_{SS} \simeq 6.2 \times 10^{34} mr_3^{-1}\dot{m}_{-2}s_{0.1}^2 + 8.9 \times 10^{33} mr_3^{-1}\alpha_{0.1}^2 s_{0.1}^4 \text{ ergs/s},$$
$$L_{BL} \simeq 1.7 \times 10^{36} mr_3^{-1}\dot{m}_{-2} \text{ ergs/s}. \quad (4)$$

Note that for sufficiently low $\dot{m}$, the luminosity is independent of $\dot{m}$ and is dominated by the settling flow. Below the radius $r_c \sim 45\alpha_{0.1}^{1/2} s_{0.1}$ Comptonization is significant (although optical depth is always smaller than unity); therefore the self-similar solution (1) is not accurate. The observed spectrum from the settling flow (assuming free-free emission) is calculated to be:

$$\nu L_\nu \simeq 1.7 \times 10^{31} m\alpha_{0.1}^2 s_{0.1}^4 \, (h\nu/[3 \text{ keV}]) \text{ ergs/s}. \quad (5)$$

which is sufficiently accurate up to $h\nu \sim kT_e(r_c) \sim 400\alpha_{0.1}^{-1/4} s_{0.1}^{-1/2}$ keV.

Stability of the flow — We demonstrate that the settling flow is convectively stable and may not have strong winds and outflows (the Bernoulli number is negative) if the adiabatic index satisfies

$$\gamma > (3/2)\left(1-s^2/2\right)/\left(1-s^2/4\right) \sim 1.5. \quad (6)$$

The flow is thermally unstable because it is cooling-dominated by free-free emission. Stabilization by thermal conduction is studied elsewhere.

REFERENCES

1. Narayan, R., and Yi, I., *Astrophys. J.* **428**, L13 (1994)
2. Narayan, R., and Yi, I., *Astrophys. J.* **452**, 710 (1995)
3. Narayan, R., Garcia, M. R., and McClintock, J. E., *Astrophys. J.* **478**, L79 (1997)
4. Medvedev, M. V., and Narayan, R., *Astrophys. J.* in press (astro-ph/0007064) (2001)
5. Popham, R., and Narayan, R., *Astrophys. J.* **370**, 604 (1991)
6. Paczyński, B., *Astrophys. J.* **370**, 597 (1991)
7. Yi, I., Wheeler, J. C., and Vishniac, E., *Astrophys. J.* **481**, L51 (1997)

Isolated Neutron Stars - Optical Non-Thermal Phenomenology

Andrew Shearer*, Aaron Golden*, Padraig O'Connor* and Raymond Butler*

*Astrophysics and Scientific Computing Group, National University of Ireland, Galway, Ireland

Abstract.
We show that observations of pulsars with pulsed optical emission indicate that the peak flux scales according to the magnetic field strength at the light cylinder. The derived relationships indicate that the emission mechanism is common across all of the observed pulsars with periods ranging from 33ms to 385 ms and ages of 1000-300,000 years. It is noted that similar trends exist for γ ray pulsars. Furthermore the model proposed by Pacini (1971) [5] [6] [7] still has validity and gives an adequate explanation of the optical phenomena.

INTRODUCTION

Many suggestions have been made concerning the non-thermal optical emission process for young and middle-aged pulsars. Despite many years of detailed theoretical studies and more recently limited numerical simulations, no convincing models have been derived which explain all of the high energy properties. There are similar problems in the radio but as the emission mechanism is radically different (being coherent) only the high energy emission will be considered here. This failure of the detailed models to explain the high energy emission has prompted this work. We have taken a phenomenological approach to test whether Pacini type scaling is still applicable. We try to restrict the effects of geometry by taking the peak luminosity as a scaling parameter rather than the total luminosity. In this regard we are removing the duty cycle term from [7].

THE PHENOMENOLOGY OF PULSED MAGNETOSPHERIC EMISSION

The three optically brightest pulsars (Crab, Vela and PSR 0540-69) are also among the youngest. However all these pulsars have very different pulse shapes resulting in a very different ratio between the integrated flux and the peak flux. In this work we will use the peak emission as the primary flux parameter. A number of definitions can exist for this and in this context we have taken the 95%-95% level of the primary pulse. To first order, this correlates well with the luminosity per pulse divided by the Full-Width at Half-Maximum (FWHM). For the Crab pulsar the cusp is effectively flat over this region [2]. The FWHM can be considered to scale with the pulsar duty cycle. Our proposition is that the peak flux represents the local power density within the emitting region with minimal effects from geometrical considerations such as observer line of sight and magnetic and rotational axis orientation.

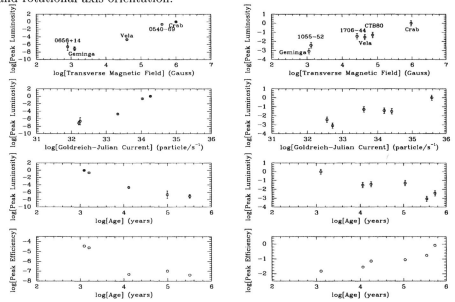

Figure 2 (left panels): Peak optical luminosity vs light cylinder field, Goldreich & Julian current [4] and canonical age. Also shown is the efficiency of the peak emission against age. The peak luminosity has been normalised to the Crab pulsar. The error bars represent both statistical errors from the pulse shape and uncertainty in the pulsar distance.

Figure 3 (right panels): Peak γ-ray luminosity vs light cylinder field, Goldreich & Julian current and canonical age. Also shown is the efficiency of the peak emission against age. The peak luminosity has been normalised to the Crab pulsar. The γ-ray peak luminosity has been inferred from various sources ([1] [10] [11]).

CONCLUSION

It seems clear, from both polarisation studies ([8] [9]) and from this work, that we expect that the optical emission zone is sited towards the outer magnetosphere. Timing studies of the size of the Crab pulse plateau indicates a restricted emission volume ($\approx$ 15 kms in lateral extent) [2]. Importantly, the simple relationships we have derived here indicate that there is no need to invoke complex models for this high energy emission. Observed variations in spectral index, pulse shape and polarisation can be understood in terms of geometrical and absorption factors rather than differences in the production mechanism.

From a combination of previous phenomenological studies [3] and our results we can begin to understand how the high energy emission process ages. Goldoni et al. compared the known spectral indices and efficiencies in both the optical and γ-ray regions. They noted that the spectral index flattened with age for the γ-ray pulsars whilst the reverse was true for the optically emitting systems. They also noted a similar trend reversal for the efficiency, with the γ-ray pulsars becoming more efficient with age. We note (see the bottom panel in Figures 2 and 3) a similar behaviour with the peak emission. From the temporal coincidence between the γ-ray and optical pulses, it seems likely that the source location is similar for both mechanisms. One explanation is that we have a position whereby from the same electron population there are two emission processes - i.e. curvature for the γ-ray photons and synchrotron for the optical ones. It seems likely that the optical photon spectrum has been further modified to produce the reversal in spectral index with age. The region over which the scattering can occur would scale with the size of the magnetosphere and hence with age. With the outer magnetosphere fields for these pulsars being $< 10^6$ G, electron cyclotron scattering is not an option. However synchrotron self-absorption could explain the observed features. In essence we would expect the most marked flattening to be for the Crab pulsar, where the outer field strengths are of order 10^6 G, and less so for the slower and older systems. These results (both optical and γ-rays) are consistent with a model where the γ and optical emission is coming from the last open-field line at some constant fraction of the light cylinder. The drop in efficiency with age for the production of optical photons points towards an absorbing process in the outer magnetosphere. Clearly more optical and γ-ray observations are needed to confirm these trends.

These results (both optical and γ-rays) are consistent with a model where the γ and optical emission is coming from the last open-field line at some constant fraction of the light cylinder. The drop in efficiency with age for the production of optical photons points towards an absorbing process in the outer magnetosphere.

BIBLIOGRAPHY

1. Fierro et al., 1998, ApJ, 494, 734
2. Golden, A. et al., 2000, A&A, 363, 617
3. Goldoni et al, 1995, A&A, 298, 535
4. Goldreich, P. & Julian, W., 1969, ApJ, 245, 267
5. Pacini, F., 1971, ApJ, 163,17
6. Pacini, F. & Salvati, M., 1983, ApJ, 274, 369
7. Pacini, F. & Salvati, M., 1987, ApJ, 321, 445
8. Romani, R. W., & Yadigaroglu, I.-A., 1995, ApJ, 438, 312
9. Smith, F. G. et al., 1988, MNRAS, 233, 305
10. Thompson, D. J. et al., 1996, ApJ, 465, 385
11. Thompson, D. J. et al., 1999, ApJ, 516, 297

'Inverse Mapping' of Non Thermal Optical Emission from Isolated Neutron Stars

P. O' Connor*, A. Shearer*, A. Golden* and S. Eikenberry[†]

*Astrophysics & Scientific Computing Group, N.U.I. Galway, Ireland
[†]Dept. of Astronomy, Cornell University, Ithaca, N.Y., USA

Abstract. The region(s) within the magnetosphere where the IR → UV emissions arise is essentially unknown. This work is an attempt to place constraints on *possible* emission regions using a computationally intensive, 'inverse mapping' approach. This approach can yield a localisation of emission regions within the magnetosphere and also gives an estimate of pulsar inclination angles. Initial results for PSR B0531+21 (the Crab) are presented.

INTRODUCTION

Our aim is to try and restrict the possible location(s) of the IR → UV emission within the magnetosphere by computationally modeling a simple, global synchrotron process [11] with no a-priori assumptions of location or pulsar orientation. Simulation output is in the form of photon emission location with phase resolved lightcurves and polarisation data. Observational data is used to constrain paramater space and also for the final selection of emission locations.

MODEL OUTLINE

- The magnetic field (**B**) is given by the Liénard-Wiechert potentials applied to a rotating magnetic dipole ($\vec{m}$):

$$\vec{B}(\vec{r},t) = \frac{3\hat{r}(\vec{m}.\vec{r}) - \hat{m}}{r^3} + \frac{3\hat{r}(\dot{\vec{m}}.\hat{r}) - \dot{\vec{m}}}{r^2} + \frac{\hat{r} \times (\hat{r} \times \ddot{\vec{m}})}{r} \quad (1)$$

- This structure is sampled numerically by imposing a cylindrical mesh on the whole magnetosphere and allowing each region to radiate independently with a local charge density given by $N_{GJ} = \mathbf{\Omega}.\mathbf{B}/(2\pi ec)$ [5].

- This population emits **synchrotron radiation**. This radiation is extremely well documented [3]. The particles emit into a tightly confined 'cone' centered on the **B** field lines. The power emitted per unit frequency and polarisation can be written in component form [1] by keeping track of the orthogonal projection of **E** onto the observers plane. Abberation effects (such as relativistic beaming and the Doppler shift, etc.) are taken into account on the emitting synchrotron cone.

- The total optical emission obeys a **power law** e.g $P(\omega) \propto \omega^{-\eta}$ [4]. The source particle distribution should then be a power law with spectral index p where $\eta = (p-1)/2$. The model particle distribution used obeys:

$$N(\gamma)d\gamma = C\gamma^{-p}d\gamma, \quad \gamma_1 < \gamma < \gamma_2 \qquad (2)$$

- The output is stored as a series of 2-D arrays redording phase resolved intensity, lightcurves and polarisation data as seen by an observer at infinity.

RESULTS

There exists some restrictions on individual pulsar magnetic inclination angle (χ) and observer line of sight angle (α) from radio data and theory [6,8] but these values do not in general agree. For a given pulsar we sample all of α - χ parameter space. In the model 'viable' χ values are those which have lightcurves with the correct peak offset relative to the radio ($\Delta\phi_R$) and also the correct relative phase separation between pulse peaks ($\Delta\phi_P$). For the Crab pulsar (see figure 1) $\Delta\phi_R \approx 0.12, \Delta\phi_P \approx 0.4$ [4].

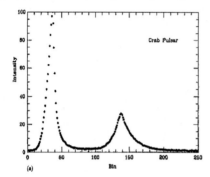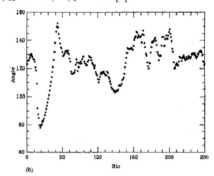

Source: Smith et.al. 1988

FIGURE 1. Observed light-curve and polarisation for the Crab

The simulation results (figure 2) reproduce a double peaked light-curve with peaks of relative intensity similar to observations. The inter-peak phase separation is $\approx 10\%$ more than observed with but with $\Delta\phi_{R_{model}} \approx \Delta\phi_{R_{obs}}$. The rise and fall time of the modeled peaks is much different for the model and observations [2]. The polarisation angle variation is reproduced to first order across the main peak with a similar angular spread.

The inverse mapping of emission regions is shown in figure 3 which represents in one dimension the scaled relative number of points emitting as a function of radial distance from the rotation axis (integrated azimuthally). This may be interpreted as a probability distribution - in this case such that there is a very high probability of emission arising beyond a weighted-average point at $\approx 0.8 \times$lightcylinder radius - which correlates with the Pacini et. al. [7] model for optical pulsar emission.

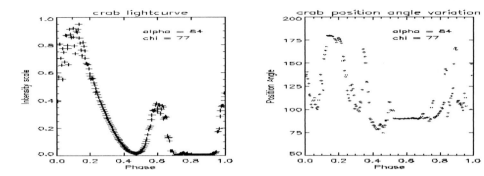

FIGURE 2. Simulated light-curve and polarisation position angle swing for Crab like parameters

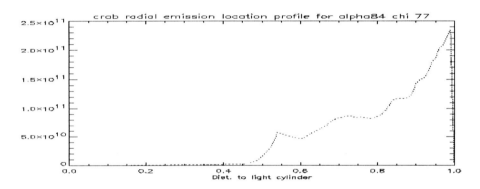

FIGURE 3. Radial plot of emission points for $\alpha = 84°$ $\chi = 77°$

DISCUSSION / CONCLUSIONS

The 'inverse mapping' approach can be used for three primary purposes: (1) the localisation of *likely* emission regions within the magnetosphere (from figure 3 emission is preferred at large radial distances which is in agreement with a recent phenomenological paper of Shearer & Golden [9]). (2) the 'unbiased' selection of pulsar α and χ values([6] estimate an α of $\approx 85°$ for the Crab). (3) Combining (1) and (2) above may aid in the proposal of detailed physical emission models.

BIBLIOGRAPHY

1. Bekefi, G., *Radiation processes in Plasmas*, J. Wiley & sons, Inc., 1966
2. Eikenberry, S. S., Fazio, G. G., Ransom, S. M., Middleditch, J., Kristian, J., Pennypacker, C.R., 1997b, Ap. J., 447, 465
3. Ginzburg, V. L., Syrovatskii, S. I.,1969, ARA&A, **7**, 375
4. Golden, A., Shearer, A., Beskin, G. M., 2000ApJ, **535**, 373
5. Goldreich, P. & Julian, H. 1969, Ap.J., **157**, 869

6. Lyne, A. G., Manchester, R. N. 1988, MNRAS **234**, 477
7. Pacini, F., Salvati, M. 1987, ApJ, **321**, 447
8. Rankin, J. M. 1990, ApJ, **352**, 247
9. Shearer, A. & Golden, A., Accepted for publication by Ap. J. 2000
10. Smith, F. G., Jones, D. H. P., Dick, J. S. B., Pike, C. D. 1988, MNRAS, **233**, 305
11. Sturrock, P. A., 1971, ApJ, **164**, 529

Magnetic Domain in Magnetar-Matters and Soft Gamma Repeaters

In-Saeng Suh and Grant J. Mathews

*Center for Astrophysics, Department of Physics, University of Notre Dame,
Notre Dame, Indiana 46556, USA*

Abstract. Magnetars have been suggested as the site for the origin of observed soft gamma repeaters (SGRs). In this work we discuss the possibility that SGRs might be observational evidence for a magnetic phase separation in magnetars. We study magnetic domain formation as a new mechanism for SGRs in which magnetar-matter separates into two phases containing different flux densities. We use relativistic Hartree theory to estimate the densities at which there is an instability in magnetic domain formation for a given magnetic field.

INTRODUCTION

By now, 4 (perhaps 5) SGRs have been observed [1]. They are believed to be a new class of γ-ray transients separate from the source of ordinary gamma-ray bursts. Recent observations [2] have confirmed the fact that these SGRs are newly born neutron stars that have very large surface magnetic fields (up to 10^{15} G) based upon measurements of the spin-down timescale. Such stars have been named magnetars [3]. AXPs also are included in a class of magnetar [4].

Whether or not magnetars are the source of SGRs or AXPs, as relics of stellar interiors, the study of the magnetic fields in and around degenerate stars should give important information on the role such fields play in star formation and stellar evolution. Indeed, the origin and evolution of stellar magnetic fields remains obscure. An upper limit to the internal field strength of a star given by the scalar virial theorem [5] is $B \lesssim 2 \times 10^8 (M/M_\odot)(R/R_\odot)^{-2}$ G for a star of size R and mass M. For a typical neutron star the maximum interior field strength could thus reach $B \sim 10^{18}$ G.

Since strong interior magnetic fields modify the nuclear equation of state for degenerate stars [6], their structure also will be changed [7]. Magnetic properties of magnetar-matter such as the magnetization and the susceptibility may possibly be coupled directly to observable consequences such as starquakes, glitches, X-ray or γ-ray emission.

MAGNETIC PROPERTIES OF MAGNETAR-MATTERS

The equation of state of neutron-star matter can be described by relativistic nuclear mean field (Hartree) theory [9,8] in which the baryons (neutrons, n, and protons, p) interact via the exchange of scalar σ and vector ω, ρ mesons. Then the effective baryon masses become $m_b^* = m_b - (g_\sigma/m_\sigma)^2(n_p^S + n_n^S)$, where g_σ and m_σ are the coupling constant and meson mass respectively. $n_{n,p}^S$ is the scalar number density. Assuming chemical equilibrium, then $\mu_n - \mu_p = \mu_e = \mu_\mu$, and charge neutrality gives $n_p = n_e + n_\mu$.

The particle number density, pressure, and energy density in a magnetic field are given by Suh & Mathews [6]. The magnetization is given by $\mathcal{M} = -(\partial\Omega/\partial B)_{\mu,T,V}$, the susceptibility $\chi = \mathcal{M}/H \simeq \mathcal{M}/B$ and the differential susceptibility η is defined by $\eta = (\partial\mathcal{M}/\partial B)_{\mu,T,V}$ [10], where μ is the chemical potential, T the temperature, and V the volume of the system.

MAGNETIC DOMAIN FORMATION

For material in a uniform magnetic field, the magnetic field H is related to the flux density B by the relation [11]: $H = B - 4\pi(1-\mathcal{D})\mathcal{M}(B)$ where $\mathcal{D}$ is the demagnetization coefficient which is fixed by the shape of the system. For example, for a neutron star crust permeated by an approximately vertical magnetic field, $\mathcal{D} \approx 0$. When the differential susceptibility η exceeds $1/4\pi$, $(\partial H/\partial B)_\mu$ becomes negative. Then thermomagnetic equilibrium becomes unstable.

We have deduced the unstable regions for magnetic domain formation [12]. We find that in the outer crust ($\rho \lesssim 10^{11}$ g cm^{-3}) of magnetars with surface magnetic fields $\sim 10^{14-15}$ G, magnetic domains can not be formed. However, within the inner crust and crust inside of the magnetar with stronger magnetic fields $B \gtrsim 10^{13}$ G, the magnetic domains become unstable to the formation of layers of domains of alternating magnetization.

The spacing between layers associated with the maximum Landau orbital is $\Delta z \sim 100(g/10^{14}\text{cm/sec}^2)(\mu_e/2)^{-1}(\epsilon_e/5)^{-1}(B/10^{15}\text{G})$ m, where g is the surface gravity and μ_e is the mean molecular weight per electron. The domain might have a

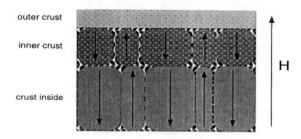

FIGURE 1. Hypothetical simple magnetic domain structure in magnetars.

horizontal scale of $\sim \Delta z$. However the actual size and shape of the domains are difficult to estimate.

SGRs MECHANISM

Under these conditions it sometimes becomes energetically favorable for the material in the inner crust to separate into two phases containing different flux densities. The stippled regions in Figure 1 indicate where magnetic field distortion increases the magnetic energy. At the boundary there also will exist regions around each wall where the magnetic field is distorted. The formation or adjustment of domain structure involves magnetic field amplitude fluctuations of a few percent which have anisotropic magnetostrictive stress $2\pi \mathcal{M}^2$ associated with the demagnetization of the field. Any sudden readjustment of the domain structure will cause a local departure from isostacy which will be relieved on the ohmic dissipation timescale ($\tau_D \sim \sigma A/4\pi c^2 \sim 10^4$ yr). The anisotropic magnetostrictive stresses may be large enough to crack the inner crust. We can estimate the cracking timescale in the inner crust to be $\tau_C \sim \Delta z/v_s \sim 10^{-3}(\theta_m/10^{-3})$ sec, where v_s is the shear velocity and θ_m the maximum allowed strain angle. Finally we obtain the released energy $\Delta E_D \sim 6 \times 10^{42} \chi^2 \theta_m (R_D/10^4 \text{cm})^2 (B/10^{15}\text{G})^2$erg, where $R_D \sim \Delta z$ is the characteristic horizontal size of the domain.

In conclusion, this cracking of the crust by magnetostrictive stress could provide a source of X-ray and γ-ray emission. The formation of magnetic domains together with sudden readjustments of the domain can produce an energy source of soft gamma-rays in SGRs.

Work supported by DoE Nuclear Theory Grant DE-FG02-95ER40934.

REFERENCES

1. Hurley, K. 2000, in *Gamma-ray burst*: Proceedings of the 5th Huntsville Gamma-Ray Symposium, ed. R. M. Kippen, R. S. Mallozzi, and V. Connaughton (AIP, New York, 2000)
2. Kouveliotiu, C., et al. 1998, Nature, 391, 235; 1999, Astrophy. J. 510, L115
3. Duncan, R. C. & Thompson, C. 1992, Astrophy. J. 392, L9
4. van Paradijs, J.,Taam, R. E., & van den Heuvel, E. P. J. 1995, A & A 299, L41
5. Chandrasekhar, S. & Fermi, E. 1953, Astrophy. J. 118, 116
6. Suh, I.-S. & Mathews, G. J. 2001, Astrophy. J., 546, 1126
7. Cardall, C. Y., Prakash, M., and Lattimer, J. M., astro-ph/0011148
8. Chakrabarty, S., Bandyopadhyay, D., & Pal, S. 1997, Phys. Rev. Lett. 78, 2898
9. Broderick, A., Prakash, M., & Lattimer, J. M. 2000, Astrophy. J. 537, 351
10. Blandford, R. D. & Hernquist, L. 1982, J. Phys. C: Solid State Phys. 15, 623
11. Pippard, A. B. 1980, in *Electrons at the Fermi surface*, ed M. Springford (Cambridge: Cambridge University press)
12. Suh, I.-S. & Mathews, G. J. in preparation

CHAPTER 12

GAMMA-RAY BURSTS

Gamma-Ray Bursts - When Theory Meets Observations

Tsvi Piran

Racah institute for Physics,
The Hebrew University, Jerusalem Israel 91904

Abstract.
Gamma-Ray Bursts (GRBs) are the brightest objects observed. They are also the most relativistic objects known so far. GRBs occur when an ultrarelativisitic ejecta is slowed down by internal shocks within the flow. Relativistic particles accelerated within these shocks emit the observed gamma-rays by a combination of synchrotron and inverse Compton emission. External shocks with the circumstellar matter slow down further the ejecta and produce the afterglow, which lasts for months. Comparison of the predictions of this fireball model with observations confirm a relativistic macroscopic motion with a Lorentz factor of $\Gamma \geq 100$. Breaks in the light curves of the afterglow indicate that GRBs are beamed with typical opening angles of a few degrees. The temporal variability of the gamma-rays signal provide us with the best indirect evidence on the nature of the "internal engine" that powers the GRBs and accelerates the relativistic ejecta, suggesting accretion of a massive disk onto a newborn black hole: GRBs are the birth cries of these black holes. Two of the most promising models: Neutron Star Mergers and Collapars lead naturally to this scenario.

INTRODUCTION

Gamma-Ray bursts - GRBs, short and intense bursts of γ-rays arriving from random directions in the sky were discovered accidentally more than thirty years ago. During the last decade two detectors, BATSE on CGRO and BeppoSAX have revolutionized our understanding of GRBs. BATSE has demonstrated [1] that GRBs originate at cosmological distances in the most energetic explosions in the Universe. BeppoSAX discovered X-ray afterglow [2]. This enabled us to pinpoint the positions of some bursts, locate optical [3] and radio [4] afterglows, identify host galaxies and measure redshifts to some bursts [5].

Since their discovery GRBs were among the prime topics of the Texas Symposia. The high energy release and the rapid time scales involved suggested immediately association with relativistic compact objects. The discoveries of BATSE and BeppoSAX confirmed these expectations. These observations have established the Fireball model demonstrating that GRBs are the most relativistic objects known so far: GRBs involve macroscopic ultrarelativistic flows with Lorentz factors $\Gamma \geq 100$. Furthermore, while the "central engines" that drive the relativistic flow and power the GRBs are hidden we have excellent evidence that they involve accretion onto a newborn black hole. GRBs are the birth cries of these black hole.

I review, here, the recent progress in our understanding of GRBs, emphasizing, as appropriate for this conference, their relativistic nature. I begin, in section I with a brief tribute to the 7th Texas symposium. This was the first Texas meeting after the discovery of GRBs was announced and GRBs were the highlight of the discussion there. I continue in section II with a brief exposition of the Fireball model (see [6,7] for details), confronting its predictions with the observations. In III I summarize the implication of the fireball model to the "inner engines". Concluding remarks, further predictions and open questions are discussed in section IV.

I THE 7TH (NEW YORK) TEXAS SYMPOSIUM

GRBs were the hightlight of the Seventh (New York) Texas symposium that took place in 1974. Five out of the 57 talks were devoted to GRBs (this record was repeated only in this symposium with 3 out of 29 talks): Two observational reviews, a theoretical review, a theoretical model and even a description of an automated system for searching for optical transients accompanying GRBs!

M. Ruderman [8] reviewed the theory[1]. emphasizing the *compactness problem*: If GRBs are cosmological then the energy budget and the time scales seem to be incompatible with the observed non thermal spectrum of the bursts. The argument is simple: the variability time scale, δt, imposes an upper limit on the size ($R \leq c\delta t$). The observed flux and the assumed (cosmological) distance determine the energy. Together these yield an extremely large lower limits on the photons density within the source and on the optical depth for pair creation by the energetic photons. Pairs would be copiously produced and the source would be optically thick. The observed optically thin spectrum is impossible. Ruderman points out, however, that relativistic effects would change this conclusion. If the source is moving relativistically the relations between the time scale and the implied distance are modified (by a factor of Γ^2). Furthermore, photons that are observed with energy below $500\Gamma \text{keV}$ have energy below 500keV in the source rest frame and could not produce pairs. These ideas lay the foundation for the current Fireball model. Recent observations have indeed confirmed ultrarelativistic motion in GRBs, showing at least in one case $\Gamma \geq 100$.

---

[1] This review enumerates more than thirty models proposed during the short time passed since the announcement of the discovery. It is remarkable (Ruderman, 1998, private communication) that today we know that none of these models is even remotely relevant.

II THE FIREBALL MODEL, PREDICTIONS AND CONFIRMATIONS

One can never prove a scientific theory. However we gain confidence in a theory when its predictions are confirmed by observations. I discuss, here, the predictions of the Fireball model (specifically of the Fireball-Internal-External shocks model) and their confirmation by numerous observations. My goal is to demonstrate the success of this model. While some of the specific observations could certainly be interpreted within other theories I strongly believe that the bulk of those observations tell us that this is the correct model.

The Fireball model asserts that GRBs are produced when the kinetic energy (or Poynting flux) of a relativistic flow is dissipated by shocks[2]. These shocks accelerate electrons and generate strong magnetic fields. The relativistic electrons emit the observed γ-rays via synchrotron or SSC. There are two variants of this model: The External Shocks model [9] assumes that the shocks are between the relativistic flow and the surrounding circumstellar matter. The Internal Shocks model [10,11] assumes that the flow is irregular and the shocks take place between faster and slower shells within the flow. According to the Internal-External shocks model [12] both kinds of shocks take place: Internal shocks are responsible for the GRB while external shocks produce the longer lasting afterglow (see Fig. 1). Both shocks occur at relatively large distances ($10^{13} - 10^{14}$cm for internal shocks and $10^{14} - 10^{16}$cm for external shocks) from the source that generates the relativistic

---

[2]) Given the low densities involved these shocks, like SNR shocks, must be collisionless.

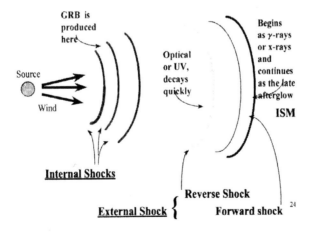

flow. The observed radiation from the GRB or from the afterglow reflects only the conditions within these shocks. We have only indirect information on the nature of the "inner engines".

Fig. 1 depicts a schematic picture of the Internal-External shocks model. An inner engine produces an irregular wind. The wind varies on a scale δt and its overall duration is T. The variability scale δt corresponds to the variability scale observed in the GRB light curve [13] thus, $\delta t \sim 1$ sec. Internal shocks take place at $R \approx \delta t \Gamma^2 \approx 3 \cdot 10^{14}$cm $(\delta t/1 \ sec)(\Gamma/100)^2$. External shocks shocks become significant and the blast wave that propagates into the circumstellar matter and produces the afterglow begins at $\sim 10^{16}$cm (see [14,6] for details). At this initial stage there is also a short lived reverse shock that propagates into the ejecta. This reverse shock is responsible to the prompt optical emission observed in GRB990123.

I review now the predictions of this model and confront them with observations.

• **Relativistic Motion - Predictions:** Relativistic motion is the key ingredient of the Fireball model. Relativistic motion arises naturally (a Fireball forms) when a large amount of energy is produced in a compact region with $E/Mc^2 \gg 1$ [15–17]. This relativistic motion overcome the compactness problem. Various estimates [18–21], of the Lorentz factor Γ based on the compactness problem lead to comparable values, $\Gamma \sim 100$, (see [21] for a critical review).

⋆ **Relativistic Motion - Observations I:** The radio afterglow observations of GRB 970508 provided the first verification of relativistic motion. The radio light curve (in 4.86Ghz) variesd strongly during the first month. These variations died out later. Even before this transition Goodman [22] interpreted these variations as scintillations. The observation of a transition after one month enabled Frail et al., [4] to estimate the size of the afterglow at this stage as $\sim 10^{17}$cm. It immediately follows that the afterglow expanded relativistically. Additionally, the source is expected to be optically thick in radio [23] leading to a ν^2 rising spectrum at these frequencies. The observed flux from the source enables us (using the black body law) to estimate the size of the source. As predicted the radio spectrum increases like ν^2. The size estimated with this method agrees [4] with the one derived by the scintillations estimate implying as well a relativistic motion.

• **The Afterglow - Predictions:** The Afterglow - lower frequency emission that follows GRBs was one of the earlier predictions of the Fireball model [24–27]. Paczynski and Rhoads [24] predicted radio afterglow on the basis of the analogy between external shocks and SNRs. Later Meszaros and Rees [25,26] performed detailed calculations of multi-wavelength afterglow. Vietri [27] predicted soft x-ray afterglow as a test for the external shocks model. These predictions were done in the context of the external shocks model. According to this model the GRB arises due to a shock between a relativistic flow and a circumstellar medium. In this case the emission observed at time t_{obs} arises when $t_{obs} \approx R/2\Gamma^2$. Later emission is related to lower Γ (and larger emission radii) and hence to lower observed frequencies. This afterglow would be a direct extrapolation of the GRB and its basic features should be strongly correlated with the properties of the corresponding GRB.

The theory of the afterglow is well understood. Blandford and McKee [28] have worked out (already in the seventies!) the theory of an adiabatic relativistic blast wave. They show that (as long as the flow is ultrarelativistic, $\Gamma \gg 1$), the blast wave is self similar, the relativistic analog of the well known Sedov-Taylor solution. Electrons are accelerated to relativistic velocities by the shocks and their interaction with the magnetic field leads to synchrotron radiation. This provides an excellent model for the observed emission [29]. Overall we have a simple theory characterized by five parameters: the total energy, E_0, the ambient density, n_0, the ratio of the electrons and magnetic fields energy density to the total energy density, ϵ_e, ϵ_B and the exponent of the electrons' energy distribution function p. An additional sixth parameters, the exponent of the circumstellar density distribution, n, arises in cases when the external matter density ($\rho \propto r^{-n}$). Most notable is $n = 2$ corresponding to a pre-GRB wind expected in some models [30]. This rather simple theory predicts a robust relations between α and β the exponents describing the flux as a function of frequency, $F_\nu \propto t^{-\alpha} \nu^{-\beta}$. At the high frequencies, above the cooling frequency, we have (for $n = 0$), $\alpha = (3p-2)/4$ and $\beta = p/2$.

⋆ **The Afterglow - Observations:** On Feb 28 1997, in a wonderful anniversary celebration for SN87A, BeppoSAX detected x-ray afterglow from GRB 970228. The exact position given by BeppoSAX led to the discovery of optical afterglow [3]. Radio afterglow was detected in GRB 970508 [4]. By now more than thirty x-ray afgterglows have been observed. About half of these have optical and radio afterglow as well and in most of those the host galaxy has been discovered.

Most x-ray and optical afterglow decay as power laws with $\alpha \sim 1.2$ and $\beta \sim 1.2$, in excellent agreement with the predictions of the simplest afterglow model: An adiabatic Blandford-McKee hydrodynamics with Synchrotron emission [29]. As α and β are determined by p the electron's distribution power law index, these observations suggest that as predicted [29,31], $p \approx 2.5$ and it is fairly invariant from one burst to another [32]. A simultaneous spectral fit for GRB980508, all the way from the radio to the x-ray also agrees with this picture [33].

• **The GRB-Afterglow Transition - Predictions:** The rapid time variability seen in most GRBs cannot be produced by external shocks [12,34]. This leaves internal shocks as the only viable model! Shortly before the discovery of the afterglow from GRB970228, Sari and Piran [12] pointed out that afterglow should arise also within the internal shocks scenario. The efficiency of the internal shocks depends on the parameters of the flow, most specifically on the variability of the Lorentz factor between different shells [13,35]. Even in the most efficient cases a significant fraction of the energy remain as kinetic energy. Sari and Piran [12] suggested that this energy would be dissipated later by interaction with the surrounding matter and produce an afterglow. Within this Internal-External shocks model the GRB is produce by internal shocks while the afterglow is produced by external shocks. The predictions of this model for the afterglow are similar to those of the External shocks model. However, a critical difference is that here the afterglow is not an extrapolation of the GRB.

The internal shocks take place at a distance $R_{IS} \sim \delta t \Gamma^2 \sim 10^{14}$cm. These shocks last as long as the inner engine is active. The typical observed time scale for this activity ∼ 50sec (for long bursts) and ∼ 0.5sec (for short ones). External shocks begin at $R_{Ex} \sim 10^{16}$cm. If $R_{Ex}/\Gamma^2 \leq T$ this happens while internal shocks are still going on and the afterglow overlaps the late part of the GRB. At the early time the afterglow emission peaks in the high x-rays contributing also to the observed γ-ray flux. We expect, therefore, a transition within the GRB from hard (pure GRB) to softer and smoother (GRB and afterglow) signal.

⋆ **The GRB - Afterglow Transition - Observations:** The extrapolation of the x-ray afterglow fluxes backwards generally does not fit the γ-ray fluxes. Moreover there is no direct correlation between the γ-ray fluxes and the x-ray (or optical) afterglow fluxes. This result is in a nice agreement with the predictions of the Internal - External shocks scenario in which the two phenomena are produced by different effects and one should not expect a simple extrapolation to work.

The expected GRB afterglow transition have been observed in several cases. The first observation took place (but was not reported until much latter) already in 1992 [36]. Recent BeppoSAX data shows a rather sharp transition in the hardness that takes place several dozen seconds after the beginning of the bursts [37]. This transition is seen clearly in the different energy bands light curves of GRB990123 and in GRB980923 [38]. Connaughton [39] have averaged the light curves of many GRBs and discovered long and soft tails: the early x-ray afterglow.

• **The Prompt Optical Flash - Predictions:** The collision between the ejecta and the surrounding medium produces two shocks. The outer forward shock propagates into the ISM. This shock develops later into the self similar Blandford-McKee blast wave that drives the afterglow. A second shock, the reverse shock, propagates into the flow. This reverse shock is short lived. It dies out when it runs out of matter as it reaches the inner edge of the flow. While it is active, it is a powerful source of energy. Comparable amounts of energy are dissipated by the forward and by the reverse shocks [14]. We expect that the system is radiative at this stage, namely most of the energy converted by the shock is radiated away.

Sari and Piran predicted at the First Rome Meeting (Oct 1998) an intense (brighter than 11th magnitude) prompt optical flash from this reverse shock [40]. Previous work [26] done prior to the discovery of the afterglow considered various possibilities and estimated the magnitude of the prompt optical flash to be anywhere from 9th to 19th magnitude. The observations of the afterglow constrained severely the relevant models and the relevant parameter space. With the new data the constrained model led to a clear prediction with a narrow range of magnitude. At that time this prediction was almost conflicting with upper limits given by systems like LOTIS and ROTSE.

One prediction that was, unfortunately, missed: prompt radio emission from the reverse shock. This radio emission should be short lived, like the burst and should have initially an optically thick component that becomes optically thin later.

⋆ **The Prompt Optical Flash - Observations:** In Jan 23 1999 just three month after this prediction ROTSE recorded six snapshots of optical emission from GRB990123 [41] . Three of those were taken while the burst was still emitting γ-rays. The other three snapshots spanned a couple of minutes after the burst. The second snapshot, taken 70sec after the onset of the burst corresponds to a 9th magnitude signal. A comparison of these optical observations with the γ-rays and x-rays light curves (see e.g. [7]) shows that the optical emission does not correlate with the γ-rays pulses. The optical photons and the γ rays are not emitted by the same photons [42,41]. The optical pulses peak some 70sec after the onset of the burst simultaneously with a late peak in the soft x-ray emission.

Radio observations of GRB990123 revealed a short lived radio pulse. This emission can be explained as coming from the reverse shock [42]. Using the parameters of the reverse shock derived from the optical flash Sari and Piran [42] estimated the magnitude of this radio emission. The theoretical curve and the observations are in excellent agreement (see e.g. [7]). Note that the theoretical curve was calculated just from the optical flash data and it was not "fitted" in any way to the observed data. While prompt optical flashes were not detected in other bursts, short lived radio flashes have been detected in GRB000926 and in GRB970828.

⋆ **Relativistic Motion II - Observations:** The radio emission from GRB970508 showed relativistic motion in its afterglow. However, the significant observations were done one month after the burst and at that time the motion was only "mildly" relativistic with a Lorentz factor of order a few. The observations of GRB990123 enabled us to obtain three independent estimates of the ultrarelativistic Lorentz factor at the time that the ejecta hits first the ISM [42]. First the time delay between the GRB and the optical flash suggests $\Gamma \sim 200$. The ratio between the emission of the forwards shock (x-rays) and the reverse shock (optical) gives another estimate of $\Gamma \sim 70$. Finally the fact that the maximal synchroton frequency of the reverse shock was below the optical band led to $\Gamma \sim 200$. The agreement between these three crude and independent estimates is reassuring.

These observations provide us also with an estimate of the position of the external shocks, $\sim 10^{15}$cm at 70 seconds after the bursts. It is an impressive measurement considering the fact that the distance to this burst is $\sim 3.5 Gpc$. The corresponding angular resolution is 10^{-13} or ~ 50nanoarcsec.

• **Jets - Predictions:** With redshift measurements it became possible to obtain exact estimate the total energies involved. While the first burst GRB970508 required a modest value of $\sim 10^{51}$ergs, the energies required by other bursts were alarming, 3×10^{53}ergs for GRB981226 and 4×10^{54}ergs for GRB990123, and unreasonable for any simple compact object model. These values suggested that the assumed isotropic emission was wrong and GRBs are beamed. Significant beaming would of course reduce, correspondingly the energy budget.

Beaming was suggested even earlier as it arose naturally in some specific models. For example the binary neutron star merger has a natural funnel along its rotation axis and one could expect that any flow would be emitted preferably along this axis. The Collapsar model also requires beaming, as only a concentrated beamed energy could drill a whole through the stellar envelope that exists in this model.

Consider a relativistic flow with an opening angle θ. As long as $\theta > \Gamma^{-1}$ the forwards moving matter doesn't "notice" the angular structure and the hydrodynamics is "locally" spherical [43]. The radiation from each point is beamed into a cone with an opening angle Γ^{-1}. It is impossible to distinguish at this stage a jet from a spherical expanding shell. When $\theta \sim \Gamma^{-1}$ the radiation starts to be beamed sideways. At the same time the hydrodynamic behaviour changes and the material starts expanding sideways. Both effects lead to a faster decrease in the observed flux, changing α, the exponent of the decay rate of the flux to: $\alpha = p/2$. Thus we expect a break in the light curve and a new relation between α and β after the break [44–46]. The magnitude of the break and the duration of the transition will change if the jet is expanding into a wind with r^{-2} density profile [47]. The break is expected to take place at $t_{jet} \approx 6.2(E_{52}/n_0)^{1/3}(\theta/0.1)^{8/3}$hr [45]. Recently numerical simulation [48] have shown that the break appears in a more realistic calculations, even though the numerical results suggest that the analytical model developed so far are probably too simple.

⋆ **Jet - Observations:** GRB980519 had unusual values for $\alpha = 2.05$ and $\beta = 1.15$. These values do not fit the "standard" spherical afterglow model[3]. However, these values are in excellent agreement with a sideway expanding jet [45]. The simplest interpretation of this data is that we observe a jet during its sideway expansion phase (with $p = 2.5$). The jet break transition from the spherical like phase to this phase took place shortly after the GRB and it was not caught in time. The light curves of GRB990123 shows, however, a break at $t \approx$ 2days [49]. This break is interpreted as a jet break, corresponding to an opening angle $\theta \sim 5°$. Another clear break was seen in GRB990510 [50,51].

The brightest bursts, GRB990123 and GRB980519 gave the first indications for jet like behaviour [45]. This suggested that their apparent high energy was due to the narrow beaming angles. A compilation of more bursts with jet breaks suggests that all bursts have a comparable energy $\sim 10^{51}$ergs and the variation in the observed energy is mostly due to the variation in the opening angles θ [52,53,32]

III THE INNER ENGINES

The Fireball model tells us how GRBs operate. However, it does not answer the most interesting astrophysical question: what produces them? which astrophysical process generates the energetic ultrarelativistic flows needed for the Fireball model? Several observational clues help us answer these questions.

• **Energy:** The total energy involved is large $\sim 10^{51}$ergs, a significant fraction of the binding energy of a stellar compact object. *The "inner engine" must be able to generate this energy and accelerate $\sim 10^{-5} M_\odot$ to relativistic velocities.*

• **Beaming:** Most GRBs are beamed with typical opening angles $0.02 < \theta < 0.2$. *The "inner engine" must be able to collimate the relativistic flow.*

• **Long and Short Bursts:** The bursts are divided to two groups according to their overall duration. Long bursts with $T > 2$sec and short ones with $T < 2$sec.

• **Rates:** GRBs take place once per $10^7(4/\theta^2)$yr per galaxy. *GRBs are very rare at about 1/1000 the rate of supernovae.*

---

[3] A possible alternative fit is to a wind (n=2) model but with a unusual high value of $p = 3.5$

The Fireball Internal-External shocks model provides us with another key clue:

- **Time Scales:** The variability time scale, δt, is at times as short as 1ms. The overall duration, T, is of the order of 50sec. According to the internal shocks model these time scales are determined by the "inner engine". $\delta t \sim$ *msec suggests a compact object. $T \sim$ 50sec is much longer than the dynamical time scale, suggesting a prolonged activity.*[4]. *This rules out any "explosive" model that release the energy in a single explosion.*

The internal shocks model requires two (or possibly three [55,56]) different time scales operating within the "inner engine". These clues, most specifically the last one suggest that GRBs arise due to accretion of a massive ($\sim 0.1 m_\odot$) disk onto a compact object, most likely a newborn black hole. A compact object is essential because of the short time scales. Accretion is needed to produce the two different time scales, and in particular the prolonged activity. A massive ($\sim 0.1 m_\odot$) disk is required because of the energetics. We expect that such a massive disk can form only simultaneously with the formation of the compact object. This leads to the conclusions that GRBs accompany the formation of black holes. This model is supported by the observations of relativistic (but not as relativistic as in GRBs) jets in AGNs, which are powered by accretion onto black holes. This system is capable of generating collimated relativistic flows even though we don't understand how.

An important alternative to accretion is Usov's model [57] in which the relativistic flow is mostly Poynting flux and it is driven by the magnetic and rotational energies of a newborn rapidly rotating neutron star. However this model seems to fall short by an order of magnitude of the energy required.

Several scenarios could lead to a black hole - massive accretion disk system. This could include mergers (NS-NS binaries [58,10], NS-BH binaries [59] WD-BH binaries [60], BH-He-star binaries [61]) and models based on "failed supernovae" or "Collapsars" [62–64]. Narayan et al. [65] have recently shown that accretion theory suggests that from all the above scenarios only Collapsars could produce long bursts and only NS-NS (or NS-BH) mergers could produce short bursts.

Additional indications arise from afterglow observations. One has to use these clues with care. Not all GRBs have afterglow (for example, so far afterglow was not detected from any short burst) and it is not clear whether these clues are relevant to the whole GRB populations. These clues seem to suggest a GRB-SN connection:

- **SN association:** Possible association of GRB980425 with SN98bw [66] and possible SN signatures in the afterglows of GRB970228 [67] and GRB980326 [68].
- **Iron lines:** have been observed in some x-ray afterglows [69]. Any model explaining them requires a significant amounts of iron at rest near those GRBs.
- **Association with Star formation:** GRBs seem to follow the star formation rate. GRB are located within star forming regions in star forming Galaxies [63,53].
- **GRB distribution:** GRBs are distributed within galaxies. There is no evidence for GRBs kicked out of their host galaxies [70,53] as would be expected for NS-NS mergers [10].

---

[4] The ratio $\delta t/T \ll 1$ for short bursts as well [54]

All these clues point out towards a SN/GRB association and towards the Collapsar model. However, the situation is not clear cut. The association of GRB980425 with SN98bw is uncertain. There are alternative explanations to the bumps in the afterglows of GRB970228 and GRB980326 [71]. Iron is produces in Supernovae. But there is no simple explanation what is iron at rest doing around the GRB (see however, [72]). The association with star formation and the distribution of GRBs within galaxies is real but all that it indicates is short lived progenitors. One cannot rule out a short lived binary NS population [73] which would mimic this behaviour. Even worse, there are some indication that seem incompatible with the SN association:

• **No Windy Afterglow:** No evidence for a wind (n=2) in any of the afterglow light curves? Furthermore, most fits for the afterglow parameters show low ambient density [32,53].

• **No Jets:** Some GRBs dont show evidence for a jet or have very wide opening angles [32,53], this would be incompatible with the Collapsar model.

IV CONCLUSIONS, PREDICTIONS AND OPEN QUESTIONS

There is an ample observational support For the Fireball model. It also has several other predictions. The most interesting ones are those concerning the very early afterglow and the GRB-afterglow transition. The early afterglow phase is radiative and a detailed look at the first hour of the afterglow should show the radiative to adiabatic transition. It should also show (mostly in radio) small bumps corresponding to refreshed shocks [74] which would enable us to learn more on the nature of the flow produced by the "inner engine". With an operational HETE II and Swift in the not too distant future we hope that this crucial phase will be explored soon. Another prediction of a ring structure of the afterglow [75] will have to wait, however, to futuristic ultrahigh resolution detectors.

We know how GRBs are produced. We are less certain what produces them. We can trace backwards the evolution at the source from the observations of the emitting regions to an accretion disk - black hole system. The traces from this point backwards are less clear. Theoretical considerations [65] suggest that only Collapsars can produce the disk-black hole systems needed for long bursts while only NS-NS (or possibly NS-BH) mergers can produce the systems needed for short bursts. These conclusions are supported by the afterglows observations that suggest SN/GRB association for the long burst population. However, the picture is far from clear yet. While the information on the location of the bursts points out towards the SN connection the physical conditions within the afterglow indicate a low circumstellar density and does not show any indication for the almost inevitable pre explosion wind. The origin of the iron lines is still mysterious and confusing.

The Fireball model has still many open questions. Some are concerned with the physics of the fireball model: How do the collisionless shocks work? How are the electrons accelerated and how are the magnetic fields amplified? How do jets expand sideways? What controls the "typical" emission to be in the soft γ-ray region? Other questions deal with more astrophysical issues like: What happens in all the cases (like WD-BH merger) in which a GRB almost form but the conditions are not exactly right? What distinguishes between the prgenitor of a GRB-"failed supernova" and the progenitor of a successful supernova with no GRB? Finally, we have, of course, the sixty four thousand dollars question: How does the "inner engine" accelerates the ejecta to relativistic velocities?.

REFERENCES

1. C. A. Meegan et al., Nature, **355**, 143 (1992).
2. E. Costa et al., Nature, **387**, 783 (1997).
3. J. van Paradijs et al., Nature, **386**, 686 (1997).
4. D. A. Frail et al., Nature, **389**, 261 (1997).
5. M. R. Metzger et al., Nature, **387**, 879 (1997).
6. T. Piran, Physics Reports, **314**, 575 (1999).
7. T. Piran, Physics Reports, **333**, 529 (2000).
8. M. Ruderman, New York Academy Sciences Annals **262**, 164 (1975).
9. P. Meszaros and M. J. Rees, Ap. J., **405**, 278 (1993).
10. R. Narayan, B. Paczynski, and T. Piran, Ap. J. Lett., **395**, L83 (1992).
11. M. J. Rees and P. Meszaros, Ap. J. Lett., **430**, L93 (1994).
12. R. Sari and T. Piran, Ap. J., **485**, 270 (1997).
13. S. Kobayashi, T. Piran, and R. Sari, Ap. J., **490**, 92 (1997).
14. R. Sari and T. Piran, Ap. J. Lett., **455**, L143 (1995).
15. J. Goodman, Ap. J. Lett., **308**, L47 (1986).
16. B. Paczynski, Ap. J. Lett., **308**, L43 (1986).
17. A. Shemi and T. Piran, Ap. J. Lett., **365**, L55 (1990).
18. E. E. Fenimore, R. I. Epstein, and C. Ho, Astron. & Astroph. Supp., **97**, 59 (1993).
19. E. Woods and A. Loeb, Ap. J., **453**, 583 (1995).
20. T. Piran, in *Some Unsoved Problems in Astrophysics*, Eds. J. N. Bahcall and J. P. Ostriker, (Princeton University Press,1996).
21. Y. Lithwick and R. Sari, astro-ph /0011508 (2000).
22. J. Goodman, New Astronomy **2**, 449 (1997).
23. J. I. Katz and T. Piran, Ap. J., **490**, 772 (1997).
24. B. Paczynski and J. E. Rhoads, Ap. J. Lett., **418**, L5 (1993).
25. P. Meszaros and M. J. Rees, Ap. J. Lett., **418**, L59 (1993).
26. P. Meszaros and M. J. Rees, Ap. J., **476**, 232 (1997).
27. M. Vietri, Ap. J. Lett., **478**, L9 (1997).
28. R. D. Blandford and C. F. McKee, Physics of Fluids **19**, 1130 (1976).
29. R. Sari, T. Piran, and R. Narayan, Ap. J. Lett., **497**, L17 (1998).
30. R. A. Chevalier and Z. Li, Ap. J., **536**, 195 (2000).
31. R. Sari, R. Narayan, and T. Piran, Ap. J., **473**, 204 (1996).
32. A. Panaitescu and P. Kumar, to appear in Ap. J., (2001).

33. R. A. M. J. Wijers and T. J. Galama, Ap. J., **523**, 177 (1999).
34. E. E. Fenimore, C. D. Madras, and S. Nayakshin, Ap. J., **473**, 998 (1996).
35. F. Daigne and R. Mochkovitch, Mon. Not. RAS, **296**, 275 (1998).
36. R. A. Burenin et al., Astron. & Astroph., **344**, L53 (1999).
37. E. Costa, Astron. & Astroph. Supp., **138**, 425 (1999).
38. T. W. Giblin et al., Ap. J. Lett., **524**, L47 (1999).
39. V. Connaughton, in *The 19th Texas Symposium on Relativistic Astrophysics and Cosmology, Eds.: J. Paul, T. Montmerle, and E. Aubourg (CEA Saclay)*. (1998).
40. R. Sari and T. Piran, Astron. & Astroph. Supp., **138**, 537 (1999).
41. C. W. Akerlofy et al., Nature, **398**, 400 (1999).
42. R. Sari and T. Piran, Ap. J., **520**, 641 (1999).
43. T. Piran, in *AIP Conf. Proc. 307: Gamma-Ray Bursts* (1994), pp. 495.
44. J. E. Rhoads, Ap. J., **525**, 737 (1999).
45. R. Sari, T. Piran, and J. P. Halpern, Ap. J. Lett., **519**, L17 (1999).
46. A. Panaitescu and P. Mészáros, Ap. J., **526**, 707 (1999).
47. A. Panaitescu and P. Kumar, Ap. J., **543**, 66 (2000).
48. J. Granot et al., astro-ph/0103038 (2001).
49. S. R. Kulkarni et al., Nature, **398**, 389 (1999).
50. F. A. Harrison et al., Ap. J. Lett., **523**, L121 (1999).
51. K. Z. Stanek et al., Ap. J. Lett., **522**, L39 (1999).
52. D. A. Frail et al., astro-ph/0102282 (2001).
53. S. R. Kulkarni, This volume (2001).
54. E. Nakar and T. Piran, astro-ph/0103192 (2001).
55. E. Ramirez-Ruiz, A. Merloni, and M. J. Rees, astro-ph/0010219 (2000).
56. E. Nakar and T. Piran, astro-ph/0103210 (2001).
57. V. V. Usov, Nature, **357**, 472 (1992).
58. D. Eichler, M. Livio, T. Piran, and D. N. Schramm, Nature, **340**, 126 (1989).
59. B. Paczynski, Acta Astronomica **41**, 257 (1991).
60. C. L. Fryer, S. E. Woosley, M. Herant, and M. B. Davies, Ap. J., **520**, 650 (1999).
61. C. L. Fryer and S. E. Woosley, Ap. J. Lett., **502**, L9 (1998).
62. S. E. Woosley, Ap. J., **405**, 273 (1993).
63. B. Paczynski, Ap. J. Lett., **494**, L45 (1998).
64. A. I. MacFadyen and S. E. Woosley, Ap. J., **524**, 262 (1999).
65. R. Narayan, T. Piran, and P. Kumar, astro-ph/0103360 (2001).
66. T. J. Galama et al., Nature, **395**, 670 (1998).
67. D. E. Reichart, Ap. J. Lett., **521**, L111 (1999).
68. B. J. S. et al., Nature, **401**, 453 (1999).
69. L. Piro et al., Science **290**, 955 (2000).
70. B. J. S., S. R. Kulkarni, and Djorgovski, astro-ph /**0010176**, (2000).
71. A. A. Esin and R. Blandford, Ap. J. Lett., **534**, L151 (2000).
72. M. Vietri et al., Ap. J. Lett., **550**, L43 (2001).
73. A. V. Tutukov and L. R. Yungelson, Mon. Not. RAS, **268**, 871 (1994).
74. P. Kumar and T. Piran, Ap. J., **532**, 286 (2000).
75. J. Granot, T. Piran, and R. Sari, Ap. J., **513**, 679 (1999).

GRB and Environment Interaction

P. Mészáros[1]

[1] *Pennsylvania State University, 525 Davey, University Park, PA 16802*

Abstract.
We discuss three aspects of the interaction between GRB and their surroundings. The illumination of the progenitor remnant and/or the surroundings by the X-ray afterglow continuum can produce substantial Fe K-alpha line and edge emission, with implications for the progenitor model. The presence of large dust column densities, capable of obscuring the GRB optical afterglow, will lead to characteristic delayed X-ray and far-IR light curve signatures. Pair production induced by the initial gamma-rays in the nearby environment will modify the initial spectrum and the afterglow light curve, and the magnitude of these changes provides a diagnostic for the external density.

I FE X-RAY LINES FROM GRB PROGENITORS

Important clues for identifying the nature of the progenitors of the long ($t \gtrsim 2$ s) GRBs may be available from the recent report at a 4.7σ level of X-ray Fe line features in the afterglow after 1.5 days of the gamma-ray burst GRB 991216 [10], as well as similar detections at the 3σ level in 5 other bursts with Beppo-SAX and ASCA. X-ray atomic edges and resonance absorption lines are theoretically expected to be detectable from the gas in the immediate environment of the GRB, and in particular from the remnants of a massive progenitor stellar system [15,14,16].

A straightforward interpretation [10] of the GRB 991216 observation would imply a mass $\gtrsim 0.1 - 1 M_\odot$ of Fe at a distance of about 1-2 light-days, possibly due to a remnant of an explosive event or supernova which occurred days or weeks prior to the gamma-ray burst itself (a 'supranova', [10,12]). The long time delay is necessary both to get the relatively massive, slow moving ejecta out to few light-day distances (to explain the line appearance at a few days with light travel arguments), and in order to get the initial Ni and Co to decay to Fe (~ 55 days). This requires a two-step process, in which an initial supernova leads to a temporarily stabilized neutron star remnant, which after weeks collapses to a black hole leading to a canonical burst ([11,12]). It is unclear whether fall-back from the supernova leading to the second collapse to a BH could occur with such a ($\sim$ weeks) long delay (e.g. [18]). Another possibility is that a massive progenitor has previously emitted a copious wind ($\dot{M} \gtrsim 10^{-4} M_\odot$/yr), which would need to be unusually Fe-rich and highly inhomogeneous ([14]; c.f. [10]).

An alternative, and perhaps less restrictive scenario for such Fe lines [17] involves an extended, possibly magnetically dominated wind from a GRB impacting the expanding envelope of a massive progenitor star. This could be due either to a spinning-down millisecond super-pulsar or to a highly-magnetised torus around a black hole (e.g. [13]), which could produce a luminosity that was still, one day after the original explosion, as high as $L_m \sim 10^{47} t_{day}^{-1.3}$ ergs. An outflow with such a dependence can also be powered by accretion of fall-back material onto a central black hole [18]. This jet luminosity may not dominate the continuum afterglow; but its impact on the outer portions of the expanding stellar envelope at distances $\lesssim 10^{13}$ cm, even with just solar abundances, can be efficiently reprocessed into an Fe line luminosity comparable to the observed value, together with a contribution to the X-ray continuum. Under this interpretation, the dominant continuum flux in the afterglow, even in the X-ray band, is still attributable to a standard decelerating blast wave.

The relativistic magnetised wind from the compact remnant would develop a stand-off shock before encountering the envelope material, and shocked relativistic plasma would be deflected along the funnel walls. Non-thermal electrons will be accelerated behind the standoff shock in the jet material; the transverse magnetic field strength (which decreases as $1/r$ in an outflowing wind) would be of order 10^4 G at 10^{13} cm – strong enough to ensure that the shock-accelerated electrons cool promptly, yielding a power- law continuum extending into the X-ray band. Some of these X-rays would escape along the funnel, but at least half (the exact proportion depending on the geometry and flow pattern) would irradiate the material in the stellar envelope. Pressure balance in the shoked envelope wall implies densities of $n_e = \alpha L_m / 6\pi r^2 ckT \sim 10^{17} \alpha L_{47} r_{13}^{-2} T_8^{-1}$ cm^{-3}, where $\alpha \sim 1$ is a geometric factor, and the recombination time for hydrogenic Fe in the funnel walls photoionized by the non-thermal continuum is $t_{rec} = 6 \times 10^{-6} T_8^{1/2} n_{17}^{-1} \sim 6 \times 10^{-6} \alpha L_{m47}^{-1} r_{13}^2 T_8^{3/2}$ s. Standard calculations of photoionization of optically-thin slabs (e.g. [7]) show that the equivalent width of the Fe K-alpha line, for solar abundances, is about 0.5 kev, or twice as strong if the Fe has ten times solar abundances. These results are applicable provided that the ionizing photons encounter a Fe ion before being scattered by free electrons i.e. provided that $\tau_T = \sigma_T d_i n_e \lesssim 1$. Under these conditions the Fe K-α photon flux is about 0.1 of the X-ray continuum [17], $\dot{N}_{LFe} \sim 10^{54} L_{47} \beta$ ph/s, where $\beta < 1$ is the ratio of ionizing to MHD luminosity. This line luminosity compares well with Fe line luminosity 6×10^{52} ph/s observed $t \sim 1.5$ day after the GRB 991216 burst by [10].

The total amount of Fe needed to explain the observed K-α line flux, arising in a thin layer of the funnel walls of a collapsar model, amounts to a very modest mass of $M_{Fe} \sim 10^{-8} M_\odot$, which could be Fe synthesized in the core. The Fe-enriched core material can easily reach a distance comparable to $r \sim 10^{13}$ cm in 1 day for an expansion velocity below the limit $v \sim 10^9$ cm s^{-1} inferred by [10] from the line widths. Even without this, a solar abundance ($10^{-5} M_\odot$ of Fe) in the envelope is sufficient to explain the observations. The initial, energetic portion of the relativistic jet, with a typical burst duration of 1 − 10 s, will rapidly expand beyond the stellar envelope, leading in the usual way to shocks and a decelerating blast wave. A continually decreasing fraction of energy, such as put out by a decaying magnetar, may continue being emitted for periods of a day or longer, and its reprocessing by the stellar envelope can be responsible for the observed Fe line emission in GRB 991216. Since the energy in this tail can decay faster than t^{-1}, the usual standard shock gamma-ray and afterglow scenario need not be affected, being determined by the first 1-10 s worth of the energy input.

II DUSTY GRB DELAYED XR/IR AFTERGLOWS

For GRB in large star forming regions, a significant fraction of the prompt X-ray emission will be scattered by dust grains. Since dust grains scatter X-rays by a small angle, time delays of the scattered x-rays will be small (minutes to days, depending on the X-ray energy and the grain size). If the dust column density is substantial, the softer part of the X-ray afterglow on the above timescales will be dominated by the dust scattering, the direct X-ray emission from the blast wave being weaker. This intermediate time, soft(er) X-ray light curve will be steeper than the unscattered X-ray afterglow.

As a specific example [6], consider a typical GRB whose unscattered X-ray light curve is parametrized as $F_0(t) = [1 + (t/100\text{s})]/[1 + (t/100\text{s})^{2.3}]$, with an arbitrary normalization depending on the X-ray energy band. This is represented by the thin line in Fig.1. We assume that the GRB occurs in a large star forming region, of typical radius R about 100pc, where the dust grain populations and optical depths are close to what is observed in our Galactic center region. Thus for numerical estimates we assume that (1) visual extinction is ~ 10, (2) X-rays are scattered preferentially by those dust grains whose size is in the range $a \sim 0.06\mu$m, (3) the optical depth to dust scattering at the X-ray energy ϵ is $\tau(\epsilon) = 3 \left(\frac{\epsilon}{1\text{keV}}\right)^{-2}$. At X-ray optical depths less than few, dust grains of size a will scatter X-rays of energy ϵ by an angle $\theta \sim 0.2\lambda/a$, where λ is the X-ray wavelength, $\theta(\epsilon) \simeq 4 \times 10^{-3} \left(\frac{a}{0.06\mu m}\right)^{-1} \left(\frac{\epsilon}{1\text{keV}}\right)^{-1}$. The time lag is $t \sim R\theta^2/2c$, or $t(\epsilon) \sim 9 \times 10^4 \text{s} \left(\frac{a}{0.06\mu m}\right)^{-2} \left(\frac{\epsilon}{1\text{keV}}\right)^{-2} \left(\frac{R}{100\text{pc}}\right)$.

At 2 keV, the optical depth is $\tau \sim 1$. The time lag is $t \sim 2 \times 10^4$s. The scattered flux is $F_s \sim \tau f/t \sim 0.03$. The unscattered flux at 2×10^4s is $F_0 \sim$

10^{-3}. In the time interval from hours to weeks, the dust scattering dominates the afterglow, and, as shown in Fig.1, the afterglow is approximately a power law $F \propto t^{-1.75}$ [6]. This is because dust grains of radius $a < 0.06\mu m$ will scatter the prompt emission with longer time lags, $t \propto a^{-2}$, and with smaller optical depths τ. To calculate τ, we take a standard dust grain size distribution where the number of grains of size of order a is $\propto a^{-2.5}$ For a scattering cross section $\propto a^4$ the optical depth is $\tau \propto a^{1.5} \propto t^{-0.75}$, so the flux $F \propto t^{-1.75}$.

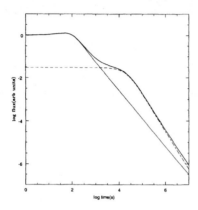

FIGURE 1. Dust-scattered X-ray afterglow. Thin line: unscattered X-ray flux. Thin dashed line: scattered X-ray flux. Thick line: total flux. The flux normalization is arbitrary, the relative fluxes correspond to the example discussed in the text for an energy of 2 keV [6].

A GRB in such a highly obscured star-forming region should lead to specific signatures in the X-ray afterglow, i.e. a bump in the X-ray light curve at energies $\epsilon \sim 2-3$ keV, hours to days after the burst [6]. This X-ray signature is expected for bursts which do not produce a detectable optical transient (OT). Such OT-less, X-ray peculiar GRBs will also lead to thermal reemission and scattering of the O/UV flux causing a delayed IR emission, as is the case also for partially absorbed bursts [8,9]. For an isotropic total burst energy $E \sim 10^{53}$ erg at a redshift $z \sim 1$ the normalization of the X-ray flux for the burst of Figure 1 would be $F_x \sim 10^{-9}$ erg cm^{-2} s^{-1} keV^{-1} for $t \lesssim 100$ s, in the usual range of X-ray afterglow fluxes detected by Beppo-SAX. The dust reradiation occurs beyond the sublimation radius $R_s \sim 10\, L_{49}^{1/2}$ pc at wavelengths $\lambda \gtrsim 2(1+z)\mu m$, where $10^{49} L_{49}$ erg/s is the early UV component of burst afterglow [8]. The time delay associated with the reradiated flux is $t_{IR} \sim (R_s/2c)\theta_j^2$ where $\theta_j = 10^{-1}\theta_{-1}$. At $z \sim 1$ the corresponding infrared flux at 2.2 μm would be $F_{2.2\mu m} \sim L_{49}\theta_j^2/[4\pi D_L^2(R_s/2c)\theta_j^2 \nu] \sim 0.3 L_{49}^{1/2}$ μJy, independent of θ_j, or $m_K \sim 23.3$ compared to Vega [6], approximately constant for a time $t_{IR} \sim 5 \times 10^6 \theta_{-1}^2 L_{49}^{1/2}$ s. Such γ-ray detected GRBs with anomalous X-ray afterglow behavior and no OT may be used as tracers of massive stellar collapses. It may thus be possible to detect star-forming regions out to redshifts larger than achievable with O/IR techniques, since typical GRB γ-ray, X-ray and IR fluxes can in principle be measured out to $z \sim 10-15$.

III PAIR PRODUCTION IN GRB ENVIRONMENTS

Gamma-ray burst sources with a high luminosity can produce $e^\pm$ pair cascades in their environment as a result of back-scattering of a seed fraction of their original hard spectrum. New pairs can be made as some of the initial energetic photons are backscattered and interact with other incoming photons. Previous work on this investigated the acceleration of new pairs for a particular fireball model [1,4], the effect of pair formation for a low compactness parameter external shock model of GRB [3], and Compton echoes produced by pairs [2]. Here we discuss a simplified analytical treatment [5] of pair effects from γ-rays arising in internal shocks in a wind; the remaining wind energy drives a blast wave which decelerates as it sweeps up the external medium, and gives rise to the afterglow emission. The γ-rays would propagate ahead of the blast wave, leading to pair production (and an associated deposition of momentum) into the external medium. The pair cascades saturate after the external (pair-enriched) medium reaches a critical bulk Lorentz factor, which is generally below that of the original relativistic wind. For external baryonic densities similar to those in molecular clouds the pairs can achieve scattering optical depths $\tau_\pm \lesssim 1$. Even for less extreme external densities the effect of the additional pairs can be substantial, increasing the radiative efficiency of the blast wave and leading to distortions of the original spectrum. This provides a potential tool for diagnosing the compactness parameter of the bursts and thus the radial distance at which shocks can occur. It also provides a tool for diagnosing the baryonic density of the external environment, and testing the association with star-forming regions.

For the maximum Lorentz factor to which an $e^\pm$ can be accelerated by scattering, and the maximum Lorentz factor at which back-scattered photons can still make new pairs, one finds two regimes defined by the effective duration of the light pulse seen by the screen of accelerated pairs. At low radii (wind regime) the effective duration is the burst duration t_w; for large radii (impulsive regime), the effective duration is $\Delta t \sim r/c\Gamma_\pm^2$. For an incident photon number index $\beta = 2$, in the former $\Gamma_\pm \propto r^{-1/3}$ and in the latter $\Gamma_\pm \propto r^{-2}$. The critical radius and Lorentz factor for the transition between the wind and the impulsive regimes are [5] $r_c = 5 \times 10^{14} L_{w50}^{2/5} t_{w1}^{3/5}$, $\Gamma_c = 3 \times 10^1 L_{w50}^{1/5} t_{w1}^{-1/5}$. The maximum radius at which pair cascades cut off is $r_\ell \sim (4r_* ct_w/3)^{1/2} \sim 4 \times 10^{15} L_{w50}^{1/2} t_{w1}^{1/2}$ cm. Before the pairs start accelerating, assuming they are held back by the environmental protons through magnetic fields, an initial cascade amplification fator $k_p \sim (m_p/m_e)$ is achieved. After the mean mass per scatterer drops to a value comparable to the electron mass, before reaching r_ℓ a further amplification $k_a \sim 2^s \sim 50$ (where $s \sim log\Gamma_c/\log 2$) is possible, so the total pair amplification factor is [5] $k_c = k_p k_a(r_c) \sim (m_p/m_e)\Gamma_c \sim 5 \times 10^4 L_{w50}^{1/5} t_{w1}^{-1/5}$. The maximum pair optical depth at r_c, which is prevented from exceeding $\tau_\pm \sim 1$ by self-shielding, is achieved for external densities $n_p \gtrsim n_{p,c}$, where $n_{p,c} \simeq 10^5 L_{w50}^{-3/5} t_{w1}^{-2/5} \text{cm}^{-3}$.

The external density and the initial Lorentz factor η determine when the outer shock and the reverse shock become important and whether this happens within the radius already polluted with pairs. If $\eta \lesssim r_l/ct_w$ the external shock responsible for the afterglow occurs beyond the region "polluted" by new pairs, and otherwise the afterglow shock may experience, after starting out in the canonical manner, a "resurgence" or second kick as its radiative efficiency is boosted by running into an $e^\pm$-enriched gas [5].

Additional effects are expected when $\tau_\pm \to 1$, for external baryon density $n_p \gtrsim n_{c,p} \sim 10^5 L_{w50}^{-2/5} t_{w1}^{-3/5} \text{cm}^{-3}$ at radii $r < r_\ell$. Such high densities could be expected if the burst is associated with a massive star in which prior mass loss led to a dense circumstellar envelope. The pair optical depth saturates to $\tau_\pm \sim 1$ and in addition to an increased efficiency and softer spectrum of the afterglow reverse shock, the original gamma-ray spectrum of the GRB will be modified as well. One of the consequences of such a critical external density leading to $\tau_\pm \sim 1$ would be the presence of an X-ray quasi-thermal pulse, whose total energy may be a few percent of the total burst energy [5].

This research was supported by NASA NAG5-9192, the Guggenheim Foundation and the Sackler Foundation. I am grateful to M.J. Rees, A. Gruzinov and E. Ramirez-Ruiz for valuable discussions on these topics.

REFERENCES

1. Madau, P & Thompson, C, 2000 ApJ, 534, 239
2. Madau, P, Blandford, R & Rees, M.J., 2000, ApJ, 541, 712.
3. Dermer, C & Böttcher, M., 2000 ApJ 534, L155
4. Thompson, C & Madau, P, 2000 ApJ, 538, 105
5. Mészáros, P, Ramirez-Ruiz, E & Rees, M, 2001, ApJ subm(astro-ph/0011284)
6. Mészáros, P & Gruzinov, A, 2000, ApJL, 543, L35 (astro-ph/0007255)
7. Young, A.J., 1999, Ph.D. thesis, Cambridge University
8. Waxman, E & Draine, B., 2000, ApJ, in press (astro-ph/9909020)
9. Esin, A & Blandford, R.D., 2000, ApJL in press (astro-ph/0003415)
10. Piro, L, et al., 2000, Science, 290, 955)
11. Vietri, M & Stella, L.A., 1998, ApJ 507, L45
12. Vietri, M, Perola, G, Piro, L & Stella, L, 2000, MNRAS 308, L29.
13. Wheeler, J.C, et al., 2000, ApJ in press (astro-ph/9909293)
14. Weth, C, Mészáros, P, Kallman, T & Rees, M.J, 2000, ApJ 534, 581
15. Mészáros, P & Rees, M.J. 1998, MNRAS, 299, L10
16. Böttcher, M & Fryer, C.L, 2000, ApJ, subm. (astro-ph/0006076)
17. Rees, M.J. & Mészáros, P, 2000, ApJ, 545, L73
18. MacFadyen, A, Woosley, S & Heger, A, 2000, ApJ(astro-ph/9910034)

A Survey of the Host Galaxies of Gamma-Ray Bursts[1]

Stephen Holland*

*Department of Physics
University of Notre Dame
Notre Dame, IN 4655–5670
U. S. A.

Abstract. We used 45 HST/STIS orbits in Cycle 9 to obtain deep images of the host galaxies of eleven gamma-ray bursts. Our goals are to study the morphologies of the host galaxies, to obtain precise locations of gamma-ray bursts within their host galaxies, and to determine star-formation rates in the hosts. We present preliminary results for GRB 980425/SN1998bw, GRB 980613, and GRB 9809703.

INTRODUCTION

We have obtained deep images of the host galaxies of eleven gamma-ray bursts (GRBs) using the Space Telescope Imaging Spectrograph (STIS) aboard the *Hubble Space Telescope* (*HST*). Data was taken using the 50CCD (clear) and F28X50LP (long pass) apertures, which peak at approximately the V and R photometric bands. Each image was taken more than one year after the burst to study the overall morphology, and the small-scale structure of each host, without contamination from the GRB's optical afterglow (OA). Our goals are to classify the morphologies of the host galaxies, identify substructure in each host, and to probe the star-formation rate at high redshifts. Combining ground-based observations of the OAs with our *HST* data allows us to determine precise positions for each GRB relative to substructure (such as bulges, spiral arms, HII regions, star-forming regions, etc.) in the host. The images will be used to compare the distribution of host morphologies to galaxies in the Hubble Deep Fields, and to search for correlations between specific types of substructure and GRBs. Morphological information, and the spectral energy distribution, will allow us estimate overall star-formation rates and the amount of dust present. We have waived all proprietary rights to this data. The reduced data are available at http://www.ifa.au.dk/~hst/grb_hosts/index.html. This paper presents some preliminary results from this project.

---

[1] Based on observations with the NASA/ESA *Hubble Space Telescope*, obtained at the Space Telescope Science Institute, which is operated by the Association of Universities for Research in Astronomy, Inc. under NASA contract No. NAS5-26555.

GRB 980425

There is growing evidence that GRBs are related to the deaths of massive stars. GRB 980425 and the Type Ib/c supernova SN1998bw coincided in time and position on the sky [5], GRB 990123 coincided with a star-forming region [8], and there is evidence for a GRB/SN connection for eleven other GRBs (see [10] for a list).

The host galaxy of GRB 980425 is ESO 184−G82, a sub-luminous, barred spiral (SBc) galaxy that is in a stage of strong star formation. The galaxy lies at a distance of 36.64 Mpc, making GRB 980425 the nearest known GRB. The host has a size, structure, luminosity, and star formation rate similar to those of the Large Magellanic Cloud. ESO 184−G82 appears to be a member of a compact group with the nearest neighbour located at a projected distance of only 11.9 kpc and shows indications of being morphologically disturbed. Therefore, this galaxy may be undergoing interaction-induced star formation.

SN1998bw/GRB 980425 occurred in an H II region approximately 300 pc in diameter. There are several bright, young stars within a projected distance of approximately 100 pc of the supernova/GRB (Fig. 1).

FIGURE 1. This Figure shows our *HST*/STIS 50CCD image of the star-forming region where GRB 980425/SN1998bw occurred. The plate scale for all the *HST*/STIS images presented in this paper is 0."0254/pixel. The circle shows the location of the supernova/GRB. Three of the stars in the star-forming region have colours that are consistent with red giants while three stars have blue colours that are consistent with their being massive main-sequence stars [4].

GRB 980613

GRB 980613 occurred near the edge of a compact blue star-forming region which may be part of a larger structure (Fig. 2). The morphology of the host is chaotic, and we find no evidence for spiral structure, or faint substructure connecting the various components. We find a star-formation rate of $\approx 3\mathcal{M}_\odot\mathrm{yr}^{-1}$ assuming no extinction in the host galaxy. The rest-frame B-band luminosity is $L_B \approx (0.2 \pm 0.1)L_B^*$ where L_B^* is the luminosity of a typical galaxy at a redshift of $z \approx 1$. The specific star-formation rate per unit blue luminosity is $\approx 20\ \mathcal{M}_\odot\mathrm{yr}^{-1}L_B^{*-1}$, the highest value of any known GRB host galaxy [7].

Djorgovski et al. [3] suggested that the host is a set of interacting galaxies. If this is the case then the lack of tidal features suggests that the system has only recently interacted. Alternatively, the colour and morphology of the host are similar to what is seen in low surface brightness (LSB) galaxies. However, the host galaxy shows strong nuclear activity. This is unlike LSB galaxies, which tend not to exhibit nuclear activity, although some of the larger LSB disk galaxies do show nuclear activity [2] similar to that seen in the host of GRB 980613. The large specific star-formation rate is unusual for an LSB. However, most of the star formation occurs in the nucleus and is not distributed through-out the galaxy. Therefore, the host may be an LSB galaxy where star formation is in the process of turning on.

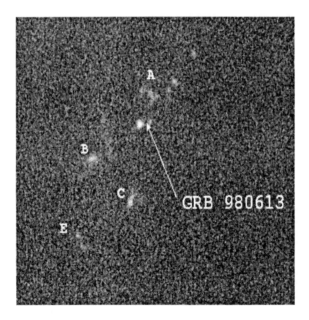

FIGURE 2. This is our STIS 50CCD (clear) image of the environment of GRB 980613. The resolution (i.e., the apparent diameter of a point source) is $0\rlap{.}{''}084$. The field of view is approximately $7\rlap{.}{''}5 \times 7\rlap{.}{''}5$ ($= 60 \times 60$ kpc assuming $(H_0, \Omega_m, \Omega_\Lambda) = (70, 0.3, 0.7)$).

GRB 980703

The host of GRB 980703 is a bright galaxy with an "egg-shaped" morphology that resembles a lop-sided barred spiral (Fig. 3a). We estimate a star-formation rate of 8–13$\mathcal{M}_\odot$yr^{-1} using Eq. 2 of [12] and assuming that there is no extinction in the host galaxy. The rest-frame B-band luminosity is $(1.6 \pm 0.4)L_B^*$. Therefore, the star-formation rate per unit luminosity is $\approx 6.5 \mathcal{M}_\odot\text{yr}^{-1}L_B^{*-1}$, which is similar to the values found for several other GRB host galaxies.

We fit two-dimensional Sersic [16] models to the HST/STIS images, after convolving with the appropriate point-spread function [10], and found a half-light radius of $0\rlap.{''}13 \pm 0\rlap.{''}01$, $n = 1.05 \pm 0.02$ ($n = 4$ corresponds to a de Vaucouleur $R^{1/4}$ profile), and an ellipticity of 0.24 ± 0.02. This is consistent with an exponential disc with a scale radius of $0\rlap.{''}21 \pm 0\rlap.{''}01$. Fig. 3b shows the host with the best-fitting model subtracted. Except for the central regions the galaxy is well fit by this model. The systematic residuals in the central few pixels suggest that there is substructure in the galaxy. The excess of light on the west side of the galaxy may be a spiral arm. The derived half-light radius is much smaller than those seen in local late and early type galaxies [11]. However, the size, colour, and spectrum of the host are similar to those of compact galaxies in the Hubble Deep Field North [13,6].

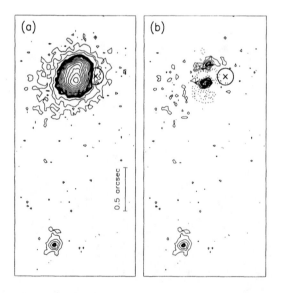

FIGURE 3. a): This Figure shows the HST/STIS 50CCD GRB 980703. The outer contours are linear while the inner contours are logarithmic to show the host over a large dynamic range. The location of GRB 980703 is marked with an "X" and our 1σ error circle. **b)**: This Figure shows the host with the best-fitting Sersic [16] model subtracted.

STAR FORMATION AND HOST MORPHOLOGY

There is evidence that redshifts of GRBs can be determined from the intrinsic properties of the bursts [15,14]. These techniques will allow the distribution of GRBs to be mapped to high redshifts solely from the observed gamma-ray pulse. However, in order to connect the GRB rate with star formation in the early Universe it is important to understand the connection between GRBs and star formation in individual galaxies. Preliminary results from our Cycle 9 observations suggest that long-duration ($T_{90} > 2$ s) GRBs are located in star-forming regions [9,10,7]. This is consistent with the growing evidence that GRBs are produced in the early stages of a supernova explosion. Our data suggests that hosts tend to be sub-luminous with high rates of star-formation per unit luminosity (see Table 1).

TABLE 1. Specific star-formation rates for several GRB host galaxies.

GRB	z	R_{host}	$\mathcal{M}_\odot \text{yr}^{-1} L_B^{*\,-1}$
970508	0.835	25.20	11.0
980613	1.096	24.56	20.0
980703	0.966	22.57	6.5
990123	1.600	24.07	11.0
990712	0.434	21.91	4.4

Fig. 4 shows images of eight GRB host galaxies. There is no single morphological type for GRB hosts, although most appear to occur in late-type and irregular galaxies (but see [1]). Our preliminary results suggest that the morphologies of GRB host galaxies are no different from other star-forming galaxies at the same redshift.

REFERENCES

1. Bloom, J. S., Kulkarni, S. R., and Djorgovski, S. G., **ApJ** (2001) submitted, astro-ph/0010176
2. Bothun, G. D., Impey, C. D., and McGaugh, S. S., *PASP*, **109**, 745 (1997)
3. Djorgovski, S. B., Bloom, J. S., and Kulkarni, S. R., *ApJL* (2001) submitted, astro-ph/0008029
4. Fynbo, J. U., et al., *ApJL* **542**, L89 (2000)
5. Galama, T. S., *Nature* **395**, 670 (1998)
6. Guzmán, R., Gallego, J., Koo, D. C., Phillips, A. C., Lowenthal, J. D., Faber, S. M., Illingworth, G. D., and Vogt, N. P., *ApJ* **489**, 559 (1997)
7. Hjorth, J., et al., (2001) in prep.
8. Holland, S., and Hjorth, J., *A&A* **344**, L67 (1999)
9. Holland, S., et al., *GCN Circ.* **704**; *GCN Circ.* **715**; *GCN Circ.* **749** (2000)
10. Holland, S., et al., *A&A* (2001) in press

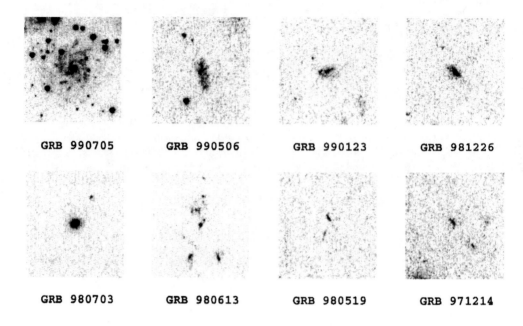

FIGURE 4. This Figure shows *HST*/STIS images of eight host GRB galaxies. Each image is in 50CCD except GRB 990123, which is in F28X50LP. The fields of view are $6''\!.5 \times 6''\!.5$.

11. Im, M., Casertano, S., Griffiths, R. E., Ratnatunga, K. U., and Tyson, J. A., *ApJ* **441**, 494 (1995)
12. Madau, P., Pozzetti, L., and Dickinson, M., *ApJ* **498**, 106 (1998)
13. Phillips, A. C., Guzmán, R., Gallego, J., Koo, D. C., Lowenthal, J. D., Vogt, N. P., Faber, S. M., and Illingworth, G. D., *ApJ* **489**, 543 (1997)
14. Reichart, D. E., Lamb, D. Q., Fenimore, E. E., Ramirez-Ruiz, E., Cline, T. L., and Hurley, K., *ApJ* (2001) in press, astro-ph/0004302
15. Schmidt, M., *ApJ* (2001) in press, astro-ph/0101163
16. Sersic, J. L., *Atlas de Galxias Astrales*, Cordoba: Observatorio Astronomico (1968)

Construction and Preliminary Application of the Variability → Luminosity Estimator

Daniel E. Reichart[1,2] and Donald Q. Lamb[3]

[1] *Department of Astronomy, California Institute of Technology, Mail Code 105-24, 1201 East California Boulevard, Pasadena, CA 91125*
[2] *Hubble Fellow*
[3] *Department of Astronomy & Astrophysics, University of Chicago, 5640 South Ellis Avenue, Chicago IL, 60637*

Abstract. We present a possible Cepheid-like luminosity estimator for the long-duration gamma-ray bursts based on the variability of their light curves. We also present a preliminary application of this luminosity estimator to 907 long-duration bursts from the BATSE catalog.

INTRODUCTION

Since gamma-ray bursts (GRBs) were first discovered [1], thousands of bursts have been detected by a wide variety of instruments, most notably, the Burst and Transient Source Experiment (BATSE) on the *Compton Gamma-Ray Observatory (CGRO)*, which detected 2704 bursts by the end of *CGRO*'s more than 9 year mission in 2000 June (see, e.g., [2]). However, the distance scale of the bursts remained uncertain until 1997, when BeppoSAX began localizing long-duration bursts to a few arcminutes on the sky, and distributing the locations to observers within hours of the bursts. This led to the discovery of X-ray [3], optical [4], and radio [5] afterglows, as well as host galaxies [6]. Subsequent observations led to the spectroscopic determination of burst redshifts, using absorption lines in the spectra of the afterglows (see, e.g., [7]), and emission lines in the spectra of the host galaxies (see, e.g., [8]). To date, redshifts have been measured for 13 bursts.

Recently, [9] (see also [10]), [11] (see also [12]), and [13] (see also [14]) have proposed trends between burst luminosity and quantities that can be measured directly from burst light curves, for the long-duration bursts. Using 1310 BATSE bursts for which peak fluxes and high resolution light curves were available, [9] have suggested that simple bursts (bursts dominated by a single, smooth pulse) are less luminous than complex bursts (bursts consisting of overlapping pulses); however, see [15]. Using a sample of 7 BATSE bursts for which spectroscopic redshifts, peak

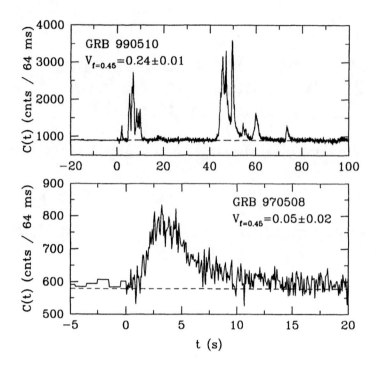

FIGURE 1. The > 25 keV light curves of the most (GRB 990510) and least (GRB 970508) variable cosmological BATSE bursts in our sample. In the case of GRB 990510 ($z = 1.619$), we find that $V = 0.24 \pm 0.01$. In the case of GRB 970508 ($z = 0.835$), we find that $V = 0.05 \pm 0.02$.

fluxes, and high resolution light curves were available, [11] have suggested that more luminous bursts have shorter spectral lags (the interval of time between the peak of the light curve in different energy bands). Using the same 7 bursts, [13] have suggested that more luminous bursts have more variable light curves. These trends between luminosity and quantities that can be measured directly from light curves raise the exciting possibility that luminosities, and hence luminosity distances, might be inferred for the long-duration bursts from their light curves alone.

In this paper (see also [15,16]), we present a possible luminosity estimator for the long-duration bursts, the construction of which was motivated by the work of [14] and [13]. We term the luminosity estimator "Cepheid-like" in that it can be used to infer luminosities and luminosity distances for the long-duration bursts from the variabilities of their light curves alone. We also present a preliminary application of this luminosity estimator to 907 long-duration bursts from the BATSE catalog.

We discuss the construction of our measure V of the variability of a burst light curve §2. In §3, we discuss our expansion of the original [14] sample of 7 bursts to include a total of 20 bursts, including 13 BATSE bursts, 5 *Wind*/KONUS bursts,

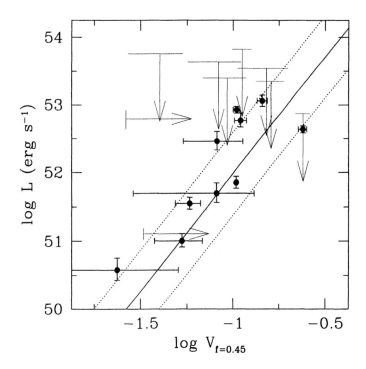

FIGURE 2. The variabilities V and isotropic-equivalent peak photon luminosities L between source-frame energies 100 and 1000 keV (see [15]) of the bursts in our sample, excluding GRB 980425. The solid and dotted lines mark the center and 1 σ widths of the best-fit model distribution of these bursts in the log L-log V plane.

1 *Ulysses*/GRB burst, and 1 NEAR/XGRS burst. Also in §3, we discuss the construction of our luminosity estimator. We present our preliminary application of this luminosity estimator in §4.

THE VARIABILITY MEASURE

Qualitatively, V is computed by taking the difference of the light curve and a smoothed version of the light curve, squaring this difference, summing the squared difference over time intervals, and appropriately normalizing the result. We rigorously construct V in [15]. We require it to have the following properties: (1) we define it in terms of physical, source-frame quantities, as opposed to measured, observer-frame quantities; (2) when converted to observer-frame quantities, all strong dependences on redshift and other difficult or impossible to measure quantities cancel out; (3) it is not biased by instrumental binning of the light curve,

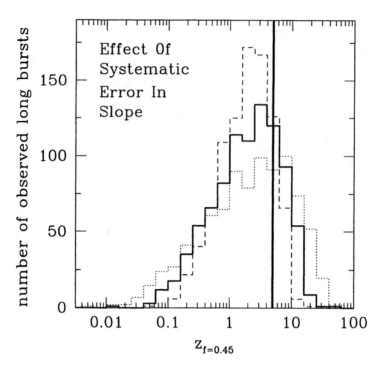

FIGURE 3. Solid histogram: The distribution of variability redshifts as determined using the best-fit luminosity estimator (Figure 2). Dotted and dashed histograms: The effect of varying the fitted slope of the luminosity estimator by $\pm$ its 1 σ statistical uncertainty. The vertical line marks $z = 5$.

despite cosmological time dilation and the narrowing of the light curve's temporal substructure at higher energies [17]; (4) it is not biased by Poisson noise, and consequently can be applied to faint bursts; and (5) it is robust; i.e., similar light curves always yield similar variabilities. Also in [15], we derive an expression for the statistical uncertainty in a light curve's measured variability, and we describe how we combine variability measurements of light curves acquired in different energy bands into a single measurement of a burst's variability. We plot the > 25 keV light curves of the most and least variable cosmological BATSE bursts in our sample in Figure 1.

THE LUMINOSITY ESTIMATOR

We list our sample of 20 bursts in Table 1 of [15]; it consists of every burst for which redshift information is currently available. Spectroscopic redshifts, peak fluxes, and high resolution light curves are available for 11 of these bursts; partial

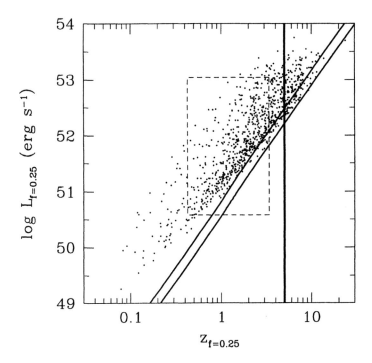

FIGURE 4. The joint redshift and luminosity distribution of the qualitatively acceptable redshift distribution of Figure 3 (the dashed histogram). The solid curves mark the 90% and 10% detection thresholds of BATSE. The dashed box marks the redshift and luminosity ranges of the bursts that were used to construct the luminosity estimator, everything outside of this box is technically an extrapolation. The vertical line marks $z = 5$.

information is available for the remaining 9 bursts. We rigorously construct the luminosity estimator in [15], applying the Bayesian inference formalism developed by [18]. We plot the data and best-fit model of the distribution of these data in the $\log L$-$\log V$ plane in Figure 2.

PRELIMINARY APPLICATION

We apply the best-fit luminosity estimator (Figure 2) to 1667 BATSE bursts, which are all of the BATSE bursts for which the necessary information was available as of the summer of 2000. To remove the short-duration bursts, we conservatively cut the bursts with durations of $T_{90} < 10$ sec from the sample, reducing the number to 907. We plot the distribution of variability redshifts in Figure 3. A rigorous analysis of how statistical and systematic errors affect this distribution will be presented elsewhere, but the largest effect is the statistical uncertainty in the fitted slope of the luminosity estimator. We plot how reasonable variations of this slope affect the distribution also in Figure 3. Although the original distribution has too many low-z and high-z bursts for comfort, reasonable variations appear to yield at least qualitatively acceptable distributions (see Figure 5 of [19]).

We plot the joint redshift and luminosity distribution of the qualitatively acceptable redshift distribution of Figure 3 (dashed histogram) in Figure 4. If systematic effects can be ruled out near BATSE's detection threshold, this distribution suggests that the luminosity distribution of the bursts is evolving, in which case no more than about 15% of bursts have redshifts greater than $z = 5$, in contrast to the results of [13] and [20].

REFERENCES

1. Klebesadel, R. W., Strong, I. B. & Olson, R. A. 1973, ApJ, 182, L85
2. Paciesas, W. S., et al. 1999, ApJ, 122, S465
3. Costa, E., et al. 1997, Nature, 387, 783
4. van Paradijs, J., et al. 1997, Nature, 386, 686
5. Frail, D. A., et al. 1997, Nature, 389, 261
6. Sahu, K. C., et al. 1997, Nature, 387, 476
7. Metzger, M. R., et al. 1997, Nature, 387, 878
8. Kulkarni, S. R., et al. 1998, Nature, 393, 35
9. Stern. B., Poutanen, J., & Svensson, R. 1999, ApJ, 510, 312
10. Stern. B., Svensson, R., & Poutanen, J. 1997, in 2d INTEGRAL Workshop: The Transparent Universe, ESA SP-382 (Paris: ESA), 473
11. Norris, J. P., Marani, G. F., Bonnell, J. T. 2000, ApJ, 534, 248
12. Norris, J. P., Marani, G. F., Bonnell, J. T. 2000, in Gamma-Ray Bursts: 5th Huntsville Symposium, AIP Conference Proceedings 526, eds. R. M. Kippen, R. S. Mallozzi, & G. J. Fishman (Melville, New York: AIP), 78
13. Fenimore, E. E., & Ramirez-Ruiz, E. 2001, ApJ, submitted (astro-ph/0004176)
14. Ramirez-Ruiz, E., & Fenimore, E. E. 1999, contributed oral presentation at the 5th Huntsville Gamma-Ray Burst Symposium
15. Reichart, D. E., et al. 2001, ApJ, in press (astro-ph/0004302)
16. Reichart, D. E., & Lamb, D. Q. 2001, in Procs. of Gamma-Ray Bursts in the Afterglow Era: 2nd Workshop, in press
17. Fenimore, E. E., et al. 1995, ApJ, 448, L101
18. Reichart, D. E. 2001, ApJ, in press (astro-ph/9912368)
19. Lamb, D. Q., and Reichart, D. E. 2000, ApJ, 536, 1, 1
20. Schaefer, B. E., Deng, M., & Band, D. L. 2001, ApJ (Letters), submitted (astro-ph/0101461)

Gamma-Ray Bursts as a Probe of Cosmology

Donald Q. Lamb* and Daniel E. Reichart[†]

*Department of Astronomy & Astrophysics, University of Chicago,
5640 South Ellis Avenue, Chicago, IL 60637
[†]Department of Astronomy, California Institute of Technology, Mail Code 105-24, 1201 East California Boulevard, Pasadena, CA 91125

Abstract. We show that, if the long GRBs are produced by the collapse of massive stars, GRBs and their afterglows may provide a powerful probe of cosmology and the early universe.

INTRODUCTION

There is increasingly strong evidence that gamma-ray bursts (GRBs) are associated with star-forming galaxies [1,2,3,4] and occur near or in the star-forming regions of these galaxies [2,3,4,5,6]. These associations provide indirect evidence that at least the long GRBs detected by BeppoSAX are a result of the collapse of massive stars. The discovery of what appear to be supernova components in the afterglows of GRBs 970228 [7,8] and 980326 [9] provides tantalizing direct evidence that at least some GRBs are related to the deaths of massive stars, as predicted by the widely-discussed collapsar model of GRBs [10,11,12,13,14]. If GRBs are indeed related to the collapse of massive stars, one expects the GRB rate to be approximately proportional to the star-formation rate (SFR).

DETECTABILITY OF GRBS AND THEIR AFTERGLOWS

We have calculated the limiting redshifts detectable by BATSE and HETE-2, and by *Swift*, for the sixteen GRBs with well-established redshifts and published peak photon number fluxes. In doing so, we have used the peak photon number fluxes given in Table 1 of [15], taken a detection threshold of 0.2 ph s^{-1} for BATSE and HETE-2 and 0.04 ph s^{-1} for *Swift*, and set $H_0 = 65$ km s^{-1} Mpc^{-1}, $\Omega_m = 0.3$, and $\Omega_\Lambda = 0.7$ (other cosmologies give similar results). Figure 1 displays the results. This figure shows that BATSE and HETE-2 would be able to detect half of these

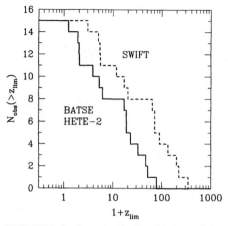
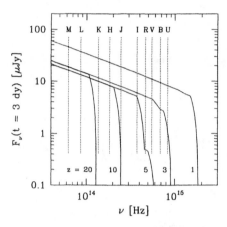

FIGURE 1. Cumulative distributions of the limiting redshifts at which the 15 GRBs with well-determined redshifts and published peak photon number fluxes would be detectable by BATSE and HETE-2, and by *Swift*.

FIGURE 2. The best-fit spectral flux distribution of the early afterglow of GRB 000131, as observed one day after the burst, after transforming it to various redshifts, and extinguishing it with a model of the Lyα forest.

GRBs out to a redshift $z = 20$ and 20% of them out to a redshift $z = 50$. *Swift* would be able to detect half of them out to redshifts $z = 70$, and 20% of them out to a redshift $z = 200$, although it is unlikely that GRBs occur at such extreme redshifts. Consequently, if GRBs occur at very high ($z > 5$) redshifts (VHRs), BATSE has probably already detected GRBs at these redshifts, and HETE-2 and *Swift* should detect them as well.

The soft X-ray, optical and infrared afterglows of GRBs are also detectable out to VHRs. The effects of distance and redshift tend to reduce the spectral flux in GRB afterglows in a given frequency band, but time dilation tends to increase it at a fixed time of observation after the GRB, since afterglow intensities tend to decrease with time. These effects combine to produce little or no decrease in the spectral energy flux F_ν of GRB afterglows in a given frequency band and at a fixed time of observation after the GRB with increasing redshift:

$$F_\nu(\nu, t) = \frac{L_\nu(\nu, t)}{4\pi D^2(z)(1+z)^{1-a+b}}, \qquad (1)$$

where $L_\nu \propto \nu^a t^b$ is the intrinsic spectral luminosity of the GRB afterglow, which we assume applies even at early times, and $D(z)$ is the comoving distance to the burst. Many afterglows fade like $b \approx -4/3$, which implies that $F_\nu(\nu, t) \propto D(z)^{-2}(1+z)^{-5/9}$ in the simplest afterglow model, where $a = 2b/3$ [16]. In addition, $D(z)$ increases very slowly with redshift at redshifts greater than a few. Consequently, there is little or no decrease in the spectral flux of GRB afterglows with increasing redshift beyond $z \approx 3$.

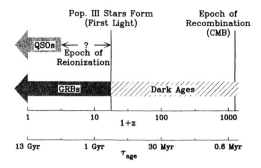

FIGURE 3. Cosmological context of VHR GRBs. Shown are the epochs of recombination, first light, and re-ionization. Also shown are the ranges of redshifts corresponding to the "dark ages," and probed by QSOs and GRBs.

In fact, in the simplest afterglow model where $a = 2b/3$, if the afterglow declines more rapidly than $b \approx 1.7$, the spectral flux actually *increases* as one moves the burst to higher redshifts! An example of this is the afterglow of GRB 000131. Its peak flux F_{peak} was in the top 5% of all BATSE bursts and the break energy E_{break} in its spectrum was 164 keV, yet it occurred at a redshift $z = 4.50$. We have calculated the best-fit spectral flux distribution of the afterglow of GRB 000131 from [17], as observed three days after the burst, transformed to various redshifts. The transformation involves (1) dimming the afterglow, (2) redshifting its spectrum, (3) time dilating its light curve, and (4) extinguishing the spectrum using a model of the Lyα forest (for details, see [15]). Finally, we have convolved the transformed spectra with a top hat smearing function of width $\Delta\nu = 0.2\nu$. This models these spectra as they would be sampled photometrically, as opposed to spectroscopically; i.e., this transforms the model spectra into model spectral flux distributions.

Figure 2 shows the resulting spectral flux distribution. The spectral flux distribution of the afterglow is cut off by the Lyα forest at progressively lower frequencies as one moves out in redshift. Thus high redshift afterglows are characterized by an optical "dropout" [4], and VHR afterglows by a near infrared "dropout." We conclude that, if GRBs occur at very high redshifts, both they and their afterglows can be easily detected.

GRBS AS A PROBE OF COSMOLOGY AND THE EARLY UNIVERSE

Theoretical calculations show that the birth rate of Pop III stars produces a peak in the SFR in the universe at redshifts $16 \lesssim z \lesssim 20$, while the birth rate of Pop II stars produces a much larger and broader peak at redshifts $2 \lesssim z \lesssim 10$ [18,19,20]. Therefore one expects GRBs to occur out to at least $z \approx 10$ and possibly $z \approx 15 - 20$, redshifts that are far larger than those expected for the most distant quasars.

Figure 3 places GRBs in a cosmological context. At recombination, which occurs at redshift $z = 1100$, the universe becomes transparent. The cosmic background radiation originates at this redshift. Shortly afterwards, the temperature of the

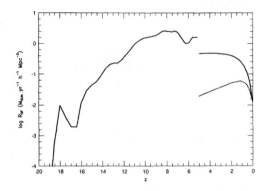

FIGURE 4. The cosmic SFR R_{SF} as a function of redshift z. The solid curve at $z < 5$ is the SFR derived by [25]; the solid curve at $z \geq 5$ is the SFR calculated by [18] (the dip in this curve at $z \approx 6$ is an artifact of their numerical simulation). The dotted curve is the SFR derived by [24].

cosmic background radiation falls below 3000 K and the universe enters the "dark ages" during which there is no visible light in the universe. "First light," which occurs at $z \approx 20$, corresponds to the epoch when the first stars form. Ultraviolet radiation from these first stars and/or from the first active galactic nuclei re-ionizes the universe. Afterward, the universe is transparent in the ultraviolet.

QSOs are currently the most powerful probes of the high redshift universe. GRBs have several advantages relative to QSOs as probes of cosmology. First, GRBs are expected to occur out to $z \approx 20$, whereas QSOs occur out to only $z \approx 5$. Second, very high redshift GRB afterglows can be 100 - 1000 times brighter at early times than are high redshift QSOs. This makes possible very sensitive high dispersion spectroscopy of the metal absorption lines and the Lyman α forest in the spectrum of the afterglows. Third, no "proximity effect" on intergalactic distances scales is expected for GRBs and their afterglows, in contrast to QSOs. Thus GRBs may be relatively "clean" probes of the intergalactic medium, the Lyman α forest, and damped Lyman α clouds, even in the vicinity of the GRBs.

The important cosmological questions that observations of GRBs and their afterglows may be able to address include the following:

• Information about the epoch of "first light" and the earliest generations of stars from merely the detection of GRBs at very high redshifts;

• Information about the growth of metallicity in the universe in the star-forming entities in which the bursts occur, in damped Lyman α clouds, and in the Lyman α forest from observations of the metal absorption line systems in the spectra of their afterglows;

• Information about the large-scale structure of the universe at VHRs from the clustering of the Lyman α forest lines and the metal absorption-line systems in the spectra of their afterglows; and

• Information about the epoch of re-ionization from the depth of the Lyman α break in the spectra of their afterglows.

Below we consider the first of these questions: the epoch of "first light" and the earliest generations of stars.

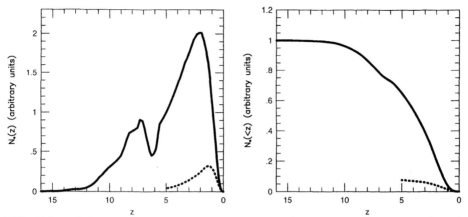

FIGURE 5. Left panel: The number N_* of stars expected as a function of redshift z (i.e., the SFR from Figure 4, weighted by the differential comoving volume, and time-dilated) assuming that $\Omega_M = 0.3$ and $\Omega_\Lambda = 0.7$. Right panel: The cumulative distribution of the number N_* of stars expected as a function of redshift z. Note that $\approx 40\%$ of all stars have redshifts $z > 5$. The solid and dashed curves in both panels have the same meanings as in Figure 4.

GRBS AS A PROBE OF STAR FORMATION

Observational estimates [21,22,23,24] indicate that the SFR in the universe was about 15 times larger at a redshift $z \approx 1$ than it is today. The data at higher redshifts from the Hubble Deep Field (HDF) in the north suggests a peak in the SFR at $z \approx 1 - 2$ [24], but the actual situation is highly uncertain.

In Figure 4, we have plotted the SFR versus redshift from a phenomenological fit [25] to the SFR derived from submillimeter, infrared, and UV data at redshifts $z < 5$, and from a numerical simulation by [18] at redshifts $z \geq 5$. The simulations done by [18] indicate that the SFR increases with increasing redshift until $z \approx 10$, at which point it levels off. The smaller peak in the SFR at $z \approx 18$ corresponds to the formation of Population III stars, brought on by cooling by molecular hydrogen. Since GRBs are detectable at these VHRs and their redshifts may be measurable from the absorption-line systems and the Lyα break in the afterglows [4], if the GRB rate is proportional to the SFR, then GRBs could provide unique information about the star-formation history of the VHR universe.

We have calculated the expected number N_* of stars as a function of z assuming (1) that the GRB rate is proportional to the SFR[1], and (2) that the SFR is that given in Figure 4 (see [15] for details). The left panel of Figure 5 shows our results for $N_*(z)$ for an assumed cosmology $\Omega_M = 0.3$ and $\Omega_\Lambda = 0.7$ (other cosmologies give similar results). The solid curve corresponds to the star-formation rate in Figure 4;

---

[1] This may underestimate the GRB rate at VHRs since it is generally thought that the initial mass function will be tilted toward a greater fraction of massive stars at VHRs because of less efficient cooling due to the lower metallicity of the universe at these early times.

the dashed curve corresponds to the star-formation rate derived by [24]. Figure 5 shows that $N_*(z)$ peaks sharply at $z \approx 2$ and then drops off fairly rapidly at higher z, with a tail that extends out to $z \approx 12$. The rapid rise in $N_*(z)$ out to $z \approx 2$ is due to the rapidly increasing volume of space. The rapid decline beyond $z \approx 2$ is due almost completely to the "edge" in the spatial distribution produced by the cosmology. In essence, the sharp peak in $N_*(z)$ at $z \approx 2$ reflects the fact that the SFR we have taken is fairly broad in z, and consequently, the behavior of $N_*(z)$ is dominated by the behavior of the co-moving volume $dV(z)/dz$; i.e., the shape of $N_*(z)$ is due almost entirely to cosmology. The right panel in Figure 5 shows the cumulative distribution $N_*(>z)$ of the number of stars expected as a function of redshift z. The solid and dashed curves have the same meaning as in the upper panel. Figure 5 shows that for the particular SFR we have assumed, $\approx 40\%$ of all stars (and therefore of all GRBs) have redshifts $z > 5$.

REFERENCES

1. Castander, F. J., & Lamb, D. Q. 1999, ApJ, 523, 593
2. Fruchter, A. S., et al. 1999, ApJ, 516, 683
3. Kulkarni, S. R., et al. 1998, Nature, 395, 663
4. Fruchter, A. S. 1999, ApJ, 516, 683
5. Sahu, K. C., et al. 1997, Nature, 387, 476
6. Kulkarni, S. R., et al. 1999, Nature, 398, 389
7. Reichart, D. E., 1999, ApJ, 521, L111
8. Galama, T. J., et al. 2000, ApJ, 536, 185
9. Bloom, J. S., et al. 1999, Nature, 401, 453
10. Woosley, S. E. 1993, ApJ, 405, 273
11. Woosley, S. E. 1996, in Gamma-Ray Bursts, eds. C. A. Meegan, R. D. Preece, & T. M. Koshut (New York: AIP), 520
12. Paczyński, B. 1998, ApJ, 494, L45
13. MacFadyen, A. I., & Woosley, S. E. 1999, ApJ, 524, 262
14. Wheeler, J. C., et al. 2000, ApJ, 537,
15. Lamb, D. Q., & Reichart, D. E., 2000, ApJ, 536, 1
16. Wijers, R. A. M. J., Rees, M. J., & Mészáros, P. 1997, MNRAS, 288, L51
17. Andersen, M. I., et al. 2000, A&A, 364, L54
18. Ostriker, J. P., & Gnedin, N. Y. 1996, ApJ, 472, L63
19. Gnedin, N. Y., & Ostriker, J. P. 1997, ApJ, 486, 581
20. Valageas, P., & Silk, J. 1999, A&A, 347, 1
21. Gallego, J. 1995, ApJ, 455, L1
22. Lilly, S. J., et al. 1996, ApJ, 460, L1
23. Connolly, A. J. 1997, ApJ, 486, L11
24. Madau, P., Pozzetti, L., & Dickinson, M. 1998, ApJ, 498, 106
25. Rowan-Robinson, M. 1999, Ap&SS, 266, 291

Gamma Ray Bursts statistical properties and limitations on the physical model

G.S.Bisnovatyi-Kogan

Space Research Institute, Profsoyuznaya 84/32, Moscow 117810, Russia

Abstract. The present common view abour GRB origin is related to cosmology. There are two evidences in favour of this interpretation. The first is connected with statistics, the second is based on measurements of the redshifts in the GRB optical afterglows. Red shifts in optical afterglows had been observed only in long GRB. Statistical errors, and possibility of galactic origin of short GRB is discussed; their connection with Soft Gamma Repeaters (SGR) is analyzed.

INTRODUCTION

Cosmological origin of GRB had been first suggested in [11] soon after their discovery. The present model of cosmological GRB based on production of gamma quanta from neutrino collisions $\nu + \bar{\nu} \to e^+ + e^-$ was first considered in [2]. The efficiency of transformation of the neutrino flux $W_{\nu\bar{\nu}} \sim 6 \times 10^{53}$ ergs into gamma quanta was estimated as $\sim 6 \times 10^{-6}$, giving a pulse $W_\gamma \sim 3 \times 10^{48}$ ergs. It could explain the cosmological GRB only at rather narrow pulse beam. In the giant GRB 990123 the isotropic energy production is very large [10] in gamma $W_\gamma \approx 2.3 \times 10^{54}$ ergs, and in optics $W_{opt} \sim 10^{51}$ ergs. Simultaneous strong beaming in gamma and optical bands is rather unplausible. Strong beaming would modify the observed smooth optical light curve in presence of a source rotation. Some problems in GRB interpretation and modeling are discussed.

STATISTICS AND RESTRICTIONS TO THE MODEL

BATSE data start to deviate from the uniform distruibution with 3/2 slope at rather large fluences, for which KONUS data are well defined. Analysis of KONUS data had been done in [9]. Taking into account selection effects, the resulting value $V/V_{max} = 0.45 \pm 0.03$ was obtained. KONUS data had been obtained in conditions of constant background. Similar analysis [14] of BATSE data, obtained in conditions of substantially variable background, gave resulting $V/V_{max} = 0.334 \pm 0.008$. These two results seems to be in contradiction, because KONUS sensitivity was only 3 times less than that of BATSE, where deviations from the uniform distribution $V/V_{max} = 0.5$ in BATSE data are still rather large [8].

In presence of a threshold deviations of V/V_{max} from its uniform Euclidean value 0.5 may be connected with the errors in determination of the burst peak luminosity or total fluence [3]. Such errors may be connected with spectral differences, variable sensitivity of detectors for bursts coming from different directions, variable background. All these reasons lead to underestimation of the burst luminosity, and decrease the slope of the curve $\log N(\log S)$. There is no angular correlation between GRB and sample of any other objects in the universe.

From the energy conservation law it follows $W < Mc^2$, where M is a mass of the source. A proper account of physical laws put much stronger restrictions to the energy output. Calculations of ns-ns collisions gave energy output in (X, γ) region not exceeding 10^{50} ergs [12], and similar results characterize ns-bh collision [13]. Magnetorotational explosion does not give larger energy output in (X, γ) region, transforming about 5% of the rotational energy into a kinetic one [1]. The problems with vaguely defined "hypernova" model had been discussed in [6].

The largest γ-ray production efficiency, close to $100\% Mc^2$ may be expected, if GRB originate from matter-antimatter star collisions. That arises a problem of antistar creation in the early universe.

Simultaneous γ, X and optical observations in GRB, accompanied by spectral and polarization experiments are very important. Search of hard X-ray lines and of annihilation 0.511 keV, line declared by KONUS, remain to be a puzzle which should be solved. Cosmological GRB explosion in a dense molecular cloud would lead to a specific optical light curve [5], which discovery would reveal conditions in the region of cosmological GRB explosion. Study of hard γ afterglow, similar to the one observed by EGRET [15] is expected in a near future.

SHORT GRB AND SGR

All afterglows had been measured only for long bursts. It cannot be excluded that short bursts could have another, may be Galactic origin. There is also a possibility, that short bursts are connected with a giant bursts observed in 3 soft gamma repeaters (from 4 known). At larger distances only giant bursts, appeared as short GRB could be observed. If we accept the present interpretation of SRG, as galactic and LMC sources at distances 10-50 kpc, than only giant bursts should be visible in the nearest galaxies as weak (about few 10^{-7}) short GRB. Taking into account that Andromeda is ~ 4 times more massive than our Galaxy [16], we should expect [4] to see about 10 short GRB in its direction during the observation time, while no one was yet observed. Another large galaxy in the local group of galaxies Maffei IC 1805 is also more massive than the Galaxy, and short GRB from it are also expected.

Presently SRG are interpreted as young neutron stars with very strong magnetic field - "magnetars" [7]. The estimation of the distance and, consequently, the luminosity is based on SGR identification with supernovae remnants (SNR), leading to large energy losses. This interpretation has several theoretical objections [4].

1. Hard gamma pulsars observed in 3 SGR have luminosities strongly exceeding the critical Eddington luminosity. At such luminosity a strong mass loss should smear out the pulses.

2. Rotational energy losses estimated from the period increase rate are much smaller than the observed gamma and X-ray luminosity even in a quiescent state. In the magnetar

model the energy comes from the annihilation of magnetic field. Such annihilation should be accompanied by creation of energetic electrons and radio-emission. The radio-emission of SGR is very weak, its discovery is very difficult, and still not firmly established.

3. Giant bursts observed in 3 SGR at present interpretation are accompanies by a huge energy production, part of which should go into particle acceleration and kinetic energy outbursts. It should influence the near-by SNR, and produce a visible changes in radio and optics, similar to those produced by pulsar glitches in the Crab nebula, when much smaller amount of energy is released. No such changes had been reported up to now, probably because they have not been present there.

Another interpretation of SGR, free of these contradictions needs a smaller distance to SGR, what is possible if the connection with SNR would not be confirmed. Note, that all SGR are situated at the very edge, or even outside of the SNR envelope, requiring very high 1000-3000 km/s speed of the neutron star. Refusing this connection and suggesting $\sim 10-30$ times smaller distance to SGR would remove the upper objections. The even smaller distances are less probable, because most SGR are situated in the galactic disk, and so should be situated at distances larger than this disk thickness. Existence of one SGR outside the galactic disk direction could indicate to its big age during which it could leave the galactic disk. Discovery of big population of neutron stars in the globular clusters and in the galactic bulge, as recycled pulsars, indicate to existence of neutron stars in the whole volume of the Galaxy.

ACKNOWLEDGMENTS

This work was partially supported by RFBR grant 99-02-18180 and INTAS-ESA grant 120. The authors are grateful to Prof. J.C.Wheeler and the Organizing Committee for support and hospitality, and to B.Schaefer for discussion.

REFERENCES

1. Ardeljan, N. V., Bisnovatyi-Kogan, G. S., Moiseenko, S. G., *A&A*, **355**, 1181-1190 (2000).
2. Berezinsky, V. S., Prilutsky, O. F., *A&A*, **175**, 309-311 (1987).
3. Bisnovatyi-Kogan, G. S., *A&A*, **324**, 573-577 (1997).
4. Bisnovatyi-Kogan, G. S., *Astro-ph/9911275*; Proc. Workshop Vulcano 1999.
5. Bisnovatyi-Kogan, G. S., Timokhin, A. N.*Astron. Zh.*, **74**, 483-496 (1997).
6. Blinnikov, S. I., Postnov, K. A., *Month. Not. R.A.S.*, **293**, L29-L32 (1998).
7. Duncan, R. C., Thompson, C., *ApJ Lett.*, **392**, L9-L13 (1992).
8. Fishman, G. J., Meegan, C. A., *Ann. Rev. A&A*, **33**, 415-458 (1995).
9. Higdon, J. C., Schmidt, M., *ApJ*, **355**, 13-17 (1990).
10. Kulkarni, S. R. et al., *Nature*, **398**, 389-399 (1999).
11. Prilutsky, O. F., Usov, V. V., *Astrophys. Sp. Sci.*, **34**, 387-393 (1974).
12. Ruffert, M., Janka, H.-T., *A&A*, **338**, 535-555 (1998).
13. Ruffert, M., Janka, H.-T., *A&A*, **344**, 573-606 (1999).
14. Schmidt, M., *ApJ Lett.*, **523**, L117-L120 (1999).
15. Schneid, E. J. et al., *A&A*, **255**, L13-L16 (1992).
16. Vorontsov-Velyaminov, B. A., *Extragalactic Astronomy*, Nauka. Moscow (1972).

Modification of relativistic jets in the presence of neutron component

E.V. Derishev*, V.V. Kocharovsky* and Vl.V. Kocharovsky*

Institute of Applied Physics RAS, 46 Ulyanov Street, 603600 Nizhny Novgorod, Russia

Abstract.
We show that free neutrons in relativistic jets from compact ultra-luminous objects alter dynamical and radiative properties of outflows. We review the phases of outflow evolution and analyze observable consequences of neutron-related effects.

The neutron-proton decoupling at the acceleration phase produces energetic photons and neutrinos. In many cases the neutrons attain a relativistic velocity dispersion either as a result of the decoupling or due to intrinsic variations of the bulk Lorentz factor. At the time of external shock formation, the interplay between primary and secondary (started by protons from neutron decay) shocks causes a reach variety of phenomena.

We consider a general case of nucleon (neutrons + protons) -loaded fireballs. In our analysis of the relative motion of neutrons and protons in the relativistic wind we pay particular attention to fireballs of Gamma-Ray Bursts (GRBs).

Specific effects of the neutron component depend on whether the final Lorentz factor of a plasma wind exceeds some critical value or not. If yes, decoupling of the neutron and proton flows takes place giving rise to an electromagnetic cascade induced by pion production in inelastic collisions of nucleons [1]. Otherwise, all nucleons in the wind behave as a single fluid. In both cases neutrons can strongly influence a GRB by changing dynamics of a shock initiated by protons in the surrounding medium [2]. Under certain conditions the decoupling leads to neutrino catastrophe (when the major part of outflow's energy is converted to neutrino emission) and furthermore to changes of the beam pattern [3].

The value of Lorentz factor corresponding to decoupling of the neutron flow is estimated to lie in the range expected for cosmological GRBs. Depending on whether neutron and proton flows decouple or not and whether neutrons decay before the shock of proton origin decelerates significantly or after that, four types of bursts are possible. Their lightcurves have different appearance and a number of lightcurve features are expected to be correlated. For example, the height, delay time and duration of the secondary pulse are related to each other. Multi-peaked lightcurves may be as common as single-peaked ones.

Effect of neutron decoupling is of particular interest in the case of cosmological GRBs because the decoupling threshold, Γ_*, falls in the expected range of fireball's Lorentz factors, 200 - 1500. Hence, a confirmation of the decoupling may give important clues to understanding physical conditions in the GRB source itself. Among proposed observational tests we find detection of energetic 100 GeV - photons by ground-based

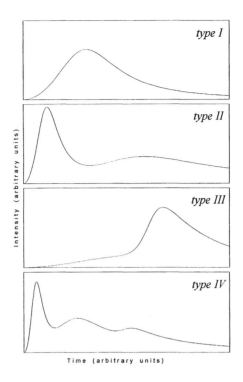

FIGURE 1. Four types of GRB bolometric lightcurves. Proportions are not exactly maintained.

telescopes to be the most promising.

The photons at the maximum of cascade spectrum, which appear blueshifted to 1-2 GeV for an observer, also carry approximately the same or even larger energy than the uprocessed quanta from pion decay. Effective area of GLAST telescope allows detection of the cascade photons from a GRB at redshift $z \sim 0.1$, provided the GRB energy was 10^{52} erg. Other fireball emission features, associated with neutron flow decoupling, such as ejection of $\sim 10^{52}$ positrons and burst of neutrinos of energy ~ 30 GeV, are too week to be detected by existing instruments. It should be noted that structured fireballs (i.e., those with internal shocks and/or sharp angular boundaries) may emit neutrinos [4] and produce pairs without decoupling.

Free neutrons may take most of the source's power, but do not contribute to the energy budget of a shock wave in surrounding medium until they decay into charged particles – protons and electrons. So, if the neutron-induced shock is visible in gamma-rays, duration of the main pulse of the burst should have a lower limit, $\sim 900/\Gamma_n$ s, contrary to the picture resulting from simple shock models [5, 6, 7].

The plasma component of a fireball pushes the surrounding medium and forms a shock at the interface, while neutrons propagate freely until they decay into charged particles. There are two independent alternatives: (i) the neutron flow may decouple from the proton one or may not, and (ii) the lifetime of a free neutron, t_n, either exceeds or is smaller than the deceleration time of the proton shock, t_p. A source satisfying the condition of decoupling will be a typical GRB representative. The apparent lifetime of free neutrons, ~ 900 s $/\Gamma_n$, also appears to be comparable with duration of an ordinary burst. Therefore, the above two alternatives produce four combinations; each one gives rise to a distinct type of lightcurves (see Fig. 1):

First type: $t_n < t_p$, no decoupling.
Second type: $t_n < t_p$, neutron flow decouples.
Third type: $t_n > t_p$, no decoupling.
Fourth type: $t_n > t_p$, neutron flow decouples.

Clearly, there are no sharp boundaries between type I and type III bursts as well as between type II and type IV. However, lightcurves of first and second class are expected only in short GRBs with duration < 10 s, while longer bursts should possess lightcurves of third and forth class.

Interference of primary and secondary (from decayed neutrons) shocks may result in rather complex GRB time histories even if the source has only one period of activity. Two separate episodes of relativistic wind ejection may already yield a lightcurve having up to seven peaks. It is clear that two-flow model of a fireball can easily account for very complicated lightcurves. Since it is decoupling that separates type I and III from type II and IV bursts, there must be a correlation between an observation of a flash of energetic quanta and the shape of the lightcurve.

The results obtained for the radiation-driven wind allow straightforward generalization for winds driven by other mechanisms, e.g., for the MHD winds.

This work has been supported by Russian Foundation for Basic Research (the project 99-02-18244).

REFERENCES

1. E.V. Derishev, V.V. Kocharovsky, Vl.V. Kocharovsky, *ApJ* **521**, 640 (1999)
2. E.V. Derishev, V.V. Kocharovsky, Vl.V. Kocharovsky, *A&A* **345**, L51 (1999)
3. A.A. Belyanin, E.V. Derishev, V.V. Kocharovsky, Vl.V. Kocharovsky, *Radiophys. Quantum Electr.* **44**, 1 (2001)
4. P. Mészáros, M.J. Rees, *ApJ* **541**, L5 (2000)
5. M.J. Rees, P. Mészáros, *ApJ* **430**, L93 (1994)
6. R. Wijers, M.J. Rees, P. Mészáros, *MNRAS* **288**, L51 (1997)
7. E. Waxman, *ApJ* **489**, L33 (1997)

Cascaded sub-TeV emission from Gamma-Ray Bursts: origin and observational tests

E.V. Derishev*, V.V. Kocharovsky* and Vl.V. Kocharovsky*

*Institute of Applied Physics RAS, 46 Ulyanov Street, 603600 Nizhny Novgorod, Russia

Abstract.
Initially, the spectra of comptonized radiation from Gamma-Ray Bursts (GRBs) have a broad maximum above 1 TeV [1], so that the VHE photons are strongly absorbed by infrared background radiation and cannot be observed from a source at cosmological distance [2]. On the other hand, we find that the comptonized radiation may be reprocessed into sub-TeV domain via electromagnetic cascade which develops after two-photon absorption of VHE photons by soft X-rays and optical quanta scattered by the surrounding gas. In this case, the VHE photons attain a large dispersion of arrival times, so that the bulk of VHE emission is delayed by a few minutes with respect to the beginning of GRB. The proposed theory can be applied to the delayed multi-GeV photons observed by EGRET as well as to the recent MILAGRO data [3, 4].

A part of radiation from a GRB source is scattered by the surrounding plasma. From the point of view of an observer in the laboratory frame, the photons emitted from the front of relativistic shock with the Lorentz factor Γ move faster than the shock itself and the velocity difference is small, $\sim c/2\Gamma^2$. Therefore, from the time when scattering occurred in some point and till the arrival of the GRB shock in the same point all scattered photons may be considered as frozen into interstellar medium. The energy density w_r of such photons can be expressed via the GRB fluence, which is equal to $E_{\rm GRB}/4\pi R^2$ at the distance R from the source (here $E_{\rm GRB}$ is GRB energy), and the density of surrounding plasma ρ:

$$w_r(R) = \frac{\sigma_T \rho}{m} \frac{E_{\rm GRB}}{4\pi R^2}, \qquad (1)$$

where m is the proton mass and σ_T Thomson cross-section. In the innermost zone of radius[1] $R_{\rm cl} \sim \left(\sigma_T E_{\rm GRB}/4\pi mc^2\right)^{1/2}$ the energy density of scattered radiation saturate at the level $\sim \rho c^2$, which is below the value given by Eq.(1). Therefore, this zone does not contribute too much to the optical depth for two-photon absorption.

On its way to an observer, the ultra-hard radiation passes through a medium filled with soft X-ray photons. This may alter significantly the GRB spectrum in the ultra-high energy domain provided the optical depth for the two-photon pair production exceeds unity at the distance $R = \Gamma^2 c t_{\rm GRB}$ (here $t_{\rm GRB}$ is duration of a burst). A projectile (photon) of energy ε has the largest cross-section of electron-positron pair production when it

---

[1] It should be noted that for a GRB with very hard spectrum the value of $R_{\rm cl}$ may increase by a factor up to 10-20 due to runaway pair creation [5], which we do not discuss here.

interacts with photons of energy $\sim 2m_e^2c^4/\varepsilon$, which make the main contribution to the opacity for the two-photon absorption. If absorbing photons are produced by the synchrotron mechanism, their spectrum is $I_\omega \propto \omega^{-1/2}$ and the effect of absorption is stronger for more energetic projectiles.

Let us introduce the new variable J, which shows what fraction of the total fluence is contained within logarithmic interval of photon energies around ε, $J(\varepsilon) = \varepsilon I_\varepsilon/I$. This is an actual observable. The two-photon absorption optical depth for quanta with energy ε is equal to $\tau_a = J(2m_e^2c^4/\varepsilon)\sigma_{pp}w_r\varepsilon/2m_e^2c^4$, where $\sigma_{pp} \simeq \sigma_T/8$ is the angle-averaged value of the pair-production cross-section. Assuming $J(\varepsilon) = \sqrt{\varepsilon/\varepsilon_m}$ one has

$$\tau_a \simeq \frac{\rho\sigma_T^2}{8m} \frac{\sqrt{\varepsilon/\varepsilon_m}}{\sqrt{2m_ec^2}} \frac{E_{GRB}}{4\pi\Gamma^2 ct_{GRB}} \sim \frac{0.3 E_{51} n \sqrt{\varepsilon_{12}}}{\Gamma^2 t_{GRB}}. \tag{2}$$

Here E_{51} is the GRB energy in units 10^{51} erg, n the electron number density in surrounding plasma (in cm^{-3}), ε_{12} the photon energy (in TeV), and t_{GRB} the GRB duration in seconds. Photons with energy above

$$\varepsilon_{2ph} \sim \frac{10\Gamma^4 t_{GRB}^2}{E_{51}^2 n^2} \text{ TeV} \tag{3}$$

are absorbed in the vicinity of GRB source and cannot be observed. Because the relativistic shock should not decelerate significantly during the burst, the density of the interstellar medium has an upper limit. Therefore, for GRBs having $\Gamma \geq 300$ a value of ε_{2ph} is larger than a characteristic energy of ultra-hard photons, i.e., there is no absorption. At the same time, for short and powerful bursts located in relatively dense galactic environment the absorption threshold ε_{2ph} may drop below 1 GeV. In these rare cases the effect of reprocessing may be observed by orbital telescopes operating in the appropriate spectral range.

Now let us follow the fate of absorbed photons. Each of them produces an electron-positron pair where the daughter particles have roughly equal energies. Electrons and positrons have larger interaction cross-section than photons do, so that if the optical depth for interaction of initial ultra-hard quanta with scattered soft photons is larger than unity the same is true for daughter electrons (positrons). The first photon scattered off by one of these energetic electrons takes away about a half of its energy. Step by step, the electromagnetic radiation becomes softer until the absorption threshold ε_{2ph} is reached. It is this energy domain where the bulk of energy initially contained in the ultra-hard radiation is concentrated. If the absorption threshold is below 1 TeV, then the reprocessed radiation may reach the Earth. For a typical burst having the total fluence $\sim 10^{-6}$ erg/cm^2 the fluence of ultra-hard photons is of the order of 10^4 TeV/ε_{2ph} km^{-2}. It is more than sufficient for detection by means of modern ground-based telescopes.

The reprocessed VHE emission is delayed with respect to GRB main pulse. This effect is purely geometrical: magnetic field bends trajectories of secondary electrons and positrons so that the emission pattern for the cascade photons becomes broader than the emission pattern of the relativistic shock itself. Now an observer is able to see a larger segment of (spherical) relativistic shock, which results in a larger difference of photon arrival times, $t_d = \phi^2 R_{sh}/2c$. Here R_{sh} is the shock radius and ϕ is an angle between electron's momentum and the line of sight. This angle grows in time until the shock catches up the electron, that limits ϕ to the value

$$\frac{\phi^2}{3} \simeq \frac{e^2 B_\perp^2}{12 \varepsilon_{2ph}^2} \ell^2 = \frac{1}{\Gamma^2}\left(1 + \frac{R_{sh}}{\ell}\right), \qquad (4)$$

where $B_\perp$ is the component of magnetic field perpendicular to electron's trajectory and ℓ the distance required for shock to catch up the electron. The delay becomes observable ($\phi > 1/\Gamma$, i.e. $\ell < R_{sh}$) if the strength of magnetic field exceeds the minimal value

$$B_{min} = 2\sqrt{6}\frac{\varepsilon_{2ph}}{e\Gamma R_{sh}} \sim \frac{5 \times 10^{-2}}{\Gamma^3}\frac{\varepsilon_{2ph}}{1\,\text{TeV}}\frac{t_{GRB}}{10\,\text{s}}\,\text{G}. \qquad (5)$$

When the magnetic field is much stronger than B_{min}, the delay of reprocessed VHE emission may be calculated via the following asymptotic expression:

$$t_d \simeq \frac{2^{4/3}}{3}\left(\frac{B_\perp}{B_{min}}\right)^{2/3}. \qquad (6)$$

The detection of delayed VHE emission and measurement of its duration t_d and peak photon energy ε_{2ph} should provide the unique information about GRB birth sites, namely, the interstellar gas density and magnetic field strength in the vicinity of GRB progenitors.

This work has been supported by Russian Foundation for Basic Research (the project 99-02-18244).

REFERENCES

1. Derishev, E.V., Kocharovsky, V.V., Kocharovsky, Vl.V., *A&A*, accepted, 2001 (astro-ph/0006239).
2. Primack, J.R., Bullock, J.S., Somerville, R.S., MacMinn, D., *Astropart. Phys.* **11**, 93 (1999)
3. Hurley, K., *Nature* **372**, 652 (1994)
4. Atkins, R., et al., *ApJ* **533**, L119 (2000)
5. Thompson, C., Madau, P., *ApJ* **538**, 105 (2000)

Effects of self-Compton cooling on the synchrotron spectrum of GRBs

Vl.V. Kocharovsky[*], E.V. Derishev[*], V.V. Kocharovsky[*] and P. Mészáros[†]

[*]*Institute of Applied Physics RAS, 46 Ulyanov Street, 603600 Nizhny Novgorod, Russia*
[†]*Department of Astronomy and Astrophysics, Pennsylvania State University, University Park, PA 16803*

Abstract.
We develop a self-consistent theory of the synchrotron-self-Compton emission for optically thin sources with a stationary injection of relativistic monoenergetic electrons. We investigate the electron distribution function, synchrotron and inverse Compton spectra and find their analytical asymptotics. The steady-state electron distribution entangled with synchrotron radiation via Compton losses is shown to produce spectral indices ranging from 1/2 to 1. Possible ways of addressing steep low-energy slopes in some bursts are pointed out.

THE EQUATION FOR SYNCHROTRON-SELF-COMPTON COOLING

The spectra of Gamma-Ray Burst (GRB) emission have been investigated theoretically by a number of authors (e.g., [1, 2, 3]) for different versions of a general fireball-shock scenario [4, 5]. However, the existing analytical models give rather coarse predictions and suffer from many uncertainties.

Below we present a refined analytical model of synchrotron-self-Compton emission, in which we assume a monoenergetic injection of electrons (with the Lorentz factor γ_{max}) and calculate their distribution over the Lorentz factors taking into account both synchrotron and inverse Compton losses. Then the synchrotron and Compton spectra are found. The second and higher cascades of Comptonization are negligible in the particular physical conditions, which we found in GRB emitting regions [6].

Let us introduce a function $\kappa(\gamma) = \mathcal{L}_s/(\mathcal{L}_s + \mathcal{L}_{ic})$, which is the ratio of synchrotron luminosity to the total luminosity (including Compton losses $\mathcal{L}_{ic}$) for an electron with the Lorentz factor γ. By definition, $0 < \kappa < 1$. The synchrotron and inverse Compton luminosity of a single electron are

$$\mathcal{L}_s = \frac{4}{3}\sigma_T \gamma^2 c \frac{B^2}{8\pi} \quad \text{and} \quad \mathcal{L}_{ic} = \frac{4}{3}\gamma^2 c \int_0^\infty \sigma(\omega) w_{sy,\omega}\, d\omega, \tag{1}$$

where ω is the photon frequency and $w_{sy,\omega}$ the energy density of the synchrotron radiation per unit frequency interval. Below we assume that synchrotron radiation of an electron is monochromatic with the frequency $\omega(\gamma) = 0.5\gamma^2(eB/m_e c)$ and $\sigma(\omega) = \sigma_T \Theta(\omega_* - \omega)$, where Θ is the step function, σ_T Thomson cross-section.

The integral in Eq. (1) is then reduced to the product of σ_T and the energy density of synchrotron radiation emitted by all electrons with Lorentz factors less than γ_*, where γ_* satisfies the equation $\omega(\gamma_*) = \omega_*$. If the electron distribution over the Lorentz factors is stationary, it can be found simply as

$$N_e(\gamma) = \frac{1}{\dot{\gamma}} = \frac{m_e c^2}{L_s + L_{ic}} \propto \frac{\kappa(\gamma)}{\gamma^2}, \tag{2}$$

and the energy density of synchrotron radiation below the Klein-Nishina cut-off is

$$w_{sy}(\omega < \omega_*) = \frac{\tau S_e m_e c^2}{V} \int_1^{\gamma_*} \kappa(\gamma') d\gamma'. \tag{3}$$

Here τ is the variability timescale and V the volume of the emitting region.

For convenience, we introduce a new variable $x \equiv \gamma/\gamma_0$, where $\gamma_0 = \gamma_*(\gamma_0)$, i.e.,

$$\gamma_0 = \left(\frac{2m_e^2 c^3}{\hbar e B}\right)^{1/3}. \tag{4}$$

For electrons with the Lorentz factor γ_0 the Klein-Nishina cut-off is at their own synchrotron frequency (in GRBs $1 \ll \gamma_0 < \gamma_{max}$). Finally, we obtain an equation

$$\frac{1}{\kappa(x)} = 1 + \mathcal{K} \int_0^{1/\sqrt{x}} \kappa(x') dx'. \tag{5}$$

It is valid for $x_{max}^{-2} \leq x \leq x_{max}$; for $x < x_{max}^{-2}$ one has $\kappa = (1 + \tau_{ic})^{-1}$, where $\tau_{ic} \simeq (4/3)\gamma_{max}^2 \sigma_T n_e L$ and L is the size of the emitting region. The parameter

$$\mathcal{K} \equiv \frac{E_r \gamma_0}{W_m \gamma_{max}} = \frac{\tau_{ic}}{\bar{\kappa}} \left(\frac{m_e c^2}{\varepsilon_p \gamma_{max}}\right)^{1/3}. \tag{6}$$

Here ε_p is photon energy at the peak of synchrotron spectrum (in the comoving frame), $W_m = (B^2/8\pi)V$ is the total energy of magnetic field in the emitting region and $\bar{\kappa}$ the average value of κ in the interval $(0, x_{max})$. The value of $\bar{\kappa}$ shows what fraction of the total energy radiated by GRB is emitted in sub-MeV spectral domain (we assume $\bar{\kappa} > 1/2$). In GRBs, the parameter $\mathcal{K}$ is within $0.1 \leq \mathcal{K} \leq 100$.

Let us introduce the reference point $x_{1/2}$ so that $\kappa(x_{1/2}) = 1/2$. In the absence of analytical solution, one can use approximate expressions

$$\kappa(x) > \frac{1}{1 + \mathcal{K}/\sqrt{x}} \quad \text{and} \quad \kappa(x) \simeq \left[1 + \frac{2}{3x^{3/4}}\right]^{-1}, \tag{7}$$

which are valid for $x \ll 1/x_{1/2}^2$ and $x \gg x_{1/2}^4$ respectively. Both asymptotics join more or less smoothly at $x_{1/2}$ if $x_{1/2} < 1$, so that Eq. (7) solves the probem.

SYNCHROTRON AND COMPTON SPECTRA

The GRB sub-MeV spectrum is just blueshifted synchrotron spectrum, and – as a function of $x \propto \sqrt{\omega}$ – it is actually given by $\kappa(x)$ in the following way:

$$I_\omega \propto \frac{\kappa(x)S(x)}{\sqrt{\omega}}, \qquad x = \left(\frac{2\hbar^2}{m_e c^3 eB}\right)^{1/6} \sqrt{\omega}. \qquad (8)$$

In particular, Eq. (8) gives the following power-law asymptotics: $\omega I_\omega \propto \omega^{3/4}$ for $1/x_{max}^2 < x \ll \min[x_{1/2}, 1/x_{1/2}^2]$, and $\omega I_\omega \propto \omega^{1/2}$ for $x < 1/x_{max}^2$ and $x \gg \max[x_{1/2}, x_{1/2}^4]$ if $\mathcal{K} \sim 1$.

The above spectrum extends down to $\omega_r = (\varepsilon_p/\hbar)(x_r/x_{max})^2$. Below this frequency I_ω reproduces the synchrotron spectrum of a single particle because the electron distribution is cut at $x = x_r$, as at smaller electron energies the cooling timescale becomes larger than the GRB variability timescale.

The spectrum of comptonized radiation is represented analytically as a convolution of syncrotron spectrum with the same spectrum but taken as a function of re-scaled frequency. As far as the resulting inverse Compton spectrum is softer than the spectrum of those synchrotron photons which undergo Comptonization, one may take an approximation assuming that the most energetic electrons produce monochromatic comptonized radiation with photon energy $\varepsilon \simeq \gamma m_e c^2/2$. For the hardest part of inverse Compton spectrum this gives $\varepsilon I_\varepsilon \propto (1-\kappa)x \propto \varepsilon^{1/4}$, where $x = x_{max}\varepsilon/\varepsilon_{ic}$, $\varepsilon_{ic} = \gamma_{max} m_e c^2$, and Eq. (7) is used to substitute $(1-\kappa)$. The inverse Compton spectrum has a very broad maximum at $\simeq \varepsilon_{ic}/2$.

In conclusion, we draw attention to the fact that the cooling distribution of electrons depends on the Compton losses and can produce a broad range of spectral indices. The low energy portion of the synchrotron spectrum can be as steep as $I_\omega = const$. Still steeper spectra with $\alpha > 1$ may be obtained in alternative ways (e.g., [7]), but this leads to low radiative efficiency of GRBs and rises concerns about the visibility of emission from fireball photosphere against the non-thermal component. Also, in a model with efficiency ≤ 0.02 and $\alpha < 1$, the blackbody emission from photosphere may account for steepening of the observed low-energy spectral indices.

REFERENCES

1. R. Sari, T. Piran, *MNRAS* **287**, 110 (1997)
2. G. Ghisellini, A. Celotti, *ApJ* **511**, L93 (1999)
3. C.D. Dermer, J. Chiang, K.E. Mitman, *ApJ* **537**, 785 (2000)
4. P. Mészáros, M.J. Rees, *MNRAS* **257**, P29 (1992)
5. P. Mészáros, M.J. Rees, *MNRAS* **269**, L41 (1994)
6. E.V. Derishev, V.V. Kocharovsky, Vl.V. Kocharovsky, *A&A*, accepted 2001 (astro-ph 0006239)
7. A. Panaitescu, P. Mészáros, *ApJ* **544**, L17 (2000)

An Analytic Model of Gamma-Ray Burst Pulse Profiles

Dan Kocevski and Edison P. Liang

Department of Physics and Astronomy
Rice University, Houston, Tx, 77054

Abstract. Gamma-ray Bursts (GRBs) have puzzled astronomers for over three decades. Recent reports of correlations between a burst's lag and variability to redshift have begun to demonstrate how the analysis of a burst's temporal structure can play an important role in our understanding of this phenomenon. Here we attempt to derive an analytic function that is independent of an emission mechanism, that can then be used to model burst light curves in a unique way. We use this function to fit 22 clean non-overlapping fast rise exponential decay (FRED) bursts in an attempt to uncover temporal patterns that may exist among bursts. Using our model, we find that there appears to be a trend in the decay phase of the FRED profile. The parameter that is associated with the asymptotic decay slope of a GRB appears to remain constant among bursts of varying duration and intensity. Although it is not immediately clear what is driving this trend, it could be due to distinct dependencies of the rise and decay portions of the light curves on the physical conditions of the GRB progenitor.

INTRODUCTION

The BATSE instrument was placed on the Compton Gamma Ray Observatory primarily as an all-sky gamma-ray transient monitor. There are two principal instruments that make up BATSE, the Large Area Detectors (LADs) and the Spectroscopy Detectors (SDs). The data sets obtained from these instruments have time resolutions of 64, 256, and 1024 ms, allowing for detailed temporal studies. The observed light curves that have been produced with BATSE show that the temporal structure of GRBs varies drastically from pulse to pulse, with no apparent pattern between bursts. It is generally believed that the fundamental constituent of GRB light curves is a fast rise, exponential decay (FRED) profile and that more complex bursts would are a result of overlapping FRED pulses of varying amplitude and intensity. Although the duration and amplitude of FREDs vary considerably, the shape is the only recurring pattern that can be distinguished. It is not clear whether the broad variety of observed temporal behaviors is due to varying physical properties of the source or other factors such as viewing angle. One of the primary questions that arises is whether there are characteristic of GRBs that can uniquely describe their temporal profiles and do these parameters have typical values for all GRBs.

Here we have attempted to derive an analytic function that it is motivated by physical first principles, which can be used to model the temporal profiles of FRED pulses in a unique way. Our approach starts with the physical assumption that the energy flux F_E of the burst has a power law dependence on the peak energy E_{pk}. This is a general form for non-thermal energy flux spectra that is largely independent of the emission mechanism. This assumption gives F_E an implicit time dependence because the peak energy evolves with time. Now if we add to this an explicit time dependence, the energy flux would be given by:

$$F_E \propto E_{pk}{}^b t^r \tag{1}$$

In the case of optically thin synchrotron emission the energy flux is given by $F_E \propto E_{pk}\tau_T B$, where b = 1, τ_T is the Thompson depth, and B is the magnetic field. In this case the explicit time dependence arises from the evolution of the Thompson depth and the magnetic field strength. Our initial assumptions give rise to a general form that should also emerge when considering other emission mechanisms. Assuming this form and using the Liang-Kargatis (1996) relation, we have been able to produce a function that has been remarkably successful in reproducing and fitting the FRED pulses. Equation 3 below gives our function in normalized form.

$$\frac{F}{F_{max}} = \frac{x^r}{[\frac{d+rx^{1+r}}{d+r}]^{\frac{d+r}{1+r}}} \tag{2}$$

Where the r and d parameters are equivalent to the asymptotic rise and decay indices as $x \to 0$ and $x \to \infty$ respectively.

LIGHT CURVE FITS

Using this function, we fit clean non-overlaping pulses in order to calculate a distribution for the rise and decay indices, with hopes that such data could be used to help work as constraints on physical models. To do this we visually searched through the concatenated 64ms BATSE LAD data for bursts containing long pulse structures that exhibited a general fast rise exponential decay profile. A total of 22 bursts met our criteria and were selected for analysis. We then performed a 5 parameter fit to each burst, including the rise, decay, time of peak flux (T_{max}), peak flux (F_{max}), and zero point offset. Two example fits for BATSE triggers 973 and 1406 are shown in figures 1 and 2 respectively.

RESULTS

After fitting all 22 bursts, we find that in our sample the rise index varies from burst to burst, but the decay index is fairly constrained. We receive a mean rise

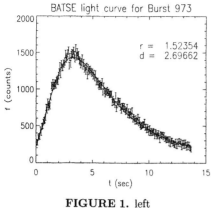 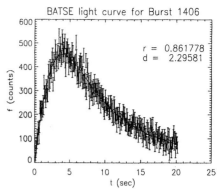

FIGURE 1. left **FIGURE 2.** right

index of r = 1.515 with a standard deviation of $\sigma = 1.458$ and d = 2.272 with a much smaller associated standard deviation of $\sigma = 0.599$. This is a direct result from the fact that our decay slope is highly dependant on the rise index, therefore bursts that do not look similar by eye, may actually have similar decay parameters. It is not immediately clear as to why this trend should exist, but it is most likely due to the rise portion of the burst being highly dependant on environmental conditions that may vary from burst to burst, such as the Thompson depth, magnetic field, etc. Whereas the cooling process may not be as heavily dependant on these parameters. This could prove to be an unexploited trend that may exist in most GRB that exhibit clean discernable peaks, with potential uses in emission model constraints and cosmological correlations with fit parameters. Work needs to be done on more bursts as well as a study of how the rise and decay slopes depend on BATSE energy bands. It is generally known that lower energies from GRBs tend to arrive later and have a broader profile, creating a soft energy lag. It would be interesting to see how these slopes and this apparent trend is effected by this broadening.

Above in equation 1 we wrote the flux-peak energy proportionality with the implicit and explicit time dependencies separated. When analyzing the spectra of a GRB, these two dependancies cannot be resolved, leading to a single paramenter that is a function of both. This value is sometimes referred to as the correlation index between a GRB's flux and energy, i.e. $F \propto E^\gamma$. The correlation index has been measured directly by Kargatis (1995) and seperately by Ryde (2000), who found that γ had a typical value of 1.5 to 1.7. It can be shown that our decay index is directly related to this correlation parameter by $\gamma = \frac{d}{d+1}$. Using the values of the decay index d that we obtained with our fits, we recieve a calculated γ of about 1.786 ± 0.13, which is in excellent agreement to the measured values. We believe that this is a self consistancy check for our approach and substantiates the trend observed in our fitted decay slopes.

GRBs from unstable Poynting dominated outflows

Maxim Lyutikov

Physics Department, McGill University, Montreal, QC, H3A 2T81 Canada [1]

Abstract. Poynting flux driven outflows from magnetized rotators are a plausible explanation for gamma-ray burst engines. We suggest a new possibility for how such outflows might transfer energy into radiating particles. We argue that in region near the rotation axis the Poynting flux drives non-linearly unstable large amplitude electromagnetic waves (LAEMW) which "break" at radii $r_t \sim 10^{14}$ cm where the MHD approximation becomes inapplicable. In the "foaming" (relativistically reconnecting) regions formed during the wave breaks the random electric fields stochastically accelerate particles to ultrarelativistic energies which then radiate in turbulent electromagnetic fields. The typical energy of the emitted photons is a fraction of the fundamental Compton energy $\epsilon \sim f\hbar c/r_e$ with $f \sim 10^{-3}$ plus additional boosting due to the bulk motion of the medium. The emission properties are similar to synchrotron radiation, with a typical cooling time $\sim 10^{-3}$ sec. During the wave break, the plasma is also bulk accelerated in the outward radial direction and at larger radii can produce afterglows due to the interactions with external medium. The near equipartition fields required by afterglow models maybe due to magnetic field regeneration in the outflowing plasma (similarly to the field generation by LAEMW of laser-plasma interactions) and mixing with the upstream plasma.

A Poynting flux driven outflows (PFDO) from a magnetized rotator is a promising paradigm for gamma-ray burst (GRB) engines and there have been various implementations of this concept (c.f. Usov 1992; Thompson 1994; Blackman et al. 1996; Mészáros & Rees 1997; Kluzniak & Ruderman 1998). Scenarios that could produce Poynting flux dominated outflows (PFDOs) require a source of magnetic fields $\sim 10^{15}$ Gauss, and rotation speeds of order $\Omega \sim 10^4$/sec. The total available energy from a compact ($R_0 \sim 10^6$ cm, $M \geq M_\odot$) object is then $M\Omega^2 R_0^2/2 \geq 10^{53}$ ergs and a dipole type luminosity $2B_0^2 R_0^6 \Omega^4/3c^3 \sim 2 \times 10^{50}$ ergs/sec.

PFDOs may differ from the conventional blast wave models in several important respects. The lepto-baryonic content of the outflow in the inner region is expected to be small by analogy to pulsar winds, and the energy carried by matter in the

---

[1] CITA National Fellow

central region will be less than the energy carried by EM fields. In the recent work (Lyutikov & Blackman 2001) we address the fundamental but unresolved issue of how Poynting flux is converted into gamma-rays and the subsequent emission characteristics (see also Smolsky & Usov 2000).

In may respects the PFDOs resemble pulsar outflows (e.g., Kennel & Coroniti (1984), Coroniti (1990)) which generically include (i) two components of magnetic field alternating and helical; alternating component changes on the scale length of $2\pi(c/\Omega) \sim 2 \times 10^7$ cm, in the equatorial region it resembles a linearly polarized wave, while in the polar region it resembles a circularly polarized wave. (ii) alternating magnetic fields are dynamically more important than the helical - this condition implicitly requires a strongly oblique rotator. (iii) along the magnetic poles a jet-like flow is produced (iv) initially Poynting energy flux dominates over the particle flux by a ratio $\sim 10^6$ - typical for pulsars.

At small distances the magnetic field of the wind, which decreases with radius as $\sim r^{-1}$,1 is frozen in the plasma. As the plasma flows out, its density decreases in proportion to r^{-2} reaching a radius $r_t \sim 10^{14}$ cm, where it becomes less than the critical charge density n_t required for applicability of MHD (Melatos & Melrose 1996). At $r > r_t$ the wind field is transformed into a Large Amplitude Electromagnetic Wave (LAEMW). The dimensionless strength parameter of the LAEMW, $\nu_0 = eE/(mc\Omega)$, near the transition point is

$$\nu_0 = \frac{\Omega R_0^3 B_0}{c^2 r_t} \sim 10^9. \tag{1}$$

This large value implies that the electromagnetic waves can drive particles to super-relativistic energies in one period of oscillation. We suggest that gamma-rays will be emitted at the point where MHD breaks down due to the *overturn instability* of LAEMWs.

The interaction of LAEMW with the plasma bears similarity to laser-plasma interactions. It is well known from laboratory experiments that interaction of strong electromagnetic radiation with plasma leads to a number of violent instabilities (e.g. Kruer 1988).

Once the EM wave overturns, it creates a broad spectrum of random EM fields. It is an electromagnetic analog of the foam of the deep water surface waves. Overturn of the initial coherent wave will create random EM fields and currents which will be strongly dissipated. In general, there will be a nonzero component of electric field along magnetic field. We will call this state relativistically strong electro-magnetic turbulence. Since the end result is the dissipation of the magnetic energy, this region may also involve relativistically strong reconnection. Particles will be accelerated in this EM turbulence to relativistic energies (in the rest frame of the foam). As the accelerated relativistic particles move through the EM foam they generate the observed GRB emission. Emission would emanate from multiple breaking regions. The separate spikes in the GRB profiles are due to separate "foaming regions" - separate regions of the wave break.

In many respects the "foaming regions" resemble optical shocks observed in the laboratory when an optical pulse steepens as it propagates in an optically active medium (Agrawal 1989). A wave overturn creates an electromagnetic shock. Interaction of such EM shocks will unavoidably lead to a creation of internal plasma discontinuities with strong dissipative effects. The "foaming regions" play the role of "internal shocks" (c.f. Piran 1999 for a review) of GRB theory. This term can be interpreted to mean "internal processes in the blastwave which generate bursts" not necessarily a specific mechanism, which we are providing here. Alternatively, the "foaming regions" may be thought of as relativistically reconnecting regions, where the energy of magnetic field is converted into particles in a relativistically turbulent manner.

When the smallest scale of the field fluctuations is larger than the radiative loss length the acceleration and emission result from quasistatic electric and magnetic fields. In this case the typical Lorentz factor of the particles and emission energy (dependent only on a combination of fundamental constants!) are

$$\gamma \sim \sqrt{\frac{r_L}{2r_e}} \sim 7 \times 10^4, \quad \epsilon \sim \frac{1}{\sqrt{2}} \frac{\hbar c}{r_e} \sim 40 \text{MeV}, \qquad (2)$$

where $r_L = c/\omega_{eff}$ is the effective Larmor radius and $r_e = e^2/mc^2$.

For the small scale turbulence, when most energy is concentrated on the skin depth scale, the average energy is $\sim (2r_e\Omega/c)^{1/14}$ times smaller.

Summarizing the main points of this work, (i) Poynting flux dominated outflows possesses internal instabilities which may also explain radiation generation, (ii) emission from particles stochastically accelerated in turbulent *electromagnetic* fields, with electric fields as important as magnetic, is a viable mechanism for the GRB emission.

REFERENCES

1. Agrawal, G. P. 1989, Nonlinear fiber optics, Boston : Academic Press
2. Blackman E.G., Yi. I., Field G.B. 1996, ApJ,
3. Coroniti, F. V. 1990, ApJ, 349, 538
4. Kluzniak & Ruderman 1998, ApJ Lett., 505, 113
5. Kruer W.L. 1988, The physics of laser plasma interactions, Redwood City, Calif. : Addison-Wesley
6. Lyutikov, M. & Blackman, E.G., MNRAS, 321, 177
7. Melatos A. & Melrose D.B., 1996, MNRAS, 279, 1168
8. Piran T., Phys Rep., 2000, 315, 575
9. Thompson, C. 1994, MNRAS, 270, 480
10. Smolsky M.V. & Usov V.V., 2000, ApJ, 531 764.
11. Usov V.V., 1992, Nature, 357, 472

Generation of Magnetic Fields and Jitter Radiation in GRBs. I. Kinetic Theory

Mikhail V. Medvedev

Canadian Institute for Theoretical Astrophysics, University of Toronto, Toronto, Ontario, M5S 3H8, Canada
Harvard-Smithsonian Center for Astrophysics, 60 Garden Street, Cambridge, MA 02138

Abstract. We present a theory of generation of strong (sub-equipartition) magnetic fields in relativistic collisionless GRB shocks. These fields produced by the kinetic two-stream instability are tangled on very small spatial scales. This has a clear signature in the otherwise synchrotron(-self-Compton) γ-ray spectrum. Second, we present an analytical theory of jitter radiation, which is emitted when the correlation length of the magnetic field is smaller then the gyration (Larmor) radius of the accelerated electrons. We demonstrate that the spectral power $P(\nu)$ for pure jitter radiation is well-described by a sharply broken power-law: $P(\nu) \propto \nu^1$ for $\nu < \nu_j$ and $P(\nu) \propto \nu^{-(p-1)/2}$ for $\nu > \nu_j$, where p is the electron power-law index and ν_j is the jitter break, which is independent of the magnetic field strength and depends on the shock energetics and kinematics. Here we mostly focus on the first problem. The radiation theory and comparison with observations will be discussed in the forthcoming publications.

INTRODUCTION

There is currently no satisfactory explanation for the origin of strong magnetic fields required in GRB shocks. Compression of the ISM magnetic field in external shocks yields a field amplitude $B \sim \gamma B_{\rm ISM} \sim 10^{-4}(\gamma/10^2)$ gauss, which is too weak and can account only for $\epsilon_B = (B/B_{eq})^2 \leq 10^{-11}$ (here γ is the Lorentz factor of the outflow). Neither a turbulent magnetic dynamo, nor the magnetic shearing (Balbus-Hawley) instability, nor any other MHD process can be so efficient to produce the required strong fields. In principle, some magnetic flux might originate at the GRB progenitor and be carried by the outflowing fireball plasma. Because of flux freezing, the field amplitude would decrease as the fireball expands. In this case, only a progenitor with a rather strong magnetic field $\sim 10^{16}$ gauss might produce sufficiently strong fields during the GRB emission. However, since the field amplitude scales as $B \propto R^{-4/3}$, even a highly magnetized plasma at $R \sim 10^7$ cm would possess only a negligible field amplitude of $\sim 10^{-2}$ gauss, or $\epsilon_B \leq 10^{-7}$, at a radius of $R \geq 10^{16}$ cm, where the afterglow radiation is emitted. Here we discuss how strong magnetic fields are generated by the kinetic relativistic two-stream instability [2] and consider their properties. We postpone the major discussion on the jitter radiation theory [3] to a forthcoming publication.

THE TWO-STREAM INSTABILITY

The non-relativistic instability was first discovered by Weibel [4]. It has been used by Moiseev and Sagdeev [5] to develop a theory of collisionless non-relativistic shocks in the interplanetary space.

Let us consider, for simplicity, the dynamics of the electrons only, and assume that the protons are at rest and provide global charge neutrality. The electrons are assumed to move along the x-axis (as illustrated in Fig. 1a) with a velocity $\mathbf{v} = \pm\hat{\mathbf{x}}v_x$ and equal particle fluxes in opposite directions along the x-axis (so that the net current is zero). Next, we add an infinitesimal magnetic field fluctuation, $\mathbf{B} = \hat{\mathbf{z}}B_z \cos(ky)$. The Lorentz force, $-e\frac{\mathbf{v}}{c} \times \mathbf{B}$, deflects the electron trajectories as shown by the dashed lines in Fig. 1a. As a result, the electrons moving to the right will concentrate in layer I, and those moving to the left – in layer II. Thus, current sheaths form which appear to *increase* the initial magnetic field fluctuation. The growth rate is $\Gamma = \omega_p v_y/c$, where $\omega_p^2 = (4\pi e^2 n/m)$ is the non-relativistic plasma frequency [6]. Similar considerations imply that perpendicular electron motions along y-axis, result in oppositely directed currents which suppress the instability. The particle motions along $\hat{\mathbf{z}}$ are insignificant as they are unaffected by the magnetic field. Thus, the instability is driven by the anisotropy of a particle distribution function and should quench for the isotropic case.

The Lorentz force deflection of particle orbits increases as the magnetic field perturbation grows in amplitude. The amplified magnetic field is *random* in the plane perpendicular to the particle motion, since it is generated from a random seed field. Thus, the Lorentz deflections result in a pitch angle scattering which makes the distribution function isotropic. The thermal energy associated with their random motions will be equal to their initial directed kinetic energy. This final state will bring the instability to saturation.

Here are the main properties of the instability and the produced magnetic fields:

- This instability is driven by the *anisotropy* of the particle distribution function and, hence, can operate in both internal and external shocks.

- The characteristic *e*-folding time in the shock frame for the instability is $\tau \simeq \gamma_{sh}^{1/2}/\omega_p$ (where γ_{sh} is the shock Lorentz factor) which is $\sim 10^{-7}$ s for internal shocks and 10^{-4} s for external shocks. This time is much shorter than the dynamical time of GRB fireballs.

- The characteristic coherence scale of the generated magnetic field is of the order of the relativistic skin depth $\lambda \simeq 2^{1/4} c \bar{\gamma}^{1/2}/\omega_p$ (where $\bar{\gamma}$ is the mean thermal Lorentz factor of particles), i.e. $\sim 10^3$ cm for internal shocks and $\sim 10^5$ cm for external shocks. This scale is much smaller than the spatial scale of the source.

- The generated magnetic field is randomly oriented in space, but always lies in the plane of the shock front.

- The instability is powerful. It saturates only by nonlinear effects when the magnetic field amplitude approaches equipartition with the electrons (and possibly with the ions). Therefore $[B^2/8\pi]/[mc^2 n(\bar\gamma - 1)] = \eta \sim 0.01 - 0.1$. This result is in excellent agreement with direct particle simulations.

- The instability isotropizes and heats the electrons and protons.

- Random fields scatter particles over pitch-angle and, thus, provide effective collisions. Therefore MHD approximation works well for the shocks. The magnetic fields communicate the momentum and pressure of the outflowing fireball plasma to the ambient medium and define the shock boundary.

- The generated small-scale fields affect the radiation processes [3] and produce non-synchrotron spectra of radiation, as shown in Fig. 1b.

REFERENCES

1. Sari, R., Narayan, R., & Piran, T., *Astrophys. J.* **473**, 204 (1996)
2. Medvedev, M. V., and Loeb, A., *Astrophys. J.* **526**, 697 (1999)
3. Medvedev, M. V., *Astrophys. J.* **540**, 704 (2000)
4. Weibel, E. S., *Phys. Rev. Lett.* **2**, 83 (1959)
5. Moiseev, S. S., and Sagdeev, R. Z., *J. Nucl. Energy C* **5**, 43 (1963)
6. Fried, B. D., *Phys. Fluids* **2**, 337 (1959)

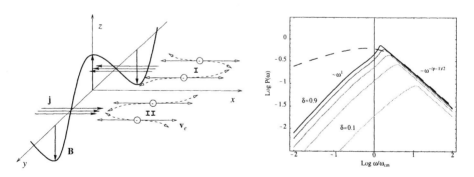

FIGURE 1. (*a*) — Illustration of the instability. A magnetic field perturbation deflects electron motion along the *x*-axis, and results in current sheets (*j*) of opposite signs in regions I and II, which in turn amplify the perturbation. The amplified field lies in the plane perpendicular to the original electron motion. (*b*) — Typical jitter spectra from small-scale magnetic fields (*dashed* curve is synchrotron, for comparison).

The Kinematics of the Lag-Luminosity Relationship

Jay D. Salmonson

Lawrence Livermore National Laboratory, Livermore, California 94550

Abstract. Herein I review the argument that kinematics, i.e. relativistic motions of the emitting source in gamma-ray bursts (GRBs), are the cause of the lag-luminosity relationship observed in bursts with known redshifts.

INTRODUCTION

Norris *et al.* [1] have discovered a relationship between the peak luminosity of gamma-ray bursts (GRB) and the pulse time lag between BATSE energy channels. In Salmonson 2000 [2] this correlation was shown to be improved when we neglect our poor knowledge of received photon energy. Thus instead we use the correlation between photon number luminosity and pulse time lag. It was found that the inferred isotropic peak number luminosity N_{pk} (photons s^{-1}) is related to the observed spectral lag between energy channels Δt by

$$N_{pk} = 8.6 \times 10^{56} \Delta t^{-0.98} . \tag{1}$$

I argue that kinematics are the origin of this relation [2]. Specifically, bursts with emitting material moving with a higher velocity toward the observer appear more luminous and have shorter observed lags (derived from an intrinsic pulse cooling timescale) between observed energy channels due to relativistic blue shift. Relativistic beaming allows one to only consider emitters moving directly toward the observer. I propose that the wide range of observed (cosmological redshift compensated) spectral lags and inferred luminosities (see Figure 1) can be explained if GRBs derive from a relativistic jet, with opening angle θ_0, in which the fastest material moves along the core of the jet and the velocity of the material monotonically decreases with increasing angle from the jet axis. The variety of observed bursts then derives from our perspective of the jet. All of the material is presumed to move relativistically ($\gamma \gg 1$) and so all of our received flux is derived from a very small $\sim 1/\gamma^2$ solid angle of the jet; much smaller than the jet opening angle ($1/\gamma \ll \theta_0$). It is from this small region that all of our information about a burst is derived.

This hypothesis that all GRBs derive from a single jet morphology viewed at various orientations then predicts that there will be GRB jets viewed at such high inclinations, and thus with such low Lorentz factors that the relativistic beaming angle will be larger than the jet opening angle ($1/\gamma > \theta_0$). The emission will no longer be consistent with isotropy and thus I predict a break in the lag-luminosity relation: $N_{pk} \propto \Delta t^{-3}$ [3]. Due to the low Lorentz factor, such bursts would be very dim. If GRB980425, which was unusual in several respects [3] and, by its association with supernova 1998bw [4], was sub-luminous, defines this break (see Figure 1), then one can calculate key features of the GRB jet. In particular, Lorentz factors range from $\gamma_{max} \sim 1000$ for bright bursts such as GRB990123, to $\gamma_{break} \sim 100$ for more middling bursts like GRB980703. The position of the break implies an opening angle of $\theta_0 \sim 1°$ and thus a total gamma-ray burst energy of $E_{tot} \sim 10^{50}$ ergs, which is much less than isotropic energy estimates (up to 10^{54} ergs [5]).

This work was performed under the auspices of the U.S. Department of Energy by University of California Lawrence Livermore National Laboratory under contract W-7405-ENG-48.

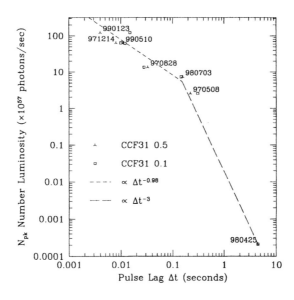

FIGURE 1. Peak photon number luminosity N_{pk} versus spectral pulse lag for six bursts with known redshifts plus GRB980425. A break is inferred by fitting a break slope $\propto \Delta t^{-3}$ to intersect GRB980425. Spectral cross-correlation function lags between BATSE channels 3 and 1 (CCF31) for regions down to 0.5 and 0.1 of peak intensity were obtained from Norris et al. (2000). The line of best fit for 0.1 (squares) is $\propto \Delta t^{-0.98}$.

REFERENCES

1. Norris J. P., Marani G. F., Bonnell J. T., *ApJ* **534**, 248 (2000).
2. Salmonson J. D., *ApJL* **544**, L115 (2000).
3. Salmonson J. D., *ApJL* **546**, L29 (2001).
4. Galama T. J. et al., *Nature* **395**, 670 (1998).
5. Galama T. J. et al., *Nature* **398**, 394 (1999).

Near-infrared polarimetric observations of the afterglow of GRB 000301C

B. Stecklum[a], O. Fischer[b], S. Klose[a], R. Mundt[c], and
C. Bailer-Jones[c, 1]

[a] *Thüringer Landessternwarte, Tautenburg, Germany*
[b] *Universitätssternwarte, Jena, Germany*
[c] *Max-Planck-Institut für Astronomie, Heidelberg, Germany*

Abstract. Based on near-infrared polarimetric observations we constrain the degree of the linear polarization of the afterglow light of GRB 000301C to less than 30% 1.8 days after the burst.

I INTRODUCTION

The question whether a GRB is always accompanied by a collimated outflow, often called a jet, is one of the key issues of current GRB research. Because of relativistic effects a collimated explosion will reduce the deduced energy release of a GRB by the beaming factor. Theoretical considerations have demonstrated that in the case of a collimated outflow the afterglow light can be partly linearly polarized and the degree of the linear polarization p should vary with time /1-4/.

In 1999 we established a Target of Opportunity (ToO) program aiming at the measurement of the linear polarization of GRB afterglows. The project is being carried out at the 3.5-m telescope at Calar Alto, Spain, utilizing the near-infrared camera Omega Cass /5/. This instrument is equipped with a 1024 × 1024 HAWAII array. Thus far, it was employed in wide field mode (0″.3/pixel), yielding a field of view of ∼ 5′ × 5′. The observations were performed in the K' band, using wiregrid polarizers to obtain images at four position angles (0, 45, 90, and 135°). The limiting magnitude of the combined images amounts to $K' \sim 19$.

The burst GRB 000301C was the first burst for which we can constrain the degree of linear polarization of the afterglow light.

---

[1] Visiting Astronomer, German-Spanish Astronomical Centre, Calar Alto, operated by the Max-Planck-Institute for Astronomy, Heidelberg, jointly with the Spanish National Commission for Astronomy

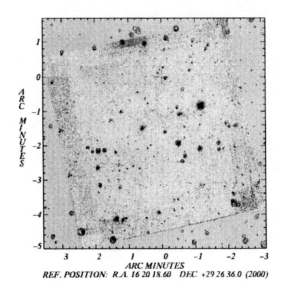

FIGURE 1. The afterglow of GRB 000301C was imaged 1.8 days after the burst with the Calar Alto 3.5-m telescope at a magnitude of $K'=17.5$ /12/. Displayed here is the combined K'-band image after adding all frames taken at four different polarization angles. The GRB afterglow is indicated by a cross. Contour lines represent the overplotted DSS-2 red image of the field.

II THE BURST GRB 000301C

The GRB 000310C was detected with *RXTE, Ulysses,* and *NEAR* on March 1, 2000, at 9:51 UTC /6/. In the high-energy band it lasted ~ 2 seconds /7/. The optical afterglow was soon detected on images taken on 2000 March 3.2 UT at $R = 20.3 \pm 0.5$ /8/. The redshift of the host galaxy turned out to be 2.04 /7/. A break in the afterglow lightcurve was seen some days after the burst suggesting that this burst was accompanied by a jet /9, 10/. Of particular interest is the possible detection of a microlensing event in the afterglow lightcurve /11/.

The polarimetric data were acquired at Calar Alto on March 3, 5:00 UT when the GRB afterglow was already at a magnitude of $K'=17.5$ (Fig. 1). The instrumental and interstellar contribution to the linear polarization were corrected by assuming a zero net polarization of all stars in the field. Based on aperture photometry of all well-isolated stars we can constrain p of the GRB afterglow to be less than 30% (Fig. 2). This result is in agreement with predictions according to which p should never exceed about 20% /1, 2/. We note, however, that our result might still be influenced by residual instrumental polarization which is currently under investigation.

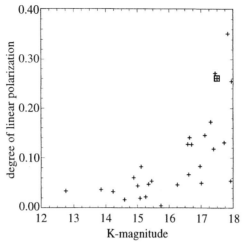

FIGURE 2. Polarization *vs.* magnitude diagram for all stars in the field ($p = 1$ corresponds to 100% polarization). The GRB afterglow is indicated by a square.

III   FUTURE WORK

The project entered its second year of operation in January 2001. Our results show that a GRB afterglow has to be brighter than $K' \sim 16$ in order to measure, or to constrain, its degree of linear polarization with an error of $\lesssim 10\%$ using Omega Cass. From *HETE* 2 we expect rapid alerts and accurate GRB locations which would allow to use the high-resolution mode of Omega Cass. The instrumental capabilities for this mode and the larger brightness of the GRB afterglows due to the shorter response time suggest that our goal of measuring p and its temporal variation will be achievable in the near future.

REFERENCES

1. Ghisellini, G. & Lazzati, D., *MNRAS* **309**, L7 (1999).
2. Sari, R., *ApJ* **524**, L43 (1999).
3. Gruzinov, A., *ApJ* **525**, L29 (1999).
4. Medvedev, M. V. & Loeb, A., *ApJ* **526**, 697 (1999).
5. Lenzen, R. et al., SPIE **3354**, 493 (1998).
6. Smith, D. A. et al., GCN #568 (2000).
7. Jensen, B. L. et al., astro-ph/0005609 (2000).
8. Fynbo, J. P. U. et al., GCN #570 (2000).
9. Rhoads, J. E. & Fruchter, A. S., *ApJ* **546**, 117 (2001).
10. Berger, E. et al., *ApJ* **545**, 56 (2000).
11. Garnavich, P. M. et al., *ApJ* **544**, L11 (2000).
12. Stecklum, B. et al., GCN #572 (2000).

The Trans-Relativistic Blast Wave Model for SN 1998bw and GRB 980425

Jonathan C. Tan[1], Christopher D. Matzner[2], and Christopher F. McKee[1,3]

[1]*Dept. of Astronomy, University of California, Berkeley, CA 94720, USA*
[2]*CITA, University of Toronto, 60 St. George Street, Toronto, Ontario M5S 3H8, Canada*
[3]*Dept. of Physics, University of California, Berkeley, CA 94720, USA*

Abstract. The spatiotemporal coincidence of supernova (SN) 1998bw and gamma-ray burst (GRB) 980425 and this supernova's unusual optical and radio properties have prompted many theoretical models that produce GRBs from supernovae. We review the salient features of our simple, spherical model in which an energetic supernova explosion shock accelerates a small fraction of the progenitor's stellar envelope to mildly relativistic velocities. This material carries sufficient energy to produce a weak GRB and a bright radio supernova through an external shock against a dense stellar wind.

Crack! A whip's tail flies supersonically through the air after a slow transverse wave has accelerated down its tapering length. In the same manner a shock wave propagating through a star may accelerate to relativistic speeds in the tenuous layers of the stellar atmosphere [1,5,8]. Furthermore, a pressure gradient, imprinted in the expanding postshock gas, greatly boosts the kinetic energy of the outer ejecta. Large energies are naturally concentrated in small amounts of matter since shock acceleration only occurs at steeply declining density gradients. Supernova-driven shock acceleration thus provides an attractive model for gamma-ray bursts (GRBs) [1], one that naturally overcomes the baryon-loading problem.

To quantify this process, we have performed a suite of numerical simulations of the acceleration of blast waves from nonrelativistic to ultrarelativistic velocities in simple, idealized density distributions [8]. One of our principal results is an analytic expression for $E_k(>\Gamma\beta)$, the kinetic energy of ejecta moving with final velocity greater than $\beta \equiv v/c$ and Lorentz factor greater than $\Gamma \equiv (1-\beta^2)^{-0.5}$, from explosions in centrally concentrated density distributions with polytropic envelopes. We applied this analysis to a more realistic, though still one dimensional, supernova progenitor, kindly provided by Stan Woosley, that was matched to SN 1998bw's light curve [9]. It involves a 2.8×10^{52} erg explosion in a Wolf-Rayet (WR) star consisting of the $\sim 7\,M_\odot$ carbon-oxygen core of an initially $\sim 25\,M_\odot$ main sequence star. We also simulated this explosion with our relativistic code. Both our analytic and numerical estimates place $\sim 2 \times 10^{48}$ ergs in ejecta with $\Gamma\beta > 1$, sufficient to account for GRB 980425, which has isotropic $E_\gamma \simeq 8.5 \times 10^{47}$ ergs.

We find that $E_k(> \Gamma\beta) \propto E_{\rm in}^{2.67\gamma_p} M_{\rm ej}^{1-2.67\gamma_p}$, where $E_{\rm in}$ and $M_{\rm ej}$ are the total kinetic energy and mass of the ejecta and $\gamma_p \equiv 1 + 1/n$, with n being the polytropic index of the outer stellar envelope that makes up the relativistic ejecta. SN 1998bw, classified as a peculiar Type Ic and thus lacking H and He, is an energetic explosion with a relatively small total ejecta mass. This explains its ability to produce relativistic ejecta. However, whether a given supernova creates a detectable GRB or not also depends on the properties of the circumstellar wind (see below).

How do the more energetic cosmic GRBs relate to this model for GRB 980425? $E_k(> \Gamma\beta)$ depends sensitively on $E_{\rm in}$, which itself depends on the gravitational energy release, $E_{\rm grav}$, and the efficiency of coupling to the ejecta, $\epsilon_{\rm in} \equiv E_{\rm in}/E_{\rm grav}$. Our model for SN 1998bw forms a 1.8 $M_\odot$ neutron star, but much more energetic explosions ("hypernovae") may result from the formation of some, more massive black holes [6]. While estimates of $\epsilon_{\rm in}$ are uncertain, explosions ejecting orders of magnitude more kinetic energy than SN 1998bw are astrophysically conceivable and need occur only rarely to account for the observed cosmic GRBs. Of course the limit $E_k \leq E_{\rm in}$ truncates the above scaling. Our preliminary analysis of cosmic GRBs in the context of shock acceleration models, suggests that asymmetric explosions, perhaps resulting from rotationally flattened progenitors, are required [8].

Fig. 1 summarizes our model of SN 1998bw and GRB 980425. The energy of the mildly relativistic ejecta is greater than the observed γ-rays, but to liberate this energy as radiation the ejecta must collide with circumstellar material; in this case a stellar wind. The ~ 35 s timescale of the GRB sets the required density of the wind since relativistic ejecta lose $\sim 1/2$ of their energy after sweeping up $1/\Gamma$ of their own mass. This wind density is high: $\dot{M}_{-4}/v_{w,8} \sim 3$, where the mass loss rate is $\dot{M}_{-4} = \dot{M}/10^{-4}\, M_\odot\, {\rm yr}^{-1}$ and the wind velocity is $v_{w,8} = v_w/10^8$ cm s^{-1}. In fact some carbon rich Wolf-Rayet stars do lose mass at these extreme rates [2].

For a given shock Γ and postshock magnetic energy fraction, the energy of synchrotron photons emitted by electrons behind the shock front increases with density. For the mildly relativistic ejecta produced in our model of SN 1998bw, the high densities implied by the short GRB timescale are also required to give the correct energies ($\sim 10 - 100$ keV) of the GRB photons. Note, however, that inverse Compton scattering may also affect the burst spectrum. Interestingly, the high wind densities imply that at the start of the interaction producing the burst, the wind is optically thick to the γ-rays, while at the end it is optically thin. The energy dependence of the Klein-Nishina cross section then predicts the early emergence of the hardest photons from the wind, as was observed [7].

Radio scintillation observations of the supernova at 12 days imply a mean shock expansion velocity $\gtrsim 0.3c$ over this period [3]. An external shock driven by ejecta from our model explosion satisfies these constraints. The shock decelerates only very slowly as the interior ejecta contain progressively more mass, while the wind becomes ever more tenuous. At ~ 20 days the radio light curve brightens, and this

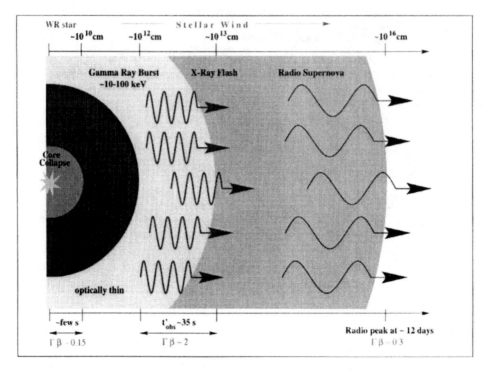

FIGURE 1. SN 1998bw - GRB 980425 schematic.

has been modeled as resulting from additional energy injection into the shock front [4]. A natural explanation is that inner mass shells of ejecta, with smaller energy per baryon but greater total energy than the outer layers, eventually catch up to the decelerating shock front. The relatively abrupt brightening may result from a density shelf initially present in the star's atmosphere, $\sim 10^{-3}\,M_\odot$ from the surface.

REFERENCES

1. Colgate, S. A., *ApJ.*, **187**, 333, (1974).
2. Koesterke, L., Hamann, W. -R., *A.&A.*, **299**, 503, (1995).
3. Kulkarni, S. R., et al., *Nat.*, **395**, 663, (1998).
4. Li, Z. Y., Chevalier, R. A., *ApJ.*, **526**, 716, (1999).
5. Matzner, C. D., McKee, C. F., *ApJ.*, **510**, 379, (1999).
6. Paczynski, B., *Gamma-Ray Bursts: 4th Huntsville Symposium*, eds. C.A. Meegan, R.D. Preece, & T.M. Koshut, Woodbury: New York, (1998).
7. Pian, E. & 27 colleagues, *ApJ.*, **536**, 778, (2000).
8. Tan, J. C., Matzner, C. D., McKee, C. F., *ApJ.*, in press, (2001).
9. Woosley, S. E., Eastman, R. G., Schmidt, B. P., *ApJ.*, **516**, 788, (1999).

CHAPTER 13

MAGNETIC ACCRETION INTO BLACK HOLES

X-rays and accretion discs as probes of the strong gravity of black holes

A.C. Fabian

Institute of Astronomy, Madingley Road, Cambridge CB3 0HA, UK

Abstract. The observations and interpretation of broad iron lines in the X-ray spectra of Seyfert 1 galaxies are reviewed. The line profiles observed from MCG–6-30-15 and NGC 3516 show extended red wings to the line explained by large gravitational redshifts. The results are consistent with the emission expected from an X-ray irradiated flat accretion disc orbiting very close to a black hole. Results from XMM-Newton and Chandra are presented and the possibility of broad oxygen lines discussed.

INTRODUCTION

The optical (e.g. Ghez et al 2000; Eckart et al 1997; Gebhardt et al 2000; Ferrarese & Merritt 2000) and radio (e.g. Miyoshi et al 1995) evidence for compact central masses in the nuclei of many nearby galaxies is now very clear. The compact objects are consistent with being supermassive black holes. The data do not however probe the strong gravity regime of these holes, since the observed stars and gas orbit at more than 40,000 gravitational radii from the centre.

Much of the X-ray emission from accreting black holes should emerge from within a few 10s of gravitational radii. Aspects of this emission can provide us with a probe of the strong gravity of black holes. In particular the rapid variability, or in some cases quasi-periodic oscillations, and soft thermal emission from an accretion disc demonstrates the small size of the X-ray emission region.

Here I concentrate on the broad iron emission lines seen in the spectra of some Seyfert 1 galaxies. The line profile and its variability can reveal the geometry of the accretion flow, and strong gravitational effects, within a few gravitational radii of the black hole. The dominant expected line is that due to iron, as shown in Fig. 1 (George & Fabian 1991; Matt, Perola & Piro 1991). The spectrum shown is the result of a Monte-Carlo simulation of the emergent spectrum when a flat surface has been irradiated with a power-law X-ray spectrum. The gas is assumed to be neutral, although the important issue here is that the metals retain their K and L-shell electrons. The major emission line is iron K-α, due to its relatively high abudance (a cosmic mix is assumed) and high fluorescent yield (which increases as Z^4). Doppler broadening of this emission line, produced by the irradiation of

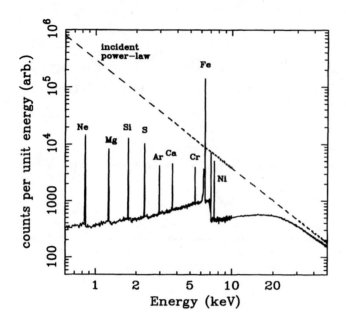

FIGURE 1. Monte Carlo simulation (by C. Reynolds) of the reflected continuum plus fluorescent line emission from a neutral slab of gas with comsic abundances. The incident power-law continuum is indicated by the dashed line. In practice the sum of the incident and reflected spectra are usually seen.

the surface of an accretion disc by hard X-rays from its corona, together with thr special relativistic effects of beaming and the transverse doppler effect and the general relativistic effect of gravitational redshift, lead to the observed profile being very broad and skew (Fabian et al 1989: Figs. 2, 3). The 'blue horn' of the line is most sensitive to the inclination angle of the disc (i.e. the importance of doppler broadening) and the 'red wing' is most sensitive to the inner radius (i.e. the gravitational redshift).

I OBSERVATIONS OF BROAD IRON LINES

A clear broad iron line was first seen with data from a 1994 ASCA observation of the Seyfert 1 galaxy MCG–6-30-15 (Tanaka et al 1995: Fig. 4). This was the result of a 4.5 day exposure on the source and is shown with the dominant power-law continuum removed. The data are well fit (Fabian et al 1995) by emission from an accretion disc stretching from about 6 gravitational radii (i.e. $6r_g = 6GM/c^2$) to about 40 r_g and an inclination of about 30 deg. The surface emissivity falls as

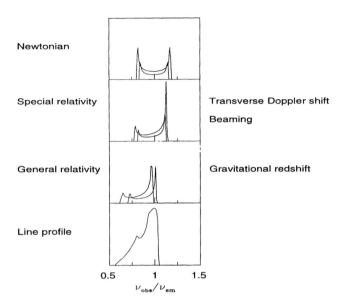

FIGURE 2. Line broadening from an intrinsically narrow line emitted from two radii in an accretion disc. The lowest panel shows the result obtained by summing many disc radii, weighted by the expected emissivity. This figure originally appeared in the Publications of the Astronomical Society of the Pacific (Fabian et al 2000, PASP 112, 1145). Copyright 2000, Astronomical Society of the Pacific; reproduced with permission of the Editors.

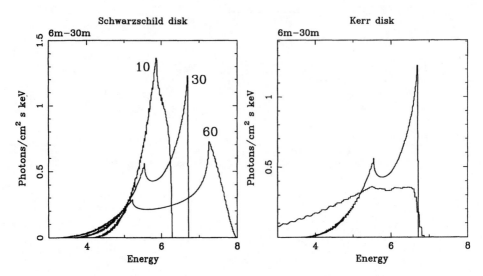

FIGURE 3. Spectra prdeicted for the iron Kα 6.4 keV line emitted from an accretion disc at radii (Left) 6 − 30r$_g$ around a non-spinning (using the model of Fabian et al 1989) and (Right) 6 − 30r$_g$ and (broader profile) 1.25 − 30r$_g$ around a maximally-spinning Kerr black hole (using the model of Laor 1991).

radius r^{-3}. A similar clear line has also been seen in NGC 3516 (Nandra et al 1999: Fig. 4). The shape of these lines is entirely consistent with the emitting region being a flat disc in Keplerian rotation close to a black hole.

K. Nandra (priv. comm.) has compiled the line spectra of several more Seyfert 1 galaxies (Fig. 5) which generally are weaker in flux, or have shorter exposures than for MCG-6-30-15. A broad red wing is seen in most. Further work on the strength and width of iron lines in Seyferts is reported by Lubiński & Zdziarski (2000).

Occasionally the iron line in MCG–6-30-15 changes its profile, as in a deep minimum (Iwasawa et al 1996) or after a flare (Iwasawa et al 1999), and appears to stretch more to the red. This is possible evidence that the black hole is spinning rapidly (Iwasawa et al 1996; Dabrowski et al 1997), although matter streaming from the innermost stable orbit around a non-spinning black hole might mimic this result (Reynolds & Fabian 1997; Young et al 1998).

XMM-Newton, which has a much larger effective area than ASCA, has now observed MCG–6-30-15 and several other Seyfert 1 galaxies. The iron line profile, kindly supplied by the PI of the observation, Joern Wilms, is shown in Fig. 8. The data in this ratio plot resemble the shape of the iron line from ASCA (Fig. 2). In the case of Mrk205 (Fig. 9) the line appears to peak around 6.7 keV, which may indicate an ionized disc (i.e. the metals have only K-shell electrons). The sharp peak at 6.4 keV may be a fluorescent line from more distant cold gas within the galaxy. Models of ionized discs have been made for constant density (e.g. Ross & Fabian 1993) and more recently for atmospheres in hydrostatic equilibrium

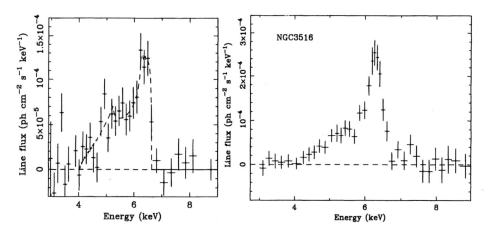

FIGURE 4. Observed profiles of broad iron lines; (Left) MCG–6-30-15 from Tanaka et al (1995), reproduced with permission from *Nature*, and (Right) NGC 3516 from Nandra et al (1999).

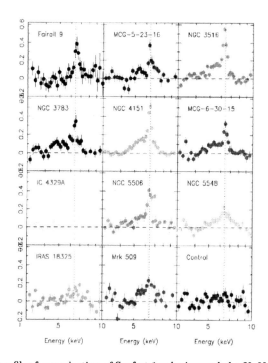

FIGURE 5. Line profiles for a selection of Seyfert 1 galaxies made by K. Nandra (priv. comm.).

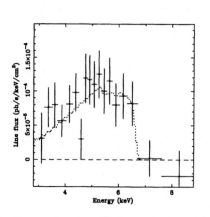 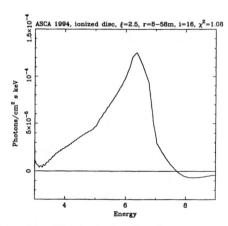

FIGURE 6. (Left) line profile during the deep minimum of the ASCA 1994 observation (from Iwasawa et al 1996, reproduced with permission from Blackwell Science Ltd.). The fitted profile is for a maximal Kerr black hole. (Right) model profile for a relativistically-broadened, constant density ionized disc model fitted to the time-averaged 1994 ASCA data. (Compare with Fig. 4, left panel).

(Nayakshin, Kazanas & Kallman 2000; Ballantyne, Ross & Fabian 2001: Fig. 7). An example of a constant density model fitted to the ASCA MCG–6-30-15 data is shown in Fig. 6. The models predict that the iron line shifts to 6.7 keV in the rest frame and oxygen and other features should be present below 1 keV. Whether these low energy features are detectable depends on their strength relative to the steep continuum.

Broad iron lines are generally not seen in objects with luminosity exceeding a few time 10^{44} erg s^{-1} in the 2–10 keV band (Iwasawa & Taniguchi 1993; Nandra et al 1997). An example from XMM-Newton of a narrow iron line from a powerful distant quasar, PKS 0537-286, is shown in Fig. 9, from Reeves et al 2001). Narrow iron lines are expected in many AGN from reflection by outer gas at parsecs and beyond (Ghisellini et al 1994) and are seen in some Seyferts (e.g. Yaqoob et al 2000).

II IRON LINE VARIABILITY

The intensity of the iron line in MCG–6-30-15 does vary (Iwasawa et al 1996; 99), but not in any obvious way in response to the continuum. A long observation made simulataneously with ASCA and RXTE in 1997 (Lee et al 2000) confirms this. Vaughan & Edelson (2000) have re-analysed the RXTE data with the results shown in Fig. 10. There is real variability in the iron line flux but it is not correlated with the continuum flux. The continuum power-law slope is correlated in the sense that the spectrum steepens as the flux increases. Further work by Matsumoto et al (2000) on a 10 day ASCA observation in 1999 extends the peculiar behaviour.

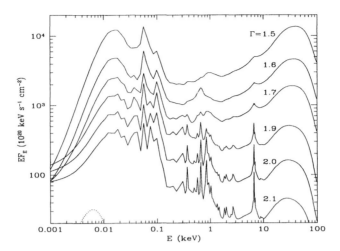

FIGURE 7. Reflection spectra from ionized accretion disc models, before relativistic blurring; see Ballantyne et al (2001) for details. The observed spectrum will usually need the addition of the power-law continuum of photon index Γ, which is not included here.

The RMS variability decreases towards high energy, particularly around the iron line (Fig. 11).

Various ideas have been mooted in order to explain the variability. Possibilities include; a) transrelativistic motion of flares (Reynolds & Fabian 1997; Beloborodov 1999) so that an observed bright continuum is associated with flux beamed away from the disc and vice versa (this has the problem that the spectral index trend with flux would be opposite to that observed); b) smoke from the coronal flares themselves (Merloni & Fabian in prep.) in which electron scattering in the corona smears out transmitted line emission when the regions are spatially large, and hence bright; and c) ionization variations (Lee et al 2000; Reynolds 2000).

The lack of any clear iron line – continuum correlation so far suggest that iron line reverberation (Reynolds et al 1999) may be complicated.

III BROAD OXYGEN LINES?

A recent development has been the observation of Seyfert galaxies at high spectral resolution, provided by the gratings on Chandra and XMM-Newton. Branduardi-Raymont et al (2001) claim from XMM-Newton RGS spectra of MCG–6-30-15 the presence of relativistically-broadened oxygen, nitrogen and carbon lines. As already noted, emission lines from these elements are expected from an ionized disc. The surprise is that they might be so strong. Moreover, the spectrum of the source below 2 keV has been previously well modelled by a variable warm absorber (Fabian et al 1994; Otani et al 1996). In their original paper, Branduardi-Raymont et al (2001)

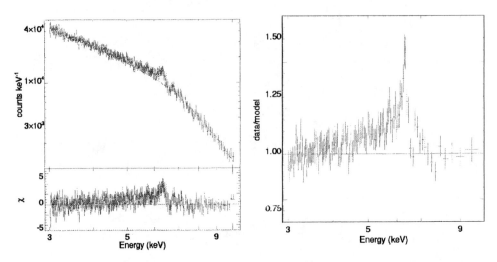

FIGURE 8. XMM-Newton EPIC-pn spectra of MCG–6-30-15. The left panel shows the whole spectrum from 3–10 keV and the right panel shows the ratio between the data and a simple power-law fitted only to the data from 3–4 and 8–10 keV. Note that the ratio plot must be multiplied by the power-law spectrum to produce a line intensity plot such as in Fig. 4.

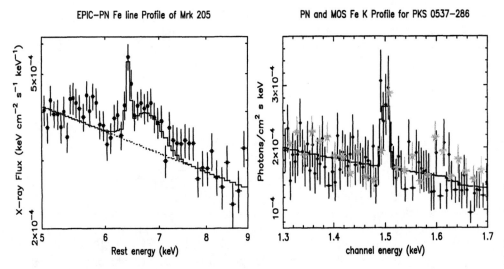

FIGURE 9. EPIC-pn spectra of the Seyfert 1 galaxy Mrk 205 (left) and the $z = 3.104$ quasar PKS 0537-286 (right), from Reeves et al (2001a,b).

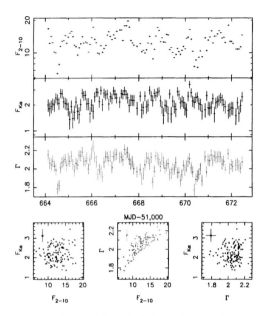

FIGURE 10. Light curves of the total 2–10 keV band flux (top), iron K-α flux (middle) and photon index (bottom) during the 1997 RXTE observation of MCG–6-30-15, from Vaughan & Edelson (2000). The lower panels show that there is no correlation between $F_{K\alpha}$ and F_{2-10}.

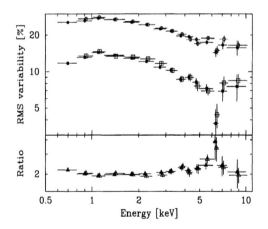

FIGURE 11. The energy dependence of the RMS variability of MCG–6-30-15, as observed with ASCA (from Matsumoto et al 2001). The 2 plots in the top panel refer to time bins of (upper) 2×10^4 s and (lower) 2×10^5 s.

rely on some unspecified mechanism for the observed spectral turndown below 2 keV and argue from the weak, broad ($2000\,\mathrm{km\,s^{-1}}$) absorption lines seen in their spectra that the column density of any warm absorber is too low to provide an edge large enough to account for the spectral jump at $\sim 0.7\,\mathrm{keV}$ (Fig. 12), moreover the jump occurs in the wrong place, requiring a redshift of $16,000\,\mathrm{km\,s^{-1}}$.

Chandra HETG spectra provide a different view of the same object (Lee et al 2001; Fig. 13). The 3 times higher spectral resolution shows that the linewidths are $< 200\,\mathrm{km\,s^{-1}}$ and, from curve of growth analysis of the OVII line series, argues for the presence of a significant warm absorber. Part of the redshift of the OVII photoelectric edge is then accounted for by the convergence of the OVII absorption line series.

The poster paper on the XMM-Newton RGS spectrum by Sako et al (2001; these proceedings) now includes a warm absorber, together with relativistically-broadened hydrogenic, C, N and O lines. Absorption by neutral iron is suspected to be a further important ingredient by the Chandra team. The neutral iron presumably being in the form of dust, which must be present to explain the high optical/UV reddening seen in the object (Reynolds et al 1997).

Since the conference, Lee et al (2001) have accounted for a major feature around the OVII edge in terms of neutral iron absorption. Of the three edges of iron-L, known as L3, L2 and L1, the absorption near threshold of Fe L3 is a deep trough in laboratory data (Kortright & Kim 2000) and in the spectrum of Cyg X-1 (Schulz et al 2001). This trough exactly matches the broad absorption feature seen in the Chandra spectrum of MCG–6-30-15 at 0.71 keV (Fig. 13) and accounts for the apparent redshift of the OVII edge (Fig. 12). Fe L2 and OVII absorption together explain the remaining features between 0.7 and 0.75 keV.

Much more detailed modelling is required, covering a wider spectral band, but it is now clear that MCG–6-30-15 does have a significant dusty warm absorber. It is plausible that there is in addition some emission from ionized C, N and O within the disc, but debatable as to whether it is strong enough to create detectable features. The relativistic line broadening parameters quoted by Branduardi-Raymont et al (2001) do not yet agree with those found for the time-averaged ASCA spectra. The ASCA results (Fig. 4) appear to be in accord with the XMM-Newton EPIC data (Fig. 8) taken simultaneously with the RGS spectra.

The Chandra spectrum does not show two spectral jumps, one of which can be explained by photoelectric absorption and the other by the blue wing of an emission line. I conclude that the large spectral jump seen around 0.7 keV is dominated by absorption features from a dusty warm absorber. Whether relativistically-broadened emission features are also present remains to be seen.

IV SUMMARY

The broad iron lines seen in MCG–6-30-15 and NGC 3516 are well explained by X-ray irradiation of an accretion disc extending very close to a black hole. The

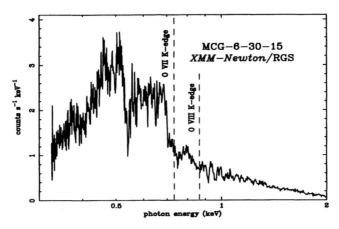

FIGURE 12. XMM-Newton RGS spectrum of MCG-6-30-15. The positions of the OVIII and OVII photoelectric absorption edges are marked (see Branduardi-Raymont et al (2001)).

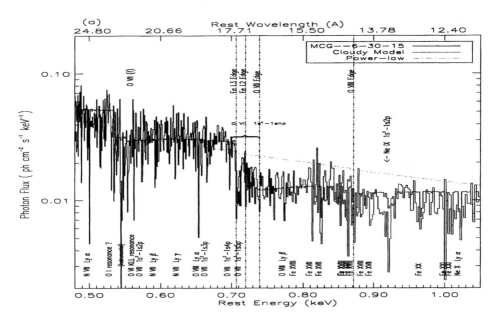

FIGURE 13. Chandra HETG spectrum of MCG–6-30-15 (Lee et al 2001). The CLOUDY model fit includes absorption due to neutral Fe-L assumed to arise from dust in the source. The large trough centred at 0.71 keV is due to Fe-L3 absorption.

skewed shape of the lines is mainly due to the strong gravitational redshift in this region.

The variability of the iron line and the lack of correlation with continuum variations is not understood. The emission region is plausibly complex with ionization, bulk motion and scattering effects likely.

The case for a broad oxygen line in MCG–6-30-15 is unclear.

V ACKNOWLEDGEMENTS

I thank David Ballantyne, Claude Canizares, Kazushi Iwasawa, Julia Lee, Raquel Morales, Patrick Ogle, Randy Ross, Norbert Schulz and Simon Vaughan for help and Chiho Matsumoto, Paul Nandra, James Reeves, Masao Sako and Joern Wilms for sending me plots in advance of publication.

REFERENCES

Ballantyne D.R., Ross R.R., Fabian A.C., 2001, MNRAS in press, (astro-ph/0102268)
Beloborodov A., 1999, ApJ, 510, 123
Branduardi-Raymont G., M. Sako, S.M. Kahn, Brinkman A.C., Kaastra J.S., Page M.J., 2001, A&A, 365, L140
Dabrowski, Y. et al., 1997, MNRAS, 288, L11
Eckart A., Genzel R., 1997, MNRAS, 284, 576
Fabian A.C., et al 1994, PASP, 46, L59
Fabian A.C., Nandra K., Reynolds C.S., Brandt W.N., Otani C., Tanaka Y., Inoue H., Iwasawa K., 1995, MNRAS, 277, L11
Fabian A.C., Rees M.J., Stella L., White N.E., 1989, MNRAS, 238,729
Fabian A.C., Reynolds C.S., Iwasawa K., Young A., 2000, PASP, 112,1145
Ferrarese L., Merritt D., 2000, ApJ, 539, L9
Gebhardt K., 2000, ApJ, 539, L13
George I.M., Fabian A.C., 1991, MNRAS, 249, 352
Ghez A.M., Morris M., Becklin E.E., Tanner A., Kremenek T., 2000, Nature 407, 349
Ghisellini G., Haardt F., Matt G., 1994, MNRAS, 267, 743
Iwasawa K., Taniguchi Y., 1993, ApJ, 413, L15
Iwasawa, K. et al., 1996, MNRAS, 282, 1038
Iwasawa K., Fabian A.C., Young A., Matsumoto C., Inoue C., 1999, MNRAS, 306, L19
Kortright J.B., Kim S.K., 2000, Phys Rev. B, 12216
Laor, A., 1991, ApJ, 376, 90
Lee J., Fabian A. C., Reynolds C. S., Brandt W. N., Iwasawa K., 2000, MNRAS, 318, 857
Lee J., Ogle P.M., Canizares C.S., Marshall H.L., Schulz N.S., Morales R.M., Fabian A. C., Iwasawa K., 2001, ApJ, 554, L13
Matsumoto C., Inoue H., Fabian A.C., Iwasawa K., 2001, PASJ, submitted
Matt, G., Perola, G.C., Piro L., 1991, AaA, 247, 25

Miyoshi M., et al 1995, Nature, 373, 127

Nandra K., George I.M., Mushotzky R.F., Turner T.J., Yaqoob Y., 1997, ApJ, 488, 91

Nandra K., George I. M., Mushotzky R. F., Turner T. J., Yaqoob T., 1999, ApJ, 523, L17

Nayaksin S., Kazanas D., Kallman T. R., 2000, ApJ, 537, 833

Otani C. et al., 1996, PASJ, 48, 211

Reeves J.N., et al 2001a, A&A, 365, L116

Reeves J.N., et al 2001a, A&A, 365, L134

Reynolds C. S., 2000, ApJ, 533, 811

Reynolds, C.S., Ward, M.J., Fabian A.C., Celotti A., 1997, MNRAS, 291, 403

Reynolds, C.S., Begelman, M.C., 1997, ApJ, 488, 109

Reynolds, C.S. Young A.J., Begelman M.C., Fabian A.C., 1998, ApJ, 514, 164

Ross, R.R., Fabian, A.C., 1993, MNRAS, 261, 74

Schulz N.S., et al 2001, in preparation

Tanaka, Y. et al., 1995, Nature, 375, 659

Vaughan S., Edelson R. 2000, ApJ, 548, 694

Yaqoob T., George I. M., Nandra K., Turner T. J., Serlemitsos P. J., Mushotzky R.F., 2000, ApJ, 546, 759

Lubiński P., Zdziarski A.A., 2001, MNRAS, 323, L37

Convection in radiatively inefficient black hole accretion flows

Igor V. Igumenshchev* and Marek A. Abramowicz[†]

*Institute of Astronomy, 48 Pyatnitskaya Ulitsa, 109017 Moscow, Russia
[†]Department of Astronomy & Astrophysics, Göteborg University & Chalmers University of Technology, 412-96 Göteborg, Sweden

Abstract. Recent numerical simulations of radiatively inefficient accretion flows onto compact objects have shown that convection is a general feature in such flows. Dissipation of rotational and gravitational energies in the accretion flows results in inward increase of entropy and development of efficient convective motions. Convection-dominated accretion flows (CDAFs) have a structure that is modified significantly in comparison with the canonical advection-dominated and Bondi-like accretion flows. The flows are characterized by the flattened radial density profiles, $\rho(R) \propto R^{-1/2}$, and have reduced mass accretion rates. Convection transports outward a significant amount of the released binding energy of the accretion flow. We discuss basic dynamical and observational properties of CDAFs using numerical models and self-similar analytical solutions.

INTRODUCTION

Observations of accreting black holes of different mass, from the stellar mass black holes to the supermassive black holes in the center of galaxies, show impressive similarities of data, which point at identical physical processes in accreting plasma. Also, black hole candidates demonstrate a great variety of physical conditions in the flows and, possibly, existence of a variety of accretion regimes. Existing theories describe different regimes of black hole accretion flows, which can be realized under different physical conditions. If matter accretes with a low specific angular momentum, $j \ll R_g c$, it forms spherical flows, which are described by Bondi solution [1] in the case of adiabatic flows. Here R_g is the gravitational radius of the black hole, and c is the speed of light. If matter has a large specific angular momentum, $j \gg R_g c$, then accretion disks are formed. The structure of accretion disks crucially depends on the efficiency of radiative cooling. If matter radiates efficiently, i.e. the radiative cooling time is shorter than the accretion time, $t_{rad} \ll t_{accr}$, the disks are geometrically thin, $H \ll R$, where H is the scale height of the flow, and R is the corresponding radius. The radial structure of such disks is described by the Shakura & Sunyaev model [2]. In the case of radiatively inefficient flows, $t_{rad} \gg t_{accr}$, the internal energy of accreting matter is close to the virial energy, and the formed disks are geometrically thick, $H \sim R$. The thick disks can be formed in two limiting cases of very high, $\dot{M} \gg L_{Edd}/c^2$, and very low, $\dot{M} \ll L_{Edd}/c^2$, accretion rates, where L_{Edd} is the Eddington luminosity of the black hole. In the former case, the optical depth of accretion flow is very large, and photons, which carry most of the internal energy, are trapped inside the inflowing matter and can not be radiated away. In the latter case, the accreting plasma is very diluted, so that different radiative mechanisms are inefficient to cool plasma on the accretion time scales. A model which describes the geometrically thick accretion disks was called the advection-dominated accretion flow (ADAF) and attracted a considerable attention during the last two decades [3–10]. The main feature of ADAFs is that most of the locally released gravitational and rotational energies of accretion flow is advected inward in the form of the gas internal energy, and the latter is finally absorbed by the black hole. The recent interest to ADAFs was generated mostly by attempts using this model to explain the low-luminosity X-ray objects of both galactic and intergalactic nature (for recent reviews see [11,12]).

ADVECTION-DOMINATED FLOWS

Properties of ADAFs could be better understood by analyzing the self-similar solution of the height-integrated hydrodynamical equations [10]. The solution depends on two parameters, the viscosity parameter α and the adiabatic index γ, and satisfies the following scalings for the density ρ, the radial velocity v_R, the angular velocity Ω, the isothermal sound speed c_s and H:

$$\rho(R) \propto R^{-3/2},$$
$$v_R(R) \propto R^{-1/2},$$
$$\Omega(R) \propto R^{-3/2}, \qquad (1)$$
$$c_s(R) \propto R^{-1/2},$$
$$H(R) = c_s/\Omega_K \propto R,$$

where Ω_K is the Keplerian angular velocity. Because of neglection of the radiative energy losses of the flow, the flow structure is independent of the mass accretion rate $\dot{M}$. The accretion rate determines only scales of density through the relation $\rho = \dot{M}/(4\pi R H v_R)$.

Two inconsistencies were found related to the model of ADAFs. The first is that ADAFs must be convectively unstable [4,5,7]. The second is connected to the positiveness of "Bernoulli parameter" in ADAFs [9],

$$Be = \frac{v_R^2}{2} + W - \frac{GM}{R} > 0,$$

where W is the specific enthalpy. A hypothesis was proposed that the latter problem could be solved by assuming the presence of powerful bipolar outflows in the accretion flows [13,14]. However, according to the later investigations [15,16], the hypothesis met some difficulties. The consequences of the earlier mentioned inconsistencies for ADAFs was understood after performing two- and three-dimensional hydrodynamic simulations of the inefficiently radiative accretion flows [17–22]. The simulations have revealed a new accretion regime of inefficiently radiated plasma, in which convection plays a dominant role determining structure and dynamics of the flow. Such a regime we shall refer as CDAF, the convection-dominated accretion flow.

CONVECTION INSTABILITY

Behavior of convective blobs in non-rotating medium depends sensitively on the superadiabatic gradient

$$\Delta \nabla c_s^2 = -c_s^2 \frac{d}{dR} \ln \left(\frac{P^{1/\gamma}}{\rho} \right),$$

which determines the Brunt-Väisälä frequency N given by

$$N^2 = -\frac{g_{eff}}{c_s^2} \Delta \nabla c_s^2,$$

where $g_{eff} = -(1/\rho)(dP/dR)$ is the radial effective gravity. When N^2 is positive, perturbations of the blobs have an oscillatory behavior with the frequency N and the medium is convectively stable. However, when N^2 is negative, perturbations have a runaway growth, leading to convection. Convection is present whenever $\Delta \nabla c_s^2$ is positive, i.e. when the entropy gradient, given by

$$T\frac{ds}{dR} = -\frac{\gamma}{\gamma - 1} \Delta \nabla c_s^2,$$

is negative. This is the well-known Schwarzschild criterion.

When there is rotation, a new frequency enters the problem and convection is no longer determined purely by the Brunt-Väisälä frequency. In this case, the effective frequency N_{eff} of convective blobs is given by

$$N_{eff}^2 = N^2 + \kappa^2,$$

where κ is the epicyclic frequency; for $\Omega \propto R^{-3/2}$, we have $\kappa = \Omega$. Again, when N_{eff}^2 is negative, the rotating flows are convectively unstable.

When convection is developed, the convective blobs carry outward some amount of thermal energy. Under certain approximations (see details in [24]) the convective energy flux can be expressed in the form,

$$F_{conv} = -\nu_{conv}\rho T \frac{ds}{dR}, \qquad (2)$$

where s is the specific entropy, T is the temperature, ν_{conv} is the diffusion coefficient, defined by $\nu_{conv} = (L_M^2/4)(-N_{eff}^2)^{1/2}$, and L_M is the characteristic mixing-length. The coefficient ν_{conv} can also be expressed in the α-parameterization form, $\nu_{conv} = \alpha_{conv} c_s^2/\Omega_K$, where α_{conv} is a dimensionless parameter that describes the strength of convective diffusion; this parameter is similar to the usual Shakura & Sunyaev α.

In the case of the self-similar ADAF solution (1), calculations lead to the conclusion that $N_{eff}^2 < 0$, i.e. ADAFs are convectively unstable and must have the outward energy fluxes defined by equation (2).

NUMERICAL RESULTS

To describe the results of numerical simulations of radiatively inefficient accretion flows, we shall mainly follow [21,22]. In these works the accretion flows were simulated by solving the nonrelativistic, time-dependent Navier-Stokes equations with thermal conduction,

$$\frac{d\rho}{dt} + \rho \nabla \mathbf{v} = 0,$$

$$\rho \frac{d\mathbf{v}}{dt} = -\nabla P + \rho \nabla \Phi + \nabla \mathbf{\Pi}, \qquad (3)$$

$$\rho \frac{de}{dt} = -P\nabla \mathbf{v} - \nabla \mathbf{q} + Q,$$

where e is the specific internal energy, Φ is the gravitational potential of the black hole, $\mathbf{\Pi}$ is the viscous stress tensor with all components included, $\mathbf{q}$ is the heat flux density due to thermal conduction and Q is the dissipation function. No radiative cooling was assumed and the ideal gas equation of state, $P = (\gamma - 1)\rho e$, was adopted. Only the shear viscosity was considered,

$$\nu = \alpha c_s^2/\Omega_K, \qquad (4)$$

where α is a constant, $0 < \alpha \leq 1$.

In the simulations, it was assumed that mass is steadily injected into the calculation domain from an equatorial torus near the outer boundary. Matter is injected with almost Keplerian angular momentum. Owing to viscous spread, a part of the injected matter moves inward and forms an accretion flow. The computations were started from an initial state in which there is a small amount of mass in the domain. After a time comparable to the viscous timescale, the accretion flow achieves a quasi-stationary behavior.

Results obtained by using two-dimensional axisymmetric simulations are summarized in Figure 1. Four types of accretion flows can be distinguished from the models with no thermal conduction (see Figure 1, left panel).

1. Convective flows. For a very small viscosity, $\alpha \lesssim 0.03$, accretion flows are convectively unstable. Axially symmetric convection transports the angular momentum *inward* rather than outward. Convection governs the flow structure, making a flattened density profile, $\rho(R) \propto R^{-1/2}$, with respect to the one for ADAFs. Convection transports a significant amount (up to $\sim 1\%$) of the dissipated binding energy outward. No powerful outflows are present.

2. Large-scale circulations. For a larger, but still small viscosity, $\alpha \sim 0.1$, accretion flows could be both stable or unstable convectively, depending on α and γ. The flow pattern consists of the large-scale ($\sim R$) meridional circulations. No powerful unbound outflows are presented. In some respect this type of flow is the limiting case of the convective flows in which the small-scale motions are suppressed by larger viscosity.

3. Pure inflows. With an increasing viscosity, $\alpha \simeq 0.3$, the convective instability dies off. Some models (with $\gamma \simeq 3/2$) are characterized by a pure inflow pattern, and agree in many respects with the self-similar ADAFs. No outflows are present.

4. Bipolar outflows. For a large viscosity, $\alpha \simeq 1$, accretion flows differ considerably from the simple self-similar models. Powerful unbound bipolar outflows are present.

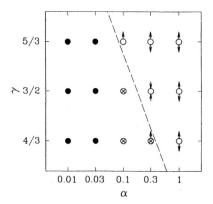 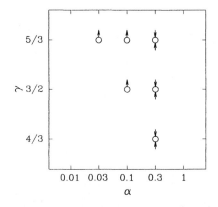

FIGURE 1. Properties of two-dimensional axisymmetric models of accretion flows. Models with no thermal conduction are shown in the left panel. Models with thermal conduction (Prandtl number **Pr** = 1) are shown in the right panel. Each circle represents a model in the (α,γ) parameter space. The empty circles correspond to laminar flows, the crossed circles represent unstable models with large-scale ($\sim R$) meridional circulations and solid circles indicate convective models. Two outward directed arrows correspond to bipolar outflows, whereas one arrow corresponds to a unipolar outflow. Two inward directed arrows correspond to a pure inflow. The dashed line on the left-hand panel approximately separates regions of convectively stable/unstable flows.

Figure 2 illustrates the density distribution and flow pattern in the low viscosity models. The models show complicated time-dependent behavior with numer-

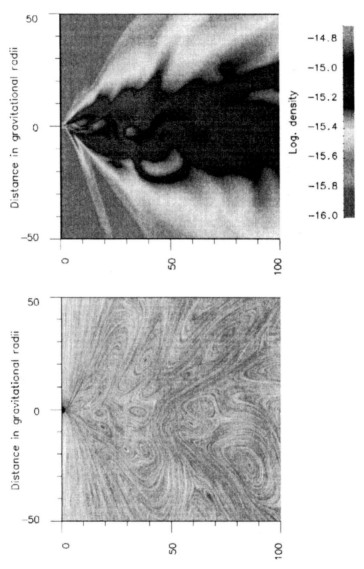

FIGURE 2. Snapshots of density distribution (upper panel) and streamlines (lower panel) from two-dimensional model of CDAF with $\alpha = 0.01$ and $\gamma = 5/3$, in the meridional cross-section. The black hole is located at the origin. The flow pattern is highly time dependent and consists of numerous vortices of different spatial scales. The variability of flow pattern is accompanied by density variations and results in variability of the accretion rate.

ous vortices and circulations, and with density fluctuations. However, the time-averaged flow patterns are smooth and do not demonstrate small-scale features. Figure 3 shows typical behavior of radial profiles of variety of time- and angle-averaged variables in convective models. Except near the inner and outer boundaries ($R < 10R_g$ and $R > 10^3 R_g$), the profiles of ρ, c_s, v and Ω show prominent power-law behaviors.

Two-dimensional models have demonstrated that the (r,φ)-component of the volume averaged Reynolds stress tensor takes negative values in the case of convective flows. This means that axisymmetric convection turbulence transports the

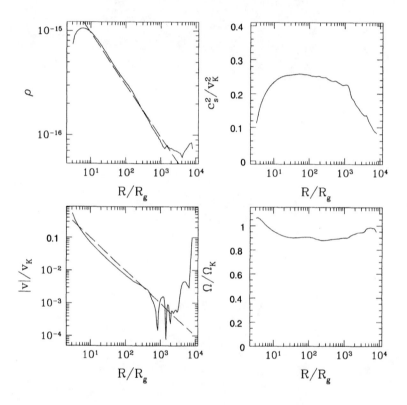

FIGURE 3. Selected properties of two-dimensional axisymmetric model with $\alpha = 0.01$ and $\gamma = 5/3$ [23]. All quantities have been averaged over polar angle θ and time. Except near the boundaries ($R < 10R_g$ and $R > 10^3 R_g$), the profiles of ρ, c_s^2/v_K^2, and Ω/Ω_K show the power-law behaviors predicted by the self-similar convective envelope solution. In the upper left panel the dashed line corresponds to the analytical scaling $\rho \propto R^{-1/2}$. Similarly, in the lower left panel the dashed line corresponds to the expected scaling $v \propto R^{-3/2}$ for CDAFs.

angular momentum inward. Indeed, if there are no azimuthal gradients of pressure, the turbulence attempts to erase the angular momentum gradient, which for the Keplerian-like angular velocity, $\Omega \propto R^{-3/2}$, means that the angular momentum transports inward. However, this is a property of axisymmetric flows. Does convection in real three-dimensional flows move angular momentum inward? Do such flows have significant azimuthal perturbations which destroy axisymmetry? The answer to this question has been found with the help of fully three-dimensional simulations, without the limiting assumption of axisymmetry [21]. The simulations have demonstrated a good qualitative and quantitative coincidence of the results of two- and three-dimension modeling. It was demonstrated that in the differentially rotating flows with nearly Keplerian angular velocities, all azimuthal gradients are efficiently washed out, leading to almost axisymmetrical structure of three-dimensional flows. Figure 4 illustrates axisymmetric structure of the low-viscosity accretion flow obtained in three-dimensional simulations.

All low-viscosity models have negative volume-averaged Be in all ranges of the

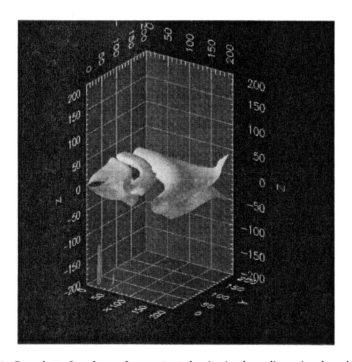

FIGURE 4. Snapshot of surfaces of a constant density in three-dimensional model of CDAF with $\alpha = 0.01$ and $\gamma = 5/3$. Only a quarter of the innermost region of the full domain is shown. Accretion matter rotates around z-axis. The black hole is located at the origin. All axes are labeled in the units of R_g. One can clearly see that density perturbations, associated with motion of convective blobs in the accretion flow, have axially symmetric form with respect to the axis of rotation.

radii. Only temporal convective blobs and narrow regions at the disk surfaces show positive Be. It seems that the outward directed convection energy transport provides an additional cooling mechanism of the flows that always results in averaged $Be < 0$, opposite to positive Be obtained in ADAFs.

In several axisymmetric simulations, effects of turbulent thermal conduction have been studied. It was found that the conduction has an important influence on the flow structure, but it does not introduce a new type of flow in addition to those already discussed (see Figure 1, right panel). The conduction leads to a suppression of small-scale convection in the low-viscosity models. In the moderate-viscosity models the thermal conduction acts as a cooling agent in the outflows, reducing or even suppressing them.

Obtained properties of CDAF can have important implications for the spectra and luminosities of accreting black holes [20,22,23]. Indeed, since $\rho \propto R^{-1/2}$ and $T \propto R^{-1}$, the bremsstrahlung cooling rate per unit volume varies as $Q_{br} \propto \rho^2 T^{-1/2} \propto R^{-3/2}$. After integration of Q_{br} over bulk of the flow one obtains the bremsstrahlung luminosity $L_{br} \propto R$. This means that CDAFs mostly radiate on the outside, whereas most of the radiative energy of ADAFs comes from the innermost region [10].

SELF-SIMILAR CONVECTIVE ENVELOPE

Numerical simulations of convective accretion flows have been understood qualitatively after construction of analytical self-similar solutions of such types of flows [24,25]. In the presence of convection, the height-integrated angular momentum and energy equations can be written as follows,

$$J_{adv} + J_{visc} + J_{conv} = 0, \tag{5}$$

$$\rho v_R T \frac{ds}{dR} + \frac{1}{R^2}\frac{d}{dR}(R^2 F_{conv}) = -(J_{visc} + J_{conv})\frac{d\Omega}{dR}, \tag{6}$$

where J_{adv}, J_{visc} and J_{conv} are the advective, viscous and convective angular momentum fluxes respectively. Solutions of equations (5) and (6), together with the equation of motion in the radial direction, crucially depend on the direction of convective angular momentum transport. There are two extreme possibilities. The first one is that convection behaves like normal viscosity,

$$J_{conv} = -\nu_{conv}\rho R^3 (d\Omega/dR). \tag{7}$$

An alternative possibility is that the convective flux is directed down the specific angular momentum gradient,

$$J_{conv} = -\nu_{conv}\rho R[d(\Omega R^2)/dR]. \tag{8}$$

For $\Omega \propto R^{-3/2}$, convective fluxes (7) and (8) correspond to outward and inward transport of angular momentum respectively. Numerical simulations have demonstrated that $J_{conv} < 0$ in CDAFs.

Choosing the convective flux (8) one can construct a *nonaccreting* solution with $v_R = 0$. We refer to it as a "convective envelope". For this solution, from equation (5) one obtains $J_{adv} = 0$ and $J_{conv} = -J_{visc}$. The latter means that in the convective envelopes the net angular momentum flux vanishes: the inward convective flux is exactly balanced by the outward viscous flux. Because of absence of advection (due to $v_R = 0$) and zero local dissipation rate (due to zero net angular momentum flux), the energy equation (6) is satisfied trivially with $F_{conv}(R) \propto R^{-2}$. Finally, the convective envelope solution has the following scalings:

$$\rho(R) \propto R^{-1/2},$$

$$v_R(R) = 0,$$

$$\Omega(R) \propto R^{-3/2}, \quad (9)$$

$$c_s(R) \propto R^{-1/2},$$

$$H(R) \propto R,$$

The outward energy flux F_{conv} corresponds to luminosity $L_{conv} = 4\pi R H F_{conv}$, which is independent of radius. Source of this energy flux is formally located at $R = 0$, but the nature of the source is not specified in the self-similar solution. In more realistic CDAFs, the mass accretion rate $\dot{M}$ is small, but it is not exactly zero. The fraction of the binding energy of the accreted mass released in the innermost region is the source of the energy required to support convection. In the case of CDAFs the convective luminosity can be expressed in the following form, $L_{conv} = \varepsilon \dot{M} c^2$, where the parameter ε has been estimated in numerical simulations, $\varepsilon \simeq 0.01$. A non-zero $\dot{M}$ leads to a finite $v_R = \dot{M}/(4\pi R H \rho) \propto R^{-3/2}$. The latter scaling and the scalings of other quantities in (9) agree quite well with the numerical results (see Figure 3).

RECENT MHD SIMULATIONS

There is good reason to believe that "viscosity" in differentially rotating accretion flows is produced by magnetic stress generated by the magneto-rotational instability (MRI) [26]. MHD simulations of accretion flows under conditions, in which CDAFs could be formed, have been recently performed ([27–32] and Matsumoto et al. in this Proceedings). A general conclusion of these studies is that MRI, and probably other instabilities of magnetized medium, leads to development of turbulence in accretion flows. The turbulence re-distributes angular momentum in the flows with an effective $\alpha_{eff} \sim 0.01 - 0.1$. The simulations show a tendency for density profiles in accretion flows to be flatter, than those in ADAFs.

In general, these results of MHD simulations do not contradict the results of hydrodynamical simulations of convective flows, which were discussed in previous sections. The MHD simulations have demonstrated the range of α_{eff}, in which hydrodynamical models are convectively unstable. The flattened density profiles in MHD models are in good agreement with those found in convective hydrodynamical models. However, a direct comparison of results of both approaches meets difficulties. The main problem is a low spatial resolution, and as a consequence, a small radial range of flow patterns studied in the discussed MHD simulations. In this case of small radial ranges the effects of boundaries become significant enough to influence the flow structure. Thus, more powerful computers and more developed numerical codes are required to obtain a comparable spatial resolution to that used in hydrodynamical simulations.

Another problem which can introduce difficulties for the comparison is that some MHD simulations have uncontrolled losses of energy due to numerical reconnection of magnetic lines. These energy losses can artificially suppress convection in the considered flows and change the flow structures, independently of the used spatial resolution. The problem could be solved by using an "artificial" resistivity in MHD schemes (see [32] for detailed discussion).

CONCLUSIONS

- Radiatively inefficient hydrodynamical black hole accretion flows with small viscosity ($\alpha \lesssim 0.1$) are always dominated by convection. Thus for low viscosity flows the concept of ADAF is unphysical.

- At present, CDAFs provide the best theoretical model for understanding of radiatively inefficient, low viscosity accretion flows.

- CDAFs are hot, their thermal energy is close to the virial energy. They have relatively reduced mass accretion rate. The density distribution in CDAFs is flattened, comparing to the one in ADAFs and Bondi-like flows ($\rho \propto R^{-1/2}$ in the former case vs. $\rho \propto R^{-3/2}$ in the latter case). Convection transports outward a significant amount (up to $\sim 1\%$) of dissipated binding energy of accretion flows. No powerful outflows are present in CDAFs.

- CDAF is a very simple model which is now undergoing a rapid development. A more detailed account of plasma, magnetic, and radiative processes is likely to change this model.

ACKNOWLEDGMENTS

The authors thank NORDITA and I.V.I. thanks Harvard-Smithsonian Center for Astrophysics for hospitality while part of this work was done. This work was supported by NSF grant PHY 9507695, the Royal Swedish Academy of Sciences, and RFBR grant 00-02-16135.

REFERENCES

1. Bondi H., *MNRAS*, **112**, 195 (1952).
2. Shakura N.I., and Sunyaev R.A., *Astron. & Astrophys.*, **24**, 337 (1973).
3. Ichimaru S., *Astrophys. Journal*, **214**, 840 (1977).
4. Gilham S., *MNRAS*, **195**, 755 (1981).
5. Begelman M.C., and Meier D.L., *Astrophys. Journal*, **253**, 873 (1982).
6. Abramowicz M.A., Czerny B., Lasota J.-P., and Szuszkiewicz E., *Astrophys. Journal*, **332**, 646 (1988).
7. Narayan R., and Yi I., *Astrophys. Journal*, **428**, L13 (1994).
8. Abramowicz M.A., Chen X., Kato S., Lasota J.-P., and Regev O., *Astrophys. Journal*, **438**, L37 (1995).
9. Narayan R., and Yi I., *Astrophys. Journal*, **444**, 231 (1995).
10. Narayan R., and Yi I., *Astrophys. Journal*, **452**, 710 (1995).
11. Kato S., Fukue J., and Mineshige S., *Black-Hole Accretion Disks*, Kyoto: Kyoto Univ. Press (1998).
12. Abramowicz M.A., Blörnsson G., and Pringle J.E., *Theory of Black Hole Accretion Disks*, Cambridge: Cambridge Univ. Press (1998).
13. Xu G., and Chen X., *Astrophys. Journal*, **489**, L29 (1997).
14. Blandford R.D., and Begelman M.C., *MNRAS*, **303**, L1 (1999).
15. Ogilvie G.I., *MNRAS*, **306**, L9 (1999).
16. Abramowicz M.A., Lasota J.-P., and Igumenshchev I.V., *MNRAS*, **314**, 775 (2000).
17. Igumenshchev I.V., Chen X., and Abramowicz M.A., *MNRAS*, **278**, 236 (1996).
18. Igumenshchev I.V., and Abramowicz M.A., *MNRAS*, **303**, 309 (1999).
19. Stone J.M., Pringle J.E., and Begelman M.C., *MNRAS*, **310**, 1002 (1999).
20. Igumenshchev I.V., *MNRAS*, **314**, 54 (2000).
21. Igumenshchev I.V., Abramowicz M.A., and Narayan R., *Astrophys. Journal*, **537**, L27 (2000).
22. Igumenshchev I.V., and Abramowicz M.A., *Astrophys. Journal Sup.*, **130**, 463 (2000).
23. Ball G., Narayan R., and Quataert E., *Astrophys. Journal*, in press, astro-ph/0007037.
24. Narayan R., Igumenshchev I.V., and Abramowicz M.A., *Astrophys. Journal*, **539**, 798 (2000).
25. Quataert E., and Gruzinov A., *Astrophys. Journal*, **539**, 809 (2000).
26. Balbus S.A., and Hawley J.F., *Astrophys. Journal*, **376**, 214 (1991).
27. Matsumoto R., and Shibata K., in *Accretion Phenomena and Related Outflows*, ed. D. Wickramsinghe, G. Bicknell & L. Ferrario, ASP Conf. Ser. 121, 1997, p. 443.
28. Matsumoto R., in *Numerical Astrophysics*, ed. S. Miyama, K. Tomisaka & T. Hanawa, Dordrecht: Kluwer, 1999, p. 195.
29. Hawley J.F., *Astrophys. Journal*, **528**, 462 (2000).
30. Machida M., Hayashi M.R., and Matsumoto R., *Astrophys. Journal*, **532**, L67 (2000).
31. Machida M., Matsumoto R., and Mineshige S., *PASJ*, in press, astro-ph/0009004.
32. Stone J.M., and Pringle J.E., *MNRAS*, in press, astro-ph/0009233.

Magnetohydrodynamic Simulations of Black Hole Accretion

Christopher S. Reynolds[1], Philip J. Armitage[2] and James Chiang[1,3]

[1] *JILA, Campus Box 440, University of Colorado, Boulder, CO 80303*
[2] *School of Physics and Astronomy, University of St Andrews, Fife, KY16 9SS, UK*
[3] *Laboratory for High Energy Astrophysics, Code 660, NASA/Goddard Space Flight Center, Greenbelt, MD 20771.*

Abstract. We discuss the results of three-dimensional magnetohydrodynamic simulations, using a pseudo-Newtonian potential, of thin disk ($h/r \approx 0.1$) accretion onto black holes We find (i) that magnetic stresses persist within $r_{\rm ms}$, the marginally stable orbit, and (ii) that the importance of those stresses for the dynamics of the flow depends upon the strength of magnetic fields in the disk outside $r_{\rm ms}$. Strong disk magnetic fields ($\alpha \gtrsim 0.1$) lead to a gross violation of the zero-torque boundary condition at $r_{\rm ms}$, while weaker fields ($\alpha \sim 10^{-2}$) produce results more akin to traditional models for thin disk accretion onto black holes. Fluctuations in the magnetic field strength in the disk could lead to changes in the radiative efficiency of the flow on short timescales.

INTRODUCTION

There is now a consensus that, in well-ionized accretion disks, magnetorotational instabilities (MRI) create turbulence that provides the 'anomalous viscosity' required to drive accretion [1]. This removes a major uncertainty that afflicted previous theoretical models for black hole accretion, and opens the possibility of using numerical simulations to study directly the structure and variability of the accretion flow. Questions that we might hope to address include:

- The magnetohydrodynamics (MHD) of the flow as it crosses $r_{\rm ms}$, the marginally stable circular orbit. Recent work has suggested that MHD effects interior to $r_{\rm ms}$ (in the 'plunging' region) could invalidate existing models of black hole accretion, with consequences that include an increase in the predicted radiative efficiency of thin disk accretion [2–4]. This suggestion remains controversial [5].

- The predicted variability in emission from the inner disk.

- The structure of the disk magnetic field and the rate of transport of magnetic flux through $r_{\rm ms}$, which together determine the efficiency of the Blandford-

TABLE 1. Summary of the simulations discussed in this article. All runs have the same sound speed, $c_s/v_\phi = 0.065$ (evaluated at $r_{\rm ms}$), spatial domain ($0.666 < r/r_{\rm ms} < 3.3$, $z/r_{\rm ms} = \pm 0.166$, $\Delta\phi = 45°$), and resolution ($n_r = 200$, $n_z = 40$, $n_\phi = 60$). The time units are such that the orbital period at the last stable orbit is $P = 7.7$.

Run	Vertical boundary conditions	Initial field	Final time	Output times
Azimuthal field	$B_z = v_z = 0$	$\beta_\phi = 100$	600	400-600
Vertical field	Periodic	$\beta_z = 5000$	250	100-250
High saturation	Periodic	$\beta_z = 500$	150	62.5-150

Znajek mechanism for extracting spin energy of the black hole [6–8].

This article presents results from simplified MHD simulations [9,10] of black hole accretion that focus on the first of these questions. We outline the numerical approach, and discuss our results and how they compare with those of other groups [11–13]. Our conclusion from work to date is that MHD effects in the plunging region *can* have an important influence on the dynamics of the inner disk flow, provided that the magnetic fields in the disk are already moderately strong.

NUMERICAL SIMULATIONS

Ideally, we would like to simulate a large volume of disk (to allow for global effects [14] and ease worries about treatment of the boundaries), for a long time period (to average out fluctuations), at high resolution (to resolve the most unstable scales of the magnetorotational instability), with the most realistic physics possible. Unfortunately, we can't, so compromises are needed. Our approach has been to simulate the simplest disk model that we believe includes the essential physical effects, while aiming for the highest resolution near and inside the last stable orbit.

We use the ZEUS MHD code [15,16] to solve the equations of ideal MHD within a restricted 'wedge' of disk in cylindrical (z, r, ϕ) geometry. The equation of state is isothermal, and a Paczynski-Wiita potential [17] is used to model the effect of a last stable orbit within the Newtonian hydrocode. To further simplify the problem, we ignore the vertical component of gravity and consider an unstratified disk model. The boundary conditions are set to outflow at the radial boundaries, are periodic in azimuth, and are either periodic or reflecting in z. The simulations begin with a stable, approximately Gaussian surface density profile outside $r_{\rm ms}$, which is threaded with a weak magnetic field. This initial seed field has a constant ratio of thermal to magnetic energy β, and is either vertical or azimuthal. We evolve this setup, which is immediately unstable to the MRI, until a significant fraction of the mass has been accreted, and plot results from timeslices towards the end of the runs when the magnetic fields have reached a saturated state.

Table 1 summarizes the parameters of three simulations, which are improved versions of those previously reported [9]. The simulated disks are 'thin' in the sense that pressure gradients are negligibly small in the disk outside $r_{\rm ms}$. Specifically, the ratio of sound speed to orbital velocity, which in a stratified disk is approximately equal to h/r, is $\lesssim 0.1$ at the last stable orbit. We have not (yet) attempted the more difficult task of simulating very thin disks, which have longer viscous timescales, and caution against extrapolating our conclusions into that regime.

Following a suggestion by Charles Gammie (personal communication), we investigated whether varying the magnetic field strength in the *disk* (i.e. at $r > r_{\rm ms}$) led to changes in the dynamics of the flow within the plunging region. To vary the field strength in the simulations, we make use of the fact that the saturation level of the MRI can be altered by varying the net flux of seed fields in the initial conditions. Local simulations [18] show that the resultant Shakura-Sunyaev α parameter [19] is approximately,

$$\alpha \sim 10^{-2} + 4\frac{\langle v_{Az}\rangle}{c_s} + \frac{1}{4}\frac{\langle v_{A\phi}\rangle}{c_s} \qquad (1)$$

where v_{Az} and $v_{A\phi}$ are the Alfven speeds for initial conditions with uniform vertical and azimuthal seed fields, and c_s is the sound speed. For our simulations, we find that the range of initial conditions (and vertical boundary conditions) shown in Table 1 leads to a variation in α between 10^{-2} and 10^{-1} in the disk.

Of course, this is a numerical trick. The *true* value of α in the disk immediately outside $r_{\rm ms}$ will probably depend upon details of the disk physics (for example, the relative importance of gas and radiation pressure), and may vary with time.

DYNAMICS OF THE FLOW CROSSING $R_{\rm MS}$

Figure 1 illustrates the geometry of the simulations. All MHD disk simulations look broadly similar, and these are no exception. We obtain a pattern of surface density fluctuations that are strongly sheared by the differential rotation, and disk magnetic fields that are predominantly azimuthal. A map of the ratio of magnetic to thermal energy, also shown in the Figure, displays clearly the predicted increase [2] in the relative importance of magnetic fields near and interior to the marginally stable orbit. Determining the influence of these fields upon the dynamics of the flow in the inner disk and in the plunging region is the main goal of the calculations.

Figure 2 shows the magnetic torque as a function of radius in the three simulations, expressed as the magnetic contribution to the equivalent Shakura-Sunyaev α parameter,

$$\alpha_{\rm mag} = \frac{2}{3}\langle\frac{-B_r B_\phi}{4\pi\rho c_s^2}\rangle. \qquad (2)$$

There are also hydrodynamic (Reynolds) stresses, which are somewhat harder to measure [13], but which are found to be substantially smaller than the magnetic stresses in local simulations.

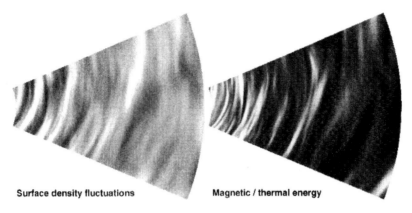

FIGURE 1. Maps showing (left panel) the surface density fluctuations and (right panel) the ratio of the energy density in magnetic fields to the thermal energy, from a simulation in which the saturation value of the magnetorotational instability in the disk was artificially boosted. The spatial domain covers $0.666 < r/r_{\rm ms} < 3.3$. A clear increase in the relative importance of magnetic fields in the inner regions of the disk, and within the marginally stable orbit, is obtained.

The three choices for the initial flux and vertical boundary conditions lead to large variations in the saturation level of the MRI and associated $\alpha_{\rm mag}$. The run with an initially azimuthal field produces an $\alpha_{\rm mag} \sim 10^{-2}$, while the run with a relatively strong initial vertical flux leads to a disk $\alpha_{\rm mag}$ that exceeds 0.1. This increase is qualitatively in agreement with local simulations [18]. For the run with the strongest field, the magnetic field energy density in the disk near $r_{\rm ms}$ is on average near equipartition with the thermal energy ($\beta \sim 1$). There are large fluctuations with time, however, including brief periods where the magnetic energy substantially exceeds the thermal energy. In all runs, the relative importance of magnetic fields compared to the thermal energy increases within $r_{\rm ms}$.

Figure 2 also shows how the torques influence the dynamics of the flow. We plot the specific angular momentum of the flow l as a function of radius. Hydrodynamic models for thin disk accretion onto black holes obtain $dl/dr = 0$ within the last stable orbit, corresponding to a zero-torque boundary condition for the disk [20]. A non-zero dl/dr implies transport of angular momentum (and implicitly energy) into the disk from within the plunging region. We find that dl/dr in the plunging region correlates with $\alpha_{\rm mag}$ in the disk. Weak disk fields lead to a small (but significant) decline in l within the last stable orbit, while strong fields lead to a steeply declining specific angular momentum profile at small radii. The former behavior is similar to that seen in our earlier simulations [9] (which also had rather weak fields), while the latter is comparable to the results obtained by Hawley [11] and Hawley and Krolik [12]. Their global simulations, which are substantially more ambitious than ours in terms of the included physics and spatial domain, did indeed generate relatively strong magnetic fields.

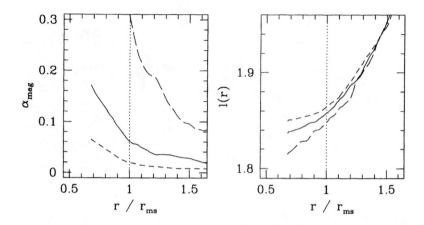

FIGURE 2. The dynamics of the flow inside the marginally stable orbit is correlated with the magnetic field strength in the disk. The left panel shows the contribution made by magnetic stresses to the Shakura-Sunyaev α parameter as a function of radius for the three runs: azimuthal initial field (short dashes), vertical initial field (solid), and strong vertical initial field (long dashes). The right panel shows the resulting specific angular momentum of the flow. All results are averaged over several independent timeslices to reduce fluctuations. Strong disk fields lead to a larger violation of the zero-torque boundary condition at r_{ms}.

DISCUSSION

The limitations of the current simulations are myriad and obvious. We are accutely aware that effects in the disk corona could be important [21,22], and that there is more to General Relativity than a pseudo-Newtonian potential [23,24]. Nonetheless, we believe that some conclusions can be drawn from existing work:

- Unstratified global simulations confirm that magnetic torques persist within the last stable orbit, but suggest that if $\alpha \sim 10^{-2}$ their influence on the dynamics of the flow is relatively modest [9,13]. By modest we mean that the gradient of the specific angular momentum, dl/dr, is non-zero but small at and inside r_{ms}. If these simulations reflect reality, existing models of black hole accretion would be a pretty good approximation for thin disks [5].

- A gross violation of the zero-torque boundary condition at r_{ms} is also possible [11,12]. This would increase the radiative efficiency of thin disk accretion above the usual $\epsilon \approx 0.1$, and have other consequences [2–4]. We believe that these strong effects *only occur* if $\alpha \gtrsim 0.1$ in the disk. This would be consistent with existing simulations [9,11], and with the results presented here.

For observations, these results suggest that the radiative efficiency of thin disk accretion may vary, both between systems, and in an individual system if the magnetic field strength in the disk varies with time.

ACKNOWLEDGMENTS

We thank the developers of ZEUS and ZEUS-MP for making these codes available as community resources. P.J.A. thanks JILA for hospitality during the course of part of this work. C.S.R. acknowledges support from Hubble Fellowship grant HF-01113.01-98A. This grant was awarded by the Space Telescope Science Institute, operated by AURA for NASA under contract NAS 5-26555. C.S.R. also thanks support from the NSF under grants AST 98-76887 and AST 95-29170. J.C. was supported by NASA/ATP grant NAGS 5-7723.

REFERENCES

1. Balbus S.A. and Hawley J.F., *Astrophys. Journal*, **376**, 214-233 (1991)
2. Krolik J.H., *Astrophys. Journal*, **515** L73-L76 (1999)
3. Gammie C.F., *Astrophys. Journal*, **522**, L57-L60 (1999)
4. Agol E. and Krolik J.H., *Astrophys. Journal*, **528**, 161-170 (2000)
5. Paczynski B., astro-ph/0004129 (2000)
6. Blandford R.D. and Znajek R.L., *MNRAS*, **179**, 433-456 (1977)
7. Ghosh P. and Abramowicz M.A., *MNRAS*, **292**, 887-895 (1997)
8. Livio M., Ogilvie G.I. and Pringle J.E., *Astrophys. Journal*, **512**, 100-104 (1999)
9. Armitage P.J., Reynolds C.S. and Chiang J., *Astrophys. Journal*, in press, astro-ph/0007042
10. Reynolds C.S. and Armitage P.J., in preparation
11. Hawley J.F., *Astrophys. Journal*, **528**, 462-479 (2000)
12. Hawley J.F. and Krolik J.H., *Astrophys. Journal*, in press, astro-ph/0006456
13. Hawley J.F., *Astrophys. Journal*, submitted, astro-ph/0011501
14. Armitage P.J., *Astrophys. Journal*, **501**, L189-L192 (1998)
15. Stone J.M. and Norman M.L., *Astrophys. Journal Sup.*, **80**, 753-790 (1992)
16. Stone J.M. and Norman M.L., *Astrophys. Journal Sup.*, **80**, 791-818 (1992)
17. Paczynski B. and Wiita P.J., *Astron. & Astrophys.*, **88**, 23-31 (1980)
18. Gammie C.F., in *Accretion Processes in Astrophysical Systems: Some Like it Hot!*, eds S.S. Holt and T.R. Kallman, New York: AIP, 99-107 (1998)
19. Shakura N.I. and Sunyaev R.A., *Astron. & Astrophys.*, **24**, 337-355 (1973)
20. Muchotrzeb B. and Paczynski B., *Acta Astronomica*, **32**, 1-11 (1982)
21. Miller K.A. and Stone J.M., *Astrophys. Journal*, **534**, 398-419 (2000)
22. Machida M., Hayashi M.R. and Matsumoto R., *Astrophys. Journal*, **532**, L67-L70 (2000)
23. Koide S., Meier D.L., Shibata K. and Kudoh T., *Astrophys. Journal*, **536**, 668-674 (2000)
24. Font J.A., *Living Rev. Relativ.*, **3**, 2000-2 (2000)

Magnetic Stresses at the Inner Edges of Accretion Disks Around Black Holes

Julian H. Krolik*

*Department of Physics and Astronomy, Johns Hopkins University, Baltimore MD 21209

Abstract.
For the past twenty-five years, nearly all analyses of accretion disk dynamics have assumed that stress inside the disk is locally proportional to pressure (the "α-model") and that this stress goes to zero at the marginally stable orbit. Recently, it has been demonstrated that MHD turbulence accounts for the bulk of internal disk stress. In contradiction with the traditional view, the stress from this MHD turbulence does not diminish near the marginally stable orbit, and the ratio of magnetic stress to pressure rises sharply there. Examples of the consequences include: an increase in accretion efficiency that may also be time- and circumstance-dependent; a decrease in the rate of black hole spin-up by accretion; and generation of disk luminosity fluctuations. Preliminary results from numerical simulations lend support to analytic estimates of these effects.

INTRODUCTION

Ever since the seminal work of Novikov and Thorne [1] and Shakura and Sunyaev [2], our understanding of the physics of accretion disks around black holes has been built on two fundamental beliefs: the energy available for radiation is equal to the binding energy of the innermost stable orbit, and is therefore determined solely by the spin of the black hole; and the (vertically-integrated) stress responsible for transporting angular momentum outwards through the disk is proportional to the local pressure. Recent work involving magnetic forces in the innermost regions of relativistic accretion disks has undercut both of these claims. In this review I will summarize the reasons for this change of heart and explain the consequences that are likely to follow if the new point of view proves correct.

THE TRADITIONAL VIEW OF THE INNER EDGES OF RELATIVISTIC ACCRETION DISKS

The basic framework for understanding the inner regions of relativistic accretion disks was laid out more than twenty-five years ago [1,3]. In the simplest possible picture (i.e., one assuming that disks are time-steady, axi-symmetric, and geometrically thin), their radial structure may be defined in terms of conservation equations that appear almost Newtonian, as the general relativistic effects can be collapsed into multiplicative correction factors. Conservation of angular momentum, for example, may be expressed through a single first-order differential equation in radius.

Thus, to specify the entire solution, all that is necessary is to choose a boundary condition for this differential equation. This boundary condition may be physically interpreted as determining the conserved outward angular momentum flux through the disk, or, equivalently, as specifying $T_{r\phi}$, the r–ϕ component of the stress exerted on the disk, at its inner edge. This inner edge is conventionally taken to lie at the radius of the marginally stable orbit, r_{ms}.

Selecting this boundary condition is the only part of the procedure in which there is any guesswork. Until very recently, it was almost universally assumed that the stress must be zero at r_{ms}. Two reasons were commonly cited for this choice. The first, due to Thorne and collaborators [1,3], was that the radial speed inside r_{ms} is so much larger than the radial speed outside r_{ms} that, assuming a constant mass accretion rate, the inertia of mass inside r_{ms} must be much smaller than the inertia of the disk proper in the region of stable orbits. Given that, it appears difficult to see how the small amount of matter in the plunging region could exert any significant force on the much heavier disk. The second argument [4] begins with the assumption that the stress carrying angular momentum outward should always be proportional to the local pressure; i.e., $T_{r\phi} = \alpha p$ [2]. If this is so, as material accelerates inward and expands, its pressure falls rapidly, so the stress must do likewise.

If there is no stress inside r_{ms}, then no forces change the energy or angular momentum of material inside that point. Matter therefore arrives at the event horizon with the energy and angular momentum it had at r_{ms}. The maximum energy per unit mass available to be radiated in the disk (the maximum radiative efficiency) is then defined simply by the binding energy of the marginally stable orbit, and the spin-up rate of the black hole is the mass accretion rate times the specific angular momentum of that orbit. Both the specific binding energy and the specific angular momentum at r_{ms} depend only on the black hole's normalized angular momentum a/M; thus, it has long been thought that both the radiative efficiency and the spin-up rate per unit accreted mass are functions only of a/M.

Interestingly, the very first work on this subject [3] acknowledged that strong magnetic fields could upset the arguments that the stress must go to zero at r_{ms}, but chose (reasonably, given the state of knowledge at the time) to ignore this possibility. However, work of the past decade (summarized in [5]) has shown that magnetic fields are essential to angular momentum transfer in accretion disks. We must therefore re-open the question of the appropriate boundary condition on $T_{r\phi}$ at r_{ms}.

MAGNETIC FIELDS IN ACCRETION DISKS

This recent work has shown that, in the main body of the disk, MHD turbulence creates a stress that, when vertically-averaged, is roughly proportional to the pressure, with an $\alpha \sim 0.01 - 0.1$ [5–7]. The magnetic stress accounts for the bulk of the angular momentum flux so long as two conditions are met: there must be some seed magnetic field in the accreting gas, but in the disk midplane the energy density of any imposed magnetic field must be smaller than the total pressure; and the matter must have high enough conductivity that the MHD approximation is valid. In the conditions surrounding black holes, these are all almost certain to apply. Weak magnetic fields are virtually ubiquitous in astrophysical environments; and near a black hole the gas temperature will very nearly always be high enough to maintain the gas in an ionized, and therefore highly conducting, state.

This turbulence is generated by an MHD instability driven by the orbital shear [8]. Its growth rate is always comparable to the orbital frequency, so it grows roughly as rapidly (relative to an orbital period) in the relativistic part of the disk as in the Newtonian part. The nonlinear amplitude of the turbulence in the Newtonian part of the disk is determined by a cascade of energy toward shorter wavelengths, where it can ultimately be dissipated. This process, too, should operate more or less in the same fashion (as viewed in the fluid frame) in the relativistic part of the disk as in the non-relativistic part, provided the local inflow time is long enough for equilibrium to be established. We may therefore reasonably conclude that the ratio $B^2/(8\pi p)$ doesn't change dramatically at radii a few times r_{ms}. Because orbital shear automatically stretches field lines in the sense that produces outward angular momentum flow ($\langle B_r B_\phi \rangle < 0$), the effective α-parameter ($= \langle B_r B_\phi \rangle/(4\pi p)$) is also unlikely to change much. But if this is so, *why should the stress diminish as the marginally stable orbit is approached?*

The only aspect of this process that changes in any qualitative way as r_{ms} is approached from the outside is the ordering of four timescales: the inflow time t_{in}, the thermal time t_{th}, the turbulent dissipation time t_{diss}, and the dynamical time t_{dyn}. In the disk body, $t_{in} \sim \alpha^{-1}(r/h)^2 t_{dyn}$ and $\beta t_{diss} \sim t_{th} \sim \alpha^{-1} t_{dyn}$, where β is the ratio of gas (+ radiation) pressure to magnetic energy density. Thus, $t_{in} \gg t_{th} > t_{diss} > t_{dyn}$. However, just outside r_{ms} the effective potential flattens (this is, of course, what it means for r_{ms} to be the location of the innermost *marginally stable* orbit), so t_{in} diminishes toward t_{dyn}. The concomitant decline in surface density likewise reduces t_{th}. However, t_{diss} (crudely $\sim h/v_A$ for disk thickness h and Alfven speed v_A) declines more slowly than t_{th}. As a result, in the vicinity of r_{ms}, the interplay of plasma dynamics and flux-freezing should be at least as important to determining the field intensity as the balance between the turbulent dynamo and turbulent dissipation.

If flux-freezing really does determine the evolution of the field as the matter plunges inside r_{ms}, the magnetic field strength in the fluid frame stays roughly constant or increases somewhat even while the fluid pressure decreases dramatically [9]. It follows that the effective α rises sharply as the inflow accelerates near and inside r_{ms}. In fact, when the radial component of the velocity becomes relativistic, it is easy to show that these assumptions imply $B^2/(8\pi) \sim \rho c^2$. That is, when the inflow is relativistic, magnetic forces become competitive with gravity, and the Alfven speed becomes relativistic. Matter may then retain a causal coupling with the disk it left behind even when it has fallen well inside r_{ms}, allowing significant transfer of energy and angular momentum [9,10].

Looking back at the old argument that the small inertia of mattter in the plunging region cannot do much to the main body of the disk, we now see that magnetic fields in effect turn this argument on its head. Magnetic connections between the disk and plunging matter act as "tethers" by which the massive disk restrains angular acceleration of the low-inertia streams inside r_{ms}. In so doing, stress can be exerted on the disk itself.

CONSEQUENCES

The existence of significant stress at the marginally stable orbit has major consequences for the most fundamental properties of accretion onto black holes. Energy taken from plunging matter and delivered to the disk is potentially available for radiation; removal of angular momentum from this matter retards the rate of black hole spin-up [12]. In fact, when the black hole is rapidly rotating, the rotational energy of the black hole itself can be tapped: frame-dragging gives plunging matter a high orbital frequency; magnetic connections from this region to the disk exert a torque; the end-result is energy drawn from black hole rotation given to matter in the disk [10,11]. In other words, the amount of energy drawn from accreting matter is no longer a quantity fixed by orbital mechanics: it is a dynamical quantity that is the product of complicated MHD dynamics, and may even be time-variable.

The fate of the work done by magnetic fields at $r \geq r_{ms}$ depends critically on the ratio between the dissipation and inflow timescales in the place where the energy is brought. If the energy can be converted into heat and radiated before the matter finds its way into the plunging region inside r_{ms}, it adds to the radiative efficiency; if, on the other hand, dissipation is slow, so that the energy stays in the gas either as organized kinetic energy or as magnetic field energy, it may end up being taken into the black hole. Given the arguments of the previous section, it is as yet unclear how this balance works out.

If additional heat is deposited in the disk, there can be substantial alterations to the disk spectrum [12]. Additional flux is radiated at high frequencies, and relativistic effects beam it strongly into the equatorial plane. In addition, because so much more radiation is produced in the most relativistic portion of the disk, the returning radiation fraction is greatly enhanced. This latter effect has implications for phenomenology as diverse as polarization of the emitted light and the synchronization of fluctuations.

However, there is an additional implication of stress at the inner edge of the disk that may be important even if the radiative efficiency is hardly altered by these mechanisms: dynamics at the marginally stable orbit can be a powerful "noise" source to the entire dynamical system. In the long run, these dynamical fluctuations may be reflected in luminosity fluctuations, which are, of course, a hallmark of all known accreting black holes. The origin of this "noise" may be seen from a simple thought experiment: Consider a small magnetized fluid element orbiting just outside r_{ms}. Imagine that half this fluid element receives a negative angular momentum perturbation from the turbulence and begins to fall inward. As it does so, its orbital frequency increases simply due to its decreasing orbital radius. The resulting shear between the two halves of the fluid element stretches the magnetic flux tube connecting them, and a magnetic tension force creates a torque that transfers angular momentum from the falling half to the half that is still orbiting stably. The stably orbiting half must then move outward, launching a compressive wave into the disk. Thus, the basic mechanics of accretion create disk "noise" when there are magnetic connections across the marginally stable orbit.

SIMULATIONS

All the arguments presented so far have been qualitative. Whether these effects are quantitatively important can only be ascertained as a result of genuine calculation. MHD turbulent dynamics are sufficiently complicated that (almost) all credible calculations are numerical simulations. Although much work remains to be done before simulations can be run with a requisite level of realism, some preliminary results have been achieved [13–15].

To date, all simulations touching on these issues have been limited in a number of regards, both physical and numerical. All have assumed Newtonian physics; general relativistic dynamics is mimicked by the use of the Paczyński-Wiita potential $U = -GM/(r - 2GM/c^2)$. This potential qualitatively reproduces motion in a Schwarzschild metric by creating a marginally stable orbit at $r = 6GM/c^2$. In addition, all simulations so far have substituted an assumed equation of state for a real energy equation. This assumption has two deleterious consequences: magnetosonic waves don't propagate at the right speed because the pressure isn't correctly computed; and, more importantly, it is impossible to use these simulations to estimate the radiative efficiency because energy taken from the plunging region to the disk falls right back in. There are also aspects of the numerical methods used in these simulations that are less than optimal. Both kinetic energy and magnetic energy can disappear if fluid from two adjacent cells with oppositely directed velocity or magnetic field components is combined. It also appears likely that no simulation to date has used a grid fine enough to resolve the magnetic field structure.

Nonetheless, putting aside these qualms, the simulational results may still be used as an indication of what may appear in more realistic simulations. In particular, angular momentum transport should be more reliably treated than energy transport because it is not subject to the problems described in the previous paragraph. The result shown in Figure 1 [14] is particularly noteworthy. Two distributions of $T_{r\phi}$

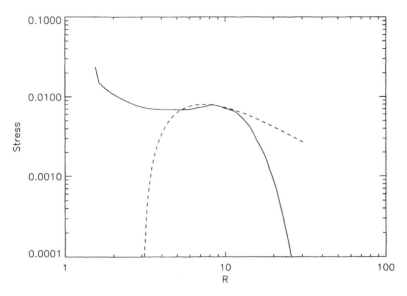

FIGURE 1. The solid curve shows the time- and azimuthal-average of the vertically-integrated r-ϕ component of the magnetic stress in a simulation by Hawley and Krolik [14]. The dashed curve is the prediction of the Novikov-Thorne model for the mean mass accretion rate found in that simulation. Note that the unit of distance is $2GM/c^2$, so the marginally stable orbit is at $R = 3$.

integrated vertically and averaged over both azimuth and time are shown: the prediction of the Novikov-Thorne model for the mean mass accretion rate found in the simulation; and the magnetic contribution to the stress. The two curves coincide very closely over the radial range from $\simeq 8 - 24GM/c^2$, demonstrating that magnetic stresses account for very nearly all the angular momentum transport in the main part of the disk. They depart in an uninteresting way at large radius—the deviation here is due to the fact that the disk in the simulation had an outer edge, whereas the Novikov-Thorne model refers to a disk that extends to very large radius. At small radius, however, the contrast is very significant—as expected on the basis of the qualitative arguments presented earlier, magnetic stress remains important across the marginally stable orbit and throughout the region where gas plunges toward the event horizon. Indeed, as a result of this continuing stress, the mean specific angular momentum of matter crossing the inner edge of the simulation (at $r = 3GM/c^2$) is about 5% smaller than the specific angular momentum at r_{ms}. Also as expected, the effective α parameter increases sharply in the innermost part of the accretion flow: between $r = 8GM/c^2$ and $r = 4GM/c^2$, it increases by an order of magnitude.

FUTURE PROSPECTS

The results of these first simulations are encouraging, but they are far from the final answer. Fortunately, significant improvements are feasible. Improved resolution can be obtained both by cleverer gridding schemes and, of course, by a few years' technological development. Real energy equations can be computed by explicitly incorporating phenemonological viscosity and resistivity and a radiative cooling function (following time-dependent radiation transfer is also possible [16], but will require somewhat greater computer power before it is feasible for this kind of simulation). Likewise, there is no fundamental impediment to writing MHD codes that work in a fixed relativistic metric. Thus, in a few years, we should be able to attach quantitative values to the effects pointed out qualitatively here.

REFERENCES

1. Novikov, I.D., and Thorne, K.S., in *Black Holes*, C. De Witt and B. De Witt, eds., New York: Gordon & Breach, 1973, p. 343
2. Shakura, N. and Sunyaev, R., *Astron. Astrop.* **24**, 337 (1973)
3. Page, D. and Thorne, K.S., *Ap. J.* **191**, 499 (1974)
4. Abramowicz, M.A. and Kato, S., *Ap. J.* **336**, 304 (1989)
5. Balbus, S.A. and Hawley, J.F., *Revs. Mod. Phys.* **70**, 1 (1998)
6. Stone, J.M., Hawley, J.F., Gammie, C.F., and Balbus, S.A. *Ap. J.* **463**, 656 (1996)
7. Brandenburg, A., Stein, R.F., Nordlund, A., and Torkelsson, U., *Ap. J. Letts.* **458**, L45 (1996)
8. Balbus, S.A. and Hawley, J.F., *Ap. J.* **376**, 214 (1991)
9. Krolik, J.H., *Ap. J. Letts.* **515**, L73 (1999)
10. Gammie, C.F., *Ap. J. Letts.* **522**, L57 (1999)
11. Agol, E. and Krolik, J.H., *Ap. J.* **528**, 161 (2000)
12. Krolik, J.H., in *Explosive Phenomena in Astrophysical Compact Objects, Proceedings of the 1st KIAS Astrophysics Workshop*, I. Yi and M. Rho, eds., New York: AIP, in press (2001)
13. Hawley, J.F., *Ap. J.* **528**, 462 (2000)
14. Hawley, J.F. and Krolik, J.H., *Ap. J.* in press (2001)
15. Armitage, P.J., Reynolds, C.S. and Chiang, J., *Ap. J.* in press (2001)
16. Turner, N.J. and Stone, J.M., *Ap. J. Suppl.* in press (2001)

Magnetohydrodynamic turbulence in warped accretion discs

Ulf Torkelsson*, Gordon I. Ogilvie[†], Axel Brandenburg[‡], James E. Pringle[†], Åke Nordlund[||], Robert F. Stein[¶]

*Chalmers University of Technology/Göteborg University, Department of Astronomy and Astrophysics, S-412 96 Gothenburg, Sweden
[†]Institute of Astronomy, Madingley Road, Cambridge CB3 0HA, United Kingdom
[‡]Department of Mathematics, University of Newcastle upon Tyne, NE1 7RU, United Kingdom, Nordita, Blegdamsvej 17, DK-2100 Copenhagen Ø, Denmark
[||]Theoretical Astrophysics Center, Juliane Maries Vej 30, DK-2100 Copenhagen Ø, Denmark, Copenhagen University Observatory, Juliane Maries Vej 30, DK-2100
[¶]Department of Physics and Astronomy, Michigan State University, East Lansing, MI 48824, USA

Abstract. Warped, precessing accretion discs appear in a range of astrophysical systems, for instance the X-ray binary Her X-1 and in the active nucleus of NGC4258. In a warped accretion disc there are horizontal pressure gradients that drive an epicyclic motion. We have studied the interaction of this epicyclic motion with the magnetohydrodynamic turbulence in numerical simulations. We find that the turbulent stress acting on the epicyclic motion is comparable in size to the stress that drives the accretion, however an important ingredient in the damping of the epicyclic motion is its parametric decay into inertial waves.

I INTRODUCTION

Warped accretion discs have been a part of the astronomical vocabulary since the discovery of the 35-day cycle of the X-ray binary Her X-1 [1–3]. Incidentally Her X-1 was the first occulting X-ray pulsar for which an optical counterpart was found as discussed by Neta Bahcall at the 6th Texas Symposium in 1972 [4]. Later on warped accretion discs have been found in a multitude of systems. In later years one of the most interesting examples has been the maser source in the active galactic nucleus of NGC 4258 [5].

While the warped accretion disc offered a simple interpretation of the observations, it was not easy to describe it theoretically at the hydrodynamic level. The problems have been both to explain the excitation mechanism of the warp and its coherence. Pringle [6] showed that the radiation pressure from the central radiation source may produce a warp in the outer disc, and Schandl & Meyer [7] described a similar mechanism in which the irradiation produces a wind, which in its turn excites a warp.

A crucial condition for any of these mechanisms to work is that the tendency of the disc to straighten itself must be sufficiently weak. There are two different forces that strive to produce a flat disc. Firstly there is the usual viscous stress due to the local turbulence, which is also driving the accretion, but in general it is insignificant compared to the hydrodynamic stress due to the epicyclic shear flow which is driven by the warp itself [8,9]. The amplitude of the epicyclic motion is inversely proportional to the ordinary turbulent viscosity. For that reason the hydrodynamic stress due to the epicyclic motion will also be inversely proportional to the turbulent viscosity, and the time scale for flattening the disc will be anomalously short compared to the ordinary viscous time scale.

The warping instability therefore requires a mechanism that can damp the epicyclic motion much more efficiently than it transports the angular momentum in the radial direction. That would for instance be the case if the turbulent viscosity was strongly anisotropic. The intention of this paper is to estimate how anisotropic the turbulent viscosity is and to check whether there are any other mechanisms that can limit the amplitude of the epicyclic motion. We describe the numerical model that we use for these estimates in Sect. 2. There are then two ways in which we have studied the interaction between the turbulence and the epicyclic motion. Firstly we have studied the free decay of an epicyclic motion (Sect. 3) and secondly we have studied the motion that results from a radial forcing (Sect. 4). We discuss and summarize our results in Sect. 5.

II THE MATHEMATICAL MODEL

We solve the magnetohydrodynamical equations in a Keplerian shearing box [10–12]. Our units are chosen such that $H = GM = \mu_0 = 1$, where G is the gravitational constant, M the mass of the central object, and H is the Gaussian scale height of the shearing box. The density distribution assuming isothermality can then be written as $\rho = \rho_0 e^{-z^2/H^2}$, where we put $\rho_0 = 1$. The physical size of the shearing box is $L_x : L_y : L_z = 1 : 2\pi : 4$, and the box is positioned such that x and z vary between $\pm\frac{1}{2}L_x$, and $\pm\frac{1}{2}L_z$, respectively, while y goes from 0 to L_y, and the distance of the origin to the central object, $R_0 = 10$. This gives the orbital period $T_0 = 199$, and the mean internal energy $e_0 = 7.4\,10^{-4}$. To stop the box from heating up we add a cooling function

$$Q = -\sigma_{\text{cool}}(e - e_0),\qquad(1)$$

where σ_{cool} is the cooling rate, which typically corresponds to a time scale of 1.5 orbital periods. Our boundary conditions are (sliding) periodic in the (x-) y-direction. In the z-direction they are impenetrable and stress-free for the velocity, and acts as a perfect conductor with respect to the magnetic field.

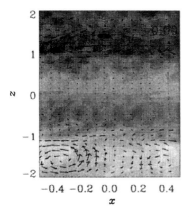 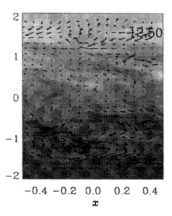

FIGURE 1. The magnetic field in a meridional cut of the simulation at the beginning of the simulation (left) and after 12.5 orbital periods (right). The toroidal field is plotted using a grey scale and the poloidal field as vectors. Note that the toroidal field has been reversed between the two images.

We start the simulations from a snapshot from a previous simulation in which the magnetohydrodynamic turbulence is already fully developed. In the first set of simulations we then add a radial velocity of the form $u_x = u_0 \sin(\pi z/L_z)$, which will have the time evolution of an epicyclic motion. In the second set of simulations, we do not modify the velocity field of the initial snapshot, but rather add a radial forcing term to the equation of motion. This forcing gives an acceleration with a harmonic time-dependence on the orbital time scale and the same z-dependence as the velocity above.

III THE FREE DECAY OF AN EPICYCLIC MOTION

In the first set of simulations we add an epicyclic motion to the initial state of the simulation, and then follow the decay of the epicyclic motion. These simulations have previously been described in [13]. With a weak epicyclic motion, its maximum Mach number is initially 0.38, it is difficult to follow the evolution of the epicyclic motion as it is comparable in size to the turbulent velocities. When the amplitude is increased to a Mach number of 3.3, we can distinguish two stages in the damping. After a brief period of essentially no damping the epicyclic velocity quickly drops by a factor of 2 to 3. This damping is followed by an extended phase of exponential decay with a time scale of 25 orbital periods. The time scale of the exponential decay can be translated to a Shakura-Sunyaev [14] α-parameter of 0.006, which is within a factor of two of the values usually derived from turbulence simulations, e.g. [10,15]. The preceding rapid damping is a new phenomenon though, which we interpret in terms of that the epicyclic motion is decaying to inertial waves

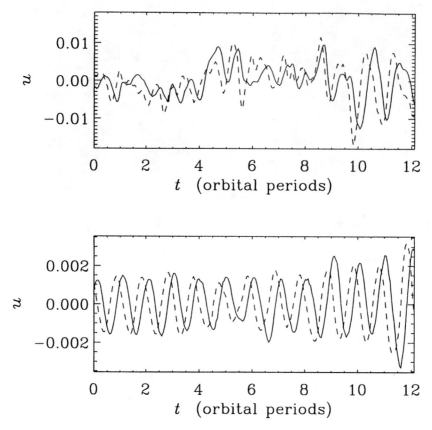

FIGURE 2. $\langle u_x \rangle$ (solid line) and $2\langle u_y \rangle$ (dashed line) as functions of time at $z = 1.17$ (top) and $z = -0.25$ (bottom).

via a parametric instability [16]. We also note that the toroidal magnetic field in the shearing box reverses its direction during the damping of the strong epicyclic motion (Fig. 1).

IV DRIVEN EPICYCLIC MOTION

The dynamics in a real accretion disc is significantly different from the case we have studied above. In reality the epicyclic motion will be driven by the radial pressure gradient that is set up by the warp. To mimic this we have carried out a new set of simulations in which we drive the epicyclic motion by adding a time-periodic radial force to the equation of motion.

We plot the horizontally averaged velocities $\langle u_x \rangle$ and $\langle u_y \rangle$ in Fig. 2. The epicyclic motion is in this case comparable to that of Run 1b in [13]. The results are similar in the two cases, and the epicyclic motion is difficult to distinguish at $z = 1.17$ due to the effect of the turbulent stresses, while it is easily distinguishable at $z = -0.29$.

V CONCLUSIONS

In this paper we have studied the dynamics of an epicyclic flow in a Keplerian shear flow. In our simulations we find two damping mechanisms for the epicyclic motion. The turbulent stresses can damp the motion in a way which can be described in terms of a turbulent viscosity comparable in strength to that driving the radial angular momentum transport, but a sufficiently fast epicyclic motion can lose significant amounts of energy by exciting inertial waves through a parametric instability.

ACKNOWLEDGMENTS

Computer resources from the National Supercomputer Centre at Linköping University are gratefully acknowledged. UT is supported by the Swedish Research Council (formerly the Natural Sciences Research Council, NFR), and RFS is supported by NASA grant NAG5-4031. This work was supported in part by the Danish National Research Foundation through its establishment of the Theoretical Astrophysics Center (ÅN).

REFERENCES

1. Tananbaum, H., Gursky, H., Kellogg, E. M., Levinson, R., Schreier, E.,, Giacconi, R., *ApJ*, **174**, L143–L149 (1972)
2. Katz, J. L., *Nat. Phys. Sci.*, **246**, 87–89 (1973)
3. Roberts, W. J., *ApJ*, **187**, 575–584 (1974)
4. Bahcall, N. A., "Optical Observations of HZ Herculis", in *Sixth Texas Symposium on Relativistic Astrophysics*, edited by D. J. Hegyi, Ann. N. Y. Acad. Sci. 224, New York, 1973, pp. 178–183
5. Miyoshi, M., Moran, J., Herrnstein, J., Greenhill, L., Nakai, N., Diamond, P.,, Inoue, M., *Nat*, **373**, 127–129 (1995)
6. Pringle, J. E., *MNRAS*, **281**, 357–361 (1996)
7. Schandl, S.,, Meyer, F., *A&A*, **289**, 149–161 (1994)
8. Papaloizou, J. C. B.,, Pringle, J. E., *MNRAS*, **202**, 1181–1194 (1983)
9. Papaloizou, J. C. B.,, Lin, D. N. C., *ApJ*, **438**, 841–851 (1995)
10. Brandenburg, A., Nordlund, Å., Stein, R. F.,, Torkelsson, U., *ApJ*, **446**, 741–754 (1995)
11. Hawley, J. F., Gammie, C. F., Balbus, S. A., *ApJ*, **440**, 743–763 (1995)
12. Miller, K. A., Stone, J. M., *ApJ*, **534**, 398–419 (2000)

13. Torkelsson, U., Ogilvie, G. I., Brandenburg, A., Pringle, J. E., Nordlund, Å., Stein, R. F., *MNRAS*, **318**, 47–57 (2000)
14. Shakura, N. I., Sunyaev, R. A., *A&A*, **24**, 337–355 (1973)
15. Stone, J. M., Hawley, J. F., Gammie, C. F., Balbus, S. A., *ApJ*, **463**, 656–673 (1996)
16. Gammie, C. F., Goodman, J., Ogilvie, G. I., *MNRAS*, 318, 1005–1016 (2000)

Formation and dynamics of neutron haloes in disk accreting black holes

A.A. Belyanin* and E.V. Derishev*

Institute of Applied Physics RAS, 46 Ulyanov Street, 603600 Nizhny Novgorod, Russia

Abstract.
  Hot plasma of ion temperature exceeding several MeV can exist in the inner accretion disks around black holes. This inevitably leads to the neutron production via dissociation of helium nuclei.
  We show that, for a broad range of accretion parameters, neutrons effectively decouple from protons and pile up in the inner disk leading to the formation of self-sustained halo. This means that new neutrons in the halo are supplied mainly by the splitting of helium nuclei in their collisions with existing neutrons. Once formed, such a halo can exist even if the proton temperature is much lower than the energy of helium dissociation.

There are many theoretical and observational indications that the hot plasma of ion temperature T_i exceeding several MeV can exist in the inner accretion disks around black holes. Formation of two-temperature region with $T_i > T_e$ in accretion disks has been predicted as long ago as in 1976 (Shapiro et al., 1976). When the energy transfer from ions to electrons is slow as compared with accretion timescale, the accretion flow becomes nearly adiabatic, and the ion temperature can reach virial values. This idea has led to the prediction of advection-dominated accretion flows which seems to find support in observations of several X-ray transients (Narayan and Yi, 1996).

The presence of hot ions in accretion disks leads to the rich variety of inelastic nuclear interactions. The most prominent of them is neutron production which primarily occurs through dissociation of helium nuclei. This process has potentially important observational implications arising e.g. from possibility of radiative neutron capture with subsequent formation of gamma-ray lines (Aharonian and Sunyaev, 1984).

Neutron production in accretion disks was considered several times (Aharonian and Sunyaev, 1984; Guessoum and Kazanas, 1989) under the assumption that neutrons are quickly thermalized in the disk and are advected to the black hole with the same rate as protons. In this case the neutron fraction in the accretion flow is inevitably small and the radiative-capture gamma-ray lines are weak and smeared out by hot rapidly rotating disk. However, the above assumption is not true if the proton density in the disk is sufficiently low. In this case, as shown below the neutron and ion components in the disk effectively decouple, allowing for the accumulation of neutrons. Such a pile-up can lead to much larger neutron number than in previously considered cases.

Let us consider bulk radial motion of neutrons which is the result of angular momentum losses caused by elastic collisions with ions (mostly protons) in the accretion disk.

One directly obtains the following expression:

$$V_n^{(r)} = -\frac{d}{dt}\left(\frac{R_g\, c^2}{2\, V_n^2}\right) = R_g \frac{c^2}{V_n^3} \frac{dV_n}{dt} = -\frac{R_g\, c^2}{2\, V_n^3} \nu_{pn}(V_n - V_d)\,, \qquad (1)$$

where V_n and $V_n^{(r)}$ are the orbital and radial velocities of neutrons, R_g the Schwarzschild radius of the black hole, and c the speed of light. The time derivative of V_n in the above expression was substituted by $-\nu_{pn}(V_n - V_d)/2$, where ν_{pn} is the proton-neutron collision rate and V_d the disk orbital velocity.

As the orbital velocity of neutrons is larger than that of the accretion disk, the excessive centrifugal force has to be balanced by the friction force, which originates from the difference of radial velocities of the neutrons and the disk. This gives another relation:

$$\frac{V_n^2 - V_d^2}{R} = \frac{\nu_{pn}}{2}\left(V_n^{(r)} - V_d^{(r)}\right). \qquad (2)$$

Here $V_d^{(r)}$ is the radial velocity of the disk and R the current radius. Assuming $V_n \simeq V_d \simeq V_0$ (V_0 is the Keplerian velocity), one finds the expression for the radial velocity of neutrons:

$$V_n^{(r)} = -\frac{1}{4}\left(\frac{R}{V_0}\nu_{pn}\right)^2 \left(V_n^{(r)} - V_d^{(r)}\right). \qquad (3)$$

The value $(R/V_0)\nu_{pn}$ in the above equation is the number of collisions a neutron undergoes as it completes one orbital revolution, divided by 2π.

The collision rate ν_{pn} may be expressed in terms of disk optical depth for proton-neutron collisions, τ_d, which is equal to the average number of collisions experienced by a neutron passing through accretion disk with Keplerian velocity directed perpendicular to the disk plane. Given the heights of ion and neutron "disks", h_d and h_n respectively, and assuming that the cross-section of proton-neutron collisions is inversely proportional to their relative velocity, one has $\nu_{pn} = \langle\sigma V\rangle_{pn} n_p (h_d/h_n) = (V_0/h_n)(\langle\sigma V\rangle_{pn}/V_0) n_p h_d = (V_0/2h_n)\tau_d$. Finally, the expression for the radial velocity of neutrons takes the following form:

$$V_n^{(r)} = \frac{(R/2h_n)^2 \tau_d^2}{4 + (R/2h_n)^2 \tau_d^2} V_d^{(r)}. \qquad (4)$$

Neutrons start to pile up in the inner parts of accretion disk when $V_n^{(r)}$ is considerably smaller than $V_d^{(r)}$, i.e., when $\tau_d \leq 4h_n/R$.

It is important that, once formed, such a halo becomes self-sustained. This means that, even if the ion temperature in the disk falls below the threshold for He dissociation by protons, the neutron production is supported by collisions of energetic neutrons from the halo with helium nuclei. The self-sustained mode implies that at least one of three neutrons resulting from helium dissociation breaks another helium nucleus before decaying or being advected into the black hole. These two possibilities provide, respectively, lower and upper bounds on the range of accretion rates in which the self-sustained neutron halo may exist.

Our analysis shows (Derishev and Belyanin, 2000) that neutron halo exists in a broad range of disk parameters and can lead to a number of important observational and dynamical effects. In particular, neutron haloes can be the natural source of relativistic electrons and positrons produced in proton-neutron and neutron-neutron collisions. In the presence of abundant soft X-rays from disk, relativistic electron-positron pairs are quickly cooled, and the snapshot of electron distribution would demonstrate that the majority of them has energies between 50-500 keV. The resulting spectrum above 100 keV has a typical comptonization pattern. Pair cascade, if developed, leads to a hard power-law tail with index around 2. Both features are frequently observed in the hard state of accreting black-hole candidates. In addition, very hard radiation should be present with photon energies up to 70 MeV. However, it is expected to contain much less energy than hard X-rays and is difficult to observe. Deuterium gamma-ray line at 2.2 MeV resulting from neutron capture is also expected at a level detectable by future INTEGRAL mission.

Furthermore, the presence of neutron halo strongly affects the dynamics of accretion and leads to the rich variety of transient dynamical regimes.

This work has been supported by Russian Foundation for Basic Research (the project 99-02-18244) and by the Russian Academy of Science through a grant for young scientists.

REFERENCES

1. S.L. Shapiro, A.P. Lightman, and D.M. Eardley, *ApJ* **204**, 187 (1976)
2. R. Narayan, J. McClintock, and I. Yi, *ApJ* **457**, 821 (1996)
3. F.A. Aharonyan and R.A. Sunyaev, *MNRAS* **210**, 257 (1984)
4. N. Guessoum and D. Kazanas, *ApJ* **345**, 356 (1989)
5. A.A. Belyanin and E.V. Derishev, *A&A*, to be submitted.

Shot Noise in USA Lightcurves of XTE J1118+480 and Cygnus X-1

Warren B. Focke*, E. D. Bloom*, B. Giebels*, G. Godfrey*, K. T. Reilly*, P. Saz Parkinson*, G. Shabad*, K. S. Wood[†], P. S. Ray[†], R. M. Bandyopadhyay[†], M. T. Wolff[†], G. Fritz[†], P. Hertz[†], M. P. Kowalski[†], M. N. Lovellette[†], D. Yentis[†] and Jeffrey D. Scargle**

*Stanford Linear Accelerator Center, Stanford University, Stanford, CA 94309
[†]E. O. Hulburt Center for Space Research, Naval Research Laboratory, Washington, DC 20375
**NASA/Ames Research Center, Moffett Field, CA 94035

Abstract. We investigate the behavior of the new transient black hole candidate XTE J1118+480 using shot noise models in the time domain.

Using light curves from the Unconventional Stellar Aspect experiment (USA), we model a portion of the emission as a superposition of shots. Individual structures in the light curve are fit to a template function. The results are compared to a similar analysis performed on light curves from Cygnus X-1. Results indicate that the characteristic timescale for XTE J1118+480 is around 3 times as long as for Cygnus X-1.

INTRODUCTION

Shot noise has long been a favored phenomenological model for describing the time variability of black hole candidates in the hard state [1]. In these models, the light curve is represented as being composed, at least in part, of the sum of many individual flashes of emission, or shots. In the simplest models, the shots are assumed to occur at random times, and to all be the same shape and size, but in general their shapes, heights and timescales may vary, and their occurrence times may be correlated [2, 3, 4].

Here we compare XTE J1118+480 and Cygnus X-1 in the context of shot noise. XTE J1118+480 is a transient black hole candidate which was active from 2000 January through 2000 July. It spent much of this outburst, and all of the time studied here, in the hard state. Cygnus X-1 is one of the oldest black hole candidates, and is a persistent source which spends most of its time in the hard state.

ANALYSIS

We apply a peak detection routine, a modification [5, 6, 7] of that used by Negoro et al. [8] to find individual shots, and fit each one to a template shape. The peak detection routine triggered on bins which were the maximum within 20 s to either side, and were at least 1.5 times the local average, which was formed for 40 s to either side.

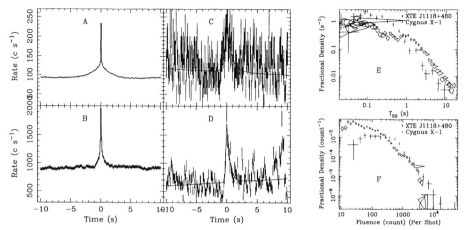

FIGURE 1. A–D: Average and example shots. The left panels are averages, those on the right are individual examples with their best-fit models. The upper panels are for XTE J1118+480, the lower ones for Cygnus X-1. The center bin in each average is much higher than the bins to either side of it, not only because of intrinsic sharpness of the shots, but is also biased because it is subject to a selection effect — since it was the highest bin in the region, it is more likely to have a positive Poisson fluctuation. The number of events and time in each curve was XTE J1118+480: 2205 shots in 62817 s (0.035 shot s^{-1}); Cygnus X-1: 446 shots in 14097 s (0.032 shot s^{-1}). The horizontal axis shows the time relative to the detected peak bin. E, F: Distributions of shot timescales (E) and fluences (F). The fluence is the integrated number of photons in the fit shape. No correction for the different count rates of the sources was applied to the fluence distribution — the correspondence of the positions of the curves is a coincidence. While we might expect the curves to have similar shapes, since the sources seem to both be BHCs in the low/hard state, we would not expect typical fluences to match, since the source rates depend on distance. When the analysis of the data was run on light curves with their rates and errors divided by 4, the curves shifted, indicating that their coincident position is not an artifact of the analysis method.

The model used had an exponential rise and fall, with different rise and fall times. The peak was constrained to be continuous. The parameters of the model were peak time, T_{50} (the time containing half the photons in the peak), the ratio of rise to fall times, and the number of photons in the peak. The model was integrated into the time bins, rather than evaluating it at a point.

Data from one observation at a time was read, and a binned lightcurve formed. The bin size used was 0.1 s. The shot detection routine was performed, the times of the detected peaks stored, and an average profile for the observation formed. this average profile was fit to the template shape plus a linear continuum. The parameters from this fit were used as the starting point for each peak in the observation, except the constant term in the continuum was set to the average rate for the interval fit, and the template amplitude was adjusted to match the peak bin. Uncertainties were found by freezing the parameter of interest and fitting the others, varying the frozen parameter until χ^2 rose by 1.0. If a new minimum was found while measuring the uncertainties, all the parameters were refit starting from that point, and the uncertainty measuring technique restarted, repeating until a stable minimum was found. Average and example shots can be seen in Figure 1, panels A–D.

To form histograms of the timescales (T_{50}, which is the same as the full width at half maximum for this shape) and fluences, bins geometrically spaced in time and fluence were formed, in a number equal to the integer part of the square root of the total number of events, covering a range from the lowest to the highest data points. The height of the histogram is the number of events which fell into a particular bin divided by the width of the bin and by the total number of events, so that the histogram estimates a probability density, and the distributions for samples with different numbers of events and bins can be compared. The horizontal errors were formed by the RMS of the uncertainties of the points that went into the bin divided by the square root of the number of events in the bin. The timescales were poorly determined when less than the lightcurve bin size. These histograms can be seen in Figure 1, panels E and F.

In order to check whether the timescale distribution in XTE J1118+480 changed during the course of its outburst, distributions were formed from the first and second halves of the data. The two were entirely consistent, except for timescales smaller than the lightcurve bin size.

SUMMARY

We have applied a shot detection and fitting procedure XTE J1118+480 and Cygnus X-1 data. Shots in XTE J1118+480 show longer characteristic timescales than those in Cygnus X-1. This can be seen both in the average shape of all detected shots and in the distribution of fit timescales. The distributions of fluences show a very similar shape for the two sources, except at the lowest values of fluence.

ACKNOWLEDGMENTS

The Office of Naval Research supports basic research in X-ray Astronomy at the Naval Research Laboratory. Work at SLAC was supported by Department of Energy contract DE-AC03-76SF00515.

REFERENCES

1. Terrel, N. J., *The Astrophysical Journal Letters*, **174**, L35–L41 (1972).
2. Miyamoto, S., and Kitamoto, S., *Nature*, **342**, 773–774 (1989).
3. Belloni, T., and Hasinger, G., *Astronomy and Astrophysics*, **227**, L33 (1990).
4. Lochner, J. C., Swank, J. H., and Szymkowiak, A. E., *The Astrophysical Journal*, **376**, 295–311 (1991).
5. Focke, W. B., and Swank, J. H., "Shot Distributions from Cygnus X-1", in *Accretion Processes in Astrophysical Systems: Some Like it Hot!*, edited by S. S. Holt and T. R. Kallman, American Institute of Physics, Woodbury, New York, USA, 1998, pp. 343–346.
6. Focke, W. B., *X-ray Timing Properties of Cygnus X-1 and Cygnus X-2*, Ph.D. thesis, University of Maryland, College Park (1998).
7. Focke, W. B., and Swank, J. H., *in preparation* (2001).
8. Negoro, H., Kitamoto, S., Takeuchi, M., and Mineshige, S., *The Astrophysical Journal Letters*, **452**, L49–L52 (1995).

Magnetic Flux through a Slightly Charged Kerr Black Hole

Chul H. Lee*, Hongsu Kim* and Hyun Kyu Lee*

Dept. of Physics, Hanyang University, Seoul 133-791, Korea

Abstract. In association with the Blanford-Znajek mechanism for the extraction of rotational energy from a Kerr black hole, it is of interest to explore how much of magnetic flux can actually penetrate the horizon. For the completely uncharged Kerr hole case, it has been known that the magnetic flux gets entirely expelled when the hole is maximally rotating. In the mean time, it is known that a rotating hole immersed in a magnetic field, when it is surounded by plasma, accretes a certain amount of electric charge. In this work, we show that, as a result of this accretion charge, small enough not to disturb the background geometry, the magnetic flux through this slightly charged Kerr black hole depends not only on the hole's angular momentum but also on the hole's charge such that it never vanishes for any value of the hole's angular momentum.

INTRODUCTION

The Blanford-Znajek process, a theoretical model mechanism for extracting rotational energy from a black hole [1], requires magnetic field lines penetrating the event horizon. However, according to the field configuration obtained by Wald [2] around a black hole immersed in an asymptotically uniform magnetic field, the magnetic flux penetrating the horizon decreases as the angular momentum of the hole increases, becoming zero when the hole is maximally rotating. This conclusion is true only when there is no electric charge accreted on the hole. If the black hole is surrounded by plasma, the hole selectively accretes charge until it builds up "equilibrium" net charge. Using the injection energy argument of Carter [3], Wald showed that this equilibrium charge is $2B_0 J$ where B_0 is the strength of the magnetic field and J is the angular momentum of the black hole. With a net electric charge accreted onto the black hole, the magnetic flux penetrating the horizon does not become zero even when the hole is maximally rotating. These arguments, explained in more detail in Sec. 2, are based on the situation where a Kerr black hole is placed in an originally uniform magnetic field aligned along the symmetry axis of the hole. In Sec. 3 we consider the case where the rotation axis of the black hole is misaligned with the asymptotic direction of the magnetic field. The solution given by Bicak and Janis [4] describes the electromagnetic field configuration with no charge accretion onto the black hole. We extend this solution to construct the field configuration in the case where the charge is accreted. Then the magnetic flux penetrating the horizon is calculated. It is shown that the magnetic flux does not vanish in any case with charge accreted onto the black hole. We use the notations of the "3+1" formalism of Ref. [5] and the Kerr metric as the background geometry around the rotating black hole.

WALD'S FIELD CONFIGURATION

Wald's construction of the solution is based on the following two statements;
(1) *The axial Killing vector $\psi = \partial/\partial\phi$ generates a stationary, axisymmetric test electromagnetic field, $F_\psi = d\psi$, which asymptotically approaches a uniform magnetic field. The source of this field has no magnetic monopole moment and has electric charge $Q = 4J$.*
(2) *The time translational Killing vector $\xi = \partial/\partial t$ generates a stationary, axisymmetric test electromagnetic field, $F_\xi = d\xi$, which vanishes asymptotically. The source of this field has no magnetic monopole moment and has electric charge $Q = -2m$ where m is the mass of the hole.*

Then the field configuration with the asymptotic magnetic field strength B_0 and the charge Q accreted onto the hole is given by

$$F = \frac{1}{2}B_0[d\psi + \frac{2J}{m}(1 - \frac{Q}{2B_0 J})d\xi]. \tag{1}$$

The magnetic flux $\Phi = \int dS \cdot B$ evaluated on the upper hemisphere (with the symmetry axis in the z-direction) is

$$\Phi = B_0 \pi r_+^2 (1 + \frac{a^2}{r_+^2})[1 - \frac{a^2}{r_+^2}(1 - \frac{Q}{B_0 J})]. \tag{2}$$

In the absence of the accreted charge($Q = 0$), the above equation shows that the magnetic flux vanishes when the black hole is maximally rotating($a \to m$ and $r_+ \to m$). With the charge, the flux becomes $\Phi = 2\pi Q$. Particularly, with the equilibrium charge $Q = 2B_0 J$, it becomes $\phi = B_0 4\pi m^2$ which is exactly the same as the flux across a Schwarzschild black hole of mass m [6], [7].

THE MAGNETIC FLUX THROUGH A MISALIGNED KERR BLACK HOLE

We now explore the general case where the asymptotically uniform stationary magnetic field happens to be "oblique", i.e., aligned at some angle to the hole's axis of rotation. Indeed the case of uncharged Kerr hole has been studied long ago by Bicak and Janis [4]. We denote their solution by F^{BJ}. With the rotation axis of the black hole in the z-direction, the asymptotic magnetic field is supposed to have, without loss of generality, the z and x components with B_0 denoting the field strength of the z component and B_1 of the x component. In order to construct the solution in the presence of the accreted charge Q, we recall statement (2) of the previous section. Then the new solution is obtained by

$$F = F^{BJ} - \frac{Q}{2m}d\xi. \tag{3}$$

The magnetic flux on a generally oriented hemisphere of the event horizon can be calculated from the field configuraton of Eq. (3). We present here only the results for some particular directions. For the hemisphere whose symmetry axis is in the z-direction, the flux is

$$\Phi = B_0 \pi r_+^2 (1 + \frac{a^2}{r_+^2})[1 - \frac{a^2}{r_+^2}(1 - \frac{Q}{B_0 J})]. \tag{4}$$

This result is the same as the one in the aligned case given by Eq. (2), and never vanishes with the sign of Q the same as that of B_0. For the hemisphere whose symmetry axis is in the xy-plane, the flux is

$$\Phi = B_1 \pi [r_+^2 \cos\beta - (r_+ + m)a \sin\beta]. \tag{5}$$

where β is the angle between the symmetry axis and the x-axis. The fact that the total flux depends only on the B_1-component is also expected since with this orientation of the hemisphere, the B_0-component contribution to the total flux is obviously zero. The flux becomes maximum for $\beta = 0$ when $\Phi = B_1 \pi r_+^2$ and minimum for $\beta = \beta_0$, with $\tan\beta_0 = [r_+^2/a(r_+ + m)]$, when $\Phi = 0$. For the extreme Kerr hole, $a = m = r_+$ and hence $\tan\beta_0 = 1/2$ or $\beta_0 = 27°$. Moreover, since we may assume $-\pi/2 \leq \beta \leq \pi/2$, $\cos\beta_0$ is always positive and hence $\Phi = 0$ occurs only if $a\sin\beta_0 > 0$. Thus if $a > 0$, then $\beta_0 > 0$ and this confirms our intuition that the field lines are bent near the hole in the same direction in which the hole rotates since the rotating hole drags field lines along.

ACKNOWLEDGMENTS

This work is supported in part by the KOSEF(Grant No. 1999-2-112-003-5) and in part by the BK21 Program of the Korean Ministry of Education.

REFERENCES

1. R. Blanford and R. Znajek, Mon. Not. R. Astron. Soc. **179**, 433 (1977) ; as good review articles, also see, D. Macdonald and K. Thorne, *ibid* **198**, 345 (1982) ; H. K. Lee, R. A. M. J. Wijers, and G. E. Brown, Phys. Rep. **325**, 83 (2000).
2. R. Wald, Phys. Rev. **D10**, 1680 (1974).
3. B. Carter, in *Black holes*, eds C. DeWitt and B. DeWitt, Gordon and Breach, New York (1973).
4. J. Bicak and V. Janis, Mon. Not. R. Astron. Soc. **212**, 899 (1985).
5. D. Macdonald and K. Thorne, *MNRAS* **198**, 345 (1982).
6. V. Dokuchaev, Sov. Phys. JETP **65**, 1079 (1987).
7. M. H. P. M. van Putten, Phys. Rev. Lett. **84**, 3752 (2000).

Relativistic Slim Disk Model for NLS1s

T. Manmoto* and S. Mineshige[†]

*Department of Physics, Chiba University, 1–33 Yayoi-cho, Inage-ku, Chiba 263-8522, Japan
[†]Department of Astronomy, Kyoto University, Sakyo-ku, Kyoto 606-8502, Japan

Abstract. Narrow-line Seyfert 1 galaxies (NLS1s) are known to be exhibit two distinct features in soft X-ray band. They are large soft X-ray excess and rapid variability. As a model to explain their relatively narrow Balmer line, [4] suggested the slim disk model where the black hole with a moderate mass $M \simeq 10^{5-7} M_\odot$ accretes nearly at the Eddington rate. In spite of the fact that the emitting region is very close to the black hole, their model was nonrelativistic. In this work, we treat everything relativistically to obtain accurate results which we show in this poster. We also examined the effect of black hole spin on the observed spectrum. We find that it is possible to tell a Kerr black hole from a Schwarzschild black hole when the mass accretion rate is less than 50 times the Eddington rate.

INTRODUCTION

NLS1s are known to exhibit large soft X-ray excess ([2], [3]). If one relates this excess to the big soft X-ray bump in Galactic black-hole candidates during the so-called soft state, the mass of a central black hole in NLS1s should be moderate, of the order of $M \simeq 10^{5-7} M_\odot$.

For such black-hole masses, the observed luminosity, $L \sim 10^{43-44}$, is near the Eddington luminosity, $L_{\rm Edd}$. [4] explored the observational appearance of a slim disk ([1]) in the context of NLS1s and found that the soft X-ray excess can be explained satisfyingly well with a superposition of blackbody spectra emitted from the surface of a slim disk.

Although the previous slim-disk model successfully explains the basic trends observed in these near-Eddington sources, its formulation is not adequate, since it is basically non-relativistic in spite of the fact that emitting region is fairly close to the event horizon of the hole in the high $\dot{M}$ accretion system ([5], [4]). In this article, we calculate a slim disk structure by treating everything relativistically to obtain the results which can directly be used for comparison with observations. Exploring how the observed spectrum is modified by changing the spin parameter is another important theme of the current article.

RELATIVISTIC SLIM DISK MODEL

First, we solve general relativistic conservation laws in the curved spacetime to obtain relativistically corrected temperature profile (figure 1), velocity field of accreting gas, and so on.

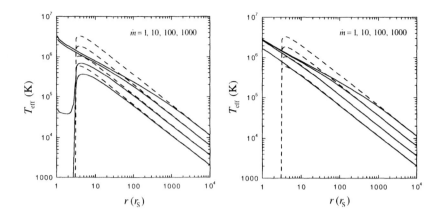

FIGURE 1. Effective temperature of the disk around a Schwarzschild hole (left panel) and a Kerr hole (right panel). The mass of the hole is $10^5 M_\odot$. The mass accretion rate is 1, 10, 100, 1000, from bottom to top, in the unit of the Eddington mass accretion rate. Dashed lines are the effective temperature of Shakura & Sunyaev's standard disks.

Then, we use ray shooting method to calculate the photon path. This method allows us to set the observer at an arbitrary inclination angle and also to take into account the effect of blocking radiation from the innermost portion of the flow by the outer disk surface. We then calculate the gravitaional and the Doppler redshift of each photons to obtain observed spectra (left panel of figure 2).

For the purpose of an easy comparison of our results with the observational data, we perform the spectrum fitting with the formula $T_{\rm eff}(r) \sim T_{\rm in}(r/r_{\rm in})^{-p}$, where $r_{\rm in}$, $T_{\rm in}$, and p are fitting parameters. We show the $\dot{m}$ dependence of $r_{\rm in}$ and $T_{\rm in}$ in the right panel of figure 2.

DISCUSSIONS

We notice that the disks around Kerr black holes ($a = 0.998$) systematically give smaller $r_{\rm in}$ and higher $T_{\rm in}$, compared with those around Schwarzschild black holes. When $\dot{m}$ is large, both cases exhibit similar spectra and hence Kerr holes and Schwarzschild holes give quite identical values of fitting parameters.

There is a range of $r_{\rm in}$ (i.e. $r_{\rm in} < 5r_{\rm S}$) which cannot be explained by the disks around Schwarzschild holes. If the spectral fitting of X-ray data indicates that the radius of the inner edge of the disk lies in this range, we can conclude that the object harbors a Kerr black hole. Note, however, that this criteria can be used only when the black hole mass

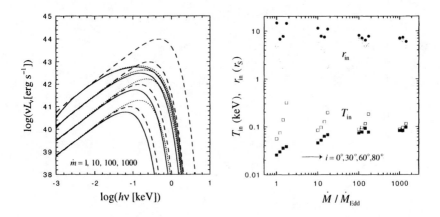

FIGURE 2. Left panel: observed X-ray spectra of the slim disk around a Schwarzschild hole (solid lines), and around a Kerr hole (dotted lines). Local spectrum is assumed to be the black body spectrum with the effective temperature shown in figure 1. Dashed lines correspond to Shakura & Sunyaev model, without any relativistic correction including redshifts. Right panel: $\dot{m}$-dependence of fitting parameters. Filled squares and circles correspond to the disk around a Schwarzschild hole, and open squares and circles the disk around a Kerr hole. Cases of four inclination angles are shown shifted sequentially for $\dot{m} = 1, 10, 100, 1000$.

is accurately known, since the unit of r_{in}, r_S, scales in proportion to the mass of the black hole. Nevertheless, it is potentially important to emphasize that we can identify Kerr holes with the help of T_{in} derived by observing the soft X-ray radiation of black hole objects.

REFERENCES

1. Abramowicz, M. A., Czerny, B., Lasota, J. P., and Szuszkiewicz, E., *ApJ*, **332**, 646 (1988).
2. Boller, Th., Brandt, W.N., and Fink, H.H., *A&A*, **305**, 53 (1996).
3. Leighly K.M., *ApJS*, **125**, 317 (1999).
4. Mineshige, S., Kawaguchi, T., Takeuchi, M., and Hayashida, K., *PASJ*, **52**, 499 (2000).
5. Watarai, K., Fukue, J., Takeuchi, M., and Mineshige, S., *PASJ*, **52**, 133 (2000).

Estimation of Relativistic Accretion Disk Parameters from Iron Line Emission

Vladimir I. Pariev*[‡◊], Benjamin C. Bromley[†], Warner A. Miller*

> *Theoretical Astrophysics Group, T6, Los Alamos National Laboratory, Los Alamos, New Mexico 87545, USA
> [†]Department of Physics, The University of Utah, 201 James Fletcher Bldg, Salt Lake City, Utah 84112, USA
> [‡]Steward Observatory, 933 N. Cherry Ave., Tucson, AZ 85721, USA
> [◊]P.N. Lebedev Physical Institute, Leninsky Prospect 53, Moscow 117924, Russia

Abstract. The observed iron Kα fluorescence lines in Seyfert I galaxies provide strong evidence for an accretion disk near a supermassive black hole as a source of the line emission. Here we present an analysis of the geometrical and kinematic properties of the disk based on the extreme frequency shifts of a line profile as determined by the measurable flux in both the red and blue wings. The edges of the line are insensitive to the distribution of the X-ray flux over the disk, and hence provide a robust alternative to profile fitting of disk parameters. Our approach yields new, strong bounds on the inclination angle of the disk and the location of the emitting region. We apply our method to interpret the observational data from a few Seyfert I nuclei.

The parameters which uniquely define the system of black hole and accretion disk are the mass of the black hole M, spin parameter of the black hole $a_* = a/M$, and the inclination of the disk relative to the observer i. We use gravitational units, $R_g = GM/c^2$, so that for a nonrotating black hole the radius of event horizon is $2R_g$, and the radius of the innermost stable orbit is $6R_g$; for the maximally rotating astrophysical black hole with $a_* = 0.998$, the radius of the event horizon decreases to $1.063R_g$ and the accretion disk becomes stabilized down to the radius $1.237R_g$. It is this extension of the stable disk down to small radii which is the most prominent effect of the rotation of the black hole.

To describe our method of analyzing the observed broad line profiles, let us first define $g_{\min} = \nu_{\min}/\nu_e$ and $g_{\max} = \nu_{\max}/\nu_e$ to be the minimum and maximum frequency shifts of a broad line profile relative to the rest-frame frequency. We then consider the emission from an infinitesimal annulus of the disk with Boyer-Lindquist radius r. For a given value of the spin parameter a_* (as well as luminosity L and viscosity parameter α for a thick disk) there exists a unique mapping $g_{\min} = g_{\min}(r, i)$, and $g_{\max} = g_{\max}(r, i)$, connecting frequency extrema of a line and the inclination angle i and radius r of the annulus.

Emission in an observed line consists of the sum of the contributions from many infinitesimal annuli. These annuli produce a curve in the g_{min}–g_{max} plane, which corresponds to a single, constant inclination angle i. Of course, a measured profile will generate a single point in this plane (with some error bars due to observational errors) which is just the location of the extreme redshift and blueshift along a curve of constant inclination i. An example of such a mapping with some data points from observations of MCG-6-30-15 is shown in Figure 1. The position of the measured point in the g_{min}–g_{max} diagram indicates that emission can come only from the quadrant of the g_{min}–g_{max} plane defined by inequalities $g(i,r) > g_{min}$ and $g(i,r) < g_{max}$. At the same time, some emission must come from the points on both sides of this quadrant, i.e. from the points having either $g(i,r) = g_{min}$ or $g(i,r) = g_{max}$. The position of this point can fix the range of possible disk inclination angles i: Since g_{min} increases with increasing r along the curve i = constant, the

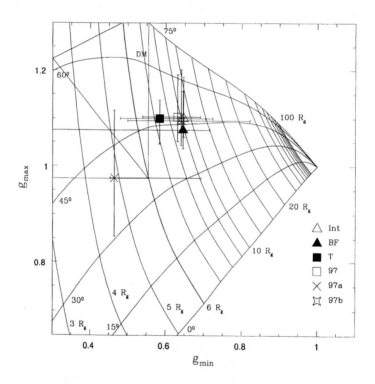

FIGURE 1. Data from MCG-6-30-15 plotted on a maximum and minimum frequency shift diagram for a thin Keplerian disk model around a Schwarzschild black hole. Asymmetric error bars indicate confidence limits at a 2σ level using the edge fitting by nonlinear "sharp-edge" model described in [1]. The point marked as T stands for the [2] data. Int, BF and DM denote intermediate, bright flare and dark minimum spectra from [3]. Points 97, 97a and 97b are for the average profile (97), bright flare subset (97a) and minimum subset (97b) from [4].

upper limit on inclination angle of the disk is given by the value of i for the curve i = constant passing through the data point. The lower limit is provided by the i = constant curve which just touches the horizontal line $g = g_{\max}$. If $g_{\max} < 1$, then the lower limit for i is 0. Furthermore, for each i within the observed bounds one can determine the outer $r_{\text{out}}(i)$ and inner $r_{\text{in}}(i)$ radii of annulus, which give a contribution to the observed profile. Generally, the middle regions of the disk can have arbitrary amounts of iron line emission, but there must always be some emission coming from points of the disk surface at radii $r = r_{\text{out}}(i)$ and $r = r_{\text{in}}(i)$.

If several line profiles are available for the same object in different phases of emissivity, then we may be able to place further constraints on the inclination angle of the disk and to check the validity of the accretion disk model (i.e. that bounds for i are not mutually exclusive). Narrower limits for i put tighter constrains on r_{in} and r_{out}. We emphasize that these constraints are independent of the emissivity law across the disk.

We applied the method described above to estimate the position of the edges of the broad iron line in MCG-6-30-15 and find that the commonly assumed inclination 30° for the accretion disk in MCG-6-30-15 is inconsistent with the position of the blue edge of the line at a 3σ level (see Figure 1, the most restrictive lower bound on i comes from Int data point). A thick turbulent disk model or the presence of highly ionized iron may reconcile the bounds on inclination from the line edges with the full line profile fits based on simple, geometrically thin disk models. The bounds on the innermost radius of disk emission indicate that the black hole in MCG-6-30-15 is rotating faster than 30 % of theoretical maximum. When applied to data from NGC 4151, our method gives bounds on the inclination angle of the X-ray emitting inner disk of $50 \pm 10°$, consistent with the presence of an ionization cone grazing the disk as proposed by [5]. The frequency extrema analysis also provides limits to the innermost disk radius in another Seyfert I galaxy, NGC 3516, and is suggestive of a thick disk model.

These studies provide the first evidence for turbulence in accretion flows around supermassive black holes. Details concerning the frequency extrema method and our results from observations of relativistic disks can be found in [1,6].

Partial support from the NASA Astrophysics Theory Program (NAG5-8277) and LDRD Program at Los Alamos National Laboratory is acknowledged.

REFERENCES

1. Pariev, V.I., Bromley, B.C., and Miller, W.A., *Astrophys. J.* **547**, 649 (2001)
2. Tanaka, Y., et al., *Nature* **375**, 659 (1995)
3. Iwasawa, K., et al., *Mon. Not. Roy. Astron. Society* **282**, 1038 (1996)
4. Iwasawa, K., Fabian, A.C., Young, A.J., Inoue, H., and Matsumoto, C., *Mon. Not. Roy. Astron. Society* **306**, L19 (1999)
5. Pedlar, A., et al., *Mon. Not. Roy. Astron. Society* **263**, 471 (1993)
6. Bromley, B.C., Miller, W.A., and Pariev, V.I., *Nature* **391**, 54 (1998)

Probing Black Holes with Constellation-X

Kimberly A. Weaver

Code 662 NASA/Goddard Space Flight Center
Greenbelt, MD 20771

Abstract. *Constellation-X* is a premiere X-ray spectroscopy mission due to launch within the next decade. With a factor of 100 increase in sensitivity over current X-ray spectroscopy missions and an excellent energy resolution of 2 eV at 6 keV, one of the prime science goals of the mission will be to observe activity near the black hole event horizon by measuring changes in the Fe Kα fluorescence emission line profile and time-linked intensity changes between the line and the continuum. Detailed variability studies with *Constellation-X* will allow us to reconstruct "images" of the accretion disk, probe the effects of strong gravity in the vicinity of black holes and measure black hole mass and spin via deconvolution of the line profile.

INTRODUCTION

X-rays that are emitted near a supermassive black hole, whether from the base of a jet, from the accretion disk corona or by some other mechanism, are excellent tools for probing the matter within a few to tens of gravitational radii ($r_g = GM/c^2$). One important product of X-rays interacting with dense matter is the Fe Kα fluorescence line at 6.4 keV. In general, it is the most prominent spectral feature in active galactic nuclei (AGN).

One likely region that produces the Fe Kα line is an accretion disk. For a geometrically thin and optically thick disk that extends inward to the event horizon, the line is predicted to be broad with a redshifted wing that can extend down to energies as low as 3 keV, depending on whether or not the black hole is rotating [1] [2]. The shape of the line also depends on the disk emissivity and the inclination of the disk to our line of sight.

The importance of the Fe Kα line for studying the accretion phenomenon in AGN was solidified by the Advanced Satellite for Cosmology and Astrophysics (ASCA), which discovered extremely broad and asymmetric lines [3] [4] (Figure 1). At the limit of ASCA's energy resolution, most of the time-averaged Fe K line profiles in Seyfert galaxies can be described by an accretion disk model.

THE INTRICACY OF THE FE Kα LINE

Standard theoretical disk models predict profiles like those shown in Figure 1b. However, there is growing evidence from *ASCA*, *RXTE* and *Chandra* that Fe K lines are neither simple nor predictable in their behavior. Approximately 70% of Seyfert 1 galaxies have variable Fe K lines [5]. While this implies that at least part of the line originates on small scales (and supports the accretion disk hypothesis), the variability does not appear to be associated with changes in the continuum flux in any predictable way [6] [7] [8]. The erratic variability patterns are a problem for reverberation studies and standard Compton reflection models. The Fe Kα lines also have complex profiles that include both narrow *and* broad components (Figure 2a). *Chandra* grating spectra have clearly detected the narrow component in NGC 5548 [9], NGC 4151 [10] and MCG$-$5-23-16 (Figure 2b). The physical widths of the narrow lines are all different, indicating that they originate from different locations (likely to be either the NLR, the BLR, or the obscuring torus).

THE ROLE OF CONSTELLATION-X

With a collecting area of 15,000 cm^2 at 1 keV and 6,000 cm^2 at 6 keV, shared between multiple satellites, *Constellation-X* will perform high throughput X-ray spectroscopy. Each satellite contains a spectroscopy X-ray telescope (SXT), covering the 0.25 and 10 keV band with a 5 to 15 arcsec angular resolution, and a hard X-ray telescope (HXT) covering the 10 to 40 keV band with a 30 to 60 arcsec angular resolution. At the focal plane of the SXT is a microcalorimeter array with 2 eV resolution at 6.4 keV and the ability to handle 1,000 counts s^{-1} pixel^{-1} (> 900 pixels). The microcalorimeter will be the workhorse instrument for Fe K studies.

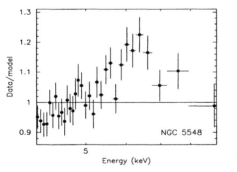 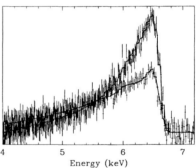

FIGURE 1. a) Ratio of the ASCA SIS data for NGC 5548 to a power-law continuum model. b) Simulated 60 ks *Constellation-X* exposure of Fe Kα from an accretion disk that extends uniformly to the event horizon and orbits a non-rotating (Schwarzschild) and maximally rotating (Kerr) black hole. The Kerr profile has a weaker blue peak and stronger red wing than the Schwarzschild profile.

By measuring Doppler shifts and the gravitationally induced increase in wavelength as matter is pulled closer to the event horizon, *Constellation-X* will allow us to reconstruct "images" of the accretion disk, as well as distinguish material much further out (Figure 2b). *Constellation-X* will probe regions of strong gravity, determine black hole mass and spin and probe accretion disk dynamics. Simulations (Figure 2c) of Fe K lines based on theoretical models of flares erupting above the disk [11] [12] [13] show that *Constellation-X* will be able to probe timescales of 1000 s, corresponding to the light crossing time of a 10^8 solar mass black hole.

REFERENCES

1. Laor, A., *ApJ*, **376**, 90 (1991)
2. Fabian, A. C., Rees, M. J., Stella, L., and White, N. E., *MNRAS*, **238**, 729 (1989)
3. Tanaka, Y., et al., *Nature*, **375**, 659 (1995)
4. Nandra, K., et al., *ApJ*, **476**, 70 (1997)
5. Weaver, K. A., Gelbord, J., and Yaqoob, T., *ApJ*, **550** (2001)
6. Chiang, J., *ApJ*, **528**, 292 (2000)
7. Lee, J. C., et al., *MNRAS*, **318**, 857 (2000)
8. Done, C., Madejski, G. M., and Zycki, P. T., *ApJ*, **536**, 213 (2000)
9. Yaqoob, T., et al., *ApJ*, **546**, 759 (2001)
10. Ogle, P. M., Marshall, H. L., Lee, J. C., and Canizares, C. R., (astro-ph/0010314)
11. Ruszkowski, M., *MNRAS*, **315**, 1 (2000)
12. Reynolds, C. S., et al., *ApJ*, **514**, 164 (1999)
13. Nayakshin, S., and Kazanas, D., (astro-ph/0101312)

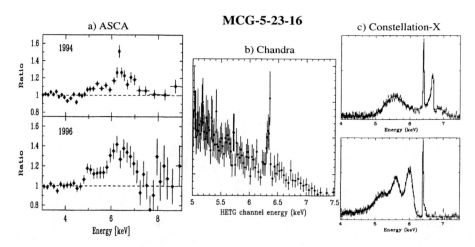

FIGURE 2. a) Ratio of the ASCA SIS data to a power-law continuum model for MCG−5-23-16. b) A *Chandra* HETG spectrum. c) A 10 ks *Constellation-X* simulation of the Fe K line based on the theoretical disk models of [11], with the narrow line also included.

CHAPTER 14

NUMERICAL RELATIVITY AND BLACK HOLE COLLISIONS

Recent developments in classical relativity

B. G. Schmidt

Max-Planck-Institute for Gravitational Physics
Albert-Einstein-Institute
14476 Golm — Germany

Abstract. In the period spanned by the Texas meetings, — the term "classical relativity" was not yet coined 40 years ago — the notions of gravitational collapse, gravitational radiation, singularities and black hole where in the center of almost all investigations and developments. 40 years ago black holes were exotic theoretical concepts far from reality. Now they seem to exist all over the univers.

In the last 40 years a scenarium describing the collaps or collision of stellar objects or BHs has formed. In my talk I want to outline this picture, tell you which parts are firmly established and where the big open questions are.

I INTRODUCTION

During the 40 years period spanned by the Texas meetings, — the term "classical relativity" was not yet coined — the notions of gravitational collapse, gravitational radiation, singularities and black hole were in the center of almost all investigations and developments. Black holes were exotic theoretical concepts, far from reality. Nowadays, as we saw this week, they seem to exist all over the universe.

In the past four decades a scenario describing the collapse or collision of stellar objects and black holes has formed. In my talk, I want to outline this picture, tell you which parts are firmly established and where the big open questions are. This will also lead us naturally to the present front of research.

It is clear from the beginning that I can only present a personal view and — even more dangerous — a view from the perspective of a mathematical relativist. I hope you will be patient and follow my line of argument.

I organized the talk in two parts: 1) collapse and equilibrium of a single object; 2) binary systems.

There are important topics of "classical relativity" which I shall not talk about: lensing, cosmology, initial value problem, exact solutions of the field equations, asymptotics, etc. Some of these topics are covered in the recent volume [42].

II COLLAPSE AND EQUILIBRIUM OF A SINGLE OBJECT

The present picture of gravitational collapse originated from the collapse of a homogeneous spherically symmetric dust cloud which can be explicitly described by an exact solution found by Oppenheimer and Snyder in 1939 [1].

The spacetime picture of this solution (Fig 1) illustrates how the world lines of the spherically symmetric homogeneous dust cloud approach each other and meet at some time at the center with infinite density, forming a singularity of infinite curvature.

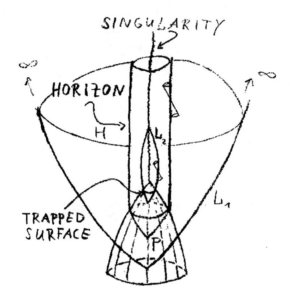

FIGURE 1. Schematic diagram of the collapse of a dust cloud.

A unique light cone of one event P at the center, the horizon H, has the property that its null geodesics do not escape to infinity. For ealier cones L_1 the photons escape, for later ones L_2 they fall back onto the singularity. The intersection of the surface of the dust cloud with the horizon is a "marginally trapped 2–surface." This is such a 2-surface that the area of the cross section of the orthogonally outgoing null hypersurface remains constant. If the outgoing as well as the ingoing congruence of null geodesics is converging — the cross section decreases - we have a "trapped surface". A spacelike singularity forms, but it is invisible from the outside for all times!

The essential concepts derived from this example are: trapped surface, horizon, singularity.

There are inhomogeneous initial data determining a spacetime with a "naked singularity", i.e. a singularity which can be seen from far away.

Inspired by this example there were far reaching generalizations — partly proofs, partly conjectures — to form the present picture of the gravitational collapse.

The Chandrasekhar limit indicates the collapse of realistic matter: stars above the Chandrasekhar mass have to explode or contract until their radius is less than the Schwarzschild radius and — as we know in the spherical case — a trapped surface has formed.

A trapped surface implies a singularity of spacetime via the famous Hawking – Penrose singularity theorems [2], the first of which was proved by Penrose in 1964. No symmetry assumptions are made in these theorems, the particular nature of the matter is irrelevant, except for some energy positivity. However, the structure of the singularities is not predicted and remains unexplored. We just know that spacetime comes to an end after some finite time, according to classical relativity.

Around 1969 the idea of cosmic censorship appeared! (R. Penrose [3]) The hope was, and still is, that the singularities are always shielded by a horizon and cannot be seen from anywhere far away. Furthermore, the only possible final state is that of a Kerr–Newman black hole.

So far the general picture. For which matter systems have we really established it?

For a spherically symmetric scalar field coupled to Einstein's field equations, beginning 1991 Christodoulou [4] has analytically confirmed the above picture in detail (1999). For weak fields the scalar field disperses, for strong fields a black hole forms. A very careful analysis shows that naked singularities are possible. However, generically cosmic censorship holds, i.e. for data with a naked singularity any arbitrarily small change of the data results in a solution in which the singularity is behind a horizon.

In 1996 Hübner [5] has numerically calculated the corresponding spacetimes globally. He finds future null infinity, the horizon and the singularity using the conformal field equations, [38].

Choptuik [6] found numerically (1993) a new phenomenon in this system. At the transition from dispersion of the scalar field to black hole formation, the solutions show some universality and discrete self–similarity.

It is still an open question whether there are also critical phenomena for pure gravitational waves at the threshold of black hole formation.

For spherically symmetric fluids we cannot expect analytic results similar to the case of a scalar field source because of the formation of hydrodynamical shocks. Numerically this case is well understood [7].

For axisymmetric stationary rotating fluids in equilibrium the first existence theorem was proved by Heilig in 1995, [8]. He generalized techniques, developed by Lichtenstein at the beginning of the last century, to the relativistic case. We have numerical results since about 1980. There are new 3–D calculations (2000) by Baumgarte, Shapiro and Shibata [9]: fully relativistically they derive mass bounds for differentially rotating neutron stars, follow oscillations of stable configurations and see the bar–mode instability of fast rotating fluid stars. The calculations seem to be good enough to display the essential behaviour of the matter. However, because the outer boundary in the numerical calculation is not far enough away, uncontrollable boundary influences do not allow to determine radiation reaction and gravitational wave forms reliably.

An overview and a more sophisticated treatment of the hydrodynamic is described by Font [7] in a recent review (2000).

In particular, it is not yet really possible to check numerically cosmic censorship for 1–body systems.

Let me describe in more detail some other interesting new results related to 1–body systems:

The **Einstein–Vlasov system** describes a self–gravitating collisionless gas of particles with mass m. The equations $G_{\alpha\beta} = 8\pi T_{\alpha\beta} = 8\pi \int p_\alpha p_\beta f dp$, while f *constant along geodesics*, determine a metric $g_{\alpha\beta}$ and a distribution function $f(x^\mu, p^\nu)$. The distribution function describes the density of particles with momentum p^ν at some spacetime point x^μ. Beginning 1992, Rein and Rendall explored this matter model [10]. There are essential differences to the case of dust. For example, in the Newtonian case the solutions are regular for all times. Similarly in Einstein's theory spherically symmetric solutions exist for all times for small data. There are also results on the Newtonian limit of relativistic solutions. A mass gap is seen at the threshold of black hole formation numerically. This matter model is a very good candidate for a generalization of the Christodoulou – Klainerman theorem to a case with sources. (In essence this theorem shows that vacuum initial data near flat data determine a solution of Einstein's equations whose global structure is like Minkowski space (1993) [43].)

The **Penrose–inequality** [11] appeared in the attempt to construct a counter example to the cosmic censorship hypothesis. In 1973 Penrose put forward the following conjecture: for asymptotically flat initial data with mass m (for Einstein's field equations) which contain a marginally trapped 2–surface of area A it holds: $16\pi^2 m \geq A$. Violation of this inequality would allow the construction of a counter example to cosmic censorship. In 1997 a proof was given by Huisken and Ilmanen [12] for the case of time symmetric initial data. The almost trapped surface then is a minimal surface. The proof used an idea of Geroch ; Bray [13] generalized the theorem to the case in which A has several components.

Non–radial linear oscillations of stars or black holes generate gravitational waves. If these waves could be observed, they would give information about internal properties of the oscillating object.

In relativity a Newtonian mode of oscillation gets a tiny imaginary part, because the mode is damped by emitting gravitational waves. New modes, called w–modes, not present in a Newtonian treatment appear, which describe the oscillation of spacetime geometry.

For rotating neutron stars things become rather complicated, even in a Newtonian treatment. Modes are difficult to calculate. In the relativistic case there are up to now only calculations for zero frequency modes (Friedman and Stergioulas

(1998) [14]).

However, a new instability, found by Chandrasekhar in 1974, occurs, the CFS (Chandrasehkar – Friedman – Schutz) instability [15]: all rotating perfect fluid stars are unstable! How can then the millisecond pulsar exist? Estimates of various damping mechanisms seem to allow just a certain window in a temperature versus period diagram in which some of the modes (f–mode, r–mode) can be unstable.

In the last years a possible further instability via the r–mode has attracted a lot of attention. r–modes are fluid motions without density changes. Instabilities can occur even for rather slow rotation of the star. It is not yet clear whether this instability is astrophysically relevant in the evolution history of a newly born neutron star. Estimates indicate that the gravitational radiation produced by the r–mode instability might be observable by LIGO II, (see the review by Andersson and Kokkotas, (2000) [16]).

There is also theoretical work on the continuous part of the spectrum of rotating stars. Its relevance has been mysterious in Newtonian theory as well as in Einstein's theory. However, if there is a continuous part of the spectrum it has to be considered in the stability analysis.

Finally, there has been some progress on the question of linear stability of the Kerr black hole. Beyer [17] has proved (2000) that linear perturbations with $e^{im\phi}$ angular behavior can grow at most linearly in time for fixed m. As the bound, however, grows with m, a general perturbation could grow still faster. Numerical simulations show no evidence for exponential growth.

III BINARY SYSTEMS

As long as general relativity has existed, the treatment of the 2–body system has been investigated. With the detection of the binary pulsar, post Newtonian calculations found their most important application. What do we know about binary systems of collapsing neutron stars or black holes? Today, such systems are of particular importance because they are possible sources of gravitational radiation to be observed by VIRGO, LIGO and GEO600 in the near future.

The following picture was developed by Hawking [18] and others about 30 years ago for the collision and merger of two collapsing neutron stars or black holes. (Fig.2)

The two legs of the trousers show the horizon H around the two neutron stars colliding after they have already collapsed. After some time the two components of the horizons merge and at late times form one horizon around the whole system, which should settle down to a Schwarzschild black hole for an axisymmetric head–on collision. Since this picture assumes cosmic censorship, an outside observer sees nothing of the collapse of the individual stars or of the violent processes which supposedly take place when the stars actually collide and merge. Inside the horizon there are trapped surfaces, first around the two single objects T_1, T_2, then around the final black hole T_3.

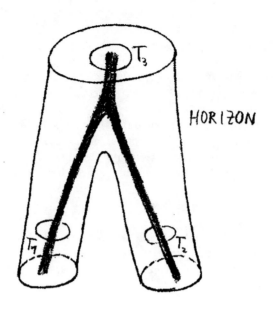

FIGURE 2. Schematic diagram of the collision of two black holes.

The following two famous axisymmetric numerical calculations gave first indications that this picture may be right!

Shapiro and Teukolsky calculated 1992 the head–on collision of two clusters of particles (Einstein–Vlasov system) [19]. Two clusters fall against each other while they collide. At some time marginally trapped surfaces begin to form, first around the individual cluster and later aroung both clusters. "Ray tracing" of null geodesics indicates the existence of a horizon.

1993 Anninios et al [20] numerically evolved time symmetric vacuum initial data constructed in 1963 by Misner [21], which have two asymptotic ends and two marginally trapped surfaces.

The evolution of the two black holes could be calculated long enough to see the individual trapped surfaces merge. The gravitational radiation and wave forms were determined; ray tracing confirmed cosmic censorship and produced the well known "green trouser" on the cover of Science in November 1995! In this black hole collision the black holes have been there always in the distant past.

For non–axisymmetric collisions the situation becomes much more complicated. One can distinguish the following phases of such a general collision: adiabatic inspiral, plunge and merger, and the final ring down.

The **adiabatic inspiral** can be calculated to leading order by the 2.5 PN–approximation, which has very successfully been applied to the binary pulsar (Blanchet et al (1995) [23]). To obtain information on the last circular orbit and the wave forms generated near it, 3PN information about the 2–body system is needed.

This enormous task — evaluation of about 100 3PN coefficients, 100000 terms had to be manipulated in intermediate calculations — was tackled and recently completed by two groups: Damour, Jaranowski, Schäfer (2000) [24] and Blanchet, Faye (2000) [25]. The post–Newtonian framework in both approaches is very different. Nevertheless, it could be proved by the first group of authors that both obtain equivalent equations of motions for the two bodies! I find this very remarkable because in particular the necessary regularisation procedures for a point–particle description seem to be rather ambiguous!

In the result a noval feature appeared: there is a coefficient which is not determined by the regularisation procedures, which indicates that, in contrast to the general belief, the structure of the bodies shows up already at this order. PN–investigation by Jaranowski and Schäfer, (1999) [26] of two different binary black hole initial–data sets found by Brill, Lindquist [22], and Misner [21] a long time ago showed that the undetermined coefficient in the 3PN expansion is different for the two initial data sets.

To obtain information about the last circular orbit, Damour, Iyer, Sathyaprakash, (2000) [27], and Damour, Jaranowski and Schäfer, (2000) [28] used resummation techniques for the badly converging PN–expansions. Various methods gave agreement for the last stable orbit for a certain value of the unknown constant.

Uryu, Shibata and Eriguichi, (2000) [29] calculated the inspiral up to the last stable orbit using neutron stars. Their results agree with point particle calculations only for rather high compactness.

For the **plunge and merger** full numerical relativity is needed. Unfortunately, we are presently not yet capable to perform such a calculation with sufficient accuracy.

Presently there are three 3d–calculations for two black holes (grazing collisions):

Brügmann (1999) [30], Brügmann and Seidel, (2000) [31] evolved initial data with two asymptotic ends. It is not clear whether such data generate similar amounts of radiation during the collision as data coming from black holes formed (from neutron stars or pure gravitational radiation) in a spacetime with one asymptotic end!

Brandt at al, (2000) [32] used succesfully "excision" techniques which "cut out the interior of marginally trapped surfaces".

Both simulations can be evolved long enough to find that marginally trapped surfaces merge, a local indication for the possible existence of a horizon. Estimates for the radiation emitted are given and agree.

Both computations cannot be evolved long enough to investigate whether there is really a horizon and whether the spacetime settles down to a Kerr black hole. The initial data used in these calculations are not yet related to PN solutions.

There is presently just one 3-D calculation of merging neutron stars by Shibata (1999) [33]. In his calculations the two stars merge and form a single differentially rotating neutron star. The calculation does not yet allow to determine radiation reaction and wave forms because the outer boundary is not far enough.

In the **ring down phase** perturbation theory was used once a single black hole has formed and spacetime is almost the Kerr spacetime at the late stages (Pullin, Price (1994) [34]). The wave forms in this phase are related to the quasi normal mode ringing of the final Kerr black hole.

Assuming that all 3 phases can be reliably calculated and have some overlap, there are attempts by various groups to join these individual calculations. There is, however, much arbitrariness at the interfaces! A proposal by Baker, Brügmann, Campanelli and Lousto (2000) [35] give interesting and promising results. Ultimately, only a full numerical calculation can demonstrate whether a combination of these methods gives good results.

Note that in this approach one gives up to investigate the issue of cosmic censorship because one assumes that the spacetime is almost Kerr at late times.

Let me finally describe two new ways to calculate black hole collisions.

Up to now there is no numerically calculated non–axisymmetric spacetime in which one could locate a horizon by tracing the null rays. Winicour, Gomez and Husa (1999) [36] performed an analysis of the structure of "stand alone horizons", which leads to interesting new insights. A horizon is a null hypersurface whose intrinsic structure (null lines, space metric on 2–dimensional sections) satisfies certain equations as a consequence of Einstein's field equations. Generic solutions of these equations were found which show that two spherical components of a horizon — the legs of the trousers — merge through a phase in which the horizon has toroidal sections. (This was also seen numerically in the calculation of rotating clusters.)

Using this approach, a characteristic initial value problem can be formulated in which the horizon is one of the boundary null hypersurfaces. Similar in spirit but technically different is the "Isolated Horizon" approach by Ashtekar (2000) [37].

The conformal approach is a method to calculate "the whole spacetime". The "conformal field equations" (Friedrich (1981) [38]) are partial differential equations for a metric $g_{\mu\nu}$ and a conformal factor Ω. On the part of the solution, where $\Omega > 0$ holds, the metric $\Omega^{-2} g_{\mu\nu}$ satisfies Einstein's equations. Points with $\Omega = 0$ are at infinity. In a numerical treatment these points have finite values on the numerical grid. This method is ideally suited to extract gravitational radiation, to determine the horizon, to detect cosmic censorship — or its violation!

Hübner(1999) [39] has developed a 3D–code for the constraints and the evolution equations based on the conformal field equations, and obtained first results. For small gravitational waves the spacetime has been calculated up to and beyond future timelike infinity. The Bondi mass and the gravitational radiation can be determined at future null infinity.

All results obtained so far for 2–body collisions show that for non–axisymmetric collisions there is presently no real evidence for the complete picture invented by theoreticians. There are, however, promising developments which should soon provide more information. The big challenges for numerical relativity still are: the calculation of gravitational radiation, exploration of numerical evidence for or against cosmic censorship, and the properties of singularities (even if they may be hidden!).

Let me finally make some remarks about properties of singularities. Even if it turns out that cosmic censorship is true and the singularities generated by collapsing objects are hidden, there is still the cosmological singularity.

Homogeneous isotropic cosmological models have singularities whose structure is well understood. For more realistic inhomogeneous models Belinskii, Kalatnikov and Lifshitz [40], in 1970, put forward an approximation scheme together with the claim to describe the generic nature of the cosmological singularity. Not much could be done to find out whether their picture was right or wrong at that time.

In 2000 Andersson and Rendall, [41] proved — at least in one particular case — that Belinskii, Khalatnikov and Lifshitz were correct. The more complicated case of singularities with an oscillating character is still open.

ACKNOWLEDGEMENTS

I want to thank J. Ehlers, J. Bičák, B. Brügman and J. Winicour for helpful discussions.

REFERENCES

[1] J. R. Oppenheimer and H. Snyder, Phys. Rev.**56**, 455 (1939)
[2] S. W. Hawking and G. F. R. Ellis, The Large–Scale Structure
  of the Universe, Cambridge University Press, (1973)
[3] R. Penrose, Nuovo Cimento, Numero Speciale **1**, 252 (1969)
[4] D. Christodoulou, Class. Quantum. Grav. **16**, A23 (1999)
  Commun. Pure Appl. Math. **XLIV**, 1131, (1991)
[5.] P. Hübner, Helv. Phys. Acta Vol. **69**, 317 (1996)
[6] M. W. Choptuik, Phys. Rev. Lett., **70**, 9, (1993)
[7] T. Font, Living Reviews in Relativity, 2000-2 (2000)
[8] U. Heilig, Commun.Math. Phys., **166**, 457 (1995)
[9] T.W. Baumgarte, S. L. Shapiro and M. Shibata,
  Ap.J. **542**, 453-463 (2000)
[10] G. Rein and A. Rendall, Math.Proc.Cambridge Phil.Soc. **128**, 363-380, (2000),
  Commun. Math. Phys. **150**, 561 (1992)
[11] R. Penrose, Ann. New York Acad. Sci.**224**, 125 (1973)
[12] G. Huisken and T. Ilmanen, Int. Math. Res. Not. **20**, 1045 (1997)
  J. Diff.Geom. (to appear)
[13] H. L. Bray, thesis, Stanford University (1997)
[14] N. Stergioulas and J. L. Friedman, Ap.J. **444**, 804 (1998)
[15] J. L. Friedman and B. F. Schutz, Ap.J. **221**, 937 (1978)
[16] N. Andersson and K. Kokkotas, gr-qc/0010102 (2000)
[17] H. Beyer, Commun. Math.Phys. (to appear), astro-ph/0008236 (2000)
[18] S. W. Hawking, article in "Black Holes", ed. C. DeWitt and
  S. DeWitt, Gordon and Breach (1973)
[19] S. L. Shapiro and S. A.Teukolski, Phys. Rev. D **45**, 2739 (1992)
[20] P. Annininos et al, Phys. Rev. Lett.,**71**,2851 (1993)
[21] C. Misner, Phys. Rev. **118**, 1110 (1960)
[22] D. Brill and R. Lindquist, Phys. Rev., **131**, 471 (1963)
[23] L. Blanchet et al, Phys. Rev. Lett., **74**, 3515 (1995)
[24] P. Jaranowski and G. Schäfer, Phys.Rev. D **57**, 7274 (1998)

T. Damour, P. Jaranowski and G. Schäfer, Phys.Rev. D**62**, 44024 (2000)
[25] L. Blanchet and G. Faye, Phys.Lett. A, **58** (2000)
[26] P. Jaranowski and G. Schäfer, Phys.Rev.D **60**,124003 (1999)
[27] T. Damour, B. R. Iyer and B. S. Sathyaprakash, gr-qc/0010009 (2000)
[28] T. Damour, P. Jaranowski and G. Schäfer, Phys.Rev.D **62** , 084011 (2000)
[29] K. Uryu, M. Shibata and Y. Eriguchi, Phys.Rev. D **62**, 104015 (2000)
[30] B. Brügmann, Int.J.Mod.Phys. D**8**, 85 (1999)
[31] B. Brügmann and E. Seidel, gr-qc/0012079 (2001)
[32] S. Brandt et al. gr-qc/0009047 (2000)
[33] M. Shibata, Phys. Rev. D **60** ,104052 (1999)
[34] R.H. Price and J. Pullin, Phys.Rev.Lett. **72**, 3297 (1994)
[35] J. Baker et al, gr-qc/0003027 (2000)
[36] J. Winicour, Prog.Theor.Phys.Supp. **136**, 57 (1999)
[37] A. Ashtekar et al. Phys.Rev.Lett. **85**, 3564 (2000)
[38] H. Friedrich, Proc.Roy. Soc. A **375**, 169 (1981),
Proc.Roy. Soc. A **378**, 401 (1981)
[39] P. Hübner, Class. Quantum Grav. **16**, 2823 (1999)
P. Hübner, gr-qc 0010052 (2000), gr-qc 0010069 (2000)
[40] V. A. Belinskii, M. Khalatnikov and E. M. Lifshitz, Adv. in Phys.**19**, 523 (1970)
[41] L. Andersson and A.D. Rendall, gr-qc/0001047 (2000)
[42] B. G. Schmidt,(ed.), Einstein's Field Equations and Their Physical Implications, LNP 540, Springer (2000)
[43] D. Christodoulou and S. Klainerman, The Global Nonlinear Stability of the Minkowski Space, Princeton University Press (1993)

Binary neutron star mergers in fully general relativistic simulations

Masaru Shibata* and Kōji Uryū[†]

*Graduate School of Arts and Sciences, University of Tokyo, Komaba, Meguro 153-8902, Japan
[†]Department of Physics, University of Wisconsin-Milwaukee, P.O. Box 413, Milwaukee, WI53201, USA

Abstract. We perform 3D numerical simulations for merger of equal mass binary neutron stars in full general relativity preparing irrotational binary neutron stars in a quasiequilibrium state as initial conditions. Simulations have been carried out for a wide range of stiffness of equations of state and compactness of neutron stars, paying particular attention to the final products and gravitational waves. We take a fixed uniform grid in Cartesian coordinates with typical grid size $(293, 293, 147)$ in (x, y, z) assuming a plane symmetry with respect to the equatorial plane. A result of one new large-scale simulation performed with grid size $(505, 505, 253)$ is also presented.

We find that the final product depends sensitively on the initial compactness of the neutron stars: In a merger between sufficiently compact neutron stars, a black hole is formed in a dynamical timescale. As the compactness is decreased, the formation timescale becomes longer and longer. For less compact cases, a differentially rotating massive neutron star is formed instead of a black hole. In the case of black hole formation, the disk mass around the black hole appears to be smaller than 1% of the total rest mass. It is also indicated that waveforms of high-frequency gravitational waves after merger depend strongly on the compactness of neutron stars.

I INTRODUCTION

Interest in the merger phase of binary neutron stars has been stimulated by the prospect of future observation of extragalactic, close binary neutron stars by gravitational wave detectors [1]. A statistical study indicates that mergers of binary neutron stars may occur at a few events per year within a distance of a few hundred Mpc [2]. Since the amplitude of gravitational waves from a binary of mass $\sim 3M_\odot$ (where $M_\odot$ denotes the solar mass) at a distance of ~ 100Mpc can be $\sim 10^{-22} - 10^{-21}$, the merger of binaries is a promising source for gravitational wave detectors. Although the frequency of gravitational waves in the merging regime will be larger than 1kHz (see Sec. IV) and hence lies beyond the upper end of an accessible frequency range of laser interferometers such as LIGO for a typical event at a distance ~ 100Mpc, it may be observed using specially designed narrow band interferometers or resonant-mass detectors [1]. Future observations, although they may not be done in near future, will provide valuable information about merger mechanisms and final products of binary neutron stars.

Interest has also been stimulated by a hypothesis about the central engine of γ-ray bursts (GRBs) [3]. Recently, some of GRBs have been found to be of cosmological origin [4]. In cosmological GRBs, the central sources must supply a large amount of the energy $\geq 10^{50}$ ergs in a very short timescale of order from a millisecond to a second. It has been suggested that the merger of binary neutron stars is a likely candidate for the powerful central source [3]. In the typical hypothetical scenario, the final product should be a system composed of a rotating black hole surrounded by a massive disk of mass $> 0.1 M_\odot$, which could supply a large amount of energy by neutrino processes or by extracting the rotational energy of the black hole [3].

To investigate merger of binary neutron stars theoretically, numerical simulation appears to be unique promising approach. Considerable effort has been made for this in the framework of Newtonian and post-Newtonian approximation (see, e.g., [5]). Although these simulations have clarified a wide variety of physical features which are important during the merger of binary neutron stars, a fully general relativistic (GR) treatment is obviously necessary for determining the final product and associated gravitational waves because GR effects are crucial.

Effort has been made for constructing a reliable code for 3D hydrodynamic simulation in full general relativity (see, e.g., [6–8]). Recently, the authors have succeeded in constructing a numerical code in which stable and fairly accurate simulations are feasible [8]. Applying the newly developed code, the authors and collaborators have been performing fully GR simulations for a wide variety of astrophysical problems [8–12]. In this paper, we present some of our recent results with regard to the merger of binary neutron stars.

The paper is organized as follows. In Sec. II, we summarize our formulation in 3D numerical relativity. In Sec. III, we briefly describe how to give initial conditions. In Sec. IV, numerical results with regard to merger of binary neutron stars are presented, paying particular attention to the final product and gravitational waveforms. Sec. V is devoted to a summary. Throughout this paper, we adopt the units $G = c = 1$ where G and c denote the gravitational constant and speed of light, respectively. We use Cartesian coordinates $x^k = (x, y, z)$ as the spatial coordinates; t denotes coordinate time. In the following, BH and NS denote black hole and neutron star, respectively.

II SUMMARY OF THE FORMULATION

We perform hydrodynamic simulations in full general relativity using (3+1) formalism. We use the same formulation and gauge conditions as in our previous works [8,13], to which the reader may refer for details and basic equations. The fundamental variables used in this paper are ρ: rest mass density, ε: specific internal energy, P: pressure, u^μ: four velocity, $v^k = u^k/u^t$, $\Omega = v^\varphi$, α: lapse function, β^k: shift vector, γ_{ij}: three metric, $\gamma = e^{12\phi} = \det(\gamma_{ij})$, $\tilde{\gamma}_{ij} = e^{-4\phi}\gamma_{ij}$, K_{ij}: extrinsic curvature. Definitions of several quantities are also found in Table 1.

Geometric variables, ϕ, $\tilde{\gamma}_{ij}$, the trace of the extrinsic curvature $K_k{}^k \equiv K_{ij}\gamma^{ij}$, a tracefree part of the extrinsic curvature $\tilde{A}_{ij} \equiv e^{-4\phi}(K_{ij} - \gamma_{ij}K_k{}^k/3)$, as well as three auxiliary functions $F_i \equiv \partial_j\tilde{\gamma}_{ij}$, where ∂_j is the partial derivative, are evolved with an unconstrained evolution code in a modified form of the ADM formalism

[14]. GR hydrodynamic equations are evolved using a van Leer scheme for the advection terms [15] adding artificial viscous terms [8]. Violations of the Hamiltonian constraint and conservation of mass and angular momentum are monitored to check numerical accuracy. Reliability of the code has been checked by several test calculations, including spherical collapse of dust, stability of spherical neutron stars, and the stable evolutions of rigidly and rapidly rotating neutron stars which have been described in [8].

We adopt a Γ-law equation of state $P = (\Gamma - 1)\rho\varepsilon$ where Γ is the adiabatic constant. For isentropic configurations the equation of state can be rewritten in the polytropic form $P = \kappa\rho^\Gamma$ and $\Gamma = 1 + 1/n$ where κ is the polytropic constant and n the polytropic index. This is the form that we use for constructing quasiequilibrium states as initial conditions. We adopt $n = 2/3, 4/5, 1$, and $5/4$ as a reasonable qualitative approximation to a moderately stiff, nuclear equation of state.

Instead of ρ and ε we numerically evolve the densities $\rho_* \equiv \rho\alpha u^0 e^{6\phi}$ and $e_* \equiv (\rho\varepsilon)^{1/\Gamma}\alpha u^0 e^{6\phi}$ as the hydrodynamic variables [8]. In our numerical method, the total rest mass of the system

$$M_* = \int d^3x \rho_*. \tag{1}$$

is automatically conserved.

The time slicing and spatial gauge conditions we use in this paper for the lapse and shift are the same as those adopted in our series of papers [8–13]; i.e. we impose an "approximate" maximal slice condition ($K_k{}^k \simeq 0$) and an "approximate" minimum distortion gauge condition ($\tilde{D}_i(\partial_t \tilde{\gamma}^{ij}) \simeq 0$ where $\tilde{D}_i$ is the covariant derivative with respect to $\tilde{\gamma}_{ij}$; see [13]).

III INITIAL CONDITION

Even just before merger, binary neutron stars are considered to be in a quasiequilibrium state because the timescale of gravitational radiation reaction $\sim 5/\{64\Omega(M_g\Omega)^{5/3}\}$ [16], where M_g and Ω denote the total mass of the system and the orbital angular velocity of the binary neutron stars, is several times longer than the orbital period (cf. Table 1). Thus, for a realistic simulation of the merger, we should prepare a quasiequilibrium state as the initial condition.

Since the viscous timescale in the neutron star is much longer than the evolution timescale associated with gravitational radiation, the vorticity of the system conserves in the late inspiraling phase of binary neutron stars [17]. Furthermore, the orbital period just before the merger is about 2msec which is much shorter than the spin period of typical neutron stars, implying that even if the spin of neutron star would exist at a distant orbit and conserve throughout the subsequent evolution, it could be negligible at the merger phase in most cases. Thus, it is quite reasonable to assume that the velocity field of neutron stars just before the merger is irrotational.

Model	C	$\bar{\rho}_{max}$	$\bar{M}_*$	$\bar{M}_g$	q	X	R_τ	C_{mass}	L/M_g	Products
(A)	0.12	0.139	0.186	0.173	1.03	0.090	5.1	0.58	37.5	NS
(B)	0.14	0.169	0.216	0.198	0.98	0.106	3.4	0.67	31.1	marginal
(C)	0.16	0.202	0.244	0.220	0.93	0.124	2.3	0.75	26.3	BH

TABLE 1. A list of quantities for initial conditions of irrotational binary neutron stars with $\Gamma = 2.25$. The compactness of each star in isolation $C \equiv (M/R)_\infty$, the maximum density $\bar{\rho}_{max} \equiv \kappa^n \rho_{max}$, total rest mass $\bar{M}_* \equiv \kappa^{-n/2} M_*$, gravitational mass at $t = 0$ $\bar{M}_g \equiv \kappa^{-n/2} M_g$, $q \equiv J/M_g^2$ where J is the total angular momentum, compactness of orbit $X \equiv (M_g \Omega)^{2/3} (\sim M_g/a$ where a is orbital separation), approximate ratio of the emission timescale of gravitational waves to the orbital period $R_\tau \equiv 5(M_g \Omega)^{-5/3}/128\pi$, ratio of the rest mass of each star to the maximum allowed mass for a spherical star $C_{mass} \equiv M_*/2M_*^{sph}{}_{max}$, location of outer boundaries L along three axes in units of M_g (i.e., L/M_g) for simulations with $(293, 293, 147)$ grid size, and final products are shown. All quantities are normalized by κ to be non-dimensional: We can rescale by appropriately choosing κ. Here, $M_*^{sph}{}_{max}$ denotes the maximum allowed mass of a spherical star ($\kappa^{-n/2} M_*^{sph}{}_{max} \simeq 0.162$ at $\bar{\rho}_{max} \simeq 0.52$).

To prepare quasiequilibrium states, we assume the existence of a helicoidal Killing vector in the form, $\ell^\mu = (1, -y\Omega, x\Omega, 0)$. For irrotational fluid, the hydrostatic equation is integrated to give a Bernoulli type equation in the presence of ℓ^μ [18], resulting in a great simplification for handling the hydrodynamic equations.

Currently, we restrict our attention to initial conditions satisfying $\tilde{\gamma}_{ij} = \delta_{ij}$, $\partial_t \tilde{\gamma}_{ij} = 0$ and $K_k{}^k = 0$. The initial conditions for geometric variables are obtained by solving the Hamiltonian and momentum constraint equations, and equations for gauge conditions. In this case, the basic equations reduce to two scalar elliptic type equations for α and $\psi (= e^\phi)$ and one vector elliptic type equation for β^k [19].

The coupled equations of Bernoulli type equation and elliptic type equations for metric are solved using the method developed by Uryū and Eriguchi [19]. Several test results and scientific results are found in [19] and [20].

IV RESULTS

We have performed simulations for $\Gamma = 1.8, 2.0, 2.25$ and 2.5. The results for $\Gamma = 2$ have been already presented in [11]. In this manuscript, we show results for $\Gamma = 2.25$ which have recently obtained. In Table 1, we list several quantities which characterize the quasiequilibrium state of irrotational binary neutron stars used as initial conditions. All the quantities are scaled with respect to κ (in units of $c = G = 1$) to be non-dimensional. We choose binaries at the innermost orbits for which the Lagrange points appear at the inner edge of neutron stars [20]. To induce merger, we reduce the angular momentum J by $\sim 2 - 3\%$ (see discussion below). The gravitational mass M_g and non-dimensional angular momentum parameter $q \equiv J/M_g^2$ listed in Table 1 are calculated from initial data sets which are recomputed by solving constraint equations after the reducing of J.

We have performed simulations using a fixed uniform grid assuming reflection symmetry with respect to the equatorial plane. The simulations were mainly performed using FACOM VPP300/16R. In this case, the typical grid size was $(293, 293, 147)$ in (x, y, z). One large-scale simulation was recently performed using FACOM VPP5000 with $(505, 505, 253)$ grid size to enlarge the computational region. In both cases, the grid spacing is identical and determined from the condition that major diameter of each star is covered with ~ 33 grid points initially. The computational time for one simulation was typically ~ 100CPU hours for $\sim 10^4$ timesteps. Test simulations were performed with $(193, 193, 97)$ grid size on FACOM VX/4R to check the convergence of numerical results.

The wavelength of gravitational waves at $t = 0$ is computed as

$$\lambda_{t=0} \equiv \frac{\pi}{\Omega} \simeq 100 M_g \left(\frac{X}{0.1}\right)^{-3/2}, \qquad (2)$$

where $X \equiv (M_g \Omega)^{2/3}$ denotes a compactness of orbits (cf. Table 1). To accurately extract the waveform near outer boundaries, $\lambda_{t=0}$ should be shorter than the size of the computational region along three axes L. However, as found in Table 1, this condition is not satisfied since taking such a large computational region is a very difficult task in the present restricted computational resources: $\lambda_{t=0}$ is typically $\sim L/3$ with $(293, 293, 147)$ grid size. Even with $(505, 505, 253)$ grid size, $\lambda_{t=0}$ is $\sim 2L/3$. This implies that gravitational waves in the early phase cannot be accurately computed. However, the wavelength of quasi-periodic waves of the merged object excited during merger (see Fig. 6) is much shorter than the wavelength of binaries in quasiequilibrium and L. Therefore, the waveforms in the late phase to which we here pay main attention can be computed fairly accurately.

As found in [20], orbits of all irrotational binaries with $\Gamma \lesssim 2.25$ are stable and the merger in reality should be triggered by radiation reaction of gravitational waves. To take into account the gravitational radiation reaction approximately, we use the following method. Using the quadrupole formula, we can estimate the angular momentum loss in one orbital period $\Delta J (> 0)$ as $4\pi M_g X^2/5$ [16] and hence can write the ratio of ΔJ to the total angular momentum as

$$\frac{\Delta J}{J} = \frac{4\pi}{5q} X^2 = 0.025 \left(\frac{1}{q}\right) \left(\frac{X}{0.1}\right)^2. \qquad (3)$$

Thus, we initially reduce the angular momentum by $\sim 2 - 3\%$.

In Figs. 1 and 2, we display the density contour lines and velocity vectors for ρ_* and v^i at selected timesteps for simulations of models (A) and (C). For (A) the final product is a massive neutron star. For (C) a black hole is formed and the apparent horizon is able to be located (thick solid circle in the last panel).

The total rest mass of the binary for model (A) is $\sim 20\%$ larger than the maximum allowed value of a spherical star in isolation. Even with such large mass, the merged object does not collapse to a black hole. As indicated in Fig. 1, the merger proceeds very mildly, because the approaching velocity at the contact of two stars is not very large. Consequently, the shock heating is not very important

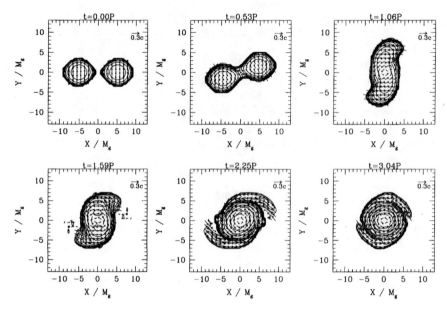

FIGURE 1. Snapshots of the density contours for ρ_* in the equatorial plane for model (A). The contour lines are drawn for $\rho_*/\rho_{*\,\mathrm{max}} = 10^{-0.3j}$, where $\rho_{*\,\mathrm{max}}$ denotes the maximum value of ρ_* at $t = 0$ (here $\bar{\rho}_{*\,\mathrm{max}} = 0.355$), for $j = 0, 1, 2, \cdots, 10$. The maximum density for ρ_* in the final panel is about 2.8 times larger than the initial value. Vectors indicate the local velocity field and the scale is as shown in the top left-hand frame. P denotes the orbital period of the initial quasiequilibrium ($P_{t=0}$). The length scale is shown in units of GM_g/c^2 where M_g is the gravitational mass at $t = 0$.

in merging. This implies that the rotational centrifugal force plays an important role for supporting the self-gravity of such supramassive outcome. To illustrate the importance of the rotation for the supramassive neutron star, the angular velocity along x and y axes and density contour lines in x-z slices are displayed in Fig. 3. It is found that the supramassive neutron star is differentially rotating and consequently has a highly flattened configuration. Fig. 3(a) shows that the magnitude of the angular velocity in the inner region is of order of the Kepler velocity, i.e., $\Omega M_g = O[(M_g/R)^{3/2}]$ where R denotes the typical radius of the merged object.

All these results are qualitatively the same as those found in simulations for $\Gamma = 2$ [11]. However, the results are quantitatively different. For $\Gamma = 2$, the maximum allowed rest mass for formation of a massive neutron star after merger is $\sim 1.5 M_{*\mathrm{max}}^{\mathrm{sph}}$ where $M_{*\mathrm{max}}^{\mathrm{sph}}$ denotes the maximum allowed rest mass of a spherical star in isolation for a fixed value of Γ. As shown here, the threshold is slightly smaller as $\sim 1.3 - 1.4 M_{*\mathrm{max}}^{\mathrm{sph}}$ for $\Gamma = 2.25$ (we determine that model (B) is located near the threshold; see below), and we have found that it is $\sim 1.6 - 1.7 M_{*\mathrm{max}}^{\mathrm{sph}}$ for $\Gamma = 1.8$. Thus, the threshold value depends on the stiffness of the equation of state.

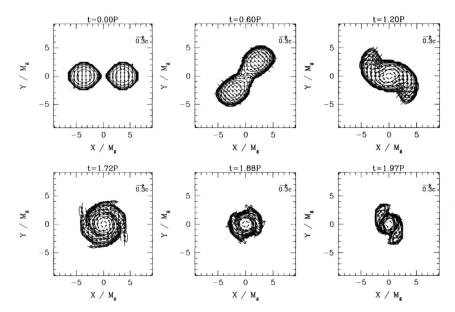

FIGURE 2. The same as 1, but for model (C). The contour lines are drawn for $\rho_*/\rho_{*\,max} = 10^{-0.3j}$, where $\bar{\rho}_{*\,max} = 0.757$, for $j = 0, 1, 2, \cdots, 10$. The maximum density for ρ_* in the final panel is about 80 times larger than the initial value. The thick solid circle in the final panel denotes the apparent horizon.

It should be noted that the supramassive neutron star formed is temporarily stable, but not forever. It will eventually collapse to a black hole as argued in [10].

For models (B) and (C) in which $C \geq 0.14$, a black hole is formed. However, the formation process is slightly different between models (B) and (C). For model (C), a black hole is quickly formed after the first contact of two stars. On the other hand, for model (B), the merged object quasi-radially oscillates for a couple of times after the first contact instead of prompt formation of black hole. Indeed, as shown in Fig. 4, the lapse function at $r = 0$ does not quickly approach to zero in this case. The collapse toward a black hole seems to occur after dissipating the angular momentum by gravitational radiation. The difference with regard to the formation process of black holes is reflected in gravitational waveforms (cf. Fig. 6).

In Fig. 4, we also show time evolution of α at $r = 0$ for simulations with (193,193,96) grid size. It is found that the results do not contradict with those with (293,293,147) grid size, and convergence is achieved fairly well. However, with lower resolution, the numerical dissipation of the angular momentum is larger so that the merger happens earlier than that for higher resolution. Also, high density peaks are captured less accurately because of larger numerical diffusion. For even lower resolution, the merger for model (B) could avoid immediate collapse and form a massive neutron star because of large diffusion. Thus, we deduce that

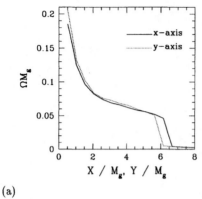
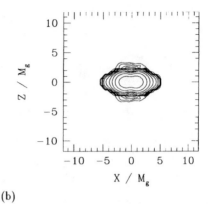

(a) (b)

FIGURE 3. (a) The angular velocity in units of M_g^{-1} along x and y axes and (b) the density contour in x-z slices at $t = 3.04P_{t=0}$ for the merged object of model (A). The contour lines are drawn in the same manner as for Fig. 1.

the total rest mass of model (B) is near a threshold for formation of black hole. To determine the threshold of the total rest mass for black hole formation more accurately, better-resolved numerical simulations are necessary.

In the outcome of model (C), the rest mass outside the apparent horizon is less than 1% of the total. This implies that the rest mass of the disk around the black hole is very small. For model (B), we were not able to determine the apparent horizon before the computation crashed. However, we found that the mass fraction outside spheres of a fixed coordinate radius (e.g., $r = 1.5$ and $3M_g$) is decreasing with time to be very small. Thus, we expect that the mass of the disk is also very small in this case. In the following, we describe the reason for these results.

In Fig. 5, we show the mass spectrum with respect to the specific angular momentum $M_*(j)/M_*$ at $t = 0$ [20] for models (B) and (C). Here, j is the specific angular momentum $(1 + \varepsilon + P/\rho)u_\varphi$ and $M_*(j)$ is defined as

$$M_*(j) = \int_{j'>j} d^3x' \rho_*(x') \quad \text{and} \quad M_*(0) = M_*. \tag{4}$$

It is found that there is no fluid element for which $j/M_g > 1.6$.

As we found in the simulations, quite a large fraction of the fluid elements are swallowed in the black hole. Gravitational radiation carries the energy from the system in particular in the early phase, but it should be less than 1% of M_g according to the quadrupole formula (see, e.g., [13]). Thus, the mass of the black holes is approximately equal to the initial value. On the other hand, the angular momentum may be dissipated by gravitational waves by about 10% of the initial value. These facts imply that $q = J/M_g^2$ should slightly decrease from the initial value to be $q \sim 0.9$ for both models. The specific angular momentum of a test particle in the innermost stable circular orbit around a Kerr black hole of mass M_g and $q = 0.9$

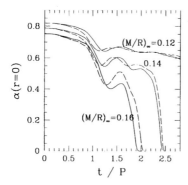
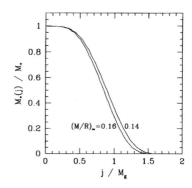

Fig. 4   Fig. 5

FIGURE 4. The lapse function α at $r = 0$ as a function of coordinate time for models (A)–(C). The solid and dashed lines denote the results for simulations with $(293, 293, 147)$ grid resolution and with $(193, 193, 97)$ grid resolution. For smaller scale simulations, the grid size is $116/96$ times larger than that for larger scale simulations. (Namely, the outer boundaries along each axis are located $116L/146$ in these cases.)

FIGURE 5. $M_*(j)/M_*$ as a function of j/M_g for quasiequilibrium configurations (B) and (C).

(0.95) is $\simeq 2.1 M_g$ $(1.9 M_g)$. Therefore, *any fluid element of irrotational binary neutron stars just before the merger does not have large specific angular momentum enough to form a disk around the formed black hole*. For the disk formation, certain transport mechanism of the angular momentum such as hydrodynamic interaction is necessary. Since the black holes are formed in the dynamical timescale of the system, the mechanism has to be very effective to transport the angular momentum by more than 30% in such short timescale. However, such rapid process is unlikely to happen as indicated in the present simulations.

To observe gravitational waveforms, we extract $+$ and $\times$ gravitational waves along z-axis defined as

$$\bar{h}_+ \equiv \frac{\tilde{\gamma}_{xx} - \tilde{\gamma}_{yy}}{2}\left(\frac{r}{M_g}\right)\left(\frac{M}{R}\right)_\infty^{-1}, \text{ and } \bar{h}_\times \equiv \tilde{\gamma}_{xy}\left(\frac{r}{M_g}\right)\left(\frac{M}{R}\right)_\infty^{-1}. \quad (5)$$

In Fig. 6 (a), we show $\bar{h}_\times$ as a function of retarded time near the outer boundary for models (A)–(C) with $(293, 293, 147)$ grid size. To illustrate the effect of the location of outer boundaries, we show $\bar{h}_+$ and $\bar{h}_\times$ for simulations with $(505, 505, 253)$ and $(293,293,147)$ grid sizes for model (C) in Fig. 6(b). Note again that outer boundaries along each axis reside inside the wave zone in the early stage of the simulation. This effect results in the underestimation of the wave amplitude for $t - z_{\rm obs} \lesssim P_{t=0}$ where $P_{t=0}$ is the orbital period at $t = 0$ as shown in Fig. 6(b) (compare results in two different grid sizes). According to a second post Newtonian study [21], the maximum wave amplitude of $\bar{h}_{+,\times}$ at $t \sim 0$ should be ~ 0.7. Thus, with $(293, 293, 147)$ grid size, the wave amplitude is underestimated by a factor of 2. However,

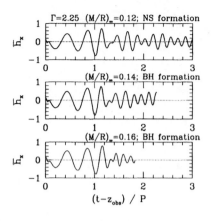 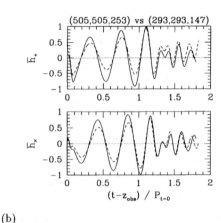

(a) (b)

FIGURE 6. (a) $\bar{h}_\times$ near the outer boundary along the z-axis as a function of retarded time (in units of P) for models (A)–(C) with $(293, 293, 147)$ grid size. P denotes $P_{t=0}$. (b) $\bar{h}_+$ and $\bar{h}_\times$ for model (C) with $(505, 505, 253)$ (solid line) and $(293, 293, 147)$ (dashed line) grid sizes.

with larger grid size in which $L/\lambda_{t=0} \sim 2/3$, the factor of the underestimation is $\sim 10\%$, indicating that fairly accurate waveform could be calculated with slightly larger grid size with $L/\lambda_{t=0} \sim 1$.

For $t - z_{\text{obs}} \gtrsim P_{t=0}$, on the other hand, L is smaller than gravitational wavelength because the characteristic wavelength becomes short after the merger starts. Therefore, the waveforms in the late phase are considered to be fairly accurate. Indeed, the wave amplitudes for two simulations of different grid sizes agree well.

In the case of massive neutron star formation, quasi-periodic gravitational waves of a fairly large amplitude, which are excited due to the non-axisymmetric oscillations of the merged object, are emitted after the merger. Since the radiation reaction timescale is much longer than the dynamical (rotational) timescale of the system, the quasi-periodic waves will be emitted for many rotational cycles.

Even for the case of black hole formation, quasi-periodic gravitational waves are excited due to the non-axisymmetric oscillation of merged objects before collapse to a black hole. Since the computation crashed soon after the formation of the apparent horizon, we cannot draw definite conclusion with regard to gravitational waves in the last phase. However, we can expect that after the formation of a black hole, its quasi-normal modes are excited and gravitational waves will damp eventually. Since the formation timescale of the black hole is different between models (B) and (C) depending on initial compactness of neutron stars, duration of the quasi-periodic waves induced by non-axisymmetric oscillations of the transient merged objects is also different. The wavelength of the quasi-periodic oscillation is ~ 3 times shorter than $\lambda_{t=0}$, and thus, the typical frequency can be estimated as

$$f \simeq 2.0\,\text{kHz}\left(\frac{2.8 M_\odot}{M_g}\right)\left(\frac{X}{0.1}\right)^{3/2}. \tag{6}$$

The amplitude of the quasi-periodic oscillation in the Fourier domain is determined by its duration and hence, depends strongly on the initial compactness of neutron stars and final product. Therefore, by observing the amplitude of this peak, we will be able to obtain information about the compactness of neutron stars before the merger, and final product. From gravitational waves emitted in the inspiraling phase with post Newtonian templates of waveforms [21], mass of two neutron stars, and hence, total mass will be determined [22]. This implies that we could constrain the maximum allowed mass of neutron stars, and hence, nuclear equations of state from the amplitude of the quasi-periodic oscillation emitted by the merged object.

Since the frequency of this Fourier peak is rather high, it will be difficult to detect by first generation, kilo-meter-size laser interferometers such as LIGO. However, the resonant-mass detectors and/or specially designed narrow band interferometers may be available in future, to detect such high frequency gravitational waves. These detectors will provide us a variety of information on neutron star physics.

V SUMMARY

We have performed fully GR simulations of merger of binary neutron stars. As demonstrated in this paper, the simulations are feasible stably and fairly accurately to yield scientific results.

One of the most interesting results found in this work is that the products after merger depend sensitively on the compactness of neutron stars before merger. If the total rest mass of the system is sufficiently (1.3–1.7 times depending on Γ) larger than the maximum rest mass of a spherical star in isolation, a black hole is formed, and otherwise, a massive neutron star is formed. It is noteworthy that the rest mass of the massive neutron star can be significantly larger than the maximum value for a spherical star of identical equation of state. The self-gravity of such high mass neutron stars can be supported by a rapid, differential rotation [11,10]. We also found that the difference of the final products is significantly reflected in the waveforms of gravitational waves, suggesting that detection of gravitational waves of high frequency could constrain the maximum allowed mass of neutron stars.

In the case of prompt black hole formation, the disk mass is found to be very small, i.e., less than 1% of the total rest mass. The main reason is that the specific angular momentum of all the fluid elements in binary neutron stars of irrotational velocity field just before the merger is too small and transport timescale of the angular momentum is not short enough to help the disk formation. It should be noted that this conclusion may hold only for binary neutron stars of equal (or nearly equal) mass. In binaries of a large mass ratio, the conclusion could be modified, because the neutron star of smaller mass may be tidally disrupted before the separation of two stars becomes small and hence, before the angular momentum of the system is not significantly dissipated by gravitational radiation. In this case, many of fluid elements in the neutron star of smaller mass may have a large angular momentum enough to form a disk during the merger. To clarify whether such scenario is promising or not, it is necessary to perform simulations for merger of binary neutron stars of unequal mass.

ACKNOWLEDGMENTS

We would like to thank J. C. Wheeler for inviting M.S. to this meeting to give an opportunity for presenting our works. We also thank T. Baumgarte, E. Gourgoulhon, T. Nakamura, K. Oohara and S. Shapiro for discussions. Numerical computations were performed on the FACOM VPP300/16R, VX/4R and VPP5000 machines in the data processing center of National Astronomical Observatory of Japan.

REFERENCES

1. For example, K. S. Thorne, in *Proceeding of Snowmass 95 Summer Study on Particle and Nuclear Astrophysics and Cosmology*, eds. E. W. Kolb and R. Peccei (World Scientific, Singapore, 1995), p. 398, and references therein.
2. E. S. Phinney, Astrophys. J. **380**, L17 (1991): R. Narayan, T. Piran and A. Shemi, Astrophys. J. **379**, L17 (1991).
3. T. Piran, this volume.
4. S. R. Kulkarni, this volume.
5. K. Oohara, T. Nakamura, and M. Shibata, Prog. Theor. Phys. Suppl. **128**, 183 (1997) for a review before 1997: For recent researches, e.g., J. A. Faber and F. A. Rasio, Phys. Rev. D **62**, 064012 (2000).
6. K. Oohara and T. Nakamura, Prog. Theor. Phys. Suppl. **136**, 270 (1999).
7. J. A. Font, M. Miller, W.-M. Suen, and M. Tobias, Phys. Rev. D **61**, 044011 (2000).
8. M. Shibata, Phys. Rev. D **60**, 104502 (1999).
9. M. Shibata, T. W. Baumgarte, and S. L. Shapiro, Phys. Rev. D **61**, 044012 (2000).
10. T. W. Baumgarte, S. L. Shapiro, and M. Shibata, Astrophys. J. Lett. **528**, L29 (2000).
11. M. Shibata and K. Uryū, Phys. Rev. D **61**, 064001 (2000).
12. M. Shibata, T. W. Baumgarte, and S. L. Shapiro, Astrophys. J. **542**, 453 (2000).
13. M. Shibata, Prog. Theor. Phys. **101**, 1199 (1999).
14. M. Shibata and T. Nakamura, Phys. Rev. D **52**, 5428 (1995).
15. B. J. van Leer, J. Comp. Phys. **23**, 276 (1977).
16. S. L. Shapiro and S. A. Teukolsky, *Black Holes, White Dwarfs, and Neutron Stars*, Wiley Interscience (New York, 1983).
17. C. S. Kochanek, Astrophys. J. **398**, 234 (1992): L. Bildsten and C. Cutler, *ibid*, **400**, 175 (1992).
18. M. Shibata, Phys. Rev. D **58**, 024012 (1998): S. A. Teukolsky, Astrophys. J. **504**, 442 (1998): See also, S. Bonazzola, E. Gourgoulhon and J.-A. Marck, Phys. Rev. D **56**, 7740 (1997): H. Asada, Phys. Rev. D **57**, 7292 (1998).
19. K. Uryū and Y. Eriguchi, Phys. Rev. D **61**, 124023 (2000): See also, E. Gourgoulhon et al., gr-qc/0007028 with regard to other approach.
20. K. Uryū, M. Shibata and Y. Eriguchi, Phys. Rev. D **62**, 104015 (2000).
21. L. Blanchet, B. R. Iyer, C. M. Will and A. G. Wiseman, Class. Quant. Grav. **13**, 575 (1996).
22. C. Cutler and E. E. Flanagan, Phys. Rev. D **49**, 2658 (1994).

Numerical Evolution of the Kruskal Spacetime using the Conformal Field Equations

B. G. Schmidt

Max-Planck-Institute for Gravitational Physics
Albert-Einstein-Institute
14476 Golm — Germany

Abstract. Initial data sets for the conformal field equations which describe spacelike hypersurfaces in the conformally extended Kruskal spacetime will be presented. These data have been evolved using the code for the conformal field equations developed by P. Hübner. The simulations will be described and analyzed.

The conformal approach is a method to calculate global spacetimes up to and beyond infinity. The congformal field equations [H. Friedrich,1,2] are PDEs for a metric g_{ik} and a conformal factor Ω. On the part of the solution where $\Omega > 0$ holds, the metric $\Omega^{-2} g_{ik}$ satisfies Einstein's equations. Points with $\Omega = 0$ are at infinity. In a numerical treatment these points have finite values on the numerical grid.

P.Hübner described in [3] how the conformal field equations can be used numerically to calculate asymptotically flat spacetimes, especially spacetimes containing black holes, without introducing artificial boundaries in the physical space time. In [4] a scheme to numerically evolve data for the conformal Einstein equations is presented and tested on A3-like spacetimes. The paper [5] contains a numerical scheme to calculate 3–dimensional initial data for the conformal field equations for the asymptotically Minkowskian as well as for the toroidal case. In [6] hyperboloidal gravitational wave data without symmetries are evolved up to and beyond timelike infinity, a regular point in the conformal spacetime as shown by H. Friedrich [7]. The radiation at future null infinity and the decay of spacetime curvature near timelike infinity is calculated with high precision.

In this paper we report how these conformal techniques have been used to calculate the evolution of initial data defined by a spacelike hypersurface intersecting both future null infinities of the conformally extended Kruskal spacetime. We take the evolution code in the form developed and take no particular measures to obtain optimal evolution times.

The Kruskal spacetime in the usual U, V coordinates shows (Fig1).

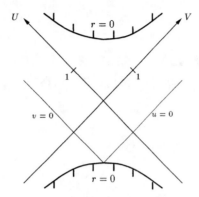

FIGURE 1. Kruskal diagram

This spacetime admits a conformal extension (Fig2).

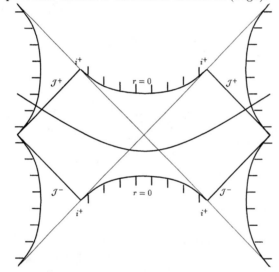

FIGURE 2. Schematic diagram of the conformally extended Kruskal spacetime.

D. Brill, et al [8] determined the inner geometry of hypersurfaces of constant mean curvature $\tilde{k}$ in the Schwarzschild spacetime. Such spacelike slices intersect both future null infinities. The inner geometry of a particular such slice in the conformally extended Kruskal spacetime (conformal factor $\Omega = r^{-1}$) is

$$h = \frac{dw^2}{2 + (2 - \frac{1}{4}w^2) + (2 - \frac{1}{4}w^2)^2} + d\Sigma^2 \qquad (1)$$

This 3-geometry is analytic for $-\infty < w < +\infty$. The "throat" is at $w = 0$,

$\frac{1}{r} = \Omega = 2 - \frac{1}{4}w^2$ and the sections of null infinity are at $w^2 = 8$, $w = \pm 2\sqrt{2}$. The second fundamental form of this hypersurface in the rescaled spacetime vanishes.

The inner geometry and the second fundamental form determine an initial data set for the conformal field equations.

Another initial data set can be constructed as follows: The rescaled Kruskal metric

$$(1 - \frac{1}{r})\frac{4}{r^2}\frac{dU\,dV}{UV} + d\Sigma^2 \tag{2}$$

becomes after a coordinate transformation for $U > -1, V > -1$,

$$U = e^{\frac{1}{\hat{u}}}(\frac{1}{\hat{u}} - 1)\,,\quad V = e^{\frac{1}{\hat{v}}}(\frac{1}{\hat{v}} - 1) \tag{3}$$

$$\frac{4l^2(1-l)}{\hat{u}^2(1-\hat{u})\hat{v}^2(1-\hat{v})}d\hat{u}\,d\hat{v} + d\Sigma^2 \tag{4}$$

with

$$UV = -e^r(r - 1) = e^{\frac{1}{\hat{u}}}(\frac{1}{\hat{u}} - 1)e^{\frac{1}{\hat{v}}}(\frac{1}{\hat{v}} - 1) \tag{5}$$

This metric is analytic at both null infinities near timelike infinity i^+. Fig.3 shows the conformally extended Kruskal spacetime in $\hat{u}, \hat{v}$ coordinates.

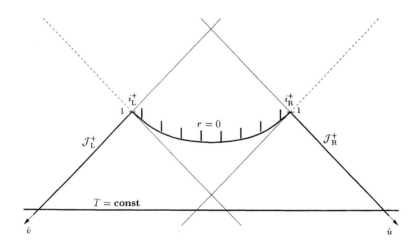

FIGURE 3. Diagram of part of extended Kruskal spacetime

The singularity $r = 0$ is tangent to the horizon in analytic double null coordinates. The hypersurfaces $r = const$ near i^+ behave for $\hat{v} \to 0$ as $\hat{u} = 1 + const\ \hat{v}e^{-\frac{1}{\hat{v}}}$. The singularity of the conformal structure at i^+ is not understood at all!

In this representation it is easy to find spacelike hypersurfaces intersecting both future null infinities. The corresponding initial data are explicit up to an inverse function which has to be calculated numerically.

The first initial data set is evolved on a 160^3 grid. We choose the gauge source functions by "intuition!" Along with the numerical solution we calculate geodesics which allows us to make the following picture. (Fig4).

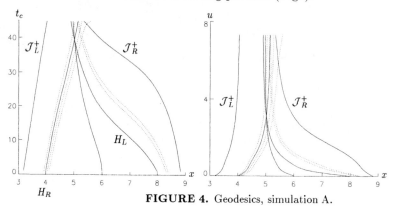

FIGURE 4. Geodesics, simulation A.

We see both null infinities and both horizons in computer coordinates. The dotted lines are null geodesics, the ones starting near 4 escaping to infinity. The Bondi time, u, at infinity was calculated. The run stops at $u = 7.3$ because of lack of resolution. Scalar invariants for the physical metric have been calculated and behave along the null geodesics as expected.

The second initial data set is evolved on a 150^3 grid. We choose the gauge source function from the representation of the exact solution we have. The chosen foliation looks as follows (Fig5)nfigure

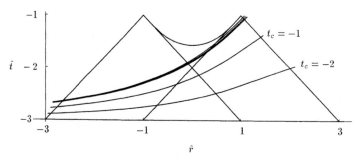

FIGURE 5. Representative leaves of the foliation chosen for simulation B.

Calculation of geodesics gives the following picture of the evolved part of the spacetime.(Fig.6)

The code crashes at Bondi time $u = 7.45$ at a point between the horizon and $Scri_R$ because of large gradients and the lack of resolution. A comparison between

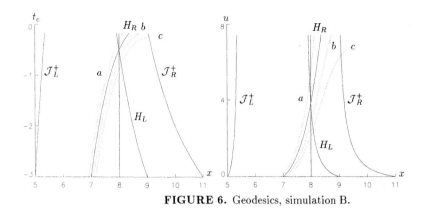

FIGURE 6. Geodesics, simulation B.

the exact solution and the numerical one, which is possible in the gauge we used for the calculation, give an error of $\sim 2\%$.

ACKNOWLEDGEMENTS

These results are largely based on the work by Peter Hübner, who has developed the 3d code for the conformal field equations used here. The simulations have been done by M. Weaver.

REFERENCES

[1] H. Friedrich, Proc.R. Soc. A 357,169,(1981)
[2] H. Friedrich, Proc.R. Soc. A 378,401, (1981)
[3] P. Hübner, Class. Quantum Grav. **16**, 2145, (1999)
[4] P. Hübner, Class. Quantum Grav. **16**, 2823, (1999)
[5] P. Hübner, gr-qc/0010052 (2000)
[6] P. Hübner, gr-qc/0010069 (2000)
[7] H. Friedrich, Commun. Math. Phys.**107**, 587 (1987)
[8] D. Brill, J.M. Cavallo and J.A. Isenberg, J. Math. Phys. **21**, 2789, (1980)

Semiglobal Numerical Calculations of Asymptotically Minkowski Spacetimes

Sascha Husa

Max-Planck-Institut für Gravitationsphysik
Albert-Einstein-Institut
D-14476 Golm, Germany

Abstract. This talk reports on recent progress toward the *semiglobal* study of asymptotically flat spacetimes within numerical relativity. The development of a 3D solver for asymptotically Minkowski-like hyperboloidal initial data has rendered possible the application of Friedrich's conformal field equations to astrophysically interesting spacetimes. As a first application, the whole future of a hyperboloidal set of weak initial data has been studied, including future null and timelike infinity. Using this example we sketch the numerical techniques employed and highlight some of the unique capabilities of the numerical code. We conclude with implications for future work.

The modern treatment of gravitating isolated systems allows to study asymptotic phenomena by local differential geometry. The principal underlying idea, pioneered by Penrose [1], is to work on an unphysical spacetime obtained from the physical spacetime by a suitable conformal compactification. Friedrich has extended this idea in a series of papers [2] to the level of the field equations by rewriting Einstein's equations in a regular way in terms of equations for geometric quantities on the *unphysical spacetime* $\mathcal{M}$. A metric g_{ab} on $\mathcal{M}$ which is a solution of the conformally rescaled equations gives rise to a physical metric $\tilde{g}_{ab} = \Omega^{-2} g_{ab}$, where the conformal factor Ω is also determined by the equations. The physical spacetime $\tilde{\mathcal{M}}$ is then given by $\tilde{\mathcal{M}} = \{p \in \mathcal{M} \,|\, \Omega(p) > 0\}$. These "conformal field equations" render possible studies of the global structure of spacetimes, e.g. reading off radiation at null infinity, by solving regular equations. It is natural to utilize everywhere spacelike slices Σ_t in $\mathcal{M}$ which cross null infinity. On $\tilde{\mathcal{M}}$ the corresponding slices $\tilde{\Sigma}_t$ are similar to the hyperboloid $t^2 - x^2 - y^2 - z^2 = k^2$ in Minkowski spacetime, and are therefore usually referred to as hyperboloidal slices. They are only Cauchy surfaces for the *future* domain of dependence of initial slice of $\tilde{\mathcal{M}}$, we therefore call our studies *semiglobal*.

The conformal field equations address a number of problems with the numerical treatment of isolated systems in general relativity: Radiation quantities can only be defined consistently at null infinity ($\mathcal{J}^+$). Artificial outer boundaries cause ambiguities and stability problems. Resolving the different length scales of radiating sources and the asymptotic falloff is numerically difficult. Our *compactified* grid is allowed to extend beyond the physical part of the unphysical spacetime, the boundary thus can not influence the physics of a simulation. No artificial cutoff at some large distance is required to keep the grid finite and there is no necessity to treat very large length scales (dominating the asymptotic falloff) along with variations on small scales. Including $\mathcal{J}^+$ in the computational domain enables straightforward extraction of radiation quantities involving only well-defined operations without ambiguities. The *symmetric hyperbolicity* of the implemented formulation of the conformal field equations guarantees a well-posed initial value formulation.

The rest of this article is organized in three parts: First the current technology of the solution of the constraints will serve as an example of how some technical problems which are particular to the conformal field equations have been solved successfully. After outlining the evolution algorithm we discuss the evolution of weak data as an example of a situation for which the usage of the conformal field equations is ideally suited: the main difficulties of the problem are directly addressed and solved by using the conformal field equations. These two topics sum up the work of Hübner on 3D numerical relativity with the conformal field equations (see [3], [4] and references cited therein). Finally, we discuss future perspectives for handling strong field situations, where a number of problems appear which are not directly addressed by the conformal field equations, but where we believe that their use will prove beneficial.

The constraints of the conformal field equations (see Eq. (14) of Ref. [5]) are regular equations on the whole conformal spacetime $\mathcal{M}$. However, they have not yet been cast into some standard type of PDE system, such as a system of elliptic PDEs. One therefore resorts to a method where one first obtains data for the Einstein equations – the first and second fundamental forms $\tilde{h}_{ab}$ and $\tilde{k}_{ab}$ induced on $\tilde{\Sigma}$ by $\tilde{g}_{ab}$ with corresponding Ricci scalar and covariant derivative denoted by $^{(3)}\tilde{R}$ and $\tilde{\nabla}_a$. After extending this *subset* of data from $\tilde{\Sigma}$ to Σ the data are then completed by using the conformal constraints. Here we restrict ourselves to a subclass of hyperboloidal slices where initially $\tilde{k}_{ab}$ is pure trace, $\tilde{k}_{ab} = \frac{1}{3}\tilde{h}_{ab}\tilde{k}$. The momentum constraint $\tilde{\nabla}^b \tilde{k}_{ab} - \tilde{\nabla}_a \tilde{k} = 0$ then implies $\tilde{k} = \text{const} \neq 0$. We always set $\tilde{k} > 0$. In order to reduce the Hamiltonian constraint

$$^{(3)}\tilde{R} + \tilde{k}^2 = \tilde{k}_{ab}\tilde{k}^{ab}$$

to *one* elliptic equation of second order, we use the standard Lichnerowicz ansatz

$$\tilde{h}_{ab} = \bar{\Omega}^{-2}\phi^4 h_{ab}.$$

The free "boundary defining" function $\bar{\Omega}$ is chosen to vanish on a 2-surface $\mathcal{S}$ – the boundary of $\bar{\Sigma}$ and initial location of $\mathcal{J}^+$ – with non-vanishing gradient on $\mathcal{S}$. The topology of $\mathcal{S}$ is chosen as spherical for asymptotically Minkowski spacetimes. Let h_{ab} be a metric on Σ, with the only restriction that its extrinsic 2-curvature induced by h_{ab} on $\mathcal{S}$ is pure trace, which is required as a smoothness condition [6].

With this ansatz $\tilde{h}_{ab}$ is singular at $\mathcal{S}$, indicating that $\mathcal{S}$ represents an infinity. The Hamiltonian constraint then reduces to the Yamabe equation for the conformal factor ϕ:

$$4\bar{\Omega}^2 \nabla^a \nabla_a \phi - 4\bar{\Omega}(\nabla^a \bar{\Omega})(\nabla_a \phi) - \left(\frac{1}{2}{}^{(3)}R\,\bar{\Omega}^2 + 2\bar{\Omega}\Delta\bar{\Omega} - 3(\nabla^a\bar{\Omega})(\nabla_a\bar{\Omega})\right)\phi = \frac{1}{3}\tilde{k}^2\phi^5.$$

This is an "elliptic" equation with a principal part which vanishes at the boundary $\mathcal{S}$ for a regular solution. This determines the boundary values as $9(\nabla^a\bar{\Omega})(\nabla_a\bar{\Omega}) = \tilde{k}^2\phi^4$. Existence and uniqueness of a positive solution to the Yamabe equation and the corresponding existence and uniqueness of regular data for the conformal field equations using the approach outlined above have been proven by Andersson, Chruściel and Friedrich [6].

The *complete* set of data for the conformal field equations is obtained from the conformal constraints via algebra and differentiation. This however involves divisions by the conformal factor $\Omega = \bar{\Omega}\phi^{-2}$, which vanishes at $\mathcal{S}$. In order to obtain a smooth error for this operation, the numerically troublesome application of l'Hospital's rule to $g = f/\Omega$ is replaced by solving an elliptic equation of the type $\nabla^a\nabla_a(\Omega^2 g - \Omega f) = 0$ for g. For the boundary values $\Omega^2 g - \Omega f = 0$, the unique solution is $g = f/\Omega$. For technical details see Hübner [3]. The Yamabe equation for ϕ and the linear elliptic equations arising from the division by Ω are solved by pseudo-spectral collocation (PSC) methods (see e. g. [7]). Fast Fourier transformations converting between the spectral and grid representations are performed using the FFTW library [8]. Nonlinearity in the Yamabe equation is dealt with by a multigrid Newton method (for details and references see [3], the resulting linear equations are solved with the AMG library [9], which implements an algebraic multigrid technique. The PSC method restricts the choice of gridpoints to be consistent with a simple choice of basis functions. Thus $\mathcal{S}$ is required to be a coordinate isosurface and the elliptic equations are solved in spherical coordinates. The polar coordinate singularities are taken care of by only using regular quantities in the computation, i.e. Cartesian tensor components, and by not letting any collocation points coincide with coordinate singularities. In order to extend the initial data to the Cartesian time evolution grid on the extended hyperboloidal slice Σ, the spectral representations are used, which define data even outside of the physical domain where the constraints have been solved. The constraints will be violated in the unphysical region, but since this region is causally disconnected from the physical interior by $\mathcal{J}^+$, the errors in the physical region converge to zero with the discretization order. For numerical purposes the coefficient functions of the spectral representation are modified in the unphysical region.

Time evolution of the conformal field equations (in particular here this is the system Eq. (13) of Ref. [5]) is carried out by a 4th order method of lines with standard 4th order Runge-Kutta time evolution. Spatial derivatives are approximated by a symmetric fourth order stencil. To ensure stability, dissipative terms of higher order were added consistently with 4th order convergence, as discussed in section

6.7 of Ref. [10]. In the unphysical region of $\mathcal{M}$ near the boundary of the grid, a "transition layer" is used to transform the conformal Einstein equations to simple advection equations. A trivial copy at the outermost gridpoint yields a simple and stable outer boundary condition.

The main result so far, obtained by Hübner, is the evolution of weak data which evolve into a regular point i^+ representing future timelike infinity. This result illustrates a theorem by Friedrich [11], who has shown that for sufficiently weak initial data there exists a regular point i^+ of $\mathcal{M}$. The complete future of (the physical part of) the initial slice was reconstructed in a finite number of computational time steps, and the point i^+ has been resolved within a *single* grid cell. For these evolutions very simple choices of the gauge source function have proven sufficient: a zero shift vector was used, the lapse was set to $N = \sqrt{\det h}$, and the scalar curvature of the unphysical spacetime was set to zero. The initial conformal metric is chosen in Cartesian coordinates as

$$h = \begin{pmatrix} 1 + \frac{1}{3}\bar{\Omega}^2(x^2 + 2y^2) & 0 & 0 \\ 0 & 1 & 0 \\ 0 & 0 & 1 \end{pmatrix}$$

and the boundary defining function as $\bar{\Omega} = \frac{1}{2}(1 - (x^2 + y^2 + z^2))$. The extraction of physics is largely based on the integration of geodesics concurrently with the evolution. This is carried out with the same 4th order Runge-Kutta scheme that is used in the method of lines. Null geodesics along $\mathcal{J}^+$ are used to construct a Bondi system at $\mathcal{J}^+$ and to compute the news function and Bondi mass. To illustrate the results we show the behavior of geodesics in the numerically generated spacetime. First, in Fig. 1 we show three timelike geodesics originating with different initial velocities at the same point $(x_0, y_0, z_0) = (\frac{1}{2\sqrt{3}}, \frac{1}{2\sqrt{3}}, \frac{1}{2\sqrt{3}})$ meeting a generator of $\mathcal{J}^+$ at i^+. Fig. 2 shows the oscillations induced by gravitational waves in the zero velocity geodesic starting out at (x_0, y_0, z_0) (Fig. (6) of Ref. [4] shows the same plot using coordinate time instead of proper time).

Future efforts will focus on developing the techniques to evolve strong data without symmetries, e.g. to describe generic black hole spacetimes. Here many problems are not yet solved, in particular issues associated with choosing the gauge source functions, the treatment of the appearance of singularities, and the limitations of computer resources for 3D calculations. These well known problems plaguing 3D numerical relativity will have to be addressed and solved in the conformal approach in order to harvest its benefits for studying the (semi)global structure of spacetimes.

A first step toward the study of black holes with the conformal field equations is to obtain suitable initial data. In the Cauchy approach topologically nontrivial data are easy to produce by compactification methods (see e.g. [12], [13] and references cited therein) where the topology of the computational grid is not influenced by the number of asymptotic regions. In the hyperboloidal case the adding of $\mathcal{J}^+$'s as suggested e.g. in [3] *does* change the computational domain and leads to technical problems, for which Hübner has suggested solutions [3]. An alternative is to use

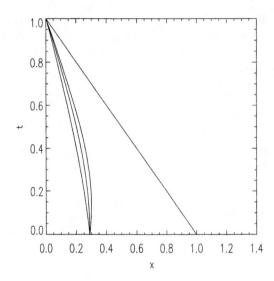

FIGURE 1. Three timelike geodesics (starting at $x = \frac{1}{2\sqrt{3}}$) meet a generator of $\mathcal{J}^+$ (starting at $x = 1$) at future timelike infinity i^+.

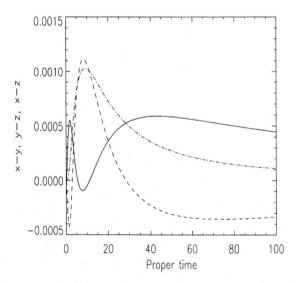

FIGURE 2. The differences $x - y$, $x - z$, and $y - z$ along a timelike geodesic show the spacetime distortion due to gravitational waves. In flat space they would vanish for symmetry reasons. The cutoff in time was chosen to better resolve the early time structure.

regular initial data containing one or more apparent horizons. Such data could be produced by parameter studies with the current code and would in some sense be more physical, since they do not require the existence of "eternal" black holes, and in principle should allow initial data which have apparent horizons and singularities in the future, but not in the past. An important question is whether such data are qualitatively different from data with "topological" black holes in the region outside of the event horizon. In order to handle the singularities inside of black holes with the code, a strategy successfully employed by Hübner in the spherically symmetric case was to handle floating point exceptions and then continue past the formation of singularities as permitted by causality [14]. However, the probably chaotic structure of the singularity and the difficulties of evolving with a time step as large as permitted by causality may turn out prohibitive in 3D. Excision of the singularity or a modification of the equations inside the black hole (analogous to the way they are changed in the unphysical region beyond $\mathcal{J}^+$) are possible alternatives.

ACKNOWLEDGEMENTS

The author thanks H. Friedrich, B. Schmidt and J. Winicour for helpful discussions, and P. Hübner and M. Weaver for letting him use their codes, explaining their results and providing general support in order to take over this project.

REFERENCES

1. R. Penrose, *Phys. Rev. Lett.* **10**, 66–68 (1963).
2. H. Friedrich, "Einstein's Equation and Geometric Asymptotics", in *Proceedings of the GR-15 conference*, edited by N. Dadhich and J. Narlikar, IUCAA, 1998.
3. P. Hübner, *Los Alamos Preprint Archive*, gr-qc/0010052.
4. P. Hübner, *Los Alamos Preprint Archive*, gr-qc/0010069.
5. P. Hübner, *Class. Quantum Grav.* **16**, 2145–2164 (1999).
6. L. Andersson, P. T. Chruściel, and H. Friedrich, *Comm. Math. Phys.* **149**, 587 (1992).
7. A. Quarteroni and A. Valli, *Numerical Approximation of Partial Differential Equations*, Springer Series in Computational Mathematics 23, Springer, Berlin, 1997.
8. M. Frigo and S. G. Johnson, "Fftw: An adaptive software architecture for the fft" in *1998 ICASSP conference proceedings, Vol. 3*, ICASSP, 1998, p. 1381.
9. J. W. Ruge and K. Stüben, "Algebraic multigrid", in *Multigrid Methods*, edited by S. F. McCormick, SIAM, Philadelphia, pp. 73–130, 1987.
10. B. Gustafsson, H.-O. Kreiss, and J. Oliger, *"Time Dependent Problems and Difference Methods"*, Pure and Applied Mathematics. Wiley, New York, 1995.
11. H. Friedrich, *Commun. Math. Phys.* **107**, 587–609 (1986).
12. S. Husa, *Los Alamos Preprint Archive*, gr-qc/9811005.
13. S. Dain, *Los Alamos Preprint Archive*, gr-qc/0012023.
14. P. Hübner, *Phys. Rev. D* **53**, 701–721 (1996).

Simulations of Black Hole Binaries: Providing Initial Data

Pedro Marronetti and Richard A. Matzner

Center for Relativity. The University of Texas at Austin, Austin TX 78712-1081

Abstract.
We present the first full numerical solutions of the initial value problem of two black holes based on a Kerr-Schild spacetime slicing. These new solutions provides more physically realistic solutions than the initial data based on conformally flat metric/maximal slicing methods. The singularity/inner boundary problems are circumvented by a new technique that allows the use of an elliptic solver on a Cartesian grid where no points are excised, simplifying enormously the numerical problem.

INTRODUCTION

The computation of gravitational wave production from the interaction and merger of compact astrophysical objects is an analytical and computational challenge. These calculations would be able to provide both a predictive and an analytical resource for the gravitational-wave interferometric detectors soon to be online. We concentrate on the case of binary black hole mergers [1].

In order to carry out such simulations, we must construct initial data sets representing binary black hole systems. These data sets should not only satisfy the Einstein constraints but also carry the desired physical content. Binary black hole initial data sets are the subject of this paper. The problem of solving the Hamiltonian and momentum constraints for two black holes has been addressed in the past by several groups (see [2] and references therein). The methods in most frequent use are based on an approach which chooses maximal spatial hypersurfaces ($K = 0$) and takes the spatial 3-metric to be conformally flat ($g_{ij} = \phi^4 \delta_{ij}$, ϕ being the conformal factor). Under these conditions, the Hamiltonian constraint can be decoupled from the momentum constraint. Analytical solutions for the extrinsic curvature K_{ij} for holes with specific linear momenta can be found [3]. This simplification leaves only one elliptic equation for ϕ, which is derived from the Hamiltonian constraint. The inner boundaries (the throats of the black holes) are usually dealt with by imposing an isometry condition between two identical asymptotically flat spatial slices, joined by an Einstein-Rosen bridge, though other boundary conditions are sometimes used. Unfortunately, the numerical solution of the equation for ϕ presents a technical challenge at the inner boundaries. Brandt and Brügmann [4] simplified this problem by compactifying the internal asymptotically flat regions to obtain a domain without inner boundaries. However, the main disadvantage of these methods is not numerical, but related to the physical interpretation of black-hole spaces described through a conformally flat 3-metric. There are no space slices for which the spatial 3-metric of a single Kerr (non-zero spin) black hole can be written in a conformally flat way [5], and recent work on sequences of initial data sets for circular orbits casts some doubt on the physical realism of conformally flat approaches to black hole binaries [6]. One way to overcome this problem is to use a Kerr-Schild slicing of spacetime as described in the next section.

SLICING AND BACKGROUND FIELDS

Under the 3+1 formulation of General Relativity, the construction of initial data requires solving the Hamiltonian and momentum constraints:

$$R + \frac{2}{3}K^2 - A_i^j A_j^i = 0$$
$$\nabla_j A_i^j - \frac{2}{3}\nabla_i K = 0 , \qquad (1)$$

respectively. Above, R is the 3-dimensional Ricci scalar constructed from the spatial 3-metric g_{ij}, and

$$K_{ij} = A_{ij} + \frac{1}{3}g_{ij}K$$

is the extrinsic curvature tensor. K and A_{ij} are the trace and trace-free parts of K_{ij}, respectively. Covariant differentiation with respect to g_{ij} is denoted by ∇_i. Spacetime indices will be denoted by Greek letters and spatial indices by Latin letters.

Our work is based on descriptions of black holes in ingoing Eddington-Finkelstein coordinates. This choice is motivated by the fact that, in these coordinates, surfaces of constant time "penetrate" the event horizon. The essence of black hole excision is the removal of the singularity while preserving the integrity of the spacetime accessible to observers outside the black hole. This is only possible if the excised region is fully contained within the event horizon, thus the need to have access to the interior of the black holes.

The 4-dimensional form of the Kerr-Schild spacetime [7]:

$$ds^2 = -dt^2 + dx^2 + dy^2 + dz^2 + 2H(x^\alpha)(l_\lambda dx^\lambda)^2 \qquad (2)$$

describes isolated single black holes. Here the scalar function H has a known form, and l_λ is an ingoing null vector congruence associated with the solution. For instance for the Schwarzschild solution, $H(x^\alpha) = M/r$, and $l_\lambda = (1; x^i/r)$, i.e. an inward pointing null vector with unit spatial part.

Initial data setting for multiple black hole spacetimes using the method described by Matzner, Huq, and Shoemaker [8] begins by specifying a conformal spatial metric which is a straightforward superposition of two Kerr-Schild single hole (spatial) metrics:

$$\tilde{g}_{ij}dx^i dx^j = \delta_{ij}dx^i dx^j + 2\,_1H(x^\alpha)(_1l_j dx^j)^2 \\ + 2\,_2H(x^\alpha)(_2l_j dx^j)^2 . \qquad (3)$$

The fields marked with the pre-index 1 (2) correspond to an isolated black hole with specific angular momentum $\mathbf{a_1}$ ($\mathbf{a_2}$) and boosted with velocity $\mathbf{v_1}$ ($\mathbf{v_2}$). The superposition of the conformal momenta is defined as follows: The extrinsic curvature for a single hole (say hole 1)

$$_1K_{ij} = (_1\partial_j\beta_i + _1\partial_i\beta_j - 2\,_1\Gamma^k_{ij}{}_1\beta_k - _1\partial_t g_{ij})/(2\,_1\alpha) \;,$$

is converted to a mixed-index object,

$$_1K_i{}^j = _1g^{nj}\,_1K_{in} \;.$$

The trace of $\tilde{K}$ is calculated as the sum of the corresponding traces:

$$\tilde{K} = _1K_i{}^i + _2K_i{}^i \;,$$

and the transverse-traceless part of the extrinsic curvature $\tilde{A}_{ij}$ as

$$\tilde{A}_{ij} = \tilde{g}_{n(i}\left(_1K^n_{j)} + _2K^n_{j)} - \frac{1}{3}\delta^n_{j)}\tilde{K}\right) , \qquad (4)$$

where the parenthesis in the subscripts denote symmetrization in i and j.

In order to further reduce the residual errors, Marronetti et al. [9] proposed a variation of the superposition method that preserves the simplicity of being analytical. Essentially, the method consists of multiplying "attenuation" functions into the recipe of the previous section. The new approximate metric $\tilde{g}^A_{ij}$, trace $\tilde{K}^A$, and tensor $\tilde{A}^A_{ij}$ take the form:

$$\begin{aligned}
\tilde{g}^A_{ij} &= \delta_{ij} + 2\,_1B\,_1H\,_1l_i\,_1l_j + 2\,_2B\,_2H\,_2l_i\,_2l_j \;,\\
\tilde{K}^A &= _1B\,_1K_i{}^i + _2B\,_2K_i{}^i \;,\\
\tilde{A}^A_{ij} &= \tilde{g}^A_{n(i}\left(_1B\,_1K^n_{j)} + _2B\,_2K^n_{j)} - \frac{1}{3}\delta^n_{j)}\tilde{K}\right) ,
\end{aligned} \qquad (5)$$

The purpose of the attenuation function B is to minimize the effects due to a given hole on the neighborhood of the other hole. For instance, the attenuation function $_1B$ is unity everywhere except in the vicinity of hole-2 where it rapidly vanishes so the metric and extrinsic curvature there are effectively that of a single black hole. The attenuation functions with this property can be constructed in a number of different ways. An example of an attenuation function is

$$_1B = 1 - e^{-r^4/\sigma^4} \;, \qquad (6)$$

where

$$r^2 = \frac{1}{2}(\rho^2 - a^2) + \sqrt{\frac{1}{4}(\rho^2 - a^2)^2 + a^2 z^2} \;, \qquad (7)$$

$$\rho = \sqrt{2\gamma^2(x - _2x)^2 + (y - _2y)^2 + (z - _2z)^2} \;. \qquad (8)$$

Here $_2x^i$, a and $_2\gamma$ denote the coordinate location, specific angular momentum and boost factor of hole-2, respectively. σ represents a free parameter of the attenuation function. An expression similar to (6) for $_2B$ is obtained by reversing the labels.

The attenuation function was chosen amongst a few different types for its simplicity and better performance. Since the constraints involve second order derivatives, it is important to pick attenuation functions that vanish up to second order derivatives, so "pure" single black hole solutions can be obtained in the neighborhood of each hole.

NUMERICAL METHOD

Following the conformal decomposition presented by York and collaborators [10], we relate the physical metric g_{ij} and the trace-free part of the extrinsic curvature A_{ij} to the background fields through a conformal factor:

$$g_{ij} = \phi^4 \tilde{g}_{ij}, \quad A^{ij} = \phi^{-10} \tilde{A}^{ij} . \tag{9}$$

In order to find a solution to the four constraint Eqs. (1), we add a longitudinal part to the extrinsic curvature A^{ij}:

$$A^{ij} \equiv \phi^{-10}(\tilde{A}^{ij} + (\tilde{l}w)^{ij}) , \tag{10}$$

where w^i is a vector potential to be solved for and

$$(\tilde{l}w)^{ij} \equiv \tilde{\nabla}^i w^j + \tilde{\nabla}^j w^i - \frac{2}{3}\tilde{g}^{ij}\tilde{\nabla}_k w^k . \tag{11}$$

Plugging Eqs. (5)-(11), into the Hamiltonian and momentum constraints (1), we obtain four coupled elliptic equations for the fields ϕ and w^i:

$$\begin{aligned}
\tilde{\nabla}^2 \phi &= (1/8)(\tilde{R}\phi + \frac{2}{3}\tilde{K}^2\phi^5 - \\
&\quad \phi^{-7}(\tilde{A}^{ij} + (\tilde{l}w)^{ij})(\tilde{A}_{ij} + (\tilde{l}w)_{ij})) \\
\tilde{\nabla}_j(\tilde{l}w)^{ij} &= \frac{2}{3}\tilde{g}^{ij}\phi^6\tilde{\nabla}_j K - \tilde{\nabla}_j \tilde{A}^{ij} .
\end{aligned} \tag{12}$$

To solve the elliptic Eqs. (12) we use an adaptation of an multigrid elliptic solver developed for the solution of the initial value problem of neutron-star binaries [11]. The four elliptic equations are solved consecutively in an iteration loop. The iteration starts with a trivial initial guess ($\phi = 1$, $w^i = 0$) and it relaxes the fields until the variation from one iteration step to the next falls below some fraction of the truncation error. The sources for these elliptic Eqs. consist mostly of the residuals of the constraint equations as defined in [9]. The use of attenuation functions eliminates the singular behavior of these sources near the ring singularities, simplifying the numerical implementation of the elliptic solver. Instead of using excision techniques around the singularities [1], we modified a standard elliptic solver to "ignore" (i.e. leave the fields with the initial values) the grid points in a small volume embedding the singularity. This "inner" region

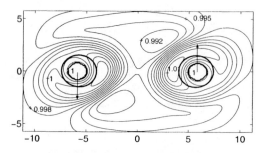

FIGURE 1. Contour plot of the conformal factor ϕ. The inner thick lines limit the "inner" region where the initial value $\phi = 1$ is preserved. The outer thick lines show the apparent horizons and the arrows the direction of the boosts.

is defined monitoring the value of the background 3-dimensional Ricci scalar, and is chosen to be smaller than the excision volume used in current numerical simulations, making the initial value data suitable for the evolutionary codes. The choice of the "inner" region values is based on the fact that arbitrarily close to the ring singularity, the background fields recover the single-hole values which are an exact solution to Eqs. (1). We use Robin conditions for the outer boundaries, guaranteeing that ϕ and w^i fall off as $constant + O(1/r)$ at infinity.

RESULTS AND CONCLUSIONS

The first results obtained using the method described above were presented by Marronetti and Matzner [12]. We studied a binary system composed of two black holes with mass m in a hyperbolic encounter configuration with parameters $\mathbf{r}_1 = (5.75\ m, 0, 0)$, $\mathbf{v}_1 = (0, 0.5\ c, 0)$, and $\mathbf{a}_1 = (0, 0, 0.5\ m)$ corresponding to black hole 1 coordinate position, velocity, and specific angular momentum. The parameters for black hole two are $\mathbf{r}_2 = -\mathbf{r}_1$, $\mathbf{v}_2 = -\mathbf{v}_1$, and $\mathbf{a}_2 = \mathbf{a}_1$. We estimated the ADM mass corresponding to the background fields and to the full numerical solution, obtaining 2.34 m (note that $(1-v^2)^{-1/2} = 1.155$) and 2.18 m respectively. This was done by integrating over the boundary surface of the computational grid. We also obtained a background total angular momentum estimation of 7.99 m^2, consistent with analytical estimates.

Figure 1 (2) is a contour plot of the conformal factor ϕ (component of the vector potential w^x). The inner thick lines limit the "inner" region where the initial value $\phi = 1$ is preserved, while the outer lines show the apparent horizons. The resulting horizon areas for the background system and the full solution were $2 \times 39.9\ m^2$ and $2 \times 38.4\ m^2$ respectively, which are consistent with the value of 45.7 m^2 obtained for a single-hole configuration.

This solution to the initial value problem presents a different approach than the conformally flat/maximal slicing methods, and allows us to specify more directly the physical content of the data. It also entails a computational simplification on the treatment of the singularities on the working grid. However, the physical aspects of these data sets can

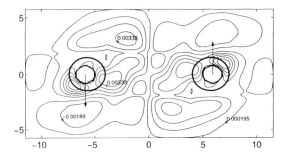

FIGURE 2. Contour plot of the x component of the vector potential w^i.

be further refined, for instance by providing background fields based on Post-Newtonian expansions or with a more astrophysically realistic initial data for the case of black holes in circular orbits, by deriving the extrinsic curvature K_{ij} from the presence of a Killing vector $\xi = \partial_t + \omega \partial_\phi$, instead of using the superposition of two boosted holes [13].

This work was supported by NSF grant PHY9800722 and PHY9800725 to the University of Texas at Austin.

REFERENCES

1. S. Brandt et al., Phys. Rev. Lett. **85**, 5496 (2000) [gr-qc/0009047].
2. G. B. Cook, "Living Reviews in Relativity" (2000) [gr-qc/0007085].
3. A. D. Kulkarni, L. C. Shepley and J. W. York Jr. Phys. Lett. **96A**, 228 (1983).
4. S. Brandt and B. Brügmann, Phys. Rev. Lett. **78**, 3606 (1997) [gr-qc/9703066].
5. A. Garat and R. H. Price, Phys. Rev. **D61**, 124011 (2000) [gr-qc/0002013].
6. H. P. Pfeiffer, S. A. Teukolsky and G. B. Cook, Phys. Rev. **D62**, 104018 (2000) [gr-qc/0006084].
7. R. P. Kerr and A. Schild, "Some Algebraically Degenerate Solutions of Einstein's Gravitational Field Equations" *Applications of Nonlinear Partial Differential Equations in Mathematical Physics*, Proc. of Symposia B Applied Math., Vol. XV11, (1965).
   "A New Class of Vacuum Solutions of the Einstein Field Equations," *Atti del Convegno Sulla Relativita Generale: Problemi Dell'Energia E Onde Gravitazionale*, G. Barbera, ed.,(1965).
8. R. A. Matzner, M. F. Huq and D. Shoemaker, Phys. Rev. **D59**, 024015 (1999) [gr-qc/9805023].
9. P. Marronetti, M. Huq, P. Laguna, L. Lehner, R. A. Matzner and D. Shoemaker, Phys. Rev. **D62**, 024017 (2000) [gr-qc/0001077].
10. J. W. York and T. Piran "The Initial Value Problem and Beyond", *Spacetime and Geometry: The Alfred Schild Lectures*, R. A. Matzner and L. C. Shepley Eds. University of Texas Press, Austin, Texas. (1982).
11. P. Marronetti, G. J. Mathews and J. R. Wilson, Phys. Rev. **D58**, 107503 (1998) [gr-qc/9803093]. P. Marronetti, G. J. Mathews and J. R. Wilson, Proceedings of 19th Texas Symposium on Relativistic Astrophysics: Texas in Paris, Paris, France. J. Paul, T. Montmerle, and E. Aubourg Eds. Nuclear Physics B, Proc. Suppl. **80** (2000). [gr-qc/9903105]. P. Marronetti, G. J. Mathews and J. R. Wilson, Phys. Rev. **D60**, 087301 (1999) [gr-qc/9906088].
12. P. Marronetti and R. A. Matzner, Phys. Rev. Lett. **85**, 5500 (2000) [gr-qc/0009044].
13. P. Marronetti and R. A. Matzner, To appear in the proceedings of 9th Marcel Grossmann Meeting on Recent Developments in Theoretical and Experimental General Relativity, Gravitation and Relativistic Field Theories (MG 9), Rome, Italy, 2-9 Jul 2000 [gr-qc/0101063].

The Lazarus project: Plunge Waveforms from Inspiralling Black Holes

J. Baker, B. Brügmann, M. Campanelli, C. Lousto, and R. Takahashi

Albert-Einstein-Institut, Max-Planck-Institut für Gravitationsphysik, Am Mühlenberg 1, D-14476 Golm, Germany

Abstract. We study the coalescence of binary black holes from the innermost stable circular orbit down to the final single rotating black hole. We use a technique that combines the full numerical approach to solve Einstein equations, applied in the truly nonlinear regime, and linearized perturbation theory around the final distorted single black hole at later times. We thus produce an estimate for the plunge radiation with a non negligible signal lasting for over $t \sim 100M$, and we obtain estimates of the total gravitational radiated energy and angular momentum during this process, plunging time, and waveforms.

Motivated by the desire to provide expectant gravitational wave observers with some long-awaited estimate of the full merger waveforms, and to prepare the arena for future, more advanced numerical simulations, we have pursued a hybrid approach to the problem, called "the Lazarus project." Our approach has recently provided the first, astrophysically plausible, theoretical estimates for the gravitational radiation waveforms and energy to be expected from binary black hole mergers based on a full non-linear treatment of the plunge dynamics [1].

The underlying idea of the Lazarus project is very simple: apply appropriate "far limit", full numerical and close limit treatments in sequence[2, 3]. In this way we can shift the finite time interval of full non-linear numerical evolution to cover the stage of the dynamics where no perturbative approach is applicable, and still derive the complete black hole ring-down and the propagation of radiation into the wave zone with a close limit perturbative treatment. The perturbative model not only allows an inexpensive and stable continuation of the evolution (which is then allowed to rise and live again like the biblical Lazarus), but also supplies a clear interpretation of the dynamics not manifest in the generic numerical simulation. Although the advantage of this approach may seem obvious, one has to face the problem of engineering a transition interface between the full numerical simulations and the close limit approach, or wait until the brute force of numerical simulation can obviate the problem. However, the continued lack of any purely numerical results at a time when theoretical information is urgently requested by the observational community has led us to tackle the transition interface problem to thereby begin building a theoretical description of coalescing binary black hole systems.

To begin getting results even before a practical interface between the post-Newtonian method and numerical simulations is available we start our simulations with initial data derived from an alternative description of the ISCO, based on effective potentials in a family of conformally flat initial data. For the equal mass, non-spinning black holes, the ISCO corresponds to a separation L of $4.9M$, and a transverse momentum for each hole of $0.335M$, implying a total angular momentum of $0.77M^2$. We then choose maximal slicing and vanishing shift, to fix the gauge in a manner practical for both the numerical simulation and the close-limit interface. The full numerical simulation of Einstein's equations then proceeds as a time succession of values on a three-dimensional numerical grid for the spatial metric and extrinsic curvature variables, updated by using the standard evolution equations applied in the Grand Challenge efforts. These simulations produce highly accurate evolutions for about 15M, but are then limited by sudden numerical instabilities.

Once the gravitational field has evolved from the ISCO stage to the point where the dynamics are dominated by the emerging final black hole a transition (on some space-like hypersurface) to the close limit perturbative evolution must be made. If the binary system has reached a regime where all further evolution can be described by the linearized Einstein equations (such as in the perturbative evolution), the final results should be independent of adjustments to the transition time T. In practice there is a limited window in which the transition can be effective. If that transition time is taken to be too early, the perturbative close-limit treatment will not yet be a good approximation for the radiative dynamics. On the other hand, if the transition time is too late, numerical inaccuracies will spoil the results. To evaluate our results we carefully study our waveforms' dependence on the transition time, and also make extensive use of the speciality invariant S, a curvature invariant that has the value 1 for the single Kerr black hole spacetime, provides an indication of the dynamical processes in a black hole forming space time. These are new tools for the application of numerical relativity technology to new physics.

In the subsequent perturbative evolution, the dynamical variables reduce to a single complex scalar field, the Teukolsky function, ψ_4 that is directly related to the outgoing gravitational radiation, and obeys a rather simple linear hyperbolic equation on a fixed Kerr background black hole. However, because the fixed background Kerr black hole is written in some preferred, analytic, coordinate system, one must in general reassign spacetime coordinates for the numerically evolved spacetime to be close to those used to identify the perturbations of a single Kerr spacetime. There is, of course, no unique way of assigning coordinates and extracting perturbations, but the Teukolsky function's first order gauge invariance makes it insensitive to the small variations in our coordinate prescription.

We have successfully tested our method on spinning Kerr black holes [3], and head on collisions [2]. We approach the ISCO data through a sequence of initial data sets constructed by keeping the black holes at a fixed ISCO separation $L = 4.9M$, but varying the magnitude of the linear momentum, from zero, corresponding to a head-on collision to its full ISCO value, $P = .335M$. The results of these "P-sequence" studies are reported in Fig. 1, which show that: (i) The minimal time of full numerical evolution T needed to switch to perturbation theory initially grows roughly linearly with increasing momentum. For the ISCO case $T = 11M$–$15M$ of numerical evolution is required. (ii) The total radiated gravitational energy grows quadratically with the momentum P for small values, as expected.

The estimate for the total energy of radiation after ISCO is 3–4% of the total (ADM) mass directed primarily along the rotation axis. This is larger than the 1.4% obtained by

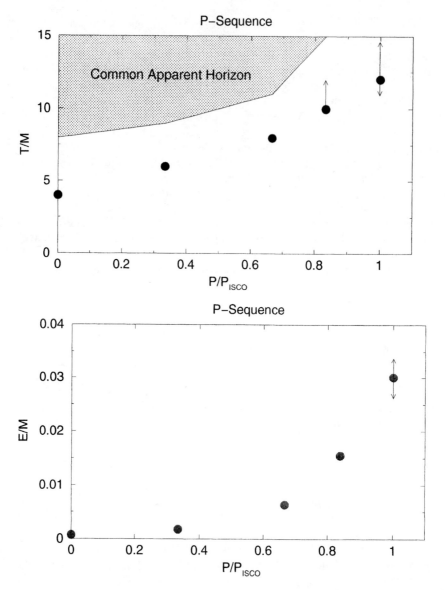

FIGURE 1. Linearization time and total radiated energy for the P-sequence. The error bars are based on transition time dependence.

extrapolating PN results and the 1–2% obtained by extrapolation of the close limit. Our results predict that 1 – 2% of the total angular momentum is lost during the plunge and ring-down. After radiation we are left with a Kerr black hole with $M_{Kerr} \approx 0.97 M_{Init}$ and $J_{Kerr} \approx 0.8 M_{Kerr}^2$.

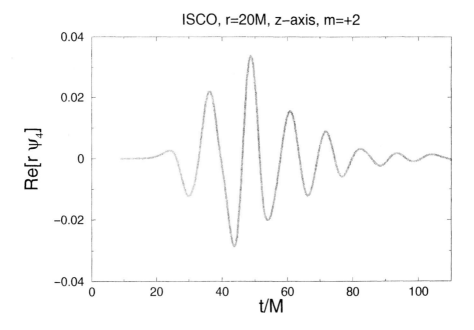

FIGURE 2. ISCO waveform for one of the two polarizations of the gravitational field, for a transition time of $T = 10M$, as seen by an observer located at a distance $r = 20M$ along the orbital axis.

Fig. 2 shows the waveform calculated for black holes mergers starting at the ISCO. Our analysis indicates that the waveform is quite robust up to $t \simeq 60M$ and probably a good estimate afterward. In the frequency decomposition of the waveform signals one can see dominating components close to the most weakly damped quasi-normal oscillation frequencies of the final Kerr black hole. For a system with total mass $35M_\odot$, the two principal frequency components correspond to frequencies of roughly $600Hz$ and $900Hz$, which are within the sensitive range of typical interferometric gravitational wave detectors.

These early results should not be seen as providing a final complete theoretical description of the binary black hole system, but rather a starting point for the astrophysical application of theoretical tools under development for many years. The key achievement is that an appropriately non-linear treatment of binary black hole coalescence has been carried out, deriving waveforms from the system and covering the dynamics from the end of its orbital epoch onward. New advances in numerics will improve the accuracy of these waveforms. Further work on the far-limit/post-Newtonian interface will reduce uncertainties about the astrophysical appropriateness of the initial data for the system at the end of inspiral. Years of further study will sketch out the astrophysical parameter space covering the variety of possible binary black hole pairs transforming interesting questions about black hole interactions from speculations to calculations. The time for applying fully non-linear numerical simulation methods to binary black hole astrophysics has begun.

ACKNOWLEDGMENTS

We also acknowledge the support of R. Price, E. Seidel, and B. Schutz, and for many helpful discussions. M.C. was also supported by the IHRP program of the European Union (Marie-Curie Fellowship HPMF-CT-1999-00334). All our numerical computations have been performed with a full scale 3D code, Cactus[4], at LRZ (Munich).

REFERENCES

1. Baker, J., Brügmann, B., Campanelli, M., Lousto, C. O., and Takahashi, R. (2001), gr-qc/0102037.
2. Baker, J., Brügmann, B., Campanelli, M., and Lousto, C. O., *Class. Quant. Grav.*, **17**, L149 (2000).
3. Baker, J., Campanelli, M., and Lousto, C. O. (2001), gr-qc/0104063.
4. Cactusweb, *http://www.cactuscode.org*.

Critical Phenomena Associated with Boson Stars

Scott H. Hawley* and Matthew W. Choptuik[†]

*Max Planck Institut für Gravitationsphysik, Albert Einstein Institut, 14476 Golm, Germany
[†]CIAR Cosmology and Gravity Program, Department of Physics and Astronomy, University of British Columbia, Vancouver, British Columbia, Canada V6T 1Z1

Here we present a synopsis of related work [1, 2] describing a study of black hole threshold phenomena for a self-gravitating massive complex scalar field in spherical symmetry.

Studies of models of gravitational collapse have revealed structure which can arise near the threshold of black hole formation. The solutions in this regime are known as "critical solutions," and their properties as "critical phenomena". These solutions can arise generically, even in simple models such a massless scalar field in spherical symmetry [3].

Critical solutions can be constructed dynamically via numerical simulations, in which one considers continuous one-parameter families of initial data with the following "interpolating" property: for sufficiently large values of the family parameter, p, the evolved data describes a spacetime containing a black hole, whereas for sufficiently small values of p, the matter-energy in the spacetime disperses to large radii at late times, and *no* black hole forms. Within this range of parameters, there will exist a critical parameter value, $p = p^\star$, which demarks the onset, or threshold, of black hole formation.

Over the past decade, numerical and closed-form studies of collapse in various matter models have enlarged the picture of critical phenomena [4, 5, 6, 7], so that we now have a more complete understanding of the relevant dynamics. (Interested readers should consult the reviews by Gundlach [8, 9] for a more comprehensive discussion of critical phenomena.) Black-hole-threshold solutions are *attractors* in the sense that they are almost completely independent of the specifics of the particular family used as a generator. Up to the current time, the only initial data dependence which has been observed in critical collapse occurs in models for which there is more than one distinct black-hole-threshold solution. Critical solutions are by construction unstable, having precisely one unstable mode [10, 11]. Thus letting $p \to p^\star$ amounts to minimizing or "tuning away" the initial amplitude of the unstable mode present in the system. These solutions also possess additional symmetry which, to date, has either been a time-translation symmetry, in which the critical solution is static or periodic, or a scale-translation symmetry (homotheticity), in which the critical solution is either continuously or discretely self-similar (CSS or DSS).

These symmetries are indicative of the two principal types of critical behavior that have been seen in black hole threshold studies (with some models exhibiting both types of behavior depending on the initial data). For Type I solutions, there is a finite minimum black hole mass which can be formed, and there exists a scaling law for the lifetime τ of the near-critical solutions such that $\tau \sim -\gamma \ln|p - p^\star|$ where γ is a model-specific exponent. In Type II critical behavior, a black hole of arbitrarily small mass can be formed, and the critical solutions are generically *self-similar*.

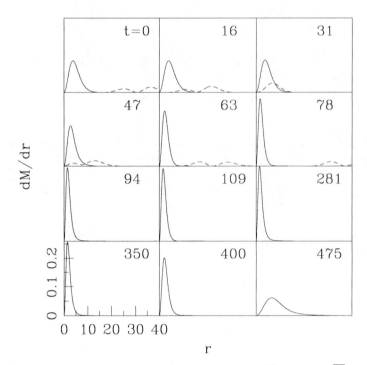

FIGURE 1. Evolution of a perturbed stable boson star with $\phi_0(0) = 0.04 \times \sqrt{4\pi}$ and mass $M_C = 0.59 M_{Pl}^2/m$. This shows contributions to $\partial M/\partial r$ due to the massive field $\phi(t,r)$ (solid line) and massless field $\phi_3(t,r)$ (dashed line). We start with a stable boson star centered at the origin, and a gaussian pulse of massless field. (We see two peaks for the massless field because it is only the gradients of ϕ_3, not ϕ_3 itself, which contribute to $\partial M/\partial R$.) In the evolution shown above, the pulse of massless field enters the region containing the bulk of the boson star ($t \simeq 15$), implodes through the origin ($t \simeq 30$) and leaves the region of the boson star ($t \simeq 50$). Shortly after the massless pulse passes through the origin, the boson star collapses into a more compact configuration, about which it oscillates for a long time before either forming a black hole or dispersing. (The case of dispersal is shown here.)

Our current interest is a critical-phenomena-inspired study of the dynamics associated with "boson stars" [12, 13, 14, 15, 16]. A boson star is given by a complex massive scalar field $\phi(t,r) = \phi_0(r)\exp(i\omega t)$, minimally coupled to gravity as given by general relativity. In this study, we dynamically construct critical solutions of the Einstein equations coupled to a massive, *complex* scalar field $\phi(t,r)$, by simulating the implosion of a spherical shell of *massless* real scalar field $\phi_3(t,r)$ around an "enclosed" boson star. The massless pulse then passes through the origin, explodes and continues to $r \to \infty$, while the massive complex (boson star) field is compressed into a state which ultimately either forms a black hole or disperses. For the massless field $\phi_3(0,r)$, we choose a gaussian of fixed width Δ and initial distance r_0 from the origin, and vary the amplitude A until the critical solution is obtained (to within machine precision).

Figure 1 shows a series of snapshots from a typical simulation in which the parameter p ($p \equiv A$) is slightly below the critical value $p^\star$, for an initial stable boson star with a mass of $M = 0.59 M_{Pl}^2/m$ (where M_{Pl} is the Planck mass). In this figure, we have

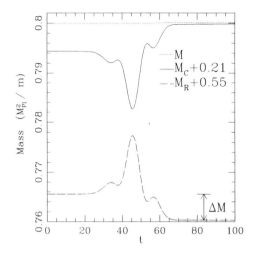

FIGURE 2. Exchange of energy between the real and complex scalar fields. The solid line shows the mass of the complex field, shifted upward by $0.21 M_{Pl}^2/m$ for graphing purposes. The long-dashed line shows the mass of the real field, shifted upward by $0.55 M_{Pl}^2/m$. The amount (and percentage) of mass transfer goes to zero as we consider boson star initial data approaching the maximum mass (the transition to instability). The dotted line near the top of the graph shows the total mass $M = M_C + M_R$, which is conserved to within a few hundredths of a percent. In all cases, we only see a net transfer of mass *from* the real field *to* to complex field, and not vice versa.

plotted the individual contributions of the complex and real fields to the total mass of the spacetime. That is, we have defined masses M_C and M_R of the complex and real fields, respectively. (Only in vacuum regions and for times at which the supports of the two fields do not overlap can M_C and M_R be interpreted as physically meaningful masses.) During this gravitational "collision," mass is is transferred from the real to the complex field, as shown in Figure 2.

The resulting critical solutions persist for some finite amount of time which depends on the fine-tuning $p - p^\star$ of the initial data. As we have shown in [1], the lifetimes τ of the near-critical solutions follow the scaling law for Type I solutions, $\tau = -\gamma \ln |p - p^\star|$. Here γ is related to the imaginary part of the growth factor σ of the unstable mode ($\sim \exp[i\sigma t]$) by $\Im(\sigma) = 1/\gamma$. In keeping with the Type I nature of these solutions, we find a finite minimum mass for the black hole formed for as we let $p \to p^\star$ (for $p > p^\star$).

The critical solutions have properties which correspond closely with those of unstable boson stars, as shown in Figure 3. To further extend the comparison between these critical solutions and boson stars, we perform a linear perturbation analysis about boson star equilibria, building on the work of Gleiser and Watkins [17]. Using the method described in [1], we find the distribution for the squared frequency σ^2 of boson star quasinormal modes with respect to $\phi_0(0)$, and we find the radial shapes of the modes. In Figure 4 we provide a comparison between unstable modes found from our simulations and corresponding results obtained via perturbation theory about a boson star which is a "best fit" to the simulation data. We find close agreement between the shapes

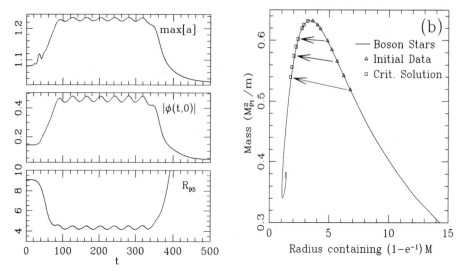

FIGURE 3. (a) Quantities describing a near-critical solution. Here we show timelike slices through the data shown in Figure 1, an evolution which ends in dispersal. Top: Maximum value of the metric function $a(t,r) = g_{rr}$. Middle: Central value $|\phi(t,0)|$ of the massive field. Bottom: Radius R_{95} containing 95% of the mass-energy in the complex field. (b) Mass vs. radius for equilibrium configurations of boson stars (solid line), initial data for the complex field (triangles), and critical solutions (squares). Arrows are given to help match initial data with the resulting critical solutions. Points on the solid line to the left of the maximum mass $M_{\max} \simeq 0.633 M_{Pl}^2/m$ correspond to unstable boson stars, whereas those to the right of the maximum correspond to stable stars. The squares show the time average of each critical solution, which exists during the oscillatory regime shown in (a). We show the radius containing $(1-e^{-1})M \simeq 0.63M$ instead of R_{95} in order to exclude the halo which forms in the critical solution (see Figure 5).

and frequencies of the unstable modes obtained by these two different methods. A comparison of the next higher (oscillatory) mode also yields favorable results.

Thus the critical solutions we obtain appear to correspond to boson stars exhibiting superpositions of stable and unstable modes. For boson stars with masses somewhat less than the maximum boson star mass $M_{\max} \simeq 0.633 M_{Pl}^2/m$ (e.g. those boson stars with masses $0.9 M_{\max}$ or less), however, we find less than complete agreement between the critical solutions and unstable boson stars of comparable mass. This is evidenced by the presence of an additional spherical shell or "halo" of matter in the critical solution, located in what would be the tail of the corresponding boson star. Interior to this halo, we find that the critical solution compares favorably with the boson star profile.

It is our contention that this halo is not part of the true critical solution, but is instead an artifact of the collision with the massless field. As one might expect, the properties of the halo are not universal, i.e. they are quite dependent on the type of initial data used. In contrast, the critical solution interior to the halo is largely independent of the form of the initial data. To demonstrate this, we use two families of initial data, given by a gaussian a "kink" ($\phi_3(0,r) = A/2[1 + \tanh[(r-r_0)/\Delta]]$). A series of snapshots from one

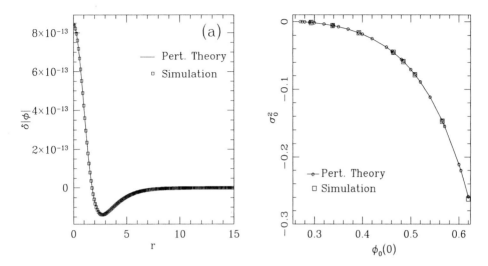

FIGURE 4. (a) Fundamental mode of unstable boson star. The solid line shows $\delta|\phi|$ obtained from the perturbation theory calculations. The squares show the difference between $|\phi|$ for two simulations for which the critical parameter p differs by 10^{-14}. Differences between the simulation data and perturbation theory results are below 1.1×10^{-15}. (b) Comparison of squared growth factors (Lyapunov exponents) for unstable modes. The circles show a subset of the perturbation theory results we obtained for unstable boson stars. The squares show the measurements of unstable growth factors in our simulations. (The solid line simply connects the circles.)

such pair of evolutions is shown in Figure 5. We suspect that the halo is radiated over time (via scalar radiation, or "gravitational cooling" [18]) for all critical solutions. We find, however, that the time scale for radiation of the halo is comparable to the time scale for dispersal or black hole formation for each (nearly) critical solution we consider. With higher numerical precision, one might be able to more finely tune out the unstable mode, allowing more time to observe the behavior of the halo before dispersal or black hole formation occur.

For the future, we consider it worthwhile to investigate similar scenarios for neutron stars. While there have bee studies regarding the explosion of neutron stars near the minimum mass (*e.g.*,[19]), we would like to see whether neutron stars of *non-minimal mass* can be driven to explode via dispersal from a critical solution.

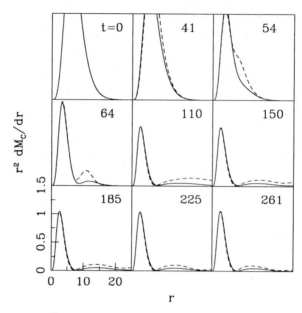

FIGURE 5. Evolution of $r^2 dM_C/dr$ for for two different sets of initial data. Both sets contain the same initial boson star, but the massless field $\phi_3(0,r)$ (not shown) for one set is given by a gaussian whereas for the other set $\phi_3(0,r)$ is given by a "kink". The amplitude of each pulse is varied (independently for each family of initial data) to obtain the critical solution. We have multiplied dM_C/dr by r^2 to highlight the dynamics of the halo; thus the main body of the solution appears to decrease in size as it moves to lower values of r. The kink data produces a larger and much more dynamical halo, but interior to the halo, the two critical solutions match closely — and also match the profile of an unstable boson star. Thus, the portion of the solution which is "universal" corresponds to an unstable boson star. One can see the additional halo of matter in the region roughly $8 \leq r \leq 23$, exterior to the bulk of the critical solution.

REFERENCES

1. S.H. Hawley and M.W. Choptuik, Phys. Rev. D **62**, 104024 (2000).
2. S.H. Hawley, Ph.D. Dissertation, University of Texas at Austin (2000).
3. M.W. Choptuik, Phys. Rev. Lett. **70**, 9 (1993).
4. D.W. Neilsen and M.W. Choptuik, Class.Quant.Grav. **17**, 761-782 (2000).
5. M.W. Choptuik, T. Chmaj and P. Bizon, Phys. Rev. Lett. **77**, 424 (1996).
6. P.R. Brady, C.M. Chambers and S.M.C.V. Conçalves, Phys. Rev. D **56**, 6057 (1997).
7. A.M. Abrahams and C.R. Evans, Phys. Rev. Lett. **70** 2980 (1993).
8. C. Gundlach, Adv. Theor. Math.Phys. **2**, 1-49 (199 8).
9. C. Gundlach, Living Reviews 1999-4 (1999).
10. C.R. Evans and J.S. Coleman, Phys. Rev. Lett. **72**, 1782 (1994).
11. T. Koike, T. Hara and S. Adachi, Phys Rev. Lett. **74**, 5170 (1995).
12. D.J. Kaup, Phys. Rev. **172**, 1331 (1968).
13. R. Ruffini and S. Bonnazzola, Phys. Rev. **187**, 1767 (1969).
14. M. Colpi, S.L. Shapiro, and I. Wasserman, Phys. Rev. Lett. **57**, 2485 (1986).
15. P. Jetzer, Phys. Rep. **220**, 163 (1992).
16. E.W. Mielke and F.E. Schunck. gr-qc/9801063.
17. M. Gleiser and R. Watkins, Nucl. Phys. B **319**, 733 (1989).
18. E. Seidel and W.-M. Suen, Phys. Rev. Lett. **72**, (1994).
19. M. Colpi, S.L. Shapiro and S.A. Teukolsky, Astrophys. J. **369**, 422 (1991).

Gravitational Waves from Rotational Core Collapse in the Conformally Flat Spacetime Approximation

H. Dimmelmeier, J.A. Font, E. Müller

*Max-Planck-Institut für Astrophysik,
Karl-Schwarzschild-Str. 1, D-85741 Garching, Germany*

Abstract. We have developed an axisymmetric general relativistic hydrodynamic code based upon an approximation scheme recently proposed by Wilson and coworkers, in which the 3-metric γ_{ij} is conformally flat, thus reducing the mathematical complexity of the Einstein metric equations to a set of 5 coupled elliptic equations. Applications of our code to rotational supernova core collapse are presented. We also discuss first results concerning the emission of gravitational waves.

According to models of core collapse supernova explosions, massive stars develop an iron core of about $1.3 - 2.2\ M_\odot$ at the end of their life [1]. This core collapses and bounces at neutron star densities, forming a shock wave which ejects the stellar envelope. It is believed that energy deposition behind the shock front by neutrino emission from the proto-neutron star plays a crucial role in driving the shock. Observed massive stars and neutron stars have significant surface rotation speeds, therefore rotation may also play an important role during core collapse. Furthermore, describing the dynamics of the collapse and the formation of the neutron star requires a general relativistic treatment.

Here we present results obtained with an axisymmetric code for simulating rotational core collapse to a neutron star. In the ADM $\{3+1\}$ formalism, the general relativistic hydrodynamic equations can be formulated as a first-order, flux-conservative hyperbolic system of conservation laws for the generalized mass density, momenta and energy density in a curved spacetime [2]. These equations are solved on a computational grid by propagating cell averages to the next time level by a conservative time-update algorithm. For this purpose we have implemented a modern flux-conservative scheme based upon the solution of approximate Riemann problems with Marquina's flux formula [3] and third-order PPM reconstruction. These high-resolution shock-capturing (HRSC) schemes converge to the physical solution, ensure the correct propagation speed of discontinuities, and sharply resolve discontinuities.

The formulation of the metric equations follows ideas of Wilson et al. [4] by approximating the 3-metric as conformally flat (conformal flatness condition, CFC hereafter): In the astrophysical scenarios we are interested in, the matter distribution does not depart too much from spherical symmetry. Therefore we assume the CFC approximation for the spatial 3-metric, $\gamma_{ij} = \phi^4 \hat{\gamma}_{ij}$, with the flat 3-metric $\hat{\gamma}_{ij}$.

In this formalism the Einstein metric equations reduce to a system of 5 coupled nonlinear elliptic equations for the conformal factor ϕ, the lapse function α, and the shift vector β^i [4]. These equations show no explicit time dependence. The system of metric equations is solved numerically using a Newton-Raphson iterative solver for a nonlinear system of coupled elliptic equations. Imposing the CFC constrains the system in such a way that gravitational waves are not present in the spacetime. To determine the gravitational wave emission by the system, we use the standard quadrupole formula.

We use a simplified equation of state [5], consisting of a polytropic and a thermal part, neglecting all other microphysics. The thermal part enforces consistent thermal heating in shock fronts.

For validation of the code, we have performed comprehensive tests: (i) Using highly relativistic shock tube tests we could to demonstrate the correct propagation and sharp resolution of shock fronts. (ii) During the evolution of rotating neutron stars in equilibrium, the density structure and angular momentum distribution are preserved to high accuracy over many rotation periods. Small oscillations, triggered by the truncation error of the scheme, are slowly damped, which proves that the code has small numerical viscosity. (iii) We have compared spherically symmetric core collapse runs with results from an independent Lagrangian artificial viscosity code, finding agreement in all state variables better than a few percent.

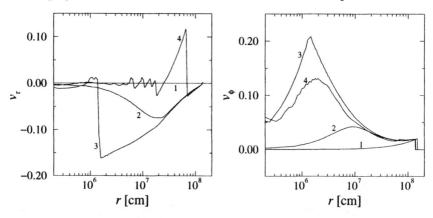

FIGURE 1. Profiles of the radial velocity (left panel) and the rotation velocity (right panel) at different evolutionary times (1: 0 ms, 2: 90 ms, 3: 94 ms, 4: 108 ms) for a typical core collapse simulation bouncing at $t \approx 93$ ms. After the formation of the neutron star, the radial velocity in the central region approaches zero, while a shock front propagates through the progenitor envelope. The rotation profile increases towards the center during infall, and then stays essentially constant within the neutron star with its maximum at the neutron star surface.

The primary goal of our study is to simulate rotational supernova core collapse to a neutron star. We construct initial data by solving the TOV equations in axisymmetry for a matter distribution in equilibrium using the self-consistent field method of [6]. The gravitational collapse is initiated by reducing the adiabatic index γ from the initial value of 4/3 by a prescribed amount, which reduces the compressibility of matter throughout the star. The unstable iron core collapses on a timescale of several 10 milliseconds, spinning up due to conservation of angular momentum, and bounces at nuclear matter density, where the equation of state stiffens (modeled by increasing the adiabatic index γ). A shock wave forms and propagates from the surface of the proto-neutron star through the outer core.

A typical evolution of the radial and angular velocity profiles of such a simulation is shown in Figure 1. The evolution of the central density, and the corresponding waveform of the emitted gravitational radiation for this collapse simulation is shown in Figure 2. The main burst of radiation is clearly associated with the core bounce, whereas the ringdown signal following the burst originates from the oscillations of the proto-neutron star, which can also be identified in the density evolution plot.

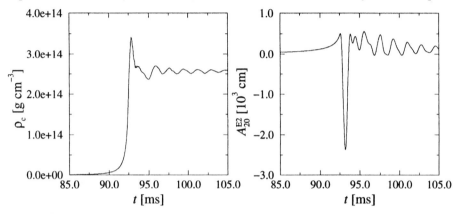

FIGURE 2. Evolution of the central density (left panel): During the infall, the density increases by more than 4 orders of magnitude until core bounce at supernuclear densities. It then settles down at a constant value $\rho \simeq 1.3\rho_{\rm nuc}$. The gravitational wave amplitude $A_{20}^{E2} \propto h_{\theta\theta}^{\rm TT} \cdot r \sin^{-2}\theta$ (right panel) was calculated using the Einstein mass quadrupole formula.

A detailed discussion of our results will be presented elsewhere [7]. For the near future we envisage to add more sophisticated microphysics in the form of a realistic equation of state for better modelling of supernova physics.

The calculations were carried out at the Rechenzentrum Garching, the MPI für Astrophysik, Garching, and the MPI für Gravitationsphysik, Golm.

REFERENCES

1. Müller, E., "Simulation of astrophysical fluid flow", in *Computational methods for astrophysical fluid flow. Saas-Fee Advanced Course 27*, edited by O. Steiner and A. Gautschy, Springer, Berlin, Germany, 1998, p. 343.
2. Banyuls, F., Font, J., Ibáñez, J., Martí, J., and Miralles, J., *Astrophys. J.*, **476**, 221–231 (1997).
3. Donat, R., and Marquina, A., *J. Comput. Phys.*, **125**, 42–58 (1996).
4. Wilson, J., Mathews, G., and Marronetti, P., *Phys. Rev. D*, **54**, 1317–1331 (1996).
5. Zwerger, T., and Müller, E., *Astron. Astrophys.*, **320**, 209–227 (1997).
6. Komatsu, H., Eriguchi, Y., and Hachisu, I., *Mon. Not. R. Astron. Soc.*, **237**, 355–379 (1989).
7. Dimmelmeier, H., Font, J., and Müller, E. (2001), in preparation.

Application of the Galerkin method to the problem of stellar stability

A.V. Dorodnitsyn* and G.S.Bisnovatyi-Kogan*

Space Research Institute, Profsoyuznaya 84/32, Moscow 117810, Russia

Abstract. Approximate method for investigating stellar stability in GR is presented. It is based on the approximate Galerkin method. Spectral Galerkin method is widely used in numerical simulations. The development of the method is based on a generalization of the well-known energetic method. In full GR treatment it results in more simple equations of the stability than in standard perturbation approach.

The Galerkin method allows to find an approximate solution of differential equations. In this method the solution of the partial or ordinary differential equation is represented in the form (Fletcher, 1984):

$$u = u_0 + \sum_{i=1}^{N} \alpha_i \varphi_i, \qquad (1)$$

where φ_i are known analytical functions, and coefficients (or functions) α_i are to be determined. Functions φ_i are assumed to be not necessary orthogonal. When substituted to the original differential equation the coefficients α_i will satisfy a set of algebraic or ordinary differential equations, obtained by minimizing the corresponding functional. Consider how the Galerkin method can be applied to the problem of the relativistic stellar stability. The functional which should be minimized, is a total energy of a star. Equating to zero variation of the total energy gives equations of the stability.
We may use the Galerkin method to approximate the system of equations in partial derivatives by the system of ordinary differential equations.
In PN approximation the density in Galerkin method is written as a following sum

$$\rho = \sum_{i=1}^{N} \alpha_i(t)\varphi_i(a), \quad \text{where} \quad \rho_c = \sum_{i}^{N} \alpha_i(t)\varphi(0). \qquad (2)$$

As for a function $\varphi_0(a)$, it is convenient to take corresponding Emden profile for one of polytropic indices. For other functions we may chose $\varphi_k = \cos\frac{1+2K}{2}\pi a$, which have increasing number of nodes. Then, satisfaction of the boundary conditions: $\varphi_i(A) = 0$, $A = a(R)$, $\varphi_i(0) = 1$ will be provided. Minimization of the energy functional for finding an equilibrium model is reduced to zero partial derivatives $\frac{\partial \varepsilon}{\partial \alpha_i} = 0$, leading in the static case of constant α_i to a set of N algebraic equations for finding equilibrium α_i^{eq}. Stability of a model is found from an evaluation of the second variation $\delta^2 \varepsilon$, which in the energetic method is reduced to the algebraic equation. In the Galerkin method with several scaling functions $\varphi_i(a)$, the second variation $\delta^2\varepsilon$ is represented by a quadratic form:

$$\delta^2\varepsilon = \sum_{i,k}^{N} \frac{\partial^2\varepsilon}{\partial\alpha_i\partial\alpha_k}\delta\alpha_i\delta\alpha_k. \tag{3}$$

In the energetic method the condition of stability reduces to one equation $\partial^2\varepsilon/\partial\alpha^2 > 0$. In the Galerkin method the stability is related to positive definiteness of the quadratic form (3), what is provided by the positiveness of the determinant $\|\frac{\partial^2\varepsilon}{\partial\alpha_i\partial\alpha_k}\| > 0$ and all its main minors. For two functions in (2) the positiveness of the main determinant, and two partial derivatives $\partial^2\varepsilon/\partial\alpha_1^2 > 0$ and $\partial^2\varepsilon/\partial\alpha_2^2 > 0$ are enough for stellar stability. Loss of stability happens close before the point where the main determinant, or one of its main minors becomes zero.

In approximate presentation of the trial function in the Galerkin method, the minimal value of the second energy variation is larger, then its value for a real trial function. So zero values of the determinant or one of its main minors, guarantees the onset of instability. Their positiveness is not an exact guarantee of the stability, but comparison of the energetic method with an exact stability analysis shows a good precision of this approximate approach in most realistic cases. Energetic method corresponds to a homologous trial function for displacement $\delta r \sim r$. In the Galerkin method the trial function may be determined with a better precision. In fact, the coefficients $\delta\alpha_i$ for the trial function of a density $\delta\rho$ are determined as an eigenvector of a set of uniform linear equations

$$\frac{\partial^2\varepsilon}{\partial\alpha_i\partial\alpha_k}\delta\alpha_k = \lambda_p\delta\alpha_i. \tag{4}$$

The eigenvector $\delta\alpha_i^e$ is used for obtaining an approximate eigenfunction, and eigenvalues λ_p are related to the square eigenfrequencies of the stellar model [1]. The positive definiteness of the quadratic form (3) coincides with the positiveness of all eigenvalues λ_p. So, Galerkin method permits to investigate a stability by finding approximate eigenvalues and eigenfunctions of a linear set of algebraic equations, instead of finding the same values from a second order differential equation exactly. In GR a barion number density n is used in (2) instead of the mass density ρ, which in this case is not presenting itself a conserved value. Inside the spherically symmetric star the Schwarzschild type metric is used. The mass-energy functional, and number of baryons a inside radius $r(a)$ for the spherically symmetrical relativistic star is written as:

$$E = e(R) = 4\pi\int_0^R T_0^0 r^2\, dr, \quad a = 4\pi\int_0^r \frac{nr^2}{\sqrt{1 - \frac{2Ge}{c^4 r}}}\, dr. \tag{5}$$

Actually when we take linear function as an eigenfunction, we take only one term in the Galerkin representation. To find it with better presision we need to take edditional terms. With the linear function the energetic method gives accuracy $R_1 = \frac{\Delta\rho}{\rho} < 10^{-5}$. In trial function $\delta r = \alpha_1\varphi_1 + \alpha_2\varphi_2$, for φ_1 we chose $\bar{r}$, what means that $r(a)$, which corresponds to the equilibrium configuration (for the cold NS it is fully determined by ρ_c), ufter simplifications,

$$\overline{\delta^2 \varepsilon} = \alpha_1^2 (J_{11} - J_{21}) + \alpha_2^2 (J_{12} - J_{22}) + 2\alpha_1 \alpha_2 (J_3 - J_4), \qquad (6)$$

where
$$J_{1i} \equiv \int_0^R e^{I_1} A_1^2(\varphi_i)\, dr, \quad J_{2i} \equiv \int_0^R e^{I_1}(a_2 + a_3)\varphi_i^2\, dr,$$

$$J_3 \equiv \int_0^R e^{I_1} A_1(\varphi_1) A_1(\varphi_2)\, dr, \quad J_4 \equiv \int_0^R e^{I_1}(a_2 + a_3)\varphi_1 \varphi_2\, dr, \qquad (7)$$

and $\overline{\delta^2 \varepsilon}$:
$$\overline{\delta^2 \varepsilon} \equiv \frac{\delta^2 \varepsilon}{e^{-I_1}}, \quad I_1 \equiv \int_0^a \left(\frac{P}{n} + \tilde{E}\right) \frac{da}{\sqrt{1 - \frac{2Ge}{c^4 r}}} \frac{G}{c^4 r}. \qquad (8)$$

Coefficients in (7) are determined by the following formulas:

$$A_1 \equiv \gamma P \left(2\varphi_i + r\frac{d\varphi_i}{dr} - \varphi_i \frac{e + 4\pi r^3 P}{1 - \frac{2Ge}{c^2 r}} \frac{G}{c^4 r}\right)^2, \qquad (9)$$

$$a_2 \equiv \frac{P + \tilde{E}n}{(1 - \frac{2Ge}{c^4 r})^2} \left(1 + \frac{4\pi r^3 P}{ec^2}\right)^2 \left(\frac{2Ge}{c^2 r}\right)^2, \qquad (10)$$

$$a_3 = \frac{P + \tilde{E}n}{1 - \frac{2Ge}{c^4 r}} \left(1 + \frac{1}{2}\frac{4\pi r^3 P}{ec^2}\right) \frac{Ge}{c^4 r}. \qquad (11)$$

Where we put $e = mc^2 \to e$. Direct numerical integration shows, that choosing a combination of liniear function with the trigonometrical polinomial gives accuracy: $R_2 = \frac{\Delta \rho_2}{\rho_2} < R_1 = 10^{-5}$.

ACKNOWLEDGMENTS

The authors are grateful to Prof. J.C.Wheeler and the Organizing Committee for support and hospitality.

REFERENCES

1. Bisnovatyi-Kogan, G. S., Dorodnitsyn, A. V., 1998, *Gravitation & Cosmology*, Vol. 4, No 3, pp. 174-182, astro-ph / 9807316

Light Cone Consistency in Bimetric General Relativity

J. Brian Pitts* and W.C. Schieve*

The Ilya Prigogine Center for Studies in Statistical Mechanics and Complex Systems
Department of Physics
The University of Texas at Austin
Austin Texas 78712

Abstract. General relativity can be formally derived as a flat spacetime theory, but the consistency of the resulting curved metric's light cone with the flat metric's null cone has not been adequately considered. If the two are inconsistent, then gravity is not just another field in flat spacetime after all. Here we discuss recent progress in describing the conditions for consistency and prospects for satisfying those conditions.

INTRODUCTION

The formulation and derivation of general relativity using a flat metric tensor $\eta_{\mu\nu}$ are well-known from the works of Rosen, Gupta, Kraichnan, Feynman, Deser, Weinberg et al. [1]. One can obtain a curved metric $g_{\mu\nu}$ by adding the gravitational potential $\gamma_{\mu\nu}$ to the flat metric $\eta_{\mu\nu}$:

$$g_{\mu\nu} = \eta_{\mu\nu} + \sqrt{32\pi G}\gamma_{\mu\nu}. \qquad (1)$$

This framework is useful [2], but is it merely formal? If general relativity can be *consistently* regarded as a special-relativistic theory, then the observable curved metric must satisfy a nontrivial consistency condition in relation to the unobservable flat background metric: the "causality principle" says that the curved metric's light cone cannot open wider than the flat metric's. The question of the relation between the cones is complicated somewhat by its gauge-variance.

PREVIOUS DISCUSSIONS OF RELATIONSHIP OF NULL CONES

While the flat spacetime field approach to general relativity has been mature since the 1950s, the question of the consistency of the effective curved metric's null cone with the original flat metric's received surprisingly little attention. In the

1970s van Nieuwenhuizen wrote: "The strategy of particle physicists has been to ignore [this problem] for the time being, in the hope that [it] will ultimately be resolved in the final theory. Consequently we will not discuss [it] any further." [3] More recently (since the late 1970s), this issue has received more sustained attention [4–7], but the treatments to date have been impaired by unnecessarily strict requirements [4,6] or lack of a general and systematic approach [5,7], as we have noted [1].

We propose to *stipulate* that the gauge be fixed in a way that the proper relation obtains, if possible. The gauge fixing can be implemented in an action principle using ineffective constraints, whose constraint forces vanish [8]. This approach does appear to be possible, because the gauge freedom allows one to choose arbitrarily g^{00} and g^{0i} (at least locally). Increasing g^{00} stretches the curved metric's null cone along the time axis, so that it becomes narrower, while adjusting g^{0i} controls the tilt of the curved null cone relative to the flat one. Stretching alone appears to be enough to satisfy the causality principle, in fact.

KINEMATIC AND DYNAMIC PROGRESS

The metric is a poor variable choice due to the many off-diagonal terms. One would like to diagonalize $g_{\mu\nu}$ and $\eta_{\mu\nu}$ simultaneously by solving the generalized eigenvalue problem $g_{\mu\nu}V^\mu = \Lambda\eta_{\mu\nu}V^\mu$, but in general that is impossible, because there is not a complete set of eigenvectors on account of the minus sign in $\eta_{\mu\nu}$ [9]. There are 4 Segré types for a real symmetric rank 2 tensor with respect to a Lorentzian metric, the several types having different numbers and sorts of eigenvectors [9]. We have recently used this technology to classify $g_{\mu\nu}$ with respect to $\eta_{\mu\nu}$. Two types are forbidden by the causality principle. One type has members that obey the causality principle, but we argue that they can be ignored. The remaining type has 4 real independent orthogonal eigenvectors, as one would hope. In that case, the causality principle is just the requirement that the temporal eigenvalue be no larger than each of the three spatial eigenvalues.

Realizing the condition $g_{\mu\nu} \to \eta_{\mu\nu}$ when the gravitational field is weak, while obeying the causality principle, is nontrivial. The causality principle puts an upper bounding surface on the temporal eigenvalue in terms of the spatial ones, and the surface is *folded*, as seen in 2 spatial dimensions in figure 1. Einstein's equations have second spatial derivatives of g^{00} (which is closely related to the temporal eigenvalue), so the fold, if not avoided, would imply Dirac delta gravitational 'forces' that make the canonical momenta jump discontinuously. On the other hand, avoiding the fold means excluding $g_{\mu\nu} = \eta_{\mu\nu}$! But why fix the temporal eigenvalue in terms of the spatial eigenvalues at the same point only (ultralocally)? It is enough to do so locally, by admitting derivatives. When the derivatives are nonzero, the fold is avoided, but as they vanish, the fold is approached. If such a partial gauge-fixing can be found, then it will facilitate interpreting the Einstein equations as describing a special-relativsitic field theory. In such a theory, one would need to consider the physical situation near the Schwarzschild radius rather carefully.

REFERENCES

1. Pitts, J. B., and Schieve, W. C., *Los Alamos Preprints*, xxx.lanl.gov, gr-qc/0101058.
2. Petrov, A.N., and Narlikar, J. V. *Found. Phys.* **26**, 1201, 1996.
3. van Nieuwenhuizen, P., in *Proceedings of the First Marcel Grossmann Meeting on General Relativity*, ed. Ruffini, R., Amsterdam: North Holland, 1977.
4. Penrose, R., in *Essays in General Relativity—A Festschrift for Abraham Taub*, ed. Tipler, F. J., New York: Academic, 1980.
5. Zel'dovich, Ya. B., and Grishchuk, L. P., *Sov. Phys. Usp.* **31**, 666 (1988).
6. Burlankov, D. E. *Sov. J. Nucl. Phys.* **50**, 174 (1989).
7. Logunov, A. A., *Theor. Math. Phys.*, trans. of *Teoreticheskaya i Matematicheskaya Fizika* **92**, 191 (1992).
8. Pons, J. M., Salisbury, D. C., and Shepley, L. C., *J. Phys. A: Math. Gen.* **32**, 419 (1999).
9. Hall, G. S., and Negm, D. A., *Int. J. Theor. Phys.* **25**, 405 (1986).

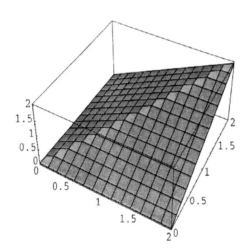

FIGURE 1. Bounding Surface for Temporal Eigenvalue as Function of Spatial Eigenvalues in 2 Dimensions

Dynamical Bar Instability in Relativistic Rotating Stars

Motoyuki Saijo [*], Masaru Shibata [*†],
Thomas W. Baumgarte [*] and Stuart L. Shapiro [*‡]

[*] *Department of Physics, University of Illinois at Urbana-Champaign,
1110 West Green Street, Urbana, Illinois 61801-3080*
[†] *Department of Earth Science and Astronomy, Graduate School of Arts and Science,
University of Tokyo, 3-8-1 Komaba, Meguro, Tokyo 153-8902, Japan*
[‡] *Department of Astronomy and NCSA, University of Illinois at Urbana-Champaign,
1002 West Green Street, Urbana, Illinois 61801*

Abstract. We study by computational means the dynamical stability against bar-mode deformation of rapidly and differentially rotating stars in a post-Newtonian approximation of general relativity. We vary the compaction of the star M/R (where M is the gravitational mass and R the equatorial circumferential radius) between 0.01 and 0.05 to isolate the influence of relativistic gravitation on the instability. For compactions in this moderate range, the critical value of $\beta = T/W$ for the onset of the dynamical instability (where T is the rotational kinetic energy and W the gravitational binding energy) slightly decreases from ~ 0.26 to ~ 0.25 with increasing compaction for our choice of the differential rotational law. Combined with our earlier findings based on simulations in full general relativity for stars with higher compaction, we conclude that relativistic gravitation enhances the dynamical bar-mode instability, i.e. the onset of instability sets in for smaller values of β in relativistic gravity than in Newtonian gravity. We also find that once a triaxial structure forms after the bar-mode perturbation saturates in dynamically unstable stars, the triaxial shape is maintained, at least for several rotational periods. To check the reliability of our numerical integrations, we verify that the general relativistic Kelvin-Helmholtz circulation is well-conserved, in addition to rest-mass and total mass-energy, linear and angular momentum. Conservation of circulation indicates that our code is not seriously affected by numerical viscosity. We determine the amplitude and frequency of the quasi-periodic gravitational waves emitted during the bar formation process using the quadrupole formula.

Stars in nature are usually rotating and subject to nonaxisymmetric rotational instabilities. An exact treatment of these instabilities exists only for incompressible equilibrium fluids in Newtonian gravity. For these configurations, global rotational instabilities arise from non-radial toroidal modes when $\beta \equiv T/W$ exceeds a certain critical value. Here T and W are the rotational kinetic and gravitational binding energies. There exist two different mechanisms and corresponding timescales for bar-mode instabilities. Uniformly rotating, incompressible stars in Newtonian theory are *secularly* unstable to bar-mode formation when $\beta \geq \beta_{\text{sec}} \simeq 0.14$. This instability can grow only in the presence of some dissipative mechanism, like viscosity or gravitational radiation. The growth time is determined by the dissipative timescale, which is usually much longer than the dynamical timescale of the system. By contrast, a *dynamical* instability to bar-mode formation occurs when $\beta \geq \beta_{\text{dyn}} \simeq 0.27$. This instability is independent of any dissipative mechanisms, and the growth time is the hydrodynamic timescale of the system.

Determining the onset of the dynamical bar-mode instability, as well as the subsequent evolution of an unstable star, requires a fully nonlinear hydrodynamic simulation. Recently, simulations in Newtonian theory [1,2] have shown that a higher degree of differential rotation enhances the onset of dynamical instability. Simulations in full general relativity [3] suggest that nonlinear gravitation has a similar effect.

The purpose of post-Newtonian simulations [4] is twofold. We verify that relativistic gravitation alone *enhances* the dynamical instability, i.e. β_{dyn} decreases with increasing compaction. We also show that in unstable configurations, the bar persists for at least several rotational periods [5,6] so that unstable stars are promising sources of quasi-periodic gravitational waves.

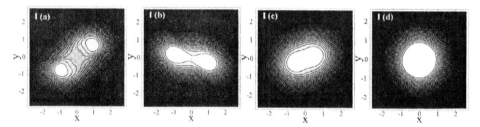

FIGURE 1. Final density contours in the equatorial plane for differentially rotating stars all of compaction $M/R = 0.05$ and varying values of β. The contour lines denote densities $\rho_* = 1.3\, i \times 10^{-3}$ ($i = 1, \ldots, 15$). Snapshots are plotted at the following times and initial β: (a) $t/P_c = 2.72$, $\beta = 0.265$ (b) $t/P_c = 3.66$, $\beta = 0.259$, (c) $t/P_c = 7.77$, $\beta = 0.249$, and (d) $t/P_c = 8.16$, $\beta = 0.238$. Here, P_c is the central rotation period.

We perform post-Newtonian simulations of rapidly and differentially rotating stars to investigate general relativistic effects on the dynamical bar-mode instability for small compactions $M/R \leq 0.05$. As a criterion for stability, we checked whether the distortion parameter follows an exponential growth. The formation of a bar is also apparent in the snapshots of density contours (See Fig. 1 for $M/R = 0.05$; Figs. 4 – 7 of Ref. [4]). These plots clearly exhibit a triaxial structure for the unstable models, while for stable models the density distribution hardly changes during the evolution. By combining these post-Newtonian results with the fully relativistic simulations [3] for configurations of higher compaction, we conclude that the critical value of $\beta = \beta_{\text{dyn}}$ decreases with increasing M/R (Fig. 2). Thus,

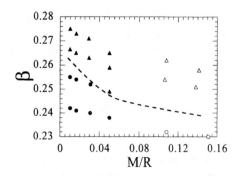

FIGURE 2. Summary of our dynamical stability analysis. All our models are plotted in a β versus M/R plane, with stable stars denoted by circles and unstable stars by triangles. The solid circles and triangles are the models studied in our post-Newtonian simulations [4]; the open circles and triangles are the models explored in full general relativity [3]. We conclude that the critical value of $\beta = \beta_{\rm dyn}$ slightly decreases with increasing compaction M/R. This trend is emphasized by the dotted line, which approximates $\beta_{\rm dyn}$ as a function of M/R.

relativistic gravitation enhances the bar-mode instability.

We also confirm that bar-structure persists at least several rotation period by checking the conservation of circulation (Fig. 8 and Table 3 of Ref. [4]). In the presence of significant numerical viscosity, long-time evolution calculations become very unreliable and may lead, for example, to erroneous evolution of a saturated bar. We have shown that in our calculations the circulation is well conserved, implying that our code is at most very weakly affected by numerical viscosity. We present a method for computing the circulation which can be applied in Newtonian, post-Newtonian and fully relativistic calculations.

Finally, we have calculated approximate gravitational waveforms in the quadrupole approximation, neglecting all post-Newtonian corrections (Fig. 10 of Ref. [4]). We found that for unstable stars a quasi-periodic oscillation with growing amplitude arises during the early bar formation. The bar-structure persists for several rotation periods, implying that bar-unstable stars are promising sources of quasi-periodic gravitational waves.

REFERENCES

1. Tohline, J. E. and Hachisu, I., *Astrophys. J.* **361**, 394 (1990).
2. Pickett, B. K., Durisen, R. H., and Davis, G. A., *Astrophys. J.* **458**, 714 (1996).
3. Shibata, M., Baumgarte, T. W., and Shapiro, S. L., *Astrophys. J.* **542**, 453 (2000).
4. Saijo, M., Shibata, M., Baumgarte, T. W., and Shapiro, S. L., *Astrophys. J.* **548**, 919 (2001).
5. New, K. C. B., Centrella, J. M., and Tohline, J. E., *Phys. Rev. D* **62**, 064019 (2000).
6. Brown, J. D., *Phys. Rev. D* **62**, 084024 (2000).

The Ultimate Future of the Universe, Black Hole Event Horizon Topologies, Holography, and the Value of the Cosmological Constant

Frank J. Tipler[1]

Department of Mathematics and Department of Physics
Tulane University
New Orleans, Louisiana 70118 USA

Abstract. Hawking has shown that if black holes were to exist in a universe that expands forever, black holes would completely evaporate, violating unitarity. Unitarity thus requires that the universe exist for only a finite future proper time. I develop this argument, showing that unitarity also requires the boundaries of all future sets to be Cauchy surfaces, and so no event horizons can exist. Thus, the null generators of the surfaces of astrophysical black holes must leave the surface in both time directions, allowing non-spherical topologies for black hole surfaces. Unitarity thus also requires the effective cosmological constant to be zero eventually, otherwise the universe would expand forever.

INTRODUCTION

Hawking showed a quarter century ago that if a black hole were to exist in a space-time that exists for infinite proper time, it would completely evaporate, destroying the information inside the BH, thereby violating unitarity. Hawking argued that this result demonstrated that unitarity was indeed violated, but since unitarity is absolutely fundamental to quantum mechanics, I shall explore the implications of assuming that Hawking's result and unitarity are BOTH correct. This assumption will be shown to imply: (1) the universe must be closed, with its future c-boundary being a single point, which means that there are no event horizons, and sets of the form $\partial I^+(p)$ for any event p are Cauchy surfaces, so the information inside a black hole both gets out in the far future and is also coded entirely on the surface of an astrophysical black hole: holography automatically holds; (2) the fact that the generators of any set $\partial I^+(p)$ eventually leave in the future direction, in contrast to event horizon null generators, means that higher genus black hole surface topologies

---

[1] e-mail address: tipler@mailhost.tcs.tulane.edu

MAY be possible; (3) the value of the cosmological constant is required to be near zero; in particular the "natural value" $\Lambda = 8\pi c^5/G\hbar$, corresponding the the Planck energy density would violate unitarity.

THE ULTIMATE FUTURE OF THE UNIVERSE

I shall now show that unitarity strongly constrains the future of the universe. Astrophysical black holes exist, but Hawking has shown that if black holes are allowed to exist for unlimited proper time, then they will completely evaporate, and unitarity will be violated. Thus unitarity requires that the universe must cease to exist after finite proper time, which implies that the universe has spatial topology S^3. (All other recollapse topologies, e.g. $S^2 \times S^1$ [9] and negative Λ universes can be eliminated as possibilities by arguments which will be published elsewhere.) The Second Law of Thermodynamics says the amount of entropy in the universe cannot decrease, but it can be shown ([1], p. 410) that the amount of entropy already in the CBR will eventually contradict the Bekenstein Bound [8] near the final singularity unless there are no event horizons, since in the presence of horizons the Bekenstein Bound implies the universal entropy $S \leq constant \times R^2$, where R is the radius of the universe, and general relativity requires $R \to 0$ at the final singularity. The absence of event horizons by definition means that the universe's future c-boundary is a single point, call it the *Omega Point*. MacCallum has shown that an S^3 closed universe with a single point future c-boundary is of measure zero in initial data space. Barrow [6] has shown that the evolution of an S^3 closed universe into its final singularity is chaotic. Yorke [7] has shown that a chaotic physical system is likely to evolve into a measure zero state if and only if its control parameters are intelligently manipulated. Thus life ($\equiv$ intelligent computers) almost certainly must be present *arbitrarily close* to the final singularity in order for the known laws of physics to be mutually consistent at all times. Misner has shown in effect that event horizon elimination requires an infinite number of distinct manipulations, so an infinite amount of information must be processed between now and the final singularity. The amount of information stored at any time diverges to infinity as the Omega Point is approached, since $S \to +\infty$ there, implying divergence of the complexity of the system that must be understood to be controlled.

THE TOPOLOGY OF BLACK HOLE EVENT HORIZONS

If event horizons do not exist, then strictly speaking neither do black holes. However, astrophysical black holes exist, and I have constructed [10] a spherically symmetric spacetime satisfying all the energy conditions which shows how this is possible: the spacetime is identical to a dust-filled closed universe with black holes in the expanding phase, but it has no event horizons. So the non-existence of event horizons does not contradict any observation on astrophysical black holes. However, this model does show that we have to take the theorems (e.g. [3], [4], [5]) proving black hole horizons to be 2-spheres with a grain of salt. These theorems must in effect make an assumption about the topology of scri [2], and use the fact that event horizon generators cannot leave the horizons in the future direction. However, if the universe is closed with the future c-boundary a single point, then it is easy to show that the boundaries of all future sets which lie in the future of some Cauchy surface must be Cauchy surfaces. Thus the null generators of an astrophysical black hole pseudo-horizon [10] must leave the black hole surface both in the future and the past, since I have argued that the universe must have S^3 Cauchy surfaces. The toroidal horizons of Hughes et al [11] are a slicing phenomenon in spacetimes with the standard scri [12], and can exist for only period $\sim M$ because in standard scri spacetimes, eventually horizon null generators must cease to enter the horizon, and once on the horizon, can never leave it. But neither is necessarily true if only pseudo-horizons exist. Thus if unitarity and standard quantum gravity are both true, higher genus black holes MAY exist.

THE VALUE OF THE COSMOLOGICAL CONSTANT

Current observations show an accelerating universe with $\Omega_\Lambda = 2/3$, and if the universe were to continue to accelerate, black holes would evaporate, violating unitarity. Hence, unitarity requires that the acceleration will eventually return to a de-acceleration, followed by a recollapse. Now Gibbons and Hawking [13] have shown that the vacuum energy in de Sitter space generates thermal Hawking radiation with a temperature of $T_H^{deS} = (\hbar/2\pi k_B)\sqrt{\Lambda/3} = 3.9397 \times 10^{-30} h\sqrt{\Omega_\Lambda}$ K, or $T_H^{deS} = 2.25 \times 10^{-30}$ degrees Kelvin with $h = 0.70$ and $\Omega_\Lambda = 2/3$. So if the cosmological constant were never to be canceled, any black hole with a Hawking temperature greater than this would eventually evaporate and violate unitarity. Thus only black holes with a mass greater than $\sim 10^{25} M_\odot$ — more mass than there is in the visible universe — could avoid evaporation.

In fact, we can use the same argument to show why the cosmological constant must be near zero, instead of being its expected value of $\Lambda = 8\pi G \rho_{vac} = 8\pi G(c^5/\hbar G^2) = 8\pi c^5/G\hbar$ given by the Planck density. If the cosmological constant were this large, in a universe that expands forever, there would be a finite (though extremely small) probability that vacuum or other density fluctuations would give rise to a black hole smaller than the de Sitter horizon $R^{deS} = c\sqrt{3/\Lambda}$, which would have a higher Hawking temperature than the de Sitter background temperature, and hence evaporate, violating unitarity. So ultimately, the cosmological constant is near zero because a large cosmological constant would violate unitarity.

REFERENCES

1. Tipler, F. J., *The Physics of Immortality* (New York: Doubleday), 1994.
2. Brill, D. R. et al, *Phys. Rev.* **D56**, 3600 (1997).
3. Galloway, G. *Comm. Math. Phys.* **151**, 53 (1993).
4. Chrusciel, P. T. and R. M. Wald, *Class. Quan. Grav.* **11**, L147 (1994).
5. Jacobson, T. and S. Venkataramani, *Class. Quan. Grav.* **12**, 1055 (1995).
6. Barrow, J. D. *Phys. Reports* **85**, (1982).
7. Yorke, J. A. et al, *Phys. Rev. Lett.* **68**, 2863 (1992).
8. Schiffer, M. and J. D. Bekenstein, *Phys. Rev.*, **D39**, 1109 (1989).
9. Barrow, J. D. and F. J. Tipler, *Mon. Not. R. Astr. Soc.*, **216**, 395 (1985).
10. Tipler, F. J. et al, *gr-qc/0003082*.
11. Hughes, S. A. et al, *Phys. Rev.*, **D49**, 4004 (1994).
12. Galloway, G. J. et al, *gr-qc/9902061*.
13. Gibbons, G. W. and S. W. Hawking, *Phys. Rev.*, **15**, 2738 (1977).
14. Weinberg, S., *Rev. Mod. Phys.*, **61**, 1 (1989).

CHAPTER 15

GRAVITY WAVE SIGNATURES

Post-Newtonian SPH Simulations of Binary Neutron Stars

Joshua A. Faber and Frederic A. Rasio

Department of Physics, M.I.T., 77 Massachusetts Ave., Cambridge, MA 02139

Abstract. Using our Post-Newtonian SPH (smoothed particle hydrodynamics) code, we study the final coalescence and merging of neutron star (NS) binaries. We find that the gravity wave signals can be computed accurately for irrotational systems in calculations of sufficient resolution, even in the presence of Kelvin-Helmholtz instabilities.

INTRODUCTION

Coalescing binary neutron stars (NS) are among the most promising sources of gravitational radiation that should be detectable by future generations of gravity wave detectors. LIGO, VIRGO, GEO, and TAMA may ultimately not only serve to test the predictions of the theory of general relativity (GR), but could also yield important information on the interior structure of neutron stars, which cannot be obtained directly in any other way.

Essentially all recent calculations agree on the basic picture that emerges during the final coalescence (see [1] and [2] for a complete list of references). As the binary approaches the dynamical stability limit, located at separations of $r = 3-4R_{NS}$, the NS plunge together rapidly and merge within a few rotation period. Mass shedding typically commences immediately after first contact, especially for initially synchronized systems. This material forms a pair of spiral arms before dissipating to form a dusty torus around the merger remnant. For stiff equations of state (EOS), the merger remnant can support a long-lived ellipsoidal (triaxial) deformation, which will radiate a significant level of gravity waves well after the merger is completed. Softer EOS relax toward spheroidal, non-radiating configurations on a dynamical timescale.

Of course, all statements about merger remnants assume that the remnant formed does not immediately collapse to a black hole (BH). Unfortunately Newtonian simulations are incapable of demonstrating such an effect. Instead, Newtonian calculations, using both Eulerian, grid based codes [3–9], and particle-based SPH methods [10–14], have studied many aspects of coalescing binaries, including the dependence of gravity wave signals on the initial spins, binary mass ratio, and NS EOS. Unfortunately, Newtonian gravity is a poor approximation to the physical situation being studied. NS have extremely deep gravitational potentials, especially in their cores, even for very stiff EOS, and reach relativistic speeds during coalescence. Thus, proper treatment of the problem requires taking into account GR effects. Perhaps the most important effect of general relativity is to change the location of the dynamical stability limit, since relativistic corrections, even when small, can greatly affect the location of the minimum equilibrium energy E_{equil} as a function of separation, altering the point at which the binary begins its final rapid plunge toward merger.

Several groups have been working on full GR calculations [15], but only preliminary results have been reported so far. Extracting waveforms from the boundaries of grids has proves to be particlarly difficult. The middle ground between Newtonian calculations and full GR lies with PN hydrodynamics treatments. We [1,16], as well as Ayal et al. [17], have constructed a PN SPH code to calculate binary mergers. Our code uses the formalism of Blanchet, Damour, and Schaefer [18], adapted to a Lagrangian SPH framework. While it has proven impossible to include full-strength first-order (1PN) relativistic corrections, since they are often of comparable magnitude to Newtonian quantities for realistic NS parameters, we have devised a formalism whereby 1PN corrections are treated at reduced strength, while radiation reaction effects (2.5PN) are included at their physical values. This formalism is described in detail in [16].

IRROTATIONAL BINARY COALESCENCE

Our most detailed calculation performed to date uses $N = 500,000$ particles per NS, corresponding to the highest spatial resolution ever for a binary coalescence calculation. The calculation was performed using an irrotational initial condition. This is generally thought to be the most realistic case since the viscous tidal locking timescale for two NS is expected to be considerably longer than the inspiral timescale [19]. We model the initial density and velocity profile of the NS as tidally stretched ellipsoids, with parameters drawn from the PN equlibrium calculations of Lombardi, Rasio, and Shapiro [20]. We choose equal mass NS, and use a $\Gamma = 3$ polytropic EOS. Particle plots showing the evolution of the equal-mass irrotational binary system are shown in Fig. 1.

Immediately prior to merger, we see that a large tidal lag angle develops, as the inner edge of each NS leads the axis connecting the respective centers of mass, with the outer regions lagging behind. This is seen in Newtonian calculations, but is greatly enhanced by the addition of 1PN terms. Unlike the standard results from synchronized calculations, we do not see significant mass shedding from the system. The rotational speed of particles on the outer half of each NS is reduced in the irrotational case with respect to the synchronized case, and such particles are never ejected. The amount of mass shed is extremely small, much less than 1% of the total mass, and the velocities of the particles ejected are not sufficient to escape the gravitational potential of the remnant.

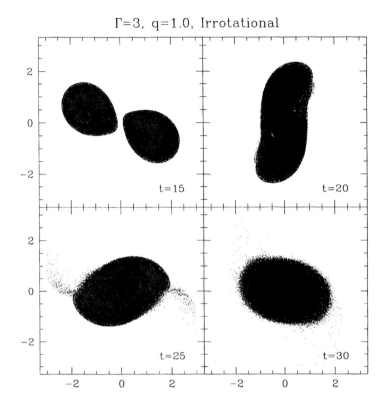

FIGURE 1. Final merger of two identical $\Gamma = 3$ polytropes with an irrotational initial condition. SPH particles are projected onto the equatorial plane of the binary. The orbital rotation is counterclockwise. Spatial coordinates are given in units of the NS radius R. Times are given in units of the dynamical timescale of the system, which here is $t_D = 0.07\text{ms} = 1$.

Density contours and velocity profiles in the equatorial plane of the binary are shown in Fig. 2. Velocities are shown in a frame corotating with the material, which highlights the Kelvin-Helmholtz unstable vortex sheet which forms along the surface of contact between the two NS. Large vortices form along this sheet, mixing material from the two NS, from $t = 20 - 25$. However, during this time the respective NS cores continue to inspiral toward the center of the forming remnant, until by $t = 30$ they have merged to form a single core, the vortices having merged together as well. This produces a characteristic differential rotation pattern, with the core spinning approximately twice as fast as the outer regions.

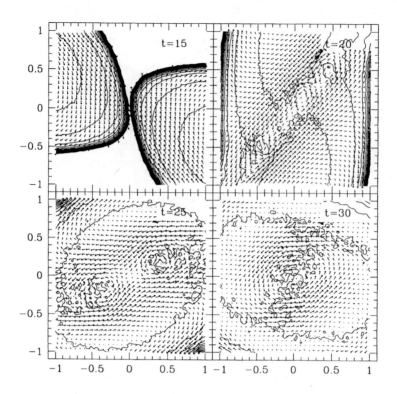

FIGURE 2. Density contours and velocities along the equatorial plane in the corotating frame of the binary, for the same times as in Fig. 1.

GRAVITY WAVE SIGNALS AND SPECTRA

We calculate the gravity wave signal for our mergers in the quadrupole approximation. The gravity wave strain h seen by an observer located a distance d from the center of mass of the system along the rotation axis is given for the two polarizations by

$$c^4(dh_+) = \ddot{Q}_{xx} - \ddot{Q}_{yy} \tag{1}$$
$$c^4(dh_\times) = 2\ddot{Q}_{xy} \tag{2}$$

where $\ddot{Q}$, the second time derivative of the quadrupole moment tensor, is given in SPH terms by

$$\ddot{Q}_{ij} = \sum_b m_b(v_i^{(b)} v_j^{(b)} + x_i^{(b)} \dot{v}_j^{(b)} + x_j^{(b)} \dot{v}_i^{(b)}) \tag{3}$$

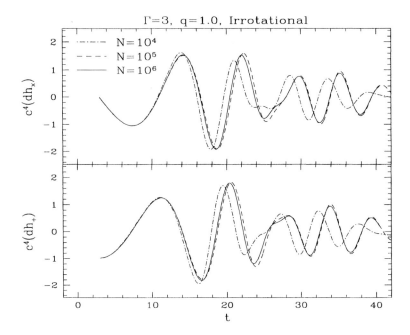

FIGURE 3. Gravity wave signals calculated for coalescences with the same initial parameters but different numerical resolutions. The solid, dashed, and dot-dashed lines correspond to runs with 10^6, 10^5, and 10^4 SPH particles, respectively.

where the summation is taken over all the particles in the calculation. In Fig. 3, we show the gravity wave signals in both polarizations, h_+ and $h_\times$, for the irrotational run described above, as well as for runs with $N = 50,000$ particles and $N = 5,000$ particles per NS, calculated as described in [10]. It is immediately apparent that the lowest resolution run shows significant discrepancies from the other two, which agree with each other quite well over the entire time history of the merger. This is a welcome result, given that the vortex sheet appearing at the contact surface is Kelvin-Helmholtz unstable on all size scales, including those much smaller than our numerical resolution. Even though differences in the exact location and size of vortices were apparent for runs of differing resolution, we found that the paths traced out by the respective NS cores, which make up the dominant contribution to the GW signal, were almost identical. The conclusion to be drawn is that numerical convergence for a given set of initial conditions and physical assumptions is possible without requiring excessive computational resources, even for this difficult problem involving small-scale instabilities.

ACKNOWLEDGMENTS

This work was supported in part by NSF Grants AST-9618116 and PHY-0070918 and NASA ATP Grant NAG5-8460. The computations were supported by the National Computational Science Alliance under grant AST980014N and utilized the NCSA SGI/CRAY Origin2000.

REFERENCES

1. Faber, J.A., Rasio, F.A., and Manor, J.B., *Phys. Rev. D* **63**, 044012 (2001).
2. Rasio, F.A., and Shapiro, S.L., *Class. Quant. Grav.* **16**, R1-R29 (1999).
3. Oohara, K., and Nakamura, T., *Prog. Theor. Phys.* **82**, 535-554 (1989); *ibid.* **83**, 906-940 (1990); Nakamura, T. and Oohara, K., *ibid.* **82**, 1066-1083 (1989); *ibid.* **86**, 73-88 (1991).
4. Shibata, M., Oohara, K., and Nakamura, T., *Prog. Theor. Phys.* **88**, 1079-1095 (1992); *ibid.* **89**, 809-819 (1993).
5. New, K.C.B., and Tohline, J.E., *Astrophys. J.* **490**, 311-327 (1997).
6. Swesty, F.D., Wang, E.Y.M., and Calder, A.C., *Astrophys. J.* **541**, 937-958 (2000).
7. Ruffert, M., Janka, H.-Th., and Schäfer, G., *Astron. Astrophys.* **311**, 532-566 (1996).
8. Ruffert, M., Janka, H.-Th., Takahashi, K., and Schäfer, G., *Astron. Astrophys.* **319**, 122-153 (1997).
9. Ruffert, M., Rampp, M., and Janka, H.-Th., *Astron. Astrophys.* **321**, 991-1006 (1997).
10. Rasio, F.A., and Shapiro, S.L., *Astrophys. J.* **401**, 226-245 (1992); *ibid.* **432**, 242-261 (1994); *ibid.* **438**, 887-903 (1995).
11. Zhuge, X., Centrella, J., and McMillan, S., *Phys. Rev. D* **50**, 6247-6261 (1994); *ibid.* **54**, 7261-7277 (1996).
12. Davies, M.B. et al., *Astrophys. J.* **431**, 742-753 (1994).
13. Rosswog, S. et al., *Astron. Astrophys.* **341**, 499-526 (1999).
14. Rosswog, S. et al., *Astron. Astrophys.* **360**, 171-184 (2000).
15. Baumgarte, T.W., Hughes, S.A., and Shapiro, S.L., *Phys. Rev. D.* **60**, 087501 (1999); Shibata, M., *Phys. Rev. D* **60**, 104052 (1999).
16. Faber, J.A. and Rasio, F.A., *Phys Rev. D* **62**, 064012 (2000).
17. Ayal, S. et al., *Astrophys. J.*, accepted, astro-ph/9910154.
18. Blanchet, L., Damour, T., and Schäfer, G., *Mon. Not. Roy. Astron. Soc.* **242**, 289-305 (1990).
19. Bildsten, L., and Cutler, C., *Astrophys. J.* **400**, 175-180 (1992).
20. Lombardi, J.C., Rasio, F.A., and Shapiro, S.L., *Phys. Rev. D* **56**, 3416-3438 (1997).

Gravitational Radiation Evolution of Accreting Neutron Stars

Robert V. Wagoner*, Joseph F. Hennawi† and Jingsong Liu*

Dept. of Physics, Stanford University, Stanford, CA 94305-4060
†*Dept. of Astrophysical Sciences, Princeton University, Princeton, NJ 08544-1001*

Abstract. The gravitational-wave and accretion driven evolution of neutron stars in low mass X-ray binaries and similar systems is analyzed, while the amplitude of the radiating perturbation (here assumed to be an r-mode) remains small. If most of the star is superfluid with mutual friction dominating the ordinary shear viscosity, the amplitude of the mode and the angular velocity of the star oscillate about their equilibrium values with a period of at least a few hundred years. The resulting oscillation of the neutron star temperature is also computed. For temperature dependent viscosity, the conditions for the equilibrium to be stable are found.

INTRODUCTION

We shall study the evolution of rapidly rotating accreting neutron stars under the influence of their emission of gravitational radiation. We modify and extend the two-component model of the star (equilibrium plus perturbation) introduced by Owen et al. [17] and also employed by Levin [11], but restrict the analysis to small perturbations. These are assumed to be in the form of r-modes [1, 7, 15, 2], which radiate mainly via Coriolis-driven velocity perturbations rather than the density perturbations of the less powerful f-modes. We must allow for large uncertainties in many of the relevant properties of neutron stars, such as the superfluid transition temperature and the effects of magnetic fields and a boundary layer.

After developing a general formalism, we shall study the evolution of the accreting neutron star under conditions in which its angular velocity remains approximately constant. One reason for this restriction is our interest in conditions in which the gravitational radiation is persistent over long time scales. This requires the existence of a stable or (possibly) overstable equilibrium. We shall see under what conditions the system can evolve toward this equilibrium, in which the rate of accretion of angular momentum from the surrounding disk is balanced by its rate of loss via gravitational radiation. If this equilibrium is achieved, the observed flux of gravitational radiation can be shown to be proportional to the observed flux of X-rays from the accretion [21, 4].

One of our longer term goals is the development of parameterized expressions describing possible time evolutions of the gravitational-wave frequency and amplitude, to facilitate detection by LIGO, VIRGO, and similar laser interferometer detectors. The brightest low mass X-ray binaries (LMXBs) are the prime targets.

DYNAMICAL AND THERMAL EVOLUTION

Consider a Newtonian neutron star in equilibrium (with equatorial radius R) which is perturbed by a nonaxisymmetric infinitesimal fluid displacement $\vec{\xi} = \vec{f}(r,\theta)e^{i(m\phi+\sigma t)} \sim \alpha R$, with $\alpha \ll 1$. Based on the work of Friedman & Schutz [8] and Levin & Ushomirsky [12], the total angular momentum J of the star can be decomposed into its equilibrium angular momentum J_* and a perturbation proportional to the canonical angular momentum J_c. That is,

$$J = J_*(M,\Omega) + (1-K_j)J_c, \qquad J_c = -K_c\alpha^2 J_*, \tag{1}$$

where M is the mass and Ω is the (uniform) angular velocity of the equilibrium star. All constants $K_{()}$ will be dimensionless, with $K_j \sim K_c \sim 1$.

In classical mechanics, the action $I = E/\omega$ of any normal mode of a set of oscillators (with frequency ω) is an adiabatic invariant. For a fluid, the analogous quantity should be $\tilde{E}_c/\omega$, where $\tilde{E}_c$ is the canonical energy of the perturbation in the corotating frame and $\omega = \sigma + m\Omega$ is its frequency in that frame. However, we also have the general relation $\tilde{E}_c = -(\omega/m)J_c$ [8]. Therefore, following Ho & Lai [10], we assume that the canonical angular momentum is also an adiabatic invariant, and should therefore be unaffected by the slow rate of mass accretion. Thus it obeys the usual relation [9]

$$dJ_c/dt = 2J_c[(F_g(M,\Omega) - F_v(M,\Omega,T)], \tag{2}$$

where F_g is the gravitational radiation growth rate and F_v is the viscous damping rate. The latter usually depends upon a spatially averaged temperature $T(t)$.

On the other hand, conservation of total angular momentum requires that

$$dJ/dt = 2J_cF_g + \dot{J}_a(t), \tag{3}$$

where $\dot{J}_a = j_a\dot{M}$ is the rate of accretion of angular momentum. The mass is accreted with specific angular momentum j_a at a rate $\dot{M}(t)$.

Combining these equations then gives the dynamical evolution relations

$$\frac{1}{\alpha}\frac{d\alpha}{dt} = F_g - F_v + [K_jF_g + (1-K_j)F_v]K_c\alpha^2 - \left(\frac{j_a}{2J_*}\right)\dot{M}(t), \tag{4}$$

$$\left(\frac{I_*}{J_*}\right)\frac{d\Omega}{dt} = -2[K_jF_g + (1-K_j)F_v]K_c\alpha^2 + \left[\frac{(j_a-j_*)}{J_*}\right]\dot{M}(t); \tag{5}$$

where $I_*(M,\Omega) = \partial J_*/\partial\Omega$ and $j_*(M,\Omega) = \partial J_*/\partial M$.

In obtaining our thermal evolution relation, the large uncertainties in some thermodynamic properties of the neutron star make it sufficient to consider slowly rotating stars. Thermal energy conservation for the entire star then gives

$$\int \frac{\partial T}{\partial t}c_v dV \equiv C(T)\frac{dT}{dt} \simeq 2\tilde{E}_cF_v(T_v) + K_n\langle\dot{M}\rangle c^2 - L_\nu(T_\nu), \tag{6}$$

where the rotating frame canonical energy $\tilde{E}_c = K_e \Omega J_* \alpha^2$. Since the main contributor to the specific heat is the degenerate relativistic electrons, its value at constant volume (c_v) is essentially the same as that at constant pressure. The emissivities are produced by viscous heating, pycnonuclear reactions and neutron emissions in the inner crust (proportional to a time-averaged mass accretion rate), and neutrino losses. The hydrogen/helium burning rate is assumed to be balanced by the surface emission of photons [20], especially at the large accretion rate $\dot{M} = 10^{-8} M_\odot$ yr^{-1} that we shall use. The mass accretion rate can be estimated from accretion energy conservation. The photon luminosity arising directly from the accretion is $L_{acc} \approx (GM/R)\dot{M}(t)$, for a slowly rotating neutron star with a negligible magnetosphere.

We are interested in the evolution of neutron stars after they have been spun up to the point where the gravitational radiation growth rate has become equal to the viscous damping rate:

$$F_g(\Omega_0, M_0) = F_v(\Omega_0, M_0, T_0) \equiv F_0 , \qquad (7)$$

so the evolution equation (2) vanishes. This equality defines our initial state. Before that time, we see from equation(2) that any intrinsic perturbation could not grow from its (infinitesimal) value α_{min}. The initial temperature T_0 is determined by the vanishing of equation (6), with the nuclear heating in the inner crust balanced by the neutrino emission [6]. Since we are only considering conditions in which $\alpha^2 \ll 1$, the properties $\Omega(t)$ and $M(t)$ evolve much slower than $\alpha(t)$ and $T(t)$. For any other property Q_* of the unperturbed star, let $Q_0 \equiv Q_*(M_0, \Omega_0)$.

From now on we shall take the perturbation to be due to the dominant $l = m = 2$ r-mode. In order to facilitate comparison with previous results, we shall adopt the neutron star model of Owen et al. [17] for numerical work. Then the gravitational radiation growth rate of this mode is

$$F_g = \tilde{\Omega}^6 / \tau_{gr} , \quad \tau_{gr} = 3.26 \text{ sec} , \quad \tilde{\Omega} \equiv \Omega(\pi G \langle \rho \rangle)^{-1/2} . \qquad (8)$$

In the temperature range of interest ($10^8 < T < 10^{10}$ K), the viscous damping rate of this mode is approximated as

$$F_v \cong \left(\frac{\tilde{\Omega}^5}{\tau_{mf}}\right) e^{-(T/T_c)^2} + \frac{1}{\tau_{sh}} \left(\frac{10^9 \text{ K}}{T}\right)^2 , \qquad (9)$$

where T_c is the superfluid transition temperature.

The first term is the contribution from the mutual friction between the neutron superfluid and the superconducting proton–relativistic electron fluid [14]. Its behavior (and that of other properties considered below) as the temperature passes through the superfluid transition temperature T_c is approximated by the exponential. Although Lindblom & Mendell [14] found that $\tau_{mf} \lesssim 10^4$ sec, we shall keep it as a free parameter because of the many uncertainties involved in the physics of this system, especially when including magnetic effects such as the interaction of the vortex lines and flux tubes [19]. In fact, we fix it to satisfy our initial condition (7), when observationally relevant values of Ω_0, M_0, T_0 are chosen. To approximately match an inferrred maximum spin rate of 330 Hz for the neutron stars in LMXBs, we shall choose $\tilde{\Omega}_0 = \Omega_0(\pi G \langle \rho \rangle)^{-1/2} = 0.25$ [11], which then fixes $\tau_{mf} \approx 13$ sec if $T \ll T_c$. The maximum rotation rate of a neutron star is $\Omega_{max} \cong (2/3)(\pi G \langle \rho \rangle)^{1/2}$.

The second term is the contribution of the ordinary shear viscosity. Lindblom & Mendell [14] obtained $\tau_{sh} = 1.0 \times 10^8$ sec for their superfluid neutron star model, dominated by electron–electron scattering in the superfluid regions and neutron–neutron scattering in the normal regions. Above the superfluid transition temperature, Lindblom, Owen & Morsink [15] obtained $\tau_{sh} = 2.5 \times 10^8$ sec. There are also large uncertainties in this contribution, due to the shear in a boundary layer between the core and crust [3, 5, 16, 18, 22] and the uncertain response of the crust to the mode [13].

In keeping with the fact that we are only working to lowest order in $\tilde{\Omega}$ (as well as the relativity parameter GM/Rc^2), we take $j_a - j_0 \cong j_a$ and $J_0 \cong I_0 \Omega_0$. We also note that $K_c = 3K_e = 0.094$. (The value of K_j is unimportant.)

Now that we have specified all properties in the equations (4) and (5) of evolution of $\alpha(t)$ and $\Omega(t)$, we can consider the thermal evolution. In what follows we shall assume that thermal conductivity timescales are short enough to give relations $T_v(T)$ and $T_\nu(T)$ between these three spatially averaged temperatures that appear in equation (6). The normal and superfluid contribution to the specific heat and the neutrino luminosity that appear in this equation are approximated by

$$C(T) = [C'_{norm} e^{-(T_c/T)^2} + C'_{super}]T \qquad (C'_{norm} \cong 20 C'_{super}), \qquad (10)$$

$$L_\nu(T) = L'_{URCA} T^8 e^{-(T_c/T)^2} + L'_{brem} T^6. \qquad (11)$$

The constants L'_{URCA} and L'_{brem} are obtained by fitting the results of Brown [6] for normal and superfluid neutron stars. We also take the nuclear heating constant $K_n = 1 \times 10^{-3}$ [6].

In Figure 1 we show the result of integrating our three coupled evolution equations (4), (5), and (6), if $T_c \gg T_0$. The main feature is the spin-down (due to loss of angular momentum in gravitational waves) and viscous heating during the time when the amplitude α of the mode exceeds its equilibrium value (denoted by the dashed line).

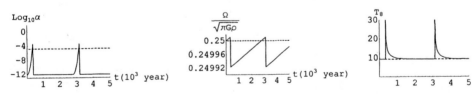

FIGURE 1. The evolution of the mode amplitude α, angular velocity $\Omega(\pi G \langle \rho \rangle)^{-1/2}$, and the temperature $T/10^8$ K; for the case $T_c \gg T_0$. Subsequent cycles are similar.

For such large values of T_c, the evolution qualitatively resembles that which occurs when $\partial F_v/\partial T = 0$ (requiring neglect of the shear viscosity). In this case equations (4) and (5) decouple from (6). In addition, $x \equiv \ln \alpha$ completely decouples, obeying the equation

$$d^2x/dt^2 - 2K_c F_0 e^{2x} dx/dt + dV/dx = 0, \quad V(x) \cong F_0(K_c F_0 e^{2x} - F_a x), \quad (12)$$

where F_a is the time-averaged rate of accretion of angular momentum. The sign of the damping term is opposite to that obtained by Levin [11], leading to overstable oscillations about the equilibrium amplitude $\alpha_e \cong (F_a/2K_c F_0)^{1/2}$ with a period $P \cong [8\ln(\alpha_{max}/\alpha_{min})/(F_0 F_a)]^{1/2} \gtrsim 300$ years (ten times shorter than in Figure 1). The fraction of the time that $\alpha > \alpha_e$ is approximately $\ln[8\ln(\alpha_{max}/\alpha_{min})]/4\ln(\alpha_{max}/\alpha_{min}) \sim 0.1$, ten times greater than in Figure 1. It is also found that α_{max} increases and α_{min} decreases on a timescale $\sim P/\alpha_{max}$. However, α_{min} is presumably limited by the intrinsic fluctuations in the neutron star, which is also why we have not allowed α to drop below its initial value in Figure 1.

On the other hand, for values of $T_c \lesssim T_0$, the temperature dependence of the viscous damping rate F_v is strong enough to produce a thermal runaway to large values of T and α (outside the range of validity of our equations), as found by Levin [11]. This result is generalized in the next section.

BEHAVIOR NEAR EQUILIBRIUM

In contrast to the initial state, the equilibrium state X_e^i of our dynamical variables

$$X^i(t) = \{\alpha, \Omega, T\} = X_e^i[1 + \zeta^i(t)], \quad |\zeta^i| \ll 1,$$

is defined by the vanishing of the evolution equation (3), in addition to (2) and (6). Employing a constant (averaged) accretion rate, the evolution equations give

$$d\zeta^i/dt = A^{ij}\zeta^j, \quad \zeta^i \propto \exp(\lambda t), \quad ||A^{ij} - \lambda \delta^{ij}|| = 0. \quad (13)$$

Assume now that $|\partial F_v/\partial T| \sim F_v/T_e$, etc.

The coefficients of the eigenvalue equation $\lambda^3 + a_2\lambda^2 + a_1\lambda + a_0 = 0$ are

$$a_2 \cong \frac{1}{C_e}\left[\left(\frac{\partial L}{\partial T}\right)_e - 2(\tilde{E}_c)_e\left(\frac{\partial F_v}{\partial T}\right)_e\right] \sim K_r \alpha_e^2 F_0, \quad (14)$$

$$a_1 \cong \frac{4(\tilde{E}_c)_e F_0}{C_e}\left(\frac{\partial F_v}{\partial T}\right)_e \sim K_r \alpha_e^2 F_0^2, \quad (15)$$

$$a_0 \cong \frac{4K_c \Omega_e \alpha_e^2 F_0}{C_e}\left[\frac{\partial(F_g - F_v)}{\partial \Omega}\right]_e \left(\frac{\partial L}{\partial T}\right)_e$$

$$- \frac{16K_c(\tilde{E}_c)_e \alpha_e^2 F_0^2}{C_e}\left(\frac{\partial F_v}{\partial T}\right)_e \sim K_r \alpha_e^4 F_0^3. \quad (16)$$

The ratio of rotational to thermal energy is $K_r \equiv 2K_e\Omega_e J_0/C_e T_e \sim 10^5$. We have used the fact that the cooling rate $F_c \equiv L_\nu(T_e)/C_e T_e \sim K_r \alpha_e^2 F_0$.

Now we also employ the inequalities $K_r \gg 1$ and (mode energy)/(thermal energy) $\sim K_r \alpha_e^2 \lesssim 10^{-4} \Rightarrow |a_1| \gg a_2^2$ to obtain the eigenvalues

$$\lambda_{1,2} \cong -a_2/2 \pm \sqrt{-a_1}, \qquad \lambda_3 \cong -a_0/a_1.$$

We have used the fact that $|\lambda_3| \sim \alpha_e^2 F_0 \ll |\lambda_1| \sim |\lambda_2|$.

Let us examine the two relevant possibilities. For cases such as dominance by superfluid shear viscosity (produced by $e-e$ scattering), with $\tilde{E}_c > 0$,

$$a_1 \propto (\tilde{E}_c)_e (\partial F_v/\partial T)_e < 0 \Longrightarrow \lambda_{1,2} \cong \pm\sqrt{-a_1} \sim K_r^{1/2} \alpha_e F_0.$$

Thus this equilibrium is unstable, with a growth rate λ_1 that is of the same magnitude as found by Levin (1999).

The other possibility $a_1 > 0 \Longrightarrow \lambda_{1,2} \cong -a_2/2 \pm i\sqrt{a_1}$. Thus stability also requires that $a_0 > 0$ and $a_2 > 0$. From their relations above, we see that this means that the variation in the cooling rate with temperature must be greater than twice the variation in the viscous heating rate with temperature (usually satisfied).

This work was supported in part by NSF grant PHY–0070935 to R.V.W. and NASA grant NAS 8-39225 to Gravity Probe B. R.V.W thanks the Aspen Center for Physics for support during a 1999 summer workshop, and benefitted from many discussions during the 2000 program on Spin and Magnetism in Young Neutron Stars at the Institute for Theoretcial Physics, U.C. Santa Barbara.

REFERENCES

1. Andersson, N. 1998, *Ap. J.*, **502**, 708
2. Andersson, N., Kokkotas, K.D. & Schutz, B.F. 1999 *Ap. J.*, **510**, 846
3. Andersson, N., Jones, D.I., Kokkotas, K.D. & Stergioulas, N. 1999, *Ap. J.*, **534**, L75
4. Bildsten, L. 1998, *Ap. J.*, **501**, L89
5. Bildsten, L. & Ushomirsky, G. 2000, *Ap. J.*, **529**, L33
6. Brown, E.F. 1999, *Ap. J.*, **531**, 988
7. Friedman, J.L. & Morsink, S.M. 1998, *Ap. J.*, **502**, 714
8. Friedman, J.L. & Schutz, B.F. 1978, *Ap. J.*, **221**, 937
9. Friedman, J.L. & Schutz, B.F. 1978, *Ap. J.*, **222**, 281
10. Ho, W.C.G. & Lai, D. 2000, *Ap. J.*, **543**, 386
11. Levin, Y. 1999, *Ap. J.*, **517**, 328
12. Levin, Y. & Ushomirsky, G. 2000, *M.N.R.A.S.*, submitted (astro-ph/9911295)
13. Levin, Y. & Ushomirsky, G. 2001, *M.N.R.A.S.*, submitted (astro-ph/0006028)
14. Lindblom, L & Mendell, G. 2000, *Phys. Rev. D*, **61**, 104003
15. Lindblom, L., Owen, B.J. & Morsink, S.M. 1998, *Phys. Rev. Lett.*, **80**, 4843
16. Lindblom, L., Owen, B.J. & Ushomirsky, G. 2000, *Phys. Rev. D*, **62**, 084030
17. Owen, B.J., Lindblom, L., Cutler, C., Schutz, B.F., Vecchio, A. & Andersson, N. 1998, *Phys. Rev. D*, **58**, 084020
18. Rieutord, M. 2000, *Ap. J.*, submitted (astro-ph/0003171)
19. Ruderman, M. 2000, private communication
20. Schatz, H., Bildsten, L., Cumming, A. & Wiescher, M. 1999, *Ap. J.*, **524**, 1014
21. Wagoner, R.V. 1984, *Ap. J.*, **278**, 345
22. Wu, Y., Matzner, C.D. & Arras, P. 2001, *Ap. J.*, **549**, 1011

Gravitational wave background from coalescing compact stars in eccentric orbits

A. G. Kuranov*, V. B. Ignatiev*, K. A. Postnov*,**,† and M. E. Prokhorov**

Physical Department, Moscow State University, 117234 Moscow, Russia
†*Max-Planck Institut für Astrophysik, 85740 Garching, German*
**Sternberg Astronomical Institute, 119899 Moscow, Russia*

Abstract.
Gravitational wave background produced by a stationary coalescing population of binary neutron stars in the Galaxy is calculated. This background is found to constitute a confusion limit within the LISA frequency band up to a limiting frequency $v_{lim} \sim 10^{-3}$ Hz, leaving the frequency window $\sim 10^{-3}$–10^{-2} Hz open for the potential detection of cosmological stochastic GW.

INTRODUCTION

The forthcoming gravitational wave (GW) interferometers (see [1] for a review) will be capable of detecting stochastic GW. Such waves (GW backgrounds, or GW noises) can be produced by a large collection of unresolved individual astrophysical sources, e.g. by binary stars [2, 3, 4] or rapidly rotating neutron stars (NS) [5]. These GW backgrounds are usually considered as unwanted additional noises to the intrinsic noises of GW detectors, since they could potentially mimic stochastic GW backgrounds of cosmological origin (primordial or relic GW) that bring a valuable information about physical processes in the early Universe [7] and references there).

Ordinary galactic binaries, in which at least one of the component is a normal main-sequence star, mostly contribute at low-frequency band ($v \sim 10^{-7}$–10^{-5} Hz) [2], where no GW detections will probably be possible in the foreseen future. Compact coalescing white dwarf (WD) binaries form a noticeable background above the LISA noise curve [3], [4], [7]. Generally, the level of the GW noise from a collection of N unresolved independent sources in terms of dimensionless amplitude $h \propto \sqrt{N}$. In case of binaries that loose the orbital angular momentum due to gravitational radiation back reaction, this level is $h \propto \sqrt{\mathcal{R}}$, where $\mathcal{R}$ is the coalescing rate of these binaries (see [7] for more detail).

Another important quantity is the upper frequency v_{lim} above which each individual source can be resolved during one-year observation time (i.e. within the frequency bin $\Delta v = 3 \times 10^{-8}$ Hz). In fact, it is this limiting frequency which mostly relates to the challenging task of detection of a relic GW background: at $v > v_{lim}$ we will be able in principle to single out individual sources and thus have prospects to register cosmological GW noise even by one interferometer [7].

In the simplest case of a collection of binary WD in circular orbits which coalesce at a constant rate $\mathcal{R}_{300} = 1/300 \text{ yr}^{-1}$ the obvious condition for v_{lim} reads $\mathcal{R}\Delta T(v_{lim}) = 1$, where $\Delta T(\nu)$ is the time each source "spends" within the frequency bin at a given frequency ν

$$\nu_{lim}(WD) \approx (1.2 \times 10^{-3} \text{Hz}) \mathcal{R}_{300}^{3/11} \left(\frac{\Delta \nu}{3 \times 10^{-8} \text{Hz}}\right)^{3/11} \left(\frac{\mathcal{M}}{0.52 M_\odot}\right)^{-5/11}$$

where $\mathcal{M}$ is the chirp mass of the binary. For merging binary neutron stars (NS) ν_{lim} calculated as above would give even a smaller value, $\nu_{lim}(NS) = 3 \times 10^{-4}$ Hz, since the binary NS coalescence rate in most optimistic scenarios is $\mathcal{R}_{NS} \sim 10^{-4}$ yr^{-1} [7, 6]. Note that the uncertainty in this rate by a factor of 3 or so is of minor importance since $\nu_{lim} \propto \mathcal{R}^{3/11}$.

An important difference between the merging binary WD and NS is that binary WD must have almost circular orbits from the very beginning since they result from a spiral-in process during the common envelope stage. Unlike them, binary NS must have (and this is what we actually observe in the known binary radio pulsars with NS companions) a significant eccentricity at birth, since they are formed after two supernova explosions with a substantial mass loss from the system. The possible asymmetry of supernova explosion leading to the natal kick velocity of newborn NS additionally affects the orbital parameters leading to higher orbital eccentricities.

It is well known [8, 9] that an eccentric binary system emits GW in a wide frequency band. This means that the high-order harmonics from an eccentric binary system should be observed at frequencies $\nu > 2\nu_K$ ($2\nu_K$ is twice the Keplerian orbital frequency, at which a circular binary radiates GW), so the entire population of galactic eccentric binary NS should contribute in a wider frequency band than analogical circular binaries would do. The effect may be not small, since the total number of NS+NS binaries in the Galaxy is of order of $\mathcal{R}_{NS} T_{gal} \sim 10^6$, where $T_{gal} = 10^{10}$ yr is the galactic age.

To calculate the GW background produced by binary NS, the following steps should be done. (1) NS binaries form with some initial distribution over orbital semi-major axes and eccentricities $F_{in}(a,e)$. This distribution can be found from evolutionary calculations using e.g. binary population synthesis method [10]. (2) Next, if orbital parameters of such binaries evolve only due to GW emission, a stationary distribution function $F_{st}(a.e)$ can be recovered [12]. (3) Knowing the GW spectrum a binary with given (a,e) [8], the GW noise from these sources can be calculated.

INITIAL BINARY NS DISTRIBUTION IN THE GALAXY

To calculate the initial binary NS distribution $F_{in}(a,e)$ in the Galaxy we used the population synthesis method [10]. It uses the simulation of evolution of a large number of binaries with initial parameters (masses of the components, semi-major axes, eccentricities etc.) distributed according to some (taken from observations or model) laws. There are also evolutionary parameters, such as efficiency of the common envelope stage, kick velocity during supernova explosion, initial spin of compact objects, etc. (see [7] for more detail), of which the kick velocity imparted to compact object (neutron star or black hole) at birth mostly impacts the distribution $F_{in}(a,e)$. We assumed a Maxwellian

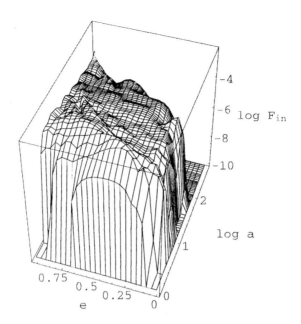

FIGURE 1. Initial binary NS distribution $F_{in}(a,e)$ for kick velocity amplitude $V_k = 200$ km/s. Initial semi-major axes a in solar units.

form of the kick velocity distribution with amplitude equal to 200 km/s. The resulting initial distribution $F_{in}(a,e)$ is shown in Figure 1.

NS binary are formed in a very broad interval of a and e, but interesting for us here will be only those that can coalesce over galactic age $T_{coal} < T_{gal}$, since only such binaries can form a stationary distribution. The integral of function in Figure 1 is equal unity.

STATIONARY BINARY NS DISTRIBUTION

GW back reaction changes orbital semi-major axis $a(t)$ and eccentricity $e(t)$ of a binary star [8] causing the distribution function $F(a,e,t)$ to evolve. In the quadrupole approximation a stationary distribution function reads [12]:

$$F_{st}(a,e) = \frac{15c^5 N_i}{304 G^3 M^3} \frac{a^4(1-e^2)^{7/2}}{e^{31/19}(1+\frac{121}{304}e^2)^{3169/2299}}$$
$$\times \int_0^1 \frac{z^{12/19}(1+\frac{121}{304}z^2)^{870/2299}}{1-z^2} F_{in}(a(z),z)\,dz$$

At the present moment only a fraction of systems from the initial distribution (with $T_{coal} < T_{gal}$) reaches stationarity. The function $F_{st}(v_K,e)$ is calculated for this part of $F_{in}(a,e)$ from Figure 1 and is shown in Figure 2. In this figure we use more useful Keplerian frequency v_K instead semi-major axes a.

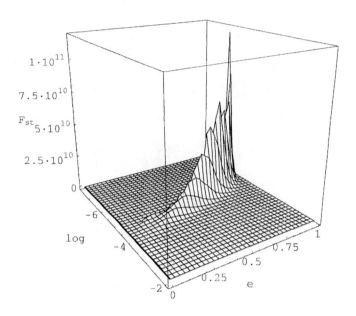

FIGURE 2. The stationary distribution $F_{st}(\nu,e)$ for F_{in} from Figure 1 for binaries with $T_{coal}(a,e) < T_{gal}$.

STATIONARY STOCHASTIC GW NOISE FROM COALESCING BINARY NS

In our calculations we shall assume all the sources to lie at one distance $d = 7.9$ kpc, which is close to the average distance to stars in our Galaxy. This simplifying assumption is unlikely to change our general conclusions.

At each frequency ν we sum up the GW flux from all the harmonics that fall within the chosen frequency bin ($\Delta \nu = 3 \times 10^{-8} Hz$) from lower-frequency non-circular systems in the calculated stationary distribution $F_{st}(\nu,e)$.

$$\frac{dE}{dtd\nu}(\nu,e) = \sum_{n=1}^{\infty} \int_0^1 \frac{dE(n,\nu)}{dt} F_{st}(\nu,e) de$$

The resulting noise curve is shown in terms of dimensional $h^2(\nu) = \frac{G}{c^3 r^2 (\pi \nu)^2} \frac{dE}{dt}$ in Figure 3. As expected, the NS+NS confusion noise lies lower than WD+WD curve due to lower $\mathcal{R}_{NS}$. High harmonics from non-circular NS binaries mostly contribute at lower frequencies, so starting from $\nu \sim 10^{-4}$ Hz the calculated noise curve practically considers with that formed by circular NS binary coalescing with the same rate $\mathcal{R}_{NS} = 10^{-4}$ yr^{-1}.

More important is the limit frequency ν_{lim}. To find it, we calculated the number of harmonics that fall within the frequency bin $\Delta \nu$ at a given ν. The number of harmonics

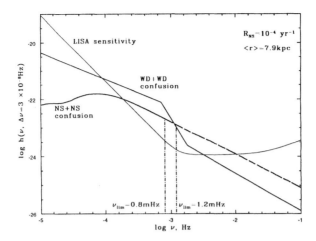

FIGURE 3. Binary neutron star GW background calculated for a model spiral galaxy with the total stellar mass $10^{11} M_\odot$. The limit frequency for binary WD ($v_{lim} \simeq 1.3$ mHz) and non-circular binary NS ($v_{lim} \simeq 4$ mHz) are plotted. Average photometric distance $7.9 kpc$ is assumed. The proposed LISA sensitivity (see [13] Fig.14). Binary WD confusion noise from [4].

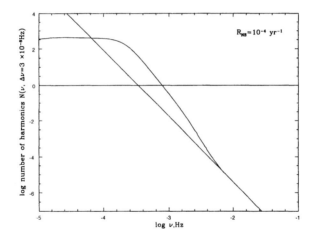

FIGURE 4. Number of harmonics from non-circular galactic NS binaries the stochastic noise in the frequency bin $\Delta v = 3 \times 10^{-8}$ Hz The straight line in the circular NS binaries.

in the frequency bin Δv is defined as a number of most powerful harmonics which are responsible for 95% of total energy of GW in the bin. The number of such harmonics as a function of frequency $N(v, \Delta v)$ is shown in Figure 4. v_{lim} is determined $N(v, \Delta v) = 1$. To within uncertainties of our calculations ($\mathcal{R}_{NS}, F_{in}(a,e), <d>$), $v_{lim} \simeq 10^{-3}$ Hz. For comparison, in the same Figure 4 we show $N(\nu,\Delta\nu)$ for stationary circular NS binaries coalescing with the same rate $\mathcal{R}_{NS} = 10^{-4}$ yr^{-1} ($\nu_{lim}(e=0) = 3 \times 10^{-4}$ Hz).

Clearly, extragalactic NS binaries would form an isotropic confusion noise, but its level is an order of magnitude lower than the galactic one and is beyond the expected LISA sensitivity. The limiting frequency ν_{lim} for extragalactic NS binaries finds from the condition for circular systems and can be as higher as 1 Hz (see also [14]).

CONCLUSIONS

We studied the effect of non-circularity of realistic coalescing NS binaries on stochastic galactic background formed by stationary NS+NS distribution $F_{st}(\nu,e)$. The limiting frequency above which we can resolve individual harmonics at the 95% level over this noise is $\nu_{lim} \approx 10^{-3}$ Hz, close to ν_{lim} for coalescing binary WD.

So our study confirms that within the frequency range 10^{-3}–10^{-2} Hz there are prospects to detect cosmological stochastic GW by means of one space interferometer LISA as suggested in [7].

ACKNOWLEDGMENTS

The authors thank L.P. Grishchuk for useful discussion. AGK thanks RFBR through travel grant 00-02-17164. KAP acknowledge MPA für Astrophysik (Garching) for hospitality.

REFERENCES

1. Braginskii, V. B., *Usp. Fiz. Nauk*, **170**, 743, (2000).
2. Mironovskij, V. N., *Soviet Astron.*, **9**, 752, (1965).
3. Lipunov, V. M., and Postnov, K. A., *Soviet Astron.*, **31**, 228, (1987).
4. Hils, D. L., Bender, P., and Webbink, R. F., *ApJ*, **360**, 75, (1997).
5. Ferrari, V., Matarrese, S., and Schneider, R. *MNRAS*, **303**, 258, (1999).
6. Kalogera, V., and Lorimer, D. R., *ApJ*, **530**, 890, (2000).
7. Grishchuk, L. P., Lipunov, V. M., Postnov, K. A., Prokhorov, M. E., and Sathyaprakash, B. S., *Usp. Fiz. Nauk*, **171**, 3, (2001).
8. Peters, P. S., and Mathews, J., *Phys. Rev.*, **131**, 435, (1963).
9. Zeldovich, Ya. B., and Novikov I. D., *Relativistic Astrophysics*, Chicago University Press, Chicago, 1971.
10. Lipunov, V. M., Postnov, K. A., and Prokhorov M. E., *Astron. Astophys.*, **310**, 489, (1996).
11. Lipunov, V. M., Postnov, K. A., and Prokhorov, M. E., *MNRAS*, **288**, 245, (1997).
12. Buitrago, J., Moreno–Garrido, C., and Mediavilla, E., *MNRAS*, **268**, 841, (1994).
13. Thorne. K. S, in *Particle and Nuclear Astrophysics and Cosmology in the Next Millenium*, edited by E. W. Kolb, and R. D. Peccei, World Scientific, Singapore, 1995, p.160.
14. Ungarelli C., and Vechido A., gr-gc/0003021.

Detecting Eccentric Globular Cluster Binaries with LISA

M. Benacquista

Montana State University-Billings, Billings, MT 59101

Abstract. The energy carried in the gravitational wave signal from an eccentric binary is spread across several harmonics of the orbital frequency. The inclusion of the harmonics in the analysis of the gravitational wave signal increases the signal-to-noise ratio of the detected signal for binaries whose fundamental frequency is below the galactic confusion-limited noise cut-off. This can allow for an improved angular resolution for sources whose orbital period is greater than 2000 s. Globular cluster sources includ possible binary black holes and neutron stars which may have high eccentricities. Cluster dynamics may also enhance the eccentricities of double white dwarf binaries and white dwarf-neutron star binaries over the galactic sources. Preliminary results of the expected signal-to-noise ratio for selected globular cluster binaries are presented.

The space-based gravitational radiation detector LISA will be sensitive to frequencies in the millihertz range. A number of interesting astrophyisical phenomena will be likely sources in this frequency band, many of interest in cosmology and as fundamental tests of relativity. LISA will also be of use as an astronomical tool for observing compact binary systems in the galaxy and globular cluster system. In the globular cluster system, LISA can test dynamical evolution nodels by detecting binary populations which are predicted by simulations but which are exceedingly difficult to observe in electromagnetic radiation. One such class of objects is a population of binary black holes with separations between 3 and 30 $R_\odot$ and a thermal distribution of eccentricities [1,2]. A Monte Carlo analysis of the detection of such signals indicate that eccentricities above 0.8 can extend the detectable systems to orbital periods on the order of 1 day.

The gravitational wave signal from an eccentric binary is given by the two polarizations [3]:

$$h_\times = \frac{\cos\vartheta}{\sqrt{2}} \sum_{n=1}^{\infty} [2h_{xy}^{(n)} \sin(2\pi nft)\cos 2\varphi - h_{x-y}^{(n)} \cos(2\pi nft)\sin 2\varphi] \quad (1)$$

$$h_+ = \frac{1}{2\sqrt{2}} \sum_{n=1}^{\infty} [(1+\cos^2\vartheta)(2h_{xy}^{(n)} \sin(2\pi nft)\sin 2\varphi + h_{x-y}^{(n)} \cos(2\pi nft)\cos 2\varphi)$$
$$- (1-\cos^2\vartheta)h_{x+y}^{(n)} \cos(2\pi nft)] \quad (2)$$

in a coordinate system centered on the binary with the x-axis along the semi-major axis and the z-axis along the angular momentum vector of the system, with the orbital frequency of the binary given by f. The angles ϑ and φ are the usual polar coordinates giving the direction to the detector. The metric components, $h^{(n)}_{xy}$ and $h^{(n)}_{x\pm y}$, share a common amplitude which depends upon the masses and the orbital frequency. In general, they are combinations of Bessel functions.

The signal detected by LISA is modified by the motion of LISA in its orbit about the sun. To describe the detected signal, we define several unit vectors. Let the orientation of the arms of the constellation of LISA spacecraft be given by the three vectors, ℓ_1, ℓ_2, ℓ_3. The direction from the sun to the source is $\hat{n}$. The orientation of the binary orbit is $\hat{L}$ (which points along the angular momentum), $\hat{a}$ (which points along the semi-major axis), and $\hat{b} = \hat{L} \times \hat{a}$. The gravitational wave propagates along $-\hat{n}$, and we define the polarization axes $\hat{p} = \hat{n} \times \hat{L}/|\hat{n} \times \hat{L}|$ and $\hat{q} = -\hat{n} \times \hat{p}$. The signal received by LISA is then $h(t) = 1/2[F^+ h_+ + F^\times h_\times]$. with:

$$F^+ h_+ + F^\times h_\times = [(p_a p_b - q_a q_b) h_+ + (p_a q_b - q_a p_b) h_\times](\ell^a_1 \ell^b_1 - \ell^a_2 \ell^b_2). \tag{3}$$

Defining

$$A_n = \frac{1}{\sqrt{2}} (2 F^\times \cos \vartheta \cos 2\varphi + F^+ (1 + \cos^2 \vartheta) \sin 2\varphi) h^{(n)}_{xy} \tag{4}$$

$$B_n = \frac{-1}{2\sqrt{2}} [(2 F^\times \cos \vartheta \sin 2\varphi - F^+ (1 + \cos^2 \vartheta) \cos \varphi) h^{(n)}_{x-y} + F^+ (1 - \cos^2 \vartheta) h^{(n)}_{x+y}], \tag{5}$$

we can write the signal in a standard phase/amplitude form:

$$h(t) = \sum \sqrt{A_n^2 + B_n^2} \cos [2\pi n f t + \phi_{np}(t) + \phi_{nD}(t)] \tag{6}$$

where the changing orientation of LISA contributes to a time-dependent phase, $\phi_{np}(t)$, due to the varying sensitivity to each polarization and the orbital motion of LISA about the sun contributes to a time-dependent phase, $\phi_{nD}(t)$, due to the Doppler shift in the received signal.

The angles ϑ and φ are determined by the source position and orientation by: $\cos \vartheta = -\hat{n} \cdot \hat{L}$ and $\tan \varphi = (\hat{n} \cdot \hat{b})/(\hat{n} \cdot \hat{a})$. The polarization phase and Doppler phase are: $\phi_{np} = \tan^{-1}(-A_n/B_n)$ and $\phi_{nD} = 2\pi n f(R/c) \sin \theta_s \cos(\phi(t) - \phi_s)$ with $R = 1\text{AU}$, and (θ_s, ϕ_s) the angular position of the source. The center-of-mass trajectory of LISA is $\phi(t) = 2\pi t/T$ with $T = 1yr$.

A second signal is generated from a linear combination of the responses from all three arms of LISA which is equivalent to the signal from a detector rotated by 45° in the plane of the initial detector [4]. These two signals are used to calculate the signal to noise ratio for eccentric binaries.

The Fourier transform of the signal is $\tilde{h}(f) = \sum \tilde{h}^{(n)}(f)$ with $\tilde{h}^{(n)}$ sharply peaked about nf. The signal to noise ratio (ρ) is found by comparing the strength of the signal at each frequency to the expected instrumental noise at that frequency. The spectral noise density $S_n(f)$ can be considered nearly constant in the small region about nf. Using Parseval's theorem and the fact that the amplitude is slowly varying with respect to the harmonic frequency, we have:

$$\rho^2 \simeq \sum_{\text{detectors}} \sum_n [S_n(nf)]^{-1} \int_0^T (A_n^2 + B_n^2) dt. \tag{7}$$

We have explored the parameter space of possible eccentric binaries using a Monte Carlo approach. The source position is assigned to be the position of a globualr cluster selected at random from the galactic globular cluster system. The source orientation is assigned with all orientations equally likely. The mass of each component of the binary as well as the eccentricity and orbital frequency are all chosen with equal likelihood from the following ranges: $8M_\odot < M < 18M_\odot$, $0 < e < 1$, and $0.5\text{mHz} < f < 0.01\text{mHz}$. Out of a total of 300 binaries, 71 were detectable by LISA with $\rho \geq 2$. Most of these had significantly higher values of ρ. The number of detectable binaries in each bin of orbital period and eccentricity are shown in Fig 1. The detection of an eccentric binary relies an multiple harmonics appearing in the data stream. The signal strengths from a simulation of 100 globular cluster binaries are also shown in Fig 1.

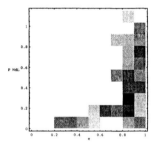
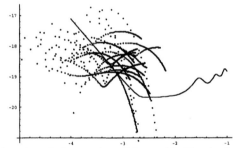

FIGURE 1. On the left, a contour plot showing the relative number of detectable binaries for a given range of orbital period (in days) and eccentricity. Darker bins contain more detectable binaries. On the right, the signal strengths from a simulation of 100 globular cluster binaries. Each track represents the harmonics for a single eccentric binary

REFERENCES

1. Portegies Zwart, S.F. and McMillan, S.L.W., ApJL 528, L17 (2000).
2. Benacquista, M.J., Portegies Zwart, S.F., and Rasio, F.A., gr-qc/0010020 (2000).
3. Pierro, V., Pinto, I.M., Spallicci, A.D., Laserra, E., and Recano, F., gr-qc/0005044 (2000).
4. Cutler, C., Phys. Rev. D 57, 7089 (1998)

Coalescing Binary Neutron Star Systems[1]

Alan C. Calder*, F. Douglas Swesty† and Edward Y. M. Wang‡

*Center for Astrophysical Thermonuclear Flashes and Department of Astronomy and Astrophysics, University of Chicago Chicago, IL 60637
†Department of Physics and Astronomy, State University of New York at Stony Brook, Stony Brook, NY 11794
‡Department of Physics and Astronomy, Northwestern University, Evanston, IL 60208

Abstract. We present numerical studies of coalescing neutron star pairs with Newtonian hydrodynamics coupled to the 2.5 Post-Newtonian radiation reaction of Blanchet, Damour, and Schäfer [1]. Our simulations evolve the Euler equations using a modification of the ZEUS 2-D algorithm [2] and use a Fast Fourier Transformation method for solving the Poisson equation for the gravitational and radiation reaction potentials. We find that the radiation reaction produces a significant effect on a neutron star pair when compared to a purely Newtonian simulation.

INTRODUCTION

General relativity predicts that binary systems of compact objects such as neutron stars (NSs) will emit energy in the form of gravitational radiation. This energy loss leads to the in-spiral and coalescence of the system, and the observed waveforms may yield information about the NS equation of state and upper bounds on NS masses. Neutron star mergers (NSMs) are a suggested source for the mysterious gamma-ray bursts observed in recent years [3,4]. NSMs are thought to release energy on the order of their gravitational binding energy $\approx 10^{53}$ erg, which may be larger than estimated gamma-ray burst energies, and simulations can test the consistency of the merger time scales and energetics with observations.

The rotating frame hydrodynamics equations we solve are

$$\frac{\partial \rho}{\partial t} + \nabla \cdot (\rho \mathbf{v}) = 0 \qquad (1)$$

$$\frac{\partial e\rho}{\partial t} + \nabla \cdot (e\rho \mathbf{v}) = -(P+Q)\nabla \cdot \mathbf{v} \qquad (2)$$

---

[1] Computational resources were provided by NCSA and PSC under Metacenter allocation #MCA975011. Funding for this research was provided by NASA under NASA ESS/HPCC contract NCCS5-153. Additional funding provided by the Department of Energy under grant No. B341495 to the Center for Astrophysical Thermonuclear Flashes at the University of Chicago.

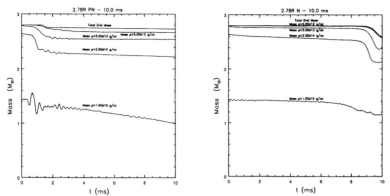

FIGURE 1. The mass at densities greater than 4 thresholds and the total mass on the grid for the Newtonian + RRXN evolution (left) and the purely Newtonian evolution (right).

$$\frac{\partial(\rho v_i)}{\partial t} + \nabla \cdot (\rho v_i \mathbf{v}) = -[\nabla(P+Q)]_i - \rho[\nabla(\Phi+\Psi)]_i \\ -2\rho(\boldsymbol{\omega} \times \mathbf{v}_r)_i - \rho[\boldsymbol{\omega} \times (\boldsymbol{\omega} \times \mathbf{r})]_i \;, \quad (3)$$

where $\mathbf{r}$, ρ, $\mathbf{v}$, e, and P are the position, density, velocity, specific internal energy, and pressure of the fluid. $\mathbf{v}_r$ is the velocity of the frame rotating at $\boldsymbol{\omega}$ relative to the lab frame, and Q is a viscous stress added to model the sub-resolution viscous processes occurring across shock fronts. Φ and Ψ are the Newtonian gravitational and 2.5 Post-Newtonian radiation reaction (RRXN) potentials, respectively. An ideal gas equation of state closes the system, and solution of the Poisson equation gives the gravitational potential Φ. Solving for Ψ requires solution of

$$\nabla^2 R = 4\pi G D^{ij} x_i \frac{\partial \rho}{\partial x^j} \quad (4)$$

where D^{ij} are third time derivatives of the mass quadrupole tensor, Q^{ij}, and the 2.5 Post-Newtonian radiation reaction potential is defined as $\Psi = \frac{2}{5}\frac{G}{c^5}\left(R - D^{ij}x_i\frac{\partial \Phi}{\partial x^j}\right)$. We use a Fast Fourier Transform to solve (4) and for the gravitational potential. The expression we use for D^{ij} is that given by Ruffert *et al.* [5]:

$$D^{ij} = \mathbf{STF}\left[2\int d^3x \left(2P\frac{\partial v_i}{\partial x_j} + \frac{\partial \Phi}{\partial x_j}(x_i \nabla \cdot (\rho \mathbf{v}) - 2\rho v_i) - \rho x_i \frac{\partial \frac{\partial \Phi}{\partial t}}{\partial x_j}\right)\right] . \quad (5)$$

Here **STF** means Symmetric Trace Free. This formulation omits 1PN corrections and assumes that the kinematic and dynamical velocities are effectively equivalent.

RESULTS

The two simulations began from identical initial conditions, an equilibrium contact binary system of 1.4 M$_\odot$ $\gamma = 2$ polytropes with a separation radius of 2.78

R_{Star} [6]. The Newtonian + RRXN simulations differed from the purely Newtonian simulations only by the inclusion of the RRXN source. The Newtonian system merged because the initial contact binary configuration was below the threshold for tidal instability [7,6]. The inclusion of the RRXN drastically affects the dynamics of simulations of symmetric contact binary polytropes as shown in Figures 1 and 2. The merger occurs on a much shorter time scale than the Newtonian merger resulting from tidal instablilites, ~1.0 ms vs. ~9.0 ms, and produced very different gravitational waveforms. Our improvement of the initial conditions and algorithms for both the Newtonian and Newtonian + RRXN evolution [6] lead to more substantial differences than we reported earlier [8]. In addition, the Newtonian + RRXN simulation lost less mass from the simulation grid than the Newtonian simulation, suggesting that NSMs will eject less mass than Newtonian simulations would predict. Currently we are comparing our numerical results to results from semi-analytic methods for determining the gravitational radiation luminosity.

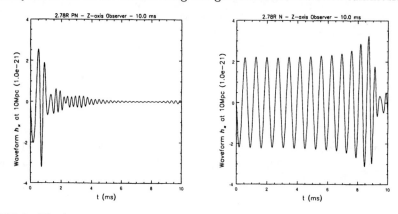

FIGURE 2. The $h_\times$ gravitational waveform produced by the Newtonian + RRXN evolution (left) and the purely Newtonian evolution (right) from the perspective of an observer at 10 Mpc along the z-axis. The waveforms are calculated from the quadrupole approximation.

REFERENCES

1. Blanchet, L., Damour, T., & Schäfer, G., *MNRAS* **242**, 289 (1990)
2. Stone, J.M. and Norman, M.L., *Ap. J. S.* **80**, 791 (1992).
3. Paczynski, B., *Ap. J.* **308**, L43 (1986)
4. Goodman, J., *Ap. J.* **308**, L47 (1986)
5. Ruffert, M., Janka, H.-T., and Schäfer, G., *Astron. Astrophys.* **311**, 532 (1996).
6. Swesty, F.D., Wang, E.Y.M., and Calder, A.C., *Ap. J.* **541**, 937 (2000).
7. Rasio, F.A., and Shapiro, S.L., *Ap. J.* **401**, 226 (1992).
8. Wang, E.Y.M., Swesty, F.D., and Calder, A.C., *Stellar Evolution, Stellar Explosions, and Galactic Chemical Evolution*, ed. A. Mezzacappa, IOP, Bristol 1998 p. 723.

Quasi-Normal Modes in Schwarzschild anti-de Sitter Spacetimes

Vitor Cardoso, José P. S. Lemos

Centro Multidisciplinar de Astrofísica-CENTRA, Departamento de Física,
Instituto Superior Técnico
Av. Rovisco Pais 1, 1096 Lisboa, Portugal.

Abstract. We study the quasinormal modes for electromagnetic perturbations of a Schwarzschild black hole in an asymptotically anti-de Sitter (AdS) spacetime.

I INTRODUCTION

The theory of black-hole perturbation was initiated by Regge and Wheeler. They separated the perturbations in an odd (axial) and even (polar) part, by using tensorial spherical harmonics, to study the Schwarzschild black hole stability. A generalized interest in perturbation theory arose after the possibility of detecting gravitational waves. Many methods have been devised to study situations of interest [1]. All these papers dealt with asymptotically flat spacetimes. Recently there has been a growing interest in asymptotically AdS spacetimes. These share with the asymptotically flat spacetimes the common property that they both have well defined charges at infinity, such as mass, angular momentum and electric charge. AdS spacetimes are also of interest to particle physics theories, such as supergravity and string theory, which require a background with a negative cosmological constant Λ.

We are going to study perturbations of Schwarzschild-AdS spacetimes. Recent work has been done on the subject [2]. These authors studied the scalar equation. We will go beyond the scalar field case and analyze the case of electromagnetic perturbations. For the three dimensional case see [3].

II ELECTROMAGNETIC PERTURBATIONS AND QUASI-NORMAL MODES

We consider the evolution of a Maxwell field in a Schwarzschild-AdS metric $ds^2 = (\frac{r^2}{R^2}+1-\frac{2M}{r})dt^2 - (\frac{r^2}{R^2}+1-\frac{2M}{r})^{-1}dr^2 - r^2(d\theta^2+\sin^2\theta d\phi^2)$, where $R \equiv \sqrt{-\frac{3}{\Lambda}}$ is the AdS radius, and M the black hole mass. The evolution is governed by

Maxwell's equations, $F^{\mu\nu}{}_{;\nu} = 0, F_{\mu\nu} = A_{\nu,\mu} - A_{\mu,\nu}$. We expand A_μ in vector spherical harmonics [4]. Puting it into Maxwell's equations we get two equations for the perturbation, one for each parity in the problem, both having the form: $\frac{\partial^2 \Psi(r)}{\partial r_*^2} + (\omega^2 - V(r))\Psi(r) = 0$, where Ψ is a linear combination of some more fundamental functions. The potential V is $V(r) = (\frac{r^2}{R^2} + 1 - \frac{2M}{r})(\frac{l(l+1)}{r^2})$, and the tortoise coordinate r_* is defined by, $\frac{\partial r}{\partial r_*} = (\frac{r^2}{R^2} + 1 - \frac{2M}{r})$. We can take $R = 1$ and measure everything in terms of R. The QNMs are defined to be solutions of the master equation which satisfy the boundary conditions: (i) only ingoing waves enter the horizon and (ii) are zero at infinity.

To find the ωs we use a method due to Horowitz and Hubeny [2]. Redefining a new wavefunction, $\Theta = e^{i\omega r_*} \Psi$, we expand Θ as $\Theta(x) = \sum_{n=0}^{\infty} a_{n(\omega)}(x-h)^n$, where $x = \frac{1}{r}$, $h = \frac{1}{r_+}$, r_+ being the horizon radius, and $a_{n(\omega)}$ is a function of ω. If we use the boundary condition $\Psi = 0$ at infinity ($x = 0$) we get: $\sum_{n=0}^{\infty} a_{n(\omega)}(-h)^n = 0$. Now we want to find a solution of the polynomial equation.

III NUMERICAL RESULTS

The roots for ω can be evaluated numerically. There are frequencies with a vanishing real part, making it also possible to use an approximation, due to Liu, to these highly damped modes [5]. We have computed the lowest frequencies for some values of the horizon radius r_+, and $l = 1$. The frequency is written as $\omega = \omega_r + i\omega_i$, where ω_r is the real part of the frequency and ω_i is its imaginary part.

For a better visualization we plot the $\omega_i \times r_+$, in figure 1.

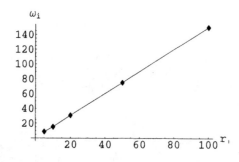

Figure 1. Lowest QNM for $l = 1$, as function of r_+.

The imaginary part of the frequency, which determines how damped is the mode, increases linearly with the cosmological constant Λ, which was expected since when one increases Λ the spacetime becomes more strained, so to speak, thereby making it more difficult for any perturbation to propagate. As one can see, the imaginary part of the frequency scales linearly with r_+, at least for large black holes, supporting the arguments given in [2]. Furthermore, there is an excellent agreement between the numerical results and analytical approximation for the strongly damped modes.

IV CONCLUSIONS

For large black holes the modes scale with the horizon [2]. The decay of the perturbation has obviously a timescale given by the inverse of the imaginary part of the frequency, $\tau = \frac{1}{\omega_i}$, i.e., for large black holes the greater the mass the less time it takes to approach equilibrium. The modes all have a negative imaginary part, which means that the black hole is stable against these perturbations, since the perturbations will decay exponentially with time (see also [6]).

Acknowledgments - We acknowledge finantial support from the portuguese Fundação para a Ciência e Tecnologia (FCT). One of us (JPSL) thanks the hospitality of Observatório Nacional do Rio de Janeiro.

REFERENCES

1. T. Regge, J. A. Wheeler *Phys. Rev.* **108** (1957) 1063; S. Chandrasekhar, S. Detweiler, *Proc. R. Soc. London, Ser.* **A344** (1975) 441; B. F. Schutz, C. M. Will , *Astrophys. J.* **291** (1985) L33; N. Fröman,P. O. Fröman, N. Andersson, A. Hökback *Phys. Rev.* **D45** (1992) 2609; H. P. Nollert, *Phys. Rev.* **D47** (1993) 5253; E. W. Leaver, *Proc. R. Soc. London, Ser. A* **402** (1985) 285; V. Ferrari, B. Mashhoon, *Phys. Rev.* **D30** (1984) 295; K. D. Kokkotas, B. G. Schmidt, gr-qc/9909058.
2. G. T. Horowitz, V. Hubeny, *Phys. Rev.* **D62**, (2000) 024027; B. Wang, C. Lin, E. Abdalla, *Phys.Lett.* **B481** (2000) 79.
3. V. Cardoso, J. P. S. Lemos, gr-qc/0101052.
4. R. Ruffini, in *Black Holes*, (Gordon and Breach Science Publishers, New York 1973).
5. H. Liu, *Class. Quantum Grav.* **12** (1995) 543.
6. V. Cardoso, J. P. S. Lemos , in preparation (2001).

Constraining Post-Newtonian Parameters With Gravitational Waves

James S. Graber*

*407 Seward Square SE, Washington, DC 20003-1113
jgraber@mailaps.org

Abstract. We re-express gravitational wave results in terms of post-Newtonian parameters. Using these expressions, and some simplifying assumptions, we compute that in a favorable case, i.e. a ten solar-mass black hole spiraling in to a at the 10% level or better a single combination of post-post-Newtonian parameters one order higher than those already constrained by solar system evidence. This significant constraint will be possible even if the signal-to-noise level is so low that the signal can only be found by matched filtering, and hence only deviations between alternate signal interpretations of order one half cycle or more can be detected.

We report here some early results of a study that tries to make an explicit connection between classical parameterized post-Newtonian (PPN) theory as developed by Nordtvedt, Will and others [1–3], and one or of the several forms of PPN analysis now being used to predict gravitational waves, particularly for the case of inspiralling binary black holes [4,5]. Classical PPN theory was used to analyze solar system tests of general relativity (GR) and compare results predicted by GR to the results predicted by any (metric-based) alternate gravity theory (AGT). Classical PPN theory, plus observational results, puts parametric constraints on the spherical geometry of a condensed body in any viable AGT. Our goal is to make the connection between the PPN analysis of AGTs and the perturbative PPN calculation of gravitational waves, and to determine the extent to which an observed inspiral can put stronger constraints on the PPN representation of a compact object in any possible AGT. In particular, we try to find an explicit expression for the second order constraints which will be imposed. It is known that very high order calculations are required to produce accurate templates for the matched filter detection of gravitational waves from binary inspirals [6,7]. This fact suggests that strong constraints will be placed on AGTs by the observation of any inspiral, but the inverse problem of deriving the constraints from the observed inspiral is not easy. This problem is clearly related to the previously studied problem of recovering the parameters describing the inspiralling binary from the observed gravitational wave information

[8–11]. Our study extends parameter estimation to include determining the second-order PPN constraints placed on AGTs by an observed binary inspiral. We report here some successful steps toward this goal. A more comprehensive summary of results achieved so far, available on the web at gr-qc/0103024, contains analytic results, including a derivation of a specific combination of second-order parameters directly constrained by gravitational wave frequency data from a binary inspiral. This short communication covers only one partly numeric study of the possibility of distinguishing one specific AGT from GR by frequency only data.

In doing a numeric study, it is necessary to choose one specific AGT. We chose the exponential metric $ds^2 = \frac{-dt^2}{e^{\frac{2M}{r}}} + e^{\frac{2M}{r}} \left(dr^2 + d\Omega^2 r^2\right)$, which is part of several theories and has been studied by many authors, including Yilmaz [12], Rosen [13], Kaniel and Itin [14], Muench, Gronwald and Hehl [15], Watt and Misner [16], and Leiter and Robertson [17]. The exponential metric agrees with the Schwarzschild metric exactly at first PPN order, and differs by 10 to 30% in the second order PPN parameters. We matched the exponential metric to the Kerr metric as well as the Schwarzschild metric. Based on the analytical work reported in the extended paper [18], three sets of two equations were derived: one set for each of the three metrics, with each set including an equation for T, time to plunge, and N, number of orbits to plunge. Using these equations, we numerically built and compared three inspiral models, called ALT, KERR, and SCHW. Based on the favorable possibilities for LISA (the proposed orbiting gravitational wave observatory) discussed in Finn and Thorne [11], where in most cases over 10,000 orbits are observable and in favorable cases over 100,000 orbits are observable, we chose our models to have 10,000 orbits from the beginning to the end of the observed inspiral.

TABLE 1. Three closely matched binary inspiral models.

Model	Number of cycles remaining before plunge: [a]								
ALT[b]	10000.00	7000.00	3000.00	1000.00	300.00	100.00	30.00	10.00	0.00
KERR[c]	10000.00	7000.28	3000.00	999.92	299.96	99.98	29.99	10.00	0.00
SCHW[d]	10000.00	7008.91	3009.61	1004.29	301.41	100.48	30.15	10.05	0.00

[a] Number of cycles remaining before plunge in the Kerr(KERR) and Schwarzschild(SCHW) models, when exactly the tabulated number of cycles remain inthe alternate(ALT) model.

[b] The ALT model has a mass of 1 arbitrary unit and no angular momentum.

[c] The KERR model has a mass of 1.012 arbitrary units and a specific angular momentum of .063 per arbitrary mass unit.

[d] The SCHW model has a mass of 1.041 arbitrary units and no angular momentum.

To test whether gravitational wave frequency information will allow us to distinguish these two alternate theories, (GR and the exponential metric theory), we try to match them as closely as possible. First, an arbitrary ALT model was computed. Since we could vary the angular momentum in the Kerr model to fit the ALT model, but not vice-versa, we froze the ALT model first. Then, the closest possible Schwarzschild solution model(SCHW) and the closest possible Kerr solution model(KERR) were fitted to the resulting gravitational wave frequency pattern from the ALT model. (Many different SCHW and KERR models were computed to find the closest possible match.) The final SCHW model has the same number of cycles as the ALT model and was fitted to match the frequency pattern of the ALT model exactly at the beginning and the end of the observed inspiral. It was fitted by adjusting the mass of the Schwarzschild black hole. The resulting fitted mass is 1.041 times the mass of the central body in the ALT model. The KERR model was fitted to the ALT model by adjusting both the mass and the angular momentum of the Kerr black hole. The KERR model matches the ALT model exactly at three points: the beginning and the end of the observed inspiral, and at one intermediate point (3000 orbits). It fits the (non-rotating) ALT pattern much more closely than does the Schwarzschild model. The final KERR model black hole has a mass 1.012 times the mass of the ALT model central body and a specific angular momentum of .063 per unit mass. The results are displayed in Table 1. The closest possible Schwarzschild match differed from the ALT model by almost 10 full orbits. This is easily detectable. The closest Kerr model differed from the ALT model by slightly more than one quarter orbit. This is marginally detectable. This model supports the possibility of constraining a second-order combination of PPN parameters at the 10 to 30% level. More extensive calculations using up to 100,000 orbits support the possibility of constraints substantially below the 10% level.

It is too early to reach conclusions, but the conclusion toward which the preliminary results of this study point is that it may be possible to constrain a single combination of second-order PPN parameters at the 10 % level or better by gravitational wave frequency data, if a favorable case is observed by LISA.

REFERENCES

1. Nordtvedt, K., and Will, C. M.,*Ap. J.* **163**, 595 (1971).
2. Damour, T., *gr-qc/9606079*.
3. Will, C. M., *gr-qc/9811036*.
4. Thorne, K. S., *gr-qc/9506084*.
5. Damour, T., *gr-qc/9606077*.
6. Finn, L. S., and Chernoff, D. F., *Phys. Rev.* **D47**, 2198 (1993).
7. Cutler, C., *et al.*, *Phys. Rev. Lett.* **70**, 1984 (1993).
8. Ryan, F. D., *Phys. Rev.* **D56**, 1845 (1997).
9. Poisson, E., and Will, C. M., *Phys. Rev.* **D52**, 848 (1995).
10. Poisson, E.,*Phys. Rev.* **D54**, 5939 (1996).
11. Finn, L. S., and Thorne, K. S., *Phys. Rev.* **D62**, 124021 (2000).
12. Yilmaz, H., *Phys. Rev.* **111**, 1417 (1958).
13. Rosen, N., *Gen. Rel. Grav.* **4**, 435 (1973).
14. Kaniel, S. and Itin, Y., *Nuov. Cim.* **113B**, 393 (1998).
15. Muench, U., Gronwald, F. and Hehl, F. W., *Gen. Rel. Grav.* **30**, 933 (1998).
16. Watt, K., and Misner, C. W., *gr-qc/9910032*.
17. Leiter, D., and Robertson, S. L., *gr-qc/0101025*.
18. Graber, J. S., *gr-qc/0103024*.

KiloHertz QPO and Gravitational Wave Emission as the Signature of the Rotation and Precession of a LMXB Neutron Star Near Breakup

J. Garrett Jernigan

Space Sciences Laboratory, University of California, Berkeley, CA 94720-7450
and Eureka Scientific Inc., 2452 Delmar St., Oakland, CA 94602-3017

Abstract. The theory of torque free precession (TFP) of the outer crust of a neutron star (NS) as the signature of the approach to NS breakup is a viable explanation of the uniform properties of kHz Quasi-periodic Oscillations (QPO) observed in X-rays emitted by Low Mass X-ray Binary (LMXB) sources. The TFP model is in strong contrast to existing models which primarily relate the kHz QPO phenomenon to the physics of gas dynamics near the inner edge of the accretion disk and the transition flow onto the surface of the NS. We suggest the possibility of the direct detection of very low frequency (~ 1 kHz) radio waves from magnetic dipole radiation and also predict kHz gravitational wave emission from the LMXB Sco X-1 that may be detectable by LIGO. The high accretion rates consistent with the predicted GW emission indicate the likely conversion of some LMXBs to maximally rotating Kerr black holes (BH) and further suggest that these systems are progenitors of some gamma-ray bursts (GRB).

I INTRODUCTION AND TORQUE FREE PRECESSION MODEL

KiloHertz (kHz) oscillations in low-mass X-ray binaries (LMXB) were first discovered ([9]; [7]) in observations with the Proportional Counter Array (PCA) onboard the *Rossi X-ray Timing Explorer* ([2]; [8]). These kHz QPO sometimes occur in pairs ([10]; and references therein) whose difference frequency is not constant and show properties that are uniform over a wide range of accretion rates ([6]).

KiloHertz oscillations are explained as a natural consequence of the spinup of a neutron star to near breakup ([5]). In this model, the fluid core of the NS spins up due to accretion torques ($\sim 300 - 800$ Hz). Further transfer of angular momentum to the NS causes the "crust" of the NS to partially decouple from the core and rotate at a higher rate (up to ~ 1100 Hz) than the fluid core. The crust of the NS both rotates and undergoes torque free precession (TFP) as a rigid body which causes a modulation of the X-ray emission from this observable outer region at a pair of frequencies in the kHz band. We designate these frequencies as ν_l and ν_u, the lower and upper kHz QPO. These oscillations are quasi-periodic because of fluctuations in the moments of inertia of the outer region of the NS.

Let $I_{c\|}$ and $I_{c\perp}$ be the parallel and perpendicular components of the moment of inertia of the crust of the NS as related to the rotational axis of the NS.

$$I_{c\|} = I_{c0}\left[1 + \frac{\nu_{cs}}{\nu_{c\|}}\right] \qquad I_{c\perp} = I_{c0}\left[1 + \frac{\nu_{cs}}{\nu_{c\perp}}\right] \qquad (1)$$

where $\nu_{c\|}$ and $\nu_{c\perp}$ are constants that characterize the response of the crust to rotation, in the directions parallel and perpendicular to the NS spin axis. We also define a dimensionless ratio ellipticity ϵ_c and its relationship to the eccentricity e_c.

$$\epsilon_c = \frac{\left[I_{c\|} - I_{c\perp}\right]}{I_{c\perp}} \qquad \epsilon_c = \frac{\nu_{cp}}{\nu_{cs}} = \frac{e_c^2}{2 - e_c^2} = \frac{\nu_l}{\nu_u} \qquad e_c^2 = 1 - \left(\frac{r_p}{r_e}\right)^2 \qquad (2)$$

Here r_e indicates the equatorial radius of the NS and r_p indicates the polar radius. The NS is assumed to have the shape of a oblate spheroid which is induced by rotation of the crust at frequency ν_{cs}. Note that in equations 2 we have made the interpretation that the observed lower kHz QPO frequency is the crust's precession frequency ($\nu_{cp} = \nu_l$) while the upper kHz QPO frequency is the crust's spin frequency ($\nu_{cs} = \nu_u$). Equations 2 are consistent with TFP of a rigid body ([4]).

The NS core is a fluid which rotates but does not precess, while the "rigid" crust rotates and precesses. This motion of the crust leads to the appearance of the pair of kHz QPO at the rotation frequency of the crust ν_{cs} ($= \nu_u$) and the precession frequency of the crust ν_{cp} ($= \nu_l$). As ν_{cs} increases to $\sim$ 1100 Hz, the crust reaches the next critical value of eccentricity $e \sim 0.9300$ which corresponds to the maximum rotation rate of the rigid crust that will allow both rotation and precession to occur.

The TFP model is confirmed by fitting the observed frequencies of the kHz QPO from Sco X-1. We can approximately model the moments of inertia of the crust ($I_{c\|}$ and $I_{c\perp}$) as a response to rotation at frequency ν_{cs} as a linear increase from an initial value at zero rotation (I_{c0}) as shown in equations 1. We use this model to fit the data for ScoX-1 and thereby empirically determine the values for $\nu_{c\|} = 1009.4 \pm 1.0$ Hz and $\nu_{c\perp} = 6435 \pm 22$ Hz. Figure 1 shows the fit to the Sco X-1 data by the TFP model (solid line) as well as the phenomenological equation of [6] (dashed line). The TFP model is the solid curve which is shown to match the Sco X-1 data and the single point from GX5-1 (near abissa 500 Hz; [11]). The point for GX5-1 has not been included in the fit for $\nu_{c\|}$ and $\nu_{c\perp}$, but is included in the figure to demonstrate that the TFP model appears to fit the data over the full observed range of ν_u (500 – 1100 Hz) with ν_l present. Horizontal dotted lines mark the critical values of the eccentricity of the crust. Also shown is one curve that correspond to a particular model of a rotating NS (EOS FPS from [3]).

REFERENCES

1. Chandrasekhar, S. 1969, Ellipsoidal Figures of Equilibrium, (New Haven:Yale University Press), 77
2. Bradt, H. V., Rothschild, R. E., & Swank, J. H. 1993, A&AS, 97, 355
3. Cook, G. B., Shapiro, S. L., & Teukolsky, S. A. 1994, ApJ, 424, 823
4. Goldstein, H. 1950, Classical Mechanics, (Reading, Ma.:Addison-Wesley), 159
5. Jernigan 2001, astro-ph/0101048
6. Psaltis, D. *et al.* 1998, ApJ, 501, L95
7. Strohmayer, T. E., Zhang, W., Swank, J. H., Smale, A., Titarchuk, L., Day. C., & Lee, U. 1996, ApJ, 469, L9
8. Swank, J.H. 1998, Nucl. Phys. B (Proc. Suppl.), 69
9. van der Klis, M., Swank, J.H., Zhang, W., Jahoda, K., Morgan, E.H., Lewin, W.H.G, Vaughan, B., & van Paradijs, J. 1996, ApJ, 469, L1
10. van der Klis 2000, in Annual Review of Astronomy and Astrophysics
11. Wijnands, R., & van der Klis, M. 1998, Nature, 394, 344

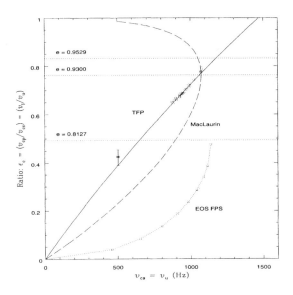

FIGURE 1. the ratio ν_l/ν_u versus ν_u. The TFP model is the solid curve; The dashed curve is the locus for a MacLaurin oblate spheroid. The locus is shown a model of a NS based EOS FPS from [3]. From the Maclaurin sequence ([1]) critical values of the ellipticity ϵ are marked by horizontal dotted lines; These values are: $\epsilon = 0.4930$; $e = 0.8127$; $\frac{T}{|W|} = 0.1375$; $\epsilon = 0.7618$; $e = 0.9300$; $\frac{T}{|W|} = 0.2379$; $\epsilon = 0.8315$; $e = 0.9529$; $\frac{T}{|W|} = 0.2738$.

A Semi-analytic Model for the Radiation Reaction Luminosity for post-Newtonian Binary Neutron Star Mergers

F. Douglas Swesty* and Alan C. Calder[†]

*Dept. of Physics and Astronomy, SUNY Stony Brook 11794 [1]
[†]ASCI Flash Center, University of Chicago, Chicago, IL 60637

Abstract. In recent years there have been numerous attempts to model the relativistic coalescence of two neutron stars in a binary system. Because of the difficulties associated with solving the Einstein equations numerically, many attempts have been made to model the coalescence via post-Newtonian (PN) numerical hydrodynamic models. Perhaps the two most popular methods have centered around the use of the slow-motion quadrupole approximation or the 2.5PN approximation of Blanchet, Damour, and Schäfer (1990). A typical result of the simulations is the gravitational luminosity as a function of time during the coalescence. However, the complexities of the numerical simulation often make it difficult to understand the PN effects. To this end we are developing a semi-analytic model of the PN driven coalescence that can help us better understand the behavior of the radiation-reaction forces. This model is based on the use of $\gamma = 2$ Newtonian polytropes to approximate the structure of more realistic neutron stars.

RADIATION REACTION (RR) MODEL

Our radiation reaction model is based on the 2.5PN contributions as calculated by Blanchet, Damour, and Schäfer (1990) (hereafter BDS). The radiation reaction potential is given by

$$\psi = U + \chi \tag{1}$$

where

$$\chi \equiv -\frac{2G}{5c^5}\left(x_j \frac{\partial \phi}{\partial x_i}\frac{\partial^3 I_{ij}}{\partial t^3}\right). \tag{2}$$

and where U is a scalar potential arising from the solution of a Poisson equation

---

[1]) This work was supported by NASA HPCC CAN S5-3099 and computing support from NCSA under allocation MCA99S010N. ACC would like to thank DOE for support under grant B341495 to the ASCI Flash Center at the University of Chicago

$$\nabla^2 U = 4\pi G \left(\frac{2G}{5c^5} x_j \frac{\partial \rho}{\partial x_i} \frac{\partial^3 \mathcal{I}_{ij}}{\partial t^3} \right) \tag{3}$$

and where $\mathcal{I}_{ij}$ is the symmetric trace free part of the mass quadrupole tensor given by

$$\mathcal{I}_{ij} \equiv \frac{1}{2}(I_{ij} + I_{ji}) - \frac{1}{3}\delta_{ij}(I_{xx} + I_{yy} + I_{zz}) \tag{4}$$

and I_{ij} is the mass quadrupole tensor defined by

$$I_{ij} \equiv \int d^3x \rho x_i x_j. \tag{5}$$

ϕ is the Newtonian potential and ρ is the density distribution of the stars.

If we assume an equation of state of the form

$$P = K\rho^\gamma \tag{6}$$

with $\gamma = 2$ the density and potential structure of an isolated neutron star is expressible analytically as a solution to the Lane-Emden equation. If we further assume a circular orbit with frequency

$$\omega = \sqrt{\frac{2GM}{a^3}} \tag{7}$$

the mass quarupole tensor is analytically expressible. The right hand side of eqs. (2) and (3) are then analytically expressible. The calculation of the RR potential only requires the numerical solution of eq. (3).

The solution of equation (3) can be accomplished by means of the James algorithm for isolated systems (James 1977). Once the radiation reaction potential ψ is evaluated then the work on on the fluid can be calculated by

$$W_{2.5PN} = -\rho \mathbf{v} \cdot (\nabla U + \nabla \chi). \tag{8}$$

The total gravitational radiation Luminosity can then be calculated by

$$L_{2.5PN} = \int W_{2.5PN} d^3x. \tag{9}$$

We can also separately track the contributions of U and χ to the luminosity which we will refer to as L_U and L_χ with

$$L_{2.5PN} = L_U + L_\chi \tag{10}$$

For comparison slow-motion quadrupole is given by

$$L_Q = \frac{G}{5c^5} \sum_{i,j} \frac{\partial^3 \mathcal{I}_{ij}}{\partial t^3} \frac{\partial^3 \mathcal{I}_{ij}}{\partial t^3} = \frac{64 G^4 M^5}{5 c^5 a^5} \tag{11}$$

and is equal to the slow-motion quadrupole point mass formula where M is the mass of a single neutron star and a is the separation distance of the two stars. A comparison of $L_{2.5PN}$ and L_Q are shown in the following figure where the binary separation is plotted in units of the polytrope radius R.

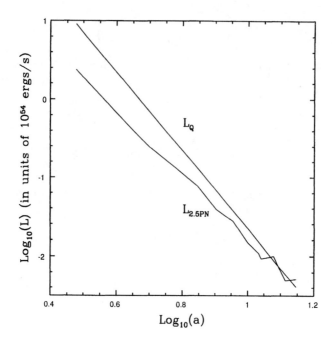

The slight jitter for large a is numerical noise as the stars near the edge of the grid.

A comparison of the 2.5PN radiation reaction term of BDS and the slow-motion quadrupole approximation reveals differences in the radiation luminosities from the two models. The 2.5PN radiation reaction luminosity shows a different scaling behavior with binary separation a than the slow-motion quadrupole would predict. Further work is needed to understand these differences.

REFERENCES

1. Blanchet, L, Damour, T., and Schäfer, G. *Mon. Not. Roy. Ast. Soc.* **242**, 289 (1990).
2. James, R. A., *J. Comp. Phys.* **25**, 71 (1977).

Detection of Gravitational Waves from Gravitationally Lensed Systems

T. Wickramasinghe* and M. Benacquista[†]

*Dept. of Physics, The College of New Jersey, Ewing, NJ 08628
[†]Dept. of Sciences, Montana State University-Billings, Billings, MT 59101

Abstract. It is accepted that quasars are powered by supermassive black holes (SMBH) with masses in the range $10^6 - 10^9 M_\odot$ in their cores. Occasionally, compact stars can plunge into SMBH. In addition, there may be a number of such compact objects circling the central SMBH in any given quasar. Both of these processes are known to emit gravitational waves. LISA has the right sensitivity to detect these waves. We show that gravitational lenses amplify the amplitudes of these gravitational waves just as they amplify the observed light of quasars. Given the geometry of the lensing configuration, this amplification can be as large as a factor of 2 to 10, allowing the waves to be above the detection threshold of LISA. We also show that waves from lensed quasars arrive with time delays which are much larger than the coherence time of the gravitational waves, making interference effects negligible. Thus, a simple geometrical optics application leads to the lensing theory of gravitational waves. In this context, we an alyze and show in this preliminary analysis that there is an enhancement of the amplitudes of gravitational radiation coming from observed lensed quasars.

It is now well accepted that the energy source of a quasar is a supermassive black hole (SMBH) [1]. The mass of the SMBH is of the order of $10^6 - 10^9 M_\odot$ [2]. In this mechanism of power generation, a compact star circling the SMBH plunges into it. Both these processes, encircling and plunging, are known to emit copious amounts of gravitational radiation [3,4]. These waves can in principle undergo gravitational lensing just as electromagnetic radiation does. We shall analyze the lensing of such waves.

A compact star plunging into the SMBH at the core of a quasar produces a burst of gravitational radiation which can be approximated as having gravitational strain h and frequency f. These two quantities are given by [5,6]:

$$h \simeq 10^{-21} \left(\frac{10 Mpc}{R}\right) \left(\frac{m}{1 M_\odot}\right) \qquad (1)$$

$$f \simeq (1.3 \times 10^{-2} Hz) \left(\frac{10^6 M_\odot}{M}\right) \qquad (2)$$

where R is the distance to the system, m is the mass of the compact object, and

M is the mass of the SMBH. LISA has enough sensitivity to detect gravitational waves with amplitudes $h \sim 10^{-20} - 10^{-21}$ and frequencies $f \sim 10^{-1} - 10^{-3} Hz$ [6].

There are a number of gravitationally lensed quasars within $z \sim 1$ [7]. The lensing object usually is a normal galaxy having a mass of the order of $10^{11} M_\odot$. The radius of curvature of the spacetime in the proximity of the lensing galaxy is $R_{curvature} \sim GM_l/c^2$, where M_l is the mass of the lensing galaxy. Eq. 2 shows that $\lambda_{GW} \sim 10^7 km$, while $R_{curvature} \sim 10^{11} km$. Thus $\lambda_{GW}/R_{curvature} \ll 1$ and we can use a simple geometrical optics approximation to analyze lensing of gravitational waves coming from lensed quasars. Gravitational waves of a lensed quasar should bend in the proximity of the lens, just as light coming from the quasar. This lensing effect will produce multiple images of the quasar due to area distortion near the lens.

The distances in Fig. 1 are Dyer and Roder angular diameter distance [8]. If the universe is filled with a smooth distribution of matter, then the angular diameter distance measured from a point at redshift z_1 to a point at z_2 is given by:

$$D_A(z_1, z_2) = \frac{c}{H_0} r(z_1, z_2) = 2 \left[\frac{1}{(1+z_2)\sqrt{1+z_1}} - \frac{1}{(1+z_2)^{3/2}} \right]. \quad (3)$$

The luminosity distance is given by: $D_l = D_A(1+z)^2$. In Eq. 1, we must replace R by D_l. An Einstein ring forms when the perfect alignment is achieved. The radius of this ring is $\xi_0 = \sqrt{4GM_l D_d D_{ds}/(c^2 D_s)}$ [9], and its angular size is $2\alpha_0$. We now define dimensionless quantities $y = \beta/\alpha_0$ and $x = \theta/\alpha_0$ so that y and x are the dimensionless positions of the source and image, respectively. With $c/H_0 =$

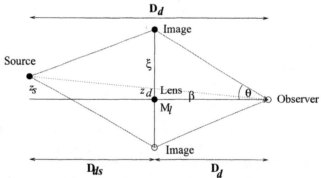

FIGURE 1. The general gravitational lensing geometry. Lens - Observer is the optic axis. β and θ are the source and image positions respectively. z_s and z_d are the redshifts of the source and the lens.

$2997.9/h_0$ Mpc, we derive the following relationships ($\alpha_0'' = \alpha_0$ in arcseconds):

$$\alpha_0'' = 1.65 \times 10^{-6} \sqrt{h_0} \sqrt{\frac{M_l}{M_\odot}} \psi(z_d, z_s) \quad (4)$$

$$\psi(z_d, z_s) = \sqrt{\frac{r(z_d, z_s)}{r(0, z_s) r(0, z_d)}} \tag{5}$$

$$\mu = \frac{y^2 + 2}{y\sqrt{y^2 + 4}} \tag{6}$$

$$c\Delta t = 1/|d\beta/d\theta| = \frac{4GM_l}{c^2}(1 + z_d) \left[\frac{1/2}{y}\sqrt{y^2 + 4} + \ln \frac{\sqrt{y^2 + 4} + y}{\sqrt{y^2 + 4} - y} \right] \tag{7}$$

Equations 6 and 7 are well-known equations in gravitational lensing theory [9]. Eq 6 gives the total amplification of images. Eq. 7 shows that there is a time delay Δt between any two images.

The equations 4 through 7 can be used to estimate the amplification of the gravitational waves coming from a lensed quasar. Eq. 7 shows that the time delay between any two images of a realistic gravitational lensing configuration of a quasar is many days. Comparison with f_{GW} shows that interference effects can safely be neglected. LISA's angular resolution scales as the inverse of the signal to noise and is a few square degrees. The separate images of a quasar are usually separated by a few arcseconds. Therefore, it is hopeless to imagine that LISA will be able to resolve images. However, it will see a composite, amplified image. The gravitational wave strain coming from this object must then be $\sim \mu h$.

For any practical astrophysical situation, $\psi(z_d, z_s) < 2$. Note that $y < 1$ for $\beta < \alpha_0$ when the source is within the Einstein ring of the lens. We note from Eq. 6 that the amplification in this case is large. It is clear also that substantially small source positions y can be expected in real astrophysical situations which implies that gravity waves coming from lensed quasars are typically amplified by a factor of 2 - 10. We see from Eq. 1 that the sensitivity of LISA is sufficient to detect gravitational waves emanating from a gravitationally lensed quasar system as far distant as 100 Mpc for an amplification as large as 10. We stress the fact that our analysis is still preliminary.

REFERENCES

1. Hoyle, F. and Fowler, W.A., Nature, 213, 373 (1963)
2. Rees, M.J. Ann. Rev. Astron. Astrophysics, 22, 471 (1984)
3. Thorne, K.S., *300 Years of Gravitation*, eds. S.W. Hawking and W. Israel, Cambridge University Press (1987)
4. Cutler, C. et al., Phys. Rev. Lett., 70, 2984 (1993)
5. Haehnelt, M.G., MNRAS, 269, 199 (1994)
6. Shibata, M., Phys. Rev. D 50, 6297 (1994)
7. Kembhavi, A.K. and Narlikar, J.V., *Quasars & Active Galactic Nuclei*, (Cambridge Univ. Press), 390 (1999)
8. Dyer, C.C. and Roder, R.C., ApJ, 180, L31 (1973)
9. Schneider, P., Ehlers, J., and Falco, E.E., *Gravitational Lenses* (Springer), 240 (1992)

CHAPTER 16

ULTRA-HIGH ENERGY COSMIC RAYS

THE ORIGIN OF ULTRA HIGH ENERGY COSMIC RAYS: WHAT WE KNOW NOW AND WHAT THE FUTURE HOLDS

A A Watson

Department of Physics and Astronomy, University of Leeds, Leeds LS2 9JT, UK

Abstract. The observational picture for cosmic rays above 10^{19} eV ($= 10$ EeV) is described and the enigma that these measurements pose is discussed. Possible ways of resolving the enigma are explained and it is concluded that the existence of particles above 10^{20} eV may come to have a major impact on our understanding of magnetic fields in intergalactic space and in possible sources. A new approach to the measurement of the mass of cosmic rays at the highest energies is demonstrated with existing data: the prospects for applying this with an instrument under construction are outlined.

INTRODUCTION

The cosmic ray spectrum is unusual: it spans over 11 decades of energy and remains almost featureless as the intensity falls through thirty-two decades. The origin of cosmic rays is not certain at any energy but most are thought to be the nuclei of the common elements, hydrogen through to iron. Because the cosmic rays are charged, magnetic fields affect their trajectories. What is observed at earth near 1 GeV is severely affected by solar wind modulation, while galactic and intergalactic magnetic fields are sufficiently strong and disordered to randomise the directions of higher energy particles. Thus it is not possible to trace back along locally observed trajectories to locate a likely source. Over the years it has become widely believed that particles above 1 GeV and less than 10^{18} eV ($= 1$ EeV) are of galactic origin while particles of greater energy are accelerated outside the galaxy. It is probably true to say that this picture is hallowed more by tradition than by incontrovertible observational data. For example energetics arguments, pressed in particular by Ginzburg from the 1950s, have pointed to supernovae as possible sources whilst since the late 1970s work on diffusive shock acceleration has provided a mechanism that may take particles to $Z \times 10^{14}$ eV at such sites. At around 3×10^{15} eV the energy spectrum steepens (the knee) quite sharply and one interpretation is that the supernovae accelerators have become exhausted. Continuity arguments suggest that the population of particles above this energy may be boosted within the galaxy through acceleration by old shocks from supernovae. Near 10^{18} eV the spectrum steepens once more (the second knee) before flattening (the ankle) and continuing in a featureless manner to 3×10^{20} eV, the highest energy yet observed. It is worth stressing that this is probably not a cut-off energy but rather a limit imposed by the exposure of detectors available so far. When it is appreciated that the rate above 100 EeV is only about 1 km^{-2} century^{-1} sr^{-1} the difficulty of coming to hard conclusions may be recognised.

At the energies where supernovae may be the source, there have been predictions of TeV gamma ray fluxes that would be detectable by current instruments - but such detections are presently lacking. At all energies a crucial observational quantity is the mass of the primaries but beyond the limits of satellite and balloon observatories (about 10^{14} eV), in which particle-by-particle mass identifications are possible, matters become very difficult. The higher energy events are only detectable because they produce giant cascades in the atmosphere that are observable at ground level but the particle amplification ($\sim 10^{11}$ at 100 EeV) and lateral spread (over several square kilometres at the same energy), while enabling detection, degrades mass sensitive information very considerably. Thus mass signatures that might be expected to be associated with the various spectral features outlined above (for example heavier particles - presumed iron - being more dominant at the second knee and protons at the highest energies) have been difficult to extract from the data.

EXPERIMENTAL DATA ON THE HIGHEST ENERGY COSMIC RAYS

Recently much attention has focused on the topic of ultra high-energy cosmic rays and the topic has been much reviewed [1-3]. Briefly it had been expected that even if the particles above about 50 EeV were produced in sources uniformly distributed throughout the Universe then the spectrum observed at earth would terminate rather abruptly near this energy. This is the well-known Greisen-Zatsepin-Kuzmin (GZK) effect [4,5] and arises from the photo-pion production that occurs when a proton propagates through the 2.7 K radiation field. Heavy primaries are similarly attenuated by photodisintegration. The results from decades of experimental effort make it highly likely that the cut-off does not exist. The published total of events above 100 EeV (an energy well beyond the GZK-cut-off that is taken as a convenient reference) is only 16 but the consensus is that the energy estimates are sufficiently accurate for it to have been established that the predicted cut-off is absent. The measurements by the AGASA group [6] are shown in figure 1. The dashed line shows the spectrum expected if the ultra high-energy cosmic rays (UHECR) are from sources distributed uniformly over all space. It is clear that the GZK cut-off is not observed, even after allowing for a 30% measurement error in the primary

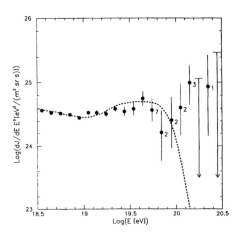

FIGURE 1. The energy spectrum observed by AGASA (Takeda et al. [6]). The dashed line represents the spectrum expected for extragalactic sources distributed uniformly in the Universe, taking into account the uncertainty in energy estimates.

energy. In figure 1 the input spectrum for the simulations was taken as E^{-3}. Olinto [7] has investigated flatter spectra and finds that even with a slope of $\gamma = -1.5$ the data points above 10^{20} eV lie above the predictions. Similarly, using the galaxy distribution from the IRAS redshift survey, an excess of observed events over expectation was found [7].

The implication of the observation of events above 100 EeV is that their sources must lie within the so-called GZK sphere which at 100 EeV is about 20 Mpc for 50% of the particles. Of course this radius depends very much on what is assumed for the intervening magnetic field strength and would be smaller if the field was very strong. It is easily shown that for an extra-galactic magnetic field of around 10^{-9} G with a structure on canonical scale of 1 Mpc, for which there is evidence (e.g. Kronberg [8]), strong anisotropies in arrival directions might be expected. For example, were the particles of galactic origin then the galactic plane would stand out in cosmic ray 'light', while if M87 was a source it might be expected to be seen rather clearly with the events so far recorded above 40 EeV. A plot of the arrival directions of 114 such events [3] is shown in figure 2: the distribution is uniform although there are a few clusters of two or three events that may occur more frequently than would be expected by chance: see [9] for a detailed discussion of clustering possibilities.

Interpretation of the data on UHECR is hampered by lack of knowledge of the mass of the incoming particles. The main method used to deduce the change of primary mass with energy is to measure the rate of change of the depth of shower maximum with energy, the elongation rate [10], and compare the result with models. The methods of deriving the shower maximum will not be described here: they are

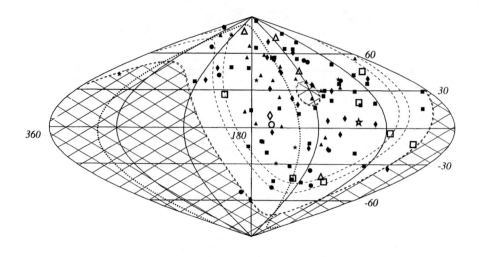

FIGURE 2. The arrival directions of events with energy $> 4 \times 10^{19}$ eV. See [3] for details.

discussed in detail in the review by Nagano and Watson [3]. A summary of data [11] from several experiments is shown in figure 3. The model-dependent nature of any conclusion is apparent. However there is some suggestion that the mean mass above 10^{18} eV is lighter than at lower energies.

There is clearly room for further work, even below 10^{19} eV, and the data are too limited to allow any strong statement about mass composition above about 10^{19} eV. However it is unlikely that the majority of the events so far detected above 10^{20} eV have photon parents as some of the largest events [12] seem to have normal numbers of muons (the tracers of hadronic primaries). Additionally the cascade profile of the most energetic fluorescence event is inconsistent with a photon primary [13]: further new evidence against photon primaries at 10^{19} eV is described below. It is also unlikely that the majority of events are created by neutrinos as the distribution of zenith angles would be different from that observed. Indeed, in all aspects so far measured, events of 10^{20} eV look like events of 10^{19} eV, but ten times larger.

Of course if the highest energy particles were Fe nuclei rather than protons the arrival direction isotropy could be more readily understood (except for the possible clustering). Similarly the magnetic field in intergalactic space might be much stronger than 10^{-9} G, as has been discussed by Farrar and Piran [14]. However that is not the only problem that the existence of trans-GZK particles poses: it is very hard to see how such particles can be accelerated to energies as high as 3×10^{20} eV in known sources as is now explained.

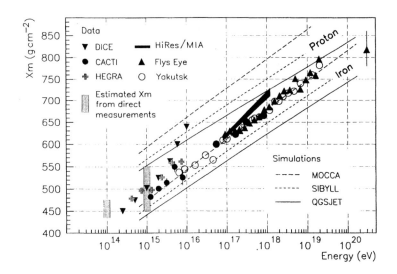

FIGURE 3. The depth of shower maximum as a function of primary energy. Data from Haverah Park, Fly's Eye and Yakutsk are shown together with the predictions for proton and iron primaries for two different interaction models, MOCCA-SIBYLL and QGSJET (after Hinton et al. [11]).

PROBLEM OF ACCELERATION OF UHECR

The question of how cosmic rays can reach such vast energies has been examined exhaustively but an early discussion [15] contains the essence of the argument. For acceleration by the diffusive shock process, or by single shot acceleration, it can be shown that the maximum energy is given by $E = kZeBR\beta c$, where k is a constant < 1 (except in a man-made accelerator where k = 1) and the other symbols have their usual meaning. This implies that the site of acceleration of a proton of 3×10^{20} eV, in which the shocks are moving with velocities close to c, must have the product $BR \sim 10^{18}$ G cm. Hillas's survey [15] of possible sites excludes all except possibly radio galaxy lobes, AGNs and galactic clusters but the potential of the latter group may be marginal and would be further reduced if a particle of even 5×10^{20} eV was discovered. It is worth noting that the field must be rather weak so that synchrotron losses are not greater than the energy gain. It was shown by Greisen [16] that the total magnetic energy in the source grows as γ^5, where γ is the Lorentz factor of the particle. For 100 EeV the energy in the magnetic field must be 10^{57} ergs with $B < 0.1$ G. Such sources are likely to be strong radio emitters with radio power $\gg 10^{41}$ ergs s^{-1}, unless hadrons are being accelerated but not electrons.

We see that the combination of an intergalactic field of 10^{-9} G and primary protons leads to a paradox. This paradox can be resolved if the intergalactic field is much stronger (perhaps 10^{-7} G) or the primary particles are iron nuclei. Iron nuclei are ostensibly easy to accelerate but are fragile against break up in the photon fields that are likely to be in the vicinity of possible accelerators. A cut-off is still expected although the cut-off energy is less well understood, as the uncertain infrared background becomes important in the estimation of the rapidity of photodisintegration. Thus protons are generally favoured (or at least most widely discussed) and strong magnetic fields are not taken very seriously. Consequently the paradox has received much attention in the last 4 or 5 years and produced many proposals for its resolution.

A detailed review of production mechanisms has been given by Bertou et al. [1] and by Olinto [2]. Mechanisms that do not involve electromagnetic acceleration have come to be known as 'top down' processes. Many of the top down mechanisms predict that the flux of photons and neutrinos above 10 EeV would be greater than the flux of protons: heavy nuclei would not be produced. One such mechanism, which I highlight because detailed predictions have been made from it, is the decay of super-heavy relic particles that may have been created in the hot phase of the early Universe following inflation. A general discussion of this idea was first advanced by Berezinsky et al. [17] and there has been extensive discussion of the crypton manifestation of such particles by Birkel and Sarkar [18] and by Rubin [19]. The idea is interesting because it is testable, or at least will be shortly, when advanced detector systems come into operation. In addition to the photon flux expected there should be measurable anisotropies, the magnitude of which will depend on the distribution of these massive particles in the halo of our galaxy (Medina Tanco and Watson [20]).

FUTURE DETECTORS

The above discussion should have made clear that there is a need for detectors with much larger apertures and with the ability to discriminate between showers produced by particles of different types. Considerable progress has been made in these directions and, in addition to the continuing operation of the 100 km^2 AGASA array, events are now being collected with the HiRes fluorescence detector. The Pierre Auger Observatory is under construction with the first events from the 55 km^2 engineering array expected in 2001. In addition a detailed accommodation study has been undertaken for the flight of a fluorescence detector on the International Space Station.

The HiRes detector is a successor to the very successful Fly's Eye instrument that operated in the Dugway desert (Utah) for many years. The latter instrument gave a near-calorimetric measure of the energy spectrum and confirmed the energy assignments made at ground arrays such as Haverah Park and AGASA. In addition, of course, the highest energy cosmic so far known was detected with it. The HiRes detector in its final form is expected to have a time-averaged aperture of 340 km^2 at 10^{19} eV and 1000 km^2 at 10^{20} eV. HiRes is a stereo system and will measure the depth of maximum of the shower to 30 g cm^{-2} on an event-by-event basis. This precision is usefully smaller than the expected difference in the mean depth of maxima for proton or Fe-initiated showers (see figure 3).

The Pierre Auger Observatory is planned eventually to have stations in the Northern and Southern hemispheres. Currently the southern observatory is under construction in western Argentina near the town of Malargue. This observatory, with an aperture of 3000 km^2, will make the first detailed exploration of the southern sky. It will use a hybrid combination of fluorescence detectors and water-Cherenkov tanks, permitting cross-calibration of energy estimates in 10% of events. The initial plans to use the muon fraction of the signal, the pulse rise time, the shower curvature and the particle lateral distribution to aid mass discrimination will be supplemented by studies of inclined showers that may also allow the aperture to be increased by about 50%. The instrument is designed to have the potential to discriminate, on a statistical basis, between showers initiated by protons and Fe nuclei and will be capable of identifying photon showers or neutrino induced events. The rate of the latter is expected to be around 5 per year if the only source is neutrinos from the decay chain associated with photo-pion production.

A very ambitious plan is underway to mount on fluorescence detector in the Columbus module of the International Space Station. This project, named EUSO for the Extreme Universe Space Observatory, is based on a line of development proposed many years ago by Benson and Linsley [21]. The time-averaged aperture of this instrument will depend upon the contribution of lightning, city lights, oceanic biofluorescence and so on to the night sky background but estimates suggest that it may be 7 to 10 times that of the Auger Observatories. This instrument will be particularly suitable for searches for neutrino and photon primaries. The launch date is planned as 2005 but progress will depend on the successful construction of the Space Station.

A NEW APPROACH TO THE MEASUREMENT OF PRIMARY MASS

The detection of neutrinos is likely to be possible with the Auger Observatory through a study of the properties of very inclined showers (zenith angle > 80°). The method envisaged is to search for events that arrive at large angles but have the characteristics of showers coming in a near vertical direction. The vast majority of showers arriving at such angles are known from earlier work Haverah Park to be characterised by having very flat particle fronts, small spreads of particle arrival times and flat distributions of densities. It has been assumed that the majority of these events are initiated by protons or heavier nuclei and that the features arise because the electronic component of the shower has been completely attenuated with only muons, produced very far from the detector, surviving. By contrast a shower initiated by a neutrino interacting in the air mass above the array will look like a vertical event with a curved shower front, a steep fall-off of densities with distance and a relatively broad spread of times over which the shower particles arrive.

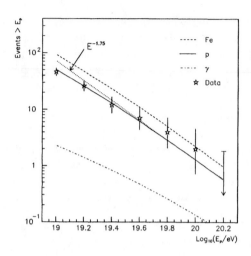

FIGURE 4. Integral rate of events above 10^{19} eV and 60° zenith angle compared with predictions for photons, protons and iron nuclei [22].

As a preliminary to the data from the Auger Observatory, the first detailed study of the inclined events recorded at Haverah Park has been made [22]. As well as demonstrating that such events can be analysed, it proved possible to set a limit to the photon flux above 10^{19} eV. It is argued that the energy spectrum in the near vertical direction is now well known from the Fly's Eye data, independent of assumptions about the particle physics or the mass at these energies. Thus this spectrum can be used with models of shower development to predict the rate of events expected for different primary masses. The results are shown in figure 4. The predicted rate assuming protons (solid line) is more than a factor of 10 above the rate expected if the primaries were photons. Indeed the rate predicted using proton primaries agrees rather well with what is observed, within the large error bars inevitably associated with only 46 events. The adopted primary spectrum is that given by Nagano and Watson [3] and, after allowing for the uncertainties in it, Ave et al. claim [22] that above 10^{19} eV less than 41% of the primaries can be photons at the 95% confidence level. This is in contradiction with the predictions of Berezinsky, Kachelreiss and Vilenkin [17], Birkel and Sarkar [18] and Rubin [19]. The later two calculations give photon fluxes that are factors of 2 and 3 respectively above the proton flux. While attenuation of the photon flux by pair production with radio photons has not been calculated in detail (Sarkar [23]) it seems likely that the new experimental data present some difficulties for the super-heavy relic model of UHECR.

The comparison between observation and prediction shown in figure 4 also allows a limit to be set to the fraction of Fe nuclei in the primary beam above 10^{19} eV to be set. At present the 95% confidence limit is 54% but this conclusion is sensitive to the shower model adopted and to the spectrum assumed. However for the calculation shown in figure 4 conservative assumptions were made such that if the predicted spectrum is higher (because of a lower intensity primary spectrum or a different model) then the photon bound will be strengthened.

Another approach to the detection of photon primaries uses the production of multiple photons in the geomagnetic field (Bertou et al. [24]). This approach may be particularly effective at 10^{20} eV where the method just described will be limited by uncertainties in the primary energy spectrum.

What is very exciting is that neither neutrino detection nor photon detection by the methods just mentioned were early drivers of the design of the Auger Observatory. The potential of the instrument has thus greatly increased and, further, the ability to find the energy of very inclined events demonstrated by Ave et al. [22] will enhance the aperture above the 7000 km^2 originally envisaged. Possibly, as with most new telescopes, the major excitement will be in discoveries that we cannot anticipate.

CONCLUSIONS

The resolution of the paradox presented by the highest energy cosmic rays clearly is some way off. If further work confirms that the photon flux is low then either there is (1) other 'new' physics [1,2] or (2) there must be an accelerator in which BR $> 10^{18}$ G-cm in a configuration that allows low synchrotron and photon losses plus magnetic fields over 20 Mpc that are strong enough (10^{-7} G) to scramble the directions. Work with the major new detector systems described above should go a long way to solving this problem. In particular after 3 years of operation with the full 3000 km^2 of the Auger Observatory it will be possible to explore the photon/proton ratio at the 10% level near 10^{19} eV.

ACKNOWLEDGEMENTS

I am very grateful to Craig Wheeler and the organising committee for inviting me to this meeting to expose the situation with regard to truly relativistic particles and for financial support. Work on ultra high-energy cosmic rays in the UK is supported by PPARC.

REFERENCES

1. Bertou, X., et al., *Int. J. Mod Phys.* **A 15**, 2181 (2000)
2. Olinto, A.V., *Physics Reports* **333 -334**, 329, (2000)
3. Nagano, M and Watson, A. A., *Rev. Mod. Phys.*, **72**, 689 (2000)
4. Greisen, K., *Phys. Rev. Lett.* **16**, 748 (1966)
5. Zatsepin, G.T. and Kuzmin, V. A., *Sov. Phys. JETP Lett.* **4**, 78 (1966)

6. Takeda, M. et al., *Phys. Rev. Lett.* **81**, 1163 (1998)
7. Olinto, A., in *Proceedings of Puebla Workshop, Mexico*, August 2000, to be published.
8. Kronberg, P. P., *Reports on Progress in Physics* **57**, 325 (1994)
9. Uchihori, Y. et al.,*Astroparticle Physics* **13**, 151 (2000)
10. Linsley J., *Proc. 15th ICRC (Plovdiv)* **12**, 89 (1977)
11. Hinton, J. A., et al., *Proc 26th ICRC (Salt Lake City)* **3**, 288 (1999)
12. Hayashida, N., et al., *Phys. Rev. Lett.* **77**, 1000 (1996)
13. Halzen, F., et al., *Astroparticle Physics* **3**, 151 (1995)
14. Farrar, G. R. and Piran, T., *Phys. Rev. Lett.* **84**, 3527 (2000)
15. Hillas, A. M., *Ann. Rev. Astron&Astrophy.* **22**, 425 (1984)
16. Greisen, K., *Proc. 9th ICCR (London)* **2**, 609 (1965)
17. Berezinsky, V., et al. 1997, *Phys. Rev. Lett.* **79**, 4302 (1997)
18. Birkel, M. and Sarkar, S., *Astroparticle Physics* **9**, 297 (1998)
19. Rubin, N. A., *M Phil Thesis, University of Cambridge*, **98** pp (1999)
20. Medina Tanco, G. and Watson, A. A., *Astroparticle Physics* **12**, 25 (1999)
21. Benson, R, and Linsley, J., *Proc. 17th ICCR (Paris)* **8**, 145 (1981), Linsley J., *USA Astronomy Survey Committee (Field Committee) Documents* (1979), Linsley J., *Proc. 19th ICCR (La Jolla)* **3**, 438 (1985)
22. Ave, M et al., *Phys. Rev. Lett.* **85**, 2244 (2000)
23. Sarkar, S., Comment at *Oxford Workshop on Radio Galaxies* (2000)
24. Bertou, X., et al., *Astroparticle Physics* **14**, 121 (2000)

Recent Results from the High Resolution Fly's Eye Detector

Douglas Bergman

Rutgers University, Piscataway, NJ 08854

Abstract. The High Resolution Fly's Eye Detector is an active, air fluorescence ultra-high energy cosmic ray detector situated in Utah. A preliminary energy spectrum is presented, along with the current status and future plans.

DESCRIPTION

The High Resolution Fly's Eye Detector (HiRes) was designed to measure the spectrum, direction and composition of ultra-high energy cosmic rays (UHECR) at the very highest energies. It uses the air fluorescence technique to observe the longitudinal development of the extensive air shower (EAS) initiated by the UHECR as it encounters the Earth's atmosphere.

The detector consists of two sites, separated by 13 km to allow for stereo viewing of the EAS. The air fluorescence light produced by the EAS is gathered by mirrors and recorded by an array of photomultiplier tubes, with each tube viewing about one square degree of the sky. Detection of the faint EAS fluorescence signals dictates the placement of the detector in the desert, with its clear, dark atmosphere. It also imposes a 10% duty cycle since the detector can only be operated on clear, moonless nights. EAS from 100 TeV UHECR can be seen from up to 50 km, which gives us the very large collection area necessary for determining the spectrum from the very low flux.

The two detector site are locate on hilltops (to get above any low haze) in the Great Salt Lake desert, within the confines of Dugway Proving Grounds in Utah. The first site to be built, HR1, consists of one ring of mirrors, viewing elevation angles from 3°–15°. It uses a sample-and-hold data acquisition system and has been taking data since June, 1997. The second site, HR2, consist of two rings of mirrors, viewing elevation angles from 3°–30°. It uses an FADC data acquisition system, which allows for better reconstruction of the longitudinal development of EAS, especially for distant events.

CALIBRATION ISSUES

The energy of the UHECR primary is directly proportional to the amount of fluorescence light produced by its EAS. Since our measurement of the energy comes primarily from detecting the amount of this light, calibration of the light collection efficiency and event geometry is vitally important to our determination of the UHECR energy. A list of calibration items includes the absolute gain of the PMTs, the tube-to-tube variations in PMT gain, filter transmission, mirror reflectivity, mirror optics, atmospheric transmission, missing energy (EAS components which do not contribute to fluorescence) and extraneous sources of light (e.g. Cerenkov light produced by EAS electrons). Of these, I will emphasize two: the absolute calibration of PMT's and the transmission of the atmosphere.

The data for the spectrum presented below used a PMT calibration based on the photon flux from a Xenon flash bulb. This flux was known to be accurate to about 30%, a significant uncertainty. A new method of calibrating the PMT gain, has been implemented which uses Poisson statistics to deduce the gain from the width of the PMT response to many flashes. The Xenon flash bulb is known to be stable to much better than 30% in continuous operation. However, complications due to the statistics of multiplication at the first dynode, delayed the implementation of this calibration. This is why I won't be showing any really new results below.

The transmission of fluorescence light through the atmosphere is actually the largest source of uncertainty in our energy measurement, due to the atmosphere's varying aerosol content. To control this uncertainty we probe the atmosphere at various points around the detector sites with lasers and Xenon flash bulbs. Light pulses from either source creates a signal similar to an EAS but moving up through the atmosphere. By measuring the amount of light scattered into our detector we can measure the amount of aerosols in the atmosphere and the extinction length of light traveling through it. This calibration data has not been implemented in the data analysis present below however, which used a constant aerosol content for all the data. This also leads to an uncertainty of about 30%.

The systematic uncertainty in energy measurement from other sources in reconstruction can be estimated by comparing the reconstructed energy of Monte Carlo (MC) generated events with their generated energy. This is shown in Figure 1. The energy resolution improves from about 30% at 1 EeV to 18% at the highest energies.

MONOCULAR SPECTRUM

UHECR events collected by HR1 between May 1997 and June 1999 are shown in Figure 2. These events have been required to pass a number of quality cuts which are detailed elsewhere [1].

Using a MC calculation of the aperture and exposure of the detector one can calculate a *preliminary* UHECR spectrum from this, which is shown in Figure 3 with the flux multiplied by E^3 to emphasize the structure. Despite the various systematic uncertainties, the position of the ankle agrees quite nicely with the old Fly's Eye experiment.

One event is displayed in illustration in Figure 4. This quite high energy event was also seen by HR2, one of the first and highest energy events to be seen in stereo.

CURRENT STATUS AND FUTURE PLANS

The second site, HR2, has been been taking quality data for almost a year now and there are a number of events which are viewed in stereo. This reduces the uncertainty in the energy due to the geometrical reconstruction of the EAS trajectory. It also allows a cross check of the atmospheric model used in reconstructing the amount of light produced in the EAS. HiRes expects to release a preliminary

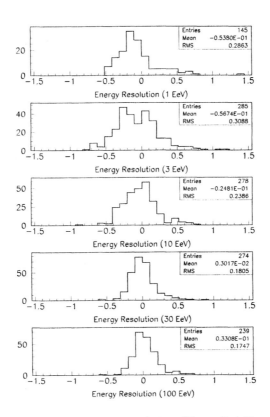

FIGURE 1. Energy resolution: $(E_{\text{out}} - E_{\text{in}})/E_{\text{in}}$.

stereo spectrum in 2001 along with an enhanced version of this monocular HR1

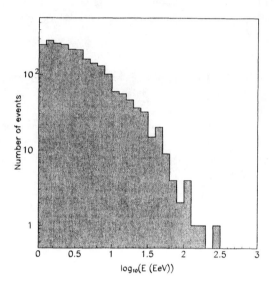

FIGURE 2. Number of HR1 events in each energy bin.

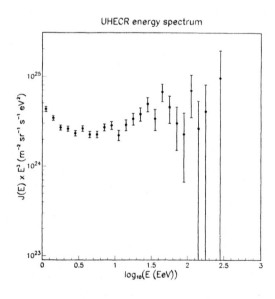

FIGURE 3. The *preliminary* UHECR spectrum as measured by HR1.

analysis with better absolute PMT gain and atmospheric calibrations.

REFERENCES

1. "The Energy Spectrum of Ultra High Energy Cosmic Rays", T. Abu-Zayyad, Ph.D. Thesis, The University of Utah (2000).

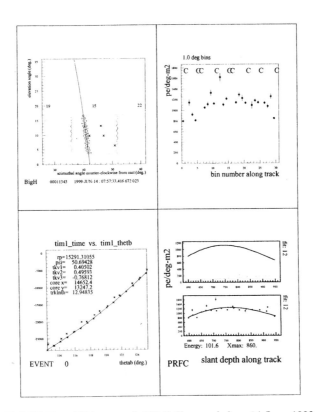

FIGURE 4. HR1 event of 100 EeV, recorded on 14 June 1999.

Probing Interactions beyond the Electroweak Scale with Ultra High Energy Cosmic Radiation

G. Sigl*

*Institut d'Astrophysique de Paris, CNRS, 98bis Boulevard Arago, 75014 Paris, France

Abstract. We demonstrate how current and future data on ultra-high energy cosmic rays and neutrinos can be used to probe new interactions at energies not or hardly accessible in accelerator experiments of the near future. As a specific example we discuss the possibility that neutrinos attain new interactions in the TeV range in particle physics models involving extra dimensions.

INTRODUCTION

Over the last few years, several giant air showers have been detected confirming the arrival of cosmic rays (CRs) with energies up to a few hundred EeV (1 EeV $\equiv 10^{18}$ eV) [1, 2, 3, 4]. The existence of such extremely high energy cosmic rays (EHECRs) pose a serious challenge for conventional theories of CR origin based on acceleration of charged particles in powerful astrophysical objects. This challenge is at least three fold.

First, it is hard to accelerate known charged particles, namely protons and heavy nuclei, up to such energies even in the most powerful astrophysical objects such as radio galaxies and active galactic nuclei. The respective problems in "bottom-up" acceleration have been well-documented in a number of studies; see, e.g., Refs. [5, 6, 7]. Second, nucleons above $\simeq 70\,\text{EeV}$ lose energy drastically due to photo-pion production on the cosmic microwave background (CMB) — the Greisen-Zatsepin-Kuzmin (GZK) effect [8] — which limits the distance to possible sources to less than $\simeq 100\,\text{Mpc}$. Heavy nuclei are photodisintegrated in the CMB within a few Mpc [9], and pair production on the CMB and the universal radio background limits propagation of photons in the hundreds EeV range to similar distances (see e.g. Ref [10]). Finally, the observed EHECR arrival directions appear quite isotropic [11] and there are no or at least not sufficiently many obvious astronomical sources within $\simeq 100$ Mpc of the Earth to explain these observations. Specifically, no obvious astrophysical source counterparts have been found in the observed EHECR arrival directions [12, 6].

The question of the origin of these EHECRs is, therefore, currently a subject of much intense debate and discussions as well as experimental efforts; see Ref. [13] for recent brief reviews, and Ref. [10] for a detailed review on theoretical work.

NEW PRIMARY PARTICLES AND NEW INTERACTIONS

A possible way around the problem of missing counterparts within acceleration scenarios is to propose primary particles whose range is not limited by interactions with the CMB. Within the Standard Model the only candidate is the neutrino, whereas in supersymmetric extensions of the Standard Model, new neutral hadronic bound states of light gluinos with quarks and gluons, so-called R-hadrons that are heavier than nucleons, and therefore have a higher GZK threshold, have been suggested [14].

In both the neutrino and new massive neutral hadron scenario the particle propagating over extragalactic distances would have to be produced as a secondary in interactions of a primary proton that is accelerated in a powerful AGN which can, in contrast to the case of EAS induced by nucleons, nuclei, or γ–rays, be located at high redshift. Consequently, these scenarios predict a correlation between primary arrival directions and high redshift sources. In fact, possible evidence for an angular correlation of the five highest energy events with compact radio quasars at redshifts between 0.3 and 2.2 was recently reported [15]. These are the kind of sources one would expect to lead to significant secondary production by the accelerated primaries. However, a new analysis with the somewhat larger data set now available does not support significant correlations [16]. This is currently disputed since another group claims to have found a correlation on the 99.9% confidence level [17]. Only a few more events could confirm or rule out the correlation hypothesis. We note, however, that the data imply a significant clustering on degree scales [11, 18], although not necessarily correlated with a known class of sources. We stress that these scenarios would require the primary proton to be accelerated up to $\gtrsim 10^{21}$ eV, demanding a very powerful astrophysical accelerator. On the other hand, a few dozen such exceptional accelerators in the visible Universe may suffice.

For the remainder we will focus on neutrinos as primary candidates since they have the advantage of being well established particles. Within the Standard Model their interaction cross section with nucleons, whose charged current part can be parametrized by [19]

$$\sigma_{\nu N}^{SM}(E) \simeq 2.36 \times 10^{-32} (E/10^{19}\,\text{eV})^{0.363}\,\text{cm}^2 \quad (10^{16}\,\text{eV} \lesssim E \lesssim 10^{21}\,\text{eV}), \quad (1)$$

falls short by about five orders of magnitude to produce ordinary air showers. However, it has been suggested that the neutrino-nucleon cross section, $\sigma_{\nu N}$, can be enhanced by new physics beyond the electroweak scale in the center of mass (CM) frame, or above about a PeV in the nucleon rest frame [20, 21, 22]. Neutrino induced air showers may therefore rather directly probe new physics beyond the electroweak scale.

The lowest partial wave contribution to the cross section of a point-like particle is constrained by unitarity to be not much larger than a typical electroweak cross section [23]. However, at least two major possibilities allowing considerably larger cross sections have been discussed in the literature for which unitarity bounds need not be violated. In the first, a broken SU(3) gauge symmetry dual to the unbroken SU(3) color gauge group of strong interaction is introduced as the "generation symmetry" such that the three generations of leptons and quarks represent the quantum numbers of this generation symmetry. In this scheme, neutrinos can have close to strong interaction cross sections with quarks. In addition, neutrinos can interact coherently with all partons in the nucleon, resulting in an effective cross section comparable to the geometrical nucleon cross section. This model lends itself to experimental verification through shower development altitude statistics [20].

The second possibility consists of a large increase in the number of degrees of freedom above the electroweak scale [24]. A specific implementation of this idea is given in theories with n additional large compact dimensions and a quantum gravity scale $M_{4+n} \sim$ TeV that has recently received much attention in the literature [25] because it provides an alternative solution (i.e., without supersymmetry) to the hierarchy problem in grand unifications of gauge interactions. The cross sections within such scenarios have not been calculated from first principles yet. Within the field theory approximation which should hold for squared CM energies $s \lesssim M_{4+n}^2$, the spin 2 character of the graviton predicts $\sigma_g \sim s^2/M_{4+n}^6$ [26] For $s \gg M_{4+n}^2$, several arguments based on unitarity within field theory have been put forward. Ref. [26] suggested

$$\sigma_g \simeq \frac{4\pi s}{M_{4+n}^4} \simeq 10^{-27} \left(\frac{M_{4+n}}{\text{TeV}}\right)^{-4} \left(\frac{E}{10^{20}\,\text{eV}}\right) \text{cm}^2, \tag{2}$$

where in the last expression we specified to a neutrino of energy E hitting a nucleon at rest. A more detailed calculation taking into account scattering on individual partons leads to similar orders of magnitude [22]. Note that a neutrino would typically start to interact in the atmosphere for $\sigma_{\nu N} \gtrsim 10^{-27}\,\text{cm}^2$, i.e. in the case of Eq. (2) for $E \gtrsim 10^{20}\,\text{eV}$, assuming $M_{4+n} \simeq 1\,\text{TeV}$. The neutrino therefore becomes a primary candidate for the observed EHECR events. However, since in a neutral current interaction the neutrino transfers only about 10% of its energy to the shower, the cross section probably has to be at least a few $10^{-26}\,\text{cm}^2$ to be consistent with observed showers which start within the first $50\,\text{g cm}^{-2}$ of the atmosphere [27, 28]. A specific signature of this scenario would be the absence of any events above the energy where σ_g grows beyond $\simeq 10^{-27}\,\text{cm}^2$ in neutrino telescopes based on ice or water as detector medium [29], and a hardening of the spectrum above this energy in atmospheric detectors such as the Pierre Auger Project [30] and the proposed space based AirWatch type detectors [31, 32, 33]. Furthermore, according to Eq. (2), the average atmospheric column depth of the first interaction point of neutrino induced EAS in this scenario is predicted to depend linearly on energy. This should be easy to distinguish from the logarithmic scaling expected for nucleons, nuclei, and $\gamma-$rays. To test such scalings one can, for example, take advantage of the fact that the atmosphere provides a detector medium whose column depth increases from $\sim 1000\,\text{g/cm}^2$ towards the zenith to $\sim 36000\,\text{g/cm}^2$ towards horizontal arrival directions. This probes cross sections in the range $\sim 10^{-29} - 10^{-27}\,\text{cm}^2$. Due to the increased column depth water/ice detectors would probe cross sections in the range $\sim 10^{-31} - 10^{-29}\,\text{cm}^2$ [34].

Within string theory, individual amplitudes are expected to be suppressed exponentially above the string scale M_s which for simplicity we assume here to be comparable to M_{4+n}. This can be interpreted as a result of the finite spatial extension of the string states. In this case, the neutrino nucleon cross section would be dominated by interactions with

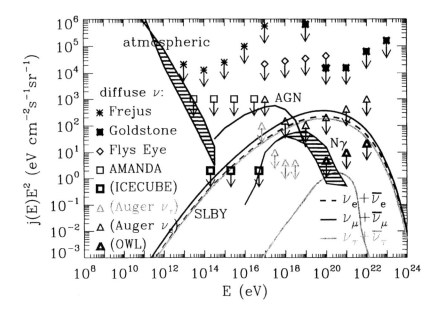

FIGURE 1. Various neutrino flux predictions and experimental upper limits or projected sensitivities. Shown are upper limits from the Frejus underground detector [35], the Fly's Eye experiment [36], the Goldstone radio telescope [37], and the Antarctic Muon and Neutrino Detector Array (AMANDA) neutrino telescope [38], as well as projected neutrino flux sensitivities of ICECUBE, the planned kilometer scale extension of AMANDA [39], the Pierre Auger Project [40] (for electron and tau neutrinos separately) and the proposed space based OWL [31] concept. Neutrino fluxes are shown for the atmospheric neutrino background (hatched region marked "atmospheric"), and neutrino flux predictions for a model of AGN optically thick to nucleons ("AGN"), for EHECR interactions with the CMB [41] ("$N\gamma$", dashed range indicating typical uncertainties for moderate source evolution), and for a "top-down" model (marked "SLBY") where EHECR and neutrinos are produced by decay of superheavy relics (see Ref. [10, 42] for more details) The top-down fluxes are shown for electron-, muon-, and tau-neutrinos separately, assuming no (lower ν_τ-curve) and maximal $\nu_\mu - \nu_\tau$ mixing (upper ν_τ-curve, which would then equal the ν_μ-flux), respectively.

the partons carrying a momentum fraction $x \sim M_s^2/s$, leading to [27]

$$\sigma_{\nu N} \simeq \frac{4\pi}{M_s^2} \ln(s/M_s^2)(s/M_s^2)^{0.363} \simeq 6 \times 10^{-29} \left(\frac{M_s}{\text{TeV}}\right)^{-4.726} \left(\frac{E}{10^{20}\,\text{eV}}\right)^{0.363}$$
$$\times \left[1 + 0.08\ln\left(\frac{E}{10^{20}\,\text{eV}}\right) - 0.16\ln\left(\frac{M_s}{\text{TeV}}\right)\right]^2 \text{cm}^2 \qquad (3)$$

This is probably too small to make neutrinos primary candidates for the highest energy showers observed, given the fact that complementary constraints from accelerator experiments result in $M_s \gtrsim 1\,\text{TeV}$ [43]. On the other hand, in the total cross section amplitude suppression may be compensated by an exponential growth of the level density [24]. It is currently unclear and it may be model dependent which effect dominates. Thus, an experimental detection of the signatures discussed in this section could lead to constraints on some string-inspired models of extra dimensions.

We note in passing that extra dimensions can have other astrophysical ramifications such as energy loss in stellar environments due to emission of real gravitons into the bulk. The strongest resulting lower limits on M_{4+n} come from the consideration of cooling of the cores of hot supernovae and read $M_6 \gtrsim 50\,\mathrm{TeV}$, $M_7 \gtrsim 4\,\mathrm{TeV}$, $M_8 \gtrsim 1\,\mathrm{TeV}$, and $M_{11} \gtrsim 0.05\,\mathrm{TeV}$ for $n = 2, 3, 4, 7$, respectively [44]. In addition, implications of extra dimensions for early Universe physics and inflation are increasingly studied in the literature.

Independent of theoretical arguments, the EHECR data can be used to put constraints on cross sections satisfying $\sigma_{\nu N}(E \gtrsim 10^{19}\,\mathrm{eV}) \lesssim 10^{-27}\,\mathrm{cm}^2$. Particles with such cross sections would give rise to horizontal air showers. The Fly's Eye experiment established an upper limit on horizontal air showers [36]. The non-observation of the neutrino flux expected from pions produced by EHECRs interacting with the CMB the results in the limit [34]

$$\begin{aligned}
\sigma_{\nu N}(10^{17}\,\mathrm{eV}) &\lesssim 1 \times 10^{-29}/\bar{y}^{1/2}\,\mathrm{cm}^2 \\
\sigma_{\nu N}(10^{18}\,\mathrm{eV}) &\lesssim 8 \times 10^{-30}/\bar{y}^{1/2}\,\mathrm{cm}^2 \\
\sigma_{\nu N}(10^{19}\,\mathrm{eV}) &\lesssim 5 \times 10^{-29}/\bar{y}^{1/2}\,\mathrm{cm}^2,
\end{aligned} \quad (4)$$

where $\bar{y}$ is the average energy fraction of the neutrino deposited into the shower ($\bar{y} = 1$ for charged current reactions and $\bar{y} \simeq 0.1$ for neutral current reactions). Neutrino fluxes predicted in various scenarios are shown in Fig. 1. The projected sensitivity of future experiments such as the Pierre Auger Observatories and the AirWatch type satellite projects indicate that the cross section limits Eq. (4) could be improved by up to four orders of magnitude, corresponding to one order of magnitude in M_s or M_{4+n}.

REFERENCES

1. See, e.g., M. A. Lawrence, R. J. O. Reid, and A. A. Watson, *J. Phys. G Nucl. Part. Phys.* **17**, 733 (1991), and references therein; see also http://ast.leeds.ac.uk/haverah/hav-home.html.
2. D. J. Bird et al., *Phys. Rev. Lett.* **71**, 3401 (1993); *Astrophys. J.* **424**, 491 (1994); ibid. **441**, 144 (1995).
3. N. Hayashida et al., *Phys. Rev. Lett.* **73**, 3491 (1994); S. Yoshida et al., *Astropart. Phys.* **3**, 105 (1995); M. Takeda et al., **Phys. Rev. Lett. 81**, 1163 (1998); see also http://icrsun.icrr.u-tokyo.ac.jp/as/project/agasa.html.
4. for a review of the data see S. Yoshida and H. Dai, *J. Phys. G* **24**, 905 (1998).
5. A. M. Hillas, *Ann. Rev. Astron. Astrophys.* **22**, 425 (1984).
6. G. Sigl, D. N. Schramm, and P. Bhattacharjee, *Astropart. Phys.* **2**, 401 (1994).
7. C. A. Norman, D. B. Melrose, and A. Achterberg, *Astrophys. J.* **454**, 60 (1995).
8. K. Greisen, *Phys. Rev. Lett.* **16**, 748 (1966); G. T. Zatsepin and V. A. Kuzmin, *Pis'ma Zh. Eksp. Teor. Fiz.* **4**, 114 (1966) [*JETP. Lett.* **4**, 78 (1966)].
9. J. L. Puget, F. W. Stecker, and J. H. Bredekamp, *Astrophys. J.* **205**, 638 (1976); L. N. Epele and E. Roulet, *Phys. Rev. Lett.* **81**, 3295 (1998); *J. High Energy Phys.* **9810**, 009 (1998); F. W. Stecker, *Phys. Rev. Lett.* **81**, 3296 (1998); F. W. Stecker and M. H. Salamon, *Astrophys. J.* **512**, 521 (1999).
10. P. Bhattacharjee and G. Sigl, *Phys. Rept.* **327**, 109 (2000).
11. N. Hayashida et al., *Phys. Rev. Lett.* **77**, 1000 (1996); M. Takeda et al., *Astrophys. J.* **522**, 225 (1999); N. Hayashida et al., e-print astro-ph/0008102.

12. J. W. Elbert, and P. Sommers, *Astrophys. J.* **441**, 151 (1995).
13. J. W. Cronin, *Rev. Mod. Phys.* **71**, S165 (1999); A. V. Olinto, *Phys. Rept.* **333-334**, 329 (2000); X. Bertou, M. Boratav, and A. Letessier-Selvon, *Int. J. Mod. Phys.* **A15**, 2181 (2000).
14. G. R. Farrar, *Phys. Rev. Lett.* **76**, 4111 (1996); D. J. H. Chung, G. R. Farrar, and E. W. Kolb, *Phys. Rev. D* **57**, 4696 (1998).
15. G. R. Farrar and P. L. Biermann, *Phys. Rev. Lett.* **81**, 3579 (1998).
16. G. Sigl, D. F. Torres, L. A. Anchordoqui, and G. E. Romero, e-print astro-ph/0008363.
17. A. Virmani et al., e-print astro-ph/0010235
18. P. G. Tinyakov and I. I. Tkachev, e-print astro-ph/0102101.
19. R. Gandhi, C. Quigg, M. H. Reno, and I. Sarcevic, *Astropart. Phys.* **5**, 81 (1996); *Phys. Rev. D* **58**, 093009 (1998).
20. J. Bordes et al., Astropart. Phys. 8 (1998) 135; in *Beyond the Standard Model. From Theory to Experiment* (Valencia, Spain, 13-17 October 1997), eds. I. Antoniadis, L. E. Ibanez, and J. W. F. Valle (World Scientific, Singapore, 1998), p. 328 (e-print hep-ph/9711438).
21. G. Sigl, *Nucl. Phys.* **B87** (Proc. Suppl.) 439 (2000), Proceedings of TAUP 99, eds.: M. Froissart, J. Dumarchez, and D. Vignaud, College de France, Paris, 6-10 September 1999.
22. P. Jain, D. W. McKay, S. Panda, and J. P. Ralston, *Phys. Lett.* **B484**, 267 (2000); J. P. Ralston, P. Jain, D. W. McKay, S. Panda, e-print hep-ph/0008153.
23. G. Burdman, F. Halzen, and R. Gandhi, *Phys. Lett. B* **417**, 107 (1998).
24. G. Domokos and S. Kovesi-Domokos, *Phys. Rev. Lett.* **82**, 1366 (1999).
25. N. Arkani-Hamed, S. Dimopoulos, and G. Dvali, *Phys. Lett.* **B 429**, 263 (1998); I. Antoniadis, N. Arkani-Hamed, S. Dimopoulos, and G. Dvali, *Phys. Lett.* **B 436**, 257 (1998); N. Arkani-Hamed, S. Dimopoulos, and G. Dvali, *Phys. Rev. D* **59** 086004 (1999).
26. S. Nussinov and R. Shrock, *Phys. Rev. D* **59**, 105002 (1999).
27. M. Kachelrieß and M. Plümacher, *Phys. Rev. D* **62**, 103006 (2000).
28. L. Anchordoqui et al., e-print hep-ph/0011097.
29. see, e.g., C. Spiering, e-print astro-ph/0012532.
30. J. W. Cronin, *Nucl. Phys. B* (Proc. Suppl.) **28B**, 213 (1992); The Pierre Auger Observatory Design Report (2nd edition), March 1997; see also http://http://www.auger.org/ and http://www-lpnhep.in2p3.fr/auger/welcome.html.
31. D. B. Cline, F. W. Stecker, OWL/AirWatch science white paper, e-print astro-ph/0003459; see also http://lheawww.gsfc.nasa.gov/docs/gamcosray/hecr/OWL/.
32. See http://www.ifcai.pa.cnr.it/Ifcai/euso.html.
33. J. Linsley, in Proc. *25th International Cosmic Ray Conference*, eds.: M. S. Potgieter et al. (Durban, 1997)., Vol. 5, 381; ibid., 385; P. Attinà et al., ibid., 389; J. Forbes et al., ibid., 273; see also http://www.ifcai.pa.cnr.it/Ifcai/airwatch.html.
34. C. Tyler, A. Olinto, and G. Sigl, *Phys. Rev. D.* **63**, 055001 (2001).
35. W. Rhode et al., *Astropart. Phys.* **4**, 217 (1996).
36. R. M. Baltrusaitis et al., *Astrophys. J.* **281**, L9 (1984); *Phys. Rev. D* **31**, 2192 (1985).
37. P. W. Gorham, K. M. Liewer, C. J. Naudet, e-print astro-ph/9906504.
38. For general information see http://amanda.berkeley.edu/; see also F. Halzen, *New Astron. Rev.* **42**, 289 (1999).
39. For general information see http://www.ps.uci.edu/ icecube/workshop.html; see also F. Halzen, *Am. Astron. Soc. Meeting* **192**, # 62 28 (1998); AMANDA collaboration, e-print astro-ph/9906205, talk presented at the *8th Int. Workshop on Neutrino Telescopes*, Venice, Feb. 1999.
40. J. J. Blanco-Pillado, R. A. Vázquez, E. Zas, *Phys. Rev. Lett.* **78**, 3614 (1997); K. S. Capelle, J. W. Cronin, G. Parente, E. Zas, *Astropart. Phys.* **8**, 321 (1998); A. Letessier-Selvon, e-print astro-ph/0009444.
41. R. J. Protheroe, P. A. Johnson, *Astropart. Phys.* **4**, 253 (1996), and erratum *ibid.* **5**, 215 (1996).
42. G. Sigl, S. Lee, P. Bhattacharjee, S. Yoshida, *Phys. Rev. D* **59**, 043504 (1999).
43. see, e.g., S. Cullen, M. Perelstein, and M. E. Peskin, *Phys. Rev.D* **62**, 055012 (2000), and references therein.
44. S. Cullen and M. Perelstein, *Phys. Rev. Lett.* **83**, 268 (1999); V. Barger, T. Han, C. Kao, and R.-J. Zhang, *Phys. Lett. B* **461**, 34 (1999).

Ultrahigh energy neutrinos as probe for weak-scale string theories?

M. Kachelrieß

TH Division, CERN, CH-1211 Geneva 23

Abstract. We discuss the potential of ultrahigh energy (UHE) neutrinos as probe for new particle physics. After briefly reviewing the proposal that the decay products from UHE neutrinos annihilations on relic neutrinos are the observed UHE primaries, we concentrate on the suggestion that UHE neutrinos can acquire cross-sections approaching hadronic size if the string scale is in the TeV range. We review the calculation of the neutrino-nucleon cross-section $\sigma_{N\nu}^{\rm KK}$ due to the exchange of Kaluza-Klein (KK) excitations of the graviton in a field theoretical framework and discuss the issue of unitarity. Assuming Regge behavior of the neutrino-nucleon interaction mediated by KK gravitons, we find that the effects of large extra dimensions are barely observable.

I INTRODUCTION

Neutrinos are the only known stable particles that can traverse extragalactic space without attenuation even at energies $E \sim 10^{20}$ eV, thus avoiding the Greisen–Zatsepin–Kuzmin (GZK) cutoff [1]. Therefore, it has been speculated that the ultrahigh energy (UHE) primaries initiating the observed air showers are not protons, nuclei or photons but neutrinos [2–4]. However, neutrinos are in the Standard Model (SM) deeply penetrating particles producing only horizontal not vertical Extensive Air Showers (EAS). Therefore, either one has to postulate new interaction that enhance the UHE neutrino-nucleon cross-section or the neutrino has to be converted locally into strongly interacting particles.

In the later scheme [5], UHE neutrinos from distant sources annihilate with relic neutrinos on the Z resonance. The fragmentation products from nearby Z decays are supposed to be the primaries responsible for the EAS above the GZK-cutoff. For energies of the primary neutrino of $E_\nu \sim 10^{23}$ eV, the mass of the relic neutrino should be $m_\nu = m_Z^2/(2E_\nu) \sim 0.1$ eV which is compatible with atmospheric neutrino data. There are, however, severe observational constraints on this model:

1. Since the Pauli principle does not allow arbitrary densely packed neutrinos, an upper limit for their number density n in the galactic halo is $n \leq 4\pi p_{\rm max}^3/3$, where $p_{\rm max} \sim m_\nu v_{\rm rot}$ and $v_{\rm rot} \sim 220$ km/s is the Galactic rotation speed. Therefore the overdensity of relic neutrinos is small and one expects in this model a rather

pronounced GZK-cutoff.

2. Since the interaction probability for a UHE neutrino in the neutrino halo is small, a large neutrino flux is needed to produce the observed UHECR. The limit on horizontal EAS set by the Fly's Eye experiment [6] limits therefore severely this model: Ref. [7] found that the neutrino spectrum has to be extremely flat, $dN/dE \propto E^{-\gamma}$ with spectral index $\gamma < 1.2$. Even if one assumes a large neutrino enhancement factor due to a lepton asymmetric Universe, the spectrum has to be much flater, $\gamma < 1.8$, than expected from astrophysical sources.

3. The observed UHECR flux implies an upper bound on the UHE neutrino flux produced in astrophysical sources which are not hidden. If UHE neutrinos annihilating on relic neutrinos contribute significantly to the observed UHECR at $\sim 10^{20}$ eV, a new class of UHE neutrino source has to be invoked which is optically thick for nucleons. The energy generation of these sources was estimated to be comparable to the total photon luminosity of the Universe [8].

II UHE NEUTRINOS AND WEAK-SCALE STRING THEORIES

Most models introducing new physics at a scale M to produce large cross-sections for UHE neutrinos fail because experiments generally constrain M to be larger than the weak scale, $M \gtrsim m_Z$, and unitarity limits cross-sections to be $\mathcal{O}(\sigma_{\rm tot}) \lesssim 1/M^2 \lesssim 1/m_Z^2$. String theories with large extra dimensions [9] are different in this respect: if the SM particles are confined to the usual $3+1$-dimensional space and only gravity propagates in the higher-dimensional space, the compactification radius R of the large extra dimensions can be large, corresponding to a small scale $1/R$ of new physics. The weakness of gravitational interactions is a consequence of the large compactification radius, since Newton's constant is then given by $G_N^{-1} = 8\pi R^\delta M_D^{\delta+2}$, where δ is the number of extra dimensions and $M_D \sim$ TeV is the fundamental mass scale. Such a scenario is naturally realized in theories of open strings, where SM particles correspond to open strings beginning and ending on D-branes, whereas gravitons correspond to closed strings which can propagate in the higher-dimensional space. From a four-dimensional point of view the higher dimensional graviton in these theories appears as an infinite tower of Kaluza-Klein (KK) excitations with mass squared $m_{\vec{n}}^2 = \vec{n}^2/R^2$. Since the weakness of the gravitational interaction is partially compensated by the large number of KK states and cross-sections of reactions mediated by spin 2 particles are increasing rapidly with energy, it has been argued in Refs. [3,4] that neutrinos could initiate the observed vertical showers at the highest energies.

In the calculations of Refs. [4,10] it was assumed that the massless four-dimensional graviton and its massive KK excitations couple with the usual gravitational strength $\overline{M}_{\rm Pl}^{-1} = \sqrt{8\pi}/M_{\rm Pl}$. Then the sum over all KK contributions to a given (t-channel) scattering amplitude,

$$M_{fi} \propto P(t) = \sum_{\vec{n}=1}^{\infty} \frac{g_{\vec{n}}}{t - m_{\vec{n}}^2}, \quad \vec{n} = (n_1, \ldots, n_\delta) \tag{1}$$

only converges in the case of one extra dimension, and for two or more extra dimensions a cutoff has to be introduced by hand. However, it has been pointed out [11] that due to brane fluctuations the effective coupling $g_{\vec{n}}$ of the level $\vec{n}$ KK mode to four-dimensional fields is suppressed exponentially,

$$g_{\vec{n}} = \frac{1}{\overline{M}_{\text{Pl}}} \exp\left(-\frac{c\, m_{\vec{n}}^2}{M_{\text{st}}^2}\right), \tag{2}$$

where c is a constant of order 1 or larger, which parameterizes the effects of a finite brane tension [11], and $M_{\text{st}} \sim M_D$ is the string scale. A similar suppression of the coupling to higher KK modes was found also in Ref. [12]. This exponential suppression thereby provides a dynamical cutoff in the sum over all KK modes. Therefore, the cross-sections obtained in Ref. [13] are smaller than those found previously, and even for $\sigma_{N\nu}^{\text{KK}} = 10$ mbarn a value of M_{st} not much above 1 TeV is required. This value of M_{st} is still marginally allowed by collider experiments, while SN 1987A gives a better limit only for the cases $\delta = 2$ and 3 [14].

The second important quantity characterizing the development of an air shower besides σ_{tot} is the energy transfer $y = (E_\nu - E'_\nu)/E_\nu$. In contrast to charged-current scattering where the electromagnetic shower initiated by the charged lepton is practically indistinguishable from a hadronic shower, only the hit nucleon can initiate an air shower in KK scattering. Therefore, even a neutrino with large σ_{tot} will behave like a penetrating particle if it does not transfer a large fraction of its energy per interaction to the shower. At energies of interest, $E_\nu \approx 10^{20}$ eV, the transferred energy fraction found in Ref. [13] is only around $y \approx 0.1$, i.e., much smaller than $y \approx 0.6$ typical of nucleon-nucleon collisions. From a more optimistic point of view, the small energy transfer of the KK-mediated interaction could be a possible signature to distinguish them from the SM interactions.

III UNITARITY LIMITS

We want to derive now a bound on $\sigma_{N\nu}^{\text{KK}}$ valid in the regime relevant for UHECR, $s \gg M_{\text{st}}^2$, where the effective field theory used above is not valid anymore. Our working hypothesis is that the Regge picture is a reasonable approximation for string theory. The Regge picture takes into account all KK modes (also from lower trajectories) and misses only genuine string modes like winding modes.

A general Regge amplitude A_R can be represented by

$$A_R(s, t) = \beta(t) \left(\frac{s}{s_0}\right)^{\alpha(t)}, \tag{3}$$

where the exponent $\alpha(t)$ is given by the Chew-Frautschi plot of the spin against the mass of the particles lying on the leading Regge trajectory contributing to the reaction. In our case, the intercept $\alpha(0)$ of this trajectory is equal to the spin j of the massless graviton, $\alpha(0) = 2$.

We first note that a Regge amplitude with intercept $\alpha(0) = 2$ gives via the optical theorem a total cross-section growing linearly with s,

$$\sigma_{\text{tot}}(s) = \frac{1}{s} \text{Im}\{A(s,0)\} \propto s^{\alpha(0)-1}. \tag{4}$$

Thus the assumed Regge-behavior alone reduces the growth of the total cross-section by one power of s compared to the naive expectation $\sigma \propto s^j = s^2$. On the other hand, the elastic cross-section

$$\sigma_{\text{el}}(s) = \frac{1}{16\pi s^2} \int_{-s}^{0} dt \, |A_R(s,t)|^2 \propto \frac{s^2}{\ln(s/s_0)} \tag{5}$$

increases even faster than the total cross-section. Therefore, elastic unitarity, $\sigma_{\text{tot}} \geq \sigma_{\text{el}}$, is above a certain energy violated.

Next, we derive the maximal total cross-section allowed for an arbitrary Regge amplitude by elastic unitarity. Following Leader [15], we rewrite $A_R(s,t)$ as

$$A_R(s,t) = \beta \left(\frac{s}{s_0}\right)^{\alpha(t)} = A_R(s,0) \left(\frac{s}{s_0}\right)^{\alpha(t)-\alpha(0)} \tag{6}$$

and expand the amplitude around $t = 0$,

$$\left(\frac{s}{s_0}\right)^{\alpha(t)-\alpha(0)} = \exp\{\alpha' t \ln(s/s_0) + O(\alpha'' t^2)\}. \tag{7}$$

Here, α' denotes the derivative of $\alpha(t)$ evaluated at $t = 0$ and we have neglected for clarity possible non-linear terms in t and the t dependence of β. Then we evaluate σ_{el},

$$\sigma_{\text{el}} = \frac{1}{16\pi s^2} \int_{-s}^{0} dt \, |A_R(s,t)|^2 = \frac{|A_R(s,0)|^2}{16\pi s^2} \frac{1}{2\alpha' \ln(s/s_0)}. \tag{8}$$

Requiring now elastic unitarity

$$\sigma_{\text{el}} \leq \sigma_{\text{tot}} = \frac{1}{s} \text{Im}\{A_R(s,0)\} < \frac{1}{s} |A_R(s,0)|, \tag{9}$$

it follows

$$\frac{1}{32\pi\alpha' \ln(s/s_0)} \frac{|A_R(s,0)|^2}{s^2} \leq \sigma_{\text{tot}} \tag{10}$$

or

$$\frac{\sigma_{\text{tot}}^2}{32\pi\alpha' \ln(s/s_0)} \leq \sigma_{\text{tot}} \tag{11}$$

and finally

$$\sigma_{\text{tot}}(s) \leq 32\pi\alpha' \ln(s/s_0). \quad (12)$$

Thus the assumption of a Regge amplitude results in a stronger bound for the total cross-section than the Froissart bound. A more general derivation of a bound that reduces for the special case of a Regge amplitude also to Eq. (12) can be found in Ref. [16].

Some remarks are now in order: First, we have always used formulae valid for $d = 4$ dimensions. This is appropriate because the main contribution to the cross-sections comes from the small t region and therefore does not probe the extra dimensions. Second, we note that this bound applies on the parton not the hadron level. Third, the bound (12) contains two parameters, the slope of the Regge trajectory α' and the unknown scale s_0, and is therefore still not useful for a *numerical* evaluation.

To proceed, we use that

$$\sigma_{\text{tot}}^{N\nu}(s) = [N(s) + \delta]\,\sigma_{\text{tot}}(s) \leq N(s)\sigma_{\text{tot}}(s), \quad (13)$$

where $N(s) \propto s^{0.4}$ [17] takes into account the increasing number of target partons in the nucleon. The term $\delta < 0$ corrects that each parton carries only a fraction $x < 1$ of the nucleon momentum, i.e. that $\ln(xs/s_0) < \ln(s/s_0)$. A numerical value for the bound (13) can now be determined by joining the field-theoretic result and the Regge result on the hadron level at that scale $s' \sim M_{\text{st}}^2$, where the field-theoretic result starts to violate s-wave unitarity on the parton level. We find that the KK contribution to the total cross-section at UHE has maximal the same order of magnitude as the SM cross-section.

Finally, we want to comment briefly on the suggestion of Ref. [3] that the exponential increase of (lepto-quark like) KK resonances in the s channel could enhance the neutrino-nucleon cross-section. Since the Horn-Schmid duality [18] connects s and t channel Regge/String amplitudes, our discussion above can be applied immediately to this case. The $n = 0$ lepto-quarks can have either spin $j = 0$ or 1. Even in the later case, the intercept will be smaller than 1 and the partonic cross-section will be asymptotically decreasing with s.

IV CONCLUSIONS

Neutrinos behave also in theories with large extra dimensions as deeply penetrating particles because of the smallness of the resulting cross-section and energy transfer per interaction. They are therefore not responsible for the observed UHECR events. If unitarity limits $\sigma_{N\nu}^{\text{KK}}$ as argued above, it will be a very difficult task to observe stringy effects in horizontal air showers even if M_{st} is in the TeV range.

ACKNOWLEDGMENTS

I am grateful to Michael Plümacher for a very pleasant collaboration and to Doug Ross for useful comments.

REFERENCES

1. For a general discussion of ultrahigh energy cosmic rays see the contribution of A. Watson, for a discussion how UHE neutrinos could reveal new interactions see the contribution of G. Sigl in these proceedings.
2. V.S. Berezinsky and G.T. Zatsepin, Phys. Lett. **B28**, 423 (1969); G. Domokos and S. Nussinov, Phys. Lett. **B187**, 372 (1987); J. Bordes, H. Chan, J. Faridani, J. Pfaudler and S.T. Tsou, Astropart. Phys. **8**, 135 (1998).
3. G. Domokos and S. Kovesi-Domokos, Phys. Rev. Lett. **82**, 1366 (1998).
4. P. Jain, D.W. McKay, S. Panda and J.P. Ralston, Phys. Lett. **B484**, 267 (2000).
5. D. Fargion, B. Mele and A. Salis, Astrophys. J. **517**, 725 (1999); T.J. Weiler, Astropart. Phys. **11**, 303 (1999).
6. R.M. Baltrusaitis *et al.*, Phys. Rev. **D31**, 2192 (1985).
7. J.J. Blanco-Pillado, R.A. Vazquez and E. Zas, Phys. Rev. **D61**, 123003 (2000).
8. E. Waxman, `astro-ph/9804028`.
9. N. Arkani-Hamed, S. Dimopoulos and G. Dvali, Phys. Lett. **B429**, 263 (1998); I. Antoniadis, N. Arkani-Hamed, S. Dimopoulos and G. Dvali, Phys. Lett. **B436**, 257 (1998), Phys. Rev. **D59**, 086004 (1999).
10. S. Nussinov and R. Shrock, Phys. Rev. **D59**, 105002 (1999).
11. M. Bando, T. Kugo, T. Noguchi and K. Yoshioka, Phys. Rev. Lett. **83**, 3601 (1999); J. Hisano and N. Okada, Phys. Rev. **D61**, 106003 (2000).
12. L. Dixon, D. Friedan, E. Martinec and S. Shenker, Nucl. Phys. **B282**, 13 (1987); S. Hamidi and C. Vafa, Nucl. Phys. **B279**, 465 (1987); I. Antoniadis and K. Benakli, Phys. Lett. **B326**, 69 (1994).
13. M. Kachelrieß and M. Plümacher, Phys. Rev. **D62**, 103006 (2000).
14. For a review of experimental bounds, see e.g. M.E. Peskin, `hep-ph/0002041`; I. Antoniadis and K. Benakli, `hep-ph/0004240`.
15. E. Leader, Phys. Lett. **5**, 75 (1963).
16. R.J. Eden, *High energy collisions of elementary particles*, Cambridge University Press 1967.
17. R. Gandhi, C. Quigg, M.H. Reno, I. Sarcevic, Astropart. Phys. **5**, 81 (1996).
18. R. Dolen, D. Horn and C. Schmid, Phys. Rev. Lett. **19**, 402 (1967).

Ultra-high-energy cosmic rays from relic topological defects

Ken D. Olum[*] and J. J. Blanco-Pillado[*]

[*]*Institute of Cosmology, Department of Physics and Astronomy, Tufts University, Medford, MA 02155*

Abstract. It is difficult for conventional sources to accelerate cosmic ray particles to the highest energies that have been observed. Topological defects such as monopoles and strings overcome this difficulty, because their natural energy scale is at or above the observed energies. Monopoles connected by strings are a particularly attractive source, because they would cluster in the galactic halo and thus explain the absence of the GZK cutoff. Heavy monopoles connected by light strings could last for the age of the universe as required. Further observations might support this model by detection of the anisotropy due to the halo, or might refute such models if strong clustering of arrival directions or correlations with known astrophysical objects are confirmed. All top-down models must contend with recent claims that the percentage of photons among the cosmic rays is smaller than such models predict.

INTRODUCTION

The observation of ultra-high-energy cosmic rays (UHECR) with energies above 10^{20} eV [1, 2] is hard to explain. First of all, is not clear that there is any site that at which particles might be accelerated to such large energies. Even if such sites exist, for example in active galactic nuclei, it is hard to explain the absence of the Greisen-Zatsepin-Kuzmin [3, 4] cutoff. Cosmic rays with energies above about $E_{GZK} = 4 \times 10^{19}$ eV will interact with cosmic microwave background photons and lose energy until they are below E_{GZK}. If the highest energy cosmic rays are produced by sources which are homogeneous in the universe (even if some of the sources are nearby), then there must be a large deficit in particles with $E > E_{GZK}$ as compared to those with $E < E_{GZK}$, because the former are converted into the latter. Such a cutoff is not observed.

In this talk, we will try to construct a relic topological defect model (see [5] for a review of such models and others) which addresses these difficulties. A relic model explains the UHECR as the decay products of some very high energy particle. As long as the progenitor has mass $M_X c^2 \gg 10^{20}$ eV, the high energies observed are trivially explained. However, to explain the UHECR, the relic must also be sufficiently long-lived to still be decaying today. This means that the progenitor lifetime must be extremely large as compared to the naive dimensional analysis value $\tau \sim \hbar/(M_X c^2)$.

To solve the GZK problem, the relics must be strongly clustered near us, and the obvious place for this is in the galactic halo. This is easily arranged, as long as the relic velocities are not too high to prevent them from being captured in the gravitational potential of the galaxy.

TOPOLOGICAL DEFECTS

Topological defects result from misalignment of fields after symmetry breaking transitions in the early universe. (For a review, see [6].) Because they are topologically stabilized, they can persist until the present time. If the scale of symmetry breaking is high, for example the grand unification scale $E \sim 10^{25}$ eV, then there is no difficulty reaching the required energies for the UHECR. However, one does need a mechanism by which the energy can be released from the defect at the necessary rate. One also needs an appropriate amount of total energy stored in defects. It must be large enough to explain the observed UHECR flux, while not being so large as to exceed bounds on the total density of the universe.

The dimensionality of the topological defects depends on the symmetry that was broken to create them. One can have monopoles (0-dimensional defects), strings (1-dimensional defects), or domain walls (2-dimensional defects). If there are multiple symmetry breaking transitions, one can have hybrid defects. For example, a high-energy transition can produce monopoles, and then a subsequent transition at a lower energy can confine the flux from the monopoles into strings. In this case, the monopoles' flux must not be the regular magnetic field, because that is not confined today.

We can now consider the various defect types as UHECR sources. Domain walls are ruled out because they contribute too much total mass to the universe. Cosmic strings would evolve into a scaling regime, so their total mass contribution is not too large. However, strings move relativistically and would not cluster in the galactic halo. Monopoles produced with a thermal abundance would also overclose the universe, but is possible that they were produced with a smaller abundance during reheating or by gravitational particle creation [7]. In this case, they are not ruled out, and they would cluster in the halo. However, it is hard to see how to get the energy out of the monopoles. The only real possibility is monopole-antimonopole annihilation [8, 9]. Unfortunately [10], monopole-antimonopole pairs would not be formed with sufficient density to explain the observed flux.

However, if after the monopoles have formed, a subsequent symmetry breaking transition connects them by strings, then every monopole will be paired with an antimonopole and there will be no problem having sufficient annihilations [10].

SCENARIO

We imagine a first symmetry breaking transition which gives monopoles of mass m_M, and a second symmetry breaking transition which connects them by strings of energy scale[1] T_s. A system of monopole and antimonopole attached by a string will oscillate with a timescale given by the acceleration of the monopole, μ/m_M, where μ is the string tension, $\mu \sim T_s^2$.

We take the monopole not to have any unconfined flux, so the loss of energy of the

---

[1] We will henceforth work in units where $k_B = \hbar = c = 1$.

system is purely in gravitational radiation. To produce a long lifetime, we want small acceleration, and thus low string tension and high monopole mass, so we take $T_s \sim 100$ GeV, and $m_M \sim 10^{14}$ GeV.

To explain the observed UHECR flux we require a sufficient density of monopoles clustered in the halo. The minimum necessary density is achieved when the decay lifetime is approximately the age of the universe. In this case, for $m_M = 10^{14}$ GeV, we found the needed present density in the universe as a whole [10],

$$N_{M\bar{M}} > 10^{-30}/\text{cm}^3. \quad (1)$$

At the time of string formation, this was

$$n_M \sim 10^{-32} T_s^3 \sim 10^{-18}/\text{cm}^3, \quad (2)$$

which gives a typical monopole separation

$$L_i \sim 10^{-6} \text{cm}. \quad (3)$$

This is much smaller than the horizon distance, $d_H \sim 3$ cm at $T \sim 100$ GeV.

When the strings are formed they may have excitations on scales smaller than the distance between monopoles, but these will propagate relativistically and thus be quickly smoothed out by gravitational radiation, leaving a straight string. The energy stored in the string is then μL_i. This is smaller than the monopole mass by the ratio

$$\frac{\mu L_i}{m_M} \sim 10^{-2} \quad (4)$$

so the monopoles will move non-relativistically.

We thus have a system of monopole and antimonopole connected by a straight string, which produces a constant force of acceleration $a = \mu/m_M$. The monopoles will move in elliptical orbits, but since their thermal velocities at the time of string formation will be small compared to the velocities they acquire during acceleration, the motion will be nearly linear, although with enough angular momentum to prevent the monopoles from colliding as they pass by.

To estimate the gravitational radiation rate, we will take the linear motion, in which a half oscillation of one monopole is parameterized by

$$x(t) = (2aL)^{1/2} t - \frac{1}{2} a t^2 \quad (5)$$

for $0 < t < (8L/a)^{1/2}$. Using the quadrupole approximation,[2] the rate of energy loss of the system is

$$\frac{dE}{dt} = \frac{288}{45} G \mu^2 \left(\frac{\mu L}{m_M} \right). \quad (6)$$

---

[2] The fully relativistic situation was considered in [11].

Now μL is just the energy in the string, so we can write $dE/dt = \mu dL/dt$, and integrate to get

$$L = L_i e^{-t/\tau_g} \tag{7}$$

with

$$\tau_g = \frac{45}{288} \frac{m_M}{G\mu^2} = \frac{45}{288} \frac{m_{pl}^2 m_M}{T_s^4}. \tag{8}$$

The monopoles thus move on smaller and smaller orbits, until they annihilate, approximately when L reaches the monopole core radius, $r_M \sim m_M^{-1}$. The system thus lives for a time about $\tau_g \ln(L_i/r_M)$. With $T_s \sim 100$ GeV and $m_M \sim 10^{14}$ GeV, Eq. (8) gives $\tau_g \sim 10^{17}$ sec, comparable with the age of the universe.

OBSERVATIONAL CONSEQUENCES

How can a model such as that presented here be verified or disproved? Unfortunately, all models which involve topological relics or relic particles decaying in the halo gave rise to the same observations, dependent essentially on one unknown parameter, the mass of the decaying primary. Thus, the specific model of monopoles bound by strings cannot be verified by cosmic ray observations. However, the low string energy scale which is necessary for long lifetimes means that the string fields might be detected in future accelerators.

Halo relic models as a class, however, do have observable consequences.

Spectrum. All relic models produce the observed cosmic ray primaries by the decay and fragmentation of super-heavy particles (produced, in this case, by monopole-antimonopole annihilation). The spectrum, thus, has little dependence on the type of defect that is decaying, but rather results primarily from the fragmentation process. Fragmentation of such high-energy particles is not completely understood, but we know that the spectrum we observe depends on the mass of the decaying particle, and in all cases it is much harder than the steeply falling spectrum of cosmic rays at lower energies. Current data does not constrain the ultra-high-energy spectrum tightly, but future experiments [12, 13] should be able to validate decaying particle models and determine the particle mass.

Particle type. Fragmentation also produces a large fraction of photons, and thus a generic prediction of relic models is that most of the observed cosmic rays will be photons. Identifying individual particles is difficult, but recent studies [14] of large zenith angle showers have found that no more than about 40% of the particles can be photons at the highest energies. If the studies are correct, then all relic models appear to be ruled out.

Anisotropy. Because the earth is not at the center of the galactic halo, cosmic rays coming from halo sources would be seen to somewhat higher degree from the direction of the galactic center. (See [15] and references therein.) The low number of observed events, combined with the lack of an observatory in the Southern Hemisphere where the galactic center is located, prevents a clear confirmation or disconfirmation of this effect. However, a statistically insignificant anisotropy is observed of generally the right form. The strongest confirmation of a halo model would be to see enhancements of the cosmic ray flux coming from the halo of M31. Unfortunately, this also must wait for future experiments.

No clustering. Any model of relic particles or monopoles will have all observed cosmic rays coming from different sources. Thus we would not expect arrival directions to be clustered into multiplets, except for an effect due to inhomogeneities in the dark matter distribution [16]. Current claims of doublets and triplets are consistent with dark matter inhomogeneity, but if further data yields greater multiplets, model such as this will be ruled out.

No correlations with known astrophysical sources. If the UHECR come from otherwise-invisible particles in the halo, there should be no correlation in arrival direction with any known object. Such correlations have been claimed [17, 18], but there is some question [19] about the correctness of these claims.

DISCUSSION

We have argued that relic topological defects have several advantages as sources of the observed ultra-high-energy cosmic rays. They naturally explain very high energies and can cluster in galactic halos and thus explain the absence of the GZK cutoff. However, most topological defect models do not have the required properties. Monopoles bound by strings seem to be a good candidate. With heavy monopoles and light strings, the required lifetime can be achieved, and because there is perfect efficiency in monopole-antimonopole binding, the required monopole density is quite small. (Necklaces — monopoles connected to two strings each [20] — also seem like a good candidate.)

Of course, hybrid topological defects are "exotic", in the sense that they involve two extra fields introduced just for this purpose. However, since conventional mechanisms do not solve the puzzle of UHECR origin, it seems reasonable to consider exotic models. Unfortunately, it appears that even exotic models don't seem to be in agreement with observation, especially the low bound on the photon fraction from recent studies [14].

ACKNOWLEDGMENTS

We would like to thank the organizers of the 20th Texas Symposium for an excellent conference, and Xavier Siemens, Gunter Sigl, Alex Vilenkin, and Alan Watson for helpful conversations. K. D. O. is grateful for support from the Symposium and from the National Science Foundation. J. J. B. P. was supported in part by the Fundación Pedro Barrie de la Maza.

REFERENCES

1. Hayashida, N., et al., *Phys. Rev. Lett.*, **73**, 3491–3494 (1994).
2. Bird, D. J., et al., *Astrophys. J.*, **424**, 491 (1994).
3. Greisen, K., *Phys. Rev. Lett.*, **16**, 748 (1966).
4. Zatsepin, G. T., and Kuzmin, V. A., *JETP Lett.*, **4**, 78–80 (1966).
5. Bhattacharjee, P., and Sigl, G., *Phys. Rept.*, **327**, 109 (2000).
6. Vilenkin, A., and Shellard, E. P. S., *Cosmic Strings and other Topological Defects*, Cambridge University Press, Cambridge, 1994.
7. Kuzmin, V. A., and Tkachev, I. I., *Phys. Rept.*, **320**, 199–221 (1999).
8. Hill, C. T., *Nucl. Phys.*, **B224**, 469 (1983).
9. Bhattacharjee, P., and Sigl, G., *Phys. Rev. D*, **51**, 4079 (1995).
10. Blanco-Pillado, J. J., and Olum, K. D., *Phys. Rev.*, **D60**, 083001 (1999).
11. Martin, X., and Vilenkin, A., *Phys. Rev. D*, **55**, 6054–6060 (1997).
12. Zavrtanik, D., *Nucl. Phys. Proc. Suppl.*, **85**, 324–331 (2000).
13. Streitmatter, R. E., "Orbiting Wide-angle Light-collectors (OWL): Observing cosmic rays from space", in *Workshop on Observing Giant Cosmic Ray Air Showers From $> 10^{20}$ eV Particles From Space*, AIP Conference Proceedings, Woodbury, New York, 1998, p. 95.
14. Ave, M., Hinton, J. A., Vazquez, R. A., Watson, A. A., and Zas, E., *Phys. Rev. Lett.*, **85**, 2244–2247 (2000).
15. Tanco, G. A. M., and Watson, A. A., *Astropart. Phys.*, **12**, 25–34 (1999).
16. Blasi, P., and Sheth, R., *Phys. Lett.*, **B486**, 233 (2000).
17. Farrar, G. R., and Biermann, P. L., *Phys. Rev. Lett.*, **81**, 3579–3582 (1998).
18. Virmani, A., et al., Correlation of ultra high energy cosmic rays with compact radio loud quasars, astro-ph/0010235.
19. Sigl, G., Torres, D. F., Anchordoqui, L. A., and Romero, G. E., Testing the correlation of ultra-high energy cosmic rays with high redshift sources, astro-ph/0008363.
20. Berezinsky, V., and Vilenkin, A., *Phys. Rev. Lett.*, **79**, 5202–5205 (1997).

Cen A as the Source of Ultrahigh Energy Cosmic Rays

Tsvi Piran[*][1] and Glennys R. Farrar[†]

[*] *Racah Institute of Physics, Hebrew University, Jerusalem, Israel*
[†] *Department of Physics, New York University, NY, NY 10003, USA*

Abstract. We argue from observed features of ultrahigh energy cosmic rays that whatever mechanism produces the events above the GZK energy, $\approx 10^{19.7}$ eV, must also account for events down to $\approx 10^{18.7}$ eV, including their isotropy and spectral smoothness. This rules out distributed sources such as topological defects and Z-bursts, and GRBs. We are lead to identify the powerful radio galaxy Cen A, at 3.4 Mpc, as a probable source of most ultrahigh energy CRs observed at Earth today, and to estimate the extragalactic magnetic field to be $\sim 0.3\mu$G.

Above an energy of about $10^{19.7}$ eV, cosmic ray protons suffer energy losses due to photopion production from the cosmic microwave background [2]. Less than 20% of protons survive with an energy above 3×10^{20} (1×10^{20}) eV for a distance of 18 (60) Mpc; ultra-high energy (UHE) nuclei and photons lose energy even more readily. Yet more than 20 cosmic rays (CRs) have been observed with nominal energies at or above $10^{20} \pm 30\%$ eV [3]. There are no apparent suitable astrophysical sources within the GZK distance to which these events point, nor is there any significant break in the spectrum in the GZK region, as would be expected if higher energy cosmic rays are attenuated.

The CR spectrum can be described by a series of power laws. At the "knee", $E \sim 10^{15.5}$ eV, the spectral index steepens from -2.7 to -3.0; at $10^{\sim 17}$ eV it steepens further to -3.3. At $E \sim 10^{18.5}$ eV the spectrum *flattens* to an index of -2.7 and is consistent, within the statistical uncertainty of the data (which is large above 10^{20} eV), with a simple extrapolation at that slope to the highest energies, possibly with a hint of a slight accumulation around $10^{19.5}$ eV. See [3] for a review.

The most straightforward interpretation of present data is that above $E \sim 10^{18.5}$ eV a new population emerges which dominates the more steeply falling galactic population, and this new population has an approximately $E^{-2.7}$ spectrum up to the highest observed energies. This interpretation is supported by AGASA's analysis of arrival directions [4]: around 10^{18} eV the angular distribution correlates with the

---

[1] talk given by T. Piran, based on [1].

galactic center and is consistent with a galactic origin, while at higher energy this anisotropy disappears. The smoothness of the observed spectrum above $10^{18.5}$ eV suggests that a single mechanism is responsible for these events. Otherwise, there is an apparently miraculous matching of spectra from different mechanisms, such that the total spectrum is smooth.

We therefore explore here the following ansatz: **(i)** *A single population of events dominates the UHECR[2] spectrum from about $10^{18.5}$ eV to the highest observed energies.* **(ii)** *The UHECR population has a smooth spectrum without a substantial discontinuity at $10^{19.7}$ eV.* The single mechanism proposition is a new and particularly powerful tool: The high statistics sub-GZK data can no longer be ignored, providing a strong constraint on models.

Above about $10^{19.7}$ eV, protons display a rapidly falling attenuation length which plateaus at about 10 Mpc for energies above $\approx 10^{20.5}$ eV (see e.g. [5]). Thus independently of the nature of the sources of UHECRs and of magnetic fields which may deflect or confine them, sub-GZK protons with pathlengths up to a few Gpc contribute to the flux at Earth, whereas super-GZK protons only reach Earth if their pathlength is of order few 10's Mpc or less. We call the ratio of the average pathlengths from source to Earth, of sub- and super-GZK protons, f_{acc} – the accumulation factor. For uniformly distributed sources active over cosmological times, $f_{acc} \approx$ few Gpc/ few 10's Mpc ≈ 100. If the UHECR sources produce a smooth spectrum, and if Earth is located in a "typical" environment, the observed spectrum should have an offset in normalization below and above the GZK region approximately equal to the accumulation factor, i.e., ≈ 100.

There are three categories of explanations for $f_{acc} \approx 1$ in the observed spectrum. **(I)** *GZK energy degradation can be circumvented by invoking new physics* such as Lorentz invariance violation or GZK-evading messengers. [3] **(II)** *The spectrum at the source could have an energy dependence complementary to the attenuation length.* However the attenuation length is practically constant in the sub-GZK region, rapidly varying in the transition region, and slowly varying above, so a very strange primary spectrum would be required to account for the data for all energies above $10^{18.5}$ eV. [4] **(III)** *The contribution to the flux on Earth from sources within the GZK distance and time ($\approx 10 - 20$ Mpc $\equiv 30 - 60$ Myr) is at least as*

---

[2]) It will be useful below to distinguish, within the UHECR population, between "sub-GZK", $10^{18.7} - 10^{19.5}$ eV, and "super-GZK", $E \geq 10^{20}$ eV, UHECRs, avoiding transition regions. For clarity we assume the UHECRs are protons.

[3]) Note that GZK-evading models do not automatically satisfy $f_{acc} \approx 1$: if the nucleon flux at the source is not lower than the messenger particle flux by at least a factor $\approx 1/f_{acc}$, the protons accompanying the messenger particles accumulate excessively.

[4]) Claims that distributed-source models can give a satisfactory accounting of the spectrum fail when the entire range including sub-GZK energies are considered. For instance models in which the spectrum is governed by hadronization of quarks (e.g., Z-burst, superheavy-relics, and topological defect models) have hard intrinsic spectra $\sim E^{-1}$ so in the GZK transition region, where the attenuation length is rapidly dropping, the attenuated spectrum fits the observed UHECR spectrum. However below the GZK energy, say at $10^{18.7}$ eV where the flux is well measured, the predicted flux is more than an order of magnitude too small.

great as the contribution from all other sources active since redshift of order 1/2 added together. Galactic sources satisfy this condition trivially, however isotropy is difficult to reconcile with a galactic explanation [6]. For extragalactic sources, this condition implies that our local sources are significantly more concentrated or more powerful than average. This is increasingly more improbable as the number of sources increases, so this scenario requires a single dominant source.

Several papers recently made use of a combination of (II) and (III), taking advantage of the overdensity of matter in the local supercluster and using an intermediate hardening the spectrum [7,8]. However the matter overdensity on the GZK scale falls far short of the f_{acc} level required to remove the offset, and these efforts fail to account for the sub-GZK spectrum below $\sim 10^{19.4}$ as can be seen from Fig. 7 of [7] where the prediction is more than a factor of 2 too low even at 10^{19} eV.

We now turn to the question of whether a single dominant source can be responsible for most UHECRs observed at Earth, (III). In this case the UHECR spectrum observed on Earth is not generic. Having a single source requires that UHECRs diffuse in the magnetic field of the local supercluster since they are observed to arrive from all directions. Diffusion not only isotropizes the flux, it also produces a time lag between the arrival of UHECRs and the arrival of photons, so UHECRs can be produced during an earlier more active phase [9]. Extragalactic magnetic fields can be of order a few tenths μG [9] rather than the $\sim nG$ fields previously assumed in discussions of UHECR propatation. These magnetic fields are required only within the local supercluster. Recent observations find magnetic fields of order a few μG at distances of 0.5Mpc from clusters [10], supporting our picture.

Magnetic fields are trapped in ionized matter whose turbulent flow leads to extensive restructuring of magnetic fields over large and small distance scales. It is plausible to assume a Kolmogorov form for the magnetic field power spectrum as a function of wavenumber, $B(k) \approx Bk^{-11/6}$ [11,12] in this regime; B denotes the magnitude of the field at scale λ, the maximum scale of coherent correlations, generally expected to be $\sim 0.1-10$ Mpc. The Larmor radius of a proton in a magnetic field B orthogonal to its motion is $R_L = 110\ E_{20}/B_{\mu G}$ kpc. Using the Kolmogorov spectrum, [11] obtained[5]: $D(E) \approx 0.05\ (E_{20}\ \lambda^2_{\text{Mpc}}/B_{\mu G})^{1/3}$ Mpc^2/Myr.

When n_0 CRs are produced by an isotropic source and propagate diffusively without energy loss their number density at a distance R from the source, and time t since emission, is: $n(r,t) = [n_0/(8\pi\ D\ t)^{3/2}]\exp[-R^2/(4Dt)]$. A diffusion front reaches R at $t = R^2/(6D)$. In order to avoid unobserved structure in the energy spectrum the diffusion front of UHECRs from the dominant source must have already reached Earth, *for all energies* $10^{18.7-20.5}$ eV. Due to the gradient in the number density with radial distance from the source, the flux at Earth will have a dipole moment $\alpha = (R/2tc)$ relative to the source.

If the source was active for a period comparable to or longer than the diffusion time we must integrate the impulsive flux over the range of possible propagation times. The minimum propagation time is $t_{min} = \text{Max}\{R^2/(3D), T_{\text{off}}\}$, where T_{off}

---

[5] We ignore here the possibility of Bohm diffusion that may be important at higher energies.

is the time since the AGN turned off its UHECR production. For emission at a constant rate n_0/τ we have: $n(R,t) \approx [2\,n_0/[(8\pi D)^{3/2}\tau t_{min}^{1/2}]$. Since $D \sim E^{1/3}$ for Kolmogorov diffusion, and shock acceleration results in a spectrum at the source of E^{-p} with p slightly greater than 2, the resultant spectral index is close to the ≈ -2.7 which is observed. The anisotropy for a continuous source is the flux-weighted average of $R/2tc$, so $\alpha \approx R/(6t_{min}c)$. The anisotropy increases with energy if t_{min} is determined by diffusion rather than the AGN turnoff.

These results allow us to constrain the source of UHECRs observed at Earth. We must demand that the number density of UHECRs from the proposed local source, n_E be equal to or greater than the accumulated number density from all UHECR sources in the rest of the Universe: $n_E \gtrsim \Gamma_s \bar{n} H^{-1}$, where $\bar{n}$ is the total number of UHECRs produced by an average source, Γ_s is the number of sources per unit volume and unit time, and $H^{-1} \approx 10^4$ Myr is the age of the Universe.

An AGN at an extragalactic scale can readily satisfy the local dominance requirement because they are rare. The local flux depends on $t_{min}^{-1/2}$, which we estimate to be $\sim 0.1 - 0.2$ Myr$^{-1/2}$. The AGN rate is $\approx \rho_{AGN}\tau_{AGN}/H^{-1}$, where $\rho_{AGN} \approx 10^{-6}$Mpc^{-3}Myr^{-1} is the number density of AGNs. The dominance condition is satisfied without assuming the UHECR output of the local source is stronger than average. It is helpful that τ_{AGN} cancels out, since it is uncertain.

Defining $D = 0.05\,D_K$ Mpc^2/Myr and requiring the diffusion time to be less than the GZK time $t_{min} < t_{max} \lesssim 50$ Myr, implies the UHECR diffusion front is closer than about $\lesssim 4D_K^{1/2}$ Mpc. Thus M87 at 18 Mpc, which has been mentioned as a possible single source of UHECRs, is too far away if we require diffusive propagation. Cen A, at 3.4 Mpc, is by far the nearest powerful radio galaxy; see ref. [13] for a comprehensive review. Already in 1978, Cavallo [15] pointed out that the size of its radio lobes and the strength of its magnetic fields satisfy the Hillas criterion [16] for acceleration of UHECRs. However with $b = 20°$ and $l = 310°$, Cen A is in the blind direction of all major operating UHECR detectors which are located in the northern hemisphere, and thus Cen A can only be the source if the UHECRs propagate diffusively. Until recently there was a prejudice that extragalactic magnetic fields are of order nG and deflection of UHECRs is negligible, so Cen A was not considered acceptable.

The observed total luminosity of Cen A is now about 10^{43} erg/s, of which about half is high energy [13], so if ϵ is the efficiency of UHECR production compared to photon production extrapolated to 10^{19} eV using equal power per decade, we estimate $E^2 dN/dE/dt \approx 3\epsilon 10^{53}$ eV/s for 10^{19} eV UHECRs. Using $t_{min}^{-1/2} = 0.1$Myr$^{-1/2}$ to be conservative, we obtain the energy-weighted flux per str at Earth $E^3 dN/dE \approx 6\epsilon 10^{25}$ eV^2 m^{-2} s^{-1} sr^{-1}, easily consistent with the observed value of $10^{24.5}$ eV^2 m^{-2} s^{-1} sr^{-1}. Furthermore, the time scale for evolution of AGNs is ≈ 10 Myr in general, and of Cen A in particular [13], so that it was likely to have been a powerful AGN within the most recent GZK time. If so, its luminosity would have been $\approx 10^{44} - 10^{45}$ erg/s. Indirect evidence suggests that Cen A was very active in the past [13], in which case the UHECR production efficiency need only be of order a percent or less.

The diffusion coefficient for $E = 10^{19}$ eV, $B_{\mu G} = 0.3$ and $\lambda_{\text{Mpc}} = 0.3$, is 0.016 Mpc^2/Myr, so the typical UHECR arrival time from Cen A is 5 Myr. The pathlength of a 10^{19} eV CR is ≈ 8 Mpc, many times the diffusion length $D/c \approx 1/4$ Mpc, so the diffusive approximation applies; the Fly's Eye event satisfies, barely, the Bohm diffusion criterion. The anisotropy $\alpha \approx R/(6t_{min}c)$ is predicted to be ≈ 0.07 for 10^{19} eV, or less if Cen A's emission of UHECRs occured more than ≈ 25 Myr ago. Analysis of AGASA data above $10^{18.7}$ eV finds isotropy down to a limiting sensitivity of $\alpha \gtrsim 10\%$ [14] Our analysis shows that the required magnetic fields within the local supercluster are a fraction of a μG. Note however, that several clusters of 2-3 events having energies from $\approx 4-30 \cdot 10^{19}$ eV have been observed [17]. If the ultra-high energy part of the cosmic ray spectrum consists of protons which experience large deflections due to few-tenth-μG fields in the local supercluster, this must be a statistical fluke. If this clustering is confirmed it would be evidence against the picture advanced here.

We have identified Cen A to be a good candidate source for UHECRs. GRBs are unlikely since similar arguments will require the "single GRB" to be within the Galaxy. Prior to concluding we consider the compatibility of various new physics explanations with conditions (i) and (ii). Two of the earliest proposals for GZK-evading messengers are (a) neutrinos produced in cosmologically distributed AGNs whose high energy cross section is anomalous so they interact strongly with nuclei in Earth's atmosphere and (b) light-gluino-containing baryons whose threshold for photo-pion production is above 3×10^{20} eV. These models are acceptable only if nucleon emission by the accelerator is $\lesssim 1/f_{acc}$ times the messenger emission. If the messenger neutrinos or hadrons are produced by electromagnetically accelerated protons, as is most conventional astrophysically, this condition demands considerable nucleon attenuation in the source. It is noteworthy that this nucleon attenuation requirement implies that, if the source were an AGN, it would necessarily display the spectral characteristics found to be associated with candidate distant sources in the analysis of [18]. If the hints of clustering of events and directional identification with powerful distant matter-enshrouded radio quasars become real signals with better data, these models would be favored.

The observed isotropy of sub-GZK events rules out all models in which the UHECR sources are proportional to the galactic DM distribution, including Z-burst models. The DM distribution is fairly well known and there is little flexibility in adjusting its typical clustering scale. AGASA compared the anisotropy in the angular distribution of 581 events above 10^{19} eV with that of two popular dark matter distributions centered on the Milky Way [4] and in both cases found poor agreement with the predicted anisotropy (reduced $\chi^2 \geq 10.0$). With the assumption of a single UHECR population, this is enough to exclude such models in spite of the fact that the statistics above 4×10^{19} eV are inadequate to do so on their own. See [19] for a numerical study which reaches the same conclusion.

Relaxing the conditions on the source distribution, as may be appropriate for a new population of astrophysical sources such as magnetars [20], topological defects, or eV neutrinos, does not allow this constraint to be evaded (except with help from magnetic diffusion). To see this, consider a simple toy model in which all sources are on a shell of radius d around the galactic center, with $d \gg 8.5$ kpc, the distance of Earth from the galactic center. This source distribution gives a dipole anisotropy of $0.34/(d/50\text{kpc})$. By increasing d the anisotropy can be decreased, however large enough values of d are unnatural; e.g., an isothermal halo with a core radius of ~ 1.5 kpc has $\langle d \rangle \sim 10$ kpc. Compatibility with $\alpha \lesssim 0.1$ requires $d \gtrsim 150$ kpc, making the local-dominance requirement problematic.

To conclude, we have deduced that Earth at this epoch is probably not exposed to a generic UHECR spectrum. The absence of a cutoff at the GZK energy probably reflects our coincidental position in space-time relative to the nearest source. The most straightforward explanation which survives the analysis here is that most UHECRs reaching Earth come from a single AGN, at a distance of at most ≈ 5 Mpc. Cen A, a powerful radio galaxy at 3.4 Mpc is an excellent candidate. This requires diffusion in a chaotic magnetic field, whose strength is of order few-tenth μG, which is consistent with observational and theoretical constraints. We predict the anisotropy of UHECRs with $E \geq 10^{19}$ eV is 7% or less.

The research of GRF was supported in part by NSF-PHY-99-96173.

REFERENCES

1. G. Farrar and T. Piran, astro-ph/0010370, 2000.
2. K. Greisen. *Phys. Rev. Lett.*, 16:748, 1966; G.T. Zatsepin and V.A. Kuzmin. *Sov. Phys.-JETP Lett.*, 4:78, 1966.
3. A. Watson, this volume.
4. N. Hayashida et al. *Astropart. Phys.*, 10:303, 1999.
5. T. Stanev et al. Phys.Rev. D62 093005, 2000.
6. See, however, G. R. Farrar in preparation and M. Giller & M. Zielińska, Ital. Phys. Soc. Conf. Ser. 57:347, 1997.
7. M. Blanton, P. Blasi, and A. Olinto. astro-ph/0009466.
8. J. Bahcall and E. Waxman. Ap., J., 542, 542, 2000.
9. G. Farrar and T. Piran. *Phys. Rev. Lett.*, 84:3527, 2000.
10. Clarke, T. E., Kronberg, P. P., & Böhringer, H., Ap. J. L., 547, L111, 2001.
11. P. Blasi and A. Olinto. *Phys. Rev.*, D59:023001, 1999.
12. A. Lemoine et al. *Astropart. Phys.*, 10:141, 1999.
13. F. P. Israel. *Astron. Astro. Rev.*, 8:237, 1998.
14. M. Takeda, M. Teshima, and G. Farrar, private communication, 2001.
15. G. Cavallo. *A & A*, 65:415, 1978.
16. A. M. Hillas. *Ann. Rev. Astron. Astrophys.*, 22:425, 1984.
17. N. Hayashida et al. *Phys. Rev. Lett.*, 77:1000, 1996.
18. G. R. Farrar and P. L. Biermann. *Phys. Rev. Lett.*, 81:3579, 1998.
19. O. E. Kalashev et al. astro-ph/0006349.
20. P. Blasi and A. Olinto. Ap. J., 533, L123, 2000. .

On a mechanism of highest-energy cosmic ray acceleration

C. Litwin* and R. Rosner*

Department of Astronomy & Astrophysics, The University of Chicago, 5640 S Ellis Avenue, Chicago, IL 60637

Abstract. A recently proposed mechanism of acceleration of highest energy cosmic rays by polarization electric fields arising in plasmoids injected into neutron star magnetospheres is discussed.

INTRODUCTION

The problem of the origin of ultra-high-energy cosmic rays (UHECR) - those with energy $\gtrsim 10^{19}$ eV - continues to pose a serious theoretical challenge (Hillas 1984, Biermann 1997, Cronin 1999, Bhattacharjee & Sigl 2000, Olinto 2000). No convincing explanation that could account for all main observables – energy, spectrum and flux – has been offered to-date. Of particular interest are cosmic rays in the highest energy range, above $\gtrsim 5 \times 10^{19}$ eV. This radiation does not exhibit any significant anisotropy connected with the galactic disk and is therefore generally assumed to be of extragalactic origin (Blandford 2000). Moreover, the seeming, albeit uncertain, change of slope of the energy spectrum at $\sim 5 \times 10^{19}$ eV is frequently taken as an indication of the appearance of a new, distinctly different (and presumably extragalactic) component of the spectrum. At the same time the UHECR do not exhibit any sign of Greisen-Zatsepin-Kuzmin (GZK) cut-off (Greisen 1966, Zatsepin & Kuzmin 1966). Thus if carried by singly charged particles, light nuclei or photons, this radiation would need to originate at distances $\lesssim 50$ Mpc. Nevertheless, the direction of the incoming radiation does not appear to be correlated with any plausible sources within this distance.

Existing theories of the UHECR generation are usually put into two general categories: the "bottom-up" scenarios, in which particles are accelerated from lower energies to the ultra-high energies; and the "top-down" scenarios in which particles are "born" with ultra-high energies in a decay of some ultra-massive X particles, usually relics of the early universe.

Top-down scenarios, in addition to relying on uncertain physics, face difficulties explaining both the flux and the energy spectrum of UHECR (see Olinto 2000). The primary difficulty for the bottom-up scenarios is the acceleration mechanism.

Acceleration scenarios are generally divided into two types (cf. Hillas 1984): (1) direct acceleration, by electric fields; or (2) statistical Fermi acceleration by shocks in magnetized plasmas.

Statistical Fermi acceleration by supernova shocks (Axford et al. 1977, Krymsky 1977, Bell 1978, Blandford & Ostriker 1978) is considered a source of the cosmic rays below the "knee" ($\sim 5 \times 10^{15}$ eV), which are believed to be of galactic origin. It gives rise to a power law energy spectrum, which combined with the energy dependence of the diffusion coefficient, as inferred from the data on secondary nuclei, yields the energy spectrum similar to the observed one. This mode of acceleration becomes inefficient at higher energies (Lagage & Cesarsky 1983).

The primary difficulty with the direct acceleration scenarios is the existence of sufficiently large voltages. Most commonly considered sources are unipolar inductors, such as rapidly spinning magnetized neutron stars or blackholes. In the case of pulsars, the rotation gives rise an emf too small to accelerate iron nuclei to the UHECR energies (Berezinskii et al. 1990, Blandford 2000). A spinning blackhole in the center of a radio-galaxy generates an emf sufficient to accelerate protons to energies $10^{19} - 10^{20}$ eV. A difficulty with this scenario, however, is the presence of a dense pair plasma and intense radiation which would cause energy losses of accelerated particles. Another argument frequently used (e.g., Hillas 1984) against direct acceleration scenarios is that it is not clear how the power-law energy spectrum, characteristic for cosmic rays, could emerge.

ACCELERATION MECHANISM

We have recently proposed (Litwin & Rosner 2001) an alternative scenario of a galactic or galactic-halo origin of the UHECR. This scenario goes some way toward overcoming some of the main difficulties described in the previous section. We describe this recent work in the present section.

We started with the observation that polarization electric fields arise in plasma "blobs" or streams injected at large angles into the magnetic field (Chandrasekhar 1960, Schmidt 1960; more recently, Litwin, Rosner & Lamb 1999 and refs. therein). The appearance of this electrostatic field allows for the plasma stream to penetrate into the magnetic field, as has been demonstrated in numerous laboratory experiments (Baker & Hammel 1965, Meade 1965, Lindberg 1978) and numerical simulations (Galvez & Borovsky 1991, Neubert et al. 1992).

Outside the plasma, the electrostatic field has an approximately dipolar character and has a nonvanishing component along the magnetic field. This field accelerates particles (electrons and ions) out of the plasma and gives rise to the plasma current that leads to the eventual halting of the plasma cross-field motion. The energy of accelerated particles is $\sim ZeU$ where Ze is the particle charge and U is one-half of the electrostatic potential difference across the plasma stream. This phenomenon of particle acceleration and the above estimate of particle energy has been verified in experiments (Lindberg 1978) and in computer simulations (Galvez & Borovsky 1991, Neubert et al. 1992).

For a sufficiently large stream width (much greater than the Larmor radius corresponding to the stream velocity), the energy of an accelerated particle can greatly exceed the kinetic energy of a bulk plasma particle. In particular, for a plasmoid of width ~ 10 km, free-falling onto a neutron star with the surface magnetic field $B \sim 10^{13}$ G, the voltage $U \sim 10^{21}$ V.

We applied the above-described basic physics to a plasmoid injected into a neutron star magnetosphere. Such plasmoids may results from planetoid impacts onto neutron stars of the type that has been previously discussed in the literature as possible sources of galactic gamma ray bursts (Colgate & Petschek 1981, Lin et al. 1991, Katz et al. 1994, Wasserman & Salpeter 1994, Colgate & Leonard 1996). The motion at distances greater than the Alfvén radius (i.e., where the ram pressure is equal the magnetic pressure) is unaffected by the magnetic field. Following Colgate & Petschek (1981) to describe the process of break-up, and subsequent compression and elongation, of an iron planetoid by tidal forces the density and the size of the impacting plasmoid at the Alfvén point is determined. During the infall the planetoid matter becomes ionized by the motional electric field. In the vicinity of the Alfvén radius it is assumed, as is customary (e.g., Lamb et al. 1973), that the magnetic field penetrates the plasma (due to, e.g., anomalous resistivity). The subsequent cross-field motion leads to plasma polarization, as described by Chandrasekhar (1960) and Schmidt (1960). The magnitude of the accelerating potential due to this polarization electric field is then determined by the energy of particles, accelerated along the magnetic field by the polarization field, from the plasmoid velocity, its size and the magnetic field at the Alfvén radius.

Subsequently we determined the energy of iron nuclei emerging from the magnetosphere, by solving numerically the particle equation of motion in the dipole magnetic field taking into account the radiation reaction force. As expected, the energy of emerging particles is much smaller than the initial energy unless they are accelerated close to the magnetic axis.

The energy of emerging particles is quite sensitive to the angle between the particle trajectory and the magnetic axis. From this, the energy spectrum of cosmic rays generated in a single impact event can be deduced (Litwin & Rosner 2001). For the range of parameters considered (magnetic field $B \sim 10^{12} - 10^{14}$ G, planetoid mass $M_p \sim 10^{22} - 10^{24}$ g), the spectrum depends weakly on the magnetic field and the planetoid mass. To a good approximation the emerging particles have a power-law energy spectrum: $dN/dE \sim E^{\mu}$. The value of the exponent $\mu \sim 2.9 - 3.0$ for magnetic fields in the considered range is within the measurement uncertainty of the value found by AGASA (Takeda et al. 1998).

The number of particles at given energy produced in a *single* event can be found (Litwin & Rosner 2001) from the energy spectrum and from the total charge carried by accelerated particles. The latter is determined by integrating the equation of motion. For an iron planetoid, with the fiducial mass of 10^{23} g, the number of particles with energies exceeding 10^{19} eV is $\sim 2 - 14 \times 10^{28}$.

MAGNETIC CONTAINMENT, ENERGY SPECTRUM AND FLUX OF UHECR

Larmor radii of iron nuclei with energies less than $\lesssim 10^{20}$ eV in a magnetic field with strength 3-10 μG, characteristic of the galactic magnetic field (Kronberg 1994), are less than ~ 1 kpc. Thus iron nuclei constituting UHECR would be confined by the galactic magnetic field with a characteristic gradient length scale of 10 kpc. Assuming a steady state, the energy spectrum observed on Earth differs from the source spectrum if the confinement time is a function of energy. It is usually believed (see, e.g., Sigl &

Bhattacharjee 2000) that confinement time decreases with increasing energy. Indeed, the chemical composition of CR in the $1-10^3$ GeV per nucleon range can be interpreted as a power law behavior of the diffusion coefficient $D(E) \sim E^\mu$, with $\mu \sim 0.3-0.7$ (see Berezinskii et al. 1990). A power law behavior of the diffusion coefficient results also in theoretical models, such as the model of Jokipii (1975) in which particles are scattered by the magnetic field fluctuations which vary only in the direction transverse to the mean field. If the particle Larmor radius is much smaller than the integral scale L_c, a power-law fluctuation spectrum, with exponent α, results in a power-law dependence of the diffusion coefficient on energy, with the exponent $\mu = (1-\alpha)/2$. In particular, for the Kolmogorov spectrum ($\alpha = -5/3$), $\mu = 4/3$.

However, the UHECR confinement time dependence on energy may be qualitatively different. First, the small Larmor radius approximation may be inapplicable to the highest energy cosmic rays. If one assumes that the integral scale of the galactic turbulence is 100 pc (Parker 1979), and that the galactic magnetic field is in the range 3-10 μG (Kronberg 1994), the Larmor radius of iron nuclei is larger than the integral scale for energies higher than $1-3 \times 10^{19}$ eV. In this regime, the dependence of the cross-field transport on energy might be significantly different. As a specific example, the previously discussed model of Jokipii (1975) yields in this regime $\mu = -1/2$. Thus the confinement times would increase with energy, assuming that the latter were determined by the cross-field diffusion. On the other hand, if the confinement time were determined by the motion along the magnetic field, it would be independent of energy on the galactic/galactic halo length scales for the ultrarelativistic, collisionless particles in the UHECR energy range.

If the confinement time is known, the rate of planetoid impact events required to generate the observed UHECR flux on Earth can be determined. The upper bound on the confinement time is given by the rate of photodisintegration on the infrared radiation background (Stecker 1998). The lower bound can be obtained from the escape time from the galactic magnetic field at the velocity of the curvature drift, assuming a 10 kpc curvature radius. For 10^{19} eV iron nuclei, this escape time is $\sim 10^{13}$ s; the escape time is longer if the field is axisymmetric and possesses closed flux surfaces. A similar estimate is found for the transit time along a spiral magnetic field twisting by 4π within radial distance of 10 kpc. Thus it is reasonable to expect that the confinement time will be in the range $\sim 10^{13}-10^{16}$ s. Then assuming that the density of neutron stars is 2×10^{-3} pc^{-3} (Shapiro & Teukolsky 1983) the observed flux of UHECR results if the impact rate is $\sim 10^{-4}-10^{-8}$ yr^{-1} on each neutron star.

One can speculate whether such an impact rate is plausible. If one assumes (Lin et al. 1991, Nakamura & Piran 1991, Colgate & Leonard 1996) that the planetoids originate in the accretion disk, formed from the estimated $10^{29}-10^{32}$ g of matter captured by the neutron star following the supernova explosion, then the rate of accretion of 10^{23} g planetoids in a galaxy cannot exceed $\sim 10^4-10^7$ yr^{-1}, assuming the neutron star birth rate to be 10^{-2} yr^{-1}. Since $\sim 10^{29}$ particles with energies higher than 10^{19} eV are generated in each impact event, the upper bound on the rate of UHECR generation is $\sim 10^{33}-10^{36}$ yr^{-1}. From the observed flux (Takeda et al. 1998) it follows that the density of particles in this energy range is 6×10^{-29} cm^{-3}. Assuming that the radiation is confined in a sphere of 10 kpc radius, this would require a confinement time exceeding 10^4-10^7 years. The estimated confinement time, mentioned in the previous paragraph, is greater than or comparable to this lower bound.

CONCLUSIONS

In this talk we reviewed a model (Litwin & Rosner 2001), which can potentially solve some of the problems, discussed in the Introduction. This model is an example of a direct acceleration process that explicitly results in a power-law energy spectrum of UHECR. The source spectrum agrees, within experimental uncertainties, with the observed spectrum. While the latter has not been calculated, it is plausible that the energy dependence of the confinement time will not result in a steepening of the spectrum in the UHECR range of energies. Also the magnitude of the observed UHECR flux results from a plausible rate of impact events; and the resulting spectrum may be only weakly anisotropic both due to the confining effect of the magnetic field and because the sources are neutron stars both in the galactic disk and in the galactic halo (cf. Bulik, Lamb & Coppi 1998).

ACKNOWLEDGMENTS

The authors thank Attilio Ferrari, Roger Hildebrand, Don Lamb, Angela Olinto and Simon Swordy for illuminating discussions. Useful comments by Pasquale Blasi, Willy Benz, Sterling Colgate, Walter Drugan, Carlo Graziani, Cole Miller, Don Rej and Eli Waxman are also gratefully acknowledged. This research has been supported by the Center for Astrophysical Thermonuclear Flashes at the University of Chicago under Department of Energy contract B341495.

REFERENCES

1. Axford, W.I., Leer, E., & Skadron, G., 1977, The acceleration of cosmic rays by shock waves, in *Proc. 15th Int. Cosmic Ray Conf. (Plovdiv)*, 11, 132 (1977)
2. Baker, D. A., Hammel, J. E., 1965, Experimental studies of the penetration of a plasma stream into a transverse magnetic field, Phys. Fluids, 8, 713
3. Bell, A.R., 1978, The acceleration of cosmic rays in shock fronts I., Mon. Not. R. Astron. Soc. 182, 147.
4. Berezinskii, V. S., Bulanov, S. V., Dogiel, V. A., Ptuskin, V. S., 1990, *Astrophysics of cosmic rays*, (Amsterdam: North-Holland), edited by Ginzburg, V.L.
5. Bhattacharjee, P., Sigl, G., 2000, Origin and propagation of extremely high-energy cosmic rays, Phys. Rep. 327, 247
6. Biermann, P. L., 1997, The origin of the highest energy cosmic rays, J. Phys. G: 23, 1
7. Blandford, R. D., Ostriker, J. F., 1978, Particle acceleration by astrophysical shocks, Astrophys. J. Lett., 221, L29
8. Blandford, R. D., 2000, Acceleration of ultra-high energy cosmic rays, Phys. Scripta T85, 191
9. Bulik, T., Lamb, D. Q., Coppi, P. S., 1998, Gamma-ray bursts from high-velocity neutron stars, 1998, Astrophys. J., 505, 666
10. Chandrasekhar, S., 1960, *Plasma Physics* (University of Chicago Press: Chicago).
11. Colgate, S.A., Petschek, A.G., 1981, Gamma ray bursts and neutron star accretion of a solid body, Astrophys. J. 248, 771.
12. Colgate, S.A., Leonard, P.J.T., 1996, Gamma-ray bursts from fast, Galactic neutron stars, AIP Conf. Proc. 366, 269 (1996)
13. Cronin, J. W., 1999, Cosmic rays: the most energetic particles in the universe, Rev. Mod. Phys., 71, 165

14. Galvez, M., Borovsky, J. E., 1991, The expansion of polarization charge layers into a magnetized vacuum: theory and computer simulations, Phys. Fluids B, 3, 1892
15. Greisen, K., 1966, End to the cosmic-ray spectrum?, Phys. Rev. Lett. 16, 748.
16. Hillas, A. M., 1984, The origin of ultra-high-energy cosmic rays, Ann. Rev. Astron. Astr. 22, 425
17. Jokipii, J. R., 1975, Motion of charged particles normal to an irregular magnetic field, Astrophys. J., 198, 727
18. Katz, J. I., Toole, H. A., Unruh, S. H., 1994, Yet another model of soft gamma repeaters, Astrophys. J., 437, 727
19. Kronberg, P. P. 1994, Extragalactic magnetic fields, Rep. Prog. Phys. 57, 325
20. Krymsky, G.F., 1977, A regular mechanism for the acceleration of charged particles on the front of a shock wave, Dokl. Acad. Nauk SSR 234, 1306 (1977)
21. Lagage, P. O.; Cesarsky, C. J., 1983, The maximum energy of cosmic rays accelerated by supernova shocks, Astron. Astrophys., 125, 249
22. Lamb, F. K., Pethick, C. J., Pines, D. A, 1973, Model for compact X-ray sources: accretion by rotating magnetic stars, Astrophys. J., 184, 271
23. Lin, D. N. C., Woosley, S. E., Bodenheimer, P. H., 1991, Formation of a planet orbiting pulsar 1829 - 10 from the debris of a supernova explosion, Nature, 353, 827
24. Lindberg, L., 1978, Plasma flow in a curved magnetic field, Astrophys. Space Science, 55, 203
25. Litwin, C., Rosner, R., Lamb, D. Q., 1999, On accretion flow penetration of magnetospheres, Mon. Not. R. Astron. Soc., 310, 324
26. Litwin, C., Rosner, R., 2001, Plasmoid impacts on neutron stars and highest energy cosmic rays, Phys. Rev. Lett., to be published
27. Meade, D. M., 1965, Experimental study of plasma motion in a toroidal octupole magnetic field, PhD thesis, University of Wisconsin
28. Nakamura, T., Piran, T., 1991, The origin of the planet around PSR 1829 - 10, Astrophys. J., 382, L81
29. Neubert, T., Miller, R. H., Buneman, O., Nishikawa, K.-I., 1992, The Dynamics of low-β plasma clouds as simulated by a three-dimensional, electromagnetic particle code, J. Geophys. Res., 97, 12057
30. Olinto, A.V., 2000, Ultra high energy cosmic rays: the theoretical challenge, Phys. Rep. 333-334, 329-348
31. Parker, E. N., 1979, *Cosmical Magnetic Fields*, (Clarendon: Oxford)
32. G. Schmidt, Plasma motion across magnetic fields, 1960, Phys. Fluids 3, 961
33. Shapiro, S. L., Tcukolsky, S. A., 1983, *Black Holes, White Dwarfs and Neutron Stars* (New York: Wiley).
34. Stecker, F. W., 1998, Origin of the highest energy cosmic rays, Phys. Rev. Lett., 80, 1816
35. Stix, M., 1975, The galactic dynamo, Astron. Astrophys., 42, 85
36. Takeda,M., Hayashida, N., Honda, K., Inoue, N., Kadota, K., Kakimoto, F.,Kamata, K., et al., 1998, Extension of the cosmic-ray energy spectrum beyond the predicted Greisen-Zatsepin-Kuzijmin cutoff, Phys. Rev. Lett., 81, 1163
37. Wasserman, I., Salpeter, E. E., 1994, Baryonic dark clusters in galactic halos and their observable consequences, Astrophys. J., 433, 670
38. Zatsepin, G.T., Kuzmin, V.A., 1966, Upper limit of the spectrum of cosmic rays, JETP. Lett. 4, 78.

Cerenkov Light from Cosmic Rays: a Comparison of Different Parametrizations

R. Barná*, A. Butkevich†, V. D'Amico*, D. De Pasquale*, A. Italiano*, A. Trifiró* and M. Trimarchi*

INFN Messina, Italy on behalf of NEMO Collaboration
†*INR Moscow, Russia*

Abstract. A detailed study regarding the different parametrizations of Cerenkov light existing in literature has been performed by comparing the various approaches used in experiments dealing with neutrino detection. Among these, we want to mention the Belyaev analytical approach, that is the only one deduced starting from theoretical formulas and the Butkevich method, based on Monte Carlo techniques. A third approach, due to Weibusch, reproduces in some way the angular distribution obtained by the Belyaev distribution.

The spatial distribution of Cerenkov light has been obtained[1-3] by using an electron angular distribution function up to π, in the approximation of pointlike source. The electron angular distribution has been chosen in the hypothesis of infinite shower energy E_0. To get the Cerenkov radiation spatial distribution (in the following, C.R.) we need the energy spectrum $N(E_0, E, z)$ and equilibrium angular distribution function $f(\vartheta)$ of the shower electrons, besides the Cerenkov angle ϑ_C.

We adopt a proper coordinate system as shown in Fig. 1, L being the radius of a sphere measured from the center of mass O of the shower, θ_0 the angle formed by the line connecting O and the P.M. detector D with respect to the shower axis. The C.R. angular distribution expression is the following:

$$F(E_0, L, \vartheta_0) = (C\rho/2\pi L t_b) \int_0^{2\pi} d\varphi \int_{z3}^{\infty} N(E_0, E_n, z') f[\theta(\vartheta_0(z'), \varphi)] dz',$$

$$C = (2\pi e^2/\hbar c) \cdot \frac{\sin^2 \vartheta_c t_b}{\rho} \cdot \left(\frac{1}{\lambda_{min}} - \frac{1}{\lambda_{max}} \right), \tag{1}$$

with $t_b = 36.4 g/cm^2$ the radiation length; ρ the water density; λ_{min} and λ_{max} the minimum and the maximum wavelength values; in this case we choose $L = 100\ m$. In his approach, Wiebusch [2] uses the pointlike source approximation, and adopts for the C.R. angular distribution the following function:

$$\frac{dn}{d\Omega} = \frac{1}{2\pi} \cdot \left[N_1 \cdot \exp\left(-\left| \frac{(\vartheta - \theta_c)}{\sigma_1 + \varepsilon_1(\vartheta - \theta_c)} \right| \right) + N_2 \cdot \exp\left(-\left| \frac{(\vartheta - \theta_c)}{\sigma_2 + \varepsilon_2(\vartheta - \theta_c)} \right|^2 \right) \right. \tag{2}$$

$$\left. + N_3 \cdot \exp\left(-\left| \frac{(\vartheta - \theta_c)}{\sigma_3 + \varepsilon_3(\vartheta - \theta_c)} \right|^3 \right) \right].$$

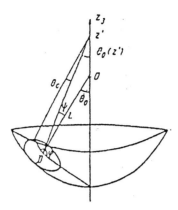

FIGURE 1.

Since non-relativistic tracks produce less Cerenkov light with a different angular distribution, Wiebusch takes into account this effect by means of an empirical formula:

$$\left(\frac{dn}{d\Omega}\right)_{corr} = \frac{dn}{d\Omega} \cdot (a + b \cdot \cos\vartheta). \tag{3}$$

a and b parameter values, obtained by means of full calculation, are 0.73 and 0.266 respectively for e.m. cascades, while they are 0.86 and 0 respectively for hadronic cascades. This calculation method is used in DADA simulation. Fig. 2 shows a comparison among different approaches, namely the Belyaev one, the Wiebusch formula with and without correction and the Hauptman one at 100 GeV electron energy.

A code for fast simulation of the C.R. from e.m. showers has been developed by means of a parametrization of the longitudinal profile [3]. In this approach, parameters have been obtained from a full Monte Carlo simulation, taking into account fluctuations and correlations among them.

A simple algorithm for parametrized showers has been used for simulation of the UA1 calorimeter [4]. The longitudinal energy profile of e.m. showers was simulated by fitting the parameters of a Γ distribution to the mean shower profile. The fast Monte Carlo code for μ tracking can be adopted for simulation of high energy event samples such as underwater neutrino detection systems like the km^3 detector. The mean longitudinal profile of e.m. showers can be satisfactorily described by the following expression:

$$f(t) = \frac{dE}{dt} = E_0 \beta \frac{(\beta t)^{\alpha-1} \exp(-\beta t)}{\Gamma(\alpha)} \tag{4}$$

t being the shower depth in units of radiation length and E_0 the shower energy;

$$\int_0^\infty f(t)dt = E_0; \frac{df(t)}{dt} = \left(\frac{\alpha-1}{t} - \beta\right) f(t).$$

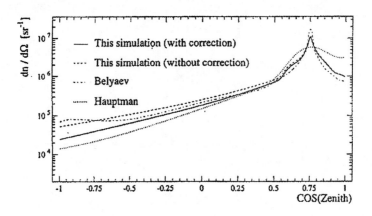

FIGURE 2.

β and $(\alpha - 1)/t$ are absorption and creation coefficients of the shower particles, respectively.

The C.R. average angular distribution emitted in the $\Delta\tau$ layer at τ shower depth and at ϑ angle with respect to the shower axis can be expressed as

$$P(\tau, \Delta\tau, \cos\vartheta) = N(\tau, \Delta\tau)\psi(\tau, \Delta\tau, \cos\vartheta)(phot/sr)$$

A fair agreement between the fast (**FLG**[5] code) and full Monte Carlo (**SIMEX**[6] code) calculations has been obtained. The peculiarity of the Butkevich approach lies on considering the shower lenght divided into several small enough intervals of amplitude $\Delta\tau$, each one associated to a point-like source τ_i with its own angular distribution depending on the shower depth considered. The flux of the C.R. from $i-th$ layer at a certain point X is $\Phi_i = \frac{P(\tau_i, \Delta\tau, \cos\vartheta)}{R_i^2}$, where R_i is the distance between the $i-th$ layer and X. The angular distribution function for different angles and depths can be obtained by the full Monte Carlo calculations. For $L >>$ *shower length* the previous approaches can be successfully applied, since we can adopt the approximation of point-like source; on the other hand, this method appears to be more exact than the other ones, and deserves further investigation.

REFERENCES

1. A. A. Belyaev, I. P. Ivanenko, V. V. Makarov, Sov. J. Nucl. Phys. **30**(1979)92, and references therein.
2. C. Wiebusch, 'The detection of faint light in deep underwater neutrino telescopes' (thesis), Aachen, unpublished.
3. L. B. Bezrukov, A. V. Butkevich, *Fast simulation of the Cerenkov light from showers*, to be published.
4. R. K. Bock *et al.*, Nucl. Instr. and Meth. **185**(1981)533.
5. A. Butkevich, *private communication*.
6. B. Stern, preprint INR P-0081, P-0082, 1978.

Pressure imbalance of FRII radio source lobes: a role of energetic proton population

M. Ostrowski[1] and M. Sikora[2]

[1] *Obserwatorium Astronomiczne, Uniwersytet Jagielloński, ul. Orla 171, 30-244 Kraków, Poland*
[2] *Centrum Astronomiczne im. Mikołaja Kopernika, ul. Bartycka 18, 00-716 Warszawa, Poland*

Abstract.
Recently Hardcastle & Worrall (2000) analyzed 63 FRII radio galaxies imbedded in the X-ray radiating gas in galaxy clusters and concluded, that pressures inside its lobes seem to be a factor of a few lower than in the surrounding gas. One of explanations of the existing 'blown up' radio lobes is the existence of invisible internal pressure component due to energetic cosmic ray nuclei (protons). Here we discuss a possible mechanism providing these particles in the acceleration processes acting at side boundaries of relativistic jets. The process can accelerate particles to ultra high energies with possibly a very hard spectrum. Its action provides also an additional viscous jet breaking mechanism. The work is still in progress.

I PARTICLE ACCELERATION AT THE JET BOUNDARY

For particles with sufficiently high energies the transition layer between the jet and the ambient medium can be approximated as a surface of discontinuous velocity change, a tangential discontinuity ('td'). If particles' gyroradia (or mean free paths normal to the jet boundary) are comparable to the actual thickness of this shear-layer interface it becomes an efficient cosmic ray acceleration site provided the considered velocity difference, U, is relativistic and the sufficient amount of turbulence is present in the medium. The problem was extensively discussed in early eighties by Berezhko with collaborators (see the review by Berezhko 1990) and in the diffusive limit by Earl et al. (1988) and Jokipii et al. (1989). The case of a relativistic jet velocity was considered by Ostrowski (1990, 1998, 2000). The simulations (Ostrowski 1990, cf. Bednarz & Ostrowski 1996 for shock acceleration) show that in favorable conditions the acceleration process acting at relativistic tangential discontinuity of the velocity field can be very rapid, with the time scale

$$\tau_{td} = \alpha \frac{r_g}{c} \quad , \tag{1}$$

where r_g is a particle gyroradius in the ambient medium and – for efficient particle scattering – the numerical factor α can be as small as ~ 10. The introduced acceleration time is coupled to the 'acceleration length' $l_{td} \sim \alpha r_g$ due to particle advection in the jet flow. In the case of a non-relativistic jet or a small velocity gradient in the boundary shear layer the acceleration process is of the second-order in velocity and a rather slow one. Then, the ordinary second-order Fermi process in the turbulent medium can play a significant, or even a dominant role in the acceleration. The acceleration time scales can be evaluated only approximately for these processes, and - for particles residing within the considered layer - we can give an acceleration time scale estimate

$$\tau_{II} = \frac{r_g}{c} \frac{c^2}{V^2 + \left(\frac{\lambda U}{D}\right)^2} \quad , \tag{2}$$

where V is a turbulence velocity ($\sim$ the Alfvén velocity for subsonic turbulence), λ is a mean particle free path normal to the jet axis and D is a shear layer thickness. The first term in the denominator represents the second-order Fermi process, while the second term is for the viscous cosmic ray acceleration. One expects that the first term dominates at low particle energies, while the second at larger energies, with τ_{II} approaching the value given in Eq. (1) for $\lambda \sim D$ and $U \sim c$.

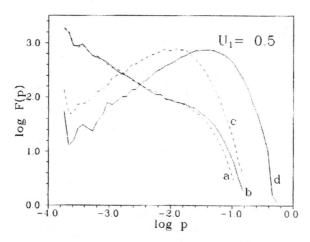

FIGURE 1. Comparison of the escaping particle spectra formed with wide (full lines 'b' and 'd') and narrow (dashed lines 'a' and 'c') turbulent cocoon surrounding the jet (cf. Ostrowski 1998). The results are presented for two possible particle injection sites: at the terminal shock (cases 'a' and 'b'), and far upstream the shock (cases 'c' and 'd'), where only the boundary acceleration is effective. Spectra of escaping particles are presented, the ones diffusively escaping from the cocoon (mostly in 'c' and 'd') and these advected downstream the terminal shock (mostly in 'a' and 'b'). Particle momentum unit is chosen in a way to give its gyroradius equal the jet radius at $p = 1$. Initial spectrum fluctuations appear near the injection momentum.

II DISCUSSION

Conditions within the large scale jets of FRII radio sources allow for acceleration of cosmic ray protons up to energies $\sim 10^{19}$ eV (e.g. Rachen & Biermann 1993, Ostrowski 1998). A characteristic feature of the boundary acceleration process in a simple considered model is formation of very flat spectrum of *escaping particles*. It is due to the on average parallel magnetic field configuration within the shear layer, limiting the low cosmic rays escape. Such particles residing within the shear layer volume can more efficiently stream to higher energies due to acting the acceleration process than to escape diffusively off the jet. At some higher energies escape becomes substantial, leading to the spectrum cut-off formation. In the considered conditions radiative losses are insignificant for protons.

Acting of the above mentioned processes can have pronounced consequences for the jet propagation if the seed particle injection at 'low energies' is efficient. Accelerated particles provide a viscous agent slowing down the jet movement. Because the jet energy is transmitted mostly to high energy nuclei, this dissipative process can occur without significant radiative effects. If a jet appearing from the central source with the Lorentz factor $\sim$ a few slows down to mildly relativistic velocities at large distances, the dissipated jet kinetic energy can be several times larger than the one available in terminal shocks. This amount is sufficient to explain additional pressure component providing stability of radio lobes against pressure of the X-ray emitting gas.

If the above interpretation is true, then some further consequences of the discussed accelerating process may arise. In particular acceleration of nuclei could be energetically inefficient in purely electron-positron jets, leading to less efficient jet breaking at large scales. Also, if the considered energetic particles escape from the radio lobes too fast, the required internal pressure component could not be sustained. A series of such problems are under study now.

The work was supported by the *Komitet Badań Naukowych* through the grant PB 258/P03/99/17 (MO) and 2 P03D 00415 (MS).

REFERENCES

1. Bednarz J., Ostrowski M., 1996, MNRAS, 283, 447
2. Berezhko E.G., 1990, Preprint "Frictional Acceleration of Cosmic Rays", The Yakut Scientific Centre, Yakutsk.
3. Earl J.A., Jokipii J.R., Morfill G., 1988, ApJ, 331, L91
4. Hardcastle M.J., Worrall D.M., 2000, MNRAS, 319, 562
5. Jokipii J.R., Kota J., Morfill G., 1989, ApJ, 345, L67
6. Ostrowski M., 1990, A&A, 238, 435
7. Ostrowski M., 1998, A&A, 335, 134
8. Ostrowski M., 2000, MNRAS, 312, 579
9. Rachen J.P., Biermann P., 1993, A&A, 272, 161

GeV γ-ray Astronomy with STACEE-64

R. A. Scalzo[1], L. M. Boone[2], C. E. Covault[1], P. Fortin[3],
D. Gingrich[4], D. S. Hanna[3], J. A. Hinton[1], R. Mukherjee[5],
R. A. Ong[6], S. Oser[1], K. Ragan[3], D. R. Schuette[6],
C. G. Theoret[3], D. A. Williams[2]

[1] *Enrico Fermi Institute, University of Chicago, Chicago, IL 60637*
[2] *SCIPP, University of California, Santa Cruz, CA 95064*
[3] *Deptartment of Physics, McGill University, Montreal, Quebec H3A 2T8 Canada*
[4] *Centre for Subatomic Research, University of Alberta, Edmonton, Alberta T6G 2N5 Canada*
[5] *Department of Physics and Astronomy, Barnard College and Columbia University, New York, NY 10027*
[6] *Department of Physics, University of California, Los Angeles, CA 90095-1562*

Abstract. The Solar Tower Atmospheric Cherenkov Effect Experiment (STACEE) is a new low-threshold atmospheric Cherenkov detector, using heliostat mirrors at a solar research facility to achieve a large collection area for Cherenkov light. The newest version of this detector, STACEE-64, should run at a threshold of 70 GeV or lower. Possible science for STACEE-64 in this energy range includes the study of AGN, supernova remnants, the extragalactic UV background, and exotic dark matter.

In recent years much effort has gone into the design of instruments capable of detecting gamma rays in the "unopened window" of energies from 10–300 GeV [1]. Future projects aim to close this gap both from orbit (GLAST) and from the ground (e.g., VERITAS, MAGIC, HESS). Both types of projects face different sorts of engineering challenges. Satellite-based experiments have small collection areas and face very low statistics at higher energies. Ground-based atmospheric Cherenkov telescopes, on the other hand, have large effective areas; but at lower gamma-ray energies, they must strain to pick out faint Cherenkov pulses from gamma-ray air showers against the ubiquitous night-sky background (NSB) light.

The STACEE experiment represents a new type of atmospheric Cherenkov detector designed to operate at low energy thresholds. It takes advantage of existing infrastructure at the National Solar Thermal Test Facility (NSTTF) at Sandia National Laboratories in Albuquerque, NM. Solar heliostat mirrors, which track the sun during the day, are used to track gamma-ray sources at night, collecting the Cherenkov light into a camera of photomultiplier tubes (PMTs). A secondary mirror is used to achieve a one-to-one mapping between heliostats and PMTs, sampling the Cherenkov wavefront at many lo-

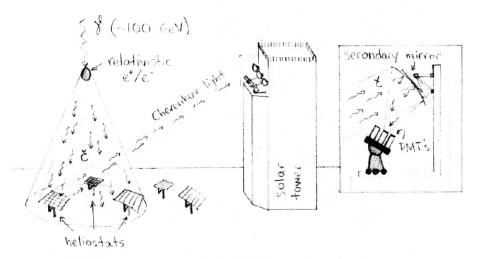

FIGURE 1. Concept drawing for the STACEE atmospheric Cherenkov telescope.

cations within the light pool. Delay electronics place Cherenkov pulses from different PMTs into coincidence, allowing formation of a high-multiplicity trigger. The arrival direction and energy of the primary particle, as well as some rejection of the hadronic background, can be recovered from the timing and intensity information recorded at each PMT. A prototype detector, STACEE-32, ran from fall 1998 to spring 1999, successfully detecting unpulsed gamma-ray emission from the Crab Nebula at an energy threshold of 190 ± 60 GeV [3].

An upgraded detector, STACEE-64, is currently under development. The upgrades can be understood via a simple scaling law for the energy threshold [2]:

$$E_{th} = \sqrt{\Phi \Omega \tau / A \epsilon},$$

where τ is the coincidence gate width, Φ is the flux of noise photons, Ω is the field of view, A is the collection area, and ϵ is an efficiency factor. For both STACEE-32 and STACEE-64, Ω is fixed at about $0.7°$ full-angle by the geometry of the heliostat field, ϵ is limited by the focusing and reflectivity of the optics and the quantum efficiency of the PMT's, and Φ is dominated by the NSB flux at the Albuquerque observing site. Careful calibration of the detector can optimize ϵ and soften the negative impact of Φ, but these are not truly "free parameters."

The parameters A and τ are more easily adjusted. It is straightforward to increase A; STACEE-64 will use 64 heliostats (twice as many as STACEE-32), totalling over 2300 m^2 of mirror area. τ is limited by the natural width of the Cherenkov wavefront for an electromagnetic air shower (about 2 ns). STACEE-32 used a combination of analog and digital electronics for its delays and its trigger; the delay units used required $\tau \geq 18$ ns. In contrast, STACEE-64 uses fast custom-made digital delay/trigger electronics which allow τ to be as small as 4 ns. The high-multiplicity triggers used by STACEE-32 and

STACEE-64 mean that the accidental coincidence rate (related to the energy threshold) actually varies as a high power of τ, making this improvement even more significant.

Improved event reconstruction will also be possible with STACEE-64. STACEE-32 used multi-hit TDC's (with a reset time of 12 ns) and charge-integrating ADCs to measure timing and intensity profiles. The STACEE-32 trigger was also relatively slow, so that the arrival of the PMT pulses had to be delayed; the analog delays used limited the ADC integration gate to 37 ns, resulting in poor signal-to-noise. In contrast, STACEE-64 features new 1 GHz flash ADCs, which will digitize PMT pulses instead of measuring times or integrated charges. This will allow dramatic improvements in timing and charge resolution on each PMT, as well as detailed information on the shapes of the Cherenkov pulses not available with STACEE-32.

Detailed Monte Carlo simulations indicate that these improvements will place STACEE-64's energy threshold at 70 GeV or lower. STACEE-64 will thus be well-suited to address many science questions requiring sensitivity at low energies, e.g.:

- *Structure of supernova remnants.* STACEE-64 can study pulsed and unpulsed gamma-ray emission from the Crab Nebula and similar objects, placing constraints on emission mechanisms and on models of pulsar magnetospheres [4,5].

- *Density of the extragalactic UV background.* Many gamma-ray AGN have been detected at MeV energies, but only a handful have been detected at TeV energies. This may be due to self-absorption in AGN, but also to absorption of gamma rays by pair production against extragalactic background light [6,7]. Such absorption has a characteristic dependence on energy and redshift, allowing it to be separated from intrinsic effects in a statistical sample of AGN observations. Observations at STACEE-64 energies would provide constraints on the density of the ultraviolet background, which is difficult to measure directly, and which is a sensitive probe of galaxy formation.

- *Exotic dark matter.* In various supersymmetric models, cold dark matter (e.g., neutralinos) is hypothesized to collect in the potential well at the galactic center. This would produce a gamma-ray continuum, and possibly annihilation lines, at energies on the order of 100 GeV [8], where STACEE-64 will be sensitive.

REFERENCES

1. Ong, R. A. *Phys. Rep.* **305**, 93–202 (1998)
2. Chantell, M. C., et al. *Nucl. Instrum. Methods A* **408**, 468 (1998)
3. Oser, S. M., et al. *ApJ* **547**, 949–958 (2001)
4. De Jager, O. C. & Harding, A. K. *ApJ* **396**, 161–172 (1992)
5. Aharonian, F. A. & Atoyan, A. M. *MNRAS* **278**, 525–541 (1996)
6. Salamon, M. H., & Stecker, F. W. *ApJ* **493**, 547–554 (1998)
7. Primack, J. R., et al., *Astropart. Phys.* **11**, 93-102 (1999)
8. Bergström, L., Ullio, P., & Buckley, J. H. *Astropart. Phys.* **493**, 547–554 (1998)

CHAPTER 17

MISCELLANEOUS TOPICS

Neutrino Factory Detector and Long Baseline Oscillations

E.J. Fenyves and R. F. Burkart

University of Texas at Dallas
Richardson, Texas

The Waste Isolation Pilot Plant (WIPP) located near Carlsbad, New Mexico, has been proposed as a favorable site to locate a neutrino detector to study neutrino beams from a neutrino factory. Numerous recent proposals have been written suggesting such a factory facility, the most notable of which is *A Feasibility Study of a Neutrino Source Based on a Muon Storage Ring* [1], a study initiated at Fermilab. A parallel study entitled *Physics at a Neutrino Factory* [2] was also prepared. At the recent workshop on *WIPP as the Next Generation US National Underground Research Facility* held in Carlsbad, New Mexico, on June 12-14, 2000, a collaboration was announced for design and development of a detector for neutrino factory beams at WIPP. Among the collaborators were representatives from UTD, UCLA, LANL, DOE/WIPP, TAMU, TAMU-Kingsville, and Princeton. Distances from the WIPP site to likely locations of neutrino factories are 1,749 km (Fermilab), 2,903 km (BNL), and 8,136 km (CERN) representing the most important ranges for neutrino oscillations.

The advantages of the neutrino factory and detector combination for unlocking new physics rest upon the neutrino factory high beam intensity, uniform composition, the energy range (10-50 GeV), and the relatively long distances possible between the factory and the detector.

The neutrino presently holds a place in the Standard Model as a neutral, massless lepton, that comes in three families, each family associating a neutrino with a corresponding charged, massive lepton. There are also three anti-neutrinos associated with the neutrinos. There is no theoretical requirement that neutrinos have non-zero mass, and a large number of GUT's predict non-zero mass. Experiments examining the tail of the beta decay spectrum of tritium claim to have established an upper limit of 3-5 eV for any electron neutrino mass [3].

In 1968 Pontecorvo [4] showed that if neutrinos have mass, then neutrino oscillations should occur if neutrino mass eigenstates do not correspond to the weak interaction eigenstates, i.e., there is neutrino flavor mixing. Numerous experiments to detect neutrino oscillations have since been carried out. These experiments have the disadvantage of not directly measuring the neutrino masses, but only the difference in mass between neutrino mass eigenstates.

Oscillation probabilities for neutrinos and anti-neutrinos will differ if CP is not conserved, providing a potential method for detecting CP violation. CP essentially acts as the particle-antiparticle conjugation [5].

Impetus for direct detection and measurement of the mass of neutrinos comes mainly from astrophysics, where solar models predict the solar neutrino flux, and cosmic ray measurements and simulations predict atmospheric neutrino fluxes. The first experiment to detect a deficit in the predicted solar neutrino flux was at Homestake Gold Mine in South Dakota in 1970 by R. Davis [6]. Standard solar model (SSM) predictions of 4.1 to 9.3 solar neutrino units (SNU) by different investigators were all higher than the average of 2.56 SNU detected at Homestake. The results of this solar neutrino puzzle were later confirmed by Kamiokande, GALLEX [7] and SAGE [8].

There have been numerous experiments for detecting neutrinos originating from both nuclear reactors and particle accelerators, all showing no neutrino oscillations with the exception of the LSND experiment (1995) at LAMPF [9]. This experiment reported the detection of muon neutrino to electron neutrino oscillations with a range of neutrino parameters excluded by atmospheric and solar experiments. A resolution of this disagreement has been proposed which postulates a fourth, sterile neutrino. The LSND experiment has not been confirmed by any other experiment, but Mini-BooNE at Fermilab is under construction to detect both v_e disappearance and appearance [11].

The most important experiment for the detection of solar and atmospheric neutrinos has been the Super-Kamiokande experiment (1998), with a large water target mass (22,500 tons) and high statistical significance [10]. This detector was sensitive to ^8B solar neutrinos, and reported a deficit in these neutrinos as compared to the SSM predictions. The experiment also reported detecting oscillations of atmospheric neutrinos evidenced by the disappearance of muon neutrinos, and going to tau neutrinos.

Three additional accelerator beam experiments utilizing pion decay neutrinos are presently operating or under development. K2K, a baseline experiment of 235 km from KEK to Super-K in Japan, is searching for v_μ disappearance and presently taking data. Two longer baseline experiments, both approximately 730 km, are in development. Minos, located in the Soudan mine in Minnesota, will operate with a neutrino beam from Fermilab, and will be capable of studying NC/CC energy distributions [12]. OPERA [13] and ICANOE [14], located at Gran Sasso, will operate with a neutrino beam from CERN. Both will be capable detecting tau appearance.

Neutrino factory beams will potentially allow all of the oscillations $v_e \rightarrow v_\mu$, $v_e \rightarrow v_\tau$, and $v_\mu \rightarrow v_\tau$ and their anti-neutrino analogs to be studied by analyzing the products of neutrino-target collisions. The number of events detected will depend upon the cross-sections of the reactions, the mass of the detector, the flux of each neutrino type in the beam, and the characteristics of the detector. The fluxes in turn will depend on the original beam intensity and the probabilities of the oscillations at the detector.

Neutrino CC interactions with the detector produce one lepton and many nucleon fragments, followed by a hadronic shower. In CC interactions the lepton will be of the same family as the incident neutrino. With a v_e and v_μ beam, this will lead to a so called *right* (no oscillations) and *wrong* (oscillations) sign of charged leptons. Tau particles detected, however, must be from oscillated neutrinos and therefore will be *wrong* sign in the same sense. It is therefore important to be able to identify and detect the charge of the primary lepton. This will require magnetized detectors in some form.

Determining the identity of the primary lepton is essential to identifying the neutrino. Muons will be the most easily detected due to their long range and unambiguous decay.

However, muons will be subject to backgrounds which may be large in comparison to the primary muon events. Electrons are readily detected by numerous methods all of which must be capable of following the electron shower to its conclusion. Otherwise, primary electrons will be easily confused with decays from other particles. Taus will be the most difficult to detect since their decays occurs within a few mm of their production and numerous decay channels are available. To see the initial tau a detector resolution of approximately 1 mm is required. For the detection of the tau by other methods, a detector capable of analyzing the hadron shower is required.

The nucleon fragments and hadronic shower can provide invaluable information on all of the neutrino interactions. The direction of the shower, the forward vs. transverse momentum of the shower, and shower reconstruction will be essential. Energy determination of both hadron and lepton showers will be necessary to determine the spectra of the incident neutrinos.

Neutrino detectors generally need to be large devices because of the small cross sections of neutrino reactions. Each detector needs to be designed with specific objectives in mind, as there is no single detector which will acquire all of the essential information needed to untangle the neutrino puzzle. Cost may dictate the choice between detectors. Specific detectors that would be suitable are magnetized steel and scintillators [15], liquid argon TPC's, nuclear emulsions, and water Cherenkov detectors.

Whatever primary neutrino detector is chosen for a factory neutrino beam, magnetic spectrometers and calorimeters should be included in the mix. It may be desirable to combine more than one type of primary detector, such as a magnetized steel/scintillator of large mass and liquid argon or nuclear emulsion detector of a smaller mass. If CP violation is to be studied, it will probably require some form of tau detection.

REFERENCES

1. Holtkamp, N., and Findley, K., ed., *A Feasibility Study of a Neutrino Source Based on a Muon Storage Ring* (2000), URL http://www.fnal.gov/projects/muon_collider/nu-factory/nu-factory.html.
2. Geer, S., and Schellman, H., ed., *Physics at a Neutrino Factory* (2000), URL http://www.fnal.gov/projects/muon-collider/nu/study/study.html.
3. Zuber, K., *On the Physics of Massive Neutrinos* (1998), hep-ph/9811267.
4. Pontecorvo, B., *Zh. Eksp Teor. Fiz.*, **33**, 548 (1957) and **34**, 247 (1958).
5. Akhmedov, E., *Neutrino Physics* (2000), hep-ph/0001264.
6. Davis, R., Jr., et al., Proc. 4th Int. Solar Neutrino Conference, ed. W. Hampel, MPIK Heidelberg, (1997).
7. Abdurashitov, J.N., et al., Proc. 4th Int. Solar Neutrino Conference, ed. W. Hampel, MPIK Heidelberg, (1997).
8. Kitsten, T., Talk presented at TAUP97, Gran Sasso (1997).
9. Athanassopoulos, C., et al., *Phys. Rev. Lett.*, **77**, 3082 (1996).
10. Fukuda, Y., et al., *Phys. Lett. B*, **433**, 9 (1998).
11. Bazarko, A., *MiniBooNE: the Booster Neutrino Experiment* (1999), hep-ex/9906003.
12. The MINOS Detectors Technical Design Report NuMI-L-337 (1998), URL http://www.hep.anl.gov/ndk/hypertext/minos_tdr.html.
13. OPERA Progress Report, LNGS-LOI19/99, URL http://www.cern,ch/opera /documents.html.
14. ICANOE Proposal, LNGS-P21/99, URL http://penometh4.cern.ch/.
15. Boyd, R., et al., Proposal to US DOE for Detector R & D Towards the Construction of OMNIS, (2000).

MAGNETIC SUPPORT AND ALFVÉN HEATING IN QUASAR BROAD-LINE REGION CLOUDS[1]

Denise R. Gonçalves[*,†], Amâncio C.S. Friaça[†] and Vera Jatenco-Pereira[†]

[*]*Instituto de Astrofísica de Canarias,*
Calle Via Lácte s/n, 38205 la Laguna, Tenerife, Spain
[†]*Instituto Astronômico e Geofísico - USP,*
Av. Miguel Stefano 4200, 04301-904 São Paulo, SP, Brazil

Abstract. Gonçalves et al. (1996) showed the relevance of Alfvén heating in order to establish the stability of quasar broad-line region clouds in the intercloud medium due to thermal instability. Although, photoionization by the continuum radiation from the central source is the most important mechanism to heat and ionize the broad-line region, Alfvenic heating being a second order process. In this article we show the results of the time-dependent calculations to follow the evolution of the forming clouds from the 10^7 K intercloud medium, calculating the UV and optical line emission associated with the clouds in order to compare them with observations.

INTRODUCTION

Understanding the broad-line regions (BLRs) of active galactic nuclei (AGN) is crucial to the study of the central engines of these objects. The pure photoionization model has some inconsistencies and an extra energy source is necessary in order to explain the observations [1]. The non-thermal process that we propose is the interaction between Alfvén waves that enables a dissipation process to heat the intercloud gas. This additional heat source, Alfvén heating (AH), when compared with the more conventional ones [2], maintains the broad-line emitting clouds and the intercloud medium in pressure equilibrium [3] in the sense of a two-phase medium, with gas at different temperatures and densities at approximately the same pressure.

---

[1] DRG acknowledges the Brazilian agency FAPESP (98/7502-0) and the Spanish (DGES PB97-1435-C02-01) grants. ACSF and VJP thank the Brazilian agency CNPq for partial support. All of us acknowledge partial support by Pronex/FINEP (41.96.0908.00).

OUR TIME-DEPENDENT MODELS AND RESULTS

In our models we suppose that the Alfvén waves can be damped by nonlinear and turbulent dissipation mechanisms, and that this AH heats the BLR intercloud medium. The onset of nonlinear heating occurs for $\beta = P_B/P_{gas} = (B^2/8\pi)/nk_BT > \beta_{on} > 1$, where β_{on} is the initial value of β when AH becomes important. In terms of AH, e.g. H_{nl} (erg cm^{-3} s^{-1}), we have: $H_{nl} = \Phi_w \Gamma_{nl}/v_A$, where Φ_w is the wave flux, Γ_{nl} is the nonlinear damping rate [4] and v_A is $B/\sqrt{4\pi\rho}$. For a spherical geometry, $B \propto A^{-1} \propto \rho^{2/3}$, where A is the cross-sectional area perpendicular to the magnetic field. As a consequence, $v_A \propto \rho^{1/6}$. Let the dependence of Φ_w on ρ be given as $\Phi_w \propto A^{-1} \propto \rho^{2/3}$. Taking $\Phi_w = \rho\langle\delta v^2\rangle v_A$, we have $\rho\langle\delta v^2\rangle \propto \rho^{1/2}$. Since $\Gamma_{nl} \propto c_s$ (the sound speed) and $c_s \propto T^{1/2}$, $H_{nl} \propto \rho^{-1/2} T^{1/2}$.

Following [5], the volumetric heating rate associated with the cascade process (e.g. with the turbulent damping of Alfvén waves) can be written as $H_{tu} = \rho\langle\delta v^2\rangle^{3/2}/L_{corr}$, where L_{corr} is a measure of the transverse correlation length of the magnetic field. Again, for a spherical geometry, $\langle\delta v^2\rangle \propto \rho^{1/2-1} \propto \rho^{-1/2}$, and adopting $L_{corr} \propto B^{-1/2} \propto \rho^{-1/3}$, we obtain: $H_{tu} \propto \rho^{7/12}$.

We investigate several representative models with magnetic fields, one with no AH and the others with AH for nonlinear and turbulent damping mechanisms and at different efficiencies, ζ, from the beginning of the calculations to $t_{t,col}$, the total collapse time (see Table 1).

TABLE 1. Properties of the Models

Model	$t_{t,col}$ (10^{7}s)	ζ	AH
I	9.88	-	none
IInl	10.10	10%	nonlinear
IItu	9.89	10%	turbulent
IIInl	10.60	30%	nonlinear
IIItu	9.93	30%	turbulent

The evolution of the cooling clouds is obtained by solving the hydrodynamic equations of mass, momentum and energy conservation [6,7]. The total pressure is the sum of the thermal and magnetic pressures. The initial density perturbations are characterized by an amplitude C and a length scale L, following $\delta\rho/\rho = C \sin(x)/x$ (here $x = 2\pi r/L$). We set $C = 1$ and $L = 5 \times 10^{16}$ cm. The unperturbed hot phase intercloud medium is characterized by $n_m = 5 \times 10^6$ cm^{-3} and $T_m = 10^7$ K, appropriate for the inner regions of an AGN, $\sim 10^{18}$ cm from the central source, in which lies the BLR.

The results of the profiles of temperature, thermal pressure and β show that at the beginning of the calculations, the magnetic pressure is much smaller than the thermal pressure. Although, with the drop in temperature, and the compression of the gas the magnetic field increases, thereby preventing collapse. In all cases (models I to IIItu) an increasing in ζ leads to an enhancement of the total collapse time that can be interpreted as the efficiency of the AH in inhibiting a strong growth of the perturbations, particularly the nonlinear one, see $t_{\rm t,col}$ in Table 1.

Eventually the gas begins to emit optical and UV lines, when $T \lesssim 5 \times 10^5$ K. We calculate the emissivity as well as the variability for the low Hβ, and high He II 1640 Å, C IV 1549 Å, N V 1240 Å and O VI 1034 Å ionization lines of the gas during this phase, and compare them with the observations of Fairall 9 [8,9] and NGC 5548 [1]. The variability pattern of the five models is very similar for all the lines studied. In addition, AH enhances the broad-line emission significantly, albeit not strongly. We found that the amplitude of variation of the Hβ line is smaller than that of the high-ionization ions, and that the time-scale of its variability is much longer, which is in first-order agreement with Fairall 9. However the comparison of the luminosities and time–lags between the Balmer and high-ionization lines show that the $L_{\rm H\beta}$ is too low and its variability too slow. The usual interpretation for this behavior is that the He II–N V emission region is located at a shorter distance from the central continuum source than the C IV region — in the case of NGC 5548, at 6 and 10 light-days, respectively [1]. The present results suggest that at least part of this trend could be due to different evolutionary time-scales for each line within a single cloud or cloud ensemble at a given position with respect to the central source, instead of different clouds giving rise predominantly to different lines at differing locations.

The variability represents the greater challenge for the stratified model for the formation of the high- and low-ionization lines, in which the more ionized lines are formed closer to the center (C IV and He II) and the less ionized further out [10]. Our results also allow for several locations for the formation of the broad lines, with the Balmer lines being produced at a more distant location, and powered mostly by photoionization, which would be responsible for most of its variability. On the other hand, the high-ionization lines would be produced, at the same location, closer to the central continuum source, however, being not only photoionized but also heated/excited by Alfvenic heating.

REFERENCES

1. Dumont, A.M., Collin–Souffrin, S. and Nazarova, L., *AA*, **331**, 11 (1998).
2. Mathews W.G. and Doane J.S., *ApJ*, **352**, 423 (1990).
3. Gonçalves D.R., Jatenco-Pereira V. and Opher R., *ApJ*, **463**, 489 (1996).
4. Lagage P.O. and Cesarsky C.J., *AA*, **125**, 249 (1983).
5. Hollweg J.V., *J. Geophys. Res.*, **91**, 4111 (1986).
6. Friaça A.C.S., *AA*, **269**, 145 (1993).
7. Gonçalves D.R., Friaça A.C.S. and Jatenco-Pereira V., *MNRAS* (2001) (submitted).
8. Rodríguez-Pascual P. M., et al., *ApJS*, **110**, 9 (1997).
9. Santos-Lleó M., et al., *ApJS*, **112**, 271 (1997).
10. Collin-Souffrin S., Hameury J. M. and Joly M., *AA*, **205**, 19 (1988).

Quantum Singularity of Spacetimes with Dislocations and Disclinations

D.A. Konkowski* and T.M. Helliwell[†]

*Department of Mathematics, U.S. Naval Academy, Annapolis, Maryland, 21402 U.S.A.
[†]Department of Physics, Harvey Mudd College, Claremont, California, 91711 U.S.A.

Abstract. A class of spacetimes with the mildest true classical singularities is shown to be quantum mechanically singular as well. These static quasiregular spacetimes with dislocations and disclinations are quantum-mechanically singular since the spatial portion of the wave operator is not essentially self-adjoint and thus the evolution of a test quantum wave packet is not uniquely determined by the initial wave function.

INTRODUCTION

A maximal spacetime is classically singular if it is geodesically incomplete or contains incomplete curves of bounded acceleration. Timelike or null incompleteness indicates that the evolution of some test particle is not defined after a finite proper time. What about the evolution of quantum wave packets? Is there a way to define their dynamics through the spacetimes alone? Is there a way to avoid having to add arbitrary boundary conditions at the classical singularity? Horowitz and Marolf [2] found that some classically singular spacetimes have repulsive barriers which shield the singularity; in these spacetimes quantum wave packets simply bounce off the barrier and never reach the singularity. Geodesics are the geometric optics limit of infinite frequency waves and only in that limit is the singularity reached.

CLASSICAL SINGULARITIES

In classical general relativity a singularity is indicated by incomplete geodesics or incomplete curves of bounded acceleration [1] in the spacetime. Since, by definition, a spacetime is smooth, all irregular points (singularities) have been excised. A singular point must therefore be a boundary point of the spacetime.

Classical singularities have been classified by Ellis and Schmidt [1] into three basic types: quasiregular, non-scalar curvature, and scalar curvature. The mildest is quasiregular and the strongest scalar curvature. A singular point q is a quasiregular singularity if all components of the Riemann tensor $R_{(abcd)}$ evaluated in an orthonormal frame parallel propagated along an incomplete geodesic ending at q are C^0 (or C^{0-}). In other words, the Riemann tensor components tend to finite limits (or are bounded); quasiregular singularities are mild, topological singularities such as dislocations and disclinations. On the other hand, a singular point q is a curvature singularity if some components are not bounded in this way. If all scalars in g_{ab}, the antisymmetric tensor η_{abcd}, and $R_{(abcd)}$ nevertheless tend to a finite limit (or are bounded), the singularity is non-scalar, but if any scalar is unbounded, the point q is a scalar curvature singularity.

QUANTUM SINGULARITIES

Horowitz and Marolf [2] have defined a static spacetime as quantum-mechanically singular if the spatial portion of the scalar wave operator is not essentially self-adjoint [4]. In this case, the evolution of a test quantum wave packet is not uniquely determined by the initial wave function. Using this definition they found that Reissner-Nordstrom, negative mass Scharzschild, and the two- dimensional cone remain singular when probed by quantum wave packets, but certain orbifold spacetimes, extreme Kaluza-Klein black holes, and some other string theory examples are nonsingular. In another context, Kay and Studer [5] have also shown that neither the 2D cone nor an thin idealized cosmic string have essentially self-adjoint wave operators.

QUANTUM SINGULARITY OF STATIC SPACETIMES WITH DISLOCATIONS AND DISCLINATIONS

A quasiregular spacetime is a spacetime with a classical quasiregular singularity. Conical singularities such as those in the 2D cone and the 4D idealized cosmic string spacetime are quasiregular singularies. Thus some of the mildest true singularities have been shown to be singular quantum mechanically as well as classically. What about other static quasiregular spacetimes?

We examined [7] a broad class of quasiregular spacetimes with dislocations and disclinations [10] including the exceedingly unusual Galtsov/Letelier/Tod spacetime [8, 9] for quantum singularities. For each spacetime we studied the spatial portion of the Klein-Gordon operator. If more than one solution for each mode was square integrable then the operator failed to be essentially self-adjoint [4] and the spacetime was quantum-mechanically singular. Static spacetimes with edge dislocations, screw dislocations, and disclinations were all found to have more than one solution for at least one mode. In fact, in certain screw-dislocation spacetimes there are an infinite number of wave modes where both solutions were square integrable. Thus we have shown a broad class of spacetimes containing disclinations and disclinations are also quantum-mechanically singular since the appropriate wave operator is not essentially self-adjoint.

DISCUSSION

The question remains: Can these quantum singularities be removed? One can, of course, add boundary conditions at the singularity [3, 5], but this is adding additional information not contained in the original spacetime. Or, one could use a Sobolev norm on the wave mode solutions instead of requiring square integrability. Ishibashi and Hosoya [6] show that this removes the ambiguity. A Sobolev norm is not, however, the usual choice to make sense of observables in ordinary quantum theory. Further study is necessary.

ACKNOWLEDGMENTS

DAK was partially funded by NSF grants PHY98-00118 and PHY99-88607 to the U.S. Naval Academy. DAK also thanks the Relativity Group at Queen Mary and Westfield College where some of this research was carried out.

REFERENCES

1. G.F.R. Ellis and B.G. Schmidt, *Gen.Rel.Grav.* **8**, 915 (1977).
2. G.T. Horowitz and D. Marolf, *Phys. Rev.* D **52**, 5670 (1995).
3. R.M. Wald, *J. Math. Phys.* **21**, 2802 (1980).
4. M. Reed and B. Simon, *Functional Analysis* (New York: Academic Press, 1972); *Fourier Analysis, Self-Adjointness* (New York: Academic Press, 1975).
5. B.S. Kay and U.M. Studer, *Commun. Math. Phys.* **139**, 103 (1991).
6. A. Ishibashi and A. Hosoya, *Phys. Rev.* D **60**, 104028 (1999).
7. T.M. Helliwell and D.A. Konkowski, *Quantum Singularity of Quasiregular Spacetimes* to appear *Gen. Rel. Grav.*.
8. D.V. Galt'sov and P.S. Letelier, *Phys. Rev.* D **47**, 4273 (1993).
9. K.P. Tod, *Class. Quantum Grav.* **11**, 1331 (1994).
10. R.A. Puntigam and H.H. Soleng, *Class. Quantum Grav.* **14**, 1129 (1997).

From Stellar Entropy to Black Hole Entropy

Manasse Mbonye*, Fred C. Adams* and Malcolm J. Perry[†]

*Physics Department, Univ. Michigan, Ann Arbor, MI 48109
[†]DAMTP, Univ. Cambridge, Cambridge, CB3 9EW, England

Abstract. We present a semi-classical argument to show how the entropy of a compact stellar object can *smoothly* approach its maximum value (that appropriate for a black hole of the same mass) as the star is compressed to form an event horizon. To have a smooth transition, the entropy must be a continuous function of the stellar radius. We show that a continuous description can be derived by considering the wave function of the star to be a quantum mechanical superposition of both classical and black hole states.

In the standard theory of black hole thermodynamics, the entropy of a black hole is specified by the area of its event horizon [1]. Associated with this entropy is a nearly thermal emission spectrum at the Hawking temperature T_H [2]. The interpretation of the black hole entropy as a particular form of the usual thermodynamic entropy is thus well established. A consequence of this formalism is that, in the absence of an event horizon, a compact stellar body contains no "black hole entropy".

This state of affairs, however, poses a conceptual problem: Consider a compact star made up of $\tilde{N}$ constituent particles. Its classical entropy is $S_{cl} = \ln \tilde{N}! \approx \tilde{N} \ln \tilde{N}$. The entropy S_{bh} of an equivalent mass black hole is $S_{bh} = 4\pi (m\tilde{N}/M_{pl})^2$, where we use units with $G = M_{pl}^{-2}$. Thus, for $N = 10^{57}$, e.g., $S_{bh} \approx 10^{76} \gg S_{cl} \approx 10^{59}$. Now consider a sequence of compact bodies, with ever smaller sizes, so that an event horizon eventually appears. At the point in the sequence where the horizon forms, the entropy must experience a step function increase from S_{cl} to S_{bh}, a jump of ~ 17 orders of magnitude in this example. According to the present theory, no matter how the stellar sequence is constructed, this troublesome step function increase in entropy remains perfectly sharp. This step function thus poses a conceptual problem which we seek to alleviate.

This work smooths the entropy problem by supposing [3] that an extended compact body of N_* particles with no event horizon must still contain small admixtures of black hole states $|BH\rangle$ in its wave function $|star\rangle$ so that

$$|star\rangle = \cos\theta |N_*\rangle + \sin\theta |BH\rangle. \qquad (1)$$

In general, the black hole portion of the wave function contains many different contributions, i.e.,

$$|\widetilde{BH}\rangle = \sin\theta |BH\rangle = \sum_j A_j |M_{bhj}\rangle, \qquad (2)$$

where $|M_{bhj}\rangle$ represents the wave function for a black hole of mass M_{bhj} and $|A_j|^2 = P_j$ is the corresponding probability of the star being in that state. The total probability of the star being in a black hole state is then given by

$$\mathcal{P} = \sin^2\theta = \sum_j |A_j|^2. \tag{3}$$

In addition to its classical entropy, $S_{cl} \approx N_\star \ln N_\star$, the star thus contains an additional gravitational contribution S_{gr} to its entropy given by $S_{gr} = \sum |A_j|^2 S_{bhj}$, where S_{bhj} is the entropy of a black hole with mass M_{bhj}.

We can estimate the probabilities $P_j = |A_j|^2$ for representative limiting cases. The total entropy $S_T = S_{cl} + S_{gr}$ includes both the classical entropy and the gravitational contribution and the sum S_T remains a continuous function of the physical variables that describe the system. As a first approximation, the star is a collection of particles of mass m with a number density n. We want the probability that N particles (mass $M_{bh} = mN$) happen to lie within their own Schwarzschild radius. The effective volume is

$$V_{bh} = 8m^3 N^3 M_{pl}^{-6}. \tag{4}$$

These N particles generally occupy a much larger volume $V_0 = N/n$. The probability p_1 that a single particle lies within this specified black hole volume is $p_1 = V_{bh}/V_0$ = $8nm^3 N^2 M_{pl}^{-6}$, and the corresponding probability p_N that N particles lie within the black hole volume is $p_N = (p_1)^N$. Since the star contains $N_\star/N$ volumes of N particles, the total probability P_N that the star contains a black hole composed of N particles is given by

$$P_N = N_\star N^{-1} (8nm^3 N^2 M_{pl}^{-6})^N = N_\star \lambda^N N^{2N-1}, \tag{5}$$

where we have defined $\lambda \equiv 8nm^3 M_{pl}^{-6}$.

The likelihood of the star containing black hole states thus depends critically on the value of the dimensionless parameter λ. Even for highly compact stellar objects, such as neutron stars which attain nuclear densities, $\lambda \approx 10^{-115} \ll 1$ and macroscopic black hole states are highly unlikely. As the density increases further, however, towards the dense configurations required for an event horizon to form, the probability of black hole states grows accordingly.

Summing over all macroscopic black hole states, we find that the black hole contribution to the entropy is specified by

$$S_{gr} = \sum_N P_N S_N = S_{bh} \sum_N \beta^{3N} (N/N_\star)^{2N+1}, \tag{6}$$

where we have defined a compression factor $\beta \equiv 2GM/R_\star < 1$. Note that $\beta \to 1$ as $R_\star \to 2GM$.

For configurations with β approaching unity the last term $S_{N_\star}$ in the series dominates. We then obtain for the entropy

$$S_{gr} \approx S_{N_\star} = S_{bh} \beta^{3N_\star} = S_{bh}(2GM/R_\star)^{3N_\star}. \tag{7}$$

In general, the true contribution to the black hole entropy will be larger than that given above, but, nonetheless, as the compact body approaches a black hole configuration, the entropy $S_{N\star}$ smoothly approaches the value S_{bh} appropriate for a black hole of mass M. In terms of the mixing angle θ introduced in equation (3), we can write

$$\mathcal{P} = \sin^2\theta \approx (2GM/R_\star)^{3N_\star}. \tag{8}$$

Notice that the extremely large exponents make the transition between a low entropy state and a high entropy state exceedingly sharp. To illustrate this sharpness, we define η to be the fractional difference between the stellar radius and the Schwarzschild radius, i.e., $R_\star/(2GM) \equiv 1 + \eta$. The function that describes the transition from stellar entropy to the black hole value of the entropy is thus given by

$$\frac{S_{N\star}}{S_{bh}} = (1+\eta)^{-3N_\star} \approx e^{-3N_\star\eta}, \tag{9}$$

where the second approximate equality is valid for $\eta \ll 1$. We thus define an exponential scale length, $L \equiv 2GM/3N_\star \sim m/M_{pl}^2$, which determines how rapidly the entropy grows as the radial extent of the compact body decreases. This scale length is exceedingly small, much less than a Planck length, so that such small deviations between the stellar radius and the Schwarzschild radius can never be realized. Although the quantum mechanical considerations of this paper are successful in formally removing any singular behavior, the classical picture remains accurate for most practical considerations.

In this paper, we have shown that the entropy of a compact star varies smoothly as the body is compressed from an extended configuration into a more compact configuration with an event horizon (a black hole). This argument considers stellar configurations to be a superposition of quantum mechanical states, which include admixtures of black hole states. The black hole states, which typically have low probabilities, have large entropies. As the stellar body becomes increasingly compact, the probability of black hole states increases, and the entropy increases accordingly. As a consequence, the increase from low classical values of entropy ($S_{cl} \approx N_\star \ln N_\star$) to large black hole values of entropy ($S_{bh} = 4\pi[mN_\star/M_{pl}]^2$) is no longer a step function when quantum fluctuations are taken into account. Although mathematically continuous, however, the transition from low entropy to high entropy remains an extremely steep and sensitive function of stellar radius. In hindsight, the fact that quantum mechanical considerations smooth out putative step function behavior is sensible and not unexpected. Here, we have also presented a semi-classical model that provides a rough mathematical description of this effect. Presumably, a more accurate treatment of this transition can be obtained once a more complete understanding of quantum gravity is available.

REFERENCES

1. J. D. Bekenstein, Phys. Rev. D **9**, 3292 (1974).
2. S. W. Hawking, Comm. Math. Phys. **43**, 199 (1974).
3. F. C. Adams et al., Phys. Rev. D **58**, 083003 (1998).

GEST

Sun Hong Rhie

Physics department, University of Notre Dame, In 46556

Abstract. Galactic Exoplanet Survey Telescope (GEST) was proposed for a discovery mission to search for microlensing terrestrial planets toward the Galactic bulge and also Kuiper Belt Objects (KBOs) that are believed to hold vital information of the early history of the solar system. Microlensing planet search method is hinged on photometric singular behavior of lensing (refraction) that is due to discontinuities in the distribution of photon paths whose phenomenon is better known in total reflection. The singularities (caustics) directly translate into potentially large planetary signals but the small size of the caustics imposes four essential requirements for earth mass planet searches: massive survey, angular resolution, temporal resolution, and continuous monitoring capability in a statistically stable observing condition. A 1-2m scale wide FOV space telescope with a large focal plane such as the GEST meets the needs.

Habitability of earth mass planets in the habitable (liquid water) zone is suspected to depend crucially on a giant planet near the snowline. The large mass of the giant planets makes the planetary caustics larger and easier to detect them in abundance. In microlensing, one needs not wait 12 years to detect a jupiter at the Jupiter orbit (5 AU) unlike in other planet search methods (doppler, astrometry, transit) because the planetary microlensing events will complete their courses within 50 days or so – 70 times shorter!

Close-in giant planets of the bulge stars will be registered as transits in the microlensing data base and will form a valuable platform for comparison studies of the planets from different detection methods and different environments.

When the Galactic bulge is behind the sun, the GEST will be ideal for high resolution mapping of the large scale structures which will include weak lensing by dark matter, quasar lensing and host galaxies, strong lensing by clusters, and a large volume of galaxies, which with selective follow-up measurements of whose redshifts and IR images will shed light on the dark stuff (matter, energy, essence, extra dimensions, topological defects, ...).

Gravitation, Relativities, Microlensing and Planets Microlensing involves both special and general relativities in a most trivial way: the spacial warp due to lensing masses is scanned in time by specal relativistic particles (photons) emitted from the lensed source and the general relativistic effect is contained in a factor 2. In microlensing, none of the 3 D's of lensing – delay, deflection, and distortion – are measured unlike in cosmological lensing because the time of flight differences and the image sizes and separations are too small. What is measured is the total magnification of the images that varies in time because of the relative motion of the lens and the lensed as seen from the observer. Microlensing light curves are essentially smooth – characteristically smooth – except at caustic crossing discontinuities. Even the discontinuities are characteristically orderly and smoothed over the size of the source (lensed) star. So, given sufficient quality of data, microlensing light curves are easily distinguished and identified against the background of variable stars and flare stars.

A microlensing planetary system will be discovered as a low multiplicity n-point lens. The interference pattern of the Newtonian potentials of the masses of the host star(s) and planets determines the angular positions, sizes, and strengths of

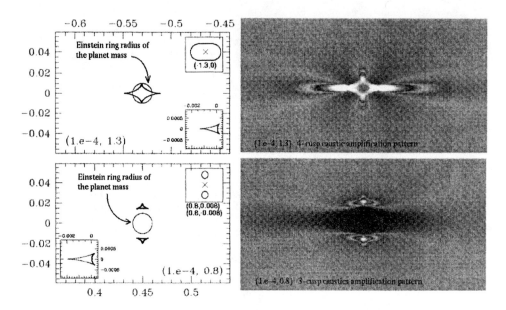

FIGURE 1. A planetary binary lens of a given mass fraction (here 10^{-4}) has two types of caustics that contribute to planet detections. When the separation (ℓ) between the star and the planet is larger than Einstein ring radius ($= 1$), the planetary caustic has 4 cusps and is located between the star and the planet. When $\ell < 1$, two 3-cusped caustics jointly define the planetary caustic region. Shown are the cases of $\ell = 1.3$ and $\ell = 0.8$. A cusp can be considered a point caustic with directionality, and the right panel exhibits varying magnification power of the cusps. Compare the sizes of the caustics and caustic regions. The Einstein ring of the planet as a single lens is commonly used as a rough estimator of the planetary caustics and is shown at the centers of the caustic planes. As a source star traverses the terrain of a planetary caustic, the light curve develops telltale "wavelets". The star is at the coordinate origin and the 4-cusped stellar caustics (resembling arrowheads) are shown in the insets. The planet positions are shown (marked by ×) in relation to the planetary critical curves which are centered around $(-1.3, 0)$ and $(0.8, \pm 0.008)$. See Bennett and Rhie (1996) for more details of the caustic regions and lensing zones.

the caustics – the planetary signal generators. The caustic curve of a planetary system lens consists of a stellar caustic and planetary caustics. See figure 1 for the case of planetary binary lenses. The stellar caustic dominates the behavior of the planetary system as a gravitational lens except in the planetary caustic regions. The planetary caustics behave like small magnifying glasses that modify the would-be single lens light curve by the host star. The "size of the planetary caustic" (circles in figure 1) of an earth mass planet with mass fraction $\epsilon = 3 \times 10^{-6}$ is ≈ 1 μas.

Ground-based Searches for Earth Mass Planets? The previous estimations of the feasibility of finding earth mass planets from the ground were based on the idea that turn-off stars (brighter than main sequence stars and smaller than gi-

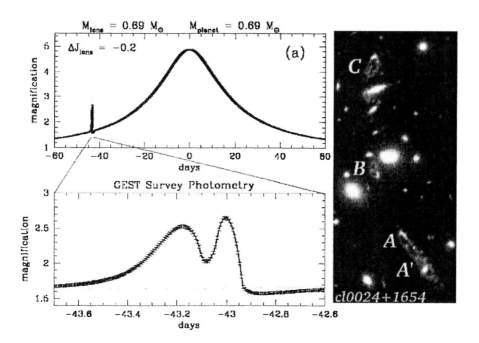

FIGURE 2. The photometrically singular nature of the caustics allows detection of earth mass planets ($\epsilon = 3 \times 10^{-6}$) with large S/N via microlensing. It is essential that the backlighting beam size (size of the source star) is sufficiently smaller than the size of the caustics in order to be able to reconstruct the magnification pattern and so the lens parameters from light curves. Here the lensed star is a main sequence star and the light curve has been sampled every 10 minutes (Bennett and Rhie 2000). The caustic singularity is associated with creation of two highly magnified images. The blue galaxy images A and A' lensed by cl0024+1654 seem to be one such pair conjoined at the critical curve which we believe is due to local mass concentrations. The image A' was reconstructed in the fit by Tyson et al. (1998) but has so far been ignored by other authors.

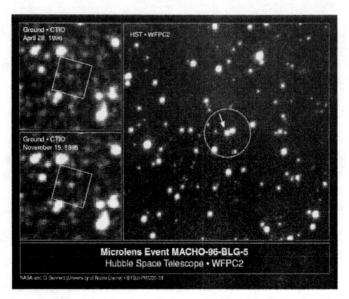

FIGURE 3. A typical Galactic bulge field is seen from the ground and space. The HST frame on the right is the image of the area inside the square boxes on the ground based images (from CTIO) on the left. The circle in the HST image indicates the seeing disk of the CTIO observation shown in the left bottom panel. In the color magnitude diagram of gound based observations, the main sequence (MS) stars inside the circle in the HST frame will show up as a "turn-off" star that is a few times brighter than a main sequence star. Since the microlensing beam - size of the Einstein ring radius – is small $\approx$ 1mas, only one of the MS stars is lensed (here arrowed) and the blended light from the unlensed MS stars adds photon noise and diminishes the effective amplification of the observed light curve. For example, a 10% signal will drop to 3.5% with 1.73 times lareger error bars.

ant stars) can be surveyed as source stars [10,1,4,7]. The caustic regions in figure 1 were calculated with the size of a turn-off star ($3R_\odot$). However, it is a false assumption that was derived from a color magnitude diagram of ground-based observations of the crowded Galactic bulge fields. The alleged turn-off stars are mostly blended main sequence stars (see figure 3). More recent HST data show that turn-off stars are indeed rare [6] as one would expect from stellar evolutions. This is detrimental to ground-based searches for earth mass planets because of the large seeing disks. It is also the case that the moon closely accompanies the bulge during the high bulge season. With a 4m scale dedicated telesope at the best site (such as dedicated VISTA in Paranal), one can detect earth mass planets, but it takes a very long time to acquire sufficient statistics. The details are being calculated and will be reported in the revised version of [2].

GEST, Planets, KBOs, and Large Scale Structures: GEST is an ongoing effort to search from space for microlensing planets in the Galactic bulge and the disk. The lack of close-in giants in the globular cluster 47 Tucanae indicates the possibility of metallicity constraints on the formation of planets or density constraints on the survival of the planets [3]. Galactic bulge is relatively metal rich ($\approx$ solar metallitcy in average) and is expected to be rich with terrestrial planets which may have been habitable when the Galaxy was younger. Gonzalez et al. [5] predicts massive terrestrial planets in the bugle, and GEST can explore the metallicity effect on the terrestrial planet mass limit. Ejection of planets is a commonality according to numerical studies, and GEST can detect the free-floating planets further constraining the planet formation and evolution scenarios. Microlensing is the only means to find mature free-floating planets.

Recent studies of habitability of planetary systems find a strong correlation between habitable terrestrial planets and jovian planets [8], and this throws a double jeopardy for most of the planet search efforts: small mass of terrestrial planets and long orbital period of jovian planets. Microlensing is a peculiar method that can handle the both challenges with ease.

In cosmological lensing, information is extracted from the spatial intensity distribution of the photons in the images. In order to reconstruct the mass distributions and extract the cosmological parameters accurately, it is crucial to find systems that offer sufficient number of constraints which in turn requires wide and deep imaging with an angular resolution of a small fraction of an arcsecond. Excellent systems can be followed up by the NGST or ground-based adpative instruements for further information.

We thank the GEST/Discovery proposal team members, especially the PI D. Bennett, and many other supporters of large format space imaging.

REFERENCES

1. Bennett, D., and Rhie, S., *Astrophys. J.* **472**, 660 (1996).
2. Bennett, D., and Rhie, S., *Astroph. J.* submitted (2000) (astro-ph/0011466).
3. Gilliland, R. et al., *Astroph. J.* **545**, L47 (2000).
4. Gould, A., *Hollywood Strategy for Microlensing Detection of Planets*, astro-ph/9608045 (1996).
5. Gonzalez, G., Brownlee, D., Ward, P., *Icarus* in press (2001) (astro-ph/0103165).
6. Holtzman, J., *Astron. Astroph.* **115**. 1946 (1998).
7. Sackett, P., *Planet Dectection via Microlensing*, ESO-SPG-VLTI-97/002 (1997).
8. Lunine, J., *Proc. Nat. Aca. Sci.* **98**, 809 (2001).
9. Tyson., A., Kochanski, G., and dell'Antonio, I., *Astroph. J. Lett.* **498**, 107 (1998).
10. Tytler, D., *A Road Map for the Exploration of Neighboring Planetary Systems (ExNPS)*, Chap. 7. (1996).

CHAPTER 18

CONFERENCE SUMMARY

Conference Summary

Steven Weinberg

Department of Physics, University of Texas at Austin
weinberg@physics.utexas.edu

Abstract. After a brief summary of some of the highlights at the conference, comments are offered on three special topics: theories with large additional spatial dimensions, the cosmological constant problems, and the analysis of fluctuations in the cosmic microwave background.

I OVERVIEW

Speaking as a particle physicist, an outsider, I have to say that my chief reaction after a week of listening to talks at this meeting is one of envy. You astrophysicists are blessed with enlightening data in an abundance that particle physicists haven't seen since the 1970s. And although you still face many mysteries, theory is increasingly converging with observation.

For instance, as discussed by Shri Kulkarni, it now seems clear that gamma ray bursters are at cosmological distances, producing over 10^{50} ergs in particle kinetic energies alone in a minute or so, making them the most spectacular objects in the sky. Tsvi Piran described a fireball model for the gamma ray bursters, in which gamma rays are produced by relativistic particles accelerated by shocks within material that is ejected ultra-relativisticaly from a central source. One can think of various mechanisms for the hidden central source, but even without a specific model for the source, the fireball model does a good job of accounting for what is observed.

According to this fireball model, gamma rays from the bursters are strongly beamed. Peter Höflich presented evidence that core collapse supernova are also highly aspherical. Both conclusions may be good news for gravitational wave astronomers — only aspherical explosions can generate gravitational waves.

Spectacular things seem to be turning up all over. Amy Barger told us how X-ray observations are revealing many active galactic nuclei in what had previously seemed like ordinary galaxies, and John Kormendy reported evidence that the events that produce galactic bulges or elliptical galaxies are the same as those that produce black holes in galactic centers.

Astrophysics is currently the beneficiary of massive surveys that are providing or will soon provide a flood of important data. We heard from John Peacock about the 2dF Galaxy Redshift Survey, from Bruce Margon about the Sloan Digital Sky Survey, and from George Ricker about the HETE x-ray and γ- ray satellite mission. Together with cosmic microwave background observations, about which more later, there seems to be a general consistency with the big bang cosmology, with about 30% of the critical mass furnished by cold dark matter, and about 70% furnished by negative-pressure vacuum energy.

This is not to say that there are no puzzles. Alan Watson reported on the long-standing puzzles of understanding how the highest energy cosmic rays are generated and how they manage to get to earth through the cosmic microwave background. There are also persistent problems in matching the cold dark matter model to observations of the mass distribution in galaxies. Ben Moore cast some doubt on whether cold dark matter really leads to the missing "cuspy cores" of galaxy haloes, and he concentrated instead on a different problem: cold dark matter models give much more matter in satellites of galaxies than is observed. He suggests that the missing satellites may really be there, and that they have not been observed because they have not formed stars. The reionization processes discussed here by Paul Shapiro may be responsible for the failure of star formation.

Our knowledge of the dark matter mass distribution within galaxies is receiving important contributions from observations of the lensing of quasar images by intervening galaxies, discussed by Genevieve Soucail. There have been hopes of using surveys of gravitational lenses to distinguish among cosmological models, but I have the impression that the study of galactic lenses will turn out to be more important in learning about the lensing galaxies themselves. Andrew Gould reported that microlensing observations have ruled out the dark matter being massive compact halo objects with masses in the range $10^{-7} M_\odot$ to $10^{-3} M_\odot$.

It would of course be a great advance if cold dark matter particles could be directly detected. We heard a lively debate about whether weakly interacting massive dark matter particles have already been detected, between Rita Bernabei (pro) and Blas Cabrera (con). It would be foolhardy for a theorist to try to judge this issue, but at least one gathers that, if the dark matter is composed of WIMPs, then they can be detected.

I have now completed my 10 minute general summary of the conference. There were other excellent plenary talks, and I have not mentioned any of the parallel talks, but what can you do in 10 minutes? In the remaining 35 minutes, I want to take up some special topics, on which I will have a few comments of my own.

II LARGE EXTRA DIMENSIONS

It is an old idea that the four spacetime dimensions in which we live are embedded in a higher dimensional spacetime, with the extra dimensions rolled up in some sort of compact manifold with radius R. This would have profound cosmological consequences: the compactification of the extra dimensions could be the most important event in the history of the universe, and such theories would contain vast numbers of new types of particle.

In the original version of this theory any field would have normal modes that would be observed in four dimensions as an infinite tower of 'Kaluza–Klein recurrences,' particles carrying the quantum numbers of the fields, with masses given by multiples of $1/R$. It had generally been supposed that R would be of the order of the Planck length, or perhaps 10 to 100 times larger, of the order of the inverse energy M at which the strong and electroweak coupling constants are unified. Even setting this preconception aside, it had seemed that in any case R would have to be smaller than 10^{-16} cm $\approx (100 \text{ GeV})^{-1}$, in order that the Kaluza–Klein recurrences of the particles of the standard model would be heavy enough to have escaped detection.

The possibilities for higher dimensional theories became much richer with the increasing attention given to the idea that the spacetime in which we live does not merely appear four-dimensional — our three-space may be a truly three-dimensional surface that is embedded in a higher dimensional space. (This is the picture of higher dimensions that was vividly described in Edwin Abbott's 1884 novel *Flatland*, and has more recently become an important part of string theory, starting with Polchinski's work on D-branes[1].) This idea opens up the possibility that some fields may depend only on position on the four-dimensional spacetime surface, while others 'live in the bulk' — that is, they depend on position in the full higher-dimensional space. Only the fields that live in the bulk would have Kaluza–Klein recurrences.

Craig Hogan here discussed the recently proposed idea that the compactification scale R may actually be much larger than 10^{-16} cm, with no Kaluza–Klein recurrences for the particles of the standard model because the standard model fields depend only on position in the four-dimensional spacetime in which we live[2]. According to this idea, it is only the gravitational field that depends on position in the higher dimensional space, and so it is only the graviton that has has Kaluza–Klein recurrences, which at ordinary energies would interact too weakly to have been observed. The long range forces produced by exchange of these massive gravitons would be small enough to have escaped detection in measurements of gravitational forces between laboratory masses as long as $R < 1$ mm. (There are stronger astrophysical and cosmological bounds on R, arising from limits on the production of graviton recurrences in supernovas[3] and in the early universe[4].)

In any such theory with large compactification radius R the Planck mass scale of the higher dimensional theory of gravitation would be very much less than the Planck mass scale in our four dimensional spacetime. In a world with $4+N$ spacetime dimensions the gravitational constant G_{4+N} (the reciprocal of the coefficient of the term $\int d^{4+N}x \sqrt{g}\, g_{\mu\nu} R^{\mu\nu}$ in the action) has dimensionality $[\text{mass}]^{-2-N}$, so we would expect it to be given in terms of some fundamental higher dimensional Planck mass scale M_* by $G_{4+N} \approx M_*^{-2-N}$. Dimensional analysis then tells us that the gravitational constant G in four spacetime dimensions must be given by

$$G \approx M_*^{-2-N} R^{-N} . \qquad (1)$$

The usual assumption in theories with extra dimensions has been that $R \approx M_*^{-1}$, in which case $G \approx M_*^{-2}$, and M_* would have to be about 10^{19} GeV. But if we take $N = 1$ and $R \approx 1$ mm, then $1/R \approx 10^{-13}$ GeV, and $M_* \approx 10^8$ GeV. With $N = 2$ and $R \approx 1$ mm, $M_* \approx 300$ GeV. This is the most attractive aspect of theories with large extra dimensions: they can reduce or eliminate what had seemed like a huge gap between the characteristic energy scale of electroweak symmetry breaking and the fundamental energy scale at which gravitation becomes a strong interaction.

Theories with large extra dimensions are very ingenious, and they may even be correct, but I am not enthusiastic about them, for they give up the one solid accomplishment of previous theories that attempt to go beyond the standard model: the renormalization group equations of the original standard model showed that there is an energy, around 10^{15} GeV, where the three independent gauge coupling constants become nearly equal[5]. In the supersymmetric version of the standard model the convergence of the couplings with each other becomes more precise[6], and the energy scale M_U of this unification moves up to about 2×10^{16} GeV [7], which is less than would be expected in string theories of gravitation by a factor of only about 20. (This is also a plausible energy scale for the violation of lepton number conservation that may be showing up in the neutrino oscillation experiments discussed here by Masayuki Nakahata.) The Kaluza–Klein tower of graviton recurrences does nothing to change the running of the strong and electroweak coupling constants, and since the higher dimensional Planck mass M_* is very much less than 10^{15} GeV in theories with large extra dimensions (this, after all, is the point of these theories), it appears that *in these theories the standard model gauge couplings are not unified at the fundamental mass scale M_**. Of course, they might be unified at some higher energy, but we have no way to calculate what happens in these theories at any energy higher than M_*.

In his talk here Hogan mentioned that Dienes, Dudas, and Gherghetta[8] have proposed a way out of this problem. I looked up their papers, and found that they modify the renormalization group equations for the gauge couplings of the standard model by allowing the gauge and Higgs fields (and perhaps some fermion fields) to depend on position in the higher dimensional space, along with the gravitational field. Of course, then they have to avoid conflict with experiment by taking $1/R$ greater than 100 GeV. The Kaluza-Klein recurrences of the gauge bosons greatly increase the rate at which the coupling constants of the standard model run, but with little change in their unification. To put this quantitatively, Dienes *et al.* find the bare (Wilsonian) couplings evaluated with a cut-off Λ are

$$\frac{4\pi}{g_i^2(\Lambda)} = \frac{4\pi}{g_i^2(m_Z)} - \frac{b_i}{2\pi} \ln \frac{\Lambda}{m_Z} + \frac{\bar{b}_i}{2\pi} \ln \Lambda R - \frac{\bar{b}_i X_N}{2\pi N} \left[(\Lambda R)^N - 1 \right] , \qquad (2)$$

where g_1 and g_2 are defined as usual in terms of the electron charge e and the electroweak mixing angle θ by $g_1^2 = e^2/\sin^2 \theta$ and $g_2^2 = 5e^2/3\cos^2 \theta$; g_3 is the coupling constant of quantum chromodynamics; and X_N is a number of order unity. The constants (b_1, b_2, b_3) are the factors $(33/5, 1, -3)$ appearing in the renormalization group equation of the supersymmetric standard model with two Higgs doublets, while the constants $(\bar{b}_1, \bar{b}_2, \bar{b}_3)$ are the corresponding factors $(3/5, -3, -6)$ in the renormalization group equations for Λ above the compactification scale $1/R$ (with a possible constant added to each of the $\bar{b}_i$, proportional to the number of chiral fermions that live in the bulk). Dienes et al. remark that the standard model couplings still come close to converging to a common value, because the ratios of the differences of the $\bar{b}_i$ are not very different from the ratios of the differences of the b_i. I would like to put this more quantitatively, by asking what value of $\sin^2 \theta$ is needed in order for the couplings to become exactly equal at some value of Λ. In the supersymmetric standard model, this is

$$\sin^2 \theta = \frac{3(b_3 - b_2) + 5(b_2 - b_1)e^2/g_3^2}{8b_3 - 3b_2 - 5b_1} = \frac{1}{5} + \frac{7}{15}\frac{e^2}{g_3^2} = 0.231 \;, \tag{3}$$

in excellent agreement with the measured value 0.23117 ± 0.00016. (Here e and g_3 are taken as measured at m_Z, in which case $e^2/4\pi = 1/128$ and $g_3^2/4\pi = 0.118$.) If all the running of the couplings were at scales greater than $1/R$, then $\sin^2 \theta$ would be given by Eq. (3), but with b_i replaced with $\bar{b}_i$:

$$\sin^2 \theta = \frac{3(\bar{b}_3 - \bar{b}_2) + 5(\bar{b}_2 - \bar{b}_1)e^2/g_3^2}{8\bar{b}_3 - 3\bar{b}_2 - 5\bar{b}_1} = \frac{3}{14} + \frac{3}{7}\frac{e^2}{g_3^2} = 0.243 \;. \tag{4}$$

This is not bad, but nevertheless outside experimental bounds. (It would be necessary to consider higher-order contributions in the renormalization group equations and threshold effects to be sure that there is really a discrepancy here.) In order not to spoil the prediction for $\sin^2 \theta$, $1/R$ would have to be considerably larger than 1 TeV, so that much of the running of the coupling constants would occur at scales below $1/R$, where the renormalization group equations are those of the supersymmetric standard model.

In any case, the running of the couplings is so rapid above the compactification scale $1/R$ that the couplings become equal (to the extent that they do become equal) at an energy not far above $1/R$. The $4 + N$ dimensional Planck scale M_* given by Eq. (1) is very much greater than this. Taking $1/R$ greater than 1 TeV, Eq. (1) would give M_* greater than 10^{13} GeV for $N = 1$. Even for $N = 7$, we would have M_* greater than 10^6 GeV. *Thus theories of this sort save the unification of couplings at the cost of reintroducing a large gap between the higher-dimensional Planck scale M_* and the electroweak scale.*

III VACUUM ENERGY

There are now two problems surrounding the energy of empty space[9]. The first is the old problem, why the vacuum energy density is so much smaller than any one of a number of individual contributions. For instance, it is smaller than the energy density in quantum fluctuations of the gravitational field at wavelengths above the Planck length by a factor of about 10^{-122} and it is smaller than the latent heat associated with the breakdown of chiral symmetry in the strong interactions by a factor about 10^{-50}. All these contributions can be cancelled by just adding an appropriate cosmological constant in the gravitational field equations; the problem is why there should be such a fantastically well-adjusted cancellation. The second, newer, problem is why the vacuum energy density that seems to be showing up in supernova studies of the redshift-distance relation (reviewed in a parallel session by Nick Suntzeff and Saul Perlmutter) is of the same order of magnitude (apparently larger by a factor about 2) as the matter density *at the present time*. There are five broad classes of attempts to solve one or both of these problems:

1) *Cancellation Mechanisms*
It has occurred to many theorists that the gravitational effect of vacuum energy might be wiped out by the dynamics of a scalar field, which automatically adjusts itself to minimize the spacetime curvature. So far, this has never worked. Some recent attempts were described by Andre Linde in a parallel session.

2) *Deep Symmetries*
There are several symmetries that could account for a vanishing vacuum energy, if they were not broken. One is scale invariance; another is supersymmetry. The problem is to see how to preserve the vanishing of the vacuum energy despite the breakdown of the symmetry. No one knows how to do this.

3) *Quintessence*
It is increasingly popular to consider the possibility that the vacuum energy is not constant, but evolves with the universe[10]. For instance, a real scalar field ϕ with Lagrangian density $-\partial_\mu \phi \partial^\mu \phi/2 - V(\phi)$ if spatially homogeneous contributes a vacuum energy density and a pressure

$$\rho = \frac{1}{2}\dot{\phi}^2 + V(\phi), \qquad p = \frac{1}{2}\dot{\phi}^2 - V(\phi), \qquad (5)$$

so the condition $\rho + 3p < 0$ for an accelerating expansion is satisfied if the field ϕ is evolving sufficiently slowly so that $\dot{\phi}^2 < V(\phi)$.

It must be said from the outset that, in themselves, quintessence theories do not help with the first problem mentioned above — they do not explain why $V(\phi)$ does not contain an additive constant of the order of $(10^{19}\text{GeV})^4$. It is true that superstring theories naturally lead to "modular" scalar fields ϕ for which $V(\phi)$ does vanish as $\phi \to \infty$, in which case the vacuum becomes supersymmetric. It might be hoped that the vacuum energy is small now, because the scalar field is well on its way toward this limit. The trouble is that the vacuum now is nowhere near supersymmetric, so that in these theories we would expect a present vacuum energy of the order of the fourth power of the supersymmetry-breaking scale, or at least $(1 \text{ TeV})^4$.

On the other hand, such theories may help with the second problem, if the quintessence energy is somehow related to the energy in matter and radiation, because the present moment is not so many e-foldings of cosmic expansion (about 10, in fact) from the turning point in cosmic history when the radiation energy density (including neutrinos) fell below the matter energy density. Paul Steinhardt here described a model in which the quintessence energy density was less than the radiation energy density by a constant factor r, as long as radiation dominated over matter[11]. (It is necessary that r be considerably less than unity, in order that quintessence should not appreciably increase the expansion rate during the era of nucleosynthesis, increasing the present helium abundance above the observed value.) Then when the radiation energy density fell below the matter energy density at a cross-over redshift $z_C \approx 3000$ the quintessence energy dropped sharply by a factor of order r^2, and has remained roughly constant since then. Since the cross-over between radiation and matter dominance the matter energy density has decreased by a factor z_C^{-3}, so the ratio of the quintessence energy density and the matter energy density now should be of order $r^2 \times r \times z_C^3 = (z_C r)^3$. For the quintessence and matter energies to be about equal now, r must be equal to about $1/z_C \approx 3 \times 10^{-4}$. Steinhardt tells me that when these calculations are done carefully, the required ratio r of quintessence to radiation energy density at early times is about 10^{-2}, rather than 3×10^{-4}. But whatever the value of r that makes the quintessence energy comparable to the matter energy density now, it requires some fairly fine tuning: changing r by a factor 10 would change the ratio of the present values of the quintessence energy density and the matter energy density by a factor 10^3.

4) Brane Solutions

Several authors have found solutions of brane theories of the Randall–Sundrum kind[2] in which our four-dimensional spacetime is flat, despite the presence of a large cosmological constant in the higher dimensional gravitational Lagrangian[12]. These solutions contained an unacceptable essential singularity off the brane, but there are models in which this can be avoided[13]. I don't believe that there is anything unique in these solutions, so that instead of having to fine tune parameters in the Lagrangian one has to fine tune initial conditions. Also, it is not clear why the effective cosmological constant has to be zero *now*, rather than before the spontaneous breakdown of the chiral symmetry of quantum chromodynamics, when the latent heat associated with this phase transition would have given the vacuum an energy density $(1 \text{ GeV})^4$.

5) Anthropic Principle

Why is the temperature on earth in the narrow range where water is liquid? One answer is that otherwise we wouldn't be here. This answer makes sense only because there are many planets in the universe, with a wide range of surface temperatures. Because there are so many planets, it is natural that some of them should have liquid water, and of course it is just these planets on which there would be anyone to wonder about the temperature. In the same way, if our big bang is just one of many big bangs, with a wide range of vacuum energies, then it is natural that some of these big bangs should have a vacuum energy in the narrow range where galaxies can form, and of course it is just these big bangs in which there could be astronomers and physicists wondering about the vacuum energy. To be specific, a constant vacuum energy if negative would have to be greater than about $-10^{-120} m_{\text{Planck}}^4$, in order for the universe not to collapse before life has had time to develop[14], and if positive it would have to be less than about $+10^{-118} m_{\text{Planck}}^4$, in order for galaxies to have had a chance to form before the matter energy density fell too far below the vacuum energy density[15]. As far as I know, this is at present the only way of understanding the small value of the vacuum energy. But of course it makes sense only if the big bang in which we live is one of an ensemble of many big bangs with a wide range of values of the cosmological constant. There are various ways that this might be realized:

(a) Wormholes or other quantum gravitational effects may cause the wave function of the universe to break up into different incoherent terms, corresponding to various possible universes with different values for what are usually called the constants of nature, perhaps including the cosmological constant[16].

(b) Various versions of "new" inflation lead to a continual production of big bangs[17], perhaps with different values of the vacuum energy. For instance, if there is a scalar field that takes different initial values in the different big bangs, and if it has a sufficiently flat potential, then its energy appears like a cosmological constant, which takes different values in different big bangs[18].

(c) As the universe evolves the vacuum energy may drop discontinuously to lower and lower discrete values. One way for this to happen is for the vacuum energy to be a function of a scalar field, with many local minima, so that as the universe evolves the vacuum energy keeps dropping discontinuously to lower and lower local minima[19]. Another possibility[20] with similar consequences is based on the introduction of an antisymmetric gauge potential $A_{\mu\nu\lambda}$, which enters in the Lagrangian density in a term proportional to $F^{\mu\nu\lambda\kappa} F_{\mu\nu\lambda\kappa}$, where $F_{\mu\nu\lambda\kappa}$ is $\partial_\kappa A_{\mu\nu\lambda}$ with antisymmetrized indices. Instead of a scalar field tunneling from one minimum of a potential to another, the vacuum energy evolves through the formation of membranes, across which there is a discontinuity in the value of Lorentz-invariant gauge fields $F_{\mu\nu\lambda\kappa} = F \epsilon_{\mu\nu\lambda\kappa}$. To allow an anthropic explanation of the smallness of the vacuum energy, it is essential that the metastable values of the vacuum energy be very close together. Several models of this sort have been proposed recently[21].

Under any of these alternatives, we have not only an upper bound[15] on the vacuum energy density, given by the matter energy density at the time of formation of the earliest galaxies, but also a plausible expectation, which Vilenkin calls the principle of mediocrity[22], that the vacuum energy density found by typical astronomers will be comparable to the mass density at the time when most galaxies condense, since any larger vacuum energy density would reduce the number of galaxies formed, and there is no reason why the vacuum energy density should be much smaller. The observed vacuum energy density is somewhat smaller than this, but not very much smaller. This can be put quantitatively[23]: under the assumption[24] that the *a priori* probability distribution of the vacuum energy is approximately constant within the narrow range within which galaxies can form, the probability that an astronomer in any of the big bangs would find a value of Ω_Λ as small as 0.7 ranges from 5% to 12%, depending on various assumptions about the initial fluctuations. In this calculation the fractional fluctuation in the cosmic mass density at recombination is assumed to take the value observed in our big bang, since the vacuum energy would have a negligible effect on physical processes at and before recombination. There are also interesting calculations along these lines in which the rms value of density fluctuations at recombination is allowed to vary independently of the vacuum energy[25].

IV COSMIC MICROWAVE BACKGROUND ANISOTROPIES

Perhaps the most remarkable improvement in cosmological knowledge over the past decade has been in studies of the cosmic microwave background. Since COBE, there is for the first time a cosmological parameter — the radiation temperature — that is known to three significant figures. More recently, since the BOOMERANG and MAXIMA experiments reviewed here by Paolo de Bernardis, our knowledge of small angular scale anisotropies has become good enough to set useful limits on other cosmological parameters, such as the present spatial curvature.

Unfortunately, this has produced a frustrating situation for those of us who are not specialists in the theory of the cosmic microwave background. We see papers in which experimental results for the strengths C_ℓ of the ℓth multipole in the temperature correlation function are compared with computer generated plots of C_ℓ versus ℓ for various values of the cosmological parameters, without the non-specialist reader being able to understand why the theoretical plots of C_ℓ versus ℓ look the way they do, or why they depend on cosmological parameters the way they do. I want to take the opportunity here to advertise a formalism[26] that I think helps in understanding the main features of the observed anisotropies, and how they depend on various cosmological assumptions.

One can show under very general assumptions that the fractional variation from the mean of the cosmic microwave background temperature observed in a direction $\hat{n}$ takes the form

$$\frac{\Delta T(\hat{n})}{T} = \int d^3k \, \epsilon_{\mathbf{k}} \, e^{id_A \mathbf{k}\cdot\hat{n}} \left[F(k) + i\hat{n}\cdot\hat{k}\, G(k) \right] , \qquad (6)$$

where d_A is the angular diameter distance of the surface of last scattering

$$d_A = \frac{1}{\Omega_C^{1/2} H_0 (1+z_L)} \sinh\left[\Omega_C^{1/2} \int_{\frac{1}{1+z_L}}^{1} \frac{dx}{\sqrt{\Omega_\Lambda x^4 + \Omega_C x^2 + \Omega_M x}}\right], \quad (7)$$

(with $z_L \simeq 1100$ and $\Omega_C \equiv 1 - \Omega_M - \Omega_V$); $k^2 \epsilon_k$ is proportional to the Fourier transform of the fluctuation in the energy density at early times (with $\mathbf{k}$ the physical wave number vector at the nominal moment of last scattering, so that $d_A \mathbf{k}$ in the argument of the exponential is essentially independent of how this moment is defined); and $F(k)$ and $G(k)$ are a pair of form factors that incorporate all relevant information about acoustic oscillations up to the time of last scattering, with $F(k)$ arising from intrinsic temperature fluctuations and the Sachs–Wolfe effect, and $G(k)$ arising from the Doppler effect. Given the form factors, one can find the coefficients C_ℓ for $\ell \gg 1$ by a single integration

$$\ell(\ell+1) C_\ell \to \frac{8\pi^2 \ell^3}{d_A^3} \int_1^\infty d\beta\, \mathcal{P}(\ell\beta/d_A) \left[\frac{\beta F^2(\ell\beta/d_A)}{\sqrt{\beta^2 - 1}} + \frac{\sqrt{\beta^2 - 1}\, G^2(\ell\beta/d_A)}{\beta}\right]. \quad (8)$$

where $\mathcal{P}(k)$ is the power spectral function, defined by

$$\langle \epsilon_\mathbf{k} \epsilon_{\mathbf{k}'} \rangle = \delta^3(\mathbf{k} + \mathbf{k}') \mathcal{P}(k). \quad (9)$$

(The first term in the square brackets in Eq. (8) appeared in a calculation by Bond and Efstathiou[27]; I think the second is new.)

As you can see from the $F^2(k)$ term in Eq. (8), for $\ell \gg 1$ the main contribution to C_ℓ of the Sachs–Wolfe effect and intrinsic temperature fluctuations comes from wave numbers close to d_A/ℓ, but this well-known result is not a good approximation for the Doppler effect form factor $G(k)$. Since it is the form factors rather than C_ℓ that really reflect what was going on before-recombination, it is important to try to measure them more directly, as for instance through interferometric measurements of the temperature correlation function, of the sort described in a parallel session by K. Y. Lo et al. and B. S. Mason et al.

The Harrison–Zel'dovich spectrum suggested by theories of new inflation[28] is $\mathcal{P}(k) = B k^{-3}$, with B a constant. In this case Eq. (8) gives a formula for C_ℓ that is valid for $\ell \gg 1$ and $\ell \ll d_A/d_H$ (where $d_H \ll d_A$ is the horizon distance at the time of last scattering):

$$\ell(\ell+1) C_\ell \to 8\pi^2 B F_0^2 \left\{1 - \frac{\ell^2}{d_A^2} \left[d^2 \left(\ln\left(\frac{\bar{d}\ell}{2d_A}\right) - C\right) - d'^2\right] + \ldots\right\}, \quad (10)$$

where C is the Euler constant $C \equiv -\Gamma'(1) = 0.57722$, and d and d' are a pair of characteristic lengths of order d_H:

$$d^2 \equiv \frac{2 F_0 F_2 + G_1^2}{F_0^2}, \qquad d'^2 \equiv \frac{3 F_0 F_2 + \frac{1}{2} G_1^2}{F_0^2}, \quad (11)$$

expressed in terms of coefficients in a power series expansion of the form factors:

$$F(k) = F_0 + F_2 k^2 + \cdots, \qquad G(k) = G_1 k + G_3 k^3 + \cdots. \qquad (12)$$

(This formula applies even when ℓ is not much larger than unity, except for $\ell = 0$ and $\ell = 1$ [29], provided we replace ℓ^2 with $\ell(\ell+1)$ and $\ln \ell$ with $\sum_{r=1}^{\ell} 1/r + C$.) The quantity $\bar{d}$ in the logarithm is another length of order d_H, this one given by a much more complicated expression involving the form factors at all wave numbers, but since $d_H \ll d_A$ the precise value of $\ln(\bar{d}/2d_A)$ does not depend sensitively on the precise value of $\bar{d}$.

One advantage of this formalism is that it provides a nice separation between the three different kinds of effect that influence the observed temperature fluctuation, that arise in three different eras: the power spectral function $\mathcal{P}(k)$ characterizes the origin of the fluctuations, perhaps in the era of inflation; the form factors $F(k)$ and $G(k)$ characterize acoustic fluctuations up to the time of last scattering; and the angular diameter distance d_A depends on the propagation of light since then. This allows us to see easily what depends on what parameters. The form factors $F(k)$ and $G(k)$ depend strongly on $\Omega_B h^2$ (through the effect of baryons on the sound speed) and more weakly on $\Omega_M h^2$ (through the effect of radiation on the expansion rate before the time of last scattering), but since the curvature and vacuum energy were negligible at and before last scattering, $F(k)$ and $G(k)$ are essentially independent of the present curvature and of Ω_Λ. The power spectral function $\mathcal{P}(k)$ is expected to be independent of all these parameters. On the other hand, d_A is affected by whatever governed the paths of light rays since the time of last scattering, so it depends strongly on Ω_M, Ω_Λ, and the spatial curvature, but it is essentially independent of Ω_B. In quintessence theories d_A would be given by a formula different from (7), but $\mathcal{P}(k)$ and the form factors would be essentially unchanged as long as the quintessence energy density was a small part of the total energy density at and before the time of last scattering. In particular, Eq. (8) shows that $\ell(\ell+1)C_\ell$ for $\ell \gg 1$ depends on ℓ and d_A only through the ratio ℓ/d_A, so changes in Ω_Λ or the introduction of quintessence would lead to a re-scaling of all the ℓ-values of the peaks in the plots of $\ell(\ell+1)C_\ell$ versus ℓ, but would have little effect on their height.

Another advantage of this formalism is that, although C_ℓ must be calculated by a numerical integration, it is possible to give approximate analytic expressions for the form factors in terms of elementary functions, at least in the approximation that the dark matter dominates the gravitational field for a significant length of time before last scattering. (There have been numerous earlier analytic calculations of the temperature fluctuations[30], and their results may all be put in the form (6), but my point here is that this form is general, not depending on the particular approximations used.) In this approximation the form factors for very small wave numbers are

$$F(k) \to 1 - 3k^2 t_L^2/2 - 3[-\xi^{-1} + \xi^2 \ln(1+\xi)]k^4 t_L^4/4 + \ldots, \qquad (13)$$
$$G(k) \to 3k t_L - 3k^3 t_L^3/2(1+\xi) + \cdots, \qquad (14)$$

while for wave numbers large enough to allow the use of the WKB approximation

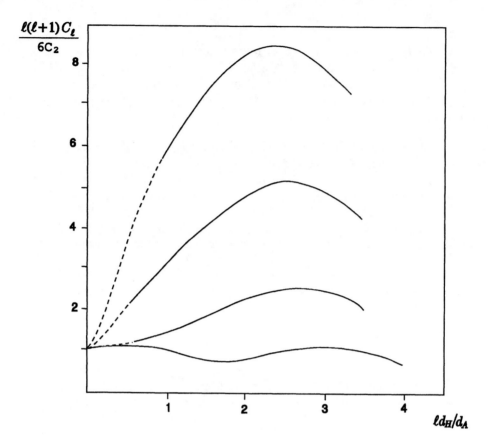

FIGURE 1. Plots of the ratio of the multipole strength parameter $\ell(\ell+1)C_\ell$ to its value at small ℓ, versus $\ell d_H/d_A$, where d_H is the horizon size at the time of last scattering and d_A is the angular diameter distance of the surface of last scattering. The curves are for $\Omega_B h^2$ ranging (from top to bottom) over the values 0.03, 0.02, 0.01, and 0, corresponding to ξ taking the values 0.81, 0.54, 0.27, and 0. The solid curves are calculated using the WKB approximation; dashed lines indicate an extrapolation to the known value at small $\ell d_H/d_A$.

the form factors are

$$F(k) = (1 + 2\xi/k^2 t_L^2)^{-1}\left[-3\xi + 2\xi/k^2 t_L^2 + (1+\xi)^{-1/4} e^{-k^2 d_\Delta^2} \cos(kd_H)\right], \quad (15)$$

and

$$G(k) = \sqrt{3}\,(1+2\xi/k^2 t_L^2)^{-1}(1+\xi)^{-3/4} e^{-k^2 d_\Delta^2} \sin(kd_H). \quad (16)$$

Here t_L is the time of last scattering; $\xi = 27\Omega_B h^2$ is 3/4 the ratio of the baryon to photon energy densities at this time; d_H is the acoustic horizon size at this time; and d_Δ is a damping length, typically less than d_H. Using these results in Eq. (8) gives the curves for $\ell(\ell+1)C_\ell/6C_2$ versus $\ell d_H/d_A$ shown in Figure 1, in the approximation that damping and the term $2\xi/k^2 t_L^2$ may be neglected near the peak. In this approximation the scalar form factor $F(k)$ has a peak at $k_1 = \pi/d_H$ for any value of $\Omega_B h^2$, but the peak in $\ell(\ell+1)C_\ell$ does *not* appear (as is often said) at $\ell = k_1 d_A = \pi d_A/d_H$; instead, $\ell d_H/d_A$ at the peak ranges from 3.0 to 2.6, depending on the value of $\Omega_B h^2$.

We see even from these crude calculations how sensitive is the height of the first peak in $\ell(\ell+1)C_\ell/6C_2$ to the baryon density parameter $\Omega_B h^2$. (The experimental value[31] for the height of this peak is about 6.) Right now, there is some worry about the fact that the value of $\Omega_B h^2$ inferred from the ratio of the heights of the second and first peaks is larger than that inferred from considerations of cosmological nucleosynthesis. Perhaps it would be worth trying to estimate $\Omega_B h^2$ by comparing theory and experiment for the ratio of $\ell(\ell+1)C_\ell$ at the first peak to its value for small ℓ, discarding the data at the second peak where the statistics are worse and complicated damping effects make the theory more complicated.[1]

ACKNOWLEDGMENTS

I am grateful to Willy Fischler, Hugo Martel, Paul Shapiro, and Craig Wheeler for their help in preparing this report. This research was supported in part by a grant from the Welch Foundation and by National Science Foundation Grants PHY 9511632 and PHY 0071512.

REFERENCES

1. For a review, see J. Polchinski, in *Fields, Strings, and Duality – TASI 1996*, eds. C. Efthimiou and B. Greene (World Scientific, Singapore, 1996): 293.

2. This was first discussed in the context of string theory by I. Antoniadis, *Phys. Lett.* **B246**, 377 (1990); I. Antoniadis, C. Muñoz, and M. Quirós, *Nucl. Phys.* **B397** 515 (1993); I. Antoniadis, K. Benakli, and M. Quirós, *Phys. Lett.* **B331**, 313 (1994); J. Lykken, *Phys. Rev.* **D54**, 3693 (1996); E. Witten, *Nucl. Phys.* **B471**, 135 (1996); and then developed in more general terms by N.

---

[1] At the meeting someone in the audience said that this has been done, but that was in the early days, not I think with the more detailed information now available.

Arkani-Hamed, S. Dimopoulos, and G. Dvali, *Phys. Lett.* **B 429**, 263 (1998); I. Antoniadis, N. Arkani-Hamed, S. Dimopoulos, and G. Dvali, *Phys. Lett.* **B 436**, 257 (1998). A different approach has been pursued by L. Randall and R. Sundrum, *Phys. Rev. Letters* **83**, 3370 (1999).

3. N. Arkani-Hamed, S. Dimopoulos, and G. Dvali, *Phys. Rev.* **59**, 086004 (1999); S. Hannestad and G. G. Raffelt, astro-ph/0103201.

4. S. Hannestad, astro-ph/0102290.

5. H. Georgi, H. Quinn, and S. Weinberg, *Phys. Rev. Lett.* **33**, 451 (1974).

6. S Dimopoulos and H. Georgi, *Nucl. Phys.* **B193**, 150 (1981); J. Ellis, S. Kelley, and D. V. Nanopoulos, *Phys. Lett.* **B260**, 131 (1991); U. Amaldi, W. de Boer, and H. Furstmann, *Phys. Lett.* **B260**, 447 (1991); C. Giunti, C. W. Kim and U. W. Lee, *Mod. Phys. Lett.* **16**, 1745 (1991); P. Langacker and M.-X. Luo, *Phys. Rev.* **D44**, 817 (1991). For other references and more recent analyses of the data, see P. Langacker and N. Polonsky, *Phys. Rev.* **D47**, 4028 (1993); **D49**, 1454 (1994); L. J. Hall and U. Sarid, *Phys. Rev. Lett.* **70**, 2673 (1993).

7. S. Dimopoulos, S. Raby, and F. Wilczek, *Phys. Rev.* **D24**, 1681 (1981).

8. K. R. Dienes, E. Dudas, and T. Ghergetta, hep-ph/9806292, 9807522.

9. For recent detailed reviews, see S. Weinberg, in *Sources and Detection of Dark Matter and Dark Energy in the Universe — Fourth International Symposium*, D. B. Cline, ed. (Springer, Berlin, 2001), p. 18; E. Witten, *ibid.*, p. 27; and J. Garriga and A. Vilenkin, hep-th/0011262.

10. K. Freese, F. C. Adams, J. A. Frieman, and E. Mottola, *Nucl. Phys.* **B287**, 797 (1987); P. J. E. Peebles and B. Ratra, *Astrophys. J.* **325**, L17 (1988); B. Ratra and P. J. E. Peebles, *Phys. Rev.* **D 37**, 3406 (1988); C. Wetterich, *Nucl. Phys.* **B302**, 668 (1988).

11. C. Armendariz-Picon, V. Mukhanov, and P. J. Steinhardt, astro-ph/0004134.

12. N. Arkani-Hamed, S. Dimopoulos, N. Kaloper, and R. Sundrum, *Phys. Lett.* **B 480**, 193 (2000); S. Kachru, M. Schulz, and E. Silverstein, *Phys. Rev.* **D62**, 045021 (2000).

13. J. E. Kim, B. Kyae, and H. M. Lee, hep-th/0011118.

14. J. D. Barrow and F. J. Tipler, *The Anthropic Cosmological Principle* (Clarendon Press, Oxford, 1986).

15. S. Weinberg, *Phys. Rev. Lett.* **59**, 2607 (1987).

16. E. Baum, Phys. Lett. **B133**, 185 (1984); S. W. Hawking, in *Shelter Island II – Proceedings of the 1983 Shelter Island Conference on Quantum Field Theory and the Fundamental Problems of Physics*, ed. by R. Jackiw et al. (MIT Press, Cambridge, 1985); *Phys. Lett.* **B134**, 403 (1984); S. Coleman, *Nucl. Phys.* **B 307**, 867 (1988).

17. A. Vilenkin, *Phys. Rev.* **D 27**, 2848 (1983); A. D. Linde, *Phys. Lett.* **B175**, 395 (1986).

18. J. Garriga and A. Vilenkin, astro-ph/9908115.

19. L. Abbott, *Phys. Lett.* **B195**, 177 (1987).

20. J. D. Brown and C. Teitelboim, *Nucl. Phys.* **279**, 787 (1988).

21. R. Buosso and J. Polchinski, JHEP 0006:006 (2000); J. L. Feng, J. March-Russel, S. Sethi, and F. Wilczek, hep-th/0005276.

22. A. Vilenkin: Phys. Rev. Lett. **74**, 846 (1995); in *Cosmological Constant and the Evolution of the Universe*, ed. by K. Sato et al. (Universal Academy Press, Tokyo, 1996).

23. H. Martel, P. Shapiro, and S. Weinberg, *Ap. J.* **492**, 29 (1998).

24. S. Weinberg, in *Critical Dialogs in Cosmology*, ed. by N. Turok (World Scientific, Singapore, 1997). Counterexamples in theories of type (b) are pointed out in reference [18], and the issue is further discussed in reference [9].

25. G. Efstathiou, *Mon. Not. Roy. Astron. Soc.* **274**, L73 (1995); M. Tegmark and M. J. Rees, *Astrophys. J.* **499**, 526 (1998), J. Garriga, M. Livio, and A. Vilenkin, *Phys. Rev.* **D61**. 023503 (2000); S. Bludman, *Nucl. Phys.* **A663-664**, 865 (2000).

26. S. Weinberg, astro-ph/0103279 and 0103281.

27. J. R. Bond and G. Efstathiou, Mon. Not. R. Astr. Soc. **226**, 655 (1987), Eq. (4.19).

28. S. Hawking, Phys. Lett. **115B**, 295 (1982); A. A. Starobinsky, Phys. Lett. 117B, 175 (1982); A. Guth and S.-Y. Pi, Phys. Rev. Lett. **49**, 1110 (1982); J. M. Bardeen, P. J. Steinhardt, and M. S. Turner, Phys. Rev. **D28**, 679 (1983); W. Fischler, B. Ratra, and L. Susskind, Nucl. Phys. **B259**, 730 (1985).

29. In Eq. (6) terms are neglected that only affect C_0 and C_1; for these terms, see A. Dimitropoulos and L.P. Grishchuk, gr-qc/0010087.

30. P. J. E. Peebles and J. T. Yu, Ap. J. **162**, 815 (1970); J. R. Bond and G. Efstathiou, Ap. J. Lett. **285**, L45 (1984); Mon. Not. Roy. Astron. Soc. **226**, 655 (1987); C-P. Ma and E. Bertschinger, Ap. J. **455**, 7 (1995); W. Hu and N. Sugiyama, Ap. J. **444**, 489 (1995); **471**, 542 (1996).

31. A. H. Jaffe et al., astro-ph/0007333.

APPENDICES

A. After Dinner Remarks

B. Reprint of the Report of the
   First Texas Symposium in Austin

C. The Formation of a Black Hole

Texas Symposium 20

Winnie Schild

Austin, Texas

In the beginning there was Alfred Schild. I am his widow, and tonight, I bear witness to the meaning and accomplishments of his life, which ended at the early age of 55. He did a lot in those 55 years and affected the lives of many people.

In 1957, Alfred left the corporate world of Westinghouse Research in Pittsburg, Pennsylvania, to take up a professorial position in the Math Department of the University of Texas, in Austin. With a great sigh of relief, he turned to his native habitat, the bracing air of a university campus. Our three kids and I tagged along, happy to see him happy after the years he had spent in exile.

There was one cloud in the otherwise blue Texan sky. There was no one in Austin with whom he could have discussions about research ideas in the Relativity Field, neither colleague nor student. So he set about remedying this emptiness by getting the Center for Relativity Theory started in the Math Department.

People who spoke Relativity began to arrive in this quiet, a bit off-the-beaten-trail, but very attractive city. Alfred succeeded in wooing them here by dangling jobs and opportunity to do research in their passion. Austin quickly became a mecca for Relativists. Within two years it was a success, and The Center became known around the world, in its own special world of Relativists who had assumed the mantel of Albert Einstein in order to further his discoveries.

Amongst those who arrived were Engelbert Schucking in Austin, and Ivor Robinson in Dallas, where Alfred found a position for him, thus widening the area being populated by the clan. One day, Schild, Schucking and Robinson decided that there needed to be a forum for research papers on the rapidly expanding and exciting field of Relativistic Astrophysics.

The Texas Symposium for Relativistic Astrophysics was born, and saw light-of-day first in Dallas, in 1963, in the shadow of the assassination of President Kennedy. The second gathering was in Austin, the following year, to play catch-up, I think, on the backlog of papers which had accumulated due to the difficulty of having a place to present them. Subsequent to that, they have been held at two-year intervals in cities around the world, as well as in Texas.

We come now to the 20th Session, in the year 2000. Appropriately, it is here, in Austin, where it all started with Alfred Schild, the lonely Relativist, planting the Relativity Tree. I take great pride in being a Founding Mother of the two institutions born in Austin, again appropriately so, because I took part in the original beginnings. Since there was no one else back then to listen to the original ideas, I was the sounding board.

As mother of you all, then, I welcome you to your explorations of our Universe in the new millenium. Long may you explore!!

20th Texas Symposium

E. L. Schucking

Department of Physics, New York University

In the summer of 1963, I visited Ivor Robinson at Lloyd Berkner's Southwest Center for Advanced Studies. One weekend my colleague, Alfred Schild, from Austin came over and we were having some of Alfred's strong Martinis while lounging in the Texas heat round a swimming pool. Larry Marshall at the Center suggested we might organize some symposium to break the boredom.

Now, what is a symposium? A symposium is a big party to which one can invite all one's friends with their airfare and liquor paid by someone else. But to have a symposium one needs a topic. I was familiar with the recent discovery of the quasars and there was the suspicion that their incredible energies had something to do with Einstein's gravity. So we invented boldly the words *relativistic astrophysics* for a non-existing topic to secure funding from government agencies. Al Schild, the Father of Texas relativity, got money from UT Austin and Ivor from the Southwest Center, especially valuable, because, unlike Austin money, it could be used for spirits.

With advice from Harlan Smith, the astronomy chair from Austin, we organized the first conference of quasars that brought together the abstract relativists with physicists, astrophysicists and quasar observers. The time of the conference was fixed around the time of the full moon when optical observers are more sociable. But we also made mistakes; we paid Hoyle an honorarium, but not the Burbidges – and they found out. Anyhow, instead of the originally planned party for twenty we had a blast for two hundred.

The next summer, Ivor visited Austin, and we decided it would be nice to do another party in December. The problem was that there wasn't that much new about quasars to fill a week. So we expanded the definition of relativistic astrophysics to include cosmic rays, cosmic neutrinos, X-ray and gamma-ray astronomy and even CP violation.

When Alfred and I traveled to the East coast to sell the idea of this new stew we found Phil Morrison at MIT already gungho for it. Not so in the West. At Caltech Ivor's friend, Yuval Ne'eman, threw a party for Alfred and me to which he invited other physicists, including Feynman. Feynman was dead set against talks about cosmic neutrinos because none had been seen yet and one should not have talks about topics without experimental results.

I reminded him of his researches on Compton scattering of gravitons with cross sections of $10^{-\text{umpteen}}$ barns and all the diagrams in quantum gravity he had calculated. Did he alone have the privilege to discuss the unobserved? He had been to the Warsaw conference on relativity and become allergic to relativists. He told me at that time – over drinks – that relativists at that conference reminded him of worms in a bottle, crawling over each other. Here in Austin, we are at least in a pretty big bottle.

Alfred invited Dirac for the symposium and took him swimming at Barton Springs. Murray Gell-Mann and Yuval Ne'eman came from Caltech, Shelley Glashow, Ray Davis, Ken Greisen, Kuzmin and Zatsepin, Bruno Rossi, Herb Friedman, Phil Morrison and all the usual suspects came to Austin. Ivor and Alfred worked hard to get Ginzburg, Schlovski and the expatriate, Bohm.

In the quasar session names for the things astronomers insisted on calling quasi-stellar objects were proposed and voted on. Ivor remembers that Hong Yee Chiu's quasar won the day with eleven to four and a Florida size undervote of more than a hundred.

The locally supported proposal to call them the *Dallas* or *Texas* stars – perhaps *Texars* would have been better – was defeated. In hindsight, the name would seem appropriate because it was in Dallas where they were first discussed and it was here in Austin that the rotating black hole, the heart of the quasar, was envisioned by Roy Kerr. To be precise, the place where this happened was the basement of the Business Economics Office Building on campus, the refuge of Alfred's relativity group. The mathematician, R. L. Moore, had threatened to machine gun the foreign relativists that Alfred had brought into the Math department if they were given offices occupied by Moore's graduate students.

Alfred suggested that Roy Kerr should reveal his discovery at the first Texas Symposium. However, none of the three summarizers took notice of it then.

Looking back thirty-six years to the second Texas Symposium, I would like to remind you of the debt we owe to Harlan Smith and Alfred Schild. Both were visionaries. Harlan had discovered the light fluctuations in 3C273 just before he went to Austin to take charge of the astronomy department, making it into one of the finest in the country. Alfred was one of the world's leading relativists. He brought Roy Kerr to Austin, Roger Penrose, Jurgen Ehlers, Ray Sachs, George Ellis, Luis Bell to begin with, and made Austin into one of the world's foremost centers for research in relativity.

Supported by his charming wife, Winnie, the home of the Schild's became the real center of relativity, where their warmth and understanding made one feel belonging to a happy family.

Both Ivor and I were brought to Texas by Alfred and we both cherish the memory of a friend, a great scientist and inventor of relativistic astrophysics.

We have to be grateful to Craig Wheeler of supernova fame and author of a beautiful book on relativistic astrophysics to have brought home a symposium that has recently travelled to England, Germany, California and France to the great state of Texas, turning it into a five-star symposium.

Relativistic Astrophysics

A report on the Second Texas Symposium

By I. Robinson, A. Schild, and E. L. Schucking

Scientists of 1965 see the universe with diverse eyes. They look with the two-hundred-inch pyrex mirror on top of Mount Palomar, with a hundred thousand gallons of cleaning fluid buried more than a mile underground, with scintillation counters flying in rockets and satellites, with a retina covering several square miles of the New Mexico desert at Volcano Ranch, with a steel bowl two hundred and ten feet across at Parkes in Australia, and with a gently swinging aluminum bar in Maryland, waiting patiently for the tremor of a gravitational wave.

Some of these eyes perceive only darkness, some have blurred and distorted vision, others detect a profusion of fine detail. Scientists have mapped the sky with meticulous care as they see it by radio waves, light, x rays, γ rays and cosmic radiation. There are some striking differences among these charts.

The finest picture is that available in the optical region. By comparison, the universe as seen in the radio spectrum is rather bare. A radio astronomer confined to strictly professional sources of information would not think of distinguishing between night and day. He discovered the sun by chance as late as 1943. He would have no reason to regard the quiet sun as an energy source of importance on the local scene; it is very faint besides Cygnus A which is 10^{13} times farther away. In a region where Palomar plates show millions of stars and thousands of galaxies, only one subject, insignificant on the photograph, may appear on the corresponding radio, x-ray, or γ-ray map.

It is only within recent years that the optical and radio maps of the skies have begun to merge. This "two-color" picture of radio and light reveals a new universe. It is not the smooth, quiet, placid world which the "black and white" map of conventional optical astronomy had shown. As staining revealed the mysteries of cell division and genes under the microscope and showed the dynamic nature of life, so the two-color maps of the sky, pinpointing hot spots of great energy events, are initiating a revolution in our view of the cosmos. Majestic galaxies, giving the impression of dull harmony, are now believed to evolve in a cataract of explosions that involve millions of solar masses.

Supernovae were the first indication that violent events play an important role in the life and death of a star. Now we see that such cataclysms occur on a galactic scale, beside which supernovae look like innocent firecrackers. High-energy particles, so difficult to obtain in laboratories on earth, are commonplace in the skies, and illuminate it with synchrotron light. Relativity theory, until recently believed to be unimportant for astronomy, provides the basic laws that govern these great events.

The new observations may give more than a new picture of galactic and cosmological evolution. The merging of the radio and light maps of the sky led to the discovery of the quasi-stellar and strong radio sources, whose tremendous energy output may well present a problem in basic physics.

At present, the other maps seem to be painfully unrelated to the optical picture. When they become sharper and more precise, when they begin to show common features, we can expect the new "many-color" view of our universe to lead to new revolutions in astronomy and in physics.

This will only be achieved by the concerted effort of scientists from many different fields. Optical astronomers will have to join forces with elementary-particle physicists, radio astronomers will have to talk to cosmic-ray experts. The Texas Symposia on Relativistic Astrophysics were con-

---

The authors of this report are I. Robinson of the Southwest Center for Advanced Studies, Dallas, and A. Schild and E. L. Schucking of The University of Texas, Austin. The three authors and J. A. Wheeler of Princeton University were the organizers of the symposium.

ceived with the deliberate purpose of overcoming the traditional segregation of specialists in their narrow cubbyholes.

The first symposium was held in Dallas in December 1963.* It dealt mainly with optical and radio observations of quasi-stellar sources and with theories which offered promise of explaining the enormous energy releases.

The second symposium was held at the University of Texas in Austin, from December 15 to 19, 1964.** It continued the discussion of quasi-stellar sources and reported on the progress made in the preceding year. Half of the symposium was devoted to the new maps of the sky, to x-ray, γ-ray, and cosmic-ray astronomy, to the search for cosmic neutrinos and to the possible large-scale implications of the breakdown of CP-invariance.

The opening session was chaired by P.A.M. Dirac of Yeshiva and Cambridge Universities. Participants were welcomed by C. H. Green of the Graduate Research Center of the Southwest, W. W. Heath of The University of Texas, John Connally, governor of the State of Texas, and E. L. Schucking, on behalf of the organizing committee.

Geoffrey Burbidge from the University of California, San Diego, was the first speaker. He reported that the strongest radio objects may be the result of events involving energies of 10^{64} erg. Such an energy, equivalent to the mass of five billion suns, is about a thousand times larger than the energy proposed a year earlier at the Dallas Symposium. Astrophysicists had then assumed that the fast-moving atomic particles in these sources were accelerated by a mechanism of almost 100 percent efficiency. Burbidge, assuming that the Lord was no better an electrical engineer than the terrestrial builders of the biggest nuclear accelerators at Brookhaven, CERN, and Berkeley, came up with an efficiency factor of only 0.03 percent for the cosmic machines.

---

* The proceedings of the first Texas symposium have been published under the titles *Quasi-Stellar Sources and Gravitational Collapse*, $10.00, and *Gravitation Theory and Gravitational Collapse*, $6.50, both by the University of Chicago Press, Chicago and London.

** Over four hundred scientists from all over the world attended the Austin symposium. It was sponsored jointly by the University of Texas and the Southwest Center for Advanced Studies, with support of the Aerospace Research Laboratories (Wright Patterson Air Force Base), Air Force Office of Scientific Research, Atomic Energy Commission, National Aeronautics and Space Administration, National Science Foundation, Office of Naval Research, American Astronomical Society, and the American Physical Society.

Not everybody in the audience agreed. It was held against the argument that electrical engineers on earth are limited by the conductivity of copper, whereas the Lord, in the vast vacuum of space, might have other means at His command. Burbidge thought that the only way of reducing the energy estimates was to bring these radio sources implausibly near to us. In conclusion, he recalled the difficulties that had beset astronomy more than thirty years ago when nuclear energy, the energy source of the stars, had yet to be discovered. He suggested that our difficulty in understanding strong radio sources might point to a gap in our knowledge of basic physics.

John Bolton of CSIRO, Sydney, one of the founders of modern radio astronomy, an art that was developed largely at the antipodes, also discussed in detail the properties of the strong radio sources. With his beautiful new instrument, the 210-foot dish at Parkes near Sydney, he found evidence that not all of the volume of the huge clouds is evenly filled with high-energy particles. He said that his observations were more indicative of shell-like structures with bright spots. This would indicate that the energy, calculated by assuming a uniform distribution of particles over the emitting volume, had been overestimated. At this stage, the chairman of the meeting, Murray Gell-Mann of the California Institute of Technology, took the microphone and asked: "Would Dr. Burbidge please defend his volume?"

He did.

Among the beautiful new results presented by Bolton were measurements of the direction of polarization of the radiation from strong sources. Many of them look like dumbbells or hourglasses hundreds of thousands of light years long, leaky magnetic bottles containing high-energy cosmic rays. Bolton's observations show that in many cases the direction of the magnetic field is perpendicular to the long axis of the dumbbell with an accuracy of one percent.

Thomas Matthews, of Caltech's Owens Valley Radio Astronomy Observatory, had shown earlier in the day that these dumbbells exhibit bright edges at both ends. In these regions, it is believed, the gas smashes its way into the ambient medium.

Since astronomers have available for interpretation only static pictures of these cosmic explosions, it is often difficult to say how evolution is occurring, and at what stage it is being observed. Frequently they can only guess. Radio astronomers had believed that the biggest objects were the older generation in the radio community, while the quasi-stellar radio sources, smaller in size, were the new born. Bolton told the audience that he thought evolution might go the other way: the strongly energetic radio sources were not on the verge of radio death, but in fact were infants which later collapsed into the less energetic smaller objects, including quasistellars. "I am very happy that somebody is following this line of approach," commented Fred Hoyle amiably, "because I think it is wrong."

Although they are not the most spectacular radio transmitters, the quasi-stellar sources are objects of enormous and bewildering brilliance, brighter than a million million suns. Maarten Schmidt of the Mount Wilson and Palomar Observatories, who had identified the first of them in 1962, reviewed carefully the evidence that these were indeed the most distant objects in the universe, and were not being confused with nearby large celestial bodies. Disposing first of the possibility that the red shift is gravitational and that these objects are as heavy as the sun, smaller than Austin, and at a distance of fourteen kilometers, Dr. Schmidt led his audience step by step to the conclusion that they are at least several million light years distant, and probably much further away. His model of a quasi-stellar source is a shell-like structure, with an outer cloud, huge but very thin, filled with atomic particles of high energy which emit radio synchrotron radiation. It may be several thousand light years across, much smaller than an ordinary galaxy. Within this radio cloud lies a thin shell of glowing rarefied gas with a diameter of some ten light years. This region emits light predominantly at a few fixed frequencies. The extreme brilliance of the quasi-stellar comes from an inner, much smaller core, a hot superstar of perhaps a hundred million solar masses with a temperature in excess of 10 000°C. Its diameter may be as small as one light year.

Allan Sandage of Mt. Wilson and Palomar reported that, in the few weeks preceding the Austin Conference, he and his colleague, P. Veron, had found fifteen new quasi-stellar sources in a systematic search with the 48-inch Palomar Schmidt telescope. This brought the total count to 34, leaving some others whose identification is not yet certain. Sandage described how he found these objects by taking survey photographs of the sky, first with a violet, and then with an ultraviolet filter. After the first filter has been removed, the photographic plates are shifted so that the stellar image seen through the second filter is slightly displaced. This makes it possible to spot these objects which are more brilliant in ultraviolet light than in violet. There may be more than 80 000 of these "interlopers," as Sandage called them, which are possible candidates for identification with quasi-stellar radio sources. One of them has the same position as the second entry in the third Cambridge catalogue of radio sources, known for short as 3C2. This "star" of 20th magnitude is very probably located in the depths of space-time, much further out than anything seen before.* Its light, which reaches us now, may have been emitted before the solar system was formed (5×10^9 B.C.). It is running away from us at something very close to the speed of light, but the precise velocity

---

* Maarten Schmidt has just measured the red shift of 3C9. It corresponds to a special relativistic recession velocity of four-fifths that of light (May 1965). See Table 1.

Table 1.* Properties of quasi-stellar sources with known redshift. The fluxes are computed under the assumption that the cosmos is a Friedman universe. The cosmological constant is assumed to be zero, the space curvature positive, the present value of the Sandage parameter $q_o = -\ddot{R}R/\dot{R}^2 = 1$, and the Hubble constant 100 km/sec/Mpc.

Source	Red shift $z = \dfrac{\lambda' - \lambda}{\lambda}$	log of intrinsic radio flux emitted at 10^9 Hz in $W(Hz)^{-1}$	log of intrinsic optical flux emitted at 10^{15} Hz in $W(Hz)^{-1}$
3C 273	0.158	26.8	23.8
3C 48	0.367	27.5	23.0
3C 47	0.425	27.0	22.5
3C 147	0.545	27.9	23.1
3C 254	0.734	27.4	22.9
3C 245	1.029	27.5	23.5
CTA 102	1.037	27.6	23.5
3C 287	1.055	27.8	23.4
3C 9	2.012	28.1	23.6

* M. Schmidt, Astrophys. Journ. 141, 1299 (1965).

The transportable 25-ft dish in Yorkshire. With Mark I at Jodrell Bank, it forms an interferometer of maximum baseline 180 000λ. (H. P. Palmer)

has not yet been measured. Sandage reported a four-fold change in brightness for 3C2 over the last two years. This is many times more than the fluctuations previously observed in quasi-stellar radio sources.

Henry Palmer of the Jodrell Bank Radio Observatory near Manchester started the hunt for quasi-stellar sources nearly a decade ago. At the symposium in Austin he reported on the observations he had made with the world's largest scientific instrument, a radio interferometer a hundred and thirty-two kilometers long. Palmer revealed that some of the quasi-stellar sources have an apparent radio diameter of only a fraction of a second of arc. He predicted that the quasi-stellar source CTA 102 may have a radio diameter as small as one hundredth of a second of arc. This source was recently reported by Russian radio astronomers to have variable radio brightness, which led some of them to suggest that it was a broadcasting station operated by extraterrestrial intelligence.

J. E. Baldwin of Cambridge University showed that the quasi-stellars are distinguished from other radio sources by their radio spectrum. W. W. Morgan, of the Yerkes Observatory, reviewed the properties of D-galaxies, supergiant systems which are predominant among the radio sources identified with optical objects.

The next sessions of the symposium were devoted to what may be called exotic astrophysics, the exploration of the skies through the observation of relativistic particles and high-energy radiation. The speakers were mostly physicists. This accounted for a quantum jump in terminology: the act of detecting the sun's visible radiation is an observation; that of observing its neutrino flux is an experiment. The first of these sessions was chaired by A. E. Chudakov of the Soviet Academy of Sciences. He presented some recent work by V. L. Ginzburg, who again was unable to attend the conference.

Bruno Rossi of the Massachusetts Institute of Technology reported on cosmic rays of the highest energy. At Volcano Ranch in New Mexico, the MIT group recorded about a dozen large air showers, each caused by a primary proton with energy larger than 10^{19} electron volts. Data from Mount Chacaltaya in Bolivia indicate that the cosmic-ray spectrum has a kink at about 10^{17} electron volts. Particles below this energy, Rossi suggested, come from our own galaxy, those above from outside. Fred Hoyle, on the other hand, believes that all the high-energy cosmic rays are emitted by strong radio sources.

Ken McCracken, of the Southwest Center for Advanced Studies, discussed the local cosmic-ray background. He said: "In the same way that an astronomer may be hindered in his work by peculiarities of his local environment, in that atmospheric dust, and man-made lights may limit the scope of his work, so is the cosmic-ray physicist limited in his ability to measure the properties

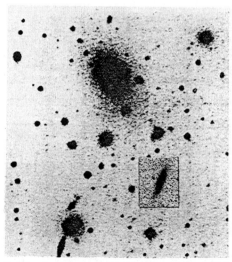

Central region of galaxy cluster Abell 2199. Supergiant is NGC 6166 = 3C 338. Insert is Andromeda Nebula M 31 reduced to linear scale of cluster. NGC 6166 is a prototype supergiant. Main body of M 31 as shown is about 24 000 pc. Cluster photo from National Geographic Society—Palomar Observatory Sky Survey; M 31 photo from 6-inch refractor plate by E. E. Barnard. (W. W. Morgan)

Cosmic x-ray experiment payload flown in October 1964 on an Aerobee rocket. (Giacconi, Gursky, Waters, Rossi, Clark, Garmire, Oda, and Wada)

of the cosmic radiation in the galaxy by 'bad seeing'—in this case, the reason for the bad seeing being the magnetic fields which pervade the solar system." He reviewed in detail the solar cosmic-ray component.

Peter Meyer of Chicago University, one of the discoverers of primary cosmic-ray electrons, discussed their origin. These electrons, together with protons, could be injected into the galaxy by supernova remnants. They may also arise, together with a larger number of positrons, from the collision of cosmic-ray protons with interstellar hydrogen. Meyer's observations on the electron-positron ratio in primary cosmic rays point to a supernoval origin.

George Clark of MIT reported on recent observations of cosmic γ rays, photons ranging in energy from 10^4 to more than 10^{15} electron volts. In a balloon experiment of July 1964, he observed a peak in the counting rates when the Crab Nebula was within the field of view of his scintillation detector, which was sensitive in the range 15 to 62 keV. He stated that the Crab Nebula, the only known source of x rays in this region of the sky, was also likely to be the source of the higher-energy radiation. This seems to be the first observation of a cosmic γ-ray source.

Philip Morrison of MIT gave a masterly review of the whole field of cosmic γ rays. He stressed the need for further observations in order to distinguish the relative importance among different production mechanisms. He pointed out that the isotropic component of the γ rays may originate from the collision of starlight with relativistic electrons. Through this inverse Compton effect, soft photons are converted into hard γ rays.

R. Giacconi, of American Science and Engineering, reported on x-ray observations with detectors flown in a rocket in October 1964. The x rays were in the range 0.5 to 15 Å. He and his coworkers resolved two new sources near the galactic equator. Each of these point sources has intensity less than 1/10 Scorpio (one Scorpio being the new intensity unit of the x-ray astronomer). It is the flux of the brightest x-ray source, 10^{-7} erg cm^{-2} sec^{-1} between 2 and 8 Å. This flux, in the visible range, would be that of a sixth magnitude star.

Herbert Friedman of the Naval Research Laboratory, with the energetic assistance of his chairman Hayakawa, wrote on the blackboard the locations of ten discrete x-ray sources in the sky, giving the latest results from observations made with Geiger counters aboard unstabilized Aerobee rockets. Astronomers in the audience took down the coordinates, and prepared to search for peculiar objects on their photographic plates. Friedman's observations showed conclusively that the x rays from the Crab Nebula in the constellation Taurus (Tau XR-1) did not have a point source. This disposed of the theory that they came from a neutron star. The x rays could be the high-energy tail of the synchrotron-radiation spectrum arising from the inner parts of the Crab cloud. The Scorpio source (Sco XR-1) has not been definitely identified with a known optical object. I. S. Shklovsky, who also was again unable to attend the conference, had suggested that it might be the remnant of a supernova which exploded about 50 000 years ago in our vicinity, some 150 light years away. Prehistoric astrologers must have greeted the sudden appearance of this object, as bright as the full moon, with something of the respect and bewilderment which the quasi-stellar objects have evoked in our own generation. Friedman concluded: "All of the x-ray sources observed lie rather close to the galactic plane and within plus and minus 90° of the galactic center. This distribution resembles that of galactic novae and suggests that all of the x-ray sources thus far observed may be associated with supernova remnants in our galaxy."

G. Wataghin of Turin University gave the results of new calculations on URCA processes and on equilibrium problems which include neutrinos at high temperatures and densities.

The neutrino session continued with a report by R. D. Davis of Brookhaven National Laboratory. He discussed the chances of catching neutrinos from the sun. The main part of his equipment will be a tank 20 ft in diameter and 48 ft long, filled

The Davis neutrino observatory in the Barberton (Ohio) mine, 2300 ft underground. The two 500-gallon tanks are filled with perchloroethylene (C_2Cl_4). Neutrino radiation transforms ^{37}Cl into radioactive ^{37}Ar.

with cleaning fluid (C_2Cl_4). This will be buried in a deep mine, under more than 4400 ft of rock. It is hoped that a few of the chlorine atoms in the fluid will undergo the reaction $\nu_e + {}^{37}Cl \rightarrow {}^{37}Ar + e^-$. Positive results would give the first direct proof that the sun is in fact a nuclear fusion device. If Davis' $600 000 neutrino eye does see solar neutrinos, it will actually be looking right into the sun's central region, which is inaccessible to all other scientific instruments.

In 1953, Frederick Reines from the Case Institute of Technology and his coworker C. L. Cowan were the first to detect neutrinos produced by a nuclear reactor. Reines described two neutrino telescopes. The first is to be buried this year 2000 feet deep in a salt mine near Cleveland. It is made of lithium and is designed to detect solar electron-neutrinos by the reaction $\nu_e + {}^7Li = {}^7Be + e^-$. The second telescope, located two miles underground in the East Rand Proprietary Gold Mines near Johannesburg, South Africa, is constructed to detect energetic muons produced by high-energy neutrinos (ν_μ) in the surrounding rock. If such neutrinos are seen, there won't be many of them. A friend told Dr. Reines: "You may possibly have a long-distance record for commuting to an experimental site, but you are one of the few under these conditions who can commute between counts." Dr. Reines remarked: "It is interesting that despite the size of this detector, it has too little sensitivity, by perhaps a factor of 10^3 or more, to detect an expected flux of true cosmic (that is, extraterrestrial) neutrinos of high energy. But we must take one step at a time and see whether, as we increase our sensitivity, nature is as we think it is or whether some surprises might not be in store for us."

John Bahcall from Caltech reviewed different possibilities for neutrino detection. He stressed that a neutrino-spectroscopic study of the solar interior should be included in the long-range program as a means of determining quantitatively conditions in the interior of the sun—in much the same way as astronomers have already studied its surface by photon spectroscopy. He discussed possible observations of neutrinos from strong radio sources and their connection with mass estimates of the hypothetical W^- boson. Bahcall also proposed that the military be persuaded to surrender their supplies of tritium to the neutrino enthusiasts to build detectors. Luis Alvarez from Berkeley, chairman of the session, expressed his doubts about the technical feasibility of this approach to nuclear disarmament. In the discussions, G. G. Zatsepin, delegate to the symposium from the Soviet Academy of Sciences, was in favor of building neutrino observatories on the moon. Hong-Yee Chiu of the Goddard Space Flight Center preferred a detector, consisting of 1000 tons of completely ionized ^{37}Cl, located at the outer limits of the solar system on the planet Pluto. He envisaged counting time of a few hundred years, and conceded that his Pluto experiment required a civilization more affluent than our own.

The next session was chaired with commendable firmness by Leopold Infeld of the Polish Academy of Sciences. R. K. Sachs of The University of Texas proposed tests which would enable astronomers to deduce the structure of the universe from their observations without begging the question by presupposing a particular cosmological model, such as a homogeneous isotropic Friedman universe. He pointed out that spherical galaxies or clusters at a great distance might all appear elliptical in shape because they are seen through ripples of gravitational waves which pervade the space-time ocean.

This gathering of physicists and astronomers was a natural environment for the discussion of the possible new long-range force proposed by J. Bernstein, T. D. Lee, N. Cabibbo, J. Bell, and J. K. Perring. A hard apple had struck the heads of these physicists a few months earlier: the CP-invariance experiment of Christenson, Cronin, Fitch, and Turlay. J. H. Christenson of Columbia University described the experiment. Gerald Feinberg, also from Columbia, reported cautiously on the hypothetical new long-range force. This fifth force, much weaker than the gravitational interaction, was invented in order to preserve time-reversal symmetry in elementary-particle processes. It would affect matter and antimatter differently

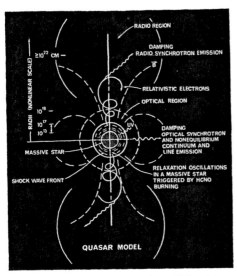

W. A. Fowler's quasar model. Relaxation oscillations between radii of 10^{13} and 10^{17} cm are energized in a massive star by HCNO burning at about 2×10^8 °K. Shock waves transmit energy to tenuous outer envelope from which relativistic particles are ejected into surrounding region. An associated dipole magnetic field channels particles into two regions ($\geq 10^{22}$ cm) where radio synchrotron emission occurs. Optical synchrotron radiation comes from region immediately surrounding star ($\sim 10^{18}$ cm). Nonequilibrium continuum and line emission are also stimulated in this region by ultraviolet radiation from star. It is this region which is visible and not star itself.

[*Physics Today*, April 1965, p. 88]. Limitations on the range and strength of the force are provided by measurements which verify Einstein's principle of equivalence. R. H. Dicke of Princeton University reviewed the Dicke-Eötvös experiments which show that the acceleration toward the sun of gold and aluminum are equal to within the impressive accuracy of one part in 10^{11}. He said: "Some idea of the required sensitivity can be obtained by noting that this requires the detecting of a relative acceleration as small as 6×10^{-12}cm/sec^2. Starting from rest a body would reach the enormous velocity of 1.2×10^{-4}cm/sec after being accelerated a whole year at this rate."

Freeman Dyson chaired the next session, devoted to general discussion and summaries. Thomas Gold of Cornell University discussed his model of a quasi-stellar source: an extremely dense cluster of stars where frequent collisions give rise to fluctuating emission of light. Harlan Smith of the University of Texas summed up the observational results on quasi-stellar sources. He proposed "stark", the astronomer's quark, as a new name for these intriguing objects. His motion was not seconded. "Quasar" received the most enthusiastic support. The vote was twenty ayes to some four hundred abstentions. Other summaries were given by S. Hayakawa of Nagoya University on x-ray and γ-ray astronomy, G. Cocconi of CERN on cosmic rays, W. A. Fowler of Caltech on neutrinos, R. Hanbury Brown of Sydney University on radio astronomy, G. Gamow of Colorado University on cosmology, J. Bjorken of Stanford University on CP-violation, and T. Page of Wesleyan University on optical astronomy.

The symposium concluded with a seminar on gravitational collapse, chaired by L. Gratton of the University of Rome and J. A. Wheeler of Princeton University. Short theoretical papers were presented by J. Bardeen and W. A. Fowler (Caltech), S. A. Colgate, M. May, and R. H. White (Livermore), C. W. Misner (Maryland), R. W. Lindquist (Texas), R. A. Schwartz (NASA and Columbia), D. H. Sharp, L. Shepley, and K. S. Thorne (Princeton), J. N. Snyder (Illinois), A. H. Taub (Berkeley), and L. Gratton (Rome). They dealt mainly with the collapse of large masses under their own gravitational weight and with the resulting release of energy which may be the origin of the strong radio sources. These papers revealed the impressive progress achieved during the preceding year in relativistic hydrodynamics applied to astronomical situations with strong gravitational fields. Fowler suggested that quasi-stellar sources consist of pulsating and rotating supermassive stars, energized by nuclear reactions, with radio and optical emissions from extended surrounding regions which the star excites with ultraviolet radiation and relativistic particles. Only after the exhaustion of nuclear energy would gravitational energy from collapse become available and the evolution of a quasi-stellar into an extended radio source become possible.

Astrophysics today draws on a wide range of talents from many countries. Scientists came to the symposium from Argentina, Australia, Brazil, Canada, England, Denmark, France, Germany, Holland, Hungary, India, Ireland, Israel, Italy, Japan, Mexico, Pakistan, Poland, Sweden, the USA, and the USSR. There was widespread regret at the absence of several distinguished Soviet scientists who are not permitted to attend meetings outside the Soviet Union, and of one well-known West-European physicist who was refused a US visa.

A third Texas Symposium on Relativistic Astrophysics is planned for December 1966.

Recipe for Black Hole Formation

Hugo Martel* and J. Craig Wheeler*

Department of Astronomy, University of Texas, Austin, TX 78712

Abstract. One of the highlights of the 20th Texas Symposium on Relativistic Astrophysics was the night out at Cedar Street, during which Black Holes were served to the participants. In this paper, we outline the basic principles behind the formation of a Black Hole.

INTRODUCTION

In the Fall of 2000, while planning the organization of the 20th Texas Symposium on Relativistic Astrophysics, it was decided that the largest meeting on relativistic astrophysics worldwide should have an official cocktail. We rapidly and unanimously agreed that the drink should be appropriately called a *Black Hole*. We wondered whether a cocktail with such a popular name already existed; however, after consulting the literature, we found recipes for cocktails such as Black Devil, Black Hawk, Blackjack, Black Magic, Black Maria, Black Russian, Black Sombrero, Blackthorn, Black Velvet, and Hole-in-One [1], but no Black Hole. Hence, we designed our own recipe.

Most cocktails are spatially homogeneous. They are made by pouring various liquors and mixers in a glass and mixing them until the concentration gradients become negligible. There are, however, a sub-class of cocktails, called *floaters*, in which the various ingredients are laid on top of each other and not mixed. This design is clearly more appropriate for a Black Hole, a highly inhomogeneous system with an event horizon that separates two distinct regions of space.

THE STABILITY CRITERIA

A floater consists of n liquors L_i, $i = 1, \ldots, n$, poured into a glass in succession. The order in which the various liquors are poured must be chosen carefully to avoid a Raleigh-Taylor instability that would lead to mixing. The stability condition is

$$\nabla \rho \cdot \nabla \phi < 0, \qquad (1)$$
$$\nabla \rho \times \nabla \phi = 0, \qquad (2)$$

where ρ is the density of the liquor, and ϕ is the gravitational potential. Equation (1) is the condition for Raleigh-Taylor stability [2]. Equation (2) simply expresses the requirement that the glass must be held straight up. This criterion assumes that the bartender is at rest with respect to the surface of the Earth and breaks down in the case of a free-falling bartender. Hence, Black Holes could not be served at cocktail parties aboard the International Space Station. This would require modifying the station's design, to create mock gravity using accelerated frames [3].

BASIC COMPONENTS

The first step is to set appropriate boundary conditions. A standard shot glass has a curved inner surface that is an approximation to the embedding diagram of a black hole. This establishes the appropriate boundary conditions for the desired curved-space metaphor. Given the strength of the liquors defined below we advise in any case against a container significantly larger than a shot glass.

The next step is to select L_1, the first liquor that goes into the glass. Since it represents the inside of the Black Hole, a region from which light cannot escape, it had to be dark. Also, its density must be higher than the density of all L_i's for $i > 1$ in order to satisfy equation (1). We note, for instance, that Black Sambuca™, which appropriately captured the spirit of the trapped surface within the event horizon, nevertheless failed to satisfy equation (1) for all sampled L_i, $i > 1$. After much deliberation, we selected Kahlua™, a popular coffee-flavored liquor.

In classical General Relativity, the surface of the black hole (i.e. the event horizon) is infinitely thin. This is not the case, however, if quantum effects are included. The event horizon is a region of finite thickness, which is not dark, since it shines by Hawking radiation. To represent this region, we used Bailey's Irish Cream™. This liquor is light, and satisfies the necessary stability condition $\rho(L_2) < \rho(L_1)$.

Finally, we needed a third liquor L_3 to represent the asymptotically flat spacetime outside the Black Hole. L_3 should be as light as possible, and have a composition that relates to the composition of the interstellar medium (ISM). We opted for a mixture of two compounds commonly found in the ISM: hydrogen oxide (H_2O) and ethyl alcohol (CH_3CH_2OH), with relative proportions of 3 to 2. This combination, commonly called "80-proof vodka," satisfies the stability condition $\rho(L_3) < \rho(L_2)$.

SUMMARY

We have designed an $n = 3$ floater, called a Black Hole, that is Raleigh-Taylor stable and symbolizes all the known physical properties of astrophysical black holes. This new design was tested on a statistically significant sample of astronomers and physicists during the Texas Symposium, and the feedback we received from this study group was quite positive. We are confident that the current design is viable, and does not require any substantial modification. We note that a singularity can arise from excess consumption. We strongly advise avoiding this singularity.

ACKNOWLEDGMENTS

We are grateful to Kim Orr for professional advice on the color, texture, and physical and fluid dynamical properties of various liquors and to Katelyn Allers for providing laboratory space and supplies that allowed many of the key experiments to be performed.

REFERENCES

1. "*Mr. Boston Official Bartender's Guide*," ed. J. Rothstein (New York: Warner Books) (1988)
2. Chandrasekhar, S., "*Hydrodynamics and Hydromagnetic Stability*," (Long Beach: Dover) (1981)
3. Clark, A. C., and Kubrick, S., "*2001 Space Odyssey*" (1968)

List of Attendees

Abramowicz, Marek — marek@fy.chalmers.se
Ahn, Kyungjin — kjahn@astro.as.utexas.edu
Airhart, Marc — mairhart@earthsky.com
Akiyama, Shizuka — shizuka@astro.as.utexas.edu
Albrecht, Andreas — albrecht@physics.ucdavis.edu
Alcubierre, Miguel — miguel@aei-potsdam.mpg.de
Alfaro, Francisco Frutos — frutos@linmpi.mpg.de
Allen, Roland — allen@tamu.edu
Allers, Katelyn — kallers@astro.as.utexas.edu
Alvarez, Marcelo — marcelo@astro.as.utexas.edu
Ambrose, Elizabeth — eba@astro.as.utexas.edu
Anderson, Sheryl — ivanova@astro.as.utexas.edu
Annis, James — annis@fnal.gov
Ansari, Reza — ansari@lal.in2p3.fr
Antilogus, Pierre — Pierre.Antilogus@cern.ch
Asztalos, Stephen — istvan@poptop.llnl.gov
Baker, John — baker@aei.mpg.de
Barger, Amy — barger@IfA.Hawaii.Edu
Barger, Vernon — ldolan@pheno.physics.wisc.edu
Barná, Renato
Bash, Frank — fnb@astro.as.utexas.edu
Battye, Richard — R.A.Battye@damtp.cam.ac.uk
Belyanin, Aleksej — belyanin@appl.sci-nnov.ru
Benacquista, Matthew — benacquista@msubillings.edu
Benne, M.
Benson, Katherine — benson@physics.emory.edu
Bergman, Douglas — dbergman@fnal.gov
Bernabei, Rita — bernabei@roma2.infn.it
Bisnovaty-Kogan, Gena — gkogan@mx.iki.rssi.ru
Bloom, Elliot — elliott@slac.stanford.edu
Bochner, Brett — bochner@prodigy.net
Borrill, Julian — borrill@cfpa.berkeley.edu
Boushaki, Ishak — ishak@hera.phy.queensu.ca
Bunner, Alan — alan.bunner@hq.nasa.gov
Burdhyuzha, Valdimir — burdyuzh@ASC.rssi.ru
Byers, Nina — Byers@physics.ucla.edu

Byrd, Deborah	dbyrd@earthsky.com
Cabrera, Blas	cabrera@cabrera.pobox.stanford.edu
Calder, Alan	calder@flash.uchicago.edu
Campbell, Alison	awc@astro.as.utexas.edu
Campanelli, Manuela	manuella@aei-potsdam.mpg.de
Campos, Antonio	toni.campos@port.ac.uk
Carmeli, Moshe	carmelim@bgumail.bgu.ac.il
Cecchini, Stefano	cecchini@bo.infn.it
Chamblin, Andrew	chamblin@mit.edu
Chatterjee, Sujit	sujit@juphys.ernet.in
Chiba, Takeshi	chiba@tap.scphys.kyoto-u.ac.jp
Chiueh, T. H.	chiuehth@phys.ntu.edu.tw
Cole, K. C.	kc.cole@latimes.com
Colgate, Stirling	colgate@lanl.gov
Coots, Lonique	lonicoots@yahoo.com
Coppi, Bruno	coppi@mit.edu
Costa, Enrico	costa@ias.rm.cnr.it
Cowan, John	cowan@phyast.nhn.ou.edu
Cowen, Ron	scinews@sciserv.org
Croley, Jr., Richard A	croley@utdallas.edu
D'Amico, Nichi	damico@bo.astro.it
Davé, Romeel	rad@as.arizona.edu
Davis, Marc	marc@deep.berkeley.edu
de Bernardis, Paolo	Paolo.DeBernardis@roma1.infn.it
Denur, Jack	
Derishev, Evgeny	derishev@appl.sci-nnov.ru
DeYoung, David	deyoung@noao.edu
Diaz, Mario	
Diener, Peter	diener@aei-potsdm.mpg.de
Dimmelmeier, Harald	harrydee@mpa-garching.mpg.de
Dolan, L.	ldolan@pheno.physics.wisc.edu
Dorodnitsyn, Anton	dora@mx.iki.rssi.ru
Duke, Charlie	duke@grinnell.edu
Duncan, Robert	duncan@astro.as.utexas.edu
Eckmann, Reinhard	eckmann@mail.desy.de
Ehlers, Juergen	ehlers@aei-potsdam.mpg.de
Eikenberry, Stephen	sse2@cornell.edu
Faber, Joshua	jfaber@mit.edu
Fabian, Andy	acf@ast.cam.ac.uk
Falk, Dan	danfalk@pathcom.com
Fargion, Daniele	daniele.fargion@roma1.infn.it
Fenimore, Ed	efenimore@lanl.gov
Fenyves, Ervin	ezbd@utdallas.edu

Ferlet, Roger	ferlet@iap.fr
Flam, Faye	fflam@phillynews.com
Flannagan, Eanna	flanagan@spacenet.tn.cornell.edu
Focke, Warren	focke@slac.stanford.edu
Fowler, James	jrf@astro.as.utexas.edu
Fragile, Patrick C.	pfragile@nd.edu
Freedman, Wendy	wendy@ociw.edu
French, Rica	rfrench@astro.as.utexas.edu
Gair, Johnathan	jgair@ast.cam.ac.uk
Gammie, Charles	gammie@uiuc.edu
Gawiser, Eric	egawiser@ucsd.edu
Gebhardt, Karl	gebhardt@astro.as.utexas.edu
Gonthier, Peter	gonthier@physics.hope.edu
Gordon, Christopher	christopher.gordon@port.ac.uk
Gotthelf, Eric	eric@astro.columbia.edu
Gould, Andrew	gould@astronomy.ohio-state.edu
Graber, James	jgra@loc.gov
Graff, David	graff.25@osu.edu
Gursky, Herbert	herbert.gursky@nrl.navy.mil
Haiman, Zoltan	zoltan@astro.princeton.edu
Hasinger, Guenther	ghasinger@aip.de
Hawley, Scott	shawley@aei-potsdam.mpg.de
Herczeg, Tibor	herczeg@mail.nhn.ou.edu
Hirotani, Kouichi	hirotani@hotaka.mtk.nao.ac.jp
Höflich, Peter	pah@hej1.as.utexas.edu
Hogan, Peter	phogan@ollamh.ucd.ie
Hogan, Craig	hogan@astro.washington.edu
Holder, Gilbert	holder@oddjob.uchicago.edu
Holland, Steven	sholland@nd.edu
Holley-Bockelmann, Kelly	Kelly@eor.astr.cwru.edu
Horacek, Kristina	khoracek@astro.as.utexas.edu
Horne, Keith	kdh1@st-and.ac.uk
Horowitz, Gary	gary@cosmic.physics.ucsb.edu
Huang, Zhihong	zhhuang@astro.as.utexas.edu
Husa, Sascha	shusa@aei-potsdam.mpg.de
Huterer, Dragan	dhuterer@sealion.uchicago.edu
Hwang, Una	hwang@milkyway.gsfc.nasa.gov
Igumenschchev, Igor	ivi@fy.chalmers.se
Iliev, Ilian T.	iliev@arcetri.astro.it
Ioannov, Zach	zac@astro.as.utexas.edu
Italiano, Antonio	antonio.italiano@ME.INFN.IT
Jackman, Kevin	jackman@astro.as.utexas.edu
Jacobson, Heather	hrj@astro.as.utexas.edu

Jaffe, Andrew	jaffe@cfpa.Berkeley.edu
Jansen, Nina	jansen@tac.dk
Jatenco-Pereira, Vera	jatenco@orion.iagusp.usp.br
Jernigan Jr., Jesse	jgj@ssl.berkeley.edu
Jha, Saurabh	sjha@cfa.harvard.edu
Kachelriess, Michael	Michael.Kachelriess@cern.ch
Kaspi, Victoria	vkaspi@physics.mcgill.com
Keros, George	photonphysics@mctttele.com
Kilic, Mukremin	kilic@astro.as.utexas.edu
Kinney, William	kinney@phys.columbia.edu
Klose, Sylvio	klose@tls-tautenburg.de
Knez, Claudia	claudia@astro.as.utexas.edu
Kocharovsky, Vitaly	vkochar@atlantic.tamu.edu
Kocharovsky, Vladimir	kochar@appl.sci-nnov.ru
Kolb, Rocky	rocky@rigoletto.fnal.gov
Konknwski, Deborah	dak@usna.edu
Koppitz, Michael	koppitz@aei-potsdam.mpg.de
Kormendy, John	kormendy@astro.as.utexas.edu
Kouveliotou, Chryssa	chryssa.kouveliotou@msfc.nasa.gov
Kravtsov, Andrey	andrey@astronomy.ohio-state.edu
Krolik, Julian	jhk@pha.jhu.edu
Kuehn, Kyler	kkuehn@uci.edu
Kulkarni, Shri	srk@astro.caltech.edu
Kuranov, Alexandr	alex@xray.sai.msu.ru
Laguna, Pablo	pablo@astro.psu.edu
Lamb, Don	lamb@oddjob.uchicago.edu
Lambert, David	dll@astro.as.utexas.edu
Lara, Juan	juan@einstein.ph.utexas.edu
Larson, Michelle	mlarson@physics.montana.edu
Larson, Shane	shane@physics.montana.edu
Lee, Chul Hoon	chlee@hepth.hanyang.ac.kr
Lee, Hyun Kyu	hklee@hepth.hanyang.ac.kr
Lee, Jounghun	taiji@asiaa.sinica.edu.tw
Lemos, Jose	lemos@kelvin.ist.utl.pt
Lenz, Mary	mary@opa.wwh.utexas.edu
Li, Chen-Biu	cbli@physics.utexas.edu
Liang, Edison	liang@spacsun.rice.edu
Liebendörfer, Matthias	liebend@quasar.physik.unibas.ch
Linde, Andrei	linde@hbar.Stanford.edu
Litwin, C.	c-litwin@chicago.edu
Lo, K.Y.	kyl@asiaa.sinica.edu.tw
Lousto, Carlos Oscar	lousto@aei-potsdam.mpg.de
Lovelace, Richard	rvl1@cornell.edu

Lyutikov, Maxim	lyutikov@cita.utoronto.ca
MacGibbon, Jane	jane.macgibbon@jsc.nasa.gov
Madsen, Jes	jesm@ifa.au.dk
Manmoto, Tadahiro	manmoto@spica.c.chiba-u.ac.jp
Mannheim, Philip	mannheim@uconnvm.uconn.edu
Maran, Steve	Stephen.P.Maran.1@gsfc.nasa.gov
Maraschi, Laura	maraschi@brera.mi.astro.it
Margon, Bruce	margon@astro.washington.edu
Marion, Howie	hman@astro.as.utexas.edu
Marronetti, Pedro	pmarrone@physics.utexas.edu
Marsh, Jasmina	jasna@astro.as.utexas.edu
Martel, Hugo	hugo@simplicio.as.utexas.edu
Mason, Brian	bsm@astro.caltech.edu
Matsumoto, Ryoji	matumoto@c.chiba-u.ac.jp
Matzner, Richard	richard@einstein.ph.utexas.edu
Mazzali, Paolo	mazzali@oat.ts.astro.it
Mbonye, Manasse	mbonye@umich.edu
Medvedev, Mikhail	medvedev@cita.utoronto.ca
Meier, David	dlm@cena.jpl.nasa.gov
Mendes, L. M.	lmendes@star.cpes.susx.ac.uk
Mészáros, Peter	nnp@astro.psu.edu
Miller, Mark	mamiller@wugrav.wustl.edu
Moiseenko, Serji	moiseenko@mx.iki.rssi.ru
Mondragon, Antonio	mondragon@tamu.edu
Montez, Rudy	rudy@astro.as.utexas.edu
Moodley, Kavilan	K.Moodley@damtp.cam.ac.uk
Moore, Ben	Ben.Moore@durham.ac.uk
Mueller, Kaisa	mueller@astro.as.utexas.edu
Nakahata, Masayuki	nakahata@suketto.icrr.u-tokyo.ac.jp
Ne'eman, Yu'val	MatildaE@tauex.tau.ac.il
Newman, Jeffrey A.	jnewman@astron.Berkeley.EDU
Nowak, Michael	mnowak@rocinante.colorado.edu
Noyola, Eva	eva@astro.as.utexas.edu
O'Connor, Padraig	padraig@itc.nuigalway.ie
Olinto, Angela	olinto@oddjob.uchicago.edu
Olum, Ken	kdo@cosmos5.phy.tufts.edu
Ostrowski, Michael	mio@oa.uj.edu.pl
Pacini, Franco	pacini@arcetri.astro.it
Pariev, Vladimir	vpariev@lanl.gov
Paulson, Diane	apodis@astro.as.utexas.edu
Peacock, John	jap@roe.ac.uk
Pen, Ue-Li	pen@cita.utoronto.ca
Perdereau, Oliver	perderos@lal.in2p3.fr

Pereira, Vera	
Perkins, Kelly	kellyp@utdallas.edu
Perlmutter, Saul	s_perlmutter@lbl.gov
Piran, Tsvi	tsvi@nikki.phys.huji.ac.il
Pitts, J. Brian	jpitts@physics.utexas.edu
Pollney, Denis	pollney@aei-potsdam.mpg.de
Poon, Wing-Chi	wcpoon@cs.utexas.edu
Premadi, Premana	premadi@bosscha.itb.ac.id
Qayum, Hisham	hqayum@hotmail.com
Ragot, Brigitte	Bragot@astro.as.utexas.edu
Reichart, Dan	der@astro.caltech.edu
Reinecke, Martin	martin@mpa-garching.mpg.de
Reynolds, Christopher	chris@rocinante.colorado.edu
Rhie, Sun Hong	srhie@condor.phys.nd.edu
Ricker, George	grr@space.mit.edu
Ricker, Paul	ricker@flash.uchicago.edu
Robinson, Edward	elr@astro.as.utexas.edu
Robinson, Ivor	robinson@utdallas.edu
Roeschel, K.	
Romanova, Marina	romanova@astrosun.tn.cornell.edu
Rosswog, Stephan	sro@star.le.ac.uk
Sadoyan, Avetis Abel	asadoyan@www.physdep.r.am
Saijo, Motoyuki	saijo@astro.physics.uiuc.edu
Sako, Masao	masao@astro.columbia.edu
Salmonson, Jay	salmonson@llnl.gov
Samuelson, Frank	fws@lanl.gov
Scalo, John	parrot@astro.as.utexas.edu
Scalzo, Richard	rscalzo@hep.uchicago.edu
Schaefer, Bradley	schaefer@astro.as.utexas.edu
Schlickeiser, R.	rsch@tp4.ruhr-uni-bochum.de
Schmid, Christopher	chschmid@itp.phys.ethz.ch
Schmidt, Bernd	bernd@aei-potsdam.mpg.de
Schmidt, Brian	brian@mso.anu.edu.au
Schwarzschild, Bertram	bschwarz@aip.org
Schwitters, Roy	schwitters@physics.utexas.edu
Seidel, Ed	eseidel@aei-potsdam.mpg.de
Selwood, Jerry	sellwood@physics.rutgers.edu
Shapiro, Maurice	shapiro@sigmanet.net
Shapiro, Paul	shapiro@galileo.as.utexas.edu
Shearer, Andy	andy.shearer@nuigalway.ie
Shepley, Lawrence	larry@einstein.ph.utexas.edu
Shibata, Masura	shibata@astro.physics.uiuc.edu
Shields, Gregory	shields@astro.as.utexas.edu

Shiromizu, Tetsuya	siromizu@utap.phys.s.u-tokyo.ac.jp
Shoemaker, Deirdre	deidre@astro.psu.edu
Shucking, Engelbert	elschucking@dellnet.com
Siegfried, Tom	tsiegfried@dallasnews.com
Sigl, Guenter	sigl@iap.fr
Silva, Eduardo C.	eduardo@slac.stanford.edu
Simmerer, Jennifer	jensim@astro.as.utexas.edu
Sincell, Mark	sincell@compuserve.com
Smith, Julia	dorothea@astro.as.utexas.edu
Sneden, Christopher	chairman@astro.as.utexas.edu
Soucail, Geneviève	soucail@ast.obs-mip.fr
Springen, Clyde	cspringen@inri.com
Steinhardt, Paul	steinh@princeton.edu
Stella, Luigi	stella@merate.mi.astro.it
Suh, In-Saeng	isuh@nd.edu
Suntzeff, Nicholas	nsuntzeff@noao.edu
Sutherland, Peter G.	deansci@mcmaster.ca
Swaters, Rob	swaters@dtm.ciw.edu
Swesty, Doug	dswesty@mail.astro.sunysb.edu
Tan, Jonathan	jt@astron.berkeley.edu
Ticona, Rolando	rticona@fiumsa.bo
Tipler, Frank J.	tipler@math.Tulane.edu
Tomita, Kenji	tomita@yukawa.kyoto-u.ac.jp
Torkelsson, Ulf	torkel@fy.chalmers.se
Truran, James	truran@nova.uchicago.edu
Tsuruta, Sachiko	uphst@gemini.oscs.montana.edu
Tucci, Marco	marco@axrialto.uni.mi.astro.it
Turner, Michael	mturner@oddjob.uchicago.edu
Tyler, Pat	tyler@lheapop.gsfc.nasa.gov
Urry, Megan	cabrera@leland.Stanford.EDU
Velden, Theresa	velden@aei.mpg.de
Venturi, Giovanni	armitage@bo.infn.it
Wagoner, Robert	wagoner@stanford.edu
Wanjek, Christopher	wanjek@sfc.nasa.gov
Watson, Alan	a.a.watson@leeds.ac.uk
Weaver, Kimberly	kweaver@cleo.gsfc.nasa.gov
Weinberg, Steven	weinberg@utaphy.ph.utexas.edu
Wheeler, J. Craig	wheel@astro.as.utexas.edu
Whelan, John T.	jtwhelan@utb1.utb.edu
Williams, Paul	pwilliams@astro.as.utexas.edu
Williams, Reva-Kay	revak@astro.ufl.edu
Wilson, David	wilsondb@coatlicue.colorado.edu
Witze, Alexandria	awitze@dallasnews.com

Woods, Peter	Peter.Woods@msfc.nasa.gov
Wu, Jiun-Huei Proty	jhpw@astron.berkeley.edu
Xia, TongSheng	xiats@astro.as.utexas.edu
Yamaguchi, Masahide	gucci@resceu.s.u-tokyo.ac.jp
Yamasaki, Tasuya	yamasaki@vega.ess.sci.osaka-u.ac.jp
Yokoyama, Junichi	yokoyama@vega.ess.sci.osaka-u.ac.jp
Yong, David	tofu@astro.as.utexas.edu
Young, Chad	cyoung@astro.as.utexas.edu
Zakarov, A.F.	zakarov@vitep5.itep.ru
Zhang, Pengjie	zhangpj@cita.utoronto.ca
Zingale, Michael	zingale@flash.uchicago.edu

Author Index

A

Abramowicz, M. A., 656
Abroe, M., 214
Abusaidi, R., 107
Adams, F. C., 882
Ade, P. A. R., 157, 214
Akerib, D. S., 107
Allen, R. E., 331
Alvarez, M., 143
Amato, M., 95
Ardeljan, N. V., 433, 439
Armitage, P. J., 668

B

Bailer-Jones, C., 635
Baker, J., 746
Balbi, A., 214
Bandyopadhyay, R. M., 690
Barbosa, D., 214
Barger, A. J., 382
Barkana, R., 136
Barná, R., 862
Barnes, Jr., P. D., 107
Bauer, D. A., 107
Baumgarte, T. W., 766
Beaver, E. A., 394
Belli, P., 95
Belyanin, A. A., 538, 687
Benacquista, M., 793, 811
Benson, K., 28
Bergman, D., 827
Bernabei, R., 95
Berrier, J., 547
Bisnovatyi-Kogan, G. S., 433, 439, 611, 760
Blanco-Pillado, J. J., 844
Bloom, E. D., 690
Bochner, B., 55
Bock, J. J., 157, 214
Bolozdynya, A., 107
Bond, J. R., 157
Boone, L. M., 868
Borrill, J., 157, 214
Boscaleri, A., 157, 214

Brandenburg, A., 681
Brink, P. L., 107
Bromley, B. C., 699
Brügmann, B., 746
Bucher, M., 196
Bunker, R., 107
Burbidge, E. M., 394
Burdyuzha, V., 58
Burkart, R. F., 873
Butkevich, A., 862
Butler, R., 562

C

Cabrera, B., 107
Calder, A. C., 484, 490, 796, 808
Caldwell, D. O., 107
Camilo, F., 526
Campanelli, M., 746
Campos, A., 34
Cardoso, V., 799
Carmeli, M., 316
Carretti, E., 184
Cartwright, J. K., 178
Castle, J. P., 107
Cecchini, S., 184
Cerulli, R., 95
Chakrabarty, D., 501
Chamandy, L., 550
Chamblin, A., 22
Chang, C., 107
Chatterjee, S., 64, 541
Chiang, J., 668
Chiba, T., 319
Chiueh, T. H., 172, 208
Cho, I., 28
Choptuik, M. W., 751
Clarke, R. M., 107
Coble, K., 157
Cohen, R. D., 394
Colgate, S. A., 259, 426
Colling, P., 107
Collins, J., 214
Cortiglioni, S., 184
Covault, C. E., 868
Cowan, J. J., 337

Crill, B. P., 157
Cristler, M. B., 107
Cummings, A., 107

D

Dai, C. J., 95
D'Amico, N., 526
D'Amico, V., 862
Da Silva, A., 107
Davies, A. K., 107
Davies, M. B., 343
de Bernardis, P., 157, 214
De Gasperis, G., 157
De Pasquale, D., 862
Derishev, E. V., 614, 617, 620, 687
De Troia, G., 157
Dimmelmeier, H., 757
Dixon, R., 107
Dorodnitsyn, A. V., 760
Dougherty, B. L., 107
Driscoll, D., 107
Duncan, R. C., 495
Dursi, L. J., 484, 490

E

Eichblatt, S., 107
Eikenberry, S., 565
Emes, J., 107

F

Fabbri, R., 184
Faber, J. A., 775
Fabian, A. C., 643
Farese, P. C., 157
Farrar, G. R., 850
Fenyves, E. J., 873
Ferreira, P. G., 157, 214
Fischer, O., 635
Focke, W. B., 690
Font, J. A., 757
Fortin, P., 868
Fragile, P. C., 544
Freiburghaus, C., 343
Friaça, A. C. S., 876

Fritz, G., 690
Frutos Alfaro, F., 262
Fryxell, B., 484, 490
Futamase, T., 268

G

Gaitskell, R. J., 107
Ganga, K., 157
Gavriil, F. P., 501
Gawiser, E., 202
Gebhardt, K., 363
Giacometti, M., 157
Giebels, B., 690
Gingrich, D., 868
Godfrey, G., 690
Golden, A., 565
Golwala, S. R., 107
Gonçalves, D. R., 876
Gonthier, P. L., 547
Gordon, A., 562
Gordon, C., 68
Gotthelf, E. V., 513
Gould, A., 119
Graber, J. S., 802

H

Haiman, Z., 136, 303, 322
Hale, D., 107
Haller, E. E., 107
Hamann, F., 394
Hanany, S., 214
Hanna, D. S., 868
Harding, A. K., 547
Hawley, S. H., 751
He, H. H., 95
Helliwell, T. M., 879
Hellmig, J., 107
Hennawi, J. F., 781
Hertz, P., 690
Hinton, J. A., 868
Hirotani, K., 445, 532
Hivon, E., 157
Hix, W. R., 472
Höflich, P., 459
Hogan, C. J., 11
Holder, G. P., 303, 322

Holland, S., 593
Holley-Bockelmann, J. K., 400
Holmgren, D., 107
Horowitz, G. T., 3
Hristov, V. V., 157, 214
Huber, M. E., 107
Husa, S., 734
Huterer, D., 297

I

Iacoangeli, A., 157
Ignatiev, V. B., 787
Ignesti, G., 95
Igumenshchev, I. V., 656
Iliev, I. T., 146
Incicchitti, A., 95
Irwin, K. D., 107
Ishak, M., 550
Italiano, A., 862

J

Jaffe, A. H., 157, 214
Jatenco-Pereira, V., 876
Jernigan, J. G., 805
Jochum, J., 107
Johnson, B., 214
Junkkarinen, V., 394

K

Kachelrieß, M., 487, 838
Kaspi, V. M., 501
Kawasaki, M., 49
Kesteven, M., 172
Khokhlov, A., 459
Kim, H., 693
Kinney, W. H., 43
Klose, S., 635
Kocevski, D., 623
Kocharovsky, V. V., 538, 614, 617, 620
Kocharovsky, Vl. V., 538, 614, 617, 620
Kohri, K., 355
Konkowski, D. A., 879
Kormendy, J., 363
Kowalski, M. P., 690

Kravtsov, A. V., 130
Krolik, J. H., 674
Kuang, H. H., 95
Kuranov, A. G., 787
Kusunose, M., 454
Kuzmenko, T., 316

L

Lackey, J. R., 501
Lake, K., 550
Lamb, D. Q., 490, 599, 605
Lange, A. E., 157, 214
Lara, J. F., 349
Larson, M. B., 553
Lee, A. T., 214
Lee, C. H., 693
Lee, C.-H., 556
Lee, H. K., 556, 693
Lemos, J. P. S., 799
Leubner, M. P., 325
Li, H., 259, 426
Liang, E. P., 623
Liang, H., 172
Liddle, A. R., 37
Liebendörfer, M., 472
Link, B., 553
Lipshultz, F. P., 107
Litwin, C., 856
Liu, J., 781
Lo, K. Y., 172
Lousto, C., 746
Lovelace, R. V. E., 426, 519
Lovellette, M. N., 690
Lu, A., 107
Lyne, A. G., 526
Lyons, R. W., 394
Lyutikov, M., 626

M

Ma, C.-J., 208
Ma, C.-P., 172
Ma, J. M., 95
MacNeice, P., 484, 490
Maeda, K., 478
Maloney, C., 107
Manchester, R. N., 526

Mandic, V., 107
Manmoto, T., 696
Mannheim, P. D., 328
Maraschi, L., 409
Marronetti, P., 740
Martel, H., 143, 265, 268, 922
Martin, R. N., 172
Martinis, J. M., 107
Martinis, L., 157
Masi, S., 157
Mason, B. S., 178
Mason, P., 157
Mathews, G. J., 544, 569
Matzner, C. D., 638
Matzner, R. A., 268, 740
Mauskopf, P. D., 157, 214
Mazzali, P. A., 478
Mbonye, M., 882
McKee, C. F., 638
Medvedev, M. V., 149, 559, 629
Meier, D. L., 420
Melchiorri, A., 43, 157
Mendes, L. E., 37
Messer, O. E. B., 472
Mészáros, P., 587, 620
Meunier, P., 107
Mezzacappa, A., 472
Miglio, L., 157
Miller, W. A., 699
Mineshige, S., 696
Mohr, J. J., 303, 322
Moiseenko, S. G., 433, 439
Mondragon, A. R., 331
Montecchia, F., 95
Montroy, T., 157
Moodley, K., 196
Moore, B., 73
Mukherjee, R., 868
Müller, E., 757
Mundt, R., 635
Muno, M. P., 501

N

Nakahata, M., 83
Nakamura, T., 319, 478
Nam, S. W., 107
Nelson, H., 107
Netterfield, C. B., 157, 214

Neuhauser, B., 107
Ng, K.-W., 172
Niemeyer, J. C., 490
Nomoto, K., 478
Nordlund, Å., 681

O

O'Brien, S., 547
O'Connor, P., 562, 565
Ogilvie, G. I., 681
Oh, S., 214
Olson, K., 484, 490
Olum, K. D., 844
Ong, R. A., 868
Oser, S., 868
Ostriker, J. P., 136
Ostrowski, M., 865
Ouelette, M. S., 547

P

Padin, S., 178
Pariev, V. I., 259, 699
Pascale, E., 157, 214
Peacock, J. A., 245
Pearson, T. J., 178
Pen, U.-L., 172, 190
Penn, M. J., 107
Perera, T. A., 107
Perillo Isaac, M. C., 107
Perry, M. J., 882
Peterson, J., 172
Piacentini, F., 157
Pierpaoli, E., 184
Piran, T., 575, 850
Pitts, J. B., 763
Pogosyan, D., 157
Polenta, G., 157
Ponomarev, Y., 58
Possenti, A., 526
Postnov, K. A., 787
Premadi, P., 268
Pringle, J. E., 681
Pritychenko, B., 107
Prokhorov, M. E., 787
Prosperi, D., 95
Prunet, S., 157

R

Rabii, B., 214
Ragan, K., 868
Rao, S., 157
Rasio, F. A., 775
Ray, P. S., 690
Readhead, A. C. S., 178
Reichart, D. E., 599, 605
Reilly, K. T., 690
Reynolds, C. S., 668
Rhie, S. H., 885
Richards, P. L., 214
Ricker, P. M., 152, 484, 490
Riotto, A., 43
Robinson, I., 915
Romanova, M. M., 426, 519
Romeo, G., 157
Rosner, R., 484, 490, 856
Ross, R. R., 107
Rosswog, S., 343
Ruhl, J. E., 157

S

Saab, T., 107
Sadoulet, B., 107
Saijo, M., 766
Salmonson, J. D., 632
Sander, J., 107
Sarazin, C. L., 152
Sarkissian, J., 526
Sault, R., 172
Saz Parkinson, P., 690
Scalzo, R. A., 868
Scaramuzzi, F., 157
Scargle, J. D., 690
Schieve, W. C., 763
Schild, A., 915
Schild, W., 911
Schmidt, B. G., 707, 729
Schnee, R. W., 107
Schucking, E. L., 913, 915
Schuette, D. R., 868
Seitz, D. N., 107
Sforna, D., 157
Shabad, G., 690
Shapiro, P. R., 143, 146, 219, 265
Shapiro, S. L., 766

Shearer, A., 562, 565
Shepherd, M., 178
Shestople, P., 107
Shibata, M., 717, 766
Shibata, S., 532
Shields, G. A., 394
Shutt, T., 107
Sievers, J., 178
Sigl, G., 832
Sikora, M., 865
Smith, A., 107
Smith, G. W., 107
Smoot, G. F., 214
Sneden, C., 337
Sonnenschein, A. H., 107
Sopuerta, C. F., 34
Soucail, G., 233
Spadafora, A. L., 107
Stecklum, B., 635
Stein, R. F., 681
Steinhardt, P. J., 279
Stockwell, W., 107
Stompor, R., 214
Subrahmanyan, R., 172
Suh, I.-S., 569
Swesty, F. D., 796, 808

T

Takahara, F., 454
Takahashi, R., 746
Tan, J. C., 638
Taylor, J. D., 107
Teter, M. A., 507
Theoret, C. G., 868
Thielemann, F.-K., 343
Timmes, F. X., 484, 490
Tipler, F. J., 769
Tomàs, R., 487
Tomita, K., 310
Torkelsson, U., 681
Toropin, Y. M., 519
Toropina, O. D., 519
Trifiró, A., 862
Trimarchi, M., 862
Truran, J. W., 337, 490
Tsuruta, S., 507
Tucci, M., 184
Tufo, H. M., 484, 490

Turner, M. S., 297
Turok, N., 196

U

Udomprasert, P., 178
Uryū, K., 717
Ustyugova, G. V., 426

V

Valle, J. W. F., 487
Vereshkov, G., 58
Vittorio, N., 157

W

Wagoner, R. V., 781
Wang, E. Y. M., 796
Wang, L., 459
Watson, A. A., 817
Weaver, K. A., 702
Weinberg, S., 893
Wheeler, J. C., 922

White, S., 107
Wickramasinghe, T., 811
Williams, D. A., 868
Williams, R. K., 448
Wilson, J. R., 544
Wilson, W., 172
Winant, C., 214
Wolff, M. T., 690
Wood, K. S., 690
Wu, J.-H. P., 211, 214, 271, 274

Y

Yamaguchi, M., 49
Yamasaki, T., 454
Yanagida, T., 49
Yellin, S., 107
Yentis, D., 690
Yi, I., 556
Yokoyama, J., 355
Young, B. A., 107

Z

Zhang, P., 190
Zingale, M., 484, 490